TO

My Late Parents,   My Wife Carolyn
Fat-Hia Hitata   and my children
Al-Sawi Fayed   Mark and Susan Otten

All of whom have given us far too much without reservation

HANDBOOK OF POWDER SCIENCE & TECHNOLOGY

SECOND EDITION

edited by
Muhammad E. Fayed
Lambert Otten

CHAPMAN & HALL

I(T)P® International Thomson Publishing

New York • Albany • Bonn • Boston • Cincinnati • Detroit • London • Madrid • Melbourne
Mexico City • Pacific Grove • Paris • San Francisco • Singapore • Tokyo • Toronto • Washington

Copyright © 1997 by Chapman & Hall, New York, NY

Printed in the United States of America

For more information contact:

Chapman & Hall
115 Fifth Avenue
New York, NY 10003

Chapman & Hall
2-6 Boundary Row
London SE1 8HN
England

Thomas Nelson Australia
102 Dodds Street
South Melbourne, 3205
Victoria, Australia

Chapman & Hall GmbH
Postfach 100 263
D-69442 Weinheim
Germany

International Thomson Editores
Campos Eliseos 385, Piso 7
Col. Polanco
11560 Mexico D.F.
Mexico

International Thomson Publishing - Japan
Hirakawacho-cho Kyowa Building, 3F
1-2-1 Hirakawacho-cho
Chiyoda-ku, 102 Tokyo
Japan

International Thomson Publishing Asia
221 Henderson Road #05-10
Henderson Building
Singapore 0315

1 2 3 4 5 6 7 8 9 XXX 01 00 99 98 97

Library of Congress Cataloging-in-Publication Data

Handbook of powder science & technology / edited by M. E. Fayed, L. Otten. -- 2nd ed.
     p.   cm.
   Rev. ed. of: Handbook of powder science and technoilogy. c1984.
   Includes bibliographical references and index.
   ISBN 0-412-99621-9 (alk. paper)
   1. Particles. 2. Powders. I. Fayed, M.E. (Muhammad E.) II. Otten, L. (Lambert)
   III. Title: Handbook of powder science and technology. IV. Handbook of powder
science and technology.
   TP156.P3H35 1997   97-3463
   620'.43--dc21   CIP

Visit Chapman & Hall on the Internet http://www.chaphall.com/chaphall.html

To order this or any other Chapman & Hall book, please contact **International Thomson Publishing, 7625 Empire Drive, Florence, KY 41042.** Phone (606) 525-6600 or 1-800-842-3636. Fax: (606) 525-7778. E-mail: order@chaphall.com.

For a complete listing of Chapman & Hall titles, send your request to **Chapman & Hall, Dept. BC, 115 Fifth Avenue, New York, NY 10003.**

CONTENTS

PREFACE TO THE SECOND EDITION

Since the publication of the first edition of *Handbook of Powder Science and Technology*, the field of powder science and technology has gained broader recognition and its various areas of interest have become more defined and focused. Research and application activities related to particle technology have increased globally in academia, industry, and research institutions. During the last decade, many groups, with various scientific, technical, and engineering backgrounds have been founded to study, apply, and promote interest in areas of powder science and technology. Many professional societies and associations have devoted sessions and chapters on areas of particle science and technology that are relevant to their members in their conferences and career development programs. Two of many references may be given in this regard; one is the recent formation of the Particle Technology Forum by the American Institute of Chemical Engineers. The second reference is the intensified effort given by the American Filtration and Separation Society to define the areas of particle and particle fluid science and technology with the objective to promote the inclusion of courses on these topics at American universities, for undergraduate and graduate circula. On the academic level, many universities in the United States, Europe, Japan, Canada, and Australia have increased teaching, research, and training activities in areas related to particle science and technology.

In addition, it is worth mentioning the many books and monographs that have been published on specific areas of particle, powder, and particle fluid by professional publishers, technical societies and university presses. Also, to date, there are many career development courses given by specialists and universities on various facets of powder science and technology.

Taking note of all these developments, the editors of this second edition faced the need for evaluating and reorganizing, as well as updating and adding to the content of the first edition. In this edition, topics are organized in a logical manner starting from particle characterization and fundamentals to the many areas of particle/powder applications. Comprehensive upgrade of many of the first edition chapters were made and three more chapters were added: namely pneumatic conveying, dust explosion, and fire hazard and health hazard of dust.

The extent to which we have succeeded may be judged from the authors contributions and the contents of this book.

THE EDITORS

ACKNOWLEDGMENTS

We wish to thank Nadeem Visanji, senior student at Ryerson Polytechnic University, for his assistance in preparing the index of this book.

We also would like to thank the Editorial and Production Staff of Chapman and Hall Publishing Co., particularly Margaret Cummins, James Geronimo, and Cindy Zadikoff for their attention and cooperation in the production of this book.

Last, but not least, we thank our families for their patience and understanding throughout the preparation of this text.

CONTRIBUTORS

Leonard G. Austin, Professor Emeritus, Department of Mineral Engineering, The Pennsylvania State University, University Park, PA. (Ch. 12).

Larry Avery, President, Avery Filter Co., Westwood, NJ. (Ch. 14).

Wu Chen, The Dow Chemical Company, Freeport, TX. (Ch. 13).

Douglas W. Cooper, Associate Professor, Department of Environmental Sciences and Physiology, School of Public Health, Harvard University, Boston, MA. (Ch. 18).

Francis A. L. Dullien, Professor Emeritus, Department of Chemical Engineering, University of Waterloo, Waterloo, ON, Canada (Ch. 3).

Norman Epstein, Professor Emeritus, Department of Chemical Engineering, The University of British Columbia, Vancouver, B.C., Canada (Ch. 10).

John R. Grace, Dean of Graduate Studies and Professor, The University of British Columbia, Vancouver, B.C., Canada (Ch. 10).

Stanley S. Grossel, President, Process Safety & Design Inc., Clifton, NJ. (Ch. 19).

Donna L. Jones, Senior Engineer, ECI Environmental Consulting & Research Co., Durham, NC. (Ch. 15).

Mark G. Jones, Senior Consulting Engineer, Centre for Industrial Bulk Solids Handling, Glasgow Caledonian University, Glasgow, Scotland, U.K. (Ch. 7).

Jacob Katz, Consultant, Coconut Creek, FL. (Ch. 16).

Brian H. Kaye, Professor, Department of Physics and Astronomy, Laurentian University, Sudbury, Ontario, Canada (Ch. 1, 2, 11, 20).

David Leith, Professor, Department of Environmental Science and Engineering, University of North Carolina, Chapel Hill, NC. (Ch. 15).

Wolfgang Pietsch, President, COMPACTCONSULT, Inc., Naples, FL. (Ch. 6).

Alan Roberts, Director and Professor, TUNRA Bulk Solids Handling Research Associates, University of New Castle, New South Wales, Australia (Ch. 5).

Keith J. Scott, (Deceased), Chemical Engineering Research Group, Council for Scientific and Industrial Research, Pretoria, South Africa (Ch. 13).

Kunio Shinohara, Chairman and Professor, Department of Chemical Process Engineering, Hokkaido University, Sapporo, Japan (Ch. 4).

Gabriel I. Tardos, Professor, Department of Chemical Engineering, The City College of The City University of New York, New York, N.Y. (Ch. 17).

Fred M. Thomson, Consultant, Bulk Solids Handling and Storage, Wilmington, DE. (Ch. 8).

Olev Trass, Professor Emeritus, Department of Chemical Engineering, University of Toronto, Toronto, Ontario, Canada (Ch. 12).

Frederick A. Zenz, Professor Emeritus, Department of Chemical Engineering, Manhattan College, Riverdale, N.Y. (Ch. 9, 17).

HANDBOOK OF POWDER SCIENCE & TECHNOLOGY

1
Particle Size Characterization

Brian H. Kaye

CONTENTS

1.1 WHAT IS THE SIZE OF A POWDER GRAIN?

It must be firmly grasped at the beginning of a discussion of techniques for characterizing the size of fineparticles that for all except spherical fineparticles there is no unique size parameter that describes an irregularly shaped fineparticle.[1,2]

When an irregular grain of powder is studied by various characterization techniques, the different methods evaluate different parameters of the fineparticle. Thus in Figure 1.1 various characteristic parameters and equivalent diameters of an irregular profile are illustrated. When selecting a parameter of the fineparticle to be evaluated, one should attempt to use a method that measures the

1

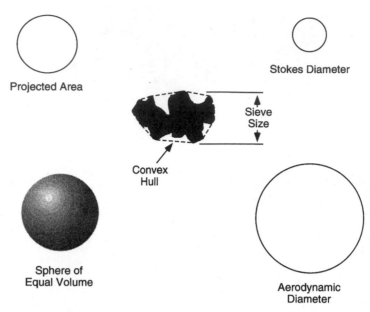

Figure 1.1. The size of a fineparticle is a complex concept for all but smooth, dense, spherical fineparticles.

parameter that is functionally important for the physical system being studied. Thus, if one is studying the sedimentation of grains of rock tailings in a settling pond one should measure the Stokes diameter of the powder grains. The Stokes diameter is defined as the size of a smooth sphere of the same density as the powder grain that has the same settling speed as the fineparticle at low Reynolds number in a viscous fluid. It is calculated by inserting the measured settling velocity of the fineparticle into the Stokes equation, which is:

$$d_{\mathrm{S}}^2 = \frac{18\eta v}{(\rho_{\mathrm{P}} - \rho_{\mathrm{L}})g}$$

where

v = the measured velocity
d_{S} = Stokes diameter
g = acceleration due to gravity
η = viscosity of the fluid
ρ_{P} = density of powder grain
ρ_{L} = density of a liquid.

On the other hand, if one is measuring the health hazard of a dust one may need to characterize the powder grains by two different methods. Thus, the movement of a fineparticle suspended in the air into and out of the mouth of a miner is governed by the aerodynamic diameter of the fineparticle. This is defined as the size of the sphere of unit density that has the same dynamic behavior as the fineparticle in low Reynolds number flow. However, when one is considering the actual health hazard caused by the dust fineparticles, one may want to look at the number of sharp edges on the fineparticle, in the case of a silocotic hazard, or the fractal dimension and surface area of the profile, in the case of a diesel exhaust fineparticle. Furthermore, if one is interested in the filtration capacity of a respirator, the actual physical dimensions of a profile may have to be measured by image analysis. In recent years there has been a great deal of development work regarding the problem of characterizing the shape and structure of fineparticles and this recent work is the subject of a separate chapter in this book.

Many methods used for characterizing fineparticles have to be calibrated using standard fineparticles. These are available from several commercial organizations.[3-6] The European technical community has evolved some standard powders for reference work.[7] Because different methods measure different parameters of irregular fineparticles the data generated by the various methods are not directly related to each other and one must establish empirical correlations when comparing the data from different characterization proceedings. From time to time we discuss this aspect of particle size analysis in this chapter.

It is useful to distinguish between direct and indirect methods of fineparticle characterization. Thus, in sedimentation methods, one directly monitors the behavior of individual fineparticles and the measurements made are directly related to the properties of the fineparticles. On the other hand, in gas adsorption and permeability methods, the interpretation of the experimental data involves several hypotheses. As a consequence, the fineness measurements should be regarded as secondary, indirect methods of generating the information on the fineness of the powdered material.

1.2 OBTAINING A REPRESENTATIVE SAMPLE

An essential step in the study of a powder system is obtaining a representative sample. Procedures have been specified for obtaining a powder sample from large tonnage material. In this chapter we concern ourselves mainly with the obtaining of a small sample for characterization purposes for a sample of powder sent to a laboratory from the plant.[1, 2, 8-10,11, 12, 13, 14, 15]

For many years the spinning riffler has been recognized as a very efficient sampling device for obtaining a representative sample. This piece of equipment is shown in Figure 1.2a. In this device a ring of containers rotates under a powder supply to be sampled. For efficient sampling the total time of flow of powder into the system divided by the time of one rotation

must be a large number. Although the spinning riffler is an efficient sampling device it has two drawbacks. First, the total supply of the powder has to be passed through the sampling device to ensure efficiency; this can sometimes be inconvenient. Second, if the powder contains very fine grains the rotary action of this sampling device can result in the fines being blown away during the sampling process. Both of these difficulties are avoided if one uses the free fall tumbler powder mixer shown in Figure 1.2b to carry out the sampling process. It has always been appreciated that if a powder could be mixed homogeneously then any snatch sample from the powder is a representative sample. However, there has been some reluctance to use this approach to sampling because of the uncertain performance of powder mixers. Recent work has shown that the device shown in Figure 1.2b is a very efficient mixer and that samples taken from a container placed in the mixer would normally constitute a representative sample.[14, 15] The mixing chamber is a small container in which the powder to be mixed or sampled is placed. In the case of the system shown in Figure 1.2b a cubic mixing chamber is used. The chamber must not be filled to capacity because this would restrict the movement of the powder grains during the chaotic tumbling that constitutes the mixing process. Usually the container should be half full. The lid of the chamber is removable and contains the sampling cup on a probe (rather like a soup ladle fixed to the top of the mixing chamber). The mixing chamber is placed inside the tumbling drum which is coated with rough-textured foam to cause the mixing chamber to tumble chaotically as the tumbling drum is rotated. This chaotic tumbling of the mixing chamber results in the complete mixing of powder grains inside the container. When the tumbling is complete the sampling cup attached to the roof of the chamber contains a representative sample. The power of the system to act as a mixer/sampler is illustrated by the data in Figure 1.3. A crushed calcium carbonate powder nominally 15 microns was sampled after tumbling a con-

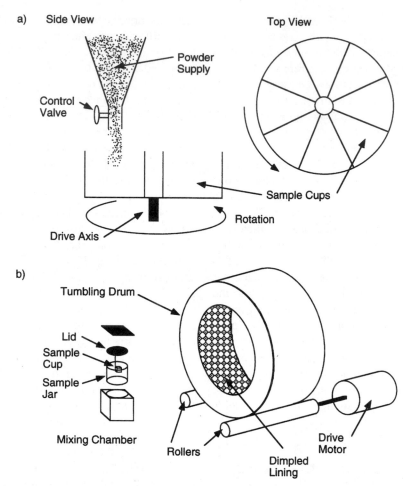

Figure 1.2. Systematic representative sampling of a powder can be achieved with a spinning riffler or chaos generating devices can be used to generate representative samples taken at random. (*a*) Side and top views of a spinning riffler. (*b*) The free-fall tumbling powder mixer can be used for powder homogenization and sampling.

tainer of the powder for 10 min. The sample was characterized by the AeroSizer®, an instrument to be described later in the text. The measured size distribution and that of the subsequent sample taken after a further 10 min are shown in Figure 1.3a. In Figure 1.3b the size distributions of a nominally 6 micron and 15 micron powder as measured by the AeroSizer are shown along with the size distribution of a mixture prepared of these two components in the proportion 25%, 6 micron powder to 75% of the 15 micron powder. In

Figure 1.3c the mathematically calculated size distribution of the mixture based on the known size distributions of the two ingredients is indistinguishable from that of the mixture as obtained from the AeroSizer after the mixture had been tumbled for 20 min in the mixer/sampler. Because the powders were not free flowing, the ability to mix these two powders so that a representative sample matched exactly the predicted structure of the mixture is a good indication of the power of the system to homogenize a powder that had segregated

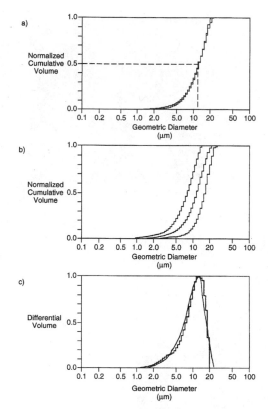

Figure 1.3. If a powder is mixed well before sampling, any snatch sample is a representative sample. (*a*) Separate samples of 15 micron calcium carbonate taken from a free-fall tumbling mixer, and characterized by the Aerosizer®, are nearly indistinguishable. (*b*) Measured size distributions of nominal 6 micron and 15 micron calcium carbonate powders, compared with a mixture of 25% of 6 the micron powder with 75% of the 15 micron powder. (*c*) The measured size distribution of the mixture in (*b*) is nearly identical to the predicted size distribution (smooth curve) calculated from the known size distributions of the constituent powders.

during previous handling.[14, 15] (See also discussion on powder mixing monitoring in Chapter 11)

Sometimes the fineparticles of interest have to be sampled from an air steam, in which case one can use several types of filters. Thus in Figure 1.4, three different types of filter are shown. The filter in Figure 1.4a is an example of a type of filter made by bombarding a plastic film with subatomic particles with subsequent etching of the pathways in the plastic. This process produces filters with very precise holes perpendicular to the surface of the plastic. This type of filter is available from the Nuclepore® Corporation and other companies.[16, 17]

When this type of filter is used to trap airborne fineparticles they remain on the surface of the filter so that they can be viewed directly for characterization by image analysis. The filter shown in Figure 1.4b is a depth filter of the same rating as that of Figure 1.4a. (The rating of the filter is the size of the fineparticle that cannot pass through the filter.) It can be seen that there are much larger holes in the membrane filter and the trapped fineparticles are often in the body of the filter and may not be readily visible. To view the fineparticle trapped by the filter, the filters may have to be dissolved with the fineparticles being deposited on a glass slide for examination. They are, however, much more robust than the Nuclepore type filter and are generally of lower cost.

The third type of filter shown in Figure 1.4c is a new type of filter known as a collimated hole sieve. These glass filter-sieves are made by a process in which a fiber optic array is assembled and then the cores are dissolved to generate orthogonal holes of closely controlled dimensions in the filter-sieving surface.[18] These glass sieves are available in several different aperture sizes and can be reused for many sampling experiments. It should be noted that when studying aerosols it is preferable to study them in situ rather than after filtering because the deposition of the fineparticles on a filter can change their nature. Thus if one is studying a cloud of fineparticles it may be better to use a diffractometer for in situ studies rather than to filter and subsequently examine the fineparticles. If one has to take a sample from a slurry stream a sampler such as the Isolock® sampler should be used.[19]

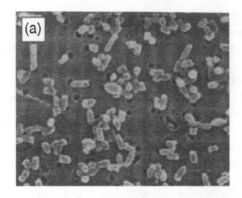

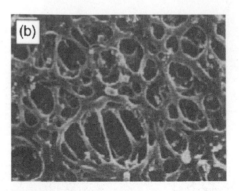

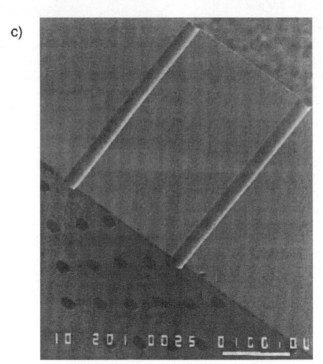

Figure 1.4. Various types of special filters are available for sampling aerosols to generate fields of view for use in image analysis procedures. (*a*) The appearance of a Nuclepore® surface filter. (*b*) Appearance of a cellulosic depth filter. (*c*) Oblique view of a 25 micron "collimated hole" sieve.[17]

Once a representative sample of a powder has been obtained, preparing the sample for experimental study is often a major problem. If one is not careful the act of preparing the sample can change its structure radically. For example, some workers recommend that when preparing a sample for microscopic examina-

tion one places the powder to be studied in a drop of mineral oil and spreads it gently with a glass rod. From the perspective of the fineparticle the glass rod is many hundreds of times bigger than itself and the pressure of the rod can crush its structure into a myriad of fragments. Other workers sometimes use ultrasonic dispersion to create a suspension of fineparticles and again such treatment can inadvertently change the structure of the fineparticle population. In general one should not use a dispersion severity that is greater than that to which the system is going to be subjected in the process of interest. Thus if a pharmaceutical powder is going to be stirred gently in a container of water then one should not use ultrasonics to disperse the fineparticles. On the other hand if the substance is a pigment such as titanium dioxide that is going to be dispersed in a medium by processing it through a triple roll mill then one should use a very severe form of shear dispersion so that agglomerates are broken down. Otherwise, a gentle dispersion technique will leave agglomerates untouched and give a false impression of the fineness of the material when dispersed in a medium. The dispersion of powders in liquids is a very difficult task and requires specialist knowledge.[20]

1.3 SIZE CHARACTERIZATION BY IMAGE ANALYSIS

It is often assumed that image analysis is the ultimate reference method because "seeing is believing." Unfortunately image analysis is often carried out in a very superficial manner to generate data of doubtful value. The first problem that one meets in image analysis is the preparation of the array of fineparticles to be inspected. If one uses a fairly dense array of fineparticles a major problem is deciding exactly what constitutes a separate fineparticle. Thus, in Figure 1.5a a simulated array of monosized fineparticles deposited at random on a field of view to achieve a 10% coverage of the field of view is shown. It can be seen that many clusters exist in the field of view. When one inspects a filter through the microscope there is no fundamental method of deciding whether a cluster viewed has formed during the filtration process or existed in the cloud of fineparticles that were filtered from the air stream. The only way that one can do this is to

repeat the sampling process at a series of dilutions. As shown in Figure 1.5b even at 3% coverage of the field of view there are three clusters that have been formed by random juxtaposition of the monosized fineparticles. If fineparticles, which are really separate entities, cluster in the field of view the loss of the smaller fineparticles is described as primary count loss due to the sampling process and the false aggregates, which are interpreted as being larger fineparticles, are called secondary count gain. (The whole question of clustering

a)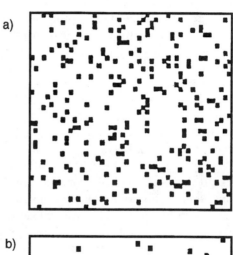

b)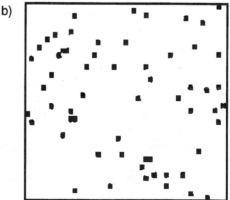

Figure 1.5. Random juxtaposition of fineparticles in a field of view can lead to false aggregates that distort the measured size distribution of the real population of fineparticles.[21-23] (a) The appearance of a simulated 10% covered field of monosized fineparticles. (b) The appearance of a simulated 3% covered field of monosized fineparticles.

in a field of view by random chance is discussed at length in Refs. 21, 22, 23.)

Many different automated computer-controlled image analysis systems have been developed for characterizing fineparticle profiles. If profiles contain indentations of the type shown by the carbonblack profile of Figure 1.6a the logic of the computer can have serious problems as the scan lines of the television camera cross the indentations. To deal with this problem many commercial image analyzers have what is known as erosion-dilation logic.[1] In the dilation logic procedure, pixels are added around the profile with subsequent filling in of the fissures of the profile as shown in Figure 1.6b. If the dilated profile is subsequently stripped down by the erosion process the resulting smoothed out profile can be evaluated more readily by the scan logic of the image analyzer. In Figure 1.6b the smoothing out of the profile by the addition of 32 layers of pixels in a series of operations is shown. Although the original purpose of the dilation followed by erosion was to create a smoothed out profile, the erosion logic can also be used to strip down an original profile to see how many components are in the original structure as shown in Figure 1.6c. The carbonblack profile of Figure 1.6a probably formed by agglomeration in the fuming process used to generate the carbonblack and the erosion strip down of the original profile suggests that it was formed by the collision of three to four original subsidiary agglomerates. Note that there is no suggestion that the agglomerates of the carbonblack were formed by deposition from the slide; in this case it probably was a real agglomerate formed in space during the fuming process.

The analyst must be very careful before using erosion dilation logic to separate juxtaposed aggregates in a field of view being evaluated by computer-aided image analysis. A major mistake made by analysts when looking at an array of fineparticles is to over count the finer fineparticles and the failure to search for the rare events represented by the larger fineparticles in the population to be evaluated.[24,25] One should always use a stratified count procedure to increase the efficiency of the evaluation process (See Exercise 9.1, pp. 411–414 of Ref. 22.)

1.4 CHARACTERIZING POWDERS BY SIEVE FRACTIONATION

In sieving characterization studies a quantity of powder is separated into two or more fractions on a set of surfaces containing holes of a specified uniform size. In spite of the development of many alternate sophisticated procedures for characterizing powders, sieving studies are still widely used and have the advantage of handling a large quantity of powder, which minimizes sampling problems. It is a relatively low-cost procedure, especially for larger free-flowing powder systems. There are many different manufacturers of sieving machines and of material from which the sieves are fabricated.[1,2] Most industrial sieves used for fractionating powders are made by weaving wire cloth to create apertures of the type shown in Figure 1.7a. For more delicate analytical work one can purchase sieve surfaces that are formed by electroforming or by other processes.

Because there is a range of aperture sizes on a sieve in which theoretically all the apertures are the same size, fractionation is never clear cut and it is necessary to calibrate the aperture range and effective cut size of any given sieve. This can be carried out either by examining the apertures directly under a microscope or by looking at near-mesh fineparticles that are trapped in the sieve surface during a sieving experiment. These near-mesh sizes are cleared from the sieve by inverting the sieve, rapping it sharply on the surface, and collecting the particles that fall out on a clean sheet of paper. In Figure 1.7b the size distribution of the apertures of a sieve as determined by direct examination of the apertures, and by examining glass beads and sand

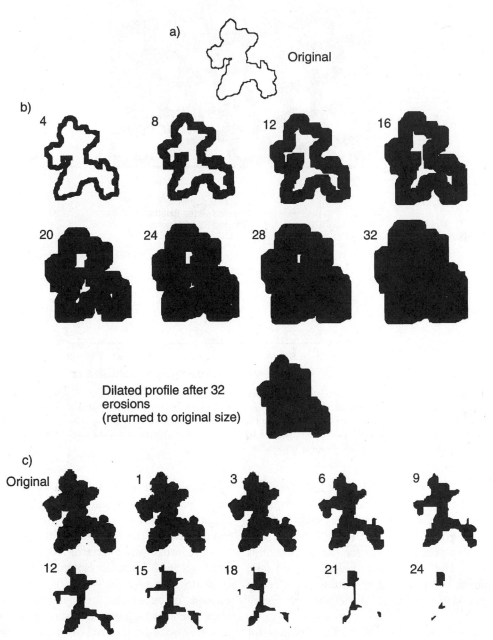

Figure 1.6. Computer-aided image analysis system routines allow routine characterization of convoluted profiles. (*a*) A typical carbonblack profile traced from a high-magnification electromicrograph. (*b*) Dilation can be used to fill internal holes and/or deep fissures in a profile being evaluated. (The number indicates the number of dilations applied to reach this stage from the original profile.) (*c*) Repeated application of the erosion routine suggests that this cluster was formed by the collision of several subagglomerates. (The number indicates the number of erosions applied to reach this stage from the original profile.)

a)

b)

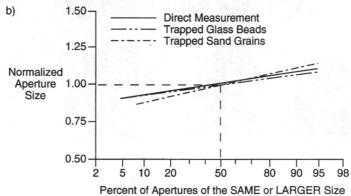

Figure 1.7. A major problem in sieve characterization of powders arises from variations in the mesh aperature size. The aperture size range increases with sieve usage. (*a*) Magnified view of the apertures of a woven wire sieve. (*b*) Variations in aperture size can be determined either by direct examination of the apertures by microscope or by examining near mesh size fineparticles that were lightly trapped in the mesh during sieving and subsequently removed by inverting the sieve and rapping it on a hard surface.

grains that were trapped in the mesh, is shown. It can be seen that the range of sizes trapped in the mesh depends on the shape of the powder grains. Thus, in Figure 1.8a a typical set of the sand grains used in the calibration is shown. The shape distribution of the sand grains as determined from a study of the grains trapped in the mesh is shown in Figure 1.8c.[26, 27]

(For a recent discussion of techniques for calibrating sieves see Ref. 28. For a discussion of the various ways in which a sieve mesh can be damaged and the subsequent changes of aperture sizes monitored see the extensive discussion given in Ref. 1.)

Apart from the uncertainty as to the exact aperture size in the surface of a sieve, another major problem when carrying out characterization by means of sieve analysis is to determine when the fractionation of the powder on a sieve with given apertures is complete. Methods have been developed to predict the ultimate residue on a sieve from the rate of passage of materials through the sieve but these techniques have not found wide acceptance. The falling cost of data processing equipment, however, will probably lead to a renewed interest in automated characterization of powders by sieve fractionation.

When carrying out a sieve fractionation study one must carefully standardize the experimental protocol and several countries have

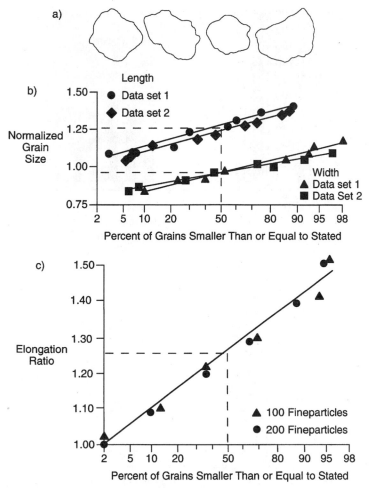

Figure 1.8. As a byproduct of calibrating a sieve mesh using trapped fineparticles, one obtains a subset of powder grains, typical of the powder being characterized, which can be used to generate a shape distribution of the powder grains. (a) Typical sand grains removed from a sieve mesh. (b) Length and width distributions of two sets of sand grains removed from a sieve mesh. (c) Distribution of the elongation ratio of two sets of sand grains removed from a sieve mesh.

prepared standard procedures for carrying out sieve characterization studies.[29] Specialist sieve equipment is available from several companies.[30-35]

Electrostatic phenomena can interfere with the progress of a sieve fractionation of a powder. Thus, in Figure 1.9 the size distributions of a plastic powder fractionated on a 30-mesh ASTM sieve are shown. (ASTM stands for the American Society for Testing of Materials; this organization has specified a whole series of tests for sieves. The mesh number refers to the number of wires per inch with the wire diameter being the same as the aperture of the sieve.) The nominal size of a 30-mesh sieve is 600 microns. When the fractionated powder was characterized by image analysis study there were considerable numbers of fineparticles less than 150 microns clinging to the coarser grains. On a mass basis, the fines do not constitute a significant fraction of the weight of powder of nominal size 600 to 1100 microns but their

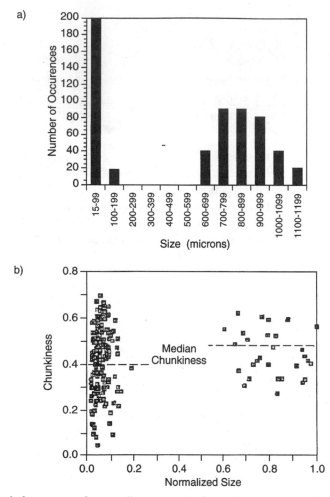

Figure 1.9. Electrostatic forces cause fines to cling to oversize fineparticles on the surface of a sieve, preventing them from passing through the sieve apertures. (*a*) Size distribution of a sieved plastic powder showing a large number of fines still contained in the oversize fraction of the powder. (*b*) Chunkiness versus size domain for the plastic powder of (*a*). (Note that chunkiness is the reciprocal of aspect ratio.)

presence could severely modify the flow and packing behavior of the powder. The fines clinging to the coarser grains had a wider range of shapes as demonstrated by the chunkiness size data domain of Figure 1.9b. Sometimes the fines of such a powder can be removed by adding a silica flow agent into the powder while sieving the powder. (For a discussion of the effect of flow agents on the behavior of a powder see the discussion in Ref. 36.)

1.5 CHARACTERIZING THE SIZE OF FINEPARTICLES BY SEDIMENTATION TECHNIQUES

As stated earlier in this chapter, in sedimentation methods for characterizing fineparticles the settling dynamics of the fineparticles in suspension are monitored and the observed data substituted into the Stokes equation to calculate what is known as the Stokes diameter of the fineparticle. During the 1960s and

1970s sedimentation methods were the dominant techniques in size characterization studies and many different instrument configurations have been described.[1,2] Several international standard protocols for using sedimentation equipment have been prepared. Recently the International Standards Organization of the European Community has prepared standards for centrifugal and gravity sedimentation methods.[37] In Figure 1.10 some of the basic instrument designs that have been used to study the sedimentation dynamics of a suspension of fineparticles are shown. In instruments known as sedimentation balances the weight of fineparticles settling onto a balance pan suspended inside the suspension, as shown in Figure 1.10a, is used to monitor the settling behavior of suspension fineparticles. This type of instrument is known as a "homogeneous suspension start" instrument. The presence of the pan in the suspension interferes with the dynamics of the settling fineparticles but this interference can be allowed for in the interpretive equations and minimized by specialized design of the equipment.

In an alternate method, the suspension of fineparticles to be studied is introduced as a layer at the top of a column of suspension. The movement of the settling fineparticles

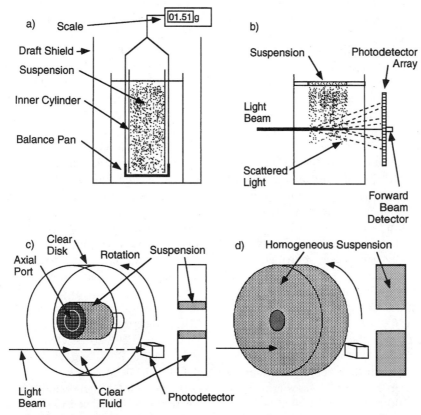

Figure 1.10. Sedimentation methods for characterizing the size distribution of powders uses the settling speed of the fineparticles in suspension and is interpreted as the size of the equivalent spheres using Stokes' law. (a) In sedimentation balances the fineparticles are weighted as they arrive at the base of the sedimentation column. (b) In a photosedimentometer, fineparticles are monitored by noting the scattering or extinction of light or X-rays passing through the suspension. (c) In the linestart centrifugal method, a thin layer of suspension is injected onto the surface of a clear fluid so that all the fineparticles start at the same distance from the wall of the disc. (d) In the homogeneous start centrifugal method the disc is filled with suspension.

down the column of clear fluid is monitored using a device such as a beam of light or a beam of X-rays as shown in Figure 1.10b. Workers started to use X-rays because of the complex diffraction pattern of irregular shaped particles and the difficult interpretation of concentration data from the measured observation of the light beam. Procedures in which a layer of suspension was floated onto a column of clear fluid are known as linestart methods. Their advantage vis a vis the homogeneous start method is the simplicity of data interpretation; however, complex interaction of the fineparticles moving in a clear fluid can cause complications in interpretation of the settling dynamics of linestart methods.

Overall, workers have preferred to work with the homogeneous start method, especially because the rapid development of low-cost data processing instrumentation facilitated the complex data manipulations required for the interpretation of homogeneous suspension sedimentation procedures.

The Micromeretics Corporation of Georgia manufactures an instrument for sedimentation studies based on X-ray evaluation of concentration changes in a settling suspension known as a Sedigraph®.[38] This instrument has been widely used, especially since some industries have written standard protocols for using the instrumentation.[2]

Accelerated sedimentation of very small fineparticles by means of centrifugal force has been the basic principle of several instruments for characterizing fineparticles. See, for example, the trade literature of the Horiba Corporation.[39]

In recent years the favored technique for doing centrifugal sedimentation studies utilizes the disc centrifuge. The basic construction of this instrument is shown in Figure 1.10c and 1.10d.[40,41] Again the analyst has the basic choice of using a homogeneous suspension at the start of the analysis or a line start system.[1,2] As with other sedimentation equipment light or X-rays can be used to monitor the sedimentation dynamics in the centrifuge.[1,2,41]

1.6 DIFFRACTOMETERS FOR CHARACTERIZING THE SIZE OF FINEPARTICLES

Advances in laser technology have made it possible to generate diffraction patterns from an array of fineparticles in a relatively simple manner. It can be shown that if one has a random array of fineparticles the resultant diffraction pattern is the same as that of the individual fineparticles times the number of fineparticles. This is shown by the diagram in Figure 1.11a. The diffraction pattern generated by a real fineparticle profile is dependent on the structure of the profile as shown by the diffraction patterns shown in Figure 1.11b. In the commercial instruments that measure size distributions from group diffraction patterns the interpretation of the data is in terms of the spherical fineparticles of the same diffracting power as the fineparticles. As can be seen from Figure 1.11b, sharp edges on the profile will diffract light further out than the smooth profile and this is interpreted by the machines as being due to the presence of smaller fineparticles rather than corresponding smooth, spherical fineparticles of the same size as the real fineparticles.[52] The basic systems of the various diffractometers are similar except that for very small fineparticles some systems study side scattered light rather than forward scattered light.[42-48]

One of the first diffractometers to become commercially available was developed by the CILAS Corporation to characterize the fineness of cement. The basic system used by the CILAS diffractometer is shown in Figure 1.12. The fineparticles to be characterized are dispersed in a fluid and circulated through a chamber in front of a laser beam. A complex diffraction pattern generated by the light passing through the suspension of fineparticles is evaluated by using a photodiode array. In essence the smaller the fineparticle the further out the diffraction pattern from the axis of the system. The optical theory of software strategies behind the evaluation of the diffraction patterns differs in complexity and sophistica-

a)

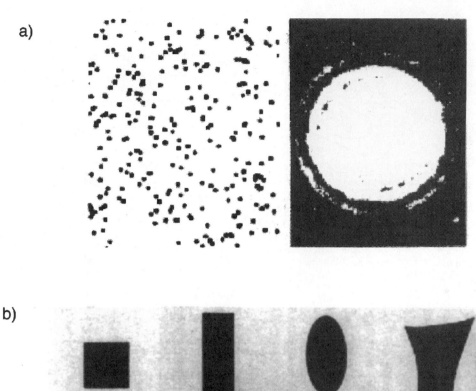

b)

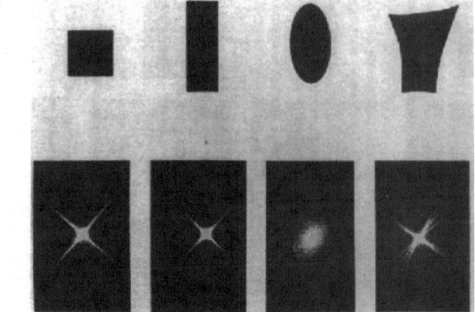

Figure 1.11. When interpreting the physical significance of the diffraction pattern data of a random array of fineparticles, one should remember that the structural features and the texture of a fineparticle affect the light scattering behavior of the fineparticle.[52] (*a*) A random array of dots and its associated diffraction pattern. (*b*) The effect of shape and sharp points on the diffraction pattern of a single profile.

tion from machine to machine, but in essence Fraunhoffer or Mie theory of diffraction pattern analysis is used to interpret the diffraction pattern. In the various presentations of the theory of the instrument, one is sometimes given the impression that the deconvolution (the mathematical term for the appropriate process) of the diffraction pattern proceeds without any basic assumptions. In practice many diffractometers take short cuts in the

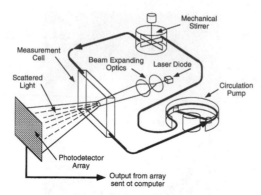

Figure 1.12. Schematic of the CILAS Corporation laser diffractometer size analyzer. In this instrument the size distribution of a random array of fineparticles is deduced from the group diffraction pattern. (Used by permission of CILAS Corporation).[43]

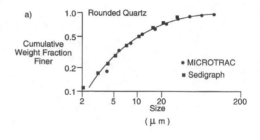

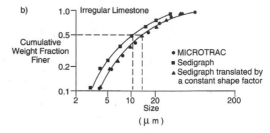

Figure 1.13. By comparing size distribution information derived from studies that evaluate different parameters of the fineparticles, one can sometimes deduce shape information factors.[53] (Microtrac is a registered trademark of Leeds and Northrup Co. and Sedigraph is a trademark of the Micromeretics Corporation.) (*a*) Sedimentation studies and diffractometer evaluations of particle size generate comparable data for spherical fineparticles. (*b*) Sedimentation and diffractometer data for angular crushed limestone can be correlated by means of an empirically determined shape factor. Thus:

$$\frac{\text{mean size by Sedigraph}}{\text{mean size by Microtrac}} = \frac{10}{7} = 1.4$$

data processing of their machines by curve fitting an anticipated distribution function to the generated diffraction pattern data. The customer should always inquire diligently as to any assumptions that are being made in the software transformations of the patterns in any particular commercial diffractometer. The fact that the shape and features, such as edges, on the fineparticles can contribute to the diffraction pattern has been used to generate shape information by comparing the data generated by diffractometer machines with other methods of particular size analysis.[49-51]

The way in which shape information can be deduced by comparing data from different methods is shown by the data summarized in Figure 1.13.[53] The type of distortion that can creep into size distribution information because of the software used in the deconvolution of a diffraction pattern is illustrated by the data of Figures 1.14 and 1.15 taken from the work of Nathier-Dufour and colleagues.[49] These workers studied the size distributions of three food powders: a pulverized wheat flour, maize flour, and a soya bean meal. When these were sized by means of a diffractometer (the Malvern size analyzer; see Ref. 44) the three distribution functions were similar as shown in Figure 1.14. All three distributions appear to be slightly bimodal, indicating that

the software being used to deconvolute the pattern was probably anticipating a bimodal distribution. When the same flours were analyzed by means of sieves the size distributions were very different as illustrated by the data of Figure 1.15. First, the wheat and maize flours did indeed appear to be slightly bimodal but did not have peaks in the positions corresponding to those calculated from the diffractometer data. Note that all three size distributions had ghost large and small fineparticles that did not exist according to the sieve characterization data. Further the peaks of the distributions did not correspond to those calculated from the diffractometer. If one is only wishing to compare a size distribution data then the fact that the diffractometer seemed

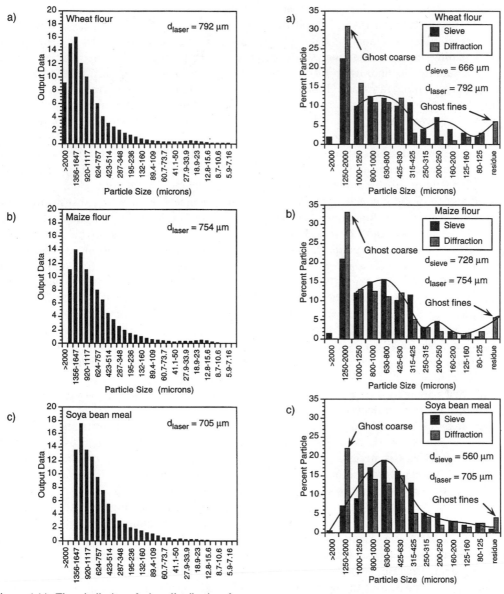

Figure 1.14. The similarity of size distribution functions for various food powders analyzed by a light diffraction size analyzer is an illusion created by insensitive software used to deconvolute the group diffraction pattern of the random array of fineparticles.[49]

Figure 1.15. The "real" size distributions of the powders in Figure 1.14, as measured by sieve analysis, is quite different from the self-similar curves generated by the software of the diffractometer. The sieve distribution data do not conform to any simple distribution function.[49]

to indicate both the presence of an excess of coarse and fines that did not match reality is not a serious problem. If, however, one is wanting to assess the performance of equipment, such as an air classification piece of

equipment, such distortions of the data are misleading. In particular, the soya bean meal appears to have a Gaussian distribution rather than a bimodal log-normal distribution. It is very difficult to fit simple distribution functions to the sieve data of Figure 1.15 and in fact the writer feels that a rush to fit distribution functions to any size analysis data can be a self-defeating process. The analyst should report the data that he finds even if it does not fit simple distribution functions. Distribution functions are of use only if one can interpret the formation dynamics in terms of the distribution function produced by a given process. (For an extensive discussion of distribution functions that have been used for size analysis, see the first edition of this book and Ref. 2. For a discussion of the physical significance of the various distribution functions that have been used see the discussion in Ref. 54.) It is not possible to give general guidance on how to interpret diffractometer data because the software used by the various companies is constantly changing. However, when reporting size characterization data generated by different diffractometers the research worker should specifically detail the year and model of the equipment being used in their studies. If possible comment on the deconvolution algorithms being used to interpret the data. In recent years many manufacturers of diffraction size analyzers have modified their equipment to be able to work with dry aerosols and/or sprays. This has necessitated the development of systems for generating aerosols from dry powder supplies prior to size analysis. This is not an easy task and the ancillary equipment for generating the necessary aerosol can be expensive. Again it is not possible to give firm figures or exact descriptions of the equipment because manufacturers are constantly modifying and changing the design of their equipment.

1.7 TIME-OF-FLIGHT INSTRUMENTS

The falling cost of data processing equipment and the ready availability of lasers have generated another family of instruments for size characterization studies that can be called "time-of-flight instruments." In the first type of instrument a narrow focused beam of laser

light explores an area of a suspension and the size of the particles in suspension is measured by the time it takes for the laser beam to track across the profile of the fineparticle. Sophisticated electronic editors are used to generate the size distribution data from the information generated by the scanning laser. The basic system of this type of instrument developed by the Galai Instruments of Israel is shown in Figure 1.16. (Note that for many years this instrument was sold in the United States by the Brinkman Instrument Company and so many publications in which this instrument is used refer to it as the Brinkman Size Analyzer). The fineparticles to be characterized are placed in the suspension and the laser is rotated by means of a rotating optical wedge. The system also incorporates a video camera for inspecting the actual fineparticles being measured. The logic of the Galai system can be manipulated to provide shape information. It also provides logic modules for advanced image analysis using the video camera data collection system.[55]

Another time-of-flight analyzer that uses a system similar to the Galai instrument is known as the Lasentech Instrument. This system is portable and has been used for online monitoring of fineparticles moving in the slurry or suspension as well as for size analysis in the laboratory.[56]

Another time-of-flight instrument, the AeroSizer®, is manufactured by Amherst Process Instruments Inc., in Massachusetts.[57] The basic system of this instrument is shown in

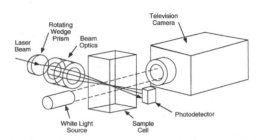

Figure 1.16. The basic layout of the Galia laser-based "time-of-flight" particle size analyzer. (Used by permission of Galia Instruments.)

Figure 1.17a. The aerosol fineparticles to be characterized are sucked into an inspection zone operating at a partial vacuum. As the air leaves the nozzle at near sonic velocity the fineparticles in the stream are accelerated across this inspection zone. It should be noted that, as the aerosol stream emerges into the inspection zone, it is surrounded by a stream of clean air that confines the aerosol stream to the measurement zone. The use of a stream of clean air to focus an aerosol stream to be characterized is a widely used technique known as hydrodynamic focussing. The term is somewhat confusing because it was originally developed with instruments employing liquid streams to examine a series of fineparticles.

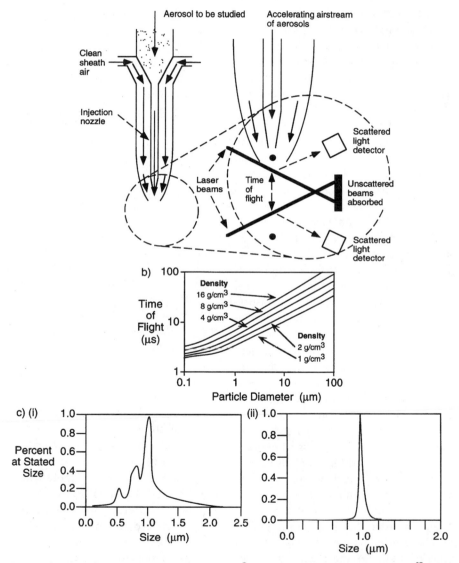

Figure 1.17. The Amherst Process Instruments Aerosizer® is a "time-of-flight" size analyzer.[57] (a) The basic layout of the AeroSizer. (b) Calibration curves for materials of various densities. (c) The AeroSizer can distinguish the various components in a mixture of standard polystyrene latex spheres. (i) Results for a mixture of 0.494 μm, 0.806 μm, and 1.037 μm latex spheres. (Density 1.05 g/cm³.) (ii) Results for 1.037 μm latex spheres alone.

Over the years the term was extended to clean gas sheaths that serve the same function to improve the efficiency of the size characterization equipment. Inherently, the instrument measures aerodynamic diameters directly; however, if the density of the fineparticles is known, the data can be converted to geometric diameters that are essentially Stokes' diameters because of the adjusted term involving the density of the fineparticle. The smaller the fineparticle, the faster the acceleration through the measurement zone. The individual fineparticles are characterized by the time they take to travel across two laser light beams. As they pass through the laser beams, they scatter light which is detected and converted into electrical signals by the two photomultipliers. A computer correlation establishes which peak from the second laser constitutes the matching peak to the initial peak as the fineparticle crosses the first beam. This cross-correlation editorial process enables the machine to operate at very high fineparticle flow densities. The equipment can measure fineparticles at a rate of 10,000/s. The instrument is calibrated using standard fineparticles as shown in Fig. 1.17b. The useful feature of the instrument is that the system used to generate the aerosol for inspection has variable shear rate dispersion force so that one can study the force needed to disperse a given material into an aerosol. In Figure 1.18, some typical data generated for a difficult cohesive powder are shown.[57] The instrument allows the information on size to be printed out either in differential or cumulative form and by number or volume. The powder data in Figure 1.18 are taken from a study of the size distribution of paint pigments. In the differential display of the data by number, the fines dominate the chart whereas if the data are presented by volume, there appears to be a small amount of agglomerated powder that may be dispersible by higher shear dispersion study. The particular sample of titanium dioxide used in this experiment had stood on a shelf for several years and it may well have agglomerated over that period. It should be noticed that pigments

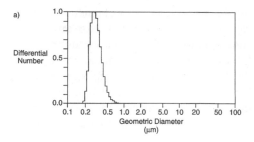

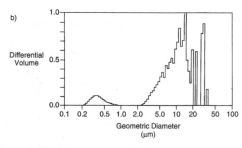

Figure 1.18. Pigments can be characterized by time-of-flight instruments by making the powder into an aerosol.[57] (a) Size distribution, by number, for a sample of titanium dioxide as obtained from the AeroSizer. (b) Size distribution, by volume, for the titanium dioxide of (a) as obtained from the AeroSizer.

such as titanium dioxide are notoriously difficult to disperse into a dry aerosol form and one needs to study the measured distribution of different shear rates before one can decide the physical significance of data such as that displayed in Figure 1.18.

Another time-of-flight instrument is manufactured by TSI Incorporated.[58] Their instrument is known as the Aerodynamic Particle Sizer (APS). This system operates at subsonic flow conditions and cannot tolerate as high a flux of fineparticles as the AeroSizer. The early models also did not have a cross-correlating editor so that one had to operate at a flow rate that permitted unique identification of a pair of light scattering peaks as the aerosol fineparticle crossed the inspection lasers. Constant developments are always underway at all the instrument companies and the reader should check with vendor companies as to the operating conditions and devices in their current equipment.

1.8 SIZE CHARACTERIZATION EQUIPMENT BASED ON THE DOPPLER EFFECT

In the time-of-flight instruments, the basic properties of a laser that are exploited are the ability of a laser system to concentrate high optical power in a narrow nonspreading beam of light. In Doppler-based methods of size characterization it is the monochromacity of the laser source that is exploited. The Doppler effect is the shift in the frequency of a wave motion caused by the relative motion between a source of the wave motion and an object reflecting those waves. It can be shown that the shift in the frequency caused by the relative motion is related to the relative velocity of the source-reflector system. In Figure 1.19, one of the basic systems used to measure the size of aerosol fineparticles by measuring the Doppler shift in light reflected off of a moving fineparticle is shown. Instruments using this type of measurement are available from several companies.[58–61]

The prime laser beam is split into two beams and sent into the interrogation zone of the equipment at different angles by the lens as shown. The aerosol fineparticles to be characterized are sent across this beam in a single stream using hydrodynamic focussing of the type discussed in the previous section with respect to the AeroSizer. In essence, the light reflected from the two different beams suffers different Doppler shifts. Thus the lower beam is heading into the direction of the air flow whereas the upper beam is angled away from the flow of aerosols. The scattered light from the two beams therefore has slightly different frequency. When the reflected light is combined in the photomultiplier tube it generates an interference frequency that is much lower than that of the laser light. This interference frequency is related to the speed at which the aerosol fineparticles are moving through the

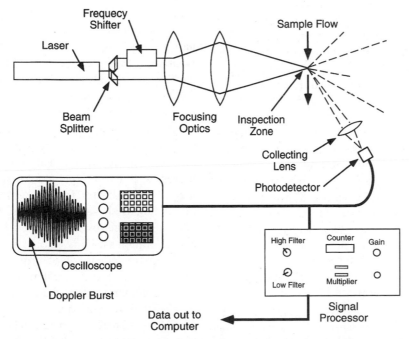

Figure 1.19. In the Doppler shaft-based technique of size characterization, the aerodynamic diameter of fineparticles is determined by accelerating the fineparticles through the inspection zone created by crossed laser beams.[1]

interrogation zone. As in the case of the time-of-flight instruments, the instrument is calibrated with fineparticles of known size. When one reads the theory of the methods such as that in Figure 1.19 it is sometimes hard to discover how the Doppler effect is involved because the interpretation of the data is sometimes phrased in terms of movement of the aerosol fineparticles through the interference fringes created by the two laser beams. In fact the optical arrangement of Figure 1.19 is identical to that used in introductory physics laboratories to generate Newton's interference fringes. The fringes constitute a series of linear fringes perpendicular to the plane of the light beam intersection and to the flow of the aerosol. Therefore to an external observer the aerosol fineparticles appear to be moving through a series of interference fringes. The speed of the aerosol fineparticle is deduced from the frequency with which the aerosol fineparticle moves past the fringes (see diagram in Ref. 1). The interpretation of the data in terms of interference fringes is not strictly correct from a physical theory point of view, but can help one to intuitively understand what is happening to the interrogation zone.

If the fact that the method used in the system outlined in Figure 1.19 involves the Doppler method is difficult to understand from published discussions, then it is even more difficult to track down the involvement of the Doppler shifts in a technique known as photon correlation spectroscopy (PCS). In this method the size of fineparticles in suspension undergoing Brownian motion are studied by looking at the Doppler shifts of laser light scattered by the wandering fineparticles. The technique is useful for fineparticles several microns in diameter downwards. In particular it is widely used to look at the size distribution of latex and colloidal fineparticles.[62]

In a recent review article Finsy makes the following comments:

Originating some 20 years ago from a research tool in a form only suitable for experts, PCS has become a routine analytical measurement for the determination of particle sizes. Its major strong point is that it is difficult to imagine a faster technique for sizing submicron particles. Average particle sizes and distribution width can be determined in a few minutes without elaborate sample preparation. However, reasonably accurate resolution of the shape of the particle size distribution requires extremely accurate measurements over a period of ten hours and more.[64]

The recent trend for this application is the development of measuring systems that allow the control of production processes by on line and in situ measurements in highly concentrated dispersions. It should be noted that the method is known by several names; thus it is sometimes referred to as quasi elastic light scattering (QELS) and DLS, standing for dynamic light scattering. Commercial equipment based on PCS is available through several companies.[63]

Another instrument that used Doppler shifts to investigate the size of airborne fineparticles is known as the E-SPART analyzer. This instrument can measure aerosol sizes in the range 0.3 to 70 microns aerodynamic diameter. This instrument was developed by Mazumder and co-workers.[65-67] The E-SPART analyzer is an acronym for the term the Electrical Single Particle Aerodynamic Relaxation Time analyzer. This instrument is used not only to measure size but also to measure the electrostatic charge of aerosol fineparticles, a parameter of importance when predicting the behavior of electrostatic copying machines and therapeutic aerosol sprays used in the pharmaceutical industry. The basic system of the instrument is shown in Figure 1.20. A knowledge of the electrostatic charge distribution on aerosol fineparticles is also of interest when studying the efficiency of crop dusting with pesticides and the electrostatic coating of objects in industry such as the automotive industry.[68] The instrument must be calibrated using particles of known particle size. Typical performance data for the E-SPART analyzer are shown in Figures 1.20b and 1.20c.

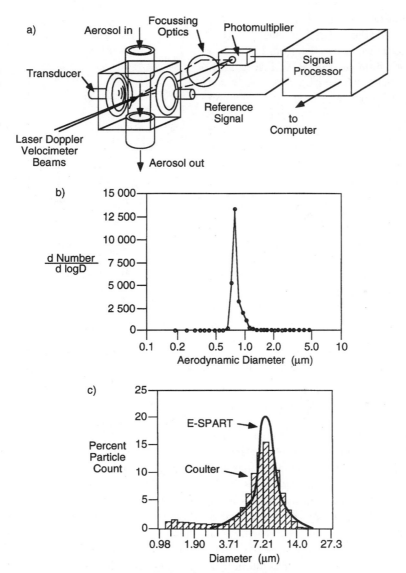

Figure 1.20. In the E-SPART analyzer, aerosols are oscillated in the inspection zone by acoustic waves. Different sized fineparticles are accelerated at different rates and laser Doppler velocimetry (LDV) is used to derive the sizes.[67,68] (*a*) The basic layout of the E-SPART analyzer. (*b*) Size distribution of 0.8 μm polystyrene latex (PSL) spheres used to calibrate the E-SPART analyzer. (*c*) Comparison of the size distributions obtained from a Coulter Multisizer and the E-SPART analyzer for dry ink powder, known as toner, used in laser printers and photocopiers.

1.9 STREAM COUNTERS

In a stream counter, as the name implies, the fineparticles to be characterized are passed in a stream through an interrogation zone where they change the physical properties of the interrogation zone.[69-73] (In instruments such as the AeroSizer the fineparticles being characterized pass through the interrogation zone in the stream but the size of the fineparticles is not monitored from the changes that they cause in the physical properties of the interrogation zone; rather, they are deduced from a measurement being made on the dynamics of the fineparticle in the zone.) Thus in the Coulter Counter the inspection zone is a cylindrical orifice between two electrodes placed in a conducting fluid as shown in Figure 1.21. The fineparticle, when it enters the zone, changes the electrical resistance of the column of electrolyte in the zone and the measured changes in the properties of the zone are used to deduce a size characteristic parameter of the fineparticle. In many discussions of the performance of the Coulter Counter it is claimed that the equipment measures the volume of the fineparticle directly. This is not so, as can be easily shown by anyone who attempts to do a mass balance on the measured size distribution and number count when looking at flakes of gold. Such fineparticles spin as they enter the zone, blocking off a larger volume of the electroyte rather than the actual volume of the flake of the material. Other steam counters such as the Climet, HIAC, and the Accusizer use optical signals to measure the size of the fineparticles in the interrogation zone. Sometimes when looking at opaque liquids sonic signals can be used to measure the size of the fineparticle.[1, 2, 69-71]

A major problem with all stream counting devices is the possibility of multiple occupancy of the interrogation zone. The fact that the number count of the small fineparticles is reduced by the loss of the identity of the two or more fineparticles in the interrogation zone is known as primary count loss. The ensemble of fineparticles in the interrogation zone is interpreted by the machine as a pseudo larger fineparticle. These false fineparticles are known as secondary count gain. Together the effects of multiple occupancy are known as coincidence effects. Some workers have attempted to deal with coincidence effects by using statistics of probability of multiple occupancy but this requires assumptions about the known size distribution being anticipated from the data, which is a somewhat dangerous way of measuring an unknown size distribution. The safe way of dealing with coincidence effects is to carry out a series of measurements at a sequence of dilutions until further dilu-

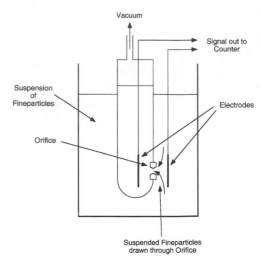

Figure 1.21. In stream counters, fineparticles to be characterized pass through an inspection zone and the size of the fineparticle is deduced from changes in the physical properties of the inspection volume. In the Coulter Counter the inspection zone is a cylindrical orifice between two electrodes. A fineparticle in the orifice changes the electrical conductivity of the column of electrolyte between the electrodes by an amount proportional to the size of the fineparticle.

tion of the suspension does not alter the measured size distribution. Some modern stream counters incorporate automated dilution technology. Many stream counters incorporate hydrodynamic focussing of the type discussed in the previous section. Some of them employ electronic editors to reject signals from fineparticles that do not travel down the center of the interrogation zone. It is sometimes claimed that the signals from stream counters can be interpreted from fundamental first principles but in practice many of them are calibrated using standard fineparticles such as latex spheres. The type of resolution available with a light inspected stream counter is illustrated by the data shown in Figure 1.22.[74]

1.10 ELUTRIATORS

The term *elutriator* comes from a Latin word meaning to wash out. In their earliest form elutriators were used to wash fine rock debris

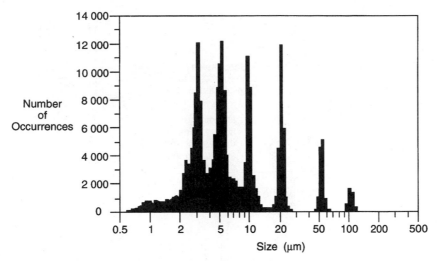

Figure 1.22. The resolution of particle size information obtainable with light inspected stream counters, also known as photozone stream counters, is illustrated by the data above, generated by the Particle Sizing Systems Accusizer® 770.[63, 74]

away from heavier gold grains that settle down to the bottom of a container of moving fluid. Today the term elutriator refers to any device in which powder fractionation is achieved by means of fluid movement with differential settling of the fineparticles. In the 1950s and 1960s elutriators were some of the first devices used to characterize fineparticles. In general they tend to have been displaced in modern technology by diffractometers and stream counters. However, they are still the basic devices for fractionating powders into different groups. For example, the preparation of various grades of fine diamond polishing powder is still achieved using liquid elutriation. (See Blythe elutriator in Ref. 1.) Various configurations of elutriator have been devised as illustrated by the systems shown in Figure 1.23. In the basic gravity elutriator, the fineparticles to be fractionated are placed in a cylinder through which a liquid is moved. The cut size of the elutriator is the size of fineparticle that cannot settle down the column but must move out of the system with the fluid as it exits from the elutriator chamber. Complex flow takes place in the elutriator body. (Not only is there a parabolic flow front because of the cylindrical structure of the chamber, but the settling fineparticles interfere with the ris-

ing stream of liquid and the entrained smaller fineparticles.) It is difficult to predict the cut size of an elutriator and the fractionation is never clean cut. (See chapter on elutriators in Ref. 1.) To accelerate the fractionation process, cyclones are often used. In a cyclone, as illustrated in Figure 1.23b, the fluid stream of fineparticles entering the main body tangentially is made to spiral around the fractionation chamber. Under the influence of centrifugal force the larger fineparticles are thrown to the walls of the vessel and slide down into the bottom of the chamber. A tube placed axially into the fractionation chamber accepts the returning fluid which is made to spiral out (vortex finder of Figure 1.23b). The fluid dynamics of cyclones is a complex subject and usually the exact cut size of the cyclone has to be established empirically. Part of the problem in predicting the performance of the cyclone is that the concentration of fineparticles in the entering fluid stream can effect the performance of the device. Personnel cyclones are widely used to fractionate industrial dusts into respirable and nonrespirable hazards (see Chapter 20). Another type of elutriator that is widely used to study aerosols and to sample aerosol systems is the impactor shown in Figure 1.23c. An air stream containing suspended fineparti-

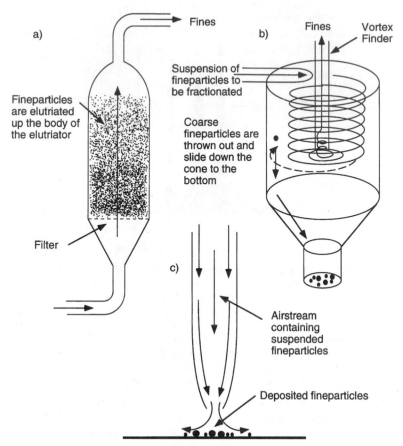

Figure 1.23. Elutriators are devices in which fineparticles are fractionated by a moving fluid. (*a*) A gravity elutriator. (*b*) A centrifugal elutriator, also known as a cyclone. (*c*) In an impactor fineparticles are centrifugally deposited on a slide as the jet of air is forced to turn by the slide.

cles is deflected by a glass slide or other collection surface. The turning fluid acts as a centrifugal system, throwing a certain size out onto the surface of the slide. The deposition of the fineparticles is controlled by the speed of the air jet and the distance from the deflecting surface.[75]

1.11 PERMEABILITY METHODS FOR CHARACTERIZING FINEPARTICLE SYSTEMS

Thus far in our discussion we have been dealing with direct methods of fineness assessment. In this section we will study what are known as permeability methods that are indirect techniques for studying the fineness of a powder. The basic concept of a permeability method for fineness assessment is that the resistance to flow offered by a compact of the powder can be used to characterize the fineness of the powder.[76] The permeability methods for assessing fineness of substances such as pyrotechnical powders, pharmaceutical powders, and cement powders was widely used for 50 years and is still a major technique in the cement industry. The techniques have tended to fall into disuse in recent years because of the availability of instruments such as the diffractometer and the time-of-flight instruments; however, they still have a role to play in quality control of the heavy industries particularly since the sampling problems associated with permeability measurements are much less severe than those for diffractometers. Thus a sample of cement for assessment with a permeability instrument can be as large as 200 g of powder as compared to the milligrams of powder used in diffractometers. It is much simpler to obtain the 200 g as a representative sample than it is to go all the way down to a few milligrams. The inherent cost of permeameters is also low. Studies aimed at the optimization of permeability equipment for quality control situations would seem to offer the potential for a renewed interest in the permeability methods. One of the work horses of industrial powder fineness measurement over the last 50 years was an instrument known as the Fisher Subsieve Sizer. The basic instrumentation of this device is shown in Figure 1.24a. The components represent 50 year old technology and newer pneumatic control devices can be used in modern pneumatic circuits.[77, 78]

It can be seen that the basic pneumatic circuit of the Fisher Subsieve Sizer is that of measuring an unknown resistance with a potentiometer and a standard resistance. The two taps, A and B, of the diagram represent different calibrated orifices for use in the comparative circuitry.

In Figure 1.24b the basic pneumatic circuitry of the Blaine fineness tester is shown. This equipment is widely used in the cement industry to assess the fineness of cement. Because of its widespread use, the fineness of cement is often referred to as its Blaine number, which is an arbitrary number derived from the performance of the equipment. The cement to be calibrated is placed as a powder plug at the top of a U-Tube Manometer. A driving pressure is established by closing tap A and opening tap B. The time required for the air to flow through the plug of cement as the manometer decays from height H_I to H_F is measured. This instrument is calibrated with cements of known fineness. It is necessary to have a strict protocol for assembling the powder plug; otherwise different operators using different pressures and assembly technique can change the apparent fineness of the powder.[79, 80]

In an alternative design of permeameter circuit suggested by Kaye and Legault, the powder plug to be used in the studies is compressed using hydrostatic pressure. The use of the hydrostatic compression technique makes it possible to automate the loading and emptying of the permeability cell. The use of the equivalent of a wheatstone bridge circuit permits automated calibration of a feedback controlled instrument. Work is underway to completely automate permeability control of cement circuits using this design. Relating the measured fineness of a cement or other powder to the constitution of the powder plug is not a simple matter because the resistance to flow of the powder plug is related not only to the fineness of the powder but also to the pore structure of the compact. This fact is illustrated by the data in Figure 1.25. If the measurement made on the permeability was directly related to the surface area, then when one mixed two powders of comparable fineness the relationship between the measured fineness of the mixture and the constitution of the mixture would be a linear relationship of the type shown for the mixture of aluminum powder and molybdedum oxide powder shown in Figure 1.25a. However, when one attempts to use the same interpretation for a mixture of an aluminum powder and a vanadium pentoxide powder there is a more complex curve which indicates that initially the vanadium pentoxide filled the interstitial spaces of the powdered aluminum, increasing its resistance to air flow which was interpreted externally as an increase in fineness as shown in Figure 1.25b. The opposite effect occurred when the aluminum powder was mixed with copper oxide powders as shown in Figure 1.25c. These curves indicate, however, that although in some cases one could not follow the constitution of a mixture from the measured permeability, the permeability richness curves indicate whether interpacking or interference with the structure

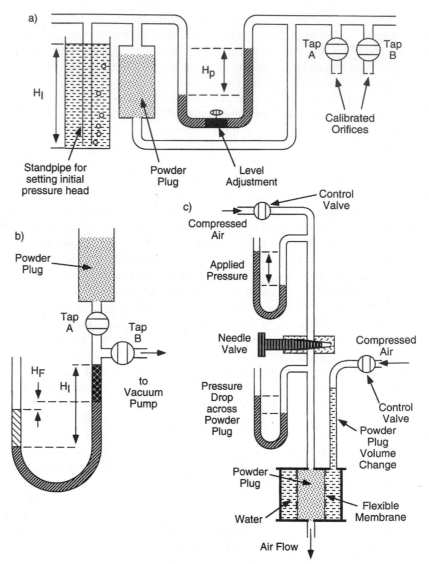

Figure 1.24. In permeability methods for characterizing fineness of a powder, the resistance of a powder plug to air flow is related to the fineness of the powder.[76] (*a*) Schematic of the Fisher Sub-Sieve Sizer. (*b*) Schematic of the Blaine Fineness Tester. (*c*) Schematic of the improved Kaye flexible wall permeameter.

of the powder plug is occurring. See discussion of these curves in Ref.[76]

1.12 SURFACE AREA BY GAS ADSORPTION STUDIES

The way in which Blaine fineness can be related to data from other size characterization techniques is illustrated by the information summarized in Figure 1.26. The surface area of a powder can be measured directly by means of gas adsorption studies. In these techniques the amount of gas or other molecular items, such as dye molecules, adsorbed onto the powder to form a monolayer is studied.[82] (See study of gas adsorption in Ref. 2.) In earlier discussions of gas adsorption (before 1977) it was stated that one of the problems with gas adsorption studies was the uncertainty in the

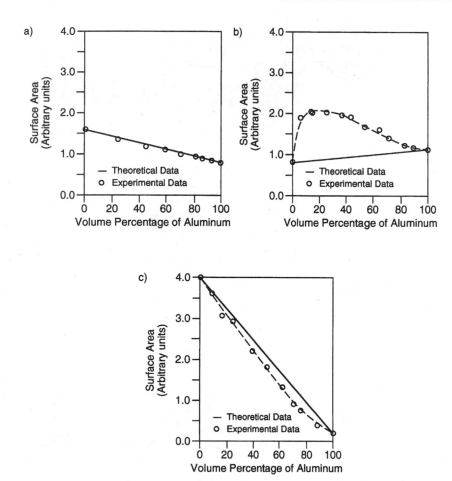

Figure 1.25. The permeability of a powder compact is related to not only the fineness of the powder, but also to the pore structure of the compact, as demonstrated by experimental data for several powders.[76] (*a*) Permeameter surface area data for a mixture of molybdenum oxide and aluminum. (*b*) Permeameter surface area data for a mixture of vanadium pentoxide and aluminum. (*c*) Permeameter surface area data for a mixture of copper oxide and aluminum.

knowledge of the cross-section area of the adsorbed molecule, which made estimates of the surface vary from gas to gas used in the adsorption studies. In recent years gas adsorption studies of surface areas are being reinterpreted from the viewpoint of fractal geometry. It has been shown that the surface area measured using a given gas depends on the accessibility of the rough surface to the particular molecule being used, as illustrated by the sketch in Figure 1.26a. Avnir and co-workers have shown that if you plot a graph of the surface area against the molecular size of the adsorbent gas or the molecule, one can draw a

straight line through the data to obtain a fractal dimension descriptive of the rough surface as shown in Figure 1.26b.[81] Neimark recently described a method for calculating the surface area and surface roughness of a powder by studying capillary condensation of a liquid on a powder.[83]

1.13 PORE SIZE DISTRIBUTION OF A PACKED POWDER BED

Sometimes when a powder has been made into a compressed structure, or other packed pow-

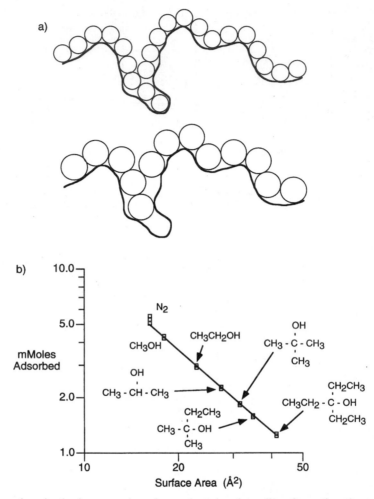

Figure 1.26. Innovations in the interpretation of gas adsorption data allow the surface fractal dimension of a substance to be deduced. (*a*) In gas adsorption the surface area is estimated from the number of a particular molecule required to cover the surface. This estimate is dependent on the size of the molecule used, smaller molecules can fit into smaller crevices. (*b*) The surface fractal dimension can be determined by using the gas adsorption technique with a series of different gases.

der bed, the pore size distribution of the interstitial spaces is of interest. These aspects of fineparticle systems can be studied by techniques known as mercury intrusion. Thus in Figure 1.27 data from the study of a powder using the mercury intrusion technique to study the structure of a bed of powder the grains of which were porous, as reported by Orr, is shown.[84] The amount of mercury entering a bed at different pressures is used to generate the data. One needs to know the contact angle of mercury with the material of the powder so that the applied pressure can be interpreted in terms of the capillary tube through which the mercury will move at that pressure. There has always been controversy as to the physical significance of mercury intrusion data because it obviously only measures access for diameter. The size of the pore behind the neck may be very different from that of the entrance capillary as shown in Figure 1.27b. (Note that interpreting mercury intrusion data in terms of

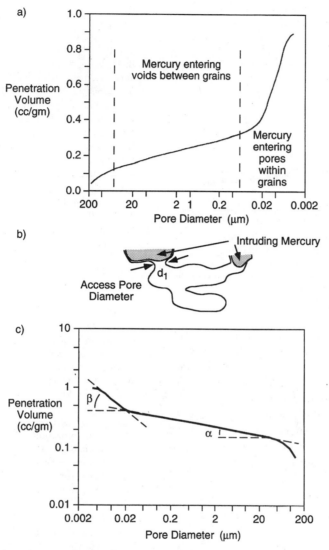

Figure 1.27. Mercury intrusion porosimetry data can be reinterpreted to obtain a pore fractal dimension of a system. (a) Mercury intrusion porosimetry data as it is traditionally presented.[84] (b) An illustration of how a "bottleneck" might cause pore volume to be underestimated. (c) A possible reinterpretation of the data of (a) leading to fractal dimensional information on the pore structure.[85]

pore diameters is often referred to as an ink bottle interpretive model.) For the data in Figure 1.27a mercury is entering the void between the grains of the powders at low pressure. When the pressure reaches approximately 2000 lb/in.^{2} the mercury starts to intrude into pores within the powder grains having accessed diameters of the order of 0.1 microns. It has recently been shown that one

can reinterpret mercury intrusion data to generate what is known as a fractal subdimension in data space. Thus in Figure 1.27c the data of Figure 1.27a have been reinterpreted in a manner that enables one to calculate the fractal dimensions description of the between-grain space and the within-grain space.[85]

In Figure 1.28 a new technique for studying the pore structure of items such as porous

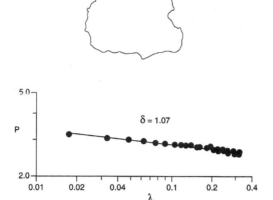

Figure 1.28. The boundary of a spreading drop on colored fluid placed on a pharmaceutical tablet is fractally structured and may be used as a quick quality control test for the pore structure of tablets.[85]

rocks and compressed powder plugs is shown. In this technique a drop of suitable colored fluid is placed on the porous body. As the fluid moves out through the pore structure, a fractal boundary is created that is related to the pore structure of the porous body. The data for Figure 1.28 was generated by observing the behavior of the fluid drop on a tablet of commercially available acetophenomen.[85]

REFERENCES

1. B. H. Kaye, *Direct Characterization of Fineparticles*, John Wiley & Sons, New York (1981). An updated book entitled *Characterizing Powders, Mists and Fineparticles Systems* is in preparation to be published by VCH 1997.
2. T. Allen, *Particle Size Analysis*, 4th ed. Chapman and Hall, London (1992).
3. Duke Scientific Corporation, 1135 D. San Antonia Road, Palo Alto, CA 94303.
4. Rhône-Poulenc, Paris Lab—Defencee, Cedex 29-F-92097, Paris, France.
5. Dyno Particles A. S., P.O. Box 160, N-2001 Lillestrom, Norway.
6. Interfacial Dynamics Corporation, 4814 N.E., 107th Avenue, Sweet Bee, Portland, OR 97220.
7. Information on the reference powders known by the initials BCR can be obtained from the Community Bureau of Reference, Commission of the European Community, Directorate General for Science Research and Development, 200 Rue De la Loi, B-1049, Brussels.
8. B. H. Kaye, "Efficient Sample Reduction of Powders by Means of a Riffler Sample." *Soc. Chem. Ind. Monographs 18*:159–163 (1964).
9. British Standards Methods for the Determination of Particle Size Powders, Part 1, "Subdivision of Gross Sample Down to 0.2 ml.," Part 1 (1961).
10. B. H. Kaye, "An Investigation Into the Relative Efficiency of Different Sampling Procedures." *Powder Metal. 9*:213–234 (1962).
11. Sampling equipment literature is available from Gustafson, 6340 LBJ Freeway, Suite 180, Dallas, TX 75240.
12. Sampling equipment literature is available from Gilson Screen Company, P.O. Box 99, Malinta, OH 43535.
13. Information on the spinning riffler system is available from Microscal Ltd., 20 Mattock Lane, Ealing, London, W5 5BH.
14. B. H. Kaye, Sampling and characterization research: Developing two tools for powder testing. *Powder Bulk Eng. 10*(2):44–56 (1996).
15. The freefall tumbler powder mixer is known as the AeroKaye® and is available commercially from Amherst Instruments, Amherst Process Instruments Inc., Mountain Farms Mall, Hadley, MA 01035.
16. Nuclepore® is the registered trademark of the Costar Corporation, 7035 Commerce Circle, Pleasanton, CA 94566. Comprehensive literature on the structure and properties of Nuclepore filters is available from the manufacturer who kindly provided the photograph reproduced in Figure 1.4.
17. The Poretics Corporation, 151 Lindbergh Avenue, Livermore, CA.
18. Collimated Holes Incorporated, 460 Division St., Campbell, CA 95008.
19. Isolock samplers are available from Bristol Engineering Company, 204 South Bridge St., Box 696, Yorkville, IL 60560.
20. G. D. Parfitt, *Dispersion of Powders in Liquids*, 2nd ed. John Wiley & Sons, New York (1973).
21. B. H. Kaye, *A Randomwalk Through Fractal Dimensions*, VCH, Weinheim, Germany (1989).
22. B. H. Kaye, *Chaos and Complexity: Discovering the Surprising Patterns of Science and Technology*, VCH Publishers, Weinheim, Germany (1993).
23. See Chapters 23 and 24 in *Particle Size Distribution II: Assessment and Characterization*, edited by T. Provder, ACS Symposium Series 472. Published by the American Chemical Society, Boston (1991).
24. B. H. Kaye and G. G. Clark, "Monte Carlo Studies of the Effect of Spatial Coincidence Errors on the

Accuracy of the Size Characterization of Respirable Dust." *Part. Syst. Charact.* 9:83–93 (1992).

25. B. H. Kaye, "Operational Protocols for Efficient Characterization of Arrays of Deposited Fineparticles by Robotic Image Analysis Systems." Chapter 23 in *Particle Size Distribution II: Assessment and Characterization*, edited by T. Provder, ACS Symposium Series 472. Published by the American Chemical Society, Boston (1991).

26. M. A. K. Yousufzai, "A Study of the Physical Parameters Affecting the Efficiency of Sieve Fractionation of Powders." M.Sc. Thesis, Laurentian University, Sudbury, Ontario (1984).

27. B. H. Kaye, M. A. K. Yousufzai, "How to Callibrate a Wire-Woven Sieve." *Powder Bulk Eng.* 6(2):29–34 (1992).

28. T. Allen, "Sieve Calibration Using Tacky Dots." *Powder Technol.* 79:61–68 (1994).

29. British Standards Institute Publication Nos. 410 (1943) and 1796 (1952).

30. Electroformed sieves are available from Buckbee-Mears Company, 245 East Sixth Street, St. Paul, MN 55101.

31. Photoetched sieves are available from Vecto-Stork International, 4925 Silabert Avenue, Charlotte, NC 28205.

32. W. S. Tyler Company, 8570 Tylor Blvd., Mentor OH 44060.

33. Cenco Instrument Corporation, 1700 Irving Park Road, Chicago, IL 60613.

34. Alpine American Corporation, 3 Michigan Drive, Natick, MA 01760.

35. Sonic Sifter available from ATM Corporation, 645 S. 94th Place, Milwaukee, WI 53214.

36. See B. H. Kaye, *Powder Mixing*, to be published by Chapman and Hall, 1996.

37. Information on the ISO Standards for particle size analysis is available from the International Organization for Standardization Secretariat, Building Division of DIN, Burggrafenstrasse 4-10.D-1000, Berlin 30. In Canada information is available from the Standards Council of Canada, 1200-45 O'Connor Street, Ottawa, Ontario, K1P 6N7. At the time of writing the ISO Council had issued draft standards for particle size by gas adsorption, photon correlation, diffraction methods, and for sedimentation procedures. (Other standards are in preparation.)

38. Micromeritics, 1 Micromeritics Drive, Norcross, GA 30093.

39. Horiba Instruments Incorporated, 17671 Armstrong Avenue, Irving, CA 92714.

40. See technical literature of Brookhaven Instruments Corporation, Brookhaven Corporate Park, 750 Bluepoint Road, Holtsville, NY 11742.

41. See H. G. Barth, *Modern Methods of Particle Size Analysis*, John Wiley & Sons, New York (1984).

42. Leeds & Northrup, A Division of MicroTrac, 351 Sumneytown Pike, North Wales, PA 19454.

43. CILAS U.S.A., agents Denver Autometrics Inc., 6235 Lookout Road, Bolder, CO 80301 Company headquarters in France Osi 47, Rue de Javel, 75015 Paris, France.

44. Malvern Instruments, Inc., 10 Southville Road, Southborough, MA 01772.

45. Coulter Corporation, Division of Scientific Instruments, P.O. Box 169015, Miami, FL 33116-9015.

46. Sympatec, Inc., Systems for Particle Technology Division, 3490 U.S., Route 1, Princeton, NJ 08540.

47. Shimadzu Scientific Instr., Inc., 7102 Riverwood Drive, Columbia, MD 21046.

48. Insitec, Inc., 2110 Omega Rd., San Ramon, CA 94583.

49. Nathier-Dufor et al., "Comparison of Sieving and Laser Diffraction for the Particle Size Measurements of Raw Materials Used in Food Stuffs." *Powder Technol.* 76:191–200 (1993).

50. J. W. Novak, Jr. and J. R. Thompson, "Extending the Use of Particle Size Instrumentation to Calculate Particle Shape Factors." *Powder Technol.* 45:159–167 (1986).

51. G. Baudet, M. Bizi, and J. P. Rona, "Estimation of the Average Aspect Ratio of Lamellae-Shaped Particles by Laser Diffractometry." *Part. Sci. Technol.* 11:73–96 (1993).

52. The data for the diffraction patterns given in this diagram are taken from a review of diffraction properties of fineparticles given in the book *A Randomwalk Through Fractal Dimensions* by B. H. Kaye, 2nd ed. VCH, Weinheim, Germany. pp. 84–89 (1994).

53. The data for Figure 1.13 are taken from the trade literature of Leeds and Northrup; see Ref. 42.

54. Brian H. Kaye, *Chaos & Complexity: Discovering the Surprising Patterns of Science and Technology*. VCH Publishers, Weinheim, Germany (1993).

55. Galai Instruments Inc., 577 Main Street, Islip, NY 11751.

56. Lasentec., 15224 NE 95th Street, Redmond, WA 98052.

57. Amherst Process Instruments, Mountain Farms Technology Park, Hadley, MA 01035.

58. TSI Incorporated, 500 Cardigan Road, P.O. Box 64394, St. Paul, MN 55164-9877.

59. Dantec Measurement Technology A-S, Tomsbakken, 16-18, Dk-740, Skovlunde, Denmark. In the United States and Canada, Dantec Measurement Technology Incorporated, Mahwah, NJ 07430.

60. For a special issue of the journal "Particle and Particle Systems Characterization" devoted to the use of Doppler shifts to measure aerosol fineparticles, see Volume 11, No. 1, February 1994. *Particle*

and Particle Systems is published by VCH, Weinheim, Germany.

61. M. Gautam, K. Car, H. Yang, K. Clifton, and J. G. Jurewicz, "LDV Measurements in Gas Solid Flows, A review." *Part. Sci. Technol. 11*:57–71 (1993).

62. For a concise introduction to the theory of photon correlation spectroscopy see *Modern Methods of Particle Size Analysis* edited by Howard G. Bart, John Wiley & Sons, Chapter 3 (1984).

63. Commercially available photon correlation equipment is available from Brookhaven Laboratories, Particle Sizing Systems, 75 Aero Camino, Santa Barbara, CA 93117.

64. R. Finsy, "Particle Sizing in the Sub Micron Range by Dynamic Light Scattering," *KONA*, No. 11, pp. 17–32 (1993). *KONA* is published by the Hosokawa Micron Corporation.

65. M. K. Mazumder, R. E. Ware, T. Yokayama, B. Rubin, and D. Kamp, "Measurement of Particle Size and Electrostatic Charge of a Single Particle Basis in the Measurement of Suspended Particles by Quasi Elastic Light Scattering." John Wiley & Sons, New York, pp. 328–341 (1982).

66. M. K. Mazumder, R. E. Ware, J. D. Wilson, R. G. Renninger, F. C. Hiller, P. C. McLeod, R. W. Raible, and M. K. Testamen, "E-SPART Analysers: Its Application to Aerodynamic Size Distribution." *J. Aerosol Sci. 10*:561–569 (1979).

67. M. K. Mazumder, "E-SPART Analyser: Its Performance and Applications to Powder and Particle Technology Processes." *KONA*, 11, pp. 105–118 (1993).

68. The E-SPART analyzer is available commercially from the Hosokawa Micron Corporation, 10 Chatham Road, Summit, NJ 07901.

69. An optical stream counter is available from the Climet Corporation, 1320 Colton Avenue, Redlands, CA 92373.

70. Royco Instruments for studying aerosols and fineparticles in liquids are available from Royco Instruments Inc., 141 Jefferson Drive, Menlo Park, CA 94025. Also, a widely used stream counter for fineparticles in fluid is the HIAC counter, HIAC Instruments, Division P.O. Box 3007, 4719 West Brooke St. Monte Claire, CA 91763.

71. Particle Measuring Systems Incorporated, 1855 South 57th Court, Boulder, CO 80301.

72. Information on the Electrozone Counter is available from Particle Data Inc., P.O. Box 265, Elmhurst, IL 60126.

73. See trade literature of Coulter Counter Electronics, 5990 West, 20th St. Hialeah, FL 33010.

74. Particle Sizing Systems, 75 Aero Camino, Santa Barbara, CA 93117.

75. See discussion of impactors in Ref. 1.

76. B. H. Kaye, "Permeability Techniques for Characterizing Fine Powders." *Powder Technol. 1*:11–21 (1967).

77. B. H. Kaye and P. E. Legault, "Real-Time Permeametry for the Monitoring of Fineparticle Systems." *Powder Technol. 23*:179–186 (1979).

78. A. D. Hoffman, "A Soft-Wall Permeameter for Online Characterization of Grinding Circuits." M.Sc. Thesis, Laurentian University, Sudbury, Ontario (1989).

79. R. L. Blaine, "A Simplified Air Permeability Fineness Apparatus." *ASTM* No. 123, 51 (1943).

80. S. Ober and K. S. Frederick, "A Study of the Blaine Fineness Tester and a Determinator of Surface Area from Air Permeability Data." *Symposium on Particle Size Measurement ASTM, Spec. Tech. Pub.*, No. 234, p. 279 (1958).

81. Dr. D. Avnir of the Hebrew University pioneered the fractal reinterpretation of gas adsorption studies. The research scientist will find an excellent review of this work in *The Fractal Approach to Heterogeneous Chemistry*, edited by D. Avnir, John Wiley & Sons, Chicester (1989).

82. Gas adsorption equipment for measuring surface areas is available from Micromeritics Instrument Corporation, 800 Goshen Springs Road, Norcross, GA 30071.

83. A. V. Neimark, "Calculating Surface Fractal Dimensions of Adsorbers." *Adsorpt. Sci. Technol. 7*(4):210–219 (1990).

84. C. Orr. "Application of Mercury Penetration in Material Analysis." *Powder Technol. 3*:117–123 (1969–1970).

85. See Chapters 7 and 8 in Ref. 21.

2
Particle Shape Characterization

Brian H. Kaye

CONTENTS

2.1 INTRODUCTION

In powder technology the size characterization problems range from determining the shape of rock fragments measuring several centimeters to the problems of characterizing the shape and structure of items such as diesel soot and industrially important pigments. In the first section of this chapter we look at the techniques for characterizing the shape and size of rock fragments that are large enough for measurements to be made directly on the items of interest. In a later section we discuss how computer-aided image analysis is being used to look at two-dimensional images of very small fineparticles. As the sophistication of computer-aided image analysis increases and the cost of processing falls many of the techniques currently limited to two-dimensional characterization will be extended to three dimensions. In the third section we look at how dynamic behavior is being used to characterize the shape of tumbling fineparticles.

2.2 DIMENSIONLESS INDICES OF FINEPARTICLE SHAPE

Shape characterization of particles large enough to be handled directly was pioneered by Heywood[1-3] and Hausner.[4] Heywood suggested using several dimensionless factors based on the measurements of length, width, and thickness of the fineparticles as defined by the sketches in Figure 2.1c. The maximum length of a profile is intuitively obvious. Once this has been fixed the width is defined as the maximum dimension at right angles to the measurement of length and the thickness is the dimension perpendicular to the plane defined by the width and the length. On the basis of these dimensions Heywood suggested using the elongation ratio defined as the length divided by the width. He also defined the flakiness as the width divided by the thickness. Another name widely used for the elongation ratio is the aspect ratio. Note that in occupational hygiene fineparticles having an aspect ratio greater than 3 are defined as being fibers. The reciprocal of the aspect ratio is known as the chunkiness factor. For graphical display of

data the chunkiness factor has the advantage that its total range of values is 0 to 1. This results in a compact display of the relevant data.[5] In Figure 2.2 the range of aspect ratio sizes in a population of nickel slag fragments similar to those of Figure 2.1 are shown.[6] It can be seen that the range of shape factors in the population of nickel slag is bimodal with two functions with Gaussian distribution. (The graph axes of Figure 2.2a are linear versus Gaussian probability.) From the graph of the chunkiness versus size domain of Figure 2.2b, there appears to be some correlation of chunkiness with size in that the larger fragments on the whole are more compact than the smaller fragments.

When it is possible to measure three dimensions of a fragment one can display the shape

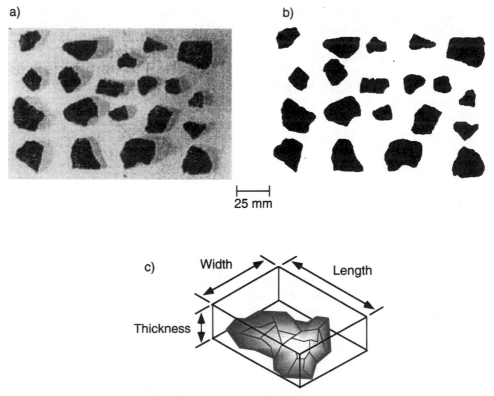

Figure 2.1. Images and silhouettes of 20 fragments of crushed nickel ore slag, such fragments may be characterized by their physical dimensions, length, width, and thickness. (*a*) Shadows cast by the slag fragments illuminated by directed light. (*b*) Silhouettes of slag fragments. (*c*) The geometric dimensions used to characterize the shape of a fragment.

information on triangular mesh graph paper as shown in Figure 2.3. The physical basis of the way in which the information on the length, width, and thickness is normalized and then plotted on the triangular mesh graph paper is illustrated in Figure 2.3a. In Figure 2.3b the triaxial display of information on the geometric shape of the fragments of Figure 2.1 is shown.[5-7] On the left-hand side the data obtained by physical measurement using calipers

are summarized whereas on the right-hand side the measurements made on the image of the profiles using the shadow cast by the illumination to retrieve the thickness of the profiles are shown. It can be seen that the two sets of data are basically the same, indicating that one can carry out triaxial measurements on very small fineparticles provided one is prepared to use shadow casting techniques. Thus one can obtain triaxial displays of shape

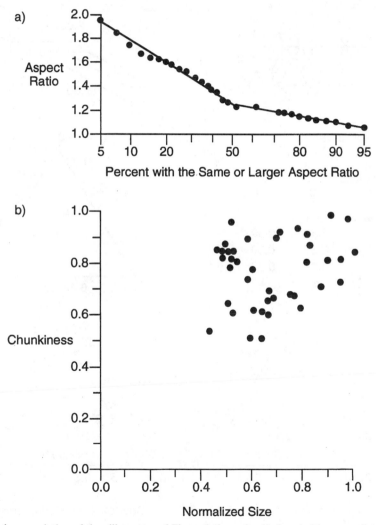

Figure 2.2. The shape variation of the silhouettes of Figure 2.1b can be displayed either as a distribution function or as a domain display.[6] (a) Distribution of the aspect ratio of the profiles of Figure 2.1b plotted on probability graph paper. (b) Domain plot of chunkiness (the reciprocal of aspect ratio) versus normalized size for the profiles of Figure 2.1b.

data for profiles that are imaged in a scanning electron microscope and are well below 1 μm.

Hausner has suggested that one way of generating dimensionless indices of shape is to construct a rectangle around the profile so that its area is at a minimum when considering other rectangles that could be drawn around the profile.[4,5] If A is the projected area of a profile and C is the actual perimeter of the profile and a and b are the lengths of the

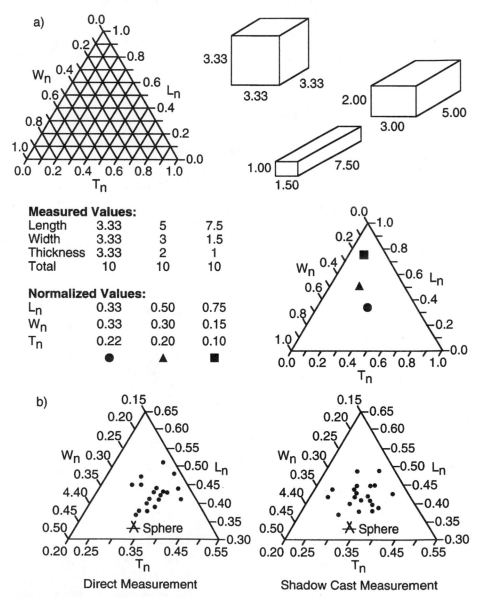

Figure 2.3. A useful way to summarize the geometric shape information for a population of fineparticles is to use triaxial graph paper.[6,7] (*a*) Illustration of how geometric dimensions are normalized and plotted on triaxial graph paper. (*b*) Shape information summaries for the slag fragments of Figure 2.1 plotted on triaxial graph paper. Note that the range of the axis is a small portion of the axis in (*a*).

sides of the minimum area embracing rectangle, then Hausner suggested the following shape descriptions:

$$\text{Elongation factor} = \frac{a}{b}$$

$$\text{Bulkiness} = \frac{A}{a \times b}$$

$$\text{Surface factor} = \frac{C^2}{12.6A}$$

These measures are known as Hausner shape indices. Medalia has also defined other measures of shape.[8]

2.3 GEOMETRIC SIGNATURE WAVEFORMS FOR CHARACTERIZING THE SHAPE OF IRREGULAR PROFILES

To be able to store information on the shape of a fineparticle profile in a computer, and to use such data for deducing characteristic shape description parameters, the basic problem is to reduce the dimensionality of the information available on the shape of the profile. One way to take a two-dimensional profile and reduce its information content to a nondimensional list is to generate what has become known as a geometric signature waveform. Various authors have suggested different reference points for carrying out the transformation of a boundary profile into a signature waveform but we shall restrict our discussion here to the use of the centroid of the profile, the profile is to be a laminar shape with mass.[9-12] The procedure for generating the geometric signature waveform using this reference point is illustrated in Figure 2.4. One plots the magnitude of the vector R against the angle θ with respect to some reference direction. It is convenient to normalize the values of this vector with respect to the largest value of R generated during the exploration of the perimeter of the profile. For many purposes it is useful to start the plotted waveform with $\theta = 0$ for the maximum value of R. This adjustment of the starting point can often be carried out

within a computer memory rather than at the beginning of the geometric exploration of the profile. The information generated by carrying out the geometric signature waveform can be stored as a list in a computer or alternatively the waveform can be subjected to Fourier analysis to generate the power spectrum of the waveform. Thus, in Figure 2.5 the geometric signature waveform of a profile generated by Flook is shown and the power spectrum of the wave is as illustrated.[11,12] Basically it has been established that the first five harmonics of the power spectrum specify the structure of the profile whereas the higher harmonics are linked to the texture of the profile. Some workers have specified shape descriptors based on the first five harmonics and then the harmonics above five.

The geometric signature waveform is quite useful for profiles of the type shown in Figure 2.4 but if there are any convolutions and fissures in the profile the physical significance of the vector crossing the fissure or the convolution becomes very difficult to interpret and the geometric signature waveforms are not normally used for such complex profiles. Again some workers have started to investigate the direct use of two-dimensional Fourier transforms for such profiles but the technique has not been widely used.[6] The technique has found applications in describing the change of the profile to be found among the grains of gravels, beach sands, and other systems of interest to the geologist.[10] However, using the geometric signature waveform on profiles such as those of Figure 2.1b would cause problems at the sharp edges, because if one attempts to do a Fourier transform of the waveform of a profile with a sharp edge one obtains a very large number of components in the power spectrum of the profile.[6]

When one attempts to characterize the shape of sharply angled fractured material such as that in Figure 2.1 one can use another technique to measure the edges on the profile and generate a three-dimensional graph using parameters such as size, chunkiness, and the number of edges on the profile. The basic

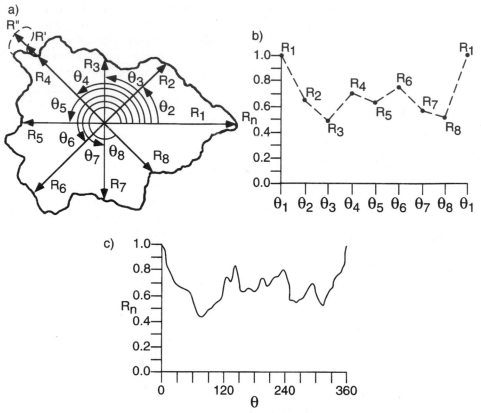

Figure 2.4. Generating the signature waveform of a fineparticle profile reduces the dimensionality of the information describing the profile to a list of information that can be processed in a computer.[12, 13] (a) Method for generating signature waveform data from a fineparticle profile. (b) A low resolution waveform generated from the data shown in (a). (c) A high-resolution waveform generated from the profile of (a). θ_n, Angle at which the nth radius vector is generated; R_n, normalized magnitude of the nth radius vector at angle θ_n.

steps used to generate information from which an edge count can be retrieved from computer-aided image analysis are illustrated in Figure 2.6. First the profile to be characterized is digitized as shown by the digitization of the rectangular profile in Figure 2.6a. One then starts to draw chords at a given number of steps around the profile as illustrated in the diagram for chords based on 10 digitized steps. A whole set of chords is generated by a process known as slip chording. Thus chord 1 is from 1 to 11, chord 2 is from 2 to 12, chord 3 is from 3 to 13, as illustrated in Figure 2.6a. The lengths of these chords are at a maximum

when they lie along a straight portion of the profile and a minimum in the series of chord lengths indicates that the slip chords have spanned a corner on the profile. Continuing the process around the profile of Figure 2.6a, what is known as the facet signature of the profile is generated. This signature is shown for the rectangular profile in Figure 2.6b. The dips in the facet signature indicate the presence of a corner. The distance between the dips generates information of the length of the edges of a profile as shown in the diagram. In Figure 2.6c the facet signatures for three profiles are shown. It can be seen that informa-

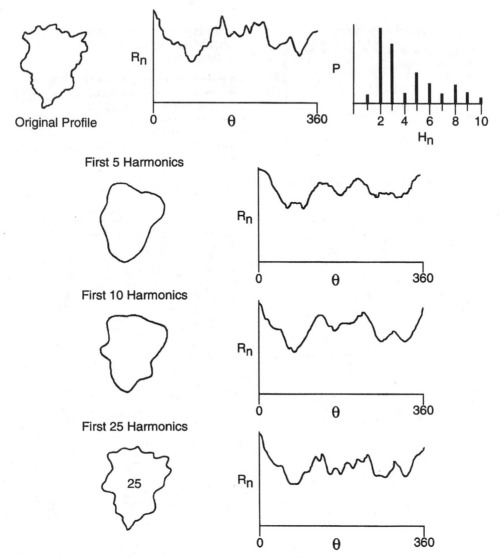

Figure 2.5. Data generated by Flook illustrates how Fourier analysis for a geometric signature waveform can be used to characterize the structure of the profile.[14] θ, angle at which the radius vector is generated; R_n, normalized magnitude of the nth radius vector at angle θ_n; H_n, harmonic number from Fourier analysis; P, relative strength of the stated harmonic.

tion in these waveforms can be used to re-trieve information on the number of edges and the sharpness of the edge (the sharper the corner the bigger the dip in the facet signa-ture). The information on the shape, size, and the number of edges of different sharpness for a set of profiles can be summarized in a three-dimensional data space graph of the type shown in Figure 2.7. The graph is generated by using normalized estimates of size and chunki-ness with a third axis that indicates the sharp-ness of any corners as a profile. To use this

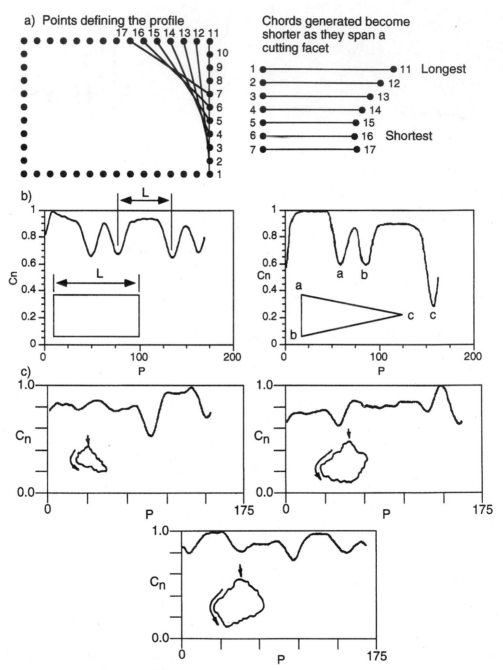

Figure 2.6. The number of sharp edges on a profile can be evaluated by generating what is known as the facet signature waveform by "slip chording" exploration of a digitized version of the profile. (*a*) An illustration of how the chord length varies for a given number of steps between ends as the chord "slips" around a sharp corner. (*b*) A demonstration of the facet signature waveform applied to two standard shapes. (*c*) The facet signature waveforms for three profiles of Figure 2.1. *P*, Position on the profile at which a particular chord was generated; C_n, normalized chord length generated at position *P*.

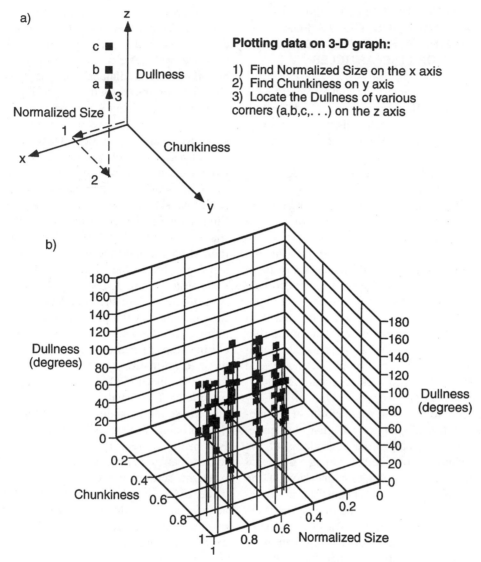

Figure 2.7. The information on the shape, size, and "edginess" of a set of profiles can be summarized in a three-dimensional data space.[6] (*a*) The method for plotting the "edginess" data for a profile on a three-dimensional graph. (*b*) Shape, size, and edge information for the profiles of Figure 2.1.

type of display, first a point on the "size-chunkiness" plane is located. Then at this point an ordinate is erected. Along this ordinate points representing the sharpness of the edges to be found on the profile are plotted. Thus for the profiles of Figure 2.1 the three-dimensional data space summary of the information on the number of edges and the

size and shape of the profiles is as shown in Figure 2.7b.[6] Another approach to characterizing the number of edges on a profile is to generate a two-dimensional Fourier transform of the profile. The sharp edges in the profile will generate high-frequency signals in the frequency domain of the two-dimensional Fourier transform.[6]

2.4 FRACTAL DIMENSIONS OF FINEPARTICLE BOUNDARIES FOR DESCRIBING STRUCTURE AND THE TEXTURE OF FINEPARTICLES

The basic theorems of fractal geometry were set out in a book by B. B. Mandelbrot in 1977.[13] Fractal geometry discusses the structure of rough surfaces and complicated space-filling boundaries. The basic theorems of fractal geometry have been applied to describing the structure and texture of fineparticles.[12,14] The basic concept of applied fractal geometry with reference to the description of the texture and structure of fineparticles can be appreciated from the systems presented in Figure 2.8a. The topological dimensions of these lines are all 1. Topology has been called rubber sheet geometry, the idea being that if you drew the lines of Figure 2.8a on a sheet of rubber one could stretch and deform the rubber so that everyone of the lines looked identical. Structures that can be pulled and twisted on rubber sheets to look identical are said to be topologically identical.[15] One of the key ideas put forward by Mandelbrot in his creation of fractal geometry is that the space-filling ability, that is the ruggedness of the lines, could be described by adding a fractional number to the topological dimension of the system. For a rugged fineparticle in three-dimensional space the fractal dimension of the object would be 2 plus the fractional number describing the ruggedness. Thus an extremely porous fumed silica fineparticle has a fractal dimension of nearly 2.99 describing its enormous surface area. One can measure the fractal dimension of objects in three-dimensional space using measurements such as light scattering and neutron scattering but a discussion of such measurement systems is beyond the scope of this introductory exploration of the use of fractal dimensions to describe rugged systems. Interested readers will find useful information in Refs. 16 and 17.

The term *fractal dimension* was used rather loosely in some of the early publications on applied fractal geometry and the type of frac-tal dimension that can be measured by light scattering and X-rays etc. should properly be called mass fractal dimensions. In this discussion we will restrict our considerations to boundary fractal dimensions. The fractal dimensions of the rugged lines of Figure 2.8a were measured using a technique that is variously known as the yardstick measure, or the structured walk, exploration of the rugged line. The physical principles employed in use of the structured walk exploration to evaluate the ruggedness of a boundary can be appreciated from the sketches presented in Figure 2.8b. First a set of dividers is opened so that they span a distance λ between the points. One then strides around the perimeter to generate a polygon fitting the profile with the sides of the polygon being λ. Usually a fractional step is required to close the boundary. To estimate the perimeter of the polygon one adds up the number of complete sides plus the fractional closing step. In many experiments it is convenient to normalize λ and the polygon perimeter estimate with respect to some convenient parameter. It is usual in applied fractal geometry to use the maximum projected length of a profile being studied as a normalizing factor. For historical reasons this projected magnitude is called Feret's diameter. One now adjusts the setting of the dividers to a smaller value of λ and the polygon construction to achieve an estimate of the perimeter of the profile is carried out again. Two stages of this exploration procedure are illustrated in the sketch.

The fractal addendum to the topological dimension to characterize the structure of a boundary can now be deduced by plotting the perimeter estimates versus λ on log–log graph paper. If the boundary is fractal in structure there will be a linear relationship displayed by the data on such a plot. This type of data plot for historic reasons is known as a Richardson plot.[18] When experiments were initially carried out to explore the possibility of describing the ruggedness of fineparticle boundaries by fractal dimensions there was some confusion in the literature because early data displays

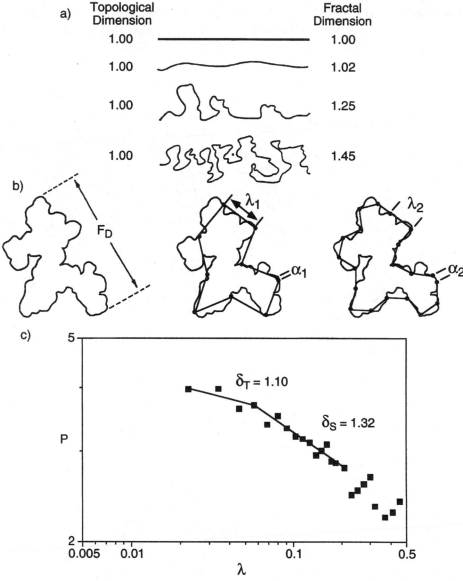

Figure 2.8. The boundary fractal dimension of a two-dimensional fineparticle profile can be deduced from data generated by a structured walk around the profile. (*a*) The fractal dimension of a line describes its ruggedness. (*b*) The structured walk exploration technique generates perimeter estimates of a boundary by constructing polygons of side length λ by walking a pair of dividers around a profile to be characterized. (*c*) The fractional addendum to the topological dimension of a profile is deduced from the slope of the best fit line on a Richardson plot of perimeter estimates against inspection resolution. λ, inspection resolution used to stride around the profile; P, perimeter estimate found with dividers set a resolution λ; F_D, Feret diameter used to normalize the values P and λ before plotting; δ_S, structural boundary fractal dimension; δ_T, textural boundary fractal dimension.

showed the presence of more than one data line instead of a single line. It is now appreciated that these two data lines represent different interaction of the multiplicity of causes creating a structure. Thus for a system such as that of Figure 2.8, which is a greatly magnified agglomerate of carbonblack spheres created by combustion in the absence of a sufficient supply of oxygen to create total combustion, the first data line of Figure 2.8c at coarse resolution inspection is known as the structural fractal dimension of the boundary. The

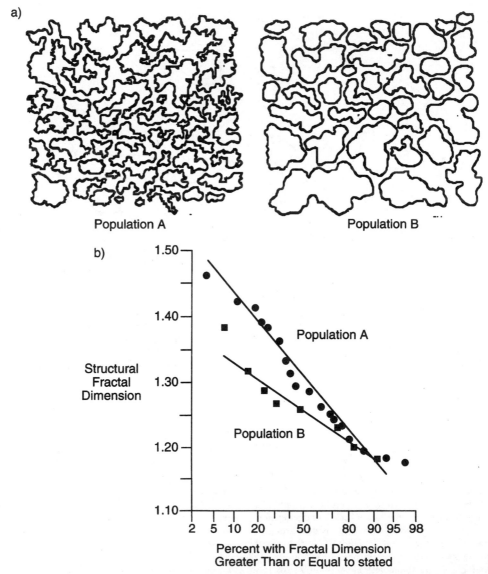

Figure 2.9. The fractal dimension contains information on the information dynamics of aerosol agglomeration.[20] (*a*) Tracing of profiles from electronmicrographs of two populations of carbonblack agglomerates produced by different methods. (*b*) The distribution function of structural fractal dimension of the profiles of (*a*) illustrated that Population A agglomerates were produced by agglomeration of primary agglomerates.

data line at high-resolution inspection (small quantities of λ) represents the packing texture of the profile and is known as the textural fractal dimension of the boundary. Agglomerates of fume produced fineparticles such as that of Figure 2.8 can exhibit more than two linear regions on a Richardson plot and in such cases the magnitude of the fractal dimensions can often be linked to the formation dynamics of the fume agglomerates.[19] Thus in Figure 2.9 two different populations of carbonblack fineparticles are shown from two different products. It can be seen by the plot of the structural fractal dimensions of the two populations that one group of fineparticles is more rugged than the others.[20] This indicates that the agglomeration process in the case of product A has been allowed to proceed to the point where agglomerates were forming from agglomerates whereas in product B the agglomerates appeared to be a simple first stage of turbulent agglomeration. In general the structured walk technique is not used in more recent strategies for computer-aided image

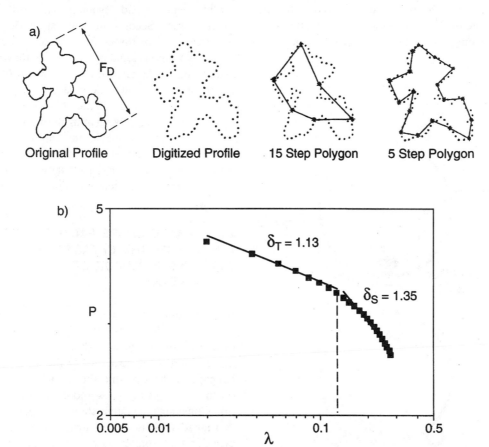

Figure 2.10. The equipaced technique for exploring the structure of a rugged boundary is convenient for use with computer-aided image analysis procedures.[23] (a) In the equipaced method for exploring the ruggedness of a profile, polygons are constructed on the digitized version of the profile by stepping out a certain number of paces along the boundary. (b) Richardson plot for data generated by the equipaced exploration of the profile of (a). λ, Inspection resolution used to stride around the profile; P, perimeter estimate found at resolution λ; F_D, Feret diameter used to normalize P and λ before plotting; δ_S, structural boundary fractal dimension, δ_T, textural boundary fractal dimension.

analysis. The method known by the name of equipaced exploration is the preferred method of computer-aided image analysis measurement of fractal dimensions of rugged lines. This method was originally developed by Schwarz and Exner.[21] In this technique the boundary to be evaluated is digitized as shown in Figure 2.10a. The polygon that becomes the perimeter estimate is now constructed by marking off a number of steps along the digitized profile as indicated in Figure 2.10a. Once the x, y coordinates of the digitized points have been stored in a computer memory the process of marking off a certain number of steps along the profile can be automated and the size of the polygon side spanning the paced out steps is calculated using the Pythagorean

theorem. Thus the Richardson plot of a series of polygons created in the computer memory can be automated to the point that once the digitized profile has been stored in the computer the entire process of calculating the data lines representing fractal dimensions can be carried out off line in the computer processing unit.

When carrying out an experimental study of the fractal dimensions of a line there are two major problems to be aware of. First, many real fineparticles have different ruggedness at different parts of their boundary. For such profiles one should split the profiles into different regions and evaluate the different ruggedness.[22] Second, in some situations the projection of a three-dimensional structure into a two-dimensional profile can result in the smoothing of the ruggedness of the profile by occlusion of the lower boundary by a projection in an upper part of the structure or vice versa. Thus when looking at agglomerates such as the carbonblack profile one can sometimes clearly see regions where there has been smoothing by occlusion, which should be taken into account as the measurements proceed.[23]

2.5 DYNAMIC SHAPE FACTORS FROM A STUDY OF THE CATASTROPHIC TUMBLING BEHAVIOR OF FINEPARTICLES

It has recently been suggested that the behavior of a tumbling rock can be used to define a dynamic shape factor that describes the three-dimensional structure of the rock.[24] The concept of the dynamic shape factor is based on the concepts of chaos and catastrophe theory. Consider the system shown in Figure 2.11. A rugged rock is placed inside a foam-lined cylinder that rotates slowly. The rock moves up the side of the cylinder until it reaches a position of instability, where it tumbles. If one studies successive tumblings of such catastrophic behavior (the term catastrophic is used here in the technical rather than in the popular sense, that is, it describes a discontinuous

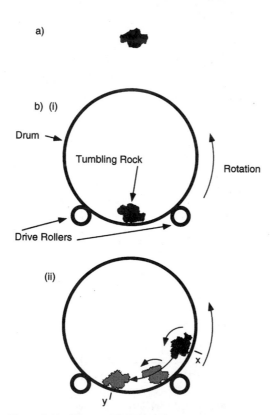

Figure 2.11. The tumbling of a rugged rock in a slowly rotating cylinder is a physical system that exhibits catastrophic behavior. (*a*) Profile of a rugged rock to be tumbled. (*b*) (i) Apparatus for tumbling experiments. (ii) One catastrophic tumbling event of a rock in the slowly rotating cylinder.

behavior when a system is subjected to smoothly evolving forces, not a colossal disaster). The subject of deterministic chaos deals with complex behavior that philosophically is governed by the laws of determinism so that its future state should be predictable from the theory of mechanics. However, the system is so dependent on the initial starting conditions and the complexity of the interactions that the outcome might as well be chaotic.[24] Unfortunately, in everyday speech the subject known as deterministic chaos has become known sim-

ply as chaos. This can be confusing to some because in Greek philosophy chaos means completely unorganized whereas in the mathematical subject now known as chaos the systems exhibit patterns of complex behavior that can be determined by empirical study. In the subject of deterministic chaos the tumbling behavior of a rock in a cylinder system such as that of Figure 2.11 is summarized graphically using a data display known as a discrete time map. The discrete time map is constructed from the time series of catastrophic events

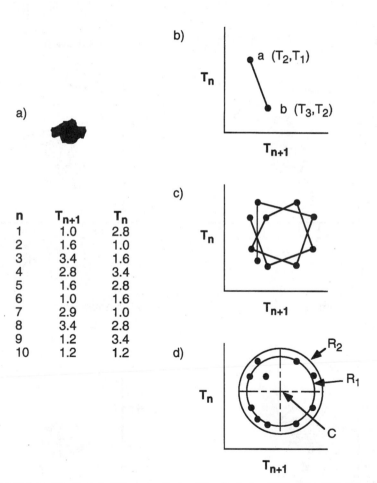

n	T_{n+1}	T_n
1	1.0	2.8
2	1.6	1.0
3	3.4	1.6
4	2.8	3.4
5	1.6	2.8
6	1.0	1.6
7	2.9	1.0
8	3.4	2.8
9	1.2	3.4
10	1.2	1.2

Figure 2.12. When studying the discrete time map of a tumbling rock, the data points delineate what is known as a strange attractor by clustering about a centroid that constitutes an equivalent oscillator. The centroid and the moments of the data points, when they are treated as unit masses, can be used to create a shape descriptor characterizing the behavior of the rock. (*a*) Time sequence for 10 tumbles of the rock. (*b*) First two data points plotted on the discrete time map. (*c*) All 10 data points define a rudimentary strange attractor. (*d*) Centroid, C, and first two moments, R_1 and R_2 for the data. (Note that this is an example only.)

(i.e., the tumbling of the rock at the height that it reaches in the rotating cylinder). In Figure 2.12 the way in which data from the time series of catastrophic tumbling are plotted as a discrete time map is shown. The two axes of this map are T_n and T_{n+1}. Points are plotted in the discrete time domain by using the pairs from the time series as illustrated. This plotting strategy generates a scattered set of points that for convenience are often shown joined in the same way that we show lines joining successive positions in Brownian mo-

tion. In the discrete time domain the data points cluster around the point that is the frequency of an equivalent oscillator which would be moving up and down the cylinder in a regular manner. The data points of the discrete time map appear to be attracted to this point. Therefore mathematicians call the pattern of points around the centroid of the data points a strange attractor. One can calculate the centroid of the data points, treated as if they were unit masses, in a data plane and then one can calculate the first and second

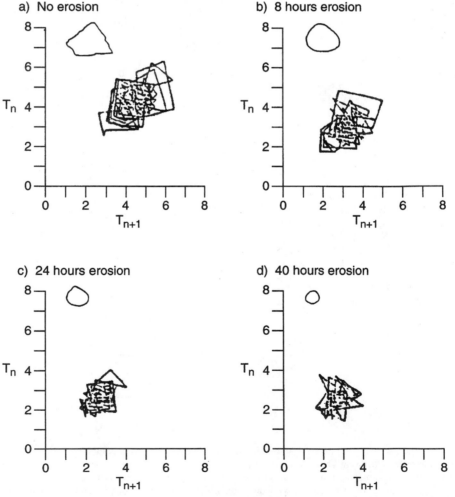

Figure 2.13. The strange attractors of the record of a tumbling rock shrink as the rock is eroded toward a spherical shape.

moments of these points around the centroid to characterize the shape of the irregular profile. The utility of this technique for describing shape is shown in Figure 2.13. The data of Figure 2.13 are taken from a study of the way in which a piece of synthetic sandstone (a fragment of a sand-lime brick) eroded in a rotating mill, simulating the tumbling of the rock in a river bed, by means of a group of sharp rocks placed in the mill with the brick fragment. When the piece of brick was freshly produced from a fragmentation process it produced the strange attractor shown in Figure 2.13a. As the rock eroded its strange attractor became more compact and the centroid representing the equivalent oscillator diminished. Similar data for three very different assembled clusters are shown in Figure 2.14 and it can be

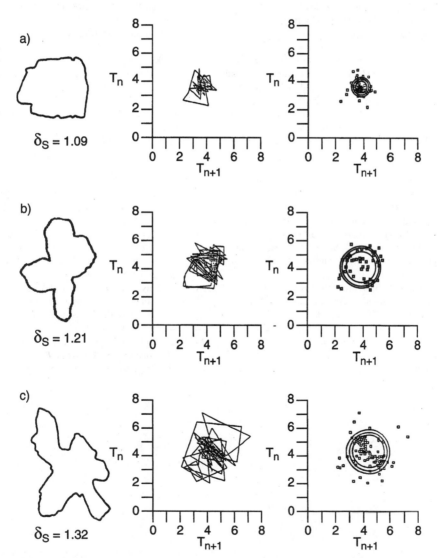

Figure 2.14. The structure of the strange attractor of the discrete time map is obviously related to the fractal dimension of the agglomerate.

shown that the properties of the strange attractor can be related to the fractal dimension of the rock.

REFERENCES

1. H. H. Heywood, "Size and Shape Distribution of Lunar Fines Sample 12057, 72," in *Proceedings of Second Lunar Science Conference*, Vol. 13, pp. 1989–2001 (1971).
2. H. Heywood, "Numerical Definitions of Particle Size and Shape." *Chem. Ind. 15*:149–154 (1937).
3. H. Heywood, "Particle Shape Coefficients." *J. Imp. Coll. Eng. Soc. 8*:25–33 (1954).
4. H. H. Hausner, "Characterization of the Powder Particle Shape," in *Proceedings of the Symposium on Particle Size Analysis*, Loughborough, England; published by the Society for Analytical Chemistry, London, England, pp. 20–77 (1967).
5. B. H. Kaye, *Direct Characterization of Fineparticles*. John Wiley & Son, New York (1981).
6. B. H. Kaye, G. G. Clark, Y. Liu, "Characterizing the Structure of Abrasive Fineparticles." *Part. Part. Syst. Charact. 9*:1–8 (1992).
7. R. Davies, "A Simple Feature Based Representation of Particle Shape." *Powder Technol. 12*:111–124 (1975).
8. A. I. Medalia, "Dynamic Shape Factors of Particles." *Powder Technol. 4*:117–138 (1970–71).
9. H. P. Schwartz and K. C. Shane, "Measurement of Particle Shape by Fourier Analysis." *Sedimentology 13*:213–231 (1969).
10. R. Ehrlich and B. Weinberg, "An Exact Method for Characterization of Grain Shape." *J. Sediment. Petrol. 40*(1):205–212 (March 1970).
11. A. G. Flook, "A Comparison of Quantitative Methods of Shape Characterization." *Acta Stereol. 3*:159–164 (1984).
12. See Chapter 15 of B. H. Kaye, *Chaos and Complexity; Discovering the Surprising Patterns of Science and Technology*, VCH, Weinheim, Germany (1993).
13. B. B. Mandelbrot, *Fractals, Form, Chance, and Dimension*, Freeman, San Francisco (1977).
14. B. H. Kaye, *A Randomwalk Through Fractal Dimensions*, VCH, Weinheim, Germany (1989).
15. I. Stewart, *Concepts of Modern Mathematics*, Penguin Books, Harmondsworth, Middlesex, England; also, 41 Steelcase Road, West Markham, Ontario, Canada (1975).
16. D. W. Schaeffer, "Fractal Models and the Structure of Materials." *Mater. Res. Soc. Bull. 13*(2):22–27 (1988).
17. D. W. Schaeffer, "Polymers, Fractal and Ceramic Materials." *Science 243*:1023–1027 (February 24, 1989).
18. See discussion of Richardson pioneering work in Ref. 13.
19. B. H. Kaye, "Characterizing the Structure of Fumed Pigments Using the Concepts of Fractal Geometry." *Part. Part. Syst. Charact. 9*:63–71 (1991).
20. B. H. Kaye and G. G. Clark, "Formation Dynamics Information; Can It be Derived from the Fractal Structure of Fumed Fineparticles?" Chapter 24 in *Particle Size Distribution II; Assessment and Characterization*, edited by T. Provder. American Chemical Society, Washington (1991).
21. H. Schwarz and H. E. Exner, "The Implementation of the Concepts of Fractal Dimensions on a Semi-Automatic Image Analyzer." *Powder Technol. 27*:207–213 (1980).
22. B. H. Kaye and G. G. Clark, "Experimental Characterization of Fineparticle Profiles Exhibiting Regions of Various Ruggedness." *Part. Part. Syst. Charact. 6*:1–12 (1989).
23. B. H. Kaye, G. G. Clark, and Y. Kydar, "Strategies for Evaluating Boundary Fractal Dimensions by Computer Aided Image Analysis." *Part. Part. Syst. Charact. 11*:411–417 (1994).
24. See Chapter 13 of B. H. Kaye, *Chaos and Complexity: Discovering the Surprising Patterns of Science and Technology*, VCH, Weinheim, Germany (1993).

3
Structural Properties of Packings of Particles

Francis A. L. Dullien
Department of Chemical Engineering,
University of Waterloo, Waterloo,
Ontario, Canada N21 3GI

CONTENTS

3.1 INTRODUCTION

There is a wealth of literature on packed beds of particles, a study of which leads to the conclusion that it is difficult to meaningfully divide the great variety of known packs into different categories. The range of complexity of the structure of beds of particles is extremely wide: at one end of the spectrum are some natural soils, containing particles of various highly different shapes and sizes that are arranged in an infinite variety of irregular and unknown configurations, whereas at the other

extreme of the scale of orderliness one finds the regular arrangements of monosized spheres. The latter are distinguished by a perfect order that facilitates a detailed description of their geometry. Owing to their great potential for analytical work, the structural properties of regular packings of equal spheres have been studied extensively. The results of these studies have been very interesting and enlightening, but sometimes they led to erroneous assumptions on the capillary and transport properties of irregular packs, containing particles of a wide range of different sizes and shapes. More recently, so-called random packs of monosized spheres have also been studied extensively, using statistical methods.

Although packs of particles of practical importance usually do not consist of arrangements of equal spheres, the amount and the significance of the work done on such systems make it a logical choice to place them in a separate category and to delegate all the rest of the packings to the other, second category of "general systems."

Of all the packs, probably soils and other unindurated sediments are of greatest significance in nature, and they also occur in the largest quantities. Other packed systems of great importance in technology include the so-called powders, usually products of crushing and grinding operations, the so-called bulk goods, such as various grains, coal, gravel, ores, etc., and the artificial packs of Raschig rings, Berl saddles, etc., used in contacting equipment of the chemical industry. In this chapter no attempt is made to discuss systematically and separately the various different systems of irregularly packed particles.

In this chapter the structures of packings, as well as certain properties that are determined by the packing structure and the void geometry, are reviewed. The so-called macroscopic structure parameters include the porosity, the permeability, the specific surface, the reduced breakthrough capillary pressure, and the resistivity factor.

3.2 MACROSCOPIC STRUCTURE PARAMETERS

All properties of packings of particles are influenced to a greater or lesser degree by the geometry of the void space which, in turn, depends on: (1) the type of packing, (2) the particle shape, and (3) the particle size distribution.[1] It is of great practical significance to identify those properties of the packings that are uniquely defined by geometry and, hence, do not depend on the nature of fluids con-

tained in the void space unless, to be sure, the geometry itself is changed as a result of the action of the fluid (or fluids) present in the voids.

3.2.1 Mean Voidage or Porosity

The most obvious and most readily measurable macroscopic structure parameter is the *void fraction* ϵ, usually called "mean voidage" in reference to packed beds and "porosity" in general. Its definition is:

$$\epsilon = \frac{\text{volume of voids in packing}}{\text{bulk volume of packing}} \quad (3.1)$$

The various experimental methods used to determine ϵ have been adequately reviewed in the technical literature.[2,3]

The mean voidage is such a convenient basic parameter that it is common practice to use it as an *independent* parameter, separated from type of packing, particle shape, and size distribution,[1] even though it *depends* on these more fundamental, but also more baffling factors.

3.2.2 Specific Surface

Another simple geometrical parameter is the *specific surface* S_0 based on solids volume, that is,

$$S_0 = \frac{\text{internal surface of packing}}{\text{volume of solids in packing}} \quad (3.2)$$

which is altogether independent of the type of packing and the mean voidage. It is completely determined, in principle, by the particle shape and size distribution. It can be readily combined with the mean voidage to give the specific surface S_v based on the bulk volume of the packing:

$$S_v = S_0(1 - \epsilon) \quad (3.3)$$

The major methods of determining specific surfaces have been reviewed in the literature.[2,3]

3.2.3 Permeability and Inertia Parameter

The *permeability* k, often called specific permeability, is defined by Darcy's law

$$Q = (kA/\mu)(\Delta\mathscr{P}/L) \qquad (3.4)$$

Q is the volumetric flow rate, or "discharge," in sufficiently slow, unidirectional, flow of a Newtonian liquid of viscosity μ, through a sample of normal cross-sectional area A and length L, in the macroscopic flow direction, under the influence of the pressure drop $\Delta\mathscr{P}$, where:

$$\mathscr{P} = P + \rho gz \qquad (3.5)$$

where P is hydrostatic pressure, ρ the liquid density, g the acceleration due to gravity, and z is the distance measured vertically upward from an arbitrarily chosen datum level.

$\mathscr{P}$ is measured, in principle, by a pipe called the piezometer (see Fig. 3.1) and is indicated as the "piezometric head" ϕ (dimension of length):

$$\phi = \mathscr{P}/\rho g = (P/\rho g) + z \qquad (3.6)$$

which is the sum of the elevation head z and the pressure head $P/\rho g$.

Darcy's law is used mostly in differential form:

$$\mathbf{V} = -(k/\mu)\nabla\mathscr{P} = -(k/\mu)(\nabla P - \mathbf{g}) \quad (3.7)$$

where $\mathbf{V} = (\delta Q/\delta A)\mathbf{n}$, the filter velocity or specific discharge, and $\mathbf{g}$ the gravitational acceleration vector. $\mathbf{n}$ is an outward unit normal vector.

When the only fluid of interest is water the *hydraulic conductivity* k_H defined as

$$k_H = k\rho g/\mu \qquad (3.8)$$

is used. Darcy's law can then be written as:

$$V = -k_H\nabla\phi \qquad (3.9)$$

In the case of gas flow, in most practical cases the elevation head may be neglected and $\Delta\mathscr{P} \approx \Delta P$. Owing to the compressibility of gases, however, both the volume flow and the filter velocity vary with pressure from one face of the sample to the other. In this case the correct integral form of Darcy's law is obtained by integrating the differential form, using the condition that at constant temperature and steady state the (pressure $\times$ velocity) product is constant throughout the sample. Thus, for gases:

$$\begin{aligned}
V_2 &= (k/\mu)(P_2^2 - P_1^2)/(2P_2L) \\
&= (k/\mu)(P_m/P_2)(\Delta P/L) \quad (3.10)
\end{aligned}$$

where P_m is the arithmetic mean pressure.

It has been found that gas permeabilities sometimes vary with P_m owing to a so-called slip effect. The equation taking the effect

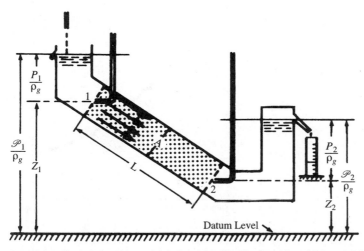

Figure 3.1. Illustration of the "piezometric head," "elevation head," and "pressure head."

of "slip" into account is named after Klinkenberg:[4]

$$V_2 P_2 L \mu / (\Delta P P_m) = k[1 + (b/P_m)] \quad (3.11)$$

where b is the constant characteristic of both the gas and the porous medium. The left side of Eq. (3.11) is plotted versus $1/P_m$, and a straight line is fitted to the data, whenever possible. The slope is bk and the intercept is k. Numerous methods have been proposed to measure permeabilities.[2,5-7]

As the filter velocity V is increased, increasing deviations from Darcy's law are observed, due to inertial effects. The appropriate modified form of Darcy's law is the so-called Forchheimer equation, written in vectorial form as follows:

$$-\nabla \mathscr{P} = \mathbf{V}(\alpha \mu + \beta \rho V) \quad (3.12)$$

where $\alpha = (1/k)$, that is, the reciprocal permeability and β the so-called inertia parameter.

The Forchheimer equation has been used mostly in its one-dimensional form, which has been rearranged[8-10] into the following dimensionless form:

$$\frac{\Delta \mathscr{P}}{L \beta \rho V^2} = \frac{\alpha \mu}{\beta \rho V} + 1 \quad (3.13)$$

All available data indicate (see Fig. 3.2) that Eq. (3.13) is a universal relationship.[18] The inertia parameter β appears to be also independent of the fluid properties and determined uniquely by the geometry of the pore space.

3.2.4 The Carman–Kozeny Equation and the Ergun Equation

S_v and ϵ are often used to calculate permeabilities based on a channel diameter characteristic of the packing by assuming[11-14] that the void space in packed beds is equivalent to a conduit, the cross-section of which has an extremely complicated shape but, on the average, a constant area. This assumption is, strictly speaking, incorrect because it ignores the facts that (1) the void space in a packing is not a single conduit but the network consisting of a multitude of conduits, (2) each conduit has a variable area along its axis because it consists of an alternating sequence of "voids" and

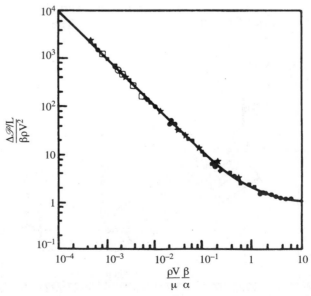

Figure 3.2. Typical fit of data to dimensionless form of Forchheimer equation.

"windows" separating adjacent voids, and (3) different conduits consist of "voids" and "windows" of different sizes. Notwithstanding its shortcomings, this assumption, called the hydraulic radius theory or the Carman–Kozeny approach, has been very useful in the case of random packings consisting of rotund particles of a narrow size distribution. Presumably the reasons for its success are that (1) in a packing of this type most of the conduits are of not very different sizes (i.e., most of the "voids" and the "windows" are contained in a narrow size range), and (2) the shapes of the conduits in different packings are not very different from each other. The effects of the axial variations in the conduit cross-section are taken care of by a (fairly constant) empirical coefficient, the so-called Kozeny constant k'. Details of this theory follow below.

In analogy with established practice in hydraulics (although valid only in fully developed turbulent flow), the channel diameter D_H governing the flow rate through the conduit of uniform cross-section is assumed to be four times the hydraulic radius, which is defined as the flow cross-sectional area divided by the wetted perimeter, that is,

$$D_H = \frac{4 \times \text{void volume of medium}}{\text{surface area of channels in medium}}$$

$$= \frac{4\epsilon}{S_0(1 - \epsilon)} \quad (3.14)$$

It is noted that sometimes the pore diameter calculated from the breakthrough capillary pressure of a nonwetting phase (such as air through a system saturated initially with water), often called the bubbling pressure (for explanations see next section), is also called hydraulic diameter,[15] although this quantity is related to the volume-to-surface ratio of a porous medium only if all the conduits are equal and of uniform cross-section (cf. Ref. 3, p. 59).

The average pore, interstitial, or seepage velocity $\bar{V}_p$ in the hypothetical single conduit

is assumed to be given by a Hagen–Poiseuille type equation for a noncircular conduit:

$$\bar{V}_p = (\Delta \mathscr{P}/L_e)(D_H^2/16k_0\mu) \quad (3.15)$$

where L_e is the length of the conduit and k_0 is a "shape factor." It is of interest to note that in a conduit of uniform cross-section Eq. (3.15) stays valid up to a Reynolds number of about 2100, whereas in a packing Darcy's law and, hence Eq. (3.15), start to break down at a Reynolds number of 0.2. This striking difference in behavior is due, in part, to variations in cross-section of the conduits present in packings.[16, 17]

The average "pore velocity" $\bar{V}_p$ and the filter velocity V [Eq. (3.7)] are assumed to be related as follows:

$$\bar{V}_p = (V/\epsilon)(L_e/L) = V_{DF}(L_e/L) \quad (3.16)$$

The division of V by ϵ is often referred to as the Dupuit–Forchheimer assumption, which defines the interstitial velocity V_{DF}, corresponding to flow through conduits of a net cross-section equal to ϵA (A is the normal cross-section of the system) in which all the microscopic streamlines are parallel to the macroscopic flow direction. The recognition of the fact that the microscopic streamlines follow a tortuous path is attributed to Carman, and it is reflected also in the fact that in Eq. (3.15) L_e was used, instead of L.

Combination of Eqs. (3.7), (3.14), (3.15), and (3.16) gives k_{CK}, the Carman–Kozeny permeability:

$$k_{CK} = \frac{\epsilon^3}{k_0\left(\dfrac{L_e}{L}\right)^2 (1 - \epsilon)^2 S_0^2} \quad (3.17)$$

The coefficient $(L_e/L)^2 \equiv T$ is usually called tortuosity factor or tortuosity. According to a great deal of data, the best empirical value of k', the Kozeny constant, that is,

$$k' = k_0(L_e/L)^2 \quad (3.18)$$

is in the neighborhood of 5.

Defining a mean particle diameter $\overline{D}_{p2}$ as the diameter of the hypothetical sphere with the same S_0 as the packing, that is,

$$\overline{D}_{p2} = 6/S_0 \qquad (3.19)$$

the following popular form of the Carman–Kozeny equation is obtained from Eqs. (3.17)–(3.19):

$$k_{CK} = \frac{\overline{D}_{p2}^2 \epsilon^3}{180(1 - \epsilon)^2} \qquad (3.20)$$

At higher filter velocity, where Darcy's law breaks down and the Forchheimer equation [Eq. (3.12)] applies, a more general form of the Carman–Kozeny equation is the so-called Ergun equation:

$$-\nabla \mathscr{P} = \mathbf{V} \left[\frac{A\mu(1 - \epsilon)^2}{\overline{D}_{p2}^2 \epsilon^3} + \frac{B\rho V(1 - \epsilon)}{\overline{D}_{p2} \epsilon^3} \right] \qquad (3.21)$$

In the original form of this equation[9] $A = 150$ and $B = 1.75$. Macdonald et al.,[18] using all available data, found that the best values are $A = 180$ and $B = 1.80$ for very smooth particles, but $B = 4.00$ for the roughest particles. Almost all experimental data lie within the $\pm 50\%$ envelope. For some of the data the following correlation holds:

$$A = 3.27 + 118.2\overline{D}_{p2}(10^{-3} \text{ ft}) \quad (3.22)$$

which improves the accuracy of Eq. (3.21) considerably.

It is customary to write the one-dimensional form of the Ergun equation in the form of a so-called friction factor versus Reynolds number correlation:

$$f_p \frac{\epsilon^3}{1 - \epsilon} = \frac{180(1 - \epsilon)}{\text{Re}_p} + 1.8 \quad (3.23)$$

where

$$f_p = \frac{\overline{D}_{p2}\Delta \mathscr{P}}{\rho V^2 L} \qquad (3.24)$$

is a friction factor, and

$$\text{Re}_p = \frac{\overline{D}_{p2} V \rho}{\mu} \qquad (3.25)$$

is the "superficial" or "particle" Reynolds number.

At the end of this section some comments are in order regarding the method of obtaining the mean particle diameter $\overline{D}_{p2}$ of the packing, which is not usually obtained by a direct application of Eq. (3.19), because S_0 of a packing is seldom measured. Instead, one calculates $\overline{D}_{p2}$ from a particle size distribution $N(D_p)$ (or, n_i particles of diameter D_{pi}) of a sample. Subject to the condition:

$$D_{pi} = 6 \frac{v_i}{s_i} \qquad (3.26)$$

there exist the following identities:

$$\overline{D}_{p2} = \frac{\Sigma_i D_{pi}^3 n_i}{\Sigma_i D_{pi}^2 n_i} = 6 \frac{\Sigma_i v_i}{\Sigma_i s_i} = \frac{1}{\Sigma_i Y_i / D_{pi}} \qquad (3.27)$$

where v_i and s_i are the volume and the surface, respectively, of a particle of diameter D_{pi}, and Y_i is the volume (or mass) fraction of the sample represented by particles of size D_{pi}.

In the case of wide particle size distributions, there is evidence[19] that $\overline{D}_{p2}$ is an inadequate mean particle diameter or, in other words, the Carman–Kozeny and the Ergun equations are inapplicable in the case of wide particle size distributions. The reason for this behavior is that even a relatively small fraction by volume of very small particles causes $\overline{D}_{p2}$ to be very small, whereas the resistance to flow of the sample is determined mainly by larger channels between larger particles.

For nonspherical particles, however, the customary methods of particle sizing and, in particular, sieve analysis do not determine particle diameters in conformity with Eq. (3.26), but yield values that need correcting to be equal to 6 v_i/s_i for the particular shape under consideration.

3.2.5 Reduced Breakthrough Capillary Pressure

The capillary pressure P_c is defined[3,7] as the pressure difference in mechanical equilibrium across an interface separating two immiscible fluids nw and w:

$$P_c = P'' - P' = P_{nw} - P_w > 0 \quad (3.28)$$

Here P'' and P' are the pressures on the concave and the convex sides of the interface, respectively, and nw and w refer to the non-wetting and the wetting fluid, respectively. *Wetting fluid* is defined as the one through which the "effective contact angle" $\theta + \phi$ is less than 90° (see Fig. 3.3). θ is the contact angle and ϕ is half of the cone angle of the tapered capillary. P_c is related to the capillary radius R, the interfacial tension σ, and the angle $\theta + \phi$, by the following form of Laplace's equation:

$$P_c = \frac{2\sigma \cos(\theta + \phi)}{R} \quad (3.29)$$

A great deal of experimental research has been done on the so-called capillary pressure curves of packings.[20-32] Details of capillary pressure curve determination may be found in these references (see also Ref. 7). Here only its most essential features will be outlined with reference to Figure 3.4. Refer to curve R_0, the so-called primary drainage curve. The experiment is started usually with the packing saturated 100% with a wetting phase, such as water. In an appropriate apparatus (there are many different types; see references) one face of the pack is contacted with a nonwetting

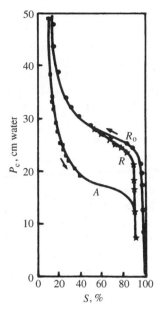

Figure 3.4. Capillary pressure curve with hysteresis loop.

phase, such as air, whereas at the other face a capillary barrier is used that permits the passage of water but prevents the penetration of air by virtue of its very fine pores. Starting at $P_c = 0$, as the value of P_c is increased in small steps, first there is penetration only into the surface irregularities of the sample. Later a value of P_c is reached (so-called entry pressure) that is high enough to force the air through the largest "windows" or "necks" at the surface of the sample and into the voids lying behind them. In regular packings of equal spheres (see section below) the same "window" size is repeated throughout the entire sample and, hence, once the entry pressure has been reached the air penetrates all across the sample and it "breaks through" at the other face. Hence, in this case, the entry pressure is also the breakthrough pressure. In random packs of equal spheres, and in general systems (see section below), however, somewhere along the path of the invading air, there always are windows that are smaller than the entry windows and, as a result, the penetration at the entry pressure is limited to a thin layer of the

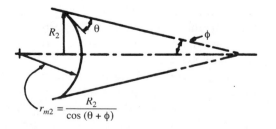

Figure 3.3. Meniscus in conical capillary.

packing at the penetration face. The value of P_c must be raised to a higher value, permitting penetration through small windows (but *not the smallest windows* in the packing), before breakthrough is achieved. Hence, both in the "random" and the "general" packings the value of the breakthrough pressure P_{cb} exceeds the entry pressure. P_{cb} is at the first elbow of the curve R_0. The value of the breakthrough pressure P_{cb} is readily determined if no capillary barrier is used, because at this pressure the air will start bubbling across the sample. This accounts for the name *bubbling pressure*. It is an important parameter, characteristic of the packing and the fluid pair (in this case, water/air) used. Using Eq. (3.29) applied to a cylindrical capillary ($\phi = 0$) the *reduced breakthrough capillary pressure P'_{cb}* is defined:

$$P'_{cb} = \frac{P_{cb}}{2\sigma \cos \theta} = \frac{1}{R_b} \qquad (3.30)$$

where R_b is the effective radius of the windows which it is necessary and sufficient to traverse to pass from one end of the system to the other. The two parameters P'_{cb} and ϵ correlate well with the permeability k and the resistivity factor F (see Ref. 15)(see below).

As the pressure is increased past P_{cb} (after replacing the capillary barrier), even smaller windows (and the voids behind them) are penetrated. The voids, however, are never completely filled because (1), as shown in Figure 3.5, there is water trapped between two touching particles in the form of so-called pendular or toroidal rings, and (2) there are also entire voids, and even systems of voids, containing water that are bypassed by the invading air,

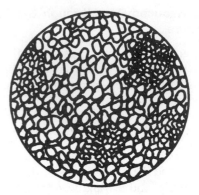

Figure 3.6. Microscopically heterogeneous distribution of grains illustrated as a monolayer. All the water between the small grains is trapped.

because they are "protected" by small windows all around. One way that this may happen is shown in Figure 3.6. As a result, even very high capillary pressures fail to reduce the water saturation below what is called the "irreducible" water saturation S_{wi}.

Reducing the capillary pressure will result, under appropriate experimental conditions, in reimbibition of water along the imbibition curve (refer to curve A in Fig. 3.4). The phenomenon that the curves R_0 and A do not coincide is called capillary hysteresis, which is due to (1) the nonuniform conduit cross-section and (2) the hysteresis of the contact angle θ. Whenever one of the two phases is perfectly wetting, that is, $\theta = 0$, such as water on clean glass and against air as the other fluid, only (1) is held responsible for capillary hysteresis.

Even if the capillary pressure is reduced back to zero there is always some air trapped in the packing in voids that have been bypassed by the invading water. The corresponding saturation, S_{nwr}, is the "residual" nonwetting phase saturation.

If P_c is increased once more, the so-called secondary drainage curve (refer to curve R in Figure 3.4) will join up with curve R_0. The loop RA is called a "permanent" hysteresis loop because it can be traced reproducibly.

Figure 3.5. Pendular rings of water trapped between touching spheres.

3.2.6 Resistivity Factor

The *resistivity factor* or formation resistivity factor F is defined as the ratio of the resistivity of the (electrically nonconductive) packing saturated with an ionic solution to the bulk resistivity of the same ionic solution. Thus:

$$F = R_0/R_w \qquad (3.31)$$

where R_0 is the resistivity of the saturated sample and R_w the resistivity of the electrolyte solution. It is evident that F, by definition, is always greater than 1. The experimental technique of measuring F has been described in the literature.[33,34]

In the absence of surface conductivity, the value of F is uniquely determined by the geometry of void space. It is important to realize, however, an important difference between k and F as far as their dependence on the pore geometry is concerned. The reason for this difference is that the electric conductivity of a conductor of uniform cross-section A_i is proportional to A_i, whereas the hydraulic conductivity in laminar flow is proportional to A_i^2. Hence, in two samples of identical mean voidage ϵ and geometrically similar void structure, but different void (pore) size, the values of F will be the same, whereas the value of k will be smaller in the sample with the smaller pore sizes. If all conduits in a packing had uniform cross-sections along their axes, F would not depend at all on the diameter of the conduits but only on their lengths and, of course, the mean voidage ϵ. Things are complicated, however, by the fact that the cross-sections of the conduits vary periodically between windows and voids. Evidently if the sizes of all the windows were decreased to zero, the value of F would increase to infinity. This effect has been taken into account by several authors.[32,38] Owing to this, as well as to the tortuosity of the conduits, F is not simply in inverse proportion to the mean voidage ϵ, as one might first expect. The first empirical relationship for F was suggested by Archie.[158]

$$F = \epsilon^{-m} \qquad (3.32)$$

where m is the cementation exponent. For random suspensions of spheres, cylinders, and sand in aqueous solutions of zinc bromide of approximately the same densities as the particles, De La Rue and Tobias[39] found Eq. (3.32) to hold with $m = 1.5$ with a high degree of precision over a range of ϵ extending from 0.25 to 0.55. Wyllie[40] used the following relationship:

$$F = X/\epsilon \qquad (3.33)$$

which defines the electric tortuosity X. One has the following relationship:

$$X = TS' \qquad (3.34)$$

where S' is a "constrictedness factor" that takes into account the axial variation of the conduit cross-sections, referred to above.

3.3 PACKING STRUCTURES OF EQUAL SPHERES

3.3.1 Regular Packings

Important literature references on this subject include Haughey and Beveridge,[41] Fejes Toth,[42] Hrubisek,[43] and Graton and Fraser.[44]

It is convenient to think in terms of rows of touching spheres as the basic unit. The rows may be assembled parallel to and touching each other to form an infinite variety of different regular two-dimensional layers. There are two limiting forms, the square pattern characterized by an angle of 90° and the equilateral triangular (rhombic, hexagonal) pattern characterized by an angle of 60°. The layers are always thought of as being arranged in the horizontal plane. Considering only the action of the force of gravity, there are three stable ways in which two square or two triangular layers may be stacked one on top of the other.

1. Each sphere in the second layer is placed with its center exactly above that of a sphere in the first layer (sequence A). The number of layers in this sequence that is necessary to obtain spatial periodicity, the so-called order of the packing, is equal to 1. If

this sequence is repeated (AA...) the simple cubic and the orthorhombic (cubic-tetrahedral) packings are obtained in the case of the square and the triangular layers, respectively.

2. Each sphere in the second layer is placed with its center exactly above the half-distance point between centers of touching spheres (i.e., the point of contact of touching spheres) in the first layer.

In the case of the square layer, depending on the direction of the displacement of the second layer relative to the first (i.e., whether this is in the x or in the y direction), there are two sequences possible (sequences B and C). If either sequence B or sequence C is repeated (BB... *or* CC...) equivalent orthorhombic packings of order 2 are obtained in which the sphere centers in layers i and $i + 2$ overlap, and there are clear vertical passages throughout the pack. When sequences B and C alternate (BCBC...), however, a different orthorhombic packing of order 3 is obtained where the sphere centers in layer $i + 2$ lie exactly above the centers of the holes between touching spheres in layer i, and there are no clear vertical passages throughout the pack (blocked passages).

In the case of the triangular layer each sphere touches six other spheres in the same layer. When placing the spheres in the second layer exactly above the points of contact of touching spheres in the first layer, the centers of two touching spheres in the second layer will lie above two of the six points of contact of a sphere in the first layer. Hence, there are three different ways of placing the second layer over the first one (sequences B, C, and D). When placing the third layer over the second one, however, there may result only two different packings. Sequences BB..., CC..., and DD... lead to equivalent tetragonal sphenoidal packings of order 2, where the centers of spheres in layers i and $i + 2$ overlap and that contain vertical passages throughout the pack. Sequences BCBC..., BDBD..., and CDCD..., on the other hand, result in

equivalent tetragonal sphenoidal packings of order 3, where the sphere centers in layer $i + 2$ lie exactly above the points of contact between touching spheres in layer i, as well as those between touching spheres in layer $i + 1$. In this packing there are no clear vertical passages (blocked passages).

3. Each sphere in the second layer is placed with its center exactly above the center of the hole formed by touching spheres in the first layer.

For the square layers only one sequence (sequence E) is possible which if repeated (sequence EE...) results in the rhombohedral (dense cubic) packing of order 2, where the centers of spheres in layers i and $i + 2$ overlap and there are no clear vertical passages (blocked passages).

In the case of the triangular layers, however, only every second hole is covered, and thus there are two possible sequences (sequences E and F). When placing the third layer over the second one, two different packings may be obtained, depending on how one proceeds. Sequences EE... and FF... result in equivalent rhombohedral (close packed hexagonal) structures of order 2, where sphere centers in layers i and $i + 2$ overlap and there are clear vertical passages. Sequence EFEF..., however, results in a different rhombohedral, that is, face-centered cubic packing of order 3 such that the sphere centers in layer $i + 2$ lie exactly above centers of holes in layer i, and there are no clear vertical passages (blocked passages).

Some important *structural properties* of regular packings of equal spheres are listed in Table 3.1.[41] *Unit cell* of the pack is defined as the smallest geometrical unit that will reproduce the pack if translated periodically, parallel to itself in the three spatial directions. The layer spacing β ($\beta = d/D_p$, where d is the distance between two consecutive parallel layers of the packing), in general, varies with the direction in the pack. The values contained in the table are the layer spacings in the vertical

Table 3.1. Structural Properties of Regular Packings of Equal-Sized Spheres

PACKING GROUP	COMMON NAME	ORDER	PACKING PATTERN SQUARE LAYER	PACKING PATTERN TRIANGULAR LAYER	UNIT CELL VOLUME PER D_p^3	UNIT CELL VOID VOLUME PER D_p^3	MEAN VOIDAGE ϵ	LAYER SPACING PER D_p' β	COORDINATION NUMBERS n	COORDINATION NUMBERS z	FRACTIONAL FREE AREA, ϵ_A MAX.	FRACTIONAL FREE AREA, ϵ_A MIN.
Cubic	Cubic, cubic No. 1	1	AA...	—	1.000	0.476	0.4764	1.000	6	6	1.000	0.21
Orthorhombic	Orthorhombic No. 4 Cubic-tetrahedral	1		AA...	0.867	0.343	0.3954	1.000	8	5	1.000	0.09
	Orthorhombic No. 2 (clear passage)	2	BB... CC...					0.866			0.635	0.21
	Orthorhombic No. 2 (blocked passage)	3	BCBC...	—								
Tetragonal-sphenoidal	Tetragonal-sphenoidal No. 5 (clear passage)	2	—	BB... CC... DD...	0.750	0.226	0.3019	0.866	10	6	0.580	0.09
	Tetragonal-sphenoidal No. 5 (blocked passage)	3		BCBC... BDBD... CDCD...								
Rhombohedral	Rhombohedral No. 3 Pyramidal, dense cubic, Cubic close packed	2	EE...		0.708	0.184	0.2595	0.707	12	5.33	0.349	0.214
	Rhombohedral No. 6 (clear passage), close packed hexagonal, Tetrahedral	2		EE... FF...				0.816			0.455	0.093
	Rhombohedral No. 6 (blocked passage) Face centered cubic	3		EFEF...								

Figure 3.7. Unit cell of simple cubic packing.

direction (cf. the description, given above, of the method of construction of regular packs). The coordination number n of a packing is the number of spheres in contact with every sphere in the pack. It is seen to increase from 6 to 12 as the packing structure becomes more compact.

The coordination number of the pack n should be carefully distinguished from the *coordination number z of the network of void space* in the pack. Groupings of neighboring spheres surround so-called voids that are separated from adjacent voids by so-called windows, access openings, or necks. The mean number of windows per void is the coordination number z of the pore or void network. From the point of view of transport phenomena taking place in the void space it is z and not n that is of importance. The value of z is 6 in the simple cubic packing, as is evident by inspecting the unit cell shown in Figure 3.7. In the rhombohedral (close packed) structures, however, the

"square" windows split into two "triangular" ones, as shown in Figure 3.8, and the rhombohedral unit cell contains two tetrahedral cells and a rhomboidal one, each of which contains a void inside. Each tetrahedral cell has four triangular windows, whereas the rhomboidal cell has eight, the average number of windows per void, that is, $z = 5.33$. Another method of calculating the coordination number of a network which sometimes yields a different value will be discussed in the section void structures.

The mean free area fraction ϵ_A represents the fractional void area in a plane section of the unit cell. In regular packs the value of ϵ_A varies both with the orientation and the position of the sectioning plane. The values of ϵ_A shown in the table are taken in the horizontal plane (cf. the above discussion of the method of construction of regular packs) and they show the maximum and the minimum values of ϵ_A as the plane is moved across the unit cell. As a corollary, one may note that it is not permissible to try to infer the bulk porosity of regular packs from measurements performed in sections of the bed.

The surface areas per unit bulk volume of the packing multiplied by D_p ($S_v \times D_p$) can easily be shown to be a function of ϵ only, that is,

$$D_p S_v = 6(1 - \epsilon) \qquad (3.35)$$

The number of spheres per unit bulk volume of the packing N_s multiplied by D_p^3 can readily be shown to be equal to:

$$D_p^3 N_s = 6(1 - \epsilon)/\pi \qquad (3.36)$$

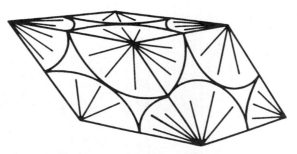

Figure 3.8. Unit cell of rhombohedral packing.

The *mathematical models* used to represent certain ranges of regular packs by Frevel and Kressley,[45] and by Mayer and Stowe[46, 47] are useful. The first of these authors used triangular (close-packed) layers, which they stacked, one on top of the other. The relative position of the two layers was varied continuously between the two limits, corresponding to cubic-tetrahedral and rhombohedral (complete close-packed) structures. These arrangements include the tetragonal sphenoidal packing, but not the simple cubic packing. Mayer and Stowe, however, used a single packing angle σ, as shown in Figure 3.9, which they varied from 90° to 60° to obtain a continuous spectrum of regular packings, ranging from the simple cubic to the rhombohedral one, but bypassing the orthorhombic and the tetragonal sphenoidal structures.

3.3.2 Random or Irregular Packings

It is easy to define a regular pack, but there does not seem to exist a completely satisfactory mathematical definition of a "random" pack.

A conceptual problem, inherent in "random" packs, is related to efforts to link the "random" distribution of particles to the mathematical abstraction of random distribution of points in a volume. Whereas points do not interfere with each other, particles do and, therefore, it may not be a fruitful idea to try to define the "random" distribution of particles in terms of the random distribution of points.

The definition of a random packing used by Debbas and Rumpf[48] is as follows: "All particles of the same size and shape have the same probability to occupy each unit volume of the mixture." The following generally accepted theorems[49] follow from this definition.

1. The mean fractional free area ϵ_A is equal to the mean voidage ϵ of the packing.
2. The probability density function of diameters of circles appearing in any sufficiently large section plane through a random bed of spheres is identical to the theoretical distribution obtained by parallel sectioning a single sphere at infinitesimal, constant intervals, perpendicular to an arbitrarily selected radial direction.

Most details of the characteristics of "random" packings have been obtained experimentally. Some of these may be summarized by the so-called radial distribution function, shown for close random packings in Figure 3.10. The concept of radial distribution function shows the difference between regular packs, on the one hand, and "random" packs, on the other, most clearly.

Choosing the center of an arbitrary sphere as the origin of a coordinate system, in a

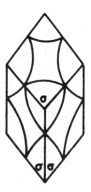

Figure 3.9. Unit cell after Mayer and Stowe.[46]

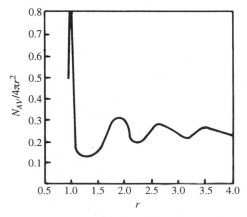

Figure 3.10. The radial distribution of the average number of sphere centers per unit area in a spherical shell.

regular pack the positions of *all the other sphere centers can be given exactly*. In a random pack, however, the position of *none of the other sphere centers can be given exactly*. In this case one can talk only about the *probability of finding sphere centers* at a given distance from the center of the base sphere. For the case of so-called uniformly random distribution one must, by definition, have the same probability density, independently of the distance. This is indeed the case in random packings of equal spheres at great distances from the center of the base sphere. Closer to the base sphere, however, it has been found experimentally[50,51] that the probability density varies with distance, as shown in Figure 3.10. In view of the fact that in the case of regular packs the radial distribution consists of a discrete set of Dirac delta functions (i.e., vertical lines) separated by gaps (i.e., regions of zero probability density). It is logical to interpret the peaks in the diagram at $R = 1$, etc. ($R = r/D_p$, where r is the radial distance measured from the center of the base sphere) as an indication of a relative order over a short range around the base sphere. As expected, if the random packing is looser (greater mean voidage), the short range order becomes less pronounced.[52-54]

For random packings, average values of the layer spacing β have been defined and expressed in dependence on the mean voidage, as follows:[55,56]

$$\beta = \sqrt{(2/3)} \; \{\pi/[3\sqrt{2} \,(1 - \epsilon)]\}^{1/3} \quad (3.37)$$

The mean coordination number n has been determined experimentally in random packs of equal spheres by several researchers.[57-61] Ridgway and Tarbuck[62] found the following correlation between n and ϵ in random packings of equal spheres:

$$\epsilon = 1.072 - 0.1193n + 0.00431n^2 \quad (3.38)$$

This correlation was based on data covering an extremely wide range of values of the parameters, that is, $3 \leq n \leq 12$ and $0.27 \leq \epsilon \leq 0.78$, whereas in all common random packs of equal spheres $0.36 \leq \epsilon \leq 0.44$, approximately.

Considering the fact that there is an infinite number of different regular packings, it is not surprising to find also different kinds of random packs. These may be divided into the following four categories:[41]

3.3.2.1 Close Random Packing

These are obtained when the bed is vibrated or vigorously shaken down, and results in mean voidages of 0.359 to 0.375,[50,51,63-71] which is considerably in excess of the mean voidage of 0.26, corresponding to hexagonal close packing.

3.3.2.2 Poured Random Packing

Pouring spheres into a container, corresponding to a common industrial practice of discharging powders and bulk goods, results in mean voidages of 0.375 to 0.391.[11,42,73-75]

3.3.2.3 Loose Random Packing

Dropping a loose mass of spheres into a container, or packing spheres individually and randomly by hand, or permitting them to roll individually into place over similarly packed spheres, results in mean voidage values of 0.40 to 0.41.[52,57,60,64-70,76-78]

3.3.2.4 Very Loose Random Packing

The fluidized bed at the minimum fluidization has a mean voidage ϵ of 0.46 to 0.47.[79]

By slowly reducing the fluid velocity to zero in a fluidized bed,[80] or by sedimentation of spheres,[81] or by inversion of a bed container[82,83] a mean voidage of 0.44 is obtained.

In the case of random packs a so-called *wall effect* exists, because the proximity of a solid surface will introduce some local order into a random packing. Thus, the particles next to the solid surface tend to form a layer of the same shape as the surface. This so-called base layer is a mixture of clusters of square and triangular units. Randomness increases with increasing distance from the base layer, with resultant disappearance of distinct layers.

Another important aspect of the *wall effect* is the existence of a region of relatively high voidage next to the wall due to the discrepancy between the radii of curvature of the wall and the particles.[44,84] A cyclic variation of the local voidage ϵ' with distance from a cylindrical wall has been measured,[62,85-90] extending some three to four particle diameters into the packing (see Figure 3.11). The local voidage ϵ' is the fractional free volume in a small or thin element of the bed dv', whereas the local mean voidage $\bar{\epsilon}$ is the mean value of ϵ' taken over a region of the bed, that is,

$$\bar{\epsilon} = \frac{1}{v'} \int_0^{v'} \epsilon' \, dv' \qquad (3.39)$$

The *wall effect* has been studied as a function of the particle diameter ratio D_p/D_T. Various empirical formulas exist to correct for this effect.[41]

3.4 PACKING STRUCTURES OF GENERAL SYSTEMS

Almost all real packings of particles fall in this category. Unfortunately, it is not possible to say anything specific about the structure of the general systems, and one must be content to state a few principles regarding this subject.

There are two major factors to consider: (1) *particle shape* and (2) *particle size distribution*. Reviews (e.g., Ref. 41) offer very little useful information on the subject of the effect of *particle shape* on packing structure. One may note only that very special packing structures arise in the case of highly anisometric particles, such as platelets and needles, which can be packed in a great variety of very different ways, resulting in systems either of very high void fractions, when the orientations of the particles are random, or of very low void fractions, when the particles are stacked with their axes aligned. Systems of this latter type are also highly anisotropic, that is, their physical properties are very different in the direction of the particle axes than in the perpendicular direction.

Particulate systems generally involve a range of particle size between 2- and 10^5-fold. When smaller particles are mixed into a bed of larger particles then the former, on one hand, tend to increase the voidage, by forcing the larger particles apart but, on the other hand, also

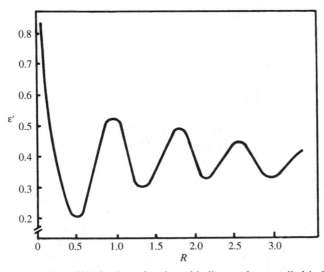

Figure 3.11. Variation of local volume fraction with distance from a cylindrical surface.

tend to decrease the voidage by filling the voids between the larger particles. The latter effect predominates for a size ratio greater than about 3:1. The net effect will depend not only on the particle size ratio (and particle shape), but also on the relative amounts of each size fraction present. Furnas,[91] and Sohn and Moreland,[92] showed that in binary mixtures of particles the bulk voidage always decreases below the values existing at both pure component ends of the composition range, and for any given ratio of particle diameters it has a minimum value at some composition.

Horsfield[93] showed that the filling of voids of a rhombohedral packing by five successive specified sizes would give a minimum voidage of 1481, which can be further reduced by the addition of even finer particles. Similar studies have been made also in other kinds of packings.[41]

A very important phenomenon that is directly related to particle size is the fact that *packings of smaller particles tend to result in relatively high voidages*, because small particles have a tendency to pack less closely than do larger particles. In beds of small particles there is frequent "arching" or "bridging" which yields larger voids.[65, 94–96] This phenomenon is related to the smaller volume-to-surface ratio of small particles, resulting in weights that are insufficient to overcome a variety of surface resistances opposing attainment by the particles of positions of a minimum local potential energy by rolling into nearby wells or holes.

The *packing structures* of general systems vary widely and are seldom known in any detail. The only practicable technique known to this author that is available to explore the packing structures of "general systems" consists of filling the voids with some colored or fluorescent plastic, or a low melting alloy, which make it possible to distinguish between the voids and the solid matrix, and study polished sections of the systems.[7] Evidently both the packing structure and the void structure may be analyzed simultaneously by this technique, but only analysis of the latter has been attempted and that only in a few instances.

3.4.1 Pore Size Distribution

Pore structure is interpreted as a characteristic "pore size," which is sometimes also called "porosity." Most generally, however, "pore structure" is identified with a so-called "pore size distribution," characteristic of the sample of the porous material. "Pore size distribution" is a poorly defined quantity, partly because it depends, sometimes very markedly, on the particular method used in its determination. The general procedure used for the determination of a pore size distribution consists of measuring some physical quantity in dependence on another physical parameter under the control of the operator and varied in the experiment. For example, in mercury porosimetry, the volume of mercury penetrating the sample is measured as a function of the pressure imposed on the mercury; in vapor sorption, the volume of gas absorbed is measured as a function of the gas pressure; the volume of liquid displaced miscibly is measured as a function of the volume of displacing liquid injected into the sample in a miscible displacement experiment, etc.

3.4.1.1 One-dimensional Pore Structure Models

The experimental data have invariably been interpreted in terms of an arbitrary model of pore structure, the most popular one consisting of a bundle of parallel capillary tubes of equal length and distributed diameters. The fact that this model may give rise to vastly different "pore size distributions" when used in conjunction with the results obtained in different types of experiments on the same sample has been demonstrated by Klinkenberg[97] for the case of mercury porosimetry and miscible displacement, as illustrated in Figure 3.12.

3.4.1.2 Dead-End Pores and Periodically Constricted Tubes

More sophisticated one-dimensional models have included dead-end pores, called also "ink-bottle" pores, "pockets," or "Turner

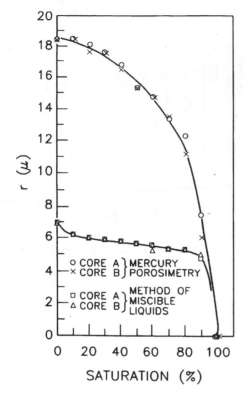

Figure 3.12. Cumulative pore size distribution curves of Bentheim sandstone. Permeability: 975×10^{-11} cm^2. Porosity: 0.289 (After Klinkenberg.[97])

structures," depending on their shape and also periodic variations of the diameter of each capillary tube, the so-called "periodically constricted tube" models. Both types of pore models introduce the concept of "pore throats," the local minima in the pore size that separate "pore bodies." These models have been reviewed in Ref. 7 and in Ref. 98. Each is designed to account only for certain properties of the porous medium under study and is quite unable to account for other properties. To account for three-dimensional flow, intrinsically one-dimensional models have often been generalized in three dimensions.[7,98]

3.4.1.3 Definition of Pore Size Distribution

The definition of pore size distribution, in the usual sense, is "the probability density function giving the distribution of pore volume by

a characteristic pore size." Whereas the volume parameter is usually measured directly, the characteristic pore size is always calculated from some measured physical parameter in terms of the arbitrary model of pore structure. Owing to the complexities of pore geometry, the characteristic pore size is often not at all characteristic of the pore volume to which it has been assigned. In mercury intrusion porosimetry, for example, the volume of the pore space penetrated through a pore throat is assigned to the size of the throat, resulting in an unrealistic picture of the real pore structure, as shown in Figure 3.13. This procedure is analogous to characterizing the size of lecture halls with a given door size by stating their combined volume and the size of the doors. The realization of this state of affairs has given rise to the custom of referring to "entry pore size."

Very recently, constant rate injection mercury porosimetry has been used to extract more information on pore structure from the fluctuations observed in the porosimetry trace representing abrupt pressure changes related to the jumps of mercury from pore throats into (larger) pore bodies (Haines jumps) with concomitant redistribution of mercury also at other locations. The lower extremes of the fluctuating pressure are related to the size of pore bodies. This method has the promise to develop into a nondestructive quantitative technique for the determination of both pore throat and pore body size distributions.

3.4.1.4 Pore Size

Another matter than needs clarification is the definition of "pore size." Only if the pores were cylindrical tubes of uniform diameter, or spherical bodies, would the pore size be unique. As neither is the case, "pore size" needs to be defined. A convenient definition of "pore size" is that it is twice the hydraulic radius, $2r_H$, which is either identical or very close to the mean radius of curvature

$$r_m = \frac{1}{2}\left(\frac{1}{r_1} + \frac{1}{r_2}\right) \qquad (3.40)$$

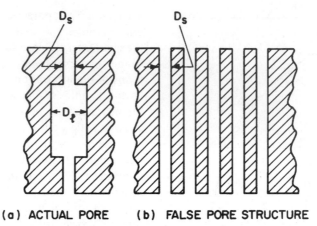

(a) ACTUAL PORE   (b) FALSE PORE STRUCTURE

Figure 3.13. Illustration of error introduced by the usual interpretation of mercury intrusion porosimetry data. (*a*) Actual pore; (*b*) false interpretation of data. (D_s = entry pore throat diameter; D_1 = pore body diameter). (After Dullien.[154])

of the interface separating two immiscible fluids in mechanical equilibrium in the pore, for the special case of zero contact angle (see Table 3.2.[99] The mean radius of curvature, r_m, can be calculated from measured capillary pressures, p_c, by Laplace's equation of capillarity:

$$p_c = \frac{2\sigma}{r_m} \qquad (3.41)$$

where σ is the interfacial tension. For the case of nonzero contact angle θ the values of the pore radius r_m in Table 3.2 must be replaced by R, according to the relation,

$$R = I_m \cos \theta \qquad (3.42)$$

The definition of hydraulic radius r_H of a capillary of uniform cross-section is:

$$r_H = \frac{\text{volume of capillary}}{\text{surface area of capillary}} \qquad (3.43)$$

For the case of a variable cross-section the above definition can be generalized for any normal cross-section of the capillary as follows:

$$r_H = \frac{\text{area of cross-section}}{\text{length of perimeter of cross-section}} \qquad (3.44)$$

Table 3.2. List of Comparative Values to Show Equivalence of the Reciprocal Hydraulic Radius ($1/r_H$) and Twice the Reciprocal Mean Radius of Curvature $2/r_m = [(1/r_1) + (1/r_2)]$ in a Capillary.[a]

CROSS-SECTION	$(1/r_1) + (1/r_2)$	$1/r_H$
Circle	$2/r$	$2/r$
Parallel plates	$1/b$	$1/b$
$a{:}b = 2{:}1$	$1.50/b$	$1.54/b$
Ellipse $a{:}b = 5{:}1$	$1.20/b$	$1.34/b$
$a{:}b = 10{:}1$	$1.10/b$	$1.30/b$
Rectangle	$1/a + 1/b$	$1/a + 1/b$
Equilateral triangle	$2/r_i$	$2/r_i$
Square	$2/r_i$	$2/r_i$

After Carman.[99]

[a] r_i is the radius of the inscribed circle.

The applicability of Eq. (3.44), however, is limited to the special case of capillaries of a rotational (axial) symmetry that have normal cross-section. For the general case of irregular capillaries the minimum value of the ratio given by the right-hand side of Eq. (3.44) must be found by varying the orientation of the sectioning plane about the same fixed point inside the capillary. The minimum value of this ratio is, by definition, the hydraulic radius r_H of the irregular capillary at the fixed point. The value of r_H of a section is assigned to the center of gravity of the section. Both definitions (i.e., r_m and r_H) are best suited to the case of pore throats that control both capillary penetration by a nonwetting fluid into the porous medium and the flow rate of fluids through the porous medium. The size of a pore body is not readily related in a unique manner to any measurable physical quantity and the problem of characterizing the size of a pore body is best dealt with by using photomicrographs of sections made through the porous medium where the pore body is made visible. The problem of defining the size of a pore body is similar to the problem of defining the size of an irregularly shaped particle.

3.4.2 Network Models of Pore Structure

3.4.2.1 Two-dimensional Network Models

The most fundamental flaw of all the simple models of pore structure is that they do not account for the fact that in permeable porous media all the conducting pores are interconnected and form a continuum of a network of pores. In a network where pores of different sizes are interconnected either in a random or in a correlated manner there is a large number of different pathways characterized by different resistances to transport. In any given porous medium the distribution of paths of different resistances depends on the nature of the particular transport phenomenon and, as a result, a variety of phenomena may occur in two- and three-dimensional networks that are impossible in one-dimensional models. Repeating an intrinsically one-dimensional model

in the three spatial directions, even if tubes are made to intersect, cannot account for these phenomena.[7] Even though there are important differences between the properties of two- and three-dimensional networks, nevertheless the first move in the direction of introducing accurate pore structure models was the pioneering work by Fatt,[100-102] who proposed random two-dimensional network models of pore structure for the first time. Whereas Fatt was primarily interested in immiscible displacement, Simon and Kelsey[103,104] used two-dimensional network models for the simulation of miscible displacement. These studies, important as they were owing to their pioneering character, were only of qualitative nature because

- the networks were two-dimensional,
- the size of the network was too small,
- the networks were regular,
- the networks consisted only of tubes, in analogy with a network of resistors,
- the geometry of the tubes and the distribution of the geometry over the network ("size distribution") did not correspond to that of any real porous medium because this was not available.

3.4.2.2 Percolation Theory: Three-dimensional Network Models

During the same time the powerful mathematical theory of percolation was developed (e.g., Refs. 105–107) and was also suitable for the rigorous treatment of immiscible displacement and two-phase flow phenomena in pore networks of infinite size.

The first published reference in which pore structure was modeled by both two- and three-dimensional networks appears to be the work of Chatzis and Dullien,[108] which was followed by a vastly improved treatment by the same authors a year later.[109] The unique features of this treatment, the final form of which was published in the English language literature only in 1985,[110] included three-dimensional network models of pore structure consisting of pore bodies situated at the nodes of the network (sites) connected by pore throats, modeled by the bonds of the network. The two three-dimensional networks used, that is, simple cubic and tetrahedral networks of coordination numbers 6 and 4, respectively, gave similar results. Different numbers were assigned randomly to the sites, indicating their relative sizes only. It was assumed that a bond can never be larger than either of the two sites connected by it. Hence the same number characterizing the smaller one of the two sites was automatically also assigned to the bond between the two sites. This resulted in bond sizes correlated with the site sizes ("bond-correlated site percolation"). The penetration of the network by a nonwetting fluid (drainage) was simulated by playing the following game. Initially all the sites and bonds were assumed "closed." The game was started by first declaring the largest sites "open," then the second largest sites, and so forth. A bond became open if and only if both sites at its two ends were "open." "Open" sites and bonds communicating with the face of the network exposed to the fluid were automatically penetrated. The rest of the faces of the network were assumed impervious. The fractional numbers of sites and bonds that were penetrated were recorded as functions of the fractional number of "open" sites. The fractional number of "open" sites P_s, at which the penetration first reaches the opposite face of the network is called the "breakthrough" value. In the case of a network of infinite size this value is the "critical percolation probability" or "percolation threshold" which is known from percolation theory. Close agreement with the published percolation theory value has been obtained by repeating the game in an 18 × 18 × 12 mesh size cubic network and taking the average breakthrough value. The game is continued until all sites and bonds are penetrated. The fractional number or probability of "open" bonds, p_b, is related to the fractional number or probability of "open" sites, p_s, as follows:

$$p_b = p_s^2 \qquad (3.45)$$

owing to the assumption that both sides at the two ends of a bond must be "open" for a bond to be also "open." As a result, relationships have been obtained for the fractional numbers of penetrated sites and bonds as functions of the fraction of open bonds, p_b. These relationships were used for predicting the mercury porosimetry curves of sandstone samples, among other things. Before one can attempt to do this, first pore sizes and volumes must be assigned to the sites and the bonds. The pore diameters, D_b for the bonds and D_s for the sites, are related to p_b and p_s, respectively, by the following relationships:

$$p_b = \int_{D_b}^{D_{bmax}} f_b(D_b)\, dD_b \qquad (3.46)$$

and

$$p_s = \int_{D_s}^{D_{smax}} f_s(D_s)\, dD_s \qquad (3.47)$$

where $f_b(D_b)$ and $f_s(D_s)$ are the bond (pore throat) and site (pore body) diameter distribution densities. The ranges of pore sizes used were consistent with photomicrographs prepared of polished sections of sandstone samples that had been previously saturated with Wood's metal, but the functions, shown in Figure 3.14, were adjusted so as to obtain the best agreement between prediction and experiment for the case of a particular Berea sandstone sample.

For each value of D_b the capillary pressure of mercury penetration, p_c, was calculated by the relation:

$$p_c = \frac{4\sigma \cos \theta}{D_b} \qquad (3.48)$$

Based on petrographic studies[111] the pores in sandstones are slit-shaped. Hence, geometrically similar, slit-shaped pore bodies and pore throats have been assumed, where the pore "diameter" (D_s or D_b) is the width of the slit and the other two dimensions of the slit are $L_1 = C_1(D)^{1/2}$ and $L_2 = C_2(D)^{1/2}$, resulting in the following expressions for the volume of a pore body V_s and a pore throat V_b, respectively:

$$V_s = D_s L_{1s} L_{2s} = D_s^2 C_{1s} C_{2s} = D_s^2 \cdot l_s \quad (3.49)$$

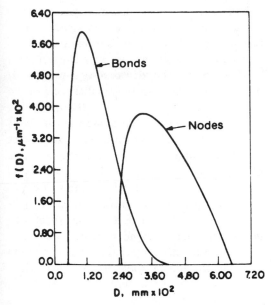

Figure 3.14. Density distributions of pore throat (= bond) and pore body (= node) diameters. (After Chatzis and Dullien.[110])

and

$$V_b = D_b L_{1b} L_{2b} = D_b^2 C_{1b} C_{2b} = D_b^2 \cdot l_b \tag{3.50}$$

where l_s and l_b are dimensional constants.

Thus the predicted volume of pores penetrated by the mercury at any given capillary pressure could be readily calculated. The assumed fraction of volume contributed by the pore throats, anywhere in the range of under 50%, had very little effect on the total volume predicted. An additional fine point considered in the calculation is that in the case of capillaries of irregular, rather than circular, cross-section penetration of the capillary by mercury does not result in complete filling of the entire cross-section because it would take a higher capillary pressure for the mercury to penetrate the corners than the pressure required to penetrate the core of the pore. Three calculations were carried out: one for circular pore cross-section, one for the case of 0.46, and another for the case of 0.65 unfilled volume fraction when the pore was first penetrated by mercury. As the capillary pressure on the mercury

is increased progressively these empty fractions keep decreasing proportionately to the square of the capillary pressure (i.e., twofold increase in the capillary pressure results in a fourfold decrease in the unfilled volume).

The results of these calculations have been compared with experimental mercury intrusion porosimetry curves obtained on 10 different sandstone samples of permeabilities ranging from 5370 md to 0.36 md. The data have been plotted in reduced form in terms of $p_c^* = P_c/P_c^0$, where P_c^0 is the breakthrough capillary pressure of the particular sample, versus mercury saturation ($S_{nm}\%$) in Figure 3.15. The fact that, with the exception of Belt Series, the data lie on the same curve indicates that the pore structures of the remaining nine sandstones are geometrically similar. The predicted curve starts predicting unrealistically high mercury saturations at higher values of the saturation. The assumption of partially unfilled pores does not eliminate this discrepancy completely which remains in the range of 5% to 10% pore volume. The most likely explanation for this is the presence of micro-

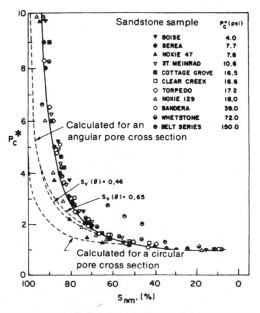

Figure 3.15. Dimensionless mercury intrusion porosimetry curve of sandstone samples. (After Chatzis and Dullien.[110])

pores in the cementing materials (clays) of the sandstones which were not penetrated by mercury under the capillary pressures used in the experiments and which were not included into the network model.

The same network model was used subsequently to model both the drainage and the imbibition oil–water capillary pressure curves in sandstones,[112] where also the trapping of each phase by the other was simulated. The experimental primary drainage curve has been predicted satisfactorily. In this case, too, lower residual wetting phase (water) saturation was predicted than the measured value (see Figure 3.16). The reason for this discrepancy is probably the water present in the micropores of the cementing materials (clays), the existence of which was not taken into account in the network model. The experimental secondary imbibition curve, however, indicates a far more gradual displacement of the nonwetting phase (oil) than the predicted trend. The reason for this is that in the simulation of imbibition the sites of the network, representing pore bodies were assumed to control the displacement process. Technically speaking, in the simulation of

imbibition first the smallest sites were allowed to be "open," then the second smallest sites, and so forth. Any "open" site in communication with the wetting fluid was automatically penetrated. The bond was also automatically penetrated along with the smallest one of the two sites connected by it. As the imbibition capillary pressure was calculated on the basis of capillary equilibrium in a pore body (site), the value obtained for breakthrough of the wetting phase turned out to be very small and after breakthrough the imbibition process was completed over a very narrow range of capillary pressures. Visual observations of imbibition in transparent capillary micromodels and other experiments have shown that imbibition is not controlled by the pore bodies because the imbibing fluid does not always advance in a pistonlike manner.[113–117] Instead, it often propagates in pore edges, wedges, corners, and surface grooves and it can pass through pore bodies while filling them only fractionally. As a result, the wetting fluid does not have to fill relatively large pore bodies in its path before it can fill pore throats and relatively small pore bodies over the entire network. Indeed much closer agreement with experiment was obtained by the author, using the same accessibility [7] data, but assuming that every bond and every site was penetrated by the wetting phase at a value of $p_c^* = 28\ \mu m/D$ where 28 μm is the "breakthrough" bond diameter in primary drainage and D is the diameter of the bond, or site, in question expressed in μm units. Bonds and sites occupied by trapped nonwetting phase were excluded. It is apparent that there has been a great improvement as a result of changing the assumption of piston-type imbibition displacement to an assumption of independent domains.

Diaz et al.[112] demonstrated the considerable effect the form of the assumed distributions of pore throat and pore body sizes has on the predicted reduced primary drainage curve, while keeping all the other parameters unchanged. It is evident from this study that accurate prediction of capillary pressure curves requires an a priori knowledge of the pore

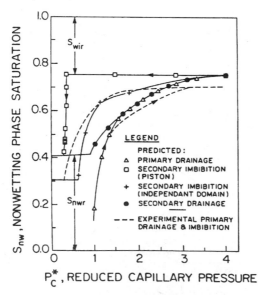

Figure 3.16. Simulated and experimental Berea sandstone capillary pressure curves. (After Diaz et al.[112])

structure. Minimum requirement consists of the density distributions of pore throat and pore body diameters, the relationship between the volume and the diameter of pore bodies, and the distribution of coordination numbers. Additional information on pore shapes and the rugosity characteristics is required for the prediction of imbibition type processes.

In addition to simulating the capillary pressure curves, the information obtained from the cubic network model on the accessibility and occupancy of the pores of the network by the wetting and the nonwetting fluids, respectively, has been used also for predicting the relative permeability curves.[110, 118, 119, 120] The same choice of density distributions of pore throat and pore body diameters that were used in the simulation of the capillary pressure curves resulted in excellent predictions of the relative permeability curves in drainage; however, the imbibition relative permeability curves could not be simulated with satisfactory accuracy (see Fig. 3.17) for the same reasons stated

under the discussion of simulation of capillary pressure curves.

The above contributions, discussed in some detail, are distinguished from the rest of the literature of network modeling and percolation theory by both the application of consistent photomicrographic pore structure data in the model calculations and the regular comparisons made between prediction and experiment on specific samples of porous media.

An incomplete list of references to (largely theoretical) contributions to the field of application, of percolation theory and network modeling to porous media is Refs. 121–133. An interesting computerized network simulation study of drainage and imbibition capillary pressure curves dependent on a variety of pore structure parameters has been published by Wardlaw et al.[134] The stated purpose of this work has been to interpret the pore structure of porous media from the capillary pressure curves and scanning loops. While a number of trends have been observed in this work, the authors did not work out a procedure leading to the determination of the density distributions of pore throat and pore body diameters, coordination numbers, the nature of correlation between neighboring pore body and pore throat sizes, and the type of pore shapes present in a particular sample. Another recent important contribution is by Lenormand et al.[135] involving both experiments and numerical modeling of immiscible displacement in networks. In contrast to Ref. 134, in this work no attention was paid to the effects of pore structure on the immiscible displacement process but instead the emphasis was on the role played by the capillary number and the viscosity ratio in determining whether there is (1) stable displacement, (2) viscous fingering, or (3) capillary fingering.

As evident from this review of the relevant literature, there are at present powerful network simulation techniques available to predict capillary pressure curves, immiscible and miscible displacement (hydrodynamic dispersion), and relative permeabilities for the case of arbitrarily assumed pore structures, mod-

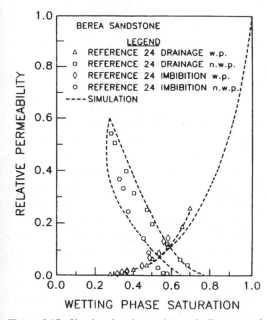

Figure 3.17. Simulated and experimental oil–water relative permeability curves for Berea sandstones. (After Kantzas and Chatzis.[118])

eled by the network. It is reasonable to expect that these simulators would be able to predict the actual behavior in samples of real porous media if the correct pore structure data of the samples were used as input. It is, therefore, of considerable practical importance to develop techniques to obtain such data. In the remaining part of this review attempts to develop a technique for the determination of the accurate pore structure of samples of real porous media are presented and discussed. All of these attempts are based on computer reconstruction of the pore structure from serial sections of the porous sample.

3.4.3 Pore Structure Determination from Serial Sections

Serial sectioning of samples has been performed for some time by metallographers for the purpose of determining the genus of a phase.[136-139] More recently this technique has been also applied to porous media.[140-144] The data used in Ref. 144 have formed the basis of a number of pore structure studies. The data were generated by first injecting, under high pressure, molten Wood's metal into a previously evacuated sandstone sample. After solidification of the metal the rock matrix was dissolved with hydrofluoric acid and replaced with clear epoxy resin. A piece of this sample of dimensions $1310 \times 1040 \times 762$ μm was encased in epoxy resin and its surface was successively ground and polished to remove layers of about 10 μm thickness. Seventy-eight layers were removed and each layer was photographed. A sample photomicrograph is shown in Figure 3.18. The white features represent the pore space as the metal is reflective and the black regions represent the rock matrix since the epoxy is clear.

3.4.3.1 Determination of Genus

The micrographs were entered into the computer, where they were digitized, and then connection between features in neighboring photomicrographs was established based on the assumption that two features are con-

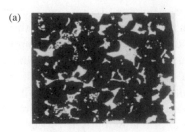

Figure 3.18. Representative serial section. (a) Photomicrograph; (b) original digitized picture; (c) digitized picture after filtering to remove nonconnected features. (After Kwiecien et al.[151])

nected if and only if they overlap. Connection of two pixels between two features present in two neighboring micrographs k and $k + 1$ is illustrated in Figure 3.19. For the purpose of topological studies, where distances and directions play no role, this kind of treatment of the data results in a branch-node chart of the sample from which the genus may be readily determined. The genus is defined as the number of nonintersecting cuts that can be made upon a surface without separating it into disconnected parts. It has been shown[145] that the genus of the enclosing surface is numerically equal to the connectivity of the branch-node network derived from that surface. The con-

FEATURE 65 'POINTS TO' FEATURE 132

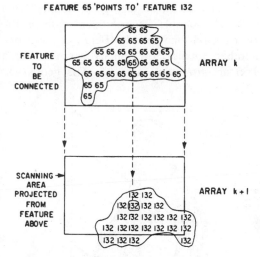

Figure 3.19. Connection of features on adjacent serial sections by the overlap criterion. (After Macdonald et al.[144])

nectivity is a measure of the number of independent paths between two points in the pore space and, hence, of the degree of interconnectedness of the pores. The genus G is given by

$$G = b - n + N \qquad (3.51)$$

where b is the number of branches, n is the number of nodes, and N is the number of separate networks in the sample. Simple examples for the calculation of genus are shown in Figure 3.20. A small portion of a serially sectioned sample is shown in Figure 3.21a. In Figure 3.21b the corresponding branch-node chart is shown. The surface nodes introduce complications because it is not known how they are connected on the outside of the sample. The surface nodes in the plane of polish are visible, but the lateral surface nodes are not and they must be obtained by means of the overlapping criterion. Much of the study[144] deals with the problem of edge effects in the construction of the branch-node chart which are very important owing to the small size of the sample. Larger samples, however, would result in much increased computer time and decreased resolution of the features. The maximum possible value of the genus G_{max} is obtained by connecting all the surface nodes to one external node. The least possible value of the genus, G_{min}, is calculated by not connecting any of the surface nodes to an external node. In the Berea sample $G_{max} = 593$ and $G_{min} = 420$. Dividing these values by the sam-

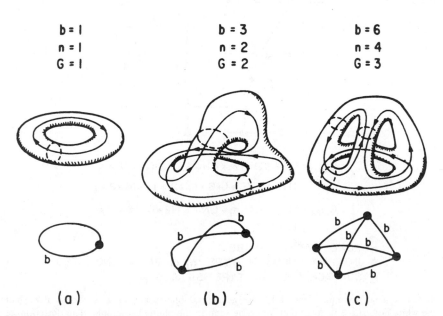

Figure 3.20. Simple shapes and their branch-node chart used to compute the genus. (After Macdonald et al.[144])

ple volume of 10.4×10^8 μm^3 the genus per unit volume can be calculated. This yields a genus of about 5×10^{-7} per μm^3, or a genus of 2 per 4×10^8 μm^3 which corresponds to a sphere of a diameter of about 200 μm^3, which is the size of an average grain in the Berea sandstone sample. The pertinent data on the Berea sandstone sample (Berea 2c) and one another smaller, preliminary sample (Berea 1xx) are listed in Table 3.3. The genus was

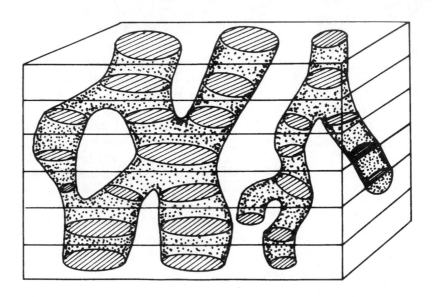

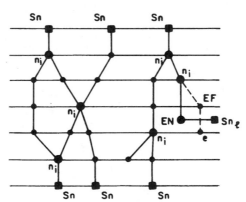

$$G_{MIN} = (b_i + b_\ell) - (n_i + EN + Sn + Sn_\ell) + N =$$
$$= (12 + 1) - (6 + 1 + 6 + 1) + 2 = 13 - 14 + 2 = 1$$
$$G_{MAX} = (b_i + b_\ell) - (n_i + EN) = (12 + 1) - (6 + 1) =$$
$$= 13 - 7 = 6$$

- • PORE SPACE FEATURE
- ● INTERNAL NODE OF TYPE n_i or EN AS INDICATED
- ■ SURFACE NODE OF TYPE Sn or Sn_ℓ

Figure 3.21. (a) A small example pore cast. The horizontal planes are serial sections. The shaded ellipses represent the white pore space features that would be seen in the photomicrographs. The dotted material is pore space between planes and not viewable. (After Macdonald et al.[144]); (b) The branch-node chart for the pore space in Figure 3.21a, showing the types of nodes and boundary features used (after Macdonald et al.[144])

Table 3.3. Berea Sandstone Samples

PROPERTY	BEREA 2C	BEREA 1XX
Cross-section	$1310 \ \mu m \times 1040 \ \mu m$ $= 1.36 \times 10^6 \ \mu m^2$	$1350 \ \mu m \times 950 \ \mu m$ $= 1.28 \times 10^6 \ \mu m^2$
Total depth	$762 \ \mu m$	$514 \ \mu m$
No. of serial sections	78	50
Avg. spacing $\Delta \bar{z}$	$9.9 \ \mu m$	$10.5 \ \mu m$
Volume	$10.4 \times 10^8 \ \mu m^3$	$6.6 \times 10^8 \ \mu m^3$
No. of features in sample	3564	2583
Avg. grain size	$\approx 200 \ \mu m$	$\approx 200 \ \mu m$

After Macdonald et al.[144]

determined versus the volume of the sample section by section with the interesting result that past a certain minimum volume the genus is a linear function of the volume (see Fig. 3.22). The slope of the line gives the best estimate of the genus per unit volume. It is logical that linearity could not exist if the pore topology of the sample had varied in the direction of grinding and polishing, that is, normal to the planes of sectioning. It is likely that topology varies less than pore geometry, for example, pore size distributions.

The work of Macdonald et al.[144] is an improvement over that of Pathak et al.,[140] who performed a manual trace and count of branches and nodes.

3.4.3.2 Pore Structure Determination Based on Computer Reconstruction

In a series of articles Lin and co-workers[141, 142, 149] presented a deterministic approach to modeling the three-dimensional pore and grain geometry and pore network topology, based on computer reconstruction of se-

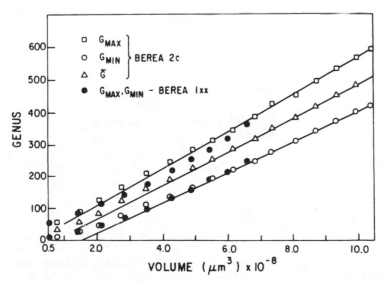

Figure 3.22. Genus versus volume of Berea sandstone samples.

rial sections. Lin and Perry[142] used a pore (or grain) surface triangulation technique as a shape descriptor, which gives the following parameters: surface area, Gaussian curvature, genus, and aspect ratio of the pore. The aspect ratio was obtained by using a spheroidal model. In their article, however, they pointed out that their method is not suitable for modeling the pore network. The method used by Lin and Cohen[141] is similar to the one described by DeHoff et al.[138] and Pathak et al.[140] In another study, Lin[149] carried out three-dimensional measurements in the pore space in the direction of the three orthogonal axes and then used these as parameters for pore models, consisting of ellipsoids, or elliptical cylinders or double elliptical cones.

The same set of 78 photomicrographs, representing serial sections through a Berea sandstone sample, that were processed by Macdonald et al.[144] were used for locating, at random, points in the digitized three-dimensional pore space and measuring, in three orthogonal directions, the length of straight lines passing through each point.[146] The set of the three orthogonal lengths measured was stored in the form of a joint distri-

bution function $f(a, b, c) \, da \, db \, dc$, with $2a$, $2b$, and $2c$ being the wall-to-wall lengths measured in the three orthogonal directions, as illustrated in Figure 3.23. In addition to the Berea sandstone sample, 80 serial sections were also prepared of a $4.3 \times 3.5 \times 1.4$ mm random glass bead pack at 15- to 20-μm increments and then photographed and digitized. The glass beads were in the 177- to 350-μm size range. Finally, three regular packings of uniform size spheres—(1) simple cubic, (2) orthorhombic, and (3) rhombohedral—were also tested. In this case there were no physical samples because the media could be described as continuous functions mathematically. They were chosen to test the method, owing to their known pore structures.

Using the number of random points generated, the sample porosities ϕ were calculated as follows:

$$\phi = \frac{\sum_{i=1}^{n} f(x, y, z)}{n} \qquad (3.52)$$

where n is the total number of points and $f(x, y, z) = (0, 1)$, where 1 represents pore space and 0 represents solid space. The results

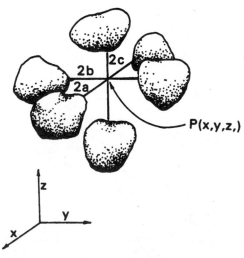

Figure 3.23. Determination of pore size in three orthogonal directions at a randomly chosen point $P(x, y, z)$. (After Yanuka et al.[146])

of the porosity determinations are given in Table 3.4.

The cumulative pore size distributions found in the samples in the x, y, and z coordinate directions, $f(2a)$, $f(2b)$, and $f(2c)$, are plotted in Figure 3.24. The z-direction is perpendicular to the plane of polish of the samples. It is apparent from the figure that the pore sizes range beyond 200 μm, in contrast with the maximum pore size of about 70 μm assumed in the network simulation studies[110, 112, 118, 119] which yielded good agreement with experimental drainage capillary pressure and relative permeability curves. The large wall-to-wall lengths measured by Yanuka et al.[146] are probably due to the presence of relatively large pore throats through which the line could pass, resulting in the combined size of several pores. Anisometric pore geometry may also contribute to this effect.

The joint distribution function was used also to obtain the minimum and the maximum harmonic mean pore radius R_{min} and R_{max} by forming the three possible combinations of pairs of the lengths (a, b), (a, c), and (b, c),

that is, $[1/2(1/a + 1/b)]^{-1}$, $[1/2(1/a + 1/c)]^{-1}$, and $[1/2(1/b + 1/c)]^{-1}$. Choosing the minimum and the maximum values of these gave frequency distribution densities $f(R_{min})$ and $f(R_{max})$, respectively. These were transformed to volume-based size distributions $V(R_{min})$ and $V(R_{max})$ by assuming pores of ellipsoidal shape. The pore size distributions of the Berea sandstone and the glass bead pack $V(R_{min})$ have been reproduced in Figure 3.25. It is evident from this figure that for the sandstone the values of R_{min} extended beyond 100 μm, consistently with the distributions shown in Figure 3.24.

The cumulative joint distribution function $F(a, b, c)$ was used to generate a model of the porous medium composed of ellipsoids distributed randomly in space. Random points were generated in a cube-shaped space which were used as centers of ellipsoids. Values of $F(a, b, c)$ between 0 and 1 were generated by a uniformly distributed random number generator and values of a, b, and c were obtained by taking the inverse of the function $F(a, b, c)$. The ellipsoids thus generated often inter-

Table 3.4. Porosities of the Different Porous Media

	TYPE OF MEDIUM				
NUMBER OF COUNTS AND REPETITIONS[a]	SIMPLE CUBIC PACKING (%)	ORTHORHOMBIC PACKING (%)	RHOMBOHEDRAL PACKING (%)	PACK OF GLASS BEADS (%)	BEREA SANDSTONE (%)
				From a total count of the digitized data	
	47.64 (exact)	39.54 (exact)	25.95 (exact)	38.4	23.6
				Experimental values obtained in bulk samples	
				38–39[b]	22–23.2[c]
100 × 10	46.30 ± 4.33	39.40 ± 4.09	23.30 ± 3.20		
1000 × 10	47.94 ± 2.44	38.91 ± 1.91	25.10 ± 1.60	38.89 ± 1.72	23.80 ± 1.40
10,000 × 10	47.55 ± 0.52	39.45 ± 0.51	25.80 ± 0.46	38.56 ± 0.35	23.69 ± 0.44
50,000 × 10	47.75 ± 0.23	39.60 ± 0.21	25.95 ± 0.16	38.69 ± 0.21	23.86 ± 0.15

After Yunuka et al.[146]

[a] Experiment repeated 10 times (e.g., an experiment of 100 points repeated 10 times).

[b] Calculated from the measured bulk density of the pack of glass beads and the density of the particles $B_d = 1.50$ to 1.53 and $p_d = 2.45$ g/cm^3, respectively.

[c] Calculated by taking the ratio of the measured volume of water filling the pore space under vacuum to the total volume of the sample.

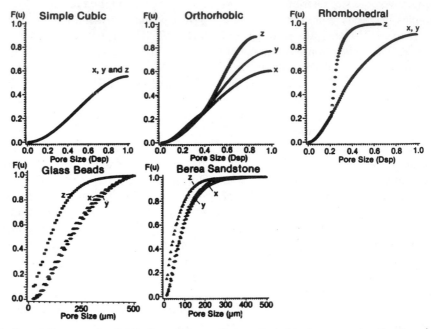

Figure 3.24. Cumulative pore size distributions of the media investigated in the x, y, and z directions ($n = 2a, 2b, 2c$). (After Yanuka et al.[146])

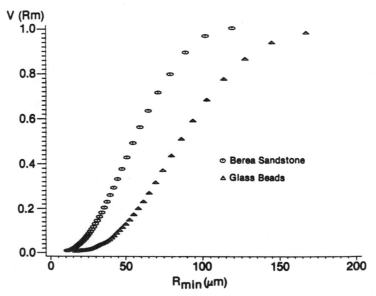

Figure 3.25. Cumulative normalized (volume-based) pore size distributions of a bead pack and a Berea sandstone sample. (After Yanuka et al.[146])

sected with each other and every time the volume of intersection was excluded. The random process of generating ellipsoids continued until the total volume of ellipsoids generated (excluding the volume of intersection) yielded the known porosity of the sample. The intersection between two ellipsoids was used to calculate the throat size between the two pores by calculating the radius of a sphere of the same volume as the volume of intersection between the two ellipsoids. The sphere radius r was assumed to represent the throat radius. The throat radius frequency distribution densities $f(r)$ of the different media are shown in Figure 3.26. For the Berea sandstone the peak of the distribution is at about 20 μm radius and the maximum radius is about 60 μm. These values are again much greater than the throat diameters, ranging from about 5 μm to about 42 μm, used in Refs. 110, 112, 118, and 119, which resulted in realistic predictions of the drainage capillary pressure and the relative permeability curves.

The average coordination number Z of the Berea sandstone sample was found to be 2.8, which is very close to the value of 2.9 calculated from the relation[147]:

$$1 - \phi = 1.072 - 0.1193Z + 0.004312Z^2 \tag{3.53}$$

Equation (3.53) has also predicted the average coordination number of the random glass bead pack (4.6 versus 4.3) and the (exact) coordination numbers of the three regular sphere packings (simple cubic: 5.8; orthorhombic: 4.7 versus 4.6; and rhombohedral: 3.1 versus 3.3). This relation, therefore, appears to be quite reliable for both regular and random structures.

The validity of the modeling approach used by Yanuka et al.[146] was checked also by comparing radii of the circles inscribed in the narrow passages of the three different regular sphere packings as calculated by Kruyer[148] with the average throat radii found in Yanuki et al. expressed in units of sphere diameter (simple cubic: 0.207 versus 0.156, orthorhombic: 0.142 versus 0.130, and rhombohedral: 0.077 versus 0.077). Evidently, this agreement is quite good and it seems to indicate that the

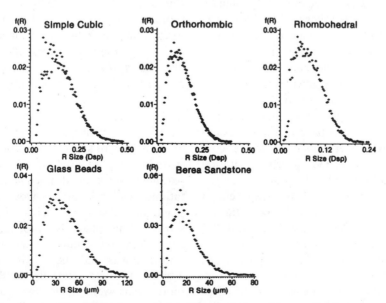

Figure 3.26. Pore throat radius distributions calculated from volumes of intersection between randomly chosen ellipsoids. (After Yanuka et al.[146])

method used in Yanuka et al.[146] may have some validity as long as the straight lines ("yardsticks") in the pore space cannot pass through pore throats as was indeed the case for the three regular sphere packs, where the throats were situated on the faces of the unit cell of the packing.

One important lesson learned from the study reported in Bernal et al.[50] is that the correct pore body sizes cannot be found in the computer reconstruction of pore structure unless first the throats are located and then partitions are erected at the throats that separate adjacent pore bodies. This procedure is analogous to closing the doors in a building that were originally wide open: as a result every room will be a separate, isolated entity whereas with the doors open one could walk freely from room to room. Similarly, with partitions erected at all the pore throats the "yardstick" used to measure pore body sizes cannot inadvertently measure the combined size of more than one pore body any more.

3.4.3.3 Method of Locating Pore Throats in Computer Reconstruction

The digitized serial sections (photomicrographs) were used as follows.[150, 151] Each pixel was assumed to be the top surface of a volume element (voxel) with a cross-section equal to the pixel area and a depth equal to the spacing between the two consecutive serial sections. When both the pixel and the one immediately below it are pore space pixels then the two two-dimensional pore space features containing these pixels were assumed to be connected. This is the same "overlap" criterion that was used by Macdonald et al.[144]

The approach followed by Kwiecien et al.[150, 151] was to first locate the pore throats and then, by symbolically closing them, define the pore bodies. A pore throat is defined as a local minimum in the "size" of pore space which thus separates two pore bodies from one another. As discussed earlier in this chapter, the most practical definition of pore radius is that it is twice the minimum value of the

ratio: area of cross-section passing through a fixed point in the pore space-to-the perimeter of this section [see Eq. (3.44)]. The ratio has, in general, different values for different orientations of the sectioning plane passing through the same fixed point and its minimum value is defined here as the "hydraulic radius" r_H of the pore at that point. Hence, pore throats, by definition, correspond to minima of the hydraulic radius r_H (minima of minima!). In principle, then, a pore throat could be located by passing series of parallel sectioning planes of all possible different orientations through the pore space, calculating the cross-section-to-perimeter ratio and keeping track of the location of each section through the pore space. Proceeding along each pore channel, for every fixed orientation of the sectioning plane there will be local minima of the cross-section-to-perimeter ratio, indicating the presence of a pore throat. The true throat size, that is, the hydraulic diameter equals four times the hydraulic radius, at a given location is found by varying the orientation of the sectioning plane over all possible angles until the least value of the cross-section-to-perimeter ratio is found near that location. The least value is, by definition, the hydraulic radius of that perpendicular throat. All the throats and their hydraulic diameters can be found, rigorously speaking, only by using parallel series of sectioning planes of all possible orientations with a sufficiently small interplanar distance.

The ideal way of locating pore throats, described previously, was replaced with the practical way of scanning the computer reconstruction of pore structure with a few sets of parallel planes of distinct, different orientations. The first and obvious plane is the plane of polish, or serial sectioning, of the sample. This plane is perpendicular to the z-axis. Next, the scanning planes perpendicular to the x-axis and the y-axis were used. In addition to these relatively simple cases, four more scans were made: two parallel to the y-axis and another two parallel to the x-axis, as illustrated in Figure 3.27. Had both the

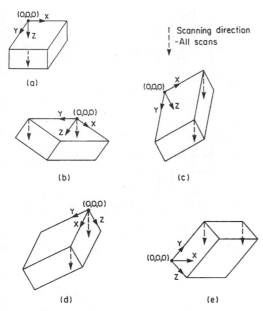

Figure 3.27. Orientations of the data matrix for various "diagonal" scans. (a) Original orientation, (b) scan with planes parallel to the y-axis (XMIN scan), (c) scan with planes parallel to the y-axis (XMAX scan), (d) scan with planes parallel to the x-axis (YMIN scan), (e) scan with planes parallel to the x-axis (YMAX scan). (After Kwiecien et al.[151])

sample and the voxels been exactly cube shaped, then all these scans would have been parallel to diagonal planes passing through two opposite edges of the cube. As the scanning was carried out in terms of pixels (or voxels) the diagonal scans are best understood by the example shown in Figure 3.28, where XMIN denotes one of the two diagonal scans parallel to the y-axis. N_x, N_y, N_z, and N_x', N_y', and N_z' denote the number of pixels in the three coordinate directions in the original arrangement and in the diagonal arrangement, respectively. Figure 3.28c shows the new overlap criterion for the diagonal scanning. (It should be noted that the pixel shape was not quadratic and was nonuniform because of the unequal spacings between consecutive serial sections. The pixel size in the x direction was 5.20 μm and in the z direction it varied from 6.5 μm to 17.8 μm.) For each scan there is a set of potential pore throats. These sets are compared to identify the true throats, using the principles outlined earlier. At the time of writing, the work of improving this technique is still in progress, because a number of throats appear to have been missed by the scanning and some other throats exhibit anomalous behavior.

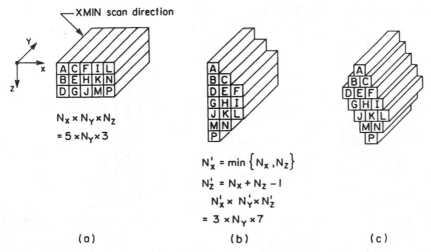

Figure 3.28. Example of data matrix transformation for "diagonal" scans. (a) Original data matrix, (b) new data matrix for XMIN scan: columns filled with zeroes are added (not shown) to fill out rectangular array, (c) overlap relationship of pixels for the data matrix in (b). (After Kwiecien et al.[151])

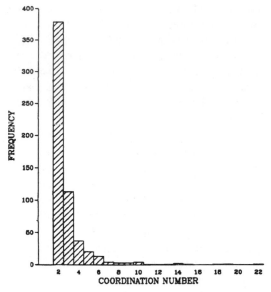

Figure 3.29. Frequency distribution of coordination numbers. (After Kwiecien.[150])

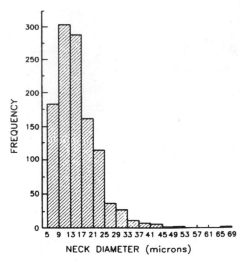

Figure 3.30. Frequency distribution of neck (pore throat) diameters. (After Kwiecien.[150])

After identifying a pore throat, a set of solid matrix voxels have been introduced in its place, thus separating the two adjacent pore bodies. The coordination number, that is, the number of throats belonging to each pore, has been determined. The volume of each pore body was directly obtained by adding up the volumes of the voxels contained in it. In addition, the dimensions of the smallest rectangular parallelepiped completely containing each pore body have also been determined. Some of the results are presented in Figures 3.29 to 3.31.

Figure 3.29 shows the frequency distribution of the coordination numbers. The average coordination number is 2.9, which is about the same as the value obtained in the ellipsoidal model in Ref. 146 and is exactly the same value as calculated by Eq. (3.53). The shape of the distribution is also very similar to the one obtained in Ref. 146.

The frequency distribution of pore throat diameters is presented in Figure 3.30. The average neck diameter was found to be 15.5 μm. The throat size distribution found is very close to the corresponding distribution

used with good results in the network simulations[110, 112, 118, 119] (see Fig. 3.14).

The frequency distribution of pore bodies, modeled as cubes, is shown in Figure 3.31. The number average pore body diameter, based on the cube model, was found to be about 29 μm. The frequency distribution of pore bodies, modeled as spheres, yields an average pore body diameter of about 36 μm. These values are very close to the distribution shown in

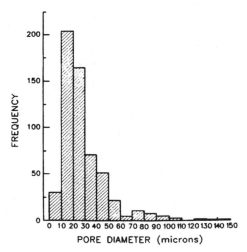

Figure 3.31. Frequency distribution of pore body diameters. Pore bodies modeled as cubes. (After Kwiecien.[150])

Figure 3.14, except for the apparent presence of a relatively small number of very large pores. These may be due to that fact that not all the pore throats were located, and, therefore, counting several distinct pore bodies as one and the same pore. The relatively small number of very large coordination numbers in Figure 3.29 is probably due to the same error.

3.4.4 Microscopic Distribution of the Wetting and the Nonwetting Phases in Immiscible Displacement

The distribution of the phases in the pore space in immiscible displacement is of great interest. It depends, in addition to the saturation, on the wettability conditions, the history (including the effect of parameters such as the capillary number, the viscosity ratio, and the individual viscosities), and, last, but not least, the pore structure. Pioneering work in this area has been reported in Ref. 152 in Berea sandstone for the special case of strong preferential wettability and quasistatic displacement (vanishingly small capillary number). The technique used consists of "phase immobilization."

A suitable pair of immiscible fluids have been used as the wetting and the nonwetting phases, one of which can be conveniently solidified in situ and the other which can be readily removed from the pore space after-

wards. The "empty pore space" permits conventional permeability measurements to be carried out instead of the usual steady-state relative permeability measurements. Resistivity index measurements can also be performed if the "empty pore space" is filled with an electrolyte solution. After these measurements are carried out the "empty pore space" is filled with another liquid of the type that can afterwards be solidified in situ. The rock matrix was also replaced with epoxy resin after etching with hydrofluoric acid and finally has either been polished or thin sections have been prepared. The following fluid pairs, representing the wetting and the nonwetting phases, respectively, have been used: System I—ethylene glycol/Wood's metal (alloy 158); System II—epoxy resin ERL 4206™/N_2 gas; System III—brine/styrene (containing benzoyl peroxide as the catalyst).

In Figure 3.32 a thin section shows the microscopic phase distribution in primary drainage obtained with the help of System II. The nonwetting phase channels were impregnated with Resin 301™, containing solvent blue dye. In Figure 3.32a a UV light source, and in Figure 3.32b normal light source was used. Figure 3.32c is a superimposition of Figures 3.32a and 3.32b, achieved through the controlled use of both light sources.

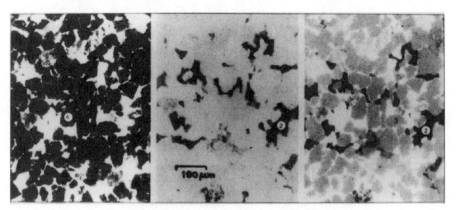

Figure 3.32. Microscopic distribution of fluids in a typical thin section of Berea sandstone at a wetting phase saturation of 53%, showing (*a*) the wetting phase only (white portion); (*b*) the nonwetting phase only (dark portion); (*c*) the wetting phase, nonwetting phase, and rock (white, black and gray portions, respectively). (After Yadav et al.[152])

In Figure 3.33 12 consecutive serial sections, prepared by the grinding and polishing procedure, are shown. System I was used in primary drainage. The wetting phase saturation is 58% pore volume. The ethylene glycol was replaced with ERL 4206.

Finally, in Figure 3.34 "relative permeability curves" obtained by conventional permeability measurements in the presence of another, immobilized phase are compared with the conventional steady-state relative permeability curves measured in a similar Berea sandstone.[120] The agreement is very good.

3.4.5 Discussion and Conclusions

Throughout the present chapter the position has been taken by the author that any model of pore structure should have as its first and foremost aim to approximate the significant features of the real pore structure of the sample of the porous medium as closely as possible and necessary. Those details of the pore structure that have no or only very little bearing on the transport properties of the medium are to be omitted, as they would unnecessarily increase the complexity of the model without any concomitant improvement in its predictive ability. The irrelevance of certain details may even lead to predictions that are at variance with experience in some cases whereas there may be other cases when a certain peculiar behavior of the medium can be explained only with the help of certain pore structure features that for most other purposes are irrelevant. As has always been the case in mathematical modeling of physical phenomena, judgment must be used in deciding what features to retain in the model and what other features to omit. While admittedly there exists a "gray zone" of uncertainty when deciding

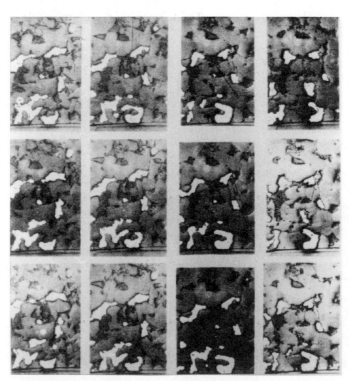

Figure 3.33. Twelve consecutive serial sections of etched Berea sandstone at about 10 μm apart, seen under normal light. The white portions are Wood's metal, representing the nonwetting phase. The dark gray portions are resin ERL 4206, replacing ethylene glycol, the wetting phase. The lighter gray areas are Buehler resin, replacing the rock that was etched away. (After Yadav et al.[152])

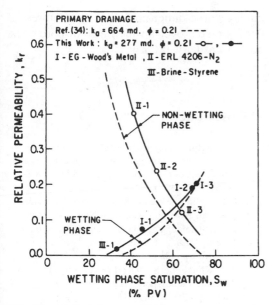

Figure 3.34. Relative permeability versus saturation curves for Berea sandstone sample obtained using the phase immobilization technique compared with the curves obtained by the usual steady-state technique.[120] (After Yadav et al.[152])

where to draw the fine line between what is kept and what is discarded as superfluous, that does not in the least put in jeopardy the requirement that *any* model of pore structure should account for the main features of the real pore structure that determine the collection of the most important transport properties of the medium. A rough comparison may be made with the blueprint of a building where all the essential constructional features are shown, however, without specifying the location of every hole to be drilled in the walls, the quality of wall surface, etc.

It is generally realized that it is possible to model transport properties of porous media without any reference to pore geometry, and merely use a large number of adjustable parameters in the model that do not have any physical meaning. In this author's opinion, models of this kind are less useful in facilitating our understanding of the observed physical phenomena than those that incorporate the basic features of pore morphology. This point

of view finds ample support in the excellent critical studies published by van Brakel[155] and van Brakel and Heertjes.[156]

A good pore structure model should simulate with the same values of the parameters the effective molecular diffusion coefficient, the absolute permeability, the dispersion coefficient (function of Peclet number), drainage and imbibition capillary pressure curves, dendritic portion of the nonflowing parts of saturations, saturation versus height of capillary rise, rate of capillary rise, relative permeabilities versus saturation (the last five are also contact angle and history dependent), formation factor, resistivity index, and drying.

At the present there is no proven model that would be able to simulate all the above properties and, therefore, there is no guarantee that the following requirements regarding a good pore structure model would be sufficient. In any event, they are most likely to be necessary to do the job:

1. A three-dimensional network of pore bodies connected by pore throats, representing the main skeleton of pore structure of the medium
2. A representative coordination number distribution and the connectivity of the network
3. Representative pore body and pore throat shapes (aspect ratios)
4. Representative pore body size and pore throat size distributions
5. A representative correlation (if present) between the pore throat sizes and the sizes of the two pore bodies connected by a throat
6. Similar properties of secondary networks of smaller (micro-) pores if such are present (e.g., cementing clays in sandstones or micropores present in the individual particles of aggregates).

An additional requirement for the purpose of predicting surface transport properties is the quality of the pore surface (rugosity).

The only way it appears possible to obtain all this body of information is by visualization of the pores and, at present, for meso- and micropores this is possible only by preparing micrographs of sections of the sample. [For macropores on the order of about 1 mm and above X-ray tomography (CAT-scanners) can do an excellent job.]

This is the reason why the author has chosen the route of three-dimensional computer reconstruction of the pore structure from serial sections of the sample. This technique, however, has the drawback of requiring lengthy and painstakingly careful sample preparation and it is restricted to pore sizes down to about 5 μm, because the precision of preparing serial sections is a few microns at best in the interplanar distance between consecutive sections. Hence there is no point in trying to prepare serial sections with a spacing of less than about 10 μm. Additional difficulties must be overcome in making the spacings as uniform and the sections as plane and parallel as possible.

Hence it would be of great advantage to be able to avoid having to make serial sections and work instead with two-dimensional sections from which the three-dimensional pore structure would be reconstructed. A major contribution to this technique has been made by Quiblier,[157] who has shown that it is possible to perform image analysis on thin sections of porous media, resulting in an autocorrelation function and a probability density function of optical densities in the micrograph analyzed, and use this information to generate a three-dimensional pore structure that may have the same morphology as the sample from which the thin section was prepared. It appears that this technique would actually be simpler to use in the case of polished sections containing Wood's metal in the pore space because of the higher contrast between pore space and solid rock matrix. The method needs much more extensive testing before it can be accepted and used routinely, but the time and money would be spent on a very worthwhile project because it has a good chance of achieving its objective, that is, to have a relatively fast and convenient, routine method of determining three-dimensional pore structures of porous media over a wide range of pore sizes, down to at least 0.1 μm.

REFERENCES

1. H. Rumpf and A. R. Gupte, "Einflüsse der Porosität and Korngrössenverteilung im Widerstandgesetz der Porenströmung," *Chem. Ing. Tech.* *43*:367–375 (1971).
2. R. E. Collins, *Flow of Fluids through Porous Materials*, Van Nostrand Reinhold, New York (1961).
3. A. E. Scheidegger, *The Physics of Flow through Porous Media*, University of Toronto Press, Toronto, Canada (1974).
4. L. J. Klinkenberg, "The Permeability of Porous Media to Liquids and Gases," *A.P.I. Drill. Prod. Pract.*, pp. 200–213 (1941).
5. J. Bear, *Dynamics of Fluids in Porous Media*, American Elsevier, New York (1972).
6. R. M. Barrer, *In The Solid-Gas Interface*, E. A. Flood (ed.), Vol. II, Dekker, New York, pp. 557–609 (1967).
7. F. A. L. Dullien, *Porous Media-Fluid Transport and Pore Structure*, Academic Press, New York (1979).
8. L. Green and P. Duwez, "Fluid Flow through Porous Media," *J. Appl. Mech. 18*:39–45 (1951).
9. S. Ergun, "Fluid Flow through Packed Columns," *Chem. Eng. Progr. 48*:89–94 (1952).
10. N. Ahmed and D. K. Sunada, "Nonlinear Flow in Porous Media," *J. Hyd. Div. Proc.* ASCE 95(HY6):1847–1857 (1967).
11. P. C. Carman, "Fluid Flow through a Granular Bed," *Trans. Int. Chem. Eng. London 15*:150–156 (1937).
12. P. C. Carman, "Determination of the Specific Surface of Powders," *J. Soc. Chem. Ind. 57*:225–234 (1938).
13. P. C. Carman, *Flow of Gases through Porous Media*, Butterworths, London (1956).
14. J. Kozeny, "Concerning Capillary Conduction of Water in the Soil (Rise, Seepage and Use in Irrigation)," Royal Academy of Science, Vienna, *Proc. Class I 136*:271–306 (1927).
15. R. B. MacMullin and G. A. Muccini, "Characteristics of Porous Beds and Structures," *AIChE J. 2*:393–403 (1956).
16. M. I. S. Azzam and F. A. L. Dullien, "Application of Numerical Solution of the Complete Navier-Stokes Equation to Modelling the Permeability of Porous Media," *IEC Fundamentals 15*:281–285 (1976).

17. M. I. S. Azzam and F. A. L. Dullien, "Flow in Tubes with Periodic Step Changes in Diameter: A Numerical Solution of the Navier-Stokes Equations," *Chem. Eng. Sci. 32*:1445–1455 (1977).

18. I. F. Macdonald, M. S. El-Sayed, K. Mow, and F. A. L. Dullien, "Flow through Porous Media—the Ergun Equation Revisited." *Ind. Eng. Chem. Fundamentals 18*:199–208 (1979).

19. A. R. Guptc, Dr.-Ing. Ph.D. dissertation, University of Karlruhe, Germany (1970).

20. N. R. Morrow, "Physics and Thermodynamics of Capillary Action in Porous Media," *Ind. Eng. Chem. 62*:32–56 (1970). Reprinted as a chapter in *Flow through Porous Media*, American Chemical Society, Washington, D.C. (1970).

21. N. R. Morrow and C. C. Harris, "Capillary Equilibrium in Porous Materials," *Soc. Petrol. Eng. J. 5*:15–24 (1965).

22. D. H. Everett, *The Solid-Gas Interface*," E. A. Flood (ed.), Vol. II, Dekker, New York, pp. 1055–1070 (1967).

23. F. A. L. Dullien and G. K. Dhawan, "Bivariate Pore Size Distribution of Some Sandstones," *J. Interface Colloid Sci. 52*:129–135 (1975).

24. G. C. Topp and E. E. Miller, "Hysteretic Moisture Characteristics and Hydraulic Conductivities for Glass-bead Media," *Soil Sci. Soc. Am. Proc. 30*:156–162 (1966).

25. A. Poulovassilis, "Hysteresis of Pore Water, an Application of the Concept of Independent Domains," *Soil Sci. 93*:405–412 (1962).

26. A. Poulovassilis, "Hysteresis of Pore Water in Granular Porous Bodies," *Soil Sci. 109*:5–12 (1970).

27. N. R. Morrow and N. Mungan, "Mouillabilité et Capillarité en Milicux Porcux," *Revue I.F.P. 26*:629–650 (1971).

28. F. F. Craig, Jr., "The Reservoir Engineering Aspects of Water-Flooding," *Society of Petroleum Engineers of AIME*, Monograph, Vol. 3, Dallas, Texas (1971).

29. N. R. Morrow, "Capillary Pressure Correlations for Uniformly Wetted Porous Media," *J. Can. Pet. Technol. 15*:49–69 (1976).

30. J. M. Dumoré and R. S. Schols, "Drainage Capillary Pressure Functions and the Influence of Connate Water," *Soc. Pet. Eng. J. 14*:437–444 (1974).

31. M. Muskat, *Physical Principles of Oil Production*, McGraw-Hill, New York (1949).

32. W. O. Smith, "Minimum Capillary Rise in an Ideal Uniform Soil," *Physics 4*:184–193 (1933).

33. J. W. Amyx, D. M. Bass Jr., and R. L. Whiting, *Petroleum Reservoir Engineering*, McGraw-Hill, New York (1960).

34. M. S. El-Sayed, Ph.D. dissertation, University of Waterloo, Canada (1978).

35. E. E. Petersen, "Diffusion in a Pore of Varying Cross Section," *AIChE J. 4*:343–345 (1958).

36. A. S. Michaels, "Diffusion in a Pore of Irregular Cross Section: A Simplified Treatment," *AIChE J. 5*:270–271 (1959).

37. J. A. Currie, "Gaseous Diffusion in Porous Media, Part 2—Dry Granular Materials, *Brit. J. Appl. Phys. 11*:318–324 (1960).

38. F. A. L. Dullien, "Prediction of 'Tortuosity Factors' from Pore Structure Data," *AIChE J. 21*:299, 820–822 (1975).

39. R. E. De La Rue and C. W. Tobias, "On the Conductivity of Dispersions," *J. Electrochem. Soc. 106*:827–833 (1959).

40. M. R. J. Wyllie, *The Fundamentals of Electrical Log Interpretation*, Academic Press, New York (1957).

41. D. P. Haughey and G. S. C. Beveridge, "Structural Properties of Packed Beds—A Review," *Can. J. Chem. Eng. 47*:130–140 (1969).

42. L. Fejes Toth, *Lagerungen in der Ebene, auf der Kugel und in Raum*, Springer, Berlin (1953).

43. J. Hrubisek, "Filtration–Geometry of Systems of Spheres," *Kolloid-Beihefte 53*:385–452 (1941).

44. L. C. Graton and H. J. Fraser, "Systematic Packing of Spheres—with Particular Relation to Porosity and Permeability," *J. Geol. 43*:785–909 (1935).

45. L. K. Frevel and L. J. Kressley, "Modifications in Mercury Porosimetry," *J. Anal. Chem. 35*: 1492–1502 (1963).

46. R. P. Mayer and R. A. Stowe, "Mercury Porosimetry—Breakthrough Pressure for Penetration Between Packed Spheres," *J. Coll. Sci. 20*:893–911 (1965).

47. R. P. Mayer and R. A. Stowe, "Mercury Porosimetry: Filling of Toroidal Void Volume Following Breakthrough Between Packed Spheres," *J. Phys. Chem. 70*:3867–3873 (1966).

48. S. Debbas and H. Rumpf, On the Randomness of Beds Packed with Spheres or Irregular Shaped Particles, *Chem. Eng. Sci. 21*:583–607 (1966).

49. E. E. Underwood, *Quantitative Stereology*, Addison-Wesley, Reading, Mass. (1970).

50. J. D. Bernal, J. Mason, and K. R. Knight, Letter, *Nature 194*:957–958 (1962).

51. G. D. Scott, "Radial Distribution of the Random Close Packing of Equal Spheres," *Nature 194*: 956–957 (1962).

52. G. D. Scott, A. M. Charlesworth, and M. K. Mak, "On the Random Packing of Spheres, *J. Chem. Phys. 40*:611–612 (1964).

53. G. Mason and W. Clark, "Distribution of Near Neighbours in a Random Packing of Spheres," *Nature 207*:512 (1965).

54. G. Mason and W. Clark, "Fine Structure in the Radial Distribution Function from a Random Packing of Spheres," *Nature 211*:957 (1966).

55. W. O. Smith, P. D. Foote, and P. F. Busang, "Capillary Rise in Sands of Uniform Spherical Grains," *Physics 1*:18–26 (1931).

56. F. B. Hill and R. H. Wilhelm, "Radiative and Conductive Heat Transfer in a Quiescent Gas-Solid Bed of Particles: Theory and Experiment," *AIChE J. 5*:486–496 (1959).

57. W. O. Smith, P. D. Foote, and P. F. Busang, "Capillary Rise in Sands of Uniform Spherical Grains," *Phys. Ref. 34*:1271 (1921).

58. E. Manegold, R. Hoffmann, and K. Solf, "Capillary Systems—XII Mathematical Treatment Ideal Sphere Packing and the Free Space of Actual Frame-like Structures," *Kolloid-Z 56*:142–159 (1931).

59. E. Manegold and W. VonEngelhardt, "Capillary Systems XII (2). The Calculation of Content of Substance of Homogeneous Frame-like Structures. I. Sphere Planes and Sphere Layers as Structure Elements of Sphere Lattices," *Kolloid-Z 62*:285–294 (1933).

60. J. D. Bernal and J. Mason, "Co-ordination of Randomly Packed Spheres," *Nature 188*:910–911 (1960).

61. W. H. Wade, "The Co-ordination Number of Small Spheres," *J. Phys. Chem. 69*:322–326 (1965).

62. K. Ridgway and K. J. Tarbuck, "The Random Packing of Spheres," *Brit. Chem. Eng. 12*:384–388 (1967).

63. W. E. Ranz, "Friction and Transfer Coefficients for Single Particles and Packed Beds," *Chem. Eng. Progr. 48*:247–253 (1952).

64. H. Susskind and W. Becker, "Random Packing of Spheres in Non-rigid Containers," *Nature 212*:1564–1565 (1966).

65. J. C. Macrae and W. A. Gray, "Significance of the Properties of Materials in the Packing of Real Spherical Particles," *Brit. J. Appl. Phys. 12*:164–172 (1961).

66. J. D. Bernal and J. L. Finney, "Random Packing of Spheres in Non-rigid Containers," *Nature 214*:265–266 (1967).

67. R. Rutgers, "Packing of Spheres, *Nature 193*:465–466 (1962).

68. G. D. Scott, "Packing of Equal Spheres," *Nature 188*:908–909 (1960).

69. G. Sonntag, "Einfluß des Lückenvolumens auf den Druckverlust in Gasdurchströmten Füllkörpersäulen," *Chem. Ing. Tech. 32*:317–329 (1960).

70. R. Jeschar, "Pressure Loss in Multiple-Size Charges Consisting of Globular Particles," *Archiv. Eisenhüttenwesen*, p. 91 (1964).

71. W. H. Denton, "The Packing and Flow of Spheres," AERE-E/R-1095, Gt. Brit., Atomic Energy Research Establishment, Harwell, England, pp. 22 (1953).

72. R. K. McGeary, "Mechanical Packing of Spherical Particles," *J. Am. Ceramic Soc. 44*:513–522 (1961).

73. J. Wadsworth, Nat. Res. Council of Canada, Mech. Eng. Report MT-41-(NRC No. 5895) (February, 1960).

74. J. Wadsworth, "An Experimental Investigation of the Local Packing and Heat Transfer Processes in Packed Beds of Homogeneous Spheres," *International Developments in Heat Transfer Proceedings*: American Society of Mechanical Engineers, pp. 760–769 (1963).

75. A. E. R. Westman and H. R. Hugill, "The Packing of Particles," *J. Am. Ceram. Soc. 13*:767–779 (1930).

76. R. J. Parsick, S. C. Jones, and L. P. Hatch, "Stability of Randomly Packed Beds of Fuel Spheres," *Nucl. Appl. 2*:221–225 (1966).

77. O. K. Rice, "On the Statistical Mechanics of Liquids, and the Gas of Hard Elastic Spheres, *J. Chem. Phys. 12*:1–18 (1944).

78. N. Epstein and M. M. Young, "Random Loose Packing of Binary Mixtures of Spheres," *Nature 196*:885–886 (1962).

79. S. Ergun and A. A. Orning, "Fluid Flow Through Randomly Packed Columns and Fluidized Beds," *Ind. Eng. Chem. 41*:1179–1184 (1949).

80. D. E. Lamb and R. H. Wilhelm, "Effects of Packed Bed Properties on Local Concentrations and Temperature Patterns," *I and EC Fund 2*:173–182 (1963).

81. H. H. Steinour, "Rate of Sedimentation, Non-flocculated Suspensions of Uniform Spheres," *Ind. Eng. Chem. 36*:618–624 (1944).
H. H. Steinour, "Rate of Sedimentation, Suspensions of Uniform-size Angular Particles," *Ind. Eng. Chem. 36*:840–847 (1944).
H. H. Steinour, "Rate of Sedimentation, Concentrated Flocculated Suspensions of Powders," *Ind. Eng. Chem. 36*:901–907 (1944).

82. J. Happel, "Pressure Drop Due to Vapor Flow Through Moving Beds," *Ind. Eng. Chem. 41*:1161–1174 (1949).

83. A. O. Oman and K. M. Watson, "Pressure Drops in Granular Beds," *Natl. Petrol. News 36*:R795–802 (1944).

84. M. Rendell, J. Ramsey, *Soc. Chem. Eng. 10*:31 (1963).

85. R. F. Benenati and C. B. Brosilow, "Void Fraction Distribution in Beds of Spheres," *AIChE 8*:359–361 (1962).

86. L. H. S. Roblee, R. M. Baird, and J. W. Tierney, "Radial Porosity Variations in Packed Beds," *AIChE 4*:460–464 (1958).

87. G. Speck, Über die Randgängigkeit in Gasdurch-strömten Füllkörperschüttungen, dissertation TH Dresden (1955).

88. J. T. Chiam, "Voidage and Fluid distribution in Packed Beds," Ph.D. thesis, Manchester College of Technology, 1962.

89. M. C. Thadani and F. N. Pebbles, "Variation of Local Void Fraction in Randomly Packed Beds of Equal Spheres," *I and EC Process Design and Development* 5:265–268 (1966).

90. M. Kimura, K. Nono, and T. Kaneda, "Distribution of Void in Packed Tubes, *Chem. Eng.* (Japan) *19*:397–400 (1955).

91. C. C. Furnas, "Grading Aggregates I—Mathematical Relations for Beds of Broken Solids of Maximum Density," *Ind. Eng. Chem.* 23:1052–1058 (1931).

92. H. Y. Sohn and C. Moreland, "The Effect of Particle Size Distribution on Packing Density," *Can. J. Chem. Eng.* 46:162–167 (1968).

93. H. T. Horsfield, "Strength of Asphalt Mixtures," *J. Soc. Chem. Ind.* 53:107–115T (1934).

94. B. S. Newman, *Flow Properties of Disperse Systems*, J. J. Hermans (ed.), *Interscience Pub.*, Chapter 10 (1953).

95. J. C. Macrae, P. C. Finlayson, and W. A. Gray, "Vibration Packing of Dry Granular Solids," *Nature* 179:1365–1366 (1957).

96. J. Kolkuszewski, "Notes on the Deposition of Sand," *Research, Lond.,* 3:478–483 (1960).

97. L. J. Klinkenberg, "Pore size distribution of porous media and displacement experiments with miscible liquids," *Pet. Trans. Am. Inst. Min. Eng.* 210, 366 (1957).

98. F. A. Dullien, "One and two phase flow in porous media and pore structure." In "Physics of Granular Media' (Daniel Bideau and John Dodds, eds.) Nova Science Publishers, Inc. (1991).

99. P. C. Carman, "Capillary rise and capillary movement of moisture in fine sands," *Soil Sci.* 52, 1. (1941).

100. I. Fatt, "The network model of porous media I. Capillary pressure characteristics," *Pet. Trans. AIME* 209, 114 (1956).

101. I. Fatt, "The network model of porous media II. Dynamic properties of singe size tube network," *Pet. Trans. AIME* 207, 160 (1956).

102. I. Fatt, "The network model of porous media III. Dynamic properties of networks with tube radius distribution," *Pet. Trans. AIME* 207, 164 (1956).

103. R. Simon and F. J. Kelsey, "The use of capillary tube networks in reservoir performance studies: I. Equal-viscosity miscible displacements," *Soc. Petroleum Engrs. J.* 11, 99 (1956).

104. R. Simon and F. J. Kelsey, "The Use of Capillary Tube Networks in Reservoir Performance Studies: II. Effect of Heterogeneity and Mobility on Misci-ble Displacement Efficiency." *Soc. Petroleum Eng. J. 12*:345 (1972).

105. S. R. Broadbent and J. M. Hammersley, "Percolation Processes. I. Crystals and Mazes," *Proc. Cambridge Philos. Soc.* 53:629 (1951).

106. V. K. S. Shante and S. Kirkpatrick, "An Introduction to Percolation Theory," *Adv. Phys.* 42:385 (1971).

107. S. Kirkpatrick, "Percolation and Conduction." *Rev. Math. Phys.* 45:574 (1973).

108. I. Chatzis and F. A. L. Dullien, "Modeling Pore Structure by 2-D and 3-D Networks with Application to Sandstones." *J. Can. Petr. T.* 16:97 (1977).

109. I. Chatzis and F. A. L. Dullien, "A Network Approach to Analyze and Model Capillary and Transport Phenomena in Porous Media." *Proceedings of the IAHR Symposium: Scale Effects in Porous Media*, Thessaloniki, Greece, Aug. 29–Sept. 1 (1978).

110. I. Chatzis and F. A. L. Dullien, "The Modelling of Mercury Porosimetry and the Relative Permeability of Mercury in Sandstones Using Percolation Theory." *ICE 25*:1, 47 (1985).

111. N. C. Wardlaw, "Pore Geometry of Carbonate Rocks as Revealed by Pore Casts and Capillary Pressure." *Am. Assoc. Pet. G. Bull. 60*(2):245 (1976).

112. C. E. Diaz, I. Chatzis, and F. A. L. Dullien, "Simulation of Capillary Pressure Curves Using Bond Correlated Site Percolation on a Simple Cubic Network." *Transp. Por. Media* 2:215 (1987).

113. R. Lenormand and C. Zarcone, "Role of Roughness and Edges During Imbibition in Square Capillaries," SPE 13264. Paper presented at the 59th Annual Technical Conference and Exhibition of the Society of Petroleum Engineers of AIME, Houston, TX, Sept. 16–19 (1984).

114. Y. Li and N. C. Wardlaw, "Mechanisms of Nonwetting Phase Trapping During Imbibition at Slow Rates," *J. Colloid Interface Sci. 109*:473 (1986).

115. I. Chatzis and F. A. L. Dullien, "Dynamic Immiscible Displacement Mechanisms in Pore Doublets: Theory versus Experiment." *J. Colloid Interface Sci. 91*:199 (1983).

116. F. A. L. Dullien, F. S. Y. Lai, and I. F. Macdonald, "Hydraulic Continuity of Residual Wetting Phase in Porous Media." *J. Colloid Interface Sci. 109*:201 (1986).

117. F. A. L. Dullien, C. Zarcone, I. F. Macdonald, A. Collins, and R. D. E. Bochard, "The Effects of Surface Roughness on the Capillary Pressure Curves and the Heights of Capillary Rise in Glass Bead Packs." *J. Colloid Interface Sci. 127*:362 (1989).

118. A. Kantzas and I. Chatzis, "Network Simulation of Relative Permeability Curves Using a Bond

Correlated-Site Percolation Model of Pore Structure." *Chem. Eng. Commun. 69*:191 (1988).

119. A. Kantzas and I. Chatzis, "Application of the Preconditioned Conjugate Gradient Method in the Simulation of Relative Permeability Properties of Porous Media." *Chem. Eng. Commun. 69*:169 (1988).

120. P. K. Shankar and F. A. L. Dullien, "Experimental Investigation of Two-Liquid Relative Permeability and Dye Adsorption Capacity versus Saturation Relationships in Water-Wet and Dry-Film-Treated Sandstone Samples," in *Third International Conference on Surface and Colloid Science: Surface Phenomena in Enhanced Oil Recovery*, edited by D. O. Sha, Plenum Press (1981).

121. P. H. Winterfeld, L. E. Scriven, and H. T. Davis, "Percolation and Conductivity of Random Two-Dimensional Composites." *J. Phys. Chem. 14*:2361 (1981).

122. R. Chandler, J. Koplik, K. Lerman, and J. F. Willemsen, "Capillary Displacement and Percolation in Porous Media." *J. Fluid Mech. 119*:249 (1982).

123. K. K. Mohanty and S. J. Salter, "Flow in Porous Media II. Pore Level Modeling," SPE paper no. 11018, presented at the 57th Annual Fall Technical Conference and Exhibition of SPE of AIME, New Orleans, LA, Sept. 26–29 (1982).

124. D. Wilkinson and J. F. Willemsen, "Invasion Percolation: A New Form of Percolation Theory." *J. Phys. A Math. Gen. 16*:3365 (1983).

125. A. A. Heiba, M. Sahimi, L. E. Scriven, and H. T. Davis, "Percolation Theory of Two Phase Relative Permeability," SPE Paper No. 11015, presented at the 57th Annual Fall Technical Conference of SPE-AIME, New Orleans, Sept. 26–29 (1982).

126. R. G. Larson, L. E. Scriven, and H. T. Davis, "Percolation Theory of Two Phase Flow in Porous Media." *Chem. Eng. Sci. 36*:75 (1981).

127. J. Koplik, C. Lin, and M. Vermette, "Conductivity and Permeability from Microgeometry." *J. Appl. Phys. 56*:3127 (1984).

128. E. Guyon, J. P. Hulin, and R. Lenormand, "Application de la Percolation a la Physique des Milieux Poreux." *Ann. Mines*, mai-juin, p. 17 (1984).

129. D. Wilkinson and M. Barsony, "Monte Carlo Study of Invasion Percolation Clusters in Two and Three Dimensions. *J. Phys. A Math. Gen. 17*:L129 (1984).

130. D. Wilkinson, "Percolation Model of Immiscible Displacement in the Presence of Buoyancy Forces." *Phys. Rev A 30*:520 (1984).

131. D. Wilkinson, "Percolation Effects in Immiscible Displacement." *Phys Rev. A 34*:1380 (1986).

132. M. M. Dias and D. Wilkinson, "Percolation with Trapping." *J. Phys. A 19*:3131 (1986).

133. L. de Arcangelis, J. Koplik, S. Redner, and D. Wilkinson, "Hydrodynamic Dispersion in Network Models of Porous Media." *Phys. Rev. Lett. 57*:996 (1986).

134. Y. Li, W. G. Laidlaw, and N. C. Wardlaw, "Sensitivity of Drainage and Imbibition to Pore Structures as Revealed by Computer Simulation of Displacement Process." *Adv. Colloid Interface Sci. 26*:1 (1986).

135. R. Lenormand, E. Touboul, and C. Zarcone, "Numerical Models and Experiments on Immiscible Displacements in Porous Media." *J. Fluid Mech. 189*:165 (1988).

136. H. F. Fischmeister, "Pore Structure and Properties of Materials," in *Proceedings of the International Symposia RILEM/UPAC*, Prague, Sept. 18–21, 1973, Part II, C435, Academic, Prague (1974).

137. R. T. DeHoff and F. N. Rhines (eds.), *Quantitative Microscopy*, McGraw-Hill, New York (1968).

138. R. T. DeHoff, E. H. Aigeltinger, and K. R. Craig, "Experimental Determination of the Topological Properties of Three-Dimensional Microstructures." *J. Microsc. 95*:69 (1972).

139. R. T. DeHoff, "Quantitative Serial Sectioning Analysis: Preview." *J. Microsc. 131*:259 (1983).

140. P. Pathak, H. T. Davis, and L. E. Scriven, "Dependence of Residual Nonwetting Liquid on Pore Topology." SPE Preprint 11016, 57th Annual SPE Conference, New Orleans (1982).

141. C. Lin and M. H. Cohen, "Quantitative Methods for Microgeometric Modeling." *J. Appl. Phys. 53*:4152 (1982).

142. C. Lin and M. J. Perry, "Shape Description Using Surface Triangularization." *Proceedings IEEE Workshop on Computer Visualization*, N.H (1982).

143. P. M. Kaufmann, F. A. L. Dullien, I. F. Macdonald, and C. S. Simpson, "Reconstruction, Visualization and Topological Analysis of Sandstone Pore Structure." *Acta Stereol. 2 (Suppl. I)*:145 (1983).

144. I. F. Macdonald, P. Kaufmann, and F. A. L. Dullien, "Quantitative Image Analysis of Finite Porous Media. I. Development of Genus and Pore Map Software." *J. Microsc. 144*:277; II. Specific Genus of Cubic Lattice Models and Berea Sandstone." Ibid. *144*:297 (1986).

145. L. K. Barrett and C. S. Yust, "Some Fundamental Ideas in Topology and Their Application to Problems in Metallography." *Metallography 3*:1 (1970).

146. M. Yanuka, F. A. L. Dullien, and D. E. Elrick, "Percolation Processes and Porous Media. I. Geometrical and Topological Model of Porous Media Using a Three-Dimensional Joint Pore Size Distribution." *J. Colloid Interface Sci. 112*:24 (1986).

147. H. L. Ridgway and K. J. Tarbuk, "The Random Packing of Spheres." *Br. Chem. Eng. 12*:384 (1967).

148. S. Kruyer, "The Penetration of Mercury and Capillary Condensation in Packed Spheres." *Trans. Faraday Soc. 54*:1758 (1958).

149. C. Lin, "Shape and Texture from Serial Contours." *J. Int. Assoc. Math. Geol. 15*:617 (1983).

150. M. J. Kwiecien, "Determination of Pore Size Distributions of Berea Sandstone Through Three-Dimensional Reconstruction." M.A.Sc. Thesis, University of Waterloo (1987).

151. M. J. Kwiecien, I. F. Macdonald, and F. A. L. Dullien, "Three-Dimensional Reconstruction of Porous Media from Serial Section Data." *J. Microsc. 159*, 343 (1990).

152. G. D. Yadav, F. A. L. Dullien, I. Chatzis, and I. F. Macdonald, "Microscopic Distribution of Wetting and Nonwetting Phases During Immiscible Displacement." *SPE Reservoir Eng. 2*:137 (1987).

153. H. H. Yuan and B. F. Swanson, "Resolving Pore-Space Characteristics by Rate-Controlled Porosimetry." *SPE Formation Evaluation 4*:17 (1989).

154. F. A. L. Dullien, "New Permeability Model of Porous Media." *AIChE J. 21*:299 (1975).

155. J. van Brakel, "Pore Space Models for Transport Phenomena in Porous Media. Review and Evaluation with Special Emphasis on Capillary Liquid Transport." *Powder Technol 11*:205 (1975).

156. J. van Brakel and P. M. Heertjes, "Capillary Rise in Porous Media. Part I: A Problem, *Powder Technol. 16*; Part II: Secondary Phenomena. Ibid. *16*:83; Part III: Role of the Contact Angle." Ibid. *16*; 91 (1977).

157. J. A. Quiblier, "A New Three-Dimensional Modeling Technique for Studying Porous Media." *J. Colloid Interface Sci. 98*:84 (1984).

158. G. E. Archie, "The Electrical Resistivity Log as an Aid in Determining Some Reservoir Characteristics." *Trans. AIME 146*:54–62 (1942).

4

Fundamental and Rheological Properties of Powders

Kunio Shinohara

CONTENTS

Powders exhibit several kinds of bulk properties such as mechanical, thermal, electrical, magnetic, optical, acoustic, and surface physico-chemical properties. Among these, the rheological property of particles is widely investigated in the applied fields. It is closely related not only to the material properties of a single particle but also to the unit operations in powder technology. Included here are the most fundamental characteristics of deformation and flow of particulate solids, that is, packing, permeability and strength, mainly in the dry system. Thus, the bulk properties are essential for describing various rheological behaviors in powder handling processes, and constitute one of the current topics in the field of particle science.

4.1 PACKING CHARACTERISTICS OF PARTICLES

Each packing particle has unique physical properties such as size, density, shape, restitution, etc., as mentioned in earlier chapters. Though a powder consists of a number of individual particles, the bulk property of a powder is not usually the simple summation of the physical properties of single particles.

General characterization of the particle has not yet been established, and it is difficult to define the location of each particle under the influence of external and/or self-exerting forces in open, as well as closed, systems. Information is, however, available on the packing structure of particle assemblage as a basis of the rheological properties of particles.[1]

4.1.1 Representative Parameters of Packing[2]

There are some fundamental and useful representations of the overall state of packing.

Void fraction or fractional voidage ϵ is defined as the interstitial void volume in the unit bulk volume of particle assemblage.

Packing density or fractional solids content ϕ_p is then defined in terms of the void fraction as:

$$\phi_p = 1 - \epsilon \qquad (4.1)$$

Bulk density ρ_p is the apparent density of a powder mass, given by the mass per bulk volume of powder. Thus, the following relationship holds among the void fraction, the particle density ρ_p, and the bulk density:

$$\rho_b = \rho_p(1 - \epsilon) \qquad (4.2)$$

Apparent specific volume V_s is the bulk volume of powder of unit mass, which is the inverse of bulk density as:

$$V_s = \frac{1}{\rho_b} = \frac{1}{\rho_p(1 - \epsilon)} \qquad (4.3)$$

Bulkiness ϕ_b indicates the bulk volume of solids in comparison with the unit volume of particles alone, that is, the inverse of packing density as:

$$\phi_b = \frac{1}{\phi_p} = \frac{1}{1 - \epsilon} \qquad (4.4)$$

Void ratio ϕ_v is written as the ratio of the void volume to the net volume of particles:

$$\phi_v = \frac{\epsilon}{1 - \epsilon} \qquad (4.5)$$

The coordinate number N_c, which is defined as the number of contact points per particle, is often considered to relate to the rheological behaviors of particle assemblage.

Specific surface is the surface area of particles per unit mass S_w, per net volume of solids S_v, or per apparent volume of powder mass S_{av}. The relationship among them is:

$$S_w = \frac{S_v}{\rho_p} = \frac{S_{av}}{\rho_p(1 - \epsilon)} = \frac{\phi_s}{\rho_p \cdot d_{p,av}} \qquad (4.6)$$

where ϕ_s is the shape factor, and $d_{p,av}$ is the average particle diameter based on the specific surface.

Tortuosity is defined as the ratio of the length of the hypothetical curved capillary consisting of voids to the thickness of a powder layer.

Other ways of representing the distribution of voids are available, as mentioned later in the section on the random packing of equal spheres.

4.1.2 Regular Packing of Spheres

Though only limited cases are known of regular packing of spheres, the geometrical arrangements will be the basis of understanding the general state of packing of particles.

4.1.2.1 Packing of Equal-Sized Spheres

There are two types of primary layers of equal spheres, that is, square and simple rhombic layers. Four central points of spheres form a square on the same plane, and three centers of spheres in contact make a triangle, respectively. Thus, six variations are considered as stable geometrical arrangements in placing such a primary layer as the bottom one, as shown in Figure 4.1.[3] It is possible to introduce the concept of a unit cell for these arrangements, as shown in Figure 4.2.[3] The corresponding packing characteristics are listed in Table 4.1. In fact, except for direction of the arrangement of voids, the second and the third types of arrangements in the square layer are the same as the fourth and the sixth types in the rhombic layer, respectively.

It could be inferred in general that the void fraction decreases and the coordination num-

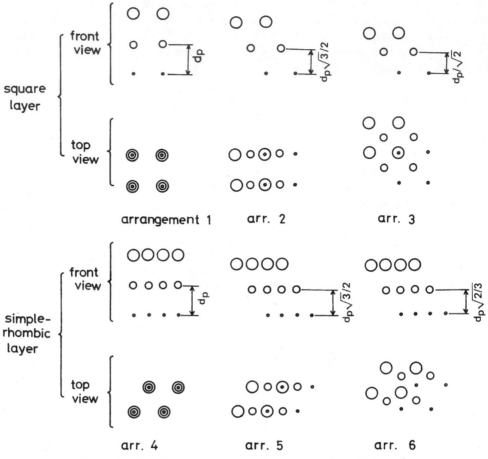

Figure 4.1. Regular arrangements of primary layers of equal spheres.[3]

ber increases with increasing degree of deformation along four different kinds of regular arrangements of equal spheres.

Of interest is that ellipsoidal particles yield the same void fractions for the stable and the unstable packing arrangements as those of the closest and loosest of the spheres, respectively.[4]

4.1.2.2 Packing of Different-Sized Spheres

Interspaces among equal spheres in regular packing can ideally be filled with smaller spheres to attain a higher density for the assemblage.

When only one small sphere is placed in each void, the diameter of which is the maximum to fit the void space, the packing characteristics are as given in Table 4.2.[5]

After the square hole among six equal spheres in the rhombohedral arrangement is filled with one secondary sphere of maximum size, the triangular hole surrounded by four primary spheres is occupied by the third largest sphere. Furthermore, the fourth and the fifth largest spheres are packed into the interspaces between the secondary and primary spheres and between the third and primary spheres, respectively. All the remaining voids are finally filled up with considerably smaller spheres of equal size in the closest rhombohedral pack-

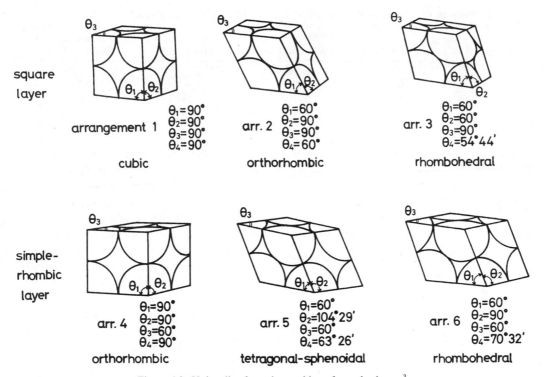

Figure 4.2. Unit cells of regular packing of equal spheres.[3]

ing. Table 4.3 presents the packing characteristics indicating the minimum voidage of 0.039. This is the so-called Horsfield packing.[6,7]

When more than one equal sphere is filled into the interspaces of the closest rhombic arrangement, the void fraction varies with the size ratio of the smaller sphere to the primary one, as tabulated in Table 4.4. It appears that the void fraction decreases with each increase in the number of smaller spheres in the square hole, but this is not always true because of the intermittent number of spheres in the triangular hole. A void fraction of 0.1130 is the minimum at the size ratio of 0.1716 on the basis of the triangular hole. This arrangement is called Hudson packing.[8]

4.1.3 Random Packing of Equal Spheres

Even in spheres of equal size the geometrical structure of random packing deviates far from that of regular packing. In other words, the

Table 4.1. Packing Properties of Unit Cell[3]

ARRANGEMENT NO.	BULK VOLUME	VOID VOLUME	VOID FRACTION	COORDINATION NO.
1	1	0.4764	0.4764	6
2	$\sqrt{3}/2$	0.3424	0.3954	8
3	$1/\sqrt{2}$	0.1834	0.2594	12
4	$\sqrt{3}/2$	0.3424	0.3954	8
5	$3/4$	0.2264	0.3019	10
6	$1/\sqrt{2}$	0.1834	0.2595	12

Table 4.2. Packing Properties of Mixture with One Largest Sphere in Each Void[5]

ARRANGEMENT	VOIDAGE	DIAMETER OF SMALL SPHERE	VOIDAGE OF MIXTURE	VOLUME RATIO OF SMALL SPHERE
Cubic	0.4764	$0.723d_p$	0.279	0.274
Orthorhombic	0.3954	$0.528d_p$	0.307	0.128
Rhombohedral	0.2595	$0.225d_p$	0.199	0.011
		$0.414d_p$		0.066

characteristics of random packing are closer to the actual ones. Thus, computer simulation of random packing is becoming popular, as mentioned below.

4.1.3.1 Overall Packing Characteristics

In reality randomness is always associated with the effects of particle properties, ways of filling, and the dimension of the container and its wall surface properties.

On the basis of the usual packing experiments under gravity alone, the overall void fraction is approximately 0.39 and the coordinate number is around 8 for relatively large spheres such as steel balls, round sand,[9] and glass beads.[5]

When spherical particles of about 3 mm diameter are poured without free fall, the datum value of voidage for the loose packing ranged from 0.393 to 0.409 for different particle densities and surface friction.[10]

Without wall effects the void fractions in the loose and close packings were 0.399 and 0.363, respectively, which were extrapolated by filling dimpled copper cylinders of various heights and diameters. For dense packing, steel balls of 3.18 mm were gently shaken down for about 2 min. Nonrigid balloons were also filled

Table 4.3. Properties of Horsfield Packing[6,7]

SPHERES	SIZE RATIO	NUMBER OF SPHERES	VOIDAGE OF MIXTURE
Primary	1.0	—	0.260
Secondary	0.414	1	0.207
Ternary	0.225	2	0.190
Quaternary	0.175	8	0.158
Quinary	0.117	8	0.149
Filler	Fines	Many	0.039

with the balls to confirm $\epsilon = 0.37$ for dense random packing with small peripheral error.[11]

On tapping vertically the same steel balls in the glass cylinder, the void fraction becomes 0.387 after 400 taps, which is close to that of the orthorhombic packing, whereas three-dimensional vigorous and prolonged shaking yielded the hexagonal close packing, whose void fraction if 0.26.[12] The structure was examined by removal of layers and arrays of balls and individual balls frozen in water as the thawing progressed.

According to experiments with spherical lead shots of 7.56 mm diameter poured into a beaker,[13] the relationship between the average coordination and the voidage could be derived by assuming that the state of packing is represented by the mixture of cubic and rhombohedral packing in between the two.[14] The void fraction is written by using the fraction of the rhombohedral packing, R_r as

$$\epsilon = 0.2595 R_r + 0.4764(1 - R_r) \quad (4.7)$$

The average coordination number N_c is then given by:

$$N_c = \frac{12\sqrt{2}\,R_r + 6(1 - R_r)}{\sqrt{2}\,R_r + (1 - R_r)} \quad (4.8)$$

Here the volumes of unit cells for cubic and rhombohedral packings are 1 and $1/\sqrt{2}$, the numbers of spheres per unit volume are 1 and $\sqrt{2}$, and the coordination numbers are 6 and 12, respectively. Thus, eliminating R_r in Eqs. (4.7) and (4.8) leads to ϵ versus N_c as

$$\epsilon = \frac{0.414 N_c - 6.527}{0.414 N_c - 10.968} \quad (4.9)$$

Table 4.4. Properties of Hudson Packing[8]

SYMMETRICAL PACKS GOVERNED BY DIMENSIONS OF THE SQUARE INTERSTICE

$n_{p,s}$	STOICHIOMETRY	SQUARE INTERSTICE			TRIANGULAR INTERSTICE		TOTAL DENSITY INCREMENT
		PACKING	d_{ps}/d_p	DENSITY INCREMENT	$n_{p,s}$	DENSITY INCREMENT	
1	Simple cube	Tight	0.4142	0.07106	0	—	0.07106
2	Along cube diagonal	Tight	0.2753	0.04170	0	—	0.04170
4	Crossed parallel face diagonals	Tight	0.2583	0.06896	0	—	0.06896
6	Centers of cube faces	Tight	0.1716	0.03028	4	0.04038	0.07066
8	Simple cube	Tight	0.2288	0.09590	0	—	0.09590
9	Body-centred cube	Tight	0.2166	0.09150	1	0.02034	0.11184
14	Face-centred cube	Slack	0.1716	0.07074	4	0.04042	0.11116
16	Concentric simple cubes	Slack	0.1693	0.07768	4	0.03882	0.11647
17	Concentric cubes, body-centred	Slack	0.1652	0.07660	4	0.03605	0.11265
21	Hopper-faced cube, body-centred	Slack	0.1782	0.11892	1	0.01132	0.13025
26	Hopper-faced cube, face-centred	Slack	0.1547	0.09626	4	0.02962	0.12588
27	Simple cube	Slack	0.13807	0.07108	5	0.02632	0.09740

SYMMETRICAL PACKS GOVERNED BY DIMENSIONS OF THE TRIANGULAR INTERSTICE

$n_{p,s}$	STOICHIOMETRY	SQUARE INTERSTICE		TRIANGULAR INTERSTICE					TOTAL DENSITY INCREMENT
		PACKING	DENSITY INCREMENT	$n_{p,s}$	$d_{p,s}/d_p$	PACKING	STOICHIOMETRY	DENSITY INCREMENT	
8	Simple cubic	Slack	0.09083	1	0.22475	Tight	Single	0.02271	0.11354
21	Hopper-faced cube, body-centred	Slack	0.10611	4	0.1716	Tight	Tetrahedral	0.04042	0.14653
26	Hopper-faced cube, face-centred	Slack	0.07457	5	0.14208	Tight	Body-centred tetrahedral	0.02868	0.10325

Based on the number density of a shell-like distribution about the central sphere, the coordination number was derived from the structure of the first-layer neighbors as:[15, 16]

$$N_c = \frac{2.812(1 - \epsilon)^{-1/3}}{(b_1/d_p)^2 + \left\{1 + (b_1/d_p)^2\right\}} \quad (4.10)$$

where b_1/d_p is obtained from

$$(1 - \epsilon)^{-1/3}$$

$$= \frac{1 + (b_1/d_p)^2}{1 + (b_1/d_p)\exp\left\{(d_p/b_1)^2\right\} \cdot \text{erfc}(d_p/b_1)} \quad (4.11)$$

The packing density of spherical particles, $1 - \epsilon$, increases with the diameter ratio of the vessel to the sphere up to about 10; above 10 the packing density becomes nearly constant, 0.62.[17, 18]

Purely by data correlation, the parabolic curve was found to fit well in the wide range of void fractions:[19]

$$\epsilon = 1.072 - 0.1193N_c + 0.00431N_c^2 \quad (4.12)$$

In the range of void fractions between 0.259 and 0.5 a model gives:[20]

$$N_c = 22.47 - 39.39\epsilon \quad (4.13)$$

4.1.3.2 Local Packing Characteristics

The point of contact between rigid spheres is classified into close and near contacts, which are distinguished by a black paint ring with a clear center and a black spot, respectively. In random close packing with a particle friction of 0.62, the model coordinate number lies between 8 and 9, the largest number having between 6 and 7 close contacts and between 1 and 2 near contacts. In random loose packing of $\phi_p = 0.6$, the modes were between 7 and 8 for the total coordination and between 5 and 6 for close contracts, the means of which were 7.1 total and 5.5 close contacts.[21]

The local voidage of randomly packed spheres fluctuates in the vessel, as illustrated in Figure 4.3, according to the results of various types of measurements.[22-25] The cyclic damping curve coincides irrespective of the sphere size, taking the distance in sphere diameters, and the wall effect disappears at a distance of about 4 or 5. Similar fluctuation curves are also observed in the radial direction with respect to the average number of sphere centers per unit area in the spherical shell, curve A,[20] and the total number of spheres cut per unit area by a spherical envelope, curve B[26] (Fig. 4.4).

Based on the distribution function of the distance from a reference particle, the number

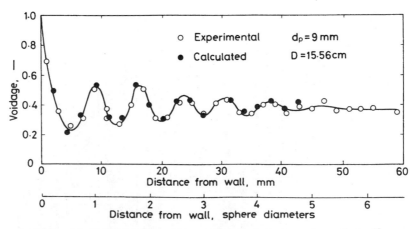

Figure 4.3. Voidage variation for randomly packed spheres in a cylinder.[25]

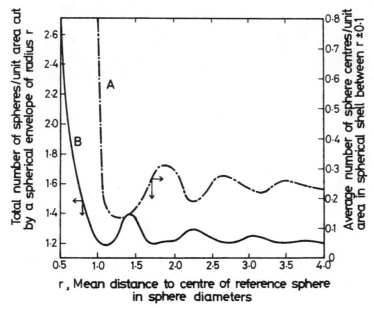

Figure 4.4. Variation of number of spheres in radial direction, A^{20} and $B.^{26}$

of spheres within the spherical shell of differential thickness is theoretically calculated[27] and compared with the packing data of an average number of neighboring particles, as shown in Figure 4.5.[28,29,30] The average number of spheres in contact is 6.0, irrespective of the packing density.[28,29]

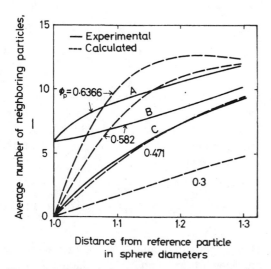

Figure 4.5. Average number of neighboring particles within a spherical shell.[27] (A,[28] B,[29] and C[30] are experimental.)

Angular distribution of contacting spheres around a sphere could represent the packing structure of a randomly packed bed, as shown in Figure 4.6, which compares the experimental curve A^{31} with the theoretical curve $B.^{32}$ Any given contact point is taken as a pole. Analogously to the liquid structure,[33] the relationship between the coordination number and the packing density is approximated by

$$\phi_p = 0.1947N_c - 0.1301N_c^2 + 0.05872N_c^3$$
$$- 0.0128N_c^4 + 1.438 \times 10^{-3}N_c^5$$
$$+ 8.058 \times 10^{-5}N_c^6 + 1.785 \times 10^{-6}N_c^7$$

$$(4.14)$$

with sufficient accuracy for $\phi_p \geq 0.15$. The local mean packing density distribution function f_p is derived on the basis of allocation of spheres to space cells according to a binomial probability mechanism:[34]

$$f_p \, d\phi_{p1} = \frac{1}{\sqrt{2\pi} \, \zeta_p}$$
$$\cdot \exp\left[-\frac{1}{2}\left(\frac{\phi_{p1} - \phi_p}{\zeta_p}\right)^2\right] d\phi_{p1}$$

$$(4.15)$$

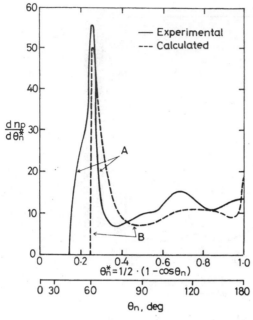

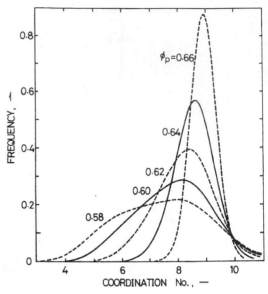

Figure 4.6. Angular distribution of contacting spheres around a sphere.[1] (A[31] is experimental and B[32] is theoretical.)

Figure 4.7. Distribution of coordination number.[33]

where ϕ_{p1} and ϕ_p are the local mean and the bulk mean packing density, respectively. The standard deviation ζ_p becomes

$$\zeta_p = 4.75(0.7405 - \phi_p)^2 \cdot \phi_{p1} \sqrt{\pi/6} \quad (4.16)$$

Putting Eqs. (4.14) and (4.16) into Eq. (4.15) gives the distributions of the coordination number, as shown in Figure 4.7, which agree with the data.[21]

4.1.3.3 Microscopic Packing Structure[35,36]

The microscopic structure of a bed of spherical particles can be expressed by the size distribution of voids among the particles in two dimensions. The size distribution of the particle sections over a cross-section of the bed composed of differently sized spheres is geometrically given in terms of the number of the particle sections, $N_A(i,j)$, per unit area as

$$N_A(i,j) = N_v(j)\Delta\left\{\sqrt{j^2 - (i-1)^2} - \sqrt{j^2 - i^2}\right\} \quad (4.17)$$

where i and j are the size-class numbers of the particle section and the particle diameter, respectively. $N_v(j)$ indicates the number of particles of size j per unit bulk volume. On the other hand, the probability, $F(a_0)$, of having no particles of any size within an inspection area, a_0, in a cross-sectional area, A, of voidage, ϵ, is written as

$$F(a_0) = \prod_{i=1}^{n}\left(1 - \frac{a_0 - a_{p,i}}{\epsilon A}\right) \quad (4.18)$$

In the general case of an elliptical inspection area of long axis, $(d_{p,i} + d_{v,l})$, and short axis, $(d_{p,i} + d_{v,s})$, a_0 is given by Eq. (4.19), and the arrangement of the regular-shaped voids is shown in Figure 4.8:

$$a_0 = \frac{\pi}{4}(d_{p,i} + d_{v,1})(d_{p,i} + d_{v,s}) \quad (4.19)$$

Hence, the probability, $P(\chi)$, of having no particles of area equivalent diameter, $d_{v,c}$, inside the elliptical space is derived from Eq.

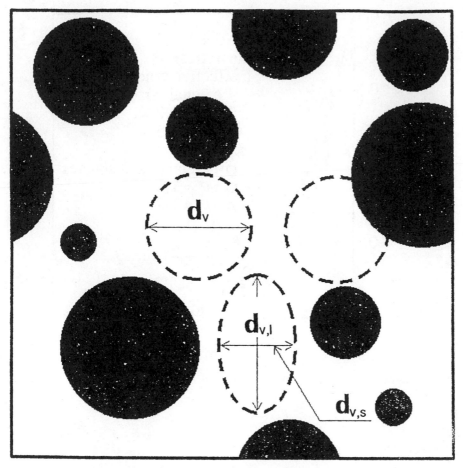

Figure 4.8. Arrangement model of regular-shaped voids.[36]

(4.18) as:

$$P(\chi) = \epsilon \prod_{i=1}^{n} \left[1 - \frac{1-\epsilon}{\epsilon} \left\{ \left(\frac{d_{p,i}}{d_{p,\max}} + \frac{d_{v,1}}{d_{p,\max}} \right) \right. \right.$$

$$\times \left(\frac{d_{p,i}}{d_{p,\max}} + \frac{d_{v,s}}{d_{p,\max}} \right)$$

$$\left. \left. - \left(\frac{d_{p,i}}{d_{p,\max}} \right)^2 \right\} \middle/ \sum_{j=1}^{n} \left(\frac{d_{p,j}}{d_{p,\max}} \right)^2 \right]$$

$$(4.20)$$

where χ is the dimensionless void diameter defined as $d_{v,c}/d_{p,\max}$, and $P(\chi = 0)$ corresponds to ϵ. Therefore, when the size distribu-

tion of particles on a number basis and the voidage are known, the particle section number of each diameter is obtained from Eq. (4.17). Thus, the existence probability of voids of χ is yielded by Eq. (4.20) as a kind of microscopic representation of the packing structure. Figure 4.9 illustrates this for spherical particles of geometric standard deviation, σ_g, and log-normal particle size distribution.

4.1.4 Packing of General Particles

Particles are not always spherical, packed regularly, or perfectly at random. Thus, the following packing characteristics are known to be useful in practice.

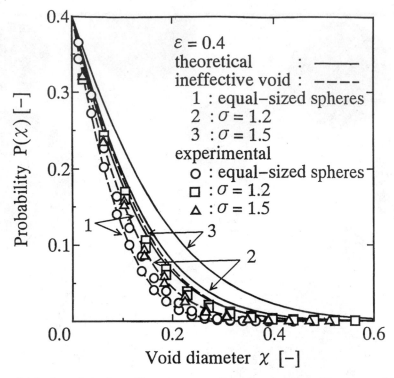

Figure 4.9. Size distribution of circular voids over cross-section of packed bed with spheres of log-normal size distribution.[36]

4.1.4.1 Overall Degree of Packing

The bulk density of general particles within a vessel decreases under gravity alone with decreasing diameter of the container and increasing height of the particle bed. Higher filling rate gives rise to smaller bulk density for coarse particles, but for fine cohesive powder such as flour, dilute feeding yields a loose packing.[37]

The void fraction increases with decreasing sphericity in general, as shown in Figure 4.10.[38] Here the sphericity is defined as the ratio of the sphere surface to the irregular particle surface of the same volume. But it is also reported that angular particles have a large void fraction in loose packing, and it is inverse in close packing.[39] Particles of higher surface roughness exhibit higher voidage, as shown in Figure 4.11.

Smaller particles have higher voidage because of cohesion between particles, contrast-

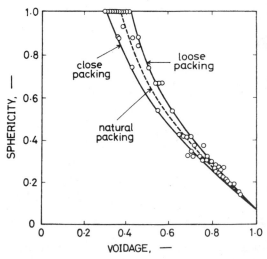

Figure 4.10. Relationship between void fraction and sphericity.[38]

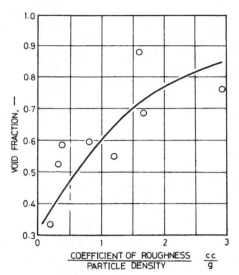

Figure 4.11. Influence of particle surface roughness on void fraction.[39]

ing with the ideal independence of particle size of void fraction.[40, 41] Thus, the apparent volume of moist powder grows larger with increasing water content.[42]

Particles of distributed size tend to yield closer packing like spheres. But, in general, it is difficult to evaluate theoretically the void fraction corresponding to the mixing ratio and the particle size ratio. An experimental relationship is available, as shown in Figure 4.12.[43]

4.1.4.2 Packing Model of Powder Mass[44, 45]

In order to express an uneven packing structure of general particles over a wide range of void fraction, a mathematical model is proposed (illustrated in Fig. 4.13). A unit cross-sectional area of powder mass is assumed to be composed of two kinds of regular packing portions of equivalent spheres, the cubic packing R_c and the rhombohedral packing R_r, and an effective void ϵ_e that is independent of the particle packing portions. These three kinds of subdivisions are mixed at random over a certain cross-section area. The void fraction ϵ is then written as the sum of the voids by:

$$\epsilon = 0.476R_c + 0.260R_r + \epsilon_e \quad (4.21)$$

Since

$$\epsilon_e = 1 - (R_c + R_r) \quad (4.22)$$

putting Eq. (4.22) into Eq. (4.21) gives

$$\epsilon = 1 - (0.524R_c + 0.740R_r) \quad (4.23)$$

The area ratio R of the loosest to the closest packing portion is assumed to increase with

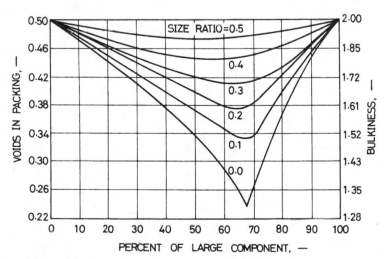

Figure 4.12. Relation between voids and size composition in two-component system of broken solids when the voids of single components are 0.5.[43]

 CUBIC PACKING PORTION, Rc

RHOMBOHEDRAL PACKING PORTION, Rr

EFFECTIVE VOID, Ee

Figure 4.13. Nonuniform packing model of powder mass.[44]

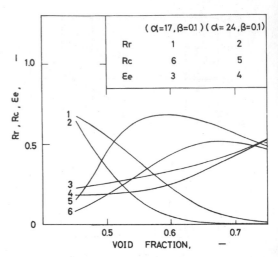

Figure 4.14. Variation of R_c, R_r, and ϵ_e with void fraction.[45]

the void fraction along consolidation of powder mass by:

$$\frac{dR}{d\epsilon} = \alpha R + \beta \qquad (4.24)$$

where

$$R = R_c/R_r \qquad (4.25)$$

and α and β are positive constants. Integrating Eq. (4.24) for the initial condition of $R = 0$ in $\epsilon = 0.260$, when the cubic packing portion disappears ($R_c = 0$) and all the portions are packed in the rhombohedral packing ($R_r = 1$), we find that

$$R = \frac{\beta}{\alpha} \{e^{\alpha(\epsilon - 0.260)} - 1\} \qquad (4.26)$$

Hence, from Eqs. (4.22), (4.23), (4.25), and (4.26), each portion is obtained as a function of the void fraction, as depicted in Figure 4.14.

$$R_c = \frac{(1 - \epsilon)\{e^{\alpha(\epsilon - 0.260)} - 1\}}{0.524\{e^{\alpha(\epsilon - 0.260)} - 1\} + 0.740\alpha/\beta} \qquad (4.27)$$

$$R_r = \frac{(1 - \epsilon)\alpha/\beta}{0.524\{e^{\alpha(\epsilon - 0.260)} - 1\} + 0.740\alpha/\beta} \qquad (4.28)$$

$$\epsilon_e = \frac{(\epsilon - 0.476)\{e^{\alpha(\epsilon - 0.260)} - 1\} + (\epsilon - 0.260)\alpha/\beta}{0.524\{e^{\alpha(\epsilon - 0.260)} - 1\} + 0.740\alpha/\beta} \qquad (4.29)$$

These results are applied to analyses of shear and tensile strength, as mentioned later.

Corresponding to the ideal maximum tensile strength, the relationship between the coordination number and the voidage is obtained from the present packing model as:

$$N_c = \frac{3}{2} \frac{e^{\alpha(\epsilon - 0.260)} - 1 + 3.99\alpha/\beta}{e^{(\epsilon - 0.260)} - 1 + 1.41\alpha/\beta} \qquad (4.30)$$

Figure 4.15 shows a comparison of Eq. (4.30) with other empirical equations and data, including those for spheres mentioned previously:[46-48]

$$N_c \epsilon = \pi \qquad (4.31)$$

$$N_c = 19.3 - 28\epsilon \qquad (4.32)$$

$$N_c = 20.0(1 - \epsilon)^{1.7} \qquad (4.33)$$

The coordination number, N_{cd}, with particles of distributed size, d_p, is derived as a function of the voidage and the median diameter, $d_{p,50}$:[48]

$$N_{cd} = \frac{8(7 - 8\epsilon)(d_p + d_{p,50})}{13d_{p,50}^2} \qquad (4.34)$$

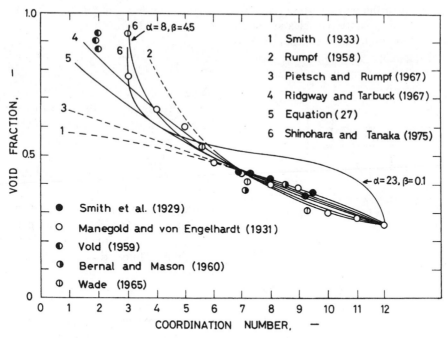

Figure 4.15. Relationship between coordination number and void fraction.[45]

4.1.4.3 Closest Packing of Different Sized Particles[49]

In a binary system of particles, interspaces among large particles are filled up with small ones to give the closest packing arrangement. The masses of large and small particles within the mixture of unit bulk volume are written, respectively, as:

$$W_1 = 1 \cdot (1 - \epsilon_1) \cdot \rho_{p1} \qquad (4.35)$$

$$W_s = 1 \cdot \epsilon_1(1 - \epsilon_s) \cdot \rho_{ps} \qquad (4.36)$$

Thus, the mass fraction f_1 of large particles is

$$f_1 = \frac{W_1}{W_1 + W_s} = \frac{(1 - \epsilon_1)\rho_{p1}}{(1 - \epsilon_1)\rho_{p1} + \epsilon_1(1 - \epsilon_s)\rho_{ps}} \qquad (4.37)$$

For the same solid material as the single components of equal voidage, that is, $\rho_{p1} = \rho_{ps}$ and $\epsilon_1 = \epsilon_s = \epsilon$, the volume fraction of large particles becomes equal to f_1 as

$$f_1 = 1/(1 + \epsilon) \qquad (4.38)$$

where small particles should be completely involved in the matrix of large ones. Then the size ratio is lower than about 0.2.[8]

In the multicomponent system of particles consisting of the same solid material, the interspaces among primary large particles are filled up with secondary small particles, the interspaces of which are packed by the tertiary small ones. Following the same way of packing by further smaller particles, the net particle volume of each component V in the bulk volume of the mixture per unit binary particle volume, $V_m = 1/(1 - \epsilon^2)$, is given as:

$$V_1 = V_m(1 - \epsilon) = f_1$$

$$V_2 = \epsilon V_m \cdot (1 - \epsilon) = 1 - f_1$$

$$V_3 = \epsilon \cdot \epsilon V_m \cdot (1 - \epsilon) = (1 - f_1)\epsilon$$

$$V_4 = \epsilon \cdot \epsilon^2 V_m \cdot (1 - \epsilon) = (1 - f_1)\epsilon^2$$

$$\vdots$$

$$V_n = (1 - f_1)\epsilon^{n-2} \qquad (4.39)$$

where the sums of the particle volumes of the primary and the secondary particles are taken as unity for computational convenience. Substituting Eq. (4.38) and summing up the volumes of all the components in Eq. (4.39) yields the following:

$$V_{tf} = \frac{1}{1 + \epsilon} + \left(1 - \frac{1}{1 + \epsilon}\right) + \left(1 - \frac{1}{1 + \epsilon}\right)\epsilon$$

$$+ \left(1 - \frac{1}{1 + \epsilon}\right)\epsilon^2 + \cdots$$

$$+ \left(1 + \frac{1}{1 + \epsilon}\right)\epsilon^{n-2}$$

$$= \frac{1 - \epsilon^n}{1 - \epsilon^2} \qquad (4.40)$$

Hence, the volume fraction of each component is obtained by dividing $V_1, V_2, V_3, \ldots, V_n$ by V_{tf}. Equation (4.40) is for the hypothetical case where each size acts as if it were infinitely small.

In the actual case of several different components uniformly mixed, the total volume of the mixed system V_{tm} is somewhat reduced as compared to the sum of the volume of separate layers of the components, V_{ts}

$$V_{tm} = \{V_{ts} - f_y(V_{ts} - f_1)\} \cdot (\rho_p / \rho_b) \qquad (4.41)$$

where ρ_b is the bulk density of each separate layer, and $-f_y(V_{ts} - f_1) \cdot (\rho_p / \rho_b)$ indicates the bulk volume decrease upon mixing and f_y is the factor ranging in value between 0 and 1.0 corresponding to separate layers of equal-sized particles and an ideal mixing with infinitely small particles. Thus, the bulk density of the mixed system is

$$\rho_{b,m} = \frac{\rho_b}{1 - \frac{f_y(V_{ts} - f_1)}{V_{ts}}} \qquad (4.42)$$

For the closest packing or the maximum bulk density, the quantity $f_y(V_{ts} - f_1)/V_{ts}$ in the denominator should be a maximum so that when it is differentiated with respect to the number of component sizes added to the

original largest single size of the system, $n' = n - 1$, the result is

$$\frac{f_1}{V_{ts}(V_{ts} - f_1)} \cdot \frac{dV_{ts}}{dn'} = -\frac{df_y}{f_y \cdot dn'} \qquad (4.43)$$

Since the diameter ratio between particles of successive sizes for the maximum density is independent of the voidage and must be constant for the entire system,

$$\frac{d_{p,2}}{d_{p,1}} = \frac{d_{p,3}}{d_{p,2}} = \frac{d_{p,n'+1}}{d_{p,n'}} = K_s^{1/n'} \qquad (4.44)$$

where K_s is defined as the size ratio of the smallest $d_{n'+1}$ to the largest particles, d_{p1}. Utilizing the experimental correlation between the total volume decrease f_y and the size ratios of binary systems,

$$f_y = 1.0 - 2.62 K_s^{1/n'} + 1.62 K_s^{2/n'} \qquad (4.45)$$

Hence, differentiating Eqs. (4.40) and (4.45) with respect to n' and putting them into Eq. (4.43) gives the relationship among ϵ, K_s and n', as

$$\frac{\epsilon^{n'} \cdot \ln \epsilon (1 - \epsilon)}{(1 - \epsilon^{n'+1})(1 + \epsilon^{n'})}$$

$$= \frac{(2.62 K_s^{1/n'} - 3.24 K_s^{2/n'})\ln K_s}{(1.0 - 2.62 K_s^{1/n'} + 1.62 K_s^{2/n'})n'^2} \qquad (4.46)$$

According to the above equations, the minimum voidage is calculated from Eq. (4.42). For example, the results for packings of two- to four-component systems are shown in Figure 4.16 and listed in Table 4.5.

For varying voidages and particles densities, it is also possible to use a similar treatment.

$$W_t = f_1 + (1 + f_1) + (1 - f_1) \cdot \left(\frac{1 - f_2}{f_2}\right)$$

$$+ (1 - f_1)\left(\frac{1 - f_2}{f_2}\right)\left(\frac{1 - f_3}{f_3}\right) + \cdots \qquad (4.47)$$

$$f_n = \frac{(1 - \epsilon_n)\rho_{p,n}}{(1 - \epsilon_n)\rho_{p,n} + \epsilon_n(1 - \epsilon_{n+1})\rho_{p,n+1}} \qquad (4.48)$$

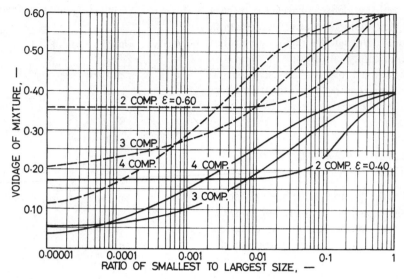

Figure 4.16. Minimum voidage for two to four component sizes for initial voidage of 0.40 and 0.60.[49]

There are quite a few models for random packings of multisized particles. Usually only the void is considered, but some refer to the relationship between the void and the coordination number.[50-53]

4.1.5 Compaction of Powders

As the essential characteristic that connects the packing structure with stress propagation within the powder mass, the compaction of powders has been extensively investigated for many years.[54]

From an operational viewpoint there are several kinds of compaction. Piston press and hydrostatic pressing are static ways of compaction, whereas tapping, vibration, hammering, and explosion belong to impact compaction. Other types of compaction are also available, such as roller pressing, vacuum pressing, multiaxis compression, and so forth.

Compaction will proceed along the free-flowing region where aggregates of particles move mutually to reduce the bulk volume, the compaction region where the aggregates are

Table 4.5. Composition of Packings for Minimum Voids[49]

INITIAL VOIDAGE IN SINGLE COMPONENT	NUMBER OF COMPONENTS	VOLUME PERCENTAGE OF EACH COMPONENT			
		$d_{p,1}$	$d_{p,2}$	$d_{p,3}$	$d_{p,4}$
0.30	2	77.0	23.0	—	—
	3	72.0	21.5	6.5	—
	4	70.7	21.1	6.3	1.9
0.40	2	71.5	28.5	—	—
	3	64.2	25.6	10.2	—
	4	61.7	24.6	9.8	3.9
0.50	2	66.7	33.3	—	—
	3	57.2	28.5	14.3	—
	4	53.3	26.7	13.3	6.7
0.60	2	62.5	37.5	—	—
	3	51.0	30.6	18.4	—
	4	46.0	27.6	16.5	9.9

broken to yield a dense phase, the region where particles undergo plastic deformation, and the pure deformation region associated with the strain of crystal lattice.[55,56]

But no systematic analysis has been carried out to derive the detailed packing structure of particles assemblages that relate to the compaction pressure.

4.1.5.1 Variation of Powder Density with Consolidating Pressure

Many empirical equations have been proposed to connect compaction pressure and powder volume for the piston press in one direction, as listed in Table 4.6.[57] These equations are rewritten and roughly classified into three types according to the differential variation of void fraction with respect to the compressive pressure:[58]

$$-\frac{d\epsilon}{dP} = c_n \cdot \epsilon^x \qquad (4.49)$$

where $x = 1$ for Athy's and $x = 2$ for Kawakita's equations.

$$-\frac{d\epsilon}{dP} = c_n \cdot \frac{(1-\epsilon)^y}{P^z} \qquad (4.50)$$

where $y = 1$ and $z \neq 1$ for Nutting's, and $y = 2$ and $z = 1$ for Terzaghi's equations.

$$-\frac{d\epsilon}{dP} = c_n \cdot \frac{\epsilon^x(1-\epsilon)^y}{P^z} \qquad (4.51)$$

where $x = 1$, $y = 1$, and $z = 0$ for Ballhausen's equation.

Some analytical equations lead to a relationship between the consolidating pressure and the volume of bulk solids. Based on the change in Mohr's circles before and after compaction, the difference in the angle of internal friction $\Delta\phi_i$ is derived as:

$$\Delta\phi_i = \frac{\sqrt{K_r}}{1 + K_r} \cdot \frac{\Delta P}{P} \qquad (4.52)$$

Table 4.6. Various Equations on Compaction of Powders.[57]

Balshin	$\ln P = -c_1(V/V_P) + c_2$
Smith	$\dfrac{1}{V} - \dfrac{1}{V_0} = c_3 P^{1/3}$
Murray	$\ln\left(\dfrac{V}{V-V_P}\right) = c_4\left(\dfrac{V_P}{V-V_P}\right)^{1/3} + c_5 P$
Ballhausen	$\ln\left(\dfrac{V_P}{V-V_P}\right) = c_6 P + \ln c_7$
Konopicky	$\ln\left(\dfrac{V}{V_0-V_P}\right) = c_8 P + \ln\left(\dfrac{V_0}{V_0-V_P}\right)$
Jones	$\ln P = -c_9\left(\dfrac{V}{V_P}\right)^2 + c_{10}$
Athy	$\dfrac{V-V_P}{V} = \dfrac{V_0-V_P}{V_0} e^{-c_{11}P}$
Nutting	$\ln\left(\dfrac{V_0}{V}\right) = c_{12} P^{c_{13}}$
Tanimoto	$\dfrac{V_0-V}{V_0} = \dfrac{c_{14}P}{V_0} + \dfrac{c_{16}P}{P+c_{15}}$
Terzaghi	$\dfrac{V-V_P}{V_P} = -c_{17}\ln(P+c_{18})$ $\qquad -c_{19}(P+c_{19}) - c_{20}P + c_{21}$
Cooper	$\dfrac{V_0-V}{V_0-V_P} = c_{22} \cdot e^{-c_{23}/P}$ $\qquad + c_{24} \cdot e^{-c_{25}/P}$
Gurnham	$P = c_{26} \cdot e^{c_{27}/V}$
Nishihara	$\ln\left(\dfrac{V_0}{V}\right) = -\left(\dfrac{P}{c_{28}}\right)^{1/c_{29}}$
Tsuwa	$\dfrac{V_0-V}{V_0} = \dfrac{V_0-V_P}{V_0} \cdot \dfrac{(1/c_{30})P}{1+(1/c_{30})P}$
Kawakita	$\dfrac{V_0-V}{V_0} = \dfrac{c_{31}c_{32}P}{1+c_{32}P}$

where K_r is the Rankine coefficient given by

$$K_r = \frac{1 - \sin\phi_i}{1 + \sin\phi_i} \qquad (4.53)$$

If $\Delta\phi_i$ is proportional to the variables representing the state of packing such as voidage, strain, bulkiness, etc., various types of empirical equations can be derived for comparison.[59] Corresponding to the linear relationship between stress P and strain γ for general solids,

$$P = C_n \cdot \gamma^{1/(1+\lambda)} \qquad (4.54)$$

was theoretically obtained for particulate matter,[60] and is comparable to Nutting's equation. By thermodynamic consideration,[61]

$$Q_d = \frac{1}{S_p} \cdot \log \frac{P - c_n}{P_0 - c_n} \qquad (4.55)$$

where Q_d and S_p are the displacement and the cross-sectional area of the piston, respectively: P_0 is the initial pressure, and c_n is the constant throughout the compaction equations above. It is similar to Terzaghi's equation.

4.1.5.2 Compaction Due to Tapping and Vibration

Tapping and vibration are often adopted as an easy way of producing the compaction of powder; however, the bulk density achieved is not as high as that with a piston press.

Experimental equations are available for the packing due to tapping:

$$\rho_{b,fn} - \rho_{b,n_{tp}} = (\rho_{b,fn} - \rho_{b,0}) \cdot \exp(-k_1 n_{tp}) \qquad (4.56)$$

where $\rho_{b,fn}$ is the final bulk density attained and is constant after many times of tapping.

$\rho_{b,0}$ is the initial bulk density before tapping, n_{tp} is the number of tappings, and k_1 is a constant.[62] $\rho_{b,fn}$ is not the material property but depends on the height of fall or the falling velocity of the tapped vessel. The straight line of the semilog plots sometimes indicates different slopes after a certain number of tappings in the series, especially with cohesive fine powders, as shown in Figure 4.17.[62, 63]

$$\frac{V_0 - n_{tp}}{V_0 - V_{nt}} = \frac{1}{k_2 k_3} + \frac{n_{tp}}{k_2} \qquad (4.57)$$

where V_0 is the initial bulk volume of powder, V_{nt} is that after n_{tp} times of tapping, and k_2, k_3 are constants.[64] For white alundum k_2 increases with decreasing diameter of particle and k_3 becomes a minimum around 3 μm in particle diameter.[65] k_3 in Eq. (4.57) is found to be in direct proportion to k_1 in Eq. (4.56).

The degree of packing under sinusoidal vibration is correlated with a measure of compaction, $\Delta v / v_0$, by[66, 67]

$$(1 - \epsilon) = (1 - \epsilon_0) + 0.073(\Delta v / v_0) \qquad (4.58)$$

where ϵ and ϵ_0 are the void fractions of a

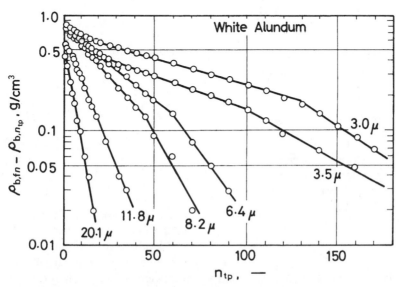

Figure 4.17. Relationship between bulk density and number of tappings.[63]

particle bed in a cylinder during and before vibration, respectively. Δv is the impact velocity or the relative velocity of a single particle and the vibrating plate, and v_0 is the initial velocity of the particles as it leaves the vibrating plate. According to the intensity of vibration, $G = a_m (2\pi f)^2 / g$, based on gravity acceleration, g,

$$G \le 1; \qquad \Delta v / v_0 \to 0$$

$$1 < G \le 3.3; \qquad \Delta v / v_0 = \frac{\sqrt{G^2 + 1 - 2G \sin 2\pi n_1} + G \cos 2\pi n_1}{\sqrt{G^2 - 1}}$$

$$3.3 < G \le 3.92; \qquad \Delta v / v_0 = \frac{\sqrt{2G(\sin 2\pi n_1 - \sin 2\pi n_2) + G^2 \cos^2 2\pi n_1} + G \cos 2\pi n_2}{\sqrt{G^2 - 1}}$$

$$(4.59)$$

where a_m and f are the amplitude and the frequency of vibration, n_0, n_1, n_2 are dimensionless times, $f \cdot t \cdot n_0$ corresponds to the point where a particle jumps from the plate, and n_1 and n_2 are the times when the particle falls on to the plate, given by

$$n_0 = \tfrac{1}{2} \sin^{-1} \frac{1}{G}$$

$$n_1 = n_0 + \frac{1}{2\pi} \left(\sqrt{G^2 + 1 - 2G \sin 2\pi n_1} + \sqrt{G^2 - 1} \right)$$

$$n_2 = n_1 + 1 \text{ for } (n_1 < n_0 + 1)$$

$$= n_1 + \frac{1}{2\pi} \left\{ \sqrt{2G(\sin 2\pi n_1 - \sin 2\pi n_2) + G^2 \cos^2 2\pi n_1} + G \cos 2\pi n_1 \right\} \text{ for } (n_1 \ge n_0 + 1)$$

$$(4.60)$$

Figure 4.18 shows the data fitted by Eq. (4.58), which indicate the maximum packing density at $G = 2.5$.

4.1.5.3 Distribution of Bulk Density

The bulk density of powder varies within a large scale for a storage vessel, or even for a small container under compression.

Assuming that the compressibility of powder is expressed by Eq. (4.61), and illustrated in Figure 4.19,[68]

$$\rho_b = aP^b \qquad (4.61)$$

the distribution of solids pressure, P_{sl} within a cylindrical vessel is described after Janssen's[69] and Shaxby's[70] derivations as

$$P_{sl} = \left\{ \left(P_0^{1-b} - \frac{aD}{4k_1 \mu_w} \right) \right.$$

$$\left. \times e^{-(1-b)4k_1 \mu_w / D \cdot h} + \frac{aD}{4k_1 \mu_w} \right\}^{1/1-b}$$

$$(4.62)$$

where P_0 is the compressive pressure acting on the upper surface of the powder bed, h is the depth from the upper surface, D is the diameter of the cylinder, μ_w is the frictional coefficient of the wall surface, and k_1 is the ratio of lateral to vertical pressure and is as-

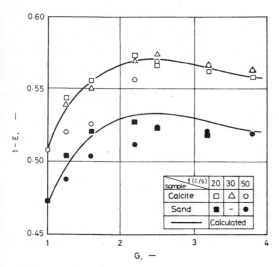

Figure 4.18. Variation of packing density with intensity of vibration.[67]

sumed to be nearly constant throughout the vessel. Finally, a and b are coefficients defined by Eq. (4.61). Hence, by putting P_{sl} from Eq. (4.62) into P in Eq. (4.61), the distribution of bulk density in the cylinder is obtained.

For a conical vessel, the density distribution is obtained in the same way by substituting Eq. (4.63) into Eq. (4.61).[68]

$c(1 - b) \neq 1$;

$$P_h = \left[P_0^{1-b} \left(\frac{Y}{h} \right)^{c(1-b)} + \frac{a(1-b)}{1 - c(1-b)} \right.$$

$$\left. \cdot Y \left\{ \left(\frac{Y}{h} \right)^{c(1-b)-1} - 1 \right\} \right]^{1/1-b}$$

$c(1 - b) = 1$;

$$P_h = \left\{ P_0^{1-b} \left(\frac{Y}{h} \right) \right.$$

$$\left. + a(1 - b)Y \log_e \left(\frac{h}{Y} \right) \right\}^{1/1-b} \qquad (4.63)$$

where Y is the distance above the apex, θ is a half of the cone angles of the conical hopper, and the coefficient c is defined as

$$c = 2\mu_w \cot \theta (k_1 \cos^2 \theta + \sin^2 \theta) \quad (4.64)$$

In the case of a bin consisting of a cylindrical silo above a conical hopper, P_0 in Eq.

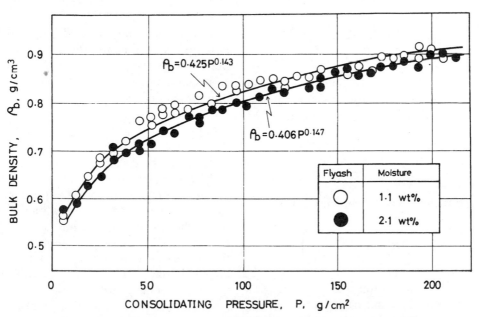

Figure 4.19. Relationship between bulk density and consolidating pressure.[68]

(4.63) is replaced by the bottom pressure in the cylinder P_{sl} given by Eq. (4.62).

For a cohesive powder the solids pressure distribution within a container is also derived under gravity alone,[71] tapping,[72] and aeration[73] in connection with the blockage criterion and a discharge rate of particles.

4.2 PERMEABILITY OF THE POWDER BED

As a result of the compaction of powder, flow of a fluid through the powder bed is governed by the uneven packing structure. Based on a microscopic packing consideration of the void-size distribution and the solids pressure distribution mentioned in the former section, the pressure drop of fluid flow can be derived as follows.

The pressure drop, Δp_a, for tubes of the same diameter is given by Ergun's equation[74] as the sum of the laminar and turbulent flow regimes:

$$\frac{\Delta p_a}{L_b} = 36 k_0 \left(\frac{L_e}{L_b}\right)^2 \frac{\mu u_{b,0}}{d_{sp}^2} \frac{(1-\epsilon_0)^2}{\epsilon_0^3}$$

$$+ 3 f_0 \left(\frac{L_e}{L_b}\right)^3 \frac{\rho u_{b,0}^2}{d_{sp}} \frac{(1-\epsilon_0)}{\epsilon_0^3} \quad (4.65)$$

In the case of bundles of tubes of different diameter, the pressure drop in laminar flow is given by Hagen-Poiseuille's equation as:

$$\Delta p_a = \frac{32 L_{e,i} \mu \langle u_{e,i} \rangle}{D_i^2} \quad (4.66)$$

and in turbulent flow by Fanning's equation as

$$\Delta p_a = 2 f_0 \rho \langle u_{e,i} \rangle^2 \left(\frac{L_{e,i}}{D_i}\right) \quad (4.67)$$

where $\langle u_{e,i} \rangle$ is the average velocity through a tube of length $L_{e,i}$, and diameter, D_i, and f_0 is the friction factor, as shown in Figure 4.20.

Here, the superficial fluid velocity, $u_{b,0}$, is obtained from the sum of flow rates through tubes of different diameter of the basis of the void-size distribution model as:

$$u_{b,0} = \frac{\sum_{i=1}^{N} n_i \frac{\pi}{4} D_i^2 \langle u_{e,i} \rangle}{A}$$

$$= \frac{\pi}{4A} \sum_{i=1}^{N} n_i D_i^2 \langle u_{e,i} \rangle \quad (4.68)$$

where n_i is the number of tubes of D_i and is given by the probability function, $\Delta P(D_i)$, as

$$n_i = -\frac{4}{\pi} \frac{\Delta P(D_i) A}{D_i^2} \quad (4.69)$$

Hence, a combination of Eqs. (4.66), (4.67), (4.68), and (4.69) after Eq. (4.65) leads to

$$\frac{\Delta p_a}{L_b} = -32 \left(\frac{L_e}{L_b}\right) \frac{\mu u_{b,0}}{\sum_{i=1}^{N} \Delta P(D_i) D_i^2}$$

$$+ 2 f_0 \left(\frac{L_e}{L_b}\right) \frac{\rho u_{b,0}^2}{\left(\sum_{i=1}^{N} \Delta P(D_i) \sqrt{D_i}\right)^2} \quad (4.70)$$

As a result, Eq. (4.70) illustrates a higher pressure drop than Ergun's for uniform tubes, as shown in Figure 4.21.[75]

In case of a powder bed prepared by a piston press from above, the voidage distribution along the axis is represented in correspondence with the solids pressure distribution derived before by[76]

$$\epsilon(x) = 1 - \frac{b_x(1-\epsilon_0)}{e^{b_x} - 1} \exp(b_x x / D_b) \quad (4.71)$$

where x_d is the distance from top surface, D_b is the bed diameter, ϵ_0 is the overall voidage, and b_x is a constant. Then the pressure drop becomes large at the same flow rate irrespective of the voidage distribution form, as shown in Figure 4.22. In case of a permeability test, the particle size is estimated to be smaller then the true value.[77] While, in case of the bed with radial distribution of voidage, the pressure drop becomes small as compared with that of uniform bed.[75,76]

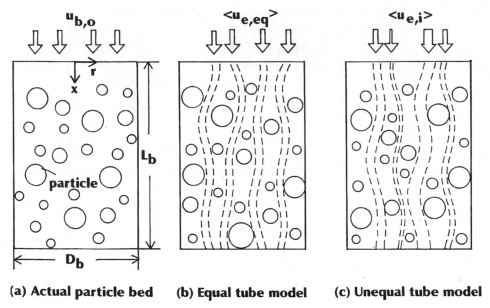

(a) Actual particle bed   **(b) Equal tube model**   **(c) Unequal tube model**

Figure 4.20. Permeation models through packed bed.[75]

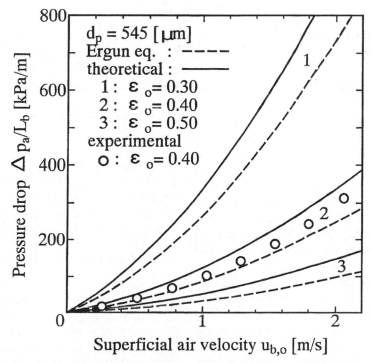

Figure 4.21. Effect of overall void fraction on pressure drop.[76, 77]

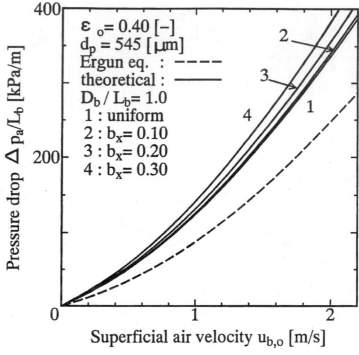

Figure 4.22. Effect of axial distribution of local voidage on pressure drop.[77]

4.3 STRENGTH OF A PARTICLE ASSEMBLAGE

The strength of powder is defined at the critical condition at which the particle assemblage initiates flow from the stationary state. Two kinds of basic factors, friction and cohesion, act in the separation of solid bodies. They correspond to two types of strength, shear and tensile, according to the breakage mechanism of particulate materials. These strengths are directly based on the packing structure of a particle assemblage through the degree of mechanical interlocking among particles and the coordination number.

It is not too much of an exaggeration to say that all the unit operations of bulk solids handling are associated with frictional and cohesive properties. They are fundamental to the interpretation of particle behavior, especially in storage, supply, transport, mixing, agglomeration, and so on. Some test devices

and methods have been proposed to evaluate the properties in a comprehensive and reproducible manner, and an analysis of data obtained is attempted on a quantitative basis.

4.3.1 Interparticle Forces at a Contact Point

In principle, the strength of powders originates from the resistant forces at a contact point between two particles. A brief review of the frictional and cohesive forces between continuous solid bodies is therefore useful.

4.3.1.1 Frictional Force Between Solid Surfaces[78]

By definition, the friction force is equivalent to the resistance exerted by one solid body against the motion of another in contact with it. This force is tangent to the contact surfaces. The coefficient of static friction is the ratio of the maximum friction force of impending motion to the corresponding normal pressure force. The coefficient of kinetic friction corresponds to the same force ratio for two surfaces moving relative to each other.

Provided Coulomb's empirical law of friction holds for comparatively dry and clean surfaces of a solid: (1) the friction force is independent of apparent area of contact, and is proportional only to the normal load on the surface; and (2) the coefficient of kinetic friction is independent of the relative sliding velocity and is less than the coefficient of static friction. Experimentally, the numerical value of the kinetic friction coefficient is found to increase gradually up to that of static friction as the velocity is decreased.

The coefficient of friction becomes large with a higher degree of vacuum, higher temperature of material, and thinner oxidation layer caused by a smaller quantity of molecules being adsorbed on or reacting with the solid surface. But the dry friction characteristics of the materials still act effectively through the boundary layer of lubricants. Coulomb's law is approximately applicable in such a wide variety of surface conditions.

4.3.1.2 Cohesive Forces Between Solids

There are various kinds of attractive forces between solid materials. Among them, the following are the most basic and often encountered in cohesion phenomena of powders:

(1) The van der Waals force F_{vw} acts between molecules of solid surfaces within the shortest distance l of about 10^{-5} cm.[79] It is said that l is equal to 4×10^{-8} cm in close contact.[80]
Between parallel planes of facing area s_f,

$$F_{vw} = \frac{A \cdot s_f}{6\pi l^3} \quad (4.72)$$

Between sphere and plane,

$$F_{vw} = \frac{A \cdot d_p}{6l^2} \quad (4.73)$$

Between different-sized spheres,[81]

$$F_{vw} = \frac{A}{12l^2} \cdot \left(\frac{d_{p1} \cdot d_{p2}}{d_{p1} + d_{p2}} \right) \quad (4.74)$$

where A is a constant inherent to the material and is usually in the order of 10^{-12} erg.[82]

The cohesive force will rapidly decrease with increasing surface roughness.[83]

(2) An electrostatic attractive force F_e for two spheres separated a distance e with electric charges positive q_1 and negative q_2 in Coulomb units, is given by

$$F_e = \frac{q_1 \cdot q_2}{d_p^2} \cdot \left(1 - 2\frac{l}{d_p} \right) \quad (4.75)$$

In a liquid phase, the electrostatic double layer causes an interparticle force between separated spheres[84] and different particles.[85]

(3) Solid bridges due to chemical reactions, sintering, melting, and recrystalization give rise to a strong bond between solids under the influence of temperature, pressure, humidity, water content, and so forth. The following is an example of one analytical approach.[86]

Based on the rates of solid dissolution and of vaporization of bonding liquid between spherical particles in contact, the radius of the narrowest portion of the solid bridge r_n is approximately related to the initial liquid volume at the contact point V_{lq} by

$$\frac{r_n}{d_p} = 1.64 \frac{c_s V_{lq}}{\rho_p \cdot d_p^3/8} \cdot X^{1/(1-X)} \quad (4.76)$$

where c_s is the saturated concentration of liquid in g/cm^3 and X is the dimensionless ratio of the rate of drying to the rate constant of dissolution in cm/s, which is a function of temperature. The bonding force F_b is then given by

$$F_b = \pi r_n^2 \cdot C_{rc} \quad (4.77)$$

where C_{rc} is the strength of the bridge material formed by recrystallization of solid constituents.

(4) Liquid bridges between solids produce the bonding force F_{lq} as the sum of the forces due to the capillary suction pressure and the surface tension T of the liquid.

Assuming constant curvature of the liquid profile and perfect wetting ($\delta = 0$), the bonding force between the different-sized spheres,

shown in Figure 4.23, is calculated at the narrowest portion of the liquid pendular ring as:[87]

$$F_{lq} = \pi r_n^2 T \left(\frac{1}{r_{lp}} - \frac{1}{r_n} \right) + 2\pi r_n T$$

$$= \pi d_p T \cdot \frac{E}{4(m + n + 1)}$$

$$\cdot \left\{ \frac{E}{4(m + n + 1)} \cdot \frac{d_p}{r_{lp}} - 1 \right\} \quad (4.78)$$

where

$$E = \sqrt{ \{4m + n(2m + n + 2)\} \cdot \left\{ 4(m + 1) \cdot \frac{2r_{lp}}{d_p} - n(n + 2m) \right\} + 2(n + 2)(n + 2m) \left\{ 8 \left(\frac{r_{lp}}{d_p} \right)^2 - n \right\} }$$

Figure 4.24 depicts variations of the dimensionless cohesive force with constant curvature; for equal spheres, $m = 1$, and for a plane, $m = $ infinity, at different separation distances.

The bonding force at the contact portion between the liquid and the equal sphere is given for the liquid volume V_{lq} with different contact angles δ, by[47]

$$F_{lq} = \pi T d_p \cos \theta \left\{ \cos(\theta - \delta) \right.$$

$$\left. + \frac{d_p}{4} \left(\frac{1}{r_{lp}} - \frac{1}{r_n} \right) \cos \theta \right\} \quad (4.79)$$

$$V_{lp} = 2\pi \left[\left\{ r_{lp}^2 + (r_{lp} + r_n)^2 \right\} r_n \sin(\theta - \delta) \right.$$

$$- \frac{r_n^3 \sin^3(\theta - \delta)}{3}$$

$$- (r_{lp} + r_n) \left\{ r_{lp}^2 \sin(\theta - \delta) \cdot \cos(\theta - \delta) \right.$$

$$+ r_{lp}^2 (\theta - \delta) \right\}$$

$$- \frac{d_p^3}{24} (2 + \sin \theta) \cdot (1 - \sin \theta)^2 \right] \quad (4.80)$$

F_{lq} at the contact portion is always greater than that at the narrowest portion, and it becomes a maximum in the close contact of solids. With the exception of the close contact case, F_{lq} increases with V_{lq} at a certain separation distance and passes through the maximum point, as shown in Figure 4.24.

In the case of a cone and a sphere, as shown in Figure 4.25, the bonding force is obtained in a similar way to the force with spheres at the narrowest portion:[87]

$$F_{lq} = \frac{\pi T d_p}{2} \cdot \left(\frac{B}{2} \cdot \frac{d_p}{r_{lp}} \cdot \cos \theta_c - 1 \right) B \cdot \cos \theta_c$$

$$(4.81)$$

where

$$B = \frac{2r_{lp}}{d_p} + \left\{ \sqrt{ 1 - n(n + 2)\tan^2 \theta_c + \frac{4}{d_p} r_{lp} \left(\frac{2}{1 - \sin \theta_c} + n \frac{\tan \theta_c}{\cos \theta_c} \right) } - (n + 1) \right\} \sin \theta_c$$

$$(4.82)$$

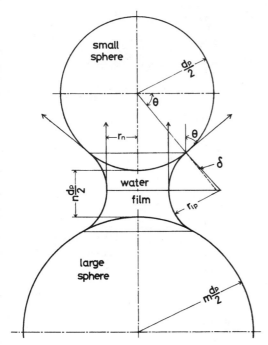

Figure 4.23. Model of cohesion due to liquid bridge between separate spheres of different size.[87]

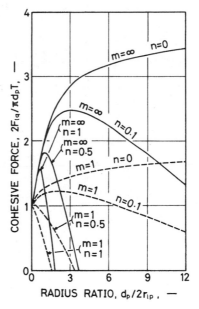

Figure 4.24. Cohesive force due to water bridge between equal spheres and between a sphere and a plane.[87]

Figure 4.26 illustrates that the cohesive force increases with a larger cone angle, shorter distance, and with greater curvature of the liquid profile. Note that d_p = infinity leads to F_{lq} for a cone and a plane as:[87, 88]

$$F_{lq} = \pi T r_n \left(\frac{r_n}{r_{lp}} + 1 \right) \quad (4.83)$$

$$V_{lq} = \pi r_{lp}^3 \left[(1 + \sin \theta_c) \left\{ \left(\frac{r_n}{r_{lp}} + 1 \right)^2 \right.\right.$$

$$+ (1 + \sin \theta_c) - \tfrac{1}{3}(1 + \sin \theta_c)^2 \Big\}$$

$$- \left(\frac{r_n}{r_{lp}} + 1 \right) \left(\sin \theta_c \cos \theta_c + \theta_c + \frac{\pi}{2} \right)$$

$$- \frac{1}{3} \left(1 + \sin \theta_c - \frac{n d_p}{2 r_{lp}} \right)^3 \tan^3 \theta_c \right]$$

$$(4.84)$$

where

$$\frac{r_n}{r_{lp}} + 1 = \frac{1 + \left(1 - \dfrac{n d_p}{2 r_{lp}} \right) \sin \theta_c}{\cos \theta_c} \quad (4.85)$$

The above conical cases assume angular particles, the surfaces of which have a conical projection at the contact point between particles.

4.3.1.3 Measurements of Cohesive Force Between Two Solids

The spring balance method is a direct way to evaluate the cohesive force between a particle and a flat plane through the displacement of the spring[89] or the elastic beam,[90] as sketched in Figure 4.27. The automatic electrobalance method is a modification where the variation of interparticle force with the distance between two sample particles is measured under various atmospheres with a sensitivity of 10^{-8} g.[91, 92] Particle diameter is usually on the order of several hundreds micrometers, and the corresponding cohesive force is on the order of several dynes.

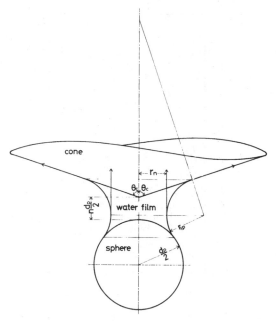

Figure 4.25. Model of cohesion due to liquid bridge between separate cone and sphere.[87]

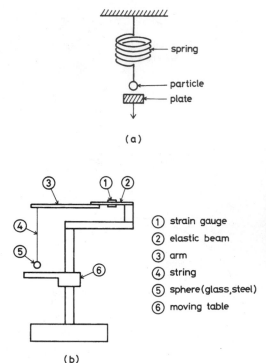

Figure 4.27. Spring balance method, (a)[89] and (b).[90]

The pendulum method is a means of separation using a gravitational component to move the wall around the center where the particle is suspended with a piece of fiber string, as shown in Figure 4.28a, or to expand two sus-

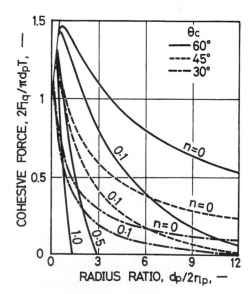

Figure 4.26. Cohesion force due to water bridge between separate cone and sphere.[87]

pending points of nylon strings, as shown in Figure 4.28b.[93] The angles or the amplitudes of the particle pendulum are measured at separation.

The centrifugal method adopts a rotating cell,[94–96] inside of which equal spheres are bonded in a line on a razor edge, as shown in Figure 4.29a,[97] or particles of distributed size are spread in a monolayer over a plane, as shown in Figure 4.29b.[98] By changing the rotational speed ω, the cohesive force between two spheres is detected at separation, and the residual percentage of particles on the wall surface ψ_r is correlated with the centrifugal force, F_0, using

$$F_c = \frac{\pi}{6} d_p^3 \cdot \rho_p \cdot r_c \omega^2 \qquad (4.86)$$

The latter result[98] indicates that the cohesive force at a contact point is distributed and that

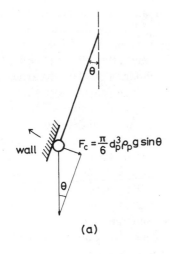

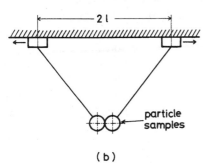

Figure 4.28. Pendulum method.[93]

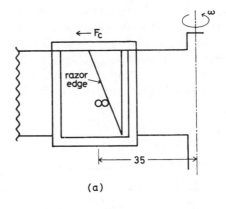

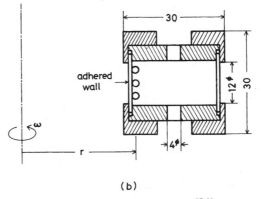

Figure 4.29. Centrifugal method.[97,98]

ψ_r versus F_c follows a log-normal function as:

$$\psi_r = \int_{\log F_c}^{\infty} \frac{1}{\sqrt{2\pi}\,\log\zeta}$$

$$\cdot \exp\left[-\frac{(\log F_c - \log F_{c,50})^2}{2\log^2\zeta}\right] d(\log F_c)$$

(4.87)

This method is applicable to a large number of finer particles around 10 μm and yields smaller values of cohesive force by 10 to 100 dynes than those obtained with the balance methods.

The cohesive force between agglomerated particles can also be measured using a high-speed gas due to the difference in re-sistant force and inertia for particles during acceleration.[99]

4.3.2 Tensile Strength of a Powder Mass

Based on the cohesive force at a contact point between particles and the number of contact points of the yield plane of a powder mass, the ultimate tensile strength is evaluated without frictional effects among particles by means of some tensile test devices.

4.3.2.1 Devices for Measuring Tensile Strength

The methods of tensile testing are classified into two kinds of direct tension and three kinds of indirect tension by their means of compaction and breakage.

1. *Vertical tensile test.* Compacted powder is subjected to a vertical tensile load, and thus the rupture plane occurs horizontally or at

right angles to the direction of compaction. Two types of testing apparatus have been devised, for example, as shown in Figure 4.30.[100, 101] Modifications of these are available, especially in the manner of application of tensile load and the detector.[101-103]

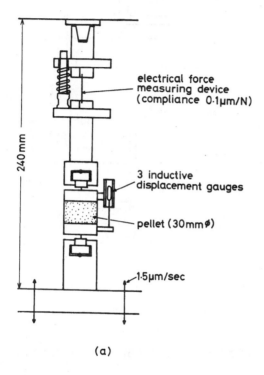

(a)

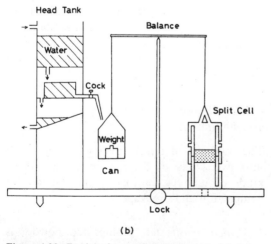

(b)

Figure 4.30. Devices for vertical tensile test. (a) Adhesive method[100] and (b) wall method.[101]

The main difference between the present apparatuses lies in the way the particle specimen is clamped.

(a) One way is the adhesive method,[102] in which a cylindrical pellet of particles prepared under high pressure (tons per cm^2) is glued to a pair of adaptors with a strong adhesive and set in the standard material-testing machine for vertical loading, as shown in Figure 4.30a.[100] The device gives the tensile stress–strain relationship at the same time. The pellet must be strong at both end planes without damage in order not to be separated from the adaptors in tension. Thus, the apparatus is not adequate for a loosely packed powder mass in the usual condition of handling.

(b) The second way of clamping the particle specimen is the wall clamping method, in which compacted powder is clamped due to friction and cohesion between particles and the walls of pistons and cylindrical cells. This method is much improved in the range of compaction pressure or voidage by employing such adaptors as shown in Figure 4.31.[101, 104] Part *a* of the figure shows the central pin inserted to increase the contact area or the resistant forces of the ring-shaped agglomerate (prepared at about 70 kg/cm^2) and to eliminate the inhomogeneous core of the cylindrical pellet.[104] Part *b* illustrates the joined cells and pistons, the internal wall surfaces of which are roughened by screw cuts to prevent the cylindrical compact (up to about $\epsilon = 0.75$) from sliding during tensile testing.[101] In contrast to the one-directional piston press in Figure 4.31a, powder is compressed in the joined cells by turning both pistons simultaneously. This produces shear loading while turning and piston pressing from both sides of the cylinder at the same time. As a result, the stress state becomes uniform and the void fraction of the sample is the largest at the joint section of the cells, that is, the yield plane is always prepared at the joint under tensile load. The void fraction over the failure plane is estimated from the mass of the powder slice at the joint section. Figure 4.32 presents typical results of measuring tensile strength by means of two kinds of wall clamping methods.

2. *Horizontal tensile test.* Powder is compacted in the shallow cylinder under vertical loads and is diametrically split into two semicircular blocks by the horizontal tensile load. The vertical fracture plane is yielded at the joint of the two half cells, one of which is fixed and the other mounted on the traction table, as shown in Figure 4.33.[105] A similar split-plate apparatus is used, in which two movable cells are attached on both sides of the central fixed plate.[106] This method is intended to measure precisely the low tensile strength of loosely packed powder. It is, however, difficult to use in practice not only because of unstable guides

for the movable half cell resting on ball bearings but also because of unavoidable inhomogeneity of stress and voidage along the powder depth. Thus, extrapolation of measured strength to zero bed height, if possible, could suggest the most appropriate value of tensile strength.[106] Figure 4.34 shows some examples of the data obtained by the horizontal tensile test.[103, 106, 107]

3. *Diametral compression test.* A discoidal or cylindrical agglomerate of particles is compressed across the diameter between two platens, as shown in Figure 4.35. In the case of ideal line loading a uniform distribution of tensile stress develops along the vertical diameter. The direction of the stress is at right angles to the vertical load. Assuming that the particle agglomerate is homogeneous and behaves like an elastic

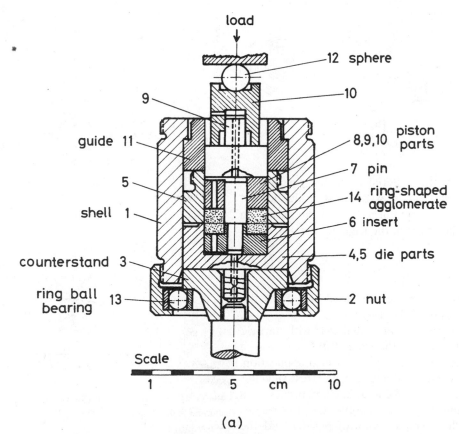

(a)

Figure 4.31. Split cells for wall clamping. (*a*) Annular cell[104] and (*b*) cylindrical cell.[101]

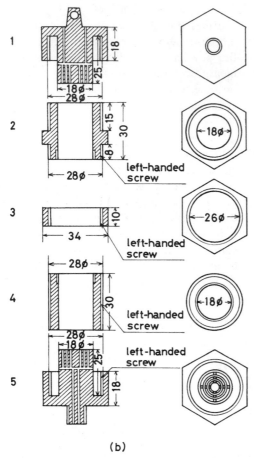

(b)

Figure 4.31. Continued

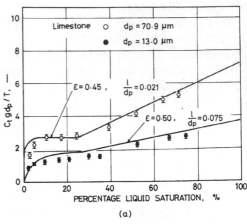

(a)

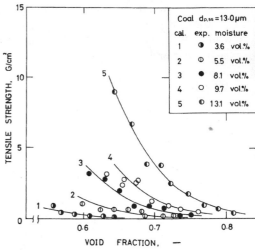

(b)

and brittle material up to the yield stress, this method gives a direct evaluation of tensile strength of powder by the maximum tensile stress in the agglomerate as:

$$\sigma_{t,\max} = \frac{2L_t}{\pi D T_k} \qquad (4.88)$$

where L_t is the total load and T_k is the thickness of the disk.

However, in an actual case with particle agglomerates, the load distributes over a finite area at the point of load application. Thus, the maximum tensile stress deviates from Eq. (4.88), and fracture occurs at a certain degree of deformation that relates to the compressibility of powder and the variation of tensile strength with consolida-

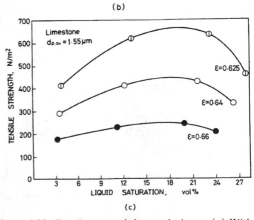

(c)

Figure 4.32. Tensile strength by vertical test. (*a*) With annular cell,[104] and (*b*)[45] and (*c*)[103] with cylindrical cell.

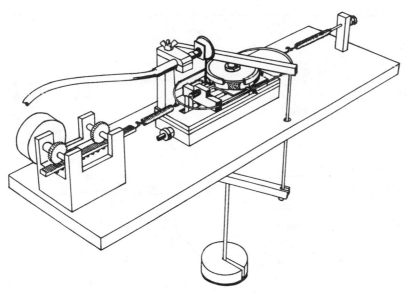

Figure 4.33. Apparatus for horizontal tensile test.[105]

tion. The general approach is given later in the analysis of tensile strength. Figure 4.36 shows one of the examples of the data.[108]

4. *Break-off test*.[109] A powder sample consolidated in a tube with a plunger is extruded in the horizontal direction until it is broken off because of its own weight, as shown in Figure 4.37.[46] The weight of the separated powder column is divided by the cross-sectional area of the cylinder to give the cohesive strength. Then a powder slice of discrete thickness is scraped off at the failure portion to determine the void fraction. Though indirect, the method is a simple way of measuring cohesiveness of powders. Figure 4.38 shows the results.

5. *The penetration method* is sometimes proposed to measure tensile strength in the liquid phase.[110] It is often unsatisfactory and attempts have been made to improve it by the hollow cylinder method.[111]

4.3.2.2 Analysis of Ultimate Tensile Strength

To investigate tensile strength, knowledge of the stress–strain behavior of particulate materials is necessary, as is the case with continuous solid bodies. But only some qualitative data are available as yet.[100, 101, 112]

On the other hand, several analytical equations have been proposed for the ultimate tensile strength of particle agglomerates in three kinds of liquid state.

1. The first is the pendular state, where only liquid bridges exist between the individual particles. Basically, the tensile strength is given as a product of the total contact points over the unit area of yield plane and the cohesive force at a contact point between two particles. The number of contact points is obtained from the number of reference particles multiplied by the coordination number of a single particle, which relates to the particle size and the packing structure of the particle assemblage. Thus, the tensile strength of a cohesive powder, C_t, is written in general as

$$C_t = k_1 \cdot \frac{1 - \epsilon}{\phi_{sp} d_p^2} \cdot N_c \cdot F \quad (4.89)$$

where k_1 is a proportionality constant, ϕ_{sp} is the shape factor to represent an effective projected area of particle, and F is the cohesive force at the contact point.

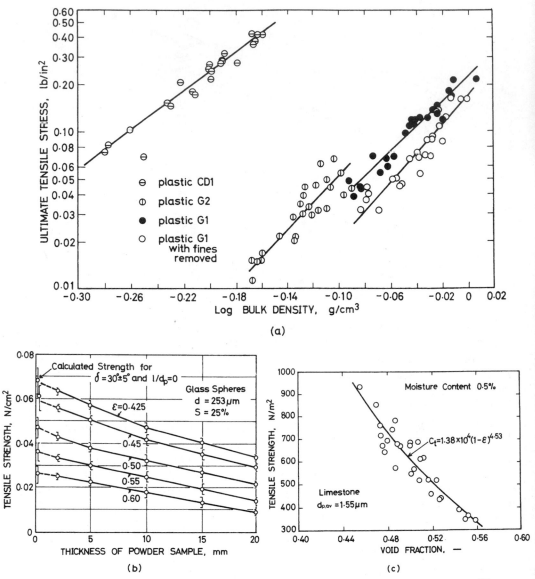

Figure 4.34. Tensile strength by horizontal test, with (a) diametral split cell,[107] (b) two split cells,[106] and (c) diametral split cell.[103]

(a) Uniform distributions of voidage and interparticle forces give rise to an ideal tensile strength that is usually the maximum and does not change much with the void fraction. The coordinate number N_c is represented as a function of the void fraction ϵ, for instance, by Eqs. (4.12), (4.13), (4.29), (4.30), and (4.31). Among them, the simple relationship of Eq. (4.12) is often adopted to give the tensile strength as

$$C_t = k_2 \cdot \frac{1 - \epsilon}{\epsilon} \cdot \frac{F}{d_p^2} \qquad (4.90)$$

where originally $k_2 = \frac{9}{8}$[109] but it is now revised to be unity.[113]

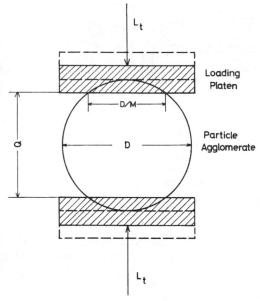

Figure 4.35. Diametral compression test.[122]

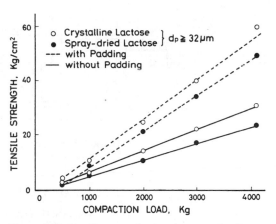

Figure 4.36. Tensile strength of agglomerate by diametral compression test.[108]

(b) An increase in interparticle force with increasing contact area due to consolidating pressure will interpret the rapid change in the tensile strength data more than that described by the voidage function $(1 - \epsilon)/\epsilon$ in Eq. (4.90), as shown in Figure 4.39.[112,114] The cohesive force over the contact area between the particle and the plane F_d is derived as the sum of the force without deformation F and that due to the pressing force on the particle, F_p.[115]

$$F_d = F + F_p \cdot \cfrac{F_{vw}}{H_p\left(1 + \cfrac{2}{3} \cdot \cfrac{s_{el}}{s_{pl}}\right)}$$

$$= F + F_p \cdot \frac{F_{vw}}{H_p} \quad \text{for } s_{el} \ll s_{pl} \quad (4.91)$$

where F_{vw} is the van der Waals force, H_p is the Hertz hardness of the particle, and s_{el} and s_{pl} are contact areas attributing to the elastic and the plastic deformations, respectively.

The deformation of spheres in the bed by the external force F_{ex} is analyzed on the contact area as:[116]

$$s = k_3 \cdot F_{ex}^q \quad (4.92)$$

where q is $\frac{2}{3}$ for elastic deformation, 1 for plastic, and lies in between $\frac{1}{2}$ and $\frac{2}{5}$ in other cases.

In principle, it is almost possible to derive the relationship between F_d and ϵ by combining these two equations. For example, the tensile strength is given by a semitheoretical equation below, taking into account the compressive force, P_{cm}, at a contact point between particles.[117] k_4 and m are constants related to the consolidation characteristics of the powder bed.

$$C_t = k_4 \frac{1 - \epsilon}{\epsilon}\left(\frac{P_{cm}}{d_p^2}\right)^m \quad (4.93)$$

(c) A nonuniform packing structure of the particle assemblage, mentioned above, results in a pronounced increase in the ultimate tensile strength with a slight decrease in the void fraction.[45] Provided that only particles in the closest packing portion govern predominantly the ultimate strength,

$$C_{t,r} = n_{p,r} \cdot 4 \cdot F \quad (4.94)$$

where 4 is the number of contact points of a single particle concerned with the break-

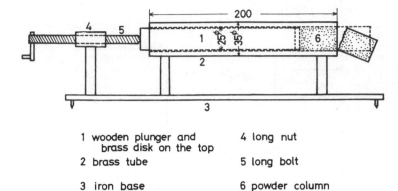

1 wooden plunger and   4 long nut
  brass disk on the top

2 brass tube   5 long bolt

3 iron base   6 powder column

Figure 4.37. Device for break-off test.[109]

age and $n_{p,r}$ is the number of particles in the rhombohedral packing portion given by

$$n_{p,r} = \frac{R_r(1 - 0.260) \cdot \left(d_p/\sqrt{2}\right)}{\pi d_p^3/6} \quad (4.95)$$

Thus, putting Eqs. (4.28) and (4.95) into Eq. (4.94), gives

$$C_{t,r} = \frac{F(1 - \epsilon)}{\pi d_p^2/6}$$

$$\cdot \frac{3.99(\alpha/\beta)}{e^{\alpha(\epsilon - 0.260)} - 1 + 1.41(\alpha/\beta)} \quad (4.96)$$

Then, the tensile strength in the cubic packing portion $C_{t,c}$ and the maximum tensile strength in both particle packing portions $C_{t,max}$ are also derived and compared with Eq. (4.96) in Figure 4.40.

$$C_{t,c} = n_{p,c} \cdot 1 \cdot F$$

$$= \frac{F(1 - \epsilon)}{\pi d_p^2/6}$$

$$\cdot \frac{e^{\alpha(\epsilon - 0.260)} - 1}{e^{\alpha(\epsilon - 0.260)} - 1 + 1.41(\alpha/\beta)} \quad (4.97)$$

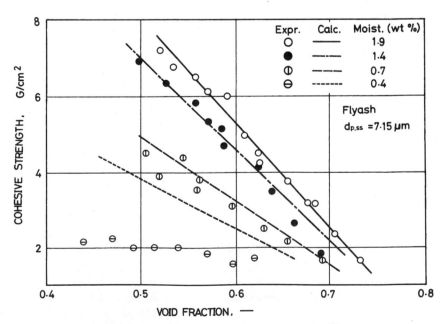

Figure 4.38. Cohesive strength by break-off test.[109]

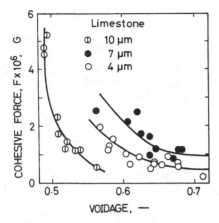

Figure 4.39. Variation of cohesive force at a contact point with void fraction.[114]

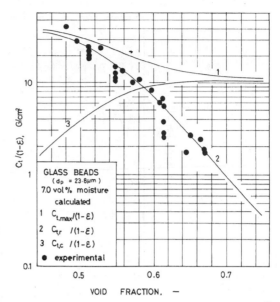

Figure 4.40. Comparison of $C_{t,max}$ and $C_{t,c}$ with $C_{t,r}$.[45]

$$C_{t,max} = n_{p,c} \cdot 1 \cdot F + n_{p,r} \cdot 4 \cdot F$$

$$= \frac{F(1-\epsilon)}{\pi d_p^2/6}$$

$$\cdot \frac{e^{\alpha(\epsilon-0.260)} - 1 + 3.99(\alpha/\beta)}{e^{\alpha(\epsilon-0.260)} - 1 + 1.41(\alpha/\beta)} \quad (4.98)$$

where

$$n_{p,c} = \frac{R_c(1-0.476)d_p}{\pi d_p^3/6} \quad (4.99)$$

All the lines in Figures 4.32b and 4.41 are calculated from Eq. (4.96).

There are other types of representation of tensile strength:

(d) For the powder compact of a single component of distributed size,[118]

$$C_t = \frac{r_a \cdot r_b \cdot N_c}{2}$$

$$\cdot S_v \cdot (1-\epsilon) \cdot F(l) \quad (4.100)$$

where r_a is the ratio of the number of particle pairs per unit area of failure to that per unit volume; r_b is the ratio of overall area of contact per particle pair to the surface area of the smaller particle of the pair, and $F(l)$ is the interparticle force per unit overall area of contact as a function of

the surface separation distance l of the particle pair, which is given by

$$l = l_{vn} - \frac{d_{p,av}}{3} \cdot \frac{\epsilon_{vn} - \epsilon}{1 - \epsilon_{vn}} \quad (4.101)$$

where l_{vn} is the effective range of the attractive interparticle force, and ϵ_{vn} is the corresponding voidage at which the tensile strength vanishes. For a binary mixture of particles, Eq. (4.100) is modified into

$$C_{t,mix} = \frac{r_a \cdot r_b}{2} N_c(1-\epsilon)$$

$$\cdot M_i \left\{ S_{v,1} \cdot f_1^2 + 2S_{v,12} \cdot f_1(1-f_1) \right.$$

$$\left. + S_{v,2}(1-f_1)^2 \right\} \cdot F(l) \quad (4.102)$$

where f_1 and f_2 are the weight fractions of components 1 and 2, respectively, $S_{v,12}$ is the specific surface for the particle pair of the mixture, and M_i is the index of mixing.[119]

(e) As an experimental correlation,

$$C_t = k_a(1-\epsilon)^{m_a} \quad (4.103)$$

where k_a and m_a are fitted parameters.[120]

2. In the capillary state,[121] where all the capillaries composed of voids in the powder are

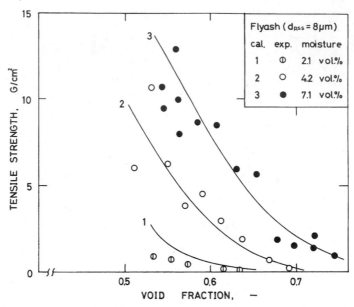

Figure 4.41. Variation of tensile strength with void fraction by wall clamping method.[45]

completely filled with the binding liquid, the tensile strength is given by the capillary pressure, p_c.

$$C_t = p_c = k_4 \cdot \frac{T}{d_{ps}} \cdot \frac{1 - \epsilon}{\epsilon} \cdot f(\delta) \quad (4.104)$$

where d_{ps} is the surface equivalent diameter, and $f(\delta)$ is the contact angle function; $f(\delta = 0°) = 1$.

3. In the funicular state,[47] $S^* < S < 100$, where both liquid bridges and capillaries filled with liquid are present,

$$C_t = C_t^* + \frac{S - S^*}{1 - S^*} \cdot (p_c - C_t^*) \quad (4.105)$$

where S is the percentage liquid saturation, and C_t^* is the tensile strength evaluated by Eq. (4.89) at the critical saturation S^*, where liquid bridges begin to touch each other, given by

$$S^* = 3 \cdot \frac{1 - \epsilon}{\epsilon} \cdot \frac{N_c}{\pi} \cdot \frac{V_{t,lq}}{d_p^3} \quad (4.106)$$

where $V_{t,lq}$ is the volume of liquid bridges.

4.3.2.3 Analysis on Diametral Compression Test

In the general case of distributed loading over the discoidal agglomerate of particles, the maximum tensile stress always occurs at the center of the disk as:[122]

$$\sigma_{mj,max} = -\frac{2L_t}{\pi D T_k} \cdot J(M, b', n) \quad (4.107)$$

where

$$J(M, b', n) = \sum_{i=1}^{n+1} \left[\frac{1 - 2^{b'} |I_n|^{b'}}{(n+1) - 2^{b'} \sum_{i=1}^{n+1} I_n^{b'}} \right.$$
$$\left. \cdot \left(\frac{16}{M^2} \cdot I_n^2 - 1 \right) \sqrt{1 - \frac{4}{M^2} I_n^2} \right]$$

$$I_n = \frac{i - 1}{n} - 0.5 \quad (4.108)$$

Here, $n + 1$ is the number of concentrated loads, and the loads distribution expressed by Eq. (4.109) is adopted.

$$L_i = -a' |x_i|^{b'} + c' \quad (4.109)$$

where a', b', and c' are constants and x_i is the x coordinate of the point of load, L_i. Thus,

$$x_i = \frac{D}{M} \cdot I_n \qquad (4.110)$$

where D/M is the contact width between the disk and the compressing plate, as denoted in Figure 4.35.

By simple geometry,

$$\frac{1}{M} = \sqrt{1 - \left(\frac{Q}{D}\right)^2} \qquad (4.111)$$

$$\rho_b = \frac{4W}{D^2 T_k}$$

$$\cdot \frac{1}{\pi - 2\cos^{-1}\dfrac{Q}{D} + 2\dfrac{Q}{D}\sqrt{1 - \left(\dfrac{Q}{D}\right)^2}} \qquad (4.112)$$

where Q is the distance between two platens and W is the weight of the agglomerate sample. Incorporating the compaction characteristics of powders mentioned above, for example, Eq. (4.61),

$$\rho_b = aP^b \qquad (4.113)$$

and equating C_1 by Eqs. (4.89), (4.103), and (4.104), and $\sigma_{mj, max}$ by Eq. (4.107) leads to the

failure condition in terms of the bulk properties of the powder as:[123]

$$\left(\frac{2L_f}{\pi D T_k}\right)_f = \frac{2P_f}{\pi} = -\frac{C_t(\epsilon_f, F)}{J_f(M_f, b', n)} \qquad (4.114)$$

where the subscript f denotes failure condition, and M_f in Eq. (4.111), ϵ_f in Eq. (4.112) and P_f in Eq. (4.113) are all represented by Q_f/D. Hence, the deformation at failure Q_f/D is obtained from Eq. (4.114) at the intersection of two curves, $\sigma_{mj, max}$ and C_f, as illustrated in Figure 4.42.

In other words, since the bulk properties of the powder give the value of Q_f/D, the ultimate tensile strength is predicted by Eq. (4.107).

4.3.3 Shear Strength of Particles

In contrast to the tensile strength in the normal direction to the failure plane, shear strength is yielded along the plane parallel to the breaking force. It arises from friction and interlocking in addition to cohesion between particles, as analyzed below.

The shear stress τ is written as a function of the normal stress σ by the following equations. For a Coulomb powder the shear stress varies as a linear relationship of the normal stress;

$$\tau = \mu_i \sigma + C_s \qquad (4.115)$$

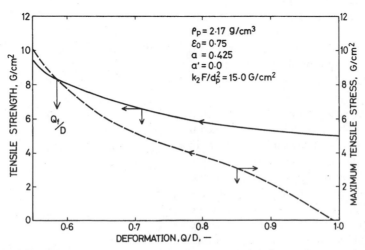

Figure 4.42. Fracture process of particle agglomerate during diametral compression.[123]

where μ_i is the coefficient of internal friction, and C_s is called the stickiness corresponding to τ at $\sigma = 0$.

For a general powder

$$\tau = \frac{\partial \tau}{\partial \sigma} \cdot \sigma + C_s(\sigma) \qquad (4.116)$$

where $\mu_i = \partial\tau/\partial\sigma$ [124]

$$\left(\frac{\tau}{C_s}\right)^{q_{id}} = \frac{\sigma - \sigma_0}{\sigma_0} \qquad (4.117)$$

where q_{id} is one sort of flowability index, and σ_0 is the normal stress σ extrapolated to $\tau = 0$ [120] or the apparent tensile strength.

$$\tau = k_6(\sigma - \sigma_0)^q \qquad (4.118)$$

where k_6 and q are fitting constants. Eq. (4.118) coincides with Eq. (4.117) by taking

$$k_6 = C_s/(-\sigma_0)^{1/q_{id}} \text{ and } q = i/q_{id}. \, [125]$$

4.3.3.1 Mohr's Stress Circle

The state of stress in the powder bed may be described by a continuum theory at static equilibrium. The stress as a reaction force per unit area can be represented by principal stresses that are not accompanied with shear stresses but are normal to principal stress planes. Among three kinds of principal stresses in three dimensions the minimum and maximum ones could be adopted at fracture and reduced to a plane stress system. Here the packing structure is assumed to be uniform throughout the bed.

To express the stress state at a certain point inside the powder bed, x- and y-axes are taken in the two-dimensional system, as shown in Figure 4.43. Thus, the force balance over a right triangle of unit thickness surrounding the point is to be considered. Then the hypotenuse length is defined as unity, and the compressive stress and the shear stress are positive in the inside and downward directions, respectively.

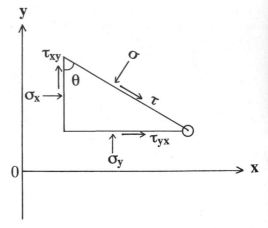

Figure 4.43. Equilibrium of powder wedge under normal and shear stresses.

The force balance in the x and y directions are written, respectively, as:

$$\sigma_x \times 1 \cdot \cos \theta + \tau_{yx} \times 1 \cdot \sin \theta$$
$$= \sigma \cos \theta - \tau \sin \theta \qquad (4.119)$$
$$\sigma_y \times 1 \cdot \sin \theta + \tau_{xy} \times 1 \cdot \cos \theta$$
$$= \sigma \sin \theta + \tau \cos \theta \qquad (4.120)$$

To get σ, multiply $\cos \theta$ and $\sin \theta$ in Eqs. (4.119) and (4.120) and sum them on both sides, respectively:

$$\sigma = \frac{\sigma_x + \sigma_y}{2} + \frac{\sigma_x - \sigma_y}{2} \cos 2\theta$$
$$+ \frac{\tau_{yx} + \tau_{xy}}{2} \sin 2\theta \qquad (4.121)$$

Similarly, subtracting these values gives τ:

$$\tau = \frac{\tau_{xy} - \tau_{yx}}{2} + \frac{\tau_{xy} + \tau_{yx}}{2} \cos 2\theta$$
$$+ \frac{\sigma_y - \sigma_x}{2} \sin 2\theta \qquad (4.122)$$

Taking a moment balance around the right corner of the triangle gives

$$\sigma \times \frac{1}{2} = \sigma_x \times \frac{\cos^2 \theta}{2} + \tau_{xy} \cdot \cos \theta$$
$$\times 1 \cdot \sin \theta + \sigma_y \times \frac{\sin^2 \theta}{2}$$

or rewriting as

$$\sigma = \frac{\sigma_x + \sigma_y}{2} + \frac{\sigma_x - \sigma_y}{2} \cos 2\theta + \tau_{xy} \sin 2\theta$$

(4.123)

Equating Eqs. (4.121) and (4.123) gives

$$\frac{\tau_{yx} + \tau_{xy}}{2} = \tau_{xy} \quad \therefore \tau_{yx} = \tau_{xy} \quad (4.124)$$

Hence, the combined stresses are yielded by substitution of Eq. (4.124) into Eqs. (4.121) and (4.122), respectively, as

$$\sigma = \frac{\sigma_x + \sigma_y}{2} + \frac{\sigma_x - \sigma_y}{2} \cos 2\theta + \tau_{xy} \sin 2\theta$$

(4.125)

$$\tau = \frac{\sigma_y - \sigma_x}{2} \sin 2\theta + \tau_{xy} \cos 2\theta \qquad (4.126)$$

Squaring and adding both sides of Eqs. (4.125) and (4.126) produces

$$\left(\sigma - \frac{\sigma_x + \sigma_y}{2}\right)^2 + \tau^2 = \left(\frac{\sigma_x - \sigma_y}{2}\right)^2 + \tau_{xy}^2$$

(4.127)

Therefore, σ and τ over the slope of θ from the y-axis will exist on a so-called Mohr's stress circle whose center is at $\{(\sigma_x + \sigma_y)/2, 0\}$ and whose radius is $[\{(\sigma_x - \sigma_y)/2\}^2 + \tau_{xy}^2]^{1/2}$,

as shown in Figure 4.44. Then, $\theta = 0$ and $\theta = \pi/2$ give opposite points through the origin on the circle and thus twice the angle of the slope corresponds to the angle of the radius on Mohr's circle.

θ for the minimum and maximum principal stresses are obtained by differentiation of Eq. (4.125) as Ψ

$$\tan 2\theta = \frac{\tau_{xy}}{(\sigma_x - \sigma_y)/2} = \tan 2\Psi \quad (4.128)$$

Thus, these stresses are generated on the plane of θ given by Eq. (4.128), which corresponds to the σ-axis on the circle. The principal stresses are yielded by substituting Eq. (4.128) into Eq. (4.125) as

$$\sigma_{1,3} = \frac{\sigma_x + \sigma_y}{2} \pm \sqrt{\frac{(\sigma_x - \sigma_y)^2}{4} + \tau_{xy}^2}$$

(4.129)

Rewriting σ and τ with the principal stresses with Eq. (4.129) gives

$$\sigma_x = \frac{\sigma_1 + \sigma_3}{2} + \frac{\sigma_1 - \sigma_3}{2} \cos 2\Psi \quad (4.130)$$

$$\sigma_y = \frac{\sigma_1 + \sigma_3}{2} - \frac{\sigma_1 - \sigma_3}{2} \cos 2\Psi \quad (4.131)$$

$$\tau_{xy} = \frac{\sigma_1 - \sigma_3}{2} \sin 2\Psi \quad (4.132)$$

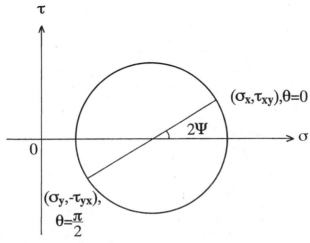

Figure 4.44. Representation of Mohr's stress circle.

Adjusting the x- and y-axis to the minimum and maximum plane, respectively, $\psi = 0$ gives $\sigma_x = \sigma_1$, $\sigma_y = \sigma_3$, and $\tau_{xy} = 0$, which leads to Eq. (4.133) from Eqs. (4.125) and (4.126).

$$\sigma = \frac{\sigma_1 + \sigma_3}{2} + \frac{\sigma_1 - \sigma_3}{2} \cos 2\theta \quad (4.133)$$

$$\tau = \frac{\sigma_3 - \sigma_1}{2} \sin 2\theta \quad (4.134)$$

As a result, the principal stress and the shear stress on the plane of θ against the maximum principal plane is represented by the minimum and maximum principal stresses as Eqs. (4.133) and (4.134), where the y-axis is taken on the maximum principal stress plane in the powder bed.

In the critical stress state, these equations describe the yield loci that are tangential to Mohr's circle, as illustrated in Figure 4.45. Usually the particle bed is not isotropic before flowing, because, in the case of a Coulomb powder, the following relation holds between the major σ_{mj} and the minor σ_{mr} principal stresses:

$$\frac{\sigma_{mn} - \sigma_0}{\sigma_{mj} - \sigma_0} \geq \frac{1 - \sin \phi_i}{1 + \sin \phi_i} \quad (4.135)$$

where it is in the elastic state up to the plastic equilibrium indicated by the equal sign.

After the failure, isotropic flow of continuous slipping is represented by the straight line through the origin with an effective angle of friction δ_e, the line is tangential to Mohr's failure circles passing through the terminal points of the yield loci, as shown in Figure 4.45.

Thus, Eq. (4.136) holds simultaneously with Eq. (4.135), and equating them gives Eq. (4.137).

$$\frac{\sigma_{mj}}{\sigma_{mn}} = \frac{1 + \sin \delta_e}{1 - \sin \delta_e} \quad (4.136)$$

$$C_s = \frac{\sigma_{mj}(\sin \delta_e - \sin \phi_i)}{(1 + \sin \delta_e)\cos \phi_i} \quad (4.137)$$

To get such fundamental flow factors as ϕ_i, C_s and δ_e, shear and compression tests are conducted to draw the yield loci.

4.3.3.2 Methods of Shear and Compression Tests

1. *Direct shear test.* A powder mass consolidated in a circular or square cell under vertical loads is subjected to horizontal shear in three ways, as shown in Figure 4.46.

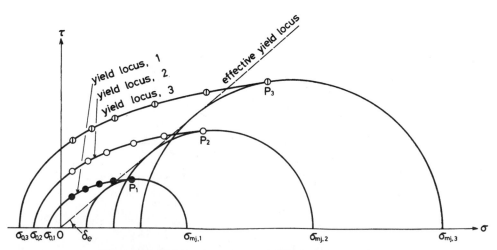

Figure 4.45. Relationship between yield loci and Mohr's circles.

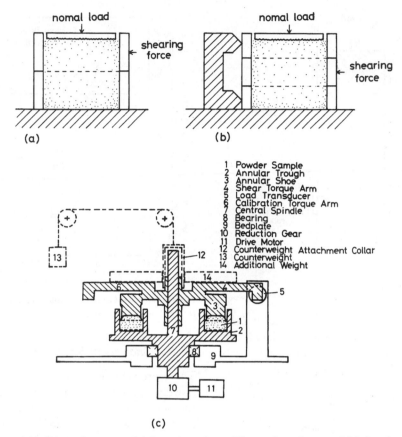

Figure 4.46. Direct shear tests. (a) One-plane shear, (b) two-plane shear, and (c) ring shear.[126]

One of the cells is horizontally moved against the fixed cells to measure the shearing force in the initial but steady state of slipping. Repeated measurements with samples of constant voidage under various loads give a linear $\sigma - \tau$ relation. As contrasted with one- or two-plane shear by the first two methods in Figure 4.46, the last method of ring shear employs annular cells.[126, 127] Its advantage lies in the constant area of shear plane during test and the possible measurement of the coefficient of dynamic friction of powder.

Some improvements and modifications were made to the one-plane shear,[128, 129] the simple shear tester,[130] the parallel-plate shear tester,[131] and the annular ring shear tester.[132-134]

2. *Compression test*. A cylindrical powder agglomerate of height of about two to three times the diameter is compressed in two ways to measure the yield strength, as schematically shown in Figure 4.47, from which the $\sigma - \tau$ relation is obtained.

The uniaxial compression test without lateral pressure gives the major principal stress at fracture in the axial direction $\sigma_{mj, f}$, and thus the Mohr's circle passes through the origin. Measuring the inclination angle of the slip plane from the vertical direction of the major principal stress θ_{uc}, the angle of internal friction ϕ_i is calculated from

$$\theta_{uc} = \frac{\pi}{4} - \frac{\phi_i}{2} \qquad (4.138)$$

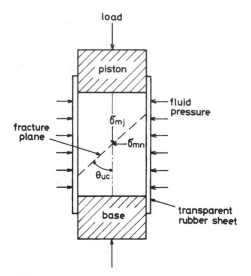

Figure 4.47. Uni- or triaxial compression test.

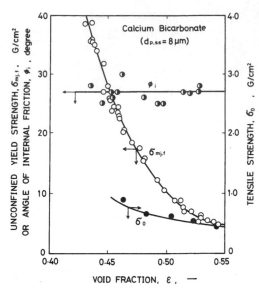

Figure 4.48. Yield strength and angle of internal friction by uniaxial compression test.[135]

As the test presents only one circle for cohesive powder ($C_s \neq 0$), the yield locus is drawn at the point on the circle whose tangent is $\tan \phi_i$. In the case of a general powder, measuring the tensile strength or σ_0, the power q_{id} in Eq. (4.117) is given as

$$q_{id} = \frac{\cos \phi_i \cdot \cot \phi_i}{(1 - \sin \phi_i) - 2\dfrac{\sigma_0}{\sigma_{mj,f}}} \quad (4.139)$$

Figure 4.48 shows an example of the data obtained by this test.[135]

The triaxial compression test employs a cylindrical powder specimen that is enveloped by a thin rubber sleeve and pressed laterally with a fluid. When one compresses the cylinder with a piston up to the point of failure, the major principal stress $\sigma_{mj,f}$ is then obtained in the vertical direction together with the minor one $\sigma_{mn,f}$ in the horizontal. Pairs of these stresses form the Mohr's circles, which are to be tangential to the yield locus or the shear strength under normal stress for constant voidage.

A new biaxial tester was designed to impose uniform principal stresses on a cubical powder compact and to measure the resulting strains in a simple manner.[136]

4.3.3.3 Analysis of Shear Strength

There are several kinds of angle properties of powders, such as the angle of internal friction, the angle of slide or wall friction, and the angle of repose. These are not primary material properties but vary with the packing conditions. Based on the nonuniform packing model mentioned above, they are analyzed as a function of void fraction.[44]

1. *Angle of internal friction ϕ_i.* The total shearing force required over a unit area of the rupture plane τ is given as the sum of the breaking forces in both particle packing portions by

$$\tau = (\mu_p \sigma_c + n_{p,c} \cdot 1 \cdot F)$$
$$+ \left(\frac{1 + \sqrt{2}\,\mu_p}{\sqrt{2} - \mu_p} \cdot \sigma_r \right.$$
$$\left. + \frac{4\sqrt{3}}{\sqrt{2} - \mu_p} \cdot n_{p,r} \cdot F \right) \quad (4.140)$$

where the first parentheses indicate the sum of the frictional force and the cohesive one in the cubic packing portion, and the term in the second parentheses is derived from the force balance in the horizontal direction of minimum shear force for a square arrangement of the rhombohedral packing. σ_c and σ_r are normal loads in each portion allotted with respect to the number of particles supporting the total load σ, as

$$\sigma_c = \frac{n_{p,c}}{n_{p,c} + n_{p,c}} \cdot \sigma, \sigma_r = \frac{n_{p,r}}{n_{p,c} + n_{p,r}} \cdot \sigma$$

$$(4.141)$$

Substituting Eq. (4.141) and Eqs. (4.27), (4.28), (4.95), and (4.99) for n_p into Eq. (4.140) leads to the shear strength of powder in the general form as

$$\tau = \left\{ \mu_p + \frac{1 + \mu_p^2}{\sqrt{2} - \mu_p} \right.$$
$$\left. \cdot \frac{(\alpha/\beta)}{e^{\alpha(\epsilon - 0.260)} - 1 + (\alpha/\beta)} \right\} \times \sigma$$
$$+ \frac{F(1 - \epsilon)}{\pi d_p^2/6}$$
$$\times \frac{e^{\alpha(\epsilon - 0.26)} - 1 + \{6.93/(\sqrt{2} - \mu_p)\}(\alpha/\beta)}{e^{\alpha(\epsilon - 0.260)} - 1 + 1.41(\alpha/\beta)}$$

$$(4.142)$$

In a comparison of Eq. (4.142) with Eq. (4.115) of Coulomb, the second term on the right-hand side indicates the physical content of the stickiness on the basis of the cohesive force between particles and the number of contact points, which increases with decreasing void fraction. The first term without σ corresponds to the coefficient of

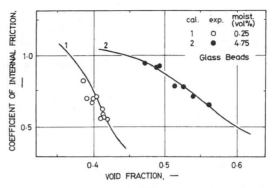

Figure 4.49. Variation of coefficient of internal friction with void fraction.[44]

internal friction as the resistant effects of solid surface friction and geometrical interlocking among particles, which decreases with increasing voidage, as shown in Figure 4.49. A similar experimental trend was also reported.[137]

2. *Angles of wall slide friction.* When the lower cell is replaced by a plane wall in the direct shear test, the shear force required per unit area of powder mass τ_w is obtained by the same considerations as is the case of τ:

$$\tau_w = \mu_w \sigma + \frac{F_w(1 - \epsilon)}{\pi d_p^2/6}$$
$$\cdot \frac{e^{(\epsilon - 0.26)} - 1 + 1.0(\alpha.\beta)}{e^{\alpha(\epsilon - 0.26)} - 1 + 1.41(\alpha/\beta)} \quad (4.143)$$

where μ_w is the coefficient of wall friction between particles and the wall, and thus $\tan^{-1} \mu_w$ is the angle of wall friction.

A force balance of the powder on the inclined plane, as shown in Figure 4.50,[44]

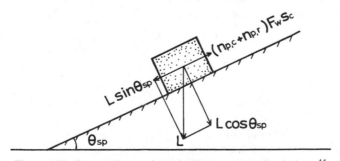

Figure 4.50. Force balance of powder block on an inclined plane.[44]

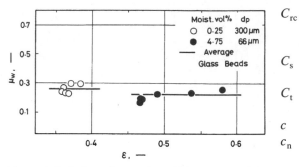

Figure 4.51. Independence of coefficient of wall friction of void fraction of powder.[44]

gives

$$L \sin \theta_{sp} = L \cos \theta_{sp} \cdot \mu_s$$
$$+ (n_{p,c} + n_{p,r}) \cdot 1 \cdot F_w \cdot s_c$$

or

$$\theta_s = \theta_{sp} - \sin^{-1}$$

$$\times \left\{ \frac{s_c}{L} \cdot \frac{F_w(1 - \epsilon)}{\left(\pi d_p^2/6\right)\sqrt{1 + \mu_s^2}} \right.$$

$$\left. \times \frac{e^{\alpha(\epsilon - 0.26)} - 1 + 1.0(\alpha/\beta)}{e^{\alpha(\epsilon - 0.26)} - 1 + 1.41(\alpha/\beta)} \right\}$$

$$(4.144)$$

where L and s_c are the weight and the cross-sectional area of powder block, θ_{sp} is the angle of inclined plane, and μ_s is the coefficient of slide friction that defines the angle of solid friction as $\theta_s = \tan^{-1} \mu_s$. For noncohesive particles, $\theta_s = \theta_{sp}$.

These two coefficients of friction, μ_w and μ_s, are independent of void fraction, as shown in Figure 4.51, and depend only on the surface properties of particle and plane wall.

LIST OF SYMBOLS

A	material constant in Eqs. (4.72) to (4.74), dyne · cm
a_m	amplitude of vibration, cm
a, b	coefficients defined by Eq. (4.61)
a', b', c'	coefficients defined by Eq. (4.109)
C_{rc}	strength of solid bridge material formed by recrystalization of constituents, dyne/cm^2
C_s	stickiness in Coulomb's equation, (4.115), G/cm^2
C_t	tensile strength of powder mass, G/cm^2
c	factor used in Eq. (4.63)
c_n	coefficients in compaction equations
c_s	saturated concentration of liquid, g/cm^3
D	diameter of cylinder of disk, cm
D_i	diameter of capillary tube, cm
d_p	particle diameter, cm
F	cohesive force at a contact point, dyne
$F(a_0)$	probability of no particles within inspection area a_0
$F(l)$	interparticle force per unit overall area of contact for surface separation l, G
F_c	centrifugal separating force, dyne
F_{ex}	external force acting on spheres in the bed, dyne
F_p	pressing force on a particle, dyne
f	frequency of vibration, s^{-1}
f_1	weight fraction of large particles
f_0	friction factor
f_p	distribution function of local mean packing density
f_y	factor between 0 and 1 used in Eq. (4.41) or total volume decrease intensity of vibration
G	acceleration due to gravity, cm^2/s
g	Hertz hardness of particle, dyne
H_p	depth from the upper surface of solids bed, cm
h	
i	integer
K_r	Rankine coefficient given by Eq. (4.53)
K_s	size ratio of smallest to largest particle in the mixture
k_a	fitting parameter in Eq. (4.103), G/cm^2
k_n	constants
k_1	ratio of lateral to vertical solids pressure

L	load, G	R_r	volume fraction of rhombohedral-packing portion of powder mass
L_0	thickness of powder bed, cm	r	radius, cm
L_e	length of the tortuous tube in powder bed, cm	r_a	ratio of number of particle pairs per unit area of failure to that per unit volume
l	shortest distance between two solid surfaces, cm	r_b	ratio of overall area of contact per particle pair to surface area of smaller particle of the pair
l_{vn}	effective range of interparticle force, cm	r_c	radius of centrifuge, cm
M	number of contact width equivalent to disk diameter	r_{lp}	radius of liquid profile, cm
		r_n	radius of narrowest portion of solid or liquid bridge, cm
M_i	mixing index	S	saturation of liquid among voids, vol%
m_a	fitting parameter in Eq. (4.103)		
m, n	integer or numerical value	S_{av}	specific surface based on apparent volume of particles, cm^2/cm^3
N_A	number of particle sections per unit area, cm^{-2}	S_v	specific surface based on net volume of particles, cm^2/cm^3
N_c	number of contact points per each particle (also called the coordination number)	S_w	surface area of particles per unit mass, cm^2/g
		s	contact area of particle, cm^2
N_{cd}	coordination number of core particle with surrounding particles of distributed size	s_c	cross-sectional area of powder block, cm^2
		s_f	facing area of parallel planes, cm^2
N_v	number of particles per unit bulk volume, cm^{-3}	T	surface tension of liquid, dyne/cm
		T_k	thickness of discoidal agglomerate, cm
n'	number of component sizes		
n_n	dimensionless time of vibration $(= ft)$	u_b	superficial velocity, cm/s
		u_e	average velocity in a tube, cm/s
n_p	number of particles	V	volume or bulk volume of particles, cm^3
$n_{t,p}$	number of tappings		
P	compaction or solids pressure, G/cm^2	V_{lq}	liquid volume at a contact point, cm^3
$P(\chi)$	probability of no particles inside elliptical space	V_p	net volume of particles, cm^3
P_0	solids pressure acting on upper surface of powder bed, G/cm^2	V_s	apparent specific volume of powder
		V_{tf}	total net particle volume of mixture
P_c	capillary pressure, G/cm^2	V_{ts}	total bulk volume of separate layers of particles, cm^3
Δp_a	fluid pressure drop across tubes of same diameter, G/cm^2		
Q	distance between two platens, cm	ΔV	impact velocity of a particle against vibrating plate, cm/s
Q_d	displacement of piston, cm		
q	exponent in Eqs. (4.92) and (4.118)	v_0	initial velocity of a particle leaving vibrating plate, cm/s
q_{id}	exponent in Eq. (4.117)		
q_1, q_2	electric charges with two spheres apart, Coulomb	W	mass of particles, g
		X	ratio of rate of drying-to-rate constant of dissolution
R	area ratio of cubic-packing portion to rhombohedral portion		
R_c	volume fraction of cubic-packing portion of powder mass	Y	distance above apex of conical hopper, cm

x, y, z	exponents in Eqs. (4.49) to (4.51)	ϕ_v	void ratio of particles
x_d	depth of powder bed, cm	χ	dimensionless void diameter
x_i	x-coordinate of loading point L_i, cm	ψ	angle of principal plane from y-axis
		ψ_r	residual percentage of particles
α, β	coefficients defined by Eq. (4.24)	ω	angular velocity, rad/s
γ	strain of particulate matters		
δ	contact angle of liquid, degree		

δ_e effective angle of friction, degree

Subscripts

ϵ void fraction of powder mass

ϵ_e effective void fraction defined by Eq. (4.22)

		av	material constant in Eqs. (4.72) to (4.74), dyne · cm
ϵ_{vn}	void fraction at which tensile strength vanishes	av	average
		b	bonding due to solid bridge
ζ_p	standard deviation of packing density given by Eq. (4.16)	c	cubic packing
		d	deformed particle
ζ_s	standard deviation of separation force due to centrifuge, dyne	e	electrostatic attractive
		el	elastic deformation
θ	angle, degree	f	at failure
θ_c	half of cone angle, degree	f_n	final
θ_s	angle of solid friction, degree	i	integer
θ_{sp}	angle of inclined plane, degree	i	index or internal
θ_{uc}	angle of fracture plane to vertical direction by uniaxial compression test, rad	l	large particles
		lq	liquid bridge
		m	mixture
θ_n	angle around a sphere, degree	max	maximum
λ	power constant in Eq. (4.54)	mj	major
μ	viscosity of fluid, g/cm · s	mn	minor
μ_i	coefficient of internal friction	0	initial or overall
μ_p	frictional coefficient of particle surface	pl	plastic deformation
		r	rhombohedral packing
μ_s	coefficient of slide friction	s	small particles
ρ_b	bulk density of powder, g/cm³	sl	cylindrical silo
ρ_p	particle density, g/cm³	sp	specific
σ	internal or normal stress, G/cm²	ss	specific surface
σ_g	geometric standard deviation of particle size distribution, cm	t	total or tensile
		tp	n_{tp} times of tapping
σ_0	normal stress extrapolated to $\tau = 0$, G/cm²	vw	van der Waals
		w	between wall and particle
τ	shear strength of powder mass, G/cm²		

ϕ_b bulkiness of particles

ϕ_i angle of internal friction of particles, rad

REFERENCES

1. D. P. Haughey and G. S. G. Beveridge, *Can. J. Chem. Eng. 47*:130 (1969).
2. K. Kubo, E. Suito, Y. Nakagawa, and S. Hayakawa, *Powder-Theory and Application*, Maruzen, Tokyo, p. 208 (1962).
3. L. C. Graton and H. J. Fraser, *J. Geol. 44*:785 (1935).

ϕ_p packing density of particles

ϕ_s shape factor of particles based on specific surface

ϕ_{sp} shape factor for effective projected area of a particle

4. T. Uematsu, K. Tsuchiya, and S. Okamura, *J. Mech. Eng. Jpn. 17*:72 (1951).

5. E. Manegold, R. Hofmann, and K. Solf, *Kolloid Z. 56*:142 (1931).

6. H. T. Horsfield, *J. Soc. Ind. 53*:108 (1934).

7. II. E. White and S. F. Walton, *J. Am. Ceramic Soc. 20*:155 (1937).

8. D. R. Hudson, *J. Appl. Phys. 20*:154 (1949).

9. A. E. R. Westman and H. R. Hugill, *J. Am. Ceramic Soc. 13*:767 (1930).

10. J. C. Macrae and W. A. Gray, *Br. J. Appl. Phys. 12*:164 (1961).

11. G. D. Scott, *Nature 188*:908 (1960).

12. T. G. Owe Berg, R. L. McDonald, and R. J. Trainor, Jr., *Powder Technol. 3*:183 (1969–1970).

13. W. O. Smith, P. D. Foote, and P. F. Busang, *Phys. Rev. 34*:1271 (1929).

14. W. O. Smith, *Physics 4*:425 (1933).

15. M. Suzuki, K. Makino, T. Tamamura, and K. Iinoya, *Kagaku Kogaku Ronbunshu 5*:616 (1979).

16. K. Gotoh, *J. Soc. Powder Technol. Jpn., 16*:709 (1979).

17. R. K. McGeary, *J. Am. Ceramic Soc. 44*:513 (1961).

18. M. Leva and M. Grummer, *Ind. Eng. Chem. 40*:415 (1948).

19. K. Ridgway and K. J. Tarbuck, *Br. Chem. Eng. 12*:384 (1967).

20. D. P. Haughey and G. G. Beveridge, *Chem. Eng. Sci. 21*:905 (1966).

21. J. D. Bernal and J. Mason, *Nature 188*:910 (1960).

22. L. H. S. Roblee and J. W. Tierney, *AIChE J. 4*:460 (1958).

23. R. F. Benenati and C. B. Brosilow, *AIChE J. 8*:359 (1962).

24. M. C. Thadani and F. N. Peebles, *Ind. Eng. Chem. Proc. Des. Dev. 5*:265 (1966).

25. K. Ridgway and K. J. Tarbuck, *Chem. Eng. Sci. 23*:1147 (1968).

26. G. D. Scott, *Nature 194*:956 (1962).

27. K. Gotoh, *J. Soc. Powder Tech. Jpn. 15*:223 (1978).

28. J. L. Finney, *Proc. R. Soc. Lond. (A) 319*:479 (1970).

29. E. M. Tory, B. H. Church, M. K. Tam, and M. Ratner, *Can. J. Chem. Eng. 51*:484 (1973).

30. J. A. Barker and D. Henderson, *Mol. Phys. 21*:187 (1971).

31. G. D. Scott and D. L. Mader, *Nature 201*:382 (1964).

32. M. M. Levine and J. Chernick, *Nature 208*:68 (1965).

33. K. Gotoh, *Nature Phys. Sci. 231*:108 (1971).

34. K. Gotoh, *Ind. Eng. Chem. Fundam. 10*:161 (1971).

35. M. Alonso, M. Satoh, and K. Miyanami, *Can. J. Chem. Eng. 70*:28 (1992).

36. K. Shinohara and T. Murai, *J. Ceramic Soc. Jpn. 101*:1369 (1993).

37. R. Aoki, *Handbook for Transport Engineering*, edited by T. Uematsu, K. Ikemori, Y, Mori, and S. Itoh, Asakura Shoten, Tokyo, p. 34 (1966).

38. G. G. Brown, *Unit Operations*, John Wiley & Sons, New York, p. 214 (1950).

39. E. J. Crosby, *Kagaku Kogaku 25*:124 (1961).

40. P. S. Roller, *Ind. Eng. Chem. 22*:1206 (1930).

41. I. Shapiro and I. M. Kolthoff, *J. Phys. Colloid Chem. 52*:1020 (1948).

42. R. Sakata, *Oyo-butsuri 21*:24 (1952).

43. C. C. Furnas, *Bur. Mines Bull. 307*:74 (1929).

44. K. Shinohara and T. Tanaka, *Kagaku Kogaku 32*:88 (1968).

45. K. Shinohara and T. Tanaka, *J. Chem. Eng. Jpn. 8*:50 (1975).

46. H. Rumpf, *Chem. Ing. Techn. 30*:144 (1958).

47. W. Pietsch and H. Rumpf, *Chem. Ing. Techn. 39*:885 (1967).

48. N. Ouchiyama and T. Tanaka, *Ing. Eng. Chem. Fundam. 19*:338 (1980).

49. C. C. Furnas, *Ind. Eng. Chem. 23*:1052 (1931).

50. N. Ouchiyama and T. Tanaka, *Ind. Eng. Chem. Fundam. 23*:490 (1984); *25*:125 (1986).

51. M. Suzuki, A. Yagi, T. Watanabe, and T. Oshima, *Kagaku Kogaku Ronbunshu 10*:721 (1984).

52. M. Suzuki and T. Oshima, *Powder Technol. 43*:147 (1985).

53. M. Cross, W. H. Douglass, and R. P. Fields, *Powder Technol. 43*:27 (1985).

54. D. Train and C. J. Lewis, *Trans. Instn. Chem. Eng. 40*:235 (1962).

55. E. Shotton and D. Grauderton, *J. Pharm. Pharmacol. 12*:87T (1960).

56. B. E. Kurtz, *Chem. Eng. Progr. 56*:67 (1960).

57. K. Kawakita and K. H. Ludde, *Powder Technol. 4*:61 (1970/71).

58. K. Kawakita and S. Taneya, *Powder Technol.*, Plant Kogaku Sha, Tokyo, p. 71 (1967).

59. T. Wakabayashi, *Powder Metal. Jpn. 10*:83 (1963).

60. A. Watanabe and K. Umeya, *Zairyo 13*:237 (1964).

61. T. Mogami, *Trans. Jpn. Soc. Civil Eng. 129*:39 (1966).

62. H. Kuno, *Proc. Fac. Eng. Keioh Univ. 11*(41):1 (1958).

63. M. Arakawa, T. Okada, and E. Suito, *Zairyo 15*:151 (1966).

64. K. Kawakita, *Zairyo 19*:574 (1970); *19*:579 (1970).

65. M. Arakawa, T. Okada, and E. Suito, *Zairyo 14*:764 (1965).

66. T. Kohata, K. Gotoh, and T. Tanaka, *Kagaku Kogaku 31*:55 (1967).

67. A. Suzuki, H. Takahashi, and T. Tanka, *Powder Technol. 2*:72 (1969).

68. K. Shinohara, K. Tamura, K. Gotoh, and T. Tanaka, *Kagaku Kogaku 31*:287 (1967).

69. H. A. Janssen, *VDI. Z 39*:1045 (1895).

70. J. H. Shaxby and J. C. Evans, *Proc. R. Instrum. 19*:742 (1910).

71. K. Shinohara and T. Tanaka, *Chem. Eng. Sci. 29*:1977 (1974).

72. K. Shinohara and T. Tanaka, *Ind. Eng. Chem. Proc. Des. Dev. 14*:1 (1975).

73. K. Shinohara and T. Tanaka, *Chem. Eng. Sci. 30*:369 (1975).

74. S. Ergun, *Chem. Eng. Progr. 48*:89 (1952).

75. K. Shinohara and T. Murai, *Kagaku Kogaku Ronbunshu 20*:198 (1994).

76. K. Shinohara and M. Kudo, *J. Soc. Mater. Eng. Res. Jpn. 5*:22 (1992).

77. K. Shinohara, *J. Soc. Mater. Eng. Res. Jpn. 3*:19 (1990).

78. N. Soda, *Friction and Lubrication*, Iwanami Shoten, Tokyo, p. 40 (1953).

79. J. Th. Overbeek, *Colloid Science*, Vol. 1, Elsevier, Amsterdam, p. 245 (1952).

80. H. Rumpf, *J. Res. Assoc. Powder Techn. Jpn. 9*:3 (Special autumn issue, 1972).

81. R. S. Bradley, *Trans. Faraday Soc. 32*:1088 (1936).

82. J. Overbeek and M. J. Sparney, *Disc. Faraday Soc. 18*:12 (1954).

83. J. Czarnecki and T. Dabros, *J. Colloid Interface Sci. 78*:25 (1980).

84. B. Derjaguin, *Kolloid Z. 69*:155 (1934).

85. R. Hogg, T. W. Healy, and D. W. Fuersteau, *Trans. Faraday Soc. 62*:1638 (1966).

86. N. Kudoh, M. Kuramae, and T. Tanaka, *Kagaku Kogaku Ronbunshu 2*:625 (1976).

87. K. Iinoya and H. Muramoto, *Zairyo 16*:70 (1967).

88. H. Tsunakawa and R. Aoki, *Kagaku Kogaku 36*:281 (1972).

89. M. Corn, *J. Air Poll. Contr. Assoc. 11*:528 (1961); *11*:566 (1961).

90. K. Hotta, K. Takeda, and K. Iinoya, *Powder Technol. 10*:231 (1974).

91. M. Arakawa and S. Yasuda, *Preprint of 37th Symposium by Tokai Branch or Society of Chemical Engineers of Japan*, p. 7 (1977).

92. M. Chikazawa, W. Nakajima, and T. Kanazawa, *J. Res. Assoc. Powder Tech. Jpn. 14*:18 (1977).

93. H. Schubert, *Chem. Ing. Techn. 40*:745 (1968).

94. M. S. Kordecki and C. Orr, *Arch. Environ. Health 1*:1 (1960).

95. G. Boehme, H. Krupp, H. Rabenhort, and G. Sandstede, *Trans. Instn. Chem. Eng. 40*:252 (1962).

96. B. V. Deryagin and A. D. Zimon, *Kolloid. Z 23*:544 (1961).

97. H. Emi, S. Endoh, C. Kanaoka, and S. Kawai, *Kagaku Kogaku Ronbunshu 3*:580 (1977).

98. S. Asakawa and G. Jimbo, *Zairyo 16*:358 (1967).

99. Y. Kousaka, S. Endoh, T. Horiuchi, and T. Araida, *Kagaku Kogaku Ronbunshu 18*:223 (1992).

100. H. Schubert, W. Herrman, and H. Rumpf, *Powder Technol 11*:121 (1975).

101. K. Shinohara and T. Takaka, *Preprint of Third Congress CHISA, Marianske Lazne*, 1969; *J. Chem. Eng. Jpn. 8*:46 (1975).

102. H. Rumpf, in *Agglomeration*, edited by W. A. Knepper, Interscience, New York, p. 379 (1962).

103. G. A. Turner, M. Balasubramanian, and L. Otten, *Powder Technol. 15*:97 (1976).

104. W. Pietsch, E. Hoffman, and H. Rumpf, *Ind. Eng. Chem. Prod. Res. Dev. 8*:58 (1969).

105. M. D. Ashton, R. Farley, and H. H. Valentin, *J. Sci. Instrum. 41*:763 (1964).

106. H. Schubert and I. W. Wibowo, *Chem. Ing. Techn. 42*:541 (1970).

107. R. Farley and F. H. H. Valentin, *Trans. Instn. Chem. Eng. 43*:T193 (1965).

108. J. T. Fell and J. M. Newton, *J. Pharm. Sci. 59*:688 (1970).

109. K. Shinohara, H. Kobayashi, K. Gotoh, and T. Tanaka, *J. Res. Assoc. Powder Tech. Jpn. 2*:352 (1965).

110. R. M. Griffith, *Chem. Eng. Sci. 20*:1015 (1965).

111. H. Schubert, *Powder Technol. 11*:107 (1975).

112. R. Aoki and H. Tsunakawa, *Zairyo 18*:497 (1969).

113. H. Rumpf, *Chem. Ing. Techn. 42*:538 (1970).

114. G. Jimbo, S. Asakawa, and N. Soga, *Zairyo 17*:540 (1968).

115. H. Schubert, K. Sommer, and H. Rumpf, *Chem. Ing. Techn. 48*:716 (1976).

116. T. Nagao, *Bull. JSME 10*:775 (1967); *Kikai Gakkai Ronbunshu (2) 33*:229 (1967).

117. J. Tsubaki and G. Jimbo, *Powder Technol. 37*:219 (1984).

118. D. C. H. Cheng, *Chem. Eng. Sci. 23*:1405 (1968).

119. S. Kocova and P. Pilpel, *Powder Technol. 7*:51 (1973).

120. M. D. Ashton, D. C. H. Cheng, R. Farley, and F. H. H. Valentin, *Rheol. Acta 4*:206 (1965).

121. W. Pietsch, *Nature 217*:736 (1968).

122. K. Shinohara and C. E. Capes, *Powder Technol. 24*:179 (1979).

123. K. Shinohara, C. E. Capes, and A. Fouda, *Powder Technol. 32*:163 (1982).

124. R. Aoki and K. Yamahuji, *J. Res. Assoc. Powder Tech. Jpn.* Special Issue, p. 33 (Feb. 1966).

125. K. Umeya, N. Kitamori, M. Araki, and H. Miwa, *Zairyo 15*:166 (1966).

126. J. F. Carr and D. M. Walker, *Powder Technol. 1*:369 (1967/68).

127. S. Kocova and N. Pilpel, *Powder Technol. 5*:329 (1971/71).

128. H. Tsunakawa and R. Aoki, *Powder Technol. 33*:249 (1982).

129. K. Matsumoto, M. Yoshida, A. Suganuma, R. Aoki, and H. Murata, *J. Jpn. Soc. Powder Powder Metal. 19*:653 (1982).

130. G. G. Enstad, *Proc. Eur. Symp. Part. Technol. B* 997 (1980).

131. M. Hirota, T. Oshima, and M. Naito, *J. Soc. Powder Technol. Jpn. 19*:337 (1982).

132. M. Suzuki, K. Makino, K. Iinoya, and K. Watanabe, *J. Soc. Powder Technol. Jpn. 17*:559 (1980).

133. A. Gotoh, M. Kawamura, H. Matsushima, and H. Tsunakawa, *J. Soc. Powder Technol. Jpn. 21*:131 (1984).

134. M. Yamada, K. Kuramitsu, and K. Makino, *Kagaku Kogaku Ronbunshu 12*:408 (1986).

135. R. Aoki, *Methods of Measuring Powder Properties*, edited by S. Hayakawa, Asakura Shoten, Tokyo, p. 85 (1973).

136. J. R. F. Arthur and G. G. Enstad, *Int. J. Bulk Solids Storage Silos 1*:7 (1985).

137. T. Ohtsubo, *J. Res. Assoc. Powder Tech. Jpn. 1*:97 (1964).

5
Vibration of Fine Powders and Its Application

A. W. Roberts

CONTENTS

5.1 INTRODUCTION

The effect of mechanical vibrations on the physical characteristics and dynamic behavior of cohesive and noncohesive powders and granular materials is a subject of broad engineering interest. For instance, it is well known that under certain conditions the application of mechanical vibrations will cause bulk powdered materials to consolidate or to compact, this process being accompanied by an increase in strength. On the other hand, given an alternative set of conditions, the application of vibrations will cause bulk materials to dilate and undergo a reduction in strength. It is important, therefore, that these two apparently diverging characteristics be understood. As is usually the case, these characteristics are

146

used to advantage to achieve certain desired objectives in powder handling and processing. However, situations may occur in practice where the presence of vibrations will have a detrimental effect on the handling and processing of bulk materials. It is important in such cases, that the full significance be understood so that adequate steps may be taken for preventative action at the design stage or for corrective action during the process or plant operation.

The industrial applications involving vibrations of powders and bulk solids (The term "bulk solid" is used to describe the general case of cohesive powder or bulk materials) are extremely wide and varied. For instance, many industries rely, to a considerable extent, on the need to handle materials in bulk form and it is important, therefore, that handling systems be designed to operate as efficiently and effectively as possible. One bulk handling function of particular significance concerns the design and operation of storage bins and their associated discharge equipment. Where it is usual to rely on gravity flow to discharge the contents of a bin, often there is a need to promote and control the flow, particularly if the storage bin has not been designed to operate under mass flow conditions. Where flow promotion is necessary, devices that impart mechanical vibrations to the bulk solid are used extensively.

The underlying principle embodied here is that the vibrations decrease the strength of the bulk material, thereby increasing its ability to flow.

Perhaps the more obvious and widely used application of vibration of powders is concerned with increasing compaction and density. This field of usage is important in many industries, such as those concerned with powder metallurgy, casting and foundry practice, ceramics, cements and silicates, and plastics. The pharmaceutical industry, with its heavy reliance on packaging and pelletizing of powders, has a dominant interest in this field of application; so too does the food industry.

The diversity in range of application is further shown in the use of vibration to fluidize granular materials and the associated application of vibrated beds in convective drying of granular materials. On other occasions, the damping characteristics of such materials as sand have been used to reduce transmitted vibration and noise.

The nature and characteristics of wave propagation through powdered or granular media are important to industries involved, in some way, with the general field of geophysics. This is true, for example, for oil exploration; it is equally true for civil engineering, where great reliance is placed on the correct design of foundations and footings for buildings, dams, and other large structures. With respect to the latter, earthquakes or tremors, blasts, traffic, and machines are all sources of vibration that may propagate through soils. It is not surprising, therefore, that the general field of soil mechanics has given much attention to this subject.

The objective of this chapter is to identify and discuss some important basic concepts of the vibration of powders and bulk solids. Following a general overview of some relevant research in this field, the body of the material presented is devoted to the study of the influence of mechanical vibrations on shear strength of consolidated bulk solids and the determination of stiffness, damping, and resonance characteristics. A vibrating shear cell apparatus, which permits these properties and characteristics to be determined, is described. A failure criterion is presented that relates the shear strength during vibration under given normal pressure with the voidage and vibration velocity on the plane of failure. The effect of vibrations in reducing friction of powders and bulk solids in contact with boundary surfaces such as the metal walls of a hopper is discussed. In view of its importance to bulk materials handling operations, the application of vibrations to promote the gravity flow of bulk solids from storage bins is briefly reviewed. Other topics discussed include the applications of random vibrations of bulk solids and some basic principles concerning the role of vibrations in compacting or consolidating powders.

5.2 LITERATURE REVIEW

The overall field of mechanical vibrations is quite extensive and has been the subject of a great deal of research in the area of vibrations of rigid and elastic bodies, machines, and structures. The general theory is now well established and thoroughly documented. The particular area of vibration of powders and bulk solids has, by comparison, received less attention, and this may be due largely to the extreme difficulty of accurately modeling and theoretically analyzing such materials under dynamic conditions. The complexities inherent in such parameters as particle shape and size distribution, moisture content, temperature, consolidation, and loading conditions have meant that much of the research to date has relied heavily on experimental investigation and, consequently, the published results are somewhat empirical.

So that some appreciation may be gained of the research associated with the vibration of powders, the following overview of the published literature is presented. This overview is by no means exhaustive, but it is based on a selection of published works deemed relevant to the present topic.

5.2.1 Developments in Bulk Solids Research

It is important to recognize the foundation work in the study of particulate solids. The origins of this area of study can be traced back to such people as Coulomb[1] and Rankine,[2] who studied the frictional behavior of sand, and Reynolds,[3,4] who observed the dilatancy effect of sand while undergoing deformation. In Reynolds' work, two areas of development are significant: that associated with the storage, flow, and handling of bulk solids and that associated with soil mechanics.

The early work associated with bulk solids was carried out in the latter part of the nineteenth century and the beginning of the present century. Of particular note are the contributions of Janssen,[5] Airy,[6] Jamieson,[7] and Ketchum,[8] who were concerned with, essentially, the behavior of granular material under static conditions.

However, it was not until the 1950s that any significant progress in this field took place. The modern developments are almost entirely due to the pioneering work of Jenike. Although Jenike and his colleague Johanson have published many papers in this field, the three University of Utah bulletins[9-11] laid the foundations for the modern theory of bulk solids storage and flow. The work of Jenike has precipitated a great deal of research in the field of bulk solids handling.

5.2.2 The Contribution of Soil Mechanics

Prior to the work of Jenike, the study of particulate solids was mainly associated with soil mechanics. Since soil mechanics is mainly concerned with retaining walls and foundation design, the internal stresses are much higher than those encountered in bulk solids handling. Furthermore, the main concern of soil mechanics is with the conditions existing within soils before failure, whereas the main interest in bulk solids handling is with the conditions under which failure and flow can occur. Nevertheless, the general similarities between the two fields of study permit some qualitative comparisons to be made.

Although the general area of soil mechanics, like the overall field of mechanical vibrations, is quite vast, only the work of certain people is particularly relevant to the present discussion and needs to be highlighted. The distinctive work of Hvorslev,[12] who established the fundamental principle of failure, is of particular importance to understanding the mechanism of failure in bulk solids as induced by mechanical vibrations. Hvorslev, who studied the stress condition in cohesive soils, showed that the peak shear stress at failure is a function of the effective normal stress on and the voids ratio (or density) in the plane of failure; this condition is independent of the stress history of the sample. The work of Hvorslev was further extended by Roscoe et al.,[13] who established the concept of a failure surface in the three-dimensional space of shear stress τ, normal stress σ, and voids ratio e. They also showed the existence of a critical void ratio boundary at which unlimited deformation could take place without change in the three variables τ, σ, and e.

The fundamental concepts developed by Hvorslev and later by Roscoe et al. stimulated further research in the same general vein. For instance, Ashton et al.[14] proposed a similar three-dimensional surface that extended into the tensile region. Their results showed a functional relationship between bulk density and strength for a given powder at constant humidity. Williams and Birks[15] proposed that the Jenike consolidation procedure was aimed at the attainment of the critical density for a given normal pressure since, for a correctly consolidated sample in the direct shear cell test, there is virtually no density change during shear.

Other researchers such as Palma,[16] Wroth and Bassett,[17] and Rowe et al.[18] have made valuable contributions by deriving stress–strain relationships for sand and clay, utilizing energy concepts. In essence, the external deformation energy was equated to the internal energy changes caused by frictional loss and stored elastic or potential energy.

In the area of soil mechanics considerable attention has been given to the study of dynamic loadings on soils and foundations. For instance, Mitchell[19] used the theory of rate processes to relate the shearing resistance of soils in triaxial compression to frictional and cohesive properties, effective stress, soil structure, rate of strain, and temperature. Lysmer and Richart[20] studied the dynamic response of footings to vertical loading. They showed that all vertically loaded footing-soil systems are strongly damped because of wave propagation

in the subsoil. Mitchell et al.[21, 22] followed the early work of Mitchell in using the fundamental theory of rate processes to study the time-dependent deformation of soils. Funston,[23] in carrying out research into footing vibrations with nonlinear subgrade support, used a simple, single degree-of-freedom model to represent the dynamic motion of a foundation. While recognizing the limitations of the model, they showed that it had considerable merit in facilitating investigation of soil properties and foundation parameters as well as permitting estimates to be made of foundation motions.

D'Appolonia[24] presented criteria for the placement and improvement of soil required to sustain dynamic loadings. Timmerman and Wu[25] studied the behavior of sands under cyclic loading. Their results led to some qualitative conclusions concerning the stress–strain relationships under dynamic loading. For the range of stress and acceleration studied, the soil deformation was primarily due to shear. In the earlier research of Greenfield and Misiaszek[26] concerning the vibration compaction of Ottawa sand, the existence of a resonant frequency at approximately 30 Hz was shown to occur. This frequency affected the maximum change in void ratio and minimum final void ratio.

5.2.3 Instrumentation and Experimental Techniques

The area of soil mechanics has also given rise to the development of instrumentation and experimental techniques for the determination of the dynamic properties of soils. For example, Suk Chae Yong[27] employed four different methods to determine certain dynamic material constants of soil. The four methods were the resonance column method, amplitude ratio method, elastic half-space method, and wave propagation method. The dynamic shear modulus and wave velocity can be determined with respect to changes in frequency, amplitude, confining pressure, and water content. Similar testing techniques were also used by

Drnevich et al.[28] and de Graft-Johnson[29] to measure the damping capacity, shear modulus, and elastic modulus of soil.

In the research of Youd[30] a direct shear apparatus was developed to investigate the effects of vibration on the shear strength and void ratio of dry granular materials. The results showed that both the critical void ratio and coefficient of internal friction are reduced considerably by the application of vibration.

5.2.4 Fundamental Studies in Powder Mechanics

A number of other researchers working in the general area of powder mechanics have contributed significantly to the study of vibrations of powders. For example, Scarlett and Eastman[31] studied the propagation of shock waves through a bed of granular materials. Their apparatus consisted of a vertical cylinder filled with granular materials, the cylinder having a centrally located loading transducer in the base. By applying single-impulse shock loading via this transducer the velocity and attenuation of the shock wave reaching the surface were measured.

It was found that the intensity of the disturbance reaching the surface was strongest vertically above the energy source and decreased along inclined lines until no disturbance was detected outside a 45° cone emanating from the energy source. The velocity through the bed was independent of the height of the bed and the size fraction of the sand used, but was a maximum in the vertical direction, while decreasing along lines of increasing angle to the vertical. The total impulse at a plane in the bed decays exponentially with the height of the bed. The velocity of propagation (around 140 m/s for dry sand) is substantially slower than that in a solid and decreases as it deviates from the original input direction.

Gray and Rhodes[32] examined the mechanism of energy transfer during the vibratory compaction of powders. They modeled the bed as plastic bodies on the one hand and as viscoelastic bodies on the other. When a powder is compacted by vibration, the final density is a function of the energy transferred to the bed from the vibrator. The authors found that at frequencies less than 150 Hz, the powder acts as a coherent mass and is projected from the container base, subsequently colliding with it. The motion is then nonsinusoidal. Both models predicted a decrease in the energy transferred to the powder as the acceleration is increased at constant frequency, but the models fail when the powder ceases to behave as a coherent mass. This occurs at frequencies above 150 Hz, accelerations greater than 10 g, or when the bed is fluidized by vibration. Within the limits of the models the viscoelastic model was shown to be superior.

The application of vertical vibrations to fluidize beds of granular materials has been the object of some research, such as that reported by Nicklin and Hopkins[33] and by Chlenov and Mikhailov.[34] Vibration frequencies between 2 and 100 Hz have been used to achieve fluidization in this way, with corresponding vibration amplitudes being such that the accelerations were in the range 0 to 10 g. The effect of the compressibility of the air in the gap between the container and the bed of granular material or the percolation of air through the material is shown to be of particular interest and importance.

This problem has been analyzed in some detail by Gutman,[35] who improved the earlier model of Kroll[36] with the inclusion of the air compressibility effect. Gutman performed a very comprehensive analytical and experimental study, and in addition to providing a more fundamental insight into the behavior of vibrated beds with air percolation, he also demonstrated the basic mechanism of heat transfer in, and energy dissipation of a vibrated bed. He showed that when a vertical plate heater is immersed in a vibrated bed, the heat transfer from the plate is enhanced by the scouring action of the gas sublayer as a result of the motion of the particles in the vertically vibrated container.

Harwood,[37] in studying the vibration segregation of lactose, used radioactive pills to trace the movements in a vertically vibrated tube. The results showed that at a critical vibration condition (100 Hz and 100 g) the powder was induced to take on a semifluidized state.

A fundamental understanding of the nature and mechanism of vibration of powders and bulk solids must ultimately be greatly assisted by a more thorough appreciation of the behavior of powders as conglomerations of discrete particles. The linking together of theories developed by such analyses with the already widely established and utilized continuum theory would be very beneficial. Some pioneering work in the mechanics of packings of discrete spherical particles was performed by Deresiewicz.[38] More recently Molerus[39] used the continuum and particulate approaches to formulate a theory on the yielding of cohesive powders. A simple model was proposed for the transmission of external stresses through a lattice of solid particles whereby contacting points on the surface of the particles were assumed to transmit force. This concept resulted in an expression relating the macroscopic externally applied stress to the microscopic interparticle force. The cohesive forces were shown to be a result of plastic deformation of particle contacts under external load. The concept underlying this behavior is important in determining the resonance frequencies of bulk granular materials.

A number of other articles of fundamental importance to the understanding of the deformation and flow of powders and bulk solids have been published. By way of example mention is made of the work of Becker and Lippman,[40] Nova and Wood,[41] Blinowski.[42] Cowin,[43] Goodman and Cowin,[44] Passman,[45,46] Passman and Thomas,[47] Nemat-Nassar,[48] and Nunziato et al.[49] In the latter article, the authors studied the behavior of one-dimensional acceleration waves in an inhomogeneous granular solid. They found that the average wave speed varied significantly for short propagation distances, but approached an equilibrium value for large distances. In general the amplitude of the waves attenuated with propagation distance. They concluded that the acceleration wave behavior was dominated by the elasticity of the granules and the dispersion caused by the initial nonuniformity of the material.

5.2.5 Vibrations of Bulk Solids

An important application of mechanical vibrations is in the promotion of flow of bulk solids from storage bins. The articles by Wahl,[50] Myers[51] and Carroll and Colijn[52] emphasized the practical applications of vibrations to the solution of flow problems. The types of vibrations and live bottom bins were discussed with some rule-of-thumb indications of appropriate frequencies and amplitudes to be used.

Shinohara, Suzuki, Tanaka, and Takahashi[53-55] studied the gravity flow of noncohesive and cohesive bulk solids from vibrating hoppers. In these processes, vibrations are applied to the whole hopper in the vertical direction and the behavior analyzed by a block flow model. Discharge occurs when impact forces caused by the vibrations, together with the weight of the materials within the block, exceed the shear force within the material and that at the hopper wall. A critical value of vibration intensity can then be determined. The experimental studies show that for the cohesive materials low frequency or high intensity of vibration gave the best flow ratio with good theoretical agreement. These results are in contrast with the results for the noncohesive materials, which show that higher frequency give the best discharge rates.

Some preliminary work reported by Arnold et al.,[56] Croft,[57] and Roberts et al.[58] referred to experiments performed on a model plane-flow hopper with a horizontally vibrating insert. Both discrete frequency sinusoidal vibrations and broad-band random vibrations were examined. The results indicated the significant influence of moisture content on the ability of the bulk material, in this case sand, to flow. The experiments also indicated the sensitivity to flow of the input frequency. Consistent performance of the bin was obtained by using broad-band random vibrations. Narrow-band vibrations had little influence in promoting flow. The need to examine the frequency and amplitude dependence of bulk solids through physical testing procedures was realized, and some preliminary work in this field was performed by Kaaden[59] and Arnold et al.[56] A vibrating shear cell apparatus was used to examine the influence of vibrations on the consolidation strength of bulk solids. Furthermore, the significant influence of mechanical vibrations on reducing hopper wall friction was shown, a result of particular importance to bin and hopper design.

The research of Roberts and Scott,[60] Li,[61] and Roberts et al.[62] extended this earlier work to show the influence of mechanical vibrations on reducing the strength of bulk solids, thus increasing their ability to flow. An important aspect of this work was the development of a special vibrating, direct shear testing apparatus. The contribution by Li[61] in establishing a theory for the failure of vibrated bulk solids when subjected to shear loading is of particular significance. The work conducted by these authors is discussed in greater detail in subsequent sections of this chapter.

The design of bulk handling equipment requires a knowledge of boundary friction characteristics between a granular material or bulk solid and an adjacent surface such as a wall, conveyer flight, or casing. Although it is common to employ the Jenike type direct shear tester for this purpose, other test equipment and associated procedures have been developed. Rademacher[63] proposed a method that permitted accurate determination of the kinetic coefficient of friction between a surface and a granular mass. Sharma et al.[64] investigated the high-frequency vibrational effects on soil–metal friction. Although this particular study was directed at the design and operation of tillage tools, the results are of interest to the field of bulk solids handling.

The application of laboratory scale experiments to full-scale operations often requires that the necessary laws for dynamic similarity be established. These laws must be kept in mind when developing laboratory models. The final application of mechanical vibrations as an aid to flow will require careful consideration of scaling laws to ensure that correct dynamic conditions are obtained in the prototype. Some progress on modeling has already been made. For example, Molerus and Schöneborn[65] have demonstrated the application of a bunker centrifuge to examine, on a model basis, the performance of actual bunkers. Johanson[66] and Sharma et al.[64] have used dimensional analysis in an attempt to generalize experimental results. Nonetheless, in the application of mechanical vibrations to flow promotion of bulk solids, no satisfactory modeling laws have yet been developed. However, the work already published may be considered as a guide to future work in this area.

5.2.6 Compaction of Powders

The application of mechanical vibrations to the compaction of powders has been studied indirectly by several authors already mentioned. For instance, Gray and Rhodes[32] and Kaaden[59] include some consideration of the influence of mechanical vibrations on the packing of powdered materials.

A comprehensive overview of the principles and methods of vibration compaction is given by Shatalova et al.,[67] McGeary,[68] Hauth,[69] and Evans and Millman.[70]

5.3 MEASUREMENT OF DYNAMIC SHEAR

5.3.1 Introductory Remarks

A knowledge of the dynamic shear properties of powders, soils, and bulk solids is of importance to particular areas of application. For example, in soil mechanics these properties will have a significant influence on the design of foundations and retaining walls. In developing testing equipment to measure these properties, it is important to give careful consideration to the manner in which consolidation and loading stresses act as well as to the way in which vibrations may be propagated.

One area where a knowledge of dynamic shear properties is of considerable importance is in the design and operation of flow promotion devices for bulk solids handling. Mechanical vibration, as an aid to flow promotion in storage bins, may be applied in several ways. In some cases the bin may be vibrated in the vertical direction, and in this respect the distribution of acceleration within the mass is important. In other cases the vibrations may be applied in the horizontal direction or more generally in an oblique plane. Much depends on the design of the flow promotion device. For instance, some flow promotion devices apply vibrations directly to the hopper walls; another type consists of a vibrating or gyrating insert as part of a live bottom bin. An important factor is the manner in which the vibration energy is transmitted through the bulk mass in a way to ensure a reduction in shear strength in the critical regions of the flow obstruction.

The theory of bin design as developed by Jenike[9, 10] is now well established and widely used. So that the application of mechanical vibrations as an aid to flow may become more effective in the design sense, there is a need for research on bulk solids vibrations, to be directed in a way that is compatible with the concepts of flowability and bin design as developed by Jenike. The research of Roberts, Li, and Scott[58, 61, 62] had this objective. Primarily the research was aimed at providing an insight into the mechanisms of shear failure and flow of bulk solids excited by vibration.

A significant aspect of the research concerned the development of a dynamic shear testing apparatus and associated measurement technique that was entirely consistent with the bulk solids testing procedures established by Jenike.[10, 71] The apparatus, which is described in the following subsection, was designed to determine the effect of vibration frequency, amplitude, and energy transfer on the shear strength of bulk solids, as well as the wall frictional characteristics.

The dynamic shear cell apparatus developed by Roberts et al. followed the earlier work of Arnold et al.[56, 72] in which a Jenike direct shear tester, fitted with vibration excitor, was used to determine the effects of vibration on the consolidation strength of bulk solids as well as wall friction. In this case the effects of vibration in reducing shear strength were not examined.

A vibrating direct shear test was used by Youd[30] in his research into the effects of vibration on the shear strength and void ratio of dry granular materials. In this case the shear cell was mounted on a shaker table, the cell being fitted with a pneumatic loading device that permitted normal pressures to be applied to the test samples in a way that could not be influenced by inertia effects due to the vertical vibrations. As will be explained in the next subsection the vertical mode of vibration employed by Youd differed from the horizontal mode used by Roberts et al.

A comment concerning the suitability of the Jenike direct shear tester for bulk solids analysis needs to be made. As is widely known, several other methods exist for determining the strength of bulk solids. These include ring shear tests, torsion shear tests, and triaxial tests. A complete review of all such methods and associated test equipment is given by Schwedes.[73] On balance, the Jenike direct shear tester is the most widely used and offers several advantages, in particular the ability to perform time consolidation tests. It is the most readily adaptable to the determination of vibration shear tests. Care needs to be taken in using the direct shear tester, since errors may occur in the measurements, as has been discussed by Rademacher and Haaker.[74]

5.3.2 Dynamic Shear Apparatus

The dynamic shear cell apparatus used by Roberts et al. is essentially a modified version of the Jenike direct shear cell apparatus with provision for the application of mechanical vibrations in the horizontal plane parallel to the plane of shear. The horizontal plane was chosen to isolate from the measurements any influence of vertical acceleration in the gravitational field. The apparatus is arranged so that either the whole cell or the top half of the cell may be vibrated. These two arrangements are shown in Figures 5.1 and 5.2. As can be observed, the vibrations are applied transversely to the direction of shear to minimize any interaction effects between the vibration excitation and the recorded shear force.

The test arrangement for vibrating the whole cell during shear is shown schematically in Figure 5.1.

The cell is located on a vibrating platform mounted on vertical leaf springs, as indicated. Provision is made for variations in shear cell size, but basically the testing is performed on a cell of 95 mm internal diameter. Vibration of the whole cell is used primarily to examine the resonance characteristics of consolidated bulk solids.

When only the top half of the cell is vibrated, the test arrangement shown in Figure 5.2 is used. Tests performed in this way permit a greater insight into the mechanism of failure, particularly with respect to vibration frequency and amplitude. The same arrangement as shown in Figure 5.2 is used in the determination of the influence of vibrations on the shear of bulk solids in contact with bin or hopper walls. In this case a sample of the hopper wall material is located in place of the lower half of the shear cell. In all cases the amplitude of vibration needs to be kept

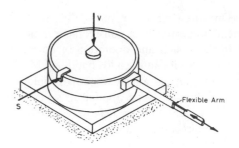

Figure 5.2. Shear cell arrangement for vibration of top half of cell.

very small to prevent premature shearing of the test sample by vibration alone.

The test instrumentation is shown in block diagram form in Figure 5.3. Provision is made for either discrete frequency sinusoidal or random vibration excitation to be applied. In the case of the latter the experimental work of Roberts et al. incorporated the use of either pseudo random binary or Gaussian signals. Vibrations are applied via the force transducer, with the associated instrumentation to provide consistent amplitude control under load.

For the tests when the whole cell is vibrated, it is necessary to tune the apparatus, so that the natural frequency and hence cut-off frequency of the vibrating cell does not influence the dynamic measurement being made. It has been found advisable to use flexible springs

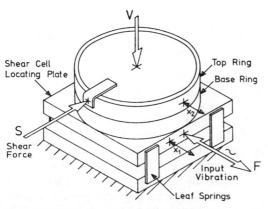

Figure 5.1. Vibrating shear cell apparatus—whole cell vibrated.

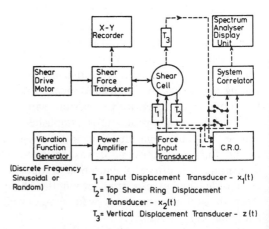

Figure 5.3. Block diagram showing test instrumentation.

for discrete frequency excitation such that the natural frequency of the cell and its mounting is well below the resonance frequencies of the bulk solid contained in the cell. On the other hand, for broad-band random vibration excitation, the spring stiffness should be sufficiently high to provide the necessary band width. That is, since the vibrating platform behaves as a mechanical low-pass filter, its cut-off frequency must be higher than the highest frequency of interest in the random vibration analysis.

Included in the instrumentation are transducers to measure the input displacement as well as the displacement of the top ring of the shear cell in both the lateral and vertical directions. As indicated in Figure 5.3, signal processing equipment such as a correlator and spectrum analyzer are desirable for analyzing the measured signals in the time domain and frequency domain, respectively. This equipment has the added advantage of permitting separate identification analysis to be performed in association with the determination of the dynamic characteristics of the particular bulk solid under test.

Since the overall objective is to determine the effect of applied vibrations on the shear strength and flow functions of bulk solids, the standard procedure as recommended by Jenike[71] for the preparation and consolidation of the samples is adopted. The samples are first preconsolidated by a twisting procedure under the applied normal load, followed by consolidation under shear to obtain the steady-state condition. Normally no vibrations are applied during the consolidation phase. Once the samples are consolidated, the vibrations are then applied during the shear test under the predetermined normal load. The apparatus does, however, permit the effects of vibration on the consolidation strength of the sample to be determined. This may be achieved by vibrating the cell under the consolidation load prior to performing the shear test. Such information is of particular value in assessing, for example, the effects on the bin performance of the vibrator being accidentally

switched on before the discharge gate is open. The information is also of relevance in determining the effects a vibratory feeder may have on the performance of a bin.

5.4 DYNAMIC SHEAR CHARACTERISTICS — SINUSOIDAL VIBRATION EXCITATION

5.4.1 Bulk Solid Strength and Flow Parameters

The purpose of the test program using the apparatus described in the previous section is to determine the relevant strength and flow properties under applied vibrations for bin design and evaluation. For a given bulk solid, the dynamic shear strength τ_f under applied sinusoidal vibrations is dependent on a number of parameters. In functional form:

$$\tau_f = f_1(\sigma_1, \bar{\sigma}, f, x, \rho, H, d, T) \quad (5.1)$$

where

σ_1 = major consolidating stress
$\bar{\sigma}$ = normal stress corresponding to shear failure
f = vibration frequency
x = amplitude of vibration
ρ = bulk density of the material
H = moisture content
d = average particle size
T = temperature of material.

The major consolidation pressure is defined for the yield locus by the applied normal consolidating process σ and corresponding shear stress τ for which consolidation in the shear cell is performed. That is, for a given set of material parameters:

$$\sigma_1 = f_2(\sigma, \tau) \quad (5.2)$$

In bin design the Flow Function, as defined by Jenike,[71] is an important parameter. For a given set of material properties the functional form of the Flow Function is given by

$$\text{FF} = f_3(\sigma_1, \sigma_{\text{cf}}) \quad (5.3)$$

where σ_{cf} = unconfined compressive stress under vibration, and is obtained from the yield function given by Eq. (5.1).

A series of tests have been performed to examine the interrelation of vibration frequency and amplitude on the shear strength for a range of material parameters. Some typical results are presented here by way of example. The results apply to pyrophyllite, a material used in the manufacture of refractories.

5.4.2 Shear Force versus Shear Deformation

It is interesting to observe the shear force versus shear deformation records obtained during an actual test. A typical set of results for the material pyrophyllite is given in Figure 5.4. Curve A_1 shows the consolidation graph during shear deformation under the applied normal stress $\tau = 8.33$ kPa, while curve A_2 is the shear versus shear deformation record for the unvibrated condition under the applied normal stress $\sigma = 2.84$ kPa.

Curve B_1 is a repeat of the consolidation condition as in A_1 for a fresh sample, while curve B_2 is the corresponding shear record for the applied normal stress $\sigma = 2.84$ kPa, this time the shear force being applied simultaneously with an impressed sinusoidal vibration of frequency $f = 200$ Hz and amplitude $X_r = 0.01$ mm measured at the shear plane. As can be seen, the sample sheared at a value of $S = 25$ N compared with a value of $S = 29.3$ N in the case of curve A_2 for the unvibrated sample.

The reduction in shear strength as a result of the impressed vibration is immediately apparent. The behavior of the second sample beyond the point b of the shear force versus shear deformation is interesting to note. At point b the vibration was switched off and, as indicated, the sample immediately reconsolidated and sheared at a value similar to that of the unvibrated sample depicted in curve A_2.

Curves A_2 and B_2, which are typical of the shear force versus deformation curves obtained for all test conditions, permit some general conclusions to be drawn concerning the dynamic characteristics of a bulk solid. During loading under the action of an applied deformation rate $\dot{x}_1$ there is an instantaneous "step" change or increase in the shear force S. This increase, denoted by F_f, is due to internal Coulomb frictional characteristics of the bulk solids. Following this initial step change the shear force then increases linearly, indicating elastic deformation. As loading further proceeds the shear force versus deformation characteristics become nonlinear as plastic flow occurs. Ultimately, the point of failure occurs, and this is accompanied by a reduction in shear force as the top half of the shear cell slides relative to the bottom half.

The behavior during unloading is interesting to observe. If at any time such as at points a and c of Figure 5.4 the load deformation input is stopped and the sample is held under constant load, there is shown to be practically no relaxation of the shear force S; apart from observed reductions over expanded periods of time of around 2% to 4%, the shear force remains substantially constant. When the shear force actuator of the testing machine described in Figure 5.1 is retracted, the shear force versus deformation characteristics are the mirror images of those obtained during loading. That is, there is an immediate "step" reduction in force due to the Coulomb frictional resistance followed by a "ramplike"

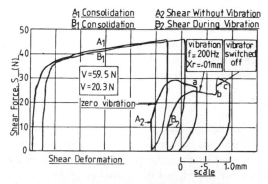

Figure 5.4. Shear force versus shear deformation for −1 mm pyrophyllite at 5% moisture content (d.b.).

elastic deformation as the elastic energy is recovered. As indicated in Figure 5.4, the slopes of the elastic deformation graphs during loading and unloading are slightly different, with the stiffness of the material being lower during unloading. This is no doubt due to the change in internal characteristics of the material resulting from the shear failure.

The characteristic behavior of the bulk material during shear is shown by Figure 5.4 to be similar for the case of shear both during and in the absence of applied vibrations. The effect of the applied vibrations is to cause a slight reduction in the internal stiffness; otherwise the general characteristics are of the same form.

5.4.3 Dynamic Shear — Whole Cell Vibrated

A set of typical results depicting the influence of consolidation stress, applied normal stress, frequency, particle size, and moisture content (m.c.) are presented in Figure 5.5. To show the magnitude of the reduction in shear stress with frequency, the shear stress ratio τ_f/τ_{f_0} is plotted against frequency in all cases, τ_{f_0} being the shear stress under zero frequency.

5.4.3.1 Effect of Consolidation Stress — Applied Normal Stress

Figure 5.5a shows τ_f/τ_{f_0} versus frequency f for the major consolidating stress $\sigma_1 = 17.4$ kPa (corresponding to $\sigma = 7.9$ kPa and $\tau = 7.6$ kPa) for -1 mm pyrophyllite at 5% moisture content (d.b.). The whole cell was vibrated with constant impressed amplitude $X_i = \pm 0.008$ mm. The shear stress τ_f was determined for the three normal stresses $\bar{\sigma} = 5.47$ kPa, $\bar{\sigma} = 4.23$ kPa, and $\bar{\sigma} = 2.98$ kPa.

The significant reduction in shear stress with increase in frequency is clearly shown. As indicated for each applied normal stress $\bar{\sigma}$, the shear stress reaches a minimum value at a particular frequency, this characteristic being indicative of a resonance effect. Similar results were obtained for other major consolidating pressures. The resonance frequency ω_n is

shown to increase with increase in major consolidating stress σ_1 (and hence σ); for each major consolidating stress the resonance frequency also increases with increase in applied normal stress $\bar{\sigma}$. These results support the results obtained in the study of X_r/X_i in the inertia model investigation. (See Section 5.4.4.)

5.4.3.2 Effect of Particle Size

Figure 5.5b shows the shear stress ratio τ_f/τ_{f_0} for four particle size ranges of pyrophyllite at 5% moisture content (d.b.). As indicated there is a well-defined resonance frequency in all cases, with the greatest reduction in shear stress occurring in the case of the largest particle size.

5.4.3.3 Effect of Moisture Content

Figure 5.5c shows the effect of moisture content on the τ_f/τ_{f_0} versus frequency characteristics for -1 mm pyrophyllite. Although there is little difference between the 5% and 10% moisture content pyrophyllites, the stress reduction at lower moisture levels is quite significant. It is clear that moisture within a bulk solid affects its dynamic characteristics, particularly with respect to stiffness and damping.

5.4.4 Amplitude Ratio — Whole Cell Vibrated

Measurements have been made of the absolute amplitude X_2 of the top half of the shear cell while the whole cell was being vibrated with input amplitude X_1 and frequency $\omega = 2\pi f$. These measurements were made both before the samples were sheared and during shear. Figure 5.6 shows the amplitude ratios X_2/X_1 for pyrophyllite for the consolidation condition $\sigma = 7.81$ kPa and for the three applied normal stresses $\bar{\sigma} = 4.69$ kPa, $\bar{\sigma} = 3.44$ kPa, and $\bar{\sigma} = 2.19$ kPa.

The effect of a resonance condition for maximum X_2/X_1 is clearly pronounced, the fundamental resonance frequency occurring at

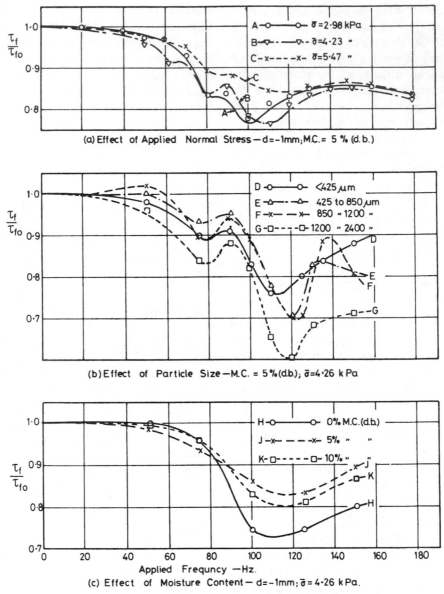

Figure 5.5. Effect of consolidation stress, applied normal stress, particle size, and moisture content on τ_f/τ_{f_0} during sinusoidal vibration of whole shear cell. Pyrophyllite; $\sigma = 7.9$ kPa, $x_i = 0.008$ mm.

values similar to that for the corresponding minimum τ_f/τ_{f_0}. Also, as indicated, there appears to be a second resonance point occurring at a frequency of approximately 200 Hz, which can be observed from the trend of the curves in Figure 5.5. It is believed that this frequency is due to the measuring equipment and not directly attributable to the bulk material.

The amplitude X_2/X_1 measured before and during shear shows an appreciable difference particularly in the region of resonance. A major influence is the effect of the loading stem (see Fig. 5.1) on the movement of the top ring.

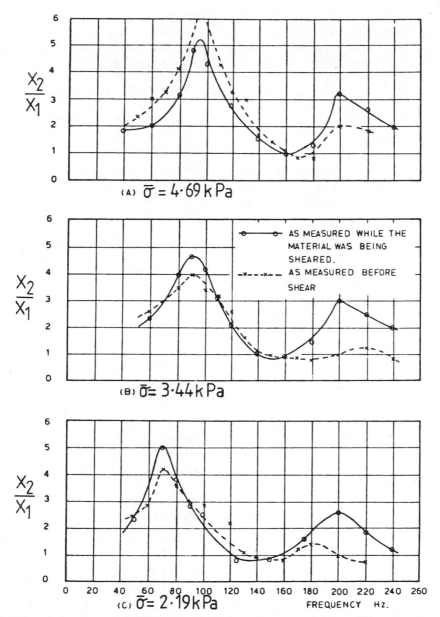

Figure 5.6. Effect of vibration on amplitude ratio X_2/X_1 for -1 mm pyrophyllite, at 5% moisture content (d.b.) $\sigma = 7.81$ kPa. $X_1 = 0.006$ mm.

Furthermore, during actual shear, the shear deformation causes the bonds between adjacent particles in contact to fail and reform in a cyclic manner. This action may also affect the dynamic properties of the material, particularly with respect to increasing the internal damping.

5.4.5 Dynamic Shear — Top Half of Cell Vibrated

When the top half of the cell is vibrated, the impressed amplitude X_1 and amplitude X_r on the shear plane are the same. Consequently there is no resonance effect, and the shear stress reduction with increase in frequency has a smooth decaying characteristic shape approaching, asymptotically, a limiting value, as illustrated in Figure 5.7.

5.4.5.1 Effect of Consolidating Stress and Applied Normal Stress

Figure 5.7a shows τ_f/τ_{f_0} versus frequency f for -1 mm pyrophyllite at 5% moisture content (d.b.). The normal consolidation stress is $\sigma = 7.9$ kPa, and the amplitude of vibration is $X_r = 0.006$ mm. Three applied normal pressures are examined; however, as the results indicate, the τ_f/τ_{f_0} variation with frequency is virtually the same for all three normal stresses.

5.4.5.2 Effect of Amplitude

Figure 5.7b shows the influence of an increase in amplitude on the reduction in shear

strength. Analysis shows that for a given frequency there is a limiting amplitude beyond which the shear stress at failure does not decrease any further.

5.4.5.3 Effect of Particle Size

The effect of particle size on the shear ratio is illustrated in Figure 5.8. The results correspond to the impressed frequency $f = 200$ Hz and amplitude $X_1 = \pm 0.01$ mm. The reduction in strength with increase in particle size confirms the results presented in Figure 5.5b for vibration of the whole cell.

5.4.5.4 Effect of Moisture Content

Figure 5.9 shows some characteristic results for the reduction in shear strength obtained by vibrating the top half of the cell at a frequency $f = 200$ Hz and amplitude $X_i = \pm 0.01$ mm. The trend of the results is similar to that indicated by Figure 5.5c for the case when the whole cell is vibrated.

5.4.6 Yield Loci and Flow Functions

From the viewpoint of assessing the effects of vibration on the strength reduction of bulk solids, the relevant information is best obtained from the yield loci and flow functions.

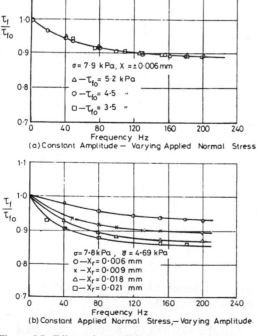

Figure 5.7. Effect of consolidation stress, applied normal stress, and amplitude on τ_f/τ_{f_0} during sinusoidal vibration of top half of shear cell for -1 mm pyrophyllite at moisture content (d.b.).

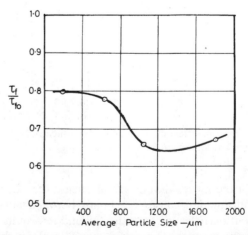

Figure 5.8. Effect of particle size on shear strength, top half of shear cell vibrated. $f = 200$ Hz, $x_i = \pm 0.01$ mm, pyrophyllite at 5% moisture content (d.b.).

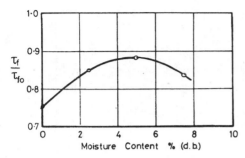

Figure 5.9. Effect of moisture content on shear strength, top half of shear cell vibrated. $f = 200$ Hz, $x_i = \pm 0.01$ mm, -1 mm pyrophyllite.

By way of illustration the yield loci for -1 mm pyrophyllite at 5% moisture content for the frequencies 0, 150, and 200 Hz and amplitude $X_1 = 0.01$ mm are plotted in Figure 5.10. These results are based on the vibrations of the whole cell, the consolidation condition being defined by the major consolidating stress $\sigma_1 = 16.0$ kPa. The reduction in shear strength is clearly evident.

The flow function of a bulk solid is a plot of the unconfined compressive strength σ_c against the major consolidating stress σ_1. For the single consolidation condition depicted by

Figure 5.10 the reduction in unconfined compressive strength as a result of applied mechanical vibrations is readily observable. To enable a complete set of flow function graphs to be drawn, a minimum of three sets of yield loci corresponding to three consolidation conditions is needed. On this basis the flow function graphs shown in Figure 5.11 have been obtained. The static zero vibration (0 Hz) or instantaneous flow function provides a reference for comparisons to be made with the vibrated flow functions. The reduction in strength under the applied vibrations is quite significant and clearly indicates the relevance of vibrations as an aid to flow.

Similar results may be obtained by vibrating the top ring of the shear cell only. The significance of this together with the implications for storage bin design is discussed in Section 5.10.

5.5 AN INERTIA MODEL FOR VIBRATION OF WHOLE SHEAR CELL

5.5.1 Introductory Remarks

In view of the complex structure of cohesive and noncohesive granular materials, difficulties are encountered when attempts are made

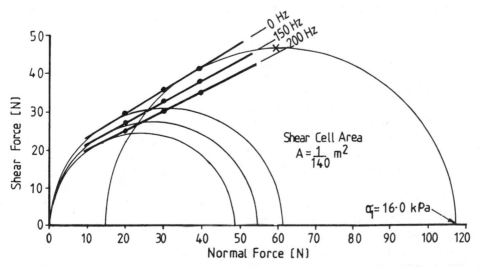

Figure 5.10. Yield loci showing influence of vibration for -1 mm pyrophyllite at 5% moisture content (d.b.), $X_1 = 0.01$ mm.

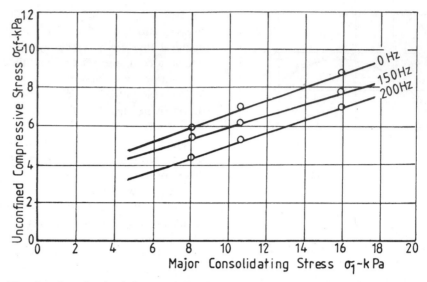

Figure 5.11. Flow functions showing influence of vibration for -1 mm pyrophyllite at 5% moisture content (d.b.), $x_1 = 0.01$ mm.

to analyze the deformation and flow characteristics of these materials. Broadly speaking, two approaches to modeling such materials may be attempted, namely the continuum approach and the particulate or discrete particle approach. Much of the work to date has been based on the continuum approach, and already a great deal of progress and development have taken place.

The continuum model assumes that the properties of the bulk materials may be represented by continuous functions of positions in both time and space. This procedure is equivalent to assuming that the material may be subdivided indefinitely into smaller components or constituents without any change in its properties. A lumped model can therefore be used to determine quantitatively the behavior of the whole aggregate or mass. So far, this approach is most commonly used principally because it lends itself to easier experimental work and subsequent analysis. It suffers from the disadvantage, in many cases, of leading to results that are often empirical in nature. However, provided the bounds on the solutions, particularly with respect to the limits of experimental and theoretical analysis, are clearly enunciated and understood, then the

continuum approach is invaluable in providing data, theories, and procedure for ready application to design and development.

In the particulate approach, a conglomeration or array of finite size particles is considered. Often it is necessary to make some simplifying assumptions, for instance, that the particles behave as rigid spheres. A physical model may be formulated by investigating the behavior of the particles resulting from their interactionary effects. On this basis attempts can then be made to deduce the behavior of the entire aggregate. This approach may provide a qualitative insight into the behavior of the material but may not be adapted to provide quantitative results. Over recent years it has become evident that the fundamental behavior of granular materials in the constitutive sense is receiving considerable attention. A selection of references[39-47,75] is included in the list at the end of this chapter. For the purpose of the present discussion only simplified lumped parameter models are considered.

5.5.2 Stiffness of Bulk Material

In the shear force versus shear deformation graphs of Figure 5.4, as indicated the characteristic "step" plus "ramp" response records are similar for the case of shear both during, and in the absence of, applied vibrations. The effect of the vibrations is to cause a slight reduction in the internal stiffness; otherwise the general behavior is the same.

The shear versus shear deformation behavior within the linear region may be represented by the simple model depicted in Figure 5.12. $\dot{x}_r$ is the applied rate of deformation and S the measured shear force. As indicated the shear force is given by

$$S = F_f + kx_r \qquad (5.4)$$

where

F_f = Coulomb force
x_r = relative deformation on shear plane
k = stiffness.

The variations in stiffness may be readily observed from the shear force versus shear deformation results obtained in the actual shear tests on the bulk solid (see Fig. 5.4). The stiffness is shown to vary with both consolidation pressure and applied normal stress as illustrated in Fig. 5.13. The stiffnesses are average values based on significant deformations in the elastic range.

5.5.3 Resonant Frequencies

As indicated in Figures 5.5 and 5.6, shear tests

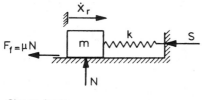

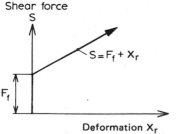

Figure 5.12. Simple Coulomb-elastic model.

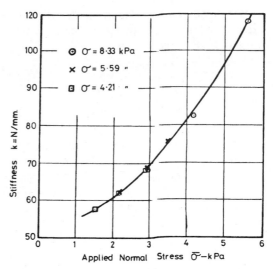

Figure 5.13. Average stiffness as a function of consolidation and applied normal stress for -1 mm pyrophyllite at 5% moisture content (d.b.).

performed on the direct shear apparatus show pronounced resonant frequencies when the whole cell is vibrated. The experimental studies indicate that the resonant frequencies vary with both the major consolidation stress σ_1, or applied normal consolidating stress σ, which acts during the consolidation phase, and the applied normal stress $\bar{\sigma}$ which acts during shear. Li,[61] in following the analysis of Molerus,[39] considered the contact forces and elastic deformation between individual particles and related the load versus deformation behavior of the particles to that of the whole mass of material in the shear cell. Li deduced that the natural frequency may be expressed by

$$\omega_n = a_0(N + C) \qquad (5.5)$$

where

a_0 = constant
N = interparticle normal force due to the external applied load
C = cohesive force generated by the external load.[57]

Molerus derived a relationship between the external compressive load and the resulting normal load. The equation is:

$$\left(\frac{1-e}{e}\right)\frac{N}{d^2} = \frac{V}{A}$$ (5.6)

where

e = void ratio
$\bar{V}$ = compressive force acting on the plane of interest (equivalent to the normal force acting on the shear cell)
A = area of the plane of interest (equivalent to area of the shear cell)
d = particle diameter.

The void ratio e depends on the degree of consolidation. However, the percentage change in e as a result of the changing consolidation is usually small. Thus, for the present discussion, it will be assumed that the bulk density, and hence voidage e, is constant. It will also be assumed that the particle diameter is constant. Hence Eq. (5.6) becomes

$$N = a_1\bar{V}$$ (5.7)

where a_1 = constant = $ed^2/(1-e)A$.

During consolidation the particle contact also receives plastic deformation and according to Molerus these deformations are responsible for the cohesive forces in bulk materials. The following relationship was derived:

$$C = a_2N_e + C_0$$ (5.8)

where

C = total cohesive force
C_0 = initial cohesive force before consolidation
N_e = interparticle normal force due to consolidation.

Using an argument similar to that embodied in Eq. (5.7) it follows that

$$N_e = a_3V$$ (5.9)

From Eqs. (5.5), (5.7), (5.8), and (5.9) the expression for ω_n, as derived by Li, becomes

$$\omega_n = a_0(a_1\bar{V} + a_2a_3V + C_0)$$ (5.10)

or in terms of stress units

$$\omega_n = c_2\sigma + c_1\bar{\sigma} + c_0$$ (5.11)

where

$$\sigma = V/A$$
$$c_0 = \frac{C_0a_0}{A}, \quad c_1 = \frac{a_0a_1}{A}, \quad c_2 = \frac{a_0a_2a_3}{A}$$
A = shear cell area.

The validity of the relationships given by Eq. (5.11) has been verified experimentally for pyrophyllite at 5% moisture content (dry basis). The result, which confirm Eq. (5.11), are plotted in Figure 5.14.

For the pyrophyllite, under the conditions examined, the constants are:

$$c_0 = 263 \qquad (s^{-1})$$
$$c_1 = 48.2 \times 10^{-3} \quad (m^2/Ns)$$
$$c_2 = 16.5 \times 10^{-3} \quad (m^2/Ns)$$

The lower bound of the resonant frequency is $\omega_n = c_0$ which occurs when both σ and $\bar{\sigma}$ equal zero. As both the consolidation stress σ and applied stress $\bar{\sigma}$ increase, the material approaches a solid condition and the resonant frequency increases.

Embodied in Eq. (5.11) are the variations in stiffness and mass that result from changes in σ and $\bar{\sigma}$. The variations in stiffness may be readily observed from the shear force versus

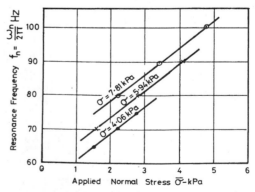

Figure 5.14. Resonance frequency for -1 mm pyrophyllite at moisture content (d.b.).

deformation results obtained in the actual shear tests on the bulk solid (see Fig. 5.4). The stiffness is shown to vary with both consolidation pressure and applied normal stress as illustrated in Figure 5.13. The stiffness and resonant frequencies will also be influenced by moisture content and particle size distribution.

5.5.4 Shear Cell Vibration Model

The graphs of Figures 5.5 and 5.6 show the existence of dominant resonance frequencies. Since these frequencies are of fundamental importance, the possibility of higher and less significant resonance frequencies will be ignored. This follows the reasoning of Li,[61] who shows that a one degree-of-freedom lumped vibration model provides a satisfactory analogy for the purposes of examining the dynamic characteristics of the shear cell test. Li assumes a linear model with viscous damping which, for the small vibration amplitudes involved, gives good correlations between the analytical predictions and experimental results.

Nevertheless, for the general case, the presence of the Coulomb resistive force cannot be ignored and for this reason the model presented by Roberts et al.[62] is proposed. This model is shown in Figure 5.15a while the corresponding forces acting on the top half of the cell are shown in Figure 5.15b. The displacement coordinates are:

x_1 = absolute displacement of the lower half of shear cell or input displacement

x_2 = absolute displacement of top ring

$x_r = x_2 - x_1$ = relative displacement between top and bottom halves of shear cell.

The force S transmitted by the shear force actuator acts perpendicular to the direction of vibration.

The formulation of the model is based on the following assumptions:

1. The material is isotropic across the shear plane.
2. The material behaves as a cohesive mass with one dominant natural frequency.

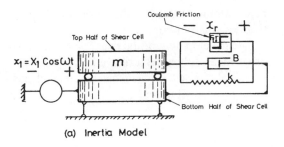

$$x_1 = X_1 \cos \omega t$$

(a) Inertia Model

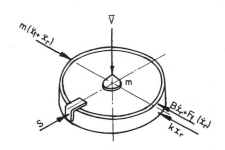

(b) Forces Acting on Top Half of Shear Cell.

Figure 5.15. Inertia model for vibration of whole shear cell.

3. During shear the properties of the material in the direction of vibration remain unchanged.
4. The applied vibration is not of sufficient intensity to shear or fluidize the material. Hence there is little loss due to friction and plastic deformation.
5. The amplitude of vibration on the shear plane is small relative to the particle size.
6. Under the influence of the applied vibration the particles in the shear cell move substantially in the direction of the applied force. Interactionary effects are negligible.

While the above assumptions are satisfactory for the present analysis it is recognized that they are not strictly correct. For instance, Molerus[39] has shown that anisotropy is inherent in the Jenike direct shear test in view of the fact that the orientation of the principal axes during consolidation may not necessarily coincide with the principal axes during incipient yield. Further, the material in the shear cell may not be completely homogeneous as a result of uneven loading and variations in particle size distribution.

The development of the dynamic model may proceed as follows: During an applied sinusoidal vibration the lower half of the cell has an impressed vibration:

$$x_1 = X_1 \cos \omega t \qquad (5.12)$$

where

X_1 = impressed amplitude (mm)
$\omega = 2\pi f$ = frequency (rad/s)
f = frequency (Hz).

The Coulomb frictional resistive force has a magnitude $F_f = \mu N$ and always acts in a direction opposing the motion. Thus the Coulomb resistive force is linked directly to the velocity but always acts in the opposite sense to the velocity. Since the top half of the shear cell vibrates relative to the bottom half with frequency ω, the Coulomb resistive force, as a function of time, will be a rectangular wave of amplitude $\pm F_f$ and frequency ω. This is illustrated in Figure 5.16. The Coulomb frictional force may be represented by the Fourier series:

$$f(t) = \frac{4F_f}{\pi}\left(\sin \omega t + \frac{1}{3} \sin 3\omega t + \cdots\right)$$

$$(5.13)$$

This is an odd function with half-wave symmetry. Hence only the sine terms and odd harmonics appear.

As in describing function analysis, a technique used in control systems analysis, a single nonlinearity such as that in the present example is approximated by the first term or fundamental frequency component of the Fourier series. Neglecting the higher harmonic terms does not introduce serious error owing to the diminishing amplitude effect associated with these terms. Thus, in this case, the nonlinear frictional resistive force may be approximated by an equivalent linear viscous drag force F_E:

$$F_E = B_E \dot{x}_r \qquad (5.14)$$

where

$$B_E = \frac{4F_f}{\pi \dot{X}_r} \qquad (5.15)$$

or

$$B_E = \frac{4F_f}{\pi X_r \omega} \qquad (5.16)$$

where

$\dot{x}_r$ = relative sliding velocity
$\dot{X}_r$ = amplitude of sliding velocity
X_r = amplitude of relative displacement.

Here B_E is the describing function for the frictional resistive force and may be regarded as the equivalent damping coefficient. Of particular interest in the study of the shear cell model is the relationship for the relative motion between the top and bottom sections of the shear cell. It may readily be shown that the differential equation is:

$$\frac{d^2 x_r}{dt^2} + 2\zeta\omega_n \frac{dx_r}{dt} + \omega_n^2 x_r = X_i \omega^2 \cos \omega t$$

$$(5.17)$$

where

$\omega_n = \sqrt{k/m}$ = natural frequency (rad/s)
ζ = damping ratio

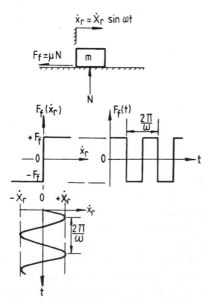

Figure 5.16. Coulomb frictional drag force during sinusoidal excitation.

x_r = relative displacement
m = total mass in motion with top half of cell
k = dynamic stiffness of bulk solid
$\omega = 2\pi f$ = impressed frequency.

The absolute motion of the top ring is also of interest since it is often easier to measure than the relative motion, particularly in the case of small impressed amplitudes. The differential equation in this case is:

$$\frac{d^2 x_2}{dt^2} + 2\zeta\omega\frac{dx_2}{dt} + \omega_n^2 x_2$$

$$= \frac{X_1\omega^2}{\sqrt{1 + (2\zeta r)^2}}\cos(\omega t + \theta) \quad (5.18)$$

where

r = frequency ratio ω/ω_n
$\theta = \tan^{-1}(2\zeta r)$

The total system damping is the sum of the actual viscous damping denoted by B and the equivalent damping B_E due to Coulomb friction.
Thus the damping ratio is:

$$\zeta = \frac{B + B_E}{B_{cr}} \quad (5.19)$$

where

$B_{cr} = 2m\,\omega_n$ = critical damping factor.

From Eqn. (5.16) it follows that when Coulomb damping is present, ζ will not be constant but will vary with the impressed frequency ω and amplitude X_r. The amplitude X_r is of particular importance in that it relates directly to the deformation on the shear plane.
Before discussing the general case of damping given by Eq. (5.19) it is important to first consider the two particular cases of viscous damping and Coulomb damping when each is present separately.

5.5.5 Viscous Damping

Here, ζ is constant and the steady-state solutions of Eqs. (5.17) and (5.18) may be expressed in terms of the amplitude and phase relationships. For relative motion:

$$\frac{X_r}{X_1} = \frac{r^2}{\sqrt{(1 - r^2)^2 + (2\zeta r)^2}} \quad (5.20)$$

The phase angle ψ_r between the relative motion and impressed motion is given by

$$\psi_1 = \tan^{-1}\frac{2\zeta r}{1 - r^2} \quad (5.21)$$

where

r = frequency ratio, ω/ω_n.

A resonance condition corresponding to maximum X_r/X_1 is obtained at the frequency

$$\omega_{re} = \frac{\omega_n}{\sqrt{1 - 2\zeta^2}} \quad (5.22)$$

For absolute motion

$$\frac{X_2}{X_1} = \sqrt{\frac{1 + (2\zeta r)^2}{(1 - r^2)^2 + (2\zeta r)^2}} \quad (5.23)$$

The phase angle ψ_a between the absolute motion and impressed motion is given by

$$\psi_a = \tan^{-1}\left[\frac{2\zeta r^3}{1 - r^2 + (2\zeta r)^2}\right] \quad (5.24)$$

5.5.6 Coulomb Damping

Using the approximated equivalent viscous damping coefficient B_E given by Eq. (5.16), it follows from Eq. (5.19) that when $B = 0$, the damping ratio is given by

$$\zeta = \frac{2F_f}{\pi m X_r \omega\omega_n}$$

or

$$\zeta = \frac{2F_f}{\pi k X_1 r} \quad (5.25)$$

Substituting for ζ in Eq. (5.20), the amplitude ratio for the relative motion becomes

$$\frac{X_r}{X_1} = \pm \sqrt{\left[\frac{r^2}{1-r^2}\right]^2 - \left[\frac{4F_f}{\pi F_s(1-r^2)}\right]^2} \quad (5.26)$$

and the phase angle ψ_1 is:

$$\psi_r = \tan^{-1}\left[\frac{\dfrac{4F_f}{\pi X_1}}{\dfrac{X_1}{X_1} F_s(1-r^2)}\right] \quad (5.27)$$

where

$F_s = kX_1$ = equivalent static restoring force.

Referring to Eq. (5.26), it can be seen that X_r has a real value only when

$$\frac{F_f}{F_s} < \frac{\pi r^2}{4} \quad (5.28)$$

When small frictional forces are involved, as is usually the case, the condition is easily satisfied. On the other hand, the condition will not be satisfied at low frequencies where $r \ll 1.0$. For all cases when Eq. (5.28) is satisfied, Eq. (5.26) shows that the amplitude X_r becomes infinite at resonance, that is when $\omega = \omega_n$. Timoshenko et al.[76] explain this condition in terms of the energy dissipated per cycle being less than the energy input.

A more rigorous solution of the Coulomb damping vibration problem with a harmonic force acting on the system has been presented by den Hartog.[77] The procedure he followed may be readily adapted to the case where the system is disturbed by a harmonic displacement, as occurs in the shear cell model of Figure 5.15. The differential equation is:

$$\frac{d^2x_r}{dt^2} + \omega_n^2 x_r = X_1\omega^2 \cos \omega t \pm \frac{F_f}{k}\omega_n^2 \quad (5.29)$$

where the + sign refers to the friction force under the condition that the mass moves in the positive direction and vice versa.

It can be shown that the amplitude ratio obtained from a solution of Eq. (5.29) is given by

$$\frac{X_r}{X_1} = \sqrt{\left[\frac{r^2}{1-r^2}\right]^2 - \left[\frac{F_f}{F_s}\right]^2 Q^2} \quad (5.30)$$

where $r = \omega/\omega_n$.

$$Q = \frac{\sin\left(\dfrac{\pi}{r}\right)}{r\left[1 + \cos\left(\dfrac{\pi}{r}\right)\right]} \quad (5.31)$$

It may be observed that Eq. (5.30) is similar in form to the approximate solution given by Eq. (5.26). In the exact solution given by Eq. (5.30) the Coulomb damping function Q, given by Eq. (5.31), varies with the frequency ratio as indicated. The phase angle from the rigorous solution is:

$$\psi_r = \tan^{-1}\left[-\frac{F_f Q}{F_s \dfrac{X_r}{X_1}}\right] \quad (5.32)$$

Diagrams of the amplitude ratio X_r/X_1 and phase angle ψ_r given by Eqs. (5.30) and (5.32), respectively, are represented in Figure 5.17. The various curves illustrate the influence of the force ratio F_f/F_s. Referring to Figure 5.17a, for comparison purposes, the plotted points indicate the curve for viscous damping with damping ratio $\zeta = 0.1$. As can be seen, this curve follows closely the results for very small Coulomb frictional forces when $F_f/F_s = 0.05$ or lower. It is readily observed that for the zero damping case, Eq. (5.30) for Coulomb damping and Eq. (5.20) for viscous damping become identical. From a practical point of view the equivalent constant viscous damping factor of $\zeta = 0.1$ provides a convenient approximation of the lower bound for the Coulomb damping case.

5.5.7 Combined Viscous and Coulomb Damping

Using the approximated equivalent viscous damping coefficient B_E for the Coulomb

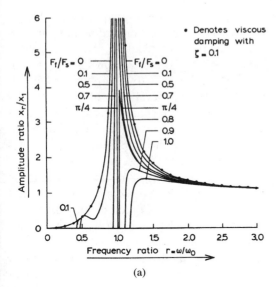

• Denotes viscous
  damping with
  $\zeta = 0.1$

(a)

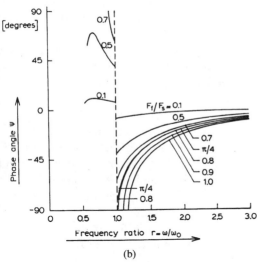

(b)

Figure 5.17. (a) Amplitude ratio. (b) Phase angle.

damping then the combined damping factor as given by Eq. (5.19) may be written in the form:

$$\zeta = \zeta_v + \frac{2F_f}{\pi X_1 kr} \qquad (5.33)$$

where

$\zeta_v = B/2m\omega_n$ = damping ratio for the viscous component of the damping.

Following a similar procedure to the approximate Coulomb damping case it can be shown

that the amplitude ratio X_r/X_1 is given by the quadratic:

$$\left(\frac{X_r}{X_1}\right)^2 \left[(1-r^2)^2 + (2r\zeta_v)^2\right] + \left(\frac{X_r}{X_1}\right)\left[\frac{16\zeta_v rF_f}{\pi F_s}\right]$$
$$+ \left[\left(\frac{4F_f}{\pi F_s}\right)^2 - r^4\right] = 0 \qquad (5.34)$$

and phase angle ψ_r is:

$$\psi_r = \tan^{-1}\left[\frac{2\zeta_v r}{1 - r^2} + \frac{4F_f}{\frac{\pi X_r}{X_1}F_s(1 - r^2)}\right]$$

$$(5.35)$$

Solutions for the real values of X_r/X_1 can be obtained from (5.34) for given values of ζ_v and F_f/F_s.

A rigorous analysis of the influence of combined viscous and Coulomb damping has been presented by den Hartog.

5.5.8 Verification of Model

The actual damping characteristics of cohesive bulk solids may be a combination of a number of factors such as interparticle friction, plastic deformation at contact points, and interfacial fluid damping. The characteristics are certainly highly nonlinear and extremely difficult to analyze in a rigorous way. However, it is reasonable to assume that the combination of viscous and Coulomb damping, as previously described, provides a satisfactory approximation for modeling purposes. As to which of the Coulomb or viscous components of the damping is dominant will depend, to some extent, on the amplitude of vibration. Certainly if the amplitude is large in relation to average particle size, then Coulomb friction will have a major influence. On the other hand, for very small amplitudes Li[61] argues that Coulomb friction is minimal and that the particles simply undergo small oscillatory motions about their pinning points or points of contact.

For the experimental work using the vibration shear cell apparatus, Li concludes that because of the small amplitudes involved, the damping is dominantly viscous. Following extensive tests he established that for pyrophyllite and iron ore the damping was not influenced significantly by the consolidation stresses and applied normal stresses during vibration. The following viscous damping factors were shown to fit the data quite well:

- For -1 mm pyrophyllite at 5% moisture content (d.b.), $\zeta = 0.1$.
- For iron ore at 5% moisture content (d.b.), $\zeta = 0.125$ (see Tables 5.2 and 5.3 for more detailed information).

As an indication of the degree of fit given by the assumption of viscous damping with $\zeta = 0.1$, the ratio of the absolute to impressed amplitude X_2/X_1 computed using Eq. (5.23) is compared with the corresponding experimentally obtained results for pyrophyllite. The two curves are shown in Figure 5.18, and the agreement is considered satisfactory. It is to be noted that the absolute amplitude rather than the relative amplitude is used, since the former was easier to obtain experimentally. Further, as previously stated, the model was developed to predict the fundamental natural frequency; no attempt has been made to analyze the presence of the second and higher natural frequency shown in the experimental results, since this frequency is of lower significance in affecting the behavior of the material during shear.

Whether the reasoning given by Li in favor of viscous damping being dominant in this case is correct, is difficult to say. Certainly the very small amplitudes used in the experimental work lend weight to his argument. On the other hand, the general form of the shear load versus deformation characteristic of Figure 5.4 favors Coulomb damping as being dominant, particularly when the amplitude is of a reasonable order. For the present results, reference to Figure 5.17 indicates that a viscous damping ratio of $\zeta = 0.1$ is equivalent to a low value of F_f/F_s in the case of Coulomb damping, that is $F_f/F_s < 0.1$. While this is feasible, it is difficult to quantify; it implies a low value of F_f and a high value of the initial stiffness k. It should be noted that the stiffness values plotted in Figure 5.13 are average values for large deformations obtained under very low (almost static) deformation rate conditions. Nonetheless, the shear force versus shear deformation graphs of Figure 5.4 indicate initial values of k substantially higher than the average values plotted in Figure 5.13.

The overriding results of this study is that the shear cell model adequately depicts the behavior of the sample during vibration. It is clear that the damping is of very low order and for Coulomb damping it is equivalent to $F_f/F_s \simeq 0.05$ for the pyrophyllite. The assumption of viscous damping with $\zeta = 0.1$ fits the data sufficiently well for practical purposes and provides a simple model for analysis.

5.5.9 Concept of Resonance

The concept of resonance in the case of a bulk solid is somewhat complex, particularly in view of the difficulty of defining, with any degree of certainty, the mass and stiffness contributions in the actual vibrating system. Even in the shear cell apparatus there are problems in establishing the actual vibrating mass, but the problems are compounded when the same exercise is attempted for a bulk solid in an actual hopper. However, as discussed later, the significant parameters affecting the dynamic shear failure are vibration frequency and rela-

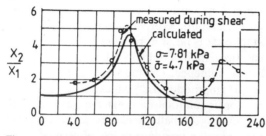

Figure 5.18. Comparison of measured and calculated amplitude ratios for -1 mm pyrophyllite at 5% moisture content (d.b.).

tive amplitude on the shear plane. Thus the dynamic properties of bulk solids are important in view of their influence on the transmission of vibrations, particularly with respect to the possible amplification of vibration amplitude.

5.6 A FAILURE CRITERION

5.6.1 Correlation of Dynamic Shear Stress with Sliding Velocity

In order that the shear cell test data be applied to the design or bin activators and flow promotion devices, consideration of the appropriate parameters for dynamic similarity is required. Such parameters need to take into account the relevant information on vibration frequency, amplitude, and energy transfer. As an initial step in this investigation, studies have been undertaken to ascertain the degree of compliance of the shear stress during vibration with the relevant parameters, notably the velocity and acceleration. Which one of these two parameters is the more appropriate will depend, to some extent, on such factors as the intensity of the vibration, the manner in which the vibration is applied, and the design features of the particular vibration system concerned. For instance, in many cases the material is vibrated in a gravitational field either in the vertical direction or at some angle to the vertical. In such cases the acceleration ratio $\omega^2/X_i/g$ is a significant parameter.

However, if the material is vibrated horizontally, as in the shear cell apparatus of Figure 5.1, then the effects of gravity and acceleration become less important.[60-62] The experimental results show that the shear stress correlates well with both the vibration velocity and acceleration. However, as shown by Li[61] the weight of evidence favors the vibration velocity (and hence energy) as being the parameter with the most significant influence on the dynamic shear behavior. The correlation with sliding velocity is shown, by way of example, with the graphs of Figure 5.19. These graphs apply to -1 mm pyrophyllite at 5% moisture content. Figure

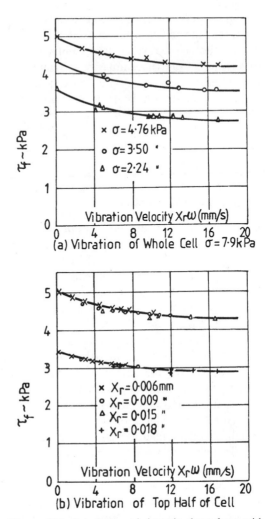

Figure 5.19. Correlation of dynamic shear force with maximum vibration velocity for -1 mm pyrophyllite at 5% moisture content (d.b.).

5.19a corresponds to the vibration of the whole cell, and it is clear that the plotting of τ_f against the maximum relative sliding velocity $2\pi f X_r$ on the shear plane has resulted in a smoothing out of the curve into a simple decaying shape with the resonance effect, as indicated in Figure 5.5, being no longer visible. This characteristic results from the fact that amplitude X_r in the maximum sliding velocity is the relative amplitude on the shear plane and is related to the impressed amplitude X_i as shown by Figure 5.6.

Figure 5.19b corresponds to the vibration of the top of the cell only. Here $X_r = X_1$ and the τ_f plot shows a decaying shape similar to that of Figure 5.19a. The similarity in shape between the two curves of the figure is clearly evident. In fact, analysis shows they are virtually identical in shape. It is clear, therefore, that resonance is meaningful only in terms of the amplitude transmitted to the plane of shear.

5.6.2 Concept of Failure

Li based his studies on the concept of failure originally developed by Hvorslev[12] and later extended by Roscoe et al.[13] Hvorslev, whose work was concerned with cohesive soils, showed that the shear stress at failure is a function of the effective normal stress on, and voids ratio (or density) in the plane of failure at the instance that failure occurs; failure, defined in this way, is independent of the stress history of the material. Rosco et al. showed the relationship between shear stress τ, normal stress σ, and void ratio e, which, for a given bulk material defines a failure surface as illustrated in Figure 5.20. The state of a sample defined by any point below the surface indicates a condition of no failure. States of stress defining a point on the surface indicate a condition of yield, and hence failure, or incipient yield. States of stress defining points above the surface are not possible. Roscoe et al. also showed the existence of the critical void ratio line (CVR) which defines a boundary of the Hvorslev failure surface. The critical void ratio indicates the limiting state of a sample at which a further increment of shear distortion will not result in any change in the void ratio. The CVR is a unique line to which all loading paths in the e–σ–τ space converge.

It is proposed that the Hvorslev failure surface applies equally well to the dynamic case as to the static. Referring to Figure 5.20, the specimen in the shear cell is originally consolidated with normal stress σ_0 and shear stress τ_0, the corresponding void ratio being e_0. The condition of the sample is given by D'. In the

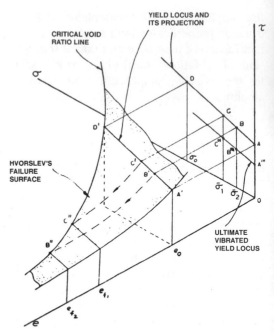

Figure 5.20. Behavior of material under vibration on the Hvorslev's failure surface.

static case, reduced normal stresses $\bar{\sigma}_1$ and $\bar{\sigma}_2$ will lead to points C' and B' on the yield surface. The projection of $B'C'D'$ onto the σ–τ plane gives the yield locus for the material for the given consolidation condition. If vibration is applied during the application of the normal stress $\bar{\sigma}_1$, the material dilates on the shear plane, causing a localized increase in the void ratio at the shear plane where relative motion exists. Elsewhere in the cell the material is in fact consolidating as the voids are reduced. The increase in voidage on the shear plane causes a change of state along the line $C'C''$. A similar argument holds for the applied stress $\bar{\sigma}_2$, where a change of state occurs along $B'B''$.

For each applied normal stress the application of mechanical vibrations may cause the void ratio to increase until a limiting value is reached. The limiting void ratios for the two stresses $\bar{\sigma}_1$ and $\bar{\sigma}_2$ are respectively e_{f_1} defined by point C'' and e_{f_2} defined by point B''; the points B'' and C'' lie on the critical void ratio line. The projection of points B''

and C'' onto the $\sigma-\tau$ plane specifies two points on the ultimate vibrated yield locus which define the lowest possible values of the shear stress. A further increase in the intensity of vibration will not cause any greater reduction in shear stress. As will be readily apparent these arguments apply only to the case of horizontally applied vibration; in the case of vertical vibrations, failure of the material may take place under conditions of high vibration acceleration.

During the application of mechanical vibrations, yielding may occur anywhere on the Hvorslev surface between the boundaries defined by the static condition and the ultimate or critical voidage condition. The condition for failure defined by points on the failure surface directly depends on the impressed velocity, frequency, and applied normal stress. For this reason the actual vibrated yield loci may not be straight or parallel lines. They may even cross each other as indicated in Figure 5.21. For a given normal stress $\bar{\sigma}$ the reduction in shear stress has been shown by Roscoe et al.[13]

to follow a negative exponential curve expressed by

$$\tau = \mu\bar{\sigma} + \beta \exp(-Be_f) \qquad (5.36)$$

where μ, β, and B are material constants and e_f is the void ratio in the plane of failure.

Referring to Figure 5.19 it is apparent that the shear strength τ_f during vibration follows a similar negative exponential curve that suggests that the vibration velocity on the shear plane is directly correlated with the void ratio e. Consequently Li proposed a decaying exponential function of the form

$$\tau_f = \tau_\infty + \beta \exp\left(\frac{-2\pi X_r f}{\gamma}\right) \qquad (5.37)$$

where $f = \omega/2\pi$ (Hz) and τ_∞, β, and γ are constants, which for a given material depend on the consolidation and normal stress during shear. The form of equation is indicated in Figure 5.22. τ_∞ represents the limiting shear stress, which is an indication of the effectiveness of the vibrations on the shear reduction.

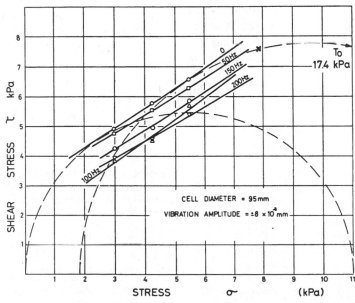

Figure 5.21. Yield loci for pyrophyllite illustrating the effect of sinusoidal vibration.

The static value of the shear stress τ_{f_0}, which occurs when the input frequency $f = 0$, is:

$$\tau_{f_0} = \tau_\infty + \beta \qquad (5.38)$$

As an alternative, Eq. (5.37) may be formulated in terms of a relationship for the shear stress ratio τ_f / τ_{f_0}. It may readily be shown that

$$\frac{\tau_f}{\tau_{f_0}} = 1 - \frac{\beta}{\tau_{f_0}} \left\{ 1 - \exp\left(-\frac{2\pi X_r f}{\gamma} \right) \right\} \qquad (5.39)$$

The constant γ in Eqs. (5.37) and (5.39) is called the vibration velocity constant and indicates the rate of decay of the shear stress. By way of example the decay in the shear stress for multiple values of γ is indicated in Table 5.1. Thus for $U = 3\gamma$, for example, the shear stress decreases by 95% of its total possible reduction. The velocity $U = 5\gamma$ causes the shear strength to decrease some 99.3% of its total possible reduction.

5.6.3 Prediction of Vibrated Shear Strength

For the purpose of this discussion attention is drawn to the various experimentally obtained parameters for pyrophyllite and iron ore listed in Tables 5.2 and 5.3, respectively. With respect to Eq. (5.37) it can be seen that the parameter β, which represents the maximum possible reduction in shear stress, depends only on the consolidation condition given by σ; for

Table 5.1. Shear Stress For Multiple Valves of γ

VIBRATION VELOCITY $U = 2\pi X_r f$	SHEAR STRESS τ_f
0	$\tau_\infty + \beta$
γ	$\tau_\infty + 0.368\beta$
2γ	$\tau_\infty + 0.135\beta$
3γ	$\tau_\infty + 0.05\beta$
4γ	$\tau_\infty + 0.015\beta$
5γ	$\tau_\infty + 0.007\beta$
10γ	$\tau_\infty + 0.00005\beta$
∞	τ_∞

Table 5.2. Typical Dynamic Properties (Material: Pyrophyllite with 5% moisture content (d.b.) input amplitude $X_i = 0.006$ mm.)

σ (kPa)	$\bar{\sigma}$ (kPa)	f_n (Hz)	ζ	τ_{f_0} (kPa)	τ_∞ (kPa)	β (kPa)	γ
7.81	4.69	100	0.11	4.86	3.92	0.94	7
7.81	3.47	90	0.10	4.21	3.28	0.94	7
7.81	2.20	80	0.095	3.36	2.42	0.94	7
5.94	4.06	90	0.105	4.07	3.28	0.80	7
5.94	2.81	80	0.10	3.42	2.62	0.80	7
5.94	1.57	70	0.10	2.59	1.79	0.80	7
4.06	2.81	75	0.10	2.94	2.24	0.70	7
4.06	2.19	70	0.10	2.59	1.89	0.70	7
4.06	1.57	65	0.10	2.22	1.52	0.70	7

each σ, β does not change with the applied normal stress $\bar{\sigma}$. The vibration velocity parameter γ is shown to be a constant for the particular material and independent of the consolidation stress and applied normal stress.

Thus, by vibrating the top ring only, an estimate of the shear strength during vibration may be obtained as follows:

1. For each consolidation condition, determine τ_{f_0} for the chosen applied normal stress.
2. For each consolidation condition and a nominal applied normal stress $\bar{\sigma}_1$, estimate the maximum possible strength reduction β. This value may be checked for other applied normal stress values. For this test it will be necessary to use a high vibration

Table 5.3. Typical Dynamic Properties (Material: Iron ore with 5% moisture content (d.b) input amplitude $X_i = 0.0075$ mm)

σ (kPa)	$\bar{\sigma}$ (kPa)	f_n (Hz)	ζ	τ_{f_0} (kPa)	τ_∞ (kPa)	β (kPa)	γ
8.54	4.41	110	0.15	5.35	4.44	0.91	10
8.54	3.04	100	0.125	4.44	3.53	0.91	10
6.44	3.72	100	0.125	4.35	3.54	0.81	10
6.44	2.35	95	0.125	3.47	2.66	0.81	10

velocity excitation. The properties of the exponentially decaying curve of Figure 5.22 will enable the values of both β and γ to be determined. As a guide, the relative amplitude X_r versus frequency f curves of Fig. 5.23 will allow various alternative values of X_r and f corresponding to $U = 5\gamma$ to be determined.

The curves are plotted for a range of possible values of γ. The values of $U = 5\gamma$ correspond to the shear stress reduction of 99.3% of β.

Where resonance effects are known to be present, such as when the whole shear cell is vibrated, the resonant frequencies may be estimated using Eq. (5.11) with the coefficients c_0, c_1, and c_2 determined for the particular bulk material. The damping characteristics need to be estimated, but for powders similar to iron ore or pyrophyllite for which the properties are known, the assumption of an equivalent viscous damping factor ζ in the range 0.1 to 0.125 would seem to be satisfactory. Using linear theory, the relative amplitude X_r as a function of impressed amplitude X_1 may be estimated. This enables the shear stress as a function of frequency for various consolidation and loading conditions to be determined.

Li[61] used this general procedure to check the validity of the parameters determined by experiment. A typical set of values for pyrophyllite is shown in Fig. 5.24. The agreement between the calculated and experimental values is considered to be quite satisfactory.

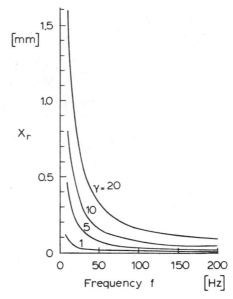

Figure 5.23. Relationship between X_r and f to obtain maximum reduction in shear strength.

5.7 BOUNDARY SHEAR AND WALL FRICTION

5.7.1 Introductory Remarks

There are many applications in industry involving powdered or bulk materials where it is necessary to obtain low frictional drag forces between the bulk material and an adjacent metal or plastic surface. For instance, in the design and operation of gravity mass-flow storage bins, a low friction coefficient at the hopper walls is essential. To achieve this condition, it is often necessary to employ polished stainless steel or an appropriate plastic coating as a lining for the hopper. The significance of this statement may be gauged from the example given in Figure 5.25, which shows the wall yield loci for a typical coal in contact with several wall surface materials. However, it has also been established by several workers,[56,58,61,64,72,79] that mechanical vibrations can significantly reduce the frictional drag between a bulk material and a solid surface.

Much work has been done to examine the effect of vibration on the friction between solid surfaces.[79-81] Although the sliding bodies

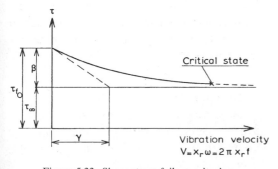

Figure 5.22. Shear stress failure criterion.

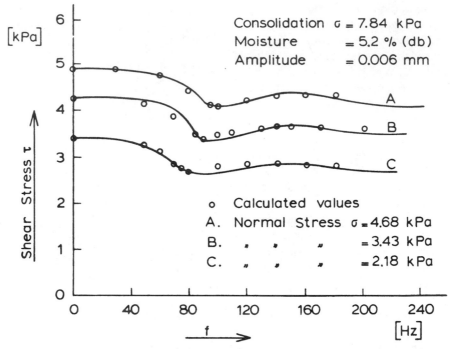

Figure 5.24. Comparison between experimental and calculated shear stress for vibration of whole shear cell for −1 mm pyrophyllite at 5% moisture content (d.b.).

in such cases may be somewhat different from that of a bulk material in contact with a solid surface, as far as the frictional drag is concerned, there are many similarities. From a macroscopic point of view, the force to overcome friction in solid contacts is equivalent to the force to shear a Coulomb powder. Microscopically both processes involve the shear deformation of asperities between adjacent contact surfaces and particles.

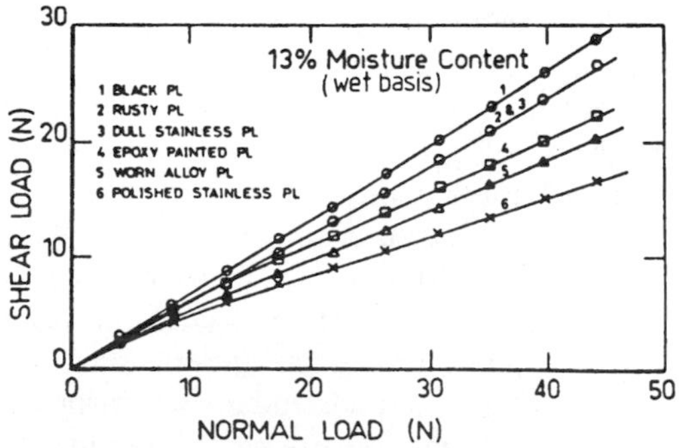

In studying the effect of vibration on the reduction of friction between metal surfaces, Lenkiewicz[79] employed vibration amplitudes which ranged from 0.005 to 0.2 mm. He showed that the sliding velocity due to the imposed vibration was a critical factor in the process although the correlations did not generally apply at large amplitudes.

Similar experiments were performed by Tolstoi,[81] who attributed the reduction in friction between metal surfaces to the increase in sliding velocity. The reasons for this behavior are summarized below:

1. The higher the velocity, the shorter is the time during which adjacent asperities compress each other, and hence the separation between contact surfaces can be maintained at a maximum.

Figure 5.25. Wall yield loci for a typical coal. (From Arnold et al.[78])

2. The increase in sliding velocity increases the upward component of the impulsive forces exerted on the asperities as they collide with those of the adjacent surfaces. In this way the amplitude of the natural vibration is increased. The effect is the same as in (1) above.

3. The higher the sliding velocity due to vibration, the shorter is the time available to squeeze the boundary lubricant between the asperities. This aids the reduction in the frictional drag force.

It is interesting to observe the similarity between the results and observations of Lenkiewicz[79] and Tolstoi[81] with the criterion for shear failure of a vibrated bulk solid presented in Section 5.6. In the case of sliding between two adjacent solid surfaces, the increase in separation between the surfaces resulting from an increase in vibrational sliding velocity is equivalent to the effect of dilation during shear of a consolidated bulk solid.

Sharma et al.[64] studied the influence of high-frequency vibration on soil–metal friction. A dimensional analysis approach was used, and several dimensionless parameters were derived. Of these, the following three are of interest to the present discussion:

$$\pi_1 = \frac{\omega X_i}{U_s}$$

$$\pi_2 = \frac{N_c \omega^2 X_i}{g}$$

$$\pi_d = \frac{F_f}{N} \qquad (5.40)$$

where

π_d = dependent term
F_f = average friction force
U_s = velocity of sliding of contact surfaces
N_c = inertial coefficient
X_i = amplitude.

The apparent coefficient of soil–metal friction given by π_d decreases as π_1 increases as illustrated in Figure 5.26. However, the varia-

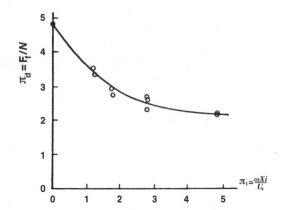

Figure 5.26. Effect of translation velocity ratio on the apparent coefficient of soil–metal friction. (From Sharma, Drew, and Nelson.[64])

tion of π_d with π_2 was within 3.07% of the mean at the 95% probability. Hence the inertia number π_2 was considered insignificant. The significance of the velocity number π_1 adds further weight to the argument that the vibrational velocity on the plane of failure being the important parameter in causing dilation or separation leading to a lowering of the frictional drag force.

5.7.2 Experiments Using Vibrating Shear Cell Apparatus

In the work of Kaaden[59] and Arnold et al.[56,72] the effect of vibration on the wall yield loci for bulk solids was determined using a modified Jenike direct shear test apparatus. Vibrations were applied in the same direction as the direction of sliding as indicated in Figure 5.27. To facilitate experimental measurements, the metal plate was vibrated, while the bulk solid

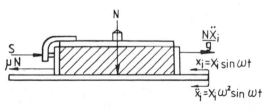

Figure 5.27. Wall yield locus analysis. (From Kaaden.[59])

sample contained within the shear cell ring was pushed relative to the plate.

Referring to the forces shown on the shear cell in Figure 5.27 and following the simplified analysis given by Kaaden, it can easily be seen that the shear force S can be reduced by an amount not exceeding

$$\Delta S = \frac{NX_i\omega^2}{g} \qquad (5.41)$$

where

N = normal force on shear plane
X_i = amplitude of vibration
ω = frequency
ΔS = maximum reduction in shear force.

Reference to Figure 5.28 will permit the reasoning behind this analysis to be seen.

Figure 5.29 shows the results obtained from tests on Mt. Newman iron ore fines at 9.3% moisture content (w.b.). For this test a constant frequency of 25 Hz and amplitude X_i = 0.23 mm was used and, as is readily observable, shear force values of from 5% to 10% of the static values have been obtained. Also, as indicated by Figure 5.29, the correlation between the measured and predicted vibrated wall yield loci is very reasonable.

To isolate the effect of the inertia force in lending assistance to the reduction in shear force as in the previous case, Roberts et al.[58, 62] and Li[61] used the vibrating shear cell apparatus described in Section 5.3 of this chapter. Vibrations were applied in the horizontal plane in a direction perpendicular to the direction of

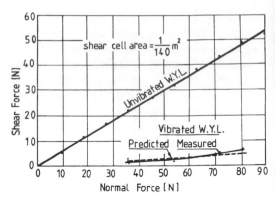

Figure 5.29. Vibrated versus normal wall yield locus, Mt. Newman fines. 9.3% moisture content (w.b.) on black steel. (Kaaden[59]–Arnold et al.[56])

shear. In this way the results obtained are consistent with those presented in Section 5.4.

By way of illustration, a set of wall yield loci curves for −1 mm pyrophyllite (5% moisture content) on mild steel is presented in Figure 5.30. A constant amplitude of X_i = 0.006 mm was used. The reduction in wall friction with increase in frequency is quite evident. The reduction in friction is considerably less than would be obtained when vibrations are applied in the direction of shear.

5.8 RANDOM VIBRATION EXCITATION

5.8.1 Application of Random Vibrations

The application of broad-band random vibrations to promote the flow of bulk solids has

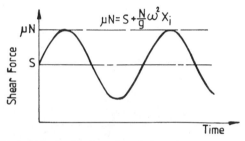

Figure 5.28. Oscillation of total shear force at limiting condition.

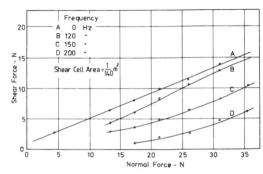

Figure 5.30. Vibrated wall yield loci for pyrophyllite on mild steel plate. X_r = ±0.006 mm; moisture content = 5% (d.b.).

been discussed by Arnold et al.[56] and Roberts et al.[58, 60] In view of the frequency dependence of many bulk solids with respect to strength reduction, broad-band random vibrations have certain advantages. Since the vibration energy is distributed uniformly across all frequencies within a given bandwidth, then it is certain that a resonant frequency of the particular material and flow promotion device would be excited, provided the resonant frequency lies within the bandwidth of the random vibration input. This will not be the case when discrete frequency sinusoidal vibration excitation is used when the input frequency differs significantly from the resonant frequency of the system.

The influence of random vibrations on the shear strength of bulk solids has been examined using the shear cell apparatus of Figure 5.1, some typical results being shown in Figure 5.31 (see Roberts and Scott[60]). Here a comparison is made between two cell diameters, namely 57 and 95 mm and two band widths 40 and 150 Hz, the bulk material is pyrophyllite. The samples were consolidated under the applied normal stresses $\sigma = 7.9$ kPa in the 95-mm cell and $\sigma = 9.5$ kPa in the 57-mm cell; in each case the samples were sheared under the normal stress of $\bar{\sigma} = 4.23$ kPa during the application of random vibrations.

Figure 5.31 shows the variation of τ_f/τ_{f_0} for pyrophyllite as a function of RMS amplitude. Curves A and B apply to the 95-mm cell, whereas curve C applies to the smaller 57-mm cell. For curves A and C the bandwidth is 90 Hz; for curve B the bandwidth is 150 Hz.

The comparison between curves A and B clearly shows the advantages, in terms of reduction in shear strength, in using the wider bandwidth. These findings support the results shown in Figure 5.5, which indicate minimum shear stresses occurring around 100 Hz under sinusoidal excitation. The comparison between curves A and C of Figure 5.31 shows that the shear stresses for the same bandwidth are lower in the smaller cell for corresponding amplitudes. This may be explained by the fact that the vibration energy input per unit mass is higher in the case of the smaller cell.

The results of this study clearly indicate that, to be most effective, the bandwidth in the case of random excitation should span the resonant frequencies. This does not happen, in this case, for the 40 Hz bandwidth.

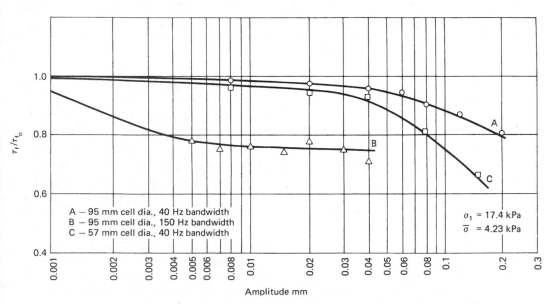

Figure 5.31. Effect of random vibration excitation on shear stress ratio τ_f/τ_{f_0} for pyrophyllite.

The results may be viewed another way. For most bulk solid storage systems, the major reduction in strength occurs at, or in the vicinity of, the resonant frequency in view of the amplitude, and hence velocity amplification, that this provides. Since the resonant frequency will vary from one material to another, then the general application of wide-band random excitation in flow promotion has some merit. The continuous range of frequencies within a given bandwidth would render this type of excitation applicable to a wide range of bulk materials.

5.8.2 Dynamic System Identification

The application of system identification by random excitation and cross-correlation and spectral analysis to bulk materials handling systems has been fully described by Roberts et al.[82-84] The method has also been used by Roberts and Scott[60] to determine the dynamic characteristics of bulk solids in the shear cell test.

The test procedure consists, essentially, of applying a pseudo-random binary coded displacement signal $x_1(t)$ to excite the whole shear cell and obtaining the cross-correlation function and cross-spectrum density of the two signals $x_1(t)$ and $x_2(t)$ where $x_2(t)$ is the dynamic displacement of the top half of the shear cell. The cross-correlation function $R_{x_1 x_2}(\lambda)$ and cross-spectral density $S_{x_1 x_2}(f)$ are, respectively, defined as:

$$R_{x_1 x_2}(\lambda) = \lim_{T \to \infty} \frac{1}{T} \int_0^T x_1(\lambda) x_2(t + \lambda)\, dt$$

$$(5.42)$$

and

$$S_{x_1 x_2}(f) = \int_{-\infty}^{\infty} R_{x_1 x_2}(\lambda) e^{-j2\pi f\lambda}\, d\lambda \quad (5.43)$$

That is $S_{x_1 x_2}(f)$ is the Fourier transform of $R_{x_1 x_2}(\lambda)$.

Under the condition that the input signal bandwidth is significantly greater than that of the system, in this case the shear cell and bulk solid, it can be shown that the impulse response or weighting function $h(\lambda)$ is given by

$$h(\lambda) \simeq K R_{x_1 x_2}(\lambda) \quad (5.44)$$

where K = constant under the same conditions.

$$H(f) \simeq K S_{x_1 x_2}(f) \quad (5.45)$$

where $H(f)$ is the system transfer function.

In performing the system identification analyses on the bulk solids, the samples were first prepared and consolidated in the normal way. The random excitation was then applied with the selected normal pressure σ applied to the shear cell, and measurements of $R_{x_1 x_2}(\lambda)$ and $S_{x_1 x_2}(f)$ were obtained.

To illustrate the application of this method, results are given for pyrophyllite and iron ore. Figure 5.32 shows the cross-correlation functions for these two materials while Figure 5.33 shows the corresponding cross-spectral densities. The results were obtained using the 95-mm diameter shear cell, the samples being consolidated under the normal stress $\sigma = 7.9$ kPa. The shear cell was subjected to pseudo-random excitation under the applied normal stress $\bar{\sigma} = 4.23$ kPa.

It is interesting to compare the two $S_{x_1 x_2}(f) \simeq H(f)$ curves of Figure 5.33 with the corresponding τ_f / τ_{f_0} curves obtained from sinusoidal excitation. These latter curves for pyrophyllite and iron ore are shown in Figure 5.34. Referring first to Figure 5.33 it may be seen that the $S_{x_1 x_2}(f)$ characteristic increases to a maximum around 90 Hz, which indicates that the peak vibration energy is concentrated at this particular frequency. On the other hand, the "flatter" $S_{x_1 x_2}(f)$ characteristic for iron ore indicates that the vibration energy is more uniformly distributed over the frequency range 0 to 100 Hz. This explains why the τ_f / τ_{f_0} graph of Fig. 5.34 for iron ore shows a greater reduction than the corresponding graph for pyrophyllite over the 0 to 100 Hz frequency range.

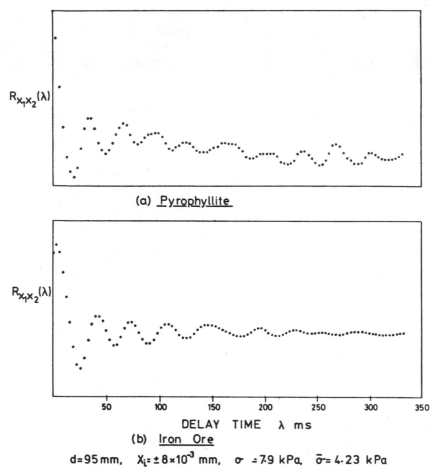

(a) <u>Pyrophyllite</u>

(b) <u>Iron Ore</u>

d=95mm, X_i=±8×10^{-3} mm, σ =7.9 kPa, $\bar{\sigma}$= 4.23 kPa

Figure 5.32. Cross-correlation functions for pyrophyllite and iron ore.

5.9 COMPACTION OF POWDERS AND BULK SOLIDS

Mechanical vibrations are employed extensively to compress and compact powders, the range of applications in this respect being wide and varied. Whereas vibrations are used to advantage in such cases, there are often situations where compaction of powders or bulk solids may have detrimental effects. This is particularly the case in bulk handling operations where increased strength due to vibratory compaction may cause flow interruptions. This section consists of a brief review of some salient aspects of the effects of vibrations on strength and bulk density.

5.9.1 Effect of Vibration on Shear Strength and Bulk Density

Kaaden[59] used a Jenike direct shear tester to examine the effect of sinusoidal vibrations on the consolidation and shear strength of bulk solids. In the standard shear test the consolidation of the sample in the shear cell comprises two phases, a preconsolidation phase involving a twisting procedure under the applied normal load followed by the consolidation under shear phase to obtain a steady-state

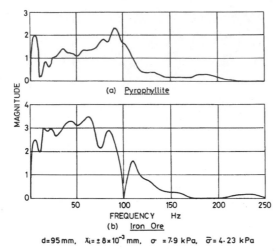

Figure 5.33. Cross-spectral densities $S_{x_1 x_2}(f) \simeq H(f)$ for pyrophyllite and iron ore.

ing applied to the top ring. For the range of frequencies examined, the increase in shear strength due to vibration of the confined samples was most affected by the increase in time of vibration. Presumably there is a limiting time of vibration consolidation when the bulk material approaches a limiting density and any further vibration would have little effect. For comparison purposes the time yield locus for a 48-hour storage time is also shown and it is readily observable that the increased strength due to vibration is far more significant than that due to time storage. The results presented in Figure 5.35 are in sharp contrast to the results given in Section 5.4; in that case the shear strength was considerably reduced when vibrations were applied during the actual shearing equation.

It is well known that a correlation exists between the shear strength and density of a bulk solid; the more a sample is consolidated, the higher will be its density and the higher its shear strength. By way of example Figure 5.36 shows the correlation between shear stress τ and bulk density for Mt. Newman iron ore fines. For curve A the complete cell was vibrated using a sieve shaker, whereas for curve B the top ring of the shear cell was vibrated in the horizontal direction using a electromagnet shaker.

condition. To assess the effect of vibration on the shear strength, Kaaden used the same consolidation procedure as in the standard test, but prior to shearing, the samples were further consolidated by vibration for a predetermined period of time.

Figure 5.35 shows a typical set of yield loci results for Mt. Newman iron ore at 9% moisture content (w.b.); in this case the samples were consolidated by horizontal vibrations be-

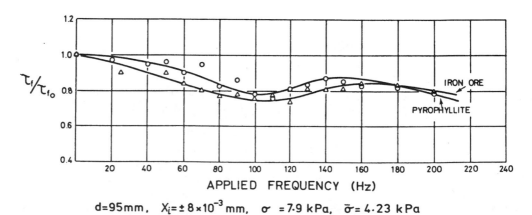

d=95mm, $X_i = \pm 8 \times 10^{-3}$ mm, σ =7.9 kPa, $\bar{\sigma}$= 4.23 kPa

Figure 5.34. Comparison of effect of sinusoidal vibration on shear ratio τ_f / τ_{f_0} for iron ore and pyrophyllite.

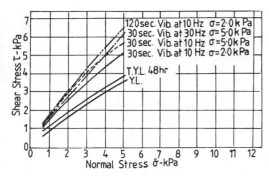

Figure 5.35. Effect of vibration consolidation on yield loci, Mt. Newman iron ore, 9% moisture content (w.b.).

The sieve shaker produced samples that were more uniformly packed and hence of greater strength than the samples produced by the electromagnetic shaker. These results indicate a 20% density increase being accompanied by an 80% to 90% increase in shear strength.

5.9.2 Vertical Compressive Deformation During Lateral Vibration

To gain a better insight into the effects of vibration on the compressibility of bulk solids, Roberts and Scott[60] examined the vertical compressive deformation of samples in the shear cell when subjected to laterally applied vibrations. The shear cell apparatus shown in Figure 5.1 was used for this test.

Samples were prepared and consolidated under shear in the usual way, the consolidating stress being $\tau = 7.9$ kPa. They were then subjected to lateral vibrations under reduced normal stress, during which records of the compressive deformation were made. For this series of tests -2 mm pyrophyllite was used instead of the -1 mm pyrophyllite, as used in the work presented in Section 5.4.

Figure 5.37 shows a set of compressive deformation versus time records obtained during sinusoidal vibration under the applied normal stress $\tau = 5.3$ kPa. The amplitude of vibration was 0.012 mm. As indicated, the major transient deformations occur within the first minute; after this the response curves approach asymptotically steady-state deformations.

The steady-state deformations as a function of frequency are shown in Figure 5.38. For the range of frequencies examined, the minimum deformation occurs around $f = 140$ Hz, with a maximum value being approached at $f = 100$

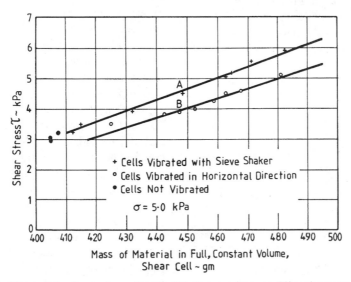

Figure 5.36. Effect of density on shear strength, Mt. Newman iron ore, 9% moisture content (w.b.).

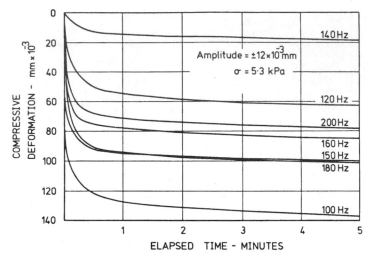

Figure 5.37. Vertical compressive deformation of pyrophyllite during lateral vibration, 95 mm shear cell.

Hz. Although the evidence is somewhat inconclusive, the maximum consolidation occurring around the value $f = 100$ Hz lends weight to the notion of this frequency being a resonance frequency similar to that depicted in the results presented in Section 5.4. A plausible argument that relates the results given here with the reduction in shear strength shown in Figure 5.5 might proceed as follows: It is apparent that the maximum reduction in shear strength occurs when the dilation, and hence voidage, on the shear plane is a maximum. This occurs at resonance. Hence, when the sample is vibrated while being sheared, the resonance condition implies that the maxi-

mum dilation on the shear plane is accompanied by a maximum compressive deformation of the material elsewhere in the shear cell.

5.9.3 Some General Remarks

The application of vibrations to the compaction of powders has received wide attention. A comprehensive review of the underlying principles and associated research in this topic is given by Shatalova et al.,[67] McGeary,[68] and Evans and Millman.[70]

It is clear that the application of vertical vibrations to powders and bulk materials held within containers provides an efficient mode of compaction, and consequently this field of application has been studied in some detail. For instance Gray and Rhodes[32] point out that the final density of vibrated powder is a function of the energy transferred to the bed from the vibrator. They modeled the bed as (1) plastic bodies and (2) viscoelastic bodies; both models predict a decrease in energy transferred to the powder as the acceleration is increased at constant frequency. For frequencies below 150 Hz and accelerations less than $10g$, the powder behaves as a coherent mass, compaction taking place as a result of the block type motion of the mass being projected

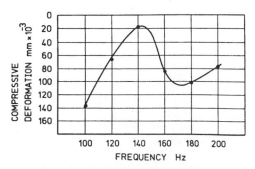

Figure 5.38. Steady-state vertical compressive deformation during lateral vibration, 95 mm shear cell.

from the base of the container and subsequently collecting with it. Above the frequency and acceleration limits indicated, the bed becomes fluidized.

Other work of relevance to vibratory compaction is that of Shinohara et al.[53] and Suzuki and Tanaka.[55] Particular attention is drawn to the work of Gutman,[35] who studied the effect of air compressed between a vertically vibrating bed and the bottom of the container.

For more details on the effect of vibrations on the compaction of powders, the reader is referred to the literature cited.

5.10 APPLICATION OF VIBRATIONS IN FLOW PROMOTION

Reference has been made several times throughout this chapter to the application of vibrations in promoting the flow of bulk solids from storage bins. When used correctly, vibrations can significantly reduce both the strength of bulk solids and the wall friction, and as a result, greatly increase the ability of the material to flow.

There are a variety of ways in which mechanical vibrations are used in practice in association with flow promotion in storage bins, and these have been reviewed in the articles by Carroll and Colijn,[52] Myers,[51] and Wahl.[50] Although a great deal of practical knowledge has been gained, the information available in the past for design purposes has been largely empirical. There has been a general lack of information concerning the relevant frequency, amplitude, and inputs that should be used for each particular bulk solid.

The manner in which the information presented in earlier sections of this chapter may be integrated with the general philosophy of bin design is discussed in this present section.

5.10.1 Bin Design Philosophy

The theory of storage bin design, as developed by Jenike,[9, 10] is well demonstrated and widely used. However, in order that the effect of

vibrations on the flow of bulk solids may be better understood, the salient aspects of the bin flow characteristics and design philosophy are briefly reviewed.

5.10.1.1 Flow Patterns

As indicated in Figure 5.39, there are two basic modes of flow, namely mass flow and funnel flow.

In mass flow the bulk material is in motion at substantially every point in the bin whenever material is drawn from the outlet. The material flows along the walls with the bin and hopper (that is, the tapered section of the bin) forming the flow channel. Mass flow is the ideal flow pattern and occurs when the hopper walls are sufficiently steep and smooth and there are no abrupt transitions or inflowing valleys.

Funnel flow, on the other hand, occurs when the material sloughs off the surface and discharges through a vertical channel that forms within the material in the bin. This mode of flow occurs when the hopper walls are rough and the slope angle α is too large. The flow is erratic, with a strong tendency to form stable pipes that obstruct the bin discharge. When flow does occur, segregation takes place, there being no remixing during flow. It is an undesirable flow pattern.

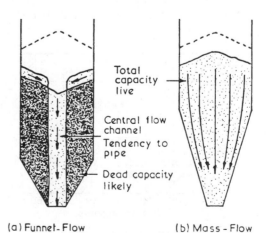

(a) Funnel-Flow   (b) Mass-Flow

Figure 5.39. Bin flow characteristics.

The limits for mass flow depend on the hopper half angle α, the wall friction angle ϕ, and the effective angle of internal friction δ. In the case of conical hoppers the limits for mass flow are clearly defined as illustrated in Figure 5.40. On the other hand plane flow or wedge-shaped hoppers have similar limits for mass flow, but these are less critical.

5.10.1.2 Hopper Geometry for Mass Flow

Basically the aim in mass flow design is to determine the hopper geometry, in particular the hopper half angle α and opening size B, so that a stable cohesive arch cannot form over the bin outlet. Two parameters are important: first, the *flow function FF* representing the strength of the material, and second, the *flow factor ff* representing the stress condition in the hopper during flow. The flow factor is given by

$$ff = \frac{\sigma_1}{\overline{\sigma}_1} \qquad (5.46)$$

where

$\overline{\sigma}_1$ = stress that can develop in an arch
σ_1 = major consolidating stress.

The flow factor is a linear function and is shown together with a typical flow function in Figure 5.41. The flow factor depends on the wall friction angle ϕ, the hopper half angle α, and the effective angle of internal friction δ.

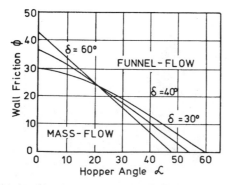

Figure 5.40. Funnel flow versus mass flow limits for conical hoppers.

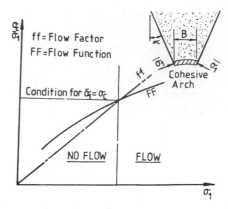

Figure 5.41. Flow/no-flow criterion for mass flow hopper design.

That is,

$$ff = f(\phi, \delta, \alpha) \qquad (5.47)$$

The determination of flow factors is described by Jenike[10] and Arnold et al.,[78] who also give the associated flow factor design charts.

With reference to Figure 5.41, the critical condition for flow is defined by $\overline{\sigma}_1 = \sigma_c$ where σ_c is the unconfined compressive strength. The minimum opening size B is defined by

$$B = \frac{\overline{\sigma}_1 H(\alpha)}{\rho g} \qquad (5.48)$$

where

ρ = bulk density
g = gravitational acceleration.

The function $H(\alpha)$ depends on the outlet shape and hopper half-angle and is plotted by Jenike[10] and Arnold et al.[78]

In practice the opening size is made larger than the minimum value given by Eq. (5.50) to ensure a satisfactory flow rate. In the case of fine powders of low permeability, the determination of opening size to achieve a prescribed flow rate is far more complex, and air permeation may be necessary.[78, 85]

5.10.2 Influence of Mechanical Vibrations

As previously discussed, two significant parameters in bin and hopper design and performance evaluation are the flow function and wall yield locus. In Figure 5.11 the flow functions from −1 mm pyrophyllite at 5% moisture content were presented for the three frequencies 0 Hz (instantaneous, static condition), 150 Hz, and 200 Hz. However, on the basis of the failure criterion presented in Section 5.6, it is really only necessary to determine, for the vibration case, the limiting flow function corresponding to the maximum vibration velocity for maximum shear strength reduction. The flow function determined in this way for the pyrophyllite is shown in Figure 5.42. This graph is based on shear tests in which the top half of the shear cell was vibrated at a frequency of 200 Hz and amplitude ±0.01 mm; the corresponding velocity on the shear plane is 12.6 mm/s, which is the value approaching that for maximum shear strength reduction as indicated in Figure 5.19. The unvibrated flow function is shown, for comparison purposes, in Figure 5.42c. The reduction in strength due to

vibration is quite significant and has a major influence on bin design.

Figures 5.42a and 5.52b show, respectively, the bulk density ρ and effective angle of internal friction. Both these parameters are required for bin design.

The vibrated wall yield loci for the pyrophyllite are presented in Figure 5.30. For the present discussion reference will be made to the instantaneous and 200 Hz curves. The reduction in wall friction due to vibration at 200 Hz will permit a significant increase in hopper half-angle α for the case of a conical hopper, as may be observed from an inspection of Figure 5.40.

The influence of the vibrations in improving the flow characteristics may be seen by reference to the mass-flow hopper proportions given in Table 5.4. Here a comparison is made between the hopper half-angle and opening sizes for conical hoppers determined on the basis of the instantaneous and the vibrated properties of pyrophyllite.

The increase in hopper half-angle α and the reduction in opening size B show the significance of using mechanical vibrations as an aid to flow. The increase in α can be even greater if the vibrations at the wall are applied in the direction of flow. The results of this study show that a funnel-flow bin can operate under mass flow giving uniform discharge, provided the vibrations are applied correctly. It is important that the bin and hopper be in a potential flow mode with the flow control gate open when the vibrations are applied; otherwise the vibrations will have the adverse effect of increasing the consolidation and strength of the bulk solid and reducing its ability to flow. This condition is indicated in Section 5.9. A flow mode will exist when an arched stress field occurs in the hopper.[78]

In order that the shear cell data and corresponding design information of the type given in Table 5.4 be applied to the design of bin activators, consideration needs to be given to the appropriate dynamic scaling parameters. Such parameters would need to translate the information on frequency and amplitude relat-

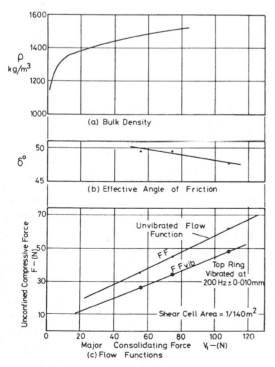

Figure 5.42. Flow properties of vibrated and unvibrated samples for −1 mm prophyllite at 5% moisture content (d.b.).

Table 5.4. Comparison of Mass Flow Conical Hoppers for Unvibrated and Vibrated Conditions [Material: -1 mm pyrophyllite, 5% moisture content (d.b.)]

HOPPER GEOMETRY	INSTANTANEOUS PROPERTIES (UNVIBRATED)	VIBRATED PROPERTIES $f = 200$ Hz $x_r = 0.006-0.01$ mm
Half-angle α	19°	33°
Opening size B	0.9 m	0.6 m

ing to the shear cell tests to equivalent data for a full-scale bin. The theory of failure presented in Section 5.6 shows that the vibration velocity on the plane of failure is the significant parameter causing local dilation and a corresponding reduction in shear strength.

Vibration velocity $X_r \omega$ is directly related to the vibration energy E_v per unit volume in the following way:

$$E_v = \rho X_r^2 \omega^2 / 2 \qquad (5.49)$$

It follows that the energy transfer characteristics of a bulk solid have a direct influence on the vibration velocity, and hence shear strength reduction, at the critical region of a flow obstruction. Based on this reasoning it seems, therefore, that the required vibration energy per unit mass at a flow obstruction needs to be the same as that determined for the shear cell.

In practice, the vibration flow promotion device may, if necessary, be located some distance from the actual flow obstruction. For this reason the vibration propagation characteristics of the bulk solid, as discussed by Scarlett and Eastman,[31] need to be taken into account. Fortunately, as already indicated, a reduction in wall friction can significantly improve the flow characteristics and, for this reason, a vibrator suitably placed on the hopper wall may be very effective. Such a device has been described by Carroll and Colijn.[52]

Another effective flow promotion device is the bin activator described by Wahl.[50] In this case the bin activator consists of a domed baffle plate located just above the outlet which, together with the rounded hopper bottom, is vibrated in the horizontal plane.

The domed baffle plate is said to eliminate overhead bridging by transmitting vibrations vertically up the hopper to cause flow to occur at least in the area vertically above the activator. An effect not mentioned is the reduction of friction on the surface of the baffle that enhances material flow off the edge of the baffle. In addition, the transmission of vibrations to the hopper walls also reduces friction and further enhances the flow.

Other types of vibratory-induced flow systems involve the vertical vibration of the whole bin or hopper. The comprehensive research of Shinohara, Suzuki, Takahashi, and Tanaka[53-55] has presented data indicating the conditions under which flow of both noncohesive and cohesive bulk solids can be best obtained by such systems. A difficulty of employing vertically vibrating hoppers in practice, particularly in large installations, is the design complexity and energy requirements to vibrate the whole hopper–bin system.

5.10.3 Performance of Bin with Vibrating Insert

Reference was made above to the vibrating baffle plate flow promotion device described by Wahl.[50] Studies performed by Croft[57] and Roberts et al.[58] have focused attention on some performance characteristics of this type of flow promotion. Salient aspects of their research are briefly reviewed.

The test rig used for this work is drawn schematically in Figure 5.43. The model bin is of rectangular cross-section and operates under plane flow, this shape and flow mode being chosen for ease of manufacture and testing. The bin is clear plastic, and the vibrating insert located inside the bin discharge section can be fitted with flat or convex baffle plates of various geometrical proportions.

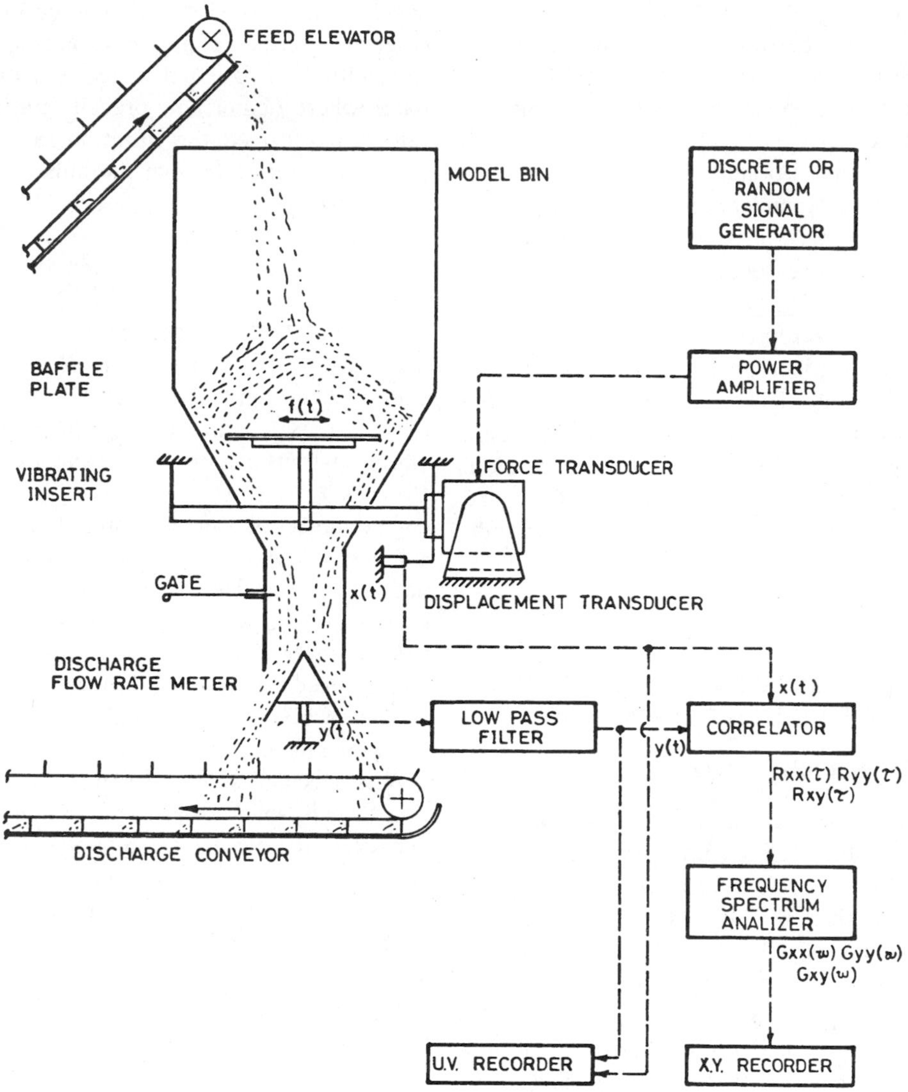

Figure 5.43. Model bin experimental test rig.

Transverse vibrations are applied to the insert by the force transducer driven by either a discrete-frequency or random-signal generator through a power amplifier. The vibrator provides the driving force and may be set at a constant force amplitude, while the input displacement of the insert and output material flow rate are recorded continuously. Signal processing instrumentation permit information on correlation functions and power spectral densities to be obtained.

By way of example, one series of tests was conducted with moist sand using a flat plate baffle. The application of sinusoidal forced excitation of varying frequencies and pseudo-random binary forced excitation of varying bandwidths were examined. In all cases the effective force amplitudes were kept approximately the same.

The results for the wet sand indicate that the low frequencies (in the order of 10 Hz) do not readily initiate flow. Quite often considerable time may elapse before flow commences, if at all. On the other hand with higher frequencies in the order of 100 to 300 Hz, flow commences almost immediately and continues at a steady rate.

The results also indicate that the excitation frequency is dependent on material properties, which confirms the results of the shear cell tests. For this reason, broad-band random excitation, in which there is an infinite number of frequencies represented in the bandwidth, will have distinct advantages over discrete-frequency sinusoidal type excitation. Table 5.5 gives a typical set of results for the wet sand, using pseudorandom binary force excitation. Although the results relate specifically to the model bin, nevertheless they provide an indication of comparative performance.

5.11 TRANSMISSION OF VIBRATION ENERGY THROUGH BULK MASS

5.11.1 General Remarks

Effective flow promotion depends on the ability of the store bulk mass to transmit vibration energy from the source of point of vibration excitation through the mass to the region of the flow blockage. In the case of a storage bin, it is usual to install the vibrator on the hopper wall; this provides an immediate benefit through the reduction in wall friction that may result. Furthermore, if the flow blockage is in the form of an arch, then vibration applied to the hopper wall at or near the outlet may cause the arch to fail and flow to occur. In this case, the vibration energy does not need to be transmitted any great distance. On the other hand, where funnel flow prevails and a stable rathole has formed, the vibration energy needs to be transmitted through the bulk mass.

The dynamic shear test described in Section 5.2, together with the theory of failure, provides information on the frequency and amplitude, and hence energy level to be applied, at the zone of the flow blockage. It then becomes necessary to determine the level of vibration energy to be applied by the flow promotion device at its location point. The problem of vibration transmission has been discussed by Roberts.[86]

The subject of wave motion in bulk granular solids is of interest to several areas of engineering. In particular, the study of seismic effects on soils and ground subsurface structures is covered in the fields of soil mechanics and geomechanics. Furthermore, the analysis of stress waves in elastic media is dealt with in the general subject area of theoretical and applied mechanics. Yet, despite this, wave motion in bulk solids storage bin systems has so far received little attention. Some salient aspects of vibration wave theory of relevance to bulk solid flow promotion are briefly discussed in the sections that follow.

5.11.2 Simplified Analysis Based on the One-Dimensional Case

The analysis of stress waves in bulk solids is exceedingly complex, particularly when consideration is given to the factors involved in the

Table 5.5. Typical Performance Results for Wet Sand using Pseudorandom Binary Force Excitation

| | ELAPSED TIME (S) | | AVERAGE | |
| | BEFORE FLOW COMMENCED | TO EMPTY BIN ONCE FLOW COMMENCED | FLOW RATE (m^3/s) | |
BANDWIDTH				comments
10 Hz	79[a]	35	0.0023	Flow not uniform
100 Hz	18	23	0.0031	Uniform flow
300 Hz	13	17	0.0042	Uniform flow

[a] This is a typical figure—much longer times have been recorded at this bandwidth.

"real" situation. Bearing in mind the difficulties that may arise in practice, factors contributing to the complexity include:

1. The problem is one of three dimensions involving uncertain boundary conditions.
2. The stored bulk solid may not be homogeneous owing to such factors as a wide variation in particle size, variation in moisture content, and variation in consolidation conditions throughout the stored bulk mass. As a result, the bulk solid is unlikely to be isotropic.
3. The damping effects within the stored bulk mass are uncertain.

The underlying principles of vibration energy transfer in relation to flow promotion may be gleaned by considering the simplified model depicted in Figure 5.44.

Figure 5.44a shows a section of a bin with vibration excitation being applied at the surface; stress waves transmitted through the stored bulk solid gives rise to planes of peak deformation as illustrated. Consider a "rod" of bulk solid material "extracted" from the bin as indicated in Figure 5.44b. Although this one-dimensional model may be somewhat unrealistic in terms of the actual bulk solid/ storage bin system, it does serve to highlight certain characteristics that aid the understanding of the objectives of flow promotion. It is assumed that the rod in Figure 5.44b is subjected to a single impulsive force at the free end, which causes a compression wave to travel along the rod and be reflected back as a tension wave.

Assuming that both Coulomb and viscous damping is present and that this combined internal damping is represented as equivalent

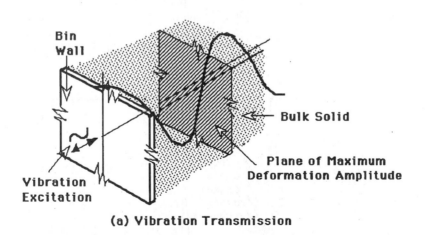

(a) Vibration Transmission

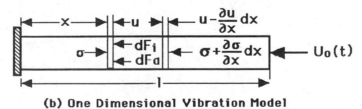

(b) One Dimensional Vibration Model

Figure 5.44. One-dimensional longitudinal vibration problem.

viscous damping, the longitudinal vibrations of the rod are governed by the damped wave equation.

$$\lambda^2 \frac{\partial^2 u}{\partial x^2} = \frac{\partial^2 u}{\partial t^2} + \omega \frac{\partial u}{\partial t} \qquad (5.50)$$

$$\lambda = \sqrt{\frac{E}{\rho}} \ \ (\text{m/s}) \qquad (5.51)$$

= velocity of wave propagation in rod

where

E = elastic modulus of bulk solids (N/m²)
u = deformation of rod at location x (m)
ρ = bulk density (kg/m³)
β = damping factor (s⁻²).

Assuming the rod is subjected to a unit impulse type displacement at the end $x = l$, the impulse response is obtained by solving Eq. (5.50).

$$h(x, t) = \sum_{n=1,3,5}^{\infty} D_n \sin\left(\frac{\omega_n x}{\lambda}\right) e^{-\zeta_n \omega_n t} \sin \omega_{dt}$$

$$(5.52)$$

where

$$D_n = \frac{4\lambda}{n\pi l \sqrt{1 - \zeta_n^2}} (-1)^{\frac{n-1}{2}} \qquad (5.53)$$

$$\omega = \frac{n\pi\lambda}{2l} \qquad f_n = \frac{n\lambda}{4l} \qquad (5.54)$$

$$\omega_{dn} = \omega_n \sqrt{1 - \zeta_n^2} \qquad (5.55)$$

$$\zeta_n = \zeta_1/n \qquad (5.56)$$

$$n = 1, 3, 5, \ldots$$

The solution $h(x, t)$ from Eqs. (5.52) to (5.56) yields damped wave responses with time. The wave form is of triangular or of truncated triangular shape.

The response to any input $U(t)$ applied to the end of the rod may be obtained by the convolution integral. Of particular interest is the response to a sinusoidal forcing function. When equation (5.52) is transformed to the frequency domain the transfer function is obtained. The steady-state amplitude ratio obtained by this process is illustrated in Figure 5.45 for the position $x = l$, for a range of damping ratios. As is evident, the higher resonant frequencies have a reducing influence on the amplitude of the forced vibration.

Figure 5.46 shows the model shapes as func-

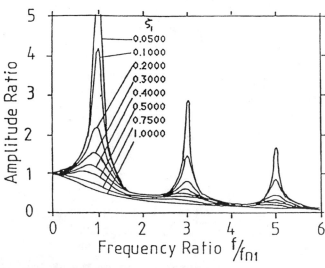

Figure 5.45. Amplitude ratio versus frequency ratio for steady-state longitudinal sinusoidal vibration of rod for $x = l$.

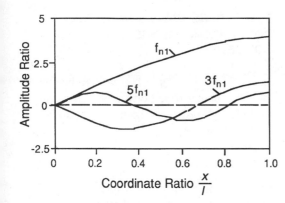

Figure 5.46. Amplitude ratio versus coordinate position x/l for steady-state sinusoidal vibration of rod-damping ratio $\zeta_1 = 0.1$.

tions of x/l for critical frequencies of excitation corresponding to the damping ratio $\zeta = 0.1$. The modal shapes for the rod are sinusoidal in shape governed by the term

$$\sin\left(\frac{\omega_n x}{\alpha}\right) = \sin\left(\frac{n\pi x}{2l}\right) \quad \text{for} \quad n = 1, 3, 5, \ldots$$

This shows that the maximum amplitudes of the steady-state vibration occur as shown for the modes $n = 1, 3, 5, \ldots$ for which the natural frequencies are $f_n = n\alpha/4l$.

For flow promotion, best results are obtained with the higher modes of vibration. The objective here is to create multiple (tensile)

planes of failure as illustrated in Figure 5.47. Research to date has indicated that excitation frequencies on the order of 100 Hz or higher are necessary. There may be a trade off in the selection of excitation frequency; the higher the frequency, the higher the mode of vibration of the bulk mass that is created and hence the greater the number of failure zones; on the other hand the vibration energy transmitted may have a higher attenuation at the higher excitation frequencies.

5.11.3 Velocity of Wave Propagation and Damping

The foregoing discussion concerning the wave motion in a one-dimensional system illustrates some basic objectives to be achieved in the use of vibrations to promote the flow of bulk solids. Where sinusoidal vibrations are applied by a flow promotion device, it is desirable that the impressed frequency corresponds to a natural frequency of the stored bulk solid. As previously stated, there are distinct advantages in exciting the bulk solid at higher natural frequencies since this will induce more zones of peak vibration amplitude and hence peak dilation. However, the attenuation of the transmitted energy due to internal damping must also be considered. The final choice of vibra-

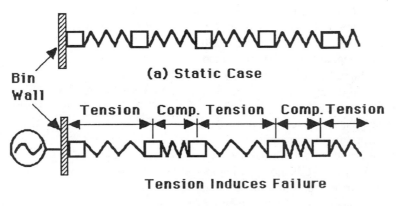

Figure 5.47. Model to illustrate formation of failure zones.

tion frequency will be a compromise based on the frequency and damping characteristics.

It is to be noted that when the simple rod analysis is extended to the two-dimensional case defined by a cylindrical coordinate system, the solution of the two-dimensional wave equation shows that peak amplitude or peak dilation surfaces result. These surfaces are cylindrical in shape with radial locations defined by Bessel type functions which depend on the order of the frequency mode of excitation. The concept of surfaces of peak dilation and possible failure zones is illustrated in Figure 5.44a.

To examine the transmission of vibration energy through stored bulk solids, it is necessary to have a knowledge of the velocity of wave propagation and the damping characteristics. Methods of determining these parameters are described in Refs. 87–90, 92. The magnitude of the wave propagation velocity, for a given bulk solid and moisture content, depends on the consolidation condition and applied load. For instance, Caldwell and Scarlett[90] measured the wave velocity for a dry, tightly packed bed of particles in which there was negligible mechanical shear or interparticle sliding. They obtained the following expression:

$$\lambda = 221 W^{0.186} \ (\text{m/s}) \qquad (5.57)$$

where W = normal load (N/m^2).

Based on the applied loads used in the measurements, the value of λ ranged from 250 to 400 m/s. Assuming, say, $\lambda = 300$ m/s, then from Eq. (5.54), the critical frequencies for the one-dimensional rod are $f_n = 75l$, $225/l$, $375l \ldots$ (Hz) for $n = 1, 3, 5 \ldots$ respectively.

The damping characteristics of bulk solids are quite complex and the assumption of viscous damping as in the model of Eq. (5.50) is a simplistic one. More detailed studies of damping are given by Hardin[89] and Snowden.[91] The latter author introduces damping into the wave equation through a complex elastic modulus term.

5.12 STRESS WAVES IN THREE DIMENSIONS — SOME BASIC CONCEPTS

Stress waves in elastic media may be divided into two categories, body waves occurring in an infinite elastic medium and boundary or surface waves by Das.[92]

5.12.1 Body Waves

For an infinite, elastic isotropic material, there are two types of waves that are of interest, compression waves and shear waves.

(a) Compression, Dilation, or P-Waves.
This type of wave is described by

$$\frac{\partial^2 \bar{\varepsilon}}{\partial t^2} = \lambda_p^2 \, \nabla^2 \bar{\varepsilon} \qquad (5.58)$$

where

$$\lambda_p = \sqrt{\frac{\Lambda + 2G}{\rho}} \qquad (5.59)$$

$$\Lambda = \frac{\nu E}{(1 - \nu)(1 - 2\nu)} \qquad (5.60)$$

ν = Poisson's ratio

E = elastic modulus

G = shear modulus = $\dfrac{E}{2(1 + \nu)}$

ρ = bulk density

$$\bar{\varepsilon} = \varepsilon_x + \varepsilon_y + \varepsilon_z. \qquad (5.61)$$

ε_x, ε_y, and ε_z are normal strain components in x, y, and z directions, respectively.

$$\nabla^2 = \frac{\partial^2}{\partial x^2} + \frac{\partial^2}{\partial y^2} + \frac{\partial^2}{\partial z^2}$$

It is to be noted that when comparing λ_p in Eq. (5.59) with $\lambda = \sqrt{E/\rho}$ given by Eq. (5.51), λ_p has the higher magnitude.

(b) Distortion, Shear, or S-Wave

This type of wave is expressed by

$$\left. \begin{aligned} \frac{\partial^2 \omega_x}{\partial t^2} &= \lambda_s^2 \, \Delta^2 \overline{\omega}_x \\[4pt] \frac{\partial^2 \omega_y}{\partial t^2} &= \lambda_s^2 \, \Delta^2 \overline{\omega}_y \\[4pt] \frac{\partial^2 \omega_z}{\partial t^2} &= \lambda_s^2 \, \Delta^2 \overline{\omega}_z \end{aligned} \right\} \quad (5.62)$$

where $\overline{\omega}_x$, $\overline{\omega}_y$, and $\overline{\omega}_z$ are components of rotation about x-, y-, and z-axes

$$\lambda_s = \sqrt{\frac{G}{\rho}} \qquad (5.63)$$

5.12.2 Rayleigh Waves

This type of wave exists near free surfaces or at the interfaces between substances of different characteristics. They are boundary type waves and, in theoretical terms, apply to elastic half spaces. They are described, for example in Ref. 92.

5.12.3 Application of Theory to Flow Promotion

Based on the discussion of wave theory presented in Ref. 92, some aspects of possible relevance to flow promotion of bulk solids may be considered. For the stored bulk solid as shown in Figure 5.48, assume that an impulse is applied at a point A as shown. It is apparent that the P and S waves will travel out with hemispherical wave fronts as shown in Figure 5.48a, while the Rayleigh waves will propagate out radially along a cylindrical wave front. At some distance from the point of the disturbance, the vertical displacement of the bulk solid will have the form indicated in Figure 5.48b. The P-waves move with the highest velocity and arrive first followed by the S-waves and then the Rayleigh waves. The latter have the highest amplitude. The amplitude of the disturbance gradually decreases with distance. The amplitude of the compression and shear waves decrease according to $1/r$ while the Rayleigh waves decrease according to $1/\sqrt{r}$. Hence the attenuation of amplitude of the Rayleigh waves is the slowest.

The loss in amplitude of the waves as they move outwards is referred to as geometrical damping. In addition, there is also loss due to absorption. Taking both damping losses into account, the vertical amplitude of the Rayleigh wave decays according to the relationship

$$y_n/y_1 = \sqrt{r_1/r_n} \, \exp[-\kappa(r_n - r_1)] \quad (5.64)$$

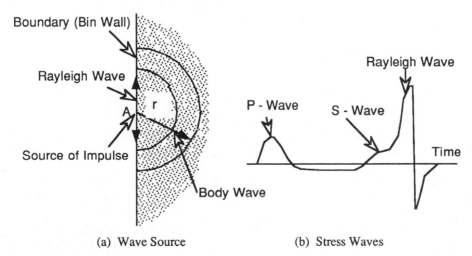

(a) Wave Source   (b) Stress Waves

Figure 5.48. Conceptual behavior of stress waves in stored bulk mass. (Based on Ref. 92.)

where y_n and y_1 are vertical amplitudes at distances r_n and r_1 respectively and κ is the absorption coefficient. For water saturated fine ground sand $\kappa = 0.1$ (m^{-1}).

5.13 CONCLUDING REMARKS

The application of mechanical vibrations with respect to powders and bulk solids is wide and varied, and a basic understanding of the effect of vibrations on strength and flow properties is of fundamental importance to all areas of application. On the basis of the research reviewed in this chapter, some general conclusions may be drawn:

1. The shear strength of a powder on bulk solid undergoing shear deformation may be reduced significantly by the application of mechanical vibrations. For a given vibration energy input, the reduction in strength is a function of the degree of consolidation, the applied normal stress during shear, and the frequency. Experimental evidence suggests that the strength reduction increases with increase in particle size and reduction in moisture content.

2. In the direct shear test using the vibrating shear cell apparatus, vibration of the whole cell during shear deformation shows the presence of critical or natural frequencies at which the shear strength is a minimum. The presence of these natural frequencies has also been observed in measurements of the amplitudes of vibration on the shear plane. The natural frequency for a particular bulk solid is shown to increase with increases in both consolidation stress and applied normal stress. While the exact nature of the damping characteristics of powders and bulk solids is difficult to determine, there is evidence to suggest the damping due to Coulomb friction has a major influence, particularly when the amplitude of vibration relative to the average particle size is not insignificant. For very small amplitudes the influence of viscous damping is believed to be more pronounced. Since the damping is of low order, it has been shown that for simplicity an equivalent viscous damping factor may be assumed. Tests on pyrophyllite and iron ore, both at 5% moisture content (d.b.), show values of $\zeta = 0.1$ and $\zeta = 0.125$, respectively, that fit the experimental data quite well.

3. In the direct shear test when the top half of the cell is vibrated, no resonance effect is observed. The shear stress reduces exponentially with frequency, approaching asymptotically a limiting value that depends on the consolidation and applied normal stress.

4. The shear failure of a vibrated powder or bulk solid may be directly related to the influence of the relative vibration velocity on the shear plane. The vibration velocity may be correlated with the voidage on the shear plane as indicated by the Hvorslev failure criterion. For a given consolidation and applied normal stress, the shear strength is a decaying exponential function of the maximum vibration velocity, approaching asymptotically a limiting value.

5. The frictional drag forces between powders or bulk solids and adjacent boundary surfaces such as the wall of a hopper may be considerably reduced by the application of mechanical vibrations. The extent of this reduction is a function of the intensity or energy level of the vibration applied. The direction of the vibration in relation to the direction of shear deformation is shown to be of importance; when these two directions coincide the greatest benefit in the lowering of the drag forces may be achieved because of the assistance rendered by the inertial forces.

6. Reductions in shear strength of powders or bulk solids may also be accomplished by the use of broad-band random vibrations. To be effective, the frequency band width should span the critical frequency or frequencies of the material.

7. System identification studies employing cross-correlation and spectral analysis may be used to obtain the dynamic characteristics of a bulk solid in both the time and frequency domains. The latter provides useful information on the energy versus frequency distribution.

8. In the application of vibrations to the compression or compaction of powders, the degree of compaction achieved is directly related to the vibration energy. Vibratory compaction greatly increases the shear strength.

9. Mechanical vibrations may be used to advantage in promoting the flow of bulk solids from storage bins. When correctly applied, the use of vibrations will significantly lower the strength of the bulk solid and its wall friction coefficient. In this way storage bins that may operate as funnel flow bins in normal circumstances can be made to operate as mass-flow bins under applied vibration, using appropriate flow promotion devices. It is most important that a flow stress field exists within the material in the bin when the vibrations are applied.

10. The effectiveness of vibrations as an aid to flow promotion, in many cases, depends on the manner in which vibration energy can be transmitted through a stored bulk mass to the region of a flow obstruction. While this is a subject of considerable complexity and one requiring further research, some general observations have been presented. It is concluded that vibration excitation corresponding to the higher modes has the advantage of inducing a greater degree of "loosening up" of the bulk solid thereby promoting flow in the most effective way.

LIST OF SYMBOLS

A	Area of shear plane (m^2)
a_0, a_1, a_2	Constants
B	Damping coefficient ($N \cdot s/mm$)
B_E	Equivalent damping coefficient ($N \cdot s/mm$)
B_{cr}	Critical damping coefficient ($N \cdot s/mm$)
C	Cohesion
C_0	Initial cohesive force before consolidation (N)
c_0, c_1, c_2	Constants
d	Average particle size (mm)
E	Constant
E_v	Energy per unit mass
e	Void ratio
f	Frequency (Hz)
f_n	Natural frequency (Hz)
FF	Flow function
F_f	Friction force (N)
F_s	Equivalent static restoring force (N)
ff	Flow factor
g	Gravitational acceleration (m/s^2)
H	Moisture content (also abbreviated as m.c.)
$H(f)$	System transfer function
$h(X)$	System impulse response or weighting function
k	Stiffness (N/mm)
K	Constant
m	Mass (kg)
N	Interparticle normal force (N)
N_c	Inertial coefficient
N_e	Interparticle normal force due to consolidation (N)
$R_{x_1 x_2}(\lambda)$	Cross-correlation function
r	Frequency ratio
Q	Coulomb damping function
S	Shear force (N)
$S_{x_1 x_2}(f)$	Cross-spectral density
T	Temperature (°C)
u	Deformation
U	Vibration velocity (mm/s)
U_s	Velocity of sliding (mm/s)
V	Normal force (N)
$\bar{V}$	Compressive force on plane of interest (N)
X	Amplitude of vibration (mm)
$x_i, \dot{x}_i$	Input displacement (mm)

X_1, X_i	Input displacement amplitude (mm)
x_r	Relative displacement (mm)
X_r	Relative displacement amplitude (mm)
x_2	Absolute displacement of top half of shear cell (mm)
X_2	Absolute displacement amplitude (mm)
α	Hopper half angle
δ	Effective angle of internal friction
ϕ	Kinematic angle friction at hopper wall
β	Reduction in shear stress (kPa)
γ	Vibration velocity constant
ρ	Bulk density
σ_1	Major consolidating stress (kPa)
σ	Normal stress during consolidation (kPa)
$\bar{\sigma}$	Applied normal stress (kPa)
σ_c	Unconfined compressive strength (kPa)
σ_{cf}	Unconfined compressive stress during vibration (kPa)
ψ	Phase angle
τ_∞	Steady state shear stress (kPa)
τ	Shear stress during consolidation (kPa)
τ_f	Shear stress plotted against frequency
$\bar{\tau}$	Measured shear stress
ω	Frequency (rad/sec)
ω_n	Natural frequency (rad/sec)
ζ	Damping ratio
ζ_v	Damping ratio, viscous component
λ	Wave velocity (m/s)
μ	Friction coefficient

REFERENCES

1. C. A. Coulomb, "Essai Sur une Application des Règles des Maximis et Minimis á Quelques Problèmes de Statique Relatifs de l'Architecture." *Mem. Math. Phys.*, pp. 343–381 (1773).

2. W. Rankine, "On the Stability of Loose Earth." *Phil. Trans.* (1857).

3. O. Reynolds, "On the Dilatancy of Media Composed of Rigid Particles in Contact." *Phil. Mag.* 20:469–481 (1885).

4. O. Reynolds, "Experiments Showing Dilatancy, a Property of Granular Material, Possibly Connected with Gravitation." *Proc. Roy. Inst. Gr. Brit.* 11:354–363 (1887).

5. H. A. Janssen, "Tests on Grain Pressure Silos." *Z. Ver. Dtsch. Ing. Beih.* 35:1045–1049 (1895).

6. W. Airy, "The Pressure of Grain." *Proc. Inst. Civ. Eng.* Paper 3049 (1897).

7. J. A. Jamieson, "Grain Pressures in Deep Bins." *Trans. Can. Soc. Civ. Eng.* 17:554–654 (1903).

8. M. S. Ketchum, *The Design of Walls, Bins and Grain Elevators*, 3rd ed., McGraw-Hill, New York (1919).

9. A. W. Jenike, "Gravity Flow of Bulk Solids." *Utah Agric. Exp. St. Bull.* 108 (1961).

10. A. W. Jenike, "Storage and Flow of Solids." *Utah Agric. Exp. St. Bull.* 123 (1964).

11. J. R. Johanson, "Stress and Velocity Fields in Gravity Flow of Bulk Solids." *Utah. Agric. Exp. St. Bull.* 116 (1962).

12. M. J. Hvorslev, "On the Physical Properties of Distributed Cohesive Soils." *Ingeniorvidensk. Skr.* 45 (1937).

13. K. H. Roscoe, A. N. Schofield, and C. P. Wroth, "On the Yielding of Soils." *Geotechnique* 8:22–53 (1958).

14. M. D. Ashton, D. C. H. Cheng, R. Farley, and F. H. H. Valentin, "Some Investigations into the Strength and Flow Properties of Powders." *Rheol. Acta.* 4(3):206–218 (1965).

15. J. C. Williams and A. H. Birks. "The Preparation of Powder Specimens for Shear Cell Testing," *Rheol. Act.* 4(3):170–180 (1965).

16. A. C. Palma, "Stress-Strain Relation for Clay, an Energy Theory." *Geotechnique* 17:348–358 (1967).

17. C. P. Wroth and R. H. Bassett, "A Stress-strain Relationship for the Shearing Behaviour of Sand." *Geotechnique* 15:32–56 (1965).

18. P. W. Rowe, L. Barden, and I. K. Lee, "Energy Components during the Triaxial Cell and Direct Shear Tests." *Geotechnique* 14:247–261 (1964).

19. J. K. Mitchell, "Shearing Resistance of Soils as a Rate Process." *Proc. Am. Soc. Civ. Eng., J. Soil Mech. and Foundations Div.*, pp. 29–61 (1964).

20. J. Lysmer and F. E. Richart, "Dynamic Response of Footings to Vertical Loading." Inl. of Soil Mechanics and Foundations Division, *Proc. Am. Soc. Civ. Eng., Soil Mech. and Foundations Div.*, pp. 65–91 (1966).

21. J. K. Mitchell, "Soil Creep as a Rate Process." *Proc. Am. Soc. Civ. Eng., J. Soil Mech. and Foundations Div.*, pp. 231–254 (1968).

22. J. K. Mitchell, A. Singh, and R. G. Campanella, "Bonding, Effective Stresses and Strength of Soils." *Proc. Am. Soc. Civ. Eng., J. Soil Mech. and Foundations Div.,* pp. 1219–1246 (1969).

23. N. E. Funston, "Footing Vibration with Non Linear Subgrade Support." *Proc. Am. Soc. Civ. Eng., J. Soil Mech. and Foundations Div.,* pp. 191–211 (1967).

24. E. D'Appolonia, "Dynamic Loadings." *Proc. Am. Soc. Civ. Eng., J. Soil Mech. and Foundations Div.* (1970).

25. D. H. Timmerman and T. H. Wu, "Behaviour of Dry Sands Under Cyclic Loading." *Proc. Am. Soc. Civ. Eng., J. Soil Mech. and Foundations Div.,* pp. 1097–1112 (1969).

26. B. J. Greenfield and E. T. Misiaszek, "Vibration-Settlement Characteristics of Four Gradations of Ottawa Sand." *Proc. Int. Symp. on Wave Propagation and Dynamic Properties of Earth Materials,* University New Mexico, pp. 787–795 (1967).

27. Suk Chae Yong, "The Material Constants of Soil as Determined from Dynamic Testing." *Proc. Int. Symp. on Wave Propagation and Dynamic Properties of Earth Materials,* University of Mexico, pp. 759–770 (1967).

28. V. P. Drnevich, J. R. Hall, and F. E. Richart, "Effect of Amplitude of Vibration on the Shear Modulus of Sand." *Proc. Int. Symp. on Wave Propagation and Dynamic Properties of Earth Materials,* University New Mexico, pp. 189–199 (1967).

29. J. W. S. de Graft-Johnson, "The Damping Capacity of Compacted Kaolinite under Low Stresses," *Proc. Int. Symp. on Wave Propagation and Dynamic Properties of Earth Materials,* University New Mexico, pp. 771–780 (1967).

30. L. T. Youd, "Reduction of Critical Void Ratio during Steady-State Vibration." *Proc. Int. Symp. on Wave Propagation and Dynamic Propagation and DynamicProperties of Earth Materials,* University of New Mexico, pp. 737–744 (1967).

31. B. Scarlett and I. E. Eastman, "Stress in Granular Materials due to Applied Vibration." *I. Chem. Eng. Symp.,* ser. 29., Inst. Chem. Eng., London, pp. 45–68 (1968).

32. W. A. Gray and G. T. Rhodes, "Energy Transfer during Vibratory Compaction of Powders." *Powder Technol.* 6:271–281 (1972).

33. D. Nicklin and W. Hopkins, "Fluidization by Vibration." Fluidization Symposium, Melbourne, Australia (1967).

34. V. Chlenov and N. Mikhailov, "Vibrofluidised Beds." *Izdatelstvo Nauka,* Moscow (1972).

35. R. G. Gutman, "Vibrated Beds of Powders." *Trans. Inst. Chem. Eng.* 54:174–183 (1976).

36. W. Kroll, "Forschung aut der Gebiete des Ingenieur wesen," 20, Ed A(1), 2 (1954).

37. C. F. Harwood, "Powder Segregation due to Vibration." *Technol. Powder Technol.* 16:51–57 (1977).

38. H. Deresiewicz, *Mechanics of Granular Matter Advances in Applied Mechanics,* vol. 5, Academic Press, New York (1958).

39. O. Molerus, "Theory of Yield of Cohesive." *Powder Technol.* 12:259–275 (1975).

40. M. Becker and H. Lippman, "Plane Plastic Flow of Granular Model Material: Experimental Set-up and Results." *Arch. Mech. Stosow.* 29(6):829–846 (1977).

41. R. Nova and D. M. Wood, "Constitutive Model for Sand in Triaxial Compression." *Int. J. Numerical and Analytical Methods in Geomechanics* 3:255–278 (1979).

42. A. Blinowski, "On the Dynamic Flow of Granular Media." *Arch. Mech. Stosow.* 30(1):27–34 (1978).

43. S. C. Cowin, "A Theory for the Flow of Granular Materials." *Powder Technol.* 9:61–69 (1974).

44. M. A. Goodman and S. C. Cowin, "A Continuum Theory for Granular Materials." *Arch. Ration. Mech. Anal.* 44:249–266 (1972).

45. S. L. Passman, "Balance Equations for Mixtures of Granular Materials." MRC Technical Summary Report No. 1390, January 1974 University of Wisconsin.

46. S. L. Passman, "Mixtures of Granular Materials." *Int. J. Eng. Sci.* 15:117–29 (1977).

47. S. L. Passman and J. P. Thomas, "On the Linear Theory of Flow of Granular Media." *Dev. Theor. Appl. Mech.* 9 (1978).

48. S. Nemat-Nasser, "On Behaviour of Granular Materials in Simple Shear." Earthquake Res. and Eng. Lab., Tech. Rep. No. 79-6-19. Dept. of Civ. Eng., Northwestern University, Evanston, Ill. (June, 1979).

49. J. W. Nunziato, J. E. Kennedy, and E. K. Walsh, "The Behaviour of One-Dimensional Acceleration Waves in an Inhomogeneous Granular Solid." *Int. J. Eng. Sci.* 16:637–648 (1978).

50. E. A. Wahl, "Bin Activators—Key to Practical Storage and Flow of Solids." Symposium on Storage and Flow of Solids, Chicago, Ill., ASME Paper 72-MH 29 (1972).

51. J. I. Myers, "Vibrating Hoppers." *Mech. Eng.,* pp. 27–31 (1970).

52. P. J. Carroll and H. Colijn, "Vibration in Solids Flow." *Chem. Eng. Prog.* 71(2):53–65 (1975).

53. K. Shinohara, A. Suzuki, and T. Tanaka, "Gravity and Vibration Effects on Flow of Cohesive Materials from Hopper." Paper presented at ASME Symp. on Flow of Solids, Boston, Mass. (October, 1968).

54. A. Suzuki, H. Takahashi, and T. Tanaka, "Behaviour of a Particle Bed in the Field Vibration, Part II: Flow of Particles through slits in the Bottom of a Vibrating Vessel." *Powder Technol.* 2:72–77 (1969).

55. A. Suzuki and T. Tanaka, "Behaviours of a Particle Bed in the Field of Vibration, Part IV: Flow of

Cohesive Solids from Vibrating Hoppers." *Powder Technol. 6*:301–308 (1972).

56. P. C. Arnold, A. S. Kaaden, and A. W. Roberts, "Effects of Vibration on the Flow of Bulk Solids." Paper presented at Symp. Am. Inst. Chem. Eng. (1976).

57. A. J. Croft, "Investigation of the Trends of Behaviour of the Flow Parameters of Cohesive Bulk Materials when Subjected to Vibration." M. Eng. Sc. thesis, University of Newcastle (1977), Australia.

58. A. W. Roberts, O. J. Scott, and Kin Wah Li, "Effects of Mechanical Vibration on the Flow of Bulk Solids." Paper presented to Conf. on Agric. Eng., Inst. Aust. Eng., Toowoomba, Australia, August (1978).

59. A. S. Kaaden, "Gravity Flow of Some Steelmaking Raw Materials with Particular Reference to the Effects of Vibration," M. Eng. Sci. thesis, University of Wollongong (1975), Australia.

60. A. W. Roberts and O. J. Scott, "An Investigation into the Effect of Sinusoidal and Random Vibrations on the Strength and Flow Properties of Bulk Solids." *Powder Technol. 21*:45–53 (1978).

61. Kin Wah Li, "Effect of Vibration on the Shear Strength and Flow Properties of Bulk Solids." Unpub. M. E. thesis, Dept. of Mech. Eng., The University of Newcastle, Australia (1978).

62. A. W. Roberts, O. J. Scott, and Kin Wah Li, "The Influence of Mechanical Vibrations on the Strength and Flow Properties of Bulk Solids." *Proc. Int. Conf. on Powder and Bulk Solids*, Philadelphia, Pa. (May, 1979).

63. F. J. C. Rademacher, "Accurate Measurement of the Kinetic Coefficient of Friction Between a Surface and a Granular Mass." *Powder Technol. 19*:65–77 (1978).

64. V. K. Sharma, L. O. Drew, and G. L. Nelson, "High Frequency Vibrational Effect on Soil Metal Friction." *Trans. ASAE 20*(1):46–51 (1977).

65. O. Molerus and P. R. Schöneborn, "Bunker Design Based on Experiments in a Bunker-Centrifuge." *Powder Technol. 16*:265–72 (1977).

66. J. R. Johanson, "Modelling Flow of Bulk Solids." *Powder Technol. 5*:93–99 (1971).

67. I. G. Shatalova, N. S. Gorbunov, and V. I. Likhtman, "Physichemical Principles of Vibratory Compacting." *Perspectives in Powder Metallurgy*, vol. 2, Vibratory Compacting, Plenum Press, New York (1967).

68. R. K. McGeary, "Mechanical Packing of Spherical Particles." *Perspectives in Powder Metallurgy*, vol. 2, *Vibratory Compacting*, Plenum Press, New York (1967).

69. J. J. Hauth, "Vibrational Compaction of Nuclear Fuels." *Perspectives in Powder Metallurgy*, vol. 2, *Vibratory Compaction*. Plenum Press, New York (1967).

70. P. E. Evans and R. S. Millman, "The Vibratory Packing of Powders." *Perspectives in Powder Metallurgy*, vol. 2, *Vibratory Compacting*. Plenum Press, New York (1967).

71. A. W. Jenike, "Determination of Flow Properties of Powders by Shear Cell Measurements." *Particulate Matter, 4*(2):11–14 (1973).

72. P. C. Arnold and A. S. Kaaden, "Reducing Hopper Wall Friction by Mechanical Vibration." *Powder Technol. 16*:63–66 (1977).

73. J. Schwedes, "Vergleichende Betrachtungen zum Einsatz von Schergeraten zur Messung von Schüttguteigenschaften" [Shear Testers for Measuring the Flow Properties of Bulk Solid], Institute fur Mechanische Verfahrenstechnik, Technische Universitat, Braunschweig (1979).

74. F. J. C. Rademacher and G. Haaker, "Analysis of the Possible Errors Caused by the Loading Mechanism of Both the Original Jenike Shear Cell and Modified Version." *Proc. Int. Conf. on Powders and Bulk Solids*, Philadelphia (May, 1979).

75. S. C. Cowin and M. A. Goodman, "A Variational Principle for Granural Materials." *Zeitschrift Fur angewandte Mathematic und Mechanik 56*:281–286 (1976).

76. S. Timoshenko, D. H. Young, and W. Weaver, *Vibration Problems in Engineering*. 8th ed., Wiley, New York (1974).

77. J. P. den Hartog, "Forced Vibrations with Combined Coulomb and Viscous Friction." *Trans. A.S.M.E. 53*:107–115 (1931).

78. P. C. Arnold, A. G. McLean, and A. W. Roberts, "Bulk Solids: Storage Flow and Handling." *TUNRA*, The University of Newcastle, (1979), 2nd Edition (1981).

79. W. Lenkiewicz, "The Sliding Friction Process— Effect of External Vibration." *Wear 13*:99–108 (1969).

80. D. Godfrey, "Vibration Reduces Metal-to-Metal Contact and causes an Apparent Reduction in Friction." *Trans. ASLE. 10*:183–192 (1967).

81. D. M. Tolstoi, "Significance of the Normal Degree of Freedom and Natural Normal Vibration in Contact Friction." *Wear 10*:199–203 (1967).

82. A. W. Roberts and W. H. Charlton, "Application of Pseudo-Random Test Signals and Cross Correlation to the Identification of Bulk Handling Plant Dynamic Characteristics," *Trans. ASME, J. Eng. Ind. 95*:31–36 (1973).

83. A. W. Roberts and G. S. Montagner, "Identification of Transient Flow Characteristics of Granular Solids in a Hopper Discharge Chute System." Paper presented at the Symp. on Solids and Slurry Flow and Handling in the Chemical Process

Industries, AICHE 77th National Meeting, Pittsburgh, Pa. (1974).

84. A. W. Roberts and W. H. Charlton, "Determination of Natural Response of Mechanical Systems Using Correlation Techniques." *Expt. Mech.,* *15*(1):17–22 (1975).

85. A. G. McLean and P. C. Arnold, "Two Phase Flow in Converging Channels." Proc. European Symp. Particle Technology Amsterdam (3–5 June 1980).

86. A. W. Roberts, "Energy Excited Gravity Flow of Particulate Solids in Silos and Channels," in *Proceedings of the Fourth World Congress of Chemical Engineering*, Karlsruhe, Germany (June 1991).

87. B. O. Hardin and F. E. Richart, "Elastic Wave Velocites in Granular Soils." *J. Soil Mech. Found. Div. ASCE, 89*(SM1):33–65 (Feb. 1963).

88. B. O. Hardin and V. P. Drnevich, "Shear Modulus and Damping in Soils: Measurement and Parameter Effects." *J. Soil Mech. Found. Div. ASCE, 89*:603–623 (1963).

89. B. O. Hardin, "The Nature of Damping in Sands." *J. Soil Mech. Found. Div. ASCE* 63–97 (Jan. 1965).

90. A. Caldwell and B. Scarlett, "The Propagation of a Mechanical Impulse in a Granular Medium," in *Proceedings of the International Symposium on Powder Technology '81*, Sept. 27–Oct. 1, Kyoto, Japan, The Society of Powder Tech. (Kyoto), pp. 165–176 (1982).

91. Snowden, "Vibration and Shock in Damped Mechanical Systems."

92. B. M. Das, *Fundamentals of Soil Dynamics*, Elsevier (1983).

6

Size Enlargement by Agglomeration*

Wolfgang Pietsch

CONTENTS

1.1 INTRODUCTION

6.1.1 Definition of Size Enlargement by Agglomeration

Size enlargement by agglomeration is a unit operation of mechanical process technology[1] (Fig. 6.1). This field deals with the transport phenomena and changes of state of particulate matter which in most cases is solid but can also be liquid (droplets) and, in a few special cases, gaseous (microencapsulated). The unit operations of mechanical process technology can be differentiated by the processes of separation and combination with and without change of particle size.

Size enlargement by agglomeration as a unit operation of mechanical process technology is characterized by the structure of the enlarged particles in which, contrary to, for example, crystals or particles obtained by solidification of melt droplets, the shape and size of the original particles are still distinguishable. This offers both advantages and disadvantages. Strength of agglomerates derives from the action of binding forces,[1] acting either at the coordination points between the particles or the interfaces between a matrix binder and the

---

*References are listed at the end of sections 6.1 through 6.6.

	Separation	Combination	
Without change of Particle Size	Mechanical Separation (Filters, Classifiers, Screens, Sifters)	Powder Mixing and Blending	Partical and Bulk Material Characterization (Size, Distribution, Shape, Volume, Surface Density, Mass, Porosity, Moisture Content, etc.)
With change of Particle Size	Size Reduction (Crushing and Grinding)	Size Enlargement by Agglomeration	
	Transport and Storage of Bulk Materials		

Figure 6.1. The unit operations of mechanical process technology and associated techniques.

particulate solids or, respectively, by the negative capillary pressure of a liquid filling the pore volume (Fig. 6.2).

6.1.2 Properties of Fine Particles

Table 6.1 shows some important characteristics of materials and disperse systems that depend on particle size.[2]

For single particles the characteristics describing quality usually improve as particle size decreases. In particular, the chemical, physical, and mineralogical homogeneity increases. Those characteristics that critically depend on uniformity of structure improve also. For example, all real solids have an imperfect structure; during loading stress concentrations occur at the structural defects that may cause

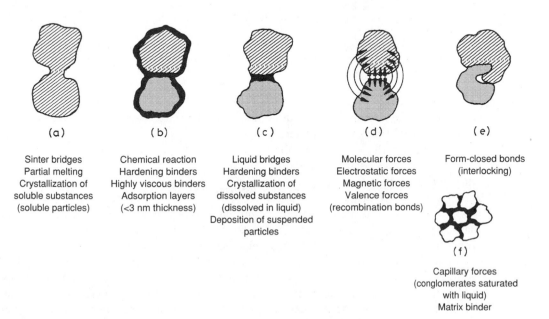

(a)	(b)	(c)	(d)	(e)
Sinter bridges Partial melting Crystallization of soluble substances (soluble particles)	Chemical reaction Hardening binders Highly viscous binders Adsorption layers (<3 nm thickness)	Liquid bridges Hardening binders Crystallization of dissolved substances (dissolved in liquid) Deposition of suspended particles	Molecular forces Electrostatic forces Magnetic forces Valence forces (recombination bonds)	Form-closed bonds (interlocking)

(f)

Capillary forces
(conglomerates saturated
with liquid)
Matrix binder

Figure 6.2. The binding mechanisms of agglomeration.

Table 6.1. Influence of Particle Size on some Important Characteristics of Materials.[2]

A.	Characteristics of Single Particle	...with decreasing particle size
A.1.	Homogeneity	Increasing...
A.2.	Elastic–plastic behavior	Increased ductility...
A.3.	a. Probability of breakage	Decreasing...
	b. Strength	Increasing...
A.4.	a. Wear	Decreasing...
	b. Resistance to mechanical surface treatment	Increasing...
A.5.	Characteristics resulting from the competition between volume and surface-related forces	Increasing...
A.6.	Vapor pressure, solubility, reactivity	Increasing...
A.7.	Optical characteristics	Increasing...
B.	Characteristics of Particle Collectives	
B.1	Bulk density (space-filling behavior)	First increasing then decreasing...
B.2.	Rheological behavior	Increasing...
B.3.	Flow characteristics, flowability (of particles)	Decreasing...
B.4.	Mixing characteristics	First increasing then decreasing...
B.5.	Separability	Decreasing...
B.6.	Wettability	Decreasing...
B.7.	Capillary pressure (system: solid/liquid)	Increasing...
B.8.	Agglomerate strength	Increasing...
B.9.	Fluid flow characteristics	
	a. Flow through pores (in particle collectives)	Decreasing...
	b. Resistance to fluid flow	Increasing...
	c. Ease of fluidization	First increasing then decreasing...
B.10.	Thermal characteristics	Increasing...
B.11.	Ignition behavior and explosiveness	Increasing...
B.12.	Taste standards	Increasing...
B.13.	Optical characteristics	Extinction, diffuse reflection

breakage. With decreasing particle size the probability of imperfections diminishes, resulting in a reduced risk of breakage and therefore higher strength. At the same time, the possibility for irreversible deformation increases with decreasing particle size. For example, limestone or quartz, with particle size of less than 10 μm and 3 μm, respectively, deforms plastically before breakage begins.

On the other hand, problems associated with mechanical processing and handling of particle systems increase with decreasing particle size mostly due to natural, undesired agglomeration including such phenomena as caking, bridging, build-up, etc.

Controlled or desired agglomeration may improve the characteristics of fine particle systems.

6.1.3 Desired and Undesired Agglomeration[3,4]

During production and processing of solid matter in disperse systems, adhesion phenomena become more and more important with decreasing particle size, causing aggregation, agglomeration, coating, caking, and build-up. The critical particle size is approx. 100 μm, but it is also possible that much coarser particulate matter may be affected if a sufficiently large fraction of finer particles is present or if specific binding mechanisms become effective.

Adhesion of finely divided material takes place during all operations of mechanical process engineering and can be either desired or undesired. Table 6.2 provides a compendium.

Adhesion during *grinding* is always undesirable because it diminishes the grinding effect, lengthens the grinding time, and increases the energy requirement. In some mills an equilibrium between size reduction and size enlargement sets in at a certain fineness and can be avoided only by the addition of dispersion agents or the application of another comminution method.

During *mechanical separation* agglomeration is undesirable if products must be classified according to particle size or composition. Only in *flotation cells* or *wet classifiers* a "selective flocculation" may be advantageous. Particle aggregation is always desirable during precipitation, thickening, filtration, and clarification, because the increased mass of agglomerates improves separation efficiency.

During analytical separation (*particle size analysis*) any agglomeration is prohibitive and must be avoided at any cost.

The quality of *mixing of solids* can be considerably impaired by undesired agglomeration. Existing or newly formed aggregates are

Table 6.2. Review of the Occurrence of Desired and Undesired Agglomeration Phenomena in Mechanical Process Engineering.

| UNIT OPERATION | PROCESS | AGGLOMERATION | |
		UNDESIRABLE	DESIRABLE
Comminution	Dry grinding	+	−
	Wet grinding	+	−
Separation	Screening, sieving	+	−
	Classifying		
	Sorting	+	(+)
	Flotation	+	(+)
	Dust precipitation	(−)	+
	Clarification, thickening	(−)	+
	Particle size analysis	+ +	−
Mixing	Dry mixing	+	−
	Wet mixing	+	+
	Stirring	+	(+)
	Suspending		
	Dispersing	+	(+)
	Fluidized bed	+	+
Particle size enlargement	Agglomerating		
	Briquetting		
	Tabletting		
	Granulating	(+)	+ +
	Pelletizing		
	Pelleting		
	Sintering		
Conveying	Vibratory conveying	+	−
	Pneumatic conveying	+	−
Storage	Silos, hoppers	+	
	Stockpile	+	−
Batching, Metering		+	−
Drying		+	+

Explanations: +, yes; −, no; (+), sometimes yes; (−), sometimes no.

normally destroyed by suitable mixing tools or by vigorous movements in the mixer. On the other hand, powder mixtures often tend to segregate during handling and storage; then, a controlled agglomeration of the final mix may be desirable prior to further processing.

Because fine powders possess a large bulk volume, generate dust, and exhibit unfavorable transport, storage, and feeding characteristics, their particle size is sometimes enlarged by *agglomeration*. In this case adhesion is desired and is systematically promoted. In some cases it is necessary to further treat the agglomerate with "anticaking" compounds to avoid clustering during storage.

Agglomeration and adhesion of fine particles are particularly annoying during *Transport*, *Storage*, and *feeding*. Conglomerates can result in clogging or feeders, prevent discharge from silos, and cause incorrect metering. The prevention or destruction of such conglomerates often requires considerable technical efforts.

Agglomeration can also play an important role in thermal unit operations. For example, if a liquid in the pores of a bulk mass contains dissolved substances that crystallize during *drying*, solid bridges may build up between the particles. Such bonding is often undesirable and must be destroyed by "deagglomeration." In other instances this method is used for "curing" a wet agglomerate, producing a stable granular material that is better suited as an intermediate product.

References

1. W. Pietsch, *Size Enlargement by Agglomeration*. John Wiley & Sons/Salle + Sauerländer, Chichester, UK/Aarau, Switzerland (1991).
2. H. Rumpf, "Mechanische Verfahrenstechnik," in *Chemische Technology*, Vol. 7, 3rd ed., edited by Winnacker-Küchler, Carl Hanser Verlag, München, Germany, and Wien, Austria (1975). English translation by F. A. Bull. "*Particle Technology*," Chapman and Hall, London, UK (1990).
3. W. Pietsch, "Das Agglomerationverhalten feiner Teilchen," Staub-Reinhalt. Luft, 27 (1967) 1, 20–33; English edition: "The Agglomerative Behavior of Fine Particles" 27 (1), 24–41 (1967).
4. W. Pietsch, "Kornvergrösserung (Agglomeration)," in *Fortschritte der Verfahrenstechnik*, Vol. 9, VDI-Verlag GmbH, Düsseldorf, Germany, pp. 831–872 (1971).

6.2 AGGLOMERATE BONDING AND STRENGTH

6.2.1 Binding Mechanisms

To obtain agglomerates from particular matter, binding forces must act between the individual particles. According to Rumpf,[1] who first published a classification, the possible mechanisms can be divided into five major groups. (Table 6.3)

6.2.1.1 Solid Bridges

At elevated temperatures, solid bridges may develop by diffusion of molecules from one particle to another at the points of contact ("sintering"). Heat can be introduced from an external, secondary source or created during agglomeration by friction and/or energy conversion. Solid bridges can also be build up by chemical reaction, crystalization of dissolved binder substances, hardening binders, and solidification of melted components.

6.2.1.2. Interfacial Forces and Capillary Pressure at Freely Movable Surfaces

Capillary pressure and interfacial forces in liquid bridges can create strong bonds that disappear if the liquid evaporates and no other binding mechanism take over.

Table 6.3. Binding Mechanisms of Agglomeration.[1]

1. Solid bridges
2. Interfacial forces and capillary pressure at freely movable liquid surfaces
3. Adhesion and cohesion forces at not freely movable binder bridges
4. Attraction forces between solid particles
5. Form-closed bonds (interlocking)

6.2.1.3 Adhesion and Cohesion Forces in Not Freely Movable Binders

Highly viscous bonding media such as tar and other highly molecular organic liquids can form bonds very similar to those of solid bridges. Thin adsorption layers are immobile and can contribute to the bonding of fine particles under certain circumstances.

6.2.1.4. Attraction Forces Between Solid Particles

The typical short-range forces of the van der Waals, electrostatic, or magnetic type can cause solid particles to stick together if they approach each other closely enough. Decreasing particle size clearly favors this mechanism. On freshly created surfaces after breakage free valence forces are momentarily present which, at certain conditions, may recombine, forming strong bonds.

6.2.1.5. Form-Closed Bonds

Fibers, little platelets, or bulky particles can interlock or fold about each other, resulting in "form-closed" bonds.

Another classification into only two groups[2] distinguishes between the presence of material bridges between the primary particles in the agglomerate and attraction forces (Fig. 6.3).

6.2.2. Theory of Agglomerate Bonding and Strength

The most important characteristic of all forms of the agglomerates is their strength. For the determination of agglomerate strength, real stresses are often simulated experimentally. In addition to the usually applied crushing, drop, and abrasion tests, methods for the determination of impact, bending, cutting, or shear strength are employed. All values obtained by these methods are strictly empirical and cannot be predicted by theory, since it is not known which stress component causes the agglomerate to fail. For the same reason, the experimental results from different methods cannot be compared with each other.

Therefore, Rumpf[1] proposed to determine the tensile strength of agglomerates. It is defined as the tensile force at failure divided by the cross-section of the agglomerate. Because with high probability failure occurs as the result of the highest tensile stress in all stressing situations, this proposal is justified. Moreover, the tensile strength can be approximated by theoretical calculations.

All binding mechanisms listed above can be described by one of three models:

1. The entire pore volume of the agglomerate is filled with a (matrix) substance that can transmit forces and thereby causes strength.
2. The pore volume of the agglomerate is entirely filled with liquid.
3. Binding forces are transmitted at the contact and coordination points of the primary particles forming the agglomerate.

6.2.2.1 Maximal Tensile Strength if the Pore Volume of the Agglomerate Is Filled with a Stress-Transmitting Substance

If the pore volume of the agglomerate is completely filled with a stress-transmitting substance, for example, a hardening binder, three strength components must be distinguished:

1. $\sigma_{t\epsilon}$ (pore volume strength) = strength of binder substance
2. σ_{ta} (grain boundary strength) = strength caused by adhesion between binder and solids
3. $\sigma_{t(1-\epsilon)}$ = strength of the solids forming the agglomerate.

The relatively lowest strength component determines the agglomerate strength. If the pore volume strength or, respectively, the strength of the solids forming the agglomerate are the determining factors and if they are everywhere the same, then the cross-section of the respective material defines the agglomerate strength. A theoretical approximation is possible using the same assumptions as described below for solid bridges between parti-

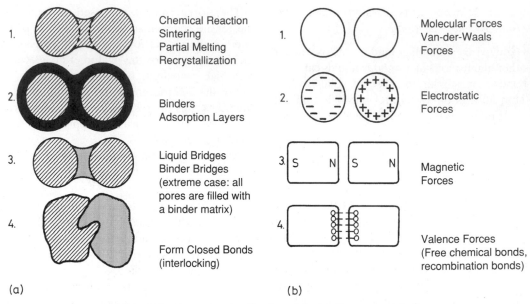

Figure 6.3. Alternative classification of the binding mechanisms.[2]

cles. If the agglomerate strength is caused by the grain boundary strength, it can be approximated by calculating the adhesion forces (see Section 6.2.2.4).

6.2.2.2 Maximal Tensile Strength if the Pore Volume of the Agglomerate Is Filled with a Liquid

If the entire pore volume of the agglomerate is filled with a liquid such that concave minisci are formed at the agglomerate surface a negative capillary pressure p_c develops in the interior. Because the membrane forces at the surface are negligibly small in relation to the capillary pressure, the tensile strength σ_{tc} of the agglomerates filled with a liquid can be approximated by the capillary pressure:

$$\sigma_{tc} \approx p_c \qquad (6.1)$$

Assuming that the pore diameter is characterized by the mean half-hydraulic radius of the pore system, further assuming complete wetting and spherical monosized particles, the following equation is obtained:

$$\sigma_{tc} \approx p_c = a' \cdot \frac{1 - \epsilon}{\epsilon} \cdot \alpha \cdot \frac{1}{x} \qquad (6.2)$$

Therefore, the maximal tensile strength of agglomerates filled with a liquid is proportional to a porosity function $(1 - \epsilon)/\epsilon$ and the surface tension of the liquid α; it is inversely proportional to the grain size x of the particles forming the agglomerate. The factor a' has a value between 6 and 8.

To correctly describe the capillary pressure, and thereby the tensile strength, a function of the wetting angle $f(\delta)$ would have to be included in the above formula. This function equals 1 if the liquid completely wets the solid.

Normally the particles forming agglomerates are not monosized and are irregularly shaped. Comparisons between experimental results and the theory showed that a mean grain size x_0, the surface equivalent diameter calculated from the specific surface area of the actual particle, describes the relations well.[3]

6.2.2.3 Maximal Tensile Strength if Binding Forces Are Transmitted at the Contact and Coordination Points

The model used for agglomerates, the strength of which is caused by solid bridges, assumes that the entire solid binder material is uniformly distributed at all contact and coordination points and, there forms bridges with constant strength σ_B. Then, the relative cross-section of that material defines the agglomerate strength. In a random packing the cross-sectional area of one component (area porosity ϵ_a) is approximately equal to the relative volume of that same component (i.e., volume porosity $\epsilon_v = \epsilon_a = \epsilon$). Thus, the tensile strength σ_{tB} of agglomerates with solid bridges can be approximated by:[4]

$$\sigma_{tB} = \frac{M_B}{M_p} \cdot \frac{\rho_p}{\rho_B} \cdot (1 - \epsilon) \cdot \sigma_B = \psi_B \cdot \epsilon \cdot \sigma_B$$

(6.3)

where M_B is the mass of the bridge building solid, M_p the mass of the particles forming the agglomerate, ρ_B and ρ_p the density of the respective solids, $(1 - \epsilon)$ the relative unit volume of the solid, σ_B the tensile strength of the bridge building solid, ϵ the specific void volume porosity of the agglomerate, and ψ_B the saturation, that is, the fraction of the void volume filled with the bridge building material.

Equation (6.3) is valid only if, in addition to the restrictive assumptions mentioned above, failure occurs only through solid bridges. In reality all these conditions are never fulfilled. Particularly the uniform distribution and the constant strength of the binder material are seldom realized.

Often, the strength is caused by adhesion forces A acting at the coordination points of the particles forming the agglomerate. Based on statistical considerations Rumpf[5] developed a general formula for the tensile strength of such agglomerates. Assuming that the particles forming the agglomerate are monosized

and spherical the tensile strength σ_t can be approximated by:

$$\sigma_t = \frac{1 - \epsilon}{\pi} \cdot k \cdot \frac{A}{x^2}$$

(6.4)

where ϵ is the specific void volume (porosity) of the agglomerate, $\pi = 3.14$, k the average coordination number and x the size of the particles forming the agglomerate.

After a small correction[5] Eq. (6.4) can also be applied for nonspherical particles. Then the estimated elementary tensile strength σ_{te} becomes:

$$\sigma_{te} = (1 - \epsilon) \cdot k \cdot A / O_p$$

(6.5)

with O_p the particle size. Equation (6.5) is valid for agglomerates formed by approximately isodisperse, convex, and monosized particles. With the third moment M_{30} of the number density distribution $n(x)$ and a shape factor f_0, a formula can be derived that is valid for distributions of similar, approximately isometric, and convex particles:

$$\sigma_{te} = \frac{1 - \epsilon}{f_0 M_{30}} \int_0^\infty k(x) \cdot A[x, n(x)] \cdot x \times n(x) \, dx$$

(6.6)

This equation can be integrated only if the relationships are known between coordination number and particle size $k(x)$ as well as between adhesion force and particle size and distribution $A[x, n(x)]$. In most instances this is not the case. To measure σ_{te}, an agglomerate free of cracks must be uniformly stressed by tensile forces. This requires very sophisticated methods and experimental care (see Section 6.2.3).

6.2.2.4 Theoretical Approximation of Adhesion Forces

Adhesion Force of a Liquid Bridge.[6,7] The maximal tensile force that can be transmitted by a liquid bridge between two monosized spheres consists of two components:

(1) An adhesion force component A_c caused by the negative capillary pressure in the bridge:

$$A_c = p_c \cdot \frac{\pi}{4} \cdot x^2 \cdot \sin^2 \beta$$

(2) An adhesion force component A_b caused by the boundary force at the contact line solid–liquid–gaseous, which is determined by the surface tension of the liquid, α:

$$A_b = \alpha \cdot x \cdot \pi \cdot \sin \beta \cdot \sin(\beta + \delta)$$

By adding the two parallel adhesion force components A_c and A_b and after introducing the dimensionless function F_A a formula for the effective adhesion force A_L of a liquid bridge is obtained:

$$A_L = \alpha \cdot x \cdot F_A = \alpha \cdot x \cdot f_1\left(\beta, \delta, \frac{a}{x}\right) \quad (6.7)$$

where α is the surface tension of the liquid, x the diameter of the spherical, monosized particles, β the angle according to Figure 6.4, and a the distance of the particle surfaces at the coordination point.

Therefore, the adhesion force of a liquid bridge is proportional to the surface tension α, the particle diameter x, and a function of the angle β, the angle of contact δ, and the dimensionless quotient a/x. β defines the size of the liquid bridge and can be substituted by ϕ, the liquid volume divided by the volume of the solid particles:

$$\phi = \frac{V_b}{2 \cdot \pi \cdot x^3/6} = \frac{3}{\pi} \cdot f_2\left(\beta, \delta, \frac{a}{x}\right)$$

with V_b the volume of the liquid bridge.

By inserting the adhesion force A_L [Eq. (6.7)] into the basic formula, Eq. (6.4), and assuming that $k \cdot \epsilon \approx \pi$,[8] the maximal tensile strength σ_{tb} of agglomerates with liquid bridges becomes:

$$\delta_{tb} = \frac{1 - \epsilon}{\epsilon} \cdot \frac{\alpha}{x} \cdot F_A \quad (6.8)$$

Adhesion due to van der Waals Forces. Depending on the geometrical model (Fig. 6.5) being used and on the theoretical approach taken, different relationships exist for the approximation of adhesion by van der Waals forces. The best-known equations are those developed by Hamaker[9] based on the microscopic theory of London–Heitler. For the model sphere/plane (Fig. 6.5a), a distance $a < 100$ nm, and a particle diameter x, the adhesion force A_{v_1} is:

$$A_{v_1} = \frac{H}{12 \cdot a^2} \cdot x \quad (6.9)$$

For the model sphere/sphere (Fig. 6.5b), and the same limitations as mentioned above, Hamaker calculates an adhesion force A_{v_2}:

$$A_{v_2} = \frac{H}{24 \cdot a^2} \cdot x \quad (6.10)$$

H, the "Hamaker Constant," which depends on the material characteristics, has values in the order of 10^{-20} to 10^{-19} J.

More recently, Krupp[10] developed a formula for the model sphere/plane (Fig. 6.5a) which is based on the macroscopic calculations of Lifshitz–Landau:

$$A'_{v_1} = \frac{\hbar \bar{\omega}}{16 \cdot \pi \cdot a^2} \cdot x \quad (6.11)$$

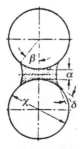

Figure 6.4. Schematic representation of a liquid bridge between two spherical, monosized particles.

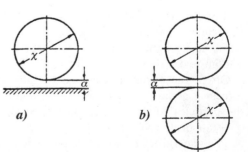

Figure 6.5. Model conceptions for the approximation of van der Waals adhesion. (*a*) Sphere/plane; (*b*) sphere/sphere.

$\hbar\bar{\omega}$ is the "Lifshitz–van der Waals Constant," which, depending on the material characteristics, has values between 0.2 and 9 eV (1 eV = $1.6 \cdot 10^{-19}$ J). All equations for the approximation of van der Waals forces differ only in the constants. The adhesion force A_v is always proportional to the particle diameter x and inversely proportional to the squared distance a:

$$A_v = c \cdot \frac{x}{a^2} \qquad (6.12)$$

By inserting Eq. (6.12) into the basic formula, Eq. (6.4), and assuming that $k \cdot \epsilon \approx \pi,$[8] the maximal tensile strength σ_{tv} of agglomerates bound by van der Waals forces becomes:

$$\sigma_{tv} = \frac{1 - \epsilon}{\epsilon} \cdot \frac{c}{a^2} \cdot \frac{1}{x} \qquad (6.13)$$

Adhesion due to Electrostatic Forces. In the case of electrostatic forces, one must distinguish between an excess charge and the electrical double layer (equilibrium).

The strength due to excess charges can be estimated if it is assumed that positively and negatively charged particles are arranged in a uniform pattern.[4] The basis for the derivation is Coulomb's formula for the attraction force between two spherical, nonconducting particles of equal size, the distance between which is much smaller than their diameter. If the charges $Q = \gamma \cdot \pi \cdot x^2$ are uniformly distributed on the surfaces the adhesion force A_c can be approximated by:

$$A_c = \frac{10\gamma^2}{(1 + a/x)^2} \cdot x^2 \qquad (6.14)$$

For quartz the maximal charge density per unit area, γ, was estimated to be $\gamma_{max}^2 \approx 0.25$ N/m^2.[1] If it is assumed that the charged particles forming an agglomerate are arranged like an ion lattice, then the attraction force between two adjacent, oppositely charged particles is approximately a factor 0.3 smaller because of the repulsion of neighboring particles with the same charge.

By inserting Eq. (6.14) into the basic formula, Eq. (6.4), and assuming that $k \cdot \epsilon \approx \pi,$[8] the maximal tensile strength σ_{tc} of agglomerated due to excess charges is:

$$\sigma_{tc} = \frac{1 - \epsilon}{\epsilon} \cdot \frac{3\gamma^2}{(1 + a/x)^2} \qquad (6.15)$$

Because of the field character of this binding mechanism, the tensile strength is independent of the particle size. Also, the strength due to excess charges is very small, and the charges tend to equalize with time. Therefore, this mechanism is most often important only for the initial formation of agglomerates.

Much more important, however, are adhesion forces due to electrical double layers. This phenomenon can develop if the particles touch each other and is permanent. According to Krupp[10] the "attraction pressure" due to electrical double layers between two semi-infinite bodies is in the order of P_{el} 10^4 to 10^7 N/m^2 (10^5 to 10^8 dyn/cm^2). In comparison, the van der Waals attraction pressure between two semi-infinite bodies is P_{vdW} 2×10^7 to 3×10^8 N/m^2 (2×10^8 to 3×10^9 dyn/cm^2).

It may seem as if the two mechanisms exclude each other. However, since P_{vdW} decreases with $1/a^3$ and P_{el} stays almost constant even over macroscopic distance, the electrical double layer will contribute to the adhesion of particles, particularly if the contact surfaces are rough.

A theoretical approximation for specific systems is still not yet possible, since little is known about the distribution of charges in different materials. The effect of magnetic particles in agglomerates corresponds to that of excess charges and is subject to the same limitations.

6.2.3 Experimental Determination of Agglomerate Bonding and Strength

The most important techniques for the experimental determination of agglomerate strength known today measure crushing, shear, and tensile strengths. Sketches A–F in Figure 6.6 show schematically the methods for measuring

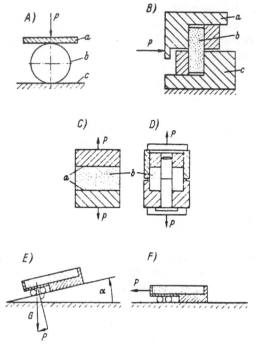

Figure 6.6. Methods for measuring the strength of agglomerates and particle conglomerates. (*A*) Determination of crushing strength: (*a*) loaded plate, (*b*) agglomerate, (*c*) support plate. (*B*) Determination of shear strength: (*a*) upper receptacle, (*b*) compact or briquette, (*c*) lower receptacle. (*C*) and (*D*) Determination of tensile strength of strong agglomerates: (*a*) adhesive, (*b*) agglomerate (eventually machined). (*E*) and (*F*) Determination of tensile strength of weak agglomerates and of particle conglomerates.

the strength of agglomerates and particle conglomerates.

Figure 6.6A shows the determination of the crushing strength.[11] This method is a very simple one. Individual agglomerates are placed between two parallel plates and loaded with a uniformly increasing force P until failure occurs. Usually the "agglomerate strength" is defined as the mean statistical force at failure of a larger number of agglomerates tested by this method. Sometimes a crushing strength is calculated by dividing the force at failure by the projection area of the agglomerate; however, from a physical point of view this is not acceptable.

The results of this test method are very rarely comparable. For spherical pellets the stressing is uniform only from test to test if all agglomerates are absolutely globular. In the case of agglomerates with flat ends the transverse expansion is blocked by friction between pellet and plate; thereby uncontrolled stress concentrations build up that can be the true cause for failure.

In Figure 6.6B, an apparatus is sketched for the investigation of shear strength. Originally this method was used in soil mechanics for the determination of shear curves of cohesive bulk solids. The "strength" of the conglomerate caused by internal friction can be determined graphically from the shear curves. The agglomerate must have two parallel surfaces, which may have to be produced by machining. The test specimen is fastened in the apparatus and stressed by the force P. The shear strength is defined by the shear force at failure divided by the shear plane.

Figure 6.6C shows in principle the "adhesive" method for the determination of tensile strength. Cylindrical agglomerates with two parallel and flat ends are centrically cemented between two so-called adaptors. To eliminate bending stresses it is necessary to machine spherical or nonsymmetrical agglomerates into cylindrical specimens using a special method (Fig. 6.7).[12] This sample is fastened to two thin wires (Fig. 6.8) and subjected to tensile forces in a conventional testing machine (Fig. 6.9a and b). The tensile strength of the agglomerate is defined as the tensile force P at failure divided by the cross section of the cylindrical specimen.

Figure 6.6D sketches the determination of the tensile strength of model agglomerates by means of the wall friction method.[7] For this test a cylindrical pellet—potentially with a central pin—is produced in a press. After removing the specimen from the press, it is stressed directly in the die shell. The tensile force is transmitted by adhesion between the end surfaces and the "pistons" as well as on the circumference and the die walls. Again, the tensile strength is defined by the quotient

Figure 6.7. Preparation of "spherical" agglomerates for the determination of tensile strength by the "adhesive" method.[12]

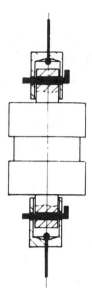

Figure 6.8. Schematic representation of the fastening method with thin wires.

rupture force P divided by the cross-section of the cylindrical or ring-shaped sample.

Figure 6.6E shows a method that is particularly suitable for the determination of low conglomerate strengths. The rupture stress is measured in a model arrangement. The powder to be investigated is filled into a flat, often rectangular split mold and densified by vibration or compaction. The upper half of the mold is fastened to a tiltable plate while the lower half is supported frictionless by rolls or spheres on the same plate. For the determination of the rupture force, the plate is slowly lifted at the end carrying the fixed mold until the particle conglomerate separates. The strength is defined as the force at failure divided by the cross-section of the agglomerate.

Another method based on the same principle uses a slowly increasing horizontal force P to pull apart the specimen (Fig. 6.6F) In this case the arrangement remains horizontal dur-

ing the entire test. The strength is defined as described above. Figure 6.10 shows a recent design of this experimental apparatus.[7] It is constructed such that two tests can be carried out on the same sample, and, if moist agglomerates are being investigated, the capillary pressure p_c can be measured simultaneously by means of a U-tube manometer. By means of inductive displacement gauges the expansion prior to rupture can be also determined.

6.2.4 Results of the Determination of Agglomerate Strength

6.2.4.1 Theoretical Approximation of Agglomerate Strength

With the theories of Section 6.2.2 the orders of magnitude of agglomerate strength can be defined for the different binding mechanisms. They often depend on the size of the particles forming the agglomerate and predict the maximum strength of conglomerates caused by a particular binding mechanism. Figure 6.11 shows the tensile strength σ_t versus particle size x in a double logarithmic plot.

The horizontal dotted line divides the entire field into two regions, I and II. These repre-

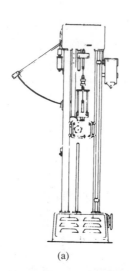

(a)

(b)

Figure 6.9. Testing machines adopted for the determination of the tensile strength of agglomerates. (*a*) Schematic overall view, (*b*) close-up during an actual tensile test.

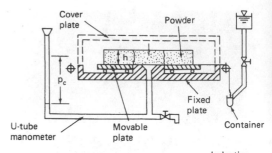

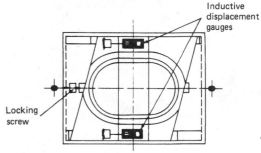

Figure 6.10. Diagrammatic representation of the split plate apparatus, according to Schubert.[13]

sent binding mechanisms that are independent of the size of the particles to be agglomerated. Region I describes high-pressure agglomeration. This technology uses high compaction pressures causing brittle disintegration as well as deformation of particles and favoring inter-

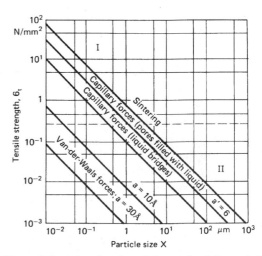

Figure 6.11. Approximation of the maximal theoretical tensile strength of agglomerates. Porosity: $\epsilon = 0.35$. Region I: for example, hardening binders. Region II: Crystallizing soluble substances, for example, salts.

particle contact. Agglomerates with highly viscous and hardening binders are also included in this region. The lower region II describes the much weaker bonds caused by recrystallization of dissolved substances.

The diagonal lines define maximal tensile strengths, which depend on the size of the particles to be agglomerated. The van der Waals lines were calculated using the model sphere/sphere and particle distances of $a = 3$ and 1 nm. Assuming a distance of $a = 0.4$ nm (equilibrium distance) and using the model sphere/plane, the line would be pushed higher, close to the one representing capillary forces. If, in addition, plastic deformation of particles is considered, still higher agglomerate strengths can be obtained.

A narrow region characterizes the effect of liquid bridges. Somewhat higher is the line for the strength of agglomerates that are completely filled with a liquid. For this diagram it was calculated assuming water and the constant $a' = 6$. The strength of agglomerates

with sinter and adhesion bridges can be expected above this line.

In each case, the predictions of Figure 6.11 are valid only for certain assumptions. In the following a few examples shall demonstrate the variability of the correlations if individual parameters are changed.

Figure 6.12a and b show salt bridges that were obtained at different drying temperatures[14] during a model experiment. The visual examination indicates that the drying temperature must play an important role in the development of agglomerate strength even if all other parameters are kept constant.

Capillary pressure and tensile strength of moist agglomerates are associated with each other. To a great degree they are influenced by the amount of liquid that is present in the pore volume of the agglomerate. Assuming that the liquid wets the solid particles ($\delta = 0$), a classification as shown in Figure 6.13 can be defined. It is valid for three-phase systems consisting of a disperse solid material and two

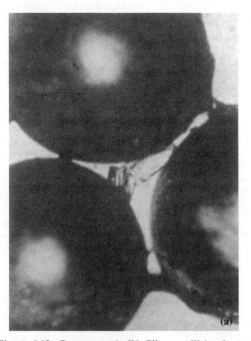 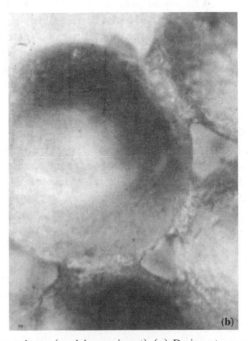

Figure 6.12. Common salt (NaCl) crystallizing between glass spheres (model experiment). (a) Drying at room temperature, (b) drying at 110°C.

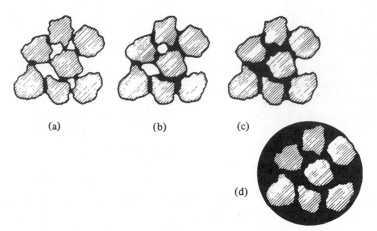

Figure 6.13. Different models of liquid distribution in moist agglomerates. (*a*) Pendular state, (*b*) funicular state, (*c*) capillary state, (*d*) liquid droplets with particles inside or at its surface.

immiscible fluid phases. The dark colored area represents the wetting fluid phase.

A small quantity of liquid causes liquid bridges between the particles forming the agglomerate (Fig. 6.13a). This region is called the pendular state. By increasing the amount of liquid, the funicular state is obtained (Fig. 6.13b) where both liquid bridges and pores filled with liquid are present. The capillary state (Fig. 6.13c) is reached when all pores are completely filled with the liquid, and concave menisci develop at the surface of the agglomerate. The last state (Fig. 6.13d), a liquid droplet with particles inside or at its surface, is an important mechanism for wet scrubbing and has relevance for agglomerate strength in spray dryer/agglomerators. Corresponding to the two patterns, Figures 6.13a and c, different models exist for the theoretical determination of agglomerate strength with a transition range in between (Fig. 6.13b).

Formerly, mathematical approximations were used for estimating the adhesion forces that can be transmitted through a liquid bridge. More recently, Schubert[15, 16] developed exact equations for all rotationally symmetric liquid bridges. In Figure 6.14 the nondimensional force $F_A = A_L / \alpha \cdot x$ [Eq. (6.7)] is plotted versus V_b/V_s for various geometric situations, where V_b is the bridge volume, V_s the volume of the solid (sphere), x the diameter of the

spheres, α the surface tension of the liquid, the wetting angle $\delta = 0$, and a the distance at the coordination point. As the value of V_b/V_s increases, the attraction forces increase for planes and cones and decrease for spheres.

Under normal atmospheric conditions and with wetting solids it must be expected that liquid bridges are developing by a capillary condensation at contact points ($a = 0$). Depending on the contact geometry involved, the attractive fores resulting from this mechanism

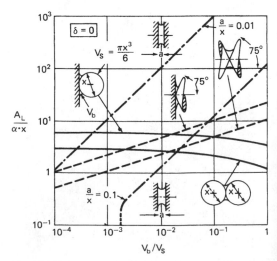

Figure 6.14. Computed adhesion forces resulting from liquid bridges for various geometric situations.[15]

may exceed the van der Waals forces. For example, Figure 6.15 shows the nondimensional force $F_A = A_L/\alpha \cdot x$ [Eq. (6.7) as a function of the dimensionless distance at the coordination point a/x for the model sphere/sphere.[17] At $a/x = 0$ and $\beta \to 0$, the maximum value $A_L = \pi \cdot \alpha \cdot x$ is obtained. If the bridge is stretched the attraction force decreases the more, the smaller the liquid volume.

Assuming complete wetting ($\sigma = 0$), the correlation between the nondimensional force $F_A = A_L/\alpha \cdot x$ [Eq. (6.7)] and the filling angle β (Fig. 6.4) at different values of a/x is presented in Figure 6.16. For spheres in contact ($a/x = 0$) the value of $F_A = A_L/\alpha \cdot x$ decreases from π (at $\beta = 0$) as the liquid content increases. For finite values of a/x, however, the curves pass through a maximum or, respectively, increase with higher liquid saturation.

The dotted curve in Figure 6.16 divides the graph into a field of capillary excess pressure ($F_{p_c} < 0$) and a field of capillary suction ($F_{p_c} > 0$). For comparison, two curves were calculated for $a/x = 0.1$, one representing the exact theory of Schubert[15,16] and the other an earlier approximation by Pietsch and Rumpf.[6]

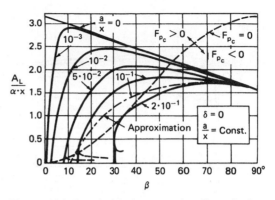

Figure 6.16. Correlation between the dimensionless adhesion force $F_A = A_L/\alpha \cdot x$ between two equal spheres and the filling angle β. Assumptions: $\delta = 0$ and $a/x = $ constant.

For the determination of the maximally transferable uniaxial tensile stress Eq. (6.8) can be rewritten:

$$\frac{\sigma_{tb} \cdot x}{\alpha} = \frac{1 - \epsilon}{\epsilon} \cdot F_A'(\beta, \delta, a/x) \quad (6.16)$$

For the transition range (Fig. 6.13b), the easily measurable liquid saturation S is introduced:

$$S = \pi \cdot \frac{1 - \epsilon}{\epsilon} \cdot \phi \quad (6.17)$$

It is defined as the ratio of liquid volume to pore volume of the agglomerate. Thus, Eq. (6.16) yields:

$$\frac{\sigma_{tb} \cdot x}{\alpha} = \frac{1 - \epsilon}{\epsilon} \cdot F_A''(\epsilon, S, \delta, a/x) \quad (6.18)$$

Equation (6.18) can be used to compute the correlation between the value $\sigma_{tb} \cdot x/\alpha$ and the liquid saturation S. Results for two different porosities ϵ and different distance ratios a/x (assuming complete wetting [$\delta = 0$]) are presented in Figure 6.17. Again, for comparison, the curves for $\epsilon = 0.35$ and $a/x = 0.1$ representing the exact theory and the approximation, respectively, have been included in Figure 6.17.

In the capillary state the strength of the agglomerate is determined by the capillary suction. Since only those pores filled with the

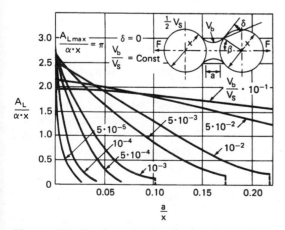

Figure 6.15. Nondimensional adhesion force $F_A = A_L/\alpha \cdot x$ as a function of the dimensionless distance at the coordination point a/x.[17]

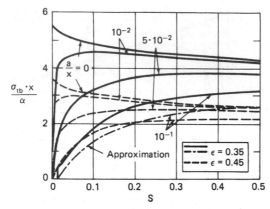

Figure 6.17. Maximally transferable uniaxial tensile stress $a_{tb} \cdot x / \alpha$ as a function of the liquid saturation S of moist agglomerates. Assumption: complete wetting ($\delta = 0$).

liquid contribute to the strength, Eq. (6.1) must be rewritten:

$$\sigma_{tc} = S \cdot p_c \qquad (6.19)$$

Equation (6.19) assumes that the liquid is uniformly distributed in the agglomerate. The product $S \cdot p_c$ can be calculated from the capillary pressure/saturation curve (Fig. 6.18b). By definition, the starting point of capillary pressure curves is $p_c = 0$ at $S = 1$. During drainage of the agglomerate the capillary pressure follows the curve marked in Figure 6.18b until it reaches a point at which only isolated capillaries exist. If, starting at that point, liquid is reintroduced into the agglomerate, $p_c = 0$ is reached at $S < 1$, as not all pores can be filled with liquid by imbibition. The remaining air pockets are blocked off by adjacent pores that are already filled. Repeated drainage/imbibition tests lead to the typical hysteresis loop shown in Figure 6.18b. It is explained by the existence of pore bulges and pore necks as well as by the contact angle hysteresis.[18]

Figure 6.18a shows schematically the maximally transferable tensile stress σ_t as a function of the liquid saturation S. The capillary pressure curve (Fig. 6.18b) is used to calculate $\sigma_{tc} = S \cdot p_c$ for $S_c \leq S \leq 1$. The capillary state ends when liquid bridges between the particles start forming ($S \leq S_c$). The funicular state ex-

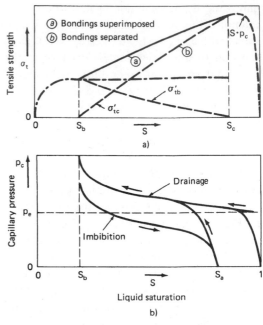

Figure 6.18. (a) Maximally transferable tensile stress σ_t and (b) capillary pressure p_c as a function of the liquid saturation S.

ists in the region $S_b \leq S \leq S_c$. For the pendular state ($S \leq S_b$) the already discussed Eq. (6.18) is valid.

In the transition range, the funicular state ($S_b \leq S \leq S_c$), in which liquid bridges coexist with liquid-filled pores, two cases can be constructed which follow a model published first by Rumpf.[1]

1. Both bonding mechanisms can be superimposed.
2. Each of the bonding mechanisms acts alone.

Assuming that the ratio of the liquid in the bridges to the total liquid diminishes linearly from 1 at S_b to 0 at S_c, one obtains the following for the individual bonding mechanisms (see Fig. 6.18a):

$$\sigma'_{tb} = \sigma_{tb}(S_c - S)/(S_c - S_b) \qquad (6.20)$$

$$\sigma'_{tc} = p_c \cdot S_c(S - S_b)/(S_c - S_b) \qquad (6.21)$$

If both mechanisms act alone curve b in Figure 6.18a represents the expected results. If the bonding mechanisms can be superimposed curve a results from the sum σ'_{tb} plus σ'_{tc}.

In all those cases where adhesion is caused by van der Waals or electrostatic attraction or by liquid bridges, surface roughness reduces the maximally transferable adhesion force. For the model sphere/plane and van der Waals attraction, Figure 6.19 shows the controlling radii for the calculation of the adhesion force according to Eq. (6.11). The shape of particles with surface roughness can be approximated by superimposing two spheres. The large radius R is considered the equivalent radius of a sphere of same volume as the particle, whereas the small radius r represents the surface roughness.[17]

Considering the model of Figure 6.19 and Eq. (6.11), the highest attraction forces A_{max} must be obtained if the adhesion partners are in contact and have smooth surfaces. Contrary to the indication of Eq. (6.11), the attraction force on contact is in reality finite. Therefore, an adjustment parameter Z_0 must be introduced:

$$A''_{v_1} = \frac{\hbar\bar{\omega}}{8\pi(a + Z_0)^2} \cdot R \qquad (6.22)$$

Krupp[10] has defined $Z_0 = 4 \cdot 10^{-8}$ cm (0.4 nm) as a measure for the atomic distance. For Figure 6.20b, A_{max} was calculated for different adhesion mechanisms using this value of Z_0, although it does not represent a true atomic distance. Rather, it is an approximate or adap-tive parameter, the accurate value of which needs still to be determined.

Figure 6.20b shows for the model sphere/ plate the correlation between the maximum adhesion force A_{max} on contact and the diameter $x = 2R$ of smooth particles for different adhesion mechanisms. The highest attraction forces are caused by liquid bridges assuming complete wetting ($\delta = 0$) and water as the liquid. Van der Waals forces are smaller by almost an order of magnitude, although a relatively high Lifshitz–van der Waals constant ($\hbar\bar{\omega} = 5$ eV $= 8 \cdot 10^{-19}$ J) was chosen. If two different materials contact, an electrostatic attraction force develops that is caused by the contact potential. The latter depends on the characteristics of the two contacting materials and their surface conditions. Again, the potential chosen ($U = 0.5$ V) represents a relatively high value. For conductors the electrostatic attraction force is higher than for nonconductors with the same contact potential because the charge is concentrated at the surface. Electrostatic attraction forces can also result from excess charges originating from friction, crushing, or electron and, respectively, ion adsorption. The highest possible excess charges are around 10^2 elementary charges $e/\mu m^2$.

Figure 6.20b indicates that for smooth spheres with sizes below 100 μm the electrostatic adhesion is negligible compared with van der Waals forces and even more so in relation to forces caused by liquid bridges.

Figure 6.20a describes the influence of surface roughness, represented by r (abscissa), on the attraction force A for different adhesion mechanisms. The curves were calculated for spheres with constant diameter $x = 2R = 10$ μm. The corresponding values of A_{max} can be determined in Figure 6.20b. Only for van der Waals forces two further curves for $R = 0.5$ μm and $R = 50$ μm were plotted since— because of their short-range character—the influence of roughness on van der Waals forces is very pronounced.

Investigating the curve for $R = 5$ μm and van der Waals attraction, the following observations can be made. At $r = 0$ (not shown in

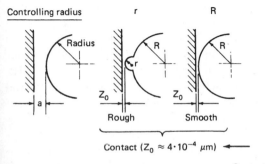

Controlling radius r R

No contact   Contact

Figure 6.19. van der Waals model sphere/plate with and without surface roughness.[17]

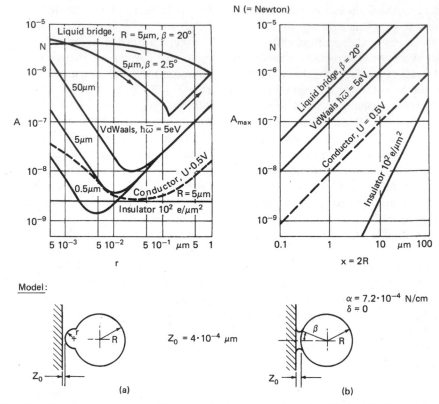

Figure 6.20. Attraction forces caused by different adhesion mechanisms for the model sphere/plate. Contact: $a = 0$, $Z_0 = 0.4$ nm.[17] (a) Influence of the roughness radius r on the attraction force A. (b) Influence of the diameter R of the smooth spheres on the attraction force A_{max}.

Fig. 6.20a) the maximum adhesion force $A = A_{max} = 10^{-6}$ N is obtained (Fig. 6.20b). With increasing r, the distance of the larger sphere from the plane grows, and the adhesion force decreases proportional to r^{-2}. Later the influence of the large sphere diminishes, and a minimum is reached at which both attraction forces act simultaneously. A_{min} is only approximately $1/250$ A_{max}. If r grows further, the influence of the large sphere disappears. Then, only the attraction force of the small sphere remains, which increases proportional to r. With growing R, r_{min} and the corresponding A_{min} increase but not at the same rate as A_{max}. At $R = 50$ μm, A_{min}/A_{max} equals $1/1000$.

Liquid bridges are much less sensitive to surface roughness. If the angle β is not too small, for example, $\beta = 20°$, the roughness is immersed in the liquid and merely increases the distance a. The smaller β, the more pronounced is this influence. At $\beta = 2.5°$ the liquid bridge breaks off from the large sphere at $\sim r = 10^{-1}$ μm and remains on the small sphere. Then, the attraction force increases proportional to r. $\beta = 2.5°$ can already correspond to capillary condensation. If β is still smaller, the transition to the line proportional to r occurs at smaller values of r, for example, for $\beta = 1°$ at $r \approx 5 \cdot 10^{-2}$ μm. In any case, the attraction force due to capillary condensation at the roughness peaks is always larger than the van der Waals force.

The electrostatic attraction forces of electrical conductors and of insulators with excess charges are smaller than the van der Waals forces. However, the influence of roughness is less pronounced and disappears completely for nonconducting particles facing a plane with an opposite charge of the same density.

For a long time the opinion existed that compared with van der Waals forces electrostatic attraction is always negligible. Based on the knowledge of the independence of roughness on electrostatic adhesion, this must now be corrected. Since the electrostatic attraction force increases with R^2, the 50 μm sphere with a charge density of 10^2 $e/\mu m^2$ would generate a higher value that could be obtained with the van der Waals model in much of the range of roughness shown in figure 6.20a. Therefore, in dry agglomerates formed by relatively large particles the electrostatic attraction forces due to excess charges—which do not depend on surface roughness—may have an important share in agglomerate strength.

The following results are generally valid:

1. For adhesion, forces caused by liquid bridges are most important and, normally, represent the highest share. Even in dry systems—due to capillary condensation—liquid bridges may be the controlling mechanism.
2. Van der Waals forces are extremely sensitive to surface roughness but should always be larger than forces caused by the contact potential.
3. Excess charges can also be a controlling factor for adhesion, particularly if relatively large particles form a dry conglomerate.
4. At distances in excess of 1 μm or, respectively, $a/x > 0.2$, only electrostatic forces are effective. They cause particle attraction before adhesion takes over and forms agglomerates.

6.2.4.2 Results of Experimental Determinations of Agglomerate Strength

In the following, some results of experimental investigations will show that if suitable model materials, agglomeration techniques, and experimental stressing methods are used, the respective theoretical expectations according to Sections 6.2.2 and 6.2.4 (see Theoretical Approximation of Agglomerate Strength) can be well approximated.

Crystallization of Dissolved Substances During Drying[14]. If drying of agglomerates that are filled with a salt solution starts at high liquid saturation S, evaporation begins at the surface of the agglomerate. The liquid flows to the surface by means of capillary suction. There, the dissolved substance crystallizes, forms a crust, and decisively controls further drying of the porous body. The crystal structure can therefore by influenced by either the drying temperature or the presence of a crust. In the latter case, the tensile strength of the agglomerate changes due to the varying drying conditions even if the strength is measured on the core after removing the crust.

A typical example is shown in Figure 6.21. The tensile strength, σ_t, is plotted versus the drying temperature, t_d. The parameter is the liquid saturation S. This diagram was obtained using a nearly saturated salt (sodium chloride) solution and a narrow limestone powder fraction. At very small liquid saturations ($S \leq$ 20%, curves a and b) no crust is built up, and the tensile strength increases with the drying temperature. This rise is caused first by the increasing crystallization velocity and second by the amount of salt forming bridges that changes almost proportionally to the liquid saturation. At a liquid saturation of 20% the formation of a thin crust starts to influence the tensile strength slightly. At the highest liquid saturations examined ($S \geq 45\%$, curves e and f), the dense crust is at all temperatures the deciding factor for drying and for the crystallization velocity. Their core tensile strength is low and remains almost constant. Such agglomerates burst at high drying temperatures much like a pressure vessel (Fig. 6.22). The high tensile strength of agglomerates with $S = 30\%$ dried at a temperature of 350°C (curve d) is caused by small hair fractures in the crust that did not cause the agglomerate to disintegrate but raised the drying

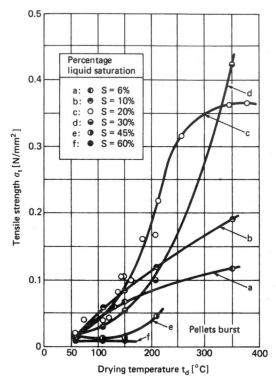

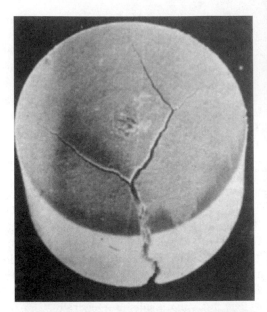

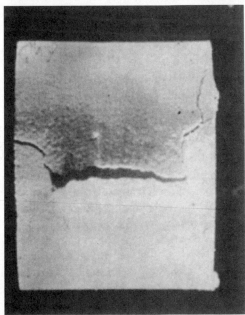

Figure 6.21. Tensile strength of σ_t of the core of agglomerates (crust removed) with salt bridges as a function of the drying temperature t_d at different liquid saturations S before drying.[14]

Figure 6.22. Cylindrical agglomerate that contained an NaCl solution and burst during drying.

rate and thus increased the tensile strength of the dry agglomerate core.[19]

Figure 6.23 shows the same set of results as shown in Figure 6.21 but plotted in a different way. This time the tensile strength σ_t is presented as a function of the liquid saturation S before drying. The parameter is the drying temperature t_d. This graph confirms that normally the highest strength is obtained at $S = 20\%$ if it is measured after removing the crust. However, an optimum drying temperature exists whereby an agglomerate dries quickly but does not build up enough inside pressure to cause cracking or disintegration. Then, the tensile strength can also be determined with crust.

Tests to investigate this optimum drying temperature were carried out in an instru-mented drying channel.[20] Figure 6.24 shows a result. Below $S = 0.2$ the strength values for agglomerates with and without crust are identical. At saturations above 0.2 the strength of agglomerates with crust increases proportionally.

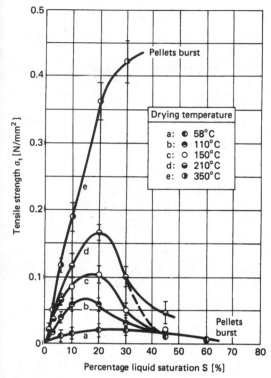

Figure 6.23. Tensile strength σ_t of the core of agglomerates (crust removed) with salt bridges as a function of the liquid saturation S before drying at different drying temperatures t_d.[14]

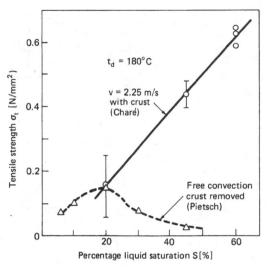

Figure 6.24. Tensile strength σ_t of agglomerates with salt bridges as a function of the liquid saturation S before drying with and without the crust.[21]

For practical applications the following conclusion can be drawn: To obtain high strength, drying should be carried out at the highest possible (without cracking) temperature using a saturated solution. Charé[20] found further that the air velocity does not substantially change the drying rate, and, therefore, the agglomerate strength.

Strength of Moist Agglomerates.

Agglomerates that are being built up by balling, that is the snowball-like forming of pellets in drums or discs, are nearly saturated with liquid. Figure 6.25 shows results of the determination of tensile strength plotted versus particle size. x_0 is the surface equivalent diameter and x_1 is the maximum of the diameter distribution. The diagonals represent the maximally transferable tensile strength calculated with Eq. (6.2) using $a' = 6$ and $a' = 8$. The diagram shows that $\sigma_t \sim 1/x$ is fulfilled. Values lower than theoretically predicted are mostly due to the fact that the agglomerates were not fully saturated when the tensile strength was determined.

The relationship $\sigma_t \sim \alpha$ was confirmed by Conway-Jones[11] with compression tests on spherical agglomerates (Fig. 6.26) and $\sigma_t x/\alpha \sim (1 - \epsilon)/\epsilon$ was checked by Schubert,[15] who confirmed this correlation, too (Fig. 6.27).

It can be assumed that up to saturations of $\approx 20\%$ to 40% the liquid in moist agglomerates is present in the form of discrete liquid bridges at the contact and coordination points between the particles forming the agglomerate. The tensile strength of such an agglomerate is predicted by Eq. (6.8). Experimentally it was investigated with the wall friction method (Figure 6.6D) using pellets made of narrowly distributed limestone powder and distilled water. In Figure 6.28 the experimental results are shown in comparison to the theory. The curves were approximated by varying the distance a (respectively, a/x). They seem to fit the experimental results well, although the parameters

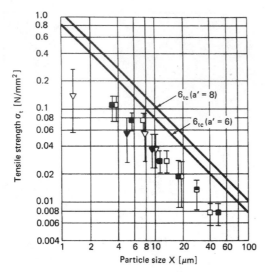

Figure 6.25. Tensile strength of moist agglomerates with high liquid saturation as a function of particle size. $\epsilon = 0.35$. Quartz powder: $\triangledown$: x_0, $\blacktriangledown$: x_1. Limestone powder: $\square$: x_0, $\blacksquare$: x_1.

a/x are purely empirical. However, considering the surface roughness of the particles the a/x values may be in the correct order of magnitude.

ALCOHOL/WATER MIXTURES

VOL. %		α	σ
ALCOHOL	WATER	(N/cm)	(N/mm²)
0	100	72.2×10^{-5}	8.31×10^{-1}
10	90	50.2×10^{-5}	6.02×10^{-1}
30	70	35.0×10^{-5}	4.53×10^{-1}
100	0	22.2×10^{-5}	2.42×10^{-1}

At higher liquid saturations more and more pores fill up and the models *liquid bridges* and *saturated pores* coexist. The theories described in Section 6.2.4 (Figs. 6.13, 6.17, and 6.18) where checked by Schubert.[15] Figure 6.29 shows the results. The tensile strength is plotted versus the liquid saturation and compared with the theory. At $S = 1$ the capillary pressure is zero. If, starting at this point, the agglomerate is drained, the capillary pressure raises steeply and then turns into a flatter curve. The point where the two tangents meet is defined as the so called entry suction pres-

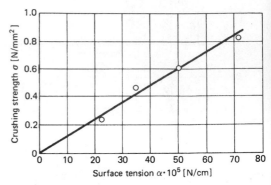

Figure 6.26. Relationship between crushing strength σ of moist agglomerates and surface tension α of the liquid.[11]

sure p_c. It is generally located near $S = 0.9$. Approximately at this point the maximum tensile strength of moist agglomerates exists. At lower and higher saturations the strength decreases. The results show that between $0.3 < S < 0.9$ both mechanisms contribute to the strength of agglomerates.

The capillary pressure and, therefore, the tensile strength are much larger if the liquid is drained than after imbibition. This knowledge can be very important for agglomeration and certain other technologies, for example, filtration (strength of filter cakes).

Strength due to van der Waals Forces. Because of the short range of van der Waals forces, particles forming an agglomerate must

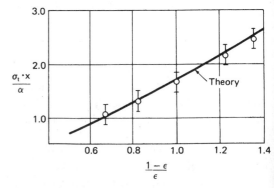

Figure 6.27. Relative tensile strength $\sigma_t \cdot x/\alpha$ of moist conglomerates formed of glass spheres plotted versus the porosity function $(1 - \epsilon)/\epsilon$. Comparison between theory [Eq. (6.8)] and experiment.[15]

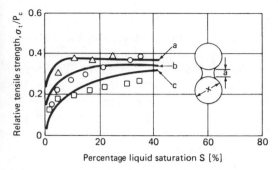

Figure 6.28. Relative tensile strength σ_t/p_c of agglomerates with liquid bridges as a function of the liquid saturation S. Limestone powder; $\triangle$: $x_0 = 71$ μm; $\bigcirc$: $x_0 = 35$ μm; $\square$: $x_0 = 13$ μm.

[a] $\epsilon = 0.45$; $a/x = 0.02$ ($a = 1.4$ μm)
[b] $\epsilon = 0.45$; $a/x = 0.04$ ($a = 1.4$ μm)
[c] $\epsilon = 0.50$; $a/x = 0.1$ ($a = 1.3$ μm)

be brought closely together to cause significant attraction. To investigate the influence of pressure and van der Waals attraction on agglomerate strength, cylindrical pellets were produced in a hydraulic press using barium sulfate as the model substance. This material excludes other binding mechanisms.

The influence of adsorption layers on agglomerate strength was demonstrated using, respectively, air-dry material and powder, which was dehydrated at a temperature of

600°C prior to pressing it into pellets at 10^{-5} mbar and room temperature. Figure 6.30 shows the results[22]. With increasing compaction pressure, that is, smaller particle distance the tensile strength increases. Pellets that were produced at atmospheric conditions from air-dry material (L) and, therefore, contain adsorbed water exhibit higher strength than those pressed at high vacuum from desorbed barite (HV). This is in general agreement with the expected influence of water adsorption on adhesion discussed above.

Herrmann[23] investigated in more detail the influence of water adsorption on the tensile and shear strength of barium sulfate briquets. Figure 6.31 shows some results. The tensile and shear strengths were determined on briquets produced and stressed in a high vacuum and at varying levels of relative humidity of the surrounding atmosphere. The normal relative humidity lies between 60% and 80%. Therefore, in the common sense, the powder must be considered dry. The strength is plotted in both parts of Figure 6.31 versus the relative water vapor pressure p/p_0 (with $p_0 =$ water vapor pressure at saturation). The following conclusions can be drawn:

1. The tensile strength σ_t increases with growing relative water vapor pressure p/p_0. Responsible for this rise is the capillary condensation. Van der Waals forces participate only to a small extent.

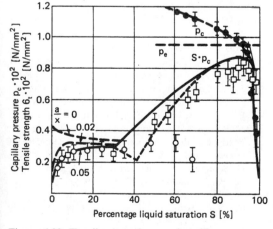

Figure 6.29. Tensile strength σ_t and capillary pressure p_c as a function of liquid saturation S. Limestone: $x_0 = 71$ μm, $\epsilon = 0.415$. $\bullet$: p_c; $\square$: σ_t drainage; $\bigcirc$: σ_t imbibition.

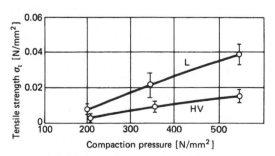

Figure 6.30. Tensile strength σ_t of barium sulfate pellets with (L) and without (HV) adsorption layers as a function of compaction pressure. Particle size of starting material: 50 to 100 μm.[22]

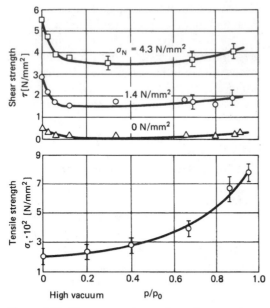

Figure 6.31. Shear strength τ and tensile strengths σ_t of barite briquettes as functions of the relative water vapor pressure p/p_0.[23] Particle size of starting material: 50 to 100 μm. Compaction pressures: 2.2×10^2 N/mm^2 (shear strength graph; 4.8×10^2 N/mm^2 (tensile strength graph).

2. Due to interparticle friction the shear strength τ is high in a high vacuum. This explains, for example, the extremely well-developed footprints that were visible during the first moon landing of man in the loose dust (Apollo II). Interparticle friction is highest at space conditions. With increasing relative humidity the shear strength τ first decreases rapidly. This is due to the fact that liquid films "lubricate" the particles and the friction decreases. Later, liquid bridges develop by capillary condensation and the strength increases again slowly.

3. At a normal load of $\sigma_N = 0$ the tensile strength σ_t of barite briquets is smaller than the shear strength τ. Therefore, the tensile strength is the critical strength, and failure occurs under tensile load.

4. The shear strength τ depends on the normal load σ_N which acts upon the specimen during shearing.

6.2.4.3 Other Investigations

A large number of other, specific investigations was carried out by various researchers, confirming still more theories of agglomerate bonding and strength. However, since this chapter is only meant to introduce some basic theoretical and experimental information, further results should be obtained from the respective scientific and technical literature.

References

1. H. Rumpf, "The strength of Granules and Agglomerates," *Agglomeration*, edited by W. A. Knepper, John Wiley, New York, pp. 379–418 (1962).
2. W. Pietsch, "Granulieren durch Kornvergrösserung," *CZ-Chemie-Technik* 1(3):116–119 (1972).
3. H. Rumpf und E. Turba, "Über die Zugfestigkeit von Agglomeraten bei verschiedenen Bindemechanismen," *Ber. dtsch, keram. Ges.* 41(2):78–84 (1964).
4. W. Pietsch, "Die Festigkeit von Agglomeraten," *Chem. Techn.* 19(5):259–266 (1967).
5. H. Rumpf, "Zur Theorie der Zugfestigkeit von Agglomeraten bei Zraftübertragung an Kontaktpunkten," *Chem-Ing. Techn.* 42(8):538–540 (1970).
6. W. Pietsch and H. Rumpf, "Haftkraft, Kapillardruck, Flüssigkeitsvolumen und Grenzwinkel einer Flüssigkeitsbrücke zwischen zwei Kugeln," *Chemie-Ing. Techn.* 39(15):885–893 (1967).
7. W. Pietsch, E. Hoffman, and H. Rumpf, "Tensile Strength of Moist Agglomerates," *I + EC Product, Research & Development* 8(3):58–62 (1969).
8. W. O. Smith, P. D. Foote, and P. F. Busang, *Phys. Rev.* 34:1271–1274 (1929).
9. H. C. Hamaker, "The London-van der Waals Attraction between Spherical Particles," *Physica* 4:1058–1072 (1937).
10. H. Krupp, "Particle Adhesion, Theory and Experiment," *Advances Colloid Interface Sci.* 1(2) (1967).
11. J. M. Conway-Jones, "An Investigation into the Mechanism of the Unit Operation of Granulation," Ph.D. thesis, University of London, 1957.
12. H. Rumpf, "Das Granulieren von Stäuben und die Festigkeit der Granulate," *Staub* 5(5):150–160 (1959).
13. H. Schubert, "Tensile Strength of Agglomerates," *Powder Technology* 11:107–119 (1975).
14. W. Pietsch, "The Strength of Agglomerates Bound by Salt Bridges," *Can. J. Chemical Engng.* 47:403–409 (1969).
15. H. Schubert, "Untersuchungen zur Ermittlung von Kapillardruck und Zugfestigkeit von feuchten Haufwerken aus körnigen Stoffen," Ph.D. thesis, University of Karlsruhe, 1972.

16. H. Schubert, "Kapillardruck und Zugfestigkeit von feuchten Haufwerken aus körnigen Stoffen," *Chemie-Ing. Technik*, 45(6):396–401 (1973); and *VDI-Bericht Nr. 190*, pp. 190–194 (1973).

17. H. Rumpf, "Die Wissenschaft des Agglomerierens," *Chemie-Ing. Technik*, 46(1):1–11 (1974).

18. N. R. Morrow, "Physics and Thermodynamics of Capillary Action in Porous Media," *Ind. Eng. Chem.* 62(6):32–65 (1970).

19. W. Pietsch, "Festigkeit und Trockungsverhalten von Agglomeraten, deren Festigkeit durch bei der Trockung auskristallisierende Salze bewirkt wird," Ph.D. thesis, Unversität (TH) Karlsruhe, 1965.

20. I. Charé, "Trocking von Agglomeraten bei Anwesenheit auskristallisierender Stoffe. Festigkeit und Struktur der durch die auskristallisierten Stoffe verfestigten Granulate," Ph.D. thesis, Universität (TH) Karlsruhe, 1976/1977.

21. H. Rumpf, "Particle Adhesion," in "Agglomeration 77," edited by K. V. S. Sastry, *Proc. 2nd Int. Symp. Agglomeration*, Atlanta, March 6–10, 1977, 1:97–129 (1977).

22. E. Turba, "Die Festigkeit von Briketts aufgrund von van der Waals Kräften und der Einfluf von Adsorptionsschichten," Ph.D. thesis, Universität (TH) Karlsruhe, 1963.

23. W. Herrmann, "Die Adsorption von Wasserdampf in Schweispat-Preßlingen und ihr Einfluß auf deren Festifkeit," Ph.D. thesis, Universität (TH) Karlsruhe, 1971/72.

6.3 SIZE ENLARGEMENT BY AGGLOMERATION IN INDUSTRY

6.3.1 Parameters of Size Enlargement by Agglomeration

A basic equation for the tensile strength of agglomerates bonded by forces transmitted at the coordination points between particles was first developed by Rumpf.[1] It is [see also Eq. (6.4)]:

$$\sigma_t = \frac{1 - \epsilon}{\pi} k \frac{\sum_{i=1}^{n} A_i(x, \dots)}{x^2} \qquad (6.23)$$

where

σ_t = tensile strength
ϵ = porosity (= relative pore space)
$(1 - \epsilon)$ = relative amount of solids
k = number of coordination points
A_i = adhesion force by mechanism i.

With

$$k \cdot \epsilon \approx \pi \qquad (6.24)$$

and taking into consideration that most adhesion forces A_i are a function of the particle size x, Eq. (6.23) becomes:

$$\sigma_t = \frac{1 - \epsilon}{\epsilon} \frac{\sum_{i=1}^{n} A_i(\cdots)}{x} \qquad (6.25)$$

Even after eliminating the influence of the particle size x the adhesion force A_i remains a function of several parameters that vary with the binding mechanism.

The particle size representing a distribution of irregularly shaped particles (= the real conditions) can be described for all models by the surface equivalent diameter x_0, the diameter of monosized spherical particles producing the same specific surface area (e.g., in m^2/g) as the actual particle size distribution.

A similar equation describes the tensile strength of agglomerates which are held together by the negative capillary pressure of a liquid filling the pore space [see also Eq. (6.2)]:

$$\sigma_t = c \frac{1 - \epsilon}{\epsilon} \alpha \frac{1}{x_0} \qquad (6.26)$$

where

c = constant between approx. 6 and 8
α = surface tension of liquid.

Equations (6.25) and (6.26) suggest that the strength of agglomerates is strongly influenced by the porosity and, respectively, the relative amount of solids and that it increases with the specific surface area (= decreasing surface equivalent diameter x_0) of the particulate matter forming the agglomerate. The latter also indicates that the presence or lack of very fine particles will favor or hinder the formation of strong agglomerates. In the case of wet agglomerates (capillary model) the surface tension participates directly in strength; it must be understood, however, that capillary forces provide only temporary bonding; post treatment will activate other binding mechanisms (Fig. 6.3) for permanent strength.

Characteristics of agglomerates from particulate solids and a matrix binder, for example, cement in concrete-like aggregates, depend on the strength of all participating materials as well as their adhesion conditions and follow different relationships (see Section 6.2.2). Strong, highly impermeable, and leach-proof agglomerates are obtained if the distribution of the particulate matter favors the formation of dense structures as shown by the model in Figure 6.32.

6.3.2 Characteristics of Agglomerated Materials

The most versatile equations describing the strength of agglomerates [Equations (6.25) and (6.26)] indicate that strength increases with decreasing particle size. Since, additionally, the representative particle size that best describes the phenomenon of agglomeration is the surface equivalent diameter x_0, the tendency of particulate matter to agglomerate increases greatly with the presence of very fine particulate matter.

The strong influence of small particles on x_0 can be easily demonstrated by the fact that a single spherical particle with density 1 g/cm^3 having a mass of 1 g (particle diameter: approx. 12.4 mm) features a surface area of approx. 4.8×10^{-4} m^2/g. If this mass of 1 g is made up by (1.9×10^{12}) monosized 1 μm spherical particles of the same material and with porosity 0.4 the specific surface area is approx. 9.3 m^2/g or more than four orders of magnitude larger (Table 6.4).

When determining the specific surface, methods must be chosen that measure only the outer particle surface and exclude accessible inner surface due to open particle porosity. Good estimates for this type of specific surface area of particulate matter can be obtained by the simple Blaine (permeability) method or the scanning of particle images.

Figure 6.33 is a dispersity scale indicating the most likely ranges for certain particulate matter (with particular reference to fines occurring in environmental control technologies) and the general applicability of agglomeration methods. Of course, the individual ranges do

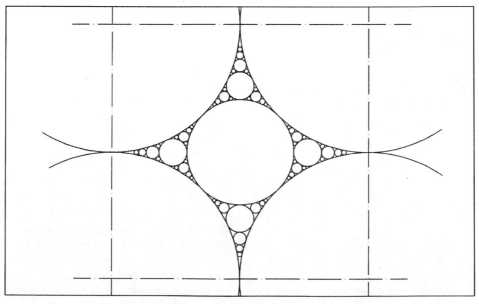

Figure 6.32. Model (cubic) depicting dense packing structure.

Table 6.4. Some Characteristics of Spherical Particles (Density of Solid: 1g/cm³).

MASS (g)	VOLUME (cm³)	DIAMETER (mm)	NUMBER (—)	SURFACE AREA (m²/g)
1	1	12.4	1	$4.8 \cdot 10^{-4}$
1	1.67	10^{-3}	$1.9 \cdot 10^{12}$	9.3

overlap and can be influenced by special conditions or processes.

In a particle size range below approx. 10 μm the natural attraction forces, such as molecular (van der Waals), magnetic, and electrostatic forces, which may be enhanced by adsorption layers, liquid films, or "binder" chemicals (e.g., flocculation agents) become significantly larger than the separating forces due to particle mass and external influences (e.g., drag and centrifugal forces) so that adhesion occurs (Fig. 6.34). Because the probability of particle-to-particle collisions, which are preconditions for adhesion, rises with concentration, the tendency of particulate matter to

naturally agglomerate increases (for example, in a fluidized bed environment) but, independent of concentration, decreases with particle size despite their greater adhesion potential. The latter is due to the fact that ultrafine particles tend to follow flow lines so that collisions do not occur as frequently.

The natural agglomeration of "submicron" particles is a reason for the relatively high efficiency of many pollution control devices that separate such solids from process effluents. The effect can be increased by forcing the particles into increased motion, for example, in the case of smokes by the application of sound.

6.3.2.1 Undesired Agglomeration

Knowing the possible binding mechanisms of agglomeration and that, with few exceptions, bonding and strength of agglomerates is strongly influenced by particle size or surface, the reasons for and potential methods for the prevention of unwanted agglomeration phenomena during processing, storage, and han-

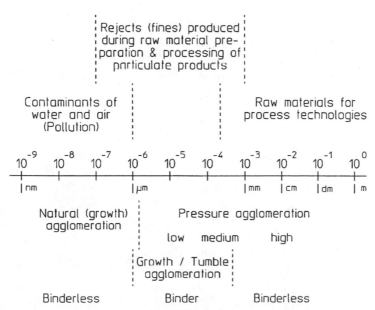

Figure 6.33. Dispersity scale relating materials and agglomeration methods to the size of particulate matter.

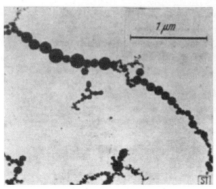
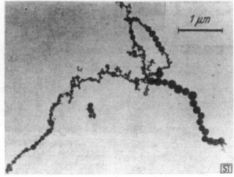

Figure 6.34. Chain-like natural agglomerates of "brown smoke" from steel converters formed by the combined effects of magnetic, electrostatic, and molecular attraction.[2]

dling of particulate solids are comprehensible. In most cases undesired agglomeration phenomena begin with the finer portion of the particle mass. In the following some examples[2] will be presented.

Comminution. During fine grinding in tube or roller mills deposits begin to form at a certain fineness, in the case of all materials, whereby two types of phenomena can be distinguished.

In the first case, the finest particles start to adhere to walls of grinding media in the mill, forming thin layers. On this basis coarser particles find excellent conditions for adhesion and massive deposits form rapidly. Experiments by Ocepek,[3] who investigated the particle size distribution across thick layers of build-up, showed that the finest particles are indeed found in the lowest layers. Figure 6.35 shows grinding balls which, after a short period of operation, are already covered with a light primary deposit, upon which additional layers will build up during extended grinding. Figure 6.36 is the photograph of the manhole cover of a ball mill, illustrating the extent of such deposits. These adhering layers produce a cushioning effect which lowers the intensity of stressing and, therefore, increases the duration of grinding.

The second phenomenon during dry fine grinding is the occurrence of agglomerates in the freely moving charge itself. Again, such agglomerates form only in the presence of a sufficiently large amount of fine particles and are frequently lamellar.

Agglomeration and adhesion in mills can be attributed to various bonding mechanisms. Since the mill housing often becomes highly charged by friction between its contents and the walls, electrostatic forces are often the cause of build-ups. This effect can be eliminated quite easily by grounding the mill. In other cases, wall deposits will begin with particles of a size that generally corresponds to that of the wall roughness. The strength of the deposited layer depends on the intensity of contact pressure which is magnified by the mill charge consisting of grinding media and mate-

Figure 6.35. Grinding balls before (*right*) and after brief grinding (*left*).[2]

Figure 6.36. Manhole cover of a ball mill before (*top*) and after grinding (*below*).[2]

rial to be crushed. Adhesion is largely affected by molecular forces; however, partial melting and sintering are also possible.

Agglomerates are formed in the freely moving charge of a tube mill by the compaction of fine particles between the grinding media. Adhesion is affected by van der Waals forces between the particles that have been compressed very tightly. Beke,[4] who determined structural changes in the agglomerated particles, went so far as to regard this mechanism as similar to cold welding. Since these agglomerates are very strong, a so-called "grinding equilibrium" is obtained which has been observed and described by many authors.[5-9] It means that, after a certain grinding time, a state of equilibrium occurs, from that point on agglomerates are crushed during further grinding and reformed so that the apparent particle size does not change.

Since every form of agglomeration decreases the efficiency of grinding and the degree of fineness obtained at the "grinding equilibrium" is not sufficient for many tasks, it is desirable to prevent or at least reduce these effects. In milling, one possibility to achieve less unwanted agglomeration is to add surface-active substances. It has long been known that small amounts of such additives may reduce the grinding time required for reaching a particular fineness by 20% to 30%.[10-14] Atoms or molecules of these substances that are present in a gas or vapor phase rapidly saturate free valences at the newly created surfaces which would otherwise give rise to recombination bonding. The effect of some of these grinding aids on the fineness of cement[15] after a specific grinding time is shown in Figure 6.37. It can be seen that, with the exception of soot, the desired effect is

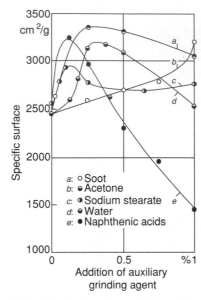

Figure 6.37. Effect of various grinding aids on the fineness of cement after a constant grinding time (2000 rev) in a rod mill (rod diameter: 25 mm, 85% critical speed). (According to Ghigi and Rabottino.[15])

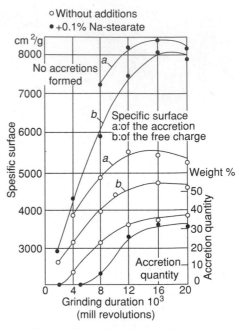

Figure 6.38. Specific surface of the build-up and of the free charge as well as amount of build-up with and without the use of a grinding aid (cement clinker, rod mill). (According to Ghigi and Rabotino.[15])

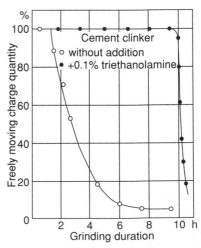

Figure 6.39. Changes in the amount of the freely moving charge during the grinding of cement in a laboratory ball mill with and without the addition of triethanolamine as a grinding aid.[4]

produced only if the amount of the grinding aid is very small. At higher concentrations the agglomeration tendency increases due to the formation of sorption layers and liquid bridges. In the case of soot a greater quantity is required because it is a solid whose molecules are not very mobile. Good results can also be obtained by merely enriching the atmosphere in the grinding chamber with certain gases or vapors that have been selected to possibly interact with the charge.[16-23]

As a rule, grinding aids also reduce caking. Figure 6.38 depicts the effect of 0.1% sodium stearate during the grinding of cement clinker. Other surface-active substances can delay build-up for longer periods or even prevent them entirely up to a certain fineness (for cement clinker, e.g., 0.1% triethanolamine,[4] Fig. 6.39). From Figure 6.38 it can also be seen that the specific surface, that is, the fineness of cement, increases when 0.1% sodium stearate is added and that the build-up consists of finer particles.

The formation of lamellas or plate-like agglomerates in tube mills has been attributed to compaction occurring between the grinding media. The same mechanism happens in all comminution processes in which the material to be crushed is subjected to stresses by two surfaces. Since the second condition for the formation of agglomerates is a sufficient fineness of the particles, the occurrence of lamellas is observed mostly in fine grinding, for example, in roller mills.

One measure for the fineness as well as the intensity of stressing—and consequently, also for the agglomerative tendency—is the so-called degree of reduction, that is, the ratio of maximum feed particle size to the gap between the rollers. Figure 6.40 shows typical agglomerates produced in a roller mill with a high degree of reduction. Since the fine material is immediately compacted, almost all free valences at the newly created surfaces participate in recombination bonding.

Consequently, the formation of agglomerates can be avoided or reduced only if a smaller degree of reduction is chosen, or by applying friction between the rollers.[24] More recently it was found[25] that the combination of a large degree of reduction in high-pressure roller mills and the desagglomeration of the conglomerates produced by this method result in a significantly lower overall energy consumption during fine grinding of brittle materials (such as cement clinker and many ores); therefore, in many cases the unavoidable agglomeration of the fine particles is not only tolerable but the technology also results in a more economical fine grinding method.

Agglomerates can also be formed during impact grinding. Figure 6.41a shows schematically the fracture lines observed during impact stressing of a glass sphere.[26] A cone of fine material is created at the impact point and is compacted by the pressure resulting from the kinetic energy of the system into an agglomerated mass (Fig. 6.41b and c). Here too, the effect of free valence forces at newly created surfaces is utilized to its almost full extent, yielding a quite strong agglomerate. During impact crushing thermoplastic materials or inorganic substances with low melting points,

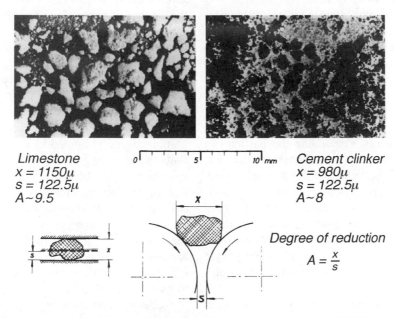

Limestone
$x = 1150\mu$
$s = 122.5\mu$
$A \sim 9.5$

Cement clinker
$x = 980\mu$
$s = 122.5\mu$
$A \sim 8$

Degree of reduction

$$A = \frac{x}{s}$$

Figure 6.40. Agglomerates produced during the grinding of limestone and cement clinker in a roller mill with a high degree of reduction.[24]

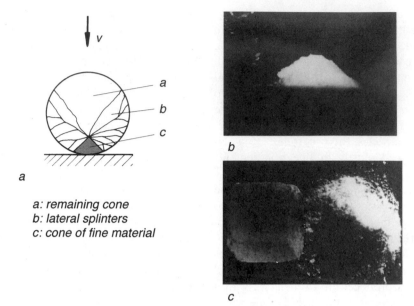

v

a

b

c

b

a

a: remaining cone
b: lateral splinters
c: cone of fine material

c

Figure 6.41. (a) Schematic representation of the fracture lines caused by impact stressing of a glass sphere.[26] (b) Agglomerated cone of fines created during the impact stressing of a glass sphere (impact velocity approx. 150 m/s; sphere diameter 8 mm).[26] (c) Agglomerated cone of fines created during the impact stressing of a sugar crystal (*left*).[27]

adhesion and agglomerate strength may further increase owing to melt bridges. It is very difficult to prevent such agglomeration; this can be affected only by reducing the impact velocity which, in turn, results in a lower degree of comminution. For glass spheres, for example, the formation of agglomerates was observed only at impact velocities exceeding 80 m/s.[26]

In wet grinding, as a rule, agglomeration is totally avoided by suspending the particles in liquid. Sometimes, the product of dry fine grinding is subjected to a brief final wet grinding to destroy the previously formed agglomerates.[50] Nevertheless, some materials also tend to flocculate in wet grinding. Since in these cases the adhesion forces are mostly electrical, the addition of a small amount of electrolyte nearly almost suffices to prevent flocculation.

Separation. During separation unwanted agglomeration can occur and needs to be avoided if a particle collective must be separated into two classes with a sharp (vertical) separation line. The so-called separation curve is a measure of the quality of separation. In this curve the degree of separation (i.e., the percentage share of particles in the coarse and fine fraction, respectively, after separation) is plotted versus the particle size interval $x_{min} \leq x \leq x_{max}$ which is to be separated. The cut size is that particle size of which half end up in the coarse fraction and half in the fine ($\phi = 50\%$). Figure 6.42 is a qualitative representation of separation curves.[2] Line (a) in Figure 6.42 represents the ideal or perfect separation at cut size x_{t1} which is possible only in theory. In industrial separation equipment curves of the type (b) are obtained. The sharpness of separation increases with a steeper slope of the curve. If the abscissa uses a logarithmic gradation, separation curves representing similar separation efficiencies at difference cut sizes are parallel to each other.

Agglomeration must be judged differently if the separation task is to remove all particles from a suspending fluid; then the cut size is x_{min}. Curve (c) in Figure 6.42 describes the

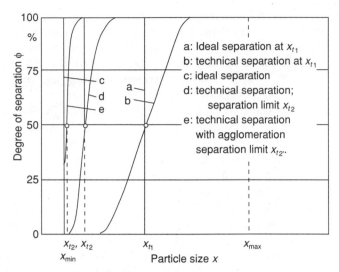

Figure 6.42. Qualitative representation of various separation curves.[2]

ideal, only theoretically possible separation curve. In reality, a certain amount of smaller particles remains and the cut size is $x_{t2} > x_{min}$ (curve d). If agglomeration occurs, the finest particles may form larger entities or attach to larger particles, changing the separation curve in Figure 6.42 to (e).

In most cases, however, agglomeration is undesired during the separation of particulate solids. Techniques include screening, sifting, classification, sorting, flotation, and, as a general analytical method, particle characterization. During screening, agglomeration is often facilitated by the motion of the material on the screen; spherical conglomerates are frequently formed from material containing fines or featuring other adhesive characteristics. Binding mechanisms are, for example: for finely divided solids, molecular forces and adsorption layers; for plastics, electrostatic forces; for iron ores, magnetic forces; for moist powders, liquid bridges and capillary forces; for fibers, interlocking; and for materials with low melting points, partial melting and sintering. In many substances several bonding mechanisms may occur simultaneously. In all cases the result of separation by screening is distorted since agglomerated fines are classified as coarse particles. Normally, the effect of

agglomeration is reduced by mechanical destruction of agglomerates with, for example, rubber cubes placed on the screen decks, the application of brushes,[29, 30] or the modification of amplitude and, respectively, frequency (ultrasonic screening)[31, 32] of vibration. Agglomerates can be also destroyed by the effect of air jets passing through the screen from below.[33]

During the screening of moist bulk materials difficulties increase with moisture content but agglomeration tendencies are almost completely eliminated during wet screening when the particles are suspended in a liquid.[34] Since in moist screening particles or agglomerates are often retained in the mesh openings by liquid bridges, the separation of such materials is facilitated by direct electric resistance heating,[35] inductive heating,[36] or by altering the wetting angle and surface tension.[37]

In air classification, typically products from dry fine grinding are separated. Particular problems arise if the material to be separated contains agglomerates that were formed during comminution. Attempts are made to destroy these agglomerates by special designs of the feeder. Destructive forces are caused, for example, by sudden changes in speed or direction of flow and by installing air jet mills in front of the classifier.[38] When classifying cement it was determined that grinding aids used during comminution also improve separation by avoiding agglomeration in the classifier.

In the classifier itself agglomerates are formed by molecular forces that may be reinforced by adsorption layers if separation is carried out in a moist atmosphere, by liquid bridges if moist materials are processed, and by electrostatic forces in a dry environment. Figure 6.43 depicts various separation curves of air classifiers.[39] With decreasing particle size the amount found in the coarse fraction increases, which is due to agglomeration whereby fine particles adhere to larger ones and conglomerates of fines behave like coarser particles. Both effects reduce the separation efficiency and can be avoided only if the causes of adhesion are removed, that is, mostly by eliminating moisture and humidity.

Sorting processes that separate materials according to particle characteristics other than size are mostly carried out in liquids. During a special technology, flotation, the relative capacity of material to float is enhanced by the addition of chemicals. Agglomeration can also reduce the separation efficiency of these processes because fine particles stick to larger ones, form conglomerates, or adhere to foam bubbles.[40] By use of modified chemicals, processing of very dilute suspensions, or multiple separation steps efficiently can be improved.

During particle size analysis, in addition to screening, sifting, and counting, sedimentation methods are often used that produce unequivocal results only if the individual particles elutriate without influencing each other. For that reason very dilute suspensions are used. Nevertheless, it is possible that agglomerates form or already present conglomerates do not disperse completely. Therefore, dispersion aids are often added that reduce particle affinity; a large number of such additives is available.[41-44] The molecules of dispersion aids attach to the particles, eliminating polarities and/or reducing interfacial tensions.

In connection with particle size analysis, the importance of correct sample preparation should be stressed. Because agglomerates always incorporate a relatively large number of the finest particles, the result of particle size analysis may be incorrect if preexisting agglomerates are not destroyed or conditions prevail during measurement that promote agglomeration.

Mixing. Many of the previously mentioned considerations apply to the formation and prevention of agglomerates during mixing. Little

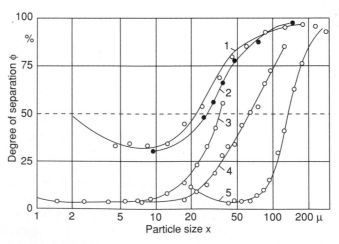

Figure 6.43. Separation of curves of various air classifiers. (According to Kayser[39].)

needs to be added concerning mixing in liquids by stirring or methods for the production of suspensions and dispersions. The addition of dispersion agents is recommended when the tendency of the solids to agglomcrate is high. Already present agglomerates or flocs can be destroyed by shear forces in the liquid. Consequently, the generation of the highest possible shear gradient is considered advantageous when selecting agitators.

During extended storage the particles in pharmaceutical suspensions often form agglomerates that can no longer be destroyed by shaking the preparation. This problem can be avoided by controlled flocculation of the solids.[45] After the addition of an electrolyte the fine particles aggregate to loose flocs that can by easily redispersed by shaking.

When mixing dry or moist bulk solids, agglomerates are formed, originating from the finest components of the mixture, which are held together by molecular and electrostatic forces as well as by capillary forces, particularly if the materials are moist.[46] These undesired agglomerates are broken up by shear or frictional stresses generated by the motion of the bulk mass or by special disintegration devices built into the blender. Figure 6.44 shows two examples. It depicts interior sections of a drum mixer with plow-like mixing elements and additional friction plates (a) or rapidly rotating cutter heads (b).

Conveying. Particulate solids, especially finely dispersed powders, tend to form agglomerates and (sometimes thick) coatings on walls during conveying. Whereas agglomerates occur mostly on vibrating or shaking conveyors, wall build-up is more common in pneumatic conveyors.[47-49] The main causes of agglomeration during conveying of fine particulate solids are molecular and electrostatic forces. However, as a result of additional mechanical and thermal energy input, other binding mechanisms can also be activated, for example, partial melting and solidification.

Although it is very difficult to avoid the formation of agglomerates on vibrating and

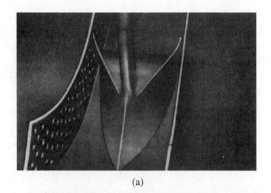

(a)

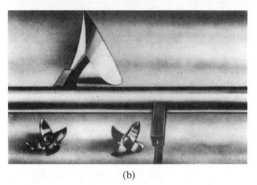

(b)

Figure 6.44. Partial views of the interior of a mixer with plow-like mixing elements. (*a*) With friction plates, (*b*) with cutter heads for desagglomeration.

shaking conveyors several possibilities exist for the prevention of wall build-up and deposits during pneumatic transport.[47] For the latter, one of the most important conditions is to provide smooth inner wall surfaces to avoid the most common reason for initial build-up, the adhesion of the smallest particles in the roughness depressions. Since high drag forces will tend to remove particles that have already adhered to the walls, high transportation velocities will reduce the danger of build-up. For the same reason deposits will start in dead or calm areas of the system; therefore, such designs must be avoided. On the other hand, sudden changes in the direction of flow will cause high-energy impacts of particles with the wall, causing build-up. Finally, friction between particulate solids and pneumatic conveyor walls can result in high electrostatic charges on both partners which depend to a large extent on whether the particles and/or walls are electrically conductive or insulators. System design must take this into consideration.

Some results, published by Möller,[50] shall be reported to illustrate typical features of pneumatic conveying systems. The investigations were carried out during the transportation of particulate matter in a horizontal tube, 58.51 m long and 0.7 m in diameter. The pressures within the system could be determined at seven locations distributed along the measured length of the tube. $\Delta p = p_1 - p_7$ is the total pressure drop in the conveying system.

In Figure 6.45 the pressures at three different locations—1, 2, and 5—are plotted versus time. Since a fan located behind the dust collector at the end of the conveyor generates a slight negative pressure in the filter housing, a small negative pressure can be measured as long as the tube is clean. After a few seconds, however, the pressure p_1 rises and the other locations follow after short delays. Part of the pressure increase is caused by loading the air with particles, but a major portion is due to depositions building up in the tube. When the tube was inspected following runs of 20 and 50 s, respectively, no deposition was found in the first case, but after the longer run deposits had

build up in the feed end portion of the system while the other parts still remained clean. Figure 6.46 depicts the total pressure drop between both ends of the tube. The lower diagram represents results of the same test as shown in Figure 6.45. After about 2.5 min the total pressure drop in the system remained almost constant. This indicates that, at least macroscopically, no further deposition takes place after this time.

The upper diagram in Figure 6.46 represents a completely different behavior. The total pressure drop increases more slowly. This is mostly due to the lower solid/fluid ratio, m_p/m_f (1.58 kg/kg as compared with 40 kg/kg) and the higher velocity (18.65 m/s versus 2.11 m/s). At rather regular time intervals, however, a high-pressure peak had been measured that was first observed at the feed end and propagated in a few seconds to the

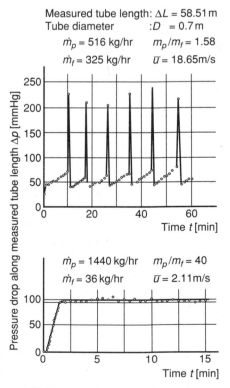

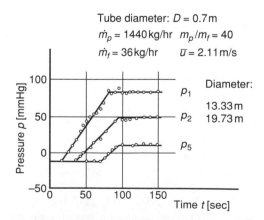

Figure 6.45. Pressure changes at three locations of an experimental pneumatic conveying system during the first 150 s of a test run.[49]

Figure 6.46. Pressure drop along the measured tube length of an experimental pneumatic conveying system as a function of time.[49]

discharge end. This, together with some other observations, indicated that deposits fell off and were carried along, thus increasing momentarily the pressure drop. When the system was opened immediately after such a pressure wave went through, the inner walls were almost completely clean. The pressure drop curve shows further that the adhesion tendency is about constant for a given material and a conveying system operated at uniform conditions.

At high conveying velocities or in vertical tubes deposits build up uniformly. Such depositions shall be called "crusts" in the following. Whereas in the upper part of a horizontal tube, for instance, bonds between particles and walls are stressed by the weight of the deposit, they are strengthened in the lower part of the tube by gravitational forces. Therefore, especially in conveying at low velocities and high solid/fluid ratios in horizontal tubes, a second type of deposit is observed that shall be called "massy" deposit. Figure 6.47 describes schematically the formation of such deposits;[50] they are affected by gravity, grow in the direction of the mass flow, and are

composed of finer particles because these particles exhibit higher adhesion tendency and because of a "sieving" or classification effect taking place in the charge by which finer components move to the bottom layer of the moving mass. The right part of Figure 6.47 is a photograph taken during a model experiment. With the exception of the formation of a crust, all other stages, including a "dune" of freely mobile particles moving over the deposits, can be distinguished clearly. Figure 6.48 is a view into a tube after pneumatically conveying a slightly moist quartz powder (particle size 50 μm) showing the heavy build-up in "crust" and "massy" deposit as well as the remainder of a "dune." Massy deposits can also be caused by the action of other forces such as centrifugal and inertial forces at an elbow.

Another important agglomeration phenomenon, which can be explained by the fact that the separation or dray forces define adhesion, is the controlled deposition yielding a more effective shape of the flow channel. Particles build up preferably in zones without flow or where the direction of flow lines is changed, such as by eddies, for example. A typical example of such deposits is shown in Figure 6.49. On the left, a partial cross-section of a "Möller pump" is presented; these pumps are used for feeding powders into a pneumatic conveying system. Powder and air enter a mixing chamber through a screw conveyor and a nozzle, respectively, and are then forced into the piping of a pneumatic conveying system. The photograph in the right part of Figure 6.49 shows a view (in direction $A-A$) of such a mixing chamber which was opened after conveying zinc oxide. Opposite the nozzle a deposit was built up forming a Venturi-like shape, which defines the most effective flow channel at this point. Similar depositions often can be found in pneumatic conveying systems that were not optimally designed and/or arranged.

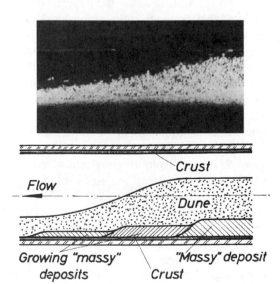

Figure 6.47. Sketch and photograph[50] of a model experiment showing different types of deposits in a horizontal pneumatic conveyor tube.[49]

Storage. Adhesion phenomena cause bridging of particulate solids in hoppers. In the case of relatively coarse materials, bridge formation

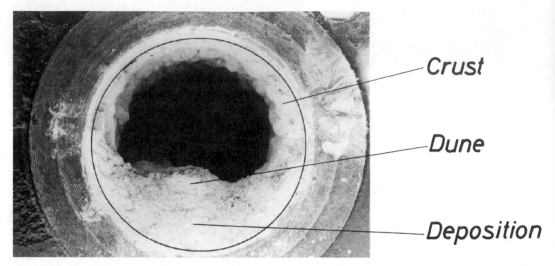

Figure 6.48. View into a tube of a horizontal pneumatic conveyor after conveying a slightly moist, finely divided quartz powder at low velocity.

is caused by building dome structures supported on the inclined walls in the lower, conical part of the bins.[51,52] With decreasing particle size, the participation of true adhesion forces in bridging and agglomeration increases. Bonding mechanisms are molecular forces and adsorption layers or liquid bridges. The latter often play an important role whereby liquid collects at the coordination points by capillary condensation.[54,55] Bridging can totally block the discharge from silos, thus causing severe operating problems. Because adhesion of finely dispersed solids cannot be avoided agglomerates and bridges must be destroyed by special devices. For this purpose, inflatable cushions are mounted in the hoppers or the material is momentarily fluidized by the injection of (pulsed) air jets. In the case of coarser solids, which tend to form domes, it is often sufficient to select a cone with steeper walls (= "mass-flow" design). Small, remaining flow problems due to adhesion can then be overcome by installing vibrators or "hammers" on the outside silo walls.

Undesirable agglomeration is often observed if the particulate material is soluble or if chemical reactions can occur between the particles, particularly in the presence of mois-

ture. These phenomena are very common in the fertilizer industry and are called caking if they occur in bulk masses or bag-set if the contents of the bags solidify.

Caking of fertilizers[49] and other soluble materials has long been and still is a great problem to producers and consumers of such materials. To get an idea about the importance and scale of this problem, three examples shall be presented at the beginning. Figure 6.50 shows the unloading of a shipload of sylvite that was expected to arrive as a free-flowing particulate mass but caked badly during transportation. Owing to the limited room in the shiphold the very costly and time-consuming method of manual unloading had to be chosen. Figure 6.51 shows the recovery of nongranular triple superphosphate from a curing pile which had to be blasted to break the so-called pile-set. This photograph, taken in 1947 by TVA, has historical value for this company because modern granular products, obtained by wanted, controlled agglomeration, no longer cake to such an extent that they require blasting. But, since especially high-nitrogen fertilizers are extremely hygroscopic, they must still be stored in bulk storage facilities with controlled, low humidity to prevent caking.

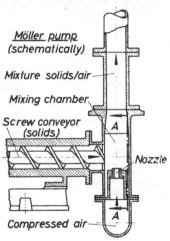

Möller pump
(schematically)

Mixture solids/air

Mixing chamber

Screw conveyor
(solids)

Nozzle

Compressed air

Figure 6.49. Sketch and photograph of a "Möller-pump." The photograph shows Venturi-like deposits after conveying of finely divided zinc oxide.[49]

The third photograph (Fig. 6.52) shows an example of the difficulties confronting the end-user. The granular fertilizer in the left bag displays the desired free-flowing behavior while the same fertilizer is caked in the other bag (bag-set). Even if such a caked mass is mechanically broken up, it will often no longer exhibit the same uniform conditions as an uncaked fertilizer and, hence, will negatively influence its uniform distribution in the field.

Figure 6.50. Manually unloading a shipload of caked sylvite.

Different materials become caked during various storage and handling procedures but caking itself is almost exclusively caused by solid bridges or, more specifically, by chemical reaction and crystallization of dissolved substances. Other binding mechanisms contribute only slightly to caking.

The rate and extent to which caking takes place depend on the moisture content, the particle size, the pressure under which the material is stored (e.g., top or bottom of the pile), the temperature and its variation during storage, as well as the time of storage. The effect of these factors changes with different materials. Figure 6.53 depicts results obtained by Adams and Ross[56] in their "caking bomb." It can be seen that the crushing strength of caked masses rises with increasing moisture content (curves a in Fig. 6.53), caking pressure, and time of storage. The influence of

Figure 6.51. Recovery of nongranular triple superphosphate from a curing pile after blasting to break "pile-set."

temperature and temperature variations depends on the solubility. Figure 6.54 shows four typical temperature–solubility curves. Whereas the solubility of sodium chloride changes little with temperature, this is not true for potassium chloride and potassium nitrate, for example. The latter especially shows a very steep curve. Some salts, such as sodium sulfate, exhibit various temperature-dependent solubility ranges.

If salts or mixtures of different salts, such as fertilizers, for example, contain only a small amount of moisture, they can cake during storage or transport even in airtight bags if they are exposed to changing temperatures. In many cases (see Fig. 6.54) more salt will be dissolved if the temperature increases; this recrystallizes and forms solid bridges between the particles when the temperature drops

Figure 6.52. Granular fertilizer treated with an anti-caking agent (*left*) and untreated control (*right*) showing severe "bag-set."

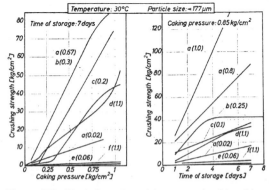

Figure 6.53. Variation of crushing strength with caking pressure (*left*) and time of storage (*right*). (*a*) NaNo$_3$, (*b*) (NH$_4$)$_2$SO$_4$, (*c*) urea, (*d*) KCl, (*e*) (NH$_4$)H$_2$PO$_4$, (*f*) superphosphate. The numbers in brackets indicate the respective moisture contents in percent.

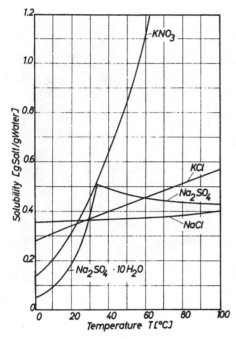

Figure 6.54. Solubility curves of four different salts.

again. Repeated cycling, for instance due to climatic changes or differences in day and night temperatures, tends to reinforce this bonding, causing bag-set.

The crushing strength of caked materials depends also on the number of bridges formed per unit volume and, therefore, decreases with increasing particle size.

In conclusion, it can be stated that the tendency for caking of a fertilizer mixture, for example, will vary with the physical and chemical properties of the components and their proportions in the mixture. It also depends on the method of mixing, the particle size after processing, and the storage conditions to which the finished products are exposed.[56-58]

The answer to what can be done to avoid or at least lessen caking is the same as in all other cases where unwanted adhesion or agglomeration must be prevented: Detect the binding mechanisms involved and the influencing parameters and then try to reduce their effect. In the following some examples shall be discussed briefly.

(a): If (unobjectional) chemical reactions between components of a mixture do occur, these components should be mixed separately until the reaction has taken place. This intermediate product can then be blended with the other components and no longer induces caking. An example for this is any mixture that contains both ammonium sulfate and superphosphate.

(b): An almost trivial precaution is very often the lowering of the moisture content. However, this is not always necessary. Different maximum moisture levels exist that depend on the materials. Figure 6.53 shows that the crushing strength of superphosphate containing 1.1% moisture is very low while the strength of some other materials is much higher although they contain considerably less water.

Silverberg et al.[58] found during microscopic studies of several types of high-analysis fertilizers that caking usually resulted from bonding by the crystals of soluble salts. These crystals often covered the entire granule in the form of a veneer or hull. Figure 6.55 shows typical granular 12–12–12 fertilizer made with an ammonia–urea solution after 3 months of storage. They were illuminated from below and photographed at a higher magnification to reveal details of the crystalline hull. Bonding-phase salts identified during the study were potassium nitrate, ammonium chloride, mono-ammonium phosphate, ammonium nitrate, and an urea–ammonium chloride complex that are all highly soluble. Those salts migrated to the surface of the granule, leaving numerous small cavities within.

This mechanism needs water and drying should, therefore, reduce caking. Figure 6.56 is a photomicrograph taken with crossed Nicol prisms. It shows the difference in hull thickness between undried and predried 12–12–12 grade fertilizer granules. The crushing strength after storage decreased correspondingly.

(c): Curve b (ammonium sulfate) in the right-hand side diagram of Figure 6.53 shows the typical behavior of materials that respond favorably to several days of bin or pile curing

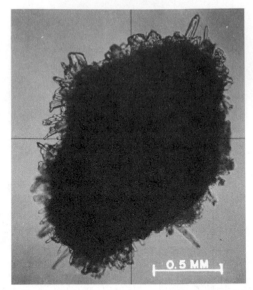

Figure 6.55. Granules of 12–12–12 fertilizer showing typical crystalline hulls of an urea–ammonium chloride complex after storage for 3 months in bags. (*left*) Uncured; (*right*) cured for 7 days prior to bagging.

prior to bagging. Such products cake in a few days to their final strength but the resulting lumps are broken up before the cured materials are finally bagged and put in storage. Curing can even accelerate hull formation as defined by Silverberg et al.[58] owing to the retention of heat and moisture in the pile. In products that respond well to curing, hull formation is apparently almost completed after curing and there is not sufficient additional development of crystals during subsequent storage to cause strong caking.

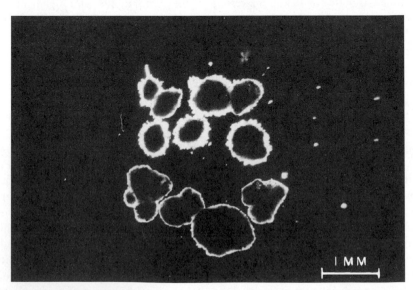

Figure 6.56. Difference in the hull thickness of undried (*top*) and predried (*bottom*) 12–12–12 granular fertilizer made with ammonia–urea solution.

Many products, however, do not improve during this type of curing. Figure 6.55 shows a comparison of uncured (left) and cured (right; 7 days prior to bagging) 12–12–12 fertilizer made from ammonia–urea solution, ammonium sulfate, superphosphate, potassium chloride, and sulfuric acid.[58] Although ammonium sulfate is present, the caking behavior of the other components dominates and both cured and uncured materials show continued growth of the hulls and caking during storage. Another curing method will be described under (e).

(d): The oldest method of "conditioning" fertilizers is the coating with a parting agent.[59, 60] Storage properties are improved after addition of up to 3% of an extremely finely divided particulate solid, such as diatomaceous earth, kaoline, vermiculite, pulverized limestone, magnesium oxide, and a variety of other inexpensive powders. Siverberg and associates' microscopical studies[58] revealed again the fundamental properties of the "conditioner," which are threefold:

1. The powder coating the granule acts as a separator between the individual fertilizer particles and prevents intergrowth of crystals during and after granule formation.
2. The hulls form beneath the coating of conditioner and crystals seldomly project beyond the layer of conditioner.
3. The moisture is distributed uniformly over the surface of the granulae due to the high sorptive capacity of the finely porous coating. Thus, the localized growth of crystals at the coordination points is prevented and the surface hulls are much finer grained, more intergrown, and more densely packed than those covering unconditional products. Such anticaking conditioning agents are usually applied by mixing them with the fertilizer in a rotary tumbler (typically a drum) prior to bagging.

(e): A modern variation of the above-mentioned conditioning process is the coating with surface-active organic chemicals. It was found, however, that not all surfactants improve the physical conditions of mixed fertilizers. Kumagi and Hardesty[62] reported that caking tendencies of mixed fertilizers were decreased by as much as 45% if nonionics were used but increased by as much as 37% with the use of anionics. Where in the process the surface-active agents were added was also found to be of decisive importance.

Typical cationic anticaking agents are fatty amines, for example, "Armoflos."[63–68] These amines, the general formula of which is $R-NH_2$ with R representing C_{16} and C_{18} chains, are believed to attach directly to the fertilizer particles with their surface-active amine group. Then, the fatty, hydrophobic part of the molecule extends outward from the surface, thus preventing hygroscopic products from attracting moisture. This is, of course, true only if a monomolecular layer covers the fertilizer particle and all amine molecules extend their hydrophobic portion outward. Therefore, too much conditioner will cause rather than prevent caking owing to the alternately hydrophobic and hydrophilic properties of additional layers.

This makes a modified curing process advantageous. The molecules of a second molecular layer, if attached, would position themselves with the amine group extending outward. These amine groups are fee to interact with other fertilizer particles, especially the phosphate portion of incompletely coated granules, to form an amine–phosphate salt. Pressure intensifies this effect. The chemical "bridge" is not as strong as the crystallized salt bridge and the "set" can be broken easily. Since, on the other hand, the amine–phosphate bond is stronger than the R–R bond, a more uniformly covered product results from a short bin cure (1 to 2 days) which is unlikely to set or cake again (see Fig. 6.52, left side).

Sometimes a combination of the two types of conditioner is used. An example for this approach is finely divided kaoline treated with a surfactant.[61]

(f): A last method, granulating is today almost obligatory, particularly for mixed fertilizers. Size-enlarged, granular fertilizers offer fewer coordination points per unit volume where solid bridges can develop. If the strength of the bridges is low anyway, as in the case of superphosphate with 1.1% moisture or monoammonium phosphate with 0.06% moisture (see Fig. 6.53), granulating alone is sufficient to prevent severe caking.

Most of the above examples data back quite some time to a period when the fundamentals of unwanted agglomeration in different industries were investigated and means to avoid these phenomena were developed. While this part of size enlargement by agglomeration is often very important, because its effects may result in considerable losses of production and profit, most of the literature deals with the methods and equipment to produce agglomerates with beneficial characteristics. Therefore, it is a most important achievement that recently a book, entitled *Cake Formation in Particulate Systems*,[69] on unwanted adhesion phenomena was published. Griffith, the author, distinguishes four major classes of particulate caking:

- Mechanical caking
- Plastic-flow caking
- Chemical caking
- Electrical caking.

In addition, several subclasses are defined whereby certain properties of components, either pure substances or part(s) of a formulation, can be expected to cause caking under certain conditions.

The chapters of the book describe the above, the chemistry of cake formation, phase behavior and cake formation, and electrically induced cake formation. Considerable emphasis is then given to laboratory techniques and test procedures that need to be considered by laboratories engaged in solving caking problems. Another chapter presents flow schemes to classify caked solids, an approach that is similar to the "old" qualitative analysis flow schematic or modern computer flow charts. On only 22 pages (out of 230 pages) the book then offers typical solutions to caking problems and concludes with a short chapter on induced cake formation, essentially a brief survey of what is called Desired Agglomeration in the context of this publication.

6.3.2.2 Desired Agglomeration

If size enlargement by agglomeration is carried out as a desired process the products resulting from this technology typically exhibit the advantages summarized in Table 6.5.[70]

Another somewhat different listing of benefits which, therefore, contained additional useful information, particularly "examples of application" was presented by C. E. Capes in Part 1 of Chapter 7 of the first edition of this book (Table 6.6).

6.3.3 Methods of Size Enlargement by Agglomeration

A common classification of methods for the size enlargement of particulate matter distinguishes between two types of processes:

- Growth/tumble agglomeration (no external forces)

Table 6.5. Advantages of Agglomerated Products.

1. No or low content of dust; therefore, increased safety during handling of, for example, toxic or explosive materials and, generally, fewer losses which may cause primary or secondary pollution
2. Freely flowing
3. Improved storage and handling characteristics
4. Improved metering and dosing capabilities
5. No segregation of co-agglomerated materials
6. Increased bulk density and lower bulk volume
7. Defined shape
8. Sometimes defined weight of each agglomerate
9. Within limits, porosity or density can be controlled; herewith dispersibility, solubility, reactivity, heat conductivity, etc. can be influenced
10. Improved product appeal
11. Increased sales value

Table 6.6. Benefits of Size Enlargement and Some Representative Applications.

BENEFIT	EXAMPLES OF APPLICATION
1. Production of useful structural forms and shapes	Pressing of intricate shapes in powder metallurgy; manufacture of spheres by planetary rolling
2. Preparation of definite quantity units	Metering, dispensing, and administering of drugs in pharmaceutical tablets
3. Reduced dusting losses	Briquetting of waste fines
4. Creation of uniform, nonsegregating blends of fine materials	Sintering of fines in steel industry
5. Better product appearance	Manufacture of fuel briquettes
6. Prevention of caking and lump formation	Granulation of fertilizers
7. Improvement of flow properties	Granulation of ceramic clay for pressing operations
8. Greater bulk density to improve storage and shipping of particulates	Pelleting of carbon black
9. Reduction of handling hazards with irritating and obnoxious materials	Flaking of caustic
10. Control of solubility	Production of instant food products
11. Control of porosity and surface-to-volume ratio	Pelleting of catalyst supports
12. Increased heat transfer rates	Agglomeration of ores and glass batch for furnace feed
13. Removal of particles from liquids	Pellet flocculation of clays in water using polymeric bridging agents
14. Fractionation of particle mixtures in liquids	Selective oil agglomeration of coal particles from dirt in water

- Pressure/agglomeration (low, medium, and high external forces)

and two techniques:

- Binderless agglomeration
- Agglomeration with the addition of binders.

The mechanism of growth/tumble agglomeration is similar to that of natural agglomeration (Fig. 6.57). Because the particles to be agglomerated are larger, the particle-to-particle adhesion must be enhanced by the addition of binders (mostly water and other liquids) and the collision probability must be increased by

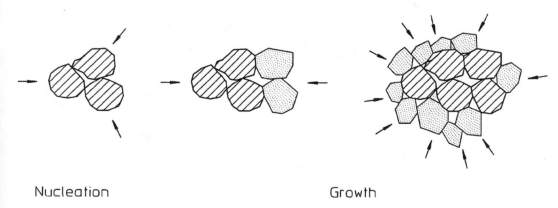

Nucleation   Growth

Coalescence   Layering

Figure 6.57. Major mechanisms of growth tumble agglomeration.

providing a high particle concentration. Such conditions can be obtained in inclined discs, rotating drums, any kind of powder mixers, and fluidized beds (Fig. 6.58). In certain cases, simple tumbling motions such as on the slope of storage piles or on other inclined surfaces are sufficient for the formation of crude agglomerates.[71]

In most instances, growth/tumble agglomeration processes yield first so-called green agglomerates after growing nuclei into larger, nearly spherical aggregates by coalescence and/or layering (Fig. 6.57). These wet agglomerates are temporarily bonded by the effects of surface tension and capillary forces of the liquid binder. While, occasionally, components within the green agglomerate naturally produce permanent bonding, for example, owing to cementitious reactions, in most cases post treatments consisting of all or some of the following processes: drying and heating, cooling, screening, adjustment of product characteristics by crushing and conditioning as well as recirculating undersized material are necessary to obtain permanent and final strength (see right-hand side of Fig. 6.58). The sometimes very large percentage of recycle must be rewetted for agglomeration and needs to pass again through the entire process, which often renders this technology uneconomical.

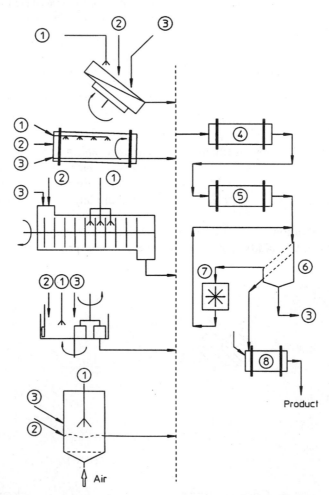

Figure 6.58. Schematic representation of typical equipment for size enlargement by growth tumble agglomeration.

With increasing size and mass of the particles to be agglomerated by growth/tumble methods, the forces trying to separate newly created bonds during agglomerate growth become larger until size enlargement by tumbling is no longer possible. Therefore, depending on the characteristics of the binder, there is a definite limitation to the coarseness of a particle size distribution which is in the range of x_0 between 200 and 300 μm. Considering the definition of x_0, the surface equivalent diameter representing the entire feed particle size distribution, it is, of course, possible to incorporate much larger particles, say up to approx. 1 mm, if a sufficient amount of finer grains is present in the mixture. Fine particles in the distribution will automatically influence x_0 through their large share in the specific surface area (see Table 6.4).

Relatively uniformly shaped and sized agglomerates can be obtained by low- to medium-pressure agglomeration whereby the feed mixture must still be made up of fine particles and binders. The moist, often sticky mass of particulate solids and a liquid binder is extruded through holes in differently shaped screens or perforated dies (Fig. 6.59). Agglomeration and shaping occur by the pressure forcing the material through the holes and by frictional forces during passage of the mass. Depending on the plasticity of the feed mix, short "crumbly," elongated "spaghetti-like," or cylindrical "green extrudates" are produced. In most cases a post treatment (typically drying and cooling) is required to obtain final, permanent strength. For the agglomeration of waste materials for recycling of beneficial use, only some of the medium-pressure "pelleting" equipment, particularly the design with flat die plate (Fig. 6.59, b.2), is applicable. As in the case of high-pressure ram extrusion (see below) this technique is particularly suitable for the agglomeration of elastic materials.

As far as applicability is concerned, high-pressure agglomeration (Fig. 6.60) is the most versatile technique for the size enlargement of particulate matter. If certain characteristics of the feed materials and conditions occurring

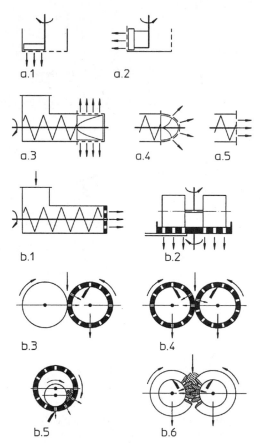

Figure 6.59. Schematic representation of equipment for low (*a*)- and medium (*b*)-pressure agglomeration.

during densification are considered during equipment selection, design, and operation, particulate materials of any kind and size (from nanometers to centimeters) can be successfully processed. Since high-pressure agglomeration is essentially a dry process, there is a limitation in regards to the highest tolerable moisture content of the feed. Typically, the products from high-pressure agglomeration feature high strength immediately after discharge from the equipment. To further increase strength, addition of small amounts of binder or application of post treatment methods are possible.

The mechanism of densification during pressure agglomeration includes as a first step, a forced rearrangment of particles requiring

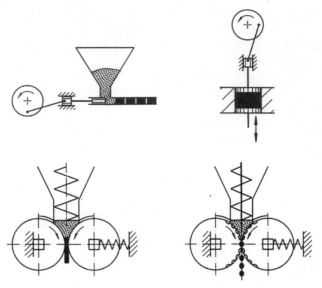

Figure 6.60. Schematic representation of equipment for high-pressure agglomeration.

little pressure followed by a steep pressure rise during which brittle particles break and malleable particles deform plastically (Fig. 6.61). Two important phenomena that limit the speed of compaction and, therefore, capacity of the equipment must be considered: compressed gas (air) in the pores and elastic spring back.

Both cause cracking and weakening or destruction of the products from pressure agglomeration.

Compressed gas can be avoided if densification occurs slowly enough so that all air from the diminishing pore space is able to escape from the particulate mass and equip-

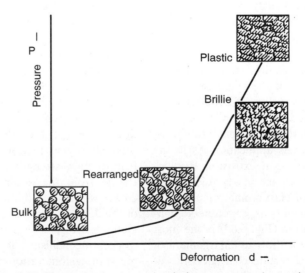

Figure 6.61. The mechanisms occurring during pressure agglomeration.

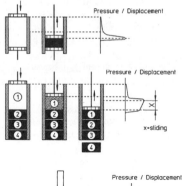

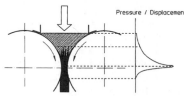

Figure 6.62. Pressure cycles in open die (ram extruder), confined volume (punch and die press), and converging die (roller press) high-pressure agglomeration.

ment. The problem becomes greater with finer particle size and requires special design features of the equipment. The effects of both phenomena, compressed air and elastic deformation, can also be reduced if the maximum pressure is held for some time (dwell time) prior to its release. Figure 6.62 shows that this is possible only with the ram extruder where all briquettes remaining in the extrusion channel are held at a certain pressure and redensified during each stroke. In punch and die presses a short dwell time can be achieved with some special drives whereas no such possibility exists with roller presses.

References

1. H. Rumpf, "The strength of granules and agglomerates," in *Agglomeration*, edited by W. A. Knepper, *Proc. 1st International Symp. Agglomeration*, Philadelphia, PA, John Wiley & Sons, New York, pp. 379–418 (1962).

2. W. Pietsch, "Das Agglomerationsverhalten feiner Teilchen," Staub-Reinhalt. Luft, *27* 1, pp. 20–33 (1967), English edition: "The agglomerative behavior of fine particles," *27* 1, pp. 24–41 (1967).

3. D. Ocepek, *Proceedings of the 2nd European Symposium on "Comminution"* (1966).

4. B. Beke and L. Opoczky, *Proceedings of the 2nd European Symposium on "Comminution"* (1966).

5. B. C. Bradshaw, *J. Chem. Phys. 19*:1057–1059 (1951).

6. G. F. Hüttig, *Dechema Monogr.* Nos. 245–268, pp. 96–115 (1952).

7. G. F. Hüttig, W. Ebersold, and H. Sales, *Radex Rdsch.* pp. 489–493 (1953).

8. B. Beke, *Rev. Matér de Construct.* No. 558, pp. 73–82; No. 559: 115–121 (1962).

9. M. Papadakis, *Rev. Matér. Construct.* No. 542: pp. 295–308 (1960).

10. H. E. Rose and R. M. Sullivan, "The Role of Additives in Milling," in *Ball, Tube and Rod Mills*, Constable, p. 236 (1958).

11. E. R. Dawley, *Pit and Quarry*, p. 7 (1939).

12. C. W. Schweitzer and A. E. Craig, *Ind. Eng. Chem. 32*(6):751–756 (1940).

13. E. R. Dawley, *Cement Lime Manufact. 17*:1–4 (1944).

14. E. von Szantho, *Erzbergbau Metallhüttenwes 2*(12):353–360 (1949).

15. G. Ghigi and L. Rabottino, *Proceedings of the 2nd European Symposium on "Comminution"* (1966).

16. J. A. Hedvall, *Z. Anorg. Allg. Chem., 283*:165–171 (1956).

17. A. Götte and E. Ziegler, *Aachener Bl.*, 5th year of publ., p. 123. (1955). Extract from: E. Ziegler, Dissertation, Aachen Technical College (1955).

18. A. Götte and E. Scherrer, *Aachen Bl.*, 8th year of publ., No. 3, pp. 77–110 (1958). Extract from: E. Scherrer, Dissertation, Aachen Technical College (1958).

19. A. Götte, and W. Wagener, *Achener Bl.*, 11th year of publ., No. 1–2, pp. 53–87 (1961). Extract from: W. Wagener, Dissertation, Aachen Technical College (1960).

20. M. Deckers, Diss. TH Aachen (1963).

21. A. Götte, *Freiberger Forschungsh. A281*:5–29 (1963).

22. W. Batel, *Chemie Ing. Techn.*, 30th year of publ., No. 9, p. 567 and No. 10, p. 651 (1958).

23. H. Börner, *Zerment Kalk Gips*, 14th year of publ., No. 6, pp. 237–253 (1961).

24. W. v.d. Ohe, Ph.D. Thesis, University (TH) Karlsruhe, Germany.

25. K. Schönert, "Method of fine and very fine comminution of materials having brittle behavior," US Patent 4,357,287 (Nov. 2, 1982).

26. J. Priemer, *Progress Reports, VDI Journal*, Series 3, No. 8, pp. 1–104 (1965). Diss. TH Karlsruhe (1964).

27. P. Kunze, Master Thesis at the Institute of Mechanical Process Engineering, TH Karlsruhe (1966).

28. H. Schubert, Contribution to discussion in (43) *Freiberger Forschungsh. A281*:27 (1963).

29. P. Schmidt, *Aufbereitungs Techn.* 7th year of publ., No. 5, pp. 265–273 (1966).

30. O. Lauer, *Measuring the Fineness of Technical Dusts.* Alpine AG, Augsbert, pp. 26–27 (1963).

31. P. Brüninghaus, *Aufbereitungs Techn.* 1st year of publ., No. 1, pp. 53–57 (1960).

32. J. Steinbusch, *Aufbereitungs Techn.* 4th year of publ., No. 11, pp. 502–506 (1963).

33. O. Lauer, *Staub, 18*(10):306–309 (1958).

34. L. Schlebusch, *Aufbereitungs Techn.* 4th year of publ., No. 11, pp. 476–481 (1963).

35. E. Burstlein, *Aufbereitungs Techn.*, 4th year of publ., No. 11, pp. 486–488 (1963).

36. T. W. Hannon and R. Sybrandy, *Aufbereitungs Techn.*, 4th year of publ., No. 11, pp. 482–485 (1963).

37. W. Batel, Research Report of the Ministry of Economics and Transport, North Rhine/Westphalia, No. 262 (1956).

38. E. Muschelknautz, Private communication (1966).

39. W. Kayser, *Proceedings of the 1st European Symposium on "Comminution,"* pp. 563–586 (1962). *Zement Kalk Gips 15*(11):469–478 (1962).

40. H. Schuber and J. Schmidt, *Bergakademie (Freiberg)*, 15th year of publ., No. 12, pp. 850–855 (1963).

41. A. H. M. Andreasen, *Staub* No. 43, pp. 5–9 (1956).

42. W. Batel, *Techniques of Particle-Size Measurement*, Springer-Verlag, New York (1960).

43. G. D. Joglekar and B. R. Marathe, *J. Sci. Ind. Res. 17A*(5):197–203 (1958).

44. VDI 2031. *Determining the Fineness of Technical Dusts.*

45. B. A. Haines, Jr. and A. N. Martin, *J. Pharmacol. Sci. 50*:228–232 (1961).

46. J. J. Fischer, *Chem. Eng.* pp. 107–128 (1960).

47. H. Rumpf, *Chemie Ing. Techn.* No. 6, pp. 317–327 (1953).

48. J. J. Fischer, *Chem. Eng. Progr. 58*(1):66–69 (1962).

49. W. Pietsch, "Adhesion and Agglomeration of Solids During Storage, Flow, and Handling—A Survey," *Trans. ASME J. Eng. Indust. Ser. B 9*(2):435–449 (1969).

50. H. Möller, Ph.D. Thesis, University (TH) Karlsruhe (1964).

51. R. Kvapil, *Aufbereitungs Techn.*, 5th year of publ. No. 3, pp. 138–144 and No. 4, pp. 183–189 (1964).

52. P. Dubach, *Aufbereitungs Techn.*, 3rd year of publ., No. 10, pp. 455–458 (1962).

53. P. Dubach, *Aufbereitungs Techn.*, 6th year of publ., No. 2, pp. 50–56 (1965).

54. J. Higuti and H. Utsugi, *Sci. Rep. (Tôhoku Univ.), 36*(1):27–36 (1952).

55. L. V. Radushkevich, *Izv. Akad. Nauk. SSR. Otdel. Khim. Hank.*, p. 1008 (1952); p. 285 (1958); p. 403 (1958).

56. J. R. Adams and W. H. Ross, (a) *I & E.C. 33*(1):121–127 (1941), (b) *Am. Fertil. 95*(2):5–8, 22–24 (1941).

57. A. L. Whynes and T. P. Dee, *J. Sci. Food Agric.* No. 10, pp. 577–591 (1957).

58. J. Silverberg, J. R. Lehr, and G. Hoffmeister, Jr., *Agric. Food Chem. 6*(6):442–448 (1958).

59. J. O. Hardesty and R. Kumagi, *Agric. Chem. 7*(2):38–39, 115, 117, 119 (1952); ibid., (3):55, 125, 127, 129.

60. J. R. Wilson, J. C. Hillyer, V. C. Vives, and R. E. Reusser, *Agric. Chem.* pp. 42, 44, 45, 116, 117 (Sept. 1962).

61. C. R. Moebus, *Proceedings 14th Annual Meeting Fertilizer Industry Round Table* (1964).

62. R. Kumagi and J. O. Hardesty, *Agric. Food Chem. 3*(1):34–38 (1955).

63. R. E. Baarson, M. R. McCorkle, and D. T. Ohlsen, Anticaking of Commercial Pelletized Fertilizers and Various Fertilizer Components with Fatty Chemicals, unpublished manuscript (1956).

64. R. E. Baarson, M. R. McCorkle, and J. R. Parks, Anticaking of Hygroscopic Salts and Multicomponent Fertilizers with Fatty Conditioning Agents, unpublished manuscript.

65. S. S. Chandler, R. E. Baarson, and J. R. Parks, Conditioning Granular Fertilizers and Fertilizer Salts with Fatty Amine Type Chemicals, unpublished manuscript (1961).

66. W. G. Sykes, S. Myers, J. R. Parks, and S. S. Chandler, Proceedings 48th National Meeting AIChE, Denver (1962).

67. S. S. Chandler, J. R. Parks, and M. R. McCorkle, Paper presented at the 145th National Meeting ASE, New York (1963).

68. J. R. Parks and J. Granok, *Farm Chem.* pp. 51, 54, 55, 57, 58, 60, 62 (Oct. 1967).

69. E. U. Griffith, *Cake Formation in Particulate Systems*. VCH, New York (1991).

70. W. Pietsch, *Size Enlargement by Agglomeration*. John Wiley & Sons/Salle + Sauerlander, Chichester, UK/Aarau, Switzerland (1991).

71. P. D. Chamberlin, "Agglomeration: Cheap Insurance for Good Recovery When Heap-Leaching Gold and Silver," *Mining Eng. 12*:1105–1109 (1986).

6.4 GROWTH / TUMBLE AGGLOMERATION METHODS— AGITATION METHODS

6.4.1 Introduction[1]

Growth/tumble agglomeration is the "most natural" of all size enlargement processes. As solid particles move in relation to each other in the relatively dense bed of a rotating or otherwise actuated containment of some kind or in a suspension with low solids density, particles will occasionally collide and, if the attraction or adhesion forces are high enough, coalesce. Theoretically, no specific piece of equipment is necessary for this phenomenon to occur; as long as the solid particles are kept in irregular, stochastic motion, the probability for collision and coalescence exists.

If in addition to this primary condition for agglomeration, the binding force remains strong enough to withstand the separating effects of all field forces and does not disappear with time without some other binding mechanism taking over, the "seed agglomerate" will survive and eventually collide with other single particles or agglomerates. At each instance of collision the above bonding criteria will be tested leading to either further growth, indifference, that is, the colliding partners will separate again and remain single, or conceivably destruction of weaker agglomerates.

For these adhesion criteria to test positive, the mass of adhering particles must be low and the specific surface high. This is equivalent to the requirement that the size of agglomerating particles must be small, typically in a range below approx. 100 to 200 μm. Micron and submicron particles (approx. <5 to 10 μm) will adhere to form an agglomerate even if they are dry. van der Waals forces are high enough to cause coalescence. Agglomeration of larger particles necessitates the addition of binders.

Drawbacks of all tumble agglomeration methods are the limitation to small dimensions of the particles forming the agglomerate and that, in most cases, only temporarily bonded conglomerates are formed. A curing step must follow to obtain permanent bonding which often also results in considerable strengthening of the agglomerate. In the green stage, the main binding mechanisms are bridges of freely movable liquids and capillary pressure at the surface of particle aggregates filled with liquid as well as adhesion caused by viscous binders; in the case of very small particles, field forces such as van der Waals, electrostatic, or magnetic attraction may also participate. After curing, agglomerate strength is achieved by solid bridges resulting from sintering, chemical reaction, partial melting and so-lidification, or crystallization of dissolved substances.

Tumble agglomeration equipment can handle large volumes of material effectively if the above criteria are fulfilled. The apparatus is simple and the design is unsophisticated. The expensive part of tumble agglomeration is normally the curing step of the process, which also contributes high operating costs. However, if very large amounts of solids must be agglomerated and the fine particulate form is also required for other reasons, for example, the concentration of valuable constituents of ores, tumble agglomeration is a preferred technology. In those cases the main binder is normally water. At capacities exceeding one million tons per year the curing facilities also become more economical and methods for, for example, heat recuperation to reduce operating costs are feasible.

Other reasons for the application of tumble agglomeration, even at small capacities, may be the high porosity of the agglomerates with other attendant beneficial characteristics such as large surface area (e.g., for catalyst carriers) and easy solubility [e.g., for food (drink) and pharmaceutical products]. These advantages may be so valuable that additional grinding costs to obtain the necessary small particle size for agglomeration will be accepted and high operating costs can be absorbed. In these cases even the agglomeration liquids (binders for the formation of green agglomerates) are sometimes so valuable that they are condensed from the dryer off-gas and recirculated.

6.4.2 Definitions[1]

With the exception of very few applications where particles are so small that they naturally agglomerate in the dry state, tumble agglomeration methods utilize binders. Even in those materials that contain the binder component inherently, this constituent of the bulk mass to be agglomerated is so obvious that one cannot classify such processes as binderless.

In this section, only those methods will be discussed in which discrete solid particles, agglomerates, and fragments of agglomerates attach themselves to each other. Other processes, such as spray dryer/granulators, use almost identical equipment as, for example, fluid bed agglomerators; however, since they utilize different growth mechanisms, they will be covered in Section 6.6.

In tumble agglomeration distinct steps can be defined in which:

1. First, green agglomerates are formed from solid particles and binder.
2. Second, green agglomerates are cured.
3. Third, if necessary, the cured agglomerates are sized (undersized material is recirculated and oversized agglomerates are crushed and rescreened or recirculated).
4. Fourth, if desired, post treatment takes place, such as the application of anticaking agents, spheronizing, etc.

Steps 3 and 4 may sometimes move in front of step 2, particularly if the post treatment involves spheronizing.

In a broad sense, equipment for tumble agglomeration itself can be divided into:

1. Apparatus producing movement of a densely dispersed bulk mass of particulate solids—dense phase agglomeration.
2. Apparatus keeping solid particulate matter suspended and loosely dispersed in a suitable fluid—suspended solids agglomeration.

In both cases, binder is sprayed into the turbulently agitated bed of particles.

If solid particles are suspended in a liquid, agglomerates may be formed in apparatus with appropriate conditions after adding a second, immiscible bridging (binder) liquid.

6.4.3. The Mechanisms of Agglomeration by Coalescence

It was previously mentioned that the basic adhesion criterion of tumble agglomeration is: two solid entities colliding with one another

coalesce and the resulting bond is stronger than the combined effects of all field forces. This principal process continues, causing size enlargement by agglomerate growth. However, as it proceeds, somewhat more complicated mechanisms evolve. Figures 6.63[2] and 6.64[3] represent almost identical sketches of the different processes. While Figure 6.63 is the more instructive presentation defining nucleation, random coalescence, abrasion transfer, as well as crushing and layering (preferential coalescence), Figure 6.64 distinguishes between size enlargement and size reduction phenomena, both of which take place simultaneously.

Nucleation, the production of microagglomerates, or, in general technical terms, of seeds, occurs when primary particles adhere to form a conglomerate. As long as more primary particles are available they tend to either adhere to each other forming more nuclei (seeds) or attach themselves to larger agglomerates. As the mass of agglomerates increases they may

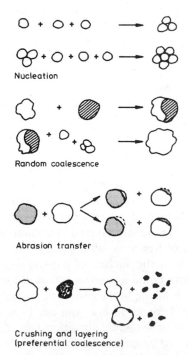

Nucleation

Random coalescence

Abrasion transfer

Crushing and layering
(preferential coalescence)

Figure 6.63. Diagram explaining the different processes taking place during tumble agglomeration.[2]

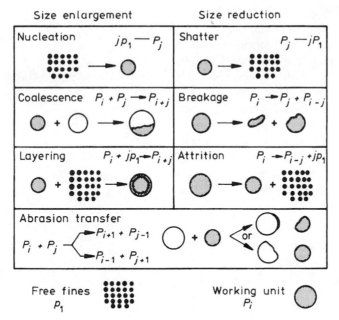

Figure 6.64. Formal representation of mechanisms of size change in size enlargement by agglomerate growth.[3]

break apart at structurally weaker areas or as a result of the force of impact. Abrasion will also take place resulting in newly liberated primary particles or small conglomerates which then try to attach themselves to entities offering better binding properties.

Depending on the density of the tumbling mass and the type of equipment causing agitation of the bed, the growth phenomena will differ.

6.4.3.1 Dense Phase Agglomeration

Agglomeration (or Balling, Granulating) Drum. Fresh material and recycle (containing "seeds") are fed together into the feed end of the drum (Fig. 6.65). Owing to natural segregation, separation takes place almost instantly whereby fines concentrate near the bottom of the kidney-shaped cross-section of the bed and the coarsest conglomerates travel near the surface. Binder (water) is sprayed onto the bed, wetting primarily the coarse particles which then pick up fines (fresh feed or fragments) as they travel through the moving bed. Since this growth takes place in "the

depth" of the bed with a certain overburden pressure and shear forces acting, only the strongest bonds will survive. Binder may be sprayed at the entire length of the drum or only during the first one third or two thirds area. While in the first case growth takes place along the entire length of the drum and oversized, wet, and relatively loose agglomerates may discharge, during the second alternative, growth toward the discharge end is limited owing to the lack of binder. In this case, an equilibrium between size enlargement and size reduction with secondary growth (bonding of fragments) occurs. The discharging material

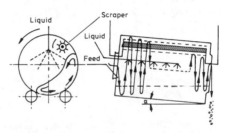

Figure 6.65. Sketch depicting the operating principle of balling drums.

lacks the oversized agglomerates and tends to be more uniform, drier, and stronger.

Since primary bonding during growth is caused by the formation of bridges of a freely moving liquid, still another mechanism may take place in a tumbling dense bed (Fig. 6.66[4]). A particle that has attached itself to the surface of an agglomerate can be moved under the influence of external forces (mostly shear) in the bed into a more favorable, permanent position without losing its primary bond. As a result of this mechanism agglomerates become stronger and denser.

Agglomeration (or Balling, Granulating) Disc (or Pan). Although similar to the balling drum, the pan agglomeration features distinct differences in growth phenomena and control equipment. The same segregation of agglomerate sizes occurs as described above in the depth of the bed. However, if, in addition, a top view of the pan is considered (Fig. 6.67) one can see that owing to the wedge-shaped configuration of the bed large pellets travel at the top near the edge of the pan, ready for discharge, while toward the center of the disc, progressively smaller agglomerate sizes are exposed. This offers an exceptional possibility for control. Since, at least in the shallow pan, feed that normally does not include recycle is also added from the top, the locations of binder

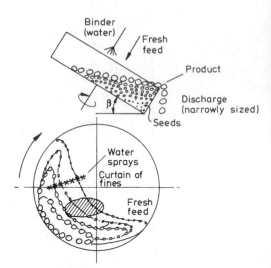

Figure 6.67. Schematic representation of the operating principle of the balling pan.

and material addition determine the overall growth mechanism.

In most cases the agglomerates are wetted first and the particulate feed is added "downstream." Depending on the relative amounts of liquid and solids and their position in relation to the pattern of movement either most of the material is picked up by the growing agglomerates with only a small amount available for the formation of new seeds or new seeds are preferentially produced with most of the growth taking place in the region of smaller agglomerates. In the first case, spray and feed are located closer to the rim of the disc, while in the second case their position is moved toward the pan center. In either case, owing to the fact that after wetting and dusting the charge, pellets move and interact without the addition of new binder and feed, rounding by means of abrasion transfer takes place but overall growth is caused by layering.

In the pan, size and strength of the agglomerates can be influenced to a certain extent by the height of the rim. A deep charge tends to produce rounder and stronger pellets. Because seed formation and early growth always happen in the lower layers it is sometimes difficult to produce enough seeds by the conventional feeding arrangement from the top without

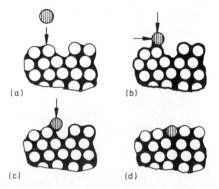

Figure 6.66. Conceptual model of incorporation of small feed particles into the surface of wet agglomerates.[4]

risking massive build-up on the pan bottom. In such cases, feeding from the bottom[5] may be advantageous.

Mixer–Agglomerators. The mechanisms of agglomeration in mixer–agglomerators is very similar to those in the drum. While industrial drum and disc agglomerators always operate continuously, mixer–agglomerators may be batch or continuous. The mechanisms are different in both cases.

Batch Mixer–Agglomerators. This type of equipment is mostly used for relatively low-capacity manufacturing in, for example, pharmaceutical applications. If batch mixer-agglomerators continue to operate for an indefinite time, a uniform agglomerate size results from the equilibrium between size enlargement and disintegration which depend largely on the amount of binder (liquid) present and/or added to the batch. More binder generally yields larger agglomerates.

In most cases the time required to reach equilibrium is too long for an economic operation and, unless very dry and relatively loose agglomerates are made, the resulting final agglomerate sizes are too large. Therefore, and in order to be able to better control agglomerate size and density, most mixer-agglomerators are equipped with mechanical disintegration tools that are called turbulizers, intensifiers, choppers, etc. Their purpose is to reduce the size of agglomerates in a controlled fashion whereupon the fragments reagglomerate with still available and freshly produced fines.

The mode of operation to achieve and control agglomeration is characterized by several consecutive steps, for example:

1. Filling
2. Mixing
3. Spraying of binder and agglomeration
4. Agglomeration without additional binder
5. "Chopping"
6. Spraying of additional binder and agglomeration

7. Repeat(s) of 4, 5, and 6
8. Discharge.

The mechanism of agglomeration is determined by alternating growth and disintegration steps whereby the final size and density can be controlled, within limits, by the time and relative duration of the individual steps.

Continuous Mixer–Agglomerators. The control mechanisms for size and density described above can be applied to only a very limited extent in continuous mixer–agglomerators. In most cases, so-called "intensive" mixers are being applied and the equipment is subdivided into different zones, that is; mixing, binder addition, and agglomeration. Turbulizers or choppers may sometimes also reduce agglomerate sizes in the second half of the mixer and provide the possibility of reagglomeration with attendant increase in density.

As compared with the product from batch mixer–agglomerators, granular products from continuously operating equipment feature wider particle size distributions, which may require screening and recirculation, and less defined, normally lower apparent density.

6.4.3.2 Agglomeration of Suspended Solids (Low Density)

Suspended solids grow according to the same basic principles. Binder is sprayed into a fluidized bed of solids and fresh feed is added simultaneously or an immiscible bridging liquid is added to a liquid suspension of solid particles. However, because much less intensive interaction occurs between agglomerates in the loose bed or liquid suspension, abrasion transfer, crushing, and layering are not as pronounced. Therefore, the porosity of the agglomerates is higher and, since a fluidized bed or a suspension of solid particles in liquid can be maintained only with relatively small and narrowly distributed particles, the capability of this type of equipment to produce larger agglomerates is very limited, particularly if nucleation and growth occur in one apparatus.

This can be overcome only if, for example, a series of fluidized beds, each feeding the other in a cascade fashion, is used. Also agglomerates produced by another method are sometimes coated by adding a layer of fresh material in fluidized beds; this allows handling of larger agglomerates.

6.4.4 Balling Drums[1]

6.4.4.1 General

Balling drums represent the most simple type of equipment for growth agglomeration by tumbling. They are typically used in industries processing large amounts of bulk solids where in the relatively crude and rough environment unsophisticated machinery performs best.

6.4.4.2 Equipment

Agglomeration and, respectively, balling drums or drum granulators are most widely used in the iron ore and fertilizer industries. The equipment consists normally of a cylindrical steel tube with a slight (typically up to 10° from the horizontal) declination toward the discharge end (Fig. 6.68). Retaining rings are often fitted to the feed and discharge ends of the drum to avoid spill-back and, respectively, increase the depth of material and/or its residence time.

Drums are often lined with cement or expanded metal to encourage build-up of material as an "autogeneous" wear liner. To control its thickness, different designs of scrapers are employed. Depending on requirements for "smoothness" and uniformity of the build-up, the scrapers may be oscillating and, therefore, are often separately driven.

In iron ore balling, where the pellets must be deposited uniformly on, for example, a travelling grate and/or to effectively feed the screen separating under- and oversized material, the discharge end of the drum is sometimes executed as a spiral.

6.4.4.3 Sizing of Balling Drums

Even though balling drums were the first large-scale industrial equipment used for tumble agglomeration, there is comparatively little information available in the literature on their sizing. The apparatus was actually developed (in spite of earlier patents) from batch type rotating drum mixers or, in the iron ore industry, from the mixing drum used in sinter plants[6] and adapted to its new function. Because of its simplicity, sizing is usually based on tests, during which drum inclination, hold-up, rpm, and liquid requirement are experimentally determined. Scale-up is achieved by using know-how from existing installations which is available from the equipment manufacturers for different materials.

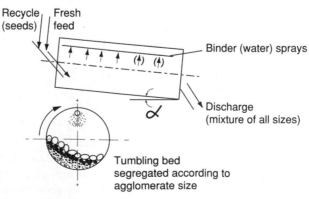

Figure 6.68. Schematic representation of a balling drum.

For actual applications the drum speed is normally kept between 25% and 40% of the critical speed which can be calculated[7] from:

$$n_{crit} = (30/\pi)\sqrt{2g/D} \text{ or } n_{crit} = const./\sqrt{D} \quad (6.27)$$

From throughput, C, and total mass in the drum, m, the average residence time t_r can be calculated:

$$t_r = m/C \quad (6.28)$$

Sommer and Herrman[8] developed mathematical relationships for estimating sizing parameters based on a simple assumption: Since the power N required to operate a drum depends on its dimensions and hold-up (mass in the drum) these authors assumed that its calculation is possible by applying the theory available for tube mills.[9] For derivation of an equation, the entire mass of material in the drum is imagined to be concentrated in its center of gravity, S (Fig. 6.69). The torque M_d necessary to keep this static "Ersatz"-mass in its excentric position is:

$$M_d = m \cdot g \cdot \cos \alpha \cdot a \quad (6.29)$$

where α is the angle of inclination of the drum's axis against the horizontal and a is the distance of the center of gravity from the vertical center line of the drum (Fig 6.69). Similarly to a parameter used for tube mills, the lifting coefficient θ is defined as

$$\theta = a(2/D) \quad (6.30)$$

If the drum rotates with circumferential speed ω or, respectively, rotational speed n, the power input N (without losses caused by motor, gear, and bearings) is:

$$N = M_d \cdot \omega = M_d \cdot 2\pi n \quad (6.31)$$

or, with Eqs. (6.29) and (6.30):

$$N = \pi \cdot \theta \cdot g \cdot m \cdot D \cdot n \cdot \cos \alpha \quad (6.32)$$

and with Eq. (6.28):

$$N = \pi \cdot \theta \cdot g \cdot C \cdot t_r \cdot D \cdot n \cdot \cos \alpha \quad (6.33)$$

The lifting coefficient θ can be estimated from results obtained for ball charges in tube mills (Fig. 6.70[10]). For the typical granulation drum loadings of $\varphi = 0.1$ to 0.3, θ is approximately constant.

Sommer and Herrman[8] also developed a model for the final size of agglomerates by assuming that the length of the bed surface, where most of the growth takes place, characterizes this parameter. The total rolling distance s_r can be estimated by:

$$s_r \sim \theta_0(\varphi) \cdot t_r \cdot n \cdot D \quad (6.34)$$

During scale-up, this characteristic must remain constant. Also, if the drum loading changes within typical limits ($\varphi = 0.1$ to 0.3), the bed surface changes only very little while the relative amount of agglomerates travelling on the surface increases inversely proportional to the drum loading φ; thus:

$$s_r \sim (1/\varphi) \cdot t_r \cdot n \cdot D$$
$$= constant \quad or \quad \varphi \sim t_r \cdot n \cdot D \quad (6.35)$$

If, as for tube mills,[9] the Froude number is kept constant during scale-up, that is, $n \sim 1/\sqrt{D}$, and one assumes that the residence time, t_r, is also kept constant, the drum loading would change according to $\varphi \sim \sqrt{D}$ [Eq. (6.35)].

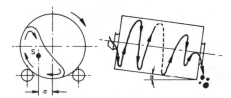

Figure 6.69. Diagram depicting the assumption for calculating the power requirement of drum granulators.[8]

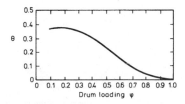

Figure 6.70. Lifting coefficient θ for the balls in a tube mill as a function of drum loading φ.[10]

This means, however, that an unwanted densification of agglomerates occurs in the deeper bed. Therefore, the drum loading is normally kept constant which requires a reduction of the residence time according to:

$$t_r \sim 1/\sqrt{D} \qquad (6.36)$$

Another possibility to scale up and keep t_r constant exists by adjusting the rotational speed and keeping the peripheral speed constant, that is:

$$n \cdot D = \text{constant} \quad \text{or} \quad n \sim 1/D \quad (6.37)$$

Then, according to Eq. (6.35) the drum loading φ is also constant.

The two operating conditions:

$n \sim 1/\sqrt{D}$, that is, constant Froude number

and

$n \sim 1/D$, that is, constant peripheral speed

are upper and lower limits. In reality, care must be exercised to guarantee a rolling movement of the bed. Figure 6.71 shows that if the rotational speed is too low the charge is sliding and if the speed is too high tumbling occurs.[6] Both conditions must be avoided.

Since, therefore, φ must be kept constant, Eq. (6.35) also determines that the term $t_r \cdot n \cdot D$ is also constant. From Eq. (6.33) follows:

$$N \sim C \quad \text{or} \quad N/C = \text{constant} \quad (6.38)$$

That means that the mass related specific energy is constant during scale-up.

The capacities C depend on whether the Froude number or the peripheral speed are kept constant. They are:

$$C \sim D^{3.5} \qquad \text{for constant Froude number} \quad (6.39)$$

$$C \sim D^3 \qquad \text{for constant peripheral speed} \quad (6.40)$$

Mathematical relationships can only provide some guidance to sizing and scale-up. Operation and capacity depend very much on the properties of the material to be pelletized and, potentially, the binders to be used. Laboratory tests always must be carried out to determine balling characteristics and basic equipment data from which scale-up can take place.

In contrast to the angle of, for example, balling discs, the inclination of drums is rather small and has almost no influence on power requirement ($\cos \alpha \approx 1$) and agglomeration behavior. It only serves to provide the required axial transport. As a result of the rather undefined movement of the charge a size classification does not take place and the discharging agglomerated mass features a rather wide particle size distribution.

As alternatives to the closed loop balling circuit there have been several proposals to achieve classification in the drum by various internal designs. For example, a multiple-cone drum pelletizer was described.[11] The disadvantage of the "classic" rotary drum granulator in regard to lacking classification also can be overcome by adopting an upward slope of the axis toward the discharge end ("Dela" drum[12]). Figure 6.72 is a sketch of the drum.

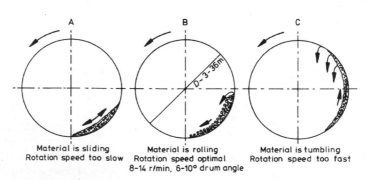

Figure 6.71. Sketches depicting different patterns of charge motion in balling drums.[6]

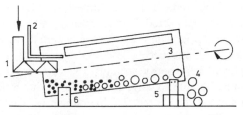

Figure 6.72. Sketch of the "Dela" drum.[12]

6.4.4.4 Balling Drum Circuits

Figure 6.73 depicts schematically a balling drum circuit for iron ore pelletizing in its simplest form.[13] It shows all major components. Feed material (concentrate), additives (e.g., limestone), binder (e.g., bentonite), and recycle are combined on a feed belt. The first three components are metered such that a constant relationship is obtained and the total amount including recycle is made up to 100% of drum capacity. The components are often premixed or somewhat homogenized with a simple "fluffer" mounted on the belt. Liquid binder is added to the tumbling mass in the first part of the drum by sprays and wall build-

up is controlled along the entire length by means of a cutter bar (scraper).

The drum is essentially an inclined, cylindrical shell with a length-to-diameter ratio of 2.5 to 3.5, a retaining ring on the feed end, and a spiral discharge to obtain uniform distribution on the screen and optimal separation. The drum dimensions are adjusted for the required throughput rate. The slope is normally very small, for example, 6° to the horizontal, and, in most cases, adjustable.

Because in the normal drum that slopes down toward the discharge end no classification occurs, the output must be screened to remove undersized material and, sometimes, oversized agglomerates. Off-specification material is recirculated.

In an ideal case it would be possible to operate the drum such that the time for agglomerate growth to the desired size is identical with the residence time. This is normally not achieved, mostly because the drum does not act as a classifier. Nevertheless, it is most important to adjust the drum speed correctly (see Fig. 6.71).

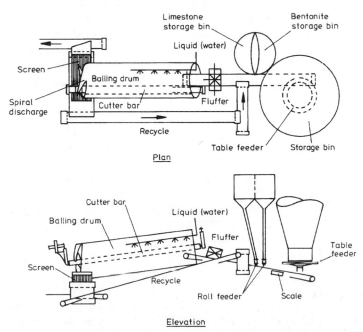

Figure 6.73. Diagram of a typical balling drum circuit for iron ore pelletizing.[13]

Drum rotation is characterized by three parameters: the speed of rotation, the depth of material held in the drum, and the time required for the desired agglomerate growth. Since part of the feed material is used for nucleation and seed growth the drum would become inordinately long if balling to the final agglomerate size would be attempted in one pass. Also, because of lacking classification, the agglomerate size distribution would be too wide. Therefore, balling drums are operated in closed circuit and ball growth occurs in more than one pass through the drum. Recycle rates are between 100% and more than 400% of the fresh feed depending on operating conditions.

Sufficient nuclei must be produced in the drum at all times to replace the pellets that are removed from the circuit, and the growth rate per pass must be such that the required production capacity of green balls is consistently maintained. Therefore, the rate of pellet production must be stabilized and the balance between material in the drum and recycle rate must be kept constant. Nevertheless, balling drum circuits tend to surging.

The green balls produced in drums are usually rather weak. They require gentle handling and curing (in most cases drying and/or sintering) to reach final strength. A trommel screen may be an integral part of the discharge end of the drum.[14]

More information on balling drums and circuits, and their design and operation can be found in the literature.[6, 13]

6.4.5 Balling Discs[1]

6.4.5.1 General

Normally, the balling disc is a simple, inclined, and shallow dish that, owing to the pattern of material motion, features a distinctive classification effect whereby only the largest pellets discharge over the rim (Fig. 6.67). To achieve special effects, modified pan designs are available.

6.4.5.2 Equipment

A typical shallow pan balling disc consists of a heavy, disc-like steel bottom to which, on one,

the lower side, a shaft mounted in roller bearings is connected in the center and, on the other, upper side, a low rim is fastened around the circumference. Disc and drive are supported in a heavy structural steel frame that must be able to carry the weight of the equipment and its charge. The pan angle is variable, normally between 40° and 60° from the horizontal. Diameter, rim height, speed, and inclination determine the capacity of a disc (Fig. 6.74).

Smaller balling discs with diameters of less than 3 m are usually driven directly with variable-speed drives; larger discs feature motor, gear reducer, and pinion/ring–gear arrangements.

As in the case of balling drums it is also necessary to control the build-up on bottom and rim. To obtain an "autogeneous wear liner," the inside is sometimes covered with expended metal to encourage build-up. In any case, a series of stationary and/or movable plows maintains a uniform layer. Depending on position of the scraper(s) and speed of the disc, operation may be such that agglomerates impact the scraper, thereby selectively destroying weak ones and further densifying/strengthening already strong pellets (Fig. 6.75). In this case the rotational speed is $n \geq n_{\mathrm{crit}}$.

As mentioned previously, location and means of feeding solids into the rotating pan together with number, distribution, and spray pattern of the liquid additions determine the performance of the disc. Therefore, feeders and spray nozzle arrangements are parts of the equipment. Their location is adjustable. Since, in most cases, dry fine powders are fed into the apparatus, dust covers are provided and are completely or partially closed during operation. To avoid selective agglomeration, which could be promoted by the pattern of movement of the charge, feeds consisting of different components, particularly if dry binders are added, should be premixed prior to the pelletizer.

While, with only few exceptions, regular balling drums (i.e., sloped downwards in direction of material flow) must be operated in a closed loop to obtain acceptably narrow prod-

(a)

(b)

Figure 6.74. Photographs of a typical shallow pan balling disc.[1]

uct particle size distributions, the balling disc is typically installed in an open circuit by selecting the fixed parameters, that is, diameter and rim height, and variation of angle of inclination, rpm, as well as feed and spray means and locations, closely sized pellets between 0.5 and 25 mm may be obtained. Of this range "micro-pelletization" (i.e., agglomerate sizes 1 to 2 mm) is mostly used in the chemical industry (e.g., agglomeration of detergents), fertilizer granulation yields products with 1 to 3 (or up to 5) mm, iron ore pelletization typically features balls with approx. 1/2 inch diameter (12 to 15 mm), and for cement raw materials

as well as special applications spherical agglomerates with dimensions of up to 1 inch (25 mm) are manufactured.

6.4.5.3 Sizing of Balling Discs

The more recent development of the balling disc and its unique pattern of movement, as well as, the possibility to control agglomerate size has triggered the interest of many researchers and scientists to investigate the balling disc theoretically and experimentally.

As with balling drums, it is most important to select the correct rotation speed of the granulation pan. It is defined as percentage of critical speed which can be calculated by (see Fig. 6.76):

$$n_{crit} = 42.3 \cdot \sqrt{\sin \beta / D}$$

or

$$n_{crit} = 42.3 \cdot \sqrt{\cos \alpha / D}$$

(6.41)

In this formula n is in rpm if the pan diagram D is introduced in m. Figure 6.77 depicts patterns of charge movement at different rotational speeds.[4] Normally,[15] balling discs are operated at approx. $n = 0.75 n_{crit}$. Klatt[16] estimated the rotational speed of granulation pans

Scraper

Direction of rotation (15 r/min)

Ore fed into this area

Water fed into this area

Figure 6.75. Schematic representation of a pan granulator operating at $n > n_{crit}$ equipped with impact plate scraper.

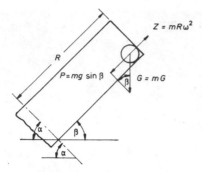

Figure 6.76. Sketch showing conditions for the determination of the critical speed of balling pans.

as a function of diameter by evaluating experimental data (Fig. 6.78) and obtained:

$$n \approx 22.5/\sqrt{D} \qquad (6.42)$$

At $\beta = 50°$ this equation represents $n \sim 0.7 n_{crit}$.

Many researchers found that, for a given material, pan design, and inclination, optimization indicates only a very small opportunity to vary speed. Therefore, it can be considered as an (adjustable) constant for given conditions. Figure 6.78 and Eq. (6.42) are directly valid for cement raw material but also can be used as a general approximation.

In addition to the angle of inclination of the pan bottom, β, the angle of repose of the material, γ, plays an important role in the

development of the pattern of charge motion in the balling disc. Bhrany[17] correlated the tilt angle of pans to the horizontal, β, and the rim height, h, with the (dynamic) angle of repose, γ, of the material (ore) to be granulated. As shown in Figure 6.79 the tilt angle β must be larger than the angle of repose γ.

The angle of repose is a measure of the "critical tilt angle." As long as the tilt angle β is smaller than the angle of repose γ the material remains stationary on the pan bottom. Only if $\beta > \gamma$ the typical pattern of charge motion develops that results in controlled pellet growth.

In practice, simple approximations are used in most cases as shown in Figures 6.78 and 6.80 and defined by the following equations:[4]

$$n \approx 22.5/\sqrt{D} \qquad (6.42)$$

$$h \approx 0.2D \qquad (6.43)$$

While Eq. (6.42) represents the data points well (Fig. 6.78), it seems (Fig. 6.80) that the rim height of larger granulating discs tends to deviate slightly from Eq. (6.43).

Equation (6.32) derived for the balling drum can also be applied for the granulation disc. In contrast to drums where the loading is very small ($\varphi_1 = 0.1$ to 0.3) Bhrany[17] found that

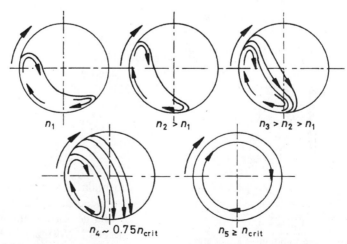

Figure 6.77. Movement of the charge in pelletizing discs at different rotational speeds.

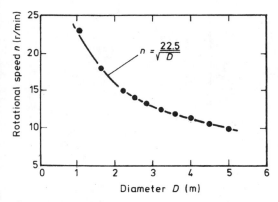

Figure 6.78. Rotational speed n of pelletizing discs as a function of pan diameter D.[4] Data points according to Klatt.[16]

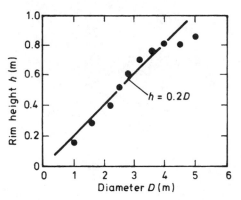

Figure 6.80. Relationship between rim height h and pan diameter D.[4] Data points according to Klatt.[16]

the mass in a pan is proportional to the square of its diameter:

$$m \sim D^2 \text{ with } h \sim D \qquad (6.44)$$

If ρ_w determines the density of the wetted mass in the pan, Eq. (6.32) can be rewritten as:

$$\varphi = m/\rho_w(\pi/4) \cdot D^2 \cdot h \qquad (6.45)$$

and with Eq. (6.44) follows:

$$\varphi \sim 1/D \qquad (6.46)$$

This means that with increasing pan diameter the loading, that is, the relative mass in a balling disc, decreases.

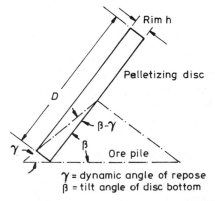

Figure 6.79. Relationship between tilt angle β, rim height h, and angle of repose γ of the material to be granulated.[17]

Test results[15,16] showed that, as in the case of drum granulators, the specific energy N/C must be kept constant during scale-up of balling discs:

$$N/C = \text{constant} \qquad (6.38)$$

The general requirement is also valid for other rotating apparatus, such as mixer agglomerators.

Specifically, for the balling disc and cement raw material Klatt[16] determined that optimum operating conditions are found only for a narrow band of parameters and concluded that the capacity, $C(t/h)$, may be calculated from the pan diameter, D(m), by:

$$C = 1.5 \cdot D^2 \qquad (6.47)$$

Figure 6.81 compares[4] the results of Klatt[16] with data published by Corney[18] and indicates that Eq. (6.47) seems to predict capacities conservatively. More generally the relationship between capacity and pan diameter can be written as:

$$C \sim D^2 \qquad (6.48)$$

Figure 6.82[19] verifies this correlation for three different materials. The actual capacity, C_a, can be calculated from Eqs. (6.47) or (6.48) by introducing a "granulator factor" Y:[7,8]

$$C_a = Y \cdot C \qquad (6.49)$$

Y must be experimentally determined during tests in a laboratory balling disc.

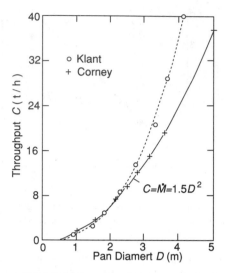

Figure 6.81. Throughput C of granulating discs as a function of pan diameter D.[4] Comparison of data obtained by Klatt[16] and Corney.[18]

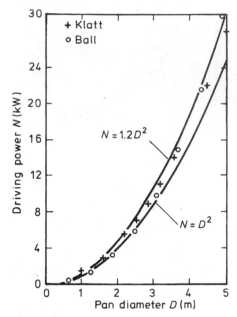

Figure 6.83. The driving power required for balling discs N as a function of pan diameter D.[4] Data points according to Klatt[16] and Ball.[20]

Because the specific energy required for granulation in a balling disc is constant [Eq. (6.38)] it follows from Eq. (6.48) that power input to the disc is also proportional to the pan diameter squared:

$$N \sim D^2 \qquad (6.50)$$

Using data of Klatt[16] and Ball,[20] Pietsch determined[4] that the proportionality factor has values between 1.0 and 1.2 (Fig. 6.83). Furthermore, from Eqs. (6.44) and (6.48) it can be deduced that the average residence time,

m/C, in a balling disc remains constant during scale-up:

$$t_r = m/C = \text{constant} \qquad (6.51)$$

The above considerations can be used as simple guidelines for sizing balling discs. Because of the rather well-defined motion and growth patterns, more complex mathematical derivations are possible resulting in relatively complicated equations.

6.4.5.4 Influence of Pan Operating Parameters on Agglomerate Quality

The well-defined motion and growth pattern in agglomeration discs allows some generalized statements in regard to agglomerate quality.

Characteristics that are of particular importance are:

- Size and shape
- Porosity
- Inner, outer, and total surface area
- Solubility
- Resistance against various stressing mechanisms.

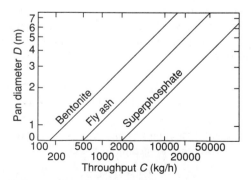

Figure 6.82. Relationship between throughput C of granulating discs and pan diameter D.[8] Data points according to Ries.[19]

Some of these depend on each other in such a way that only certain correlations exist. For example, normally, high porosity results in low strength and high solubility whereas high strength requires low porosity with an attendant low solubility.

As in the case of balling drums and, to a certain degree, in mixers, agglomerate shape, size, and quality depend on the growth mechanism taking place in the granulating disc which, in turn, is influenced by pan inclination, rim height, pan speed, as well as locations of feed and liquid binder additions. To further modify these conditions, a number of modified disc configurations have been proposed and some are being used to achieve special effects.[1]

6.4.6 Mixer–Agglomerator[21]

6.4.6.1 General

Virtually all solids mixers are capable of forming agglomerates when processing fine powders mixed with a wetting liquid. The agitation methods of size enlargement, most often used for large tonnage applications, make use of the tumbling, rolling, cascading action produced in disc, drum, and cone devices. In this section, alternative methods of mixer agglomeration are considered.

Mixer agglomeration can be broadly classified into two major groupings according to the size, density, and state of wetting of the agglomerates produced. In the first group, the agglomerates are similar in physical characteristics to those produced by tumbling. Dense, capillary-state agglomerates are formed by using agitator internals within the mixing vessel to provide a positive rubbing and shearing action. Hignett[22] claims that the agglomerates made in this way are harder and stronger than those produced by tumbling. Among other advantages over tumbling methods is the ability to process plastic, sticky materials and greater tolerance in accommodating variations in operating conditions. Less wetting phase can be used in a mixer than in a tumbling device.[23] Disadvantages include generally higher maintenance and power requirements and an irregular product that may require further shaping, as in a tumbling dryer.

In the second group of mixer–agglomerators powders are moistened to a lesser degree than in the wet capillary state. The product is in the form of weak clusters and the technique is suitable, for example, to produce "instantized" food products.

Specialized equipment has been developed for each of these two major groupings. Some mixers, however, are suitable for both methods. Some of the most common equipment in mixer agglomeration are discussed in the following examples.

6.4.6.2 Pan Mixers

The horizontal pan mixer shown in Figure 6.84 was used primarily as a batch mixer–granulator in the early development of fertilizer granulation. A typical pan might be 2.3 m in diameter and 0.5 m deep and contain a 0.5- to 0.9-Mg ($M \equiv 10^6$) batch of material. Mixing blades rotate in a direction opposite to the rotation of the pan, maintaining the charge in a constant state of agitation. If sufficient water is added, a certain degree of plasticity is created, and agglomerates are formed in 2 to 3 min of agitation. In modern fertilizer technology, horizontal pan mixers have been replaced by rotary cylinders that are better suited to continuous processing.

6.4.6.3 Paddle Mixers

These devices, also known as pugmills, pug mixers, and blungers, consist of a horizontal trough containing a mixing shaft with attached mixing blades of various designs. Single- or double-trough designs are used, although the latter type is most popular (see Fig. 6.85). Twin shafts rotate in opposite directions, throwing the materials forward and to the center of the machine as the pitched blades on the shaft pass through the charge. Incoming material may be added at various locations along the length of the mixer to ensure that the entire mixing length is used and to add versatility to the processing. Table 6.7 gives

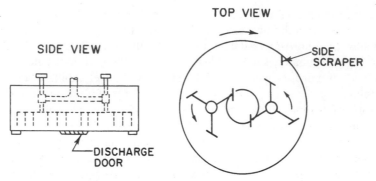

Figure 6.84. Schematic of a horizontal pan mixer.[24]

the general characteristics of the range of pug mixers offered by one manufacturer for fertilizer granulation.

6.4.6.4 High-Speed Mixers

Shaft mixers operating at high rotational speeds provide a more intensive mixing-granulating action than that obtained with conventional paddle mixers. These machines are generally single-shaft devices that may be operated either vertically or horizontally. They find application in granulating extreme fines that may be highly aerated when dry and plastic or sticky when wet. The intensive mixing action may achieve agglomeration with short residence times, leading to very compact continuous flow-through designs.

Typical examples of high-speed mixer–agglomerators are the peg granulator[25] used to treat ceramic clays in the china clay industry (see Fig. 6.86) and the pin mixer[26] used to densify carbon black into pellets (see Fig. 6.87). These machines are similar in design, consist-

Figure 6.85. Double trough pugmill for fertilizer granulation. (Courtesy of Edw. Renneburg & Sons Co.)

Table 6.7. Characteristics of Pug Mixers for Fertilizer Granulation.

MODEL	MATERIAL BULK DENSITY LB / FT^3	APPROXIMATE CAPACITY (TONS / H)	SIZE (WIDTH × LENGTH) (FT)	PLATE THICKNESS (IN.)	SHAFT DIAMETER (IN.)	SPEED (rpm)	DRIVE (hp)
A	25	8	2 × 8	1/4	3	56	15
	50	15	2 × 8	1/4	3	56	20
	75	22	2 × 8	1/4	3	56	25
	100	30	2 × 8	1/4	3	56	30
B	25	30	4 × 8	3/8	4	56	30
	50	60	4 × 8	3/8	4	56	50
	75	90	4 × 8	3/8	4	56	75
	100	120	4 × 8	3/8	5	56	100
C	25	30	4 × 12	3/8	5	56	50
	50	60	4 × 12	3/8	5	56	100
	75	90	4 × 12	3/8	6	56	150
	100	120	4 × 12	3/8	6	56	200
	125	180	4 × 12	3/8	7	56	300

Courtesy of Feeco International, Inc.

ing of a metal cylinder housing a rotating shaft carrying a number of pins or pegs arranged in a helix. Wet feed or dy feed, which is immediately moistened, enters the machine at one end and emerges as pellets at the opposite end.

As illustrated in Figure 6.87, the pelletizing of carbon black in a pin mixer is considered[26] to occur in three stages:

1. The mixing zone is roughly 15% to 20% of the total length. In this stage small droplets of binder are brought into intimate contact with the powder by the interaction of mechanical and aerodynamic forces produced by the agitator.

2. Agglomeration begins in the pelletizing zone, which is roughly 35% of the effective length. Moist solid particles introduced into the pelletizing zone are eventually combined into a number of nuclei granules and grow into spheriodal pellets fairly uniform in size and density.

3. The densifying zone comprises the final 50% of the effective machine length. The granules formed in the previous zones require

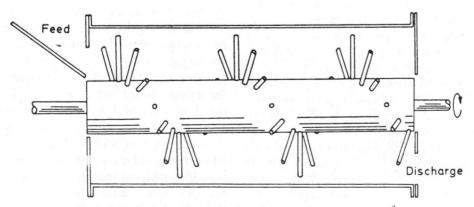

Figure 6.86. Horizontal peg granulator for ceramic clay preparation.[3]

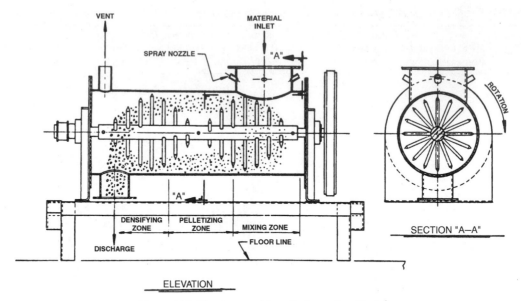

Figure 6.87. Pinmixer used in pelleting carbon black.[4]

very little additional mass but are hardened, densified, and polished through the action of the pins and interaction with each other. Table 6.8 shows pelletization test results using the pinmixer with a furnace oil carbon black.

6.4.6.5 Powder Blenders and Mixers

In applications such as the preparation of tableting feed and the manufacture of detergent powders, the aim is to produce small agglomerates (usually 2 mm diameter and less) with improved flow, wetting, dispersing, or dissolution properties. Agglomeration takes place by wetting the feed powders in a relatively dry state in standard or specialized powder mixers.

In the standard wet-granulating method used to produce tablet feed in the pharmaceutical industry, sigma blade or heavy-duty planetary mixers are often employed.[27] These machines may handle 100- or 200-kg batches and employ 5- to 7.5-kW drives to knead and mass the moistened charge. Mixing times from 15 min to an hour may be necessary, depending on the formation. The mass is then wet screened or milled, dried, and rescreened to

the required size that is dictated by the size of the tablets to be produced.

The time-consuming wet-milling step can be omitted and the agglomerates sent directly to drying, provided an appropriate granular texture can be formed in the mixer. This can be achieved by the use of specialized intensive powder mixers such as the Littleford–Lödige unit shown in Figure 6.88. Powder is fed through the filler-opening at the top of the mixer while the product is discharged through a contour door at the bottom. The working level is normally 50% of the total volume, and cleaning is easily accomplished through the two wide-access doors at the front. The material is subjected to a dry mix cycle to eliminate any lumps that might have formed during storage. The granulating solution is introduced to the mixer through liquid injectors mounted over high-speed blending choppers. Spray nozzles are not needed since the high-speed blending choppers quickly disperse the granulating liquid. Plows intermingle the powder and drive material into the high-speed choppers, which are independently powered. The choppers also control the upper size of lumps

Table 6.8. Pinmixer Used for Pelleting Carbon Black. Test Results with a Furnace Oil Carbon Black Using a 0.67 × 2.54 m Stainless Steel Unit.[26]

Carbon black feed	
Rate, Mg/day	26.3[a]
Bulk density, kg/m^3	51.3
Pellets produced	
Wet basis	
Production rate, kg/h	2108.3
Bulk density, kg/m^3	562.3
Density ratio	11.0
Dry basis	
Production rate, kg/h	1096.3
Bulk density, kg/m^3	394.1
Binder	
Specific gravity	1.05
Injection rate, kg/h	1011.5
Use ratio, weight of binder to weight of wet pellets	0.92
Power consumption[b]	
Rate, kW	18.5
Per Mg of wet pellets, kWh	15.0[c]
Production quality	
Rotap test (5 min), %	1.4 (avg. of 45 samples)
Crushing strength, g	25 (avg. of 73 samples)

[a] Average from 5-day test, plus subsequent production.
[b] Ammeter readings.
[c] Cold shell.

and agglomerates formed. Standard mixers with working capacities up to 4.8 m^3 are available in this design. Relatively short batch granulation cycles of less than 10 min are claimed for this equipment.

6.4.6.6 Other Cluster-Type Agglomeration Processes

Two other applications requiring small, cluster-type agglomerates with improved flow, wetting, dispersing, and/or dissolution properties are the manufacture of home dishwashing detergents and "instantized" powdered food products.

Powdered detergent ingredients can be gathered together into a homogeneous granular product by the application of a liquid silicate spray as the bonding agent. A unique design of agglomerator has been developed,[28, 29] for this application in which the liquid spray is applied to a falling curtain of powder ingredients of constant thickness. The curtain is generated in a rotary drum contain-

ing an internal cage of bars separated from the drum walls by a spiral ribbon (Fig. 6.89). The cage, together with inertial and centrifugal forces, holds the powder bed against the shell until it falls through the cage to form a constant density curtain. The spiral serves to

Figure 6.88. Littleford–Lödige mixer–granulator for tablet feed preparation. (Courtesy of Littleford Bros., Inc.)

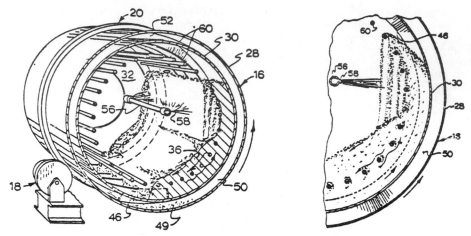

Figure 6.89. Constant-density falling-curtain agglomerator.[29]

recirculate fine material toward the feed end. The curtain of powder absorbs the liquid spray before it can impinge on internal agglomerator surfaces while the free-floating action of the internals keeps all surfaces free of build-up, both of which prevent lump formation and encourage uniform granulation. An agglomerator 1.5 m in diameter by 4.9 m long typically produces 4.5 Mg of dishwasher detergent per hour.

In the food industry, continuous-flow mixing systems are used to bring together powder and moistening liquid to form clustered products with "instant" properties. Several types of moistening–agglomerating devices are possible,[30-32] including rotating cones, powder funnels and vortex tube mixers. An illustrative example of this type of system is the Blaw–Knox Instantizer Agglomerator[33] depicted in Figure 6.90. Feed powder, at a rate controlled by a rotary valve, is introduced to the wetting section via a pneumatic conveying line. The powder falls as a narrow stream between two jet tubes that inject the wetting liquid in a highly dispersed state. Steam is often used but water, other solvents, or a combination of these may be used. Air at ambient temperature is introduced through radial wall slots in the moistening chamber to produce a vortex motion. The resulting lower particle temperature condenses fluid onto

the particles while the vortex motion enhances particle–particle collisions. The clustered material then drops through an air-heated chamber onto a conditioning conveyor where it is allowed sufficient time to reach a uniform moisture distribution. The material then passes to an after-drying, cooler, and sifter followed by bagging of the selected product.

6.4.7 Fluidized Bed / Spray Agglomerators[34]

6.4.7.1 General

In these methods of size enlargement, feed in a liquid or semiliquid form is sprayed into a gas to produce granular solids through heat and/or mass transfer. A variety of process equipment may be used, including spray dryers, spouted or fluidized beds and pneumatic conveying (flash) dryers. Agglomerates are formed by the direct conversion of feed droplets into solid particles, the layering of solids deposited from the feed onto the existing nuclei and/or the sticking together of small particles into aggregates by deposition of binding solids from the spray.

Features common to all these spray and dispersion techniques include the following:

1. The feed liquid must be pumpable and dispersible.

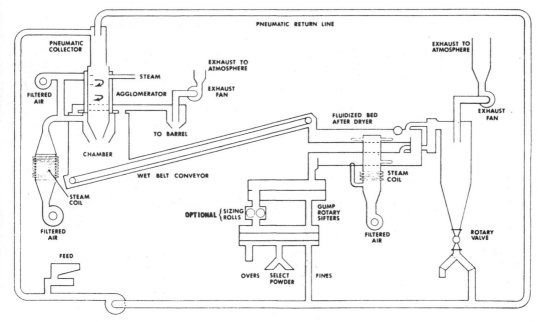

Figure 6.90. Flow diagram of the Blaw–Know Instantizer. (Trademark of Blaw–Knox Food & Chemical Equipment, Inc.)

2. The processes are usually amenable to continuous, automated large-scale operation.
3. Attraction and fines carryover are often a problem, and the systems are designed to recover and/or recycle them.
4. Product size is limited to about 5-mm diameter particles and is often much smaller.

6.4.7.2 Spray Drying[35]

In this process, feed material is dispersed in droplet form into a drying chamber where it contacts a large volume of hot gas. The liquid carrier is evaporated, and the dry product is recovered. Control of the operating variables can lead to rounded product particles varying from quite fine powders to relatively coarse granular materials (see Fig. 6.91).

Spray drying represents an attractive alternative to traditional granulation and feed preparation methods used, for example, in ceramics and pharmaceutical industries. This procedure is illustrated in Figure 6.92, where the unit operations associated with the conventional wet preparation of ceramic tilebod-

ies is compared with the spray-drying alternative. The latter method eliminates a number of processing steps between slip preparation and finished pressbody.

The four fundamental unit processes involved in spray drying are shown in Fig. 6.93. The liquid feed is dispersed into droplets in the first stage, mixed with the gas stream, and then introduced to the drying chamber. The moisture is evaporated from the droplets, which form solid granules. The dried particles are separated from the gas stream in the fourth stage. Control of the properties of spray dried products requires close attention to the design of each of these four unit processes.

Atomization of the liquid feed and contacting the spray with air are the critical features of spray dryers. Dispersion of the feed into droplets is accomplished with either rotary devices or with nozzles. In rotary atomization (Fig. 6.94a), feed is introduced centrally to a wheel (with vanes or bushings) or a disc (vaneless plates, cups, inverted bowls) and is flung off at the periphery where it disintegrates into droplets. Nozzles used can be either single-

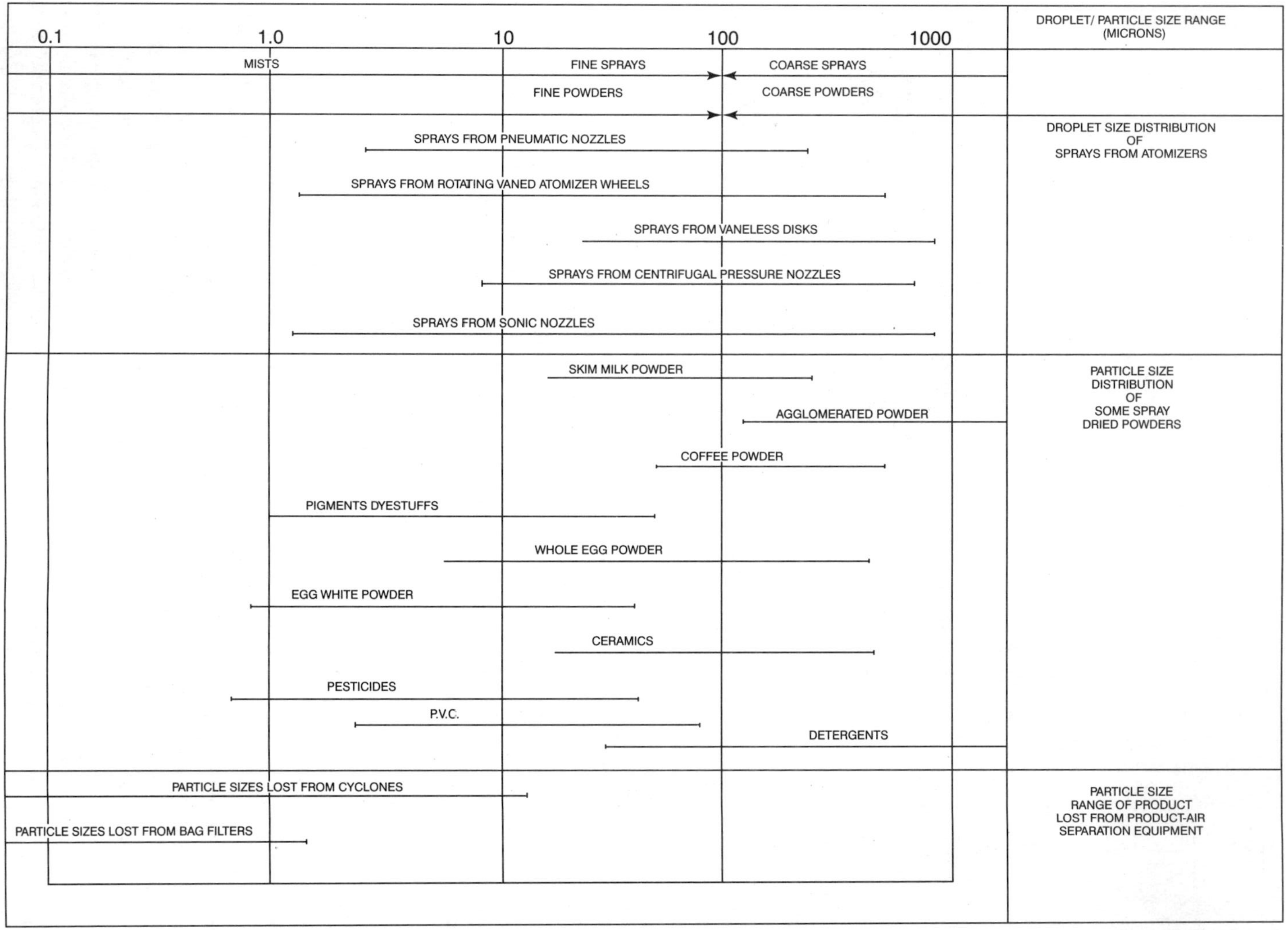

Figure 6.91. Particle size range of spray-dried products.[35]

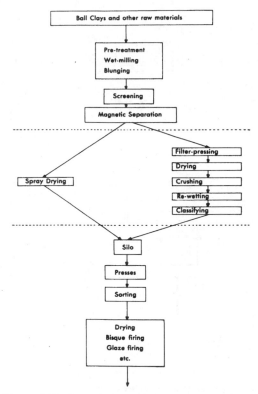

Figure 6.92. Operations used in wet preparation as compared with spray drying of ceramic press feeds. (Courtesy of Niro Atomizer, Inc.)

fluid pressure (Fig. 6.94b) or two-fluid pneumatic (Fig. 6.94c). Thus atomization of the feed can use centrifugal, pressure, or kinetic energy.

Other design features include the solid–gas flow system and the chamber shape. Solid–gas flow can be concurrent, countercurrent, or a combination of these. Chamber shape is chosen to accommodate the type of atomization. Because of the narrow cone pattern produced, a tall tower is required when nozzles are used, whereas chambers of smaller height-to-diameter ratios are suitable when droplets spun horizontally from a centrifugal atomizer are used.

Spray dryers are available with water evaporative capacities up to 40,000 lb/h (18,100 kg/h) or more. Details of the theoretical and empirical design of spray dryers is given elsewhere.[35]

Course Products by Spray Drying. Many different properties of spray-dried products (e.g., density, friability, reactivity, etc.) are of interest depending on the application under discussion. When size enlargement to a coarse granular structure is a primary objective, however, particle size and its distribution are of greatest concern.

Although the variables of spray dryer design and operation all interact to influence product characteristics, a number of these have important effects on product size and size distribution. Decreased intensity of atomization and of spray-air contact and lower exit temperatures from the dryer all tend to increase the particle size obtained. Higher liquid feed viscosity and feed rate as well as the presence of natural or added binders that lend tackiness also favor larger product size.

The flowsheet in Figure 6.95 gives one example of a system designed to yield agglomerated products. Coarse spray-dried food powders with "instant" properties are produced directly from liquid in this system. Two stages of agglomeration are involved. The initial stage occurs in the atomization zone where relatively cool air is passed to retard the evaporation rate and enhance the agglomeration of fines. Further agglomeration is achieved by operating the spray dryer so that the powder is still moist on leaving the drying chamber. The agglomerated powder passes out of the bottom of the drying chamber to a vibrating fluid bed, where drying is completed, then into a second fluid bed for cooling.

6.4.7.3 Fluid Bed Granulators

In this process, simultaneous drying and particle forming are carried out by spraying liquid feed onto a fluidized layer of essentially dry particles. Particle growth occurs either by particle coalescence or by layering of solids from the feed liquid onto the surface of bed particles.

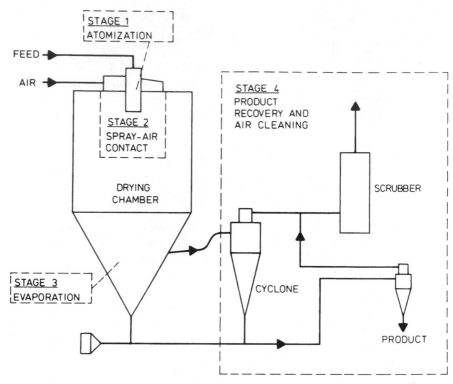

Figure 6.93. The four fundamental unit processes associated with spray drying.[35]

Because of their ability to deposit multiple layers of solids on a given particle or cluster of particles, fluidized bed (and spouted bed) systems can produce larger granules than spray dryers. The product is thus less dusty, and the longer residence times possible mean that larger dryer loads with more dilute feed liquors can be handled. Since the drying particles are less dispersed in fluid beds, smaller equipment is needed.[37]

A typical fluid bed spray granulation unit is shown in Figure 9.96. The fluidizing gas is heated externally and introduced to the base of the unit through a suitable distributor plate. In addition to product support, the distributor ensures a uniform distribution of the fluidizing

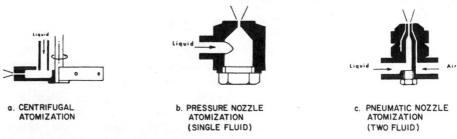

Figure 6.94. Feed atomization methods used in spray drying. (Courtesy of Anhydro. Inc.)

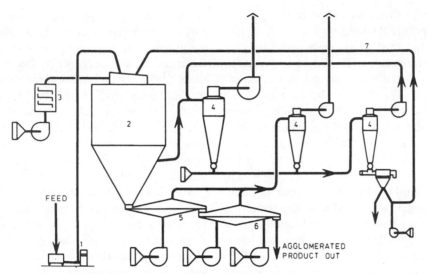

Figure 6.95. Flow sheet for production of coarse food powders with "instant" properties by spray drying:[36] (1) liquid feed system; (2) spray drying chamber; (3) drying air heater; (4) cyclones for fines recovery; (5) vibrofluidizer as after-dryer; (6) vibrofluidizer as after-cooler; and (7) fines return to drying chamber.

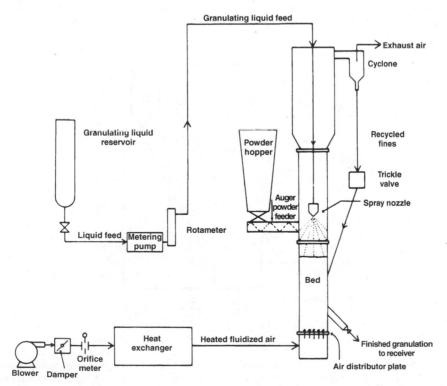

Figure 6.96. A typical fluid bed spray granulation unit. (Reproduced from Scott et al.,[38] with permission.)

medium over the cross-section of the granulator. Any poorly fluidized region which is subjected to the feed spray might cause the formation of large lumps. The liquid is sprayed by an atomization nozzle centered in the expansion section. The solids to be granulated are fed into the unit below the expansion section.

Air leaving the fluidized bed passes through a cyclone collector, which removes the entrained solids. The solids are returned to the fluidization section and the air is passed through a scrubber for further cleaning. Granulated product is removed near the bottom of the bed through an outlet pipe located slightly above the distributor plate. Pressure drop measurements indicate the weight of solids in the bed and can be used to control the rate of product removal.

A number of important design factors should be emphasized. Often the fluidization chamber is conical in shape, so that the gas velocity is highest near the distributor. In this way the larger granules which tend to segregate to the bottom of the bed are kept in motion, and overheating is prevented. Fluidizing velocity in

general should be selected so that the bed surface, where the feed spray is deposited, is maintained in vigorous movement. Under these conditions, carryover of the smaller particles can occur and a de-entrainment section in the upper part of the bed is necessary. Product discharge takes place through an opening below the bed surface, often relatively close to the distributor. In this way buildup of larger granules and lumps at the distributor plate is avoided.

The fluidizing chamber can consist of more than one compartment (see Fig. 6.97). This provides different process conditions (e.g., temperature, moisture level, gas velocity, etc.) as material flows through the bed and encourages conditions closer to plug flow for the granular solids, leading to a more uniform product size distribution.[39]

Control Parameters[40,41]. As in all size enlargement processes, the control of granule nucleation is essential to stable operation. In continuous operations such as those in Figures 6.96 and 6.97, the rate of production of stable

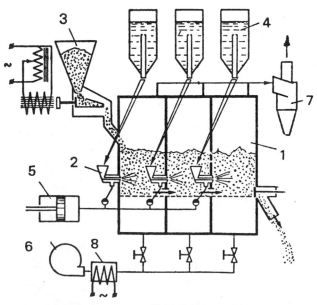

Figure 6.97. A multicompartment fluid bed granulator:[39] (1) fluid beds; (2) compressed air-operated injectors for introducing solution into the fluid bed; (3) vibratory feeder for introducing the solid phase; (4) solution tanks; (5) compressor; (6) blower; (7) cyclone; and (8) heater.

new seeds must equal the rate of production of product size granules. New seeds are generated by a number of mechanisms, including drying of liquid feed to solid before contacting the bed, by attrition and fracture of bed particles, by recycling of crushed oversize product, and by introducing new solid particles as part of the feed.

Some general guidelines on the effect of various operating parameters can be given, but these require experimental verification for each spray granulation application.

Increase in the rate of liquid feed addition and in its solids or binder content generally produces larger, stronger, and more dense granules. In some cases, as the solids content of the feed increases, the feed may tend to spray dry in the space above the bed, forming new seed particles and smaller particle size in the bed.

Large agglomerates can be obtained by decreasing the intensity of feed atomization. This effect is lessened as the granule/droplet size ratio increases. Increase in the fluidizing gas rate and bed temperature decreases the ability of the spray to penetrate and wet the bed material and hence smaller particle size is obtained. The geometry of the spray plays an important part in the product size. For example, a narrower, more concentrated spray angle wets a smaller fraction of the bed material and would be expected to yield larger granules.

It is often found that the rate of agglomeration increases as the gas velocity decreases. This is due to a less rapid exchange of particles within the wetted zone of the fluidized bed. The extent to which the gas velocity can be decreased is limited by the formation of lumps and eventually by termination of the fluidization process.

As noted above, recycled particles are an important source of new seeds for larger granules. The extent to which recycled particles are milled has a profound effect on granule size. For a constant rate of spray addition, increased grinding of recycled material in-

creases the seeding effect and reduces the size of the bed material.

Performance Data.   Performance data for two industrial versions of the fluid bed spray granulation technique are given in Tables 6.9 and 6.10. Corresponding equipment diagrams are found in Figure 6.98 and 6.99, respectively. In Table 6.9, data are listed for a range of batch spray granulators available for the production of tablet granulations in the pharmaceutical industry. In this application, the fluidized bed granulator combines into one step several of the individual operations (e.g., size control, drying, blending) normally used in other gran-

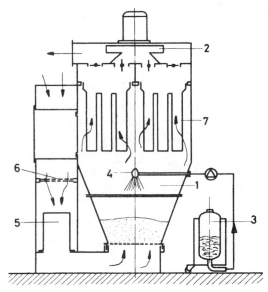

Figure 6.98. Batch fluid bed spray granulator used to produce tablet granulations in the pharmaceutical industry. Air flow necessary for fluidization is generated by a suction fan (2) mounted in the top portion of the unit, directly driven by an electric motor. The air being used is heated to the desired temperature by an air heater (5). Prefilters remove all impurities at the air inlet (6). The material to be processed has been loaded into the material container (1). The container bottom consists of a perforated plate above which a fine mesh stainless steel retaining screen is fitted. Exhaust filters (7) mounted above the product container retain fines and dust. The granulating liquid (3) is sprayed as a fine mist through a mechanical or pneumatically actuated nozzle onto the finely dispersed, fluidized material to form the desired agglomerates. (Courtesy of Aeromatic AG.)

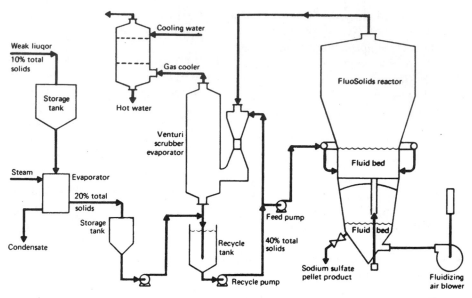

Figure 6.99. Flowsheet of fluid bed incinerator used to treat paper mill waste liquor.[42]

ulation techniques. Table 6.10 contains data on the fluid bed incineration process. Although the main objective of this process is disposal of waste sludges, the granular ash product may often by a salable chemical byproduct. In this secondary aspect, fluid bed incineration can be considered as a size enlargement process.

Spouted Bed Granulation. This technique differs from the fluidized bed process in the method used to agitate the growth bed particles. As shown in Figure 6.100 hot spouting gas is injected as a single jet into the conical base of the granulation chamber, causing the bed material to circulate much like a water fountain. Particles are carried up the central spout as a dilute phase until they lose their momentum and fall back onto the top of the bed around the outer periphery. They recirculate back down the column as a dense moving bed and are directed back into the gas stream

Table 6.9. Characteristics of Batch Fluid Bed Spray Granulators to Produce Tablet Granulations in the Pharmaceutical Industry. (Flowsheet Given in Figure 6.98)

	APPROXIMATE RANGE
Batch load, dry basis, lb	20–400[a]
Volume of container for static bed, ft^3	2–15
Fluidizing air fan, hp	5–25
Air (stream) heating capacity, Btu/h	70,000–600,000
Drying air temperature, °C	40–80
Granulating liquid spray[b]	Two fluid nozzle
Air volume	$\frac{1}{2}$–2 SCFM
Liquid volume	500–1500 cm^3/min
Batch processing time, min	30–50
Average granule size	24–8 mesh

[a] Batch capacity exceeds 1500 lb in the largest modern units.
[b] Typical granulating liquids are gelatin or sodium carboxymethyl cellulose solutions.

Table 6.10. Granular Products from Fluidized Bed Incineration.[42]

TYPE OF SLUDGE	INCINERATOR SIZE	BED TEMPERATURE	CAPACITY	GRANULAR PRODUCT COMPOSITION
Oil refinery waste sludge (85–95% water)	40 ft high; 20 ft ID at base increasing at 28 ft at top	1330°F	31×10^3 lb/h of sludge	Start-up material was silica sand; replaced by nodules of various ash components such as $CaSO_4$, Na, Ca, Mg silicates, Al_2O_3 after operation of incinerator.
Paper mill waste liquor[a] (40% solids)	20 ft ID at top	1350°F	31×10^3 ib/h	Sulfur added to produce 90–95% Na_2SO_4 and some Na_2CO_3

[a] Flowsheet, Figure 6.99.

by the conical base of the apparatus. Liquid feed, injected as a spray into the base together with the hot spouting gas, deposits a thin layer of liquid onto the recirculating seeds. Feed solids deposit by drying from the liquid onto the particles as they cycle up the spout and down the annulus.

The gas–solids contacting efficiency of fluidized systems becomes impaired at particle sizes larger than, say, 1 mm diameter when more and more gas bypasses the solids in the form of large bubbles. Spouted beds avoid this problem and are thus suited to the formation of larger granules than those produced in fluid beds.

In their recent book, Mathur and Epstein[43] have noted other advantages of spouted beds when compared with fluid beds:

1. Higher permissible inlet gas temperatures since the spray liquid rapidly cools the gas when injected into the high-velocity region at the base of the spout.
2. Layer-by-layer growth mechanism favors well-rounded and uniform granules.
3. A classification effect at the top of the bed allows the largest particles to be removed through the outlet pipe, yielding a narrow product size distribution.
4. There is no gas distribution plate to become scaled and plugged.

Performance data for the spouted bed granulation of some agricultural products are given in Table 6.11.

6.4.8 Agglomeration in Liquid Systems[45]

6.4.8.1 General

Although fine *dry* powders present difficulties such as dusting losses and other handling hazards, finely divided materials *in liquids* are also difficult to deal with. The size of the individual particles is often so small that methods to capture them (such as filtration) are difficult unless some form of size enlargement is applied.

Traditional procedures for agglomerating fine particles in liquids, such as flocculation

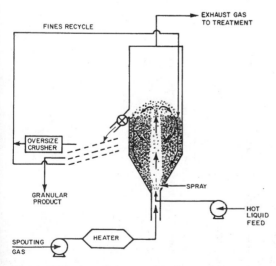

Figure 6.100. A typical spouted bed granulator.

Table 6.11. Spouted Bed Granulation Data for Some Agricultural Chemicals.[43]

	FEED SOLUTION		PRODUCT		AIR TEMPERATURE (°C)		AIR FLOW RATE (M³/S)	CAPACITY (mg/h OF PRODUCT)	WEIGHT OF BED OF SEED GRANULES (mg)
MATERIAL	MOISTURE (%)	TEMPERATURE (°C)	SIZE (MM)	MOISTURE (%)	INLET	OUTLET			
Complex fertilizer (nitro phosphorus)	27	Cold	3–3.5 (90%)	2.4	170	60	13.9	4	—
Potassium chloride	68	Cold	4–5 (oversize < 5%)	—	200	60–75	13.9	1	1
Ammonium nitrate	4	175	2.5–4	0.2	Cold	55	13.9	9.5	1.5
Sulfur	0	135	2–5	0	Cold	—	0.011[a]	0.04	0.008

Performance data reported by Berquin.[44]

[a] Injecting 1 liter/h water as spray into the spouting air reduced the air requirement to 0.007 m³/s for the same product output.

rely on relatively small interparticle bonding forces to form rather weak cluster-type agglomerates which occupy a large volume. Often the objective is simply to remove the fines from the liquid medium. In contrast, the present discussion deals with those techniques in which stronger bonding and specialized equipment are used to form generally larger and more permanent agglomerates in liquid suspensions. In addition to separation of particles from suspensions, these latter methods have other broad objectives as shown in Table 6.12, including production of granular (often spherical) materials, maximum displacement of suspending liquid from the product, and the selective agglomeration of one or more components of a multiparticle mixture.

6.4.8.2 Mechanisms of Liquid-Phase Agglomeration

In the growth of agglomeration technology in liquid systems, three broad types of processes have evolved. Two of these rely on different bridging mechanisms to pull suspended particles together into larger agglomerates. The third involves conversion of the solid feed into a liquid form, which is then dispersed as droplets in a second liquid, followed by solidification to the final particulate product.

Wetting by Immiscible Liquids[46]. Fine particles in liquid suspension can readily be formed into large dense agglomerates of considerable integrity by adding suitable amounts of a second or bridging liquid under appropriate agitation conditions. This second liquid must be immiscible with the suspending liquid and must wet preferentially the solid particles that are to be agglomerated. A simple example is the addition of oil to an aqueous suspension of fine coal. The oil readily adsorbs preferentially on the carbon particles and forms liquid bridges between these particles by coalescence during collisions under the agitation conditions. Inorganic impurity (ash) particles are not wetted by the oil and remain in unagglomerated form in the aqueous slurry.

The agglomeration phenomena that occur as progressively larger amounts of bridging liquid are added to a solids suspension are depicted in Figure 6.101. The general relationships shown are not specific to a given system. Figure 6.101 relates equally well, for example,

Table 6.12. Some Important Agglomeration Processes Carried Out in Liquid Systems.

PROCESS OBJECTIVE	MATERIAL TREATED AND PROCESS USED	REFERENCES
Sphere formation and production of coarse granular products	Nuclear fuel and metal powder production by sol–gel processes	55, 56
	Manufacturing of small spheres from refractory and high-melting-point solids (e.g., tungsten carbide) by immiscible liquid wetting	52, 54
Removal and recovery of fine solids from liquids	Removal of soot from refinery waters by wetting with oil	63
	Recovery of fine coal from preparation plant streams to allow recycling of water	58, 59
Displacement of suspending liquid	Dewatering of various sludges by flocculation followed by mechanical drainage on filter belts, in revolving drum, etc.	49, 61
	Displacement of moisture from fine coal by wetting with oil	58, 59
Selective separation of some components in a mixture of particles	Removal of ash-forming impurities from coal and from tar sands by selective agglomeration	59, 64

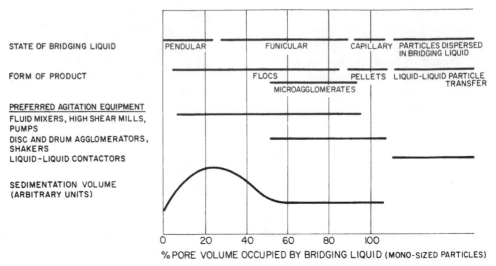

Figure 6.101. "Phase diagram" for agglomeration by immiscible liquid wetting.[46] The effect of increasing amounts of bridging liquid on the process.

to siliceous particles suspended in oil and collected with water, or to coal particles suspended in water and agglomerated with oil.

At low levels of bridging liquid, only pendular bridges can form between the particles, with the result that an unconsolidated floc structure exists. As seen in the lower part of Figure 6.101, a loose settled mass of volume larger than that of the unfloccated particles results. As the funicular region of bridging-liquid levels is reached, the flocs consolidate somewhat, and lower settled volumes are recorded. Some compacted agglomerates appear and increase in number, until about midway in the funicular region the whole system has been formed into "microagglomerates." As the amount of bridging liquid is increased, the agglomerates grow in size and reach a peak of strength and sphericity near the capillary region. Beyond this region the agglomerates exist as pasty lumps; the solids are then essentially dispersed in the bridging liquid. Figure 6.102 shows these different stages in the development of coal agglomerates.

Bridging with Polymeric Flocculants. A wide range of polymeric agents[47,48] (e.g., polyacrylamides) is available today to aid the aggrega-

tion and subsequent removal of fine particles from water. These polymeric flocculants, due to their large molecular size, cause aggregation of particles by a bridging mechanism in which several particles are united by adsorption onto one molecule of flocculant. The agglomeration of particles into a floc structure results in faster settling of the suspension and allows the supernatant liquid to be recovered more quickly. Flocculated particles, however, tend to stick to each other as they settle and form a loose, bulky layer. Although the pores in the settled layer are relatively large and its filtration and drainage rate is thus enhanced, the high porosity of the settled layer means that a larger proportion of suspending medium is often retained by the flocculated material than is the case with the unflocculated particles.

Techniques have been developed to form more compact sediments (agglomerates) of reduced liquid content in flocculated systems. These techniques, sometimes known as "pellets flocculation,"[49] combine relatively large amounts (a few pounds per ton of solids) of polymeric flocculants with gentle agitation, such as a rolling tumbling action, to reduce the moisture content of the separated solids.

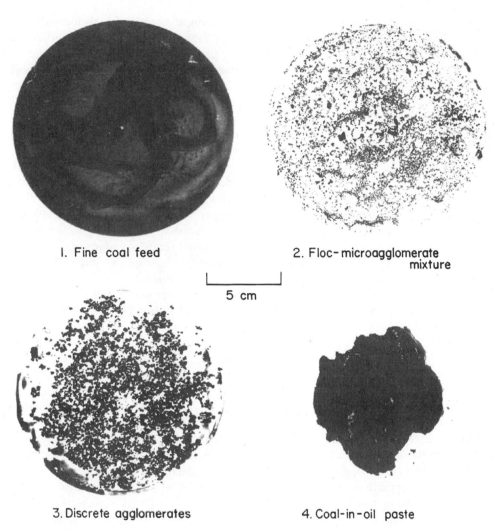

1. Fine coal feed

2. Floc-microagglomerate
mixture

5 cm

3. Discrete agglomerates

4. Coal-in-oil paste

Figure 6.102. Coal agglomerators formed with increasing amounts of bridging oil.

The agglomerates thus formed contain more interparticle bridges than with lower polymer levels, are able to grow to a larger size permitting easier separation from the liquid phase, and are strong and pliable enough to allow entrapped liquid to be squeezed out under mechanical working.

Dispersion in Liquids. A number of processes exist in which solid materials are converted into a liquid form, dispersed as droplets in a second liquid by some suitable means, and

solidified to a final particulate product. When the starting material is a massive solid, the process is then one of size reduction.[50] When a powder feed is used, size enlargement results.

Many variations are possible depending on the method used to disperse the liquid phase and on the procedure used to harden the droplets. For example, the feed liquid can be prilled into a quiescent column of the second liquid or dispersed by mechanical agitation. Hardening the droplets can be accomplished by chemical reaction, by cooling, by solvent extraction, by evaporation, or by combinations of these methods.

6.4.8.3 Processes and Equipment

A number of specific processes will now be described in which the various mechanisms of liquid-phase agglomeration discussed above are utilized.

Spherical (Immiscible Liquid) Agglomeration Processes. Spherical shapes are required for a variety of applications, many of which are associated with the field of powder metallurgy. While it is relatively easy to produce spheres by conventional techniques such as shot or prilling towers, refractory solids in general and high melting point metals do not readily respond to these methods. If the solid is available in powder form, however, various techniques are available to agglomerate the powder into highly spherical shapes.

One attractive method uses immiscible liquid wetting to pull the powder together into agglomerates while suspended in a second fluid under highly energetic agitation. Such methods are part of the family of techniques generally known as "Spherical Agglomeration Processes."[51] In this operation, the interparticle bonds formed by the immiscible liquid bridges, especially in the capillary region, are very rugged and are readily replaced if they become dislocated. Consequently, intense mechanical energy may be used to produce dense, highly spherical agglomerates. Compaction and rounding is facilitated by multitudinous collisions between the agglomerates themselves and the container walls. Where unit batch size is small, high-energy shaking devices may be utilized to optimize sphericity, size distribution, and density of the spherical agglomerates. The advantage of such a process is that a finishing operation, such as lapping and grinding after a preliminary sintering step, is much reduced when compared to that necessary for spherical powder compacts made by other techniques, for example by press molding.

A typical example of this technique[52] is the manufacture of small tungsten carbide spheres for use as blanks in the manufacture of ball-point pens. The closely sized spheres, 1 mm diameter, are prepared by agitating tungsten carbide and cobalt powders, in a closed Teflon container with hemispherical ends, on a high-speed reciprocating shaker (see Fig. 6.103). Halogenated solvents are used as the carrier liquid and water as the bridging liquid. The addition of about 6% cobalt to the tungsten carbide powder is required to reduce sintering temperatures to more acceptable levels. Coagglomeration of the mixed powders from liquid suspension tends to reduce segregation of the powders, ensuring a more homogeneous spherical product.

Energy-intensive batch agitators such as shown in Figure 6.103 are most suited to producing small spheres (1 to 2 mm diameter) since a large number can be made in a single small batch. Where comparatively high production rates are required, rotating drums and disc agglomerators are better suited to the process. In these tumbling agglomerators, the presence of a liquid slurry of feed is useful in reducing the dust nuisance that may be a problem, especially with toxic powders, when conventional dry granulation is used. The blanket of suspending liquid is helpful in at least two additional aspects. First, the phenomenon of "snowballing" is much reduced since the suspending liquid helps to disperse the bridging liquid uniformly throughout the agglomerating mass and the liquid turbulence opposes rapid agglomeration, allowing particles to layer into larger entities in a controlled manner. The liquid blanket also aids in developing a desirable tumbling, cascading motion in the equipment since the charge is more voluminous and better interparticle lubrication prevails than would be the case if no suspending liquid were present. The solids tend to be carried with the fluid. Indeed, this semi-fluid nature of the agglomerating mass has been used[54] to operate successfully a balling drum with an internal screen classifier. As shown in Figure 6.104, the horizontal, rotating

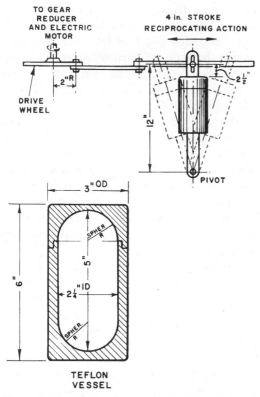

TO GEAR
REDUCER
AND ELECTRIC
MOTOR

4 in. STROKE
RECIPROCATING ACTION

DRIVE
WHEEL

PIVOT

3" OD

SPHER. R

5"

2¼" ID

6"

SPHER. R

TEFLON
VESSEL

Figure 6.103. Teflon cylinder with hemispherical ends mounted in reciprocating shaker used to form small spheres by the Spherical Agglomeration Process.[52,53] Typical conditions include 75 cm^3 carbon tetrachloride containing 200 g of tungsten carbine powder, which is agglomerated with about 10 cm^3 water. Shaking speed 300 cpm for 5 to 7 min. (Reprinted with permission from Ref. 53. Copyright 1967 by the American Chemical Society.)

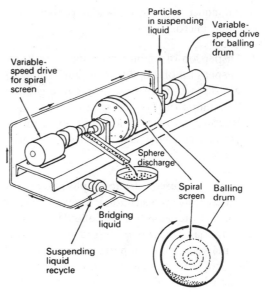

Figure 6.104. Drum agglomerator with internal screen classifier for formation of uniform spheres by immiscible liquid wetting.[54]

drum contains a spiral screen rotating at a slower speed. This screen continually passes through the pellet charge. Undersize pellets return to the drum through the screen, while the onsize material progresses along the spiral until it reaches the axis of the drum. A hollow tube at the axis then directs the pellets to a discharge point outside the drum. Highly spherical products with very uniform size distributions have been produced in this equipment. For example, carbonyl iron spheres with a median diameter of 3.9 mm and a size spread of ±0.3 mm about this median were produced

at a rate of about 40 g/min using a drum 280 mm diameter by 280 mm long. A light petroleum solvent was used to suspend the 10 μm powder and water at a rate of about 4 cm^3/min was added as the agglomerating agent. Drum speed was in the range 60 to 100 rpm while the spiral screen revolved in the same direction at 4 to 6 rpm.

Sol–Gel Processes.   In agglomeration by immiscible liquid wetting, small amounts of a bridging phase adsorbed on the particles coalesce to draw the particles into larger entities. In the sol–gel process, fine particles are initially suspended in an excess of a bridging phase; the suspension is formed into spherical droplets, and the excess bridging phase is removed to solidify the droplets into a particulate product.

The sol–gel process has been actively developed[55,56] for the preparation of spherical oxide fuel particles up to about 1000 μm diameter for nuclear reactors. The following operations are involved in converting the initial aqueous sol of solloidal particles into calcined microspheres:

1. Dispersion of sol into droplets
2. Suspension of droplets in an immiscible liquid that will extract water to cause gelation
3. Separation of gel microspheres
4. Recovery of immiscible liquid for reuse
5. Drying, calcining, and sintering of microspheres

Equipment used[55] to accomplish steps 1 to 4 in a continuous operation is shown in Figure 6.105. The aqueous sol of colloidal particles (e.g., thoria, ThO_2) is dispersed into drops at the enlarged top of a tapered forming column. The droplets are fluidized by an upflowing stream of the immiscible water-extracting fluid, such as 2-ethyl-1-hexanol. Interfacial tension holds the drops in a spherical shape, but the maximum size is limited since large drops are more susceptible to distortion. A surfactant is dissolved in the immiscible liquid to prevent coalescence of the sol droplets with each other, their adhesion to the walls of the vessel, and/or sticking together of partially dyhydrated drops. As the water is removed and the sol is converted to a gel, the particles become denser and their settling velocity increases. Column design and flow rates are controlled so that the densified particles drop out continuously to the product receiver, while new sol droplets are added to the top of the column. The extracting liquid is separated from the product and a portion of it is sent continuously to the distillation recovery system for purification to maintain a sufficiently low water concentration in the fluidizing liquid. The purified extracting liquid is then recycled to the column.

Typical capacity of a 76 mm ID (minimum) column is 9 kg/h of sintered oxide spheres using concentrated sols.

Agitation in baffled vessels can also be used to disperse and suspend the sol drops in the extracting liquid. Compared with the fluidized system described above, this more vigorous agitation produces smaller microspheres less than 100 μm in diameter.

Other Liquid-Phase Dispersion Processes. Many other granulation methods based on liquid-phase dispersion are possible depending on the way in which the feed material is con-

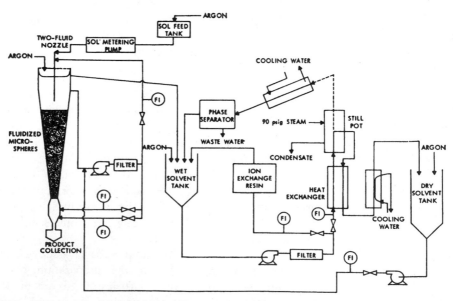

Figure 6.105. A flow diagram for microsphere formation by the sol–gel process. (Reprinted with permission from Ref. 55. Copyright 1966 by the American Chemical Society.)

verted into liquid form, the method used to disperse it into droplets, and the procedure used to harden the droplets.

One variant involves agitation of powders in nonsolvent liquid above their melting temperatures to form droplets that are then cooled below their melting temperatures to produce solid enlarged particles. For example, small spheres of naphthalene required in preparation of hollow metallic spheres[57] can be produced by agitating naphthalene in water at about 80°C to form emulsion droplets, which are then quenched at lower temperatures to yield the solid form.

A further example involves dissolution of appropriate particulate feeds in sufficient organic solvent to make them fluid, and dispersion of the liquid into water by agitation or spray followed by steam-distillation of the solvent to yield solid enlarged particles.[50]

Fine Coal Cleaning Using Oil Agglomeration[58,59].

Selective agglomeration is readily accomplished with processes based on wetting by immiscible liquids. One or more components of a complex solids mixture can be selectively agglomerated and removed from suspension (for example, by screening) while other components not wetted by the bridging phase remain in suspension. Where the natural wetting properties of a particulate component do not allow its separation from a suspending liquid and/or from other particles of a mixture, surface conditioning agents may be used to modify its surface properties and allow the desired separation.

The recovery of fine coals from aqueous waste suspensions is a problem of great current interest and will be used to illustrate the selective agglomeration process. Increasing quantities of fines in water suspension must be processed today during coal preparation. These fines result from natural degradation, increasingly mechanized mining methods, and the finer griding necessary to liberate impurities from lower-quality coals. Coal particles are readily agglomerted and recovered from aqueous suspension upon agitation with many different oils as collecting liquids. Inorganic or ash-forming constituents remain in suspension and are thus rejected.

A simplified flow diagram for the recovery of fine coal by selective oil agglomeration is given in Figure 6.106. Standard equipment well known in the chemical and mineral industries can be applied in the process. Agitation serves initially to disperse the bridging oil and secondly to contact the oil droplets and coal particles so that bonds are formed between oil-coated particles. The coal agglomerates thus formed are readily recovered on a screen of suitable mesh size while the impurity particles pass through the screen to waste. The agglomerates recovered in this way are typically in the diameter range 0.5 to 1 mm and may be suitable in some cases for direct shipment with the coarser coal products from a preparation plant. Alternatively, further pro-

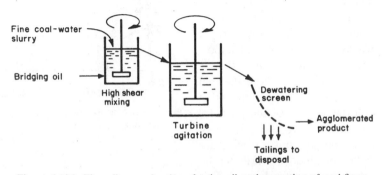

Figure 6.106. Flow diagram for the selective oil agglomeration of coal fines.

cessing may be required such as centrifugal dewatering or balling with binder in a disc or drum pelletizer.

Oil-coal contact is the most critical step in the oil agglomeration process. The required intensity and duration of mixing are determined by the oil and coal characteristics and by the solids concentration and oil usage. Predispersion of the oil as an emulsion appears to be helpful. The range of operating conditions that have been used[59] in the oil agglomeration of fine coals includes:

Coal-water feed slurry	
Wt% solids	3–50
Particle size	typically, minus 200 mesh
Ash content, wt% dry basis	10–50
Oil usage (light fuel oil), % of solids wt	2–30
Turbine agitator	
Tip speed, m/s	~ 10–30
Mixing time	30 s to several min
Power consumption, kW/m^3	~ 10–40
Product agglomerates	
Wt% recovery of solid combustible matter	> 90
Ash content, wt% dry basis	5–10

A most important characteristic of the oil agglomeration technique is its ability to recover extremely fine coal particles (for example, a few micrometers in diameter) even in the presence of clay slimes. A second important benefit of the oil agglomeration method is its ability to dewater fine coal without thermal drying. During agglomeration, the collecting oil is adsorbed on the surface of the particles and displaces the moisture to the surface of the agglomerates. The amount of moisture held by the coal then depends primarily on the surface area of the agglomerates formed. This is illustrated in Figure 6.107, in which the surface area is represented by the reciprocal of the agglomerate diameter. The moisture contents shown relate to the simple gravity drainage of the agglomerates on a 100-mesh screen.

It is evident that all the data in Figure 6.107 lie approximately on one line and as the diameter of the agglomerates or the coal particles increases, the moisture content decreases. These data indicate that the moisture content may be reduced to less than 10% by agglomeration and drainage on a screen without the need for thermal drying, provided that the size of the agglomerates is larger than about 2 mm.

The amount of oil necessary to form these large agglomerates that will drain to low moisture levels may be prohibitive in cases where extreme fines are being treated. When small agglomerates (less than 1 mm) are produced with lower oil levels, mechanically assisted dewatering in a centrifuge can be employed to

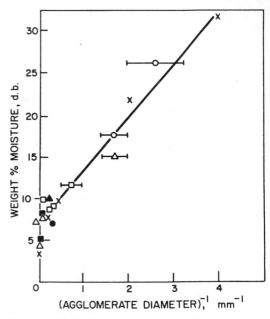

Figure 6.107. Moisture content for coal agglomerates and for unagglomerated coal as a function of the reciprocal of diameter.[60] (Data refer to gravity drainage on screens. × refers to unagglomerated coal of various sizes; △, □, ○, ▲ ■, ● refer to coal fines wetted by oil to form agglomerates of various sizes.

attain low moisture levels without thermal drying. Some results are given in Figure 6.108 for a nominal minus 28-mesh metallurgical coal. With this particular feed material, the small agglomerates formed with only a small percentage of oil dewater to low levels in a centrifuge.

Recovery of fine waste coals from preparation plants obviously reduces the load on the tailing handling system. For example, if a tailings slurry containing 70% coal is agglomerated prior to thickening and 90% of the coal is recovered, then the solids feed to the thickener and the tailings pond decreases by 63%. Thus, not only is the settling rate improved because of the reduced solids concentration, but tailings pond life almost triples. Agglomeration of coal in existing tailings ponds recovers lost coal values, extends pond life, and in some cases may eliminate the need for new ponds.

Pellet Flocculation. This technique combines relatively large amounts of polymeric flocculants with gentle rolling mixing to consolidate settled flocs into compact agglomerate-like sludges of low liquid content.

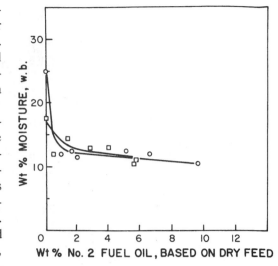

Figure 6.108. Effect of oil content on the moisture content of centrifuged (160 G) agglomerates formed from two different samples of minus-28-mesh coal fines.

Choice of flocculant is of prime importance in these processes. Criteria for choosing a floculant include the degree of floc formation and effect on water clarification, the amount of water contained in the settled and dewa-

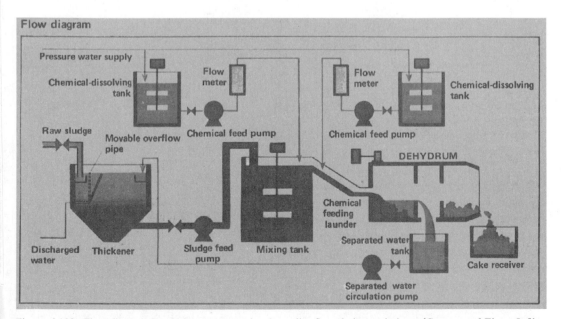

Figure 6.109. Flow diagram for sludge treatment by the pellet flocculation technique. (Courtesy of Ebara-Infilco Co., Ltd., Tokyo.)

Table 6.13. Performance Data for Treatment of Various Sludges Using "Dehydrum" of Figure 6.109.[61]

TYPE OF SLUDGE		WATER WORKS SLUDGE	BENTONITE SLURRY SLUDGE IN SHIELD TUNNELING PROCESS	SLUDGE IN MIXED WASTE EFFLUENTS FROM AUTOMOBILE SHOPS (CONTAINING OILS AND ACTIVATED SLUDGE)	SLUDGE IN GRAVEL-WASHING WASTE WATER	SEWAGE SLUDGE (MIXED RAW SLUDGE)	SLUDGE FROM DREDGING WASTE WATER
Raw sludge	Solid concentration (g/liter)	30–40	150–300	50–70	100–200	40–75	200–250
	Ignition loss in solid matter (%)	15–30	—	40–45	—	45–50	12–21
	Oil content in solid matter (%)	—	—	7–10	—	—	—
Chemicals	Amount of polymeric flocculant/amount of solid (%)	0.1–0.2	0.07–0.1	0.25–0.3	0.04–0.05	0.25–0.35	0.05–0.1
	$Ca(OH)_2$ (%)	0	3	0	0	0	2–4
Moisture content in cake (%)		65–80	46–48	82–86	38–47	79–82	46–64
Turbidity of separated water (ppm)		50	50	100	100	100	100

tered flocs, and the dosage required in terms of cost per unit weight of dry solids. Organic polyelectrolytes provide the best results with many materials, and it can be anticipated that a cationic flocculant will be most useful with organic sludges, whereas an anionic or nonionic flocculant will be best for inorganic and mineral sludges.[61]

One effective piece of equipment to accomplish "pellet flocculation" has been developed in Japan and is depicted in the flow diagram of Figure 6.109. This process makes use of a slowly revolving (1 m/min peripheral speed) horizontal drum (called a "Dehydrum") to dewater sludge.[49] The drum interior is made up of three sections for successively pelletizing, decanting, and consolidating the solids. Polymeric flocculant is added to the suspension upstream of the drum, together with auxillary agglomerating agents such as calcium hydroxide or sodium silicate. Voluminous flocs formed ahead of the drum are rolled into denser sediment in the pelletizing section. These are then pushed into the decanting section by a guide baffle where water is removed through intermittent slits in the drum wall. In the final consolidating section, the agglomerates are gently tumbled and rolled into a denser form and water again escapes through wall slits. Product solids then discharge as a low-water-content cake. These cylindrical vessels are available[62] in standard sizes up to 3.4 m diameter by 9.2 m long with a 5.5 kW drive. Typical sludge treating capacities for a 2.4 m diameter unit are 6 to 9 Mg/h for gravel waste sludge, 1.4 to 2.2 Mg/h for a dredged mud sludge, and 0.4 Mg/h for a mixed waste sludge from an automobile factory. Table 6.13 provides performance data for the treatment of a number of suspensions by this technique.

References

1. W. Pietsch, *Size Enlargement by Agglomeration.* John Wiley & Sons/Salle + Sauerländer, Chichester, UK/Aarau, Switzerland (1991).

2. P. T. Cardew and R. Oliver, "Kinetics and Mechanics in Multi-phase Agglomeration Systems," in Notes of the Waterloo Intensive Course on Ag-glomeration Fundamentals, University of Waterloo, Ont., Canada (1985).

3. K. V. S. Sastry and D. W. Fuerstenau, "Kinetic and Process Analysis of the Agglomeration of Particulate Materials by Green Pelletization," in *Agglomeration 77* Vols. 1 and 2, edited by K. V. S. Sastry, *Proc. 2nd International Symp. Agglomeration*, Atlanta, GA, AIME, New York, pp. 381–402 (1977).

4. W. Pietsch, "Die Beeinflussungsmöglichkeiten des Granuliertellerbetriebes und ihre Auswirkungen auf die Granulateigenschaften." *Aufbereitungs Technik* 7:177–191 (1966).

5. C. R. Harbison, "Pelletizer," US Patent 3 802 822 (1974).

6. K. Meyer, *Pelletizing of Iron Ores*. Springer-Verlag, Berlin, and Verlag Stahleisen GmbH, Düsseldorf, Germany (1980).

7. W. Pietsch, "Wet Grinding Experiments in Torque Ball Mill," in *Zerkleinern, Proc. International Symp.* Cannes, France (1971). *Dechema Monographien*, Vol. 69, Verlag Chemie GmbH, Weinheim, Germany, pp. 751–779 (1972).

8. K. Sommer and W. Herrman, "Auslegung von Granulierteller und Granuliertrommel." *Chemie Ingenieur Technik* 50:518–524 (1978).

9. R. Manz, "Beitrag zur Berechnung der Antriebsleistung von Rohrmühlen." *Zement Kalk Gips* 23:407–412 (1970).

10. H. E. Rose and R. M. E. Sullivan, Ball, Tube, and Rod Mills. Constable, London (1957).

11. H. T. Sterling, "Advances in Balling and Pelletizing," in *Agglomeration*, edited by W. A. Knepper, *Proc. 1st International Symp. Agglomeration*, Philadelphia, PA, John Wiley & Sons, New York, pp. 177–206 (1962).

12. G. Heinze, "Novel Rotary Drum for (the Agglomeration of) Finely Divided Dispersed Material." *Aufbereitungs Technik* 28:404–409 (1987).

13. D. F. Ball, J. Dartnell, J. Davison, A. Grieve, and R. Wild, *Agglomeration of Iron Ores*. American Elsevier, New York (1973).

14. F. P. Morawski, *Mining Eng.* 15(5):48–52 (1963).

15. M. Papadakis and J. P. Bombled, "La Granulation des Matières Premières de Cimenterie." *Rev. Matér. Construct.* 549 289–299 (1961).

16. H. Klatt, "Die betriebliche Einstellung von Granuliertellern." *Zement Kalk Gips* 11(4):144–154 (1958).

17. U. N. Bhrany, "Entwurf und Betrieb von Pelletiertellern." *Aufbereitungs Technik* 18(12):641–647 (1977).

18. J. D. Corney, "Disc Granulation in the Chemical Industry." *Br. Chem. Eng.* 10(9):405–407 (1965).

19. R. B. Ries, "Granulaterzeugung in Mischgranulatoren und Granuliertellern." *Aufbereitungs Technik* 16(12):639–646 (1975).

20. F. D. Ball, "Pelletizing before Sintering: Some Experiments with a Disc." *J. Iron Steel Inst.* pp. 40–55 (1959).

21. C. E. Capes and A. E. Fouda, "Agitation Methods," in *Handbook of Powder Science and Technology*, edited by M. E. Fayed and L. Otten, Van Nostrand Reinhold, New York, pp. 286–194 (1983).

22. T. P. Hignett, "Manufacture of Granular Mixed Fertilizers," in *Chemistry and Technology of Fertilizers*, edited by V. Sanchells, Reinhold, New York (1960).

23. P. J. Sherrington, The Granulation of Sand as an Aid to Understanding Fertilizer Granulation. *Chemie. Eng. (London) No. 220*, CE 201–CE 215 (1968).

24. J. O. Hardesty, "Granulation." In *Superphosphate: Its History, Chemistry and Manufacture*, U.S. Dept. of Agriculture, Washington, 1964.

25. R. E. Brociner, "The Peg Granulator," *Chem. Eng. (London) No. 220*, CE 227–CE 231 (1968).

26. J. A. Frye, W. C. Newton, and W. C. Engelleitner, The Pinmixer—a Novel Agglomeration Device. *Proc. Inst. Briquet. Agglom. Bien. Conf. 14*, pp. 207–217 (1975).

27. L. Lachman, H. A. Lieberman, and J. L. Kanig (eds), *The Theory and Practice of Industrial Pharmacy*, Lea and Febiger, Philadelphia (1970).

28. C. A. Sumner, "Agglomeration of Dishwater Detergents," *Soap Chem. Spec.* (July, 1975).

29. C. A. Sumner and E. O'Brien, "Constant Density Falling Curtain Agglomeration of Detergents and Other Materials," in *Agglomeration 77*, edited by K. V. S. Sastry, AIME, New York (1977).

30. J. D. Jensen, "Some Recent Advances in Agglomerating, Instantizing and Spray Drying." *Food Technol.*, Chicago, pp. 60–71 (June, 1975).

31. R. Wood, "Getting to Grips with Granulation." *Mfg. Chem. Aerosol News*, pp. 23–27 (June, 1975).

32. K. Masters and A. Stoltze, "Agglomeration Advances." *Food Eng.*, pp. 64–67 (February, 1973).

33. J. G. Moore, W. E. Hesler, M. W. Vincent, and E. C. Dubbels, "Agglomeration of Dried Materials." *Chem. Eng. Prog, 60*(5):63–66 (1964).

34. C. E. Capes and A. E. Fouda, "Prilling and Other Spray Methods," in *Handbook of Powder Science and Technology*, edited by M. E. Fayed and L. Otten, Van Nostrand Reinhold, New York, pp. 294–307 (1983).

35. K. Masters, *Spray Drying Handbook*, 3d ed., George Godwin London; Halsted Press, New York (1979).

36. K. Masters and A. Stoltze, "Agglomeration Advances." *Food Eng.*, pp. 64–67 (February, 1973).

37. J. W. Pictor, "Solids from Solutions in One Step." *Process Eng.*, pp. 66–67 (June, 1974).

38. M. W. Scott, H. A. Lieberman, A. D. Rankell, and J. V. Battista, "Continuous Production of Tablet Granulations in a Fluidized Bed." *J. Pharm. Sci. 53*(3):314–320 (1964).

39. N. A. Shakhova, B. G. Yevdokimov, and N. M. Ragozina, "An Investigation of a Multi-Compartment Fluid-Bed Granuator." *Process Technol. Int. 17*:946–947 (1972).

40. W. L. Davies and W. L. Goor, Batch Production of Pharmaceutical Granulations in a Fluidized Bed. *J. Pharm. Sci. 60*(12):1869–1874 (1971); ibid. *61*:618–622 (1972).

41. S. Mortensen and S. Hovmand, "Particle Formation and Agglomeration in a Spray Granulator," in *Fluidization Technology*, edited by D. L. Keairns, Hemisphere Pub. Corp., Washington (1976).

42. C. J. Wall, J. T. Graves, and E. J. Roberts, "How to Burn Salty Sludges." *Chem. Eng. 82*(8):77–82 (1975).

43. K. B. Mathur and N. Epstein, *Spouted Beds*, Academic Press, New York (1974).

44. Y. F. Berquin, *Method and Apparatus for Granulating Melted Solid and Hardenable Fluid Products*. U.S. Patent 3, 231, 413 (January 25, 1966).

45. C. E. Capes and A. E. Fouda, "Agglomeration in Liquid Systems," in *Handbook of Powder Science and Technology*, edited by M. E. Fayed and L. Otten, Van Nostand Reinhold Co., New York, pp. 331–344 (1983).

46. C. E. Capes, A. E. McIlhinney, and A. F. Sirianni, "Agglomeration from Liquid Suspension—Research and Applications," in *Agglomeration 77*, edited by K. V. S. Sastry, AIME, New York (1977).

47. R. Akers, *Flocculation*, Inst. Chem. Engrs., London (1975).

48. J. A. Kitchener, "Principles of Action of Polymeric Flocculants." *Br. Polym. J. 4*:217–229 (1972).

49. M. Yusa, H. Suzuki, and S. Tanaka, "Separating Liquids from Solids by Pellet Flocculation." *J. Am. Water Works Assoc. 67*:397–402 (1975).

50. R. H. Perry and C. H. Chilton, (eds.), *Chemical Engineers' Handbook*, 5th ed., section 8, McGraw-Hill, New York (1973).

51. J. R. Farnand, H. M. Smith, and I. E. Puddington, "Spherical Agglomeration of Solids in Liquid Suspension." *Can. J. Chem. Eng. 39*:94–97 (1961).

52. A. F. Sirianni and I. E. Puddington, "Forming Balls from Powder." U.S. Patent 3,368,004 (Feb. 6, 1968).

53. C. E. Capes and J. P. Sutherland, "Formation of Spheres from Finely Divided Solids in Liquid Suspension." *Ind. Eng. Chem. Process Design Develop. 6*:146–154 (1967).

54. C. E. Capes, R. D. Coleman, and W. L. Thayer, "The Production of Uniformly Sized Spherical Agglomerates in Balling Drums and Discs." *Int. Conf. Compact. and Consolid. of Part. Matter, Proc.*, 1st, London (1972).

55. P. A. Haas and S. D. Clinton, "Preparation of Thoria and Mixed Oxide Microspheres," *Ind. Eng. Chem. Product Res. Dev.* 5(3):236–246 (1966).

56. M. E. A. Hermans, "Sol-gel Processes—A Curiosity or a Technique?" *Powder Met. Int.* 5(3):137–140 (1973).

57. J. R. Farnand and A. F. Sirianni, Hollow Article Production. U.S. Patent 3,528,809 (Sept. 15, 1970).

58. C. E. Capes, A. E. McIlhinney, R. E. McKeever, and L. Messer, "Application of Spherical Agglomeration to Coal Preparation." *Int. Coal Prep. Conf. Proc.*, 7th, Sydney, Australia (1976).

59. C. E. Capes and R. L. Germain, "Selective Oil Agglomeration in Fine Coal Beneficiation, in *Physical Cleaning of Coal*, edited by Y. A. Liu, Markel Dekker, New York (1982).

60. C. E. Capes, A. E. McIlhinney, A. F. Sirianni, and I. E. Puddington, "Agglomeration in Coal Preparation." *Proc. Inst. Briquet. Agglom.* 12:53–65 (1971).

61. *Flocpress*, Bull. DB845, Infilco Degremont Inc. (Sept. 1976).

62. *Dehydrum Continuous Pelletizing Dehydrator*, Ebara-Infilco Co., Ltd., Tokyo, Japan.

6.5 PRESSURE AGGLOMERATION METHODS

6.5.1 Introduction

Pressure or press agglomeration using tabletting machines and other piston presses, roller presses, isostatic pressing equipment, and extrusion machinery, as well as some lesser known equipment, represents a large share among commercial applications of size enlargement by agglomeration. This technology is largely independent of feed particle size and the forces acting upon the particulate feed may be very high with certain equipment. Therefore, it constitutes the most versatile group of size enlargement processes by agglomeration. Because of the relative complexity of the equipment and its comparatively small capacity per unit, these techniques find their largest field of use in low to medium capacity applications (approx. 1 to 50 t/h). In addition, specialty products, such as those in the pharmaceutical industry, may be processed in very small and sophisticated machinery, handling only a few kilograms per hour, while certain high-tonnage bulk materials, for example, some fertilizers and refractory materials, are briquetted or compacted in large facilities employing multiple units.

Other advantages of pressure agglomeration are that, in most cases, essentially dry solids are processed which do not tend to set up and that the amount of material in the system is relatively small. Therefore, pressure agglomeration methods lend themselves particularly well to batch or shift operations and to applications in which several products must be manufactured from different feed mixtures. At the end of a production run, the system can be easily and completely emptied in a relatively short period of time.

In general, if several million tons per year of always the same feed composition must be agglomerated, such as in ore or minerals mining and concentrating, pressure agglomeration will normally not be the preferred first choice. In all other cases, one of the different methods of pressure agglomeration should be considered.

6.5.2 Mechanisms of Compaction[2]

The production of a powder tablet, compact or briquet can be carried out by a number of techniques, the purpose of which is usually to form the powder into a more or less well-defined shape. Within each method many routes are possible, each resulting in the manufacture of different types of products with respect to size, shape, and physical properties. However, all have in common a basic compaction mechanism.

When a particulate solid is placed into a die and pressure is applied, a reduction in volume will occur due to the following mechanisms (Fig. 6.110):

1. At low pressure, rearrangement of the particles takes place, leading to a closer packing. At this stage, energy is dissipated mainly in overcoming particle friction, and the magnitude of the effect depends on the coefficient of interparticle friction. In the case of fine powders, cohesive arches may collapse at this stage.

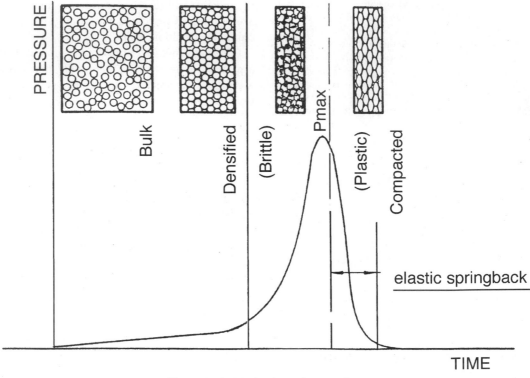

Figure 6.110. Mechanisms of compaction.

2. At higher pressures, elastic and plastic deformation of the particles may occur, causing particles to flow into void spaces and increasing the area of interparticle contact. Interlocking of particles may also occur. For materials of low thermal conductivity and low melting point, the heat generated at points of contact may be sufficient to raise the local temperatures to a point where increased plasticity and even melting facilitate particle deformation.

With brittle materials, the stress applied at interparticle contacts may cause particle fracture followed by rearrangement of the fragment to give a reduced volume.

3. High pressure continues until the compact density approaches the true density of the material. Elastic compression of the particles and entrapped air will be present at all stages of the compaction process.

The mechanisms discussed may occur simultaneously. The relative importance of the various mechanisms and the order in which they occur depend on the properties of the particles and on the speed of pressing.

The aim of compaction is to bring small particles into sufficiently close contact so that the forces acting between them are large enough to produce a product that has sufficient strength to withstand subsequent handling. Therefore, it is often necessary to carry the compaction into the bulk compression stage, in which the stressing is hydrostatic in character. Broken or deformed particles are no longer able to change position because of the few remaining cavities, and a certain amount of interparticle conformity has been achieved. With increasing pressure the apparent density will gradually approach that of the theoretical density. The rate of this approach depends on the yield point of the material. Brittle materials are more difficult to densify to a high degree by pressure only because fragmentation decreases due to the hydrostatic pressure conditions and higher strength of smaller particles. When porosity becomes fully disconnected, the isolated pores may set up considerable internal gas pressures which, together with stored elastic energy, can contribute to the disintegration of compacts if the pressure is released too quickly.

If a particulate solid were compacted in a cylindrical die with frictionless walls, it is expected that the pressure exerted by the piston would be transmitted throughout the material giving uniform pressure and, therefore, uniform density throughout the compact. In practice, the presence of frictional shear forces at the wall leads to a nonuniform pressure distribution causing variations in the density of the compact (Fig. 6.111). These variations are present in products from all pressure agglomeration techniques and lead to weakening of the compact. If a sintering step follows, distortion is possible owing to differences in the amount of contraction occurring at the positions of different density.

Figure 6.111 shows density distribution curves in tablets produced in a cylindrical die

with stationary bottom after one-directional compression (punch moves from the top into the die).[3] The individual tablets were obtained from identical bulk volumes after applying the indicated compaction forces. In such tablets the highest density is at the top edge of the compact and the lowest density at the bottom edge. A region of high density occurs near the axis a short distance above the bottom of the compact. In some cases the density in this position is higher than that observed near the axis at the top of the compact.

The general conclusion from investigations into the effects of operating conditions of pressure agglomeration equipment are that density variation:

- increases with the applied pressure and with the height of the specimen for constant diameter,
- decreases with increasing diameter even at constant height-to-diameter ratio,
- is slightly reduced by the addition of a lubricant to the powder, and

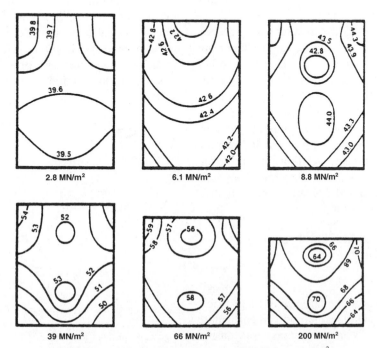

Figure 6.111. Density distributions in cylindrical compacts.[3]

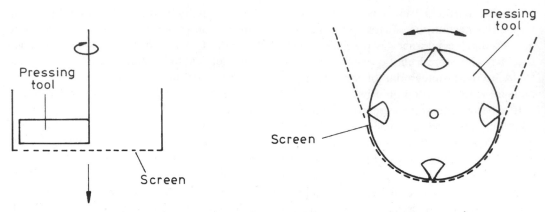

Figure 6.112. Schematic representation of two typical low-pressure agglomerators.[1]

• is considerably reduced by lubricating the die walls or tools.

Segregation during feeding and filling also leads to density variations owing to local changes in size distribution and, in the case of mixtures, to differences in the plasticity and friability of the component materials. Since there is evidence that radial flow of powder during compaction is negligible, it is expected that variations in density before compaction have an appreciable effect on the uniformity and quality of the compact.

A knowledge of the relationship between compacting pressure and density is important because pressure or force, more than any other factor, controls the attainment of high density, high strength, and low porosity in green compacts and markedly influences the same properties in the final product. A number of empirical formulas has been proposed to describe the pressure–porosity relationship; however, none of these formulas is universally applicable, giving acceptable results over a limited range of pressures only.

6.5.3 Low- and Medium-Pressure Agglomerators

6.5.3.1 General

Low-pressure agglomeration is most probably the oldest granulation method for particulate matter. Originally, a moist mass was passed through a sieve by the eminence of the hand, a spatula, specially designed handtools, or a brush. Later, this procedure was simulated by mechanization. Figure 6.112 depicts schematically two typical low-pressure (screen) agglomerators. The size of the screen openings depends on the moisture content of the mass to be agglomerated.

In most cases, the material pressed through the screen must be scraped off with suitable tools (knife blades). The green product is collected and dried. A typical system with continuous drying is shown in Figure 6.113. If necessary, all or only larger granules may be crushed to the desired size in a mill. The shape of the final, dry agglomerates produced by low-pressure agglomeration is slightly elongated

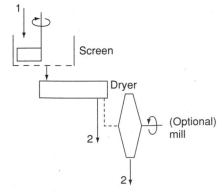

Figure 6.113. Low-pressure agglomeration system with screen, dryer, and (optional) mill. 1: Powder + binder, 2: granular product and fines.[1]

but generally irregular and density is low (high porosity and solubility).

As far as porosity, solubility, and the possibility to introduce microdoses of active ingredients with the agglomeration liquid are concerned, for example, in the pharmaceutical industry, products from low-pressure agglomeration are similar to those obtained in tumble agglomeration. The main differences are that the particle shape is more irregular, particularly if all or part of the dried material is milled to adjust particle size, and that the steps of mixing, agglomeration, as well as drying are carried out in separate process equipment. The latter may be an advantage (better control of each step) or a disadvantage (possibilities of material losses, contamination, etc.) or both.

A modern machine that may, alternatively, apply low or medium pressure is a screw extruder[4] which can optionally be used as a peripheral axial, or dome discharge, low-pressure screen agglomerator (see Fig. 6.59, a3 to a5) or, for a denser extrudate, employ a medium pressure axial die plate (see Fig. 6.59, b1). Single or twin screws convey the damp formulation from the feed hopper to the extrusion zone. In case of low-pressure extrusion tapered rotors with longitudinal blades expel the material through a screen (Fig. 6.114), which is easily replaced or changed for different extrudate diameters. Screen openings as small as 0.5 mm are possible for many materials. For medium-pressure applications the peripheral discharge attachment is replaced with axial die plates.

Medium-pressure agglomerators use extrusion for the formation of agglomerates. In this respect the mechanism is similar to screen agglomeration in low-pressure agglomeration. To achieve higher densification, forces are created in thicker dies by friction of the material sliding through mostly cylindrical extrusion channels or bores. In agglomeration, this technology is called pelleting.

Schematic representations of the machines are shown in Figure 6.59, b1 to b6. The most commonly utilized equipment features differently arranged press rollers and perforated dies (see Fig. 6.59, b2 to b6). If the extrusion bores are long and without relief counterbores, relatively high densification can be achieved. On exiting, the extrudates are scraped off by knives and form cylindrical agglomerates with defined diameter and variable length (Fig. 6.115).

To render materials suitable for pelleting or extrusion, they must have inherent binding characteristics or contain binders and feature a certain lubricity. Therefore, most medium-pressure agglomeration techniques use moist mixtures, that are prepared in a mixing step prior to pelleting.

An important advantage of medium-pressure agglomeration is that, in comparison with tumble or low-pressure granulation, only

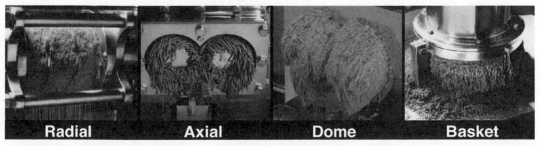

Figure 6.114. Photographs of low-pressure agglomerates exiting from screw extruders with radial, axial, and domed discharge screens. The basket-type (extruder) granulator is also shown.[4]

Figure 6.115. Typical products manufactured with a pelleting machine.

one half to one third of the agglomeration liquid is required. Therefore, drying takes place quicker and with less energy.

For mechanical reasons it is not easily possible to equip the dies with bores of less than 1 mm diameter. This is why agglomerates formed by medium pressure (extrusion) are normally dried and then "crumbled" by crushing if a finer granular product is desired. Fines may be screened out and recycled to the mixer for renewed agglomeration.

6.5.3.2 Equipment

Continuous Extrusion. The phenomenon of movement caused by the flights of rotating screws in more or less tightly fitting housings can be used to continuously produce the necessary pressure to overcome the friction in open-ended dies. These so-called screw extruders offer advantages compared with, for example, the noncontinuous ram extruders (see below) because capacity limitations due to the reciprocating movement of the plunger with its acceleration and deceleration phases do not exist. Feed and product move continuously, thus avoiding static friction, and addi-

tional work, such as plasticizing or even melting and deaeration or degassing can be performed by specially designed screws.

Screw extruders may feature single or twin screws. While most of the modern machines are used in the plastics industry to produce granular master compounds with complex equipment design,[5-8] relatively simple presses are utilized for agglomeration by extrusion of plastic and pasty materials such as clays, lightweight aggregate mixtures, building material mixtures, coal or carbon products with binders, etc.,[9] and of powders mixed with liquid binders and, sometimes, lubricants or plasticizers.

In general, the extrusion rate dm/dt of a screw extruder is determined by the combined influence of screw transport and die resistance. The operating point, defining pressure and capacity, is obtained in a mass flow/ pressure diagram at the point of intersection between the lines characterizing the screw and die performance, respectively (Fig. 6.116). Because of the influence of both characteristics, the theory of screw extruders is rather complex. The actual operating condition results from the superposition of two extremes, of

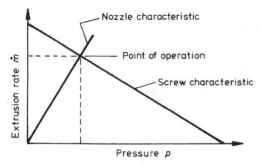

Figure 6.116. Extrusion rate $m = dm/dt$ of a screw extruder as a function of pressure of the mass to be extruded.[5]

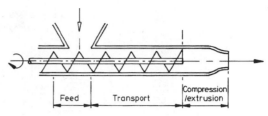

Figure 6.117. Operating conditions in a simple, axial, single-screw extruder.

screw conveying with no backpressure and pumping/mixing against a completely closed end.

The above-mentioned difficulty in describing mathematically the conditions in a screw extruder becomes even more complicated if special kneading, densification, and deaeration sections are included in the design. As shown in Figure 6.117 the simplest single screw extruder features already three distinct zones: feed, transport, and compression/extrusion zones.

In some units, a conditioning mechanism is located in the feed zone so that liquid can be introduced followed by kneading of the wetted powder mass into a moist, homogeneous mass. Some mixing of different powders may also be accomplished.

The auger-like screws then transport the material into the compression zone, where air or gases are forced from the interstitial voids as particle matter is compacted.

Screw designs vary in accordance with how much pressure is needed to obtain sufficient densification and to overcome the friction in the die. In the space between the end of the screw and the die, densification is controlled by rheological properties of the material. Less compression and less dense extrudates are obtained if this gap becomes smaller and vice versa.

Extruders that rely solely on the pressure developed by the rotating screws employ hy-drostatic pressure as the transport mechanism and are generally high-pressure extruders. Those extruders that utilize dragging or rolling motion feature a localized "drag flow" transport mechanism and, consequently, the rate of work performed and internal pressure developed are lower.

Two fundamentally different mechanisms for screw extrusion are possible: axial (Figs. 6.117 and 6.118a) and radial (Fig. 6.118b). Both machines may be equipped with either one or two screws.

While most of the axial screw extruders operate solely according to the hydrostatic pressure principle (Fig. 6.117) several other types use an extrusion blade to additionally create a wiping effect at the die plate (Fig. 6.118a). This blade looks like and performs in a fashion somewhat similar to a propeller. Nevertheless, the material discharges axially from the bores at the end of the extruder barrel.

In radial discharge extruders the extrusion blades are formed as shown in Figure 6.119. Material is extruded circumferentially through openings in the barrel wall and the direction of extrudate flow is perpendicular to the screw axis. In many cases, the barrel wall in the extrusion zone consists of a screen. Because of the extremely short length of the extrusion openings in such equipment, low-energy input and low densification prevail. Extrudates formed by this mechanism are very plastic and are normally treated in a second step, for example, achieve final shape and density.

As with all pressure agglomeration techniques, air or gases are squeezed from the

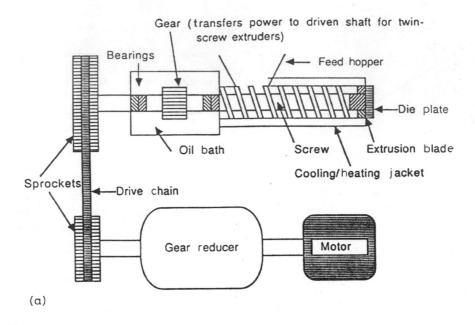

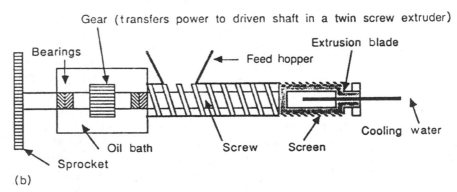

Figure 6.118. Schematic representation of (*a*) axial and (*b*) radial screw extruders.[10]

particle interstices during densification. The complete and reliable removal of this air or gas from the equipment is most important for good product quality. Because forward flow into the denser compression area and through the die opening is very restricted, air must normally flow in opposite direction of the flow of material and escape at the feed opening.

The product shape is defined by the shape and length of the opening in the die or screen. If a denser product is desired, a thicker die plate or screen is required to increase back-pressure. If feasible in regard to product size, a similar effect can be obtained by reducing the diameter of the die opening. The lower unit of this dimension is defined by the decreasing economics of manufacturing the holes and the increasing backpressure due to a reduction of the free area (relatively higher amount of land area between the holes is required for structural reasons). The upper limit on hole size is determined by the flow properties of the particular formulation, the extrusion rate, and the ability of the extruder screws to transport and compress sufficient material so that a consistent extrudate is obtained. At the same time, relatively thick die plates are necessary.

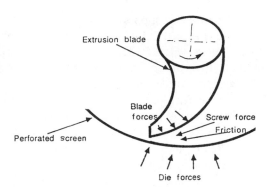

Figure 6.119. Extrusion blade and forces in radial screw extrusion.[10]

The difference between a screen and a die plate extruder is quite substantial. While die plates are 2 to 30 mm thick, screens feature usually the same thickness as the hole diameter; screens are rarely thicker than 1 mm. For extruders with the same barrel diameter, a radial discharge with screen will have more than 6 times the open extrusion area than an axial discharge with die plate.[10] This has consequences for screw design but will also generally translate into higher specific capacity and cost advantages for the radial discharge extruder if the low density and small product diameter can be tolerated.

Another continuous extrusion press that finds increasing but specialized application, mostly in the pharmaceutical industry, is the basket type extruder (Fig. 6.120).[10] This type of equipment is similar to the radial discharge extruder except that material is fed into the extrusion zone by gravity rather than screws. The perforated cylinder sits upright so that feed material falls into the basket and in front of rotating or oscillating extrusion blades with vertical axis of the rotor. The material is compressed in the nip between blade and screen and forced through the holes in the screen; the extrudates are transported to a discharge chute by a slowly rotating horizontal table. Forces developed in basket type machines are similar to those described for screw extruders except that the additional compressive force of the screw(s) is not present. These devices generally result in the least compaction of all extrusion apparatus and, therefore, the number of applications is rather limited. Attractive features for the pharmaceutical industry are: low energy input coupled with minimal temperature rise in the mass, high porosity, and quick dissolution of the product.

Power consumption, equipment geometry, wear rate, as well as capital and operating costs are all directly correlated to the internal working pressure. Therefore, there are good reasons to consider lower extrusion pressures obtained in peripheral or radial extruders and pelleting machines.

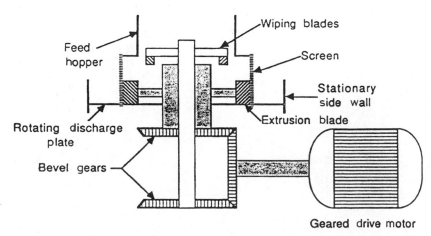

Figure 6.120. Sectional drawing of a basket type granulator with vertical rotor axis.[10]

Pelleting Machines.[11, 12] Another group of quasi-continuous extrusion machines comprises so-called pelleting machines (see Fig. 6.59b). Although, if part of a process, such equipment operates continuously, featuring uniform feed and production rates, extrusion itself is discontinuous and resembles more the process taking place in the reciprocating ram extruder. Material is first densified and then, after stationary friction in the due holes is overcome, transported or extruded. Because of design considerations, forces exerted on the mass to be pelleted (extruded) are relatively low. Therefore, binders play an important role for the technology and the product is not normally highly densified.

Figure 6.121 depicts the basic principle of pelleting. A cylindrical pressing tool (1) rolls over a layer of material depositing on a perforated (only a few holes are shown) support.[2] In the wedge-shaped nip, material is first densified and then extruded through the holes (between 3 and 4). At the point of closest approach (5) a gap remains between pressing tool and die to later obtain improved bonding between feed layers as well as better predensification and to avoid damage by metallic contact. Because materials to be pelleted normally feature considerable elasticity the residual layer expands elastically (between 5 and 6). Curve 3−m−6 represents a typical profile of the forces acting on the material in the nip and expansion zones.

The perforated support (die) can be either a flat disc (Fig. 6.122a) or concave (Fig. 6.122b) and convex (Fig. 6.122c) rings. Either the pressing tools or the die or both may be driven.

Machines with concave die rings offer advantages. Particularly, if elastic materials with a certain behavior must be pelleted, compaction force in the longer and more slender nip increases more slowly which allows for a more complete conversion of temporary elastic into permanent plastic deformation. Figures 6.123a and b show the conditions in a pelleting machine with concave die. Figure 6.123a depicts the mechanisms of compression and extrusion in the "work area," the material volume wedged in between press roller and die. Figure 6.123 explains the phenomenon. Feed deposited in a layer on the die is pulled into the space between roller and die and compressed. Neither the roll force nor the force from the die resisting extrusion (flow) is constant. The roll force increases with progressing densification while the flow resisting force remains constant until a threshold pressure defined by the static friction in the die holes, is surpassed. After extrusion (movement in the die holes) has started both the resisting and the roll forces decrease.

Friction between roller, die, and material as well as interparticle friction in the mass to be pelleted are responsible for the pull of feed into the nip region and for densification.

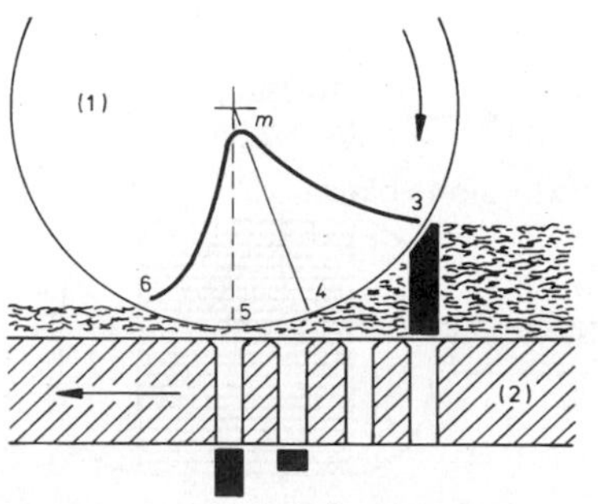

Figure 6.121. Basic principle of pelleting (for explanation see text).

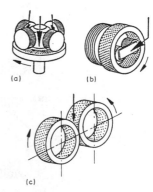

Figure 6.122. Schematic representation of the three major die designs of pelleting machines.

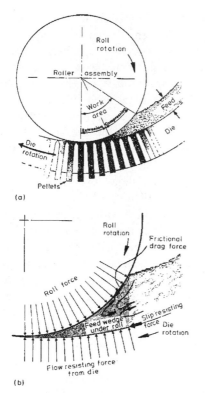

(a)

(b)

Figure 6.123. (a) Concave die and roll assembly; (b) movements of roll, die, and material and forces acting on the material.

Smooth surfaces may result in slip and low interparticle resistance to flow will result in a more or less pronounced tendency of the mass to avoid the squeeze (back flow), thus reducing densification and potentially choking the machine. While the first problem can be reduced by increasing the friction between roller and/or die and material (most press rollers feature rough or axially corrugated surfaces), the second one cannot be easily overcome unless specific machine designs are used.

As in all extrusion presses the force resisting extrusion is of great importance for the quality of pelleted products. Since in most cases the die holes are cylindrical bores, the relationship $1/d$ (bore length/diameter) determines the resistance of the die and thus the work done during compression of the material. The die, in spite of the perforations, must be structurally sound to withstand the forces that

are developing. In many cases the masses to be pelleted are organic materials that may contain fibers and feature a certain elasticity. Therefore, the holes in the die are rarely straight. Figure 6.124 shows six different die bore designs and Figure 6.125 explains the die hole characteristics. With the exception of elastic recovery, d represents the pellet diameter and L is the effective length actually performing work on the material during extrusion. T is the total, overall thickness of the die which relates to the stresses within the pellet mill. X is the counterbore depth; it reduces the die thickness T to the effective length L. The counterbore may feature a tapered bottom with angle B to obtain a gradual elastic expansion during extrusion and avoid structural defects in the pellet. Other counterbores have a square bottom; this design or straight holes with no counterbore can be used for plastic materials with no or negligible elastic expansion. The tapered inlet, from diameter D to pellet diameter d with angle ϕ, is required to either increase the open area without sacrificing die strength or to obtain additional compression according to D^2/d^2. The first effect is particularly important for fibrous materials that may produce matting and ultimately clogging of the die if too large land areas exist between the holes, and the second reason for a tapered inlet may be dictated by the need for high overall densification of feeds with very low bulk density.

While the press rollers can be hardfaced and may be rebuilt in case of wear by overlay welding, the dies deteriorate by "washing-out" the inlet and discharge cones, if applicable,

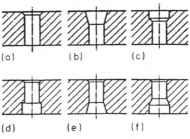

(a)   (b)   (c)

(d)   (e)   (f)

Figure 6.124. Six typical die bore designs.[11]

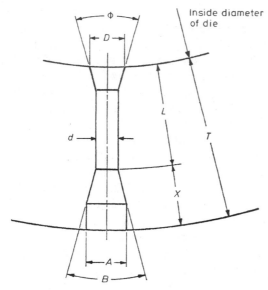

Figure 6.125. Die hole characteristics.[12]

and increasing the bore diameter d. Die life is determined by the increase of d to such dimensions that either the product size is no longer acceptable or the backpressure becomes too low (reduced compression and, therefore, inadequate product density and/or strength). If neither characteristic is critical, the limiting die life is defined by the remaining structural integrity of the die.

Particularly if pellets with high density or strength and small diameter must be produced, the necessary die thickness and effective hole length may be rather incompatible. In such cases replaceable insert plates with short bores may be used (Fig. 6.126a). Other inserts may be used as replacements in case of wear and to salvage the overall die body (Fig. 126b)[13] or to fulfill process requirements, such as cooling of the dies.[14]

Figure 6.127 shows two typical designs of pellet mills. Figure 6.127a is a partial cut through a machine with concave die rings and Figure 6.127b is a photograph of the working parts of a flat die pellet mill. For structural and process reasons the perforated concave or flat rings cannot be very wide (Fig. 6.128); therefore, to increase the capacity of a given press and more uniformly distribute the load, up to three rollers are installed in concave die presses (Fig. 6.129) and up to five rollers are used in flat die presses (Figs. 6.127b and 6.128b). Adjustable plows direct the feed in front of each press roller (Figs. 6.127b and 6.129), thus approximately increasing the capacity by the number of rollers used. Because in flat die pellet presses additional, potentially unwanted shear develops between cylindrical rollers and the die plate, relatively narrow rollers and perforated die areas are used. Press

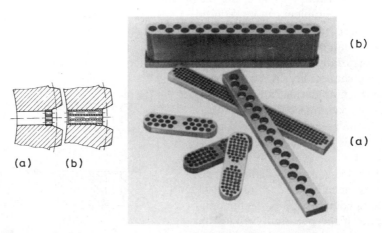

Figure 6.126. Replacement insert plates with (*a*) short and (*b*) long bores (BEPEX/Hutt, Leingarten, Germany).

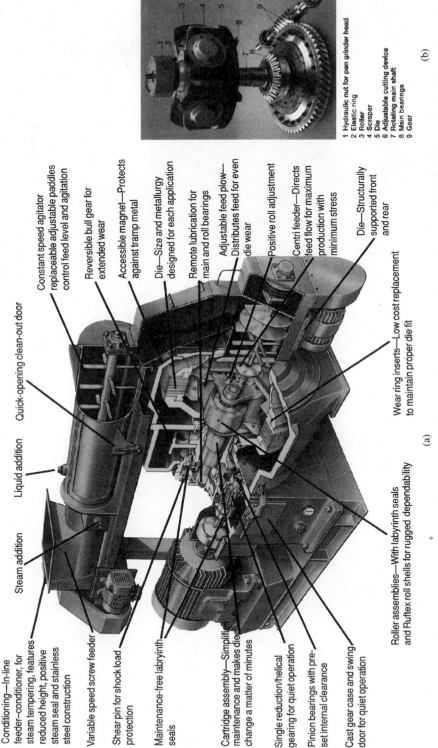

Conditioning—In-line feeder–conditioner, for steam tempering, features reduced height, positive steam seal and stainless steel construction

Steam addition

Liquid addition

Quick-opening clean-out door

Constant speed agitator replaceable adjustable paddles control feed level and agitation

Reversible bull gear for extended wear

Accessible magnet—Protects against tramp metal

Die—Size and metallurgy designed for each application

Remote lubrication for main and roll bearings

Adjustable feed plow— Distributes feed for even die wear

Positive roll adjustment

Centri feeder—Directs feed flow for maximum production with minimum stress

Die—Structurally supported front and rear

Variable speed screw feeder

Shear pin for shock load protection

Maintenance-free labyrinth seals

Cartridge assembly—Simplifies maintenance and makes die change a matter of minutes

Single reduction/helical gearing for quiet operation

Pinion bearings with pre-set internal clearance

Cast gear case and swing door for quiet operation

Roller assemblies—With labyrinth seals and Ruftex roll shells for rugged dependability

Wear ring inserts—Low cost replacement to maintain proper die fit

(a)

1 Hydraulic nut for pan grinder head
2 Elastic ring
3 Roller
4 Scraper
5 Die
6 Adjustable cutting device
7 Rotating main shaft
8 Main bearings
9 Gear

(b)

Figure 6.127. (a) Partial cut through a pelleting machine with concave die rings (Andritz Sprout-Bauer, Muncy, PA, USA); (b) photograph of the working parts of a flat die pellet mill (Amandus Kahl Nachf., Reinbek, Germany).

(a)

(b)

Figure 6.128. (*a*) Typical concave pelleting die rings (Andritz Sprout-Bauer, Muncy, PA, USA); (*b*) flat dies and roller assemblies (Amandus Kahl Nachf., Reinbek, Germany).

rollers should be conical if a larger area of the die plate shall be utilized.[15] As with other pressure agglomeration methods, density and strength of pellets can be improved if two machines operate in series.

A particular advantage of flat die pellet presses is their applicability for very wet pastes or sticky materials. If the drive mechanism is moved to the top of the machine nothing interferes with the unobstructed discharge of

Figure 6.129. Three-roll assembly and feed plows in a concave pelleting machine (Andritz Sprout-Bauer, Muncy, PA, USA).

pellets from the flat die (Fig. 6.130). Pellets from presses with top drive (Fig. 6.130b) either fall as discrete particles onto a fast-moving conveyor where they remain separate entities or discharge directly into a dryer or cooler where they are immediately flushed by air and, therefore, do not stick together.

6.5.3.3 Peripheral Equipment

Conditioning and Product Treatment. Because suitable feed for extrusion equipment must feature specific characteristics, particularly some plasticity and lubricity, many material need conditioning by heating and moistening or steaming as well as mixing with solid and/or liquid additives. Conditioners may be

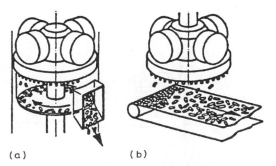

(a)   (b)

Figure 6.130. Diagrams showing two different designs of flat die presses.[11] (a) Bottom drive, (b) top drive for very wet pastes or sticky materials.

paddle and screw type mixers which can be an integral part of the extruder (see, e.g., Fig. 6.127a) or, in those cases where longer conditioning times are necessary, are separate pieces of equipment.

Figure 6.131 shows three typical mixer conditioners commonly used with pelleting machines.[11] The simple screw type machine (Fig. 6.131a) offers only limited mixing capabilities but is best suited for long, fibrous, and bulky materials. The unit shown in Figure 6.131b combines, in-line, a metering screw and a paddle mixer while the design of Figure 6.131c features a separate metering screw feeding the paddle mixer. Because in the latter arrangement, metering screw and paddle mixer are

driven separately, intensive mixing can be achieved at any feed rate. Consequently, relatively large amounts of liquid and/or solid additives can be introduced.

In many cases it is preferential to use steam for heating and moistening; this technique commonly results in higher extrusion rate (capacity), increased die life, decreased power consumption, and improved quality of the extrudate. These characteristics are most reliably obtained if conditioning takes place in separate machines in which residence times of 5 to 30 min can be achieved. Figure 6.132 shows schematically the conditioner of such a system in which material is constantly moved with slowly rotating scrapers and transported from deck to deck while steam is injected and other additives, such as molasses or fat, are incorporated.

Depending on the amount of moisture and/or heat added prior to the extruder, the

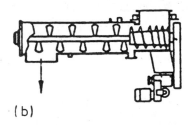

(a)

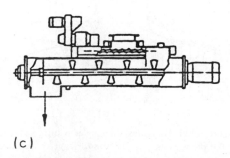

(b)

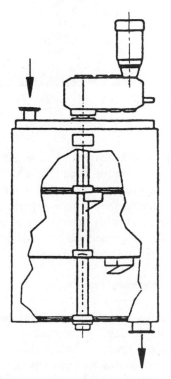

(c)

Figure 6.131. Diagrams of three different paddle- and screw-type conditioners.[11]

Figure 6.132. Schematic representation of a vertical conditioner with long dwell time.[11]

product must be dried and/or cooled. Although sometimes sophisticated equipment is necessary for these tasks, simple louvered, vertical pellet coolers with gravity flow and application of ambient air for product cooling and drying (Fig. 6.133) are commonly used in the animal feed industry, which is the single largest application of pelleting.

Spheronizing. As mentioned previously, feeds suitable for extrusion must be somewhat plastic. In fact, many wet mixtures that, during compaction, become too pasty for use in any other agglomeration equipment can be successfully densified in extruders and shaped into discrete agglomerates. Such products are still easily formable and, therefore, can be further treated in so-called spheronizing equipment to yield uniform round particles.[16]

Spheronization was developed in the 1950/60s,[17] primarily for the pharmaceutical industry where rounded particles are needed for more uniform coating.

Spheronization begins with wet extrudates obtained from one of the previously described extruders, preferentially the low-pressure type machines. Because very often small spherical particles are desired, the extrudates tend to be relatively long and thin.

A spheronizer consists of a vertical hollow cylinder (bowl) with a horizontal rotating disc (friction plate) located inside (Fig. 6.134). The spaghetti-like extrudates are charged onto the rotating plate and break almost instanta-

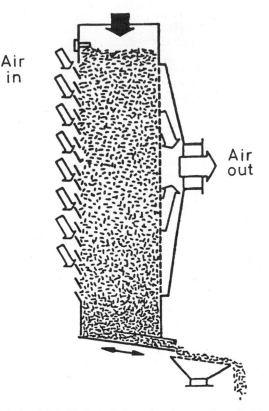

Air in

Air out

Figure 6.133. Typical design of a louvered, vertical pellet cooler.

neously into short segments of uniform length. The friction plate surface has a variety of textures designed for specific purposes. Often, a grid is applied,[16] the pattern of which is related to the desired particle size (Fig. 6.135).

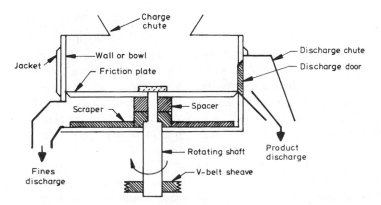

Figure 6.134. Diagram depicting the typical design of a spheronizer.[16]

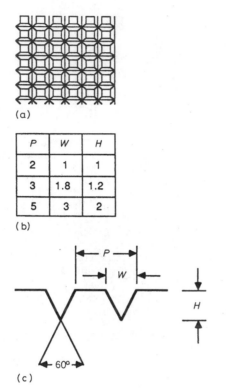

P	W	H
2	1	1
3	1.8	1.2
5	3	2

Figure 6.135. Common grid patterns of the friction plate of a spheronizer.[16] For explanations see text.

For example, a 1-mm granule would be processed on a friction plate with 50% to 100% larger groove openings, that is, 2 mm. The wider groove allows the extrudate to fall into the opening so that the leading edge of the peak will fracture the pellet into pieces with a length-to-diameter ratio of 1.0 to 1.2.

The still plastic pellet segments are being worked by further contact with the friction plate as well as by collisions between particles and with the wall. Mechanical energy is transformed into kinetic energy and the mass of particles rotates in a torus-shaped ring in the apparatus. Continued processing will cause a gradual deformation into spherical shape.

During deformation and further densification excess moisture may migrate to the surface or the mass can exhibit thixotropic behavior. In such cases, a slight dusting by means of a suitable powder dispenser reduces the likelihood of particles sticking together. Other spe-

cial features may include cooling or heating of the bowl through a jacket and cleaning of the friction plate with brushes.

Spheronization equipment principally operates batchwise. Quasi-continuous operation is possible by means of multiple batches or cascade flow. Either one of these methods used two or more spheronizers. Multiple batch operation, for example, using two spheronizers, is sequenced such that one unit discharges while the other is in the middle of the spheronizing cycle. A reversing belt can be used to alternatively feed each machine. In cascade operation two or more units are linked in series to extend the total residence time. Feed is continuously charged into the first spheronizer and continuously overflows into the next one(s).

6.5.4 High-Pressure Agglomerations

6.5.4.1 Die Pressing[1]

Die presses for compacting powder are the oldest pressure agglomeration machines. They are used by numerous industries for a wide variety of purposes. The largest user is most probably the pharmaceutical industry (see Section 6.5.4.2). However, they are also widely used in the ceramic, powder metal, confectionary, catalyst, and, to an increasing extent, the general chemical industries. The machines can be divided into two main categories: reciprocating or single-stroke machines and rotary machines.

Reciprocating Machines. Reciprocating presses operate with one upper and one lower punch in a single die (see Figure 6.60). They are mainly used for complex shapes where high pressure and/or low outputs are required (less than 100 compressions per minute).

Reciprocating machines can be subdivided into two types: ejection presses and withdrawal presses.

Ejection Presses. Ejection presses are built as very simple hand-operated units and as highly complex machines operating at up to about

$1000 \ MN/m^2$ pressure and producing compacts with a very high degree of accuracy.

The hand-operating press incorporates basic features common to all ejection presses. The die is mounted in a fixed plate and the upper and lower punches are attached to moving rams. The lower punch descends to allow the die to fill. All the compression is carried out by the upper punch moving toward the stationary lower one. Later, the lower piston ejects the compact upward from the die.

Hand operating machines are very limited in performance. They are only capable of exerting a pressure of 8 to 16 MN/m^2 and the output, obviously, depends on the operator. It is extremely difficult to predict the behavior of particulate matter at high pressure in a rotary press from data obtained by using a hand-operated machine.

A range of mechanical or hydraulic presses has been developed from the hand machine. They vary in the size of compacts that can be produced and the amount of pressure that can be exerted to form the tablet. The smaller machines are used in the pharmaceutical industry for products in which only limited output is required and, to a certain extent, for development work. Larger machines are mainly applied by the powder metal and ceramic industries, but even there, the use is limited in most cases to compacts that feature no change in cross-section, such as washers and short bushings.

The disadvantage of the machines in this category is that they produce a compact that varies considerably in density from top to bottom because the pressure is exerted only by the top punch (see Fig. 6.111). This is not particularly important in the pharmaceutical industry, although in extreme cases it could produce a tablet that disintegrates more rapidly on one side than the other. This disadvantage is of much greater consequence to the ceramic and powder metal industries, where the difference in density will cause uneven shrinkage during sintering. To overcome this problem, some ejection presses are built with "double pressure," that is, the pressure is applied equally to the upper and lower punches.

Withdrawal Presses. Withdrawal presses operate with two cams. The top cam controls the movement of the upper punch and, in turn, the lower cam controls the movement of the die. Whereas the majority of ejection presses are mechanically operated, both mechanical and hydraulic drives are common for the withdrawal type.

In a withdrawal press, compaction and ejection take place with a continuous downward movement of the upper punch and the die (Fig. 6.136).

At the beginning of the pressing cycle, the die is positioned on top of the lower punch to produce the required depth of fill. In fact, the material to be compressed is fed to the die during the return move to avoid the necessity to replace air with the solid feed. The upper punch then descends to compress the material and the die also moves downward during the compression to maintain uniform density in the compressed material. At the end of the pressure stroke, the die continues to move downward until it has been completely removed. During ejection, the compact is supported by the lower punch.

Tooling for this type of press is much more expensive and complex than that required for ejection presses. It consists of a complete die set that is removable from the machine as a complete unit. This has the advantage that the tooling is interchangeable between presses. Further advantages lie mainly in its adaptability to the production of complex components. It is also possible to obtain greater accuracy. Compacts can be made on this type of tooling with dimensional tolerances of less than 4×10^{-5} mm.

In practical terms, apart from the output, effectiveness of mechanical and hydraulic pressure systems is equal. The cycle time of the hydraulic press varies with the stroke. The low-pressure stroke can be made quite fast by using a multistage pump but as the higher pressures cut in, the remainder of the stroke

Movement of
upper punch

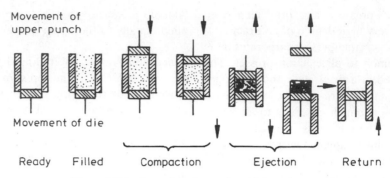

Movement of die

Ready   Filled   Compaction   Ejection   Return

Figure 6.136. Operating phases of a withdrawal press.

becomes progressively slower. The length of the high-pressure stroke depends directly on the thickness of the piece being pressed. In addition, the pumping system of the hydraulic press can rarely achieve a cycle time comparable with the mechanical press when it is used near maximum pressure. Therefore, the use of hydraulic presses is restricted to the ceramic and powder metal industries because of low output and compacts requiring very high pressures. It is also applied in the recycling industry for the reproduction of large, cylindrical compacts from, for example, metal-bearing wastes.

Rotary Machines. Rotary machines were developed to meet the demand for higher outputs of relatively small tablets, primarily in the pharmaceutical industry. Their basic principle of operation is similar to that for hand-operated machines with the exception that the dies are mounted in a rotating table and pass, in turn, under a feed position (Fig. 6.137). The tooling design resembles the one used on the simpler ejection presses whereby the punches are moved by a series of cams. The design of tooling limits the shape of compacts to those that can also be produced on the simpler type of single-stroke machines.

The feed is supplied to the die table by an open frame. The lower punch is pulled down by a cam to the lowest position while the die fills with powder. It then rises up an adjustable ramp, ejecting excess powder from the die. The surplus is scraped off flush with the top of the die table at the highest point of the "weight adjusting ramp," leaving the desired volume of material to be compacted. It is common practice for the lower punch to drop slightly after the surplus material has been scraped off and

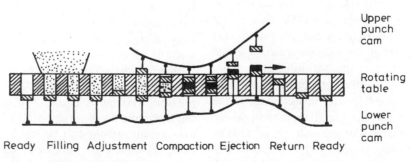

Upper
punch
cam

Rotating
table

Lower
punch
cam

Ready Filling Adjustment Compaction Ejection Return Ready

Figure 6.137. Operating schematic of rotary tabletting machines.

before the upper punch enters the die. This is done to prevent the upper punch displacing material from the die as it enters. The material is compressed by the two punches passing between two rolls, one or both of which are spring loaded. This produces the effect of double pressure. Therefore, the problem of making a compact with uneven density is not very pronounced in rotary machines. Finally, the upper punch is lifted out of the die by a cam and the lower punch travels up another cam to eject the compact from the die.

The simplest type of rotary machines is "single-sided" (one feed location); one tablet is produced from each station (die) per revolution. Therefore, the output of rotary machines depends on the number of stations in the turret (table) and the speed of the turret. It is usually in the region of 300 to 800 tablets per minute. A further increase in output is possible by using a "double-sided" machine. In this case the stations are filled twice on opposite sides of the rotating table; two compressions are carried out in each die per revolution of the turret. Outputs of up to 3000 tablets per minute can be obtained from the "double-sided" machine. Still further capacity increases can be obtained by dual or multiple tooling (two or more dies) per station.

Special Design Features of Die Presses. (Many of the schematic drawings used in this section are reproduced from the "Powder Metallurgy Equipment Manual"[18] with permission of the Metal Powder Industries Federation. Special die presses for the pharmaceutical industry are described in Section 6.5.4.2).

Shapes. The original and still most common shape of die pressed agglomerates is a more or less cylindrical "tablet" (Fig. 6.138).[19] Included in this description are flat, faceted, and crowned compacts. For these shapes, simple die and punch configurations are applicable. Structured shapes can be necessary in Powder Metallury (P/M) where a classification of I through IV characterizes the complexity of part design.[18]

One-level, relatively thin tablets or parts with any contour (Class I of PM, Fig. 6.139) can be pressed with a single punch and force may be applied from one side. The maximum dimension A (Fig. 6.139) depends on the particulate feed and the shape of the compact. Thicker parts (Class II of P/M, Fig. 6.140), while still requiring only simple tooling, must be pressed from two directions. Holes are obtained by the installation of mandrels or core rods.

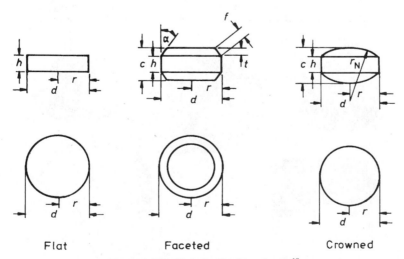

Flat   Faceted   Crowned

Figure 6.138. "Standard" tablet shapes.[19]

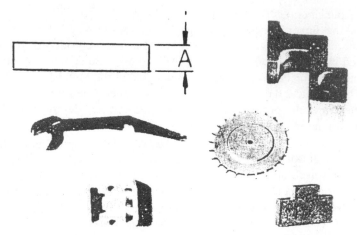

Figure 6.139. P/M classification: Class I parts.[18]

Because, owing to interparticle friction there is little or no hydrodynamic flow of particulate solids during compaction, each level of more complicated parts must be supported with a separate punch or die member to maintain reasonably uniform density throughout the green pressed part (Class III and IV of P/M, Figs. 6.141 and 6.142).

Drives. The above product shapes are usually made in mechanically operated die presses. Advantages of mechanical presses are: high production rates, low power requirements, and a large range of applicable pressing forces. The most common mechanical drives are: eccentric or crank, toggle, cam, and rotary arrangements.

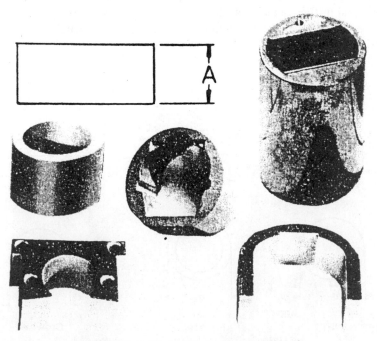

Figure 6.140. P/M classification: Class II parts.[18]

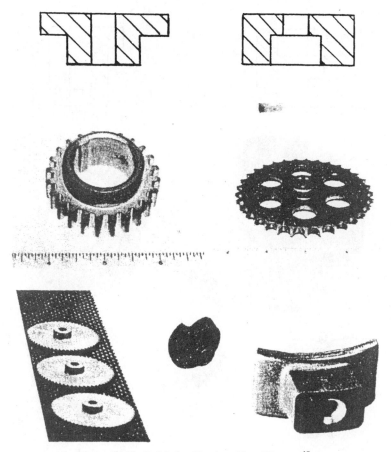

Figure 6.141. P/M classification: Class III parts.[18]

Figure 6.143 represents eccentric or crank type drives which convert rotary motion to linear, reciprocating movement. The mechanisms feature small final rate of pressing speed (approaching bottom dead center) and high loading with low torque at maximum compression (at bottom dead center). The stroke can be adjusted on the eccentric cam or "Pitman" link. Normally, this method is used when force is applied from only one side and, typically, it drives the top punch.

Another common drives mechanism is the toggle (or knuckle) type (Fig. 6.144). Actuation is normally accomplished by eccentric or crank arrangements that alternatively straighten and bend a jointed arm or lever. If one end of this lever is fixed, the other—if guided properly—will produce a reciprocating motion. The stroke can be adjusted as mentioned previously. Final pressure will be even higher and pressing speed near the end of compression is minimal.

Figure 6.145 depicts schematically the cam drive. Cam and lever arrangements are used to convert rotary motion to linear movement. Pressing speed, timing, and motion are adjustable by changing the contours of the cams or cam inserts.

The cam drive is mostly used for rotary die presses which feature a series of punches and dies arranged in a common, rotating, tool holding table (turret) (see also Section 6.5.4.2 and Fig. 6.163). The stationary axis around which the turret rotates provides a fixed reference point for mounting the press cams and pressure rolls.

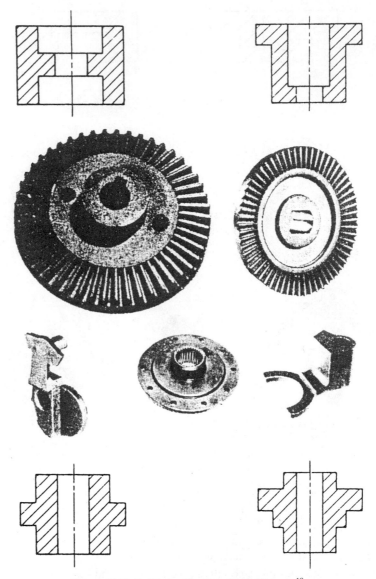

Figure 6.142. P/M classification: Class IV parts.[18]

The disadvantage of all mechanical punch drives is that, while compression speed becomes smaller as the eccentric connection of the rotating drive member approaches dead center and cam drives may follow curves that allow a certain "dwell-time" at maximum compression, compaction takes place very quickly with a sudden release of force after reaching the maximum. This is a particular problem if the material to be compacted has elastic prop-

erties. Such products reach sufficient permanent (plastic) deformation and strength only after remaining under pressure for some time. Premature pressure release results in excessive elastic spring-back which may destroy the structural integrity of the compact and result in well-known failure modes (e.g., capping, lamination, etc.) indicating "overpressing."

The only reliable way to overcome this problem in die presses is to employ hydraulic

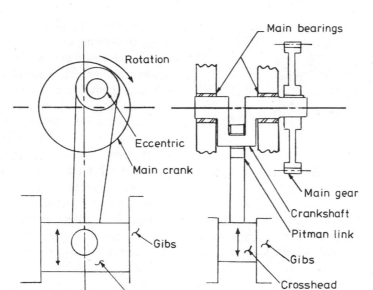

Figure 6.143. Eccentric or crank drive arrangement.[18]

actuation of the punch(es) (Fig. 6.146). The timing of the punch strokes as well as the rate of increasing or decreasing pressure and the "dwell-time" can be easily adjusted. In addition, hydraulic presses typically feature overload protection by means of gas filled accumulators and allow the densification of larger amounts of feed even with low initial bulk density. Because there is no physical limit to the length of the stroke, densification ratios can be very high; and, since pressure rise can be slow, final pressure high, and "dwell-time" adjustable without limiting constraints (other than capacity), elastic materials, such as organic refuse or other organic materials and, for example, steel turnings can be successfully compacted. Figure 6.147 is the sketch of a large, hydraulic, horizontally oriented high-pressure press.

More conventional presses feature vertical design (Fig. 6.148). They can be highly automated and, with multiple tooling, producing several compacts per stroke, as well as auto-

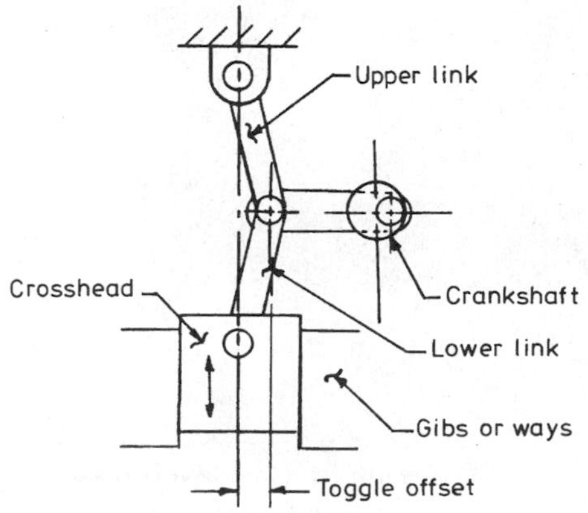

Figure 6.144. Toggle or knuckle drive mechanism.[18]

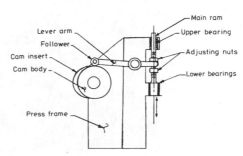

Figure 6.145. Schematic representation of the cam drive principle.[18]

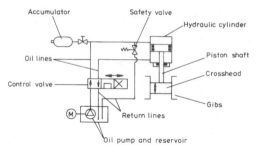

Figure 6.146. Schematic representation of a hydraulically driven press.

matic feeding and product handling systems, can have considerable capacities. Typical applications are in the refractory industry for making brick. But many other uses are conceivable as demonstrated in Figure 6.149, which shows a selection of parts.

Press Feeders. To obtain parts with high accuracy of volume and density it is necessary to employ automatic powder feeders. The design of such mechanisms is simpler for the otherwise more complicated rotary presses then for the much less complex machines with fixed tool carrier table. Rotary presses employ a stationary filling shoe (see Section 6.5.4.2, Fig. 6.162); because of the high rotational speed of the turret the feed must be free flowing and, therefore, is often preagglomerated (granulated). To further improve feeding and guarantee uniform filling at high rotating speeds, "power feeders" are employed. Their design is such that they can be easily removed and opened for cleaning.

Feeders for presses with stationary tool holders can be divided into direct shuttle, metered shuttle, and arc type feeders.[18] The di-

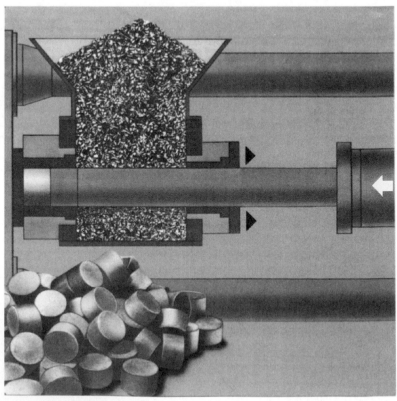

Figure 6.147. Diagram showing the principle design of a large hydraulic press with horizontal punch movement (Lindemann, Düsseldorf, Germany).

Figure 6.148. Typical vertical hydraulic press for the manufacture of refractory brick (Horn, Worms, Germany).

rect shuttle feeder (Fig. 6.150) may also be used on moving die tables. It provides a straight in-line reciprocating action over the die with the feed shoe connected directly to the supply hopper. The motion of the metered shuttle feeder (Fig. 6.151) is the same as that of the in-line system (Fig. 6.150) and may also be applied on moving die tables. It does not have a direct connection with the supply hopper. At fill, it moves with a metered amount of material from a position under the hopper to the die cavity. This system supplies the same amount of material on each press stroke. The arc type feeder (Fig. 6.152) is normally applied only on mechanical presses with a stationary table. It uses a pivoting action of the feed shoe over the die area.

Control of the lower punch and the feed shoe is typically such that material is transported to the die area when the punch is still in or near the ejection (highest) position. This avoids the cavity filling with air which must be replaced by feed and finally expelled during compaction. Particularly with high-speed

Figure 6.149. Selection of different products made with vertical hydraulic presses (Horn, Worms, Germany).

presses, sufficient deaeration may pose a major problem and compressed pockets of air can be an important cause of tablet failure (e.g., capping).

Tooling Design.   Since particulate solids do not flow under pressure, friction within the mass and on the tool walls absorbs part of the force applied by the punch(es). The "neutral axis" is the low-density zone approximately perpendicular to the direction of pressing. Control of the location of this zone in the compacted part is often important (e.g., to avoid distortion of P/M parts during sintering) and is achieved by the relative tooling motions. Under pressure, particulate matter will also not flow from one part level to another. Therefore, when parts of more than one level are pressed,

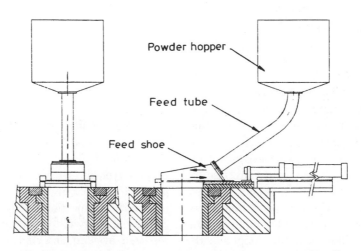

Figure 6.150. Schematic representation of the direct shuttle feeder.[18]

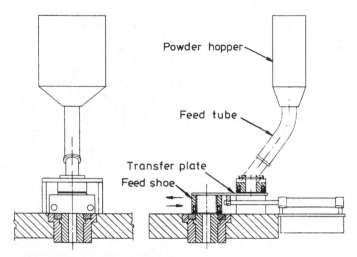

Figure 6.151. Diagram of the metered shuttle feeder.[18]

separate pressing forces must be applied simultaneously for each level. As a result, there will be a neutral axis for each part level (Fig. 6.153).

Figure 6.154 demonstrates how the location of the neutral axis of a simple, one-level part can be controlled in a die press with upper punch pressing and controlled withdrawal die (see also below and Fig. 6.159).

As far as variety of applications, complexity of shapes, and accuracy of parts are concerned, die pressing is the most versatile agglomeration method. To achieve this versatility, the basic principle of die pressing is often modified. The most important methods, reflecting the significance of the technology, are reviewed in the following.

Single-Motion Pressing

This is the simplest method and is usually limited to compacting relatively thin parts with or without through holes obtained by the installation of core rods. Only one part of the tooling is moved during compression.

Figure 6.155 depicts schematically the three stages of upper punch pressing. Other single-motion pressing designs are sketched in Figure 6.156. During sliding anvil pressing (Fig. 6.156a) the lower punch movement accomplishes filling, compaction, and ejection. Normally, powder feed, anvil, and pick-up are three separate components brought in place by a "positioner." Figure 6.156b shows the "Pentronix unitized anvil" in which all three functions are combined into one assemblage which is always in contact with the die plate. Powder spillage and blow-out are reduced to practically zero, making this design ideal for, for example, the processing of toxic materials. In anvil withdrawal pressing (Fig. 6.156c) the lower punch remains stationary while the die table is moved into positions for filling, compaction (with anvil in place), and ejection.

Double-Motion Die Pressing

This method will produce parts with more uniform density. Double-motion pressing provides force to the particulate mass to be compacted simultaneously from top and bottom through movement of two parts of the tooling, for example, the upper and lower punches (Fig. 6.157). A similar effect can be obtained by upper punch pressing with floating die (Fig. 6.158) whereby the die table moves if the frictional forces overcome the supporting or counterbalancing force holding the die. This die travel has the same effect during compaction as an active lower punch. Ejection can

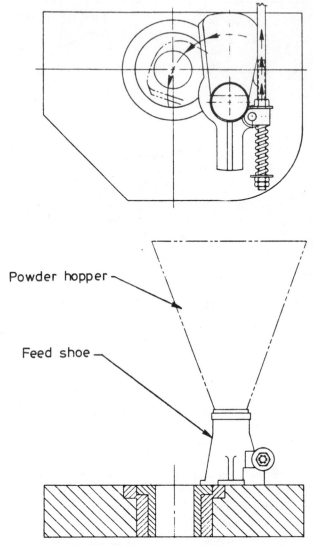

Figure 6.152. Diagram depicting the arc-type feeder.[18]

be accomplished by movement of either the lower punch (Fig. 6.158a) or the table (Fig. 6.158b, upper punch pressing, lower fixed punch, floating withdrawal die). Potential disadvantages of this system are that compacted

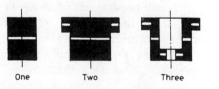

Figure 6.153. "Neutral axes" in single- and multilevel parts.[18]

parts may have density variations that are determined by the size of the supporting force of the die and that the neutral axis may not be located in the center of the part.

In upper punch pressing with controlled withdrawal die (Fig. 6.159) adjustment of timing of die travel provides positive control over the position of the part's neutral axis (see also Fig. 6.154).

Multiple-Action Pressing

Multiple-action pressing systems are those that support and compact each level of multilevel

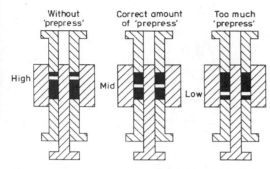

Figure 6.154. Possibilities to influence the "neutral axis" position in upper punch pressing with controlled withdrawal die.[18]

parts with a separate punch or tool member. Such tooling is used to minimize density gradients in complex compacts. Therefore, all of these more sophisticated machines also use one or the other method of double-motion pressing as demonstrated in Fig. 6.160. With descriptions the figures are self explanatory.

In the foregoing, the need to minimize density variations was mentioned several times. Excessive density gradients may cause destruction of compacts (capping, laminating, cracking, etc.) and deterioration of parts during finishing (firing, sintering, etc.). In addition to double-motion pressing and multiple-action tooling, it is sometimes necessary to decrease friction by the addition of lubricants. Because in most cases lubricants are impurities and costly it is desirable to keep their amount as low as possible. Lubricants can reduce interparticle and/or die wall and tooling friction.

In most cases, however, the lubrication is required only on the die walls and tooling. In fact, if lubricants are blended into the mixture to be compressed, the normally hydrophobic additives may reduce product quality. Therefore, new developments are directed toward the lubrication of only the tool surfaces.

Tooling design, tolerance, and finish are of utmost importance for die pressing and the quality of compacted parts. The die holder (table, turret, etc.) normally has larger holes into which die inserts are mounted. Whereas for simple, cylindrical contours sleeves can be clamped or shrunk into the openings, designs and mounting of noncylindrical die configurations require considerable know-how and skill. The problem is aggravated by the need to produce dies from abrasion-resistance material (e.g., carbides) and to provide tight tolerances with high-quality surface finish. Often, dies must be made up of different parts as shown, for example, in Fig. 6.161.

For improved deaeration and release of the compacted part, die walls and core rods are often slightly tapered. However, clearances must be small enough to retain the particulate solids in the compression chamber. Die cavities and core rods must has a high-quality surface finish (polished, lapped, etc.) and strong supports must be provided to avoid distortion under pressure. In multiple tooling arrangements, some punches must also partially serve as a die. In such a case, the punch

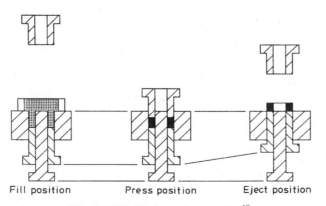

Figure 6.155. Single-motion pressing.[18]

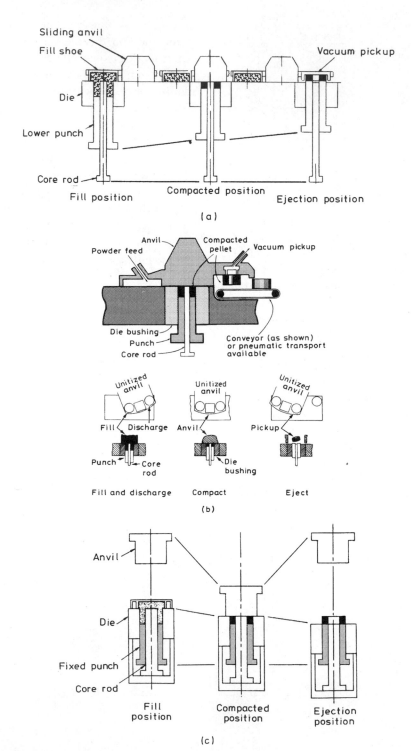

Figure 6.156. Different sketches representing anvil pressing. (*a*) Sliding anvil,[18] (*b*) unitized anvil (Pentronix, Lincoln Park, MI, USA), (*c*) anvil withdrawal pressing.[18]

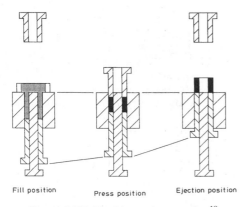

Figure 6.157. Double-motion pressing.[18]

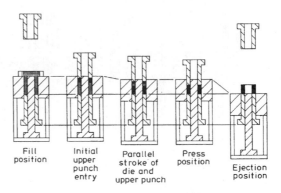

Figure 6.159. Upper punch pressing, controlled withdrawal die.[18]

must be backed up for the full length of the compact to provide rigidity.

In addition to these basic designs and general requirements there are large numbers of supplemental machine and tooling options. Requirements for nonstandard equipment or process characteristics must be determined for each particular application and discussed with

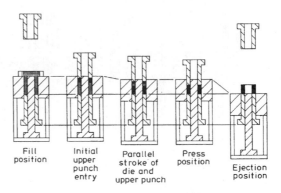

Figure 6.158. Sketches depicting presses with floating die.[18] For explanation see text.

the machine manufacturer. As usual in the field of agglomeration, suppliers of die presses maintain well-equipped technical centers in which special requirements can be tested and machine modifications are developed as necessary.

Today, the main thrust in new developments for die presses in in the area of machine data measurement and control.[20] Techniques have become available to accurately measure the parameters during a press cycle that may last only a few hundred milliseconds. Based on such information, production machines can be programmed and automatically controlled.

6.5.4.2 Tabletting in the Pharmaceutical Industry[21]

Compressed tablets are the most common pharmaceutical dosage form. The reasons for this are: (1) they are convenient, compact, easy to carry and ship, and (2) they are usually more chemically stable than other dosage forms, since most drugs decompose by hydrolysis.

This was the first paragraph of Chapter 7, Part 3 of the first edition of this book in which Carstensen[21] described in much detail "machines, manufacturing procedures, formulation parameters, and basic principles on which formulation principles are based." He also included an extensive list of machine specifications, "since such a compilation has been

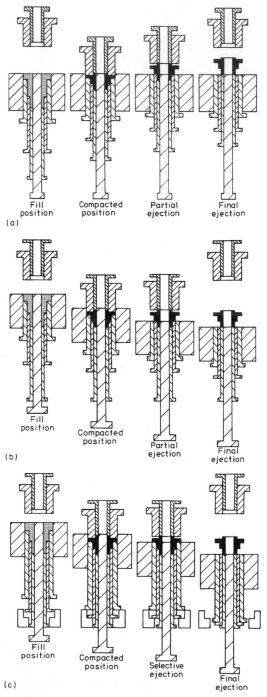

Figure 6.160. Schematic representation of three multiple-action tooling systems.[18] (*a*) Upper and lower punches pressing, die stationary; (*b*) upper punches pressing, floating withdrawal die; (*c*) upper punches pressing, controlled withdrawal of die and lower punches.

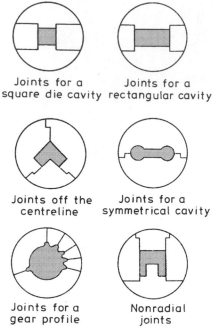

Figure 6.161. Six examples of die inserts with preferred location of joints for noncylindrical cross-sections of parts.[18] The shaded area is the die cavity.

deemed useful for both the reader and the practitioner."

Because this topic is very specialized but of considerable interest to a large industrial segment using pressure agglomeration, a much shortened version will follow. In particular the "extensive lists of machine specifications" are not given because they are no longer current and can be obtained readily from the manufacturers of tabletting equipment if desired.

Tablet Machines.   The first tablet machines were introduced in the nineteenth century, and have by now been developed into sophisticated, high-precision tools.

They may be either single-punch (eccentric) machines (Fig. 6.162) or rotary presses (Fig. 6.163, see also Fig. 6.137).

In the *eccentric machine*, powder flows from the shoe into the die in position 1. The shoe then swings away, and the upper punch is lowered to compress and powder (position 2). Both punches then are raised (position 3),

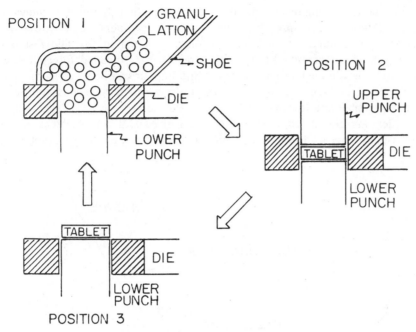

Figure 6.162. Steps in the formation of a tablet on a single punch (eccentric) machine.

lifting the tablet out of the die, and the hopper then comes back into its original position (and knocks the ejected tablet onto the discharge chute). The powder level in the shoe is maintained by gravity feed from the hopper.

The fill weight can be adjusted by the (low) position of the lower punch. The lower it is, the higher the fill weight. The fill weight is also a function of the apparent density and the flow rate of the powder. The compression pressure (and hence the tablet hardness and porosity) can be adjusted by the (low) position of the upper punch.

In a *rotary press* there is a series of dies positioned circularly on a die table (Fig. 6.163a). The upper and lower punches glide on cams (Figs. 6.163b, c, d). An evoluted picture is shown in Figure 6.163b. The filling takes place between points A and B, that is, under the feed frame. This in turn is fed by the hopper. The powder is leveled (scraped) at point B, so that the fill is a function of the level of the lower punch at this point. As the table rotates (goes from right to left in Figure 6.163b), the die passes the feed frame, and the

lower punch drops a small amount. With the pressure wheels, the upper punch is brought down and the lower punch raised to form the tablet. Both are then raised (by the cam contour), and the tablet is ejected. Point A' corresponds to A (the back end of the feed frame which serves as an ejection bar for the tablet). It is obvious from the drawing that tablet weight can be adjusted by screw E, ejection by screw F (where the ejected tablet must be flush with the table) and compression pressure by the relative position of the pressure wheels.

In the simplest case of a rotary machine there is one hopper and a certain number of "stations" (as few as four) on the die table. In other words, one rotation produces the number of tablets given by the number of dies (and punch sets) on the machine.

Expulsion of entrapped air from a granulation or (particularly) a powder mix is important since it reduces lamination and capping of the produced tablets. High-speed machines are equipped with a precompression feature. Solids for tableting are of three types: (1) noncompressible powders, (2) compressible powders possessing poor flow, (3) compressible powders possessing good flow. Noncompressible powders are either wet granulated (which adds a binder, making them compressible) or (if they are of sufficiently low dosage level) mixed with a powder excipient of type (3), so that the mixture is compressible and free flowing.[22]

When powders are granulated, flow characteristics are usually superior to those of naturally free-flowing powders; hence direct compression powders (i.e., mixtures of type 3) are usually aided in the filling step of artificial means, the so-called forced feeders. Flow in the hopper can be of concern also (and if not uniform will cause inconsistent tablet weights).

Powders of type (2) (especially if moisture sensitive) will form tablets, but because of inconsistent flow they cannot be compressed

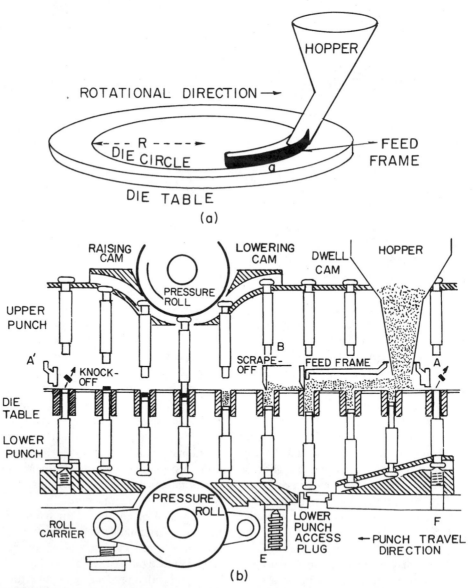

Figure 6.163. (*a*) Schematic of a rotary machine. (*b*) Path of punches during tableting on a rotary machine in evoluted presentation.[21]

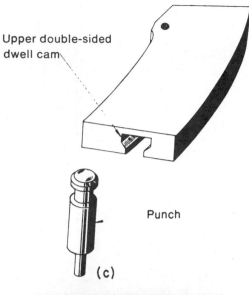

Upper double-sided
dwell cam

Punch

(c)

(d)

Figure 6.163. (c) Double-sided upper cam.[21] (d) Photograph of punch being installed on a tableting press.[21]

ture in sheets with roller presses. These slugs or sheets are broken up by milling through a suitable screen, to from fragments of a larger particle size than the parent powders. Hence flow is better, and tablets can be produced that have satisfactory weight variation.

In should be noted that a fair amount of development of tablet formulas is done at a stage where only small amounts of drug are available (the so-called stages I and II in the clinical progression of drug development), and that scale-up difficulties into high-speed equipment can be anticipated. Because of government requirements that eventual production formulas be identical to those tested in the clinic, severe pilot problems always exist in the pharmaceutical industry.

In many cases there are incompatibilities[23] among drugs, and such solids are kept apart from one another by special means, notably by double- or triple-layer tablets or by compression-coated tablets (tablet within a tablet). In the triple-layer tablet, compression takes place in several stages, requiring a special press. There are three hoppers, 120° apart on the die table. Filling takes place in three steps. In the first stage, the low position of the lower punch dictates the fill weight (of the first layer); in the next stage the "bottom" of the die is the top layer of the first filled powder; and in the last stage, the surface of the second layer is the "bottom" of the die. Intermediate "tamping" is possible, and this, for instance, improves the precision of the fill of each layer.

In the case of compression coating, a tablet is first manufactured on one press (which constitutes one half of the total press assembly), and then transferred into the half-filled larger die, with the "outer" granulation on the other half of the machine. This is then filled to the top and compressed. In both compression coating and multiple-layer tablets, the intergranulation and layer bonding and the amount of moisture are exceedingly important parameters.

In a triple-layer table, the precision of fill is less than in a conventional tablet, as far as the

directly to produce tablets with uniform tablet weights. The flow in these cases is often improved by particle-size enlargement effected by first making large tablets (slugs, boluses) on a heavy-duty machine or compacting the mix-

individual layers are concerned. Defects are primarily (1) insufficient interlayer bonding, giving rise to separation of layers, (2) unevenness of the layers (which can be seen directly if multicolor schemes are employed). In the compression-coated tablet the defects are (1) missing core, (2) poorly centered ore, which can be seen from the "outside" of the tablet, and (3) splitting caused by inadequate bonding in the outer layer. The formulation of these types of products is difficult.

Tablet Formulations. To formulate a tablet one must first know the desired size as well as shape and approximate thickness. In this manner one may estimate the approximate weight. The sum of all ingredients is, of course, the tablet weight, and estimates are then made of the required amounts of necessary ingredients. The amount is then brought to the desired weight (q.s.) with filler. A list of ingredients and approximate concentration ranges is shown below:

DRUG	EXAMPLE	RANGE
Disintegrant	Cornstarch	0–8%
Lubricant	Magnesium stearate	0–2%
Glidant	Talc	0–1%
Binder	Cornstarch	0–5%
Filler	Lactose	q.s.

Except when placebo tablets are made, the drug is present, and in an amount dictated by its nature. When a tablet is administered to a patient, it must *disintegrate* in the gastric (and intestinal) fluids. On contact with biological fluids, swelling substances such as starch, certain resins, alginic acid, and modified polyvinylpyrrolidone will expand sufficiently to "blow" apart the tablet.

When a tablet is compressed in the die, a residual force exists against the die wall (Fig. 6.164). This force P is perpendicular to the ejection force E, exerted by the lower punch during the ejection phase of the tabletting. The two forces are related by the frictional

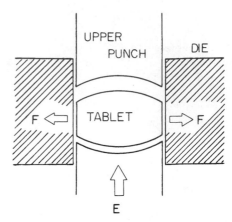

Figure 6.164. Residual die wall force F and ejection force E.

coefficient F. The function of the lubricant is to reduce the value of F. Improperly lubricated formulations will, in milder cases, give rise to tablets that are prone to cap (or that actually do cap), that is, the crown separates from the rest of the tablet. Hairline cracks in the walls of the tablet are usually indicative of this condition. In more severe cases the formulation will "bind up" in the die, and the tablet machine will stop operating. There are formulation reasons for capping as well; for instance, a too large quantity of fines will give rise to capping. The actual capping often occurs as the tablet is being ejected (i.e., actually outside the die), because at this point the tablet expands. Lubrication, machine speed, and reduction of fines are usually the remedies employed in the case of "capped tablets."

To obtain a good tablet the powder of granulation must flow wall. *Glidants* are sometimes added to improve flow, but most frequently flow is controlled by particle size and surface. The property affected by poor flow is the consistency of the tablet weight. The United States Pharmacopeia XVI states the following requirements for weight: of 20 individually weighted tablets only two may differ from the mean by more then the stated percentage, and no tablet may differ by more than twice the stated percentage:

Tablets weighing 13 mg or less	15%
Tablets between 13 and 130 mg	10%
Tablets between 130 and 324 mg	7.5%
Tablets more than 324 mg	5%

Binders are added to tablet formulations to produce granules or powders that will bind together to make a good compact in the tablet die. To describe binders, it is necessary to briefly classify manufacturing methods. These are (1) wet granulation, in which the binder is added to a paste (i.e., water is added to the granulation in the process), and (2) dry methods, in which powders are blended and compressed (direct compression); or compressed, reground, and recompressed. The pastes used in wet granulation[28] are mostly: Cornstarch paste (0 to 10%), sucrose (usually added dry, water being the granulating liquid), povidone (polyvinylpyrrolidone) (10% alcoholic solution), acacia (10% aqueous solution), and gelatin (5 to 13% aqueous solution).

Fillers are usually sugars, sugar alcohols, or inorganic substances. Lactose, dicalcium phosphate, sucrose, and mannitol are common tablet fillers. All nondrug substances in a tablet are denoted excipients.

Factors Affecting Flow and Compression. Flow rates of powders affect tableting in two ways: the flow from the hopper to the feed frame must be adequate, the flow from the feed frame to the die must be adequate.

Powder flow is a function of

1. Particle size
2. Particle shape
3. Roughness of surface
4. The chemical nature of the compound (cohesion)
5. Moisture.

In general, flow versus *particle diameter* is a parabolic function, such as shown in Figure 6.165. The maximum (d_m, W_m), where d and W are diameter and flow rate, respectively, occurs at fairly large diameters (400 to 1000 μm), so that flow problems associated with fineness and cohesiveness of powders can usually be solved by particle enlargement. The general methods employed are either wet or dry granulation or slugging.

The effect of the *particle shape* has been described by Ridgway and Rupp.[25] They define a quantity for describing particle shape (shape factor) in the following fashion: If d denotes the projected mean diameter of the particle, it is possible to express the surface A

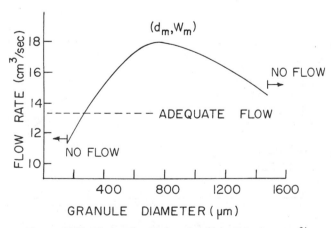

Figure 6.165. Flow rates as a function of particle diameter.[24]

and the volume V of the particle as $A = q_1 d^2$ and $V = q_2 d^3$, and the shape factor is then $G = q_1/q_2$. In general the effect of the shape factor on flow amounts to a 20% drop in flow rate with a doubling (from, e.g., 7.5 to 15) of shape factor.

The effect of orifice diameter on flow is described by the Brown and Richards Equation[26]

$$(4W/\pi\rho g)^{0.4} = mD + C \qquad (6.52)$$

The effect of the *addition of fines* to a monodisperse powder has been described for instance by Danish and Parrott.[27] The general effect of this step is shown in Figure 6.166. The amount of material that can be filled into a tablet die is the apparent density ρ' (g/cm³) multiplied by the volume V (cc) of the die cavity. If the contact time between the die and the feed frame of length a (cm) is t (seconds), then on a die table of radius R (cm) and rotational speed Ω (rotations per second),

$$t = \frac{a}{\Omega 2\pi R} \qquad (6.53)$$

In general, as long as the flow rate has a value over a critical value W' given by:

$$W' = V\rho'/t = D\Omega 2\pi/R/a \qquad (6.54)$$

the fill weight will be D (g). However, for values of $W < W'$, this is not the case, and here the tablet becomes a function of flow rate:

$$D = Wa/(\Omega 2\pi R) \qquad (6.55)$$

These relations are shown in Figure 6.167. There is no sharp break between the two linear portions predicted by Eqs. (6.54) and (6.55), and on high-speed machines, the situation is frequently in the transitional region (the curve in Fig. 6.167).

The thickness h (cm) and the hardness H (kg) of a tablet are functions of the pressure P (Pascals) applied in the formation of the tablet. This, of course, is a function of the relative distance between the two punches at their closest point of approach. The thickness h follows the Fell–Newton law:[28]

$$\ln \frac{h - h_\infty}{h_0 - h_\infty} = -k(P - P_i) \qquad (6.56)$$

and this relation is shown (in linear fashion) in Figure 6.168. h_∞ is a function of the true density of the tablet ρ (g/cm³), in that the (nonporous) mass of the compact is given by:

$$D = h_\infty \pi (D/4)^2 \rho \qquad (6.57)$$

h_0 is given by the apparent density (ρ') in a similar expression:

$$D = h_0 \pi (D/4)^2 \rho' \qquad (6.58)$$

Equation (6.56) applies only to the last steps of compaction and hence P_i somehow relates

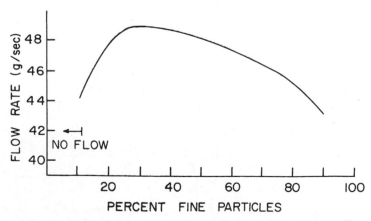

Figure 6.166. Fow rate as a function of percent fines in a granulation.[27]

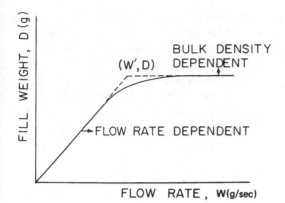

Figure 6.167. Fill weight as a function of flow rate of a granulation or powder. D is dose and W' is the critical flow rate.

to the elastic limit beyond which deformation no longer gives rise to the same shape or size of the particle when the pressure is released.

The rate at which a powder or a granulation can consolidate may be critical when high-speed machines are used, and therefore, consolidation rates play a part in compression physics.

Leigh et al,[29] have treated the pressure relations in the compression cycle by comparing the tablet with a solid (a Mohr's body), and the cycle in Figure 6.169 is suggested. Here the radial stress σ is plotted as a function of the axial stress τ. The point B is interpreted as the value where elastic recovery has its limit, and plastic flow prevails. In a manner of speaking this corresponds to the point P_i in

Figure 6.168, but it should be stressed that the analogy is but a similarity, because the tablet mass is not nonporous.

The residual stress (AE) is the pressure exerted by the tablet on the die wall after removal of the upper punch. It follows that an equation holds

$$\tau = \mu\sigma \qquad (6.59)$$

where stresses replace forces and where μ is the frictional coefficient. One of the functions of a lubricant in a tablet is to reduce the value of μ.

The lubricant also manifests itself in how well the compression pressure is propagated through the solid mass. Tablet machines are frequently instrumented[30,31] by strain gauges or piezoelectric cells, so that the pressure exerted on the upper punch P_u and the lower punch P_1 can be monitored. The closer to unity the ratio P_1/P_u is the better the tablet is lubricated. This is obviously partly a kinetic problem, since its severity is increased with increasing speed of the tablet punches. The consolidation rate plays a part in the process, and if the time for complete consolidation does not exist, then fragmentation will take place in a structure that is not completely closely packed, and consolidation, fragmentation, and fusion will occur simultaneously.

Tablet Durability. The tablet produced must have the desired physical durability to with-

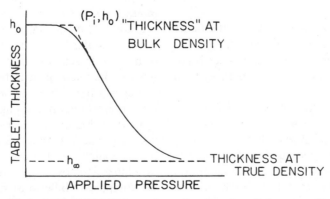

Figure 6.168. Tablet thickness h as a function of applied pressure P.

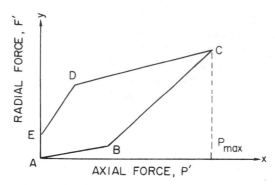

Figure 6.169. Radial force (or stress) as a function of axial force (or stress).

stand the vicissitudes of packaging, handling, and transportation. In these aspects, *hardness* is the most important quality. This is usually measured by means of a diametral hardness test: The tablet is placed (diametrally) between two anvils, and the force necessary to cause mechanical failure (breaking) is measured. This can be measured in Newton or in arbitrary units.

The systematic investigation of the diametral compression test for pharmaceutical tablets is in great part based on the studies of Newton and co-workers.[32-35] With line loading under ideal circumstances, the values of compressive, tensile, and shear stresses can be calculated by elastic theory (assuming the tablet to be a nonporous solid). The derived maximal tensile stress σ' is:[33]

$$\sigma' = \frac{2F}{\pi dh} \qquad (6.60)$$

where F is the load (Newton), d is the tablet diameter (cm), and h is the tablet thickness (cm). Some authors note that the tablet is not a nonporous solid and include a porosity term in Eq. (6.60).

The test gives different types of failure,[34] and there is a sizable scatter in results. Fell and Newton[33] have shown that the fracture strength of tablets made under identical conditions can give rise to either tensile failure (in which case the tablet splits cleanly in two parts) or to shear or compressive failure (in

which case the tablet falls into many smaller parts). Newton and Stanley[34] have shown that, if limited to tensile failure, the scatter, statistically, adheres to a Weibull function.

If Eq. (6.60) were correct, then a plot of σ' versus P would be linear through the origin. The data of Fell and Newton[34] when plotted this way are quite linear, but require a small adjustment due to the nonzero intercept.

6.5.4.3 Isostatic Pressing[36]

General and History. Isostatic or hydrostatic pressing is a compaction of a powdered material into predetermined shapes by the application of pressure via a fluid through a flexible mold. The arrangement may be such that the flexible tool contracts or dilates by the application of the pressure. Isostatic pressing covers liquids and gases as the pressure transmitting medium, whereas hydrostatic pressing is best reserved for liquids. However, the two terms are used freely to cover both aspects. Depending on whether the flexible tool is an integral part of the press or removed from the pressure vessel after each compaction cycle one distinguishes between the "dry" and "wet bag" process.

Isostatic pressing using gases as pressure-transmitting medium is still in development and practiced by only a few. It is particularly attractive at high temperatures where compaction and sintering are combined into one operation, that is, isostatic hot pressing. The preform produced by cold isostatic pressing is in most cases further consolidated by sintering, forging, extrusion, rolling, etc. When the economics of isostatic pressing are considered, it must be in relation to the final product and not for the shaping operation alone. The advantages often lie in a better product and reduction in final machining requirement.

In 1913, Madden first described an isostatic pressing technique in a U.S. patent assigned to the Westinghouse Lamp Co; the method was developed to overcome the limitations of die-compacted billets. Madden claimed that isostatically pressed billets were uniformly compacted, devoid of strata, and possessed sufficient green strength to permit handling. Further patents were taken out on the isostatic pressing of refractory metal powders by Coolridge in 1917 (for tubes of tungsten and molybdenum), and by Pfanstiehl in 1919; Fehse described the wet bag isostatic pressing of tungsten tubes in 1928. Little further interest was shown in isostatic pressing until the 1930s and early 1940s, when a series of isostatic techniques was described by Jeffery (1932–1942) and Daubenmayer (1934) in patents assigned to the Champion Spark Plug Company.

During the same period, Fessler and Russell patented a technique for pressing spark plug insulators by direct compression isostatic pressing. These workers cited the low number of rejects, rapidity, and the need for only a limited amount of equipment as economic advantages of isostatic pressing.

By 1942, most of the advantages of isostatic pressing had been recognized, and the basic principles in common use today had been established, that is,

- The wet bag pressing of large or complex shapes in which the flexible tool is filled externally and subsequently immersed in the fluid,
- The dry bag pressing of smaller, regular shapes in which the tool forms an integral part of the pressure vessel.
- The use of rigid formers to produce accurate internal or external surfaces, and
- Pressurized by pumped systems or by direct compression with punches in a die.

Materials that had been pressed included ceramics, metals, and cermets.

In recent years, fully automatic dry bag presses for producing small ceramic components have been developed, while semiautomatric wet bag presses are used to manufacture large and sometimes complex components with reasonable dimensional accuracy and requiring only minor trimming to produce the final form. The size of pressure vessels has increased greatly. Additional materials that are now isostatically pressed include plastics (particularly PTFE), explosives, and chemicals. Isostatic pressing is also being developed for the food and pharmaceutical industries.

Hot isostatic pressing, including so-called gas pressure bonding, was developed during the last 30 years. This technique has been developed for two main research applications: the solid-state diffusion bonding of components of various metals and cermets, and the hot compaction of metal, ceramic, and cermet powders. However, hot isostatic pressing has remained confined to special applications for which the high operating costs and low rates of production are acceptable.

Isostatic Pressing Equipment. Isostatic powder compaction equipment consists of a pressure vessel, pumps to generate the necessary hydraulic pressure, and related equipment to enable effective and safe machine operation.

The time to reach the required pressure depends on a number of factors, that is, volume of the cavity, volume and compaction ratio of the powder and tool, compressibility of the fluid, and delivery rate of the pumping system. To speed up pumping, it is possible to use a number of pumps in parallel. Alternatively, a pump system using different types of pumps to reach different pressure levels may be designed.

Air-driven and hydraulically driven pumps can be built easily in a variety of modules for various demands. It is simple, therefore, to change the pumping requirements by changing the intensifier (Fig. 6.170) or increasing the number of intensifiers.

Most isostatic presses operate satisfactorily up to 400 MN/m^2 on an oil/water emulsion or hydraulic oil; for higher pressures special fluids may have to be used, but the tools used must be compatible with these liquids. Problems can arise when it is necessary to dispose of contaminated fluid after each pressing operation. Such contamination may originate

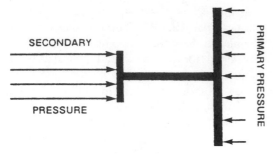

SECONDARY

PRESSURE

PRIMARY PRESSURE

Figure 6.170. The principle of intensifiers.

from powder adhering to the external tool walls or from a tool bag failure.

To be effective, an isostatic press must be joined with equipment that fulfils some or all of the following functions: filling and consolidating the powder in the tool; loading and unloading the tool set into the vessel; handling, that is, insertion and removal of the vessel's closure; controlling the pressure in the vessel; and stripping the compact from the tool.

The difference between wet bag and dry bag pressing is illustrated in Figure 6.171. In the dry bag process, the flexible tool is fixed in the

pressure vessel, and the powder can be loaded without the need to remove the tool from the vessel. The tool thus forms a membrane between fluid and powder; optionally, the tool can be placed inside a primary diaphragm so that it never comes into contact with the fluid.

Dry bag tooling is used for the production of small components at a fast rate. It is common to make provisions for loading the powder automatically into the tool by dispensing an accurately premeasured quantity. The automatic filling, the permanent location of the tool, and the smaller fluid volume result in faster operation. Dry bag tooling has also the advantage that the fluid cannot be contaminated with powder. However, because the tool has to stand up to many pressing cycles and since tool changing is time consuming, it has to be made of a very durable material.

Where mass production of simple small powder compacts (e.g., spark plug insulator blanks, grinding media, carbide tools, electrical insulators) is required, the equipment usually takes the form of a battery of small presses generally similar to and operationally having a

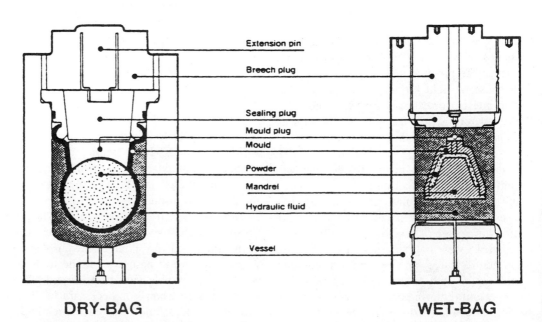

Extension pin

Breech plug

Sealing plug

Mould plug

Mould

Powder

Mandrel

Hydraulic fluid

Vessel

DRY-BAG   **WET-BAG**

Figure 6.171. Schematic representation of the difference between dry bag and wet bag pressing.

great deal in common with a conventional hydraulic press.

It has proved relatively easy to make such a machine for automatic operation with production rates up to 90 components per minute. The compaction, ejection, and filling is demonstrated on a spark plug insulator in Figure 6.172. The production rates depend on maximum pressure, size of component, powder properties, and number of tool cavities.

The larger automatic units that have been developed include a rotating pressure chamber system where loading of powder, pressurizing, depressurizing, and unloading of the compact are carried out automatically at various work stations during the complete cycle. It has also proved possible to automatically operate certain functions on large isostatic presses. For instance, loading and unloading of the tooling, insertion and closure of the breech, pressurization/depressurization of the vessel, all have been carried out automatically by ingenious arrangements of mechanisms and controls. Figure 6.173 shows the operational sequence of a three-station rotary automatic press capable of producing parts at rates of up to 300 parts per hour.

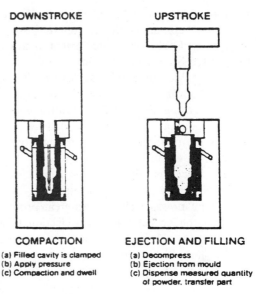

DOWNSTROKE   **UPSTROKE**

COMPACTION
(a) Filled cavity is clamped
(b) Apply pressure
(c) Compaction and dwell

EJECTION AND FILLING
(a) Decompress
(b) Ejection from mould
(c) Dispense measured quantity of powder, transfer part

Figure 6.172. Operational sequence of a "densomatic" press (Olin Energy Systems Ltd.).

In general, the development of isostatic pressing has been comparatively slow, particularly for metal powders, and even today the technique is still regarded only as an alternative to be used when the technical limitations of conventional methods are too restrictive.

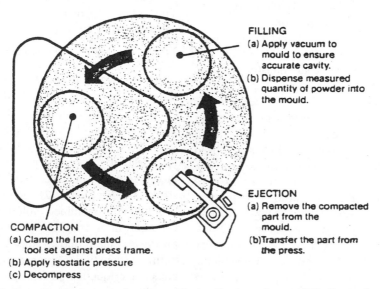

FILLING
(a) Apply vacuum to mould to ensure accurate cavity.
(b) Dispense measured quantity of powder into the mould.

EJECTION
(a) Remove the compacted part from the mould.
(b) Transfer the part from the press.

COMPACTION
(a) Clamp the integrated tool set against press frame.
(b) Apply isostatic pressure
(c) Decompress

Figure 6.173. Operational sequence of an automatic, rotating isostatic press (Olin Energy Systems, Ltd).

Until recently, mostly the ceramic manufacturers have commercially exploited isostatic pressing and only to a limited extent mainly in the United States. In addition, isostatic pressing is suitable for producing high-purity ceramics and long ceramic tubes, for which there is an increasing demand. In contrast, the common metals can be readily formed by long-established methods such as casting, rolling, forging, or extrusion and only recently have metal fabricators begun to look more closely at the feasibility of isostatic pressing.

Isostatic pressing also shows great promise of becoming an established production technique for the fabrication of components from PTFE and high-molecular-weight polyethylene. PTFE, for instance, although a thermoplastic, has a very high melt viscosity, which precludes satisfactory processing by established injection moulding and extrusion techniques. This has led to the adoption of techniques used in powder metallurgy, which involve initial cold compaction and subsequent sintering, of which isostatic moulding is the latest.

6.5.4.4 Discontinuous High-Pressure Extrusion Presses

General. To illustrate discontinuous extrusion compaction of soft, formable materials with inherent or added binding characteristics, the "extrusion briquetting" process as employed by the brown coal industry shall be discussed as a typical example.

Figure 6.174 depicts the sequence of events during a briquetting cycle in a ram extrusion press.[37] The reciprocating motion is produced by, for example, an eccenter drive symbolized by the circular representation on the left. The diagram on the right indicates the progress of force exerted on the material to be briquetted. The figure is self-explanatory. Only a few important operating stages shall be pointed out.

At (3) the force exerted by the ram has reached a level that is sufficient to overcome the friction of all briquettes in the pressing channel and the backpressure caused by the

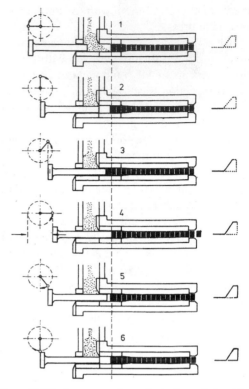

Figure 6.174. Sequence of events during a briquetting cycle in a ram extrusion press.[37]

column of briquettes in the cooling channel. The entire line of briquettes moves forward, with the force remaining approximately constant, and a new briquette emerges from the "mouth" of the press (4).

At the beginning of the backstroke (when the eccenter drive passes position 4) at first the ram face does not separate from the briquette because of considerable elastic expansion of the briquette. It is important, however, to note that the surface produced by the ram face is so highly densified that, during the next stroke and for phases (2) and (3), it acts as the bottom of a confined volume densification chamber until friction is overcome and the product column moves forward; during the entire production sequence the surfaces of adjacent briquettes do not develop significant bonding; therefore, on discharge from the cooling channel, the product will separate into single briquettes.

Equally important is that at a typical rotational speed of the eccenter drive of 90 rpm the duration of the compression phase, during which the primary briquette is compacted, is only approx. 0.4 s.[37] Because brown coal is very elastic and the time is too short to achieve conversion of elastic into plastic volume change, the elastic recovery during the backstroke is high. Without the condition that during each compression stroke all briquettes in the pressing (extrusion) channel are again loaded and compacted, whereby more and more permanent plastic deformation is obtained, successful briquetting or organic material with high elasticity would not be possible. This is an important difference from, for example, roller presses (see also Section 6.3.3 and Fig. 6.62). That all briquettes up to the point of narrowest cross-section in the extrusion channel participate in the densification and expansion was shown by Metzner[38] and Schenke.[39]

To accomplish the above, the design of a ram extrusion press must provide a relatively long extrusion channel. However, there are physical limits to this parameter because friction and drive power as well as overall stressing of the equipment increase with channel length. Briquettes may retain a certain elastic deformation which, if suddenly released, will damage or destroy the product. Therefore, in most applications, a gradual release is provided in the channel prior to product discharge.

Figure 6.175 shows cross-sections through relatively modern ram extrusion or Exter presses. The upper channel wall is adjustable such that different release angles can be obtained. In addition, a flexible support system at this point serves as a safety device to avoid excess loading due to tramp material in the feed or overcompaction. During the backstroke the energy of the drive is stored in a fly wheel (Fig. 6.175b) and again made available during compaction.

In a closed mold, the development of a predetermined pressure presents no difficulty, but in extrusion presses the situation is complicated. The peak pressure developed at each stroke depends not only on the power exerted by the ram but also on the resistance to the forward movement of the material to be briquetted. The latter is influenced by many factors: the shape and length of the channel, die or bore, the changes in cross-section in relation to length, the smoothness of the tool walls, the nature of the material to be processed including parameters such as temperature, structure, plasticity, etc., and the type and length of the curing channel if applicable.

The rate of pressure increase is also important; it depends on stroke frequency and length and on the rather complicated relationship between movement of the ram and magnitude of the resisting frictional force between extrudate and die as well as the force caused by the column of already compressed product being pushed forward. These forces change with both state of compaction and rate of movement.

Sizing of Discontinuous Extrusion Presses. As for all pressure agglomeration methods, the most important design parameter is the compaction pressure acting upon the material to be compressed and extruded. In a machine with "parallel-wall die channel," that is, a die with constant cross-section, and without curing channel, this pressure, which is necessary to produce compacts of good quality, is determined by the static frictional resistance. It depends on the radial pressure p_r acting on the die wall, the coefficient of static friction u, and the length of the channel (Fig. 6.176).[40] Since the radial pressure and the coefficient of static friction are practically constant for a given set of conditions, channel length is the only variable for obtaining the desired compaction pressure p_K.

Recently[40] earlier theories of noncontinuous extrusion were corrected by taking into account the two distinctly different phases, that is, compression and extrusion or transport (see Fig. 6.174). As long as the compaction

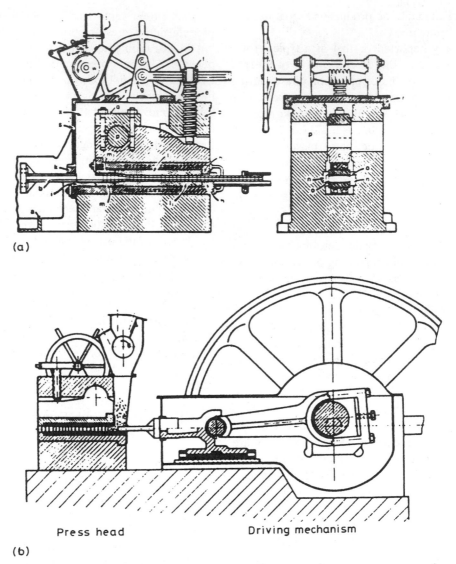

(a)

Press head   Driving mechanism

(b)

Figure 6.175. Cross-sections through relatively modern ram extrusion or "Exter" presses.[1]

pressure has not overcome the static friction of the column of already compressed compacts, the mechanism of pressure agglomeration in a ram extrusion press is the same as experienced in confined volume punch (die) presses (see Section 6.5.2). Later, during the transportation or extrusion phase, all previously densified compacts in the die are, to a certain degree, densified again while being pushed forward.

Figure 6.177 illustrates these conditions.[40]

Figure 6.177a depicts the relationship between axial (compaction) pressure p_K and radial pressure p_r. $p_{r,r}$ is the residual radial pressure after separation of the ram from the elastically recovering column of compacted material in the channel during the backstroke. Figures 6.177 b.1 and b.2 show the axial and, respectively, radial pressure distributions along the length of the channel. While the axial compaction pressure drops to "zero" each time the ram retracts, a residual radial pressure always

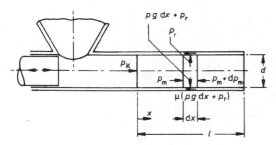

Figure 6.176. Diagram showing the development of compaction pressure in a ram extrusion press featuring a channel die with constant cross-section.[40]

This parameter describes the nonisotropic character of bulk particulate solids which results in the fact that pressures in the direction of loading are higher than those perpendicular to it.[41] The ratio [Eq. (6.62)] is well known from soil mechanics; is is always larger than 0 and smaller than 1. If the angle of internal friction φ and coefficient of cohesion C of the bulk material are known, λ can be calculated with:

$$\sigma_r = [\sigma_m(1 - \sin \varphi) - 2C \cos \varphi]/[1 + \sin \varphi] \quad (6.63)$$

and the necessary length of precompressed compacts results from:

$$1 = (d/4\mu A) \ln([Ap_G/p_{r,r}] + 1) \quad (6.64)$$

A is the slope of the de- and recompression lines in Mohr's stress diagram.[40] According to Figure 6.177b the total channel length L is $1 + H + H^* +$ densification prior to the forward movement of the column of compacts in the die, where H^* is the thickness of the new

remains which is primarily responsible for the back pressure p_G in the channel (Fig. 6.178) necessary to accomplish the compaction phase during the next stroke.

p_G can be calculated by:

$$p_G = p_K \cdot e^{-4\lambda\mu H/d} \quad (6.61)$$

λ is the ratio of radial to axial pressure (Fig. 6.179):

$$\lambda = p_r/p_m = \sigma_r/\sigma_m \quad (6.62)$$

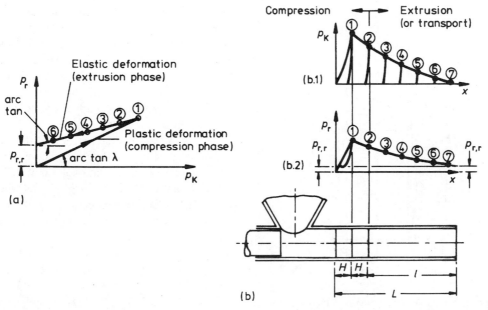

Figure 6.177. Axial and radial pressures of a compact as it moves through a channel die with parallel walls.[40]

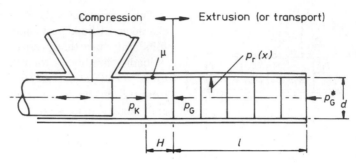

Figure 6.178. Sketch depicting the pressure acting on the particulate material in a ram extrusion press.

compact at the beginning of forward movement and H is the thickness at the dead-center turnaround point of the ram (beginning of the backstroke). Experimental investigations[40] proved that there is excellent agreement between actual data and theory.

If in additional counter pressure p_G^* acts at the press mouth onto the end of the column of compacts (Fig. 6.178), for example, because of a line of curing briquettes or a control baffle (see below) Eq. (6.64) becomes:

$$1 = (d/4\mu A) \ln([p_{r,r} + Ap_G]/[p_{r,r} + Ap_G^*]) \tag{6.65}$$

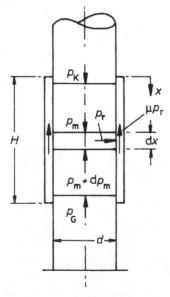

Figure 6.179. Model describing conditions in the particulate matter during the compaction phase.

From the equations a number of dimensionless parameters can be obtained that characterize the noncontinuous compression in an extrusion from open-ended dies. If these parameters are all plotted in one diagram, they can be correlated graphically which provides a method to size an extrusion press with "parallel-wall die channel."[40]

In reality, the conditions are not as simple and uniform. In most cases, the die cross-section decreases somewhat to enhance the compression phase of the method. Since this results in nonlinear differential equations, solution is not easy. Furthermore, to avoid damage of the extrudate by sudden elastic recovery when it emerges from the "press mouth" (die end), the channel walls are set at a slight taper, opening toward the discharge end, to provide for a slow and controlled release of elastic deformation. With these design features the preconditions for the above theory are no longer valid and the results can be taken to determine only approximate order of magnitude parameters.

The material characteristics are also not as constant as assumed. Relatively small inhomogeneities in the particulate solid may result in invariations in backpressure p_G as well as residual radial pressure $p_{r,r}$ and, consequently, in compaction pressure p_R as well as density or strength of the extrudate. To demonstrate the extent of variations in material characteristics, Figure 6.180 shows the compressibility presented as pressure/densification graphs of 15 lignite samples, most from the same mine and all subjected to iden-

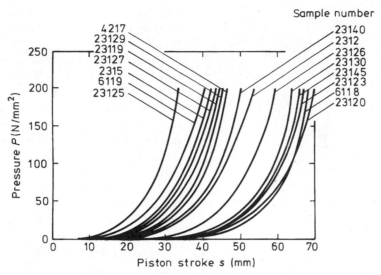

Figure 6.180. Pressure/densification graphs of 15 different lignite samples (laboratory evaluation).[42]

tical feed preparation.[42] The samples having constant weight were compacted with a maximum pressure of 200 N/mm². The large differences in compaction behavior are characterized by the piston stroke length at maximum pressure which varies from less than 35 mm to 70 mm.

There are important parameters that influence the extrusion of particulate matter. To obtain reproducible results, as many of these parameters as possible must be kept constant. The need to cool the die is rather unique for binderless briquetting of lignites in ram extrusion presses. In this application, if the die heats up, the coefficient of friction between lignite and die wall changes such that movement occurs at lower pressures, which results in less densification and inferior strength. The speed of densification, as in other high-pressure agglomeration methods, influences the amount of elastic springback. Slower speed allows conversion of a larger portion of elastic energy into plastic deformation; on the other hand, capacity is reduced by this measure.

6.5.4.5 Roll Pressing

Double Roll Presses. The most widely used roller presses are double roll presses which achieve compaction by squeezing material between two countercurrently rotating rollers (Fig. 6.181), much in the same manner as the operation of rolling mills.[43] Pockets or indentations, which have been cut into working surfaces of the rollers,[44] form briquettes or compacts.

Between smooth, fluted, corrugated, or waffled rollers, material is compacted into dense sheets. Normally, these sheets are crushed and then screened to yield a granular product.

If rows of identical pockets are machined into the working surface and the rollers are timed such that the pocket halves exactly match, so-called briquettes are formed. Roller presses do not produce compacts with the same fine detail and uniformity as those made by tabletting machines or other die presses. The flashing or web, caused by the "land areas" around each briquette pocket, which is usually found on the edges of all briquettes from roller presses cannot be removed completely and reliably and, therefore, may also be objectionable.

Because of these characteristics, roller presses find their natural field of application where relatively low investment and operating costs are more important than the absolute

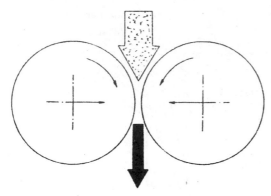

Figure 6.181. The basic principle of double roll pressing.

uniformity of the product. Double roll pressing of particulate matter is traditionally of greatest interest for all industries in which large quantities of finely divided solids, both valuable and worthless (wastes), must be handled. Originally developed as an economic method to agglomerate coal fines, today, this size enlargement technology is applied for a large number of materials in the chemical, pharmaceutical, food processing, mining, minerals, and metallurgical industries. This versatile technology lends itself to such different uses as computation and granulation of highly heat- and pressure-sensitive pharmaceutical materials, for example, pancreatin or penicillin, briquetting of extremely corrosive and poisonous materials, for example, sodium cyanide, compaction and granulation of large tonnage materials, for example, fertilizers, or briquetting of crude, hot materials, for example, metal chips and turnings, ores, or "sponge iron" at temperatures of up to 1000°C. An important, newly emerging application is the vast field of environmental control where sometimes micron or submicron sized particulate solids must be enlarged for recycling or disposal.

In the early machines and for many applications today, the particulate matter to be compacted or briquetted is fed by gravity into the nip of the rollers. Feed control is performed by adjustable tongues and distribution across

the width of the roller is achieved by simple, rotating devices mounted on top of the press (Fig. 6.182).[45]

To obtain positive feed pressure and provide a more versatile means of control, screw feeders are installed for many modern applications (Fig. 6.183).[46]

The process occurring during compaction of particulate matter in roll presses is described and interpreted by different authors in a rather similar way. The feed mechanism is characterized by the pressure caused by gravity or a force fed system and the friction between material and roll surface. Compaction between two rolls may be explained by dividing the roll nip area into two zones: the feed zone and the compaction zone.

As depicted in Figure 6.184, showing a smooth roll press, the feed zone is defined by the two angles α'_E and α_E. In the feed zone, the material is drawn into the nip by friction

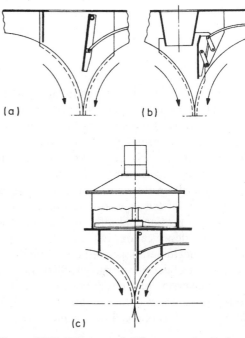

(a)   (b)

(c)

Figure 6.182. Diagrams of different gravity feed controls. (a) Standard tongue, (b) tongue with parallel movement, (c) mechanical distribution with standard tongue.[1]

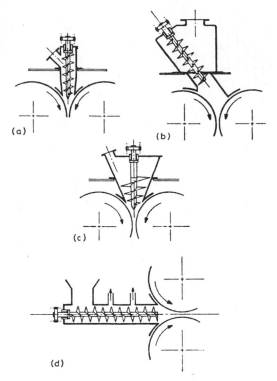

Figure 6.183. Schematic representation of some typical force (screw) feeders.[46] (a) Vertical straight or slightly tapered screw feeder, (b) inclined straight screw feeder, (c) vertical tapered (conical) screw feeder, (d) horizontal straight screw feeder.

on the roll surface. Densification is solely due to the rearrangement of particles (Fig. 6.110). The density of the feed is characterized by the bulk density ρ_0 and reaches the tap density τ_t at the point α_E. The peripheral speed w of the rolls is higher in this zone than the velocity u of the material to be compacted. α_0 is the so-called angle of delivery which is defined by he width h_0 of the feed opening above the rolls as well as the material (flowability) and feeder characteristics.

The compaction zone follows after the heavy solid line (Fig. 6.184). α_E is the angle of rolling, the gripping angle, or angle of compaction. In the compaction zone the pressing force becomes effective and the powder particles deform plastically and/or break (Fig. 6.110). α_g is the neutral angle where the sign

(direction) of the friction force changes. At this point, the pressure in the material and the density have their highest values.

α_v is the angle of elastic compression of the rolls that determines the thickness h_s of the compacted sheet. α_v becomes zero and the sheet thickness h_A if the elastic deformation of the rolls can be ignored. However, in most cases the strip is even thicker than h_s owing to elastic recovery of the compacted material. The angle corresponding to this actual outlet plane is called angle of release α_R.

During compaction between essentially smooth rollers a third zone can be defined: the extrusion zone. When the direction of the friction force changes at the neutral angle α_g, the material may "accelerate" and, in respect to the roller speed, attain a higher velocity resulting in an "extrusion" through the roller gap. This phenomenon assists in the release of the compacted material from the rollers.

In the case of briquetting, the gap between the roller approaches zero and the pockets, which were cut into the roller surface and define the briquette shape, do considerably influence and change the above compaction process. Figure 6.185 depicts the mechanism of briquetting in roller presses. Of interest is only the final compaction phase. It begins when the lower axial land area passes through the line connecting the centers of the rollers. At this point, the pocket forming the briquette is practically closed at the leading (lower) edge while the trailing (upper) edge is still open and connected with the feed in the nip. Immediately following this condition the formerly closed leading edge of the pocket opens while now the upper (trailing) edge closes and compaction of the briquette is completed. Owing to "interlocking" between material in the nip and the pocketed roller surface, the previously defined feed and compaction zones are less clearly defined and determined only by inter-particle friction. They no longer depend on friction between material and roller surface. However, as the leading edge of the pocket

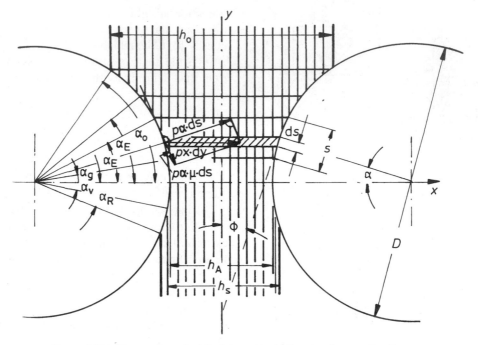

Figure 6.184. Compaction of particulate matter in the nip of a smooth roll press.

opens the force acting vertically to the line connecting the roller centers tries to "extrude" the briquette, thus assisting in the release of the briquette from the pocket, provided the shape is correctly designed.

Much of this knowledge is still phenomenological in character. A comprehensive theory of densification of particulate matter between counterrotating rollers is not yet available even though many similarities exist with the much better investigated and defined deformation of metals in rolling mills.[43]

Ring Roll Presses.[47]   In the ring roll press, an alternative to the double roll press have been developed for high-pressure work. The particulate matter, normally powdered coal, is pressed between a roll and the inner surface of a ring (Fig. 6.186). Thus, a very narrow angle of entry is achieved, and with it, of

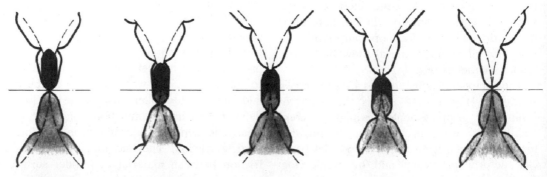

Figure 6.185. Five successive momentary conditions of briquetting between two countercurrently rotating rollers with matching pockets.[46]

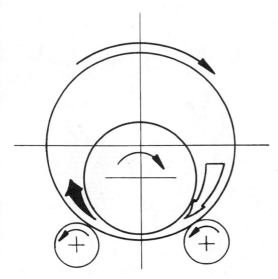

Figure 6.186. Operating principle of the ring roll press.

Sizing of Roller Presses

Theory of Rolling. The basic principle of compaction of particulate solids between two countercurrently rotating rollers (Fig. 6.187) is similar to that used in calenders for plastic foils or in rolling mills for metals. The first can be adjusted to extremely narrow gap tolerances across press rollers with face widths of up to 2 m and production speeds of approx. 100 m/min; in the latter enormous pressing forces can handle ingots of more than 35 tons weights.

While roll pressing of particulate solids is still an art rather than a science, fundamental perception and technical knowledge exist in the above mentioned fields because they were developed and investigated in modern times. Therefore, several authors concluded that it must be possible to use this knowledge and translate it into corresponding theories for roll pressing. Specifically, the basic equation obtained for rolling steel can be used to gain an understanding of roll pressing.[43, 44]

course, considerable drag, which obviates forcible feeding of the powdered coal. Such a system has many advantages, but also some disadvantages that have not yet been completely overcome.

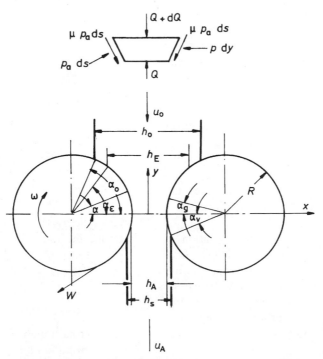

Figure 6.187. Strip model: Geometry of rolling and forces acting on a volume element.[43]

Particulate Matter.   Since pressure agglomeration between countercurrently rotating rollers deals with particulate matter, results of a theory based on homogeneous and isotropic solid material is applicable only in a general way. If smooth rollers are used, a very close correlation can be obtained. Today, however, roller surfaces for agglomeration are in most cases equipped with some sort of profile to improve the "bite" on the material, which is a never ending problem because of the noncontinuity of particular matter. In case of profiled surfaces the operating zones of the "elementary theory" described by Siebel and v. Kármán[43] defining deformation and, respectively, densification cannot freely develop owing to interlocking between material and roller surface. This is most pronounced for briquetting.

The inability of the material to develop the relative speed conditions predicted by the strip model and the relatively short densification time result in considerable elastic deformation which also modifies the pressure curve as shown, for example, in Figure 6.188.[48] In the case demonstrated, a special pocket design (Koppern) is used with alternating shallow and deep cavities. Shallow pockets of one roller dip into the corresponding deep pockets in the opposite roller, much like a piston and die arrangement. The diagram at the right illustrates compaction and expansion actions and times as well as the pressure curve. The effect caused by the elastic recovery (expansion) of the briquette during release is clearly visible. This phenomenon of pressure agglomeration is very important and often determines the quality of briquetted or compacted products.

Capacity, Throughput.   For the calculation of equipment capacity the macroscopic phenomenon of material passing through the nip of the roller is utilized and theoretical or actual characteristics are neglected. Therefore, the throughput C_c of a roller compactor can be determined as:

$$C_c = \pi \cdot D \cdot l \cdot h_A \cdot n \cdot 60 \cdot \gamma \quad (6.66)$$

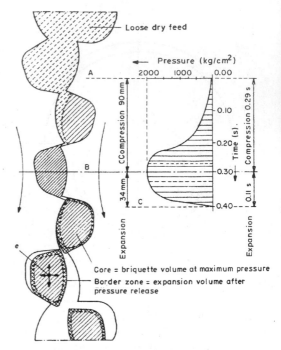

Figure 6.188. Schematic representation of the compaction process in a roll-type briquetting press[48] (1 kg/cm² = 9.81 N/m²).

where

D = roller diameter (cm)
l = roller length, working width (cm)
h_A = gap width between the rollers, sheet thickness (cm)
n = roller speed, revolutions per minute (1/min)
γ = apparent sheet density (kg/cm³)
then: C_c = throughput of the roller compactor (kg/h).

Correspondingly, the throughput of a roll type briquetting machine C_b is:

$$C_b = z \cdot V \cdot n \cdot 60 \cdot \gamma \quad (6.67)$$

where

z = total number of pockets per roller, total number of briquettes per revolution

V = volume of each briquette (cm³)

n = roller speed, revolutions per minute (1/min)

γ = apparent briquette density (kg/cm³)

then: C_0 = throughput of the roll type briquetting machine (kg/h)

Because of leakage at the sides of the rollers and, in case of roll-type briquetting machines, the flashings or webs around the briquettes, the actual throughput of and the feed to roller presses are somewhat higher (approx. 5% to 15%).

Roll Diameter. One of the most important criteria for the design of roller presses, which also determines the physical size of the entire machine, is the roll diameter, D. It is also one of the few parameters that is fixed in a given machine and cannot be adjusted to different operating conditions.

Referring to Figure 6.189 it is obvious that the sizes of the feed and compaction zones depend on the roll diameter. Under the (almost correct and therefore acceptable) assumption that the gripping angle α_E changes only slightly with roll diameter, the conditions of Figure 6.189 are obtained[49] for the nip areas between two pairs of rollers with different diameters, D_1 and D_2, and identical gap,

h_A. If the peripheral speed of both pairs of rollers is the same (i.e., theoretically, both machines yield identical volumetric output) compaction takes place more gradually in the case of the larger roll diameter. At the same time, a larger volume element is pulled into the nip, resulting in a higher density of the compacted product (i.e., potentially, a larger gravimetric output is obtained).

For smooth roll compactors a formula can be derived that correlates roller diameter and gap. With the definitions of Figure 6.184 and the restrictions imposed by the modified strip model [i.e., beginning at the line $h_E(\alpha_E)$ horizontal increments move with the peripheral speed of the rolls (no slip) and remain absolutely horizontal (no distortion)], the following equation for the porosity ϵ_{min} at the narrowest point ($\alpha = 0$) is obtained:[49]

$$\epsilon_{min} = 1 - \gamma_t[D(1 - \cos \alpha_E) + h_A]/\gamma h_A \tag{6.68}$$

Since ϵ_{min} cannot become negative, it follows:

$$\gamma_t[D(1 - \cos \alpha_E) + h_A] \le \gamma h_A \tag{6.69}$$

or:

$$h_A \ge D[1 - \cos \alpha_E]/[(\gamma/\gamma_t) - 1] \tag{6.70}$$

In Eqs. (6.68) to (6.70) γ_t is the tap density which is assumed to be equal to the density at point α_E (see Fig. 6.184).

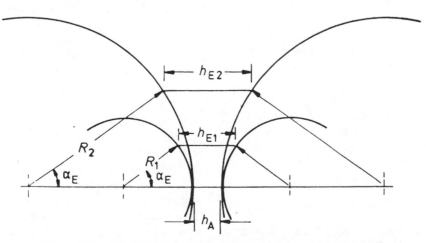

Figure 6.189. Influence of the roll diameter D on the compression zone.[49]

For materials requiring relatively little densification during compaction, the classic theory (strip model) can be used to determine the minimum roller diameter needed to form a dense sheet or briquette. Equation (6.68) can be rewritten as follows:

$$D = h_A/(1 - \cos \alpha_E)[\gamma_0/\gamma_t(1 - \epsilon) - 1] \quad (6.71)$$

where ϵ characterizes the remaining porosity at $\alpha = 0$ (disregarding elastic recovery). In the case of briquetting, an equivalent gap width h'_A must be calculated from the briquette volume and web thickness combined.

If the rollers are larger than necessary to achieve $\epsilon(\alpha = 0)$, control can be applied by restricting the flow of feed to the roll nip (see Fig. 6.182).

With increasing densification ratio the necessary roller diameter becomes larger. However, there are economic advantages in reducing the roller diameter to below the minimum diameter if materials needing high densification must be processes. Then a force feeder system must be used (see Fig. 6.183). In such a case, the above criterion can be applied to choose a diameter that is less than the dimension calculated with Eq. (6.71). The selected roller diameter should be always sufficiently smaller than the calculated minimum to allow density control by as large an adjustment of the force-feeding system as possible.

Another criterion for selection of the roller diameter, particularly of briquetting machines, is the release mechanism from the pockets (see Fig. 6.185).

Roll Gap. There is a close correlation between strip thickness h_s and theoretical roll gap h_A. Since with a given roll diameter the gap defines the compaction ratio, the strip density, γ, also depends on the gap as indicated by rewritten Eq. (6.69):

$$\gamma \sim \gamma_t[D/h_A(1 - \cos \alpha_E) + 1] \quad (6.72)$$

As a rough approximation it can be assumed that strip thickness equals roll gap. In reality,

however, h_s is always greater than h_A for the following two main reasons:

1. Under load the roll gap changes because of (a) clearance in the roll shaft bearings and frame members, (b) elasticity of the machine frame, (c) deflection of rolls and shafts, and (d) elastic deformation of the roller surface.
2. After pressure release the strip expands because of (a) recovery of elastically deformed particles and (b) expansion of compressed air trapped in pores of the compact.

The extent of the roll gap change depend only on machine design and is constant for a given compaction pressure.

Expansion of the strip after pressure release is influenced by the physical characteristics of the material to be compacted (plasticity, brittleness, particle size and distribution, particle shape, etc.), the roll diameter, the speed of rotation, and the surface configuration of the rollers. With increasing roll diameter and/or decreasing speed the expansion of compacted material is reduced owing to better deaeration during densification and a more complete conversion of elastic into permanent, plastic deformation.

The smallest theoretical roll gap can be calculated using Eq. (6.70). However, because of the mechanical deformations discussed above, it is possible to roll a strip with finite thickness even if the static (= no load) gap is set at zero. This means that, in reality, the dynamic roll gap, which develops under load, must be considered. The largest acceptable roll gap results from the need to obtain a coherent compact, that is, the compaction ratio; this is influenced by the roll diameter as well as the amount and predensification of feed; the latter is characterized by the density γ_t at $h_E(\alpha_E)$. In addition, because of pressure and density gradients in the particulate mass during compaction, it is possible that the center of strip or sheet has insufficient strength if too large a thickness is desired.

For briquetting presses the correct relationship between feed rate and volume of compacts is of special importance. To avoid thick "flashings" or "webs" on the edges of briquettes, it is necessary to use strong machines, to prevent flexing, with rigid response, and a static roll gap of close to zero.

Roll Speed. For most considerations and approximations it is assumed that the peripheral speed of the rollers and the speed of the particulate matter are identical in the entire compaction zone. In reality this is not true; throughput does not increase proportionately with roll speed. The maximum speed is determined by two effects; starved conditions in the compaction zone develop if (1) too much slip occurs between rolls and material in the feed zone and/or (2) air squeezed from the particulate mass flows upward and fluidizes the feed thus reducing the supply of material to the nip.

In the first case, compaction is not high enough to form a stable compact, and intermittent operation, accompanied by sometimes severe chattering and potential equipment failure, occurs in the second case.

The minimum speed for smooth rolls is reached if the mass flow rate M_p of the free flowing powder is higher than the mass flow rate $\dot{M}_s$ of the compacted strip. Determination of minimum speed is important only if strips with tightly controlled thickness are to be made between smooth rollers, for example, in powder metallurgy. In other applications, for example, the compaction of fertilizers, the problem of minimum speed may be circumvented by selecting a narrow static gap and adjusting the hydraulic pressure such that, when feed is introduced into the nip, the clearance increases to the operating gap and, at the same time, the pressure rises to the operating level.

A completely different situation exists if the rollers are pocketed or corrugated because the flow of material is stopped when the land areas between the pockets or the ridges of the corrugations are in close proximity. For such rollers virtually no minimum speed exists. As the filling of the cups is the controlling factor and the disintegrating forces due to the release of residual elastic deformation and compressed air trapped in the pores diminish with reduced speed of compaction, briquette quality improves in most cases if the rollers are slowed down.

Roll Feeding. The simplest form of feeding roller presses is by gravity (choke feeding). A mass flow hopper with rectangular feed opening to the nip between the rollers should be used for this purpose.

The feeder dimension h_0 (see Fig. 6.184) is characterized by the angle α_0 and depends on the roller diameter, D, the gap h_A or, respectively, the surface configuration of the briquetting roll. To make use of the full transport capability of the rolls, the feed angle or angle of delivery, α_0, should be greater than the gripping angle, α_E.

In many applications the degree of compaction necessary to produce a satisfactory agglomerate is so small that the combination of commercially and conveniently sized rollers (as well as pockets, if applicable) provides too much densification if choke feeding is used. Then, the flow of material to the nip between the rollers must be deliberately restricted to avoid overcompaction (see Fig. 6.182).

In contrast, the briquetting or compacting of some other materials demands a degree of compaction that cannot be achieved by a single pass in a choke-fed roller press, irrespective of the ratio pocket size (or gap width) to roll diameter. In addition, redistribution of material (which may be extensive) from the nip against the flow of material or from the rear of cups into following cups, for example, owing to the flow of displaced air, may further reduce the efficiency of compaction. In these cases the use of force feeders (see Fig. 6.183) is required.

Roll Pressure and Torque. After determining roller diameter, width, and gap or briquette size and shape as well as roller speed using throughput capacity and product density as input, roll force and torque as well as feed pressure must be determined. The requirements on these design parameters of a roll-type press are:

1. The press must be capable of safely supporting the roll force and sustaining the torque necessary to make a good sheet or briquette, and
2. the press with associated feed mechanism must allow development of the torque and force required to make a good product at the required throughput rate.

These parameters relate to the flow properties of the solid to be compacted.[46]

Figure 6.190 depicts schematically a typical compressibility diagram (density versus force) of a particulate solid. In a log/log plot the curve can be approximated by five straight-line segments. The first occurs at low pressures where density essentially does not change. The second range, during which density increases slowly, applies to positive force-feed systems (gravity chutes, screw feeders, etc.). The third represents the high-pressure nip region between the rollers. The compressibility factor K of the solid is characterized by the slope of the curve in this range. In the fourth segment of the curve density again remains constant; this operating condition is normally outside of the desirable working range of roller presses. In the fifth region, residual elastic deformation in the compacted solid springs back when the pressure is released.

Even though bench scale densification tests do not reliably predict the performance of a roller press, results can provide valuable information on the relative behavior of different feed materials.

The solids pressure p_{max} will be influenced by the "precompaction" pressure of the feeder, p_0. Reductions in roll force and diameter accompanying the increase in precompaction pressure lower size, weight, and cost of roller presses. In contrast, roll drive requirements remain almost unchanged[50] if the production rate is kept constant. Feed screw precompaction pressures up to and exceeding $2.8 \cdot 10^6$ Nm^{-2} have been reported. In normal operation the pressures are probably in the range of 10^4 to 10^6 Nm^{-2}.

Feed screws are axial flow type compressors whose power requirements increase with the compression ratio and also with larger frictional forces between material and screw occurring at the higher pressures provided, however, that the permeability of the densifying bulk mass remains high enough to allow unrestricted flow of the gas that is expelled during compaction. The total power requirement of the roller press with screw feeder is the sum of both drive energies. Figure 6.191 illustrates schematically the correlation between total drive energy and precompaction pressure for

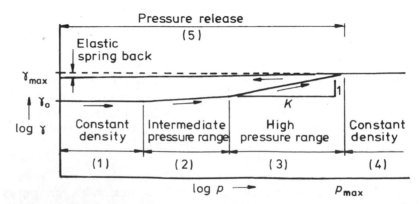

Figure 6.190. Typical compressibility diagram (density versus force) of a particulate solid.[46]

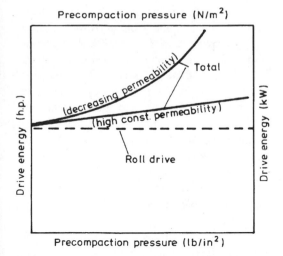

Figure 6.191. Drive energy of roller presses with screw feeder as a function of precompaction pressure and influence of permeability during compaction.

an always highly permeable particulate solid and for a material with decreasing permeability during compaction. Normally, the share of feed screw power in relation to total drive power is the range of 1% to 20%.

Optimum precompaction pressures, p_0, and corresponding feed screw designs vary widely with physical and chemical properties of the material and also with the desired quality and shape of the material. Because of the many different applications and the numerous variables, optimum precompaction pressure and feed screw design are determined during tests with a sample of the actual feed material whereby roller presses with large roll diameters are often used to avoid scale-up problems and alternative feeder designs are applied. Actual plant conditions are simulated by adding the proper amount of "recycle" to the feed.

Scale-Up Considerations. In addition to the above considerations, there are some simple relationships between roll diameter, D, force or pressure, p, and gap width, h_A, which can be applied for scaling-up or -down. An equivalent gap width may be calculated for briquetting machines and used as approximation.

These relationships are:

$$\sqrt{D_2/D_1} = p_2/p_1 \qquad (6.73)$$

$$\sqrt{h_{A2}/h_{A1}} = p_2/p_1 \qquad (6.74)$$

Although it follows from Eq. (6.73) = Eq. (6.74) that the sheet thickness (i.e., gap width h_A) can be larger with increasing roller diameter, experience teaches that the prediction $h_{A2} = (D_2/D_1) \cdot h_{A1}$ can normally not be achieved. Depending on the characteristics of the material to be compacted the minimum sheet thickness may be estimated during scale-up by:

$$h_{A1}(D_2/D_1) > h_{A2} > \sqrt{D_2/D_1}\, h_{A1} \quad (6.75)$$

Special Characteristics of Roller Presses

Phenomenology of Roll Compaction. Controlled and complete removal of gas (normally air) compaction that is expelled during densification is an important consideration for all pressure agglomeration methods. Correct and sufficient venting becomes most critical for roller presses handling large bulk volumes. For example, during roll compaction, densification ratios are typically 2:1. In the case of potash compaction, a common high-capacity application of roller presses, the bulk density of the feed is approx. 1 t/m³; it increases to an apparent density of the compacted sheet of nearly 2 t/m³; therefore, approx. 0.5 m³ of air per ton of salt must be removed during compaction. Since modern, large-scale equipment is capable of handling approx. 80 to 100 t/h, 40 to 50 m³/h of air is to be vented without disrupting uniform operation of the press.

In many applications the simple smooth, cylindrical roll surface design is used. Particularly with smaller roller diameters the gripping angle of compaction, α_E, becomes very small, resulting in reduced compaction ratio, and, therefore, a lower throughput. Especially if fine powders are to be compacted, a force feeder is necessary to overcome these shortcomings. A rough surface will increase the gripping angle and improve the situation; however, because of inevitable wear, which will eventually polish the surface, this measure is only of limited value.

Other surface "irregularities" include different types of corrugations or shallow pockets that produce "waffled" sheets.[51] The latter seem to improve deaeration.[52]

While profiled rollers are acceptable for some small-capacity applications and for materials with low abrasivity, large, high-capacity machines processing abrasive solids require the inherent advantages of smooth rollers, which are: rugged design, easy manufacturing, the possibility of refacing work rolls, and lower price. To improve the angle of grip, weld beads may be applied to the surface; these welds can be replaced from time to time as required.

A relatively homogeneous sheet is most easily formed across almost the entire width of the rollers with small roller diameter or narrow roll gap (sheet thickness) and low circumferential speed (both resulting in small capacity) as well as small compaction ratio. Then, only a relatively small amount of air is expelled which can escape partly to the top and partly to the sides of the rollers. Production of a homogeneous sheet will be further facilitated by coarser feed and correspondingly larger pore diameters (permeability).

Figure 6.192 illustrates the forces at work in a roll press when powders are compacted.[53] The rotating rollers magnify the small contact pressure between the solid particles from a value correspondingly to p_0 at the feed point

of the press to the maximum pressure, characterized by p_{max} near the narrowest point of the nip (part A of Fig. 6.192). The increase may be calculated as a function of roll diameter, sheet thickness, and roll friction coefficient as well as the material's compactibility, interparticle friction, and permeability. On discharge, the pressure is released and initial strength is caused by binding mechanisms that have been activated during compaction. Residual elastic deformation at the point of discharge is relieved by expansion of the sheet which may result in a weakening of the binding forces.

Part B of Figure 6.192 shows the corresponding increase in apparent density which is typically approx. two times higher after compaction but may reach values of up to three times the feed density depending on material as well as roll press size and design. As can be seen, densification occurs very rapidly (in 1 s or less) in the narrow part of the nip. Ideally the compact density remains constant after discharge but may become somewhat lower due to elastic expansion (dotted curve).

Part C of Figure 6.192 shows that, depending on permeability of the particulate solids being processed and its change during compaction, respectively, air pressure in the material may increase to different levels. If residual porosity (permeability) remains high enough during densification and in the compacted sheet, this air pressure equalizes by gas flow

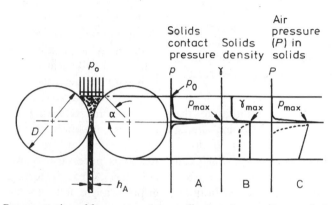

Figure 6.192. Representation of forces at work in a roll press when powders are being compacted.[53]

and venting both during and after compaction. If permeability is or becomes low, air pressure increases to very high levels and the sheet expands on release similarly to the effect of elastic recovery mentioned above. However, because it is not a relaxation of the material itself but an "explosive relief" of compressed air located in the interstices (pores) between the solids, expansion of the sheet by this mechanism always results in some, often serious destruction (bursting) of the compact. Bursting is frequently associated with a popping noise. Sheets may break into slivers or irregularly shaped pieces and sometimes disintegrate to powder.

The damage that is done by the expanding gas depends on the strength of the compacted material (Fig. 6.193) and its remaining open porosity but there is always some reduction of quality associated with the process. Therefore, it is most desirable to remove the air while it is expelled from the densifying material. If initially relatively narrow rollers, arranged side by side, are considered, three types of problems created by the removal of entrained air can be identified.

First, gas flowing countercurrently to the feed material in the nip causes particulate solids to alternatively fluidize and flow. These process conditions repeat in a cyclic manner. This operating condition is not acceptable because not only does the yield of good product drop considerably, thus reducing process economics, but equally important is that large fluctuations in pressure and torque are experienced that may result in serious damage to rolls, bearings, gear reduces, and drives (chattering).

Second, the gas flowing into higher levels of a gravity feeder hinders the free flow into the nip and reduces the roller press capacity. This problem can be overcome by installation of a feeder (e.g., screw feeder) which forces material into the nip.

Third, the conditions described above are most pronounced for fine feed materials featuring low permeability. It is possible that the problems caused by entrained air cannot be solved by simple and economical (i.e., sufficient roller speed and, thus, capacity) means without changing the feed characteristics by coarsening the particle size and, therefore, increasing permeability. One rather simple method to achieve this is to recirculate a certain amount of crushed compacted material with a particle size distribution that must be determined by experimentation.

Figure 6.194 shows the effects of roller speed and permeability on air pressure in the compacted sheet.[53] To the original graph a line has been added that characterizes the theoretical strength of the compacted sheet (i.e., prior to decompression); it decreases with increasing roll speed because of the shorter time available for the development of binding forces. Comparison of this line with the curves for air pressure shows that air entrainment does not limit roller speed for coarse granular

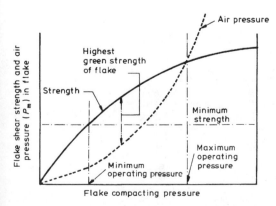

Figure 6.193. Effect of entrained air on compacted sheet strength.[53]

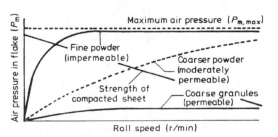

Figure 6.194. Effects of roller speed and permeability on air pressure in the compacted sheet.[53]

material, and only insignificantly influences the choice of roller speeds for "moderately permeable" coarser powder, but leaves only a small range of very low speeds for "impermeable" fine powders.

In most cases, the feed of roller presses does not consist of the coarse granular material with no limitation to roller speed. Consequently, if equipment with large capacity is required, roller width must be increased. Figure 6.195 reiterates[52] that air can escape from the nip countercurrently to the flow of material into the feeder arrangement, over the top of the rollers, and sideways between the cheek plates sealing the roller nip against excessive leakage of solids. The first portion, which causes limitations of free flow of feed to the rollers, grows with increasing roller width. While wide rollers (with working widths in excess of 1000 mm) operate without problems in high-capacity applications if materials with "high permeability" are handled,[52] decreasing feed permeabilities will reduce acceptibility of wide rollers, even if force feeders are applied.

Generally, the same phenomenon as discussed above occur during briquetting with roller presses. Differences are, that it is more difficult to vent the gas that is being squeezed out from a pocketed roll, particularly during the last stages of compaction when the pockets close (see Fig. 6.185) and essentially seal remaining air within the briquette. Since this compression of residual air cannot be completely avoided even at low roller speed and high permeability of the particulate solids and, on the other hand, during briquetting a final product is to be made, the disruptive effects of entrained air are even more critical during and after release from the rollers than in the case of compaction.

Phenomenology of Roll Briquetting.   During roll briquetting individual pieces with defined shape are generated but are not compacted simultaneously all over; rather, pressing takes place at varying rates and reaches different maxima at different times in separated points within the briquette. Only in the relatively rare case of materials with a very high intrinsic bond strength caused by compaction and requiring a low degree of densification can the product of roll briquetting be described as fault-free. Even in these cases the compact is not a perfect match to the pockets. The generative process of rolling always produces a compact that is longer than the circumferential length of the cup. This process, together with expansion due to elastic recovery and/or compressed air make the briquettes larger than the combined pocket volumes. If other materials are briquetted, especially those requiring high densification, imperfections and faults do arise that may not occur in every compact and, often, very similar problems can arise for en-

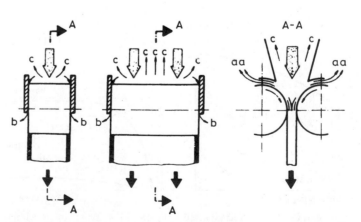

Figure 6.195. Schematic representation of deaeration in a roller press.[52]

tirely different reasons. Moreover, the precise causes of some of these faults are still unknown.

One of the most easily recognized and probably the best understood of the various faults is a narrow, broken band of material around the plane dividing the two briquette halves. This is commonly known as "flash" or "web" and results from the fact that the rollers are not in contact during operation. The web can become excessively thick owing to either stretch of the press frame or misalignment of the rollers during the setting-up procedure; in that case, briquettes are joined together and, particularly in the case of multirow presses, may have the appearance of a chocolate bar. In addition to distracting from the appearance of the product, special equipment is necessary to separate the briquettes which may also cause damage to the structure.

Another fault, equally as common as that of thick flash but probably less understood, is that in which the compacts open up along the plane of pocket contact. In the vast majority of cases, this opening is at the trailing (last compacted) edge of the briquettes but, occasionally, opening at the leading (first compacted) edge has been described. These faults are known as "clam-shelling," "oyster-mouthing," or "duck-billing."

The most common explanation of the above, especially with low-plasticity materials, is that, in attempting to achieve adequate compaction at the leading edge, the trailing edge is subjected to excessive pressure and also contains most of the compressed air; therefore, it splits as a result of elastic recovery and expansion of air when the briquette is released. However, as the phenomenon has also been observed for very plastic materials in which even forward extrusion has occurred, it is likely that other mechanisms participate in producing this fault. Breaking away the flash may be a source of cracks which could lead to splitting along the central plane. This would also provide a satisfactory explanation for clam-shelling, at the front and, occasionally, the sides.

Because the trailing edge of briquettes does not receive its final pressing until the front ends of the pockets have separated (see Fig. 6.185), compacts are not homogeneous in density and, in general, using a symmetrical cup shape, the trailing end is distinctly denser than the leading end. This may suggest that the rear end undergoes higher rolling load than does the front; this, however, is not always the case. The difference in density is least when the material is plastic because it will flow, both in part and in whole, and may even extrude forward when the cups open at the leading edge. Such flow can also result in a highly polished surface of some finished briquettes.

A near uniform state of stress and strain within a briquette is more difficult to achieve with a roll press than with uniaxial compaction presses (either closed mold or extrusion) because of the more complicated geometry of the "pressing chamber" (nip plus briquette pockets). Homogeneity (but not necessarily isotropy) could be attained if either:

1. A cup could be designed that would apply equal strain increments to all elements of the material without gross movement of the materials within the cups, or
2. the material is deliberately made sufficiently plastic (either by previous processing or the addition of a plasticizing constituent) to allow equalization of strain throughout the material during compaction.

Neither of these extreme situations is feasible. For case (1) no practically conceivable cup shape can produce equal strain increments; and in case (2) a material with the necessary degree of plasticity will normally be incompatible with a potential need to develop adequate pressing load because the material could be extruded from between the pockets at relatively low pressure. Alternatively, the product specification may exclude modification of the material or it is impossible to remove the plasticizing constituents after briquetting if they are inacceptable in the product. However, a combination of rational pocket design with a material featuring maximum plasticity commensurate with required pressing load and product specification is likely to give optimum briquette equality.

One factor that may contribute to the nonuniformity of strain is an increase in roller speed. During the main stage of the compaction process, the strain rate in volume elements varies from point to point in a cup and with cup position. In the simplest geometrical estimation of these strains, their rates of change will be directly proportional to roller speed. Therefore, it is likely that operating a briquetting roller press at the slowest possible speed consistent with economic throughput would be advantageous in reducing stress differences during compaction. Moreover a slower roller speed will allow more time for any time-dependent recovery to attain equilibrium and plastic flow to reduce high stress concentrations.

Extraction Considerations in Optimizing Pocket Design. Equally as important as designing a pocket shape to achieve stress-free compaction is the requirement to obtain stress-free extraction. Even if the briquette experiences a fairly uniform stress distribution at the point of minimum volume (owing to a combination of optimum pocket shape and good material characteristics) and is, at this point, relatively fault-free, it can be damaged during its release. Although the release portion of the cycle is geometrically the same as the compaction portion, the material has changed from a deformable particulate solid to a coherent mass that is often under considerable elastic deformation. Consequently, the principal release problems are associated with changing stress distribution within the compact.

Because the trailing edge of the briquette must ultimately attain a near closed shape, with the lands at the rear of the pockets almost touching, the rolls will continue to apply pressure until the land between successive cups passes the plane of roll axes. During this phase, the forward cup space is already increasing in volume and the constraints to the

leading part of the briquette are released while the back is still being compacted (Fig. 6.185).

The effect of this mechanism will be between two extremes: one for a highly elastic low modulus material and the other for a completely inelastic (or very high modulus) material.

Briquettes made from elastic materials can always expand sufficiently at their leading end to support the rear stress during the critical period and, except in the unlikely case that the new stress distribution is so distorted that briquette strength is exceeded at some point, the compact will remain undamaged. In contrast, inelastic briquettes cannot follow the receding pocket surfaces by expansion; therefore, it moves forward until the front edge protrudes beyond the plane containing the receding edges of the cup and very high stresses can be generated at the line or point contacts with the compact. Some damage to the briquette is almost inevitable. Furthermore, the trailing edge of the compact may remain comparatively weak because not enough material is contained to fill the now larger briquette volume. If the material can deform plastically extrusion of a "tongue" through the opening gap into the rear of the preceding compact may occur.

Secondary release problems arise from various adhesive forces between briquette and cup. Obviously, pockets cannot contain any reentrant surface because, as the pockets part, the briquette would get caught and tend to split in half. Similar forces can be caused by friction between briquette and cup and on surfaces nearly parallel to the roller radius (Fig. 6.196, left).

Generally, three factors must be considered in optimizing the pocket shape for easy release of briquettes:

1. The overall release geometry. This is governed mainly by the ratio "roll diameter/pocket length." If this ratio is large enough, the trailing edges of the cups will close before the leading edges have separated sufficiently to cause damage or extrusion.

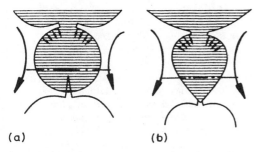

(a)   (b)

Figure 6.196. Schematic representation of two extreme pocket shapes. (*a*) Half circle, no briquette production possible owing to release difficulties; (*b*) "rationally shaped" pocket (tear drop).

2. The detailed release geometry. This is governed by the pocket shape. A "pillow shape" with the axis of its partial cylinder across the rollers and conical sides of wide angle will probably be the best. It is suggested[46] that at no point on the cup surface the normal to the surface should differ in the direction from the roll radius by more than 65°.

3. The properties of the material as briquetted. Pocket design will be more critical for inelastic than for low-modulus compacted material. However, the design will be less critical if the compacted material features high shear strength. If the front part of the briquette can survive the high stress resulting from the rear load because of its inherent strength, then any sophisticated cup shape compensation is unnecessary.

In many actual cases, compact shape must conform to commercial requirements that are unrelated to the production process (e.g., a distinctive shape may be desired for a proprietary fuel or special identifying marks may have to be applied). Consequently, the cup shape used may not necessarily be the optimum design for the material to the compacted.

The Difference in Behavior Between Single with Multirow Briquetting Presses. The different behavior of wide roll compactors as compared to narrow rollers has been discussed above.

Roll type briquetting presses feature even more pronounced differences if single row designs are compared with multirow applications (i.e., two or more pockets across the face of rollers).

For most materials, the throughput of a single row press can be increased by placing two or more rows of pockets side by side on the (correspondingly wider) rolls and enclosing the space with a single pair of cheek plates. Theoretically, the limitations of this method of increasing throughput are only in the need to provide adequately sized bearings to support the increasing roll load and in distributing feed uniformly between all rows.

Although briquettes made in a single row pilot plant may be of excellent quality, for some materials performance of a commercial multirow press may be unsatisfactory. Such problems are normally encountered with materials demanding high compaction ratios.

Three main reasons may explain the operating difference between single and multirow presses:

1. It has been noted that proportionately more work is done in the precompaction stage of a single row press. This extra work may result in a general degradation of feed material, change of the position at which compaction begins, increase of the bulk density at the start of compaction, even a difference in the adhesive properties of material's surface.

2. In the case of multirow presses it may be impossible to achieve an adequately uniform distribution of the feed on the rollers. Part of the maldistribution may be due to uneven gas backflow, particularly in the center of wide rolls. The influence of uneven distribution becomes more critical as the briquette volume decreases. With very small pockets it becomes almost impossible to produce briquettes of uniform quality in multirow presses.

3. In single row presses the cheek plates may cause substantial differences in the distri-

bution of material within the cups. The distribution within the pockets is more critical in systems requiring high compaction ratios. The effect of cheek plates is less pronounced or absent in multirow presses.

Entrainment of Material by Roller Presses. The mechanisms that control the entrainment and subsequent movement during densification are not yet fully understood. However, a number of theoretical approaches have been successful in predicting the behavior of roller systems, particularly if small changes in density are involved.

Originally, most workers considered a horizontal volume element of material in a roll press with rollers arranged side by side and assumed that it remains horizontal and retains constant thickness as it moves through the nip between the rolls. This is a gross oversimplification and leads to the prediction of excessively large changes in density for a given roll system if the material is "entrained" at the angle of friction.

Therefore, later research concluded[54-58] that material is entrained at some other angle —the "true angle of nip"—which is smaller than the angle of friction and must be determined experimentally. The use of an empirical "angle of entrainment" makes allowances for the "upward movement" of material avoiding the squeeze after compaction has commenced.

For additional information on roller presses, particularly special design features, instrumentation, and control, as well as peripheral equipment for systems with roller presses, the available literature should beconsulted.[1, 43, 46, 59]

References

1. W. Pietsch, *Size Enlargement by Agglomeration*, John Wiley & Sons/Sall + Sauerländer, Chichester, UK/Aarau, Switzerland (1991).
2. W. Pietsch, "Pressure Agglomeration—State of the Art," in *Agglomeration '77*, Vols. 1 and 2, edited by K. V. S. Sastry, *Proceedings of the Second International Symposium on Agglomeration*, Atlanta, GA, AIME, New York, pp. 649–677 (1977).
3. D. Train, "Transmission of Forces Through a Powder Mass During the Process of Pelleting." *Trans. Inst. Chem. Eng. 35*(4):258–266 (1957).
4. D. C. Hicks, Private Communication, LCI Corp., Charlotte, NC (1993).
5. G. Schenkel, *Schneckenpresse für Kunststoffe*, Carl Hanser Verlag, München, Germany (1959).
6. Anonymous, *Schneckenmaschinen*, *Mitteilungen de Verfahrenstechnischen Versuchsgruppe der BASF*, Ludwigshafen/Rh., Germany (1960).
7. G. Menges, *Einführung in die Kunststoffverarbeitung*, Carl Hanser Verlag, München, Germany (1975).
8. K. F. Mauch, "Compounding and Pelletizing of Plastic Materials with Twin-Screw Extruders," Unpublished report, Werner and Pfleiderer, Stuttgart, Germany (1986).
9. J. C. Steele, Jr. and K. A. Hanafey, "Agglomeration via Auger Extrusion," in *Proceedings of Sixteenth Biennial Conference*, IBA, pp. 287–95 (1979).
10. D. C. Hicks, "Extrusion, Spheronizing, and High-Speed Mixing/Granulation Equipment," Unpublished manuscript, LCI Corp., Charlotte, NC (1988).
11. G. Frank, "Pelletizing with Horizontal Dies," Unpublished manuscript, Amandus Kahl Nachf., Reinbek/Hamberg, Germany (1984).
12. R. H. Leaver, "The pelleting process," Unpublished manuscript, Koppers Co., Inc. (1982) (Currently Sprout-Bauer, Inc., Muncy, PA).
13. Anonymous, "Matrize für eine Pelletisiermaschine," German Patent Application OS 3 342 658 (1985).
14. Anonymous, "Pelletisiermatrize," German Patent Application OS 3 342 659 (1985).
15. Anonymous, "Flachbettpresse," German Utility Model CM 8 310 601 (1987).
16. D. C. Hicks, "Extrusion and Spheronizing Equipment," Unpublished manuscript, Luwa Corp., Charlotte, NC (1988).
17. N. Nakahara, "Method and Apparatus for Making Spherical Granules," US. Patent 3 277 520 (1966).
18. S. Bradbury (ed.), *Powder Metallurgy Equipment Manual*, 3rd ed., Metal Powder Industries Federation, Princeton, NJ (1986).
19. R. Voigt, *Lehrbuch der Pharmazeutischen Technologie*, 6th ed., VEB Verlag Volk und Gesundheit, Berlin, DDR, and VCH, Weinheim, FRG, and Deerfied Beach, FL (1987).
20. R. Ridgeway-Watt, *Tablet Machine Instrumentation in Pharmaceutics—Principles and Practice*, Ellis Horwood Series in Pharmaceutical Technology, John Wiley & Sons, New York (1988).
21. J. T. Carstensen, "Tabletting and Pelletization in the Pharmaceutical Industry," in *Handbook of Powder Science and Technology*, edited by M. E. Fayed and L. Otten, Van Nostrand Reinhold Co., New York, pp. 262–269 (1983).

22. J. T. Carstensen, *Pharmaceuticals of Solids and Solid Dosage Forms*, John Wiley & Sons, New York, p. 161 (1977).

23. J. T. Carstensen, J. B. Johnson, W. Valentine, and J. J. Vance, *J. Pharm. Sci. 53*:1050 (1964).

24. J. T. Carstensen and P. Chan, *J. Pharm. Sci. 66*:1235 (1977).

25. K. Ridgeway and R. Rupp, *J. Pharma. Pharmacol.* 21:305 (1969).

26. R. L. Brown and J. C. Richards, *Trans. Inst. Chem. Ing. 38*:243 (1960).

27. F. Q. Danish and E. L. Parrott, *J. Pharm. Sci. 60*:550 (1971).

28. J. T. Fell and J. M. Newton, *J. Pharm. Sci. 60*:1428, 1868 (1971).

29. S. Leigh, J. R. Carless, and B. W. Burt, *J. Pharm. Sci. 56*:888 (1967).

30. T. Higuchi, E. Nelson, and L. W. Busse, *J. Am. Pharm. Assoc. 43*:345 (1954).

31. E. Shotton, J. J. Deer, and D. Ganderton, *J. Pharm. Pharmacol. 15*:106T (1963).

32. J. M. Newton, P. Stanley, and C. S. Tan, *J. Pharm. Pharmacol. 29*:40P (1977).

33. J. T. Fell and J. M. Newton, *J. Pharm. Sci. 59*:688 (1970).

34. J. M. Newton and P. Stanley, *J. Pharm. Pharmacol. 26*:60P (1974).

35. J. M. Newton and D. J. W. Grant, *Powder Technol.* 9:295–297 (1974).

36. P. Popper, "Isostatic Pressing," in *Monographs in Powder Science and Technology*, edited by A. S. Goldberg, Heyden & Sons Ltd., London (1976).

37. E. Rammler, "Uber die Theorien der Braunkohlenbrikettentstehung. Sitzungsberichte der Sächsischen Akademie der Wissenschaften zu Leipzig." *Mathematisch naturwissenschaftliche Klass,* Vol. 109(1), Akademie Verlag, Berlin, Germany, 38 pp. (1970).

38. H. Metzner, "Untersuchung des Pressvorganges in Strangpressen mit Hilfe von Pressdruckmessungen unter besonderer Berücksichtigung schnellaufender Zweigelenk Pressen," Ph.D. Thesis, Bergakademie Freiberg, Germany (1962).

39. K. Schenke, "Uber die Veränderungen der Briketts beim Durchgang durch den Formkanal der Strangpressen und sich daraus ergebende Erkenntnisse über den Pressvorgang, insbesondere bei der Feinstkornbrikettierung von Braunkohle," Ph.D. Thesis, Bergakademie Freiberg, Germany (1968).

40. W. Horrighs, "Determining the Dimensions of Extrusion Presses with Parallel-Wall Die Channel for the Compaction and Conveying of Bulk Solids. *Aufbereitungs Technik, 26*(12): 724–732 (1985).

41. K. Schneider, "Druckausbreitung und Druckverteilung in Schüttgütern." *Chem. Ing. Techn. 41*(1/2): 51–55 (1969).

42. R. Kurtz, "Important Parameters for Briquetting Soft Lignite in Extrusion Presses." *Aufbereitungs Technik 27*(6): 307–316 (1986).

43. H. Herrmann, *Das Verdichten von Pulvern zwischen zwei Walzen*, Verlag Chemie GmbH, Weinheim, Germany (1973).

44. W. Pietsch, "Roll Designs for Briquetting-Compacting Machines," in *Proceedings of Eleventh Biennial Conference*, IBA, pp. 145–163 (1969).

45. G. Franke, *Handbuch der Brikettbereitung*, Vol. 1, Die Brikettbereitung aus Steinkohlen, Braunkohlen und Sonstigen Brennstoffen, Verlag Ferdinand Enke, Stuttgart, Germany (1909).

46. W. Pietsch, "Roll Pressing," in *Monographs in Powder Science and Technology*, edited by A. S. Goldberg, Heyden and Son, London (1987).

47. W. John, "Brikettieren," in *Ullmann's Enzyklopädie der Technischen Chemie*, 4th ed., Vol. 2, Allgemeine Grundlagen der Verfahrens und Reaktionstechnik. Brikettieren, Verlag Chemie GmbH, Weinheim/Bergstr., Germany, pp. 315–320 (1972).

48. K. Kegel, *Aufbereitung und Brikettierung*, Vol. 4, Part I: *Brikettierung der Braunkohle*, Wilhelm Knapp Verlag, Halle/Saale, Germany (1948).

49. W. Pietsch, "Agglomerieren problemlos—Kompaktiervorgang in Wälzdruckbrikettier—und Kompaktiermaschinen." *Maschinenmarkt MM Industriejournal 78*(88):2036–2040 (1972).

50. J. R. Johanson, "A Rolling Theory for Granular Solids." *Trans. ASME J. Appl. Mechanics, Ser. E, 32*:842–848 (1965).

51. R. Zisselmar, "Kompaktiergranulieren mit Walzenpressen." *Chem. Ing. Techn. 59*(10):779–787 (1987).

52. W. Pietsch, "Modern Equipment and Plants for Potash Granulation," in *Potash Technology*, edited by R. M. McKercher, *Proceedings of First International Potash Technology Conference*, Saskatoon, Sask., Canada, Pergamon Press Canada, pp. 661–669 (1983).

53. J. R. Johanson, "Reducing Air Entrainment Problems in Your Roll Press." *Powder Bulk Eng. 2*:43–46 (1989).

54. B. E. Kurtz and A. J. Barduhn, "Compacting Granular Solids." *Chem. Eng. Progr. 56*:67 (1960).

55. Anonymous, A study of the compression in tangential roll briquetting presses, Sahut, Conraur and Cie., Varrangeville, France (1950).

56. J. H. Blake, R. G. Minet, and W. P. Steen, "Pressure Developed in a Roll Press," in *Proceedings of Eighth Biennial Conf.*, IBA, pp. 38–48 (1963).

57. F. S. Novikov, "Calculating of Roll Briquetting Presses." *Ugol. (Russ.), 38*:50 (1963).

58. B. Atkinson, "Compaction of Powders and Pastes in Double Roll Presses." NCB/CRE/Solid Products Dept. Report No. 108 (Feb. 1964).

59. Z. Drzymala, *Industrial Briquetting, Fundamentals and Methods*, Vol. 13 of Studies in Mechanical Engineering, Elsevier, Amsterdam, NL/PWN Polish Scientific Publishers, Warszawa, PL (1993).

6.6 OTHER AGGLOMERATION METHODS

6.6.1 General

Agglomeration methods are defined and controlled by binding mechanisms. Different techniques use different binding mechanisms and the equipment applied to accomplish agglomeration is characterized by suitable handling and treatment of particulate matter to bring about the desired effect. For example, in tumble agglomeration, the particulate solids are subjected to movement that is irregular, often turbulent, and controllable, resulting in collisions between particles, development of bonds, and growth of agglomerates. In pressure agglomeration a more or less stationary bed of particles is consolidated by pressure bringing about various binding mechanisms.

Therefore, the basis of all agglomeration methods can be found in the availability and/or selection of binding mechanisms. The technique or equipment used is only the "vehicle" to obtain the agglomerated product of desired shape, size, strength, density, etc.

Consequently, "other" agglomeration methods still employ similar effects and mechanisms as mentioned before in the two main groups: tumble (Section 6.4) and pressure (Section 6.5) agglomeration. Most of the examples that will be discussed in the following are intended to show that for special applications and tasks knowledge of the binding mechanisms as well as creativity in regard to techniques to be used may result in special new methods for solving a particular problem more economically or conveniently than currently available through existing technologies.

6.6.2 Agglomeration Heat

Agglomeration by heat uses primarily the binding mechanisms sinter (or mineral) bridges and partial melting (Fig. 6.2). It is frequently called "sintering."

In the first edition of this book Limons[1] covered the sintering of iron ores in much detail; this treatice is recommended as a ready reference.

Often, agglomeration by heat is a second step (curing) in an agglomeration process, whereby in the first stage size enlargement to discrete agglomerates occurs by means of tumble or pressure agglomeration methods with or without binders and, in the second stage, hardening and development of permanent bonds is achieved by heat.

The largest application of such two-stage agglomeration procedures is the pelletization of iron ores.[2-8] Figure 6.197 shows schematically the three main induration methods used in this industry.[2] They are the vertical shaft furnace (a), the straight or sometimes circular (traveling) grate or strand machine (b), and (c) the combination of straight grate and rotary kiln ("grate-kiln"). In a complete pelletizing system these induration methods are combined with tumble agglomeration in drums or discs.

The final, often very high strength of agglomerates is obtained by development of solid bridges between the ore particles at elevated, so-called "sintering" temperatures. In the first, the tumble agglomeration stage, nearly spherical pellets are produced. These "green" agglomerates are held together by surface tension and capillary forces. During induration the pellets must be first dried and preheated before, at approx. two thirds of the melting temperature, migration of atoms and molecules sets in at solid/solid interfaces and solid bridges are formed. The problem with this and many similar processes is that, after drying, the original binding mechanism of the green agglomerates (capillary forces and surface tension) has disappeared but sintering has not yet begun. Therefore, there is a time during the process at which the agglomerates exhibit almost no strength. Theoretically, only the traveling grate may introduce low enough stresses into the essentially stationary bed of

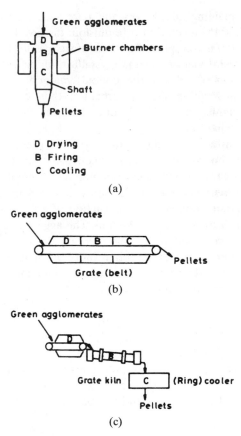

D Drying
B Firing
C Cooling

(a)

(b)

(c)

Figure 6.197. Schematic representation of the three major induration methods used in iron ore pelletization. (a) Shaft furnace, (b) grate, (c) "grate-kiln."

pellets that the latter can survive this phase. In reality, even these machines, because of their relatively crude design, have vibrations and other dynamic forces that endanger survival of the weak agglomerates. To overcome this problem, additives are used during tumble agglomeration that retain some bonding characteristics in the dry state and improve the change of survival until sintering begins. In iron ore pelletizing, this additive is traditionally bentonite, a natural montmorillonite clay.[9] In the wet stage this material imparts plasticity and in the dry stage some, but sufficient strength. Unfortunately, the addition of bentonite not only increases the cost of pelletizing but also introduces impurities (slag components in iron making) into the product. There-

fore, recent efforts to optimize the process have come up with organic additives[10,11] that do retain strength in the dry stage but burn out during sintering, thus preventing unwanted contamination.

The principle of first forming and then indurating agglomerates is also applied for other materials, particularly nonferrous ores and metal bearing recycled or reclaimed wastes.[6]

Sintering as a process of solidifying and densifying powders is very often used in modern powder metallurgy and for manufacturing of high-quality technical ceramics as well as composite materials, for example, cermets. Because of the need for good control of the process and extreme final quality of the products, a theory of sintering has been developed for these applications and extensive research has been carried out.[12-16] During sintering, shrinkage takes place that is correlated to the density of the "green" (preagglomerated) part. Since most "preforms" are produced by pressure agglomeration, density gradients (see Fig. 6.111) can cause distortion during sintering. It is therefore most important to select the correct tooling (see Section 6.5.4.1). To obtain small density variations and little distortion, isostatic pressing may be used for the production of agglomerated preforms (see Section 6.5.4.3).

6.6.3 Spray Solidification

Several methods are known that convert droplets, formed from a melt, into solid granular products by cooling. These processes are called prilling, spray cooling, spray solidification, spray congealing, as well as shoe or pastille forming.[17] Although these methods are often mentioned in connection with agglomeration, this technology is not part of the unit operation "Size Enlargement by Agglomeration."

Spray drying (Fig. 6.198) on the other hand, is a true agglomeration process. Feed material is either a solution, an emulsion, a suspension, or a slurry. While, in the first case, particles are forming by crystallization as the solution becomes supersaturated during drying, solid particles are already present in the other liquids. The feed is pumped to an atomizer, either nozzle or rotating wheel; the resulting droplets are immediately contacted by hot gas that has entered the drying chamber through a specially designed air dispenser. Hot gas and droplets move con- or countercurrently, producing excellent heat and mass transfer. Owing to the large surface area of many small droplets, rapid evaporation takes place. At the same time, heat evaporation is removed which actually results in cooling. This effect is very advantageous as it prevents overheating of the product while allowing for a relatively high inlet air temperature, thus improving economy.

During drying the drop becomes smaller and the newly formed (e.g., by crystallization) or concentrating particles are compacted within the diminishing droplets because of forces caused by surface tension. Van der Waals forces may develop. At a certain point in time a small, almost spherical wet agglomerate has formed and further drying removes

remaining liquid from the pore space. If the liquid is a solution or emulsion, dissolved material is transported to the surface and forms a crust. Final drying and crystallization or deposition of solid (often colloid) material takes place within the agglomerate, thus causing bonding by solid bridges and/or binding mechanisms.

In a single-stage spray dryer the process is finished when most of the moisture in the pore space has dried. The agglomerates accumulate in the lower part of the spray drying chamber and are removed by the suction of a fan driving a dust collection system. The agglomerates are collected in a cyclone while dust is collected in a wet scrubber (not shown in Fig. 6.198). Material-laden scrubber water may be recirculated and mixed with the liquid feed.

Since the resulting agglomerates are rather small and light (from solutions, hollow spheres are obtained), further development of the technology was directed to additional size enlargement and, potentially, increase of product density. One possibility is to treat the spray-dried material in fluidized bed whereby addi-

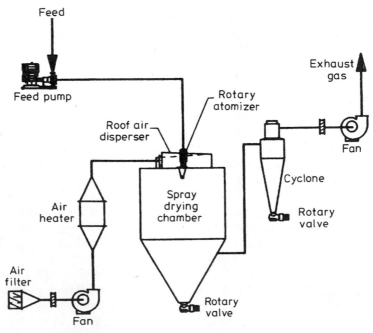

Figure 6.198. Flow sheet of a single-stage spray dryer.[18]

tional drying, cooling, and/or agglomeration can take place.[19] For the latter to happen, the product is slightly rewetted with solvent or liquid feed material, to make it sticky enough for agglomeration, and dried.

Spray and fluid bed drying technologies can be combined into one multistage process to accomplish the tasks discussed above. Figure 6.199 shows as an example the two-stage arrangement of a spray dryer and a vibrating fluid bed. Further agglomeration (size enlargement and/or densification), drying, and cooling may take place in one or more fluidized beds installed in-line with the equipment shown in Figure 6.199. Also, different spray drying and tumble agglomeration (not only low-density, fluidized bed) technologies can be combined in a similar way.

Recently, a new fluidized spray dryer–agglomerator was introduced that accomplishes both tasks in one unit.[18] Figure 6.200 shows the flow sheet. The spray dries essentially as described before but the particles now collect in a fluidized bed at the bottom of the

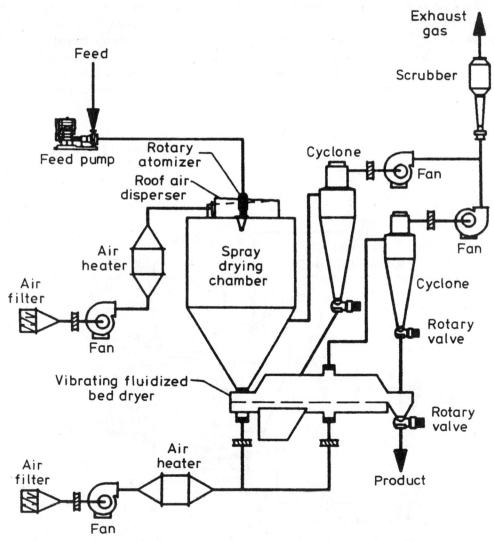

Figure 6.199. Flow sheet of a two-stage spray and fluidized bed dryer.[18]

chamber. Primary hot air for the process enters the dryer at the top through an air dispenser surrounding the atomizer and is used for the spray drying part of the process. Secondary air, about 25% to 40% of the total process air, is introduced from the plenum at the bottom through a perforated distributor plate to fluidize the fluid bed portion of the dryer. This air may be hot, warm, or cold depending on process requirements.

Because the residence time of particles in the fluidized bed can be minutes as compared to seconds in a normal spray dryer, lower air temperatures can be used for the same amount of liquid to be evaporated. Particularly, the slow, last drying stage (removal of moisture from pores of the agglomerate) will occur only in the fluidized bed and, therefore, the intermediate moisture content at the interface between spray and fluid bed areas can be relatively high and is adjustable. This moisture content in and on the surface of the fluidized particles is a major contributor to the agglomeration process. In addition, smaller particles are propelled from the fluidized bed area into the spray zone. This upward blowing of particle-laden air against the downward flow of drying gas creates a very turbulent environment causing most of the relatively dry fines to interact with the wetter particles coming from the spray and agglomerate.

The combined process air exists through outlet openings in the top of the chamber. Particles still entrained in the gases are separated in a cyclone. The material collected in the cyclone can either be removed from the system for direct use or it is recirculated into the fluidized bed for further agglomeration.

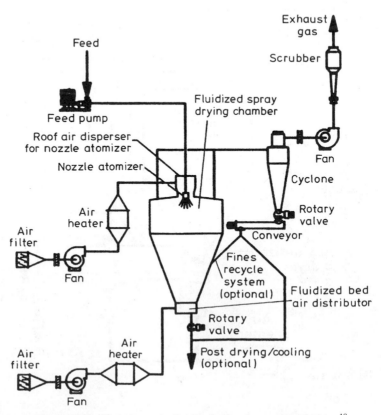

Figure 6.200. Flow sheet of a fluidized spray dryer–agglomerator.[18]

The main differences in product characteristics between spray-dried and fluid bed agglomerated materials are:

Spray-dried: Small, light (often hollow), spherical with relatively smooth surface

Fluid bed agglomerated: Larger, denser, irregularly formed with relatively rough surface.

The difference in particle size is shown in the graph of Figure 6.201. In the recently introduced fluidized spray dryer–agglomerator[18] spray-dried powders are directly agglomerated whereby differently sized spherical particles are bonded together.

6.6.4 Alternate Sources for Particle Movement

As indicated in the introduction of this section, many different techniques may be applied to induce irregular movement which will cause collisions and, if sufficiently high adhesion forces are present, bonding (agglomeration). In addition to rotating discs, drums, mixers of all kinds, fluidized beds, vibrating and shaking conveyors, etc., many other methods to produce turbulent stochastic particle movement are possible.

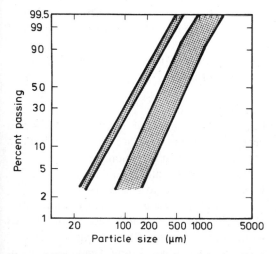

Figure 6.201. Comparison of typical particle size distributions obtained from spray drying (*left*) and fluidized bed agglomeration (*right*).

Two such technologies shall be discussed as further examples. One represents a relatively sophisticated approach, the other an extremely low-cost application.

The efficiency of currently used techniques for the removal of particulates from gases drops off sharply if the particle size becomes smaller than 1 μm. On the other hand, the human pulmonary system is most efficient in absorbing and retaining particles in the micron and submicron range. These particles are then the primary cause of respiratory ailments. While such suspended particles in emissions from, for example, stacks of power plants, are invisible, measurements have revealed that approx. 50% of the particulates suspended in the air of urban regions are smaller than 1 μm.[20]

Micron and submicron particles can be effectively removed from aerosols if they are first converted into agglomerates with a size of, say, 5 to 20 μm. To accomplish this, acoustic agglomerators can be used. As described in a historic review,[21] accelerated agglomeration of particles in sound fields is, per se, not a new idea.

Movement of micron and submicron particles in a carrier gas can be due to Brownian movement, caused by the collision of thermally agitated gas molecules with solid particles, and by convection currents or turbulence. In addition, an acoustic field would impose acoustic pressure and velocity. For a typical acoustic sound pressure of 160 dB the acoustic velocity will be about 5 m/s and a typical acoustic frequency of 2000 Hz might cause a fully entrained particle to flit back and forth 2000 times a second over a distance of about 600 μm.[21] Particle entrainment is defined by an entrainment factor η_p:

$$\eta_p = 1/(1 + \omega^2\tau^2)^{1/2} \qquad (6.76)$$

with ω the acoustic frequency and $\tau = \rho_p d_p/18$ μm the particle relaxation time. ρ_p is the particle density, d_p the particle diameter, and μ the gas dynamic viscosity. $\eta_p = 1$ characterizes full entrainment and for $\eta_p = 0$ no entrainment occurs; the latter means that the particle is not affected by the acoustic field and, in respect to particles moved by the acoustic pressure, stands still. Figure 6.202 depicts the particle entrainment factor as a function of particle size for five sound frequencies.[21] For each of the frequencies a particle size exists below which particles are almost fully entrained ($\eta_p \geq 0.5$). For example, in the case of a sound frequency of 2000 Hz, this "cut size" is approx. 4.5 μm. The larger particles, compared to the cut size, are essentially still while the smaller particles are moving through large displacements, colliding with and then adhering to the large particles because of high van der Waals forces.

In the hot gas clean-up system of a coal burning fluidized bed power plant acoustic agglomeration could be installed after the first cleaning cyclones and followed by a high-efficiency cyclone. The power required to operate the acoustic agglomerator is about 0.02% to 0.5% of the power plant output. This means that for a 250 MW power plant several hundred kilowatts of acoustic power are needed. Compared with the acoustic power output of a four-engine jet aircraft on take-off of approx. 36 kW, these are very large acoustic powers. Therefore, while the theory of acoustic agglomeration is well understood, large-scale application still requires considerable development. However, the need for micron and sub-micron particle removal from breathing air warrants large expenditures for research and commercial installations.

On the other end of the scale of sophistication are agglomeration methods needed for low-cost applications in the field of recovery of small amounts of valuable materials by leaching and waste processing for disposal. Many finely divided particulate wastes cannot be deposited in landfills or similar open storage facilities because of the danger of recontamination by wind and water. Because, in this case, agglomeration is only an additional cost, the cheapest possible method must be selected.

Such cheap solutions will be described in the following. They were developed for the low-cost heap leaching technology applied for low-concentration gold and silver ores and tailings that could not be economically processed by conventional methods. The technology relies on the ability of a liquid to contact the entire surface of a particulate mass and leach out the valuable component. The main reason for agglomeration in heap leaching is to prevent percolation problems in the heap caused by the segregation of coarse and fine particles during heap construction.[22] This segregation creates areas with significantly lower permeability because, there, fines fill the void volume between coarse particles. Consequently, the leach solutions follow the path of least resistance through "open" areas and bypass or barely wet the areas containing large amounts of fines. This results in lower extraction, longer leach time, and higher reagent consumption. In addition, after the heap has been built, fines are washed into pockets and layers of the pile and thereby further impede uniform flow of the solution.

Percolation problems can be minimized if fines are attached to the coarser particles by agglomeration. When the heap is built, fines are uniformly distributed and, if the bonds within the agglomerate are strong enough and do not deteriorate during leaching, remain immobile. Because the heap leaching technology is a low-cost process and the amount of

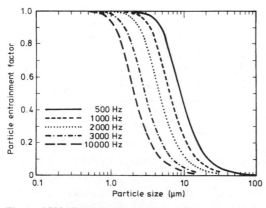

Figure 6.202. Particle entrainment factor versus particle size at different sound frequencies.[21]

recoverable metals is small, high agglomeration costs are not warranted. While any of the previously mentioned tumble or pressure agglomeration methods could be adapted for use, the investment cost for the equipment alone would be prohibitive. Therefore, two requirements have to be met:

1. Selection of a suitable binder that is cheap, easily available, effective, and produces a permanent bond
2. Development of an agglomeration method that requires minimum investment.

Figure 6.203 describes "stockpile agglomeration." An inclined conveyor discharges ore mixed with proper amounts of CaO and cement (dry binders) at a point 5 to 6 m above ground and builds a stockpile. The stream of material falling from the conveyor is wetted with coarse sprays (liquid binder). Below the sprays and suspended in the falling curtain of material are several heavy dispersion bars that act as a simple, stationary mixer. The wetted mass of ore and binder then tumbles down the slopes of the pile and agglomerates. At the foot of the pile, a front-end loader picks up the agglomerates and transfers them into a dump truck or directly onto the heap.

Another simple agglomeration method is "belt conveyor agglomeration" (Fig. 6.204). It is really a modified "stockpile agglomeration" system with additional sprays and mixing at each transfer point from one conveyor to the other. The number of transfer points depends

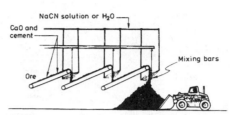

Figure 6.204. Belt conveyor agglomeration.[22]

on the amount of fines in the ore that must be bonded onto larger particles or agglomerates.

Figure 6.205 depicts the vibrating deck agglomeration. Dry binder is added prior to the vibrating conveyor and liquid binder is sprayed onto the bed at the beginning of the vibrating deck. The inclined deck is equipped with several steps over which the ore must tumble. Mixing and agglomeration take place.

A final low-cost agglomeration method uses the "reversed belt" principle (Fig. 6.206). It is a steeply inclined conveyor belt to which ore and dry binder are fed at the upper end. Liquid binder sprays are located in the upper third of the steeply inclined belt. The belt movement is such that it attempts to convey the mass to the top of the equipment but the steep inclination causes material to roll down against the direction of transport. Depending on the angle of inclination and the speed of the conveyor the ore can be retained on the belt long enough to provide adequate mixing and agglomeration.

6.6.5 Coating Techniques

Agglomeration can be applied also for coating. The technology used for this task is tumble agglomeration (Section 6.4). Very often, layering (preferential coalescence) occurs during

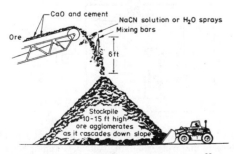

Figure 6.203. Stockpile agglomeration.[22]

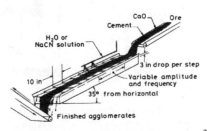

Figure 6.205. Vibrating deck agglomeration.[22]

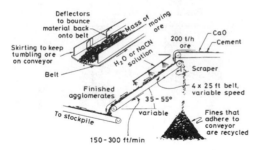

Figure 6.206. Reversed belt agglomeration.[22]

growth. During "straight agglomeration" this growth mechanism takes place continuously from nucleation until the finished agglomerate is removed from the process. In coating, nuclei are provided from elsewhere and layering occurs in turbulently moving beds of relatively large nuclei and coating powders. In most cases a liquid binder is added to assist in layer formation. Often, coating materials are also brought in by means of atomized solutions or suspensions.

Nuclei are typically agglomerates themselves. The largest applications are in the pharmaceutical and food industries where uniformity of size and shape as well as customer appeal are very important. The most common nuclei are, therefore, tabletted or spheronized particles.

Associated with the coating apparatus are four major support functions. Figure 6.207 is a sketch depicting schematically the flow sheet whereby the coating apparatus itself is a rotating drum.[23] Coaters operate batchwise, and it is most important to apply strict process control to obtain uniformly coated particles. To accomplish this, the nuclei (e.g., tablets) must tumble in the drum, the liquid sprays must cover the entire length of the particle bed, and the flow of warm or hot air must be directed such that each particle is instantaneously and sufficiently dried to guarantee the production of a smooth surface. Correct particle movement is achieved by installation of baffles or lifters or by using polygonally shaped drums. Spray systems have become very sophisticated whereby the stainless steel spray arms with nozzles are often telescopic and can be extracted through the front door for cleaning.[23]. In the case of slurries, spraying is air assisted for automatic cleaning of the nozzle.

Depending on the application, flow of drying air may be directed in different ways to obtain specific effects. In drum coaters some or all of the wall panels are double walled and perforated to allow air inlet and exhaust, controlled by specially designed valve assemblies.

The coating of smaller particles, either irregular or spheronized, is typically carried out in specially designed fluid beds.[24,25] The heart of the fluid bed process is again the liquid delivery system. Three methods of spraying

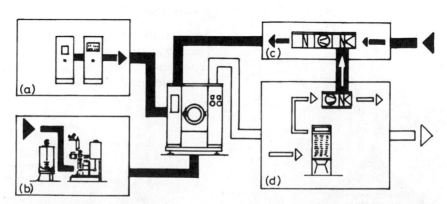

Figure 6.207. Diagram depicting schematically the flow sheet of a typical (film) coating facility.[23] (a) PLC (programmable controller), (b) storage tanks for spray liquid(s) and metering/pumping system, (c) equipment for air supply and processing, (d) air cleaning and exhaust system.

are available: top, tangential, or bottom spraying (Fig. 6.208). The nozzles used are often binary, that is, liquid is supplied at low pressure through an orifice and is atomized by air. Such pneumatic nozzles produce smaller droplets, an advantage when coating finer particles. However, it is an important principle of coating that the solution or suspension droplets impact the nuclei and uniformly distribute on the surface before the liquid is dried off (film coating). Since very fine droplets start evaporating liquid quickly as they travel from the nozzle to the fluid bed, solids concentration and viscosity increase. Therefore, droplets may contact the substrate surface and fail to spread uniformly, leaving an imperfect film. This drying of the coating spray is most severe in top-spray coaters (Fig. 6.208a) in which particle movement is the most random and liquid is sprayed against the drying air flow. Nevertheless, a sizable amount of coating is performed

in top-spray equipment because larger amounts per batch can be processed and the equipment design is simpler.

The rotating disk coater (Fig. 6.208b) combines centrifugal, high-intensity mixing with the efficiency of fluid bed drying. One major advantage of this method is its ability to layer large amounts of coating materials onto nuclei consisting of either robust granules, crystals, or nonpareil seeds. Because of the unit's high drying rate, relatively large grains in product weight can be achieved in short periods of time. Another advantage is the possibility to layer dry powders onto nuclei wetted with binder solution. Because the liquid spray nozzle(s) is (are) located below the fluid bed surface the above mentioned problems with early drying are not experienced.

The same is true of the Wurster process for bottom-spray coating (Fig. 6.208c). This is the only fluid bed coating method that is applicable for tablets, pellets, and coarse granules as well as fine powders. The Wurster coating chamber is cylindrical and contains normally a concentric inner partition with approximately half the diameter of the outer chamber. At the base of the apparatus is a perforated plate that features larger holes underneath the inner partition. The liquid spray nozzle is located in the center of the orifice plate and the partition is positioned above the plate to allow movement of material from the outside to the higher velocity air stream inside the partition. This design creates a very organized flow of product. Material moves upward in the partition, where coating and highly efficient drying occur, into an expansion area and then down as near weightless suspension in a bed of particles outside the partition. Design variations include different configurations for use in coating tablets, coarse granules, or fine powders. If larger sizes must be treated, the outer vessel diameter and the number—rather than the size—of inner partitions increase. For example, a Wurster coater with 1200 mm outer diameter for an approximate batch size of 400 to 575 kg will contain a total of seven

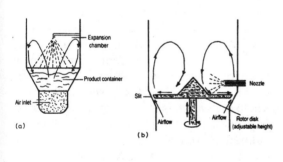

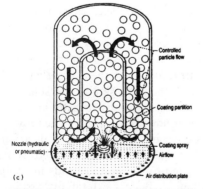

Figure 6.208. Schematic representations of the product handling sections of three fluidized bed coaters.[24] (a) Top spray, (b) tangential spray (rotary fluid bed coater), (c) bottom spray (Wurster coating system).

partition tubes (size in a circle and one in the center).[25]

6.6.6 Flocculation in Gases and Liquids

Flocculation of fine particles in gases or liquids plays an important role in industrial environmental control systems. Solid particulate contaminants are often so small that their removal from liquid or gaseous effluents is not economically possible. The agglomeration of these solids into sometimes rather loose larger "flocs," conglomerates, or "strings" of particles facilitates removal with conventional, economic environmental control devices.

Agglomeration may take place naturally or require support by forces or movements introduced from the outside or by the addition of binders. Natural aggregation has been observed and used in the precipitation of so-called, brown smoke from steel mills. The primary particles, mostly Fe_2O_3, are ferromagnetic and form dipoles that attach to each other forming string-like agglomerates (see Fig. 6.34) which can be separated from the flue gases in conventional dust collection systems. Similar but artificially induced effects take place in electrostatic precipitators. In an electrostatic field the naturally produced agglomerates of brown smoke grow into dendritic structures, thus further facilitating precipitation.

Similar aggregation takes place in liquids. If contaminated water is stirred, flocs form naturally, the size and shape of which depend on circumferential speed of the propeller and the processing time. Figure 6.209 shows that the flocs will be larger if the shear forces are low and the processing time is short. However, further investigation revealed that higher propeller speed and/or longer duration of stirring result in denser and more stable flocs.

For quite some time it has been known that polymers added to colloidal systems can have a dramatic influence on particle interaction.[27, 28] There are two ways in which polymers can promote aggregation: by making particles more susceptible to salts or by flocculating the system without aid of electrolytes. These processes are known as sensitization and adsorption flocculation, respectively. The second is the more common. To create aggregates, the polymer adsorbs on various particles simultaneously which is best accomplished by using substances with high molecular weight and a strong affinity to the particles to be agglomerates. Figure 6.210 is a sketch of a flocculate. This is how commercial flocculants that are used extensively in practice work, for instance in water purification.[29-32] By influencing the affinity of the flocculant, it is also possible to obtain selective agglomeration. This method is used in the upgrading of certain minerals.

Less well known is the fact that, more often than not, solids and droplets dispersed in aqueous solution are electrically charged owing to preferential adsorption of certain ion species, charged organics, and/or dissociation of surface groups.[27] Depending on such variables as nature of the material, its pretreatment, pH, and composition of the solution, these charges can be either positive or negative. Since the surface charges on particles are compensated by an equal but opposite countercharge surrounding them an electrical double layer develops that, even though as a whole the system is electrically neutral, results in repulsion of the particles. On addition of indifferent (nonadsorbing) electrolyte, the double layers become less active and, as a consequence, the particles can approach each other more closely before repulsion sets in. If enough salt is added, the particles may eventually come so near to each other than van der Waals attraction binds them together. This is, in principle, the expansion of the sensitivity of colloids and suspensions to salts and may, in other environments, be used to destroy stable colloids or suspensions and cause flocculation.

For technical applications, electrocoagulators are used to charge the solids in contaminated liquid effluents. Metal hydroxides are produced by a system of soluble electrodes (anodes) that, in suitable electrolytes, cause coagulation of particles into larger flocs.[33]

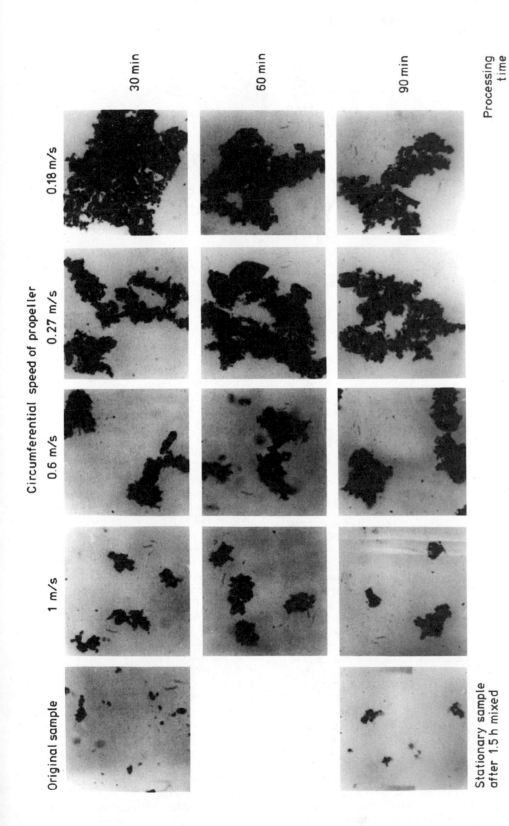

Figure 6.209. Natural flocculation of solid contaminants in river water. Parameters are circumferential speed of the stirrer and processing time.[26]

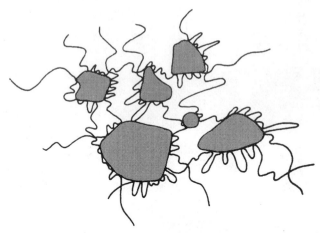

Figure 6.210. Flocculation of particles by polymers.[27]

References

1. R. A. Limons, "Sintering—Iron Ore," in *Handbook of Powder Science and Technology*, edited by M. E. Fayed and L. Otten, Van Nostrand Reinhold, New York, pp. 307–331 (1983).

2. W. Pietsch, "Stand der Welt-Eisenerzpelletierung (Pelletizing of Iron Ore, Worldwide)." *Aufbereitungs Technik 9*(5):201–214 (1968).

3. D. F. Ball, J. Dartnell, J. Davison, A. Grieve, and R. Wild, *Agglomeration of Iron Ores*, American Elsevier, New York (1973).

4. K. Meyer, *Pelletizing of Iron Ores*, Springer-Verlag, Berlin, and Verlag Stahleisen GmbH, Düsseldorf, Germany (1980).

5. K. Meyer, "Uberblick über neuere Granulierverfahren und ihre Anwendungsmöglichkeiten in der Zementindustrie," *Zement Kalk Gips, 6* (1952).

6. J. Srb and Z. Růžičková, "Pelletization of Fines," in *Developments in Mineral Processing* (advisory editor for D. W. Fuerstenau), Vol. 7, Elsevier Science, Amsterdam (1988).

7. R. L. Lappin and F. B. Traice, "A Survey of Modern Iron Ore Pelletizing Processes." British Steel Corp. PB 225 693, GS/OPER/446/1/73C, Distr.: NT1S-US Department of Commerce (1973).

8. Anonymous, "Pelletizing—a Process for the Agglomeration of Very Fine-Grained Raw Materials." Lurgi Express Info. C 1187/3.76, Frankfurt/M., Germany (1976).

9. H. Kortmann and A. Mai, "Untersuchungen über die Eignung verschiedener Bentonite für den Einsatz bei der Eisenerzpelletierung." *Aufbereitungs Technik, 11*(5):251–256 (1970).

10. F. L. Shusterich, "Production of Peridur Pellets at Minorca." *Skillings' Mining Rev. 74*(28):6–10 (1985).

11. H. A. Kortmann et al., "Peridur: a Way to Improve Acid and Fluxed Taconite Pellets. *Skillings' Mining Rev. 76*(1):4–8 (1987).

12. H. Hausner, *Bibliography on the Compaction of Metal Powders*, Hoeganaes Corp., USA (1967).

13. M. B. Waldron and B. L. Daniele, "Sintering," in *Monographs in Powder Science and Technology*, edited by A. S. Goldberg, Heyden and Son Ltd., London (1978).

14. P. J. James, "Powder Metallurgy Review 5: Fundamental Aspects of the Consolidation of Powders." *Powder Metal. Int. 4*(2) -; (3), pp. 145–149; (4), pp. 193–199 (1979).

15. S. Pejovnik et al., "Statistical Analysis of the Validity of Sintering Equations." *Powder Metal. Int. 11*(1):22–23 (1979).

16. H. Schreiner and R. Tusche, "Description of Solid State Sintering Processes Based on Changes in Length of Compacts Made from Different Metal Powders." *Powder Metal. Int. 11*(2):52–56 (1979).

17. C. E. Capes and A. E. Fouda, "Prilling and Other Spray Methods," in *Handbook of Powder Science and Technology*, edited by M. E. Fayed and L. Otten, Van Nostrand Reinhold, New York, 294–307 (1983).

18. M. M. Ball, "Revolutionary New Concept Produces Agglomerated Products While It Spray Dries," in *Proceedings of the 20th Biennial Conference*, IBA, pp. 81–96 (1987).

19. S. Mortensen and S. Hovmand, *Chem. Eng. Progr. 4*(37) (1983).

20. R. N. Davies, *Dust is Dangerous*, Faker and Faker Ltd., London (1953).

21. G. Reethof, "Acoustic Agglomeration of Power Plant Fly Ash for Environmental Clean-up," in *Proceedings of the 10th Annual Powder and Bulk*

Solids Conference, Rosemont, IL, pp. 299–312 (1985).

22. P. D. Chamberlin, "Agglomeration: Cheap Insurance for Good Recovery When Heap Leaching Gold and Silver." *Mining Eng.* 12:1105–1109 (1986).

23. Anonymous, "DRIACOATER," Prospectus DRIAM Metallprodukt GmbH and Co.KG., Eriskirch, Germany.

24. D. M. Jones, "Factors to Consider Fluid-Bed Processing." *Pharmacut. Technol. 4* (1985).

25. K. W. Olsen, "Batch Fluid-Bed Processing Equipment. A Design Overview: Part II. *Pharmaceutical Technol. 6*:39–50 (1989).

26. W. Pietsch, "Das Agglomerationsverhalten feiner Teilchen," Staub-Reinhalt. *Luft. 27*(1):20–33 (1967); English edition (The Agglomerative Behavior of Fine Particles), 27(1):24–41 (1967).

27. J. Lyklema, "The Colloidal Background of Agglomeration," in *Agglomeration '85*, edited by C. E. Capes, *Proceedings of the 4th International Symposium on Agglomeration*, Toronto, Canada, The Iron and Steel Society, Inc., Warrendale, PA, pp. 23–36 (1985).

28. B. M. Moudgil and A. McCombs, "Physical Simulation of the Flocculation Process. *Minerals Metal. Proc. 8*:151–155 (1987).

29. H. Burkert and H. Horacek, "Anwendung von Flockungsmitteln bei der mechanischen Flüssig-keitsabtrennung," *Chem. Ing. Tech. 58*(4):279–286 (1986).

30. R. Hogg, R. C. Klimpel, and D. T. Ray, "Agglomerate Structure in Flocculated suspensions and Its Effect on Sedimentation and Dewatering." *Minerals Metal. Proc. 5*:108–114 (1987).

31. L. A. Glasgow, "Effects of the Physiochemical Environment on Floc Properties," *Chemie. Eng. Proc. 85*(8):51–55 (1989).

32. B. M. Moudgil and T. V. Vasudevan, "Evaluation of Floc Properties for Dewatering Fine Particle Suspensions." *Mineral Metal. Proc. 8*:142–145 (1989).

33. M. M. Nazarian et al., "Electrocoagulator," German Patent PS 34 90 677 (1988).

6.7 ACKNOWLEDGMENTS

For the first edition of the *Handbook of Powder Science and Technology*, C. E. Capes of the National Research Council of Canada (NRC), Ottawa, Ontario, Canada, coordinated the contents of Chapter 7, entitled "Size Enlargement Methods and Equipment."

Chapter 7 in the first edition was subdivided into eight parts:

Title	Author(s), Affiliation
Part 1: Introduction	C. E. Capes, NRC
Part 2: Agglomeration Bonding and Strength	W. B. Pietsch, KOPPERN
Part 3: Tabletting and Pelletization in the Pharmaceutical Industry	J. T. Carstensen, Univ. of Wisconsin, Madison, WI
Part 4 Roll Pressing, Isostatic Pressing and Extrusion	W. B. Pietsch, KOPPERN
Part 5: Agitation Methods	C. E. Capes and A. E. Fouda, NRC
Part 6: Prilling and Other Spray Methods	C. E. Capes and A. E. Fouda, NRC
Part 7: Sintering—Iron Ore	R. A. Limons, Bethlehem Steel
Part 8: Agglomeration in Liquid Systems	C. E. Capes and A. E. Fouda, NRC

The editors decided to ask W. B. Pietsch to write Chapter 6 for the second edition using those parts of the first edition that fitted completely or partially into the current chapter 6.

For the new chapter the author used a classification that was presented in his recent book *Size Enlargement by Agglomeration*, published in 1991 by John Wiley & Sons, Ltd, Chichester, West Sussex, England, in co-operation with Salle + Sauerländer, Aarau, Switzerland, and Frankfurt/Main, Germany.

Many of the new parts of this second edition are exerpts from the above mentioned book which are presented with permission of the publishers. In addition some of the original texts were used after editing. The respective authors are acknowledged in the references.

7
Pneumatic Conveying

Mark Jones

CONTENTS

7.1 INTRODUCTION

In most applications, the major requirement of a pneumatic conveying system is to reliably convey a bulk material at a given transfer rate from one point to another. Although there are many factors that influence the specification of hardware and other aspects of detailed design, the fundamental parameters that must be considered are:

- Conveying distance and pipeline geometry:
  The solids mass flow rate for a given conveying line pressure drop will depend on the conveying distance and on the routing of the pipeline. In most cases, the number and position of bends are just as important as the length of the horizontal and vertical sections of straight pipe.

- Air supply:
  The combination of air flow rate and supply pressure must be matched to the conveying distance and bore of the pipeline. In all pneumatic conveying applications there will be a minimum velocity below which material transfer will cease.[1-4]

If these two major areas are given adequate consideration, the prospect of a successful system is almost assured. However, the apparent simplicity belies the complex relationship between these two major parameters and the extensive range of variables involved. All too often the requirements for a given system conflict, requiring intelligent compromises to be implemented.

7.2 RELATIONSHIP BETWEEN MAJOR PIPELINE VARIABLES

7.2.1 The System Operating Point

The three major variables that specify the operating point of a pneumatic conveying system are the:

- solids mass flow rate
- gas mass flow rate
- pressure gradient (pressure drop per unit length).

The relationship between these three variables is best illustrated in graphical form as shown in Figure 7.1. The zero solids mass flow rate line represents the pressure gradient required to drive the air alone through the pipeline. For a given air velocity (or flow rate), an increase in pressure gradient, above that required for the air alone, will allow some material to be conveyed. If a constant pressure gradient is available, it can be seen from Figure 7.1 that, as the air velocity is increased, the mass flow rate of material that can be conveyed decreases.

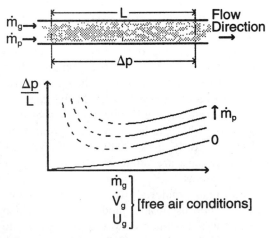

Figure 7.1. The relationship between major pipeline variables.

The pressure gradient required is a square law relationship with velocity; thus using an unnecessarily high air velocity will have an adverse effect on the solids mass flow rate.

At high conveying velocities, typically above 15 m/s (3000 ft/min), material is suspended in the conveying gas by the aerodynamic drag force on the particles. However, as the velocity of the gas is reduced, material will begin to fall out of suspension. The exact velocity at which this occurs will depend on the conveying medium and the product being conveyed. For some materials, this will lead quickly to material forming a plug, usually at a bend, that is sufficiently impermeable to block the pipeline. For other materials, conveying may continue but in a nonsuspension mode of flow. The exact nature of the nonsuspension flow regime will depend on the characteristics of the material being conveyed.[5]

7.2.2 Modes of Flow

Many different flow patterns can be observed in the pipeline of a pneumatic conveying system.[6-8] These flow patterns will vary according to the velocities of the gas and particles, and the properties of the bulk material. Owing to expansion of the conveying gas, the velocity of the gas–solids mixture increases from inlet to outlet; hence the flow patterns observed will be dependent on the location of the viewing point. In general these flow patterns can be divided into two groups:

- Suspension flow:
  In this mode of flow the majority of the particles that comprise the bulk material are suspended in the conveying gas. Systems employing this mode of flow are commonly referred to as *dilute phase* systems.
- Nonsuspension flow:
  In this mode of flow the majority of the particles that comprise the bulk material are conveyed out of suspension along the bottom of the pipe in a horizontal section. Systems employing this mode of flow are

commonly referred to as *dense phase* systems.

The link between the modes of flow that a bulk material can achieve and its properties is discussed in subsequent sections.

7.2.3 Conveying Characteristics

An alternative way of presenting the three major pipeline variables is to plot solids mass flow rate against the mass flow rate of gas as shown in Figure 7.2. This graphical form is referred to as the conveying characteristic, or performance map. A conveying characteristic applies to a particular:

- bulk material
- pipeline

In this form, the third variable, conveying line pressure drop, is presented as a set of curves. Each curve represents a line of constant conveying line pressure drop. The shape of these curves varies and depends on the conveying capability of the particular material. Bulk materials can be classified according to the modes of flow that they can achieve in the pipeline of a pneumatic conveying system. From a comparison of different conveying characteristics it can be seen that the shape of the curves is governed by the mode of conveying, which itself is determined by the physical properties of the material being conveyed.

The extent of the performance envelope for a conveying characteristic is bounded by four limits:

- The lower limit due to the air only pressure drop for the pipeline.
- The right-hand limit which is governed by the volumetric capacity of the air mover. This could be increased simply by using a larger capacity machine. However, there is no advantage in most applications for increasing the air velocity, since this simply limits the rate at which material can be conveyed.
- The upper limit can be due to either the pressure rating of the air mover, or the maximum rating of the solids feed device. In the cases shown, the maximum pressure rating of the air mover was 7 bar_g(105 psi_g); thus the upper limit is due to the solids feed device.
- The limit to the left-hand side of the characteristic is normally the most important since this marks the boundary between flow and no flow. For a system to operate without possibility of a blockage the operating point must be to the right of this boundary.

Some materials possess physical characteristics that prohibit conveying in nonsuspension modes of flow in conventional pipelines. In such cases, the limit of the pressure drop curves to the left-hand side of the graph corresponds to a minimum velocity. In this case the material remains predominantly in suspension. Typically, this minimum velocity would be about 15 to 18 m/s (3000 to 3600 ft/min). These systems are often referred to as *dilute phase* systems.

For many materials conveying is possible with a nonsuspension mode of flow, which results in a minimum conveying velocity in the range of 1 to 5 m/s (200 to 1000 ft/min). These systems are often referred to as *dense phase* systems. Figures 7.3 and 7.4 provide

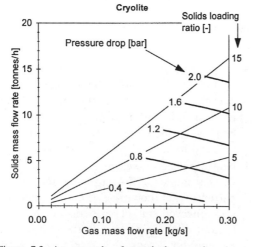

Figure 7.2. An example of a typical conveying characteristic.

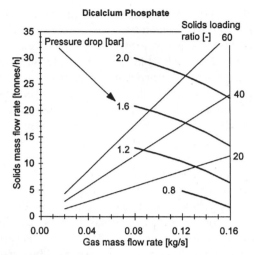

Figure 7.3. Example of a conveying characteristic for a moving-bed flow material.

more detailed examples of conveying characteristics for the two modes of *dense phase* pneumatic conveying.

7.2.4 Performance Variation

The relationship between the three major parameters (material flow rate, air flow rate and conveying line pressure drop) is unique to an individual material. For design purposes it is

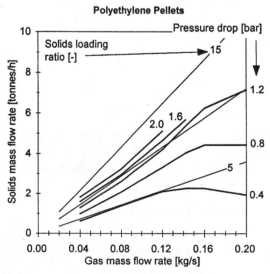

Figure 7.4. Example of a conveying characteristic for a plug flow material.

important to be aware of how much this relationship can vary from one material to another. Two bar charts are presented in Figures 7.5 and 7.6, which show graphically the variation that can occur.[9] In both cases, all the materials presented were conveyed through exactly the same pipeline using the same air mass flow rate and conveying line pressure drop. It can be seen that significant variations in material mass flow rates were achieved.

7.3 BASICS OF SYSTEM DESIGN

The basic pipeline model is shown in Figure 7.7. The two positions along the pipeline of particular interest are the point where the gas and solids are:

- mixed, referred to as the pick-up, or inlet point
- separated, referred to as the delivery, or outlet point.

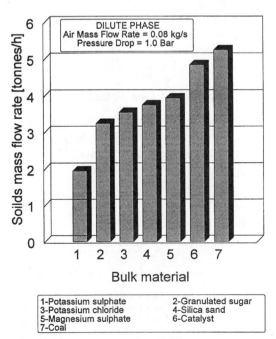

Figure 7.5. A comparison of dilute phase conveying performance.

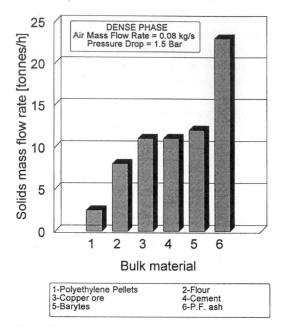

Figure 7.6. A comparison of dense phase conveying performance.

At the inlet the static pressure will be a maximum and the air velocity a minimum. Conversely, at the outlet point the pressure will be a minimum and the velocity maximum. This is due to the fact that the conveying gas expands along the pipeline as the pressure reduces. This change in velocity leads to a more complex design problem.

7.3.1 Superficial Gas Velocity

The selection of the correct gas velocity is critical to successful design. If the gas velocity is too low the pipeline will block. If the gas velocity is too high this will lead to excessive

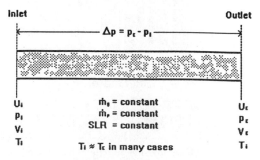

Figure 7.7. Basic pipeline model.

air only pressure drop and a reduction in the rate at which material can be conveyed. In addition, a gas velocity that is too high can lead to unacceptable damage to the material particles,[10,11] or problems of high plant wear particularly at bends.[12-14] The determination of minimum conveying velocity is dependent on the physical properties of the material to be conveyed and is difficult to predict. Often, the most reliable way of determining the conveying capability of a bulk material is to conduct a set of conveying trials.

The actual gas velocity is difficult to determine precisely in gas–solid flow, since the cross-sectional area of the flow channel is variable depending on the area occupied by particles at a given instant. To overcome this difficulty a superficial gas velocity is used that is based on the empty cross-sectional area of the pipe, but computed using the static pressure measured due to the gas–solid mixture.

$$U_g = \frac{\dot{V}_g}{A} = \frac{\dot{m}_g}{A \rho_g} \qquad (7.1)$$

7.3.2 Solids Velocity

The relationship between the gas velocity and the solids velocity will depend on the mode of flow. In *dilute phase* conveying the solids velocity is normally between about 70% and 99% of the gas velocity depending predominantly on the size and density of the particles. In *dense phase* conveying where the material is conveyed out of suspension the relationship between gas and solid velocities is less obvious and more complex in nature.

7.3.3 Gas Mass Flow Rate

The gas mass flow rate remains constant throughout the pipeline, provided no air injection system is used. This provides a useful datum for reference.

7.3.4 Solids Loading Ratio

The solids loading ratio (SLR) is the ratio of the mass flow rates of material and gas:

$$SLR = \frac{\dot{m}_s}{\dot{m}_g} \qquad (7.2)$$

The SLR gives an indication of the concentration of solids (by mass). It is constant along the pipeline and relatively easy to determine. The value of the SLR shows how the concentration of solids in the flow changes for different operating points. However, values of SLR achieved with different materials cannot be readily compared.

7.4 SPECIFICATION OF AIR REQUIREMENTS

Air movers are specified according to two major parameters:

- volumetric flow rate of gas
- supply pressure

The combination of these two parameters dictates the gas velocity at the pick-up point where the gas–solids mixture is formed. Therefore, one of the most critical decisions in the whole design process is the selection of the pick-up velocity. Having selected a value for pick-up velocity, the overall pressure drop for the system needs to be estimated.

7.4.1 Pressure Drop Considerations

The total system pressure drop is made up of a number of important elements:

- Air-only resistance in the air supply lines:
  In both positive pressure and vacuum systems the air mover can be located some distance from the conveying system. This is especially true when the air is supplied from a central source. In such cases, the pressure drop associated with the air supply line can be significant.
- Conveying line pressure drop:
  In most cases the conveying line pressure drop will be the most significant element. In many situations, the conveying line pressure drop is so dominant that the other elements can be neglected, especially where high overall system pressure drops are utilized for long-distance conveying, or for high transfer rates over short distances.

- Pressure drop across the gas–solids separation system:
  There will always be a pressure loss associated with gas–solid separation; however, its significance will depend on the gas flow rate and the size of the pressure loss compared with the total system pressure drop. For most cyclones and bag filters information can be obtained to estimate the likely pressure loss.

The pressure drop elements making up the total system pressure drop are illustrated for a positive pressure system in Figure 7.8.

7.4.2 Conveying Line Pressure Drop

In many ways, this is the most critical parameter to determine and the most difficult to obtain. In practice, the most common method of determining the pressure drop is either based on past experience of handling the same material or by undertaking pilot scale tests to obtain the relationship between the major three variables, that is, air flow rate, solids flow rate, and conveying line pressure drop.

At present, pilot testing of a material is the most reliable method of determining this relationship. It also has the advantage that tests can be carried out over a wide range of conditions including both dilute and dense phase conveying.

It would, obviously, be desirable to be able to calculate this relationship. Many academic and industrial researchers have attempted, with varying degrees of success, to model both dilute and dense phase pneumatic conveying.[15-22] To date, there is no universally accepted model even for dilute phase conveying.

7.4.3 Velocity Considerations

The ideal velocity profile would be a constant velocity throughout the pipeline. In hydraulic conveying this can almost be achieved. This situation allows the designer to specify a velocity that will fulfill the minimum conditions with a margin of safety added, without being

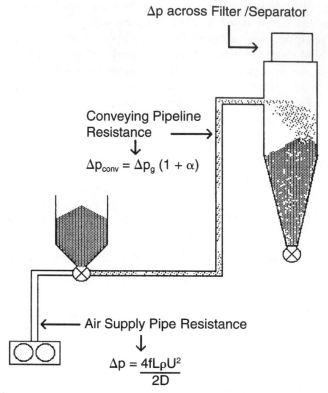

Figure 7.8. Pressure drop elements for a positive pressure system.

concerned about the problems of high velocities leading to pipeline wear, or particle degradation. However, air (or other conveying gas such as nitrogen) is compressible which means that expansion of the gas along the pipeline is inevitable. Figure 7.9 shows graphically the percentage expansion that will occur for a range of conveying line pressure drops.

The variation of gas velocity along the pipeline can be calculated using the ideal gas law:

$$pV̇_g = ṁ_g RT \qquad (7.3)$$

where p is the absolute pressure and T the absolute temperature. Substituting this expression into the equation for the superfical gas velocity:

$$U_g = \frac{V̇_g}{A} = \frac{RT}{p}\frac{ṁ_g}{A} \qquad (7.4)$$

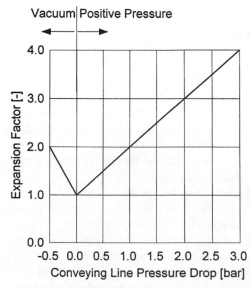

Figure 7.9. Conveying gas expansion factors for a range of pressure drops.

In many applications m_g, A, R, and T can be regarded as constant. Therefore:

$$\frac{RT\dot{m}_g}{A} = p_iU_{g,i} = p_oU_{g,o} = \text{constant} \quad (7.5)$$

where subscripts i and o refer to conditions at inlet and outlet of the pipeline respectively. This can be rearranged as follows:

$$U_{g,i} = \frac{p_i}{p_o} U_{g,o} \quad (7.6)$$

This expression relates the superficial gas velocity at the end of the pipeline to that at the pick-up.

7.4.4 Specification of Free Air Volumetric Flow Rate

As stated earlier, the free air volumetric flow rate must be specified based on the superficial gas velocity required at the inlet, or pick-up point. The relationship between these points is illustrated in Figure 7.10. Therefore, the pick-up velocity required, static pressure at the inlet, and the absolute temperature of the gas at inlet are all required to calculate the free air volumetric flow rate required. From the ideal gas law:

$$\dot{m}_g R = \frac{p_i}{T_i} \dot{V}_{g,i} = \frac{p_o}{T_o} \dot{V}_{g,o} \quad (7.7)$$

where i refers to conditions at the pick-up point and o refers to free air or normal conditions:

$$\dot{V}_{g,o} = \frac{p_i}{p_o} \frac{T_o}{T_i} \dot{V}_{g,i} = \frac{p_i}{p_o} \frac{T_o}{T_i} A U_{g,i} \quad (7.8)$$

For example, consider the flow in a 0.08 m (3 in) pipe where the pick-up velocity is 20 m/s (4000 ft/min) and the pressure drop is 1 bar (14.5 psi):

$$\dot{V}_{g,o} = \left(\frac{1.01325 \text{ bar}_a + 1 \text{ bar}}{1.01325 \text{ bar}_a} \right)$$
$$\times \left(\frac{273.15 \text{ K} + 20 \text{ C}}{273.15 \text{ K} + 20 \text{ C}} \right)$$
$$\times \left(\frac{\pi}{4}(0.08 \text{ m})^2 \right)(20 \text{ m/s}) \quad (7.9)$$

The free air volume flow rate is 0.2 m³/s (424 ft³/min) assuming that the gas temperature is constant at 20°C (68°F).

7.4.5 Air Mover Characteristics

An example of an air mover characteristic for a positive displacement twin rotary lobe blower is provided in Figure 7.11. This diagram shows that:

• The volumetric flow rate increases linearly with blower speed.

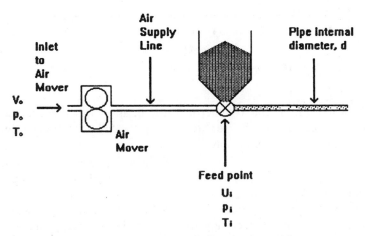

Figure 7.10. Specification of free air volumetric flow rate for positive pressure systems.

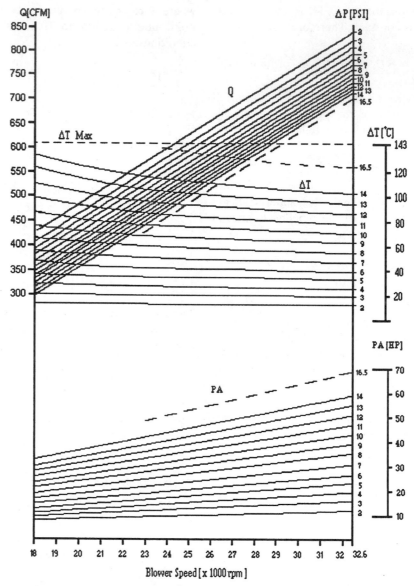

Figure 7.11. The characteristic for a twin lobe positive displacement blower.

- For a fixed blower speed, the volumetric flow rate at the inlet to the blower decreases as the delivery pressure increases.

Choosing an operating point in the middle of the characteristic provides the greatest flexibility for the pneumatic conveying system.

Once a blower with the range necessary to satisfy the system requirements has been se-

lected, the blower speed can be calculated using the following procedure:

- From the volume flow rate axis (top left) draw a horizontal line that intersects with the pressure lines.
- Find the point at which the volume flow rate line intersects the pressure line corresponding to the required pressure and draw

a line vertically down that intersects the blower speed axis.

There is a second set of pressure lines just above the blower speed axis, which can be used to calculate the power rating of the blower's motor. To calculate the power requirement:

- Find the point at which the blower speed line intersects the pressure line corresponding to the required pressure and draw a line horizontally that intersects the motor power axis (bottom right).

The third set of curves is provided (in the middle of the graph) to estimate the air temperature rise across the blower.

7.4.6 Determination of Air Requirements for Vacuum Systems

The determination of the air-mover specification for a vacuum system is similar to that for a positive pressure system in that:

- The gas velocity at the solids feed point must be specified.

- The pressure drop along the pipeline must be known.
- The flow rate of gas into the air-mover must be calculated.

The difference between this case and the positive pressure case is that the calculation of the free air volumetric flow rate is relatively simple, since the conditions at the feed point can be regarded as similar to those of free air. The sum of the pressure drops:

- along the pipeline
- through the filter
- in the pipe to the air mover

allows the pressure at the inlet to the air-mover to be calculated. This can be used to calculate the gas volume flow rate into the air-mover from the free air flow rate. A characteristic similar to that for a positive pressure system can then be used to find the necessary operating condition for the air-mover to satisfy the air requirements of the pneumatic conveying system.

NOMENCLATURE

Variables

A	Area	m^2	1 ft^2	= 0.0929 m^2
D	Diameter	m	1 in	= 0.0254 m
f	Friction Factor			
g	Gravitational acceleration	m/s^2	1 ft/s^2	= 0.3048 m/s^2
L	Length	m	1 ft	= 0.3048 m
$\dot{m}$	Mass flow rate	kg/s	1 ton/h	= 0.252 kg/s
p	Pressure	Pa	1 psi	= 6895 Pa
R	Gas constant	J/kg K	1 ft lb$_f$/lb R	= 5.381 J/kg K
T	Temperature	K	1 R	= 0.5556 K
u	Velocity	m/s	1 ft/min	= 0.00508 m/s
U_g	Superficial gas velocity	m/s		
$\dot{V}$	Volume flow rate	m^3/s	1 ft^3/min	= 0.00047195 m^3/s
Δp	Pressure drop	Pa	1 in$_{H_2O}$	= 249.083 Pa
ρ	Density	kg/m^3	1 lb/ft^3	= 16.02 kg/m^3
SLR	Solids loading ratio			

Constants

$$g = 9.81 \text{ m/s}^2 \quad = 32.19 \text{ ft/s}^2$$
$$R_{air} = 287.1 \text{ J/kg K} \quad = 53.35 \text{ ft lbf/lb } R$$
$$P_{std} = 101325 \text{ Pa}_a \quad = 14.695 \text{ psi}_a$$
$$T_{std} = 293.15 \text{ K} \quad = 527.67 \text{ } R$$

REFERENCES

1. P. A. Johnson, M. G. Jones, and D. Mills, "A Practical Assessment of Minimum Conveying Conditions," in *Proc. Powder and Bulk Solids Conf.*, pp. 221–233, Rosemont, IL (1994).
2. F. J. Cabrejos and G. E. Klinzing, "Pickup and Saltation Mechanisms of Solid Particles in Horizontal Pneumatic Transport," *Powder Technol.* 79:173–186 (1994).
3. F. Rizk, *Proc. Pneumotransport 3,* paper D4, Bath, UK (1976).
4. S. Matsumoto et al., *J. Chem. Eng. Jpn.,* 7(6): (1974).
5. M. G. Jones, "The Influence of Bulk Particulate Properties on Pneumatic Conveying Performance," PhD Thesis, Thames Polytechnic (now University of Greenwich), London (1988).
6. C. Y. Wen, "Flow Characteristics in Solids-Gas Transportation Systems," US Dept. of the Interior, Bureau of Mines, Pennsylvania, IC 8314, pp. 62–72 (1959).
7. R. G. Boothroyd, *Flowing Gas-Solids Suspensions,* Chapman and Hall, London p. 138 (1971).
8. D. J. Mason, "A Study of the Modes of Gas-Solids Flow in Pipelines," PhD Thesis, Thames Polytechnic (now University of Greenwich), London (1991).
9. M. G. Jones and D. Mills, "Some Cautionary Notes on Product Testing for Pneumatic Conveying System Design," in *Proc. Powder and Bulk Solids Conf.,* pp. 145–158, Rosemont, IL (1991).
10. A. D. Salman et al., "The Design of Pneumatic Conveying Systems to Minimise Product Degradation," in *Proc. 13th Powder and Bulk Solids Conf.,* pp. 351–362, Rosemont, IL (1988).
11. British Materials Handling Board Particle Attrition—State of the Art Review, *Trans Tech. Publications* (1987).
12. G. P. Tilley, "Erosion Caused by Airborne Particles," *Wear 14*:63–79 (1969).
13. D. Mills and J. S. Mason, "The Interaction of Particle Concentration and Conveying Velocity on the Erosive Wear of Pipe Bends in Pneumatic Conveying Lines." in *Proc. 1st Powder and Bulk Solids Conf.,* Rosemont, IL pp. 26 (1976).
14. D. Mills and J. S. Mason, "The Significance of Penetrative Wear in Pipe Bend Erosion," *Proc Int. Conf. on Optimum Resource Utilisation Through Tribology and Maintenance Management*, Indian Institute of Technology, Delhi (1988).
15. B. L. Hinkle, "Acceleration of Particles and Pressure Drops Encountered in Horizontal Pneumatic Conveying," PhD Thesis, Georgia Institute of Technology (1953).
16. K. E. Wirth and O. Molerus, "Prediction of Pressure Drop with Pneumatic Conveying of Solids in Horizontal Pipes," in *Proc. Pneumatech 1,* Powder Advisory Centre, Stratford-upon-Avon (1982).
17. F. A. Zenz and D. F. Othmer, "Fluidisation and Fluid-Particle Systems," in *Reinhold Chem. Eng. Series*, Reinhold, New York (1960).
18. J. S. Mason "Pressure Drop and Flow Characteristics for the Pneumatic Transport of Fine Particles Through Curved and Straight Circular Pipes," PhD Thesis, CNAA, Liverpool Polytechnic (1972).
19. H. E. Rose and H. E. Barnacle, "Flow of Suspensions of Non-cohesive Spherical Particles in Pipes," *Engineer 203*(5290) (1957).
20. R. G. Boothroyd, *Flowing Gas–Solid Suspensions* Chapman and Hall, (1971).
21. E. Muschelknautz and W. Krambrock, "Vereinfachte Berechnung Horizontaler Pneumatischer Forderleitungen bei Hoher Gutbeladungen mit Feinkornigen Produkten" (Simplified Calculations on Horizontal Pneumatic Conveying Feed Pipes at High Solids Loading with Finely Divided Granular Products.) Chemie-Ing-Techn, Vol. 41, Jahrg, No 21 (1969).
22. P. Marjanovic, "An Investigation of the Behaviour of Gas-Solid Mixture Flow Properties for Vertical Pneumatic Conveying in Pipelines," PhD Thesis, Thames Polytechnic (now University of Greenwich), London (1984).

8
Storage and Flow of Particulate Solids

Fred M. Thomson

CONTENTS

8.1 INTRODUCTION

This chapter is concerned with measuring the flow properties of bulk solids, and how to use this information for the functional design of storage vessels.

Quantitative measurements of the properties of bulk solids that affect their behavior when stored and discharged from bins can be made. This information can then be used for specifying the proper bin geometry for a specific application. In a mass flow bin, the solids flow channel is predictable and defined; the solids slide on the wall during discharge. In a funnel flow bin, the geometry of the flow channel is not well defined; the solids flow to the outlet through a channel formed in stagnant material.

Stresses imposed on the bin walls by the stored material are less understood. They are affected by the location of the filling point, the configuration of the flow channel, and by any deviations in the bin geometry produced during manufacture. Most published information on wall stresses deals with axisymmetric filling and discharge of a bin. It is well known that wall stresses are higher during eccentric filling and eccentric discharge and they require special consideration.

An important consideration, often overlooked, is the required rate of flow from the outlet. Flow-regulating devices at the bin outlet must be properly configured to produce the desired solids flow pattern in the area of the outlet without arching or ratholing. The air permeability of powders will vary with the consolidating pressures as they flow through bins. This can cause an erratic or restricted flow from the bin outlet. Air injection at specific points may be necessary to balance the interstitial air pressures in order to maintain a required flow rate.

8.2 DEFINITIONS

The following definitions are commonly accepted and will be used in this chapter:

Bin: Any upright container for storing bulk solids.

Silo: A tall bin, where $H > 1.5D$ (H is the height of the vertical and D is the diameter of a round bin or the dimension of the short side of a rectangular bin). A tall bin is described in some structural engineering texts as a bin where the "plane of rupture" of the contained material, determined by Coulomb's theory, intersects the side walls. There is disagreement among engineers regarding the actual location of this plane of rupture in bins having hopper bottoms and this definition is becoming less used.

Bunker: A shallow bin, where $H < 1.5D$ or, as above, where the "plane of rupture" intersects the top surface of the stored solids.

Hopper: A converging sloping wall section attached to the bottom of a silo. If a converging section stands alone as an independent bin, it is called a bunker.

Solids Flow Patterns: As solids flow from a bin, the boundaries between flowing and nonflowing regions define the flow pattern. Three common patterns—funnel flow, mass flow, and expanded flow—are defined in Section 8.4.

Flow Obstructions: It is assumed that interruption of solids flow in a bin can be caused by either of two types of obstructions: an arch (sometimes called a bridge) formed across a flow channel or bin opening, or a rathole (sometimes called a pipe) formed when the flow channel empties, leaving the surrounding stagnant material in place. These obstructions are defined in more detail in Section 8.4.

8.3 TYPES OF BIN CONSTRUCTION

Bins and silos can be categorized as either agricultural or industrial-type construction. The general descriptions that follow apply to either type.

8.3.1 Metal Construction

8.3.1.1 Shop-Welded

These are welded as a complete assembly in the shop (with roof in place) and then shipped as a complete unit to the site. The maximum width or diameter is normally limited to 3.6 to 4.0 m to accommodate rail and road clearances encountered during transportation. The maximum volume accommodated by shop-welded bins is about 1700 m^3.

8.3.1.2 Field Assembly by Welding

Preformed parts are shipped to the site, fitted together, and assembled by welding. Elevated silos and bins with hopper bottoms have been built up to 15 m diameter. Flat-bottom silos and bins resting on concrete slabs have been built up to 48 m diameter. Shop- or field-welded bins can be fabricated of any desired material. Carbon steel, aluminum, and stainless steel are all commonly used. A wide variety of wall coatings are available to protect carbon steel surfaces although sophisticated coatings that require curing, heating, etc., are best applied on shop-fabricated parts.

8.3.1.3 Field Assembly by Bolting

Bolted cylindrical bins are formed with rolled steel staves, normally 2.5 m high with flanged ends and sides (Fig. 8.1). These are gasketed and bolted together to form body rings (Figs. 8.2 and 8.3) and then stacked to form the bin or silo cylinder. The bin is anchored to a concrete slab with stirrups (Fig. 8.4).

Elevated cylindrical bolted bins with conical hopper bottoms supported on a steel structure are available up to about 8 m diameter. Flat-bottom bins supported on a concrete slab are available up to about 17 m diameter. These bins are usually fabricated of carbon steel with various paint, epoxy, or glass coatings.

A common method of assembly of bolted circular bins is shown in Figure 8.5. In the case shown, a glass-coated steel bin is being assembled by forming the plates or staves into a ring on the foundation and jacking each section vertically as it is completed. Details of the bolted seam of a glass-coated steel bin are shown in Figure 8.6.

Bins are also formed from preformed flanged plates that can be assembled into rectangular or octagonal cross-sections. These can be grouped to form large-capacity storage systems (Fig. 8.7). Details of a typical rectangular storage structure are shown in Figure 8.8.

Very large silos with capacities of 1000 to 60,000 m^3 have been built by Eurosilo.[1] These are constructed as a vertical structure steel frame supporting an outer and inner shell. The bottom is flat. Loading and discharge is accomplished by a special rotating device that levels the material while filling and draws the material to the center when unloading to underground conveyors. Other bolted agricultural silo designs are described by Reimbert.[2]

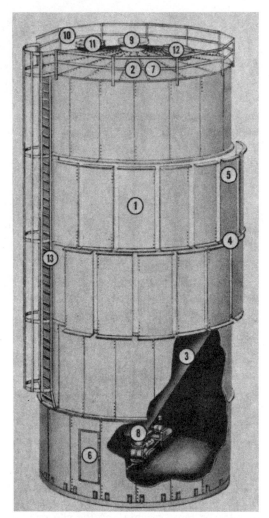

Figure 8.1. Assembly of bolted cylindrical steel bin. (1) Prefabricated cylindrical shell, (2) circular deck, (3) conical hopper, (4) flanges joining shell rings, (5) rings formed by bolting sections (staves), (6) access door in silo skirt, (7) preshaped deck plates, (8) hopper formed by joining hopper sections that are equipped with compression bars that attach to the interior of the shell for support, (9) fill opening, (10) guard rail, (11) ventilator, (12) deck manhole, (13) ladder. (Courtesy of Peabody Tec Tank, Inc., South Industrial Park, Parsons, KS.)

8.3.1.4 Spiral-Wound Coil Construction

With this unique fabrication method, aluminum or steel strip is unwound from a coil into a circular roll-forming machine that joins the edges of the strip in a continuous double-

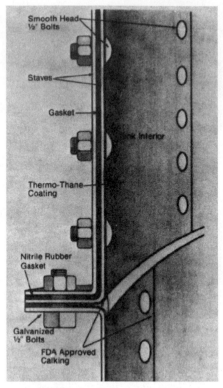

Figure 8.2. Detail of vertical and horizontal seams. Courtesy of Peabody Tec Tank, Inc., South Industrial Park, Parsons, KS.)

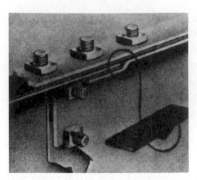

Figure 8.3. Detail of chime-lap gasket at sectional points. (Courtesy of Peabody Tec Tank, Inc., South Industrial Park, Parsons, KS.)

crimped spiraling seam. This generates a continuous rigid water tight cylinder (Fig. 8.9). Field-assembled bins are made by mounting coil and forming machine on the silo pad and continuously unwinding and seaming (Fig. 8.10), producing a continuous vertical cylinder. A detail of the double-crimped seal is shown in Figure 8.11. Prefabricated roof sections are put in place close to ground level after several revolutions have formed the initial cylinder; then unwinding and seaming is continued until the desired height is reached. These bins can also be shop-fabricated horizontally, then shipped to the site for erection.

8.3.2 Concrete Construction

Concrete bins become economically competitive with metal structures when diameters exceed 3.5 to 4.5 m. Concrete bins have been built up to 30 m diameter and larger units up to 46 m diameter are being designed. These bins can be constructed as a single circular or rectangular storage cell or in multiple cells. Common concrete bin constructions are shown in Figure 8.12. Various techniques used in concrete silo construction include the following.

8.3.2.1 Precast Construction

8.3.2.1.1 Concrete Staves. The bin is formed by assembling the staves in a circle and stacking them to form a cylinder. Circumferential steel hoops, usually consisting of three or more rods connected by steel or malleable iron lugs, are spaced at intervals along the outside of the staves and post-tensioned to

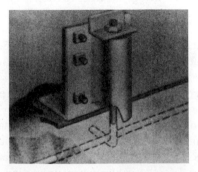

Figure 8.4. Detail of stirrup bolted to bottom ring and hook-end to foundation. (Courtesy of Peabody Tec Tank, Inc., South Industrial Park, Parsons, KS.)

Figure 8.5. Assembling a glass-coated steel bin. (Reprinted with permission of Koppers Co., Inc., Sprout Waldron Div.)

place the staves in compression (Fig. 8.13). Staves are usually about 250 to 300 mm wide by 500 to 750 mm long and 60 to 100 mm thick (Fig. 8.14). They are pressed in forms and then cured in high-pressure steam kilns. Staves can be made with hollow cores or can be made of lightweight aggregate to provide a measure of insulation to protect solids from the effects of sudden temperature changes and/or to mini-

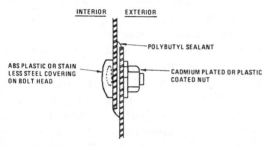

Figure 8.6. Details of bolted seam. (Reprinted with permission of Koppers Co., Inc., Sprout Waldron Div.)

mize moisture condensation on the inner walls. Solid staves can be used for less critical applications and can be cast in heavier duty construction for storage of high-density solids. The staves are fitted together with a tongue-and-groove fit. Each stave has a tongue cast on top and one side with grooves cast on the bottom and opposite side.

The exterior of the structure is coated with either of several coatings as required by the application. These include sand and cement slurry coatings, waterproof agents combined with the slurry coatings, or special paint or epoxy coatings. The coatings provide weather protection, a water drip over the hoops, and improve the appearance of the silo. The interior of the structure can be finished to provide a smooth monolithic appearance using several coats of a brush and trowelled cement plaster or several coats of epoxy coatings applied with spray equipment or trowels.

Figure 8.7. Exterior view of a bolted preformed rectangular steel bin. (Reprinted with permission of Leach Manufacturing Co., Inc. [Lemanco].)

A flat-bottom silo mounted on a concrete slab at ground level will discharge from the side. When steel hoppers are used for discharge, they are fabricated with compression ring girders and supported on steel columns from grade. The silo roof normally consists of a reinforced concrete slab, sometimes mounted on a bar joist or structural steel beam support.

Where applicable, concrete stave silos are the lowest cost concrete construction. They have been built up to 12 m diameter and 30 m high. Specifications and standards for concrete stave silos have been published by the American Concrete Institute.[16]

8.3.2.1.2 Post-tensioned Rings. These structures are assembled by stacking circular precast concrete sections vertically on an ele-

vated slab, capping with a precast roof slab, and then vertically post-tensioning them together with wire rope. Details of the design of a group of 4.9 m diameter × 9 m deep bins are described in Ref. 2.

8.3.2.1.3 Prefabricated Reimbert Silo. This construction is described in detail by the Reimberts.[3] Shaped, precast reinforced concrete slabs about 4.5 m long by 0.5 m high are used as the basic structural element. These are stacked in a horizontal position with each end fastened to vertical concrete posts. These units can be used to form storage cells of any shape —rectangular, hexagonal, etc. Storage of up to 30,000 tons of grain and other agricultural products has been reported.

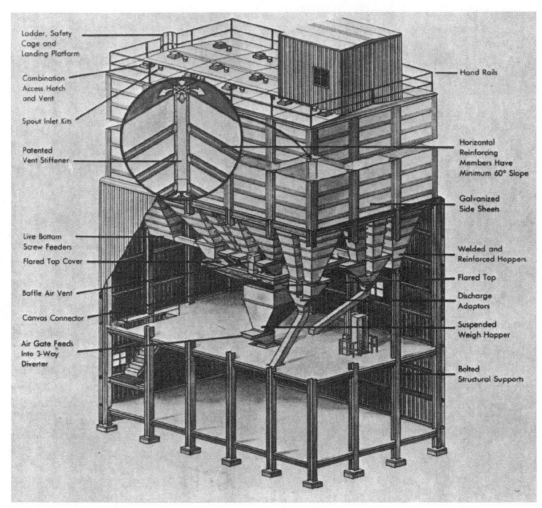

Figure 8.8. Typical construction details of a bolted preformed rectangular steel bin. (Reprinted with permission of Leach Manufacturing Co., Inc. [Lemanco].)

8.3.2.2 Cast in Place

Concrete silos or bunkers can be cast in stationary forms, slip forms, or jump forms in various configurations (Fig. 8.15). Detailed descriptions of these silos constructions are given by Safarian and Harris[5] and the Reimberts.[2] Slip forms and jump forms are well suited to silo fabrication and are the most commonly used method for large silos.

8.3.2.2.1 Slip Form. In this method, forms erected on the silo foundation are continuously raised by hydraulic screw jacks spaced

about the periphery of the wall, and guided in the lift by vertical rods (Fig. 8.16).

Concrete is cast into the forms on a continuous basis. The speed of the upward movement of the form is determined by the setting time of the concrete. Continuous pouring assures that the concrete does not set before the following layer is cast, thus providing a monolithic structure. The slip form moves from 0.2 to 0.38 m/h (with 0.3 m/h a good average) and continues around the clock until the walls are complete. Steel reinforcing is placed in the forms as they reach predetermined positions in the pour. Conventional reinforcing bars or

Figure 8.9. Spiral-wound bin. (Reprinted with permission of Conair, Inc.)

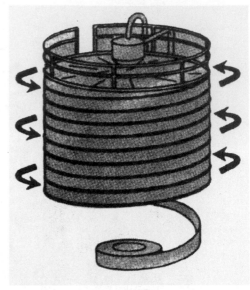

Figure 8.10. Forming a continuous spiraling seam to construct a spiral bin. (Reprinted with permission of Conair, Inc.)

post-tensioned steel strands or wire tendons may be used for reinforcing.

Slip form structures can provide greater loadbearing capacities than precast storage structures, provide smooth monolithic wall construction, and can be built in a variety of single-cell geometries, as well as grouped cells (Fig. 8.17). Slip form construction becomes most economical when silo diameters are about 12 m and larger. They have been built up to 37 m diameter and 52 m high. There are proposals for larger diameters on the drawing boards.

8.3.2.2.2 Jump Form. With this method, the forms are made up of sections about 1.8 m long by 1.3 m high, fastened together to form a continuous circular form. After two vertical layers have been cast and set, jump forming takes place. The forms from the previous pour are removed and hoisted into place above the top form and a new pour is made. Pouring need not be continuous and thus can be done on a day shift only, in contrast to slip forming, where casting must continue until the silo walls are completed. Jump form construction provides a storage system intermediate in size and cost compared to precast and slip form structures and is used mostly for cylindrical silos from about 9 to 18 m diameter and up to 46 m high.

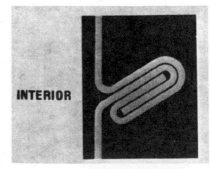

Figure 8.11. Detail of the double crimped seam. (Reprinted with permission of Conair, Inc.)

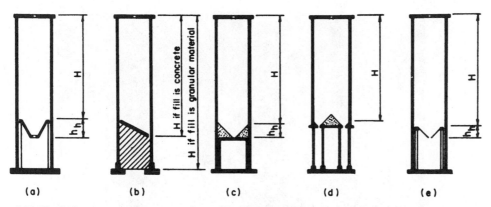

Figure 8.12. Typical concrete silo constructions: (a) Silo on raft foundation, independent hopper resting on pilasters attached to wall; (b) silo with wall footings and independent bottom slab supported on fill; (c) silo with hopper-forming fill and bottom slab supported by thickened lower walls; (d) silo with multiple discharge openings and hopper-forming fill resting on bottom slab, all supported by columns. Raft foundation has stiffening ribs on top surface; (e) silo on raft foundation, with hopper independently supported by a ring-beam and column system. (Reprinted from Ref. 16 with permission of American Concrete Institute.)

8.4 FLOW PATTERNS IN BINS AND HOPPERS

A knowledge of flow patterns occurring in a bin is fundamental to any understanding of the forces acting on the material or on the bin walls. Wall pressures are determined not only by frictional forces caused by sliding of solids along the wall, but also by the flow patterns that develop during filling and withdrawal.

8.4.1 Types of Flow Patterns

Three basic flow patterns have been identified:[55]

8.4.1.1 Funnel Flow

This is sometimes also called "core flow." It occurs in bins with a flat bottom or with a hopper having slopes too shallow or too rough to allow solids to slide along the walls during flow. Solids flow to the outlet through a channel within a stagnant mass of material. This channel is usually conical in shape, with its lower diameter approximately equal to the largest dimension of the active area of the outlet. It usually increases in size as it extends from the outlet, up into the bin (Fig. 8.18). Serious flow problems can occur if the mate-

rial compacts and exhibits poor flow properties when consolidated under solids head pressures. Material in the stagnant areas may gain strength with time and remain in place when the active flow channel empties out, forming a rathole or pipe (Fig. 8.19). In severe cases, the material can form a bridge or arch over the discharge opening (Fig. 8.20).

The flow channel may not be well defined. It may follow a serpentine path through the bin, particularly if particle segregation has occurred. Material surrounding the channel may be unstable, and in this condition will cause stop-and-start flowing, pulsating, or "jerky" flow. High pressures within the channel are often muffled by the stagnant material and may not reach the walls. At high discharge rates, however, these pulsations could lead to structural damage. As the bin is emptied (assuming the material does not compact to form a stable rathole) solids continually slough off the top surface into the channel. If solids are simultaneously charged into the top and withdrawn from the bottom, the incoming solids will pass immediately through the channel to the outlet.

In tall bins or silos, the channel boundaries may expand to intersect the cylinder walls at a

Figure 8.13. Concrete stave silos. (Reprinted with permission of First Colony Corp. and the Nicholson Co.)

point defined as the effective transition (Fig. 8.21). Material above this intersection may move in plug flow and stresses developed within the flow channel reach the bin walls.

A storage bin having a funnel flow pattern is the most common in industry and many have been designed to provide a certain volume for storage without considering that the actual discharge capacity may be much less owing to accumulation of stagnant material.

The funnel flow bin is usually the least costly design. However, it has several disadvantages when handling certain materials:

- Flow rate from the discharge opening can be erratic because arches tend to form and break and the flow channel becomes unstable. Powder density at discharge will vary widely because of the varying stresses in the flow channel. This can render ineffective any volumetric feeder installed at the silo discharge.

- Fine powders can become *aerated and flush uncontrollably when arches or ratholes collapse.* Positive sealing-type discharge devices or feeders are *mandatory* when these conditions exist.

- Solids can degrade or cake solid when left under consolidating stresses in the stagnant areas.

- A stable rathole or pipe can form if the stagnant material gains sufficient strength to remain in place after the flow channel drains out.

- Indicators mounted along the length of the bin wall to detect bin level will remain sub-

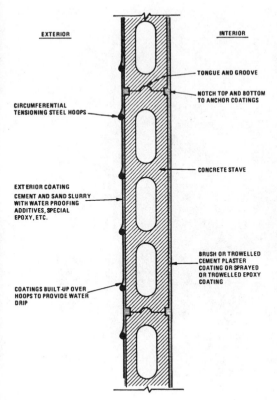

EXTERIOR

INTERIOR

TONGUE AND GROOVE

NOTCH TOP AND BOTTOM
TO ANCHOR COATINGS

CIRCUMFERENTIAL
TENSIONING STEEL HOOPS

CONCRETE STAVE

EXTERIOR COATING
CEMENT AND SAND SLURRY
WITH WATER PROOFING
ADDITIVES, SPECIAL
EPOXY, ETC.

BRUSH OR TROWELLED
CEMENT PLASTER
COATING OR SPRAYED
OR TROWELLED EPOXY
COATING

COATINGS BUILT-UP OVER
HOOPS TO PROVIDE WATER
DRIP

Figure 8.14. Detail of stave-wall. (Reprinted with permission of First Colony Corp. and the Nicholson Co.)

merged in the stagnant areas and will not correctly signal solids level in the lower regions of the bin.

Funnel flow bins are entirely adequate for the many cases where noncaking or nondegradable materials are to be stored and the discharge openings are adequately sized to prevent bridging or ratholing. Many commercial, mechanical, and aerating devices, described in Sections 8.13 and 8.14 are available for promoting flow. However, if these devices are considered for new installations, their investment and maintenance costs, as well as the probability of success in maintaining reliable flow, should be balanced against the cost of a mass flow design.

8.4.1.2 Mass Flow

This occurs in bins having sufficiently steep and smooth hoppers, and where material discharges from the entire area of the outlet (the outlet *must* be fully active for mass flow to occur). In mass flow, the flow channel coincides with the bin and hopper walls; all material is in motion and sliding against the walls

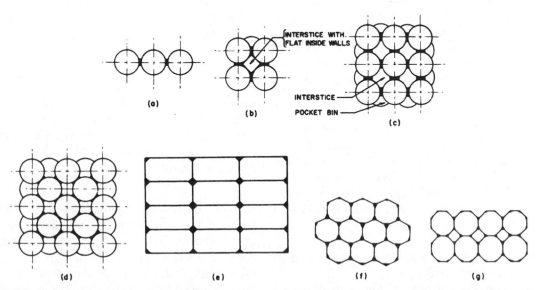

INTERSTICE WITH
FLAT INSIDE WALLS

INTERSTICE

POCKET BIN

(a)

(b)

(c)

(d)

(e)

(f)

(g)

Figure 8.15. Typical silo or bunker grouping. (Reprinted from Ref. 5, with permission of Van Nostrand Reinhold Co.)

Figure 8.16. Slip-form concrete silo under construction. (Permission First Colony Corp. and the Nicholson Co.)

of the vertical section and the converging hopper (Fig. 8.22). Material in the vertical part moves down in plug flow as long as the level is above some critical distance above the hopper–cylinder transition. If the level drops below that point, the material in the center of the channel will flow faster than the material at the walls. The height of this critical level has not been exactly defined but it is apparently a function of material angle of internal friction, material-wall friction, and hopper slope. The height shown in Figure 8.22 is approximate for many materials. In mass flow, stresses caused by the flow act on the entire wall surface of the hopper and vertical part.

Mass flow offers significant advantages over funnel flow:

Erratic flow, channeling, and flooding of powders are avoided.

Stagnant regions within the silo are eliminated.

First-in first-out flow occurs, minimizing the problem of caking, degrading, or segregation during storage.

Particle segregation is considerably reduced or eliminated.

The material in the silo can act as a gas seal.

Flow is uniform at the hopper outlet: bulk solids density is unaffected by the solids head in the upper part of the hopper. As a result, volumetric as well as gravimetric solids feeders can regulate flow from the outlet with a high degree of control.

Since flow is well controlled, pressures will be predictable and relatively uniform across any horizontal cross-section. Flow channel boundaries will be predictable and, therefore, the analysis based on steady-state flow conditions described in Section 8.6 can be used with a high degree of confidence.

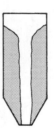

Figure 8.19. Rathole, formed when stagnant material gains sufficient strength to remain in place as flow channel empties.

Expanded flow is used where a uniform discharge is desired, but where space or cost restrictions rule out a fully mass flow bin. This arrangement can be used to modify existing funnel flow bins to correct flow problems. Multiple mass flow hoppers are sometimes mounted under a large funnel flow silo, as shown in Figure 8.24.

8.4.2 Studies of Flow Patterns

A number of techniques have been used to study flow patterns in model bins.[27-30] A comprehensive summary of these techniques is given by Resnick.[26] Included are:

1. Observing the passage of tracer layers before, during, and after flow in transparent wall "thin slice" models.
2. Immobilizing the entire model bin contents with molten wax or polyester casting resin, then slicing the model longitudinally to study the flow patterns shown by tracer layers.

Figure 8.17. Aerial view—group of slip-form silos with silos in foreground under construction. (Permission First Colony Corp. and the Nicholson Co.)

8.4.1.3 Expanded Flow

Expanded flow is a term used to describe flow in a vessel that combines a funnel flow converging hopper with a mass flow hopper attached below it, as shown in Figure 8.23. The mass flow hopper section ensures a uniform, controlled flow from the outlet. Its upper diameter is sized such that no stable pipe can form in the funnel flow hopper portion above it.

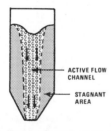

Figure 8.18. Funnel flow through an entire bin.

ACTIVE FLOW CHANNEL

STAGNANT AREA

Figure 8.20. Arch or bridge, formed across a flow channel.

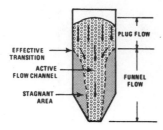

Figure 8.21. Funnel flow below an effective transition.

3. Sequentially photographing and tracking the passage of particles as viewed through a hopper with a transparent wall.
4. Photographing particles through a transparent wall using stereophotogrammetric techniques. In this method developed by Butterfield et al.,[28] photographs of moving particles are taken by a single camera. Successive photographs are viewed as a stereo pair with each eye viewing one of the photographs. An optical, three-dimensional model can then be provided and from that isovelocity contour maps can be constructed to reproduce the displacement field. An example is given in Ref. 29.
5. Measuring gamma radiation absorption in flowing beds to determine variation in porosity.

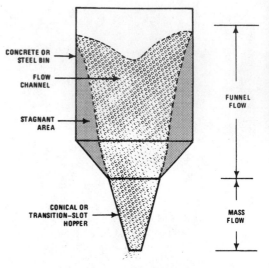

Figure 8.23. Expanded flow through a single outlet.

6. Measuring X-ray densities during and after flow in a model hopper.
7. Tracing the flow with radioactive or colored markers deposited on various parts of the bed during flow.
8. Using tracer layers in a model that can be separated longitudinally. After the flow pattern is developed, the model is laid on its side, the top half removed, and excess material brushed off to reveal the tracer layer patterns on a plane across the center.

Sketches of flow patterns observed during model tests have been given in a number of papers. Deutsch and Clyde[12] and Deutsch and

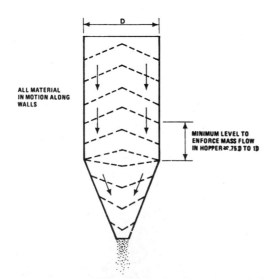

Figure 8.22. Mass flow.

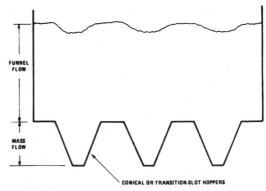

Figure 8.24. Expanded flow through multiple outlets.

Schmidt[13] presented patterns observed during flow of granular material in semicylindrical 305 mm model bins having a height varying up to 1500 mm (Figs. 8.25 and 8.26). Sugita[33] showed similar patterns occurring with 177 to 250 μm beads in a 200 mm diameter $\times$ 800 mm high semicylindrical model (Figs. 8.27 and 8.28). He also observed flow patterns into an eccentric discharge opening (Fig. 8.29). Lenczner[35] and McCabe[34] studied the flow of sand in model bins. McCabe tested sand in a 2450 mm high model with diameters varying from 80 mm to 460 mm. Data from his test is shown in Figure 8.30. All these tests show that the flow patterns are very sensitive to degree of consolidation of the material. Nguyen et al.[36,37] studied the flow of dry granular sand, polystyrene, glass beads, and rice in model conical and wedge-shaped hoppers and showed that the height of the free surface of the material in the vertical part of the bin can have a significant effect on flow pattern.

These studies showed that the boundaries of funnel flow cannot yet be predicted with certainty. They are a function not only of the hopper configuration, but also of the solids level in the vertical portion of the bin, and the angle of internal friction of the material, which in turn is sensitive to changes in bulk density

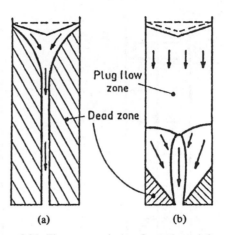

Figure 8.26. Flow zones during flow of sand from a model bin after compaction of vibration; (a) Initial flow pattern: central pipe extends to surface forming conical crater, (b) steady flow. (From Ref. 12.)

caused by changes in consolidating pressures during filling and discharge.

Blair-Fish and Bransby[27] observed variations in density in flowing material in mass and funnel flow using radiographic techniques with lead shot tracers in a model bunker. The velocity fields they detected were similar to those reported by Deutsch, and their photographs clearly showed flow direction and rupture surfaces.

Chatlynne and Resnick[29] used gamma ray absorption in a flowing bed and found a dilation wave that moves up the bed as flow as initiated, similar to that shown in Figure 8.30. They reported that the porosity of their test bed was 0.41 at rest. It increased to 0.47 during flow, which was also the value they obtained at minimum fluidization conditions.

Studies by Giunta,[38] Van Zanten et al.,[87] and Johanson[39] defined the flow channels observed on flat-bottom models in terms of the measured frictional properties of the bulk solids.

Giunta[38] used fine powders (starch, pulverized coal, and iron concentrate) with spaced layers of marking material in a 457 mm diameter $\times$ 610 mm high, split-model bin, similar to Johanson's.

He found the flow pattern to be a function of effective angle of internal friction, opening

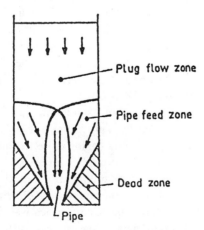

Figure 8.25. Flow zones during steady flow of sand from a model bin after loose filling, with no compaction. (From Ref. 12.)

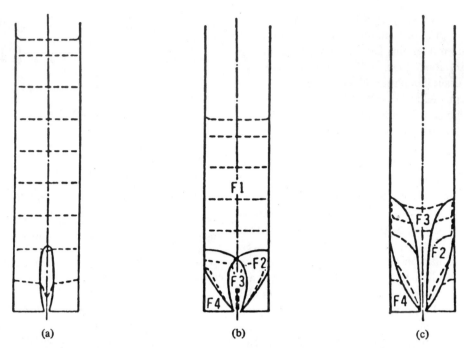

(a) (b) (c)

Figure 8.27. Flow zones during free discharge of glass beads from a model bin after loose filling with no compaction; (*a*) Immediately after discharge begins, the zone of flow extends vertically above the orifice. Height varies with orifice diameter and material properties. (*b*) During steady-state flow, the following zones appear: *F1*, Material sinks uniformly with steady velocity. *F2*, Material entering this zone then flows into zone F_3 with radial velocity. Material at boundary of *F1* and *F2* reaches state of failure and flows plastically. *F3*, Material falls vertically with a high radial velocity. *F4*, stagnant zone. (*c*) In the final state of discharge when the free surface moves down to a certain height, the boundary between *F1* and *F2* makes a slow ascent. Eventually this boundary rises up to the falling free surface, *F1* disappears, and a crater is formed. (From Ref. 33.)

size, and head of material in the bin. He proposed the following equation for determining the boundaries as shown in Figure 8.31:

$$2\bar{Y} = D + 2\tan\bar{\theta}\,\frac{H - A\dfrac{D}{2}}{1 + A\tan\bar{\theta}} \quad (8.1)$$

where

D = diameter of discharge opening (ft) (where D is large enough to prevent arching or ratholing)

$\bar{\theta}$ = angle of flow pattern boundary at edge of opening (degrees)

H = head of material in bin (ft)

Y = maximum radius of boundary between flowing and nonflowing material (ft).

Angle $\bar{\theta}$ and factor A are dependent on the angle of internal friction as given in Figure 8.32. This angle is defined in Section 8.6.

Giunta states that Equation (8.1) is valid only if $H > AD/2$. If $H \le AD/2$, the diameter of the flow boundary will remain the same as the opening ($2\bar{Y} = D$).

A study of PVC and sand flow in a 1.5 m bunker reported by Van Zanten et al.[87] confirmed the central flow region close to Giunta's prediction but also found a large cylindrical slow flow zone surrounding the central "fast flow" core. They identified several flow zones (Fig. 8.33a). The number of zones was found to be different depending on flow properties of the material (Fig. 8.33b) and the configuration of the bin. In some sloped hoppers, only a

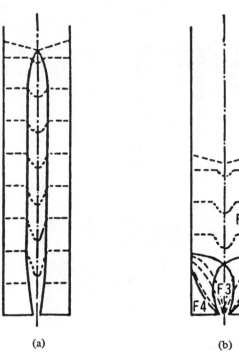

(a)   (b)

Figure 8.28. Flow zones during free discharge of glass beads from a model bin after compaction by tapping after filling; (a) Immediately after discharge begins, the zone of flow extends from the orifice to the free surface of the material. A smaller crater is formed at the surface. (b) Steady-state flow (see Fig. 8.30 for identification of zones). (From Ref. 33.)

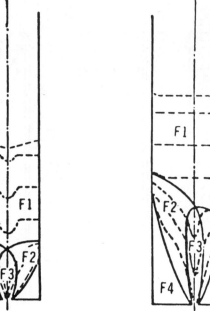

Figure 8.29. Flow zones during steady-state flow of glass beads from an eccentric opening (see Fig. 8.30 for identification of zones). (From Ref. 33.)

pulsations or jerky flow that can lead to vibration or possible structural failure.

8.5 STRESSES ON BIN WALLS

8.5.1 Static and Dynamic Conditions

Stresses on the bin walls are caused by combinations of static and dynamic conditions that occur during filling and discharge of a bin. Extensive bibliographies on this subject are presented in Refs. 11 and 101. Experimental measurements on models and industrial size bins have shown that the distribution of wall stresses changes significantly when flow begins after the initial filling, and these stresses remain after the outlet is closed.

8.5.1.1 Initial Filling: Mass Flow

When a bin is initially filled, the solids contract mostly vertically in the cylinder and hop-

conical and cylindrical fast flow, with stagnant zone were present. The angle O_T was found to be close to that predicted by Jenike[54] and shown in Figure 8.33c.

Johanson[39] studied the zones in funnel flow. He identified the steady flow zones (Fig. 8.34) as a function of angle of internal friction as in Jenike, but he further defined the surrounding region of unsteady flow that occurs with free flowing, noncohesive, frictional solids (Fig. 8.34). He attributes formation of this secondary channel to pressure changes along the steady flow channel walls during flow that causes loosening of the adjacent material and causes the outer region to become unstable. These unstable zones in funnel flow cause

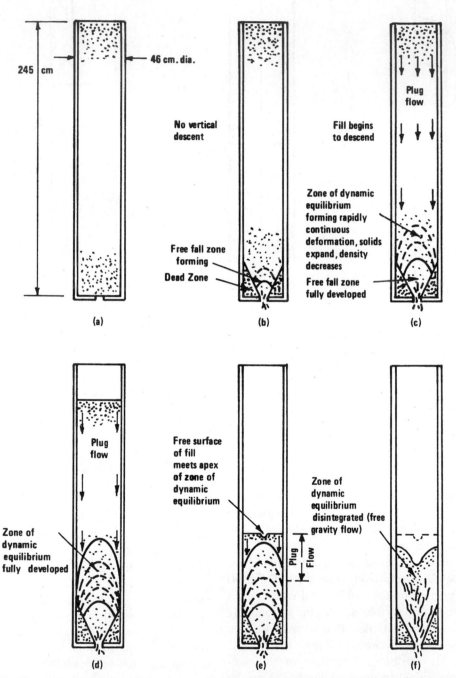

Figure 8.30. Flow zones during free discharge of sand from a model bin; (*a*) Bin is full, discharge port closed. (*b*) Discharge port opens, discharge begins. (*c*)–(*f*) Discharge continues. (From Ref. 34, with permission of the Institute of Civil Engineers.)

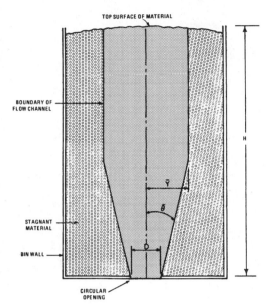

Figure 8.31. Giunta's predicted flow boundary in funnel flow. (From Ref. 38.)

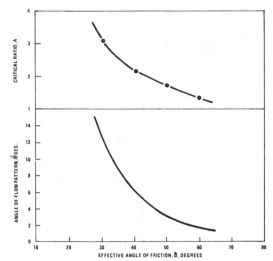

Figure 8.32. A and $\bar{\theta}$ versus δ by Giunta. (From Ref. 38.)

per sections, under the pressure of the solids head. The major principal stresses are assumed to be aligned along this near vertical direction, and they form what is defined as an active stress field, as shown in Figure 8.35a for a mass flow bin. As the solids settle slightly, slip occurs along the wall and a frictional stress develops.

8.5.1.2 Flow Conditions: Mass Flow

When the outlet gate is opened after initial filling, the unsupported solids expand downwards and contract laterally as they move in a flow channel that converges downward toward the outlet. This causes the stress field in that region to change from an active to a passive stress field transferring some of the load to the converging hopper walls, as shown in Figure 8.35b.

8.5.1.3 Switch Pressures

Nanninga[97] observed that at the transition level between active and passive fields, equilibrium of the mass requires an overpressure to occur. Jenike and Johanson[101] and Walters[98, 99] pos-

tulated that such a transient overpressure develops in the area of the outlet, at the boundary between active and passive stress fields, when flow is initiated in a bin that has been filled without having any solids withdrawn. As flow continues, the interface between the two fields moves upward to a point where the flow channel intersects the vertical section of the bin, and remains fixed in that area as shown in Figure 8.35b. Above the transition, the solids are still in an active state where initial pressures prevail. Below, the solids are in a passive state, and the smaller flow pressures have developed. The solids at the transition between the stress fields are no longer supported by the flowing material below and the equilibrium of forces results in an additional stress, or overpressure, on the hopper walls. The authors called this a "switch pressure."

8.5.1.4 Funnel Flow

The stress field in a funnel flow silo during initial filling is similar to that of a mass flow silo. When flow is initiated, an active stress field is created within the flow channel. How and if any of these stresses are transmitted through the stagnant solids to the bin wall has

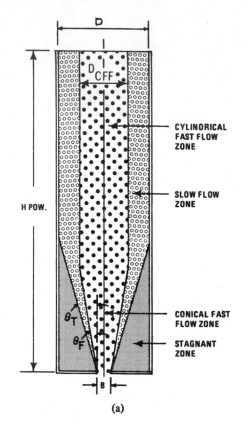

(a)

Properties of materials

Property	PVC grade X	PVC grade Y	Sand
ρ (kg/m³)	495	595	1525
δ (deg)	39.5	35.0	37.5
ϕ' steel/zinc compound (deg)	25.0	24.0	23.5
ϕ' aluminum (deg)	27.0	24.0	18.0

(b)

Summary of measured and predicted flow patterns in funnel flow

Material	Angle of cone to vertical, deg	D, m	B, m	$\dfrac{H_{pow}}{D}$	θ_T, deg	θ_F, deg	$D_{CFF,m}$	θ_F (Jenike [b]) deg	θ_F (Giunta [c]) deg	$D_{CFF,m}$ (c) [4])
PVC grade X	25	1.50	0.15	4.70	8.5	6	0.90	8	6	1.33
	25	1.50	0.15	3.63	9	5	0.70	8	6	1.05
	25	1.50	0.10	4.73	9	6	0.85	8	6	1.34
	20	2.00	0.15	2.57	<10	5	— (d)	8	6	1.00
PVC grade Y	15/25	1.50	0.10	5.00	11–12.5	7	0.80–0.90	11	8.5	1.76
	20	2.00	0.15	2.77	<9.5	6	— (d)	11	8.5	1.32
	90ª	1.50	0.15	4.00	13.5	10	1.05	11	8.5	1.42
	90ª	1.50	0.15	2.67	12	9.5	0.90	11	8.5	0.98
	90ª	1.50	0.10	4.00	9–12.5	9	1.00	11	8.5	1.38
Sand	90ª	1.50	0.10	4.00	12.5	—	0.75	9.5	8	1.26

(c)

Figure 8.33. Funnel flow patterns determined from test on model bins: (*a*) flow zones, (*b*) properties of the materials used in tests, (*c*) summary of measured and predicted flow patterns. (From Ref. 88.)

not been well defined. In the case of tall silos, the channel may expand sufficiently to intersect the cylinder wall as shown in Figure 8.36. This point of intersection has been called the "effective" transition.[101] Solids above this intersection will move down the cylinder walls and a pressure peak will occur at the point where the solids converge into the flow channel. The location of the effective transition will change as fill level in the silo changes.

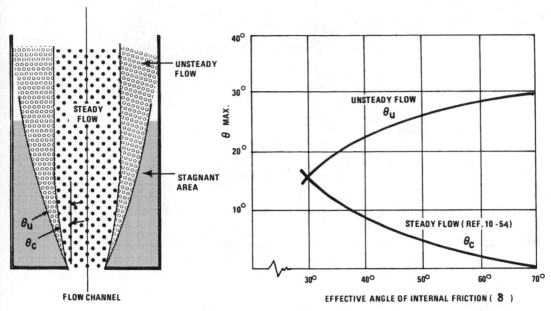

Figure 8.34. Predicted flow zones with free-flowing noncohesive frictional solids. (From Ref. 39) (Permission *Chemical Engineering*).

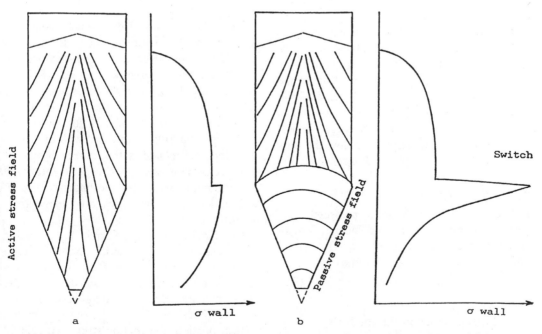

Figure 8.35. Stress field and profile of stress σ, normal to wall in mass flow. (*a*) Initial filling; (*b*) flow.

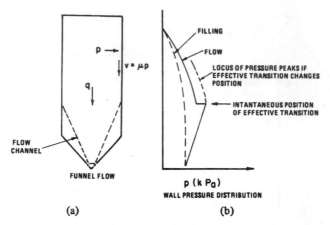

Figure 8.36. Wall pressure distribution—funnel flow: (a) Funnel blow bin, (b) wall pressure distribution.

Therefore, for structural design purposes, a locus of pressure peaks is assumed, with a distribution profile as shown in Figure 8.36.

8.5.2 Janssen's Method of Computing Stresses

Janssen's method[6] has been used for many years as a model for computing the stresses on silo walls, caused by stored solids. A similar method developed by the Reimberts[3] is also used.

Janssen evaluated the equilibrium of forces acting on an elemental horizontal slice of bulk solids, in the vertical, cylinder portion of a silo. He assumed that the vertical pressures in the solids are uniform over any horizontal bin cross-section, and these pressures vary only in the vertical direction. The average vertical static pressure q, at depth Z, below the bulk solids surface is given by the Janssen equation as:

$$q = \gamma g R / \mu K [1 - e^{-\mu k(z/R)}] \quad (8.2)$$

The ratio of vertical to lateral pressure, K, is assumed to be constant, and independent of the magnitude of the pressure. The lateral static pressure is therefore given as:

$$p = \gamma g R / \mu [1 - e^{-\mu k(z/R)}] \quad (8.3)$$

where

q = vertical pressure in the solids (N/m²)
p = lateral pressure, normal to silo wall (N/m²)
γ = solids bulk density (kg/m³)

g = gravity acceleration (m/s²)
K = ratio of lateral to vertical pressure
R = hydraulic radius: cross-section area of the cylinder/circumference. $R = D/4$ for a round silo
μ = coefficient of sliding friction—solids on wall
Z = vertical distance measured downward from the centroid of the heap of solids at the top of the cylinder, m.

The assumptions from which Janssen's equations were derived are known to be incorrect. Pressures over the cross-section, although poorly understood, are not uniform, and the bulk density and the pressure ratio are not constant throughout the silo. However, when used with the measured flow properties of the specific bulk solids, and with appropriate safety factors added, the Janssen equations agree reasonably well with experimental wall measurements made on the cylinder portion of silos under static conditions, after axisymmetric initial filling.

The Janssen model is correct in predicting that pressures in the cylinder do not increase proportionally with depth. Some of the pressures from the material are transferred to the walls through friction, adding vertical compressive (buckling) forces to the walls. In the vertical portion, the friction stress v is related to lateral pressure by:

$$v = \mu p \quad (8.4)$$

As depth Z increases, the lateral pressure on the wall approaches asymptotically the limiting value:

$$p = \gamma g R / \mu \qquad (8.5)$$

8.5.2.1 Pressure Ratio, K

Janssen determined a value for K from measurements on a silo model. Although it has a significant impact on the pressures as calculated by Janssen, there is no agreement on how to measure a value for K experimentally on a solids sample. As a result, a number of equations, most of which relate the value to the measured angle of internal friction ϕ of the solids, have been proposed and are in use. These include, for example;

$$K = 1 - \sin \phi / 1 + \sin \phi \qquad (8.6)$$

$$K = 1 + \sin \phi / 1 - \sin \phi \qquad (8.7)$$

$$K = 1 + \sin^2 \phi / 1 - \sin^2 \phi \qquad (8.8)$$

$$K = 1 - \sin \phi \qquad (8.9)$$

For powders, the effective angle of friction δ is often used in place of ϕ.

Lohnes[149] reviewed these and other equations that have appeared in the literature. He pointed out that Eqs. (8.6) and (8.7) can be derived from the Mohr circle, and are valid for smooth walls and horizontal and vertical stresses that are principal stresses. He further stated that, since the Janssen equation assumes that the load is transferred from the solids to the wall through wall friction, the horizontal and vertical stresses are not principal stresses.

Lohnes described two experimental devices to measure the lateral stress ratio: a modified low-stress triaxial test apparatus to measure K_0, the ratio of minor to major principal stress at zero lateral strain, and a confined, rigid wall compression test apparatus to measure K, the ratio of horizontal to vertical stress at failure. Tests on a variety of solids samples gave K_0 values ranging from 0.22 to 0.56, and K values ranging from 0.17 to 0.45.

Carson and Jenkyn[150] suggested that the pressure ratio is more silo-dependent than solid-dependent and therefore attempts to measure its value for a given solid are inappropriate.

Current industrial design practice is tending toward using numerical values for K. Jenike[109] states that the numerical Janssen value for K ranges from 0.30 for soft powders to 0.60 for hard particles. In most of the new drafts of national silo design codes, a variety of industrial solids are tabulated, with suggested numerical values for K, to be used in Janssen-type equations. The given K values range from 0.25 to 0.6, although the basis for this is not given. Some codes list upper and lower bound values for K and wall friction μ for each solid, and recommend that they be used in the combination that maximizes the computed horizontal and vertical pressures, and the vertical frictional wall pressures.

8.5.2.2 Stress Theories

Stress conditions within flowing solids in a bin are not well understood. All useful stress measurements published to date have been made at the bin walls. Test data from model or full-size bins that confirm the presence of varying wall stresses during filling and discharge are given in Refs. 83 through 95. There is yet no general agreement on a theory that best describes the stress conditions in bins under all conditions of flow. A number of models to describe stress distributions have been proposed. These include those by Walker,[96] Walters,[98, 99] Clague,[100] Jenike et al.,[101-104] Enstad,[78, 79] and Takahashi.[105, 106] A summary of the first five is given by Arnold and Roberts.[107] A review of stress distributions is also given by Shamlou.[151]

Janssen's slice method has been extended by others to evaluate pressures in converging mass flow hoppers including Schulz.[152] Schwedes[153] compared results by Schulz with those of others, and reported that Schulz's model gave the best agreement to hopper wall pressures measured on an experimental 0.6 m diameter silo hopper.

The theories of Walker, Walters, Jenike, and Johansen have been the most widely quoted. Details can be found in the references. Their published information is mostly concerned with *axisymmetric filling and discharge*. It is recognized that much of the silo overstressing and failures have been caused by *eccentric* flow patterns induced by eccentric single or multiple discharge openings. These conditions impose severe, unbalanced lateral stresses along the horizontal cross sections of a bin, particularly where the flow channel intercepts the wall. The points of interception are often difficult to determine. This area of study remains the least explored and requires the most caution on the part of bin designers.

Walker and Walters. Walker proposed an approximate theory to describe stresses and arching in hoppers and bins, and presented a considerable amount of data derived from tests with wet and dry coal in a 1.8 m diameter and 1.8 m square bins. His data confirmed that the stress fields developed during filling and discharge are very different; withdrawing a small amount of material while filling a bin significantly reduces the high initial pressures found near the hopper apex; during loading with no withdrawal, initial pressures increased with depth of fill; flow pressures in the lower region of a mass flow bin were linearly proportional to height above the hopper apex and independent of depth of fill (evidence of a radial stress field). Flow pressures were independent of flow rate and once established by withdrawal they remain, even when withdrawal is interrupted for a period of time.

Walters extended the Walker theory to distinguish between the stresses developed during filling and flow in conical hoppers and in conical hoppers with vertical sections above. He also proposed an approximate method of calculating the "switch stress." Clague extended it to plane flow bins. Arnold and Roberts[107, 108] integrated these theories with Jannsen to propose a generalized theory for predicting wall stresses in mass flow bins.

Jenike, Johanson, and Carson. The authors measured wall stresses on 300 mm diameter model bins handling sand and coke. They found widely varying pressure fluctuations in the cylinder portion during flow, which they attributed to very slight deviations from perfect uniformity in the shape of the bin cross-section. (These results and conclusions were confirmed by Van Zanten et al.).[87, 88] Wall pressures were measured on models with non-diverging cylinders, with cylinders having surface imperfections caused by weld shrinkage on girth seams, with continuously converging cylinders, and with cylinders having internal ledges or constrictions. The wall pressure profiles measured with dry sand and with coke in a 0.152 m model bin are shown in Figures 8.37 and 8.38.

The authors concluded that wall boundary layers tend to form, dissolve, and reform as solids move through a cylinder having these imperfections. They proposed that the flow pressures that occur in the region of these boundary layers be determined by assuming that the elastic strain energy within the flowing solids tends toward a minimum. Since the locations of these boundary layers are indeterminate, the bound enclosing all possible pressure peaks should be determined. Design charts for this purpose are given in Refs. 103 and 107. These bounds are shown in Figures 8.37 and 8.38 for the experimental models.

In a later article, Jenike[109] presented a simplified method of computing the upper bounds of the cylinder wall pressures, using the Janssen equation, with lower and upper bound values for K and μ, as shown in the following example.

Mass Flow. For initial fill pressure, use Janssen Eq. 8.2 with $K = 0.4$. Some convergence and divergence is assumed to occur along the length of an industrial silo. During flow through the cylinder, wall pressure p increases to contract the solids laterally when a layer passes a convergence, and decreases when a layer passes a divergence. This results in a varying value for Janssen's pressure ratio

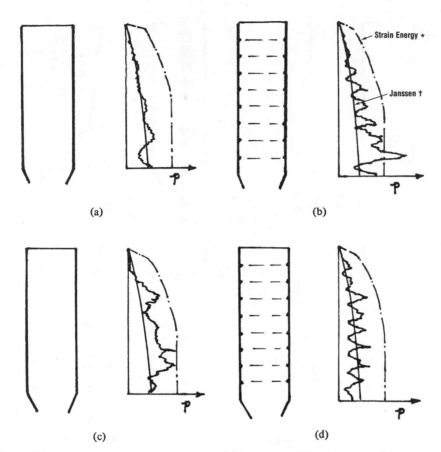

Figure 8.37. Wall pressures on model bin handling sand: (*a*) Diverging $1/2°$ with no ledges (*b*) diverging $1/2°$ with ledges, (*c*) converging $1/2°$ with no ledges, (*d*) converging $1/2°$ with ledges. *Pressures calculated by Jenike, Johanson, and Carson strain energy theory. †Pressures calculated by Janssen theory. (From Ref. 103.)

K. Wall friction, as well, may vary between kinematic (μ_k) or static conditions (μ_t). Jenike therefore suggested using the Janssen pressure distribution for the cylinder, with the following bounds:

$$0.25 > k > 0.6$$
$$(\mu_k - 0.05) \leq \mu \leq (\mu_t + 0.05)$$

For design purposes, the minimum product of $K\mu = 0.25$ ($\mu_k - 0.05$) gives the maximum value of q, and a value of $K = 0.6$ gives a maximum value of p.

Funnel Flow. In the stagnant areas, cylinder deviations have a minimum effect on the walls as long as there is no sliding on the walls. For

that case, Jenike suggests cylinder wall pressures computed from Janssen, using $K = 0.4$.

If the flow channel intersects the cylinder wall, an "effective transition" is formed, and the overpressures must be considered.

8.5.3 Simplified Calculation Procedure with Axisymmetric Flow in Silo

It is known from experimental work with models and with industrial silos that the distribution of wall pressure in mass flow silos is closely approximated by the profiles shown in Figure 8.35. The proposed constitutive models for calculating stresses on solids and silo walls are complex and the experiments or the experimental equipment required to determine spe-

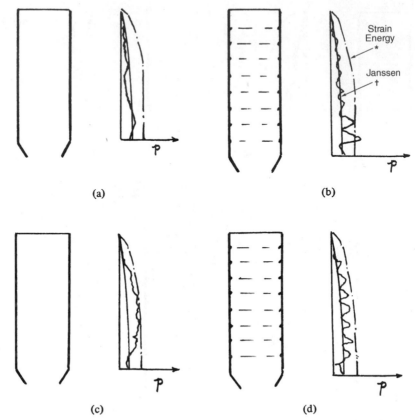

Figure 8.38. Wall pressures on model bins handling coke: (*a*) diverging $1/2°$ no ledges, (*b*) diverging $1/2°$ with ledges, (*c*) converging $1/2°$ no ledges, (*d*) converging $1/2°$ with ledges. *Pressures calculated by Jenike, Johanson, and Carson strain energy theory. †Pressures calculated by Janssen theory. (From Ref. 102)

cific bulk solids characteristics needed for the particular models have not yet been clearly defined.[153] Despite the disagreement on design equations, it is universally accepted that the silo cylinder and converging hopper should be analyzed individually, for both filling and flow conditions.

At present, industrial practice is to make use of selected portions of equations given in models appearing in the open literature, and then add safety factors to these design equations to allow for the not well understood stresses that result from nonsymmetric filling and discharge.

The following example, for estimating wall stresses, for the simple case of a round silo with central filling and discharge, and, with no allowance for vibration or shock loads that might be imposed by the flow of solids, has been suggested by Carson and Jenkyn.[150] Note: all values in Eqs. (8.10) to (8.17) are given in English units, as used in the original reference.

8.5.3.1 Mass or Funnel Flow: Cylinder— Initial Filling

Use Janssen's equation for a round cylinder ($R = D/4$)

$$p = \frac{\gamma D}{4\mu}[1 - e^{-4\mu Kz/D}] \qquad (8.10)$$

where

p = pressure acting normal to silo wall (lb/ft^2)
D = cylinder diameter (ft)
γ = solids bulk density (lb/ft^3)

Z = vertical distance measured downward from the centroid of the heap of solids at the top of the cylinder (ft)

K = a value of 0.4 is suggested.

Although it is a matter of interest, the initial filling conditions represent the lower bound on the pressures in the cylinder, and therefore are not used for silo structural design.

8.5.3.2 Mass or Funnel Flow: Converging Hopper — Initial Filling

$$p = \gamma\left[\frac{h-z}{n_i} + \left(\frac{q}{\gamma} - \frac{h}{n_i}\right)\left(1 - \frac{z}{h}\right)^{n_i+1}\right]$$

$$(8.11)$$

$$n_i = 2\left(1 + \frac{\tan\phi'}{\tan\theta_c}\right) - 3 \qquad (8.12)$$

where

h = hopper height (ft)

z = vertical distance measured downward from the top of the hopper (ft)

p = calculated from Janssen horizontal pressure at the bottom of the cylinder, divided by K ($K = 0.4$) (ft)

ϕ' = angle of wall friction

θ_c = hopper wall slope.

8.5.3.3 Mass Flow: Cylinder — Discharge

As pointed out earlier by Jenike et al.,[103] deviations in shape and concentricity of vertical walls, and the presence of girth seams and other protrusions and ledges on interior walls of an industrial silo are not unusual. These irregularities as well as nonuniformity in solids density and flow properties caused by segregation will cause changes in the stress field, and actual wall pressures during flow will be higher than those predicted by Jannsen's equation.

Carson and Jenkyn[150] suggest a simplified method to account for these conditions: use the Janssen equation, selecting the values for wall friction and pressure ratio to maximize

the design pressures:

$$0.25 \le K \le 0.6 \qquad (8.13)$$

$$\phi'_{calc.} = \phi'_{meas.} \pm 5° \qquad (8.14)$$

The plus sign is used only when calculating the maximum shear stresses (buckling stresses) on the cylinder wall.

8.5.3.4 Mass Flow: Converging Hopper — Discharge

Carson and Jenkyn[150] proposed the following simplified equations for flow pressures in a mass flow hopper:

$$p = \gamma K_f\left[\frac{h-z}{n_f} + \left(\frac{q}{\gamma} - \frac{h}{n_f}\right)\left(1 - \frac{z}{h}\right)^{n_f+1}\right]$$

$$(8.15)$$

$$K_f = \cfrac{1}{\left[\frac{2}{3}\left(1 + \frac{\tan\phi'}{\tan\theta_c}\right) - \frac{1}{6(\sigma'/\gamma B)\tan\theta_c}\right]}$$

$$(8.16)$$

$$n_f = 2K_f\left(1 + \frac{\tan\phi'}{\tan\theta_c}\right) - 3 \qquad (8.17)$$

where

q is computed by Janssen's horizontal pressure p, at the bottom of the cylinder, divided by K. For conservative design purposes, the minimum value of K is suggested.

z is vertical distance measured downward from the top of the hopper (ft).

$(\sigma'/\gamma B)\tan\theta_c$ is a function of δ, presented as design charts for conical and plane flow channels in Ref. 55.

The authors suggest that, for design purposes, peak pressures due to the switch pressures be distributed for a short distance along the bottom of the cylinder wall as shown in Figure 8.35. Details can be found in the reference.

8.5.3.5 Funnel Flow

In funnel flow, the boundary between active and stagnant material is often unstable and, particularly with powders, is very often not axisymmetric with the silo discharge opening. Limited studies to predict the shape of flow channels have been concerned with fairly free-flowing solids. On short silos (ht/diameter ≤ 2) the flow channel seldom expands sufficiently to intersect the cylinder walls, and in those cases the wall pressures during flow are assumed to be the same as those during initial fill.

In tall silos where the flow channel is likely to intersect the silo wall (effective transition), an overpressure must be considered in a manner similar to the switch pressure encountered in mass flow. Carson et al.[150] suggest that this overpressure be calculated as in a mass flow hopper, substituting an estimated flow channel angle for the hopper angle, the solids internal friction for the wall friction, and distributing the stress over a region at the intersection. Since the location of the intersection will vary with solids level, a locus of pressure peaks is assumed as an upper bound on wall pressures along the cylinder wall, within the range of the boundaries of the effective transition.

8.5.4 Silo Design Codes

At the time of this writing, there is no agreement on a general theory, suitable for use in a national silo design code, for quantifying the stresses imposed on silos by stored solids, under all the conditions that are typical of industrial situations. A number of national silo structural design codes[16, 154–156] are being currently revised to take into consideration the present understanding of the effect that the frictional properties of solids, geometry of the flow channels, and eccentricity of filling and discharge have on stresses. In most cases, these codes represent mandatory minimum requirements for design, but they are not all equal in their depth of coverage of all storage and flow situations. With respect to wall stresses imposed by the stored material, the current drafts generally recommend Janssen-type pressure models, with a variety of added safety factors to compensate for the not well understood effects of eccentric filling and discharge, and the dynamic forces imposed during discharge. These safety factors generally

take the form of overpressure multipliers for specific flow conditions and hopper geometries, and suggested upper and lower boundary limits for pressure ratio K and wall friction μ, to maximize the computed stresses.

Most codes include a tabular listing of experimentally determined properties for a range of "typical" bulk solids. However, it is recognized that, wherever possible, the flow properties of specific bulk solids to be stored should be determined by tests, rather than by reference to the listed properties of a "similar, generic material," which may or may not actually replicate the solids to be handled. The effects of moisture, particle size distribution, temperature, and chemical activity on flow and storage properties cannot be adequately described in a tabular listing. This is also true for the wall friction values for the many available types of wall surface finishes. Most of the Codes will include suggestions for testing solids as described in Section 8.6.

8.6 SOLIDS FLOW ANALYSIS AND TESTING

Attempts to develop an analytical model for predicting the behavior of bulk solids during flow have been based on either the *particulate* or the *continuum* approach. As defined by Goodman and Cowin,[50] in the particulate approach, the properties of discrete particles of finite size (idealized rigid or elastic spheres) are used to deduce the laws governing the behavior of the entire mass. In the continuum approach, the properties of the mass are assumed to be a continuous function, and the mass may be divided indefinitely without losing any of its defining properties. The discrete particles' properties are not considered.

A considerable amount of work has been underway to develop a theory of flow based on particle properties. But particulate granular and powder solids are nonhomogeneous, and may have an infinite combination of particle sizes, shapes, and interstitial voids. To date the continuum approach, being less complex, has yielded the most useful information for purposes of engineering design and has been responsible for accelerating the development of the field of powder mechanics as it evolved from the soil mechanics work of Coulomb, Rankine, and others. There are, however, important differences between soil and powder mechanics.[55] Cohesion is usually not important in soil mechanics but it is in powders. Stresses in powders stored in bins can be up to 1000 times smaller than those normally encountered in soil and are not detectable in mechanics tests of soils; boundary conditions in powder mechanics are usually not the same as in soil mechanics, since powders are usually stored in bins; powders can be subjected to much larger deformations than is common in soil mechanics.

Continuum plasticity-type models for powders have been proposed by a number of workers,[51,52,53] including Jenike and Shields. Jenike was the first to use the concepts of plastic failure with the Mohr–Coulomb failure criteria in analyzing the flow of solids in bins and hoppers to develop the concept of a flow–no-flow criterion. This has produced an extremely useful quantitative method for designing storage bins for gravity flow of solids. Since this method has been proven in engineering practice, the information that follows is based on Jenike's original work.[54,55,60,61]

Jenike assumed that a bulk solid can be closely approximated by a rigid-plastic Coulomb solid. From soil mechanics, such a solid is characterized by a yield locus that defines the limiting shear strength under any normal stress (Fig. 8.39). A Coulomb solid has a linear yield locus. Plotting shear stress τ and normal stress σ, the yield locus for a Coulomb powder intersects the τ axis at a value of τ defined as cohesion C at an angle ϕ, defined as the angle of internal friction.

After many experimental measurements, Jenike found that with real bulk solids, at low pressures, the locus deviates from a straight line (Fig. 8.40); the locus does not increase indefinitely with increasing values of σ but terminates at some point E; and the position

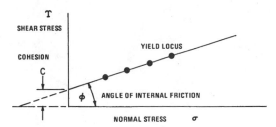

Figure 8.39. Yield locus of a Coulomb solid.

of the locus is a function of the degree of consolidation of the material. During flow the stresses in the plastic regions of the solid are continuously defined by point E.

The yield locus for a cohesive solid is shown in Figure 8.40. The yield locus for a free-flowing material such as dry sand would have a locus as shown in Figure 8.41.

In his analysis, Jenike assumed that in the plastic region, solids properties at a point are the same in all directions (isotropic), and frictional, cohesive, and compressible. During incipient failure, the bulk solid expands and during steady-state flow it can either expand or contract, stress at any point does not change with time, and stresses are not significantly affected by velocity changes.

8.6.1 Stress–Strength Relationships

As an element of material flows through a channel, the major consolidating stress σ_1 and minor consolidating stress σ_2 on the element change (Fig. 8.42) and continuous shear deformation occurs, causing slip planes as the elements slide on one another or on the bin wall. During flow the "strength" (resistance to shear

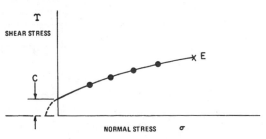

Figure 8.40. Yield locus of a cohesive solid.

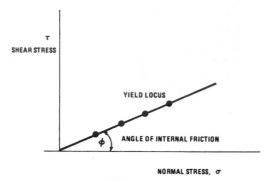

Figure 8.41. Yield locus of free-flowing sand.

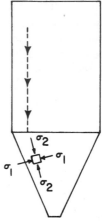

Figure 8.42. Stresses on an element flowing through a channel in a bin.

failure) and density are a function of the last set of stresses and when flow stops, it is assumed these stresses remain. As the material remains stationary under these stresses. it may gain in strength, and resist flow when the bin outlet is reopened.

8.6.2 The Jenike Direct Shear Cell

Shear testers of various types have been used to determine the stress/strength relationships of bulk solids. To date, Jenike's direct shear cell tester and his proven procedure for design of bins for flow has become a bench mark in research and in industrial practice. The shear cell is described below. The design procedure

is given in Sect. 8.9. Other testing devices are described in Sect. 8.11.1.

The Jenike shear cell assembly is shown in Fig. 8.43. It consists of a shear ring, base and cover. The ring most frequently used has an inside diameter of 95 mm (3.75 in.). A 65 mm (2.5 in.) ring is sometimes used when high consolidating forces are required. Very large rings are used for special applications. The bottom of the cover and inside of the base are roughened to increase solids adhesion. A bracket is attached to the top of the cover.

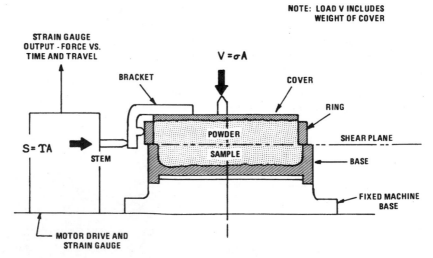

Figure 8.43. Jenike shear cell.

The base and ring are filled with powder and the cover put in place. A vertical force is applied to the cover. A horizontal shearing force is applied to the bracket by a motor-driven stem. Part of the shearing force is transferred to the ring by a loading pin attached to the cover bracket. This helps to ensure more uniform distribution of shearing force across the cell during shear. With 60 Hz electrical supply, the shearing force is applied at a constant rate of 0.91 mm/min (0.036 in./min) in older machines, and 2.7 mm/min (0.108 in./min) in newer machines. With 50 Hz electrical supply, the rates are 0.76 mm/min (0.03 in./min) and 2.3 mm/min (0.09 in./min). The shearing force is transmitted through the pin to a load cell and displayed as shear force versus time and displacement.

The flowability and yield strength of a mixture of coarse and fine particles are most dependent on the properties of the fine fraction since shear occurs across the fine fraction during flow. Therefore, when testing such a mixture, particles greater than about 3 mm are usually screened and removed from the shear test sample.

8.6.3 Determining the Yield Locus with the Jenike Shear Cell

Section 8.6.1 describes the change in stresses that act on an element of material as it flows through a bin. The Jenike test sequence is intended to simulate these conditions. The test is accomplished in three steps. The first, called preconsolidation, is to ensure uniformity between samples. The second, called consolidation, reproduces flow with a given stress, under steady-state conditions. In the third step, the sample is sheared to measure shear stress at failure.

8.6.3.1 Test Procedure

The procedure for testing with the Jenike Cell is depicted in Fig. 8.44 and briefly described below. Detailed procedures are given in Refs. 55 and 159.

Preconsolidation (Fig. 8.44a). With a packing ring in place, the cell is filled, a twisting top is placed on the sample, a force V_t is applied to the top while it is given a number of oscillating twists. The twisting top and force are then removed and the powder surface scraped level with the shear ring.

Consolidation (Fig. 8.44c). A shear cover is placed over the powder sample and a selected normal force V is applied. A shear force is then continuously applied until it reaches a steady-state value indicating plastic flow. The shear force is then interrupted and the stem retracted. The measured steady state stress is point E on the yield locus (Fig. 8.45).

Shear. The normal force V is replaced by a smaller force $\overline{V}$ and the shearing force is re-applied until the stress/strain peaks and falls off, indicating a failure plane in the sample, and a point on the yield locus. This procedure is repeated several times with fresh samples, each consolidated as above but sheared with a progressively smaller normal force.

Time Yield Locus. When the steps in the procedure just described are performed without interruption, the results are characteristic of solids placed in a bin and discharged almost immediately. The yield locus determined this way is usually referred to as the instantaneous yield locus. Solids that remain stationary in a bin, under a consolidating stress, may gain strength and resist flow as described in Section 8.6.1. To describe these conditions, a time yield locus must be determined. After preconsolidation and consolidation are completed as above, the sample is placed under a consolidating stress, V_1, and left undisturbed for a period of time equal to the expected storage time (Fig. 8.44e). The value for force V_1 is determined from the stress σ_1 at the intersection of the Mohr semicircle through the instanteous yield locus as shown in Fig. 8.46. After the time interval is complete, the sample is removed and sheared under the same $\overline{V}$ forces used for the instantaneous yield locus

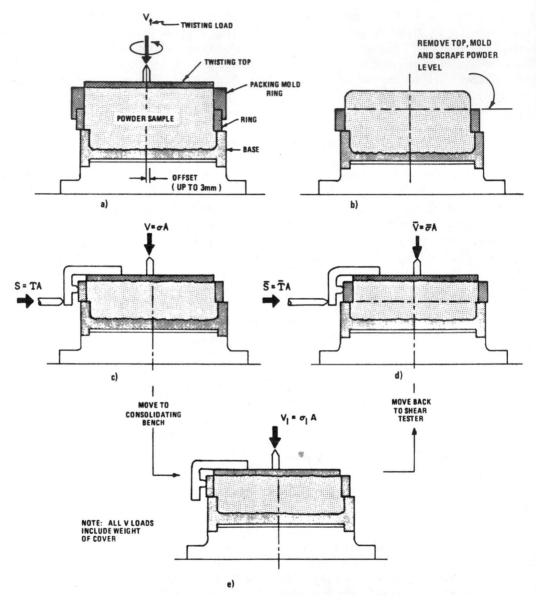

Figure 8.44. Jenike shear test sequence: (*a*) preconsolidation, (*b*) Removal of twisting top and packing mold ring, (*c*) consolidation, (*d*) shear, (*e*) time consolidation.

(Fig. 8.44d). A time yield locus is then constructed as shown in Fig. 8.47.

Mohr Stress Semicircle. Mohr stress semicircles are used to identify the frictional and strength properties of the sample from the yield locus as shown in Figure 8.46. The state of stress on any plane within the bulk solid can be represented by a Mohr circle. For any stress condition represented by a Mohr semicircle tangent to the yield locus, the bulk solids will be at yield, and the major principal stress σ_1 and minor principal stress σ_2 at this condition will be defined by the intersection of the semicircle with the σ axis. The yield locus terminates at the point of tangency of the

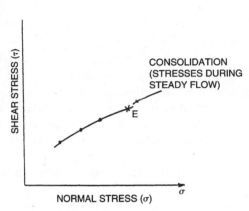

Figure 8.45. Yield locus.

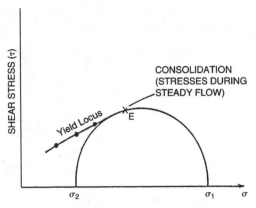

Figure 8.46. Mohr stress semicircle.

Mohr semicircle through Point E. This circle intersects the σ axis at the principal stresses σ_1 and σ_2.

8.6.4 Solids Characteristics Derived from the Yield Loci

The following characteristics, determined from shear tests and the yield loci of a bulk solid, are used in the analysis of flowability, and for specifying the geometry of a mass flow bin (refer to Fig. 8.47).

8.6.4.1 Effective Yield Locus (EYL)

From the results of many shear tests,[55] it has been found that the Mohr circles representing steady-flow stress are approximately tangential to a straight line through the point of zero stress. The envelope of Mohr circles defined in this manner is called the effective yield locus (EYL). This locus is tangent to the Mohr semicircle that defines the major and minor

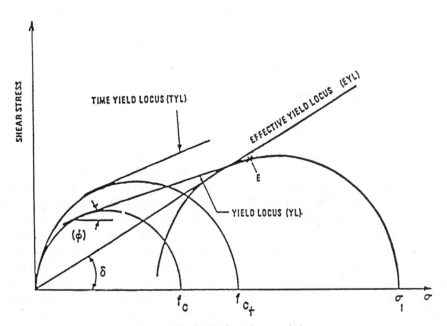

Figure 8.47. Solids flow characteristics.

principal stress σ_1 and σ_2. The effective yield locus (EYL) can be defined by:

$$\sin \sigma = \sigma_1 - \sigma_2 / \sigma_1 + \sigma_2 \quad (8.18)$$

8.6.4.2 Effective Angle of Friction (δ)

The angle δ is called the *effective angle of friction* of the solids, a measure of resistance of the solids to flow while they are in a steady-flow condition. Higher values of δ indicate lower flowability. With a given solid, it usually increases slightly with decreasing stress. The values for δ range from 25 to 70° for most materials that have been tested.

The ratio of major principal consolidating stress σ_1 and minor principal consolidating stress σ_2 during steady flow can be expressed by the effective yield function:

$$\sigma_1 / \sigma_2 = 1 + \sin \delta / 1 - \sin \delta \quad (8.19)$$

8.6.4.3 Unconfined Yield Strength (f_c)

At a free surface formed on the bottom surface of an arch, the minor consolidating stress σ_2 acting normal to the surface is equal to zero, and the major stress σ_1 is tangent to the surface. Therefore, a Mohr circle through the origin, tangent to the yield locus, defines the largest stress σ_c that the solids can withstand at a free, unsupported surface. The value of σ_c defines the unconfined yield strength f_c. For each value of consolidating stress, there is a corresponding value of f_c, and as the consolidating stress increases, f_c increases.

8.6.4.4 Flow Function (FF)

The flow function, sometimes called the failure function, characterizes the "flowability" of a bulk solid. The unconfined yield strength is a function of the major consolidating stress σ_1 and for a value of σ_1 the corresponding value of f_c can be found from the yield locus. Therefore if a family of yield loci is constructed as shown in Fig. 8.48a, the corresponding values for σ_1 and f_c for each family member can be plotted to produce a flow

function as shown in Fig. 8.48b. The flow function for a bulk solid can then be defined as:

$$FF = \sigma_1 / f_c \quad (8.20)$$

Instantaneous flow functions are determined under conditions of zero consolidation time. Time flow functions are determined under conditions occurring during time consolidation. Many materials gain strength with time consolidation. The upper line in Figure 8.48 represents the greater strength and the greater ability to support an arch.

8.6.4.5 Angle of Internal Friction (ϕ)

The slope of the yield locus at the point tangential to the Mohr circle passing through the origin defines ϕ, the angle of internal friction of the solids. This is also called the kinematic angle of internal friction since it is determined by the instantaneous yield locus. Fine and dry solids have lower values of ϕ (and δ). Coarse and wet solids and cohesive solids have higher values.

8.6.4.6 Static Angle of Internal Friction (ϕ_t)

The slope of the time yield locus at the point tangential to the Mohr circle passing through the origin defines ϕ_t, the static angle of internal friction of the solids. This is a value used in the analysis of funnel flow.

8.6.4.7 Cohesion

Cohesion is the sticking together of the particles in a bulk solid. A relative measure of the cohesion of a bulk solid sample can be determined from the intercept of a straight line extended from the solids yield locus, across the low-stress region, to the shear stress axis. Cohesion increases with decreasing particle size. With wet solids that do not absorb water, higher moisture increases cohesion. It has been reported that this increase in cohesion with moisture is more pronounced for coarse particles than for fine particles.[153] Cohesion values obtained by extending the yield locus as described are only rough estimates and are not used in this chapter for analyzing flow in a bin. The results of a study of shear testers for

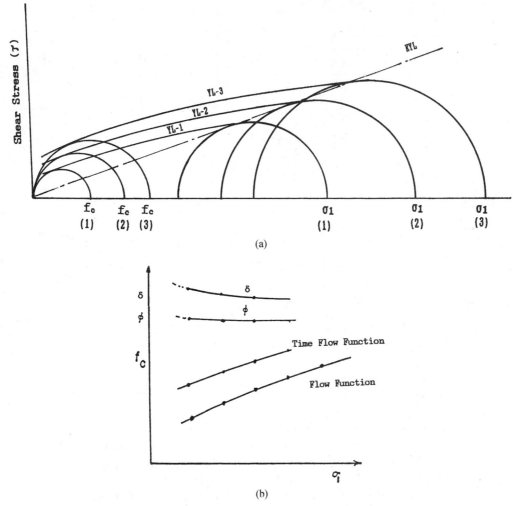

Figure 8.48. Flow function.

measuring cohesion in powders and granular materials are given in Ref. 153.

8.6.5 Wall Yield Locus (WYL)

8.6.5.1 Kinematic Angle of Friction Between a Solid and Wall Surface (ϕ')

Solids flow along slip lines that form boundaries between flowing and static solids described by the yield locus, or they can flow along rigid bin walls. Stresses along the wall during this type of flow lie along the wall yield locus (WYL).

This locus is determined by substituting, for the base in the Jenike tester, a sample plate of the same material to be used for the silo hopper wall, and measuring the shear force required to slide the exposed solids along the plate under a range of normal loads as shown in Figure 8.49.

The normal stresses and corresponding shear stresses are plotted to produce the wall yield locus (Fig. 8.50). The angle ϕ' is the angle of wall friction or, as it is also called, the kinematic angle of wall friction, since it represents continuous flow along the wall surface. Tangent ϕ' is the coefficient of friction, μ, between solids and wall.

The WYL may be a straight line (Fig. 8.50a) or convex shaped (Fig. 8.50b). A straight line

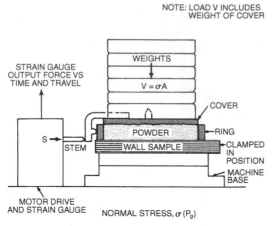

Figure 8.49. Jenike wall friction test.

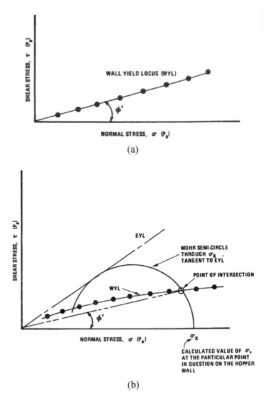

Figure 8.50. (*a*) Linear wall yield locus. (*b*) Convex wall yield locus.

locus through the origin indicates the friction is independent of wall pressure. A convex line locus indicates that the friction is pressure dependent. In that case, ϕ' must be determined for stress conditions at the hopper wall at the particular area of interest. This is done by extending a straight line through the origin to the intersection of the WYL and the Mohr circle representing stress conditions at the point in question and computing ϕ' (Fig. 8.50b), as described in Section 8.6.5. With radial stress, a low-stress region exists near the outlet of a mass flow bin, so solids having a pressure-dependent wall friction (higher friction value at lower stresses) will require a steeper sloped hopper at the lower region near the outlet.

8.6.5.2 Adhesion or Static Angle of Friction (ϕ_t) Between a Solid and Wall Surface

In some cases, solids will stick or adhere to a wall surface if allowed to remain at rest, in wall contact, under a consolidating load. When this happens, a higher stress will be required to initiate flow and restore the WYL to steady flow conditions, after the bin outlet is opened.

This can be predicted by a wall adhesion test. Steady-state shear across a plate is established as described for the WYL test. The shear is then interrupted for a specified time, and reapplied. The force required to initiate

movement is compared with the steady-state value. If higher, there is adhesion, and a value for ϕ_t can be determined.

McLean[161] critically examined the increase in wall friction with decreasing major consolidating stress displayed by many solids, particularly those having adhesion tendencies and concluded that in certain cases the definition of wall friction angle becomes meaningless below a certain critical consolidation stress—the solid will tend to slip within itself in preference to the wall.

8.7 BULK DENSITY AND COMPRESSIBILITY

Bulk density of a solid is a function of the consolidation stress, and during flow, it changes as the stresses change. Bulk density as a func-

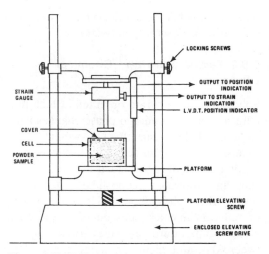

Figure 8.51. Bulk-density compression test apparatus.

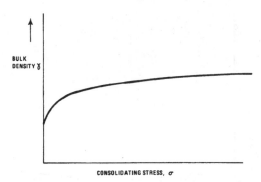

Figure 8.52. Typical plot of bulk density versus consolidating stress.

σ = major consolidating stress (kPa)

σ_0 = arbitrary chosen base value (kPa).

A large value for b implies a large compressibility.

8.8 OTHER FACTORS AFFECTING FLOW PROPERTIES DURING STORAGE

8.8.1 Impact During Loading

When a silo is initially filled, the stresses on the material at the point of impact may be higher than those that occur during flow. As-

tion of these stresses can be determined by applying vertical loads to a bulk solids sample of known mass and recording compression of the sample, with a dial indicator, scale, or electronic position indicating equipment as shown in Figure 8.51. Since the mass, consolidating load, and volume are known, the relationship can be plotted as shown in Figure 8.52. To minimize wall effects, the cylinder used for the consolidating test should have a length-to-diameter ratio (L/D) not exceeding 1. Jenike and Johanson[62] have shown that powders can be characterized by a compressibility constant b, as a function of consolidating stress. This is measured with the device shown in Figure 8.53. A solids sample of known mass is consolidated in a cylindrical cell, under a range of consolidating pressures. The change in sample volume is recorded and the bulk density computed for each consolidating pressure. The data are plotted on a logarithmic scale and a straight line fitted to the data points as shown in Figure 8.54. The compressibility constant b for the powder is defined by:

$$\gamma = \gamma_0(\sigma/\sigma_0)^b \qquad (8.21)$$

where

γ = bulk density (kg/m^3)

γ_0 = bulk density at major consolidating stress (σ_0)

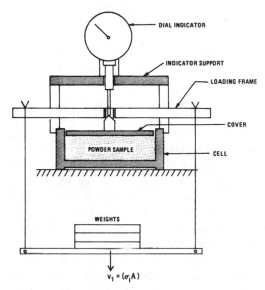

Figure 8.53. Jenike and Johanson Inc. compressibility tester. (From Ref. 61.)

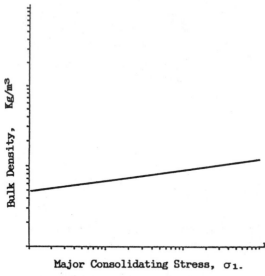

Figure 8.54. Compressibility.

suming initial velocity of the solids to be zero, this impact stress σ_p may be estimated as:

$$\sigma_p = W(2gh)^{0.5}/gA \qquad (8.22)$$

where

W = weight flow rate into bin
g = gravitational constant
A = estimated area of impact
h = height of fall.

If a tall bin is to be designed, the yield strength f_c of the solid developed under the impact stress $(\sigma_p = \sigma_1)$ should be checked to make sure it is less than f_c during time consolidation or flow. If not, the dimensions of the discharge opening should be calculated on the basis of the higher solid strength.

The impact stresses can be reduced by installing a deflector plate at the bin inlet to distribute the material over a wider impact area, directing the incoming solids stream at the bin wall, or by maintaining a minimum solids head in the hopper so that impact occurs well above the solids outlet. Stresses at the outlet can also be reduced by withdrawing material (at a low rate if necessary) for a short time during initial filling. This will cause a

passive stress field to develop immediately over the outlet as described in Section 8.5.1.

8.8.2 Temperature and Chemical Changes

Solids may agglomerate or soften at high temperature or may undergo phase changes when cooled. All these can affect flowability. Temperature changes can occur, for instance, when solids are dried, then loaded hot into storage silos and allowed to consolidate and cool at rest or when solids are loaded into silos, trucks, or rail cars that are subsequently exposed to cyclic temperature extremes occurring between day and night. In many cases, these conditions have been found to significantly increase solids strength (f_c).

Flow properties of these solids should be determined by duplicating *these conditions* (including cyclic changes) in the consolidating bench. Many startup problems can be avoided by such testing. Testing may indicate the need to cool solids before storage or during storage or to insulate storage bins or rail cars to protect the material from ambient temperature changes.

8.8.3 Moisture

Moisture in a bulk solid may vary owing to changes in operation of a process dryer or from exposure to weather when stored in outside piles. Changes also occur during storage if moisture enters the bin through open atmospheric vents, is blown in during pneumatic conveying of hot moist materials and condenses on cold bin walls, or migrates from the stored material. Moisture can effect yield strength (f_c) and wall friction ϕ' and can cause solid-wall adhesion (ϕ_t'). To properly access these effects, the expected storage conditions must be duplicated in the shear test. If the flow properties are time dependent they can be determined only by making time consolidation tests. Migration of moisture from large quantities of material in a storage bin, particularly if it is accompanied by chemical changes, may not be detected in shear tests since the sample size is small. Several useful studies of caking due to these mechanisms are described in Refs. 66 to 70. The use of flow conditioning agents to improve flow are reviewed in Refs. 71 and 72.

8.8.4 Particle Size

Flowability usually decreases, and wall friction ϕ' tends to increase as particles become finer. Permeability also decreases, resulting in an increasing potential for flooding or flow limitation from the hopper outlet.

If materials are to be handled in equipment that will cause particle attrition before storage (pneumatic conveyors for instance) the samples used for shear testing must be representative of material after handling.

Future possibilities of producing and storing finer particles should be anticipated in the initial design. Coarse particles may break and generate finer particles during storage and flow. Run crush tests to determine potential breakage with the anticipated consolidating stresses in the bin. If breakage cannot be tolerated, consider storage in smaller bins where the stresses will be lower. If breakage will occur and can be tolerated, shear test the minus 3 mm fraction since the flowability of the mixture will be most affected by these particles. Particles larger than this are considered coarse granular and are usually free flowing.

8.8.5 Vibration

Vibration can be induced into storage structures by nearby moving equipment; by transport in over-the-road vehicles, and by vibrating devices used to promote flow.

Many bulk solids, particularly those containing fine particles, will tend to compact very rapidly when vibrated. Those solids that have high instantaneous flow functions are particularly susceptible to flow stoppages caused by compaction during vibration. Use of vibrators to assist flow should be restricted to *only* the time that material is flowing in the hopper. Vibration should be stopped immediately when the silo outlet is closed or the discharge feeder stopped.

The application of vibration to assist flow is strictly empirical at this time. Many manufacturers have developed rules of thumb regarding proper locations for their particular vibrator on a particular hopper geometry.

If the shear test data (or other experience) indicate that vibration will be required to start flow (i.e., reduce the time flow function to the instantaneous flow function), the silo outlet should be designed with a safety factor to allow for possible solids over compaction due to vibration (see Section 8.6). Roberts describes the effects of vibration in more detail in Chapter 5.

8.9 DESIGN OF BINS FOR FLOW

8.9.1 Mass Flow

Mass flow bins can have a variety of shapes, but they are all characterized by steep hoppers and, usually, also by the absence of in-flowing valleys and sharp transitions. Some commonly used shapes are shown in Figure 8.55. In this chapter, all hopper slope angles, θ_c' for conical flow, and θ_p' for plane flow, are measured from the vertical, as shown in Figure 8.55. The diameter of a circular discharge opening in a mass flow hopper will be designated as B_c. The dimension of a rectangular (slot) opening in a plane flow hopper will be designated by its width B_p, and its length L. The length of the slot opening should be at least three times the width to avoid end wall effects and ensure mass flow.

For mass flow to occur in a bin, the hopper slope angle, wall frictional surface, and the size of the discharge opening must be compatible with the measured flow properties of the stored solids. In addition, any discharge device at the hopper outlet must withdraw solids from the entire cross-section. If the device or connecting chute causes the solids to flow preferentially from a portion of the outlet, the mass flow pattern within the bin will be destroyed and funnel flow will result.

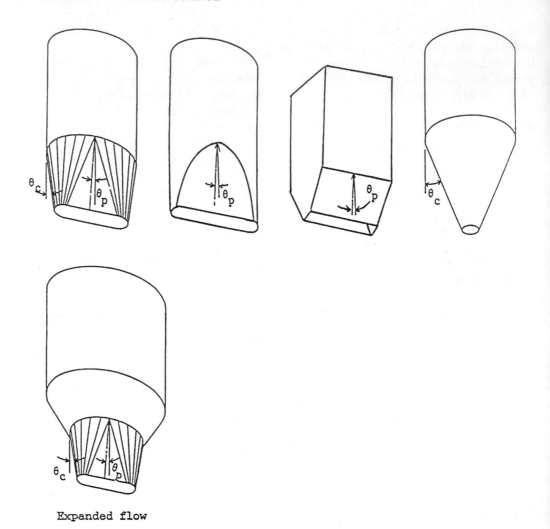

Expanded flow

Figure 8.55. Mass flow silo geometries hopper slopes: θ_c, conical; θ_p, plane flow.

Flow properties are influenced by the stresses imposed on them as they move through a bin. Figure 8.56 illustrates the approximate distribution of stresses on an element of solids as it flows along the wall of a mass flow bin. The major consolidating stress, σ_1, increases exponentially with depth (as predicted by Janssen), abruptly increases at the transition, then decreases toward zero at the vertex (area of radial stress as described in the following). The solids develop a yield strength f_c (resistance to shear failure) that changes in response to the consolidating stress. The stress $\bar{\sigma}_1$ acts at the abutments of any arch that

tends to form, and is proportional to the span B, and therefore will vary as shown. The flow–nonflow criterion[53] states that a cohesive arch will form in a hopper when the yield strength f_c exceeds the stress $\bar{\sigma}_1$, tending to break it. In a hopper, this will occur below the point of intersection in Figure 8.56 where the critical value is:

$$f_c = \bar{\sigma}_1. \qquad (8.23)$$

Jenike and Leser[73] analyzed the equilibrium of forces acting on an arch in a converging hopper at the point of collapse and obtained

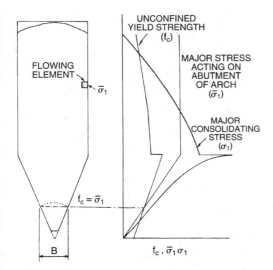

Figure 8.56. Stress on an element flowing through a mass flow bin.

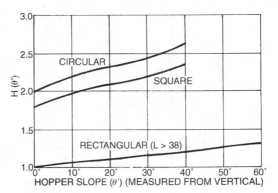

Figure 8.57. Function $H(\theta')$. (From Ref. 55).

the following expression for calculating the arch span, B:

$$B = \bar{\sigma}_1 H(\theta')/\gamma g \qquad (8.24)$$

where

B = diameter B_c, of a circular opening, or the width, B_p of a rectangular opening, m

$\bar{\sigma}_1$ = stress acting on the abutment of an arch

$H(\theta')$ = a function of the hopper slope angle and geometry; computed by Jenike and presented as a design chart in Figure 8.57[55]

γ = bulk density (kg/m^3)

g = gravity acceleration (m/s^2)

8.9.1.1 Radial Stress Field in Mass Flow

Jenike showed that solutions describing stress and velocity fields in a channel converge to a radial stress field at the vertex of the hopper.[54] That is, the stresses in the lower part of the hopper increase almost linearly with distance from the vertex. Therefore, the stresses near the outlet in a mass flow hopper, where an arch or flow obstruction is likely to occur, can be predicted without considering the solids

head above the outlet and the geometry of the walls in the upper region. In further studies, Johanson showed that radial stresses were closely approximated in regions farther from the outlet.[30, 31] Radial stress distributions in mass flow hoppers have since been confirmed by other researchers.

Jenike[54] solved the stress equations needed to satisfy this condition, and determined the boundaries for mass flow in conical and plane flow hoppers as a function of angle of wall friction ϕ', hopper slope θ, and effective angle of friction δ, as shown in Figures 8.58 and 8.59. The values for ϕ and δ are determined as described in Section 8.6. Hopper slopes required for mass flow can then be found from these figures. In the regions marked funnel flow, the boundary conditions are not solvable for radial stress, and solids will not flow along the hopper wall.

8.9.1.2 Arching Dimension

Jenike's procedure for finding the critical value that satisfies Eq. (8.22), to enable the calculation of the arch dimension B_c and B_p in Eq. (8.33), requires that the *flow function* (*FF*), describing the flowability of a bulk solids, and the *flow factor* (*ff*), describing the flowability of the channel in the hopper, be determined. The flow function *FF* has been defined earlier by Eq. (8.20). The flow factor ff is defined as:

$$ff = \sigma_1/\bar{\sigma}_1 \qquad (8.25)$$

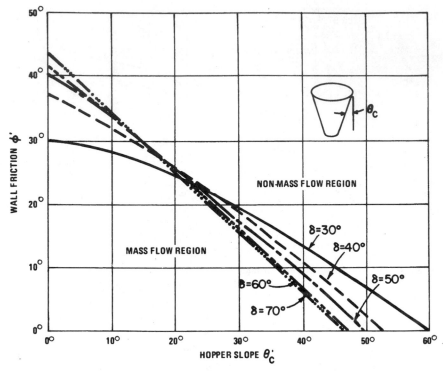

Figure 8.58. Critical wall slope for conical mass flow hoppers for δ values from 30 to 60°.

Jenike solved the equations for this relationship, as a function of angle of wall friction ϕ, hopper slope θ'_c, and angle of internal friction δ, and presented the solutions in the form of design charts for conical and plane flow (slot opening) hoppers.[55] Figures 8.60 and 8.61 are examples of the charts for conical and plane flow for solids having a δ value of 40°. The limiting hopper slope θ_c for conical flow, or θ_p for plane flow, and the flow factor, for mass flow are determined by entering the measured value of angle of wall friction ϕ' on the proper chart, and moving right to intersect the boundary. At the intersection read the flow factor ff, and then move down to read the required hopper slope. In practice, the slope angle θ_c is reduced 3 to 5 degrees from that value read from the chart to allow for the instability of conical channels in the region of convergence from the cylinder to the hopper. No reduction in the value for θ_p is made because the limits

on the region of mass flow in a plane flow hopper are wider than conical flow.

The flow factor is a constant, and plots as a linear function through zero. When it is superimposed on the flow function, the critical stress value σ_1, for determining the minimum hopper discharge opening (Eq. 8.24), is found from the point of intersection, as shown in Figure 8.62a.

If the FF and the ff do not intersect, and FF lies completely below ff, the minimum hopper opening is very small and cannot be determined with this flow analysis. Opening size will be limited only by the possibility of mechanical interlocking of particles, or by the required solids discharge rate.

If the instantaneous FF lies below ff, and the time FF lies above, then it is usually possible to use vibration or other means to start flow after time consolidation, and thus return the solids to the instantaneous flow condition.

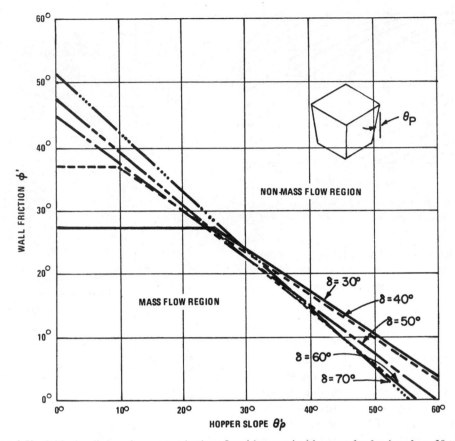

Figure 8.59. Critical wall slope for symmetric plane flow (slot opening) hoppers for δ values from 30 to 60°.

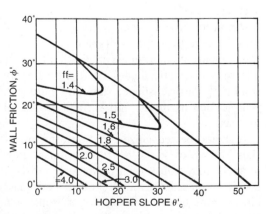

Figure 8.60. Flow factors (ff) for conical mass flow hoppers where $\delta = 40°$. (From Ref. 55).

If there is no intersection, and the FF lies above ff, unassisted gravity flow is not feasible, and mechanical flow aids must be considered.

If vibration is used to initiate flow, a safety factor equal to[25]

$$\sigma_1 = 1.5f_c \qquad (8.26)$$

is used to compute the value for B.

If the wall yield locus is convex downward, as described earlier, then ϕ' will be higher in the lower regions of the hopper, where in accordance with the radial stress field, the wall pressures are the lowest. In that case, the value for ϕ' at the region near the hopper opening should be used to determine the flow factor and wall slope for that region.

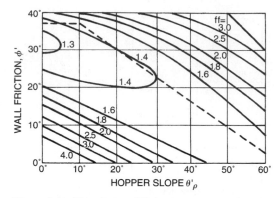

Figure 8.61. Flow factors (ff) for symmetric plane flow (slot opening) hoppers where $\delta = 40°$. (From Ref. 55).

To determine ϕ' at the hopper opening, first estimate a value for B, the discharge opening, and calculate a value for $\bar{\sigma}_1$, from Eq. (8.24).

$$\bar{\sigma}_1 = \frac{B_{\gamma g}}{H(\theta')} \qquad (8.27)$$

From the intersection of $\bar{\sigma}_1$ and an estimated ff line, determine σ_1 and estimate δ (Fig. 8.62b). Construct the EYL, note point σ_1, construct the Mohr semicircle through this point tangent to the EYL (Fig. 8.50b). The line through the origin and the point of intersection of the WYL and the Mohr semicircle determines the value of ϕ'. A flow factor is determined using this value. If it varies by more than about 10% from the assumed value, a new estimate for B is made and the calculation repeated.

8.9.1.3 Surface Finish of Hopper Wall

A low solid-wall friction ϕ' is preferred for the *hopper* section of a mass flow bin. It permits larger hopper slope angles θ' and thereby reduces the overall height of the hopper.

A variety of hopper wall lining materials can be used to reduce wall friction and adhesion. These include Teflon®, glass, various epoxy paints, smooth finished stainless steel, and ultra-high molecular weight (UHMW) polyethylene. The last four are the most commonly used.

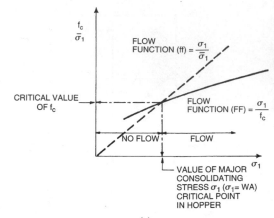

(a)

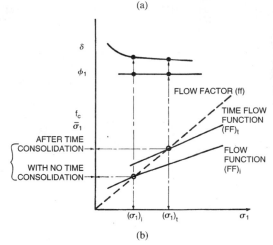

(b)

Figure 8.62. (*a*) Flow factor-flow function relation. (*b*) Flow factor with instantaneous and time-flow function.

It is not possible to make generalizations such as use of "smoother" surface will always result in lower solids-wall friction. Each solid must be tested. Hopper wall materials with the same apparent smoothness may exhibit widely differing kinematic angles of friction on dry solids.[2,74] Particle shape, size, and hardness, wall surface hardness, and surface profile all interact.

It should be recognized that abrasive materials sliding over the surface of mass flow hoppers may cause wear. Abrasive-resistant plates, replaceable metal liners, and glass blocks have been used for this reason when handling abrasive ore, minerals, coal, and so on.

Care should be taken to ensure that a wall finish specified in the design is reproduced in the actual bin.

Do not substitute "similar" paint or coated surfaces unless samples have been submitted and tested to determine wall friction.

- When using unpainted carbon steel, use the WYL for rust-coated steel unless precautions are taken to prevent rusting. Mass flow may not start if the rusting is sufficient to prevent sliding of the material during initial use.
- Specify surface finish of stainless steel sheets, and plates can be obtained in several different surface finishes (smoothness). The standard commercial finish usually described as a No. 1 finish has a surface profile ranging from about 150 to 500 μAA (micro inches, arithmetic average) and is standard on most sheet and plate. Sheets up to $\frac{3}{16}$ inch thickness can be furnished in a 2B finish, having a 5 to 15 μAA profile. Plates can sometimes be furnished with a 2D surface (40 to 60 μAA profile). Flat sheets and plates can also be polished to any desired surface finish before forming and fabrication.
- Construct the hopper so no ledges are presented to the flowing material. With lap welded construction, overlap plates in the direction of flow. Grind circumferential welds flush. Fasten any interior liners with countersunk or shallow-head fasteners.

8.9.1.4 Surface Finish of Vertical Section

Smooth walls on the vertical part of a silo may *not* be desirable. As the solids-wall friction in the vertical part of the silo is reduced, more of the consolidating stress from the stored material is transmitted directly to the material in the converging hopper below. This could cause arching across the silo at the transition between cylinder and hopper.

An inspection of Figures 8.60, 8.61 and 8.57 will show that as θ_c' decreases to zero (vertical wall) the function $H(\theta')$ decreases and the flow factor (ff) increases, indicating a decreasing flowability of the channel. The vertical wall flow factor can be superimposed on the flow function as before to determine if the material will gain sufficient strength to arch across the bin at the transition (the value of ϕ' appropriate to this area of the bin must be determined if the WYL is pressure sensitive).

If the time flow function FF_t continues to rise steeply at high pressures, even though it lies below the flow function (ff) (and indicates a small or zero arching diameter) arching may still be possible at the cylinder cone transition. This condition is described in Ref. 109.

If it is suspected that smooth walls in the vertical portion of a silo will cause arching at the transition, the vertical wall specification should call for a rough surface, for a distance of about one diameter above the transition.

Borg[147] presented an interesting paper summarizing a statistical evaluation of 500 shear tests on (unidentified) solids commonly used in the chemical industry. Hopper wall friction with various wall surfaces, hopper slopes, and critical outlet diameters required for mass flow were calculated for more than 200 bulk solids having varying degrees of cohesiveness. The critical outlet diameter exceeded 1.2 m for 35% of all products. The percentage of products for which mass flow could be achieved with the wall surface having the lowest friction (ϕ') was plotted against the hopper angle θ_c for which mass flow will occur. Eighty percent of these products would mass flow at $\theta_c = 10°$; only 25% would mass flow at $\theta_c = 30°$. The curve between these points was almost linear. Confirming comments made earlier in this chapter, the author reported that polishing a wall surface does not always reduce solid wall friction.

8.9.2 Funnel Flow

To ensure funnel gravity flow from a funnel flow bin, the discharge opening must be large enough to prevent a rathole or arch from forming. The critical opening dimensions D_f, for preventing a rathole or arch from forming over a circular, square, or rectangular opening, are shown in Figure 8.63.

8.9.2.1 Ratholing

The bin or hopper opening must be larger than the critical rathole (piping) dimension D_f. At this critical dimension, the stress imposed on the material will exceed the yield strength, and any rathole that tends to form will continually collapse.

The critical flow properties of materials determined from the shear test as described so far are based on steady-state flow conditions. As pointed out earlier, and in Section 8.11, initial pressures caused by filling a bin, without withdrawing solids, result in an active stress field and higher consolidating pressures on the material near the outlet. The material at the bottom of the bin, therefore, gains greater strength during initial filling than during steady flow.

In a mass flow bin, any obstructions to flow caused by initial filling will fail, and the critical flow properties are determined from steady-state conditions. This is not true in a nonmass flow bin. Therefore, the strength of the solids, the ability to support a rathole, and the minimum rathole diameter must be calculated on the basis of initial filling as well as steady-state flow.[56]

Under initial filling conditions (with no withdrawal) the critical rathole diameter D is determined first calculating the consolidation stresses on the material at the hopper opening. In bins where the height to diameter ratio exceeds 1, use Janssen's equation (Eq. 8.1) to calculate q, using Janssen's k value = 0.4 and assume:

$$q = \sigma_1 \qquad (8.28)$$

Transposing this value of σ_1 to the time flow function chart, determine a value of $\bar{\sigma}_1$ from the intersection of σ_1 and FF_t. Calculate the critical piping diameter D_f(m) as:

$$D_f = \bar{\sigma}_1 G(\phi_t)/\gamma \qquad (8.29)$$

where

$G(\phi_t)$ = function of ϕ_t as given in Figure 8.64.

If the bin height-to-diameter ratio is 1 or less, use:

$$q = H_{\gamma g} \qquad (8.30)$$

where H = height of material surface above the opening.

Since the major consolidating stress is lower in the upper part of the bin, the solids strength is correspondingly lower. Therefore, it is possible to have material flow in the upper part without piping but when the level drops to a certain height, a stable pipe can form as shown in Figure 8.65. The height at which this occurs can be estimated as follows. Restate Eq. (8.23):

$$\bar{\sigma}_1 = D_f \gamma_g / G(\phi_t) \qquad (8.31)$$

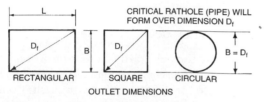

Figure 8.63. Outlet dimensions used in the analysis of funnel flow: (*a*) Rectangular, (*b*) square, (*c*) circular.

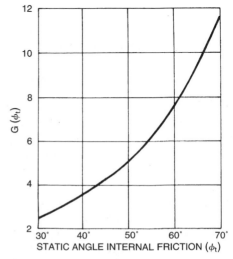

Figure 8.64. Function $G(\phi_t)$. (From Ref. 55.)

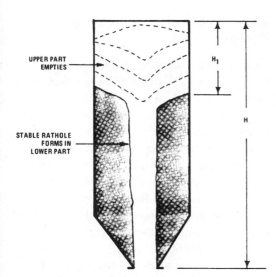

UPPER PART
EMPTIES →

STABLE RATHOLE
FORMS IN →
LOWER PART

H_1

H

Figure 8.65. Formation of a stable rathole after upper part of bin empties to a critical level.

Enter $\bar{\sigma}_1$ on the time flow function chart and determine the corresponding value of σ_1. Use this value of σ_1 to calculate H_1, the distance measured from the top of the material over which flow will occur without ratholing:

$$H_1 = \frac{\sigma_1}{\gamma_g} \qquad (8.32)$$

When the top surface of the solids drops below this level, a stable rathole may be in place.

In the case where the solids are being continuously removed from the flow channel, while it is being simultaneously filled, the surrounding solids are not subjected to the high initial filling pressures and the critical dimensions of the opening to prevent a rathole will be less than that calculated as just described. Jenike provides a design chart of no-piping flow factors for this situation, based on continuous flow and assuming that the outlet pressures are not affected by the solids head.[55] However, to ensure against ratholing, the outlet dimension, D_f should be calculated on the basis of the largest rathole that is likely to occur: that caused by *filling conditions* as described above, rather than steady flow conditions.

8.9.2.2 Arching

Arching will not occur over circular or a square outlet with a dimension, D_f, sufficiently large enough to prevent ratholing. This is not true for a rectangular opening. The width, B, must be sufficiently large to prevent the formation of an arch. This minimum dimension is computed by:

$$B \geq 1.15\sigma_1/\gamma g \qquad (8.33)$$

where σ_1 is determined from the intersection of the flow function, and a flow factor having a value of 1.7.

8.9.3 Expanded Flow

The major diameter of the mass flow section must be at least larger than the critical piping diameter for the material; the outlet must be larger than the minimum arching dimension.

Expanded flow can be achieved with an increased slope at the lower end of a hopper, but it is not the only way. A funnel flow hopper can be converted to expanded flow by applying a low-friction surface material, epoxy coating, plastic linings, glass coating, polished stainless steel, etc., to the lower wall surfaces. This technique can be particularly useful where ϕ' is pressure sensitive. The proper wall coating is selected by evaluating the WYL for the particular surface and the stresses expected in the low pressure regions near the outlet where the coating will be used.

When a funnel flow bin is converted to mass flow, the effect of higher wall stresses must be considered. The stable or unstable portions of the flow channel may expand to reach the walls in the vertical portion and impose higher wall stresses, as described in Section 8.5.

8.9.4 Summary

A number of workers have evaluated the Jenike method with model bins. Wright,[75, 76] testing a number of different iron ores in wedge-shaped hoppers with various slopes, reported that this method provided a sound basis for functional design of mass flow bunkers handling ore under dynamic conditions and provided a reasonable safety factor for engineering design. The predicted critical outlet size agreed closely with experimental results, but the predicted hopper slopes were 5 to 10° steeper than those found by test to be required for mass flow. The same conclusion regarding plane flow hopper slopes was reported by Eckhoff and Leversen.[77] These results are consistent with theory, since it is known that while the regions of mass flow are quite restrictive in conical-flow channels, they are much wider in plane-flow channels. Accordingly, the Jenike flow factor charts for plane flow contain a safety factor to allow for variations in solid head in the vertical portion of the bin. Wright also confirmed that consolidating stresses caused by impact during initial fill must be considered in sizing the discharge opening and reported that arching at the vertical, sloping wall transition can occur.

Richards[86] tested wet and dry sand in symmetrical conical hoppers and reported that the critical hopper slope and outlet for mass flow and the critical outlet for funnel flow determined by test agreed very closely with values predicted by the Jenike method.

Enstad[78, 79] pointed out that if the flow function is determined in a region of high stresses where it may be linear and this linear function is extrapolated to regions of low stress, it will intersect the flow factor at a higher, incorrect value. This overestimates the strength of the material in the region of the hopper outlet and will predict an opening considerably larger than that required. Eckhoff and Leversen[77] report similar results when the yield locus was extrapolated into a region of very low normal stress. Jenike, however, suggests that at low normal stresses, the powder in the cell is exposed to tensile stress components that produce measurements at shear stresses at failure that are actually too low. This is avoided by following a test procedure whereby normal stresses imposed on the cell during shear are greater than $\frac{1}{3}$ of the normal stress used for consolidation.

8.9.5 Arching of Large Granular Particles

Noncohesive large granular particles will bridge or arch by mechanical interlocking of particles. They develop very little, if any, unconfined yield strength and cannot be analyzed by existing powder mechanic's theory. Minimum hopper opening size to prevent bridging has been based mainly on rules-of-thumb. Orifice tests with noncohesive granular material reported by Reisner[80] indicate a theoretical *minimum* limit of 3 for the ratio of hopper opening diameter/maximum particle dimension (D/D_p).

The most commonly quoted minimum ratio for design is 5. Schwedes[81] recommends a ratio of 10 to provide a larger factor of safety to assure no mechanical block in a hopper outlet. Peschl[82] studied the flow of coarse granular materials in model bins. He described the flow as being characterized by constant formation and collapse of successive arches, which he termed "dynamical" arches. This is similar to observations reported with flow of granular material by early experimenters. Peschl concluded that the probability of arch formation with coarse granular materials cannot be predicted theoretically, but can be predicted by making a small number of repetitive tests with model hoppers, and statistically, analyzing the data.

8.10 EFFECT OF THE GAS PHASE

The previous sections on flow in bins were based on single-phase flow, under gravity forces, and the effect of the gas phase was not considered.

During loading and subsequent settling, the gas entrapped within the solid bed can have a significant influence on wall pressures and flow behavior, a fact recognized in the national silo design codes. Entrapped gas in fine powders can be retained for an appreciable time. Suggestions for estimating the settlement time of powders in industrial bins are given in Refs. 62 and 110.

In a funnel flow bin, powders move to the outlet through narrow, unstable channels. Often the residence time is not sufficient to allow powders that have become aerated during filling to deaerate before discharging. The problem then, with these bins, is how to regulate an unpredictable, relatively high flow of fluidized powder.

As mass flow bins have come into widespread use, it has been found that the discharge of powders from these bins can become flow rate limited. Interstitial gas pressures within a powder bed change during flow, and this influences the rate of discharge from a mass flow hopper.[11,112] Wlodarski and

Pfeffer[113] demonstrated this principle. They showed that air pressures within a plug-flowing bed in a 90 mm tube exhibited a significant axial gradient. In their test, the solids head was maintained constant and the air pressures along the axis of the tube were measured through hypodermic needles inserted through the tube wall. They found that the interstitial gas pressure above the discharge orifice was below atmospheric, and this pressure was dependent on the particle size, orifice size, and the rate of discharge as shown in Figure 8.66. When the lower needle was vented to the atmosphere, the powder flow rate increased, as air entered to eliminate the pressure gradient at the outlet. A similar phenomenon was pointed out by Bruff and Jenike[114] in describing a design of a mass flow

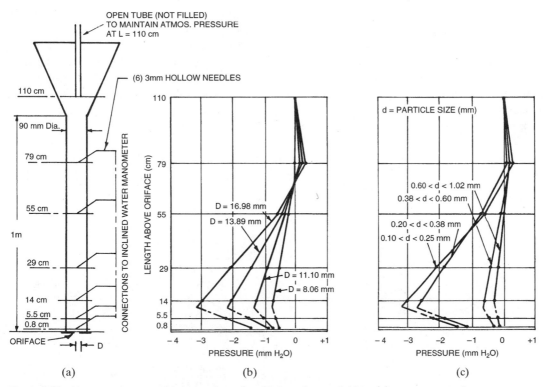

Figure 8.66. Air pressure measured during flow of sand through a model bin: (a) arrangement of test apparatus, (b) air pressure measured (in center) at varying heights, with sand particles having a size distribution d of 0.10 < 0.25 mm, and orifice diameters D varying from 8.06 mm to 16.98 mm, (c) air pressure measured (in center) at varying heights, with an orifice diameter D = 16.98 mm and various sand particle size distributions d. (From Ref. 113.)

hopper for ground anthracite, in which gas injection was found to be needed to overcome a flow rate limitation at the hopper discharge.

The permeability of powders is a defining parameter that influences the rate of discharge. The importance of permeability can be seen by the following example of a fine powder discharging from a mass flow silo, as shown in Figure 8.67.

As an element of powder moves through a mass flow silo to the outlet, the consolidating pressure on the element changes as described earlier. Initially, as the element is compressed, the voidage is reduced and interstitial air is squeezed out through the top surface. As it moves through the hopper the consolidating pressures on the element decrease, the element expands, and the voidage increases. If the powder has a low permeability to air flow, the interstitial pressure in the lower region of the hopper can decrease to below atmospheric pressure. The resulting pressure gradient will cause an influx of air from the hopper outlet that will retard the solids flow.

It has been found that injection of a *small* amount of well-distributed air at appropriate locations in the powder bed provides some support for the solids in the vertical portion of the bin, thereby reducing the compaction and subsequent expansion in the lower regions of the hopper, and it supplies gas to the interstices, reducing or eliminating the pressure gradient at the outlet. This can substantially increase the rate of powder discharge from a mass flow hopper. Experimental work and examples of air injection or permeation are given in Refs. 115–118, 162. The injection or permeating air flows used for solids flow rate enhancement in industrial silos are in the range of 0.03 to 0.3 m^3/min, much less than that which will cause the powders to become fluidized.

8.10.1 Permeability Constant

Jenike and Johanson[62] have proposed a permeability constant a to characterize powders, using the device shown in Figure 8.68. A powder sample of known mass is placed under a range of consolidating loads. At each load, the column height is recorded, and a measured flow of dry gas is permeated through the sample. The gas pressure gradient is measured, and the consolidating stress and solids bulk density calculated for each applied consolidating load.

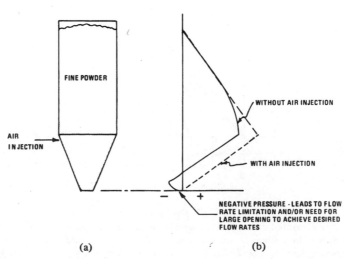

Figure 8.67. Air pressure gradient that may occur in a mass flow bin: (*a*) mass flow bin, (*b*) air pressure in flowing mass.

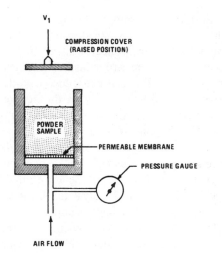

Figure 8.68. Jenike and Johanson Inc. permeability tester. (From Ref. 62.)

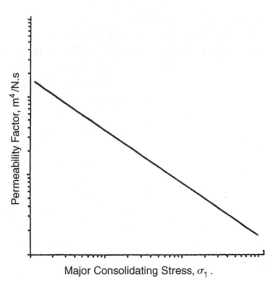

Major Consolidating Stress, σ_1.

Figure 8.69. Permeability.

A permeability factor C for the sample is determined from a form of the Darcy equation for laminar flow:

$$C = -v/dp/dx \qquad (8.34)$$

where

v = superficial air velocity
dp/dx = pressure gradient across sample column.

Computed values for the permeability factors, when plotted as shown in Figure 8.69, closely approximate a straight line. The relationship between C and the consolidating stress can be expressed by:

$$C = C_0[\sigma/\sigma_0]^{-a} \qquad (8.35)$$

where

σ = consolidation stress
C_0 = value of C, arbitrary value, corresponding to chosen value of σ_0
a = permeability constant for the powder, $m^4/N \cdot s$

The relationship of C with bulk density γ can be expressed in a similar manner, as:

$$\gamma = \gamma_0[\sigma/\sigma_0]^{-a} \qquad (8.36)$$

and a is expressed in m/s.

Gu et al.[163-166] reported on their extensive research on powder permeability and flow rates in mass flow silos, and reviewed the work of others. They concluded that using permeability as a parameter to delineate coarse and fine solids is more useful than using a single particle size. Their studies showed that the critical permeability necessary to produce a significant effect of interstitial air on the flow rate from a mass flow silo is dependent not only on the powder flow properties, but also on the hopper geometry and the outlet size. They suggest that critical permeability be determined in relation to the outlet size since it affects the flow rate as well as the interstitial pressure gradient.

Because of the complexity of the powder flow regemi, no acceptable model for predicting the limiting flow rate of fine powders from mass flow bins is yet available in the open literature. A proprietary mathematical model that includes the compressibility and permeability factors has been developed by Jenike and Johansen, Inc.[159]

8.11 OTHER METHODS FOR CHARACTERIZING BULK SOLIDS RELEVANT TO STORAGE AND FLOW

Consolidation stresses are always present during storage and during flow of bulk solids in bins, hoppers, and containers and processing equipment. Flowability is a function of these stresses. Therefore, a sample must first be preconsolidated to a predetermined level of stress in order to obtain a quantitative measure of its flowability (or yield strength).

In many pharmaceutical, bulk powder processing, and handling operations, quantitative bin design information is not needed but a reproducible, easily measured flowability or relative flowability index is highly desired for routine quality control operations.

8.11.1 Commercial Test Devices that Preconsolidate the Sample

8.11.1.1 Peschl Rotational Split Level Shear Tester

The applied shear strain is limited in the translational type shear cell of Jenike as can be seen from an inspection of Figure 8.43. The rotational, split level, shear tester was developed by Peschl[59] to overcome the shear strain limitation and to reduce the needed operator skill and the time required to complete a shear test.

As shown in Figure 8.70, the material sample is sheared in a rotary motion that has almost unlimited travel. The cell containing the material sample is clamped to a turntable that rotates at about 0.050 rpm. A consolidating load is applied to the cover, which is kept stationary through a vertical shaft attached to the cover. The stationary shaft allows the application of the loads to be automated if desired. The shaft is mounted on an air bearing to minimize friction and to hold the cover parallel with the cell base. The torque applied to the cell cover during cell rotation is transmitted through a torque arm attached to the cover, to a strain gauge load cell. The vertical movement of the cover that occurs during expansion of the sample during shear is transmitted through a vertical rod to a vertical displacement transducer. Shear force and cover displacement are measured and recorded for each test. The test procedure includes initial preconsolidation, as with the Jenike cell. However, with rotational shear, greater shear strain is possible, so multiple shear tests can be made on the same sample to obtain the complete yield locus. Usually a maximum of three to five yield loci (depending on the reproducibility of the steady-state shear values) can be made with a single sample.

Two tester models are available: a manual machine where the consolidating loads are placed manually by the operator, and an automatic machine. With the automatic machine, the operator selects the values for the consolidating loads, but the placement of the weights on the cell is done by programmed electromagnets, and the sequencing of the consolidating and shearing procedures and acquisition and evaluation of the test data are controlled by a programmed microprocessor. The resulting data, yield locus, flowability index, bulk density, and angle of internal friction can be printed out, displayed, or stored on tape or floppy disc. Time-consolidated yield loci are more time consuming because of the need to interrupt operating sequence, to time consolidate the sample each time, before proceeding with the shear.

There is disagreement as to the precise location of the shear profile within the rotational shear cell, and what effect the repeated shear of the same sample for the construction of the yield locus has on the final results.

8.11.1.2 Johanson Bulk Solids Indicizer® System

JR Johanson Inc. have developed three automated testing devices, described below, for measuring the primary characteristics of powders that affect their performance in handling and storage. With each tester, a powder sample is prepared in a cell configured for that particular test. The cell is then inserted into

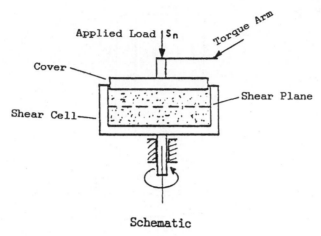

Schematic

Figure 8.70. Peschl rotational shear tester.

the tester, the information required for the particular test is entered through a keypad, and the test, guided by an on-board computor, is performed automatically.

Johanson Hang-Up Indicizer.®   This is a uniaxial shear device as shown in Figure 8.71. In the test procedure, the test cell is filled with a known weight of powder, and inserted into the tester. The sample weight and desired indice is entered on the keypad. The top disc and cylinder lower with a vertical force to consolidate the sample to a pressure approximating the condition at a hopper outlet. When consolidation is complete, the top cylinder and disc are withdrawn and the bottom disc is lowered, leaving the consolidated sample supported on the horizontal ledge within the cell. The top disc, smaller in diameter than the supporting inner ledge, is then lowered and the force required to fail the sample is measured. The top disk then retracts. From this force, an unconfined yield strength is calculated.

Two powder indices are determined with the Hang-Up Indicizer: an Arching Index (AI), a relative measure of the propensity to arch over a hopper opening, and a Ratholing Index (RI), a measure of the propensity to rathole in a bin. The Hang-Up Indicizer tester has been reported to be highly repeatable across a wide range of solids. It operates rapidly, and requires minimum operator training for its operation (168). The validity of the testers for design of bins by an unskilled technician is being debated in the literature. In one study it was reported that hopper openings required to prevent an arch and rathole from forming, as determined by the Hang-Up Indicizer, were less than that predicted by the Jenike procedure. This may be due to the fixed values assumed for the stress functions and the computation method imbedded in the program, as well as the lack of understanding of the actual stress distribution throughout the sample. (167)

Hopper Indicizer.®   This device measures the angle of slide of a powder sample, constrained within a ring, on the surface of a wall sample mounted on a tilted platform within the tester. The powder is subjected to two sliding tests, each with a different predetermined consolidating load applied through the ring cover. The measured static surface friction angles ϕ_t' (after any adhesion is broken) are interpolated to the conditions at the outlet. A conservative value of 60° is assumed for the effective angle of friction δ, and a recommended conical Hopper Index (HI) predicts hopper slope required to cause flow at the walls. A second test determines a Chute Index (HI), the minimum angle of slope of a chute, having the same

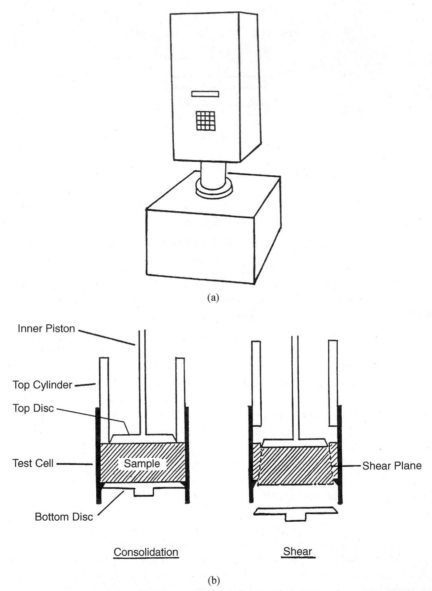

(a)

(b)

Figure 8.71. (*a*) Typical Indicizer® arrangement. (*b*) Schematic arrangement: Hang-up Indicizer.®

surface characteristics as the wall sample, after the solids impact the wall at a pressure equivalent to about 100 psf.

Flow Rate Indicizer.® The test cell is similar to the Hang-Up Indicizer® cell except that it has provisions for introducing a controlled and measured air flow through the bottom. First, air permeability of a sample is measured, followed by compressibility. A Bin Density Index (BI) and a Flow Rate Index (FRI) is calculated using a proprietary procedure. The FRI is stated as the limiting flow rate for unassisted gravity flow of fully deaerated solids through a 12 in. diameter outlet (or the diameter specified by the operator), in a mass flow bin.

8.11.1.3 Jenike & Johanson Quality Control Tester

This device is intended for routine measurements of the relative flowability of a bulk solids, mainly for quality control applications, where rapid off-line measurements can provide guidance in recognizing and diagnosing problems in solids processing control.

A solids sample is placed in a sample container configured with a cylindrical section above a converging hoppers section, as shown schematically in Figure 8.72. A perforated slide gate covers the hopper opening when the solids sample is gently filled into the container. After loading, the container top is sealed, and the container pressurized to a predetermined pressure for about 30 s to consolidate the sample as air permeates out through the screen opening at the bottom of the cone. After 30 s, the pressure is reduced to zero, the screen

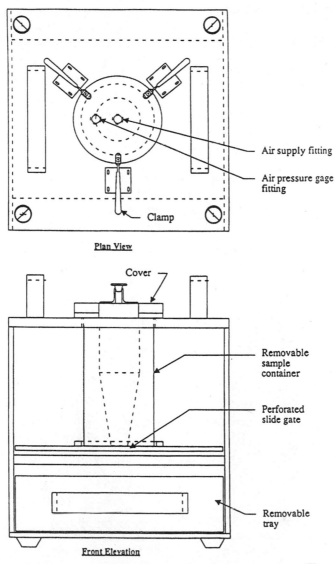

Figure 8.72. Jenike & Johanson Inc. Quality Control Tester.[170]

slide gate is removed, and the pressure reapplied until the arch at the outlet breaks and solids flow from the container. The peak pressure is recorded by a digital pressure indicator.

The consolidating and failure procedures are repeated several times and an average peak pressure value is calculated. This peak pressure value, corresponding to the "strength" of the material, can be compared to a reference value to establish relative flowability of the sample[170]. Three different size containers are available with the tester, to accommodate a range of particle sizes.

8.11.1.4 POSTEC-Research Uniaxial Tester

Scientific testers like biaxial or modified triaxial testers are indirect shear testers where the shear zone is independent of the design of apparatus. The data from these testers can define the stress-strain relationships and flow functions directly but are too complex for routine industrial use.

POSTEC-Research [171] has developed an interesting UnIaxial Tester that shows a potential for a rapid and direct method of measuring the unconfined yield strength (f_c) of a powder. In theory, the unconfined yield stress of a powder can be determined by consolidating a supported column of powder under a stress σ_1, then removing the support and applying a vertical stress on the unsupported column until it fails (at f_c). The obvious problem with this test is how to maintain an unsupported powder column. This is handled in the Postec device as shown in Fig. 8.73.

The die and piston are aligned and fixed in position. A flexible rubber membrane is fitted to the edge of the piston and the lower end of the die. The membrane is stretched, so that it will contract as the piston moves downward, and, with lubricant between membrane and die, sliding and wall friction will be reduced to a minimum. The die is filled upside down, with the bottom plate removed. The assembly is then mounted upright, and the sample consolidated by moving the piston slowly downward until a predetermined stress σ_1 has been reached. After a period of time for stabilization, the compaction stress σ_1 is reduced to a minimum value and the die is pulled up allowing the sample to stand by itself. The piston then moves slowly downward until the value of

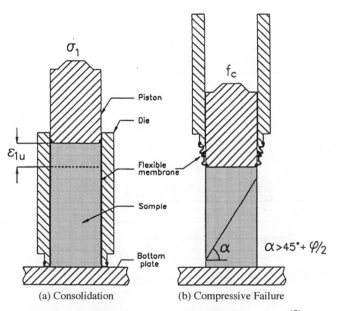

(a) Consolidation   (b) Compressive Failure

Figure 8.73. POSTEC-Research Uniaxial Shear Tester.[171]

f_c, at failure of the sample, is measured. The shear plane at failure will fall close to the angle α indicated in Fig. 8.73b. The test is repeated at several different consolidating stresses and the yield strengths are plotted to determine a flow function as described for the Jenike test. The authors report that in some cases the scatter in test results was higher than what they considered acceptable, but there was not an unreasonable agreement with the Jenike tester results. Further refinement of this patented tester is reported to be underway.

8.11.2 Other Test Procedures

Numerous empirical tests have been devised to measure and characterize the properties of bulk solids that affect their behavior in storage and handling. Most do not produce quantitative design data. However, lacking that data, the information from these tests can be useful for comparing certain characteristics with known or reference solids.

A compilation of methods of measuring physical properties of bulk solids, taken from existing trade and research literature, is available in the *Powder Testing Guide* published on behalf of the British Materials Handling Board. [172]

The American Society for Testing Materials (ASTM) [180], Subcommittee D18.24 Characterization and Handling of Powders and Bulk Solids, is embarked on a comprehensive program to accumulate, develop, and publish a series of procedures for testing powders and bulk solids. At the time of this writing, the first standard is being prepared for publication.

8.11.2.1 CEMA

The Conveyor Equipment Manufacturers' Association (CEMA) in the United States has published a guide to the Classification and Definition of Bulk Materials.[40] A similar guide is published by the British Materials Handling Board. These widely used guides attempt to establish a terminology for describing the various properties and characteristics of bulk materials that affect the design of materials handling equipment.

Two general descriptive categories are used in CEMA. First are those physical characteristics that can be determined by simple benchtop tests. Eighteen tests are described. A bulk material is assigned an alphanumeric code designation corresponding to the measured or observed results. The second category describes 20 specific properties that are difficult to quantify. These are classified as *hazards affecting conveyorability*. These are also assigned an alphanumeric code designation. Tables 8.1 and 8.2 show typical code designations. The results of all these classifications are combined into the widely used *CEMA Material Classification Code*, shown in Table 8.3. The Definition and Test Reference column shown in Table 8.2 refers to the first

Table 8.1. CEMA Factors (Reprinted with permission of Conveyor Equipment Manufacturers' Association Ref. 40).

	MOHRS NO.	CEMA FACTOR
Hardness	1	1
	2	4
	3	9
	4	16
	5	25
	6	36
	7	49
	8	64
	9	81
	10	100
	LB / CU FT	
Density	0–60	1.0
	61–120	1.1
	121–180	1.2
	181–240	1.3
	241–300	1.4
	TYPE	
Shape	Rounded	1.0
	Subround or Subangular (approach rounded or angular shape but well-rounded edges)	1.5
	Angular	2.0

Table 8.2. CEMA Abrasive Index. (Reprinted with permission of Conveyor Manufacturers' Association, Ref. 40)

CHARACTERISTICS	CEMA CODE NUMBER	ABRASIVE INDEX RANGE
Mildly abrasive	5	1–17
Moderately abrasive	6	18–67
Extremely abrasive	7	68–416

category tests by the prefix A, the second by the prefix B. As an example of this coding system, alumina, having a bulk density of 50 to 65 lb/ft^3 fine particle size less than no. 6 sieve, free flowing, extremely abrasive, can become aerated, windswept, and dusty, is assigned the material code designation of 58B$_6$27MY.

8.11.2.2 Carr's Method of Classification

Carr devised a system to characterize bulk solids with respect to what he defined as Flowability and Floodability.[43,44] With Carr's procedure, a series of tests are made and each test result is assigned a numerical value that is based on Carr's past experience in observing flow of powders and granules through hoppers and feeders. The numerical values are summed to give a "Flowability Index" and a "Floodability Index," the relative values of which indicate the level of flowability and the potential for the solids to become aerated and flood when discharged into or from a hopper. The solids are not consolidated before or during the tests. Carr's procedures have been incorporated into a testing machine, manufactured by Hosokawa Iron Works, Osaka, Japan (Figure 8.74) and Micron Powder Systems.

8.12 PARTICLE SEGREGATION DURING STORAGE AND FLOW

Whenever particulate solids are moved, deposited on piles, or withdrawn from silos, there is a tendency toward segregation. In many cases, this is undesirable, and the storage and handling system must be designed to minimize the possibility of its occurring.

Segregation occurs most frequently in free-flowing granular materials having a wide size distribution and seldom in fine powders where particle size is about 70 μm or less.

Cohesive powders usually do not segregate during handling. Powders containing cohesive and noncohesive components can segregate. The more cohesive components tend to move together in relatively thick unsegregated layers or patches when sliding in a chute or on a pile and will form rivulets of nonsegregated cohesive material extending down the face of the chute or pile.

A review of segregation of particulate materials is given by Williams[120] and Johanson.[121] Particle properties that cause segregation are due to differences in particle:

- Size
- Density
- Shape
- Resilience
- Angle of repose
- Cohesiveness

Many workers[122–127] have confirmed that differences in particle size is by far the most important cause of segregation, with differences in particle density and shape (assuming not gross shape difference) being comparatively unimportant.

8.12.1 Mechanisms

The mechanisms leading to segregation of noncohesive particles include:

8.12.1.1 Percolation of Fine Particles

Fine particles can percolate through the voids of larger particles as they rearrange themselves during a disturbance. This can occur, for example, during shear induced by stirring, shaking, or pouring the particles in a heap, or during flow through a silo.

Table 8.3. CEMC Material Classification Code Chart. (Reprinted with permission of Conveyor Manufacturers' Association, Ref. 40)

MAJOR CLASS	MATERIAL CHARACTERISTICS INCLUDED		DEFINITION AND TEST REFERENCE	CODE DESIGNATION
DENSITY	BULK DENSITY, LOOSE		A-8	ACTUAL lbs/cf
Size	Very fine	No. 200 sieve (.0029") and under		A_{200}
		No. 100 sieve (.0059") and under		A_{100}
		No. 40 sieve (.016") and under		A_{40}
	Fine	No. 6 sieve (.132") and under		B_6
		$\frac{1}{2}''$ and under		$C_{1/2}$
	Granular	3" and under	A-17	D_3
		7" and under		D_7
		16" and under		D_{16}
	Lumpy	Over 16" to be specified X-actual maximum size		D_X
	Irregular	Stringy, fibrous, cylindrical, slabs, etc.		E
Flowability	Very free-flowing-flow function > 10			1
	Free-flowing-flow function > 4 but < 10		A-12	2
	Average flowability-flow function > 2 but < 4			3
	Sluggish-flow function < 2			4
Abrasiveness	Mildly abrasive—Index 1–17			5
	Moderately abrasive—Index 18–67		A-1	6
	Extremely abrasive—Index 68–416			7
Miscellaneous Properties of Hazards	Builds up and hardens		B-3	F
	Generates static electricity		B-5	G
	Decomposes—deteriorates in storage		B-7	H
	Flammability		B-11	J
	Becomes plastic or tends to soften		B-2	K
	Very dusty		B-8	L
	Aerates and becomes fluid		B-1	M
	Explosiveness		B-10	N
	Stickiness-adhesion		B-18	O
	Contaminable, affecting use		B-19	P
	Degradable, affecting use		B-6	Q
	Gives off harmful or toxic gas or fumes		B-12	R
	Highly corrosive		B-4	S
	Mildly corrosive		B-4	T
	Hygroscopic		B-13	U
	Interlocks, mats or agglomerates		B-14	V
	Oils present		B-15.	W
	Packs under pressure		B-16	X
	Very light and fluffy—may be windswept		B-20	Y
	Elevated temperature		A-11	Z

A well-known example of surface percolation occurs during filling of a silo (see Fig. 8.75a). The particles striking the heap form a thin layer of rapidly moving material. The finer particles in the moving layer percolate into the stationary layer below and become locked in position. The large particles do not penetrate and continue to roll or slide to the outside perimeter of the heap. This has been compared to a sieving or screening mecha-

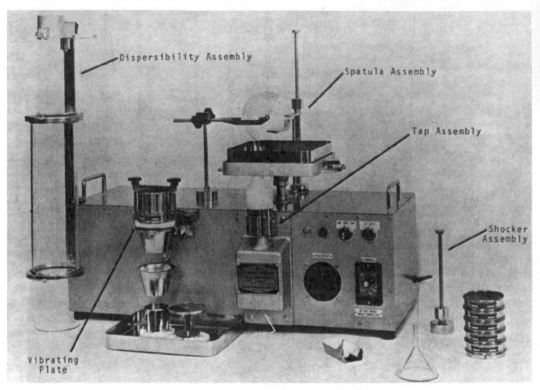

Figure 8.74. Hosokawa powder characteristics tester. (Permission VibraScrew Corp.)

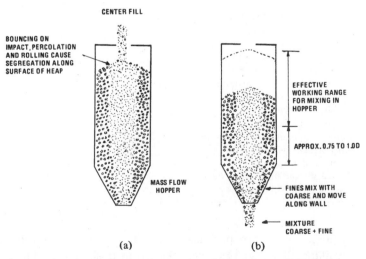

Figure 8.75. Typical segregation and mixing during mass flow: (*a*) Bin filling with no discharge, (*b*) discharge.

nism. The result is considerable radial inhomogeneity.

When the silo discharges, particle rearrangement again take place. In a *mass* flow silo, remixing occurs as the segregated material leaves the vertical section and enters the mass flow hopper (Fig. 8.75b) where the fine fraction mixes with the coarse fraction.[121]

In a *funnel* flow bin, particle segregation occurs during filling as the particles fall onto a heap (Fig. 8.76a). A central core of finer material is deposited during filling just as it does in the mass flow hopper. However, the mixture

ratio exiting the hopper can vary, depending on the rate of refill as shown in Figures 8.76b through 8.76g. If the hopper is drained, the last material to exit the hopper will be mostly coarse. If the level is lowered and then refill is begun, a short-term increase in coarse fraction will be noted at the discharge, until the new incoming material has reestablished the central core flow.

If refill continues at the same rate as discharge and a narrow flow channel has formed, segregation at the outlet will be reduced. This condition will continue until a change in silo

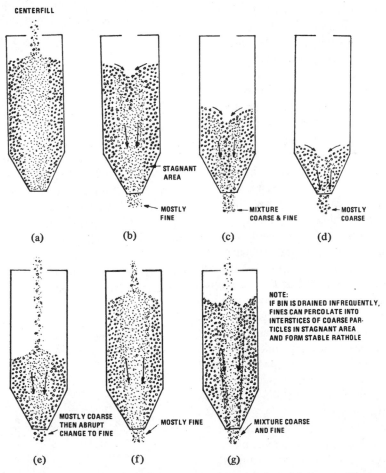

Figure 8.76. Typical segregation and discharge patterns during funnel flow; (a) Center filling, no discharge, (b) discharge begins; (c) discharge continues, level in bin dropping; (d) level continues to drop, heel discharging; (e) start to refill before heel is completely discharged; (f) level rising; (g) discharging at the same rate as filling, level remains unchanged. Note: Flow patterns shown are typical for a funnel flow bin when any free flowing (segregating or nonsegregating) material is stored.

level takes place. In time, if the silo is not emptied, fines can percolate into the coarse fraction in the stagnant region and this could cause stable ratholes to form.

Percolation also occurs when a mixture of particles are vibrated or agitated during conveying. This effect can be noticed in vibrating conveyors and chutes and small hoppers that are vibrated to promote flow.

8.12.1.2 Vibration

Williams[120] describes a condition other than percolation that can cause even a single large particle in a vibrated bed to rise to the surface. Each vertical movement of the bed allows fines to run in under the large particle. As the fine material accumulates and compacts, it supports the large particle, causing it to rise to the surface.

Ahmad and Smalley[125] studied the movement of a single 12,700 μm diameter lead ball in a vibrated bed of 500 to 600 μm dry sand particles. They reported that at a constant frequency of vibration, segregation increased as acceleration increased, but at a constant acceleration, segregation was reduced as frequency increased. Acceleration was the most critical variable affecting segregation.

Harwood[122] studied the behavior of cohesive and noncohesive powders subjected to vertical vibration using tracer powders comparable in size to the powder bed to determine segregation. He reported that particle size was the major controlling factor for segregation. In a binary system of free flowing and cohesive powders, segregation was very limited once the powder bed had become compacted, but if vibrational energy was sufficiently high to induce a semifluidized state in the bed, segregation was significantly increased.

Storage silos do not usually experience vibration with sufficient intensity to cause segregation, but small feed hoppers and chutes can.

8.12.1.3 Trajectory of Falling Particles

Material projected from a conveyor or chute onto a heap can segregate before impact due to differences in particle size, particle density and occasionally because of air drag effect.

Usually, if the material has segregating tendencies they have already occurred on the incoming conveyor or chute because of the mechanisms described above, and the trajectory of discharge serves to preserve this separation (Fig. 8.77).

Guidance on estimating the trajectory of material discharged from the head pulley of a belt conveyor can be found in Ref. 128. Calculated trajectories of a single fine particle usually have no practical significance in this case since it is not possible to account for the effects of air turbulence and particle to particle contact in a dense falling stream of material.

8.12.1.4 Impact on a Heap

After impacting on a pile, large coarse particles will tend to roll or slide over smaller coarse particles to concentrate on the outside. The more resilient larger particles will tend to bounce and also concentrate along the outside of the pile, while the smaller, less resilient particles will tend to concentrate in the center.

If the mixture contains sufficient moisture, the fine fraction will tend to stick on impact, and large particles will roll or bounce away.

If a mixture of fine powder and coarse particles is impacted directly onto a heap after discharge from an air slide,® or pneumatic

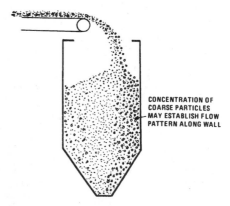

CONCENTRATION OF
COARSE PARTICLES
MAY ESTABLISH FLOW
PATTERN ALONG WALL

Figure 8.77. Trajectory segregation.

conveyor, the fines fraction can become aerated during a free-fall, and, on impact, will assume a very shallow or zero angle of repose. The heavier, coarse particles will concentrate in the impact area.

If the silo is loaded with a pneumatic conveyor impacting on the bed, the fine material can remain fluidized or can remain entrained in the moving air stream above the bed until loading is completed and the pneumatic conveyor shut down. If the silo is then completely emptied, the fine fraction that has settled at the top of the silo will discharge in a mass.

8.12.1.5 Angle of Repose

When a mixture of uniformly sized granular particles consisting of components with different angles of repose is poured on a heap, the particles having a steeper angle of repose tend to concentrate in the center of the heap.

8.12.2 Theoretical Analysis

Segregation is usually studied by sampling from a model bin or from a full-size bin and by reporting the results on a statistical basis. No theoretical basis for analyzing segregation mechanisms has yet been formulated, although some work in this area is beginning. Theoretical models to describe segregation by percolation have been proposed by Shinohara et al.[129, 130] and by particle size and density by Tanaka.[131] Matthee[132] has proposed an approach to modeling all aspects of segregation.

8.12.3 Minimizing Segregation

Particle rearrangement and segregation will occur each time a material is dumped onto a conveyor, or a chute at the loading or transfer points. Knowing the likely segregation mechanisms that will be present, the probable distribution of coarse and fine particles across and along the length of the conveyor coming into the silo can be predicted with reasonable certainty. This incoming stream then must be redistributed or mixed in the silo if a nonsegregated mixture is required at the silo discharge.

As long as the solids level in a mass flow bin remains above the transition, a distance equivalent to about three quarters the diameter, the material moves down the vertical section in plug flow with radial segregation relatively unchanged. However, radial mixing will occur in the hopper section before discharge, as noted earlier in Figure 8.75.

Mixing will not occur in a funnel flow bin. To minimize segregation in these bins, some means of redistributing the incoming material and/or changing the internal flow pattern are required. The BINSERT®[173] is such a device, and represents the newest and most important advance in the design of inserts and hopper geometry for the purpose of reducing particle segregation in storage bins. This is described in Section 8.13. A moving fill spout (Fig. 8.78a), a fixed deflector, flow spitter, or multiple loading spouts (Fig. 8.78b) have been used to distribute incoming material on the heap. A patented rotating device for this purpose is shown in Figure 8.79.

Devices to reduce segregation by changing the flow pattern are, in essence, designed to simulate mass flow as much as possible. An insert mounted high in the hopper section can widen the flow channel and assist in remixing (Fig. 8.80). Multiple discharge pipes (Fig. 8.80b) have been used to extract material from different segregated areas of the bin and recom-

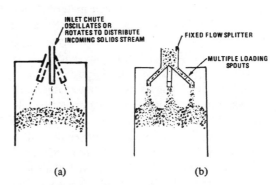

Figure 8.78. Devices to minimize segregation during filling of a bin; (a) Moving fill spout, (b) flow splitter or spreader-deflector.

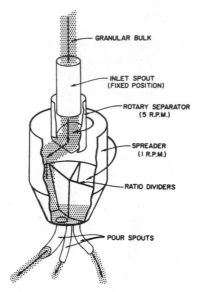

Figure 8.79. Rotary spreader (U.S. Patent 3, 285, 438). (From Ref. 134.)

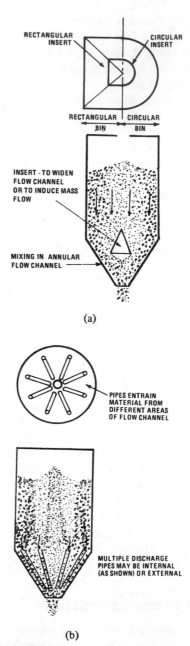

(a)

(b)

Figure 8.80. Devices to assist in mixing during discharge; (*a*) Inserts, (*b*) multiple discharge ducts.

bine them at the discharge point. A patented device for similar use is shown in Figure 8.81. A device patented by Fisher[133] is used for such a purpose. The Stock conical distribution chute is used to feed coal to stokers. The chute remains full at all times as the coal drains from a bin above to a spreader stoker below. Under these conditions, this device will produce very little segregation.

Van Denberg and Bauer[123] reported on studies of segregation of granular particles in model bins as the bins flowed from full to empty. They obtained quantitative data by filling the models with well-mixed material, then discharging and sampling continuously as the bin emptied. Each sample was analyzed with conventional sieving techniques and the results plotted as shown in Figure 8.82 with sample screen analysis as ordinate values versus sample order in terms of percent weight removed as abscissa values. Figure 8.82a shows segregation patterns typical of center filled funnel flow bins. Figures 8.82b and 8.82c show potential improvements with a well-placed insert or with multiple point fill. Although flow properties of the materials are not reported in the Van Denberg and Bauer article, it is ap-

parent that the bin geometry and discharge arrangement shown in Figure 8.82d producing mass flow and segregation at discharge was markedly reduced. Figure 8.82e shows the serious segregation effects that can be induced by filling and discharging near a vertical baffle.

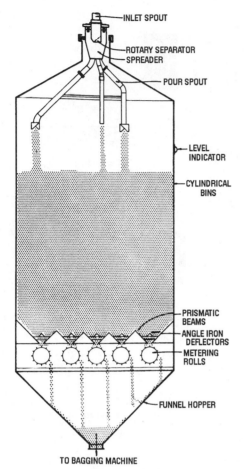

Figure 8.81. Metering rolls (U.S. patent 3, 285, 438). (From Ref. 134.)

8.13 STATIC DEVICES TO PROMOTE GRAVITY FLOW FROM BINS

8.13.1 Binsert®

The Binsert®[175] is formed by positioning a mass flow hopper in a funnel flow hopper as shown schematically in Figure 8.83. The inner hopper is configured for mass flow using the design procedures described earlier. It has been found that flow will occur in the inner hopper, as well as in the annular space between hoppers, when the slope angle of the (outer) funnel flow hopper is up to a maximum of twice the slope angle θ_c, of the mass flow insert hopper. Binserts have also been constructed for plane flow (wedge-shaped) hop-

pers. The positioning of the inner hopper and the configuration of the outlet of both hoppers determine the flow pattern in the bin. It can be made to cause mass flow to minimize segregation and provide a controlled flow, or it can be made to provide a velocity gradient between the inner and outer flow channels. With a velocity gradient (center channel moving faster), in-bin blending is possible. A Binsert® bin requires less vertical space than a conventional mass flow bin.

Granular particles and powders, free flowing and cohesive, have been handled in these in bins. A Binsert® can be retrofitted into an existing funnel flow hopper, but the higher stresses imposed on the structure by the change to mass flow, and the internal support for the inner hopper, must be considered in the retrofit.

8.13.2 Other Inserts

It has been known for a number of years that correctly placed inserts can solve flow-related problems in silos. Newton,[137] in 1945, described the use of perforated trays and inclined pipes to provide even distribution (mass flow) of a granular catayst in a moving bed. Morse[138] described the sizing and placement of inserts placed on the vertical axis of a vessel, near the junction of a cone and a vertical shell, to cause mass flow in a moving bed when shallow hoppers are used.

Sizing and placement of inserts in bins have generally been based on rules-of-thumb, or have been found by trial and error. Johanson[139–141] proposed a method of sizing and placement based on the bulk solids flow properties and hopper geometry. This work predates the more recent invention of the Binsert®. It is summarized in the following paragraphs, and is the most specific guide to insert placement that has appeared in the literature.

Johanson reasoned that since an insert forms an annular opening that approaches a long slot opening (Fig. 8.84), a plain-strain wedge-shaped hopper is closely approximated

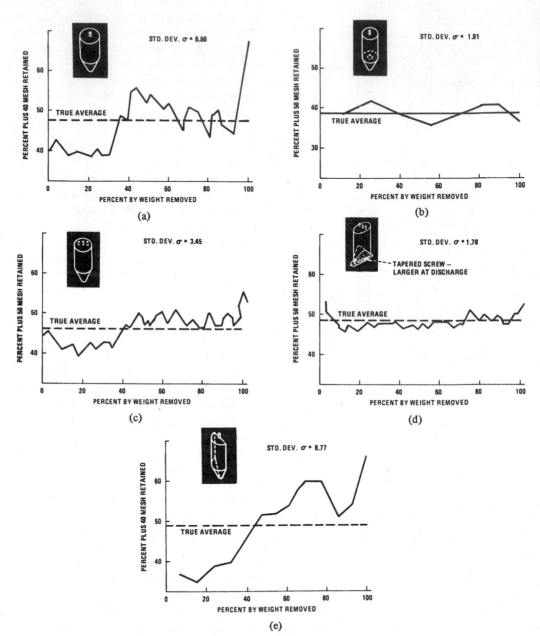

Figure 8.82. Segregation patterns in model bins handling granular material: (*a*) cylindrical unit with 60° cone bottom; axially filled, then axially discharged completely, (*b*) cylindrical unit with 60° cone bottom, well-sized and located insert, axially filled and then axially discharged completely, (*c*) cylindrical unit with 60° cone bottom, filled through three points, then axially discharged completely, (*d*) cylindrical unit with symmetrical wedge bottom, filled through three points, then discharged completely by uniform withdrawal across the slot discharge opening, (*e*) cylindrical unit with 60° bottom and vertical partition, filled and discharged completely through openings adjacent to the partition. (From Ref. 123) (Excepted by special permission from *Chemical Engineering*, copyright © 1964 by McGraw-Hill, Inc., New York, N.Y.)

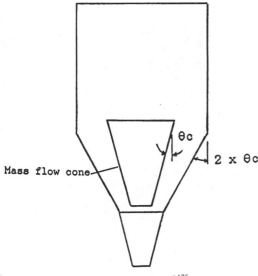

Figure 8.83. Binsert.[®175]

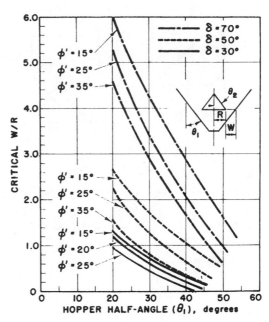

Figure 8.85. Approximate critical W/R for inserts and hoppers having the same slope.

in the area around the insert. Since flow will occur along the walls of wedge-shaped hoppers at relatively shallow slopes, an insert can change a funnel flow pattern to flow along the walls in the region influenced by the insert as shown in Figure 8.84. The critical ratio of the dimensions W/R given by Johanson and shown graphically in Figure 8.85 was calculated for the case where $\theta_2 = \theta_1$. Johanson states that the values shown in the figure will give a good approximation for any insert slope angle. The

angle between the insert wall AB and the line AC to the point at which flow occurs along the hopper wall is presented in graphical form (Fig. 8.86) as a function of the total included angle $\beta = \theta_1 + \theta_2$, assuming a symmetric channel. (The value of α is approximately the same for nonsymmetrical channels where θ_1 and θ_2 are not equal.)

8.13.2.1 Inserts to Minimize Segregation

An insert designed to eliminate segregation during withdrawal from a funnel flow bin must be placed high enough in the bin near the junction of vertical wall and cone so as to cause the entire mass in the vertical section to move uniformly.

Such an insert is shown in Figure 8.87. With materials that will not arch or rathole, the insert is designed by Johanson's method as follows:[87]

1. Select an insertslope angle θ_2. A horizontal flat plate can be used if cleanout is not a consideration.

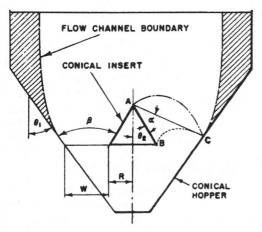

Figure 8.84. Insert geometry and placement. (From Ref. 139.)

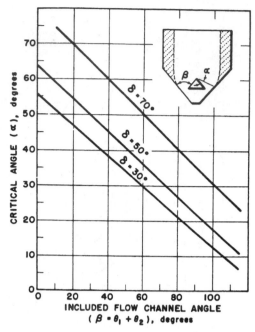

Figure 8.86. Approximate angle α to determine limit of flow along hopper walls.

2. Determine critical W/R and α values using solids flow properties and Figures 8.85 and 8.86. Johanson suggests adding a safety factor to the design by reducing the critical W/R by 10%.
3. On a sketch of the silo, draw the line AB having as its slope the angle $(\pi/2 - \alpha - \theta_2)$ from the horizontal. Draw line CD through the vertex at angle a, where $\tan a = \tan \theta_1/(1 + WR)$. Points on this line represent critical values of W/R.
4. Draw line BE at slope angle θ_2 to determine point E, the bottom of the insert.

8.13.2.2 Insert to Widen the Flow Channel

The same procedure can be used to determine insert size and placement low in the hopper to widen the flow channel and reduce stagnant areas in a funnel flow silo (assuming the materials will not bridge or rathole). The diameter of the desired flow channel locates the approximate point C (Fig. 8.84).

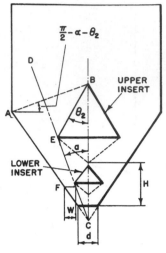

Figure 8.87. Placement of inserts.

8.12.1.3 Inserts with Cohesive Materials

An insert can be used to prevent ratholing by providing a vertical flow channel greater than the critical rathole diameter. To design this insert, first determine the minimum opening required to prevent an arch forming over a circular opening in a funnel flow bin, as described earlier. The critical dimensions W should be no less than three fourths of this minimum opening. If this insert is placed high in the hopper, a rathole may form below if the angle of repose of the material allows a depth of material above the hopper opening greater than the diameter of the opening. Johanson suggests a second, lower insert may be necessary to prevent this (Fig. 8.87).

Inserts placed in the hopper section usually do not cause higher overpressures (stresses) on the hopper wall. Inserts that project up into or are located in the cylinder portion can cause overpressure (stresses) on the cylinder walls because of the presence of the flow channel transition at the insert.

8.13.3 BCR Easy-Flo Bin

Large discharge outlets are usually required to prevent bridging over the outlet in bins used to store fine coal, particularly when surface moisture is present. Bituminous Coal Research Inc. (BCR) has developed the Easy-Flo

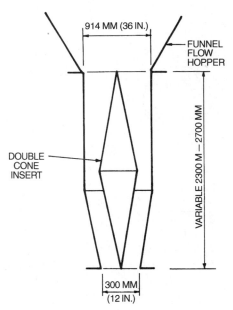

Figure 8.88. BCR Easy-Flow® cone. (Ref. 136.) (Reprinted with permission of Bituminous Coal Research Association.)

bin[136] for promoting flow of this material (Fig. 8.88). BCR reports that the interior double cone insert in this device controls pressures and flows such that the solids can be made to converge from the large silo opening to the small opening on the bottom. By reducing the coal outlet, the size and cost of associated

valves and feeders can be reduced. The silo hopper opening to which such a device is to be attached must be larger than the critical opening for arching or ratholing.

8.13.4 Reimbert "Antidynamic Tube"

This is a vertical perforated tube developed by the Reimberts and is mounted above the discharge opening in a bin (Fig. 8.89). During solids discharge, only the top layer of material moves, sliding down the surface into the top perforation. Since the tube is full, the material below this point is prevented from entering through the lower perforations and thus remains stationary. As discharge continues, each top layer is successively discharged as another tube perforation is exposed. This device, of course, can be used only with noncohesive materials that will flow freely through the perforations.

The antidynamic tube was developed as a response to structural failures of bins caused by large overpressures for which they were not designed. These overpressures were thought to be caused by mass flow conditions that were not anticipated and/or not understood by the silo designer. These include "effective transitions" occurring during flow, and off-center discharge. The Reimbert tube enforces a fun-

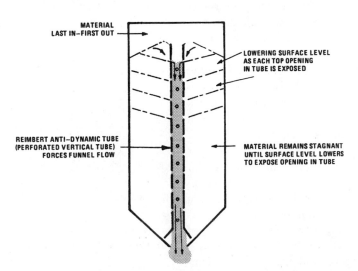

Figure 8.89. Reimbert antidynamic tube. (From. Ref. 2.)

nel flow pattern so that most of the dynamic flow pressures will not extend to the silo walls. The tube has been installed as part of the repair of damaged bins and in new installations to ensure funnel flow.

Other benefits have been found with the use of the tube. When solids are introduced into a bin through the tube and allowed to flow laterally through the tube openings, instead of falling and impacting onto a heap, particle segregation is reduced. It has also been found that with the tube installed, the vibration effects that occur due to unstable flow channels (Section 8.4) are reduced.

The tube concept, with entry ports modified to allow simultaneous flow from several levels, is used for blending in silos. Schulze and Schwedes[176] experimented with similar tubes in model bins and reported that the tube can be applied to increase mass flow discharge rates.

8.13.5 Special Hopper Geometries

8.13.5.1 Diamondback Hopper ®

The JR Johanson Inc. Diamondback Hopper® is constructed with a unique, patented[178] geometry, designed to prevent ratholing or arching in a conveying flow channel. The basic hopper unit is formed by assembling two or more bin sections, of similar shape, in a telescoping arrangement. The linear dimension of each succeeding section increases so that the bottom of each section fits the top of the one below it, with the smallest forming the hopper bottom. The Diamondback® hoppers are configured for specific applications, and the geometry is determined by the measured frictional properties of the solids.

The Arch-Breaking Diamondback® hopper (Fig. 8.90a) is for mounting under a mass or funnel flow hopper, to converge the flow channel, in mass flow. For this application, each hopper section is configured for one-dimensional convergence, with one set of opposite walls in each section that is vertical or slightly diverging, and the other set converging in a circular format. For example, with

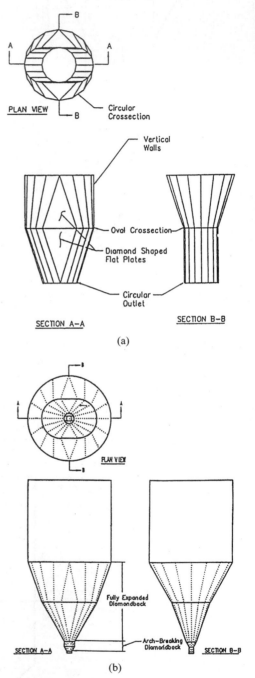

Figure 8.90. (a) Arch-Breaking Diamondback® hopper. (b) Expanded Flow Diamondback® hopper.

a two-section hopper, the upper section would have a circular inlet, transitioning to an oval cross-section where it joins the lower section. The matching oval inlet on the bottom section would transition to a circular discharge opening. The flow channel therefore changes from circular to oval to circular again. The manufacturer states that this configuration makes it possible to significantly reduce the size of the discharge opening, compared to that required for no arching or ratholing in a conventional mass flow hopper.

The fully expanded version of the Diamond back is based on the same principle as the arch-breaker version, except that all surfaces are converging, in accordance with mass flow principles similar to those of the transition hopper. This reduces the overall height that otherwise would be required for a conical mass flow hopper.

8.13.5.2 Concrete Bins

Theimer[142] describes a variety of silo and hopper geometries, mostly for concrete silos, developed through trial and error, that have proven useful for promoting gravity flow of bulk solids. Most of the examples cited refer to storage of poor-flowing grain and food products in large concrete silos. In these large concrete silos, mass flow hoppers in many cases are prohibitively expensive. However, by taking into account the flow properties of the solids and by judicious shaping and proportioning of the silo bottoms, the structures described in this article are successfully storing and discharging poor-flowing materials.

The design criteria for shaping bin and hopper geometry to improve flow include the following:

1. When handling poor-flowing powders in hoppers having a rectangular or square cross-section, avoid sloping walls that intersect to form a valley angle. These materials will not flow in the region of the valley angles and will cause ratholes to form.

Design so that inclined hopper surfaces intersect with vertical wall surfaces.
2. To promote flow of cohesive materials in large silos provide a design that reduces the consolidating pressures and allows expansion of the material as it flows through the lower hopper area. This can be accomplished by the inserts described previously, by pressure relief "noses," or by expansion of the hopper cross-section at the junction of hopper and vertical section. Other hopper geometries are reviewed by Reisner and Eisenhart.[143]

8.14 FLOW-PROMOTING DEVICES AND FEEDERS FOR REGULATING FLOW

Selection and design of feeders or other flow control devices to be installed at a bin outlet must be considered to be an integral part of the storage bin design. Feeders should be designed to withdraw material uniformly from the entire area of the discharge opening. This will ensure the largest possible flow channel in a funnel flow bin. It is a mandatory requirement for a mass flow bin. If the entire opening is not active in such a bin, mass flow will not occur.

The minimum opening size and shape required to ensure flow from a bin must be determined before selecting a discharge device or feeder. It is not always correct to select the feeder and then match the hopper opening to it. Feeders are usually rated by manufacturers on the basis of volumetric capacity. If the feeder selected on this basis has an inlet smaller than the minimum required hopper opening size, it is unacceptable. Too small an opening could result in bridging, ratholing, and erratic flow. Selecting the feeder on the basis of opening size, therefore, may require a unit that is considerably "oversized," and operate at low speed.

8.14.1 Basic Feeder Types

The most commonly used types of feeders are described below. An important selection criterion for any of these devices that are to be used at the discharge of a hopper or bin is that the device and any connecting hopper be configured such that solids are withdrawn from across the entire hopper opening. Other solids feeders are described in Ref. 143.

8.14.1.1 Rotary Feeders (Also Called Rotary Vane Feeder or Star Valve)

These can be used as a volumetric feeder and/or a gas pressure seal (air lock) to pass solids from one pressure environment to another (Fig. 8.91). It can be used under a circular, rectangular, or slot opening (Fig. 8.92). These valves are well suited for feeding materials that tend to flush or aerate in funnel flow bins, since they can be machined with close clearances between rotor and housing. The pockets of a rotary valve fill on the rising side of the rotor when the rotor is under a head of solids. Therefore, when the valve is mounted under a bin, withdrawal across the bin opening can be made more uniform by either of two arrangements. Separate the valve from the bin opening by a connecting spout having at least the same cross-section of the opening. The length of the spout should be at least two times the bin opening as viewed along the axis of the valve rotor, to allow the solids flow channel into the rotor to diverge upward to meet the full opening of the bin. An alternative arrangement would be to direct the material to the rising side of the rotor, similar to that shown for slot openings in Figure 8.92.

Where the material properties dictate a large bin opening to prevent arching or ratholing, a shallow-pocket (filled-pocket) rotor is often used instead of the standard rotor (Fig.

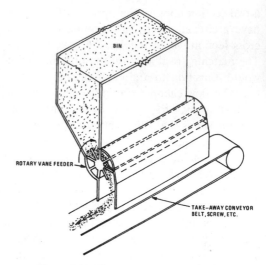

Figure 8.92. Slot-type rotary vane feeder.

8.93). When the valve opening is oversized to meet material property requirements, the volumetric capacity of the standard rotor becomes so large that very high drive speed reductions are required. In most cases, it is less expensive to reduce the capacity of the rotor with the shallow-pocket design and operate at speeds that require a more moderate, less costly drive.

Where a rotary valve is to be used to feed into a pneumatic conveyor there are additional considerations. Close clearances between rotor and body are required. Pellets or granular material can jam in these clearances. This can be prevented by using a side-entry (pellet) valve (Fig. 8.94) or a pellet shield with a flow control gate to meter the material and

Figure 8.91. Drop-through rotary feeder.

Figure 9.93. Drop-through rotary feeder with filled rotor.

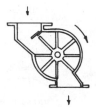

Figure 8.94. Side-entry rotary feeder.

prevent filling of the rotor pockets. When feeding material into pressure pneumatic conveyors, the gas leakage past the rotor clearances will greatly exceed the pocket displacement and will pass up into the incoming material. In some cases, this gas "fluffs" the material in the bin and assists flow. In most cases, however, this gas impedes solids flow and must be *vented*. It can be vented through a connection to the inlet feed section or through a connection in the valve housing.

8.14.1.2 Screw Feeders

Screw feeders handle a wide range of materials from lumps to powder, are relatively inexpensive, are easily enclosed to be dust tight, and easily accommodate slot openings. They will not seal against an uncontrolled flow of "flooding" fine powders and normally operate with a zero or low-pressure differential between outlet and inlet. Special designs have been made for feeding certain materials at pressure differentials up to 100 kPa.

If the required bin discharge opening determined from solids flow properties is very large, it may be necessary to use several parallel screws in a slot opening. No matter how many screws are used, they must be designed to promote uniform withdrawal from the hopper above. Nonuniform withdrawal can lead to solids arching, ratholing, or, if a rathole collapses, to *flushing through the screw*.

A standard screw feeder has a pitch-to-diameter ratio of 1. This ratio is satisfactory only for withdrawing uniformly from openings where the maximum dimension does not exceed 1 to $1\frac{1}{2}$ pitches. If the hopper opening exceeds this, solids flow will occur at the back

end of the screw, causing a channel or funnel flow to occur at that point (Fig. 8.95). There are several screw configurations that can be used to promote uniform withdrawal from slot openings.

Increasing Pitch (Fig. 8.96a). Each pitch progressively increases in the direction of flow, until the flight is out into the conveying section, away from the hopper opening. On long slot openings, exceeding about 5 to 6 screw diameters in length, successive increases in pitch may not be completely effective in achieving uniform withdrawal along the slot.

Increasing Flight Diameter (Fig. 8.96b). The diameter of the screw flight increases in the direction of flow. Some manufacturers offer this as a standard preengineered single or twin screw assembly to reduce cost. It can be very effective, but the powder properties must be such that they will not bridge over the smaller opening at the rear end of the screw.

Increasing Pitch with Decreasing Shaft Diameter (Fig. 8.96c). The flight pitch increases in the direction of flow, while the shaft diameter decreases in the direction of flow. This design is effective over long slot opening.

8.14.1.3 Vibrating Feeders

Vibrating feeders provide precise feed control, handle material gently, are self-cleaning, and can handle hot materials. They normally operate at frequencies from 12 to 60 cps and strokes to about 10 mm. There are two general

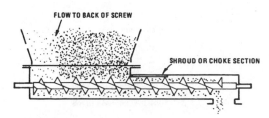

Figure 8.95. Screw feeder—uniform pitch—produces a poor flow pattern under a slot opening.

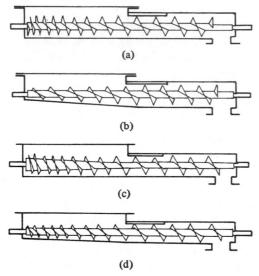

Figure 8.96. Various screw feeder geometries producing improved flow patterns under slot openings. (*a*) Increasing flight pitch, (*b*) increasing flight diameter, (*c*) increasing pitch with decreasing shaft diameter.

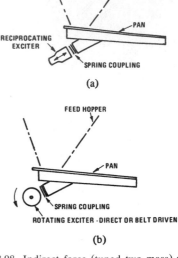

Figure 8.98. Indirect force (tuned two mass) vibrating feeders; (*a*) Electromagnetic feeder, (*b*) electromechanical feeder.

types of feeders: the direct force (single mass) machine (Fig. 8.97) and the indirect force (tuned two-mass) machine (Fig. 8.98).

A rotating counterweight or reciprocating piston causes the motion in a direct force feeder. Essentially a constant rate machine, it is low cost and can handle a wide range of particle sizes from lumps to damp fines, but does not provide precise flow control.

The vibrating forces from an exciter mass are amplified by a spring mass system to vibrate the trough of an indirect force feeder. This design is most commonly used since it provides the best control of solids flow, uses the least power, and normally requires less maintenance than the direct force machine.

Two excitation systems are in common use: electromagnetic (Fig. 8.98a) and electromechanical (Fig. 8.98b).

In the electromagnetic feeders, an alternating or pulsating direct current drives a vibrator that is coupled to the pan through metal or fiberglass leaf springs. These machines have a short stroke (approx. 0.1 mm) and high frequency (50 to 60 Hz). Feed rate is adjusted by voltage control using a rectifier and rheostat, or variable voltage transformers. Very precise control from 0% to 100% and almost instantaneous shut-off of solids is possible with this machine.

In electromechanical feeders, an electric-motor-driven eccentric weight, coupled through mechanical, elastomer, or pneumatic springs, drives the pan. These machines can have strokes up to 0.10 mm and run at frequencies varying from 12 to 17 cps. Feed control can be accomplished by varying speed of the motor through a variable voltage control circuit, automatic changing of eccentric weight loading, or varying air pressure in pneumatic couplings. Some of these feeders can control

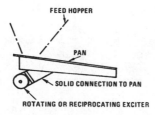

Figure 8.97. Direct force (single mass) vibrating feeder.

solid flow from 20% to 100% of rated capacity; other models are capable of 5% to 100%.

Hoppers above vibrating feeders must be designed correctly to properly deliver the solids to the feeder trough. Improperly designed hoppers can put unnecessary loads on the feeder pan and can cause compaction of the material in the hopper opening, all of which can significantly reduce feeder capacity. Suggested methods of hopper outlet design for delivering to vibrating feeders are given in Refs. 144 and 145, Figure 8.99 for circular openings, and Figure 8.100 for rectangular or slot openings.

The performance of a vibrating feeder is more sensitive to particle properties than any of the feeders discussed. Special consideration should be given to fine powders and powders that tend to aerate. These powders often move at very low rates on vibrating pans. They can deaerate on vibrating surfaces and only the top layer of material will move. They can also flush through an improperly designed inlet hopper. It would be prudent to test fine powders on a vibrating feeder before specifying their use.

8.14.1.4 Belt Feeders

Belt feeders can withdraw material from very long slot openings in bins, can be designed to take very heavy impact and solids loads from

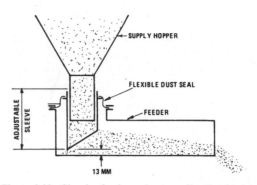

Figure 8.99. Circular feed opening to a vibrating feeder.

the hopper above, can be combined with weigh decks or weigh idlers to gravimetrically meter flow and will handle practically any solid. The belt is usually a fabric or elastomeric covered fabric reinforced band, riding on a slider bed or rollers.

Improperly designed feed hoppers over the belt can cause solids compaction, belt wear, and high horsepower demand (Fig. 8.101). Belt feeders having hoppers designed so that the opening diverges in the direction of flow have proven successful for handling a variety of granular and powdered material through long slots (Fig. 8.102).

When handling very abrasive materials or large lumps, an apron feeder may be used in place of a belt feeder. In this device, the carrying surface is made up of over-lapping

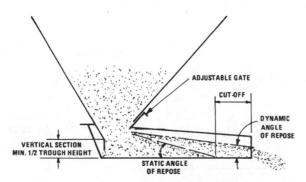

Figure 8.100. Rectangular or slot feed opening to a vibrating feeder.

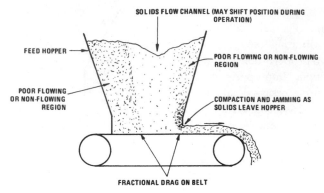

Figure 8.101. Poorly designed feed hopper over a belt feeder.

metal pans, supported on each side by a driven roller chain, riding on steel tracks.

8.14.2 Feeder / Flow-Promoting Devices

A number of flow-promoting devices and special feeders have been developed for specific applications. Several commonly used devices classified by the principal method used to induce flow are described below.

8.14.2.1 Vibratory-Type Devices

Vibrating Bin Bottom or Bin Discharger (Fig. 8.103). This is a conical hopper mounted beneath the opening in a bin or hopper and suspended from the bin or hopper by elastomeric-bushed links. Elastomeric bands connect and seal the inlet to the bin above and to the feed device or chute below. Motor-driven eccentric weights, mounted on the vibrating hopper, cause it to gyrate in an elliptical path on a horizontal plane. The frequency is fixed by the rotational speed of the weights: amplitude is varied by positioning of the weights. Frequencies vary from 15 to 50 Hz, but 15 to 30 Hz are most commonly used. Weight positioning (amplitude) is determined by solids flow characteristics, density, and amount of material in the bin, and is based on experience

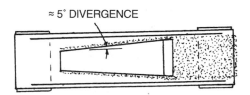

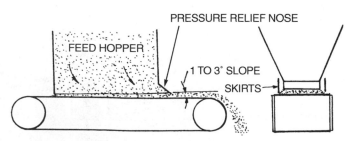

Figure 8.102. Well-designed feed hopper over a belt feeder.

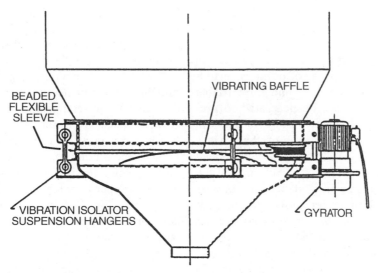

BEADED
FLEXIBLE
SLEEVE

VIBRATING BAFFLE

VIBRATION ISOLATOR
SUSPENSION HANGERS

GYRATOR

Figure 8.103. Vibrating hopper (Reprinted with permission of VibraScrew Corp.)

with similar materials or from tests on small hoppers.

A pressure cone or baffle mounted axially within the unit vibrates with the hopper and serves two purposes: It reflects the vibratory motion up into the material in the bin above and prevents solids compaction at the outlet by shielding the outlet from direct pressure from the solids.

In the Whirlpool® vibrating hopper configuration (Fig. 8.104), two motor-driven vibrators with their axis of rotation inclined to the horizontal plane are mounted 180° apart on the hopper. The action of the vibrators impart

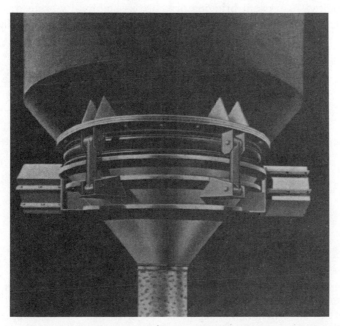

Figure 8.104. Whirlpool vibrating hopper. (Reprinted with permission of Carman Mfg. Co.)

a twisting and lifting motion that can be effective in inducing flow of very sticky or cohesive materials.

Vibrating hoppers can be very effective in promoting flow of a variety of powders, including those that agglomerate and form friable lumps, those that must be deaerated to prevent flooding, and cohesive powders that will not flow by gravity. Because of their heavy rugged design, they will accept very high head loads from material in the bin. Selection of the hopper inlet diameter is dependent on the solids flow pattern desired in the bin above. If mass flow is required, the vibrating hopper can be sized to match the full cylinder diameter, or it can be mounted at the discharge of a conical mass flow hopper. This latter arrangement is useful where a mass flow hopper would require inordinately large openings for gravity flow. The vibrating hopper in that case can be used to converge this flow to a smaller outlet.

If funnel flow in the bin above is acceptable, the vibrating hopper can be sized large enough to expand the flow channel to the desired size, and converge to a small discharge opening.

The inlet dimension of a vibrating hopper must be sized to be larger than the minimum hopper outlet required to prevent bridging or ratholing in the hopper or bin under which it is mounted. This dimension can be determined by first making allowances for possible compaction due to vibration and using the techniques described in Section 8.8.

If vibrating hoppers are improperly applied, bridging can occur above the hopper or powders can be overcompacted and flow at very reduced rates from the outlet. Conversely, flooding can occur if ratholes form and collapse in the bin above, or if hopper flow rate, discharge nozzle size, and hopper amplitude are not properly matched. The Metalfab bin discharger (Fig. 8.105) features a secondary adjustable baffle designed to prevent overcompaction at the outlet.

Since it is a vibrating device, care should be taken to prevent transmission of the vibration into building structures when designing supports for bins having large bin dischargers.

An additional feed device must be installed at the vibrating hopper outlet to achieve accurate flow control.

Vibrating Screw Feeders (Fig. 8.106). In this feeder, a screw and trough assembly are mounted on an elastomeric isolation system. Motor-driven eccentric weights cause the entire assembly to vibrate or oscillate in a rocking motion. This keeps the material in the

Figure 8.105. Metal Fab bin activator, vibrating hopper for attachment to hopper opening. (Reprinted with permission of Metal Fab Inc.)

Figure 8.106. VibraScrew® feeder. (Reprinted with permission of VibraScrew Corp.)

hopper section above the screw in motion, preventing bridging or channeling, and a more consistent solids density is achieved as the material flows into the vibrating screw. This action also permits feeding into very small screw feeders at low rates. The rate is controlled by screw size and speed. Frequency of vibration is fixed and amplitude, determined by test or experience, is set by eccentric weight positioning.

Vibrating Louver-type Discharger-Feeders.
There are two general types of these devices.

In the Silleta® and Superfeeder® design, a feed tray is suspended from a frame fastened to a silo outlet as shown schematically in Figure 8.107a. A row of fixed position, inclined blades, mounted in a feed tray, divides the flow area into a series of powder feed slots. The feed section reciprocates in response to an electromagnetic or electromechanical vibrator to provide, in theory, an infinite variability in feed rate. The fixed blade dimensions, inclination, and spacing are determined by the powder tests, to ensure that powder will flow during vibration, and stop when the vibration stops. These devices combine the function of a bin discharger, and a feeder to regulate the flow. Because they extract solids from the entire cross section of an opening, they can be used at the outlet of mass flow silos. They are fabricated to accommodate round or square openings, ranging from 0.15 m to 1.5 m in diameter or width.

The Hogan® discharger (Fig. 8.107b) is similar to those described above, except that in addition to varying the vibrator stroke, the blade positions can be adjusted to any position between closed (zero flow) to fully open (maximum flow), by manual, electric, or pneumatic actuators, while the unit is operating.

Thayer "Bridge Breaker" (Bin Discharger) (Fig. 8.108).
Expanded metal or perforated metalscreens in this device are positioned inside the hopper and parallel to the walls. They are attached by studs to externally mounted, low-frequency, high-amplitude air vibrators.

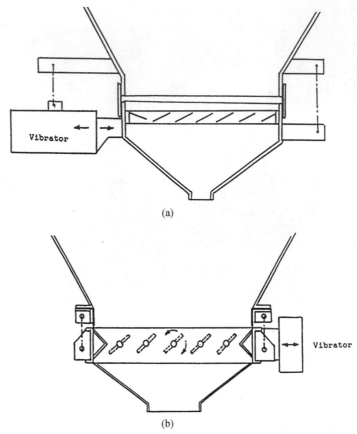

(a)

(b)

Figure 8.107. Vibrating louver-type discharger feeder. (*a*) Silleta,® and Superfeeder® fixed blade tray. (*b*) Hogan® adjustable blade.

The studs pass through and are supported by resilient elastomeric wall mounts. When activated, the vibrators agitate the screens in a reciprocating motion almost parallel to the plane of the hopper wall. Since this motion puts most of the energy directly into the material instead of the hopper walls, this device uses less energy and makes less noise, compared to standard bin vibrators.

Matcon-Buls® Discharger Valve. This device, shown in Figure 8.109, is in the form of an inverted cone, mounted on a pneumatic spring-actuator in a truncated hopper body, bolted to a bin outlet. The cone is raised in the hopper section, by pressurizing the pneumatic spring. This opens an annular gap be-tween the hopper wall and cone, allowing solids to flow. Solids flow rate is regulated by the positioning of the cone. To promote flow, a pneumatic piston vibrator, mounted inside the cone, is actuated while the cone is in the raised position.

The hopper units range from 1 to 10 ft in diameter.

8.14.2.2 Agitation-Type Devices

Acrison Bin Discharger (Fig. 8.110). In this unit, helical agitators turn at 1 to 2 rpm to prevent consolidation and maintain the solids in a flowable condition. Solids discharge from one or more openings on the bottom, with no control of rate. A variety of feeders can be

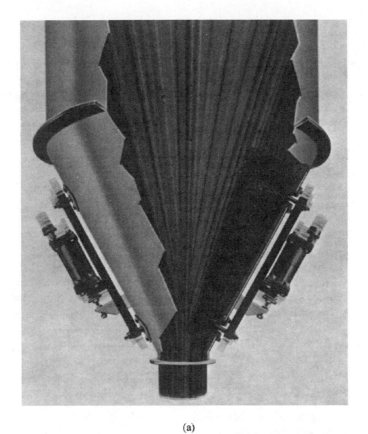

(a)

(b)

Figure 8.108. Thayer Bridge Breaker: (*a*) assembly of two units on a conical hopper, (*b*) internal view into conical hopper. (Permission Thayer Scale Co., Hyer Industries, Inc.)

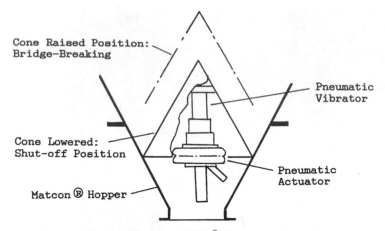

Cone Raised Position:
Bridge-Breaking

Pneumatic
Vibrator

Cone Lowered:
Shut-off Position

Matcon ® Hopper

Pneumatic
Actuator

Figure 8.109. Matcon-Buls® discharger valve.

mounted at the discharge to control feed rate. The size of the unit is governed by the maximum opening required to prevent bridging in the hopper opening under which it is mounted.

Acrison Bin Discharger Feeder (Fig. 8.111). This device combines the fixed-speed bin discharger to induce flow to a variable-speed screw. Feed rate is controlled by screw speed.

Acrison Feeder (Fig. 8.112). A slow-moving concentric ribbon (or agitator) "condi-tions" the solids as they enter the feed screw. This controlled agitation maintains density at a consistent level, reduces the tendency to arch or bridge over the feed opening, and permits feeding into very small screw feeders at low rates. Rate is controlled by screw speed and/or screw size.

Metal Fab Feeder (Fig. 8.113). Specially configured agitators mounted on the feed screw loosen the material in the feed hopper, prevent arching, and induce a consistent flow

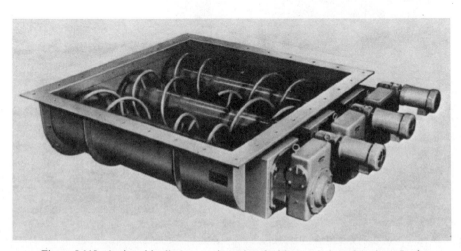

Figure 8.110. Acrison bin discharger. (Reprinted with permission of Acrison, Inc.)

Figure 8.111. Acrison bin discharger-feeder. (Reprinted with permission of Acrison, Inc.)

through the feed screw. This also permits feeding into small screws at low rates.

K-Tron Twin Screw Feeder (Fig. 8.114). The feeder uses a mechanical agitator to prevent

bridging as solids flow into wiped-surface corotating screws. Specially designed flights on these screws intermesh in close proximity to provide a wiping action that aids in discharge of sticky or cohesive materials. The feeder

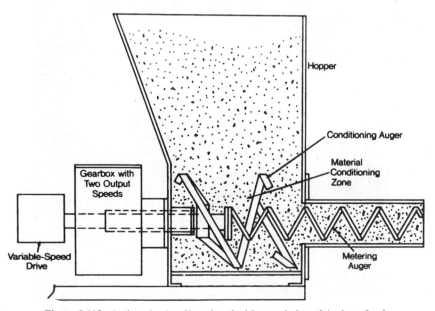

Figure 8.112. Acrison feeder. (Reprinted with permission of Acrison, Inc.)

Figure 8.113. Metal Fab feeder. (Reprinted with permission of Metal Fab, Inc.)

offers good metering capability with a variety of cohesive and noncohesive powders.

Agitated Bin Unloader (Fig. 8.115). A single shaft-mounted ribbon or agitator can be used to prevent bridging and to promote flow from a bin outlet. The agitator can be installed in the hopper, or it can be mounted in a separate housing that is installed at the bin outlet. This device can handle a variety of powders, flake and fibrous materials; it can be mounted on bins that are susceptible to arching because process needs dictate the use of a small discharge opening. Control of flow requires adjustable gates or feed devices be installed at the outlet.

8.14.2.3 Force-Extraction Devices

Stephens-Adamson Circular Bin Discharger with Arch Breaker (Fig. 8.116). In the discharger section, rotating fingers extract a layer of material and move it toward a gravity discharge chute. A centrally mounted arch breaker extends up through the discharger into the conical portion of the bin.

The arch breaker, driven from below through a universal joint, rotates slowly on its

Figure 8.114. K-Tron twin-screw feeder. (Reprinted with permission of K-Tron Inc.)

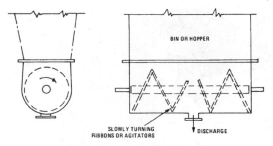

Figure 8.115. Agitated bin discharger.

axis to break up arches that may occur, and also induces flow to the discharger. This device works well on many powdery granular, flaky, and fibrous materials, as long as bridging does not occur in the cylindrical section of the bin above the arch breaker. This unit can operate with the discharge completely full since material not removed will recycle through the discharger. Multiple discharges can be furnished. This is a combination bin unloader

and rough feeder and can be equipped with a variable speed drive.

Sweep-Arm Unloader (Fig. 8.117). The sweep arm, a chain equipped with teeth, rotates to sweep the flat bottom of a bin, dragging material to a center opening where it discharges to a second traveling chain that transports it to a discharge chute. This combination bin unloader and rough feeder has been used with powders and granular and flaky materials. It can be damaged by abrasive materials and flaky materials that tend to build up between chain and sprocket. It can be equipped with a variable speed drive, although it is not intended for precise feed control.

Some manufacturers offer a horizontal screw (in place of the chain) that rotates about the bin centerline to sweep the flat bottom and draw material into a central discharge opening (Fig. 8.117). These sweep-type unloaders are

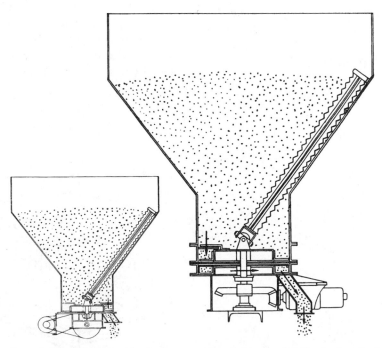

Figure 8.116. Stephens-Adamson circular bin discharger with arch breaker, (a) Single-stage feeder, (b) two-stage feeder. (Reprinted with permission of Stephens-Adamson Div., Allis Chalmers Co., Aurora, IL.)

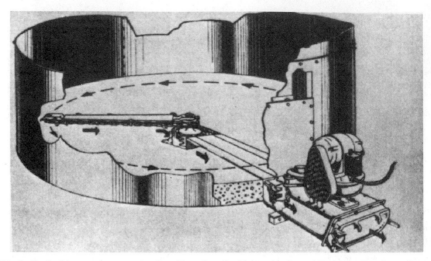

Figure 8.117. A. O. Smith sweep-arm unloader. (Reprinted with permission of Koppers Co., Inc., Sprout Waldron Div.)

used primarily with solids that tend to bridge, or to gain increased volumetric capacity at reduced cost by allowing a flat-bottom in place of a conical-bottom hopper.

Multiscrew Unloader / Feeder (Fig. 8.118). A series of parallel screws, *proportioned for proper slot flow*, as described earlier, can be used to provide a large, fully active discharge. This device is particularly useful for feeding

and discharging sticky, very cohesive or compacting-type solids that require very large hopper openings, or those that require vertical or negatively sloped hopper walls.

Rotary Table Feeder (Fig. 8.119). The table, a circular plate, rotates at about 2 to 10 rpm below a hopper opening. Material flowing onto the plate is discharged over the edge of a fixed plow. Flow can be regulated by changing the

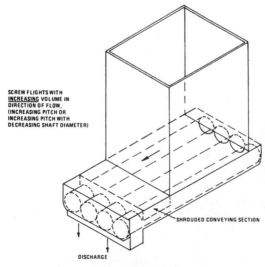

SCREW FLIGHTS WITH INCREASING VOLUME IN DIRECTION OF FLOW, (INCREASING PITCH OR INCREASING PITCH WITH DECREASING SHAFT DIAMETER)

SHROUDED CONVEYING SECTION

DISCHARGE

Figure 8.118. Multiscrew unloader.

Figure 8.119. Rotary table feeder.

height of an adjustable feed collar or the speed of the plate.

For best flow control and to keep the hopper opening fully active, the feed collar should be high enough to allow flow from under the entire perimeter of the collar onto the rotating plate. The device is not intended for precise feed control. The table feeder is selected for materials that require large bin openings to eliminate arching, such as wood chips, sticky or wet granular materials, and for abrasive materials such as minerals and sand.

Com-Bin Feeder (Fig. 8.120).   This feeder is designed for a variety of solids and is particularly effective for damp, oily, or sticky materials. It resembles a table feeder except that the shell and contained solids rotate with the plate. A stationary plow strips off solids from a gap between rotating shell and plate. Flow rate can be controlled by gap height and rotational speed.

Flow Star Feeder (Fig. 8.121).   In this device, specially configured wiping blades draw material from an annular space formed by a hopper wall and a stationary flow cone into a discharge opening. This action promotes mass flow in the vicinity of the hopper opening. This unit can be classified as a combination discharger and feeder. Feed control is achieved by varying drive-motor speed.

Disc Feeder.   This is a small-scale version of the table feeder. The table is grooved to extract a fixed volumetric amount of material. It is used for very low feed rates (about 1 to 2 cu ft/h), with fine free-flowing or cohesive powders.

Rotary Plow Feeders.   There are two general types: one in which a rotating plow is moved horizontally and one in which the plow is stationary, and coaxially mounted in a hopper.

An example of the first is shown in Figure 8.122. A self-propelled carriage, supporting a rotating plow, travels parallel to a slot opening in a bin, and above a conveyor. Solids are plowed from a continuous shelf onto the con-

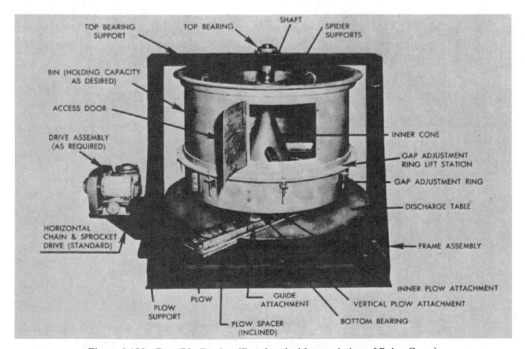

Figure 8.120. Com-Bin Feeder. (Reprinted with permission of Pulva Corp.)

Figure 8.121. Flow star feeder. (Reprinted with permission of Merrick Scale Mfg. Co., Passaic, N.J.)

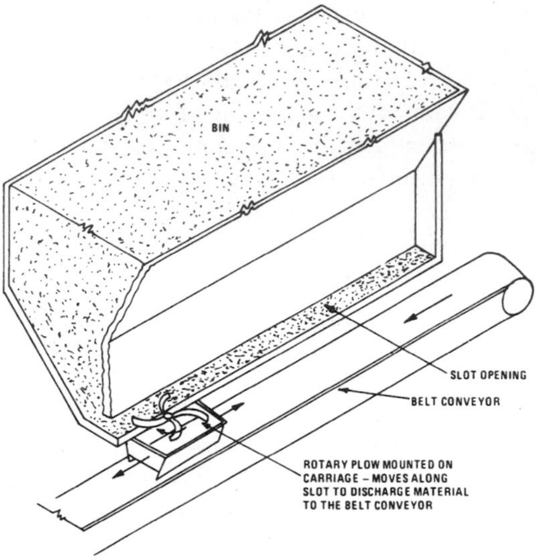

Figure 8.122. Rotary plow feeder.

veyor below. This arrangement produces a very long, fully active slot opening in a bin or hopper, requiring a minimum of overhead space.

An example of a stationary rotary plow is shown in Figure 8.123. These are particularly useful for discharging poor flowing, wet, or sticky solids. The curved sweep-arm plow rotates around the bin axis, to withdraw and sweep solids from the hopper outlet into a central discharge chute. A fixed pressure relief cone is mounted above the plow. This cone prevents accumulation of solids, and is positioned to provide an annular slot that allows the plow to withdraw solids uniformly with each revolution. Schafer et al.[179] described the design and performance of a plow feeder, successfully discharging moist limestone from a 10,000 ton mass flow bin.

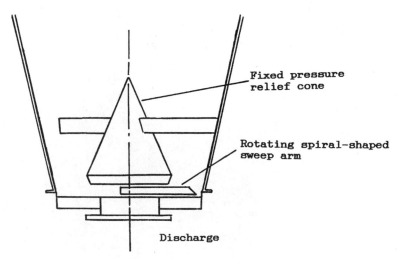

Figure 8.123. Rotary plow discharger-feeder.

8.14.2.4 Flexible Wall Devices

Wall Panels (Fig. 8.124). These elastomeric panels are fastened along the inner wall of hoppers. Periodic inflation with air expands the panel and forces solids into the flow channel. Pressure control of the air prevents over-inflation. The panels are sized and spaced to suit the storage hopper geometry. Since they are elastomeric, they are temperature limited. The sequence and timing of inflation of single or multiple panels is determined by material characteristics and flow rate. Inflation is most effective in promoting flow if there is a void to accept the displaced material. Inflation will be ineffective if it packs cohesive material into a filled channel or if the cohesive material forms a void around the panel. The hopper walls must be sufficient to withstand the reaction forces generated during inflation. The panels are useful for powders as well as sticky materi-als and are for aiding flow, but provide no control of flow rate.

Accu-Rate Feeder (Fig. 8.125). This is a low-cost volumetric feeder combining a flexi-ble wall hopper for loosening material in the hopper with a variable speed screw feeder. Motor-driven mechanical agitators distort or agitate the walls of a one-piece molded flexible-vinyl hopper during operation to pre-vent bridging or ratholing and to provide a constant solids feed to the screw.

8.14.2.5 Aeration-Type Devices

Certain solids can be aerated easily by con-trolled gas injection and are readily discharged from hoppers or fed to process by a variety of aeration devices. These devices operate at low noise levels, require little maintenance, are relatively low cost, and can handle large vol-umes of solids with low gas flows. If well

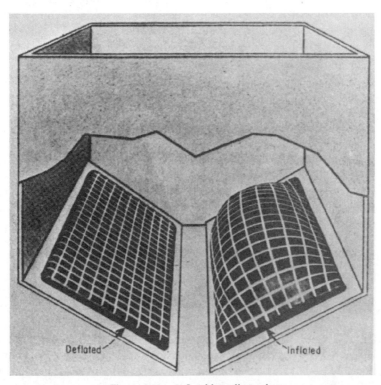

Figure 8.124. Inflatable wall panels.

Figure 8.125. Accu-Rate feeder. (Reprinted with permission of Accu-Rate Div., Moksnes Mfg. Inc.)

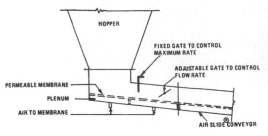

Figure 8.126. Air feeder. (Reprinted with permission of Air Slide Conveyor Reg. Trademark Fuller Co.)

distributed and controlled, most of the injected gas will exit with the powder. At low solids heads, however, more of the gas may exit up through the top of the bin.

For reliable operation, it is necessary to control dust, using dust collectors as necessary, and to provide dry, clean air to prevent fouling of membranes and prevent entry of moisture into the material, which could reduce flowability.

The following are the major types of aeration-type flow-promoting equipment. There are a variety of aeration jets, impulse tank jets, and pads that are not discussed here.

Air Feeder (Fig. 8.126). Air introduced into the solids through an inclined permeable membrane causes solids flow. Flow rate is controlled by an adjustable gate.

Air Hopper (Fig. 8.127). This can take the form of an air feeder under a rectangular or transition slot hopper, a flanged, dished head surrounding a conical membrane, or a conical or rectangular hopper fully lined with a permeable aerating membrane, or a hopper having individual spaced aerating panels or nozzles. Air (or other gas) is introduced through the membrane in sufficient quantities only as

required to reduce the particle–particle and particle–wall friction in the immediate area of the membrane wall surface and exits with the solids. The upper dimension of the aerated portion of the unit is determined by the maximum opening required to prevent arching or ratholing in the hopper above. In determining the opening dimension it may be necessary to add a safety factor to account for the supporting forces caused by air passage up through the material that subtract from the consolidating forces tending to collapse an arch. These devices are essentially bin dischargers. Precise feed control requires a valve or feeder at the outlet.

Aerated Bin Discharge Cone (Fig. 8.128). In this device, air is directed to the solids, under an elastomeric conical insert in a steel hopper. Pulsed air flow can be used to agitate and loosen material. Currently manufactured only as a 762 mm (30 in.) diameter flanged cone it has been effective in promoting flow from bins into small (100 to 300 mm) diameter discharge openings. It has also been used for aeration during pressure differential unloading of bulk trucks.

Air Blasters (Fig. 8.129). In this device, a volume of compressed air is stored in a tank with its exhaust port sealed by a linear or spherical piston. When air pressure on one side of the piston is exhausted through a quick acting valve, the compressed air in the tank is released almost instantaneously into a mass of stored solids. Location of these devices on a

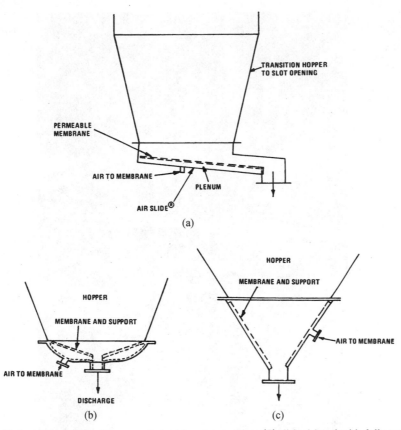

Figure 8.127. Air hoppers; (*a*) Airslide mounted under slot opening; (*b*) dished head with fully aerated interior surface, mounted under circular bin opening; (*c*) conical hopper with fully aerated interior surface, with radial aeration strips or individual aeration pads.

silo, and the orientation of the air release nozzle is determined by the probable location of flow obstructions. The expanding air pocket can break down bridges or ratholes in the material. Single or multiple units can be acti-vated sequentially at time intervals through electrical controls, to maintain the solids in a flowable condition.

Caution is advised if using blasters where serious bridging or ratholing may occur in

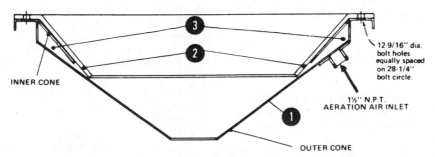

Figure 8.128. Aerated bin discharge cone; (1) Steel discharge hopper, (2) neoprene inner seal hopper, (3) aeration plenum. (Reprinted with permission of Monitor Mfg. Co.)

Figure 8.129. Air Blasters shown attached to conical hopper. (Reprinted with permission of Martin Engineering Co.)

large silos. The silo must be capable of withstanding the stresses caused by a sudden collapse of these flow obstructions, and the discharge device must be capable of sealing against a sudden rush of solids.

REFERENCES

1. Eurosilo Holland, *Wormerveer*, The Netherlands.
2. M. Reimbert and A. Reimbert, *Silos—Theory and Practice*, Clausthal, Germany: Trans. Tech. Publications (1976).
3. M. Reimbert and A. Reimbert, *Silos—Traité Théoretique et Pratique*, Editions Eyrolls, Paris (1961).
4. J. M. Haeger and S. S. Safarian, "A New Concept of Storage Bin Construction," *9 Proceed V 64: J. Am. Conc. Inst.* 9:575–597 (Sept. 1967).
5. S. S. Safarian and E. C. Harris, "Silos and Bunkers," M. Fintel (ed), *Handbook of Concrete Engineering*, Van Nostrand Reinhold Company, New York (1974).
6. H. A. Janssen, *Versuche Über Getreidruck in Silozellen*, VD1 Zeitschrift, Düsseldorf, 39:1045–1049 (1895).
7. O. F. Theimer, "Failures of Reinforced Concrete Silos," *Trans. ASME, J. Eng. Ind. (B)91(2)*:460–477 (May 1969).
8. J. E. Sadler, "More Research Needed in Coal Silo Technology," *Coal Mining and Processing*, pp. 70–72 (May, 1976).
9. J. E. Sadler, "Silo Problems," *International Confer-*

ence on Design of Silos for Strength and Flow, Univ. of Lancaster, U.K. Powder Advisory Center, London (Sept. 1980).

10. A. W. Jenike, "Denting of Circular Bins with Eccentric Draw Points," *J. Struct. Div., Am. Soc. Civil Eng.* 93(ST1):27–35 (1967).

11. F. Turitzin, "Dynamic Pressure of Granular Material in Deep Bins," *J. Struct. Div., Am. Soc. Civil Eng.* 89(ST2):49–73 (1963).

12. G. P. Deutsch and D. H. Clyde, "Flow and Pressure of Granular Material in Silos," *J. Struct. Div., Am. Soc. Civil Eng.* (EMC):103–123 (1967).

13. G. P. Deutsch and L. C. Schmidt, "Pressures on Silo Walls," *Trans. ASME, J. Eng. Ind.* (B)91(2):450–459 (1969).

14. V. Sundaram and S. C. Cowin, "A Reassessment of Static Bin Pressure Experiments," *Powder Technol.* 22:22–32 (1979).

15. D. F. Bagster, "A Note on the Pressure Ratio in the Janssen Equation," *Powder Technol.* 4:235–237 (1970/71).

16. "Recommended Practice for Design and Construction of Concrete Bins, Silos and Bunkers for Storing Granular Materials" (AC1-313-77) *J. Am. Conc. Inst.* 10:529–548 (1975).

17. "Régles de Conception et de Calcul des Silos en Beton," *Syndicat National du Beton Arme et des Techniques Industrialisées*, no. 189 (1975).

18. *Soviet Concrete and Reinforced Concrete Code*, Silo Code SN-302-65, Moscow (1965).

19. "Lastanahmen für Bauten-Lasten in Silozellen," DIN 1055 Blatt 6 (1964), *Erganzende Bestimmungen ZU*. DIN 1055 Teil 6 (1977).

20. G. Garfinkel, "Reinforced Concrete Bunkers and Silos: Steel Tanks," R. S. Wozniak, E. H. Gaylord and C. N. Gaylord (eds). *Structural Engineering Handbook*, 2nd ed., McGraw-Hill Company, New York (1979).

21. F. A. Wenzel, "A Critical Comparison of Different Standards and Methods of Computation of Silo Loads," *Am. Soc. Chem. Eng.*, 70th Annual Meeting, New York, Paper No. 71b (1977).

22. G. P. Deutsch, "Bin Wall Design and Construction, Discussion," vol. 65, *J. Am. Conc. Inst.* 3:211–218 (1969).

23. "Commentary on Recommended Practice for Design and Construction of Concrete Bins, Silos and Bunkers for Storing Granular Materials," *J. Am. Conc. Inst.* 10:549–565 (1975).

24. "Discussion of Proposed AC1 Standard, Recommended Practice for Design and Construction of Concrete Bins, Silos and Bunkers for Storing Granular Materials." V. 73: *J. Am. Conc. Inst.* 6:345–361 (1976).

25. R. L. Brown and J. C. Richards, *Principles of Powder Mechanics*, Pergamon Press, London (1970).

26. W. Resnick, "Flow Visualization Inside Storage Equipment," *Intern. Conference on Bulk Solids Storage, Flow and Handling*, Stratford-Upon-Avon, England, Powders Advisory Center, London (1976).

27. P. M. Blair-Fish and P. L. Bransby, "Flow Patterns and Wall Stresses in a Mass Flow Bunker," *Trans. ASME, J. Eng. Ind.* (B)95(1):17–26 (1973).

28. R. Butterfield, R. M. Harkness, and K. Andrews, "A Stereo-Photogrammetric Method for Measuring Displacement Fields," *Géotechnique* 20(3):308–314 (1970).

29. C. J. Chatlynne and W. Resnick, "Determination of Flow Patterns for Unsteady-State Flow of Granular Solids," *Powder Technol.* 8:177–182 (1973).

30. J. R. Johanson, "Stress and Velocity Fields in Gravity Flow of Bulk Solids," Ph.D. thesis, University of Utah, Mech. Engrg. Dept. (1962).

31. J. R. Johanson, "Stress and Velocity Fields in the Gravity Flow of Bulk Solids," *Trans. ASME, J. Appl. Mech.* (E)31(3):499–605 (1964).

32. M. G. Perry and H. A. S. Jangda, "Pressures in Flowing and Static Sand in Model Bunkers," *Powder Technol.* 4:89–96 (1970-1971).

33. M. Sugita, "Flow and Pressures in Non-cohesive Granular Materials in Funnel Flow Bins," ASME, *2nd Symposium on Storage and Flow of Solids*, Chicago, Ill. Paper no. 72-MH-20 (1972).

34. R. P. McCabe, "Flow Patterns in Granular Materials in Circular Silos," *Géotechnique* 24(1):45–62 (1974).

35. D. Lenczner, "An Investigation into the Behavior of Sand in a Model Silo," *Struct. Eng.* 41(12):389–398 (1963) and (7):243–246 (1963).

36. T. V. Nguyen, C. E. Brennen, and R. H. Sabersky, "Gravity Flow of Granular Material in Conical Hoppers," *Trans. ASME, J. App. Mech.* 46 (3):529–535 (Sept. 1974).

37. T. V. Nguyen, C. E. Brennen, and R. H. Sabersky, "Funnel Flow in Hoppers," *Trans. ASME J. App. Mech.* 47:(4):729–735 (Dec. 1980).

38. J. S. Giunta, "Flow Patterns of Granular Materials in Flat Bottom Bins," *Trans. ASME, J. Eng. Ind.* (B)91(2):406–413 (1969).

39. J. R. Johanson, "Know Your Material—How to Predict and Use the Properties of Bulk Solids," *Chem. Eng.* 85(24):9–17 (Oct. 1978).

40. *Classification and Definitions of Bulk Materials*, Conveyor Equipment Manufacturers Association, Washington, Book no. 550 (1970).

41. S. Frydman, "The Angle of Repose of Potash Pellets," *Powder Technol.* 10:9–12 (1974).

42. R. L. Brown and J. C. Richards, *Principles of Powder Mechanics*, Pergamon Press, London (1970).

43. R. L. Carr, "Evaluating Flow Properties of Solids," *Chem. Engrg. 72*(2):163–168 (Jan. 1965).

44. R. L. Carr, "Particle Behavior Storage and Flow," *Brit. Chem. Engrg. 15*(12) (Dec. 1970).

45. R. L. Carr, "Classifying Flow Properties of Solids," *Chem. Engrg. 72*(3):68–72 (Feb. 1965).

46. R. E. Tobin, "Flow Cone Sand Tests," *J. Am. Conc. Inst. 1*:13–21 (1978).

47. H. M. Sutton and R. A. Richmond, "Improving the Storage Conditions of Fine Powders by Aeration," *Trans. Instn. Chem. Engrs. 51*:97–104 (1973).

48. W. Bruff, "Some Characteristic Qualities of Powder Materials," *Trans. ASME, J. Engrg. Ind.* (B)*91*(2):323–328 (1969).

49. F. A. Zenz and D. F. Othmer, *Fluidization and Fluid-Particle Systems*, Reinhold Publishing Corp., New York (1970).

50. M. A. Goodman and S. C. Cowin, "Two Problems in the Gravity Flow of Granular Materials," *Trans. ASME J. Fluid Mech. 45*(part 2):321–339 (1971).

51. K. Terzaghi, *Theoretical Soil Mechanics*, John Wiley, New York (1943).

52. D. C. Drucker, R. E. Gibson, and D. J. Henkel, "Soil Mechanics and Work Hardening Theories of Plasticity," *Trans. ASCE 122*:338–346 (1957).

53. A. W. Jenike and R. T. Shield, "On the Plastic Flow of Coulomb Solids Beyond Original Failure," *Trans. ASME, J. Appl. Mech.* (E)*26*(4):599–602 (Dec. 1959).

54. A. W. Jenike, *Gravity Flow of Bulk Solids*, Bul. 108, University of Utah, Utah Engineering Station (1961).

55. A. W. Jenike, *Storage and Flow of Solids*, Bul. 123, University of Utah, Utah Engineering Station (1964) (revised 1976).

56. J. R. Johanson, "Effect of Initial Pressures on Flowability of Bins," *Trans. ASME. J. Engrg. Ind.* (B)*91*(2):395–399 (1969).

57. J. R. Johanson and H. Colijn, "New Design Criteria for Hoppers and Bins," *Iron and Steel Engineer.* (Oct. 1964).

58. J. F. Carr and D. M. Walker, "An Annular Shear Cell for Granular Materials," *Powder Technol. 1*:369–373 (1967/68).

59. M. U. Eiserhart-Rothe and I. A. S. Peschl, "Powder Testing Techniques for Solving Industrial Problems," *Chem. Engrg. 84*(7):97–102 (1977).

60. A. W. Jenike, P. J. Elsey, and R. H. Woolley, "Flow Properties of Bulk Solids," *Proc. ASTM 60*:1168–1181 (1960).

61. A. W. Jenike, *Flow Properties of Bulk Solids*, Bul. 95, University of Utah, Utah Engineering Station (1958).

62. A. W. Jenike and J. R. Johanson, "Settlement of Powders in Vertical Channels Caused by Gas Settlement," *Trans. ASCE J. Applied Mech.* (E)*39*(4):863–868 (Dec. 1972).

63. J. R. Adams and W. Ross, "Relative Caking Tendency of Fertilizers," *Ind. and Eng. Chem. 33*(1):121–127 (1941).

64. J. Whetstone, "Solution to the Caking Problem of Ammonium Nitrate and Ammonium Nitrate Explosives," *Ind. Eng. Chem. 44*(11):2663–2667 (Nov. 1952).

65. A. L. Whynes and T. P. Dee, "The Caking of Granular Fertilizers: An Investigation on a Laboratory Scale," *J. Sci. Food Agric., 8*:577–590 (Oct. 1957).

66. W. B. Pietsch, "Adhesion and Agglomeration of Solids During Storage, Flow and Handling—A Survey." *Trans. ASME, J. Eng. Ind.* (B)*91*(2):435–449 (May 1969).

67. W. B. Pietsch, "The Strength of Agglomerates Bound by Salt Bridges," *Can. J. of Chem. Eng. 47*:403–409 (Aug. 1969).

68. J. Silverberg, J. R. Lehr, and G. Hoffmeister, "Microscopic Study of the Mechanisms of Caking and Its Prevention in Some Granular Fertilizers," *Agric. Food Chem. 6*(8):442–448 (1958).

69. R. R. Irani, H. L. Vandersall, and W. W. Mergenthaler, "Water Vapor Sorption in Flow Conditioning and Cake Inhibition," *Ind. and Eng. Chem. 53*(2):141–142 (1961).

70. R. R. Irani, C. F. Callis, and T. Liv, "How to Select Flow Conditioning and Anti-Caking Agents," *Ind. and Eng. Chem. 51*(10):1285–1288 (1959).

71. M. Peleg and C. H. Mannheim, "Effect of Conditioners on the Flow Properties of Powdered Sucrose," *Powder Technol. 7*:45–50 (1973).

72. M. Peleg, C. H. Mannheim, and Passy, "Flow Properties of Some Food Powders," *N. J. of Food Sci. 38*:959–964 (1973).

73. A. W. Jenike and T. Leser, "A Flow-No-Flow Criterion in the Gravity Flow of Powders in Converging Channels," *Fourth International Congress on Rheology*, Brown University, Providence, R.I. (1963).

74. P. L. Bernache, "Flow of Dry Bulk Solids on Bin Walls," *Trans. ASME, J. Eng. Ind.* (B)*91*(2):489–496 (1969).

75. H. Wright, "An Evaluation of the Jenike Bunker Design Method," *Trans ASME, J. Eng. Ind.* (B)*95*(1):48–54 (1973).

76. H. Wright, "Bunker Design for Iron Ores," Ph.D. thesis, Dept. of Chem. Engrg, University of Bradford (1964).

77. R. K. Eckhoff and P. G. Leversen, "A Further Contribution to the Evaluation of the Jenike Method for Design of Mass Flow Hoppers," *Powder Technol. 10*:51–78 (1974).

78. G. Enstad, "On the Theory of Arching in Mass Flow Hoppers," *Chem. Eng. Sci. 30*:1273–1283 (1975).

79. G. Enstad, A. "Note on Stresses and Dome Formation in Axially Symmetric Mass Flow Hoppers," *Chem. Eng. Sci. 32*:337–339 (1977).

80. W. Reisner, "Behavior of Granular Materials in Flow Out of Hoppers," *Powder Technol. 1*:257–264 (1967/68).

81. Schwedes, Jr., "Dimensionierung Von Bunkern" (Dimensioning of Bins), *Aufber Technol. 10*:535–541 (1969).

82. L. A. S. Z. Peschl, "The Theory of Formation of Arches in Bins," *Trans. ASME, J. Eng. Ind.* (B)*91*(2):423–434 (May 1969).

83. D. M. Walker and M. H. Blanchard, "Pressures in Experimental Coal Hoppers," *Chem. Eng. Sci. 22*:1713–1754 (1967).

84. R. Aoki and H. Tsunakaw, "The Pressures in a Granular Material at the Wall of Bins and Hoppers," *J. of Chem. Eng.* (Japan) *2*(1):126–129 (1969).

85. H. Tsunakaw and R. Aoki, "Lateral Pressures of Granular Materials in a Model Bin, *J. of Chem. Eng.* (Japan) *7*(2):131–134 (1974).

86. P. C. Richards, "Bunker Design, Part 1: Bunker Outlet Design and Initial Measurement of Wall Pressures," *Trans. ASME, J. Eng. Ind.* (B)*99*(4):809–813 (1977).

87. D. C. Van Zanten and A. Mooij, "Bunker Design, Part 2: Wall Pressures in Mass Flow," *Trans. ASME, J. Eng. Ind.* (B)*99*(4):814–818 (1977).

88. D. C. Van Zanten, P. C. Richards, and A. Mooij, "Bunker Design Part 3: Wall Pressures and Flow Patterns in Funnel Flow," *Trans. ASME, J. Eng. Ind.* (B)*99*(4):819–823 (1977).

89. R. Everts, D. C Van Zanten, and P. C. Richards, "Bunker Design, Part 4: Recommendations," *Trans. ASME, J. Eng. Ind.* (B)*99*(4):824–827 (1977).

90. R. Moriyama and T. Jotaki, "An Investigation of Wall Pressures in Flowing and Static Bulk Materials in Model Bins," *Intern. Conf. on Design of Silos for Strength and Flow*, Univ. of Lancaster, U.K., Powder Advisory Center, London (Sept. 1980).

91. C. G. Tattersall and L. C. Schmidt, "Model Studies of a Plane Converging Hopper," *Intern. Conf. on Design of Silos for Strength and Flow*, Univ. of Lancaster, U.K. Powder Advisory Center, London (Sept. 1980).

92. P. G. Murfitt and P. L. Bransby, "Pressures in Hoppers Filled with Fine Powders," *Inter. Conf. on Design of Silos for Strength and Flow*, Univ. of Lancaster, U.K. Powder Advisory Center, London (Sept., 1980).

93. U. S. Mukhopadhyah and K. N. Srivastava, "Static and Dynamic Pressure Distribution in Steel Storage Bins Handling Fertilizer Raw Material," *Intern. Conf. on Design of Silos For Strength and Flow*, Univ. of Lancaster, U.K. Powder Advisory Center, London (Sept. 1980).

94. G. E. Blight and D. Midgley, "Pressure Measured in a 20 m Diameter Coal Load-out Bin," *Intern. Conf. on Design of Silos for Strength and Flow*, Univ. of Lancaster, U.K. Powder Advisory Center, London (Sept. 1980).

95. W. S. Patterson, "Measurement of Pressures in Hoppers and Silos," *Inter. Conf. on Design of Silos for Strength and Flow*, Univ. of Lancaster, U.K. Powder Advisory Center, London (Sept. 1980).

96. D. M. Walker, "An Approximate Theory for Pressures and Arching in Hoppers," *Chem. Eng. Sci 21*:975–997 (1960).

97. N. Nanninga, "Gigt die übliche Berechtungsart der Drücke auf die Wände und den Boden von Silobauten sichere Ergenbnisse?" *DeIngenieur 68* (Nov. 1956).

98. J. K. Walters, "A Theoretical Analysis of Stresses in Silos with Vertical Walls," *Chem. Eng. Sci. 28*:13–21 (1973).

99. J. K. Walters, "A Theoretical Analysis of Stresses in Axially Symmetric Hoppers and Bunkers," *Chem. Eng. Sci. 28*:779–789 (1973).

100. K. Clague, "The Effect of Stresses in Bunkers," Ph.D. thesis, Univ. of Nottingham (1973).

101. A. W. Jenike and J. R. Johanson, "Bin Loads," *J. Struct. Div. Am. Soc. Civil Eng. 94*(ST4): 1011–1041 (1968).

102. A. W. Jenike, J. R. Johanson, and J. W. Carson, "Bin Loads, Part 2: Concepts," *Trans. ASME, J. Eng. Ind. 95*(1):1–5 (1973).

103. A. W. Jenike, J. R. Johanson, and J. W. Carson, "Bin Loads, Part 3: Mass-Flow Bins," *Trans. ASM, J. Eng. Ind. 95*(1):6–12 (1973).

104. A. W. Jenike, J. R. Johanson, and J. W. Carson, "Bin Loads, Part 4: Funnel Flow Bins," *Trans. ASME, J. Eng. Ind. 95*(1):13–16 (1973).

105. H. Takahashi, H. Yanai, and T. Tanaka, "An Approximate Theory for Dynamic Pressure of Solids in Mass Flow Bins," *J. of Chem. Eng. Japan 12*(5):369–375 (1979).

106. H. Takahashi, H. Yanai, and T. Tanaka, "An Approximate Theory for Dynamic Pressures of Solids in Funnel Flow Bins," *J. of Chem. Eng. of Japan 12*(5):376–382 (1979).

107. P. C. Arnold, A. G. McLean, and A. W. Roberts, *Bulk Solids: Storage Flow and Handling*, The Univ. of Newcastle, New South Wales, Tunra Limited (1979).

108. P. C. Arnold and A. W. Roberts, "A Useful Procedure for Predicting Stresses of the Walls of Mass-Flow Bins," *AICHE 80th National Meeting*, Boston. Paper No. 49B (1975).

109. A. W. Jenike, "Effect of Solids Flow Properties and Hopper Configuration on Silo Loads," in *Unit*

and Bulk Materials Handling, F. J. Loeffler and C. R. Proctor (ed.). Presented at the Materials Handling Conference, ASME Century 2, Emerging Technology Conferences, San Francisco, Calif. (August, 1980).

110. P. G. Murfitt and P. L. Bransby, "Deaeration of Powders in Hoppers," *Powder Technol.* 27:149–162 (1980).

111. J. E. P. Miles, C. Schofield, and F. H. H. Valentin, "The Rate of Discharge of Powders from Hoppers," *Inst. Chem. Eng. Symp. Ser.* 29 (1968).

112. B. J. Crewdson, A. L. Ormond, and R. M. Nedderman, "Air Impeded Discharge of Fine Particles from a Hopper," *Powder Technol.* 16:197–207 (1977).

113. A. Wlodarski and A. Pfeffer, "Air Pressure in the Bulk Granular Solid Discharge from a Bin," *Trans. ASME, J. Eng. Ind.* (B)91(2):382–384 (1969).

114. W. Bruff and A. W. Jenike, "A Silo for Ground Anthracite," *Powder Technol.* 1:252–256 (1967/68).

115. A. W. Jenike and J. R. Johanson, U.S. Patent 3, 797, 707, *Bins For Storage and Flow of Bulk Solids* (March 19, 1974).

116. G. E. Reed and J. R. Johanson, "Feeding Calcine Dust with a Belt Feeder at Falconbridge," *Trans. ASME, J. Eng. Ind.* (B)95(1):72–74 (Feb. 1973).

117. L. G. Laszlo, L. Williams, J. W. Carson, "Brookfield Solves Fine Limestone Feed Problem," *Proc. Int. Bulk Solids Handling and Proc. Conf.*, Phila., Pa., International and Scientific Conference Management, Inc. (May 1979).

118. M. Turco, C. Gaffney, and J. R. Johanson, "Feeding Dry Fly Ash Without Flooding and Flushing," *Proceedings Inten. Proc. Int. Bulk Solids Handling and Proc. Conf.*, Phila., Pa., International and Scientific Conference Management, Inc. (May 1979).

119. J. R. Johanson, "Two-Phase Flow Effects in Solids Processing and Handling," *Chem. Eng.* 86(1):77–86 (Jan. 1979).

120. J. C. Williams, "The Segregation of Particulate Materials: A Review," *Powder Technol.* 15:245–251 (1976).

121. J. R. Johanson, "Particle Segregation and What to Do About It," *Chem. Eng.* 85(2):183–188 (May 1978).

122. C. F. Harwood, "Powder Segregation Due to Vibration," *Powder Technol.* 16:51–57 (1977).

123. J. F. Van Denburg and W. C. Bauer, "Segregation of Particles in Storage of Materials," *Chemical Engrg.* 71(2):135–142 (Sept. 1964).

124. J. F. G. Harris and A. M. Hildon, "Reducing Segregation in Binary Powder Mixtures with Particular Reference to Oxygenated Washing Powders," *Ind. Eng. Chem., Proc. Des. Devel.* 9(3):363–367 (1970).

125. K. Ahmad and I. J. Smalley, "Observations of Particle Segregation in Vibrated Granular Systems," *Powder Technol.* 8:69–75 (1973).

126. "Matching Size Eliminates Fertilizer Segregation," *Chem. & Eng. News* (Sept. 24, 1962).

127. K. Clague and H. Wright, "Minimizing Segregation in Bunkers," *Trans. ASME, J. Eng. Ind.* (B)95(1):81–85 (1973).

128. *Belt Conveyors for Bulk Materials* (CEMA), Cahners Publishing Company, Boston (1979).

129. K. Shinohara, K. Shojik, and T. Tanaka, "Mechanism of Size Segregation of Particles in Filling a Hopper," *Ind. Eng. Chem., Proc. Des. Devel.* 11(3):369–376 (1972).

130. K. Shinohara, K. Shojik, and T. Tanka, "Mechanism of Segregation and Blending of Particles Flowing Out of Mass Flow Hoppers," *Ind. Eng. Chem., Process Des. Dev.* 9(2):174–180 (1970).

131. T. Tanaka, "Segregation Models of Solid Mixtures Composed of Different Densities and Particle Sizes," *Ind. Eng. Chem. Proc. Des. Dev.* 3:332–340 (1971).

132. H. Matthee, "Segregation Phenomena Relating to Bunkering of Bulk Materials: Theoretical Considerations and Experimental Investigations," *Powder Technol.* 1:265–271 (1967/68).

133. G. W. Fisher, U.S. Patent 3, 575, 321, *Solid Particulate Material Blender* (April 20, 1971).

134. C. A. Lawler, "New Method Controls Particle Segregation," *Materials Handling Eng.*, pp. 105–108 (Nov. 1968).

135. F. J. Loeffler and C. R. Proctor (ed.), "Unit and Bulk Materials Handling," presented at the Materials Handling Conference, ASME Century 2, Emerging Technology Conferences, San Francisco, Calif. (August 1980).

136. Bituminous Coal Research Institute, Pittsburg, Pennsylvania.

137. R. H. Newton, G. S. Dunham, and T. P. Simpson "The TCC Catalytic Cracking Process for Motor Gasoline Production," *Trans. A.I.ChE.* 41:215–18 (1945).

138. H. H. Morse, U.S. Patent 2, 255, 052, *Method of Effecting Contract in a Pebble Heater* (May 29, 1951).

139. J. R. Johanson, "The Use of Flow Corrective Inserts in Bins," *Trans. ASME, J. Eng. Ind.* 88:224–230 (1966).

140. J. R. Johanson and W. K. Kleysteuber, "Flow Corrective Inserts in Bins," *Chem. Eng. Prog.* 62(11):79–83 (Nov. 1966).

141. J. R. Johanson, "The Placement of Inserts to Correct Flow Problems," *Powder Technol.* 1:328–333 (1967/68).

142. O. F. Theimer, "Ablauf fördernde Trichterkonstruktion von Silozellen" (Discharge—Prompting

Hopper Construction of Silos), *Powder Technol.* 3:253–248 (1969/70).

143. W. Reisner and M. Eisenhart Rothe, *Bins and Bunkers for Handling Bulk Materials*, Trans. Tech. Publications, Clausthal, Germany (1971).

144. F. M. Thomson, "Smoothing the Flow of Materials Through the Plant: Feeders," *Chem. Eng.* 85(24):77–87 (Oct. 1978).

145. P. J. Carroll and H. Colijn, "Vibrations in Solids Flow," *Chem. Eng. Progress* 71(2):53–65 (Feb. 1975).

146. P. J. Carroll, "Hopper Designs with Vibratory Feeders," *Chem. Eng. Progress* 66(6):44–49 (June 1970).

147. Lambertus ter Borg, "Evaluation of Shear Test Results on Bulk Solids in Chemical Industry," *German Chem. Eng.* 5:59–63 (1982).

148. H. Tsunakawa, "The Use of Partition Plates and Circular Cones to Reduce Stresses on Particulate Solids in Hoppers," *Intern. Chem. Eng.* 22(2):280–286 (April 1982).

149. R. A. Lohnes, "Lateral Stress Ratios for Particulate 5(4): Materials," *Powder Handl. Proc.* 331–336 (1993).

150. J. W. Carson and R. T. Jenkyn, "Load Development and Structural Considerations in Silo Design," in *RELPOWFLOW II*, EFChE Pub. Ser. 96, Oslo, Norway, 237–282 (1993).

151. P. A. Shamlou, *Handling of Bulk Solids*, *Theory and Practice*, Butterworths, London (1988).

152. D. Schulz, Dissertation, TU Braunschweig (1991).

153. J. Schwedes and H. Feise, "Modelling of Pressures and Flow in Silos," in *Proceedings*, *RELPOWFLOW II*, EFChE Pub. Ser. 96, Oslo, Norway, 193–215 (1993).

154. German Standard, DIN 1055, Part 6 Loads in Silos (1964). New draft in preparation.

155. Draft Code of Practice for the Design of Silos, Bins, Bunkers and Hoppers, British Materials Handling Board (1985).

156. Loads Due to Bulk Materials (Draft) ISO Working Group TC98/SC3/WG5 (1991).

157. Draft Australian Standard, Loads on Bulk Solids Containers.

158. G. E. Blight, "Comparison of Measured Pressures in Silos with Code Recommendations," *Bulk Solids Handling* 8(2):145–153 (1988).

159. Standard Shear Testing Procedure for Particulate Solids Using the Jenike Shear Cell, Institute of Chemical Engineers, England (1989).

160. J. Y. Ooi, W. C. Soh, Z. Zhong, and J. M. Rotter, "Bulk Mechanical Properties of Some Dry Granular Solids," in *RELPOWFLOW II*, EFChE Pub. Ser. 96, 75–86 (1993).

161. A. G. McLean, "A Closer Examination of the Variation of Wall Friction Angle with Major Con-

solidating Stress," *Bulk Solids Handling* 8(4):404–411 (1998).

162. A. T. Royal and J. W. Carson, "Fine Powder Flow Phenomena in Bins, Hoppers and Processing Vessels," *Conference: Bulk 2000: Bulk Material Handling Toward the Year 2000*, London (1991).

163. Z. H. Gu and P. C. Arnold, "Critical Permeability for Significant Effect of Air on the Flowrate from a Mass Flow Bin," in *RELPOWFLO II*, EFChE Publication Series (96), Oslo, Norway, 169–178 (1993).

164. P. C. Arnold and Z. H. Gu, "The Effect of Permeability on the Flowrate of Bulk Solids from Mass Flow Bins," *Powder Handling Proc.* 2(3):229–233 (1993).

165. Z. H. Gu, P. C. Arnold, and A. G. McLean, "A Simplified Model for Predicting the Particle Flow Rate from Mass Flow Bins," *Powder Technol.* 74(2):153–158 (1993).

166. Z. H. Gu, P. C. Arnold, and A. G. McLean, "The Influence of Surcharge Level on the Flowrate of Bulk Solids from Mass Flow Bins," *Powder Technol.* 74(2):141–151 (1993).

167. J. R. Johanson, "The Johanson Indicizer System vs. the Jenike Shear Tester," *Bulk Solids Handling* 12(2):237–240 (1992).

168. T. A. Bell, B. J. Ennis, R. J. Grygo, W. J. F. Scholten, and M. M. Shenkel, "Practical Evaluation of the Johanson Hang-up Indizer," in *RELPOWFLO II*, EFChE Publication Ser. 96, Oslo, Norway, 117–137 (1993).

169. G. G. Enstad and L. P. Maltby, "Flow Property Testing of Particulate Solids," *Bulk Solids Handling* 12(3):451–456 (1992).

170. J. W. Carson and D. A. Ploof, "Quality Control Tester to Measure Relative Flowability of Powders," in *RELPOWFLO II*, EFChE Publication Ser. 96, Oslo, Norway, 117–137 (1993).

171. C. C. Goelema, L. P. Maltby, and C. G. Enstad, "Use of a Uniaxial Tester for the Determination of Instantaneous and Time Consolidated Flow Properties of Powders," *RELPOWFLO II*, EFChE Publication Ser. 96, Oslo, Norway, 139–152, 1993.

172. L. Savarovsky, *Powder Testing Guide*, Elsevier Applied Science, London (1987).

173. J. W. Carson, T. A. Royal, and D. J. Goodwill, "Understanding and Eliminating Particle Segregation Problems," *Bulk Solids Handling* 6(1):139–144 (1986).

174. J. R. Johanson, "Controlling Flow Patterns in Bins by Use of an Insert," *Bulk Solids Handling* 2(3):495–498 (1982).

175. BINSERT®, U.S. Patent 4, 286, 883.

176. D. Schulze and J. Schwedes, "Tests on the Application of Discharge Tubes," *Bulk Solids Handling* 12(1):33–39 (1992).

177. M. Terziovski and P. C. Arnold, "Effective Sizing and Placement of Air Blasters," *Bulk Solids Handling 10*(2):181–185 (1990).
178. U.S. Patent 4,958,741.
179. R. Schafer, H. Schroer, and J. Schwedes, "Silo for Storage of 10,000 Tons of Moist Limestone," in *Reliable Flow of Particulate Solids*, Bergen, Norway, EFCE Publication Ser. 49 (1985).
180. American Society for Testing Materials: PA, USA, Herts, England.

9
Fluidization Phenomena and Fluidized Bed Technology

Frederick A. Zenz

CONTENTS

The term *fluidization* is used to designate the gas-solid contacting process in which a bed of finely divided solid particles is lifted and agitated by a rising stream of process gas. At the lower end of the velocity range, the amount of lifting is slight, the bed behaving like a boiling liquid (hence the term *boiling bed*). At the other extreme, the particles are fully suspended in the gas stream and are carried along with it; the terms *suspension, suspensoid,* and *entrainment* contact have all been used to designate this action.

9.1 HISTORICAL DEVELOPMENT

The fluidized technique as it is known today was born from the pioneering work of

Standard Oil Development Co., The M. W. Kellogg Co., and Standard Oil of Indiana in their efforts to find a better catalytic cracking process than the fixed-bed process that was introduced commercially in 1937. The fixed-bed process was a major improvement over the earlier thermal cracking methods. It yielded more gasoline of higher octane rating and less low-value heavy fuel oil byproduct.

Initial experimentation in developing a still superior process began along the lines of the fixed-bed method. Oil vapor was passed through one of a pair of beds until the catalyst became fouled with carbon formed in the reaction; then the oil vapor was fed to an adjacent fresh bed while air passed through the fouled material to burn off the carbon and regenerate the catalyst. It was soon appreciated that some innovation would be desirable to avoid the complexity and cost of such intermittent operations.

Placing the regenerating and reacting beds in series and continuously moving the catalyst mechanically from one to the other appeared to be an obvious method of approach. Initial experiments indicated that such a system might suffer considerable loss from catalyst attrition unless pneumatic rather than mechanical conveying methods were adopted. Thus, experimentation turned to studies of pneumatic transport of catalysts.

It was soon discovered that, in order to avoid severe erosion as well as attrition, relatively lower gas velocities would be required. This led to the investigation of powder-form catalysts and eventually to the observation that dense beds of powder could be maintained with relatively low carryover losses even at superficial gas velocities that were orders of magnitude greater than the calculated settling velocity of the individual particles making up the bed. At these gas velocities the particles were observed to be considerably agitated, as gas bubbles passed upward through the bed in a manner analogous to the boiling of liquids. Simultaneously, it was observed that the pressure drop through such a boiling or fluidized bed was equal to the weight of the bed charge; the bed was in effect heterogeneously buoyed by the gas stream and thus took on effective flowing properties similar to liquids. These simple experiments gave birth to the present-day fluid-bed concepts.

Before such processing techniques could be applied commercially, considerable further work had to be done to develop satisfactory solids recovery systems, proper aeration techniques, instrumentation, methods of line sizing, minimizing of erosion problems, reactor conversion correlations, heat-transfer data, regeneration rates, and numerous other matters. The first commercial fluid-bed catalytic cracking plant was put in operation in 1942 followed by 31 additional plants during the war years alone.

During the past succeeding 35 years the application of the fluidized technique has spread rapidly to metallurgical ore roasting, limestone calcination, synthetic gasoline, petrochemicals, and even to the design of nuclear reactors. A realization of the scope of applications and number of organizations with a vested interest in fluidization is some indication of its importance and its rate of growth. Table 9.1 gives a representative sampling of fluid-bed applications that have been investigated over the past 40 years. This list is far from exhaustive; yet it records substantial evidence of a lively pace of interest and activity.

Each application listed in Table 9.1 may represent a host of operating units. Table 9.2, for example, lists a number of metallurgical processing installations and Table 9.3 indicates the level of activity in fluidized bed combustion. When it is realized that none of these tables is exhaustive, that equal if not greater activity exists in fluidized bed gasification than in fluidized bed combustion, and that nearly 200 fluidized bed catalytic cracking plants are in operation, it becomes apparent that fluidization as a unit operation has touched almost every process industry and every related corporate body.

Though the commercial development of the fluidized technique was a direct outcome of the work of the major petroleum process development companies, scattered references bearing on the fluidized technique can be found as far back as 1878.

In all processes using the fluidized-solid technique, it is common to handle the solid material in one or more stages or steps and to transfer it from step to step through pipe lines in much the same manner as with a liquid. To raise the material to a higher level, it is carried as a suspension in a gas stream; to take it to a lower level or to a region of higher pressure, the settled material is allowed to flow by gravity down a pipe line to the desired

Table 9.1. Some Applications of Fluidization

APPLICATION	ORGANIZATION	REFERENCE
Acetone recovery	British Celanese Ltd. Wrexham, Wales	*C. & E.N.*, 10/25/65, p. 56
Acetylation of poly formaldehyde	Sci. Res. Inst., Sofia, Bulgaria	*Int. Chem. Eng.* No. 1 415 (1965)
Acrylonitrile from C_3H_6 and NH_3	E. I. duPont de Nemours & Co., Beaumont, TX; Standard Oil of Ohio; Montedison	U.S. Pat. 2, 736, 739 (Feb. 28, 1956). *Chem. Eng. 68*, 122 (Jan. 23, 1961); *Chem. Week 88*, 40 (Jan. 28, 1961); *Hydr. Proc* 144–146 (11/72)
Activated charcoal manufacture	A. Godel	*Chem. Eng. 55*(7) 110 (July 1948)
Activated carbon regeneration	Battelle Columbus Labs. (OH)	*Env. Sci. Tech.* *4*, No. 5, 432–437 (May 1970)
Adsorption with fluid char	L. D. Etherington, et al.	*Chem. Eng. Prog. 52* (7), 274 (1956)
Adsorption in fluid beds	Food Machinery & Chemical Corp.	*Chem. Eng. 68* (11) 87 (May 29, 1961). *Chem Tech. 11*, 647 (1964)
Adsorption separation of gases in a fluid bed	D. L. Campbell et al.	U.S. Pat. 2,446,076 (July 27, 1948)
Agglomerates from fines	The Pillsbury Co. Krimo-Ko Corp., CA and Hawaiian Sugar Refin. Corp.	*Chem. Week*, 6/27/64 p. 96
Alkylation and dehydrogenation of aromatic hydro carbons	Mamedaliev Petro- chemical Inst. ASSR	*Int. Chem. Eng. 5*, No. 3, 467 (1965)
Production of AlF_3	Montedison-Lurgi	*C.E.P. 67*, No. 2, 58–63 (1971)
AlF_3 from HF plus aluminum hydroxide	Kaiser Aluminum and Chem. Corp., Gramercy, LA	*Chem Eng.*, 1/18/65, p. 92
Calcining aluminum hydroxide	Metals Research Institute, Budapest Hungary	*Br. Chem. Eng.* *10*, No. 10, 710 (1965)
Ammonium chloride and similar con- densations of sub- limable matl's	Ivanov Chem. Engrg. Inst. (Russia)	*Int. Chem. Eng. 8*, No. 4, 651–653, Oct. 1968; 2, No. 1, 105–108 (Jan. 1962)
Ammoniation of superphosphate	Inorg. Chem. Res. Inst., Czech Republic	*Br. Chem. Eng. 10*, No. 11, 756 (1965) *Chem. Eng. 68* (13), 74 (June 26, 1961)
Aniline via fluid-bed process	American Cyanamid Co.	*Chem. Week 85* (13), 68 (Sept. 26, 1959): U.S. 2, 891, 094
Aniline from notrobenzene	M. S. Murthy et al.	*Chem. Age India 14*, No. 9, 653 (Dec., 1963)

Table 9.1. (*Continued*)

APPLICATION	ORGANIZATION	REFERENCE
Roasting of arsenopyrites	Piritas Espanolas Auxini S.A., Madrid	*I.E.C. Proc. Des. Dev.* 2, No. 3, 214 (1963)
BaCl$_2$ from Cl$_2$ + BaSO$_4$	Central Salt & Marine Chemicals Res. Inst., Bhavnagar, India	"Fluidization and Related Processes," CSIR, New Delhi, India
Barium oxide by reduction of BaCO$_3$	Barium Reduction Co., South Charleston, West VA	*Chem. Eng.* 67 (9), 107 (May 2, 1960)
Benzotrichloride from toluene	Western NY Nuclear Res. Ctr.	*Nucl. Technol.* 18, 29–45 (1973)
Bisulfite acid from SO$_2$ and limestone	Univ. of Naples, Italy	*Pulp & Paper Mag. of Canada* (Oct. 1965)
Blending solids in a fluid bed	Fuller Co.; Wilmot Castle Co.; General Electric NC	Fuller Co. Bull. B-1 Catasaugua, PA; *Chem. Week 85* (19), 73 (Nov. 7, 1959)
Conversion of ethanol to butadiene	Indian Inst. of Tech., Kharagpur	*I.E.C. Proc. Des. Dev* 2, No. 1, 45 (1963)
Calcination of phosphate rock	San Francisco Chemical Co. Montpelier, Idaho	*Chem. Eng. Prog 55* (12), 77 (Dec. 1959)
Calcination of phosphates, limestone, and magnesite	Krupp; J. R. Simplot; Door-Oliver	*Br. Chem. Eng. Proc. Tech.*, p. 381, May 1972 *Chem. Proc.* p. 78, 8/77
Coke calcination	Alum. Co. of Canada	*Chem. Week*, p. 69, 70, 10/19/63, *Can. J. Chem. Eng*, 94–96 (April 1965); 146–149 (June 1965)
Calcium carbide nitration	Süddeutsche Kalkstick-stoffwerke A.G., Trostberg, Bavaria	*Chem. Eng.* 67 (3), 66 (Feb. 8, 1960)
Spent carbon recovery	Westvaco Corp.	*Chem. Eng.*, p. 97 9/12/77 *Envirom. Sci. Tech.* 10 No. 5, 454–456 (1976)
Carbon disulfide production	R. P. Ferguson Kurashiki Rayon Co. Japan	U.S. Pat. 2, 443, 854 (June 22, 1948); U.S. Pat. 3, 402, 021
CS$_2$ adsorption on carbon in 38 ft diameter fluid bed	Courtaulds Ltd., Holywell, Wales	*Chem. Eng.*, p. 92, 4/15/63; *Br. Chem. Eng.*, 8, No. 3, 180 (1963)
Carbontetrachloride from char	C.S.I.R.O., Clayton Victoria, Australia	*Br. Chem. Eng.* 17, No. 4, 319–322 (April 1972)
Carbonization and drying by fluidization	V. Charvat	*Paliva 34*, 179 (1954)

Table 9.1. (*Continued*)

APPLICATION	ORGANIZATION	REFERENCE
Cement via fluid-bed process	R. Pyzel	*Chem. Week 80* (7), 108 (Feb. 16, 1957)
MFC fluidized cement calciner	Mitsubishi Hvy. Ind's Ltd.	*Chem. Eng.*, pp. 102–106, 6/24/74 C.E.P., pp. 36–44, (August 1977)
CCl_3F and CCl_2F_2 production	Montedison, Italy	*Chem. Eng.*, p. 40 2/23/70; pp. 75–77, (6/14/71)
Trichloroethane (methychloroform)	Ethyl Corp.	*Chem. Week*, pp. 30–31, (8/9/69)
Trichlor and perchlorethylene	PPG Industries	*Chem. Eng.*, pp. 90–91, (12/1/69)
Chloromethane from HCl, natural gas, and oxygen	E. I. duPont de Nemours & Co., Inc. Orange, TX	*Chem. Eng. 68* (15) 126 (July 24, 1961) and *68* (3), 33 (Feb. 6, 1961)
Reactivation of clays	Shir Ram Inst. for Ind. Res., Delhi, India	"Fluidization and Related Processes" CSIR, New Delhi, India
Coal refinery	W. F. Coxon	*Gas World 144*, 148 (1956)
Coal gasification	Philadelphia and Reading Corp., NY; Hydrocarbon Research, Inc. Inst. of Gas Tech., FMC Corp., M. W. Kellogg Co., Consolidation Coal Co. etc.	*C.E.N.* 4/28/66, p. 68; *C.E.P.* 60, No. 6, 35, 58, 69 (1964); *Chem Eng.*, 12/6/65, p. 78
Fluid-bed boiler	CEGB, England	*Elec. Rev. 176*, 39, 1/8/65
Coal combustion	Babcock and Wilcox (see also Table 2)	*Chem. Eng.*, p. 114–127, 8/14/78
Coating drug tablets	Abbott Laboratories	*C.E.P. 62*, No. 6, 107 (1966)
Coating particulates	D. E. Marshall	U.S. Pat. 2, 579, 944 (Dec. 25, 1951)
Coating particles by the Wurster process	Abbott Laboratories; Merck Co.; Smith, Kline & French Laboratories	*Chem. Eng. 66* (15), 55 (July 27, 1959)
Coating in fluid beds	Knapsack-Griesheim A.G., Cologne; Whirlclad Div., The Polymer Corp.; General Electric Co.; Rockwell Mfg. Co.	*Chem. Week 87* (10) 56 (Sept. 3, 1960); *Chem. Week 89* (2), 48 (July 8, 1961); *Chem. Eng. Prog. 56* (7), 75 (1960); *Chem. Eng. 66* (28), 100 (Dec. 28, 1959) *C.E.N.*, p. 37, 1/25/71; *Chem. Eng.*, pp. 36–38, 7/12/71

Table 9.1. (*Continued*)

APPLICATION	ORGANIZATION	REFERENCE
Coating particles with ceramics	Battelle Memorial Institute; American Metal Products Co.; Mallinckrodt Nuclear Corp.; 3M Co.; High Temperature Materials (Union Carbide Corp.)	*Chem. Week 87* (19), 59 (Nov. 5, 1960); *Chem. Eng. News*, p. 41 (June 12, 1961); p. 25 (Nov. 21, 1960)
Coating Mb and Cb with silicon in 1900°F fluid bed	Boeing Co., The Pfaudler Co.	*Chem. Eng. 6/24/63* p. 40; *A.I.Ch.E.*, paper 4E, Dallas Mtg. 2/6-9/66
Fluid coking of coal	Atlantic Refining Co.	*Chem. Eng. 8/31/64*, p. 22
Coking process	Esso Research & Eng. Co.	*Chem. Eng. 67* (8), 79 (April 18, 1960); *67* (10), 112–115 (May 16, 1960)
Coking of pelleted coal fines	U.S. Fuel Co. (U.S. Smelting, Refining & Mining Co.)	*Chem. Eng. 67* (2) 43 (Jan. 25, 1960)
Columbium chloride reduction	Battelle Memorial Institute	*Chem. Eng. 67* (13), 77 (June 25, 1960)
Condenser using fluidized solids	C. B. Beck et al.	M.S. Thesis, M.I.T. (June, 1952)
Cornstarch depolymerization to dextrins	R. Frederickson (A. E. Staley Mfg. Co.)	*Chem. Eng. 66* (18) 80 (Sept. 7, 1959)
Cracking with fluidized sand	Erdölchemie, GmbH, Dormagen; Lurgi, Frankfurt; Ruhrgas, Essen; Bayer, Leverkusen, Germany	*Chem. Eng. 66* (17) 66 (Aug. 24, 1959); *World Petrol. 30* (6) 62 (June 1959); *Petrol. Refiner 40* (10), 137 (1961)
Cristobalite from quartz in 2800°F fluid bed	A. D. Little, Inc.; Wedron Silica Co. Chicago	*A.I.Ch.E. Symp. Ser.* No. 62, Vol. 62, 56 (1966)
Cumene dealkylation	J. F. Mathis and C. C. Watson	*A.I.Ch.E.J. 2*, 518 (1956)
Oil decontamination of sand	Univ. of California, Santa Barbara, CA	*Chem. Eng.*, p. 58 8/10/70
Desulfurization of coke-oven aromatics	U.S. Industrial Chemicals Co.	*Chem. Eng. 68* (31), 72 (June 26, 1961); *Chem. Week 88* (23), 64 (June 10, 1961)
Desulfurization of petroleum coke	Institute of Petroleum, Zagreb, Yugoslavia	*Chem. Eng. 68* (2) 126 (Jan, 23, 1961)
Dimethyl terephthalate from toluene	Bergwerksverband A.G.; Toyokoatsu Ind., Japan	*Hydrocarb. Proc. 44*, No. 11, 275 (1965)
Distillation with fluidized solids	R. W. Krebs and C. N. Kimberlin	U.S. Pat. 2, 758, 073 (Aug. 7, 1956)
Drying of fluidized solids	W. W. Niven	U.S. Pat. 2, 715, 282 (Aug. 16, 1955)
Drying with a plurality of fluid beds	W. N. Lindsay	U.S. Pat. 2, 676, 668 (April 27, 1954)

Table 9.1. (*Continued*)

APPLICATION	ORGANIZATION	REFERENCE
Paper and textile drying	The Shirley Institute Dunlop Textiles Man-Made Fibers Res. Association	*Br. Chem. Eng.* *17*, No. 5, p. 383 (1972); *C.E.N.*, p. 37, 12/6/65; *Chem. Eng.*, p. 74, (4/30/62)
Synthetic fiber heat stretching and heat setting	Dunlop Textiles; Courtaulds Ltd. England	*C.E.N.*, 12/6/65 p. 37
Drying of heat-sensitive materials	Door-Oliver, Inc.	*Chem. Eng. 68* (15), 57 (July 24, 1961)
Electrolysis	Akzo Zout Chem. Netherlands	*Chem. Eng.*, p. 72, 77 (8/14/78)
Plating and similar electrochemistry	Centre Nat. de La Rech. Sci., Nancy, France; Central Elec- trochem. Res. Inst., India	*Ind. Eng. Chem.* *61*, No. 10, 8–17, Oct. 1969, *Ind. Eng. Chem.* *Prod. Des. Dev.* *9*, No. 4, 563–567 (1970)
Electrostatic ore beneficiation	Univ. of Western Ontario, London, ON, Canada	*C.E. Sym. Ser. 66*, No. 105, 236–242 (1970); *Revue de L'industrie* *Minerale*, pp. 442–449 (June 1967)
Electrothermal fluidized beds for Zr, P, etc.	Battelle Columbus Labs	*C.E.P. 61*, No. 2, 63–68 (1965) *Battelle Tech. Review.* *13*, No. 11, 3–9, (Nov. 1964)
Ethylene manufacture	R. P. Cahn	U.S. Pat. 2, 752, 407 (June 26, 1956)
Polyethylene, LD & HD	Union Carbide Corp.	*Hydrocarb. Proc.* p. 130–136 (Nov., 1972), *Chem. Proc.*, p. 18 (February 1978) *Chem. Eng.*, p. 72– 73, 11/26/73; p. 25–27, 1/2/78
Fuel oil from polypropylene residue	Procedyne, New Brunswick, N.J.	*Chem. Eng.*, p. 57, (12/4/78)
Ethylene oxide from ethylene	National Research Council, Ottawa, ON, Canada	*Can. J. Chem. Eng. 38* (4), 108 (Aug. 1960)
Extraction counter-currently in fluidized beds	D. E. Weiss and E. A. Swinton	U.S. Pat. 2, 765, 913 (Oct. 9, 1956)
Feeder for solids	Hanna Furnace Corp. (National Steel Corp.)	*Chem. Eng. 68* (19) 64 (Sept. 18, 1961); *Chem. Week 89* (11), 126 (Sept. 9, 1961)

Table 9.1. (*Continued*)

APPLICATION	ORGANIZATION	REFERENCE
Fertilizers from oxidation and ammoniation of coal	Central Fuel Res. Inst. of India	*Br. Chem. Eng.* (August 1966), p. 799
For protection with fluidized solids	Anonymous	*Chem. Eng. 58* (2), 160 (Feb., 1951)
Fischer–Tropsch and related processes	H. C. Anderson et. al.	*U.S. Bur. Mines Bull.* No. 544 (1955)
Flowmeter using fluid bed	C. F. Gerald	*Ind. Eng. Chem. 44*, 233 (1952)
Freezing fruits and vegetables; grilling meats and potatoes	Swedish Food Processor	*Br. Chem. Eng. 8*, No. 12, 800 (1963); ibid, *11*, No. 1, 2 (1966); *Food Proc.* (Nov. 1963)
Roasting chocolate beans	General Foods Corp.	*Chem. Week* (8/31/63), p. 37
Formaldehyde from methane	Bergbau A.G.	*Chem. Week 85* (13), 76 (Sept. 26, 1959)
Gaseous diffusion	P. W. Garbo	U.S. Pat. 2, 637, 625 (May 5, 1953)
Gasoline from methanol	Mobil Oil Corp.	*C.E.N.*, pp. 26, 28, (1/30/78) *Chem. Week*, pp. 35–37 (1/25/78)
Grain seed inoculation	Northrup King & Co.	*A.I.Ch.E.*, paper 16e, 57th Nat'l Mtg., Minneapolis (9/26-29/65)
Granulation and drying	Aeromatic, Inc.; Niro Atomizer	*Chem. Proc.*, p. 53, Mid-Nov. (1978) *Chem. Eng.*, p. 39 (11/6/78) *Br. Chem. Eng.*, p. 811, *12*, No. 6, (June 1967)
Glycol production	D. F. Othmer and M. S. Thakar	ACS meeting, New York (Sept. 8–13, 1957)
Grinding	Southwestern Eng. Co.	*Chem. Proc.* (Chicago) p. 132 (Sept. 1961)
Heat treating and stress relieving	Boeing Co., Seattle, Wash.; The Electric Furnace Co.	*Chem. Eng. News*, p. 80 (April 10, 1961); *Missiles Rockets*, p. 28 (Oct. 23, 1961)
High-temperature bath	C. E. Adams et al.	*Ind. Eng. Chem. 46*, 2458 (1954)
Hot air for peak shaving	Stal-Laval (Gt. Brit.) Ltd.	*C.E.N.*, p. 33, 34 (9/3/73)
Humic acid by fluidized oxidation of Tandur coal	D. P. Agarwal and M. S. Iyengar	*J. Sci. Ind. Res.* (India) *12B*, 443 (1953)
Hydrogen production	Pittsburgh Consolidation Coal Co.	Br. Pat. 673, 332 (June 4, 1952)
Hydrogen from steam	Inst. of Gas Tech.	*Popular Science*, p. 91–94 (Jan. 1977)

Table 9.1. (*Continued*)

APPLICATION	ORGANIZATION	REFERENCE
HCl and Fe_2O_3 from waste pickling liquors	Nittetu Chem. Eng. Ltd. American Lurgi Corp.	*Chem. Eng.*, p. 107 (8/14/78); p. 102–103 (11/13/72)
Regenerating HCl pickle liquor	Lurgi, GMBH; Hilgers AG at Rhein-Brohl, Germ.	*Chem. Eng.*, p. 32 (8/29/66)
HCN from C_3H_3 + NH_3; "Fluohmic" processes for titanium tetrachloride, CS_2, and desulfurizing coke	Shawinigan Products, Ltd., Montreal	*Chem. Eng. 68* (19), 72 (Sept. 18, 1961), *Chem. Week 89* (12) 104 (Sept. 16, 1961) *Ind. Chem. Eng. 53* (1), 19A (1961); *Chem. Eng. News*, p. 55 (Nov. 21, 1960)
HCl oxidation	A. J. Johnson and A. J. Cherniavsky	U.S. Pat. 2, 644, 846 (July 7, 1953)
Hydrogen sulfide removal	L. Jequier	Br. Pat. 708, 972 (May 12, 1954)
Hydrogenating residual petroleum oils	Hydrocarbon Research, Inc.; British Gas Council, Solihull, England	*Chem. Eng. 67* (19) 69 (Sept. 24, 1960), *Petrol. Refiner 39* (10), 151 (1960); *Chem. Eng. 67* (9), 115 (May 2, 1960)
H-Coal and H-Oil (Liquid fluidized beds)	Hydrocarbon Res., Inc. Cities Service Oil, Co.	*C.E.P. 67*, No. 8, 81–85, 8/71; *Br. Chem. Eng. 16*, No. 12, 1117–1119, (Dec. 1971); *Hydrocarb. Proc. 45*, No. 5, 153–158 (May, 1966)
Imenite chlorination	L. K. Doraiswamy et al.	*Chem. Eng. Prog.* 55 (10), 80 (1959)
Incineration of spent caustic and other refinery wastes	Amoco-Dorr-Oliver; Nichols Eng.	*Chem. Proc.*, p. 8, 9 (Sept. 1973) *Chem. Eng.* p. 87–94 (1/2/78); p. 60 (8/14/78); p. 71, 72 (10/4/71)
Econ-Abator catalytic incineration of organic emissions	Air Resources, Inc. Harshaw Chem. Co. Nipro, Inc.; Dutch State Mines	*C.E.P.*, p. 31–35 (August 1977) *Chem. Proc.* p. 13 (August 1976)
Indole and benzofuran	SNAM Progetti, Italy	*Chem. Eng.*, p. 78 (5/17/71)
Ionic mass transfer	G. J. V. J. Raju	D.Sc. thesis, Andhra Univ., Waltair, India (1959)
Ion exchange	Himsley Eng. Ltd., Toronto; Liquitech, Inc., Houston; Farbenfabriken Bayer (Mobay)	*C.E.N.*, p. 23, 24 (Aug. 2, 1976) *Chem. Week*, p. 37 (Aug. 2, 1972) *Chem. Eng.*, p. 60–62 (Jan. 8, 1973)

Table 9.1. (*Continued*)

APPLICATION	ORGANIZATION	REFERENCE
Chlorination of pyrites cinder to FeO	Montedison S.p.A.- A. G. McKee	*Chem. Eng.*, p. 45 (4/26/76)
FeS to iron oxide in 14′ diam. bed at 1900°F	Lurgi, GMBH; Outokumpu Oy, Finland	*Chem. Eng.* (2/14/66) pp. 122–124
Iron ore reducer (Wicke process)	Phoenix-Rheinrohr A.G.	*Chem. Week 85* (17), 42 (Oct. 24, 1959)
Iron ore reduction (multistage)	Arthur D. Little, Inc.	*Trans. Met. Soc. AIME*, *218* (1), 12 (Feb. 1960)
Iron ore reduction in a self-agglomerating fluid bed	Battelle Memorial Institute	Blast Furnace, Coke Oven and Raw Materials Conference., Apr. 4–6, 1960, Chicago, Ill.
Iron ore reduction by "Nu-Iron" process	U.S. Steel Corp.	*J. Metals 12*, 317 (Apr. 1960); *Chem. Eng. 67* (7), 64 (Apr. 4, 1960)
Iron ore reduction (FIOR)	Exxon Res. & Eng.; A. G. McKee	*Chem. Week*, p. 32 (9/5/73) p. 66 (3/7/64); *Chem. Eng.*, p. 49 (5/28/73)
Iron ore reduction by the "H-Iron" process	Hydrocarbon Res., Inc.; Bethlehem Steel Corp.	*Chem. Eng. 67* (3), 96 (Feb. 8, 1960)
Magnetic Fe roasting	French Iron and Steel Res. Inst.	*A.I.Ch.E. Symp. Ser.* No. 62, Vol. 62, 15 (1966)
Isomerization of paraffin hydrocarbons	P. Tristmans	*Het Ingrablad Techn. Wetenschapp. Tijdschr.* 22, 269 (1953)
Isophthalonitrile	Badger (Boston); Mitsubishi (Japan)	*Chem. Week*, p. 39 (5/3/72)
Reforming and desulfurization in a magnetically stabilized bed	Exxon Res. & Eng. Co.	*Chem. Eng.*, p. 95 (10/23/78)
Maleic anhydride	Shakhtakhtinskü et al.; Mitsubishi Chem. Industries	*Azerb. Khim. Zh.*, No. 2, 91–94 (1965) *Br. Chem. Eng. Proc. Tech.*, p. 13, Aug. 1974, *Chem. Eng.*, pp. 107–109 (9/20/71)
Maleic anhydride from benzene and butylenes	Inst. for Gen'l Chem., Warsaw, Poland, Petrochem. Processes Inst. ASSR Acad. of Science	*Br. Chem. Eng. 10*, No. 10, 710 (1965); *Int. Chem. Eng.* 6, No. 4, 674 (1966)
Manganese ore chloridization	M. L. Skow et al.	U.S. Bureau Mines Rept. Invest. No. 5271 (1956)

Table 9.1. (*Continued*)

APPLICATION	ORGANIZATION	REFERENCE
Fluidized mattress	Medical Univ. of South Carolina	*Reader's Digest* (1970) Dr's Thos. Hargest and C. P. Artz
Melamine from urea	BASF, Ludwigshafen, W. Germany U.S. Steel, Pitt. Coke Co., Chemico	*C.E.N.*, p. 62–63, 9/22/69; *Chem. Eng.* 101–103 (10/19/70) Cdn. Patents 737, 475; 737, 476,
Melamine from urea and ammonia	Osterreichische Stickstoffwerke, AG; Power Gas Corp.; Uhde, GMBH	*Chem. Week*, 3/19/66, p. 87; *Chem. Eng.* (10/11/65), p. 180–182
Mercury adsorption	S. F. Yavorovskaya et al.	*Khim. Prom.* (1955), 91
Metal oxide reductions	H. G. McGrath and L. C. Rubin	U.S. Pat. 2, 671, 765 (March 9, 1954)
Metal powders of high purity	J. E. Drapeau and R. J. Halsted	U.S. Pat. 2, 758, 021 (Aug. 7, 1956)
Reclamation of scrap metals	USI (Eng.) Ltd.	*Br. Chem. Eng.*, *14*, No. 8, p. 1041, Aug. 1969
Metal carbide and nitrides	AERE (Gt. Brit.)	*Chem. Eng.*, 134–136, 11/11/63
Methane reduction with cupric oxide	W. K. Lewis et. al.	*Ind. Eng. Chem. 41*, 1227 (1949)
LPM process for fluidized bed methanation of H_2 + CO	Chem. Systems, Inc.	*C.E.N.*, p. 30, 1/16/78
Catalytic cracking of methylcyclohexane	S. Tone et al.	*J. Chem. Eng. Jpn.* 7, No. 1, p. 44 (1974)
Molybdenum trioxide reduction	J. M. Dunoyer	*Proc. Int. Symp. Reactivity Solids*, Gothenburg (1952), p. 411
Recovery of metals from dilute solutions	Elec. Council Res. Ctr., Capenhurst, U.K.; Rockwell Intl.; Constructors John Brown	*Chem. Eng.*, p. 44 (12/18/78); p. 78, 5/11/73, *Br. Chem. Eng. 15*, No. 9, 1191 (Sept. 1970)
Condensation of naphthalene	J. Ciborowski	*Int. Chem. Eng.* 2, No. 1, 105 (1962)
Neutralization of liquids	H. W. Gehm and L. T. Purcell	U.S. Pat. 2, 642, 393 (June 16, 1953)
Nickel oxide reduction	A. Kivnick and N. Hixson	*Chem. Eng. Prog.* 48 (8), 394 (1952)
NiO and nickel chloride	Falconbridge Nickel Mines, Quebec	*Chem. Week*, p. 52 (2/10/71); *Can. Min. Met. Bull.* (August, 1961), p. 601

Table 9.1. (*Continued*)

APPLICATION	ORGANIZATION	REFERENCE
Niobium pentachloride reduction with hydrogen	Battelle Memorial Inst.; National Steel Corp.; Nova Beaucage Mines, Ltd., Lake Nipissing, ON, Canada	*Intern. Cong. Chem. Eng.*, June 19–22, 1960, Mexico City, Mexico; *Chem Eng. 68* (2), 124 (Jan. 23, 1961); *Chem. Eng. News*, p. 51 (July 4, 1960)
Nitric acid by the "Wisconsin process"	Food Machinery & Chemical Corp.	*Chem. Eng. Prog. 52* (11), 483 (1956)
Nuclear liquid fluidized-bed reactor	Martin Co.	*Chem. Week 85* (18) 59 (Oct. 31, 1959)
Numerous processes	Society of Chem. Industry, England	*Fluidisation*, (1964)
Olefin polymerization	J. A. Carver	U.S. Pat. 2, 686, 110 (Oct. 10, 1954)
Olefins from crude oil on fluidized coke	BASF, Ludwigshafen, Germany	*Chem. Eng. News 37* (45), 50 (Nov. 9, 1959); *Chem. Eng.*, p. 17 (11/30/70)
Oxidation of aromatic hydrocarbons in fluid beds	H. K. Pargal	Ph.D. Thesis, Univ. of Colorado (1954)
Oxygen production	P. J. Gaylor	*Petrol. Proc. 5,* 1211 (1950)
Generation of ozone	Iowa State Univ.	*I.E.C. 59*, No. 3 64 (1967)
Packaging of solids	St. Regis Paper Co.	*Chem. Proc. 21,* 182 (March 1958)
Paper mill black liquor recovery	Container Corp. of America; Copeland Process Corp.; Green Bay Packaging Inc.; Dorr-Oliver	*Tappi 47*, No. 6, 175A, 1964; *Chem. Week* (7/31/65) p. 31
Continuous particle size separation	Baskakov et al.; SWECO, Inc.	*Khim. Prom.*, No. 6, p. 59 (1974); *Chem. Processing*, p. 60 (mid-Nov. 1978)
Perchlorethylene	Diamond Alkali; Columbia-Southern Chemical Corp. (Pittsburgh, Plate Glass Co.)	*Hydrocarb. Proc. 46,* No. 11, 210, 1967; U.S. Pats. 2, 914, 575 2, 914, 576, (Nov. 24, 1959); 2, 951, 103 (Aug. 30, 1960); 2, 952, 714 (Sept. 13, 1960); 2, 957, 924 (Oct. 25, 1960)
Phenanthraquinone	U.S. Steel Corp.	Br. Pat. 771, 085 (March 27, 1957)
Production of phosphorus	Goldberger, W. M. (Battelle)	*C.E.P. 61*, No. 2, p. 63 (February 1965)

Table 9.1. (Continued)

APPLICATION	ORGANIZATION	REFERENCE
CaO and phosphorus from tricalcium phosphate in a plasma bed at 2000°F	Battelle Memorial Inst.	*A.I.Ch.E. Symp. Ser.* No. 62, Vol. *62*, 42 (1966)
Phthalic anhydride by air oxidation of phenathrene in a fluid bed	Central Fuel Research Institute of India	*Ind. Eng. Chem. 53* (7), 14A (1961)
Phthalic anhydride	American Cyanamid Co.	*Chem. Eng. 66* (25), 78 (Dec. 14, 1959); *Chem. Week 85* (3), 37 (July 18, 1959)
Phthalic anhydride from o-xylene	Badger (Cambridge, MA)	*Chem. Week* (11/28/64) p. 63; *C.E.P.*, *66*, No. 9, 49–58 (1970)
Prilling fertilizers	Fisons Fertilizers, Ltd.	Br. Pat. 1, 187, 372 *Br. Chem. Eng. 15*, No. 5, 585 (May 1970)
Fluidized solids propellants	Bell Aerospace Div. of Textron Corp.	*Chem. Eng.*, p. 54 (5/12/72)
Radioactive waste solidification	Newport News Ind. Corp.; Aerojet Energy Corp.; Atlantic-Richfield Hanford Co.	*C.E.N.*, p. 21 (10/16/78); *Chem. Eng.*, p. 39 (5/3/71)
Hydrocarbon reforming		*Br. Chem. Eng.*, p. 27, (Dec. 1971)
Hydrogenation of waste rubber tires	Hydrocarbon Res., Trenton, NJ	Paper presented at 165th ACS Nat'l Mtg. Dallas, Texas (4/9-13/1973)
Scouring in sea water desalination	Brookhaven Nat'l Lab.	*C.E.N.* (3/29/1965), p. 42
Scrubbing of gases	Aerotec Industries; Aluminum Co. of Canada Ltd.	*Chem. Eng. 66* (5), 106 (Dec. 14, 1959)
Shale preheater	Oil Shale Corp. (TOSCO)	*C.E.N.*, p. 24 (5/14/73)
Silicon-organic compounds	S. Nagata et al.	*Chem. Eng.* (Japan) *16*, 301 (1952)
Si and Zr tetrachlorides	Stauffer Chemical Co.	*Chem. Eng.*, p. 90 (9/3/62)
$HSiCl_3$ from Si + HCl	K. A. Andrianov, Russian Acad. of Sciences	*Br. Chem. Eng. 11* No. 9, 927 (1966)
Styrenes from aromatic hydrocarbons	Mamedaliev Inst. ASSR	*Int. Chem. Eng. 4*, No. 3, 382 (1964)
Sugar, process liquor treating	Spreckles Sugar Co.; American Sugar Co.; Socony Mobil	*Sugar Azucar 56* (5), 33 (1961); *Chem. Week 86* (6), 59 (February 6, 1960)
Combustion of S to SO_2	Celleco	*Br. Chem. Eng. 8*, No. 6, 414 (1963)

Table 9.1. (*Continued*)

APPLICATION	ORGANIZATION	REFERENCE
SO$_2$ removal from gases	R. J. Best and J. G. Yates, Exxon-Esso, Abingdon, England; Mitsubishi, Japan; Westvaco Corp., NY	*I.E.C. Proc Des. Dev. 16*, No. 3, 347–352 (1977); *Ind. Res.*, p. 23, May 1971; *Chem. Week*, p. 55, 4/15/67; *C.E.N.*, p. 85 (2/15/71)
Sulfur and NH$_3$ recovery	Chemetics Intn'l, Can. Inds., Ltd.	*Chem. Week*, p. 68 (11/22/72)
Sulfur recovery from coke-oven gas	Appleby-Frodingham Steel, England	*Chem. Eng. 67* (9), 109 (May 2, 1960); *J. Inst. Fuel 42*, 319 (July 1958); *Chem. Eng. 65* (21), 74 (Oct. 20, 1958)
Terephthalonitrile	Lummus (Bloomfield, NJ)	*Chem. Week*, p. 27 (4/11/73)
Tetrafluorohydrazine from NF$_3$ and carbon	Stauffer Chemical Co.	*Chem. Eng. 68* (2), 124 (Jan. 23, 1961); *Chem. Eng. News*, p. 85 (Sept. 19, 1960)
Titanium dioxide production	LaPorte Titanium Ltd. (London)	*Chem. Week 88* (24), 74 (June 17 1961); *Br. Chem. Eng. 16*, No. 1, 17 (Jan. 1971)
Titanium tetrachloride	Fabriques De Produits Chimiques, Belgium	B.P. No. 1, 184, 199; *Br. Chem. Eng., 15*, No. 6, 735 (June 1970)
Chlorinating rutile to TiCl$_4$ at 1800°F	Chemical and Metallurgical Research, Inc.	*Chem. Week*, p. 139 (9/15/62)
Tungsten from H$_2$ reduction of WF$_6$	Battelle Mem. Inst.; Allied Chem. Co.	*Nucl. Applic. 1*, 567 (1965)
H$_2$ reduction of Nb, Ta and W halides	E. I. duPont de Nemours	U.S. Pat. 3, 020, 148
Uranium from fluid-bed roasting of lignite	International Resources Corp., Custer, S. Dakota	*Chem. Eng. 67* (19), 113 (May 2, 1960)
Uranium from spent metal by halogenation	Brookhaven National Laboratory	*Chem. Eng. Prog. 56* (3), 96 (1960)
Uranium oxides and fluorides	General Chemical Div., Allied Chemical Corp.	*Chem. Eng. 67* (14), 70 (July 11, 1960); *Chem. Week 87* (2), 41 (July 9, 1960)
Fluorination of U to UF$_6$	General Electric Co. Union Carbide Corp.	*Business Week*, p. 76, 11/16/74; Preprint 42D, 60th annual A.I.Ch.E. Mtg. N.Y (11/26-30/67)
Direct conversion of UF$_6$ to UO$_2$ and UO$_2$ to UF$_6$	Argonne Nat'l Lab., E. I. duPont de Nemours	*Nuc. Sci. Eng. 20*, 259 (1964); *I.E.C. Proc. Des. Dev. 4*, No. 3, 338 (1965)

Table 9.1. (*Continued*)

APPLICATION	ORGANIZATION	REFERENCE
Fluorination of UF_4 to UF_6 at 2000°F	Union Carbide Nuclear Co., Paducah, KY	*C.E.N.*, p. 42 8/20/62
Uranium dioxide by reduction of UO_2	Union Carbide Corp., Nuclear Div.	*Chem. Eng. News*, p. 40 (Mar. 23, 1959); *Chem. Eng. Prog. 56* (3), 4 (1960); *Chem. Eng. 66* (2), 140 (Mar. 23, 1959)
UO_3 from denitration of uranyl nitrate	Mallinckrodt Chemical Works	*Chem. Eng. 67* (6), 80 (March 21, 1960)
UO_2 from UF_6	Royal Inst. of Tech., Stockholm	*Nuclear Technol.*, *18*, 177–184 (May, 1973)
UC and UN from UO_2	AERE, Harwell, England	*Chem. Eng.* (11/11/63), p.134
UC from U plus hydrocarbons	Argonne Nat'l Lab.; Ill. Inst. of Tech.	*C.E.P. Symp. Ser.* No. 67, Vol. 62, p. 76 (1966)
Vinyl acetate production	K. Kawamichi et al.	Jap. Pat. 1863 (April 30, 1953)
Vinyl chloride production	M. G. Geiger, Jr. Pechiney-Saint-Gobain	Ph.D. Thesis, Purdue Univ. 1953–1954; C.E.N., p. 39 (5/11/70)
Oxychlorination to vinyl chloride	B. F. Goodrich; Ethyl Corp.	*Chem. Week*, p. 93–107 8/22/64, *Chem. Eng.*, p. 105, 10/18/71; *C.E.P.*, *61*, No. 1, 21–26 (Jan. 1965)
Vibratory fluid bed cooler	Rexnord, Inc. Alfa-Lavalthermal, Inc.	*Chemical Proc.*, p. 47 (mid-Nov. 1978); Chem. Eng., p. 113 (4/10/78)
Vulcanization of rubber	Rubber and Plastics Research Assoc. of Gt. Britain	*C.E.N.*, p. 41 (8/13/62); *Chem. Eng.*, p. 60 (8/20/62)
Hy-Flo process for treating ind. and municipal waste water	Ecolotrol-Dorr Oliver	*Chem. Week*, p. 40, 9/1/76; *Chem. Eng.*, p. 51 (2/13/78)
Fluid bed distillation of wood	Ga. Inst. of Tech.	*I.E.C. Proc. Des. Dev. 2*, No. 2, 148 (1963)
Wood distillation	M. S. Dimitri et al.	*Chem. Eng. 55* (12), 124 (1948)
Wood distillation	L. W. Morgan et al.	*Chem. Eng. Prog. 49* (2), 98 (1953)

Table 9.1. (*Continued*)

APPLICATION	ORGANIZATION	REFERENCE
Wood pulping	Univ. of Florida	*Chem. Week 86* (1), 34 (Jan. 2, 1960)
Regenerating spent NH_3 base pulping liquor	Copeland Systems, Inc.	U.S. 3, 927, 174
Zinc ore roasting for contact acid	T. T. Anderson and R. Bolduc	*Chem. Eng. Prog. 49* (10), 527 (1953)
Zinc production	P. W. Garbo	U.S. Pat. 2, 475–607 (July 12, 1949)
Zinc ore roasting	Societé des Mines et Fonderies de la Vieille-Montagne, Belgium	*Chem. Week 87* (13), 66 (Sept. 30, 1961)
Recovery of zinc chloride	Conoco Coal Dev. Co.	*Chem. Week*, p. 37, 38 (12/21/77); *I.E.C. Proc. Des. Dev. 8*, No. 4, 552–558 (1969)

point, the weight of material in the pipe more than equaling the differential pressure.

9.2 ADVANTAGES AND DISADVANTAGES OF THE FLUIDIZED TECHNIQUE

9.2.1 Advantages

9.2.1.1 Temperature Control

The ability of the fluidized-solid bed to approach isothermal conditions is the outstanding advantage of this method over other methods of carrying out reactions. This factor is vital to nearly all applications, the other advantages generally being of lesser importance.

Close control of reaction variables is well known to be important in obtaining maximum yields of desired products. Of the several variables, temperature is one of the most important, for reaction rates change exponentially with temperature (often doubling for a 10°C change). In the common case in which several competing reactions may occur, a temperature change of a few degrees may shift the balance between the several rates from a favorable one to an unfavorable one.

The relatively close control of temperature that is possible in a fluidized-solid bed is due

to a combination of the following three factors (listed in the order of their importance):

1. Turbulent agitation within the fluidized mass, which breaks and disperses any hot or cold spots throughout the bed before they grow to significant size. It should not be inferred from this statement that the temperature of every solid particle in a given fluidized-catalyst bed is the same. The catalytic activity will differ somewhat from particle to particle, and those with greater activity will accelerate the reaction in their neighborhood to a greater extent. As a consequence, their temperature will be different from that of the surrounding particles of lower activity. However, the departure of the individual particle temperature from the mean value for the bed will be much less than in a fixed-bed converter because of the turbulent mixing, the high heat-transfer rates, and the high bed heat capacity.

2. High heat capacity of the bed relative to the gas within it. This factor stabilizes the temperature of the bed, permitting it to absorb relatively large heat surges with only small temperature change. For example, a bed of ordinary sand, fluidized with air at a solids concentration of about 70 lb/ft^3

Table 9.2. Some Commercial Fluid Bed Installations in the Mining and Metallurgical Industries

ROASTING PYRITE OR PYRRHOTITE FOR SULFURIC ACID MANUFACTURE

	LOCATION	NO. UNITS	TONS / DAY FEED	SIZE UNIT (FT)	TONS H_2SO_4 DAY	REMARKS
1) West Rand	S. Africa	1	22	14	35	
2) West Rand	S. Africa	1	22	14	35	
3) Daggafontein	S. Africa	3	235	20	250	
4) Western Keefs	S. Africa	3	235	20	250	
5) Bethlehem Steel	USA	3	180	18	250	Sulfating cobalt
6) Anaconda	USA	4	600	18	450	Low grade sulfur
7) Maria Cristina	P.I.	1	120	18	150	
8) Dowa Okayama	Japan	1	130	20	130	Sulfating Cu
9) Randfontein	S. Africa	3	235	20	250	
10) Albuffoe	The Netherlands	1	50	12	67	Gas recycle Flotation conc.
11) Rumianca Pieve V.	Italy	1	70	20	90	Flotation conc. Gas recycle
12) Rumianca Pieve V.	Italy	1	70	10	90	Massive pyrite Gas recycle
13) Rumianca Asenza	Italy	1	70	3	90	Massive pyrite Gas recycle
14) Virginia mines	S. Africa	3	300	26	350	
15) Stilfontein	S. Africa	1	55	18	65	
16) Stilfontein	S. Africa	2	135	20	150	
17) Sumitomo	Japan	2	240	16	300	
18) Rico Argentine	USA	1	115	20	150	
19) Electrochemica Surden	Italy	1	10	5	13	Massive and fle. conc.
20) Chemische Werke Albert	Germany	1	35	7	50	
21) Soc. Interconzorsiale Romanagnola	Italy	1	100	10	125	Massive and flot. conc.
22) Soc. Montecatini Spinetta M.	Italy	2	155	9	200	Gas recycle
23) Shin Nippon	Japan	1	100	22	125	
24) Dowa Okayama	Japan	1	125	20	110	
25) Illyvooruitricht	S. Africa	2	130	20	150	
26) Lorado	Canada	1	60	14	50	
27) Atlas Fertilizer	P.I.	1	105	18	120	
28) St. Gebain St. Fons	France	2	235	10	300	Gas recycle
29) Kennecott	USA	1	130	22	100	Calcine for sponge iron
30) Anaconda	USA	1	175	22	150	Dry feed
31) Chile Exploration	Chile	1	90	22	100	(10,000′ elev.)
32) Tokal Ryuan	Japan	2	176	22	150	
33) Felli Nutti	Italy	1	25	6		H_2SO_4 and liquid SO_2
34) Buffelsfontein	S. Africa	1	85	24	100	
35) Harmony Gold Mine	S. Africa	2	100	18	120	
36) Buffelsfontein Est.	S. Africa	2	130	20	150	
37) British Titan Prod.	England	3	340 TPD $FeSO_4$ 18 H_2O	16	180	Decomposition of ferrous sulfate monohydrate for iron and sulfuric acid recovery

Table 9.2. (*Continued*)

ROASTING PYRITE OR PYRRHOTITE FOR SULFURIC ACID MANUFACTURE

	LOCATION	NO. UNITS	TONS / DAY FEED	SIZE UNIT (FT)	TONS H_2SO_4 DAY	REMARKS
38) Calvo Sutelo	Spain	2	160	10	200	
39) Stauffer Chemical	USA	1	120	10	150	
40) Transvaal Gold	S. Africa	1	15	8	20	Also gold and copper
41) Abonos Sevdia	Spain	2	150	9	200	Massive pyrite coils in bed
42) Kaohsiung	Taiwan	1	250	20	300	Coils in bed
43) Repesa	Spain	2	360	10′ and 18′	500	Massive pyrite two-stage roaster
44) Albatros	The Netherlands	2	250	22	350	Coils in bed Flot. conc.
45) Sa. Montecatins Follonica	Italy	2	1,000	18.5	1,000	Massive pyrite 36% S-cooling coils in bed
46) Kimoshima Chemical	Japan	1	94 MT/D	14	Chamber acid	Pyrite
47) Kowa Seiko	Japan	1	440	26		Pyrite (Vanahara)
48) Kowa Seiko	Japan	2		28	600	Pyrite (Hanaoka)
49) Consolidated Mining & Smelting Co.	Canada	1	350	26		
50) Nehanga	S. Africa	1	500	22		Low-grade copper concentrate
51) Hartebeest-fontein	S. Africa	1	12.9 ST/D	10		Pyrite roaster
52) Kosovaka Mitrovica	Yugoslavia (former)	2	370 MT/D	18		Pyrrhotite Flot. conc.
53) Cinkarna Celje	Yugoslavia (former)	1	140 MT/D	20		Flot. conc.
54) Eseo	Philippines	2	540 T/D Pyrite + 25 T/D of H_2S	30	750 T/D	Pyrite roaster

ZINC-H_2SO_4 FOR ACID

	LOCATION	NO. UNITS	TONS / DAY FEED	SIZE UNIT (FT)	TONS H_2SO_4
1) Mitsubishi (Talhei)	Japan	1	55	14	55
2) National Zinc	USA	2	500	18	200
3) Anaconda	USA	1	155	22	155
4) Alcan	Canada	1	150	22	150
5) Dowa Kosaka	Japan	1	20	10	20
6) Nippon Soda	Japan	1	55	14	55
7) Mitsui Mining & Smelt	Japan	1	140	24	140
8) Cinkarna Celje	Yugoslavia (former)	2	110	15	
9) Mitsubishi Metal, Akita	Japan	1	140 MT/D	24	120
10) Mitsubishi Mining, Hosokura	Japan	1	96 T/D	18	70
11) Zorba Sabac	Yugoslavia (former)	1	50 MT/D	11	
12) Toho Zinc	Japan	1	80	18	70

Table 9.2. (*Continued*)

	LOCATION	UNITS	TONS / DAY / FEED	SIZE UNIT (FT)	MATERIAL TREATED
ROASTING OF MISCELLANEOUS SULFIDES FOR METALLURGICAL PURPOSES					
1) Cochenour Williams	Canada	1	20	6	Arsenic removal for gold
2) Golden Cycle	USA	1	65	14	Telurium removal for gold
3) Campbell Red Lake	Canada	1	65	14	Arsenic removal for gold
4) New Dickenson	Canada	1	20	6	Arsenic removal for gold
5) Giant Yellow-knife	Canada	1	55	14	Arsenic removal for gold
6) Negus Mines	Canada	1	30	10	Arsenic removal for gold
7) Dowa Mining Co.	Japan	1	85	20	Copper zinc sulfating
8) Falcon Mines	Zimbabwe	1	50	16	Arsenic removal for gold
9) St. John del Rey	Brazil	1	55	14	Arsenic removal for gold
10) Campbell Red Lake	Canada	1	45	12	Arsenic removal for gold
11) Kilembe	Uganda	1	70	16	Dead roast copper
12) Union Miniere	Zaire	1	85	14	Sulfate roast copper and cobalt
13) Anaconda	USA	1	110	10	Sulfate removal for sponge iron
14) Kerr-Addison	Canada	1	115	22	Sulfate removal for gold
15) Macalder-Nyanza	Kenya	1	75	20	Copper-zinc sulfate roast
16) Anaconda	USA	1	175	22	Calcines for sponge iron
17) Kerr-Addison	Canada	1	115	22	Sulfate removal for gold
18) Ndola Copper	Zambia	1	45	17	Copper cobalt sulfating
19) Mitsubishi Metal	Japan	1	50	14	Dead roast copper-zinc conc.
20) La Luz	Nicaragua	1	50	16	Copper dead roast
21) Sherritt Gordon	Canada	1	30	3	Nickel-copper concentrate
22) Giant Yellow-knife	Canada	2	150	14' & 16'	Arsenic removal for gold
23) Ndola Copper Refinery	Zambia	1	65	20	Copper-cobalt sulfating
24) Rhokana Corp.	Zambia	1	130	20	Sulfate roast copper and cobalt
25) Union Miniere	Zaire	1	150	16	Copper-cobalt sulfating
26) SGM d'Hoboken	Belgium	1	45	18	Cobalt sulfating
27) Union Miniere	Zaire	1	150	16	Copper-cobalt sulfating
28) International Nickel Co. (Thompson)	Canada	3	900	15	Partial roasting sulfide conc.
29) International Nickel Copper Cliff	Canada	3	500	12	Nickel sulfide roasts
30) Tennessee Copper Co.	USA	1	300	12	Partial roasting of sulfide
31) Getchell Mines	USA (Nevada)	1	1800	16	Gold ore
32) Associated Lead Mfgs.	England (Newcastle)	1	4–11	5	Oxidizing roasts of residue mattes for tin and copper recovery
33) Johnson Matthey & Co.	England (Brimsdown)	3	3–9	$4'-1\frac{1}{2}''$	Oxidizing roast of residue matter for nickel and other recoveries
34) Johnson & Sons Smelting Works Limited	England	1	280–840 T/h	$4'-1\frac{1}{2}''$	Oxidation of minerals semiprecious mineral recovery
35) Phelps-Dodge	USA	1	66 T/h	22	Copper conc. partial roast

Table 9.2. (*Continued*)

| | | | | SIZE | |
	LOCATION	NO. UNITS	TONS / H	UNIT (FT)	MATERIAL
DRYING AND SIZING					
1) Nelco-Canaan	USA	1	50	6	Dolomite dryer
2) Nelco-Adams	USA	1	110	8	Limestone dryer
3) National Gypsum	USA	1	30	9	Dolomite dryer
4) Marquette Cement	USA	1	40	4	Cement rock dryer
5) Columbia-Southern	USA	1	50	5	Limestone dryer
6) Peerless Cement	USA	1	35	8	Blast furnace slag
7) Wyandotte Chemical	USA	1	32	8	Blast furnace slag
8) Lone Star Cement	USA	1	40	6	Oyster shell
9) Universal Atlas Cement NY	USA	1	40	8	Blast furnace slag
10) Umgababa Minerals	S. Africa	1	30	12	Ilmenite conc.
11) Monsanto Chemical	USA	1	530 lbs	5	Chlorinated hydrate dryer
12) National Lime & Stone	USA	1	150	12	Limestone dryer
13) Universal Atlas (Ala.)	USA	1	40	8	Blast furnace slag
14) J. G. Stein Co.	Scotland	2	10	6	Clay dryer and iron carbonate to magnetite converter
15) Associated Cement	India	1	30	7	Slag dryer
16) Guardite Company	USA	1	15	3	Sand dryer
17) Victor Chemical	USA	1	1.14	5	Chemicals
18) Detergent Co.	USA	1	10	14	Detergent
19) Detergent Co.	USA	1	10	14	Detergent
20) Iron Ore Co.	Canada	1	10	3	Iron ore dryer
21) Quebec Cartier	Canada	2	500	12	Iron ore dryer
22) Dorr-Oliver-Long	Canada	1	15	3	Sand dryer
23) Victor Chemical	USA	1	2.75	$9\frac{1}{2}$	Chemicals
24) Victor Chemical	USA	1	0.75	$9\frac{1}{2}$	Chemicals
25) Huron Portland	USA	1	500	11	Limestone dryer
26) Chemical Lime Inc.	USA	1	167	6	Limestone dryer
27) Anglo Alpha Cement	S. Africa	1	20	4'–4"	Slag dryer (suction system) (Conversion of competitive dryer)
28) Wabash Mines Ltd.	Canada	3	370	14	Iron ore dryer
29) IMC	USA	1	323 LT/h	14	Phosphate rock conc.
30) Victorville	USA	1	150	5	Limestone dryer
31) Glidden	USA	1	20	5'–8"	Ilmenite dryer
32) Chevrolet Motors	USA	1	20	4'–9"	Foundry sand
33) Victor Chemical	USA	1	6	15	Chemicals
34) African Metals	S. Africa	2			Phosphate conc. dryer
35) Montana Phosphate	USA	1	130 T/D	9	Phosphate rock
36) U.S. Reduction	USA	1	4,000 f/h	3	Alumina compounds

Table 9.2. (*Continued*)

	LOCATION	NO. UNITS	SIZE (FT)	TONS / DAY	MATERIAL
		CALCINING SYSTEMS			
1) New England Lime Co.	USA	1	12	100 (CaO)	Limestone (5-compt.)
2) Wright Construction	USA	1	7.12 14.11	360	Chrome ore
3) City of Lansing, Mich.	USA	1	6	30	Lime sludge
4) New England Lime Co.	USA	1	12	100	Limestone
5) J. G. Stein Co.	Scotland	1	9	175	Fire clay (4 compt.)
6) Central Farmers	USA	1	14	500	Phosphate rock (3 compt.)
7) Corn Products	USA	1	7	25	Carbon reactivator (3 compt.)
8) Cleveland-Cliffs	USA	1	3	24	Iron ore pilot plant
9) San Francisco Chemical	USA	2	14	1000	Phosphate rock (3 compt.)
10) Anglo Lautaro	Chile	1	7	180	Caliche (2 compt.)
11) Caroline Tufflite Co.	USA	1	8	960	Stone preheater (3 compt.)
12) J. G. Stein Co.	Scotland	1	9	175	Fire clay (4 compt.)
13) J. R. Simplot Co.	USA	1	15	1000	Phosphate rock calciner
14) Chemical Lime Co.	USA	1	12	200 (CuO)	Limestone calciner
15) Djebel Onk	Algeria	3	23	3000	3 compartment with aftercooler. Phosphate rock
16) Soc. Montecatini Follonica	Italy	2	10	750	3 compartment magnetic roast
17) Bay State Abrasives	USA	1	3	24	SiC grit
18) W. S. Moore	USA	1	5	120 LT/D	Pilot plant-iron ore
19) GAFSA	Tunisia	1	4	50 LT/D	Phosphate rock
20) Djebel Onk	Algeria	1	4	50 LT/D	Phosphate rock
21) OCP	Morocco	1	4	50 LT/D	Phosphate rock
22) International Minerals	USA	1	4	25 LT/D	Phosphate rock
23) Smith Douglas	USA	1	4	25 LT/D	Phosphate rock
24) S. D. Warren Co.	USA	1	10	70 T/D	Lime mud reburning
25) Kimberley-Clark	USA	1	9	50	1 compartment— paper mill lime sludge
26) Billiton	The Netherlands	1	3		Pilot (2 compt.) tin volenbering
27) Anchor Minerals	USA	1	12	220 T/D	Limestone calciner
28) J. R. Simplot	USA	1	15	1000 T/D	Phosphate rock
29) Chas. Pfizer (NELCO)	USA	1	12'	200 ST/D	Limestone
30) Phelps-Dodge	USA	1	13'	220 T/D	Lime calciner
31) Ford Motor Co.	USA	1	9'	15 T/h	Foundry sand
32) Ford Motor Co.	Mexico	1	9'–6"	2 T/h	Foundry sand
33) Gainesville	USA	1	$2\frac{1}{2}''$	4 T/D	Lime mud reburning

Table 9.3. Fluidized Bed Coal Combustion Developments among Major U.S. Interests.

1) Argonne Nat'l Lab. (Argonne, IL)	PFBC	a) Designing a 24,000–94,000 lb/h steam boiler as a CTIU to fire 6 to 18 tons of coal/day, expected completion 1981; Stearns-Roger doing the actual design; Bed vel. ≈ 7 ft/s, press, 3–12 atm, size $3' \times 3'$ to $4' \times 4'$.
		b) Operated a 6″ diameter PFBC at up to 10 atm pressure.
2) American Electric Power Co. (NY)	PFBC	a) Sponsored feasibility study of combined cycle plant (105 MW_e steam turbine and 65–70 MW_e gas turbine) with Stal-Laval and Woodall–Duckham; sponsoring pilot scale tests at CRE in Leatherhead.
3) Babcock Contractors (subsidiary of Babcock & Wilcox Ltd.)	AFBC	a) Installing a 60,000 lb/h boiler in an Ohio State mental hospital (at a cost of $4,300,000).
4) Babcock & Wilcox (Alliance, OH)	AFBC	a) Operating a 20,000 lb/h steam $6' \times 6'$ boiler under EPRI contract firing 22 tons/day of coal.
		b) Operates a $39'' \times 39''$ unit under EPRI contracts to study sorbent utilization firing 500 lbs/h of coal with 8 ft/s fluidizing velocity.
		c) Operates a $1' \times 1'$ unit also studying sorbent and fuel characteristics.
5) Babcock & Wilcox (North Canton, OH)	AFBC	a) Has designed a product line of FBC boilers producing steam in the 50,000 to 300,000 lb/h range.
6) Battelle (Columbus, OH)	AFBC	a) Has operated a 6″ diameter MS-FBC bench-scale unit with a combustor superficial velocity of 30 ft/s.
		b) Is designing (with Foster-Wheeler and A. G. McKee) a 40,000 lb/h steam boiler prototype to serve Battelle's facility wherein the combustor will be 6′–0″ I.D. and the external heat exchanged $5' \times 10'$.
7) Burns & Roe Inc. (Oradell, NJ)	AFBC	a) Designing a 200–300 megawatt utility power plant under a DOE $1,300,000 study contract.
		b) Designed a 570 MW_e under a DOE contract.
	PFBC	c) Designed a 583 MW_e FBC combined cycle plant in association with Babcock & Wilcox and Pratt & Whiting div. of United Technologies, Inc.; unit will have 5 beds $71' \times 24'$.
8) Cleaver Brooks Div. of Aqua-Chem. Inc. (Milwaukee, WI)	AFBC	a) A mfr. of packaged boilers supplied the Alexandria VA test facility of Pope, Evans & Robbins.
9) Combustion Engineering (Windsor Locks, CT)	AFBC	a) Designing a 50,000 lb/h steam generator at the Great Lakes Naval Training Center scheduled to be in operation before the end of 1980.

Table 9.3. (*Continued*)

		b) Operating a $3' \times 3'$ test unit delivering 2300 lbs/h steam; also have a Plexiglas cold flow model of this unit.
		c) Made a design study for a retrofit FBC unit for Con. Edison's 500 MW$_e$ Arthur Kill #3 unit on Staten Island under EPRI and NYSERDA sponsorship.
10) Combustion Power Corp. (Menlo Park, CA)	PFBC	a) Operated a 40 ft^2 bed ($\sim 7'$ I.D.) at 4 atm press. Burning Illinois No. 6 coal (4% sulfur).
	APBC	b) Operated a 2.2 ft^2 combustor principally for corrosion testing.
11) Curtiss-Wright Corp. (Woodridge, NJ) with Stone & Webster (Boston) and Dorr-Oliver (Stamford, CT)	PFBC	a) Constructing a combustor capable of producing 130,000 lbs/h of steam from combustion of 109 tons/day of coal.
		b) Operating a bench scale unit $\frac{1}{4}$ the diameter of the above 13 MW$_e$ pilot plant to provide design data.
12) The Ducon Co. (Mineola, NY)	AFBC or PFBC	a) Has patented a modular combustor incorporating conceptual improvements in mechanical design, control, and operation.
13) Energy Products of Idaho, a subsidiary of Energy, Inc. (Idaho Falls)	AFBC	a) Has sold a dozen FB incinerators (Trade named Fluid Flame) in sizes producing 10,000 to 30,000 lbs/h steam burning wood wastes, corn cobs, fruit pits etc.
14) Energy Resources Co. (Cambridge, MA)	AFBC	a) Has operated a $2' \times 2'$ FBC unit at 4–16 ft/s testing 16 different coals.
		b) Is building a $6' \times 6'$ unit for further test work and to establish credibility.
		c) Offering to design and build industrial size AFBC boilers producing up to 500,000 lbs/h of steam.
15) Exxon Res. & Engr'g Co. (Florham Park, NJ)	AFBC	a) Operates a Plexiglas cold flow model $12'' \times 90''$ in cross-section and 12' high at super ficial air velocities up to 15 ft/s using horizontal simulated tubes 2" to 6" in diameter at various C–C spacings.
16) Exxon Res. & Engr'g Co. (Linden, NJ)	PFBC	a) Operates a 12" I.D. miniplant burning 380 lbs of coal/h with a max. operating press of 10 atm at superficial velocities up to 10 ft/s.
		b) Operates a 4.5" I.D. bench scale unit burning 30 lbs coal/h.
17) Fluidyne Eng. Corp. (Minneapolis)	AFBC	a) Constructing two 5×11 ft beds (capable of producing up to 28,000,000 MTU/h) at the Owatonna Tool Co. in Minnesota to supply high temp. air for paint-curing ovens, plating lines, and space heat.
		b) Are offering 15,000,000 BTU/h modules on the open market.
		c) Have 3 test units in operation for demonstration and customers' systems studies.
18) Fluidyne Eng. Co., The City of Wilkes-Barre, and the Shamokin Area Industrial Corp.	AFBC	a) Have submitted proposals to DOE for construction of units to burn 910,000,000 yds of culm or anthracite wastes (equivalent to 250,000,000 tons of coal) scattered in the northeastern parts of Pennsylvania.

Table 9.3. (*Continued*)

19) Foster Wheeler Corp. (Livingston, NJ)	AFBC	a) Constructing a 100,000 lb/h steam unit at Georgetown Univ.; this is a 2 bed unit with over bed feed of coal and limestone to operate an 8 ft/s fluidizing velocity burning 125 tons/day of coal.
		b) Operated a $6' \times 6''$ cold model to study scale up correlations regarding tube spacing, solids distribution, etc.
		c) Operates two test facilities in Livingston, one is $1.75' \times 1.75'$ and the other a 36 ft^2 bed.
		d) Offering commercial warranties on FBC boilers up to 600,000 lbs/h of steam
		e) Operated an 18″ diameter unit burning 100 lbs/h coal; this unit now at METC.
		f) Designed a 70,000 lb/h steam AFBC boiler for Ford Motor Co.
20) The General Electric Co. (Schenectady, NY)	AFBC & PFBC	a) Operate a $1' \times 1'$ and a $2' \times 2'$ test bed to explore effects on in-bed exchanger tubes.
21) General Electric and TVA	AFBC & PFBC	a) Completed conceptual design and cost comparison study of a 750–925 MW$_e$ FBC utility power plant.
22) Grand Forks Energy Research Center (Grand Forks, ND)	AFBC	a) This DOE laboratory has been testing various coals and lignites for bed sulfur retention in a 6″ diameter combustor.
23) International Boiler Works, subsidiary of Comb. Equip. Assoc. Inc. (East Stroudburg, PA)	AFBV	a) Manufacturing an FBC from an Energy Resources Co. design.
24) Johnston Boiler Co. (Ferrysburg, MI)	AFBC	a) Operates a 10,000 lb/h FBC steam generator at its Michigan offices.
		b) Offering 2500–50,000 lb steam/h units (under licence from Britains Fluidfire Ltd. through CSL).
		c) Constructing a 23,500,000 BTU/h boiler for the Ohio Center Convention Complex in Columbus, based on a design developed by Britain's CSL.
25) John Zink Co. (Tulsa, OH)	AFBC	a) Is reported to be fabricating an in-house FBC.
26) Morgantown Energy Technology Center (Morgantown, WV)	AFBC	a) Designing a 60,000 lb/h steam capacity unit to burn 10 to 90 tons of coal/day as a Component Test and Integration Unit with two beds each $6' \times 6'$ stacked vertically surmounted by a $3' \times 6'$ carbon burnup cell. The beds will operate at superficial velocities of 7 to 15 ft/s through beds 2–8 ft deep; this CTIU will be sited on the Medical Center Campus of West Virginia Univ. and is scheduled for operation end of 1979.
		b) Have an 18″ diameter unit for test purposes.
27) NASA-Lewis Res. Center (Cleveland, OH)	PFBC	a) Operates a bench scale PFBC for technical support studies; burns 10–80 lbs/h of coal at bed press of 25–200 psia in an 8.9″ to 20″ conical bed.

Table 9.3. (*Continued*)

28) Oak Ridge National Lab. (TN)	AFBC	a) Carried out conceptual design of a unit geared to large apartment complexes. b) Operated a $4' \times 4''$ cold flow model to observe bubble flow past tube arrays. c) Has developed a concept for an atmospheric fluidized bed coal combustor for cogeneration (ccc) to produce 300–500 kW of electricity plus some 2,500,000 BTU/h of useful heat.
29) Oregon State University (Corvalis, OR)	AFBC	a) Operate a $3' \times 3' \times 35'$ high cold flow model to study tube spacing effect, mining, etc.
30) Pope, Evans and Robbins (NY)	AFBC	a) Operating a 0.5 MW_e (5000 lb/h steam) pilot plant at Alexandria, VA), $1\frac{1}{2}' \times 6'$ in cross-section. b) Operating a 30 MW_e plant at Rivesville comprising 3 modules each $10' \times 12'$ rated at 30 MW_e plus a smaller "carbon burnup cell;" this unit produces 300,000 lbs steam/h burning 330 tons of coal/day. c) Has licensed Mitsubishi and IHI of Japan to build a 300,000 lb/h AFBC boiler to burn "coke breeze" waste.
31) Process Equipment Modelling and Mfg. Co. (Nelsonville, NY)	AFBC	a) Designed and operated a Plexiglas cold model of a 6" combustor and $3'' \times 14''$ external heat exchanger for Battelle's MS-FBC development.
32) Stone & Webster Eng. Co. (Boston)	AFBC	a) Designing a 500–600 megawatt utility power plant under a DOE $1,350,000 study contract.
33) The Trane Co. (La Crosse, WI)	AFBC	a) Has under development a proprietary design of FBC packaged boilers.
34) Univ. of West Va. (Morgantown)	AFBC	a) Operates a $2' \times 2'$ cold model for mixing studies and has under construction a $2' \times 2'$ FBC boiler.
35) Wormser Eng. Inc. (Lynn, MA)	AFBC	a) Operating a demonstration unit in a Lynn, MA, factory where it is used for space heating. b) Expects to offer a line of FBC boilers in the 2–10 MW_{Th} range.

would contain only about 0.05 lb of air/ft^3, corresponding to a mass ratio of 1400 to 1.

3. High heat-transfer rates, which are possible because of the large amount of transfer surface per unit volume of the fluidized bed. This permits rapid leveling of any temperature surges either from the incoming gas or from reaction within the bed. Although the heat-transfer coefficients are not usually high, the amount of surface per unit volume is very large; for example, the surface area of a bed of ordinary sand would be in the range of 1000 to 5000 ft^2/ft^3 of bed.

The remarkable uniformity of temperature in a well-fluidized bed has been noted in many references. Temperature traverses in large fluidized-catalyst beds indicate that the point-to-point variation is less than 10°C when the feed-gas temperature is not greatly different from that of the bed and particularly if the inlet gas is carefully distributed.

9.2.1.2 Continuity of Operation

The ability to handle the fluidized solid like a liquid permits the technique to be easily adapted to many continuous operations, thereby obtaining the advantages of lower labor requirements, precise and automatic control of process variables, and uniformity of product quality.

9.2.1.3 Heat Transfer

The fluidized-solid technique is a convenient method for transferring heat, either alone or in conjunction with other operations, such as catalysis, gas–solid reactions, and transport of solids and fluids incidental to these operations.

The advantages of the fluidized-solid technique for heat transfer are as follows:

1. The possibility of combining heat transfer with other operations.
2. For large heat-transfer units, the equipment volume can be smaller than with conventional heat exchangers, because so much transfer surface is available in the fluidized bed and there is a high rate of heat transfer between the bed and an external heat-transfer surface.
3. Corrosion resistance and extreme-temperature resistance can be easily obtained by using ceramic materials for vessels and for granular solids.
4. The transfer can be effected in two stages with the fluidized solid acting as the heat reservoir to carry the heat from one fluid to the other. The stages may be physically close together or far apart.
5. The transfer may be affected with extreme rapidity because of the large surface available; this is important when undesirable reactions would occur at intermediate temperatures.
6. Similarly, a liquid may be heated, vaporized, and dispersed in a small fraction of a

second by direct contact with hot fluidized solids.

9.2.1.4 Catalysis

Fluidized-solid technique is particularly adaptable to the contacting of free-flowing, non-sticky, granular solids with gases. It may therefore be applied in catalytic gas reactions in which solid catalysts are used. The technique has been most widely applied to the catalytic cracking of petroleum because of the unique combination of advantages that are inherent in fluidized-solid processing:

1. Control of reaction temperature
2. Maintenance of uniform catalyst activity (as well as continuous catalyst regeneration)
3. Continuous removal of solid byproducts
4. Supply of heat to endothermic reaction
5. Simple equipment with few moving parts
6. Continuous operation with automatic control.

Most catalysts gradually lose their activity during use because of poisoning or coating of the active surface with byproducts. Replacement or regeneration is therefore eventually required. As the activity decreases, operating conditions in conventional catalyst contactors must be altered in order to maintain the operating rate. Use of a higher temperature is one expedient, but this may increase the cost of the product. Lowering the operating rate is another expedient, but this also results in increased costs through higher investment because the plant must be large enough to manufacture at an average rate sufficient to compensate for the low-rate period.

In contrast, the fluidized-solid technique makes possible the maintenance of a definite level of catalyst activity because partially spent catalyst can be continuously withdrawn and fresh catalyst added. The level of activity is determined by the proportion between the rate of loss of catalyst activity and the rate of withdrawal of the catalyst. A stable catalyst will require a lower catalyst withdrawal rate than would one which loses activity rapidly; correspondingly, a higher level of activity would require a larger withdrawal rate. The activity in any actual bed must be a compromise based on economic considerations, taking into account the following factors:

1. Relationship between yields and catalyst activity
2. Value of incremental yield
3. Rate of catalyst degradation
4. Costs of handling and regenerating the catalyst
5. Catalyst losses during handling and regenerating
6. Cost of catalyst.

9.2.1.5 Gas–Solid Reactions

Advantages of the fluidized-solid technique for carrying out gas–solid reactions are the isothermal reaction bed, the easily varied time of contact, the effective contact (as compared with rotary kilns or tray-type reactors), the simple methods of handling solids (no moving parts) and transferring heat, and the ease of continuous, automatic operation.

9.2.2 Disadvantages

Solids that do not flow freely or tend to agglomerate cannot be processed in a fluidized-solid reactor; rotary kilns and tray-pipe reactors are not thus limited. As the reaction proceeds, fine solid particles may be formed that will become entrained in the gas leaving the fluidized bed; recovery means must usually be included in the design. The pressure drop in the gas system of a fluidized-solid boiling-bed-type reactor is larger than in kilns or tray-type reactors because the gas supports and fluidizes the solid; this pressure drop may sometimes be a serious objection because of the larger compressors required.

Occasionally it is desirable to obtain a temperature gradient in a catalytic process; for example, a higher temperature may be desired at the upper part of a reactor in order to effect a clean-up of residual reactant. In the simple boiling-bed fluidized-solid reactor this is not possible; however, by proper design, including baffles or other staging or zoning means plus internal heat exchangers or by the use of series reactors, temperature gradients may be secured.

The pressure drop through a fluidized bed may be large as compared with that through an ordinary heat exchanger. If the fluidized bed is used solely for heat transfer, this high pressure drop might be detrimental unless circumvented by use of shallow beds.

The pressure drops must be balanced throughout the system in such a manner that gases will not flow to undesirable parts of the system; this may necessitate providing a gas purge at several points.

The observation that gas bubbles rising through the fluidized solids contain but little of the solids indicates that the efficiency of contact in a boiling bed may be much less (in terms of the availability of the active surface of the solids) than in a fixed-bed reactor; this disadvantage is at least partially overcome by the fact that the fluidized catalyst can be of much smaller particle size (and therefore much greater surface) than than used in fixed-bed reactors.

Since 1960, a great deal of experimental and theoretical work has been carried out in studying this bubble phenomenon in fluidized solids. When finally resolved it should be possible to predict the degree of reaction in any such gas–solids contactor and, it is hoped, to indicate how bubbles might be controlled and contact maximized. To date, these studies have succeeded in rationalizing the scale-up of pilot plant results to commercial reactor designs. Without experimental reaction rates and conversions from an operating fluidized bed pilot plant, it is still a nebulous procedure to predict what might occur in a commercial scale plant; fixed-bed reaction kinetics are not as yet, with any degree of certainty, transferable to a bubbling bed.

An appreciation of the operating limits and operating characteristics of a fluidized bed of particles is likely best gleaned from an outline of the steps involved in the design procedure.

9.3 OPERATING CHARACTERISTICS AND DESIGN PROCEDURES

9.3.1 Choice of Operating Gas Velocity

Fluidization occurs in a bed of particulates when an upward flow of fluid through the interstices of the bed attains a frictional resistance equal to the weight of the bed. At this point, an infinitesimal increase in the fluid rate will lift or support the particles. Hence, the particles are envisioned as barely touching, or as "floating" on a film of fluid.

To avoid ambiguities, this condition would better be considered as incipient buoyancy, as at such a fluid rate the particles are still so close as to have essentially no mobility—whereas the usual desire in fluidization is to create bed homogeneity. Such homogeneity can be achieved only by violent mixing. This is brought about by increasing the fluid velocity to the point of blowing "bubbles" or voids into the bed, which mix it as they rise. The increased fluid velocity at which bubbles first form is referred to as the incipient-bubbling velocity.

Fine powders can exist over a wide range of bulk densities and therefore exhibit substantial differences between incipient buoyancy and incipient bubbling, as illustrated qualitatively in Figure 9.1.

There is no way to predict precisely a powder's range of bulk densities. However, this is a relatively simple physical measurement. As an example, FCC (fluidized catalytic cracking) catalysts with particle-size distributions from about $10-200$ μm in diameter exhibit incipient-buoyancy velocities in ambient air in the order of 0.01 to 0.03 ft/s, and incipient-bubbling velocities of about 0.1 to 0.3 ft/s. Many catalysts used in other processes consist of silica-alumina carriers (such as are used in FCC catalysts) impregnated with desired catalytic agents. Hence, these velocities are broadly representative. From a practical point, the incipient-bubbling velocity is the more significant one in reactor design.

The terms *particulate* and *aggregative* were coined[32] in the 1940s to differentiate between bubbling beds (aggregative) and nonbubbling

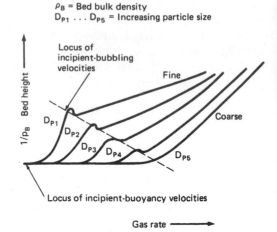

ρ_B = Bed bulk density
$D_{P1} \ldots D_{P5}$ = Increasing particle size

Figure 9.1. Particle size affects incipient-bouyancy and incipient-bubbling velocities.

beds (particulate). In general, liquid fluidized beds were nonbubbling, whereas gas fluidized beds bubbled. It is presently recognized that bubbling is related to fluid and particle properties in a manner permitting the prediction of a system's maximum attainable bubble size,[27] which if negligible leads to the observation of so-called particulate fluidization. Rather than employ the terms *aggregative* and *particulate*, it is more correct to refer to the maximum stable-bubble size for a particular system.

The "optimum" operating fluid velocity is not likely to have narrow limits. It generally represents physical compromises with entrainment, attrition, pressure drop, and economics. The lower limit is the bed's incipient bubbling velocity, and the upper limit usually approaches the terminal or free-fall velocity of the largest particle in the bed. There exists a nearly identical analogy to the optimum physical dimensions of a distillation or absorption tower.

The composition and physical properties of the fluidizing gas are fixed by the process conditions. If the density and size distribution of the particles to be fluidized are not also predetermined, then they must be obtained either from an incorporation of the cost of various degrees of grinding, or, in the case of a high-conversion gas-catalytic reaction, be taken as the size that optimally[46] leads to an incipient bubbling velocity between 0.1 and 0.3 ft/s under reaction conditions. Most frequently the optimum superficial fluidizing velocity will be based on pilot plant experience.

9.3.2 Calculation of Incipient Bubbling and Terminal Velocities

The superficial upward fluid velocity through a bed of particles that will initiate incipient bubbling is definable as the minimum flow that in passing through the interstices of the bed at its loosest bulk density suffers a particle-to-fluid frictional resistance just equalling the weight of the bed. Any additional fluid will therefore lift the bed (which must be unrestrained on its surface) and pass upwards in the form of a so-called bubble. Mechanistically it simply pushes the bed upward at its point of entry through the supporting grid port and thereby creates a void or hole into which the bed solids can slide and thereby displace[39] the bubble-form void in the upward direction. The calculation of the incipient bubbling velocity therefore constitutes the substitution of the bulk density for the pressure drop in the conventional correlations for flow through packed beds and the determination of the attendant

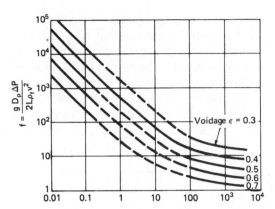

Figure 9.2. Friction factor for pressure drop through packed beads of uniform particles.

velocity. Correlations such as those of Ergun,[10] or of others[2, 5, 21] as suggested in Figure 9.2, are far more reliable than many published equations[11, 18, 29] for direct calculation of incipient fluidization velocities, as they empirically account for the voidage or loosest bulk density, which can have a very significant effect, as evidenced by the spread between the curves in Figure 9.2.

The superficial velocity at which the largest particle in the bed is able to be blown out of the reactor is frequently the practical or near-

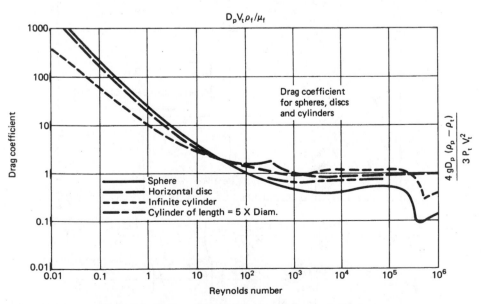

Figure 9.3. Drag coefficient for spheres, discs, and cylinders.

optimum operating velocity. This is simply calculable from the conventional drag coefficient versus Reynolds number curves[17] for the free-fall, or terminal, velocity of equivalent single spheres, as represented in Figure 9.3.

It is obvious from Figures 9.2 and 9.3 that the drag coefficient and friction factor differ solely in the constants $\frac{4}{3}$ and $\frac{1}{2}$ and are therefore interchangeable. Also, because both velocity and particle diameter appear in the abscissas as well as the ordinates, the prediction of incipient bubbling velocity or of terminal velocity involves a trial and error procedure. This can be circumvented[35] by converting friction factor to drag coefficient ($C_D = (\frac{8}{3})f$) and then plotting $(C_D Re^2)^{1/3}$ versus $(Re/C_D)^{1/3}$, as shown in Figure 9.4, where now the denominators of the abscissa and the ordinate repre-

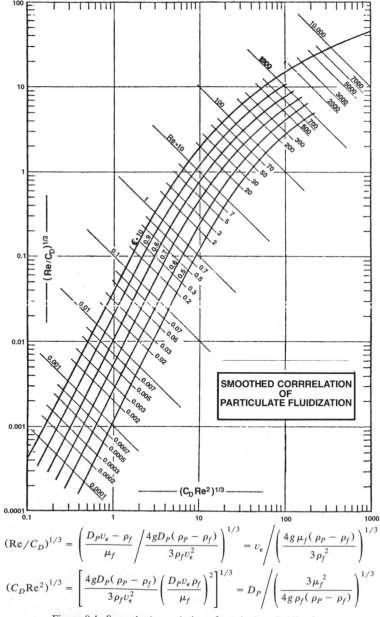

$$(Re/C_D)^{1/3} = \left(\frac{D_P v_\epsilon - \rho_f}{\mu_f} \middle/ \frac{4gD_P(\rho_P - \rho_f)}{3\rho_f v_\epsilon^2} \right)^{1/3} = v_\epsilon \middle/ \left(\frac{4g\mu_f(\rho_P - \rho_f)}{3\rho_f^2} \right)^{1/3}$$

$$(C_D Re^2)^{1/3} = \left[\frac{4gD_P(\rho_P - \rho_f)}{3\rho_f v_\epsilon^2} \left(\frac{D_P v_\epsilon \rho_f}{\mu_f} \right)^2 \right]^{1/3} = D_P \middle/ \left(\frac{3\mu_f^2}{4g\rho_f(\rho_P - \rho_f)} \right)^{1/3}$$

Figure 9.4. Smoothed correlation of particulate fluidization.

sent composites of the physicochemical properties of the fluidizing medium and the solids (which composites are constants for any given system), so that Figure 9.4 is effectively a generalized correlation of velocity versus particle diameter at any known or desired bed density or void fraction. The incipient bubbling velocity is found from the value of the ordinate corresponding to an abscissa determined by the geometric weight mean particle diameter of the bed and the bed's void fraction at its loosest bulk density. At a value of the abscissa representing the largest particle in the bed, the ordinate determined from the $\epsilon = 1.0$ curve yields the velocity that would blow the largest particle out of the bed. This is not necessarily an upper limit to bed operation because the particles entrained at any operating velocity must be separated from the exit gases and returned to the bed in order to maintain the bed. This particle recovery system usually consists of one or more cyclones[1]

in series whose diplegs are submerged in the bed to discharge as near the grid level as feasible. The greater the superficial operating velocity, the greater the entrainment rate and the greater the need to design efficient cyclone recovery systems.

9.3.3 Predicting the Particle Entrainment Rate

The entrainment of particles from the surface of a bubbling bed is directly analogous to the entrainment of liquid droplets from a boiling or bubbling pool of liquid as occurs on a distillation tray. Because solid particles cannot coalesce in the manner of liquid droplets, the entrainment above a fluidized bed of solids, though declining with increasing height, inevitably reaches a constant rate representing all the bed particles with a free-fall velocity less than the operating superficial velocity, which were ejected from the bed surface. The

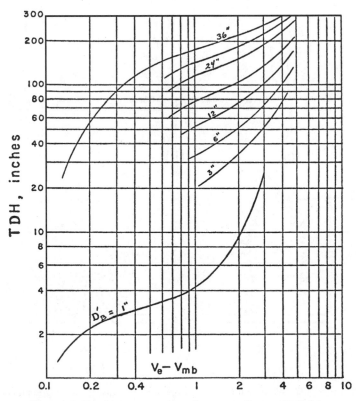

Figure 9.5. Empirical correlation of transport disengaging height.

force or mechanism of ejection is a function of the size and frequency of the bubbles erupting on the bed surface. Because the bursting bubbles represent intermittent locally higher-than-superficial velocity profiles, which dissipate with distance or so-called particle disengaging height above the bed, it is of interest to be able to predict at what height above the bed the velocity profile will eventually be stabilized to the superficial velocity, with an attendant constant entrainment rate thereafter. This so-called Transport disengaging height[36] has been empirically correlated as shown in Figure 9.5. Increasing the reactor height to more than the TDH serves no purpose in decreasing the entrainment rate. The abscissa of Figure 9.5 is related to the frequency of surface bubble eruptions and the parameter to the physical size of the surface bubbles. As discussed subsequently, the size of the surface bubbles is calculable from the grid characteristics, the bubble merger rate, and

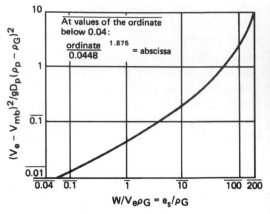

Figure 9.6. Maximum dilute phase entrainment in vertical gas–solids upflow.

the maximum stable bubble size for the particle–fluid system under consideration.

The particle entrainment rate, at and above the TDH, can be determined from the empirical correlation[36] given in Figure 9.6. The detailed design of the cyclone recovery system

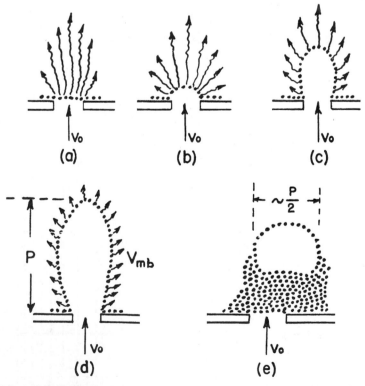

Figure 9.7. Bubble formation from bed-penetrating gas jets at the grid points.

can best be found in Chapter 11 of the *A.P.I. Emissions Control Manual.*[1]

9.3.4 Grid Design and Initial Bubble Size

Bubbles form at the grid ports when fluidizing gas enters the bed. They form simply because the velocity at the interface of the bed just above the hole represents a gas input rate in excess of what can pass through the interstices with a frictional resistance less than the bed weight, and hence the layers of solids above the holes are pushed aside until they represent a void through whose porous surface the gas can enter at the incipient fluidization velocity.[41] If the void attempts to grow larger, the interface velocity becomes insufficient to hold back the walls of the void and hence they cave in from the sides,[39] cutting off the void and presenting a new interface to the incoming gas. This sequence is illustrated in Figure 9.7. The depth of penetration of the grid gas jets has been correlated empirically,[31, 34, 40] as shown in Figure 9.8, and the diameter of the initial bubble resulting from a detached void has been observed, within experimental error, to be about half the penetration depth. Because the grid is the source of the bubbles, its design is relatively critical. The holes should be as small as is reasonable (considering cost, plate strength, and possible pluggage) and of such total area that the pressure drop of the fluidizing gas passing through the holes is sufficient to ensure gas distribution[14, 20, 26, 43] to all the holes. As in the case of a perforated plate distillation tray, bed weepage flow down

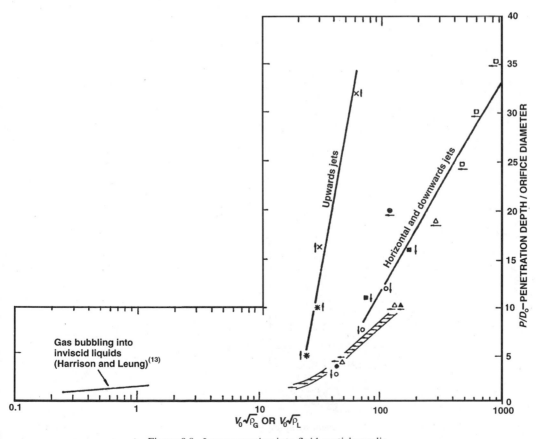

Figure 9.8. Jet penetration into fluid-particle media.

through the holes is to be avoided. It has been demonstrated that the pressure drop through a flat plate grid necessary to ensure that all the holes are bubbling must be at least 30% of the pressure drop through the bed atop the grid. This criterion establishes the grid hole velocity and, in conjunction with the lowest anticipated fluidizing gas volume, determines the total hole area. The number of holes is then dependent on the designer's choice of the hole diameter which simultaneously also determines the initial bubble size. In passing up through the bed these bubbles inevitably merge when they meet and hence their fluid mechanics must be understood in order to relate to the gas–solids contact occurring within the bed and to the size of the bubbles bursting on the bed surface.

9.3.5 Fluid Mechanics of Bubble Flow

Bubbles or "gas voids" rise in a fluidized bed by being displaced with an inflow of solids from their perimeter.[3,39] Because free-flowing or incipiently fluidized bulk solids have shallow angles of repose, their walls cannot stand at 90° and hence the solids slide down the bubble's walls into its bottom where all the peripheral streams collide to form a so-called wake as illustrated in Figure 9.9. Observations of this downflow of solids in a "shell" around the bubble have shown it to occupy an annular thickness of $\frac{1}{4}$ of the bubble diameter so that the overall diameter within which a bubble can rise "freely," as it would in a bed of infinite diameter, can be defined as 1.5 D_B.

Because the peripheral surface of the bubble is simply a layer of particles, it is at first

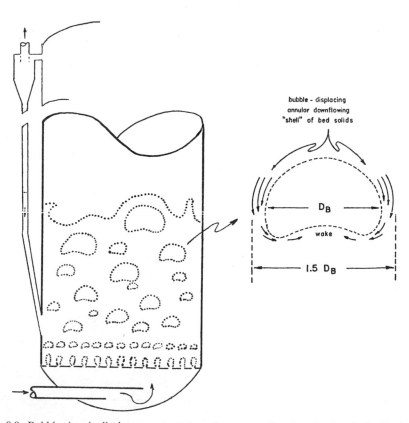

Figure 9.9. Bubble rise via displacement by inflow of a surrounding downflowing shell of bed solids.

difficult to understand why the particles do not fall from its roof and annihilate the bubble. Danckwert's[7] simple bed support experiments, illustrated in Figure 9.10, provide the physical demonstration and Rowe and Henwood's[24] experiments the classical approach.

In Figure 9.10a the air rate is raised to the point of incipient fluidization and in Figure 9.10b through 9.10f this same gas rate is passed through the bed in the opposite direction. Note that in position (d) the solids do not slide to their angle of repose but instead are held at 90° and that on reaching (f) the bed is held up without solids falling from what is now its lower side or conversely the upper surface of a bubble in a fluidized bed. When the surface of a bed is traversed by an incipiently fluidizing flow the particles cannot separate from each other. This not only explains the bubble's surface stability but also the integrity of the walls of a bed-penetrating jet, as in Figure 9.7.

Rowe and Henwood carried out classic drag measurements that revealed that the drag on a downstream particle is reduced because of the presence of an adjacent upstream particle. This simply means that a particle cannot fall from the roof of a bubble because if it did, then it would immediately be followed by the particle above it, and that by the particle still farther above, etc., so that the entire mass or bed above the bubble would have to collapse as a unit. For this to occur, the excess gas could not be passed through the bed unless the bed were physically held down or restrained at its upper surface.

The velocity at which bubbles rise in a gas fluidized bed has been measured photographically by several investigators. The results are in excellent agreement with what would be predicted for gas bubbles in liquids from the drag coefficient versus Reynolds number correlations of such investigators as Van Krewelen and Hofttijzer[33] illustrated in Figure 9.11. Over the range of Reynolds numbers corresponding to reasonable size bubbles the drag coefficient is essentially a constant so that simple substitution shows that if gas density is small relative to the bed density:

$$C_D = \frac{4gD_B(\rho_B - \rho_G)}{3\rho_B V_B^2}.$$

$$\therefore \ V_B = \sqrt{\frac{4gD_B(\rho_B - \rho_G)}{3C_D \rho_B}}$$

or

$$V_B = 4.01\sqrt{D_B}$$

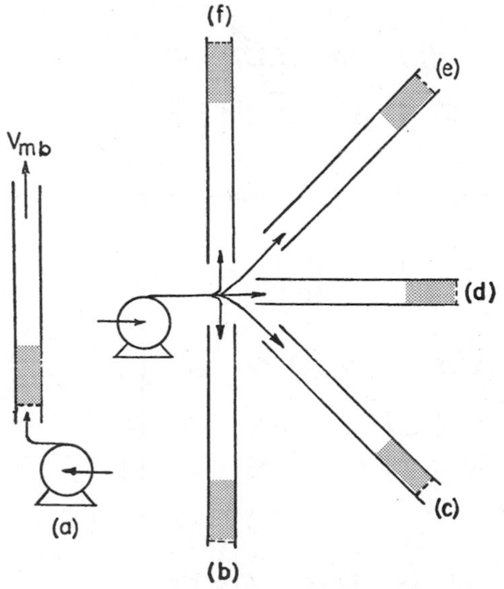

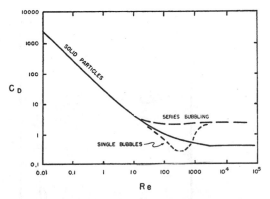

Figure 9.11. Rate of rise of gas bubbles in liquids.

Figure 9.10. Bed support experiments of P. V. Danckwerts.

This has been corroborated in experiments with freely bubbling beds.

Matsen and Tarmy[19] have shown that in slugging beds the full width of the downflowing solids shell (Figure 9.9) is restricted and the velocity of bubble rise then approximately $\frac{1}{2}$ that in a freely bubbling bed. In most instances this becomes a matter for consideration only in the design or scale-up of small pilot or bench scale fluidized bed processes.

9.3.6 Rate of Bubble Growth by Merger

That bubbles must grow by merger as they rise through the bed is obvious from the large and less frequent surface eruptions relative to a much higher frequency of small voids initiated from a usual multitude of grid ports. Growth by simple gas expansion resulting from the pressure reduction between bottom and top

of a fluidized bed is generally relatively insignificant.

From the solids inflow model of Figure 9.9 it is obvious that a bed must be exceptionally homogeneous to expect the shell of downflowing solids around a bubble to be flowing at an equal rate in every plane. Any bed nonuniformity can cause a shift in the bubble shape or position. Merely the prior passage of another bubble could alter local densities or distributions so as to make bed solids in one local area more readily flowable in a given direction than the bed solids in an adjacent area. The solids inflow model therefore obviates a simple mechanism of bubble merger. If two bubbles get close enough that their shells of downflowing solids begin to interact, the touching shells will represent a local downflowing stream of solids faced with more than one path to the nearest void. The stream could

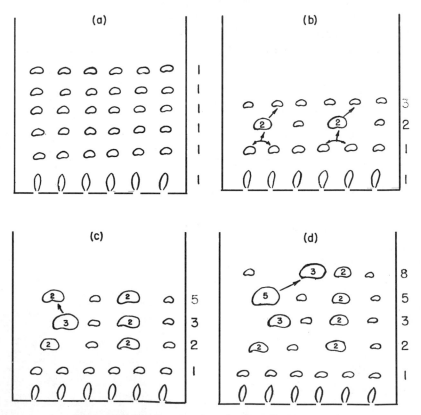

Figure 9.12. The "catch-up" mechanism of bubble growth.

be squeezed to the point of being insufficient to satisfy both bubbles and thereby drain off, leaving no wall between the voids and hence the appearance of a single somewhat larger bubble whose volume is the sum of the volumes of the two merged bubbles.

It is therefore readily acceptable that the idealized bubbling of Figure 9.12a will lead to a situation as in (b) where two bubbles of unit initial volume can merge into bubbles of twice this volume. Because larger bubbles rise more rapidly, these double volume bubbles will catch up and merge with other unit volume bubbles to yield bubbles of thrice the initial bubble volume. These newer bubbles will rise even more rapidly and catch up with bubbles of 1 or 2 times the volume of the initial bubble resulting in bubbles of at most 5 times the volume of the initial bubble. The bubble of five-fold volume can now catch up with bubbles of 1, 2, or 3 times the volume of the initial bubble, resulting in bubbles of at most 8 times the volume of the initial bubble as illustrated in Figures 9.12c and 12d. Carrying on this process of overtaking bubbles results in a sequence of maximum multiples of the initial bubble volume in which each multiple is the sum of the two previous bubbles. This sequence, illustrated in Figure 9.13, is the well-known "Fibonacci" series.[42]

Because the levels at which the maxima exist represent the summation of the diameters of their forebearers and because their diameters are proportional to the cube root of their volumes, it follows that the ratio of merged bubble diameter to initial bubble diameter is equal to the cube root of the number of initial bubbles consumed in the merger, and also that the level at which the merged bubbles exist relative to the height (or diameter) of the initial bubble is equal to the summation of the cube root of the number of initial bubbles consumed in the merged bubble.[41] For the case of the maximum size of merged bubble this is illustrated analytically in Figure 9.13 and shown graphically in Figure 9.14.

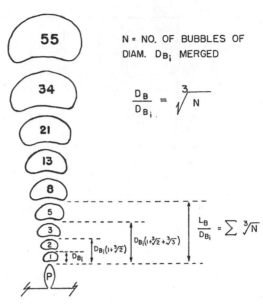

Figure 9.13. Maximum bubble growth by the "catch-up" mechanism resulting in a Fibonacci sequence.

That the mechanism of Figures 9.13 and 9.14 appears in good agreement with experimental observations is illustrated in Figure 9.15 where the empirical bubble growth relationships proposed by Chavarie and Grace,[6] Werther,[30] and Rowe[23] are superimposed on the curve representing the Fibonacci series. In using Figure 9.15 to determine the maximum bubble diameter D_B at any bed level L_B above the grid, it is necessary to determine the initial bubble diameter D_{Bi}, which could exist at the grid level as a result of individual ($P/2$) or merged jets. Figure 9.15 must also not be extrapolated beyond the maximum attainable stable bubble size.

9.3.7 Maximum Stable Bubble Size

Danckwert's bed support experiments (Figure 9.10) and those of Rowe and Henwood based on particle drag force measurements demonstrated that a bed interface (and hence a bubble) should be fundamentally stable against collapse as long as it is traversed by a superficial velocity equal to its incipient fluidization rate. Because the inflowing solids shell volume usually far exceeds the incipient fluidization

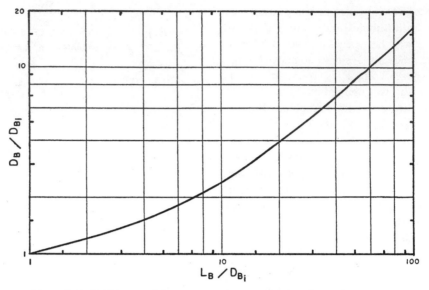

Figure 9.14. Bubble growth by merger represented by the Fibonacci sequence.

rate, there would appear to be no limit to the attainable bubble size, or dome, apt to collapse. Presumably, if the dome cannot collapse amid free-flowing bed solids then as the bubble grows it could only be limited by particles leaving the shell and being entrained into the bubble void. Such entrainment, or particle pick-up, would be most likely to occur from the bubble walls as the result of the relative velocity between gas and surface particles at

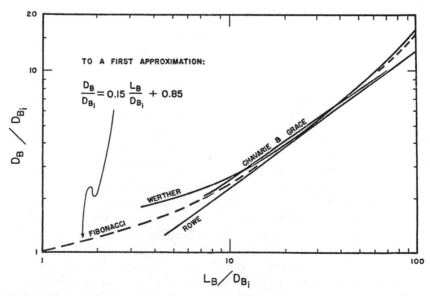

Figure 9.15. Comparison between empirical bubble growth correlations and the "catch-up" mechanism represented by the Fibonacci sequence.

the interface.[27] Because against the downward velocity of bulk solids the bubble fluid (whether gas or liquid) rises at approximately an equal velocity, the relative flow of fluid past the particles at the bubble wall is twice the shell or bubble velocity. Equating twice the bubble velocity to the particle pick-up velocity allows calculation of the minimum bubble size necessary to stir up the solids interface and thus thwart bubble appearance or growth. Because pick-up velocity is approximately twice saltation,[15,37] this is equivalent to equating bubble velocity to saltation velocity. This procedure has given results in reasonable agreement with a broad range of observations reported to date.[41] For example 80 μm particles of sand fluidized with air could sustain a maximum bubble diameter of the order of 24 in., whereas when fluidized with water the maximum bubble size would be indiscernible. Sand particles 600 μm in diameter when fluidized with water would permit a maximum stable bubble size of only $\frac{1}{4}$ in., and 3000 μm lead particles a water bubble of 7 in.

9.3.8 Gas – Solids Contact

From grid design, operating superficial velocity, and fluid particle properties, it is possible to calculate the initial bubble size at the grid, the maximum stable size, and the bed depth over which the bubbles may grow from their initial to their stable diameter. Once having reached their maximum stable diameter, any further unlikely mergers would also lead to collapse, so that bubble diameter may be considered constant once having reached the stable size. Because the bubbles represent a flow superimposed on the superficial incipient bubbling velocity passing up through the bed, they are in effect being continuously purged as they rise. Because their local size, velocity, and residence times are calculable from grid to bed surface, it is also possible to calculate the degree to which they are purged before bursting at the surface and hence to make certain that no bubble gas bypasses contact with the bed solids. It may be assumed that the mini-

mum bed depth required to avoid any feed gas breakthrough (e.g., 100% bubble purging) represents the minimum bed depth required for the desired reaction.[44] This is represented graphically in Figure 9.16 for a freely bubbling bed.

In choosing a pilot plant such as to entirely avoid scaleup considerations, superficial velocity and grid details must be identical to those anticipated in the commercial reactor; in addition, the diameter of the shell of downflowing bed solids surrounding the rising bubbles represents the minimum pilot reactor diameter necessary to simulate free bubbling and avoid approach to slugging. In addition to simulating free bubbling hydrodynamically, it may be argued that gas permeation from bubble into a surrounding bed should also be equalled. This only becomes significant or controlling with coarse and easily permeated beds having a high incipient fluidization velocity. The gas permeation or "cloud" diameter[25] is calculable from the depth of gas flow at incipient fluidization velocity over the time interval required for the bubble to rise a distance of one bubble diameter. Because the bubble rises at a velocity equal to 4 times the square root of its diameter it follows that:

$$\frac{\text{Thickness of gas penetrated "cloud"}}{\text{Thickness of downflowing solids "shell"}}$$

$$= \frac{V_{\text{mb}}}{\sqrt{D_{\text{B}}}}$$

or because

$$\text{"Shell" O.D.} = 1.5 D_{\text{B}}$$

$$\text{"Cloud" O.D.} = D_{\text{B}} + 0.5\sqrt{D_{\text{B}}}\, V_{\text{mb}}$$

In applying free shell or cloud criteria in scaleup or scaledown, the relationship between bubble diameter and bed depth is obtainable from Figure 9.15 with the limitation of the system's maximum stable bubble size.

An unquestionably conservative approach to a minimal risk pilot plant reactor free of scaleup considerations would suggest it equal the larger of either "cloud" or "shell" diame-

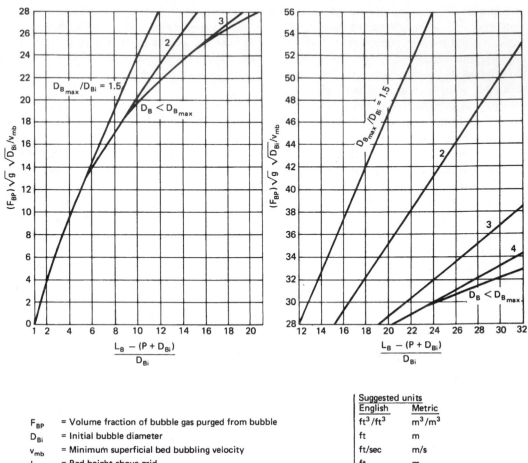

	Suggested units	
	English	Metric
F_{BP} = Volume fraction of bubble gas purged from bubble	ft^3/ft^3	m^3/m^3
D_{Bi} = Initial bubble diameter	ft	m
v_{mb} = Minimum superficial bed bubbling velocity	ft/sec	m/s
L_B = Bed height above grid	ft	m
P = Gas jet penetration at grid port	ft	m
$D_{B_{max}}$ = Maximum stable bubble diameter	ft	m
g = Gravitational acceleration	32.2 ft/sec^2	9.71 m/s^2

Note: Ordinate, abscissa, and parameter are dimensionless

Figure 9.16. Degree of bubble gas purging during its rise through a fluidized bed.

ter surrounding the system's maximum stable bubble.

9.3.9 Solids Mixing and Heat Transfer

Because rising gas bubbles are replaced with bed solids it is evident that the superficial gas velocity minus the incipient bubbling velocity also approximates the volumetric bulk solids movement across any unit bed cross section per unit time. This amounts to a relatively substantial mass movement, and hence it is not surprising that a fluidized bed exhibits

reasonably uniform particle size distribution and bed temperature throughout its volume. Reasonable quantitative estimates of such local solids mixing rates[16, 25, 45] are of major importance principally in determining allowable solids feed rates, since to avoid accumulation at any feed pipe location the bed mixing rate must be able to remove the feed material as rapidly as it enters. Such solids mixing rates have been reasonably well correlated by Talmor and Benenati.[28]

The substantial heat capacity of the bed solids relative to the gas inventory represents an enormous "flywheel" which, coupled with the high solids mixing rate, leads to a rather uniform temperature throughout the bed. The heat transfer between gas and solids is nearly instantaneous, primarily because of the high particle surface area per unit of bed volume. However, the transfer of heat between the bulk bed and the vessel walls, or any other heat transfer surface, represents a composite of mechanisms such that the average may range from 10 to 100 BTU/h $\times$ ft^2 $\times$ °F depending on particle size, bubble size, fluid properties, and superficial fluidizing velocity. The homogeneity and relative uniformity of bed temperature make fluidized beds an attractive vehicle in which to conduct exothermic as well as endothermic reactions controlled by immersed boiler tubes, exchangers, platecoils, or other heat transfer surfaces when the bed walls do not offer sufficient area for cooling or heating via a fluid circulated through a surrounding jacket.

The transfer of heat from the bed to an immersed or wall surface depends instantaneously on whether the surface is bathed in stationary solids, in moving solids, or in a bubble void[38] as illustrated at points A, B, and C in Figure 9.17. Overall coefficients averaging these local mechanisms are summa-

rized in Figure 9.18. Conductive transfer to or from stationary solids as at A in Figure 9.17, as well as transfer between bubble gas and metal surface as at C, are nearly negligible relative to the rate occurring when the surface is "wiped" by the shell of solids flowing down around a bubble passing within a distance of a quarter of its diameter from the transfer surface, as at B in Figure 9.17.

Though advantageously exploitable in industrial installations only under relatively rare circumstances, a number of experimental investigations[4, 8, 9, 12, 22] of the heat transfer coefficients under conditions such as at B in Figure 9.17 have also been correlated in terms of Nusselt and Reynolds numbers, as shown in Figure 9.19. The cross hatched area in Figure 9.19 encompasses the overall heat transfer data of Figure 9.18.

9.3.10 Bed Internals

Bed internals in the form of vertical tubes have no effect on bed hydraulics other than to slow down the rise velocity of bubbles. They do not break up bubbles or "cage" them to limit their size or growth. They simply represent impediments to the rate of inflow of surrounding bulk solids that cause the bubbles to rise by displacement. Thus the bubbles' longer residence time affords greater opportunity for their gaseous content to be purged, thereby enhancing gas–catalyst contact. The increased contact time, or reduced bubble rise velocity is derivable from the rate of rise of the equivalent or hydraulic bubble size[47] calculable as:

$$V = 4(D_h)^{1/2}$$

where

V = rate of bubble rise (ft/s)

D_h = 6 (free internal volume)/(internal surface)

Internal volume = $\pi D_B^3/6$ minus the volume occupied by the penetrating tubes

D_B = actual bubble diameter (ft)

Internal surface = bubble surface, πD_B^2 minus surface of penetrating tubes within the bubble (ft^2)

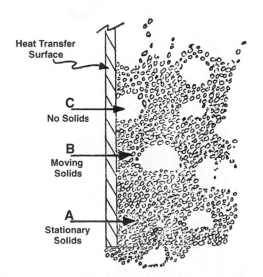

Figure 9.17. Comparison between overall bed coefficients and local flowing dense phase coefficients.

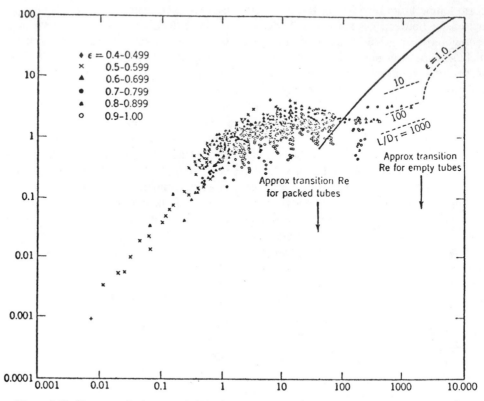

Figure 9.18. Heat transfer between fluid beds and tube walls (correlation of overall coefficients).

Type of data	Value of Re	Value of Nu
Fixed- and fluid-bed	$D_P v \rho_f / \mu_f$	$h D_P / K_f$
Empty tube pipe flow	$D_T v_\epsilon \rho_f / \mu_f$	$h D_T / K_f$

Curves

(-) data for flow of fluid through fixed beds (M. Leva, *Ind. Eng. Chem.* 39, 857–862 (1947); see also D. A. Plautz and H. F. Johnstone, *A.I.Ch.E.J* 1, 193–200 (1955)).

(---) correlation for flow of fluids through pipe (W. H. McAdams, *Heat Transmission*, 2nd edit., McGraw-Hill, New York, 1942).

Sources of Data

R. N. Bartholomew, Ph.D. Thesis, Univ. of Michigan, 1950; *Chem. Eng. Progr. Symp. Ser. 48* (4), 3–10 (1952).

L. H. Collins, M.S. Thesis, Massachusetts Institute of Technology (MIT), Cambridge, MA, 1946.

W. M. Dow, Ph.D. Thesis, Illinois Institute of Technology, 1949; *Chem. Eng. Prog. 47*, 637–648 (1951).

H. Fischer and E. F. Dillon, B.S. Thesis, MIT, 1947.

W. Lazor and S. A. Murray, M.S. Thesis, MIT, 1947.

M. Leva, M. Weintraub, and M. Grummer, *Chem. Eng. Prog. 45*, 563–572 (1949).

H. S. Mickley and C. A. Trilling, *Ind. Eng. Chem. 41*, 1135–1147 (1949).

W. H. Millick and A. S. Humphrey, M.S. Thesis, MIT, 1948.

R. V. Trense, Ph.D. Thesis, Northwestern University, 1954.

R. W. Urie, M.S. Thesis, MIT 1948.

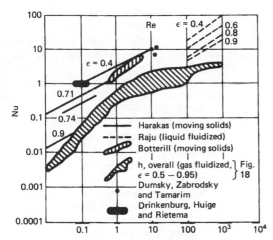

Figure 9.19. Comparison between overall bed coefficients and local flowing dense phase coefficients.

Bed internals in the form of horizontal baffles or perforated grids have no effect on bed hydraulics other than to retard the top-to-bottom mixing of the dense phase bulk solids. By proper design such structured packing can create a plug flow condition, particularly desirable in instances where gas and catalyst flow through a reactor or regenerator countercurrently. This aspect has recently received a substantial degree of investigation by AIMS, an industrial research consortium, and by Snamprogetti, and will appear in publications early in 1995.

9.3.11 Scale-up

Scale-up of fluidized bed reactors from pilot plants as small as 2 in. in diameter to industrial units as large as 40 ft I.D. has followed from the concept of bubble purging and the contact effectiveness factor.[48] In most instances this can be shown to reduce to a relationship easily solved by trial and error:

$$L_{Ind}/L_{pp} = 1.68(D_{Bh}/D_T)^{0.75}$$

where

D_{Bh} = hydraulic diameter of average rising bubble

D_T = pilot plant reactor I.D.

L_{pp} = minimum bed depth of catalyst in pilot plant producing satisfactory yield and conversion

L_{Ind} = depth of bubbling industrial reactor yielding performance equaling pilot plant results

LIST OF SYMBOLS

C_D	Drag coefficient, dimensionless (see Figs. 9.3 and 9.4)
D_B	Bubble diameter
D'_B	Bubble diameter bursting at bed surface
D_{B_1}	Initial bubble diameter at grid level
D_0	Grid hole diameter
$D_{B_{max}}$	Maximum stable, or attainable, bubble diameter
D_P	Particle diameter
ε_j	*Maximum entrainable lbs of solids/ft³ of gas*
F_{BP}	Volume fraction of bubble gas purged
f	Friction factor (see Fig. 9.2)
g	Gravitational constant
h	Heat transfer coefficient
K	Thermal conductivity of fluidizing medium
L_B	Bed depth
N_u	Nusselt number, dimensionless (see Figs. 9.18 and 9.19)
P	Jet penetration depth
$\Delta P/L$	Pressure drop per unit length, lbs/ft² × ft
Re	Reynolds number, dimensionless
TDH	Transport disengaging height
v	Superficial fluidizing medium velocity
v_ϵ	Superficial fluidizing medium velocity when bed voidage is ϵ
V_B	Bubble rise velocity
V_e	Effective superficial velocity governing rate of entrainment
V_{mb}	Superficial velocity at point of incipient bubbling
V_0	Fluidizing medium velocity through grid hole
V_t	Particle terminal or free fall velocity

W Weight rate of solids entrained, lbs/s $\times$ ft^2 of vessel area

ϵ Fractional void volume in bed

ρ_B Bed or bulk density

ρ_f Fluid or medium density

ρ_G Gas density

ρ_L Liquid density

ρ_p Apparent particle density

μ_f Viscosity of fluidizing medium

REFERENCES

1. Amer. Petrol. Inst., "Cyclone Separators," in *Emissions Control Manual*, Pub. no. 931 (May, 1975).
2. B. A. Bakameteff and N. V. Feodoroff, *J. Appl. Mechanics 4*:A97 (1937).
3. J. S. M. Botterill, J. S. George, and H. Besford, *Chem. Eng. Prog. Symp. Ser. 62*:7 (1966).
4. J. S. M. Botterill and J. R. Williams, *Trans. Instn. Chem. Engrs.* (London) *41*:217 (1963); idem, *Fluid Bed Heat Transfer,* Academic Press, New York (1975).
5. P. C. Carman, *Trans. Instn. Chem. Eng.* (London) *15*:150 (1937); idem, *J. Soc. Chem. Ind.* (London), *57*, 225 (1938).
6. C. Chavarie, J. R. Grace, *Ind. Eng. Chem. Fund.*, pp. 75–78 (May, 1975).
7. P. V. Danckwerts, Symp. on Fluidization held at AERE, Harwell, England (Oct. 5, 1959).
8. A. A. H. Drinkenburg, N. J. J. Huige, and K. Rietema, *Proc. Third Internat'l Heat Transfer Conf.*, Vol. IV, p. 271–279, A.I.Ch.E., New York (1966).
9. V. D. Dunsky, S. S. Zabrodsky, A. I. Tamarin, *Proc. Third Internat'l Heat Transfer Conf.*, Vol. IV, p. 293–297, A.I.Ch.E., N.Y. (1966).
10. S. Ergun, *Chem. Eng. Prog. 48*(89) (1952); idem, *Ind. Eng. Chem. 41*:1179 (1949).
11. J. F. Frantz, *Chem. Eng. Prog. Symp. Ser. 62*(62):21–31 (1966).
12. N. K. Harakas and K. O. Beatty, *Chem. Eng. Prog. Symp. Ser. 59*(41):122 (1963).
13. D. Harrison and L. S. Leung, *Trans. Instn. Chem. Eng.* (London) *39*:409 (1961).
14. J. W. Higby, *Chemie. Ingr. Tech. 36*:228 (1964).
15. P. J. Jones and L. S. Leung, *Ind. Eng. Chem. Proc. Des. Dev.*, *17*(4):571–575 (1978).
16. S. Katz and F. A. Zenz, *Petrol. Refiner 33*(5):203–204 (1954).
17. C. E. Lapple and C. B. Shepherd, *Ind. Eng. Chem. 32*:605 (1940).
18. M. Leva, Shirai Takashi, and C. Y. Wen, *Genie Chim. 75*(2):33–42 (1956).
19. J. M. Matsen and B. L. Tarmy, *Chem. Eng. Prog. Symp. Ser. 66*(101):1–7 (1970); idem, *Chem. Eng. Sci. 24*:1743–1754 (1969).
20. Nat'l Petrol. Refr's Assoc., *Proceedings of the Question and Answer Session on Refining Technology*, p. 86 (1970).
21. A. O. Oman and K. M. Watson, *Natl. Petrol. News 36*:R795 (1944).
22. R. Raju, Ph.D. thesis, Andhra Univ., Waltair, India (1959).
23. P. N. Rowe, International Fluidization Conf., Asilomar, Cal. (June 15–20, 1975).
24. P. N. Rowe and G. A. Henwood, *Trans. Instn. Chem. Eng.* (London), *39*, 43 (1961).
25. P. N. Rowe, B. A. Partridge, and E. Lyall, At. Energy Res. Estab. (Gt. Brit.), Repts. R-3777, R-3846, "Particle Movement Caused by Bubbles in a Fluidized Bed." (Oct. 1961); R-4108, "Gas Flow through Bubbles in a Fluidised Bed," (Jan. 1963); R-4543, "Cloud Formation around Bubbles in Gas Fluidised Beds," (Feb. 1964); *Chem. Eng. Sci. 18*:973 (1964); *Chem. Eng. Prog. 60*:75 (March 1964); *Fluidisation*, Society of Chemical Industry, London (1964).
26. P. N. Rowe and F. A. Zenz, *The School Sci. Rev.* (Pub. by the Instn. of Chem. Engrs., London), *53*(182):94–102 (Sept. 1971).
27. A. M. Squires, Paper delivered at the 54th Annual A.I.Ch.E. Meeting, New York (Dec. 6, 1961); *Chem. Eng. Prog. Symp. Ser. 58*(38):57 (1962).
28. E. Talmor and R. F. Benenati, *A.I.Ch.E. J9*(4):536–540 (1963).
29. C. Y. Wen and Y. H. Yu, *Chem. Eng. Prog. Symp. Ser. 62*(62):100–111 (1966).
30. J. Werther, International Fluidization Conference, Asilomar, Cal. (June 15–20, 1975); idem, *Fluidization Technology*, edited by D. L. Keairns, Vol. 1, Hemisphere Pub. Co., Wash pp. 215–235 (1976).
31. J. Werther, *Fluidization*, edited by J. F. Davidson and D. L. Keairns, Cambridge Univ. Press, pp. 7–12 (1978).
32. R. H. Wilhelm and M. Kuauk, *Chem. Eng. Prog. 44*:201 (1948).
33. D. W. Van Krewelen and P. J. Hoftijzer, *Chem. Eng. Prog. 44*:529 (1948).
34. W. C. Yang and D. L. Keairns, *Fluidization*, edited by J. F. Davidson and D. L. Keairns, Cambridge Univ. Press, pp. 208–214 (1978).
35. F. A. Zenz, *Petroleum Refiner, 36*(8):147–155 (1957).
36. F. A. Zenz and N. A. Weil, *A.I.Ch.E.J. 4*:472 (1958); idem, *Hydrocarbon Processing*, pp. 119–124 (April, 1974).
37. F. A. Zenz, *Ind. Eng. Chem. Fund. 3*(1):65–76 (1964).
38. F. A. Zenz, *Proc. of the Third International Heat Transfer Conf.*, Vol. VI, A.I.Ch.E., p. 311–313 (1966).

39. F. A. Zenz, *Hydrocarbon Processing* 46(4):171–175 (April, 1967).
40. F. A. Zenz, *Instn. of Chem. Eng.* (London), *Symp. Ser.*, no. 30, pp. 136–139 (1968).
41. F. A. Zenz, *Chem. Eng.*, pp. 81–91 (Dec. 19, 1977).
42. F. A. Zenz, *The Fibonacci Quarterly* 16(2):171–183 (April, 1978).
43. F. A. Zenz and D. F. Othmer, *Fluidization and Fluid-Particle Systems*, Reinhold, New York, p. 171 (1960).
44. Ibid., Chapter 8.
45. Ibid., Chapter 9.
46. F. A. Zenz and D. F. Othmer, *Fluidization and Fluid Particle-Systems*, annotated 1966 edition, p. 281, orig. pub. by Reinhold, New York (1960).
47. F. A. Zenz, "Fluidization and Fluid-Particle Systems," Vol. II, 1989, PEMM-Corp Pub., Rte. 1, Box 130A, Cold Spring Harbor, NY 10516.
48. F. A. Zenz, *Hydrocarb. Proc.*, pp. 155–156, January 1982.

ADDITIONAL READING

The editors recommend the following publications for additional reading.

L. S. Fan, "Gas-Liquid-Solid Fluidization Engineering," Butterworths Series in Chem. Eng. (1989).

L. S. Fan (ed.), "Fluidization and Fluid Particle Systems: Fundamentals and Applications," *A.I.Ch.E. Symp. Ser.* No. 270, Vol. 85 (1989).

L. S. Fan (ed.), "Advances in Fluidization Engineering," *A.I.Ch.E. Symp. Ser.* No. 276, Vol. 86 (1990).

D. Gidaspow, *Multiphase Flow and Fluidization*, Academic Press, San Diego (1994).

J. R. Grace, L. W. Shemilt, and M. A. Bergougnou (eds.), "Fluidization VI," in Proceedings of the International Conference on Fluidization, Engineering Foundation (1989).

G. Hetsroni (ed.), *Handbook of Multiphase Systems*, Hemisphere Publishing, New York (1982).

D. Kunii and O. Levenspiel, *Fluidization Engineering*, 2nd edit., Butterworth-Heinemann Series in Chemical Engineering (1991).

K. Ostergaard and A. Sorensen (eds.), "Fluidization V," in Proceedings of the Fifth Engineering Foundation Conference on Fluidization (1986).

A. W. Weimer (ed.), "Advances in Fluidized Systems," *A.I.Ch.E. Symp. Ser.* No. 281, Vol. 87 (1991).

W. C. Yang (ed.), "New Developments in Fluidization and Fluid-Particle Systems," A.I.Ch.E. *Symp. Ser.* No. 255, Vol. 83 (1987).

W. C. Yang (ed.), "Fluidization Engineering: Fundamentals and Applications," *A.I.Ch.E. Symp. Ser.* No. 262, Vol. 84 (1988).

F. A. Zenz, "Fluidization and Fluid-Particle Systems," Vol. II Draft, Pemm-Corp Publications (1989).

10
Spouting of Particulate Solids

Norman Epstein and John R. Grace

CONTENTS

10.1 INTRODUCTION

The spouted bed technique is an alternative to fluidization for handling particulate solids that are too coarse and uniform in size for good fluidization. Although the areas of application of spouted beds overlap with those of fluidized beds, the flow mechanisms in the two cases are very different. Agitation of particles in a spouted bed is caused by a steady axial jet and is regular and cyclic, as distinct from the more random and complex particle flow patterns in fluidized beds.

Figure 10.1 illustrates a spouted bed schematically and photographically. Fluid, usually a gas, is injected vertically through a centrally located small opening at the base of a conical, cylindrical, or conical-cylindrical (as in Fig. 10.1) vessel containing relatively coarse particulate solids (e.g., $d \gtrsim 1$ mm). If the fluid injection rate is high enough, the resulting jet causes a stream of particles to rise rapidly

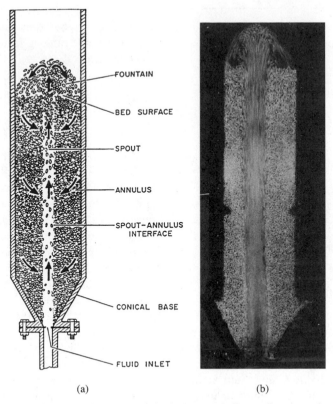

(a)   (b)

Figure 10.1. (*a*) Schematic diagram of a spouted bed. Arrows indicate direction of solids movement. (*b*) Photograph of air-spouted wheat bed in half-cylindrical column.

through a hollowed central core, or *spout*, within the bed of solids. These particles, after rising to a height above the surface of the surrounding packed bed, or *annulus*, rain back as a *fountain* onto the annulus, where they slowly move downward and, to some extent, inward as a loosely packed bed. Fluid from the spout leaks into the annulus and percolates through the moving packed solids there. These solids are reentrained into the spout over the entire bed height. The overall system thereby includes a centrally located dilute-phase cocurrent-upward transport region and a surrounding dense-phase moving packed bed through which fluid percolates countercurrently. A systematic cyclic pattern of *solids movement* is thus established, with effective contact between fluid and solids, and with unique hydrodynamics.[1]

The spouted bed regime, which occurs over a limited range of fluid velocity, is bracketed by fixed packed bed (i.e., static bed) operation at the lower velocities and by bubbling or slugging fluidized bed operation at the higher. For a given combination of fluid, solids, and vessel configuration, the transitions between regimes can best be represented quantitatively by plots of bed depth versus fluid velocity. An example of such a flow regime map is given in Figure 10.2. The demarcation line obtained by decreasing the fluid velocity until the spout collapses to give a static bed in its random loose-packed condition represents the *minimum spouting velocity*, U_{ms}, at various bed depths. The horizontal transition line separating spouting and bubbling represents the *maximum spoutable bed depth*, H_m, for the given system. Above some critical value of the inlet

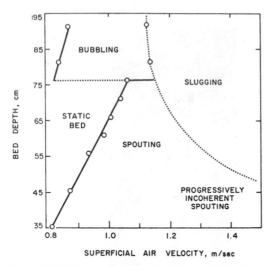

Figure 10.2. Flow regime map for wheat particles (prolate spheroids: 3.2 mm × 6.4 mm, ρ_p = 1376 kg/m³). D = 152 mm, D_i = 12.5 mm. Fluid is ambient air.[1,3]

nozzle to column diameter ratio, D_i/D, there is no spouting regime. Instead, the bed changes directly from the fixed to the aggregatively fluidized state with increasing fluid velocity. This critical value is well approximated[2] at $H = H_m$ by $(U_{mf}/U_t)^{1/2}$ or $\epsilon_{mf}^{n/2}$. The same approximation can be safely applied at $H < H_m$. The value decreases from 0.35 for gas spouting of coarse spheres (for which $\epsilon_{mf} \simeq 0.42$, $Re_T > 500$, $n = 2.39$) to 0.1 for finer particles[1] owing to the accompanying increase of n, and can be expected to increase with decreasing particle sphericity owing to the accompanying increase of ϵ_{mf}. Another critical diameter ratio is D_i/d, which must not exceed about 25 to 30[4,5] if stable nonpulsatile spouting is to be achieved. The included angle of the *conical base* is a less critical parameter and need only exceed about 40° for stable spouting of most solid materials.[1]

10.2 MINIMUM SPOUTING VELOCITY

For cylindrical vessels up to about 0.5 m in diameter, with or without a conical base, the Mathur–Gishler[3] equation continues to be the simplest predictor (within ± 15%) of the mini-

mum spouting velocity for a wide variety of solid materials, bed dimensions, nozzle diameters, and fluids ranging from air to water. The correlation is:

$$U_{ms} = \left(\frac{d}{D}\right)\left(\frac{D_i}{D}\right)^{1/3} \sqrt{\frac{2gH(\rho_p - \rho)}{\rho}}$$

(10.1)

where the particle diameter, d, is taken as the arithmetic average of bracketing screen apertures for closely sized near-spherical particles and as the volume-surface mean diameter for mixed sizes, using the equivolume sphere diameter d_p for nonspherical particles. An exception is the case of particles, such as prolate spheroids, that align themselves vertically in the spout, for which prediction by Eq. (10.1) is best when d is taken as the horizontally projected diameter (i.e., the smaller of the two principal dimensions).

For bed diameters exceeding 0.5 m, Eq. (10.1), which can be rationalized qualitatively by jet-to-particle momentum transfer considerations,[4] increasingly underestimates U_{ms}; a rough working approximation for such large bed diameters is that U_{ms} is $2.0D_c$ times the value given by Eq. (10.1), with D_c in meters.[6] Recent studies[7,8] also indicate that the effect of changing bed temperature is inadequately accounted for by Eq. (10.1), due to the omission of fluid viscosity.

For conical beds or conical-cylindrical columns in which the bed height barely exceeds that of the conical base, the minimum spouting flow rate is no longer proportional to the square root of H as in Eq. (10.1), but is approximately proportional to H.[1,9,10] It should also be noted that flat-bottomed cylindrical spouted beds usually show large dead spaces in the annular region near the bottom, so that effectively they too behave either like conical or conical-cylindrical beds, depending on their height.[11]

The value of U_{ms} at the maximum spoutable bed depth for a given solid material in a given vessel is termed U_m, the *maximum value of the minimum spouting velocity*. In general, U_m is closely related to the minimum fluidization velocity, U_{mf}, for the given material, that is,

$$U_m/U_{mf} = b = 0.9 \text{ to } 1.5 \qquad (10.2)$$

The value of b depends partly on the solid material spouted and partly on the spouting vessel geometry, its value decreasing toward unity as vessel diameter is increased for a fixed ratio of D/D_i, and increasing toward 1.5 as D_i is increased for a fixed value of D.[1] It also decreases toward 0.9 with increasing spouting gas temperature.[7]

10.3 MAXIMUM SPOUTABLE BED DEPTH

For irregularly shaped particles at the maximum spoutable bed depth, Eq. (10.1) becomes

$$U_m = \left(\frac{d_p}{D}\right)\left(\frac{D_i}{D}\right)^{1/3}\sqrt{\frac{2gH_m(\rho_p - \rho)}{\rho}}$$

$$(10.1a)$$

For flow through a packed bed at the condition of minimum fluidization, the pressure gradient may be obtained from the Ergun[12] equation and is balanced by the buoyed weight of the bed per unit volume. Hence,

$$\left(\frac{-dP}{dz}\right)_{mf} = (\rho_p - \rho)(1 - \epsilon_{mf})g$$

$$= \frac{150\mu(1 - \epsilon_{mf})^2 U_{mf}}{(\phi d_p)^2 \epsilon_{mf}^3}$$

$$+ \frac{1.75\rho(1 - \epsilon_{mf})U_{mf}^2}{\phi d_p \epsilon_{mf}^3} \qquad (10.3)$$

With substitution of the empirical approximation of Wen and Yu,[13] that is, $1/\phi\epsilon_{mf}^3 = 14$ and $(1 - \epsilon_{mf})/\phi^2\epsilon_{mf}^3 = 11$, Eq. (10.3) can be solved to yield

$$Re_{mf} = \frac{d_p U_{mf} \rho}{\mu}$$

$$= 33.7\left(\sqrt{1 + 35.9 \times 10^{-6}Ar} - 1\right)$$

$$(10.4)$$

where

$$Ar = d_p^3(\rho_p - \rho)g\rho/\mu^2 \qquad (10.5)$$

When Eqs. (10.1a), (10.2), and (10.4) are combined to eliminate U_m and U_{mf}, the result is

$$H_m = \frac{D^2}{d_p}\left(\frac{D}{D_i}\right)^{2/3}\frac{568b^2}{Ar}$$

$$\times\left(\sqrt{1 + 35.9 \times 10^{-6}Ar} - 1\right)^2 \quad (10.6)$$

McNab and Bridgwater[14] found that Eq. (10.6) with $b = 1.11$ gave the best fit to prior experimental data for H_m in gas-spouted beds at room temperature, despite considerable scatter. Subsequent investigations[7,8] show $b = 0.9$ to be a more reliably conservative value at elevated temperatures. More empirical equations for H_m are available in Refs. 1 and 2.

If Eq. (10.6) is differentiated with respect to Ar, after substituting for d_p from Eq. (10.5), and $dH_m/d(Ar)$ is set equal to zero, it is found that there is a critical value, $Ar = 223,000$, or

$$(d_p)_{crit} = 60.6\left(\frac{\mu^2}{g(\rho_p - \rho)\rho}\right)^{1/3} \quad (10.7)$$

below which H_m increases with d_p and above which H_m decreases as d_p increases. For gas spouting, the resulting value of $(d_p)_{crit}$ is typically in the range 1.0 to 1.5 mm, a result that agrees with experimental observations.[1,2,14] Substitution of $Ar = 223,000$ into Eq. (10.4) gives a corresponding critical $Re_{mf} = 67$, which is also in close agreement with experiment for gas spouting.[1] It is noteworthy that these critical values are independent of vessel geometry, though H_m itself, including its maximum value at $Ar = 223,000$, varies as $(D^4/D_i)^{2/3}$ according to Eq. (10.6).

A value of b in Eq. (10.2) close to unity is consistent with the most frequently assumed mechanism for the termination of spouting, namely, fluidization of the upper solids layer in the annulus by the increasing annular fluid flow at the higher bed levels. A small excess of b over unity is attributable to the persistence of a higher superficial velocity in the spout than in the annulus, even at $z = H = H_m$. Other postulated termination mechanisms are the onset of slugging (or "choking") in the spout owing to the particle flow exceeding its conveying capacity, and the growth of surface instability waves at the spout–annulus interface.[1] Experimental studies have shown the spout-slugging termination mechanism to prevail in gas spouting of relatively small particles at room temperature[2,5] and of larger particles at elevated temperatures[15] where $Ar <$ 223,000, while termination is due to fluidization at the top of the annulus for larger particles at room temperature,[2,15] where $Ar <$ 223,000. The critical diameter given by Eq. (10.7) therefore also appears to represent the transition between these two termination mechanisms for gas spouting.

Equation (10.7) appears to be inapplicable to liquid spouting. Liquid spouted beds are characterized by a decrease of H_m as d_p increases for all values of d_p,[16] by the onset of fluidization in the annulus at $z = H = H_m$, and by persistence of spouting to a depth of H_m even when $H > H_m$, the spouted bed of height H_m then being capped by a particulately fluidized bed of height $(H - H_m)$.[17]

10.4 FLOW DISTRIBUTION OF FLUID

For a bed height H_m, Mamuro and Hattori[18] considered a simplified force balance over a differential dz of the annulus. Based on the assumption that Darcy's law applies to the vertical component of flow through the annulus and on the boundary condition that the annular solids are incipiently fluidized at $z = H_m$, they derived the following expression for the superficial fluid velocity, U_a, in the annulus at height z in a cylindrical column:

$$\frac{U_a}{U_{mf}} = 1 - \left(1 - \frac{z}{H_m}\right)^3 \quad (10.8)$$

Grbavčić et al.,[17] among others,[19] have shown that for given vessel geometry, given spouting fluid and given solids material, U_a at any level, z, is independent of total bed depth, H. Hence, Eq. (10.8) should apply for $H \leq H_m$, and this is borne out by experiment,[17,19] especially when $U \sim U_{ms}$. There is evidence that Eq. (10.8) works well even where the annulus Reynolds number, $\rho d_p U_a / \mu$, is one or two orders of magnitude greater than the upper limit for Darcy's law. This insensitivity of fluid flow distribution to deviation from Darcy's law has been explained theoretically.[19]

If the *spout diameter* at any bed level, labelled D_s, is known, continuity at that level yields

$$U_{sz} D_s^2 + U_a(D^2 - D_s^2) = U D^2 \quad (10.9)$$

Hence, the fraction of the total fluid flow that passes through the spout at any level, for a given superficial spouting velocity, U, is simply $U_{sz} D_s^2 / U^2 D^2$. At minimum spouting, the superficial velocity, $U = U_{ms}$, can be estimated from Eq. (10.1); operating velocities for gas spouting are typically 10% to 50% above U_{ms}. As a first approximation, the additional gas flow above that required for minimum spouting may be assumed to pass through the spout, while the gas flow through the annulus is constant once $U \geq U_{ms}$. Figure 10.3 indicates that, assuming ϵ_a is invariant with respect to U, increasing the spouting velocity above U_{ms} actually results in some *decrease* in the net gas flow through the annulus.[19] This is caused by the increased spout diameter and the increased solids downflow in the annulus.[20] The same effect is responsible for decreasing U_a at $z = H = H_m$ from U_{mf}, as given by Eq. (10.8), to about $0.9U_{mf}$ at $U = 1.1U_{ms}$.[19] Eq. (10.8) must then be modified by substituting $U_{aHm}(\sim 0.9U_{mf})$ for U_{mf}.

Typical *gas streamlines* in the annulus are shown in Figure 10.4. There is considerable evidence[22-24] that, below the outermost streamline shown, that is, immediately adjacent to the gas inlet, the gas reverses itself and flows downward and radially inward from the annulus to the spout, especially at $U >$

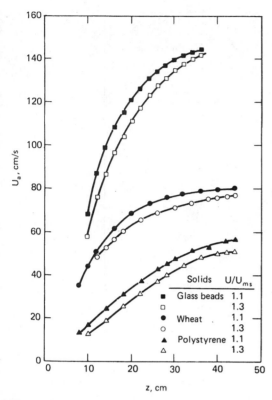

Figure 10.3. Effect of spouting air velocity on upward air velocity in the annulus. $D = 152$ mm, $D_i = 19.0$ mm, $\alpha = 60°$.[19]

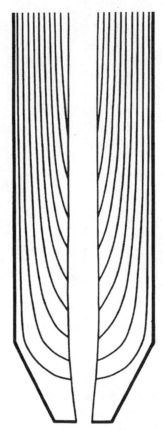

Figure 10.4. Calculated gas streamlines for a 0.24 m diameter × 0.72 m deep bed of polystyrene pellets ($d_p = 2.93$ mm, $D_i/D = 0.12$, $U/U_{ms} = 1.1$), in substantial agreement with experimental observations.[21]

$1.1U_{ms}$.[25] This gas recirculation, which is caused by the accelerated downward solids motion in the cone and the venturi effect experienced by the gas above the inlet nozzle,[24] does not, however, appear to affect the applicability of Eq. (10.8) farther up the bed. The fluid in each streamtube is in dispersed plug flow in the streamwise direction,[21] as in a moving packed bed, while that in the spout is essentially in plug flow. However, because of the large difference in gas velocities between spout and annulus, the residence time distribution of gas in the bed as a whole differs substantially from both plug flow and perfect mixing.[1]

10.5 PRESSURE DROP

Figure 10.5 shows typical plots of $(-\Delta P)$ versus U for the transition from flow through a fixed bed to a spouted bed. Before arriving at the fully spouting condition, the bed passes through a peak pressure drop, $-\Delta P_M$, associated with the energy required by the jet to disrupt the packing. As a first approximation, $-\Delta P_M$, which is bed-history-dependent, can be assumed to be equal to the buoyed weight per unit area of the initial packed bed, that is,

$$-\Delta P_M \simeq H(\rho_p - \rho)(1 - \epsilon) \quad (10.10)$$

Once the fully spouting condition is reached, the pressure drop stabilizes to a value, $-\Delta P_s$, that is essentially independent of superficial velocity.

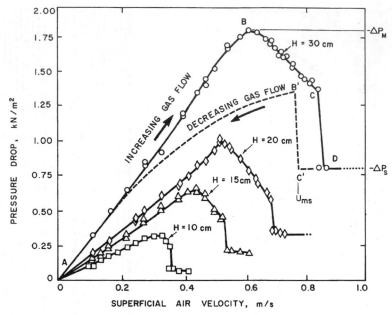

Figure 10.5. Typical pressure drop versus flow rate curves for the onset of wheat spouting. $d_p = 3.6$ mm, $D = 152$ mm, $D_i = 12.7$ mm, $\alpha = 60°$.[1,26]

The longitudinal pressure gradient in the annulus of a fully spouted bed at any level is given by

$$-dP/dz = K_1 U_a - K_2 U_a^2 \quad (10.11)$$

where K_1 and K_2 can be estimated as the coefficients of U_{mf} and U_{mf}^2, respectively, in Eq. (10.3), and U_a may be evaluated from Eq. (10.8). The pressure drop, $-\Delta P_f/H$, per unit length across a fluidized bed is:

$$-\Delta P_f/H = (-dP/dz)_{mf} = K_1 U_{mf} + K_2 U_{mf}^2 \quad (10.3a)$$

Combining Eqs. (10.11) and (10.3a), we obtain

$$\frac{-dP/dz}{-\Delta P_f/H} = \frac{2(\beta - 2)y + 3y^2}{2\beta - 1} \quad (10.12)$$

where $\beta = 2 + (3K_1/2K_2 U_{mf})$ and $y = U_a/U_{mf}$. Substitution for y in Eq. (10.12) from Eq. (10.8) and integration between limits (z, P) and (H, P_H) yields the following equation for the longitudinal pressure distribution in the annulus:

$$\frac{P - P_H}{-\Delta P_f} = \frac{1}{h(2\beta - 1)} \cdot [2(\beta - 2)$$

$$\times \{1.5(h^2 - x^2) - (h^3 - x^3)$$

$$+ 0.25(h^4 - x^4)\} + 3\{3(h^3 - x^3)$$

$$- 4.5(h^4 - x^4) + 3(h^5 + x^5)$$

$$- (h^6 - x^6) + 0.143(h^7 - x^7)\}] \quad (10.13)$$

where $h = H/H_m$ and $x = z/H_m$. The total pressure drop, $-\Delta P_s = P_0 - P_H$, across the spouted bed is obtained by putting $x = 0$ in Eq. (10.13), that is,

$$\frac{-\Delta P_s}{-\Delta P_f} = \frac{2 - (4/\beta)}{2 - (1/\beta)}(1.5h - h^2 + 0.25h^3)$$

$$+ \frac{3}{2\beta - 1}(3h^2 - 4.5h^3$$

$$+ 3h^4 - h^5 + 0.143h^6) \quad (10.14)$$

Note that the final term of Eq. (10.14) disappears when Darcy's law prevails in the annulus ($\beta \to \infty$) while the first term on the right-hand side drops out for the opposite extreme of inviscid flow ($\beta \to 2$). For a bed of maximum spoutable depth (i.e., $h = 1$), Eq. (10.14) shows that $(\Delta P_s/\Delta P_f)_{max} = 0.75$ and 0.643 for the Darcy and inviscid regimes, respectively.[1,19]

A simpler empirical relation for vertical pressure distribution was proposed by Lefroy and Davidson:[27]

$$\frac{P - P_H}{-\Delta P_s} = \cos\left(\frac{\pi z}{2H}\right) \qquad (10.15)$$

Differentiating Eq. (10.15) with respect to z yields

$$\frac{dP}{dz} = \frac{\Delta P_s}{2H} \pi \sin\left(\frac{\pi z}{2H}\right) \qquad (10.16)$$

If incipient fluidization of the annulus is assumed at $z = H = H_m$, then

$$\frac{\Delta P_f}{H_m} = \left(\frac{dP}{dz}\right)_{mf} = \frac{(\Delta P_s)_{max}\pi}{2H_m} \qquad (10.17)$$

that is, $(\Delta P_s/\Delta P_f)_{max} = 2/\pi = 0.637$, in excellent agreement with the inviscid value above, but below the more realistic Darcy value. In view of the fact that all the above equations neglect radial pressure gradients, which in the vicinity of the fluid inlet may be sufficient to raise the total pressure drop by some 25% for deep beds and considerably more for shallow beds, the more conservative Eqs. (10.13) and (10.14) are preferred over Eq. (10.15) and its corollaries.[19]

10.6 PARTICLE MOTION

The gas streamlines shown in Figure 10.4, if reversed in direction, also represent with little change the streamlines for the downward and inward flow of solids in the annulus. Because of the progressive crossflow of solids from the annulus into the spout, downward particle ve-

locities decrease with decreasing level in the column. As shown in Figure 10.6, the mass flow of solids, based on velocities measured at the wall, increases approximately linearly with height in the cylindrical part of the column, at least for $z \geq D$.[29] If the sometimes appreciable radial variation of downward solids velocity[30] is neglected, the slope of the plot at any level,

$$dW/dz = \rho_p(1 - \epsilon_a)d(v_w A_a)/dz \qquad (10.18)$$

is a measure of the crossflow rate at that level. Here $\epsilon_a(\simeq \epsilon_{mf})$ represents the constant *annulus voidage*.

Individual particles in the spout are accelerated by the surrounding fluid from a vertical velocity of essentially zero, when they first enter, to some maximum value, after which the particles decelerate to achieve zero velocity again at the top of the fountain. The parti-

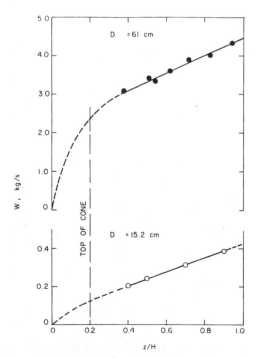

Figure 10.6. Solids flow in annulus, air-spouted wheat beds. $D/D_i = 6$, $H/D = 3$, $\alpha = 60°$, $U/U_{ms} = 1.1$,[1,3,28] $W = \rho_p(1 - \epsilon_a)A_a v_w$.

cles that enter the spout from the annulus rapidly become indistinguishable from the particles already in the spout at the same level. An experimental longitudinal profile of particle velocity, v_{sc}, along the spout axis appears in Figure 10.7. The point at the top of the *fountain* where $v_{sc} = 0$ is also shown. At any horizontal level within the spout, the variation of upward particle velocity, v_s, with radial distance, r, from the spout axis may be represented by

$$\frac{v_s}{v_{sc}} = 1 - \left(\frac{r}{r_s}\right)^m \qquad (10.19)$$

with $1.3 \le m \le 2.2$.[29]

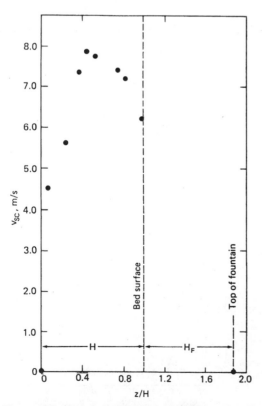

Figure 10.7. Experimental values of particle velocities along spout axis for 0.61 m diameter × 1.22 m deep bed of 3.2 mm × 6.4 mm wheat particles. $\rho_p = 1376$ kg/m³, $D_i = 102$ mm, $\overline{D}_s = 81$ mm, $U = 0.68$ m/s.[1,28]

A momentum balance on the spout particles over a differential height dz yields

$$\frac{2v_s dv_s}{dz} - \frac{v_s^2}{1 - \epsilon_s}\frac{d\epsilon_s}{dz} + \frac{v_s^2}{D_s^2}\frac{d(D_s^2)}{dz}$$
$$= \frac{3\rho(u_s - v_s)|u_s - v_s|d^2 C_D}{4 d_p^3 \rho_p}$$
$$- \frac{(\rho_p - \rho)g}{\rho_p} \qquad (10.20)$$

The drag coefficient C_D can be evaluated[31,32] as:

$$C_D = C_{DT}/\epsilon_s^{2(n-1)} \qquad (10.21)$$

where for gas spouting of spheres the terminal drag coefficient, C_{DT}, usually assumes the Newton's law value of 0.44 and the corresponding Richardson–Zaki[31] index $n = 2.39$. The upward interstitial fluid velocity in the spout, u_s, at any level is related to the corresponding superficial velocity, U_{sz}, by

$$u_s = U_{sz}/\epsilon_s \qquad (10.22)$$

Equations (10.20) to (10.22) and (10.9), in conjunction with experimental data for U_a or a relationship for U_a, such as Eq. (10.8), may be solved[29] to yield vertical profiles of v_s, ϵ_s and u_s, provided an auxiliary relationship involving at least one of these dependent variables is available. The boundary conditions at $z = 0$ are $v_s = 0$, $\epsilon_s = 1.0$ and $u_s = UD^2/D_i^2$.

If downward particle velocities at the vessel wall have been measured, then the auxiliary relationship that can be used is simply solids continuity at any level, neglecting any radial variation of particle velocity and voidage in the spout, and of particle velocity in the annulus,

$$W = \rho_p A_s(1 - \epsilon_s)v_s = \rho_p A_a(1 - \epsilon_a)v_w \qquad (10.23)$$

or its differential form, Eq. (10.18). The solution using Eq. (10.23) or (10.18) has been referred to as *Model* I.[29] Alternatively, an energy balance over a differential height of spout may be used, and this has been referred to as *Model* II.[29] Lim and Mathur[29] used a coefficient of unity instead of 2 in the first term of Eq. (10.20), and their results for v_s by both models compared to experimental data for wheat spouting are shown in Figure 10.8. Model I gives good agreement with experiment over the entire bed height, but depends on measurements of W or dW/dz, for which no generalized correlations exist. Model II, which in theory requires no experimental input except D_s as a function of z, becomes unstable at $z/H < 0.2$: reasonable values can be obtained only for $z/H > 0.2$ by starting with the experimentally measured values of v_s and ϵ_s at $z/H = 0.2$.

Recently, Krzywanski et al.[20, 24] have developed a more rigorous axially symmetric two-dimensional fluid-particle *model* of a spouted bed that predicts radial variations of pressure, gas velocity, particle velocity, and voidage.

A one-dimensional analysis similar to that used in the spout, but without the necessity of an auxiliary equation, has been applied[32] for particle motion in the fountain core. It is assumed that there is no crossflow of solids and that the interstitial gas velocity is approximated by U/ϵ_s. The boundary conditions at $z = H$ are taken as $\epsilon_s = \epsilon_{sH}$ and $v_s = v_{sH} = (v_{sc})_H \epsilon_{sH}^{0.93}$, the last relationship having been obtained empirically.[32] Equation (10.20), the left-hand side of which reduces for the present

conditions to $v_s dv_s/dz$, can then be solved in conjunction with Eqs. (10.21) and (10.22) for u_s, ϵ_s, and v_s, with $U_{sz} = U$ and

$$v_s(1 - \epsilon_s) = v_{sH}(1 - \epsilon_{sH}) \quad (10.24a)$$

when

$$v_s > v_{sH}(1 - \epsilon_{sH})/(1 - \epsilon_a)$$

or

$$\epsilon_s = \epsilon_a \quad (10.24b)$$

when

$$v_s \le v_{sH}(1 - \epsilon_{sH})/(1 - \epsilon_a)$$

The *height* of the *fountain*, H_F, can also be predicted with little error by ignoring drag in the simplified Eq. (10.20). Integration of this equation with the upper boundary condition, $z = H + H_F$, $v_s = 0$, gives

$$H_F = \frac{v_{sH}^2 \rho_p}{2g(\rho_p - \rho)} \quad (10.25)$$

where empirically v_{sH} is taken[32] as $(v_{sc})_H \epsilon_{sH}^{0.73}$; the decrease of the index on ϵ_{sH} from 0.93 to 0.73 arises from the neglect of drag.

Because of solids *crossflow* from the annulus into the spout over the entire annulus height and because of the showering effect of the fountain, a spouted bed is a good solids *mixer* when a single species of solids is used. For most practical purposes, assuming that the solids feed and discharge ports are located to preclude any obvious short-circuiting, that the cone angle is sufficiently small to prevent any dead solids zones at the base, and that the mean residence time of the solids exceeds some minimum value in the order of minutes, perfect mixing of the species is a good approximation for a continuously fed spouted bed. This is illustrated by Figure 10.9, in which the perfect mixing line is given by

$$I(\theta) = \exp[-\theta] \quad (10.26a)$$

and the nearby regression line by

$$I(\theta) = \exp[-(1/0.92)(\theta - 0.10)] \quad (10.26b)$$

When more than one species of solid material is used, for example, particles of different

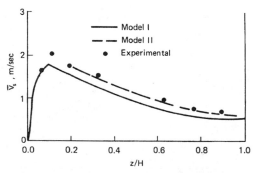

Figure 10.8. Radial-average particle velocity profile for air-spouting of 2.82 mm $\times$ 5.14 mm wheat particles. $\rho_p = 1240$ kg/m^3, $D = 152$ mm, $D_i = 19$ mm, $H/D = 3$, $U/U_{ms} = 1.1$.[29]

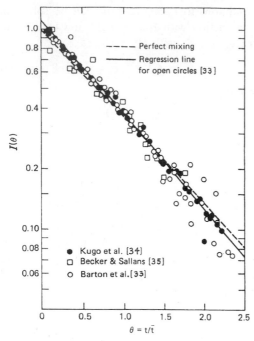

Figure 10.9. Solids mixing data correlated as internal age distribution function versus dimensionless time.[1]

size and/or density, considerable *segregation* occurs, especially at $U \simeq U_{ms}$. The heavier and/or larger particles concentrate in the upper inside part of the annulus, and, for continuously fed systems, the concentration of these particles in the bed becomes significantly higher than in the feed and discharge.[36] The primary cause of this segregation is the lower radial velocity imparted to the heavier particles when particle–particle collisions occur in the fountain region.[37] Segregation may therefore be largely countered if deflecting *baffles* are placed in the *fountain* region, or if U is increased so that the fountain particles strike the outer wall and bounce back toward the center of the bed surface.

10.7 VOIDAGE DISTRIBUTION

The *annulus* of a spouted bed near minimum spouting is a loose-packed bed of solids with its voidage, ϵ_a, very nearly the same as ϵ_{mf}.

This voidage has a value of about 0.42 for closely sized smooth spheres; somewhat higher values are found for nonspherical particles.[19] Some negative deviations from ϵ_{mf} occur in certain regions of the annulus as a result of local variations in fluid percolation and solids flow rates,[38] while some positive deviations occur at velocities well in excess of U_{ms}, but for most purposes these deviations can be ignored.

The voidage variation in the *spout* is roughly linear with height, as exemplified in Figure 10.10. These data are for the same spouted bed as in Figure 10.8. The predictions of Lim and Mathur[29] for both particle circulation models described above are also plotted on this figure. As in the case of particle velocity, there is reasonable agreement with the experimental data—over the entire range of z/H for Model I, and starting at $z/H = 0.2$ for Model II.[29] Above the bed surface, the *fountain* analysis summarized above shows a continuous decrease in voidage from ϵ_{sH} to ϵ_a in the core of the fountain.[32]

10.8 SPOUT DIAMETER

The diameter of the spout is an important parameter for determining the flow distribution between spout and annulus via Eq. (10.9),

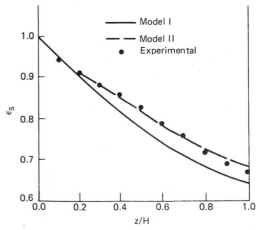

Figure 10.10. Spout voidage profile for system of Figure 10.8.[29]

profiles of v_s and ϵ_s via the one- or two-dimensional models mentioned above, and predictions of fluid-particle heat transfer, mass transfer, and chemical reaction via the two-zone model discussed below. Longitudinal variations of spout diameter that have been observed in conical-cylindrical columns are illustrated in Figure 10.11. *Shape* (a), the most common, tends to give way to (b) as column diameter increases, to (c) as particle size decreases, and to (d) for large inlet diameters.[1,39]

The variation of spout diameter with bed level for shapes (a) and (b) is predictable in good approximation by soil mechanics principles combined with variational analysis and knowledge of the longitudinal average spout diameter, $\overline{D}_s$.[40] The latter has been correlated empirically by the dimensional equation,[41]

$$\overline{D}_s = 2.00 G^{0.49} D^{0.68} / \rho_b^{0.41} \pm 5.6\% \quad (10.27)$$

over a wide range of experimental data at room temperature, where SI units are re-quired. It is better represented at both high and low temperatures by the dimensionally consistent semiempirical equation,[15]

$$\overline{D}_s = 5.61 \left[\frac{G^{0.433} D^{0.583} \mu^{0.133}}{(\rho_b \rho_g g)^{0.283}} \right] \pm 5\% \quad (10.28)$$

which, however, has been tested only for $D = 1.56$ mm and $\rho_b \sim 1500$ kg/m^3.

10.9 HEAT TRANSFER

Transfer of heat between the fluid and the solid particles in a spouted bed is most accurately described by means of the two-region *model* discussed below. A more conservative approach, based on the use of a fluid-particle heat transfer coefficient for a loose-packed bed, has also been employed.[1] In the annulus, unlike the spout, thermal equilibrium between fluid and particles is achieved even in a shal-

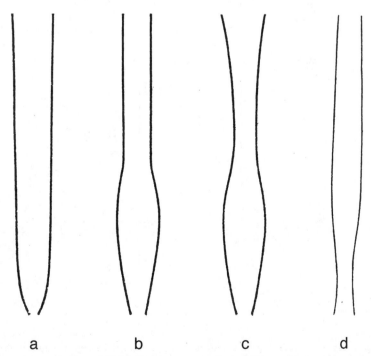

a   b   c   d

Figure 10.11. Observed spout shapes.[1,39] (*a*) Diverges continuously; (*b*)expands, then tapers or remains constant in diameter; (*c*) expands, necks, and then diverges; (*d*) necks, expands, then tapers slightly.

low bed. For the relatively large particles used in spouted beds, intraparticle heat transfer must be considered.[1]

Transfer between the bed and the wall is characterized by the development of a thermal boundary layer in the annulus, as exemplified by Figure 10.12 for gas spouting. For liquid spouting this boundary layer extends all the way to the spout. Over the range of conditions for which wall-to-bed heat transfer in gas-spouted beds has been studied,[1] the bed-to-

wall heat transfer coefficient, h_w, can be predicted from the empirical equation of Malek and Lu:[42]

$$\frac{h_w d_p}{k_g} = 0.54\left(\frac{d_p}{H}\right)^{0.17}\left(\frac{d_p^3 \rho_g^2 g}{\mu^2}\right)^{0.52}$$
$$\times \left(\frac{\rho_b c_{pp}}{\rho_g c_{pg}}\right)^{0.45}\left(\frac{\rho_g}{\rho_b}\right)^{0.08} \quad (10.29)$$

An alternative theoretical approach, based on a two-dimensional penetration model,[1] results in:

$$h_w = 1.129\left[v_w \rho_b c_{pp} k_b/(H-z)\right]^{1/2} \quad (10.30)$$

where the heat transfer surface extends over a length $(H - z)$. The mean coefficient given by Eq. (10.30) is twice the local coefficient at level z. Equation (10.30) tends to overpredict h_w somewhat, owing to the higher voidage at the wall than in the bulk of the annulus.[1]

A heating or cooling element submerged in the bed is a more efficient heater or cooler than a jacket around the column wall. Typical radial profiles of the immersed heat transfer coefficient, h_s, for a vertically aligned cylindrical heater are shown in Figure 10.13. It is seen

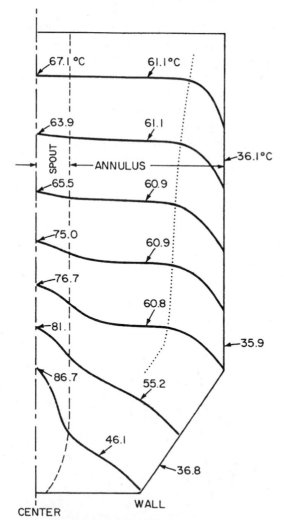

Figure 10.12. Local gas temperature profiles for a wall-cooled spouted-bed,[42] with thermal boundary layer profile (dotted line) added.[1]

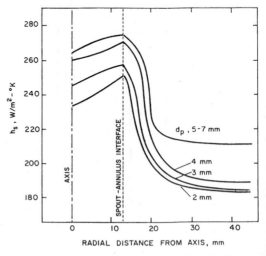

Figure 10.13. Radial profiles of submerged object-to-bed heat transfer coefficient in upper half of bed measured with a vertically aligned cylindrical heater, 4 mm diameter × 35 mm long,[43] air-spouted silica gel, $U =$ 0.945 m/s, $D =$ 94 mm, $D_i =$ 15 mm, $H =$ 100 mm.[1]

that h_s reaches a maximum at the spout–annulus interface and increases with d_p.

Typical vertical profiles of h_s in the *spout* for a horizontally aligned cylindrical heater are shown in Figure 10.14. It is seen that h_s decreases with increasing z, sharply near the bottom of the bed and then gradually farther up, and that it increases with increasing spouting velocity. Profiles of h_s within and around the fountain have also been reported for a spherical probe.[1,44] Values of h_s obtained in the annulus are generally similar to those for objects submerged in moving packed beds; values at the bottom of the spout are like those for the pure fluid flowing past the submerged object at comparable velocities, while coefficients higher in the spout and in the fountain are similar to those for objects submerged in a dense-phase fluidized bed.

10.10 MASS TRANSFER

As in the case of fluid-particle heat transfer, mass transfer between the fluid and the surface of the particles is best treated by the

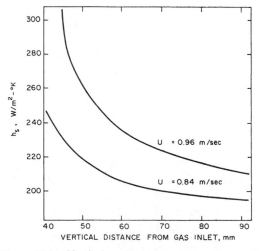

Figure 10.14. Vertical profiles of centrally submerged object-to-bed heat transfer coefficient in the spout measured with a horizontally aligned cylindrical heater, 10 mm diameter × 17 mm long,[43] air-spouted silica gel, $d_p = 2$ mm, $D = 94$ mm, $D_i = 15$ mm, $H = 100$ mm.[1]

two-region *model* described below. Again, a more conservative approach for fluid-particle mass transfer under conditions of external control, for example, constant rate drying can be based on the use of a mass transfer coefficient for the loose-packed bed conditions that prevail in the annulus.[1]

However, *drying* of such materials as agricultural products and fertilizer granules, for which a hot air spouted bed has proved to be most effective, is often carried out over ranges of moisture content that are well within the falling rate period. Moisture diffusion within the particles then controls the overall drying process. For such internal mass transfer control, the oft justified assumption that the bed is deep enough for the outlet gas to be in thermal equilibrium with the well mixed spouted solids precludes the need for heat transfer rate considerations. An overall mass balance, overall energy balance, and particle moisture diffusion equation, combined with moisture desorption isotherms for the given solids and a knowledge of particle moisture diffusivity, $\mathscr{D}$, as a function of temperature and local composition, can then be solved numerically to give good prediction of the temperature and uniform mean moisture content, $\overline{m}$, of batch-dried particles as a function of uniform drying time, t.[45] For steady continuous spouted bed drying, the residence time, t, of individual particles tends to differ from the mean residence time $\bar{t}$. Therefore, the average moisture content of the continuous solids product is given by

$$\overline{\overline{m}} = \int_0^\infty \overline{m}(\theta)E(\theta)\,d\theta \qquad (10.31)$$

where $\overline{m}(\theta)$ is the average particle moisture content for the corresponding isothermal batch process of duration $\theta = t/\bar{t}$ and the exit age distribution function, $E(\theta)$, is related to the internal age distribution function, $I(\theta)$, by[46]

$$E(\theta) = -dI(\theta)/d\theta \qquad (10.32)$$

Substitution of Eq. (10.32) into Eq. (10.31) with the appropriate change in integration limits results in

$$\bar{\bar{m}} = \int_0^1 \bar{m}(\theta)\, dI(\theta) \qquad (10.33)$$

which was derived more directly by Becker and Sallans.[35] For continuous grain drying, Eq. (10.26b) leads to even more accurate prediction of particle temperature and $\bar{\bar{m}}$ than Eq. (10.26a).[45]

The analytical simplification, used for wheat drying,[35] which results when the surface moisture content of the particles is assumed constant, and other shortcut or empirical methods, are summarized elsewhere.[1]

10.11 CHEMICAL REACTION: TWO-REGION MODELS

Spouted beds share some of the principal advantages of fluidized beds as chemical reactors —solids mobility, relatively uniform temperature and, to some extent, favorable bed-to-surface heat transfer. Shared disadvantages between spouted and fluidized bed reactors are *bypassing* of gas, back*mixing* of solids, particle *entrainment*, and *attrition*. Spouted beds give more reproducible flow patterns and have fewer *flow regimes* than fluidized beds, but their ranges of application in terms of mean particle size and vessel diameter are much more limited.

Bypassing in spouted beds is caused by fluid elements in the central spout travelling more quickly and with a much higher voidage than in the annulus. For a catalytic gas-phase reaction, it is essential to distinguish between the two regions, since reaction is much more favorable in the annulus, where gas elements are in intimate contact with the solids, than in the spout. Similar considerations apply when spouted beds are used for heat transfer between fluid and particles or for an analogous mass transfer process, for example, adsorption of a component from a gas.

The earliest and simplest representation of a spouted bed for these purposes is in terms of

a *one-dimensional model*[47] in which radial gradients within each region are ignored. For a first-order reaction in an isothermal spouted bed reactor, steady-state mass flow balances for an element of height dz of each region then result in:[48]

$$Q_s \frac{dC_s}{dz} + k_{sa}\pi D_s(C_s - C_a)$$
$$+ K_r \frac{\pi}{4} D_s^2(1 - \epsilon_s)C_s = 0 \qquad (10.34)$$

and

$$Q_a \frac{dC_a}{dz} + \frac{dQ_a}{dz}(C_a - C_s) + k_{sa}\pi D_s(C_a - C_s)$$
$$+ K_r(1 - \epsilon_a)A_a C_a = 0 \qquad (10.35)$$

for the spout and annulus, respectively. The first and last terms in each of these equations are due to convection and chemical reaction, respectively. Plug flow of fluid is assumed to prevail in each region, and the reaction rate in each region is assumed to be controlled by chemical kinetics. The middle terms arise from inter-region mass transfer, the second term in Eq. (10.35) being due to net outflow from the spout into the annulus, as discussed above; the terms involving k_{sa} account for any additional transfer. The flow rates Q_a and Q_s through the two regions and the derivative dQ_a/dz can be obtained as functions of height from Eqs. (10.8) and (10.9). The *spout diameter*, D_s, can be estimated from Eqs. (10.27) or (10.28) or measured in a *half-column*, while A_α can be obtained from the geometry of the column. The rate constant, K_r, should be determined separately under isothermal conditions in a reactor whose hydrodynamics are well understood, for example, in a packed bed or spinning basket reactor. For non-first-order kinetic rate expressions, the final terms in Eqs. (10.34) and (10.35) must be replaced by the appropriate rate expressions. There is no reliable method of estimating k_{sa}, but values are typically less than 0.1 m/s and $k_{sa} \simeq 0$ appears to be a reasonable assumption when $d > 1$ mm.[48]

The boundary condition required for solution of Eqs. (10.34) and (10.35) is $C_s = C_{in}$ at $z = 0$. The equations can be integrated numerically from $z = 0$ to $z = H$. The exit concentration is then evaluated from the overall mass balance:

$$U \frac{\pi}{4} D^2 C_{exit} = [Q_s C_s + Q_a C_a]_H \quad (10.36)$$

For the case where a component is removed from a gas stream in a spouted bed, for example, for collecting aerosol particles on the bed particles,[49] the same one-dimensional model can be applied, but with the reaction terms replaced by the respective adsorption rate per unit volume in that phase. For the annulus, the adsorption rate can be based on correlations for mass transfer between particles and fluid in packed beds. For the spout, *mass transfer* between the spouting fluid and particles can be estimated from the high-voidage correlation of Rowe and Claxton.[50] These equations should also be used when reaction rates within the individual regions are mass-transfer controlled.

Analogous equations can be developed for *heat transfer* when a hot gas enters a bed of cold particles or vice versa. Let us assume constant properties and spherical particles and neglect any interphase transfer, aside from that associated with crossflow of gas. In view of the rapid mixing of solids in spouted beds and the fact that the volumetric heat capacity of the solids, $\rho_p c_{pp}$, is much larger than that of the gas, $\rho_g c_{pg}$, we may, as a first approximation, treat the particles at any instant as being of uniform temperature, T_p. Then energy balances for gas in each of the regions yield

$$Q_s \rho_g c_{pg} \frac{dT_{gs}}{dz}$$
$$= (h_{pg})_s \frac{3\pi}{2d_p} D_s^2 (1 - \epsilon_s)(T_p - T_{gs}) \quad (10.37)$$

and

$$\rho_g c_{pg} \left[Q_a \frac{dT_{ga}}{dz} + \frac{dQ_a}{dz}(T_{ga} - T_{gs}) \right]$$
$$= (h_{pg})_a \frac{6A_a}{d_p}(1 - \epsilon_a)(T_p - T_{ga}) \quad (10.38)$$

for the spout and annulus, respectively. The particle-gas heat transfer coefficients, $(h_{pg})_s$ and $(h_{pg})_a$, for the spout and the annulus may again be obtained from correlations for dilute suspensions and packed beds, respectively. Gas entering the annulus equilibrates with the solids temperature within a small distance. To obtain the change of particle temperature for a batch process or the steady-state particle temperature for a continuous process where solids are fed at a different temperature from the gas, a heat balance is also required for the solids. Equations (10.37) and (10.38) can be solved numerically with the boundary conditions $T_{gs} = T_{ga} = T_{gi}$ at $z = 0$. The outlet gas temperature is obtained from an energy balance, that is,

$$U \frac{\pi}{4} D^2 (T_g)_{exit} = [Q_s T_{gs} + Q_a T_{ga}]_H \quad (10.39)$$

The *one-dimensional model* has been extended to *spout-fluid beds* (see section 10.13) by Hadžismajlović et al.[51] These workers also allowed for variation of ϵ_s with z rather than adopting an average value.

An alternative model, the *streamtube model*, has been applied[48] to the case of a first-order gas phase reaction in an isothermal spouted bed. The model was first used[21] to describe gas residence time distributions in spouted beds. Whereas the one-dimensional model implicitly assumes perfect radial *mixing* of gas elements in the annulus, the streamtube model is based on a physical picture, shown in Figure 10.15, in which the gas entering the annulus fans outward and upward in a finite number of streamtubes. The coordinates of the streamlines bounding each of these streamtubes are calculated on the assumption that the vertical component of gas velocity is radially uniform at each section of the annulus. Streamwise dispersion is ignored in each of the streamtubes. Any inter-region mass transfer, aside from the bulk flow obtained from Eqs. (10.8) and (10.9), is also ignored. Plug flow of gas is again assumed in the spout region.

With these assumptions, a mass balance in the spout phase gives Eq. (10.34) with $k_{sa} = 0$.

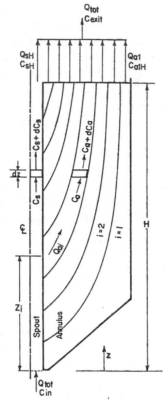

Figure 10.15. Vertical section through spouted bed showing flow distribution assumed in the streamtube model.

The concentration of gas leaving the top of the ith streamtube is given[48] by

$$C_{aiH} = C_s(z_i)\exp(-K_r(1 - \epsilon_a)\tau_i/\epsilon_a)$$
$$(10.40)$$

where z_i is the midpoint entry height (i.e., the mean height of intersection of the bounding streamtubes with the spout–annulus interface) and τ_i is the mean residence time of gas within the streamtube. The exit concentration is again obtained by performing an overall mass balance at the top of the reactor:

$$U\frac{\pi}{4}D^2 C_{exit} = \left[Q_s C_s + \sum_{i=1}^{N} Q_{ai}C_{ai}\right]_H$$
$$(10.41)$$

Piccinini et al.[48] found that 20 streamtubes gave a good compromise between accuracy and computational effort when applied to the streamtube model.

Experimental reaction data obtained in a spouted bed of catalyst pellets using the ozone decomposition reaction have been compared with both the one-dimensional model and the streamtube model.[48] Predicted concentration profiles from the latter model for one experimental run are shown in Figure 10.16. Agreement between the experimental conversions and the predictions of both models was excellent. While it can be shown theoretically that for a first-order reaction and the same flow distributions between spout and annulus, the two models predict identical overall conversions,[52] the streamline model gives a more accurate representation of the actual flow patterns and concentration profiles, especially for beds of large diameter.

Neither of the models presented above makes allowance for the additional contacting between gas and solids that occurs in the *fountain* region above the bed surface. A procedure for including the fountain (the contri-

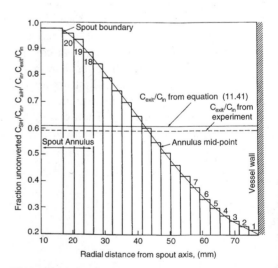

Figure 10.16. Predicted radial concentration profile at the bed surface for the streamtube model: $d = 1.48$ mm, $H = 0.41$ m, $D = 0.155$ m, $U = 1.07$ m/s = $1.1U_{ms}$, $K_r = 4.2$ s^{-1}, $\rho_p = 2330$ kg/m^3, $\epsilon_a = 0.48$, $\epsilon_s = 0.85$. Predicted and experimental overall exit concentrations are also shown.

bution of which to the overall conversion is usually small) is presented by Hook et al.,[53] whose comprehensive *streamtube model* is based on a set of relationships developed by Littman, Morgan, and their co-workers.

The *one-dimensional* and *streamtube* models described above can be used to predict the performance of gas-phase solid-catalyzed chemical reactions. In principle, they can also be applied to gas–solid heterogeneous reactions, in much the same way that two-phase reactor models for fluidized beds have been extended to the case of heterogeneous reactions.[54,55] Foong et al.[56] used the one-dimensional model to describe conversion in a spouted bed coal gasifier. However, since the kinetics of the reaction were unknown, the reaction was treated like a gas phase reaction to yield an effective rate constant, and this value was then used to predict the influence of bed height, bed composition, and column diameter. It is noteworthy that conversion is predicted to increase with increasing reactor diameter, in contrast to the case of fluidized beds where conversion almost always decreases as a reactor is scaled up. The improved performance with increasing D arises because the spout occupies a smaller fraction of the cross-sectional area of the spouted bed as the reactor is scaled up. The same trend has been predicted for spout bed reactors previously,[1,47] but has been contradicted experimentally over a limited range, $D = 0.15$ to 0.22 m.[57]

Several complications arise in applying the reactor models to the more general case of gas–solid heterogeneous reactions. The kinetic rate expression must account for the way in which particles react, for example, by assuming a shrinking core, surface reaction, or homogeneous reaction throughout the particles.[54] The physical properties (size, density, and shape) of the particles may change during their residence in the bed as a result of reaction, *attrition*, or *agglomeration*. Solids residence time distributions (commonly approximated by perfect *mixing*) must be considered, since the extent of reaction of each particle depends on its residence time in the reactor.

Population balances may be required to account for different sizes of particles as reaction, attrition, and entrainment proceed. Since heterogeneous reactions are often highly exothermic or endothermic, energy balances may also be required. Finally, the extent of reaction of gaseous components must be linked to that of the solids by means of the stoichiometry of the reactions. Models complex enough to cope with all these factors have not yet been developed.

10.12 APPLICATIONS

Originally developed for wheat drying[53] (Fig. 10.17), gas-spouted beds have since been applied to a wide variety of operations[1] involving coarse (e.g., 1 to 5 mm) solid particles. These operations rely on one or more of the following features of the technique:

1. Good solids *mixing* coupled with satisfactory gas-particle contact, thereby accomplishing for coarse solids what a fluidized bed does for fine solids.

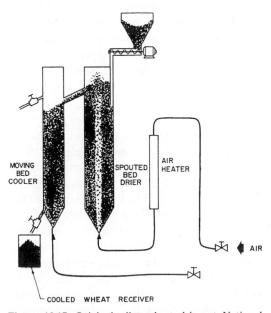

Figure 10.17. Original pilot wheat drier at National Research Council of Canada.[1,58]

2. Higher gas velocities and correspondingly lower gas residence times than for low-voidage fluidized beds of fine solids.
3. Systematic cyclic *movement of solids*, compared with the more random particle movement in fluidized beds or rotary drums.
4. Solids *attrition* and *deagglomeration* caused by high-velocity interparticle collisions in the spout.
5. The absence of a distributor plate, in contrast to the case of a fixed or fluidized bed.
6. Countercurrent heat transfer between ascending gas and descending annular solids.

Good solids mixing, together with *effective gas-particle contact*, is the basis for spouted bed *drying* of noncaking granular solids.[1] The method is particularly suitable for heat-sensitive materials such as agricultural products or polymer granules, since the rapid agitation of the solids permits the use of higher temperature gas than in nonagitated driers, without the risk of thermal damage to the particles. Commercial driers of 0.6 m diameter with a bed depth of about 2 m are capable of safely drying up to 2 Mg/h of peas through an 8% moisture range, dry basis, using about 3 Mg/h of air at temperatures up to 557 K.[59] The layout of such an industrial unit for drying peas, lentils, and flax is shown in Figure 10.18. Many other agricultural products have been successfully dried in spouted beds.[60]

Sensible heating or cooling of coarse solids in spouted beds also makes use of the favorable gas–solid contacting, but the good solids mixing is more important in heating than in cooling. In the use of a spouted bed for *blending* of solids, the intimate gas-particle contact is incidental, and only the good solids mixing is of importance. Multistage spouted bed preheating of coal feed to coke ovens has been successfully piloted, while commercial-scale rectangular (4.9 m × 1.8 m) two-stage multiple-spout fertilizer coolers with capacities up to 30 Mg/h and thermal efficiencies exceeding 85% have been developed by Fisons Ltd.; single-spout circular units of equal size have been operated by I.C.I. Fibres Ltd. for

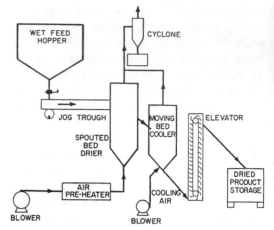

Figure 10.18. Layout of industrial drier for agricultural products.[1,59]

blending polyester polymer chips in batches exceeding 57 m³.[1]

The relatively *high gas velocities* and correspondingly low gas residence times associated with spouting of coarse particles are the basis for the bench-scale development at Hokkaido University of a dual-spouted reactor–regenerator combination for thermal *cracking* of petroleum feedstocks.[1] A similar combination has been developed by the same investigators for catalytic desulfurization of residual fuel oil, using steam at 923 K plus the fuel oil as the spouting fluid in the reactor, and air in the regenerator.[61] High gas throughput per unit cross-section and high gas–solids relative velocity also make the use of a spouted bed of coarse solids attractive for *gas cleaning* purposes, especially since high efficiencies at minimum spouting velocities have been measured for the bench-scale collection of liquid and electrified-solid aerosols from a gas in spouted beds of inert solids,[49] as well as for the chemical reduction of dilute SO_2 gas by a spouted bed of activated charcoal.[62] However, for both these gas cleaning processes, operation at velocities above minimum spouting sharply reduces the respective efficiencies, undoubtedly as a result of excessive gas bypassing through the spout, in addition to the lowering of gas residence time.

The highly systematic *cyclic movement of solids* in a spouted bed has proved to be a key advantage in such processes as granulation and particle coating. In granulation, a melt or solution is atomized into a bed containing seed granules spouted by hot gas. These granules build up layer by layer as they cycle in the bed, and yield a final product that is well-rounded and uniform in structure.[63] It is also possible to build up granules by feeding dust with the hot spouting gas, which softens the surface of the seeds.[64] Continuous operation requires that oversize product be crushed and recycled to the spouted bed together with undersize product,[65] as illustrated in Figure 10.19. This has been applied commercially by PEC Engineering of France to a number of 16 Mg/h sulfur granulators ("Perlomatic" system). Spouted bed granulation of fertilizers[66, 67] has also been applied on an industrial scale, the largest known unit—for mixed fertilizers, in Sicily—expanding upward to 3.5 m diameter and having a capacity of 500 to 700 Mg/day.[68]

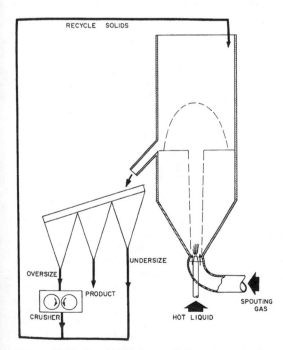

RECYCLE SOLIDS

OVERSIZE

UNDERSIZE

PRODUCT

CRUSHER

SPOUTING GAS

HOT LIQUID

Figure 10.19. Spouted-bed granulation system, after Berquin.[1, 65]

The *coating* of pharmaceutical tablets in a spouted bed is a well-established commercial operation, a typical batch being 100 kg.[1] The kinetics of this process have been studied.[69] Batch[70] or continuous[71] spouted bed coating of urea granules with sulfur to produce a slow-release fertilizer has been investigated extensively on a pilot scale. The thermal-chemical coating of pyrolytic carbon and/or silicon carbide onto submillimeter nuclear fuel kernels of uranium oxide or carbide has been standardized in spouted bed furnaces of 75 to 125 mm internal diameter, with kernel loads of about 1 kg for each coating operation.[72] Essentially the same technique has been applied to the pyrolytic coating of prosthetic devices.[73]

The *solids attrition* caused by the particle collisions in the spout is a liability for some spouted bed operations (e.g., granulation, tablet coating), but an asset for several others. The most successful of these, developed at the Leningrad Institute of Technology, is the drying of slurries and solutions by atomizing them into the lower region of a hot gas-spouted bed of inert particles.[74] The slurry or solution coats these particles and dries during the particle downward movement in the annulus. The fine product is broken away by interparticle collisions in the spout and collected from the overhead gas. Materials that lend themselves to this method of drying include organic dyes, dye intermediates, lacquers, salt and sugar solutions, several chemical reagents, animal blood,[75] and wastewater sludge.[76] Both capital and operating costs compare favorably with respect to spray drying.[77] The more conventional spouted bed drying of granular materials with caking tendencies, for example, ammonium nitrate, has also been industrially successful (where fluidized bed drying has failed), owing to the breakdown of embryonic agglomerates in the high-velocity spout.[1] Bench or pilot scale spouted bed developments for which this property has been important include simultaneous drying and comminution of particulate solids,[78] iron ore reduction,[1] shale pyrolysis,[1, 79] and coal *car-*

bonization (pyrolysis). The last of these has been conducted with coals of various caking tendencies at temperatures up to 925 K in Australia,[33] up to 815 K in India,[80] and up to 913 K in Canada.[81] The latter gave tar yields up to 31% by weight on a moisture- and ash-free basis. Indian noncaking coals have been similarly gasified with air and steam.[82] A more impressive development is the gasification at atmospheric pressure and temperatures up to 1200 K of 0.8 to 3.6 mm highly caking coals from western Canada in a 0.15 m diameter spouted bed containing a proportion of silica particles in the same size range.[56] This was achieved without the cumbersome and expensive procedures required for gasifying caking coals in a fluidized bed. Scale-up to 0.30 m has led to improved performance.[83] The larger unit (Fig. 10.20) has been used to study the effect of oxygen enrichment,[85] as well as to gasify 1.5 mm oil sand coke.[86] The process has also been successfully operated at elevated pressures[87] and modeled.[88]

The *absence of a distributor plate* in a spouted bed is a definite advantage in many of the above operations—especially in granulation and coating, in *drying* of solutions, slurries and sticky solids, and in *carbonization* or *gasification* of caking coals. It is also an important consideration in a high-temperature (1300 to 1800 K) industrial process for making granular activated carbon,[89] in bench-scale spouted-bed calcination of limestone,[90] and for production of cement clinker from decarbonated cement granules.[1] The decarbonization itself has been successfully accomplished in a system designated Kawasaki Spouted Bed and Vortex Chamber or KSV.[91] At least five cement plants with capacities of 350 Mg/day and one with 8500 Mg/day, using KSV calcining furnaces, were in satisfactory operation by 1975.[92]

Because of *countercurrent heat transfer* between the downwardly recirculating hot annulus solids and the ascending cold inlet gas, a spouted bed of inert particles is capable of sustaining the *combustion* of leaner mixtures or lower grades of gaseous,[93,94] liquid,[95,96] and solid[97,98] fuels[99] than more conventional burners. Even for the high ash solid fuels tested, combustion efficiencies exceeded 90% pro-

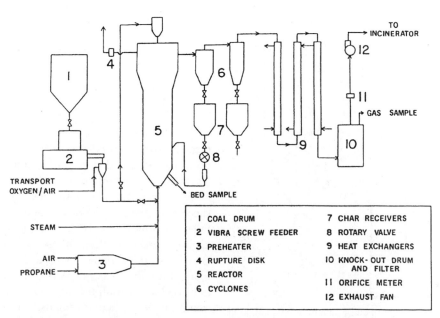

Figure 10.20. Schematic diagram of a spouted bed coal gasifier.[84]

vided that the bed temperatures were above 870°C and that the fines captured in the primary cyclone were recycled to the bed.[98]

The applications cited in this section relate, for the most part, to the "standard," "classical" or "conventional" spouted bed (CSB) described in the earlier sections. Many significant modifications to the standard geometry and/or mode of operations have, however, been made over the years. Such modifications, and a few applications thereof, are discussed next.

10.13 MODIFIED SPOUTED BEDS

The following modifications of the CSB, many of which have been detailed elsewhere,[1] are worthy of note.

10.13.1 Multiple Spouts

There is a practical limit to the vessel diameter that can be served by a single fluid inlet. Since $H/D > 1$ for stable spouting, the large bed heights required for large column diameters would give rise to excessive pressure drops. In addition, the long times spent by particles in the annulus of a large bed over the course of a single cycle, especially along the outer streamlines similar to those shown in Figure 10.4, could be a distinct disadvantage for certain processes, for example, for particle *drying* where excessive time in the hot region of the bed could cause thermal damage to the particles. One way of overcoming these difficulties is by using several fluid inlet nozzles in parallel, that is, multiple spouting.

Figure 10.21 shows a schematic of a multiple spouted bed with a flat base. Multiple *cone bases* have also been used.[1] Although the spouting fluid may originate from a single manifold, the flow to each inlet nozzle must be controlled separately. Even with such control, spouting stability problems arise when the inlet nozzles are too closely spaced, when the bed height is increased excessively (but still

below H_m for a single nozzle), and when the ratio D_i/d is less than about 8.[99a] Bed *stability* can be improved by using fluid inlet nozzles that project (e.g., about 3 mm) above the bed floor (as in Fig. 10.21), by installing vertical partitions that cut off lateral flow fluid between spouting cells,[1] and by fixing inverted-funnel spout deflectors above the fountains[1] to minimize interference between adjacent cells by spout wandering.

Foong et al.[99a] have shown that, if one takes the diameter of each spouting cell as that of a circular cylinder having the same cross-sectional area, then both the minimum spouting velocity and the pressure drop across the spouted bed can, for a *stable* multispout bed, be predicted by relationships applicable to a CSB. For handling equal inventories of solids, a multispout bed requires considerably more fluid than a single-spout unit, but results in faster solids turnover.[1]

For given solids, fluid, column configurations, and bed depth, there exists a maximum superficial velocity beyond which steady spouting gives way to chaotic fluidization.[100] CERCHAR of France utilizes a multiple cone base as an efficient distributor to a fluidized bed. In these units, the static bed height must exceed the maximum spoutable height or the gas velocity must exceed the maximum spouting velocity.[101]

10.13.2 Draft Tube

Crossflow of both fluid and solids between the spout and the annulus can be eliminated over most of a spouted bed's height by inserting in the spout region, starting at some distance (in excess of 10 d_p) above the fluid inlet nozzle, an open draft tube with walls that are impervious to both phases. The draft tube diameter is usually chosen to be similar to that of the spout without a draft tube, and is equal to or larger than the inlet nozzle diameter. The draft tube is aligned vertically with its axis collinear with the axis of the column. One result is that the bed can now function at depths greater than H_m. Other consequences

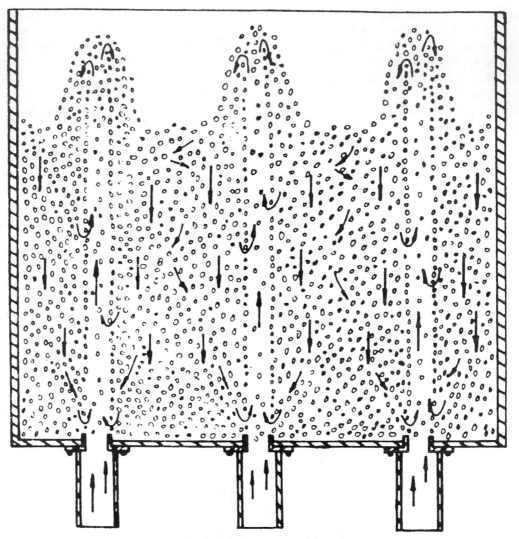

Figure 10.21. Schematic diagram of a multiple spouted bed.[99a]

are a large reduction in the fluid flow requirement for spouting, an even larger reduction in the solids circulation rate,[102] and considerably reduced solids mixing.[1] These changes are advantageous for granulation and particle coating, where plug flow of solids increases the uniformity of the product. The method also allows smaller solids to be successfully spouted.[103] The characteristics of a draft tube spouted bed grain drier have been detailed.[104] More recently, a similar unit has been shown to be viable for simultaneously *drying* and

removing by attrition the valuable pigment from the seeds of a tropical shrub.[105]

If the draft tube is permeable, for example, made of metallic screen, it can allow fluid exchange but remain impervious to the solids. Such a screen is well suited to applications in which it is desirable that all particles spend the maximum possible time in the annulus without curtailing annular fluid flow.[1] The hydrodynamic characteristics of a porous draft tube, intended for thermal disinfestation of grains, have been described quantitatively.[106]

10.13.3 Top-Sealed Vessel

By sealing the top of the spouting vessel and providing an alternative fluid outlet, either at the bottom of the bed[1] or part way up, fluid is forced to travel downward through the annulus. This results in a narrower fluid residence time distribution than otherwise. The residence time distribution can be further narrowed if, in addition, an inner draft tube is used[107] (see Fig. 10.22). The combination of side outlet and draft tube appears to give high gas conversions and versatility in operable particle sizes.[103]

10.13.4 Slotted Two-Dimensional Spouted Bed

The rectangular cross-section slot spouted bed first described by Romankov and Rashkovskaya[74] has, during the past decade, been extensively elaborated and investigated by Mujumdar and co-workers, with particular reference to grain *drying*.[111] Draft plates can be added to perform the same function as a draft tube in a conical-cylindrical spouted bed. The gas entry slot and the draft plates span the full thickness of the column (see Fig. 10.23). Scale-up of such a two-dimensional vessel with

a small thickness can then be affected, according to Mujumdar,[111] by simply increasing this dimension, without changing the width, the resulting performance differing from that in the smaller unit only by virtue of the reduced wall effect caused by the front and back plates. The practical problem of introducing gas uniformly into an ever-increasing slot length has, however, not been addressed.

Instabilities in spouted bed behavior may arise from even slight dissymmetries in alignment, horizontally or vertically, of the draft plates or of the gas entry slot. This problem can be avoided by several deliberately asymmetric alternative designs of two-dimensional spouted beds.[112]

10.13.5 Spout-Fluid Bed

If, in addition to supplying spouting fluid through a central inlet nozzle, extra fluid is also supplied through either a flat (Fig. 10.24) or a conical (Fig. 10.25) distributor to the annular region, the result is a "spout-fluid" bed. This modification of a conventional spouted bed enhances fluid–particle *heat* and *mass transfer*, and counteracts any tendency

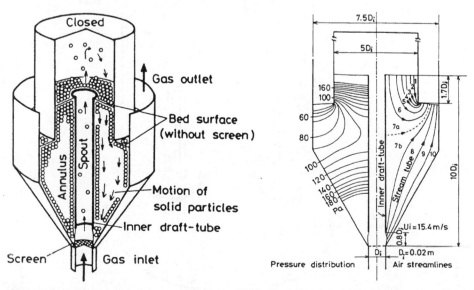

Figure 10.22. Schematic diagram of a top-sealed spouted bed with draft tube and dual surface gas outlet, together with isobars and gas streamlines in the annulus.[108]

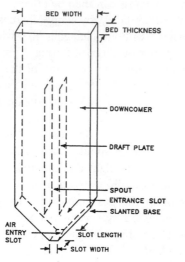

Figure 10.23. Schematic diagram of a slotted two-dimensional spouted bed with two draft plates.[110]

for particles to agglomerate in the annulus. From the standpoint of those concerned with improving the performance of a fluidized bed, addition of a central jet to the distributor promotes better circulation and *mixing* of the solids which, in the case of an exothermic reaction such as combustion, results in greater temperature uniformity and increased bed-to-surface heat transfer.[98] Thus, both spouted bed and fluidized bed designers have recognized the virtues of the spout-fluid bed hybrid, which has therefore received considerable attention during the past 15 years.

If the additional fluid fed to the annulus is insufficient to fluidize the annular solids, the total flow required to maintain the bed in the spouted condition is greater than that for spouting without the auxiliary fluid, but less than that required to fluidize the same

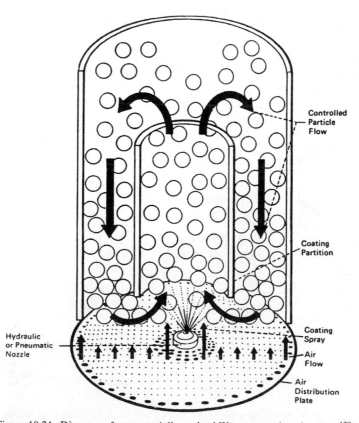

Figure 10.24. Diagram of commercially evolved Wurster coating chamber.[120a]

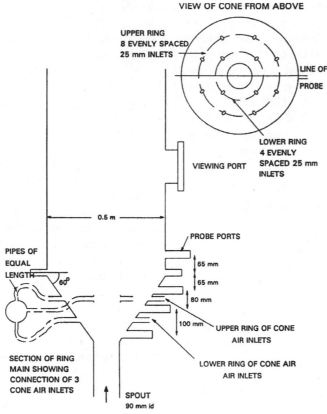

Figure 10.25. Base design for a commercial scale gasifier.[126]

solids.[113, 114] This *regime* of a spout-fluid bed has been applied to liquid contacting of an ion-exchange resin.[115]

If sufficient auxiliary fluid is supplied to completely fluidize the annular solids while maintaining penetration of the spout to a fountain above the bed, the total fluid flow requirement for such "spout-fluidization" exceeds that for either spouting or fluidizing the bed.[1] In addition to spout-fluidization and spouting with gentle aeration, other spout-fluid regimes also exist and have been mapped.[9, 116, 117]

For a spouted bed with a *draft tube*, aeration of the annulus tends to counteract the reduction in solids circulation rate caused by the draft tube.[102] For applications requiring small contact times of the spout gas with the circulating solids, as in the pyrolysis of hydro-

carbons,[118] diversion of some of the inlet spout gas to the annulus may be eliminated by reducing to zero the clearance between the draft tube and the central gas inlet, that is, by substituting a riser for the draft tube. Transfer of solids from the annulus to the riser is then effected by orifices in the wall of the riser near its base.[119]

The original "air-suspension" technique for coating pharmaceutical tablets initiated by Wurster[120] has since matured industrially into a spout-fluid bed with a draft tube, illustrated in Figure 10.24. Another interesting application of a spout-fluid bed is in the *blending* and/or *drying* of tobacco and similar fibrous masses. In this case, gas jets, introduced at relatively high velocity through the sloping sides of the distributor, are required to disentangle the fibers before they can be mobilized and circulated. Additional gas flow introduced from the central gas inlet serves only to reduce the flow requirement of the mobilizing gas, but cannot by itself produce any circulation of the fibers.[121] Current designs of spout-fluid coal gasifiers in several industrialized countries, including the United States,[122-124] Japan,[125] and the United Kingdom,[126] bear a striking resemblance to each other. The preferred British Coal design is shown in Figure 10.25.

10.13.6 Three-Phase Spouting

In countercurrent gas–liquid spouting, low-density solid spheres are spouted by an upward flow of gas and irrigated by a downward flow of liquid.[127] In its performance and applications (e.g., gas absorption, dust removal) this type of operation is comparable to that of a turbulent bed (or "mobile bed" or "fluidized packing") contactor, where low-density spheres are fluidized by an upward continuous-phase gas flow counter to a downward trickle of liquid.[128] Three-phase spouted bed operation is characterized by a higher *pressure drop* than the three-phase fluidized bed, while the latter is characterized by a greater tendency to slugging and bed nonuniformity.[129] The disadvantages of both can be overcome by using a countercurrent gas–liquid spout-fluid bed, in which a portion of the gas is introduced via a centrally located nozzle and the rest through a surrounding gas distributor.[130]

In a cocurrent gas–liquid spouted bed, gas is used to atomize the liquid feedstock through the inside of the entry nozzle and is additionally introduced around the periphery of this nozzle (as in Fig. 10.19). As in the case of gas–liquid fluidized beds, liquid phase volumetric mass transfer coefficients for the air–water system in the presence of particles exceeding 3 mm are greater than in their absence, but this situation is reversed for smaller particles.[131] The introduction of a *draft tube* gives rise to higher gas holdups.[132] In the case of a three-phase spout-fluid bed, additional liquid is introduced through a conical

distributor. Candidates for this type of vigorous gas–liquid–solid contacting include *cracking* of heavy hydrocarbons, *gasification* of residual oils, and production of adiponitrile from adipic acid using a B_2O_3 catalyst.[133]

10.13.7 Dilute-Phase Spouting

If shallow beds of solids (e.g., $H/D_i = 2$ to 5) in cone-based columns (e.g., $\alpha = 30°$ to 50°) are subjected to upward gas velocities greater than about two to four times their minimum spouting velocities,[134] the result is what has been misnamed a "jet-spouted bed"[135] (Fig. 10.26). It is a misnomer because it implies that a conventional spouted bed, unlike this nonconventional one, is actuated by something other than a jet. The main difference is that, because these beds are initially much shallower and subject to considerably higher operating velocities than a CSB, their final annulus voidage is well in excess of 0.9, in comparison with a typical value of 0.4 for the annulus voidage of a CSB.[136] A more appropriate name is therefore *dilute-phase spouting*, in contrast to conventional or dense-phase spouting. Because of the much lower solids holdups and therefore lower solids residence times, as well

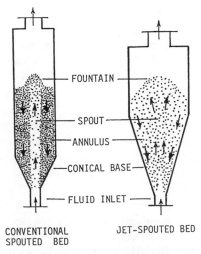

Figure 10.26. Diagrammatic representation of dense-phase or conventional spouted bed and dilute-phase or "jet-spouted" bed.[136] Arrows depict particle movement.

as higher gas-particle relative velocities and correspondingly higher heat transfer coefficients, dilute-phase spouted beds have been reported to give superior performance in *drying* paste-like materials, slurries, and solutions of heat-sensitive materials, especially bioproducts,[137] on inert solids.[138] In some cases, the dried product had an even narrower particle size distribution than the corresponding product from a spray drier.[139]

The above regime of dilute-phase spouting should be distinguished from the "spouted bed-type 2" regime observed by Littman and Morgan[140] for beds deeper than the maximum spoutable and velocities well in excess of U_m. This regime is similar in appearance to fast fluidization.

10.13.8 Other Modifications

Spouted beds can be countercurrently staged,[1,92,141] directly *vibrated*,[142,143] or subjected to flow *pulsations*.[1] The advantages of these modifications must in each case be weighed against corresponding increased costs. Fluid, instead of entering via a centrally located nozzle or slot, can be introduced through concentric rings or tangential slots.[1,92] In a "swirled spouted bed" both the fluid stream and the solid particles are subjected to a helical motion, leading to more intensive *heat* and *mass transfer* between the phases.[144]

10.14 PRACTICAL CONSIDERATIONS

We present here a number of practical suggestions, based on experience, to help designers of spouted bed processes and equipment.

10.14.1 Particle Properties

Conventional spouted beds operate best with dry, closely sized, rounded particles having a surface-to-volume mean diameter in the range of 1 to 8 mm. Even when these conditions are met, and especially when they are not, it is best to test the spouting behavior in a small column. If the solids are sticky or cohesive or

if a liquid phase is to be present in addition to gas and solids, the possibility of a *spout-filled bed* should be considered.

10.14.2 Test Column

Tests of spouting behavior for a given material should be carried out in a transparent column, not less than about 0.1 m in diameter. The column should have a *conical base* of included angle ~ 60°, constructed in such a way that the inlet orifice diameter can be varied, but no larger than $D/3$. Among the features that can be investigated in the test column are:

- Ease with which the material undergoes spouting (spoutability)
- Tendency of the material to undergo *attrition*
- *Minimum spouting velocity, maximum spoutable bed depth*, and the agreement of these measured values with the principal correlations
- Other hydrodynamic features such as the downward particle velocity at the wall and the fountain height.

Additional features can be observed and measured if a *half-column* (semicylindrical vessel) is employed:

- *Spout shape* and *diameter* and their agreement with the correlations discussed herein
- Tendency for solids *segregation* to occur
- *Dead zones*, if any, within the column.

10.14.3 Fluid Inlet

A straight vertical approach section of 10 to 12 pipe diameters should generally precede the inlet orifice. This approach section may be of the same diameter as the orifice or of larger diameter, narrowing gradually to the entry diameter. A bundle of straightening tubes is sometimes fixed inside this approach section. Sometimes an abrupt orifice plate is placed at the entry, which leads to increased spouting stability at the expense of increased pressure drop. A coarse screen or a special inlet valve can be used to prevent dumping of solids when the bed is shut down.[1] Unless special startup measures[1] are taken to avoid the peak pressure drop, $-\Delta P_M$ (see Fig. 10.5), the blower or compressor must be sized to provide this *pressure drop*, in addition to the drop across the screen, entry section and upstream pipes, valves, and fittings. If the bed is to operate as a *spout-fluid bed*, the additional fluid should enter through orifices or nozzles on the conical lower section[133] or in the flat annular base if there is no lower cone. In either case the flow of auxiliary fluid should be controlled separately from the main spouting flow to allow the ratio of auxiliary to spouting fluid to be varied.

10.14.4 Solids Feeding

The simplest way to feed solids is to deliver them via gravity from a hopper to the bed surface. Agglomerating solids should be pneumatically conveyed into the column by the spouting gas. Bottom feeding actually increases the *maximum spoutable bed height* and decreases the *minimum spouting velocity*.[145] A third means of adding solids is from the side, using the suction created by the fluid jet entering the bed.[1]

10.14.5 Solids Discharge and Entrainment

Solid material can be discharged from a spouted bed, like liquid from an orifice in the side of a container, using the difference between the local pressure inside the vessel and that on the outside. Hence, solids efflux will be more rapid the lower the discharge port. For orifice-to-particle diameter ratios of about 30 or more, the discharge coefficient is expected to be about 0.5, as for solids discharging from fluidized beds.[146] The exit pipe should slope down at an angle of $\sim 45°$, and it should be on the opposite side from any overhead solids feeder to prevent short-circuiting. For segregating solids, the position of the discharge port strongly affects the steady-state bed com-

position,[147] and this should be taken into consideration.

For friable solids, columns with little freeboard space, or materials with significant fines contents, entrained solids leaving the column should be captured by one or more cyclones which may be followed by filters or other collection devices. Solids captured in cyclones may be returned to the annulus region by means of a dipleg that enters obliquely through the wall of the spouted bed vessel.

10.14.6 Baffles

Concave axisymmetric fountain deflectors (e.g., an inverted funnel) are sometimes used[1] to restrain the fountain, prevent flowover of solids during startup, and induce greater symmetry and less wandering of the spout. For segregating solids a convex axisymmetric shape (e.g., a cone with its apex at the lowermost point) positioned near the top of the *fountain* can help to prevent *segregation*[37] by deflecting the heavier particles to the outside of the vessel. Either of these types of baffles in the fountain may, however, promote *attrition*.

ACKNOWLEDGMENT

The continuing financial support of the Natural Sciences and Engineering Research Council of Canada is gratefully acknowledged.

LIST OF SYMBOLS

A	$[\rho/(\rho_p - \rho)][U_T U_{mf}/gD_i]$
A_a	Cross-sectional area of annulus at any level
A_s	Cross-sectional area of spout at any level
Ar	Archimedes No.
	$= gd_p^3(\rho_p - \rho)\rho/\mu^2$
b	U_m/U_{mf}
C_a	Species gas-phase concentration in the annulus
C_D	Drag coefficient for particle in fluid $= F/(\pi d^2/4)(\rho/2)(u_s - v_s)^2$

C_{DT}	Drag coefficient under terminal settling conditions		that have been there for time θ or greater
C_{exit}	Exit species gas-phase concentration	i	Integer
C_{in}	Inlet species gas-phase concentration	K_1	Viscous coefficient in Ergun equation = $150\ \mu(1-\epsilon)^2/\phi^2 d_p^2 \epsilon^3$
C_s	Species gas-phase concentration in the spout	K_2	Inertial coefficient in Ergun equation = $1.75\rho(1-\epsilon)/\phi d_p \epsilon^3$
c_{pg}	Specific heat capacity of gas	K_r	First-order reaction rate constant referred to volume of solids
c_{pp}	Specific heat capacity of solid particles	k_b	Effective thermal conductivity of loose-packed bed
$\mathscr{D}$	Moisture diffusivity within particles	k_g	Thermal conductivity of gas
D	Column diameter	k_{sa}	Spout-annulus inter-region mass transfer coefficient
D_i	Fluid inlet diameter		
D_s	Spout diameter at any level	m	Index in Eq. (10.19)
$\bar{D}_s$	Longitudinal average spout diameter	$\bar{m}$	Final moisture content of batch solids, dry basis
d	Particle diameter; horizontally projected particle diameter; reciprocal mean particle diameter	$\bar{\bar{m}}$	Moisture content of continuous solids product, dry basis
d_p	Diameter of sphere with same volume as particle	N	Total number of streamtubes
		n	Richardson–Zaki[31] index
$E(\theta)$	Exit age distribution function[46] = fraction of particles leaving bed that have been in bed for time θ or greater	P	Fluid pressure
		$-\Delta P_f$	Pressure drop for fluidized bed of height H
		$-\Delta P_M$	Peak pressure drop
F	Drag force on particle	$-\Delta P_s$	Spouted bed pressure drop
G	Superficial mass flux of spouting fluid = ρU	Q_a	Fluid flow rate through annulus = $U_a A_a$
g	Acceleration due to gravity	Q_{ai}	Flow rate of fluid through ith streamtube in the annulus
H	Height of loose-packed static bed; height of annulus, measured from fluid inlet orifice	Q_s	Flow fluid rate through spout = $U_{sz} A_s$
H_F	Fountain height measured from bed surface	Q_{tot}	Total fluid flow rate
		Re	Particle Reynolds number $= d_p U \rho/\mu$
H_m	Maximum spoutable bed height		
h	H/H_m	Re_T	Terminal particle Reynolds number $= d_p U_T \rho/\mu$
$(h_{pg})_a$	Gas-to-particle heat transfer coefficient in the annulus	r	Radial distance from spout axis
$(h_{pg})_s$	Gas-to-particle heat transfer coefficient in the spout	r_s	Spout radius = $D_s/2$
h_s	Heat transfer coefficient between submerged object and bed	T_{ga}	Local temperature of gas in the annulus
h_w	Heat transfer coefficient between wall and bed, surface-mean value	T_{gs}	Local temperature of gas in the spout
		T_p	Temperature of particles
$I(\theta)$	Internal age distribution function[46] = fraction of particles within bed	t	Time
		$\bar{t}$	Mean residence time of solids

U	Superficial fluid velocity
U_a	Superficial fluid velocity in annulus at any level
U_i	Fluid inlet velocity
U_m	U_{ms} at maximum spoutable bed height
U_{ms}	Minimum superficial fluid velocity for spouting
U_{sz}	Superficial upward fluid velocity in spout or fountain core at any level
U_T	Terminal settling velocity of isolated particle in spouting fluid
u_s	Interstitial upward fluid velocity in spout or fountain core at any level
v_s	Local upward particle velocity in spout or fountain core at any level
$\bar{v}_s$	Average upward particle velocity in spout or fountain core at any level
v_{sc}	Upward particle velocity on spout axis at any level
v_w	Downward particle velocity at column wall
W	Mass downflow rate of solids in annulus at any level = mass upflow rate of solids in spout at same level
x	z/H_m
y	U_a/U_{mf} or U_a/U_{aH_m}
z	Height coordinate measured from fluid inlet orifice
z_i	Mean entry level of ith streamtube
α	Included angle of cone
β	$2 + (3K_1/2K_2U_{mf})$
ϵ	Voidage = 1 − solids volumetric fraction
ϵ_a	Voidage in annulus
ϵ_s	Voidage in spout or fountain core at any level
θ	Dimensionless time = $t/\bar{t}$
μ	Fluid viscosity
ρ	Fluid density
ρ_b	Bulk density of loose-packed solids = $\rho_p(1 - \epsilon_{mf})$
ρ_g	Gas density
ρ_p	Density of solid particles
τ_i	Mean residence time of gas in the ith streamtube

ϕ	Sphericity = surface area of equivolume sphere/surface area of particle

Subscripts

a	annulus
crit	critical
H	at $z = H$
H_m	at $z = H_m$
i	ith streamtube
max	at max spoutable bed depth
mf	at minimum fluidization
0	at $z = 0$
s	spout

REFERENCES

1. K. B. Mathur and N. Epstein, *Spouted Beds*, Academic Press, New York (1974).
2. H. Littman, M. H. Morgan III, D. V. Vuković, F. K. Zdanski, and Z. B. Grbavčić, "Prediction of the Maximum Spoutable Height and the Average Spout to Inlet Tube Diameter Ratio in Spouted Beds of Spherical Particles," *Can. J. Chem. Eng.* 57:684–687 (1979).
3. K. B. Mathur and P. E. Gishler, "A Technique for Contacting Gases with Coarse Solid Particles," *AIChE J.* 1:157–164 (1955).
4. B. Ghosh, "A Study on the Spouted Bed—A Theoretical Analysis," *Indian Chem. Eng.* 7:16–19 (1965).
5. P. P. Chandnani and N. Epstein, "Spoutability and Spout Destabilization of Fine Particles with a Gas," in *Fluidization V*, edited by K. Ostergaard and K. Sorensen, Engineering Foundation, pp. 233–240 (1986).
6. A. G. Fane and R. A. Mitchell, "Minimum Spouting Velocity of Scaled-up Beds," *Can. J. Chem. Eng.* 62:437–439 (1984).
7. B. Ye, C. J. Lim, and J. R. Grace, "Hydrodynamics of Spouted and Spout-Fluidized Beds at High Temperature," *Can. J. Chem. Eng.* 70:840–847 (1992).
8. Y. Li, "Spouted Bed Hydrodynamics at Temperatures up to 580°C," M.A.Sc. Thesis, University of British Columbia, Vancouver, Canada (1992).
9. Y.-L. He, C. J. Lim, and J. R. Grace, "Spouted Bed and Spout-Fluid Bed Behaviour in a Column of Diameter 0.91 m," *Can. J. Chem. Eng.* 70:848–857 (1992).

10. M. Choi and A. Meissen, "Hydrodynamics of Shallow, Conical Spouted Beds," *Can. J. Chem. Eng. 70*:916–924 (1992).

11. C. J. Lim and J. R. Grace, "Spouted Bed Hydrodynamics in a 0.91 m Diameter Vessel," *Can. J. Chem. Eng. 65*:366–372 (1987).

12. S. Ergun, "Fluid Flow through Packed Columns," *Chem. Eng. Progr. 48*(2):89–94 (1952).

13. C. Y. Wen and Y. H. Yu, "A Generalized Method for Predicting the Minimum Fluidization Velocity," *AIChE J. 12*:610–612 (1966).

14. G. S. McNab and J. Bridgwater, "Spouted Beds —Estimation of Spouting Pressure Drop and the Particle Size for Deepest Bed," in *Proc. European Congress on Particle Technology*, Nuremberg (1977).

15. S. W. M. Wu, C. J. Lim, and N. Epstein, "Hydrodynamics of Spouted Beds at Elevated Temperatures," *Chem. Eng. Commun. 62*:251–268 (1987).

16. H. Littman, M. H. Morgan III, D. V. Vuković, F. K. Zdanski, and Z. B. Grbavčić, "A Theory for Predicting the Maximum Spoutable Height in a Spouted Bed," *Can. J. Chem. Eng. 55*:497–501 (1977).

17. Z. B. Grbavčić, D. V. Vuković, F. K. Zdanski, and H. Littman, "Fluid Flow Pattern, Minimum Spouting Velocity and Pressure Drop in Spouted Beds," *Can. J. Chem. Eng. 54*:33–42 (1976).

18. T. Mamuro and H. Hattori, "Flow Pattern of Fluid in Spouted Beds," *J. Chem. Eng. Jpn. 1*:1–5 (1968).

19. N. Epstein, C. J. Lim, and K. B. Mathur, "Data and Models for Flow Distribution and Pressure Drop in Spouted Beds," *Can. J. Chem. Eng. 56*:436–447 (1978).

20. R. S. Krzywanski, N. Epstein, and B. D. Bowen, "Spouting: Parametric Study Using Multi-Dimensional Model," in *Fluidization VII*, edited by D. E. Potter and D. J. Nicklin, Engineering Foundation, pp. 353–360 (1992).

21. C. J. Lim and K. B. Mathur, "A Flow Model for Gas Movement in Spouted Beds," *AIChE J. 32*:674–680 (1976).

22. D. Van Velzen, H. J. Flamm, and H. Langenkamp, "Gas Flow Patterns in Spouted Beds," *Can. J. Chem. Eng. 52*:145–149 (1974).

23. G. Rovero, C. M. H. Brereton, N. Epstein, J. R. Grace, L. Casalegno, and N. Piccinini, "Gas Flow Distribution in Conical Based Spouted Beds," *Can. J. Chem. Eng. 61*:289–296 (1983).

24. R. S. Krzywanski, N. Epstein, and B. D. Bowen, "Multi-Dimensional Model of a Spouted Bed," *Can. J. Chem. Eng. 70*:858–872 (1992).

25. C. J. Lim and K. B. Mathur, "Residence Time Distribution of Gas in Spouted Beds," *Can. J. Chem. Eng. 52*:150–155 (1974).

26. L. A. Madonna, R. F. Lama, and W. L. Brisson, "Solids-Air Jets," *Br. Chem. Eng. 6*:524–528 (1961).

27. G. A. Lefroy and J. F. Davidson, "The Mechanics of Spouted Beds," *Trans. Instn. Chem. Eng. 47*:T120–T128 (1969).

28. B. Thorley, J. B. Saunby, K. B. Mathur, and G. L. Osberg, "An Analysis of Air and Solid Flow in a Spouted Wheat Bed," *Can. J. Chem. Eng. 37*:184–192 (1959).

29. C. J. Lim and K. B. Mathur, "Modeling of Particle Movement in Spouted Beds," in *Fluidization. Proceedings of the Second Engineering Foundation Conference*, edited by J. F. Davidson and D. L. Keairns, Cambridge University Press, pp. 104–109 (1978).

30. G. Rovero, N. Piccinini, and A. Lupo, "Vitesses des Particules dans les Lits à Jet Tridimensionnels et Semi-cylindriques," *Entropie 124*:43–49 (1985).

31. J. F. Richardson and W. N. Zaki, "Sedimentation and Fluidisation: Part I," *Trans Instn. Chem. Eng. 32*:37–53 (1954).

32. J. R. Grace and K. B. Mathur, "Height and Structure of the Fountain Region above Spouted Beds," *Can. J. Chem. Eng. 56*:533–537 (1978).

33. R. K. Barton, G. R. Rigby, and J. S. Ratcliffe, "The Use of a Spouted Bed for the Low Temperature Carbonization of Coal," *Mech. Chem. Eng. Trans. 4*:105–112 (1968).

34. M. Kugo, N. Watanabe, O. Uemaki, and T. Shibata, "Drying of Wheat by Spouting Bed," *Bull Hokkaido Univ. Sapporo Jpn. 39*:95–120 (1965).

35. H. A. Becker and H. R. Sallans, "On the Continuous Moisture Diffusion-Controlled Drying of Solid Particles in a Well-Mixed Isothermal Bed," *Chem. Eng. Sci. 13*:97–112 (1961).

36. N. Piccini, A. Bernhard, P. Campagna, and F. Vallana, "Segregation Phenomenon in Spouted Beds," *Can. J. Chem. Eng. 55*:122–125 (1977).

37. E. Kutluoglu, J. R. Grace, K. W. Murchie, and P. H. Cavanagh, "Particle Segregation in Spouted Beds," *Can. J. Chem. Eng. 61*:308–316 (1983).

38. Y. Eljas, "Contribution to the Study of Spouted Beds with Particular Emphasis on Porosity," Paper No. 58d, *68th Annual AIChE Meeting*, Los Angeles (November, 1975).

39. C. J. Lim, "Gas Residence Time Distribution and Related Flow Patterns in Spouted Beds," Ph.D. Thesis, Univ. of British Columbia (1975).

40. R. S. Krzywanski, N. Epstein, and B. D. Bowen, "Spout Diameter Variation in Two-Dimensional and Cylindrical Spouted Beds: A Theoretical Model and Its Verification," *Chem. Eng. Sci. 44*:1617–1626 (1989).

41. G. S. McNab, "Prediction of Spout Diameter," *Brit. Chem. Eng. Proc. Tech. 17*:532 (1972).

42. M. A. Malek and B. C. Y. Lu, "Heat Transfer in Spouted Beds," *Can. J. Chem. Eng. 42*:14–20 (1964).

43. S. S. Zabrodsky and V. D. Mikhailik, "The Heat Exchange of the Spouting Bed with a Submerged Heating Surface." Collected papers on "Intensification of Transfer of Heat and Mass in Drying and Thermal Processes," *Nauka Tekhnika BSSR*, Minsk, pp. 130–137 (1967).

44. Y. G. Klimenko, V. G. Karpenko, and M. I. Rabinovich, "Heat Exchange Between the Spouting Bed and the Surface of a Spherical Probe Element," in *Heat Physics and Heat Technology*, No. 15, Ukrainian SSR Academy of Science, Kiev, pp. 81–84 (1969).

45. A. H. Zahed and N. Epstein, "Batch and Continuous Spouted Bed Drying of Cereal Grains: The Thermal Equilibrium Model," *Can. J. Chem. Eng. 70*:945–953 (1992).

46. P. V. Danckwerts, "Continuous Flow Systems: Distribution of Residence Times," *Chem. Eng. Sci. 2*:1–13 (1953).

47. K. B. Mathur and C. J. Lim, "Vapor Phase Chemical Reaction in Spouted Beds: A Theoretical Model," *Chem. Eng. Sci. 29*:789–797 (1974).

48. N. Piccinini, J. R. Grace, and K. B. Mathur, "Vapor Phase Chemical Reaction in Spouted Beds: Verification of Theory," *Chem. Eng. Sci. 34*:1257–1263 (1979).

49. M. Balasubramanian, A. Meisen, and K. B. Mathur, "Spouted Bed Collection of Solid Aerosols in the Presence of Electrical Effects," *Can. J. Chem. Eng. 56*:297–303 (1978).

50. P. N. Rowe and K. T. Claxton, "Heat and Mass Transfer from a Single Sphere to Fluid Flowing through an Array," *Trans. Inst. Chem. Eng. 43*:321–331 (1965).

51. Dz. E. Hadžismajlović, D. V. Vuković, F. K. Zdanski, Z. B. Grbavčić, and H. Littman, "A Theoretical Model of the First Order, Isothermic Catalytic Reaction in Spouted and Spout-Fluid Beds," Paper C4.5, *6th CHISA Conference*, Prague (1978).

52. C. M. H. Brereton and J. R. Grace, "A Note on Comparison of Spouted Bed Reactor Models," *Chem. Eng. Sci. 39*:1315–1317 (1984).

53. B. D. Hook, H. Littman, M. H. Morgan III, and Y. Arkun, "A Priori Modelling of an Adiabatic Spouted Bed Catalytic Reactor," *Can. J. Chem. Eng. 70*:966–982 (1992).

54. D. Kunii and O. Levenspiel, *Fluidization Engineering*. John Wiley & Sons, New York (1969).

55. J. R. Grace, "Fluid Beds as Chemical Reactors," in *Fluid Bed Technology*, edited by D. Geldart, John Wiley & Sons, New York (1981).

56. S.-K. Foong, C. J. Lim, and A. P. Watkinson, "Coal Gasification in a Spouted Bed," *Can. J. Chem. Eng. 58*:84–91 (1980).

57. G. Rovero, N. Piccinini, J. R. Grace, N. Epstein, and C. M. H. Brereton, "Gas-phase Solid Catalysed Chemical Reaction in Spouted Beds," *Chem. Eng. Sci. 38*:557–566 (1983).

58. K. B. Mathur and P. E. Gishler, "A Study of the Application of the Spouted Bed Technique to Wheat Drying," *J. Appl. Chem. 5*:624–636 (1955).

59. W. S. Peterson, "Spouted Bed Drier," *Can. J. Chem. Eng. 40*:226–230 (1962).

60. G. Massarani, *Secagem de Produtos Agrícolas: Coletânea de Trabalhos*, Vol. 2, Editora Universidade Federal do Rio de Janeiro, 136 pages (1987).

61. O. Uemaki, M. Fujikawa, and M. Kugo, "Cracking of Residual Oil by Use of a Dual Spouted Bed Reactor with Three Chambers," *Sekiyu Gakkai Shi 20*(5):410–415 (1977).

62. S.-K. Foong, R. K. Barton, and J. S. Ratcliffe, "Reduction of Sulphur Dioxide with Carbon in a Spouted Bed," *Chem. Eng. in Australia*, Instn. Engrs. Aust., *CEI*(1/2):1–8 (1976).

63. T. Robinson and B. Waldie, "Dependency of Growth on Granule Size in a Spouted Bed Granulator," *Trans. Inst. Chem. Eng. 57*:121–127 (1979).

64. A. F. Dolidovich and V. S. Efremtsev, "Experimental Study of the Thermal Granulation Process in a Spouted Bed," *Can. J. Chem. Eng. 61*:454–459 (1983).

65. Y. F. Berquin, "Method and Apparatus for Granulating Melted Solid and Hardenable Fluid Products," U.S. Patent No. 3,231,413 (1966).

66. O. Uemaki and K. B. Mathur, "Granulation of Ammonium Sulfate Fertilizer in a Spouted Bed," *Ind. Eng. Chem. Proc. Des. Dev. 15*:504–508 (1976).

67. Y. F. Berquin, "Prospects for Full-Scale Development of Spouting Beds in Fertilizer Granulation," Paper 23, *First International Conference on Fertilizers*, London (November, 1977).

68. G. Brusasco, R. Monaldi, V. de Lucia, and A. Barbera, "Production of NPK Fertilizers with a Spouted Bed Granulator-Dryer." Presented at *I.F.A. Conference*, Port El Kantooni, Tunisia (1986).

69. J. Kucharski and A. Kmiec, "Kinetics of Granulation Process During Coating of Tablets in a Spouted Bed," *Chem. Eng. Sci. 44*:1627–1636 (1989).

70. P. J. Weiss and A. Meisen, "Laboratory Studies on Sulphur-Coating Urea by the Spouted Bed Process," *Can. J. Chem. Eng. 61*:440–447 (1983).

71. M. Choi, "Sulfur Coating of Urea in Shallow Spouted Beds," Ph.D. Thesis, University of British Columbia, Vancouver, Canada (1993).

72. N. Piccinini, "Coated Nuclear Fuel Particles," *Adv. Nucl. Sci. Technol.* 8:255–341 (1975).

73. E. H. Voice, "Coatings of Pyrocarbon and Silicon Carbide by Chemical Vapour Deposition," *Chem. Eng.* pp. 785–792 (December, 1974).

74. P. G. Romankov and N. B. Rashkovskaya, *Drying in a Suspended State*, 2nd edit. (in Russian), Chem. Publ. House, Leningrad Branch (1968).

75. Q. T. Pham, "Behaviour of Conical Spouted-Bed Dryer for Animal Blood," *Can. J. Chem. Eng.* 61:426–434 (1983).

76. C. Brereton and C. J. Lim, "Spouted Bed Drying of Sludge from Metals Finishing Industries Wastewater Treatment Plants," *Drying Technol.* 11:389–399 (1993).

77. A. G. Fane, T. R. Stevenson, C. J. Lloyd, and M. Dunn, "The Spouted Bed Drier—An Alternative to Spray Drying," in *8th National Chemical Engineering Conference*, Melbourne, Australia (August, 1980).

78. G. K. Khoe, S. L. Sun, C. J. Lim, and N. Epstein," Simultaneous Drying and Comminution of Coal in a Spouted Bed," *Drying Technol.* 9:1051–1066 (1991).

79. A. C. L. Lisboa and A. P. Watkinson, "Pyrolysis with Partial Combustion of Oil Shale Fines in a Spouted Bed," *Can. J. Chem. Eng.* 70:983–990 (1992).

80. T. B. Ray and S. Sarkar, "Kinetics of Coal Pyrolysis in Spouted Bed," *Ind. Chem. Eng.* 18(2):11–19 (April–June, 1976).

81. A. Jarallah and A. P. Watkinson, "Pyrolysis of Western Canadian Coals in a Spouted Bed," *Can. J. Chem. Eng.* 63:227–236 (1985).

82. A. N. Ingle and S. Sarkar, "Gasification of Coal in Spouted Bed," *Indian J. Technol.* 14:515–516 (1976).

83. J. Foong, G. Cheng, and A. P. Watkinson, "Spouted Bed Gasification of Western Canadian Coals," *Can. J. Chem. Eng.* 59:625–630 (1981).

84. Z. Haji-Sulaiman, C. J. Lim, and A. P. Watkinson, "Gas Composition and Temperature Profiles in a Spouted Bed Coal Gasifier," *Can. J. Chem. Eng.* 64:125–132 (1986).

85. A. P. Watkinson, G. Cheng, and C. J. Lim, "Oxygen-Steam Gasification of Coals in a Spouted Bed," *Can. J. Chem. Eng.* 65:791–798 (1987).

86. A. P. Watkinson, G. Cheng, and D. P. C. Fung, "Gasification of Oil Sand Coke," *Fuel* 68:4–10 (1989).

87. T. A. Sue-A-Quan, A. P. Watkinson, R. P. Gaikwad, C. J. Lim, and B. R. Ferris, "Steam Gasification in a Pressurized Spouted Bed Reactor," *Fuel Proc. Technol.* 27:67–81 (1991).

88. C. J. Lim, J. P. Lucas, M. Haji-Sulaiman, and A. P. Watkinson, "A Mathematical Model of a Spouted Bed Gasifier," *Can. J. Chem. Eng.* 69:596–606 (1991).

89. C. Dumitrescu and D. Ionescu, "Contributions to the Spouted-Bed Studies as an Aspect of Fluidization." Paper E5.10, *4th CHISA Conference*, Prague (1972).

90. A. V. Golubkovich, "Study of the Limestone Calcination Regimes in Spouted-bed Kilns," *Khimicheskoe i Neftyanoe Machinostroenie*, no. 4, pp. 19–21 (April, 1976).

91. N. R. Iammartino, "Cement's Changing Scene," *Chem. Eng.* 81(13):102–104 (June 24, 1974).

92. D. V. Vuković, F. K. Zdanski, and H. Littman, "Present Status of the Theory and Application of Spouted Bed Technique." Paper D2.20 *5th CHISA Conference*, Prague (1975).

93. M. Khoshnoodi and F. J. Weinberg, "Combustion in Spouted Beds," *Combust. Flame* 33:11–21 (1978).

94. H. A. Arbib, R. F. Sawyer, and F. J. Weinberg, "The Combustion Characteristics of Spouted Beds," in *18th Symposium (International) on Combustion*, The Combustion Institute, pp. 233–241 (1981).

95. H. A. Arbib and A. Levy, "Combustion of Low Heating Value Fuels and Wastes in the Spouted Bed," *Can. J. Chem. Eng.* 60:528–531 (1982).

96. E. R. Altwicker, R. K. N. V. Konduri, and M. S. Mulligan, "Spouted Bed Combustor for the Study of Heterogeneous Hazardous Waste Incineration." Paper No. 82.6, *AIChE National Meeting*, Philadelphia, PA (August 1989).

97. C. J. Lim, S. K. Barua, N. Epstein, J. R. Grace, and A. P. Watkinson, "Spouted Bed and Spout-Fluid Bed Combustion of Solid Fuels," in *Fluidised Combustion: Is it Achieving its Promise?* pp. 72–79, Institute of Energy, London (1984).

98. C. J. Lim, A. P. Watkinson, G. K. Khoe, S. Low, N. Epstein, and J. R. Grace, "Spouted, Fluidized and Spout-Fluid Bed Combustion of Bituminous Coals," *Fuel* 67:1211–1217 (1988).

99. M. Murphy and E. Cox, "Application of the Spouted Bed Combustor to the Burning of Low Heating Value Fuels." Report to U.S. Environmental Protection Agency, Battelle, Columbus, OH (Sept. 20, 1983).

99a. S.-K. Foong, R. K. Barton, and J. S. Ratcliffe, "Characteristics of Multiple Spouted Beds," *Mech. Chem. Eng. Trans. MCII* (1,2):7–12, Instn. Engrs. Aust. (1975).

100. D. V. R. Murthy and P. N. Singh, "Dynamics of Multiple Spouted Beds." Distributed at *Third International Symposium on Spouted Beds*, Vancouver, B.C., Canada (October, 1991).

101. B. Taha and A. Koniuta, "Hydrodynamics and

Segregation from the CERCHAR FCB Fluidization Grid." Free Forum, *7th International Fluidization Conference*, Banff, Alberta, Canada, Engineering Foundation (May, 1989).

102. J. R. Muir, F. Berruti, and L. A. Behie, "Solids Circulation in Spouted and Spout-Fluid Beds with Draft-tubes," *Chem. Eng. Commun. 88*:153–171 (1990).

103. H. Hattori and K. Takeda, "Side-Outlet Spouted Bed with Inner Draft-Tube for Small-Sized Solid Particles," *J. Chem. Eng. Jpn. 11*(2):125–129 (1978).

104. G. K. Khoe and J. van Brakel, "Drying Characteristics of a Draft Tube Spouted Bed," *Can. J. Chem. Eng. 61*:411–418 (1983).

105. G. Massarani, M. L. Passos, and D. W. Barreto, "Production of Annatto Concentrates in Spouted Beds," *Can. J. Chem. Eng. 70*:954–959 (1992).

106. J. K. Claflin and A. J. Fane, "Spouting with a Porous Draft Tube," *Can. J. Chem. Eng. 61*:356–363 (1983).

107. H. Hattori and K. Takeda, "Modified Spouted Beds with the Gas Outlet Located in the Side Wall Surrounding the Annual Dense Bed." *J. Fac. Text. Sci. & Technol.*, Shinshu Univ., no. 70, ser. B, Engineering no. *12*:1–13 (1976).

108. H. Hattori, A. Kobayashi, I. Aiba, and T. Koda, "Modification of the Gas Outlet Structure on the Spouted Bed with Inner Draft-Tube," *J. Chem. Eng. Jpn. 17*(1):102–103 (1984).

109. M. I. Kalwar and G. S. V. Raghavan, "Batch Drying of Shelled Corn in Two-Dimensional Spouted Beds with Draft Plates," *Drying Technol. 11*:339–354 (1993).

110. M. I. Kalwar, G. S. V. Raghavan, and A. S. Mujumdar, "Spouting of Two-Dimensional Beds with Draft Plates," *Can. J. Chem. Eng. 70*:887–894 (1992).

111. A. S. Mujumdar, "Spouted Bed Technology—A Brief Review," in *Drying '84*, pp. 1–7, Hemisphere, New York (1984).

112. T. Kudra, "Novel Drying Technologies for Particulates, Slurries and Pastes," in *Drying '92*, pp. 224–239, Elsevier, New York (1992).

113. H. Littman, D. V. Vuković, F. K. Zdanski, and Z. B. Grbavčić, "Basic Relations for the Liquid Phase Spout-Fluid Bed at the Minimum Spout-Fluid Flowrate," in *Fluidization Technology*, Vol. 1, edited by D. L. Keairns, pp. 373–386, Hemisphere, Washington (1976).

114. C. Dumitrescu, "The Hydrodynamical Aspects of a Spouted Bed Modified by the Introduction of an Additional Flow," *Rev. Chim.* (Roumania) *28*(8):746–754 (1977).

115. Dz. E. Hadžismajlović, D. V. Vuković, F. K. Zdanski, Z. B. Grbavčić, and H. Littman, "Mass Transfer in Liquid Spout-fluid Beds of Ion Exchange Resin," *Chem. Eng. J. 17*:227–236 (1979).

116. W. Sutanto, N. Epstein, and J. R. Grace, "Hydrodynamics of Spout-Fluid Beds," *Powder Technol. 44*:205–212 (1985).

117. J. Zhao, C. J. Lim, and J. R. Grace, "Flow Regimes and Combustion Behaviour in Coal-Burning Spouted and Spout-Fluid Beds," *Chem. Eng. Sci. 42*:2865–2875 (1987).

118. R. K. Stocker, J. H. Eng, W. Y. Svrcek, and L. A. Behie, "Ultrapyrolysis of Propane in a Spouted-bed Reactor with a Draft tube," *AIChE J. 35*:1617–1624 (1989).

119. B. J. Milne, F. Berruti, L. A. Behie, and T. J. W. de Bruijn, "The Internally Circulating Fluidized Bed (ICFB): A Novel Solution to Gas Bypassing in Spouted Beds," *Can. J. Chem. Eng. 70*:910–915 (1992).

120. D. E. Wurster, "Air-Suspension Technique of Coating Drug Particles," *J. Am. Pharmac. Assoc. 48*:451–454 (1959).

120a. D. M. Jones, "Value of Laboratory Testing and Scaleup," *Pharm. Tech. Conference '84 Proceedings*, Aster Publishing, Springfield, OR, pp. 317–331 (1984).

121. R. Legros, C. A. Millington, and R. Clift, "A Mobile Bed Process for Fibrous Materials," in *Fluidization V*, edited by K. Ostergaard and K. Sorenson, Engineering Foundation, pp. 225–232 (1986).

122. A. Rehmat and A. Goyal, "Fluidization Behavior in U-Gas Ash Agglomerating Gasifier," in *Fluidization. Proc. 4th Internat. Conf. on Fluidization*, edited by D. Kunii and R. Toei, Engineering Foundation, pp. 647–654 (1983).

123. D. A. Lewandowski, J. Weldon, and G. B. Haldipur, "Application of the KRW Coal Gasification Hot Gas Cleanup Technology to Combined Cycle Electric Power Generation," Presented at *AIChE National Meeting*, Boston (August, 1986).

124. F. W. Shirley and R. D. Litt, "Advanced Spouted-Fluidized Bed Combustion Concept," in *Proc. 9th Internat. Conf. on Fluidized Bed Combustion 2*:1066–1073 (1987).

125. K. Kikuchi, A. Suzuki, T. Mochizuki, S. Endo, E. Imai, and Y. Tanji, "Ash-Agglomerating Gasification of Coal in a Spouted Bed Reactor," *Fuel 64*:368–372 (1985).

126. M. St. J. Arnold, J. J. Gale, and M. K. Laughlin, "The British Coal Spouted Fluidised Bed Gasification Process," *Can. J. Chem. Eng. 70*:991–997 (1992).

127. D. V. Vuković, F. K. Zdanski, G. V. Vunjak, and Z. B. Grbavčić, "Pressure Drop, Bed Expansion

and Liquid Holdup in a Three Phase Spouted Bed Contactor," *Can. J. Chem. Eng.* 52:180–184 (1974).

128. L. S. Fan, *Gas-Liquid-Solid Fluidization Engineering*, Chapter 5, Butterworths, Boston (1989).

129. G. Vunjak-Novaković, D. V. Vuković, F. K. Zdanski, and H. Littman, "Comparative Hydrodynamical Characteristics Relevant for Mass Transfer in Three-Phase Fluidized and Spouted Bed Contactors," Paper C2.7, *6th CHISA Conference*, Prague (1978).

130. D. V. Vuković and G. V. Vunjak-Novaković, "The Three-Phase Spout-Fluid Bed—A Novel Gas-Liquid Contacting System." Paper C3.11, *6th CHISA Conference*, Prague (1978).

131. M. Nishikawa, K. Kosaka, and K. Hashimoto, "Gas Absorption in Gas-Liquid or Solid-Gas-Liquid Spouted Vessel," in *Proc. 2nd Pacific Chem. Eng. Cong. (Pachec '77)* II:1389–1396, *AIChE* (1977).

132. L. S. Fan, S. J. Hwang, and A. Matsuura, "Hydrodynamic Behaviour of a Draft Tube Gas–Liquid–Solid Spouted Bed," *Chem. Eng. Sci.* 39:1677–1688 (1984).

133. H. Kono, "A New Concept for Three Phase Fluidized Beds," *Hydrocarbon Proc.* pp. 123–129 (January, 1980).

134. M. Olazar, M. J. San José, A. T. Aguayo, J. M. Arandes, and J. Bilbao, "Stable Operation Conditions for Gas–Solid Contact Regimes in Conical Spouted Beds," *Ind. Eng. Chem. Res.* 31:1784–1792 (1992).

135. A. Markowski and W. Kaminski, "Hydrodynamic Characteristics of Jet-Spouted Beds," *Can. J. Chem. Eng.* 61:377–381 (1983).

136. O. Uemaki and T. Tsuji, "Particle Velocity and Solids Circulation Rate in a Jet-Spouted Bed," *Can. J. Chem. Eng.* 70:925–929 (1992).

137. A. S. Markowski, "Quality Interaction in a Jet Spouted Bed Dryer for Bio-Products," *Drying Technol.* 11:369–387 (1993).

138. A. S. Markowski, "Drying Characteristics in a Jet-Spouted Bed Dryer," *Can. J. Chem. Eng.* 70:938–944 (1992).

139. S. Grabowski, A. S. Mujumdar, H. S. Ramaswamy, and C. Strumillo, "Particle Size Distribution of *l*-Lysine Dried in Jet-Spouted Bed," in *Drying '92*, pp. 1940–1946, Elsevier, New York (1992).

140. H. Littman and M. H. Morgan III, "A New Spouting Regime in Beds of Coarse Particles Deeper than the Maximum Spoutable Height," *Can. J. Chem. Eng.* 64:505–508 (1986).

141. G. Rovero and A. P. Watkinson, "A Two-Stage Spouted Bed Process for Autothermal Pyrolysis or Retorting," *Fuel Proc. Technol.* 26:221–238 (1990).

142. Gy. Rátkai, "Particle Flow and Mixing in Vertically Vibrated Beds," *Powder Technol.* 15:187–192 (1976).

143. J. R. D. Finzer and T. G. Kieckbusch, "Performance of an Experimental Vibro-Spouted Bed Dryer," in *Drying '92*, pp. 762–772, Elsevier, New York (1992).

144. A. F. Dolidovich, "Hydrodynamics and Interphase Heat Transfer in a Swirled Spouted Bed," *Can. J. Chem. Eng.* 70:930–937 (1992).

145. A. G. Fane, A. E. Firek, and C. W. P. Wong, "Spouting with a Solids-Laden Gas Stream." Chemeca '85, Perth, Australia (1985).

146. L. Massimilla, "Flow Properties of the Fluidized Dense Phase" in *Fluidization*, edited by J. F. Davidson and D. Harrison, Academic Press, London (1971).

147. N. Piccinini, "Particle Segregation in Continuously Operating Spouted Bed," in *Fluidization*, edited by J. R. Grace and J. M. Matsen, Plenum Press, New York, pp. 279–286 (1980).

11
Mixing of Powders

Brian H. Kaye

CONTENTS

11.1 BASIC CONCEPTS OF POWDER MIXING

As soon as one begins a study of powder mixing theory and practice one finds that there is considerable confusion as to what constitutes a good mixture. In fact the term good is meaningless in the context of individual powder technology and one should use the term satisfactory, with the exact meaning of this term being interpreted within the context of the industrial process. Most mixers are designed to achieve a random mixture of the ingredients. By definition a random mixture is one such that if the position in a mixture of a given fineparticle is x_1, y_1 and z_1 at the beginning of a mixing process, then its final position x_2, y_2, and z_2 is completely independent of its initial starting point. Unfortunately many people have no concept as to what constitutes a random mix. If they are shown a series of randomized mixtures of black and white

fineparticles they are surprised by the amount of clustering that persists in a random mixture. Thus in Figure 11.1 a series of simulated black and white fineparticles at a richness level of 5% by volume is shown. When these were shown to people at a workshop many of the participants felt that the systems were inadequately mixed and that a better mixture could be achieved if more effort were expended on creating a mixture. The participants were surprised at the variation that can exist in a mixture of this kind. (A legal variation in a random mix is one that can arise by chance.) For some industrial purposes a randomized mixture is not sufficiently well dispersed. One then has to move to create what are known as structured mixtures by using strategies such as microencapsulation or arranging for the ingredients to be mixed under conditions in which one powder will coat another to give at least transient microencapsulation until the mixture is used in the process for which it was de-

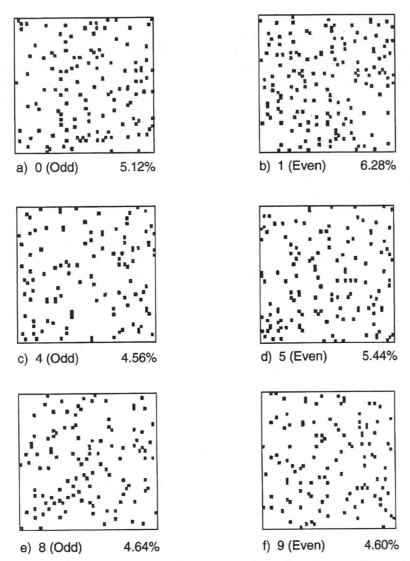

Figure 11.1. The constitution of a sample of a mixture can vary by random chance as illustrated by the above simulated samples of a 5% mixture of a monosized powder dispersed in a continuous matrix.[1]

signed.[1] To help appreciate the variations that can occur in a random mix Kaye has devised an expert system that can simulate and display mixtures of ingredients at various specified levels.[2]

In a recent review of powder mixing technology the statement was made that powder mixing is an important but academically unfashionable subject in the United States.[3] In the book that was being reviewed there was a chapter by Dr. J. C. Williams, who had studied powder mixing for many years. At the beginning of that chapter Dr. Williams made the statement that during the past 30 years there has been much work done at universities in the study of solids mixing but the results of this effort have not yet been applied in industrial practice.[4]

Opinions differ as to why the industrial community is apparently unwilling to learn from academic research into powder mixing. Some scientists have said that the reason is that the university investigations have been too abstract and of little utility to the working scientist. In my opinion the reasons why academic knowledge has failed to have much impact on the industrial operations of powder mixing arises from two factors. The first is that much of the language used by the academic scientist is inaccessible to the working technologist, because it involves advanced manipulation of data and the use of decision making concepts not normally developed in the background of people who have the responsibility of the day-to-day mixing of industrial powders. Second, much of the powder mixing research undertaken in the academic world has been based on the assumption that powder mixing

studies would eventually be organized along classic scientific lines with clear-cut deterministic equations that can be applied to the powder systems. This perspective on powder mixing is changing. It is now becoming obvious that powder mixing is not amenable to the classic investigation techniques in which one seeks to first understand the basic mechanisms with the ultimate synthesis of such understanding into a comprehensive theory of performance. Thus a great deal of time has been spent on studying the mixture of red and white glass beads of the same size. Such studies only indicate the efficiency of randomization in a given mixing system but it is of absolutely no use in predicting the mixing efficiency of a set of powders of different physical properties and different sizes. For example, in the case of

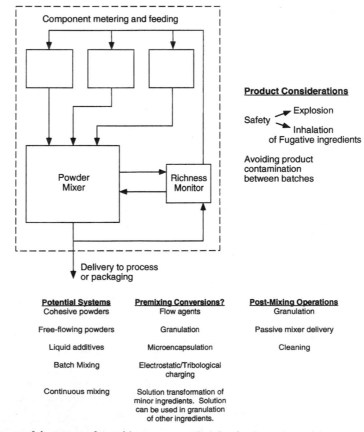

Figure 11.2. A successful strategy for achieving a specified level of powder mixing must take into account many factors.

many cohesive powders electrostatic effects are very important whereas in the study of the intermingling of glass beads, forces are of little consequence. It is now becoming clear that the number of factors that interact during the operation of a powder mixing system are so varied, and their interaction so complex, that powder mixing should properly be regarded as a branch of mechanics to which, in the last few years, the name deterministic chaos or simply chaos has been applied.[5-11]

The discipline of deterministic chaos has emerged over the last 15 years as a study of systems that, although in essence are deterministic (predictable), the process of prediction is so complex and the progress of a system so sensitive to initial conditions, that in practice the exact behavior of a system cannot be predicted with any high accuracy. One can only predict probable behavior within a broad range of expectations. When discussing fluid mixing Oldshue, an expert in the area of fluid mixing, states:

"Mixing processes are so complex that it is not possible to define process requirements using parameters that involve fluid mechanics." [12]

L. T. Fan, one of the leading experts on powder mixing theory and practice, states that:

"Various mathematical models for powder mixing have been proposed and numerous mathematical expressions for the rates of powder mixing based on these mechanisms have been developed. While many of the models and expressions are deterministic or microscopic, some resort to stochastic approaches. This may be attributed to the difficulties in delineating the inherently complex nature of solids mixing processes by means of deterministic approaches. Our understanding of solids mixing processes, the design of mixers for powders, has mainly been carried out heuristically." [13]

Fan then states:

"Due to the complexity of powder mixing behaviour, describable only by a large number of parameters, the experience gained with a pilot scale mixer may not be reliable for scale up, therefore an effective design procedure employing both heuristics and algorithms needs to be developed." [13]

The perspective of this chapter is that one needs to adopt a holistic approach to powder mixing. The assembly of the ingredients prior to the operation of the mixer and the subsequent handling of the mixture are all part of the problem of achieving a satisfactory mixture of different ingredients. Knowing the performance characteristics of mixing equipment must be accompanied by the technologist having a broad-based knowledge of the powder systems and their behavior before one can hope to achieve a satisfactory process.

At the beginning of any planning session concerned with the production of mixtures of powders one should use the chart shown in

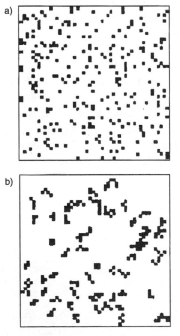

Figure 11.3. A mixture that has the required richness of the components may not have an adequate internal structure for purposes such as color consistency and penetration pathways.[1] (a) Completely randomized 10% mixture of a black, monosized ingredient in a white matrix. (b) A randomized 10% mixture of a black, monosized ingredient in which some agglomeration of the black ingredient persists.

Figure 11.2 as a focus in a protocol planning session to see if one has thought of all the variables and arranged for all the information retrieval that one needs within a given process. In the evolution of mixing strategies one should remember that one can make complex machines to achieve rapid mixing. However, the cost of cleaning the equipment between batches if more than one mixture is to be handled is an important aspect of the cost effectiveness of any mixing procedure. Thus, sometimes one can have a very efficient mixer but the cost of cleaning it between batches of different drug products is prohibitively expensive and one must look for an alternate mixing strategy. In the course of a powder mixing investigation the type of information that will be needed for planning an optimum strategy will include the flow properties of the powder, particle size distribution of the powder, and whether it is dry or not. Sometimes a powder can look as if it is dry but may contain up to 10% to 15% of moisture. This moisture, when the powder is tumbled, can initiate a spontaneous agglomeration of the grains that interferes with the mixing process. Sometimes a powder has also been treated with a surface conditioning material before it is delivered to the factory. Thus, many pigment powders have been treated with stearate to promote the flow of the powder system and the presence of such

stearate can cause agglomeration during the mixing process. Sometimes a powder ingredient will be unsatisfactory in a powder mixture and it may be that the manufacturer of the powder can change the shape or size of the powder to facilitate the mixing process. For example, some pharmaceutical powders are spray dried and others are precipitated and dried. The two different processes result in powders having the same size specification but very different physical properties.

I have been involved in a situation where a mixer, which had been performing satisfactorily, suddenly started to malfunction. The malfunction was finally traced back to the fact that the maker of the powder had changed from ball milling to attrition milling in the manufacture of the powder. This had changed the shape characteristics of the powder even though the powder met the specifications imposed upon the vendor by the purchaser. Thus, in any mixing situation one should keep a catalogue of the size, shape, and even manufacturing processes of the powders being delivered.

In many situations one needs to be able to sample powder from a mixer to see if it is performing satisfactorily and all of the precautions with regard to the efficient sampling outlined in Chapter 1 should be followed to make sure that the sample ultimately studied

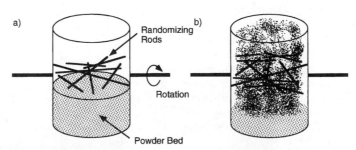

Figure 11.4. In a tumbling mixer the position of the grains of the powder ingredients are randomized by the randomizing rods and the turbulence in the transient fluidized bed created by the falling powder. This type of mixer can sometimes create a mixture in which the properties of the ingredients are at the desired level but the dispersion of the ingredients within the sample structure is not at the desired level because there are no internal shear forces to disperse local pockets of high concentration of a particular ingredient. (*a*) Appearance of a simple tumbling mixer. (*b*) When the mixer is inverted, the falling powder interacts with the upward displaced air and the randomizing rods to create a transient, turbulent, fluidized bed in which the mixing occurs.

in the laboratory is a representative sample of the material taken from the mixer. Recently there has been a renewed interest in the possibility of using fiber optic probes to monitor the internal structure of a powder mixture but such systems are not yet available commercially.[14]

In general the technology for monitoring the progress of a mixing process is poorly developed although recently different workers have started to use sophisticated methods to track particle trajectories in the powder mixing equipment.[1,6]

It is useful to distinguish between the richness of a mixture and the intimacy of the mixture ingredient. Thus a mixture may have the required richness in a sample of mixture taken from the mixer but for some purposes the intimacy of the mix may not be adequate. Consider, for example, the two systems shown in Figures 11.3a and 11.3b. Both simulated mixtures represent 10% by volume of the ingredient represented by the black squares, but the simulated fineparticles of Figure 11.3a are randomly dispersed whereas some clustering of the fineparticles has occurred in Figure

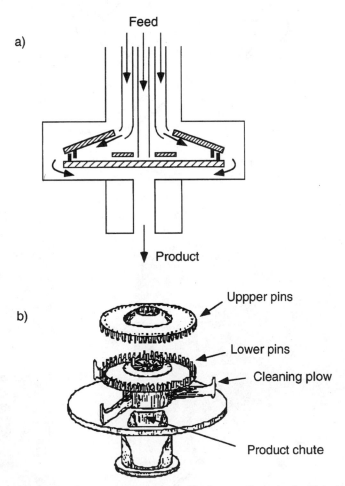

Figure 11.5. The Centri-flow mixer is a continuous mixer with vigorous dispersion by high shear forces generated between a rapidly rotating set of pins and a stationary set of pins.[15] (*a*) Schematic of feed blending systems used in the Centri-flow mixer. (*b*) Exploded view of the Centri-flow mixer showing the two sets of pins and the cleaning plows.

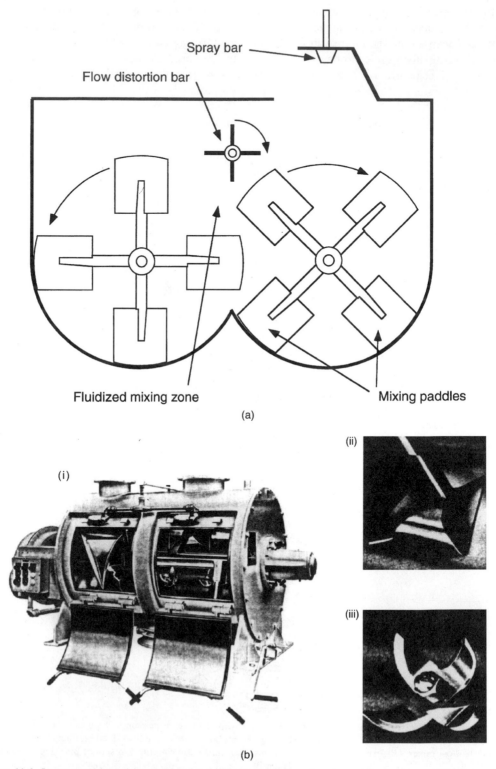

Figure 11.6. In some active mixing equipment, high-speed paddles of various shape, rotate rapidly to create a fluidized, turbulent zone in which the powder mixing takes place. (*a*) Schematic of the Forberg mixer.[16] (*b*) The Littleford mixer showing (i) the overall system, (ii) plow paddles, and (iii) intensifier choppers.[17,18]

11.3b. If the black squares represent drug fineparticles in a starch matrix both samples represent adequate mixtures for the purposes of drug delivery via a tablet, but if the samples represent a pigmented plastic mixture then the color appearance of the two samples would be different. (In fact color consistency of dispersed powder mixtures in technologies such as cosmetics manufacture is a major problem limiting the quality control capacity of the industry). To improve the quality of the mixture such as that depicted in Figure 11.3b usually requires the use of high-shear force over not more than several diameters of the fineparticles to be dispersed. Many industrial mixing machines have no component structure capable of applying such shear forces and so cannot improve the intimacy of a powder mix. Consider, for example, the operation of a tumbler mixer of the type shown in Figure 11.4a.

Continuous Ribbon agitator: standard construction for all Day Ribbon Blenders, has continuous inner and outer ribbons, and may be arranged for center or end discharge.

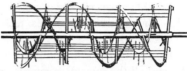

Cut-it-in agitator construction is used for cutting fats, oils and shortenings into flours and powders. Essentially a continuous agitator with cutting bars added to inner ribbon, and cutting wires mounted through ribbon arms. Leather or Tygon wipers are furnished.

Cut-out agitator construction is used when products to be mixed are heavier than normal. Basically the same as the continuous type, the cut-out agitator has alternate sections of the outer ribbon removed, efficiently mixes heavier batches.

Dry Color agitator construction is used for mixing extremely heavy crystalline or abrasive materials. A series of 'T' shaped paddles are spaced 90° apart on the shaft. All except center assembly have one long and one short arm and paddle, effectively circulate material at outer and inner areas of tank; center assembly consists of two long arms to facilitate discharge.

(a)

(b)

Figure 11.7. In ribbon blenders, complicated, extended paddle systems rotate relatively slowly to move the powder ingredients back and forth and intermingle them.[17, 20, 21, 29, 30] (a) Four different types of randomizing ribbon mixers are made by J. H. Day Company.[17] (b) An overview of a ribbon mixer.

When the tumbler is inverted the powder cascades down the body of the mixer, creating a transient fluidized bed. The internal turbulence of this bed is the main mechanism creating the powder mixture. (Note: sometimes the particle kinetics in the transient fluidized bed generates electrostatic forces that enhance the mixing action of mixing machines.) The randomized action may create an adequate mixture from the aspect of the gross properties of a sample taken from the mixer but the internal structure of the powder mixture may not be adequate and there are no shear-creating elements within the mixer. Sometimes technologists will operate simple mixers such as those shown in Figure 11.4 long after the mixer has achieved all the intermingling of the ingredients that the machine is capable of. This is done in the hope of improving the internal structure of the mixture. A far better strategy is to split the mixing process into two stages. The ingredients are first randomized in a mixer such as that shown in Figure 11.4 and then emptied through a high-shear disperser of the type shown in Figure 11.5. (An ordinary pin mill can also be used as high-intensity shearing dispersion equipment.) The failure of a powder mixing process can often be traced to the lack of adequate shear forces in the internal structure of the powder mixer.

11.2 DIFFERENT MIXING MACHINES

It is useful to classify powder mixing machines into two main groups: active and passive. The active process uses certain moving parts to assist in the randomization of the ingredients or the mixer machine moves about physically in the mixing process. In a passive mixer system the randomization of the ingredients is achieved by directed flow of the powder streams by baffles, etc., as they move through the mixing system.

In one type of active mixer rapidly rotating paddles whip up the air and the ingredients to create a fluidized zone in which intense turbulence intermingles the ingredients of the powder mixture. The Forberg mixer contains a twin paddle system as shown in Figure 11.6a. Very rapid mixing of materials such as dry soap mixes and other food products has been successfully accomplished with this type of equipment. For some purposes it has been found useful to mount ancillary rods traversing the mixer as shown in Figure 11.6a so that impacting ingredients stirred up by the paddles can be deagglomerated. The item labeled a flow distortion bar in this diagram is also known as an intensifier because it increases the efficiency of the mixing process. It will be noted that this system also incorporates a device for adding liquid to the powder ingredi-

SC style Helicone® Mixer

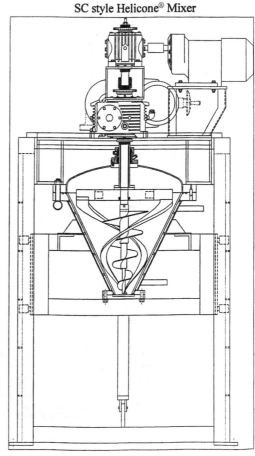

Figure 11.8. The Helicone® mixer creates turbulent, convective mixing currents by using counter-rotating lifting screw elements.[24]

ents. The addition of liquid is sometimes an integral part of an ultimate mixture; in other cases the liquid is added to stabilize the mixture to prevent segregation when the system is emptied from the mixer. Another type of mixer in which the rotation of the dispersing paddles is so intense that the ingredients are suspended in the moving air to create a turbulent mixture is the Littleford mixer shown in Figure 11.6b. (Note this mixture is known in Europe by the term Lodige mixer.) The practice of changing the name of equipment when it is licensed to a North American distributor from Europe and vice versa is confusing because the relationship between the different names of the same mixer style is not always obvious.[17,18] To increase the rate of mixing and to disperse agglomerates that can exist in the ingredients, so-called intensifier choppers are mounted in the side of the mixture and

driven independently of the movement of the high-speed paddles. It will be noted that the end of the paddles in the Littleford mixer are plow shaped so that the walls of the mixing chamber are continually cleaned by the rotation of the paddles. These plows do create some shearing action but there may not be sufficient shearing action if an infinite mix of cohesive powders is required.

In a different type of active mixer, instead of sets of paddles rotating to create the mixing action, a long complicated single paddle in the form of a mounted ribbon of material is used to disperse the ingredients of the mixture. This type of mixer is known as a ribbon mixer and the details of the different types of ribbon paddles that are available for these types of mixes are shown in Figure 11.7a. Ribbon blenders of the type shown in Figure 11.7 move much more slowly than the high-speed

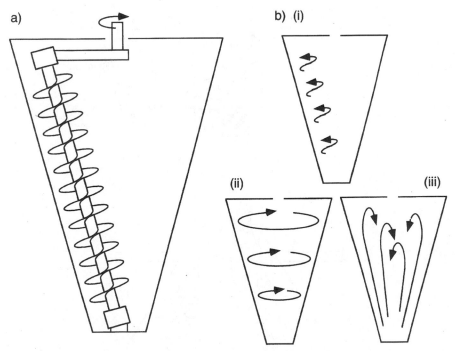

Figure 11.9. The Nauta® mixer uses a single convective lift screw that also rotates around the conical blending chamber.[25] (a) Schematic of the Nauta® mixer. (b) Three types of currents are created in the Nauta® mixer: (i) motion around the screw, (ii) motion around the mixing chamber, and (iii) convection currents from the bottom to the top of the chamber.

paddle mixers shown in Figure 11.6. They are widely used in the food industry and the pharmaceutical industry. Unless the ribbons are carefully designed, ribbon blenders can have dead pockets in some parts of the mixer, especially near the ends of the mixer close to the axis of rotation.[20–23, 32]

A vertical type of ribbon mixer is the Helicone® mixer shown in Fig 11.8.[24, 19] As mixers become more complicated in their internal structure they become more difficult to clean, especially if trace contamination from one mixing product to another is important and sometimes it is recommended that mixers as complicated as the Helicone® mixer shown in

Figure 11.8 and the Nauta® mixer shown in Figure 11.9 should be dedicated to a given product line.[19, 24, 25] All of the mixers discussed so far are available in different volume capacity and different screw paddle, etc., configurations.

In the Nauta® mixer the ingredients to be intermingled are placed in a large conical vessel. A lift screw rotates around the conical chamber creating convection and lift currents as illustrated in Figure 11.9.[26, 27] Another type of mixer making use of lift screws is manufactured by Prater Industries Inc.

Other types of active mixers are not equipped with internal moving parts but the

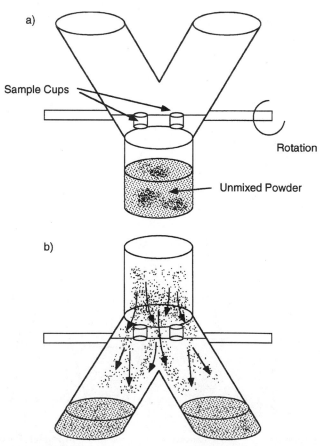

Figure 11.10. In Y and V mixers, mixing occurs when the powder divides and flows turbulently as the mixer is turned back and forth.[29–31] (*a*) Y mixer at the start of the mixing process. (*b*) When inverted the mixer is said to be in the Lambda (λ) position.

whole mixing chamber is moved to achieve mixing. Thus in the Patterson-Kelley twin shell or V Blender the materials to be mixed are loaded into the container similar to that shown in Figure 11.10a. The system should not be filled to more than half capacity to allow freedom of movement of the powder as the mixer operates. As the mixer is inverted the powder falls and splits into two streams, with each stream being turbulently mixed by the upward moving air. Better mixing is achieved if the mixer is inverted quickly with a pause to allow the powder to move down the system. In many industrial situations, however, the system moves relatively slowly and as a consequence rather longer times are required for mixing. Often a so-called intensifier bar is placed across the mixer to increase the turbulence of the falling powder stream. It is probable that the use of an intensifier bar was empirically discovered when hollow tubes were placed in the position of the intensifier bar of Figure 11.11a so that liquid could be sprayed into the falling powders to achieve granulation. Variations on the Y and V mixers are double cone mixers and so-called zig-zag mixers.[29, 31]

Again it will be noted that there are no high shear zones in the mixing system and sometimes if the intimacy of the mix needs to be improved one can drop it into a mulling device of the type shown in Figure 11.12. In this

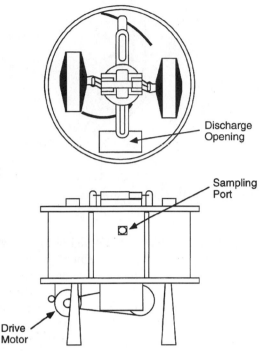

Figure 11.11. The intimacy of a powder mixture can be improved by subjecting the mixture to the high shear forces in a mulling machine of the type shown above.[33]

system the two large wheels move around the pan of the mixer and the movement of the system is deliberately designed to cause the two larger wheels to skid sideways as they roll around the mixing chamber and this skid-

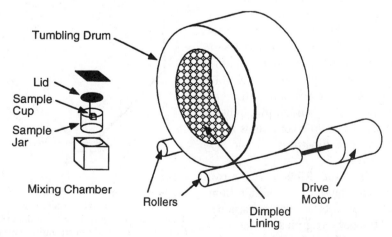

Figure 11.12. In the AeroKaye® mixer the mixing chamber tumbles randomly in a rotating drum to intermingle the ingredients.[35]

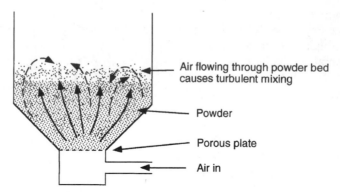

Air flowing through powder bed
causes turbulent mixing

Powder

Porous plate

Air in

Figure 11.13. The operation of a simple fluidized bed leads to turbulent mixing of the powder forming the bed.

ding and rolling effect applies high shear forces to the mixture.[33] In the Turbula® system a container is turned into a mixing device by mounting it in a cradle which then moves the container through a complex sequence of movements to create randomization of the positions of the powder grains inside the mixer. Again the mixer should be filled to only half capacity to permit freedom of motion of the grains as the container is moved through the series of positions.[34] A relative newcomer to the field of powder mixing that so far has been used only on a laboratory scale is the AeroKaye® mixer shown in Figure 11.13. The mixing chamber in Figure 11.12 is a cube but many other different kinds of mixing chambers can be used. This is half-filled with the ingredients to be mixed and then the container is placed in the large cylinder. The inside of the cylinder is covered with foam so that as the cylinder turns the chamber is lifted up until a point of instability is reached, then it tumbles randomly down to a new position at the bottom of the cylinder. This random tumbling of the container creates chaotic conditions inside the mixing chamber, leading to very rapid mixing of the ingredients.[35]

In mixers such as the Ribbon Met mixer, one of the problems hindering rapid intermingling of the ingredients is the constrained movements of the powder grains. As has been pointed out in many successful mixing systems the actual mixing takes place in a transient

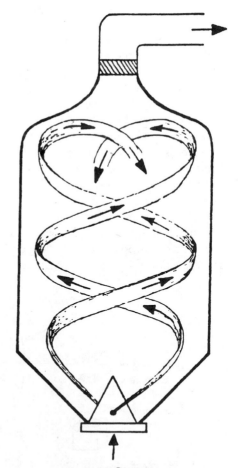

Figure 11.14. The Airmix® mixer fluidizes the powder to be mixed and creates intense turbulence to achieve randomization of position.[38] Spiraling ribbons of air are created by jets in the distribution cone at the base of the mixer. In the diagram only two ribbons are shown for clarity.

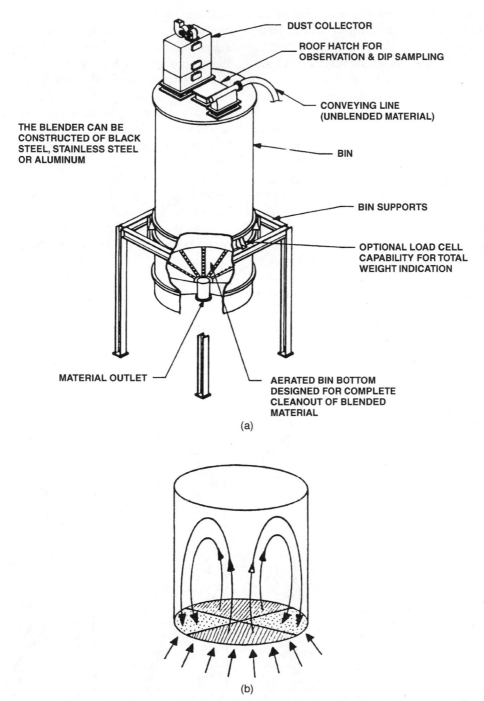

DUST COLLECTOR

ROOF HATCH FOR
OBSERVATION & DIP SAMPLING

CONVEYING LINE
(UNBLENDED MATERIAL)

THE BLENDER CAN BE
CONSTRUCTED OF BLACK
STEEL, STAINLESS STEEL
OR ALUMINUM

BIN

BIN SUPPORTS

OPTIONAL LOAD CELL
CAPABILITY FOR TOTAL
WEIGHT INDICATION

MATERIAL OUTLET

AERATED BIN BOTTOM
DESIGNED FOR COMPLETE
CLEANOUT OF BLENDED
MATERIAL

(a)

(b)

Figure 11.15. The Airmerge® blender and homogenizing silo both employ air fluidization to achieve mixing of powders. (a) The Airmerge® system manufactured by Fuller-Kovako Corp. has a completely fabric-covered fluidized bottom divided into four quadrants that are fluidized in sequence to achieve strong, varying convection currents. (b) The homogenizing silo system uses aeration pads on the silo floor divided into eight segmented areas. This silo also fluidizes the segments in sequence resulting in turbulent convection currents.

fluidized bed where there is rapid random free movement of the turbulently suspended grains. The logical extension of this fact is that fluidized bed mixing should be very efficient mixing and some fluidized bed mixers have been developed of the type illustrated in Figure 11.13.[36, 37] When describing systems such as those shown in Figure 11.13 Harnby states that fluidization is caused by the passage of a gas through a bed of particles. In such a system the bulk density of the powder is reduced and the mobility of the individual particles is increased. If the gas flow is sufficiently large there will be considerable turbulence within the bed and the combination of turbulence and particle mobility can produce excellent mixing. A constant danger in the fluidized mixer is that if the turbulence is not complete then the constituent particles can readily segregate owing to variable settling. He goes on to state that very few commercially available

fluidized mixing units exist. Because of the diversity of its application the bed is usually designed for a specific process and is not available as a standard line product. Several fluidized beds are purposely built for the pharmaceutical industry.[37] In Figure 11.14 the Airmix® mixer is shown. This equipment makes use of intermittent fluidization created by pulsed air jets at the base of the mixing chamber.[38, 40]

Similar mixers known as Dynamic Air Blendcon are available from Dynamic Air Conveying Systems.[39, 41] Another mixer that employs fluidization to achieve mixing is the Airmerge System®, manufactured by the Fuller Company.[42]

The basic operational principles of the Airmerge® system are illustrated in Figure 11.15a. The various quadrants at the base of the mixing chamber are alternately the source of fluidizing air, with the variation in these air

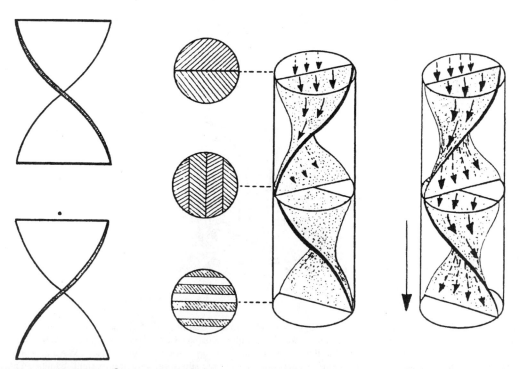

Figure 11.16. The Kenics® Static Mixer uses right- and left-handed "butterfly twisters" to achieve a structured, total processing intermingling of initially totally segregated feed components.

currents creating turbulence and freedom of motion to achieve rapid mixing.[42] A very similar piece of equipment used on a larger scale to homogenize large supplies of material such as cement and flour has been described by Harnby and it is shown in Figure 11.15b.[37]

Fluidization mixing is basically of use only when mixing relatively free flowing powders because it is not easy to fluidize cohesive powders. One should also be concerned with the potential for dust explosions when operating fluidized bed systems.

In passive mixers the ingredients to be intermingled are brought in contact with each other by passing the material through a series of randomizing veins. Passive mixers have not

been particularly successful in the mixing of powder systems, their main utility having been in the area of liquid mixing.[37,43,45] The typical passive mixer manufactured by Chenincer Inc. is shown in Figure 11.16. In this mixer randomization of the moving powder is achieved by a series of left- and right-handed butterfly twists opposed to each other in sequence as illustrated in the figure.[44] Other passive mixers using different randomizing elements placed in the system in sequence are available from several manufacturers.[46-49,51,52]

A different type of passive mixing system used on a large scale with free-flowing powders is the system known as a gravity blender. In this type of mixer, material from different

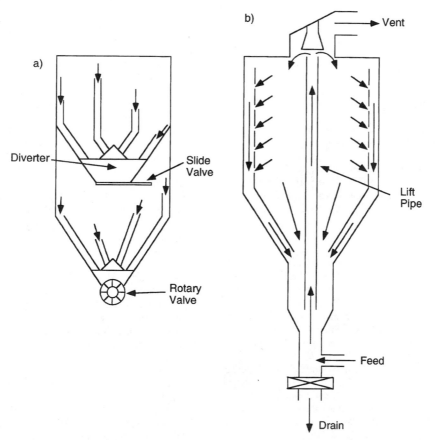

Figure 11.17. Gravity bin mixers are used in the processing industry to homogenize bin contents flowing into an industrial process.[63] (a) Young's bin mixing system.[58,63] (b) Fluidized bin mixer described by Stein using passive mixing from diverter pipes, moving down in the sketch, and pneumatic recirculation currents.[55]

parts of a storage hopper are drawn by feed pipes into a central area where they mingle to produce a mixture suitable for industrial processing as it comes out of the exit portion of the storage device.[53-63] In Figure 11.17a a proprietary design of a bin mixer patented by Young is shown.[58] This type of mixer is primarily used in industry for homogenization of the contents of a bin going to an industrial process and is not useful for mixing intimately cohesive powders. Some installations are hybrid mixers using the principles of gravity bin mixers with pneumatic recirculation of the contents to promote better homogenization. Thus in Figure 11.17b a bin mixer with pneumatically activated recirculation of the contents is shown. This type of mixing system has been extensively reviewed by de Silva and colleagues.[9,63]

REFERENCES

1. B. H. Kaye, *Powder Mixing*. Chapman & Hall, London (1996).
2. B. H. Kaye, "Using an Expert System to Monitor Mixer Performance," *Powder Bulk Eng*. Vol. 5, No. 1, 36–40.
3. H. L. Toor, Book review, in *Am. Sci. 75*:594 (1987).
4. J. C. Williams, "Mixing, Theory and Practice," in *Mixing of Particulate Solids*, Vol. 3, edited by V. W. Uhl and J. B. Gray, p. 314, Academic Press, San Diego (1986).
5. Y. Tsuji, "Discrete Particle Simulation of Gas-Solid Flows," *Kona, 11*:57–68 (1993).
6. C. J. Broadbent, J. Bridgwater, D. J. Parker, S. T. Keningley, and P. Knight, "A Phenomenological Study of a Batch Mixer Using a Positron Camera," *Powder Technol. 76*:317–329 (1993).
7. J. A. C. Gallas, J. J. Herrmann, and S. Sokolowski, "Molecular Dynamics Simulation of Powder Fluidization in Two Dimensions," *Physica A 189*:437–446 (1992).
8. G. C. Barker, *Computer Simulations of Granular Materials in Granular Matter, An Interdisciplinary Approach*, edited by A. Mehta, Springer-Verlag, New York, pp. 35–83. In this communication segregation in a powder mixture is simulated on a computer.
9. M. R. Stein, "Gravity Blenders: Storing and Blending in One Step," *Powder Bulk Eng*., pp. 32–36 (1990).
10. B. H. Kaye, *Chaos and Complexity. Discovering the Surprising Patterns of Science and Technology*. VCH Publishers, Weinheim, Germany (1993).
11. B. H. Kaye, *A Randomwalk Through Fractal Dimensions*. VCH Publishers, Weinheim, Germany (1989).
12. J. Y. Oldshue, "Mixing," *Ind. Eng. Chem. 60*(11):24–35 (1968).
13. L. T. Fan and Yi-M. Chen, "Recent Developments in Solid Mixing," *Powder Technol. 61*:255–287 (1990).
14. B. H. Kaye, 1991. "Optical Methods for Measuring the Performance of Powder Mixing Equipment," Presented at the Bulk Powder Solids Conference, Rosemont, May 6–9, 1991. Proceedings published by Cahners Exposition, Cahners Plaza, 1350 East Touhy Ave., P.O. Box 5060, Des Plaines, IL, 60019–9593.
15. Centriflow disc mixer is available from J. H. Day & Company; see Ref. 19.
16. The Forberg mixer was manufactured by Halvor Forberg A.S., Hegdal, N3261, Larvik, Norway. It is no longer being manufactured.
17. Littleford Day Inc., 7451 Empire Drive, Florence, KY 41042.
18. Lodige Mixer available from Geruber Lodige, GmbH, Elenser Strasser P 0A790 Paderborn 1, Germany.
19. Conical mixers are available from J. H. Day & Company, 4932 Beech Street, Cincinnati, OH 45212.
20. Ribbon mixers are also manufactured by several companies including Beardsley and Piper Process, Equipment Division, 5501 W. Grand Avenue, Chicago, IL 60639. Every year the May issue of the controlled circulation magazine *Powder and Bulk Engineering* is dedicated to powder mixing and this issue has a comprehensive listing of the manufacturers of powder mixing equipment.
21. Ribbon blenders are manufactured by SCOH Equipment Company, 605 Fourth Avenue N.W., New Prague, MN 56071.
22. Ribbon and other mixers available from Teledyne-Redco, 901 South Richland Avenue, P.O. Box M-552, York, PA 17405.
23. Koch Engineering Company, Static Mixing Division, 161 East 42nd Street, New York, NY 10017. The Koch mixing unit is manufactured under License from Sulzer Chemtech Mixing and Reaction Technology Ltd., CH, 8401 Winterthur, Switzerland.
24. A Helicone™ mixer is available from Design Integrated Technology Inc., 100 E Franklin Street, Warrenton, VA 22186.
25. Nauta® mixers are available from Hosokawa Micron Group, 10 Chatham Road, Summit, NJ 07901. Nauta® is a registered trademark of Hosokawa Micron International Inc.
26. L. Hixon and J. Ruschmann, "Using a Conical Screw Mixer for More than Mixing," *Powder Bulk Eng. 6*(1) (1992).

27. W. J. B. van der Bergh, B. Scarlett, and Z. I. Kollar, "Computer Simulation Model of a Nauta® Mixer," *Powder Technol.* 77:19–30 (1993).

28. Prater Industries Inc., 1515 South 55 Court, Chicago, IL 60650.

29. V. Mixers are available from Patterson-Kelley Co., Division of Harsco Corp., East Stroudsberg, PA 18301.

30. V Mixers and Double Cone Mixers are available from the General Machine Company of New Jersey, Inc. (GEMCO), 55 Evergreen Avenue, Newark, NJ 07114.

31. V Mixers and Ribbon mixers available from O'Hara Manufacturing Ltd., 65 Skagway Avenue, Toronto, Canada, M1M 3T9.

32. Ribbon and other mixer systems available from Munsun Machine Company Inc., 210 Seward Avenue, Utica, NY 13503.

33. Mulling equipment is available from National Engineering Company, 20 North Wacker Drive, Chicago, IL 60606.

34. The Turbula® system was developed by Willy A. Bachofen A.G., Maschin en fabrik, C.H. 4005 Basel, Utengasse 15, Switzerland. Available in North America from Glen Mills Inc., 395 Allwood Avenue Road, Clifton, NJ 07012.

35. The AeroKaye® mixer is manufactured by Amherst Process Instruments Inc., Mountain Farms, Technology Park, Hadley, MA 01035-9547.

36. L. T. Fan and Y-M. Chen, "Recent Developments in Solid Mixing," *Powder Technol.* 61:255–287 (1990).

37. N. Harnby, M. F. Edwards, and A. W. Nienow, *Mixing in the Process Industries*, 2nd edit. Butterworth, London (1992).

38. Air mixers are available from Andritz Sprout-Bauer Inc., Muncy, PA 17756. This equipment is manufactured in the United States under license from Gebruder Grunkg Lissberg, Germany.

39. Dynamic Air Conveying Systems, 1125 Walters Blvd., St. Paul, MN 55110.

40. V. A. Fauver and A. E. Hodel, "Pulsed Air Blender Produces Uniform 15 Ton Lots in 20 Minutes," *Chem. Proc.* (1986).

41. Blendicon is available in Canada from Ward Iron Works, Ltd., 1223 Victoria Street, P.O. Box 511, Welland, Ontario, L3B 5R3.

42. Fuller-Kovako Corporation, 3225 Schoeperville Road, P.O. Box 805, Bethlehem, PA 18016-0805.

43. J. M. Ottino, "The Mixing of Fluids," *Sci. Am.* 56–67, Vol. 260, No. 1 (1989).

44. Chemineer Inc. manufactures a passive mixer known by the trade name Kenics® Static Mixer. 125 Flagship Drive, North Andover, MA 01845.

45. L. T. Fan, S. J. Chen, N. D. Eckhoff, and C. A. Watson, "Evaluation of a Motionless Mixer Using a Radioactive Tracer Technique," *Powder Technol.* 4:345–350 (1970–71).

46. KOMAX Systems, Inc., 1947 E. 223rd Street, Long Beach, CA 90810.

47. Charles Ross & Son Company, 710 Old Willets Path, Hauppauge, NY 11787.

48. Toray Industries Inc., 3 to 3 Nakanoshima Kita-ku, Osaka 530, Japan.

49. Lightning Mixer Equipment Co. Inc., 128 Mount Road Blvd., Rochester, NY 14603.

50. R. H. Nielsen, N. Harnby, and T. D. Wheelock, "Mixing and Circulation in Fluidized Beds of Flour," *Powder Technol.* 32:71–86 (1982) describes the use of Cabosil added to the flour to facilitate fluidization and minimum fluidization velocity.

51. Statitec Mixing Systems, EMI Inc., P.O. Box 912, Clinton, CT 06413. The passive mixer available from Statitec is known as the Statiflo mixer.

52. D. A. Pattison, "Motionless Inline Mixers Stir Up Broad Interest, *Chem. Eng.* 11:94 (1969).

53. D. J. Cassidy, B. G. Scribens, and E. E. Michaelides, "An Experimental Study of the Blending of Granular Materials," *Powder Technol.* 72:177–182 (1992).

54. J. R. Johanson, "In Bin Blending," *Chem. Eng. Prog.* 66(6):50–55 (1970).

55. M. R. Stein, "Gravity Blenders: Storing and Blending in One Step," *Powder Bulk Eng.*, Vol 4, No. 1, pp. 32–36 (1990).

56. A. W. Roberts, "Storage and Discharge of Bulk Solids from Silos with Special Reference to the Use of Inserts," POSTEC-Research Report, May 1990.

57. A. W. Roberts, "Design of Bins and Feeders for Anti-segregation and Blending," in Proceedings of the Institute of Mechanical Engineers, Bulk Materials Handling—Towards the Year 2000, London 1991.

58. H. T. Young, Apparatus for Gravity Blending of Particulate Solids, U.S. Patent No. 4,353,652, October 12, 1982.

59. C. E. Roth, Blending System for Dry Solids, U.S. Patent 4,358,207, November 9, 1982.

60. I. A. S. A. Peschl, "Universal Blender—A Blending and Mixing for Cohesive and Free Flowing Powders," *Bulk Solids Hand.* 6(3) (1986).

61. H. Wilms, "Blending Silos. An Overview," *Powder Hand. Proc.* 4(3) (1992).

62. J. W. Carson and T. A. Royal, 1991. "Techniques of In-Bin Blending," in International Conference on Bulk Materials Handling—Towards 2000, 1 Mech. E., London.

63. K. S. Manjunath, S. R. de Silva, A. W. Roberts, and S. Ballestad, "Determination of the Performance of Gravity Blenders with Emphasis on Plane Symmetric Designs. POSTEC-Research Report 921600-2, June 1992. Available from POSTEC Research A/S, Kjolues Ring, Porsgrunn, Norway.

12
Size Reduction of Solids Crushing and Grinding Equipment

L. G. Austin and O. Trass

CONTENTS

12.1 INTRODUCTION

The unit operation of the size reduction or comminution of solids by crushers and mills is a very important industrial operation involving many aspects of powder technology. It is estimated that mechanical size reduction of rocks, ores, coals, cement, plastics, grains, etc. involves at least a billion tons of material per year in the United States alone. The operation ranges in scale, for a single device, from a few kilograms per hour for speciality products to hundreds of tons per hour for metallurgical extractive purposes. In this chapter, the funda-mental aspects are emphasized rather than mechanical or process engineering aspects, to form a background for intelligent decision-making in the choice and analysis of size reduction systems.

In many operations, a material must be reduced from lumps of up to a meter in size to a fine powder, sometimes a powder essentially less than 100 μm in size. It is clear that size reduction over many orders of magnitude in size cannot be efficiently achieved in a single machine and a sequence of different types of machine is used, each machine designed for efficient operation on a particular feed size.

Machines for breakage of large lumps are called crushers and machines for smaller sizes are called mills, with a range of overlap where either a fine crusher or a coarse mill can be used. The operation of crushing normally does not give problems because the energy consumption and capital cost per ton per hour is not high. The principal requirement for crushers is a mechanical requirement—they must be very robust because of the high stress required to crush a large lump. On the other hand, fine grinding consumes a great deal of energy and may lead to high abrasive wear, so the major scientific and technical problems are concerned with fine grinding and most current research is focused on these problems.

Before discussing the various types of comminution equipment in detail, it is invaluable to have a clear idea of the fundamental physical laws involved in size reduction. These involve the areas of fracture mechanics, particle-fluid dynamics, agglomerative forces (dry and wet), and powder flow. The last four topics are covered elsewhere in this book and are mentioned here only as they arise. Fracture mechanics are discussed in some detail. Since the objective of size reduction is to obtain a suitable product size, the accurate measurement of powder size distributions is a basic feature of the process; this is also covered in detail elsewhere. However, the prediction of size distributions and how they change with mill operation is dealt with in depth.

12.2 A BRIEF REVIEW OF FRACTURE MECHANICS

12.2.1 Stress, Strain, and Energy

To produce size reduction the lumps of solids must be fractured, and they must be stressed to produce fracture. Quantitative theoretical analysis is possible only for relatively simple states of stress, but the concepts that emerge are qualitatively useful for the complex stressing conditions of industrial crushers and mills. Materials are divided into two broad cate-

gories, elastic and ductile, with the corresponding failure under stress termed brittle or nonbrittle fracture, respectively. Consider a simple tensile stress, as illustrated in Figure 12.1. Stress is defined as $\sigma = F/A$, and Figure 12.2 shows the characteristics of elastic and ductile materials. An elastic material can be stressed, producing elongation, and the material returns to its original shape when the stress is removed. However, if the solid is stretched too far, catastrophic failure occurs and the solid fractures at a stress termed the tensile strength. Ductile materials undergo a partially irreversible stretching before failure occurs.

Elastic materials fail at small strain so $\sigma \approx \sigma_0$ and the strain–stress relation up to where failure occurs is the empirical Hooke's law:

$$\sigma = Y\epsilon = Y \frac{x}{L_0} \qquad (12.1)$$

where Y is Young's modulus, ε is strain. For a perfect crystal Y depends on the orientation of the stress, but most brittle solids are polycrystalline with a random arrangement of crystallites, so Y is an effective isotropic elastic constant. The work done on the solid to go from zero external stress to a stressed state by slowly

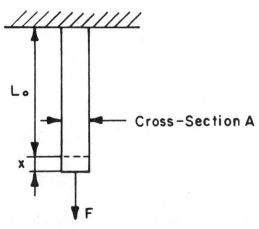

Figure 12.1. Simple tensile stress.

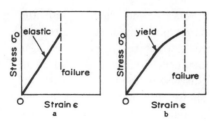

Figure 12.2. Illustration of stress–strain curves for simple tensile tests; σ_0 = force/original cross-section; strain $\epsilon = x/L_0$.

increasing F up to a final stress of σ is $\int_{L_0}^{L} F\,dx$ and using Hooke's law:

$$\text{Work per unit volume} = Y\epsilon^2/2 = \sigma^2/2Y \tag{12.2}$$

This reversible strain energy is stored in the solid. If the solid is immediately loaded to σ, the work done is $\sigma A \epsilon L$, which is σ^2/Y per unit volume. Half of this is strain energy and the other half will accelerate the solid and cause it to oscillate until frictional damping converts the kinetic energy to heat. Similarly, if a solid suddenly expands at a constant σ, the work done per unit volume is σ^2/Y and again only half is reversible strain energy.

More generally, consider a stressed solid at equilibrium. At a differential plane at any point in the solid there is no net force (since there is no movement of one part of the solid with respect to another), as illustrated in Figure 12.3, and the force of material A acting on material B must equal the force of B acting on A. The force per unit area of A acting on B is called stress, and equals B on A, so stress is a force transmission through the solid. The stress

at the point can be resolved into two components, the normal component perpendicular to the plane and the shear component in the plane. The normal stress tends to pull A away from B (tension) or force A into B (compression), whereas the shear stress tends to make A slip sideways with respect to B. From a molecular aspect, a solid consists of an array of atoms, molecules, or ions at rest (although vibrating) with respect to one another, so that the attractive and repulsive forces between them are exactly balanced. Viewing these forces as acting like springs, Figure 12.5 illustrates the three stress states. Obviously, uneven compression or tension across a solid must produce shear stress.

Drawing an arbitrary set of axes through the point that defines x, y, z directions (see Figure 12.3), the shear stress can be resolved into the components τ_{xy}, τ_{xz}. The sign convention is that material $-x$ is dragging material at $+x$ in the y direction with a force per unit area of τ_{xy}, when the sign convention for normal stress is positive for compression, negative for tension. Taking moments about a point it is readily shown that $\tau_{xy} = \tau_{yx}$, $\tau_{yz} = \tau_{zy}$, $\tau_{zx} = \tau_{xz}$ (see Fig. 12.4).

12.2.2 Directions of Normal and Shear Stress

To describe the process of fracture it is necessary to know the normal and shear stresses and their directions in the solid. The relations between stress and direction can be readily developed for a planar solid (two-dimensional) as follows. Consider an arbitrary direction de-

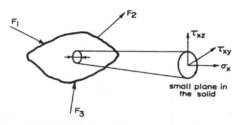

Figure 12.3. Illustration of stress through a point in a stressed solid at equilibrium.

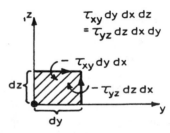

Figure 12.4. Moments about a point in the zy plane: material outside square acts on material inside.

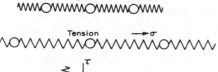

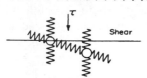

Figure 12.5. Illustration of states of stress on a molecular basis.

fined by α in Figure 12.6a, and imagine the shaded differential element of solid at equilibrium to be acted on by forces from the outside material, as shown. Because the element is a differential element, the forces are uniform over the small lengths of side and represent the forces at a point in the solid. The relative lengths along x, y, and the hypotenuse are $\cos \alpha : \sin \alpha : 1$, and since $\tau_{xy} = \tau_{yx}$ a force balance gives

$$\sigma = \sigma_x \cos^2 \alpha + \sigma_y \sin^2 \alpha + 2\tau_{xy} \cos \alpha \sin \alpha \quad (12.3)$$

$$\tau = -\left(\frac{\sigma_x - \sigma_y}{2}\right)\sin 2\alpha + \tau_{xy} \cos 2\alpha \quad (12.4)$$

The force balance will apply for any other α (see Fig. 12.6b), with a different σ and τ, of course, but the same σ_x, σ_y, τ_{xy}.

From Eq. (12.4), a particular value of α, that is $\bar{\alpha}$, can be obtained that makes $\tau = 0$.

$$\tau_{xy}/(\sigma_x - \sigma_y) = \tfrac{1}{2} \tan^2 \bar{\alpha} \quad (12.5)$$

New $\bar{x}, \bar{y}$ axes are defined along this direction, as shown in Figure 12.6c; then $\tau_{\bar{x}\bar{y}} = \tau_{\bar{y}\bar{x}} = 0$, and Eqs. (12.3) and (12.4) become

$$\sigma = \sigma_{\bar{x}} \cos^2 \beta + \sigma_{\bar{y}} \sin^2 \beta \quad (12.6)$$

$$\tau = -\left(\frac{\sigma_{\bar{x}} - \sigma_{\bar{y}}}{2}\right)\sin^2 \beta \quad (12.7)$$

where β is now a general direction variable (angle) measured from the new axes and σ, τ are the stresses at angle β (at angle $\alpha = \beta + \bar{\alpha}$; see Figure 12.6d). These axes are called the directions of principal stress and $\sigma_{\bar{x}}, \sigma_{\bar{y}}$ are the *principal* stresses.

Eliminating β between Eqs. (12.6) and (12.7) gives the equation of a circle, so the relation between τ and σ at any angle β can be represented by the *Mohr* circle as shown in Figure 12.7. The maximum shear stress occurs in a direction of $\beta = 45° (= 135°)$ and

$$\tau_{\max} = |\sigma_{\bar{y}} - \sigma_{\bar{x}}|/2 = \sqrt{\left(\frac{\sigma_x - \sigma_y}{2}\right)^2 + \tau_{xy}^2} \quad (12.8)$$

$$\sigma \text{ at } \tau_{\max} = (\sigma_{\bar{x}} + \sigma_{\bar{y}})/2 \quad (12.9)$$

The maximum normal stress is clearly the larger of the principal stress values. Also, it is readily shown that the principal stresses are

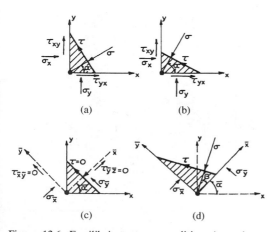

Figure 12.6. Equilibrium stress conditions in a planar element.

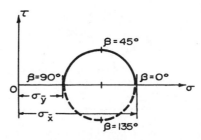

Figure 12.7. A Mohr stress circle for a planar system.

related to the normal stresses in the original coordinates by

$$\sigma_{\bar{x}},\ \sigma_{\bar{y}} = \frac{\sigma_x + \sigma_y}{2} \pm \tau_{max} \qquad (12.10)$$

Thus, knowing $\sigma_x, \sigma_y, \tau_{xy}$ at any point in the solid, the direction and magnitudes of the maximum shear stress, tensile stress, and compressive stress are readily calculated.

A similar treatment[1] in three dimensions, considering the six stress components, leads to Mohr circles for the three planes of principal stress as illustrated in Figure 12.8, where $\sigma_3, \sigma_2, \sigma_1$ are principal stresses ranked in order of magnitude. It is concluded that the maximum tensile stress has the magnitude and direction of the largest negative value of the three principal stresses and the maximum shear stress occurs at 45° between the σ_1, σ_3 directions, with a magnitude given by Eq. (12.8).

12.2.3 Differential Stress–Strain Equations

The second step is to find the values of $\sigma_x, \sigma_y, \tau_{xy}$ at all points in a solid, since these can be converted to maximum stresses and directions. For planar stress, a differential force balance of a rectangular differential element at position x, y in the solid gives

$$0 = \frac{\partial \sigma_x}{\partial x} + \frac{\partial \tau_{yx}}{\partial y} + X \qquad (12.11)$$

$$0 = \frac{\partial \sigma_y}{\partial y} + \frac{\partial \tau_{xy}}{\partial x} + Y \qquad (12.12)$$

where X, Y are the body forces in the x and y directions at the point.

The differential strains at point x, y are defined by $\epsilon_x = \partial u / \partial x$, $\epsilon_y = \partial v / \partial y$ for the linear strains, where u is the change in x dimension from the nonstressed state at point x, y; v is change in y dimension. The differential planar shear strain γ_{xy} is illustrated in Figure 12.9 and is defined by γ_{xy} = angular deformation $\theta_1 + \theta_2$. Clearly $\theta_1 = (\partial u / \partial y)\, dy / dy$ and $\gamma_{xy} = \gamma_{yx} = \partial u / \partial y + \partial v / \partial x$. The empirical physical laws relating stress and strain are Hooke's law, $\epsilon_x = \sigma_x / Y$, and the fact that a strain in the x direction causes a proportional dimensional change in the y direction (stretching in x gives a contraction in y, compression an expansion). Thus ϵ_y due to ϵ_x equals $-\nu\epsilon_x$, where ν is Poisson's ratio (≈ 0.25). For small elastic planar deformations the total strains are:

$$\epsilon_x = \frac{\sigma_x}{Y} + \left(-\nu \frac{\sigma_y}{Y}\right) \qquad (12.13a)$$

$$\epsilon_y = \frac{\sigma_y}{Y} + \left(-\nu \frac{\sigma_x}{Y}\right) \qquad (12.13b)$$

Defining a modulus of rigidity $G = Y/2(1 + \nu)$, it can be shown from Hooke's law that:

$$\gamma_{xy} = \tau_{xy}/G = \tau_{xy}\frac{2(1 + \nu)}{Y} \qquad (2.14)$$

Using the definitions of strain

$$\frac{\partial^2 \epsilon_x}{\partial y^2} + \frac{\partial^2 \epsilon_y}{\partial x^2} = \frac{\partial^2 \gamma_{xy}}{\partial x \partial y}$$

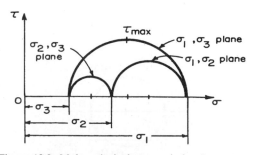

Figure 12.8. Mohr principal stress circles for a three-dimensional solid.

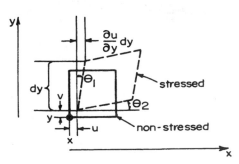

Figure 12.9. Illustration of differential strains at a point x, y in a planar solid.

and from Eqs. (12.13) and (12.14)

$$\frac{\partial^2 \sigma_s}{\partial y^2} - v\frac{\partial^2 \sigma_y}{\partial y^2} + \frac{\partial^2 \sigma_y}{\partial x^2} - v\frac{\partial^2 \sigma_x}{\partial x^2}$$

$$= \frac{2(1 + v)}{Y} = \frac{\partial^2 \tau_{xy}}{\partial x \partial y} \quad (12.15)$$

If the body forces are known, Eqs. (12.11), (12.12), and (12.15) are three simultaneous differential equations in the unknowns $\sigma_x, \sigma_y, \tau_{xy}$. They are solved using the stress and/or strain *boundary conditions*, that is, the stress–strains imposed on the solid from external action. For negligible X, Y the solution procedure is to define the Airy stress function $F(x, y)$ such that $\sigma_x = \partial^2 F/\partial y^2$ and $\sigma_y = \partial^2 F/\partial x^2$, for then $\tau_{xy} = -\partial^2 F/\partial x \partial y$ and from Eqs. (12.13) and (12.15) $(\partial^4 F/\partial y^4) + (\partial^4 F/\partial x^4) + 2(\partial^4 F/\partial x^2 \partial y^2) = 0$. Solving this equation with the transformed boundary conditions gives $F(x, y)$ and $\sigma_x, \sigma_y, \tau_{xy}$ follow by double differentiation. Equivalent but more complex equations exist for three dimensions. The strain energy above the nonstressed state is calculated from

$$\frac{1}{2} \iiint_V (\sigma_x \epsilon_x + \sigma_y \epsilon_y + \sigma_z \epsilon_z + \tau_{xy} \gamma_{xy}$$

$$+ \tau_{xz} \gamma_{xz} + \tau_{yz} \gamma_{yz}) \, dx \, dy \, dz \quad (12.16)$$

12.2.4 Ideal Strength, Stress Concentration, and the Griffith Crack Theory

The concept of *ideal strength* can be illustrated by considering an ideal solid made up of planes of molecules, subjected to simple one-dimensional tension. The tension stretches the bonds between the molecules, as illustrated in Figure 12.10, where the arrows indicate intermolecular attractive–repulsive forces. In the stretched state, any molecule still has a balance of forces on it but, as Figure 12.10b shows, the movement away from the non-stressed equilibrium against attractive forces requires addition of energy (integral of force × distance) and the solid reaches a new equilibrium at a higher energy state (stored strain energy). The maximum attractive force that the solid can exert on the surface layer is the inflection point of the potential energy curve since force = d(energy)/d(separation distance), and an external tension that exceeds this maximum causes an unbalance of forces and acceleration of one plane of molecules away from another. The solid would catastrophically disintegrate at all planes in the solid. Assuming Hooke's law to apply up to the inflection point, the strain energy per unit volume of solid is, from Eq. (12.2), $\sigma^2/2Y$. The area produced per unit volume is $2N$ where N is the number of planes per unit length; N equals $1/d$ where d is the interplanar spacing. Thus,

$$\sigma_{\text{ideal failure}} \approx \sqrt{\frac{4Y_\gamma}{d}} \quad (12.17)$$

where γ is the surface energy defined as the work required to create a unit area of surface from the unstressed solid. Equation (12.17)

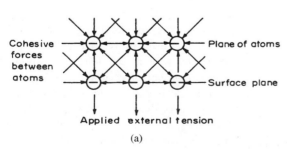

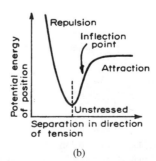

(a) (b)

Figure 12.10. Illustration of forces between molecules in a solid. (*a*) Cohesive forces; (*b*) energy of position.

must underestimate the ideal strength since Hooke's law underestimates the force required to reach the inflection point. Since γ is known for simple solids, it is readily shown that the tensile force for real fracture is orders of magnitude *less* than ideal.

The concept of stress concentration or stress intensity factor can be illustrated by considering a planar solid containing a small hole, under a uniform externally applied tensile stress of S in the x direction and zero in the y direction. Without the hole, the solution is intuitively obvious as $\sigma_x = S$, $\sigma_y = 0$, $\tau_{xy} = 0$ for all values of x and y. With a small hole of radius a present (see Fig. 12.11), the added boundary condition is

$$\sigma_r(a, \theta) = 0, \quad \tau_{r\theta}(a, \theta) = 0$$

since there is no external stress inside the hole, and the solution is:

$$\sigma_x(r, \theta) = \frac{S}{2}\left(1 + \frac{a^2}{r^2}\right) - \frac{S}{2}\left(1 + \frac{3a^4}{r^4}\right)\cos 2\theta \tag{12.18}$$

which gives a maximum stress of $3S$ in the x direction at $\theta = 90°$ and $270°$. Since a crack will open up under tension it is reasonable to expect that the solid will fail by cracks starting at the top and bottom of the hole and progressing in the $\pm y$ direction. The solution for a small elliptical hole is more complex but gives a maximum stress of

$$\sigma_{\max}/S = 1 + \frac{2a}{b} \tag{12.19}$$

where a is the ellipse axis in the y direction, b in the x direction. For an elliptical hole with its long axis perpendicular to the stress direction, a is greater than b, and stress concentration can be very high if $a \gg b$.

Griffith[2,3] argued that real solids contain many minute *flaws* corresponding to the three-dimensional equivalent of the elliptical holes discussed above and that these points of weakness initiate cracks at stress levels much below ideal. He made four basic assumptions: (1) that stress concentration occurs at the tip of the flaw, (2) that the solid is stressed to where the intermolecular bonds at the tip are stretched to breaking point, (3) that the stress state is reproduced at the tip for an infinitesimal expansion of the flaw and, (4) that energy for expanding the flaw as a propagating crack is available because the solid cannot immediately relax from its externally applied stressed or strained state. The solution of the stress–strain equations for a long ellipse gives the extra strain energy due to the presence of the ellipse as $\Delta z \, \pi c^2 \sigma^2 / Y$ where c is the long half-axis, that is, half the crack length, and Δz is the crack width. Thus, $dw_1/dc = \Delta z \, 2\pi c \sigma^2 / Y$. A sudden irreversible change from c to $c + dc$ at the instant of fracture is like a loaded solid suddenly expanding dc at constant load, so that the work done is twice the (reversible) strain energy, $dw_3/dc = 2\Delta z \, 2\pi c \sigma^2 / Y$. The energy necessary to break bonds is $4\gamma c \, \Delta z$ for a crack of half-length c, so $dw_2/dc = 4\gamma \, \Delta z$. Using the principle of virtual work, $dw_3 = dw_1 + dw_2$ at crack initiation and the critical tensile strength is

$$T_0 = -\sigma_c = \sqrt{\frac{2\gamma Y}{\pi c}} \tag{12.20}$$

Comparing Eq. (12.20) with Eq. (12.17), values of d are no more than a few Angstroms, so a flaw with a half length of hundreds of Angstroms can give orders of magnitude reduction in tensile strength T_0 from the ideal strength. As the crack progresses after initiation, $dw_3/dc > (dw_1/dc) + (dw_2/dc)$ and extra energy is available to accelerate the crack tip. The system is unstable and the crack

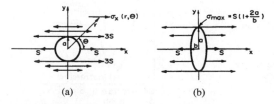

Figure 12.11. Illustration of stress concentration in a plane due to a circular or elliptical hole; s = applied tensile stress.

rapidly expands, accelerating to high velocities. The strength is lower than ideal because the bulk stress does not have to be sufficient to break all the bonding forces at once, since only the bonds around the crack tip are breaking at any instance of time. In addition, Eq. (12.20) is valid for a single flaw whereas the presence of many flaws close together will give further reductions in strength.

Obviously, pure compressive stress does not cause the flaw to open and will not cause crack propagation, so tensile stress is necessary for brittle failure. It might be thought that tensile stress will not exist under conditions of simple one-dimensional compression. However, a more detailed analysis considering all possible orientations of the flaws shows that tensile stresses are produced at the tip of an ellipse at a suitable orientation even under conditions of bulk compression. The result for a planar system with bulk normal stresses σ_1 and σ_2 and flaws of a size that would give a tensile strength of T_0 under one-dimensional tension (with the crack axis perpendicular to the stress) is shown in Figure 12.12. The compressive strength under one-dimensional compression is $8T_0$, that is, compressive strengths of brittle materials, are about an order of magnitude higher than tensile strengths.

Under combined stressed conditions the crack will propagate in a direction perpendicular to the local tensile stress conditions and

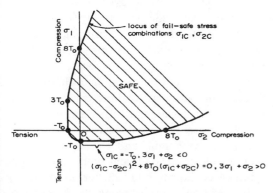

Figure 12.12. Illustration of effect of combined stress on failure from Griffith Flaws with simple tensile strength of T_0: equations are equations of locus.

may run into a region of compression that prevents further crack growth. Also, solutions of the stress–strain equations for simple compression of discs, cylinders in the "Brazilian" radial mode of testing, and spheres, show that tensile stresses are present, with maximum values along the loaded axis. Even for cubes and cylinders loaded along the axis, friction between the loading platen and the sample leads to nonuniform compressive stress and regions of tensile stress. Thus compressive loading of irregularly shaped lumps or particles will certainly produce local regions of tensile stress and, hence, brittle fracture.

Ductile materials, on the other hand, undergo plastic deformation due to sliding of planes of solid over one another, with the fundamental mechanism being that of movement of dislocations under stress gradients. In this type of movement, the bonding forces between planes are not broken all at once, but only enough bonds are broken to allow the dislocation to move to the next position, the bonds reform behind the dislocation, and so on, thus leading to slip of one plane over another by a series of low-energy steps. We have already seen that the maximum shear force occurs at 45° to the direction of principal stress, so plasticity and failure by shear will appear as illustrated in Figure 12.13. The slip process appears as the region of yielding in Figure 12.1, and is quite unlike the unstable initiation of brittle failure. Slip may initiate from a suitably oriented flaw that gives stress concentration, but there is no opening of a crack comparable to that under tensile stress.

However, other factors come into play once plastic yield has commenced. The plastic slip may cause part of the solid to act as a wedge, thus creating tensile forces that then propagate brittle fracture, as illustrated in Figure 12.13. Also, the movement of dislocations can pile up dislocations at a grain boundary, thus leading to a small hole that can nucleate a Griffith crack. Highly ductile materials under simple one-dimensional tensile loading will neck down, giving increased stress at the neck and, eventually, complete slip failure with pos-

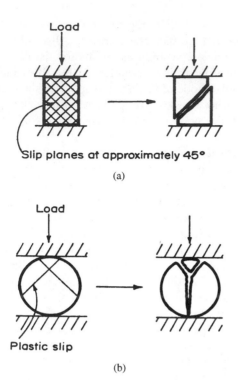

Figure 12.13. Illustration of failure by shear: slip leads to brittle fracture.

sible cleavage along crystallographic planes of weakness.

The Griffith treatment is extended[4] to allow for plasticity by including a term, dw_4, for the energy required for plastic deformation caused by the moving stress field around the crack tip. Then the initiation condition is $dw_3 - dw_1 = dw_1 \geq dw_2 + dw_4$. The value of dw_4 depends on the size and density of dislocations in the solid and dominates over bond energy dw_2 for ductile materials. Thus, ductile materials are stronger than purely brittle materials. Once fracture commences, however, the term for plastic energy may decrease because the crack moves at high velocity in relation to the time scale for movement of the dislocations that give plasticity.

Some polymeric materials have the ability to deform to high strain without fracture, for example, rubber, and the description of their failure can be considered as a separate problem. They are difficult to break by compres-

sion and induced tension because a large strain is required to make the solid reach a highly stressed state, and because dw_4 is large. Rubber materials have this property because of the shape and flexibility of the long molecules, which can coil and uncoil, bend, and straighten. A high degree of crosslinking bonds will reduce the flexibility, so these materials are weakly crosslinked, which means they are weak to shear stress. Thus, the best conditions for failure are tensile strain which straightens out the tangled and coiled molecules into a parallel array like a crystal, with superimposed shear that breaks the few crosslinking bonds.

12.2.5 Qualitative Applications of Fracture Theory: Grinding Energy

Rocks, ores, and coals being broken in size reduction machines will normally undergo brittle fracture via preexisting Griffith flaws. The strength or grindability of these materials will correlate only roughly with the hardness or chemical bond strength, because the number, size, and orientations of the flaws are additional variables. The materials are stronger in compression than tension. To calculate the strength of a lump or particle being subjected to stress, from an *a priori* theory of fracture mechanics, it would be necessary to: (1) solve the stress–strain equations for the geometry and conditions of applied stress; (2) convert the results to the *local* magnitude and direction of the principal stresses at all points in the solid; (3) consider the density (number per unit volume) distribution of sizes, and orientation (possibly random) of flaws in the solid; and (4) determine the places where local *tensile* stress can activate the flaws to the point of fracture initiation, with failure commencing at the weakest location. Such a calculation is clearly impossibly complicated for most real conditions in a mill, and can be attempted only for idealized solids and simple stress conditions [see Equation (12.23) for an example].

In addition, most grinding machines have some degree of impactive stress that propagates stress waves through the solid, activating flaws to tensile fracture in the process. The size distribution of the suite of fragments produced on fracture is as important as the fracture itself (see later), and there exists no known theory for its prediction. Theory predicts, and experiment confirms, that a fracture propagating under local tensile stress rapidly reaches high velocity (unless it reaches a zone of local compressive stress), of the order of the velocity of sound in the solid. This leads to a stress wave that propagates from the crack tip and this stress wave in turn initiates more fracture at flaws in the path of the crack. This leads to bifurcation of the crack, with bifurcation of each of the new arms, and so on, to give a "tree" of cracks through the solid (see Fig. 12.14). The energy associated with the rapidly moving stress wave is normally sufficient to pass the crack through grain boundaries and through regions of bulk compressive stress.

Ductile materials fail by initial shear, and it is again necessary to find the magnitude and direction of shear at all points through the solid. The Mohr–Coulomb criterion is that failure occurs when shear stress reaches the yield point given by

$$\tau_{y \cdot p} = \tau_0 + \mu\sigma \qquad (12.21)$$

where τ_0 is the yield shear stress under conditions of zero tensile or compressive stress perpendicular to the shear stress plane and μ is called the *coefficient of internal friction*. Equation (12.21) states that a high compressive stress perpendicular to the shear plane will tend to prevent slip, thus requiring a higher τ

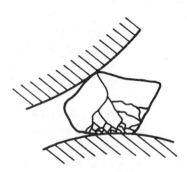

Figure 12.14. Tree of cracks in brittle failure.

to give yield, and vice versa. The slip surface is now along the direction of $\tau - \mu\sigma \geq \tau_0$. The value of μ is normally small so that the tensile strength is fairly close to the compressive strength, and slip surfaces tend to lie fairly close to 45° to the principal stress directions. From Eqs. (12.6) and (12.7) it is readily shown that

$$\mu = \frac{C_0 - T_0}{2\sqrt{C_0 T_0}} \qquad (12.22)$$

where C_0, T_0 are the magnitudes of simple one-dimensional compressive and tensile stresses required to give yield. It will be remembered that the maximum shear stress for principal stresses of $\sigma_{\bar{x}}, \sigma_{\bar{y}}$ in two dimensions is $|\sigma_y - \sigma_x|^2$, so slip is aided by a combination of compressive and tensile stresses.

A comparison between the failure of brittle and ductile materials shows the following major features:

1. Pure brittle failure is almost independent of temperature, but as temperature increases to where dislocations are more mobile, the failure may change to slip, and, hence, lower strengths. Pure ductile failure gives decrease of strength with increase of temperature owing to greater mobility of dislocations. For brittle failure with a significant plastic energy term, strength increases with temperature owing to the increase of the plastic zone around the tip, then decreases as failure changes to slip.

2. For failure from Griffith cracks, a smaller particle has a smaller probability of containing a large flaw and will be relatively stronger. Put another way, as brittle materials break, the remaining fragments are stronger because the larger flaws have broken out. On the other hand, failure by yield is not very size-sensitive because the dislocations are very small compared to lumps or particle sizes.

3. The rate of stress application is more important with ductile materials than with purely brittle materials, because a high rate

of stress application may give brittle failure whereas the same stress reached by slow steps would give time for ductile behavior.

4. Ductile materials demonstrate work hardening, that is, initial deformation produces movement and pile-up of dislocations and further deformation is more difficult. They also demonstrate stress fatigue, again owing to the gradual accumulation of dislocations on repeated cycles of stress.

5. Loading of brittle materials with uniform triaxial compressive stress, hydrostatically for example, leads to greatly increased strength by reducing local tensile forces and preventing cracks from opening.

In the case of tough, rubbery materials, the best stress application for size reduction is the scissors type of action, that is, a cutting action. This has three main features: (1) a large component of shear stress, (2) a high strain and stress caused by two forces applied in opposite directions by the blades (or stator and rotor), and (3) the creation of a surface flaw by the very high local stress of a sharp blade penetrating the material. These features are illustrated in Figure 12.15. For rubbery polymers with a substantial degree of crosslinking, which gives high shear strength, cooling the material to a low temperature can convert it to a brittle material, which can then be broken like other brittle materials. The action of the cooling is to reduce the flexibility (ability to rotate and bend) of the bonds joining the groups making up the polymer chains; it is normally necessary

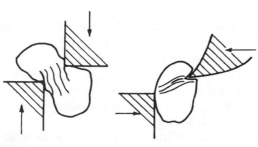

Figure 12.15. Illustration of shear-cutting actions.

to cool to very low temperatures, using liquid nitrogen (77 K).

There has been a great deal of misconception in the grinding literature concerning grinding energy. The previous discussions show that a strong solid must be raised to a higher state of stress for fracture to proceed, especially from applied compressive forces. Once the fracture has initiated, only a fraction of the *local* stored strain energy around the propagating cracks is used to break bonds (the γ term). The fragments of solid are removed from external stress when the solid disintegrates, and the rest of the strain energy stored in the solid is converted to heat and sound. Experiments on mills show that the fraction of the electric power input to the mill that is used directly to break bonding forces is very small ($< 1\%$), usually less than the errors involved in the measurement of the energy balance. Rittinger's law,[5] that the "energy of size reduction is proportional to the new surface produced," has no correct theoretical base.

To make size reduction more energy efficient it is necessary to: (1) match the machine to the particles being broken, so that mill energy is efficiently transferred to stressing the particle; (2) get nonuniform stress conditions in the particles, because nonuniform stress generates local tensile stress to activate flaws to the point where fracture can initiate; and (3) generate the right type of stress to match the failure characteristics of the material. The specific energy consumption per unit of area produced, for example, Joules/m^2, can be used as a comparative guide to efficiency, because a higher value is certainly an index of more size reduction per unit of energy input. It will not necessarily be constant for a given machine and material because it may increase or decrease with a greater degree of size reduction. On the other hand, in many cases, the production of extra fine material is undesirable, and then the specific surface area of the product is obviously not a good guide to mill efficiency, because the specific surface area is contributed largely by the extra fine sizes.

12.2.6 Property Changes and Reactions

It is known that prolonged treatment of materials with repeated stress, by batch ball milling for long periods, for example, can cause massive changes in the properties of the materials. Rose[6] showed that quartz underwent phase change from one form to another during ball milling and the topic has been reviewed,[7] giving many examples. It has been suggested that shear stress will cause nucleation and growth of one phase from crystallites of another in a particle. In ball milling, tough organic polymers can undergo a delay period in which they hardly break at all, followed by breakage. Presumably the pounding by the balls makes the material weaker by causing some degree of crystallization (molecular alignment). It is known that repeated light taps on a friable coal create or extend cracks in the coals, so that it eventually fails. Coal is a brittle polymer with planes of weakness caused by the geological process of laying down the material, but presumably other materials could show the same effect.

Benjamin[8] has discussed the formation of solid solutions of ductile metals by prolonged ball milling of a mixture of powders of the components, and the similar creation of a fine dispersion of a brittle material in a ductile matrix. The mechanism appears to be cold-welding of clean surfaces produced by fracture or flattening, so that size reduction and size growth occur simultaneously. In this case, the mill action must be such as to force particles together as well as fracture particles. It is known that organometallic compounds can be formed by ball milling chromium and nickel in organic liquids, accompanied by rearrangement of the organic molecule to other organic molecules with some H_2, CH_4, and CO_2 evolution. Similarly, reactions such as $Cr(s) + 3 TiCl_4(l) \rightarrow CrCl_3(s) + 3 TiCl_3(l)$ occur in anhydrous liquids. Again, the cause is undoubtedly the high reactivity of freshly fractured clean surfaces.

12.2.7 Abrasion

Abrasion is a special type of fracture—the tearing out of small pieces of material from the surfaces of the components used to apply stress to the material being fractured. It is obvious, of course, that these grinding components must be strong enough to stress the material being comminuted without bulk fracture themselves, but this is no guarantee that their surface will necessarily be abrasion resistant. The fracture mechanics of abrasion is not well developed theoretically, but it certainly involves high local surface stresses owing to asperities in the rock and in the grinding surface, plus local surface microflaw structure, ductility, friction, and possibly high local temperatures. High rates of surface stressing caused by high relative speed between stressing and stressed agents undoubtedly assist abrasive fracture.

The chemical environment at the surface can play a significant role, by two mechanisms. The first and most obvious is that an environment that attacks the grinding surfaces will cause surface flaws and weakness and accelerate abrasion. This effect is well recognized in wet grinding. Second, there is some evidence[9,10] that an environment can change the bond strength and ductility of material close to the surface, by strong chemical adsorption onto the surface. Such an effect will not change the bulk strength (unless conditions are such that fracture commences at surface flaws) but it can change the abrasive comminution.

The terms "hard" and "soft" are often indiscriminately used to characterize both the bulk strength or resistance to comminution of a material and its ability to penetrate or wear another material. It would be better to use the terms "strong" and "weak" for bulk comminution properties, and reserve "hard" or "soft" for the characteristics measured by one of the usual hardness tests such as the Rockwell or Vickers tests. For example, coal can be considered to be a weak rock, but certain coals are abrasive due to inclusions of quartz. Again, a tough plastic such as Teflon is not hard or abrasive but it can be very difficult to comminute.

12.3 SIZE REDUCTION MACHINES

12.3.1 Crushers

There are many different types of machines for size reduction, and almost every method of breaking lumps that one might think of has been incorporated into a crusher or grinder. Figures 12.16, 12.17, and 12.18 show common types of industrial crushers that are available in a wide range of sizes. The reciprocating action of the movable jaw in a jaw crusher strains lumps of feed to the point of fracture, as does the nonsymmetric movement of the rotating mantle in a cone or gyratory crusher (the nonsymmetric movement is produced by the bottom end of the mantle shaft being set in an off-center, eccentric bearing). The size reduction ratio, defined approximately as the largest feed size divided by the largest product size, is of the order of 10 and is varied by the adjustable gap setting. The basic action is that entering brittle material is crushed, the broken products fall under gravity into a narrower space, and bigger fragments are crushed as the metal–lump–metal space closes again, with

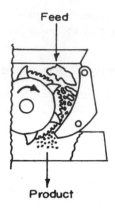

Figure 12.17. Toothed single-roll crusher.

material moving down until all of it falls through the gap. The crusher "capacity," that is, the kg/s passed, is determined by the area available for this mass flow. Feed or product fragments less than the gap setting pass out of the breakage zone and cannot be overground, so these devices can be referred to as *once-through* machines.

The machines are applying nonuniform compressive stress and the mill power must be sufficient to compress all the large pieces of rock to the fracture point when the crusher is full of large lumps. The stronger the rock, the larger the power required. Solution of the equilibrium stress–strain equations for ideal diametral loading of spheres of brittle material

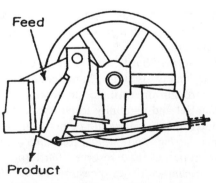

Figure 12.16. One type of jaw crusher.

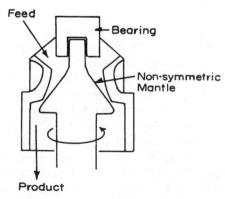

Figure 12.18. Gyratory crusher.

gives the relation between force, diameter, and maximum tensile stress σ as

$$\sigma_x = \sigma_y = -\frac{P}{\pi r^2} \frac{21}{(28 + 20\nu)} \propto -\frac{P}{r^2} \tag{12.23}$$

where P is the force, r the sphere radius, ν is Poisson's ratio, and tensile stress is negative, or

$$\sigma_c = -\frac{P^*}{r^2} \tag{12.23a}$$

where P^* is the force that causes fracture and σ_c is called the *compressive resistance*. Thus, the force to produce the tensile stress for fracture is roughly proportional to r^2. The crushing surfaces of jaw and roll crushers are often ribbed or toothed (1) to help prevent slippage of the rock as it is compressed, thus ensuring additional shear stress and (2) to give higher local stress at the surface of the material, thus activating or even producing local flaws. An important point to remember is that the same mass flow is occurring through an ever-decreasing area, so the broken material consolidates to a bed of lower porosity. It is known that a highly consolidated bed of low porosity is difficult to compress further (it has a high Young's modulus) so that high stresses can be produced on the metal surfaces squeezing the bed near the gap. This is generally avoided by controlling the rate of feed to a crusher to prevent excessive consolidation at the gap. As discussed later, compression of a bed of particles (bed compression) can produce fine material, which is often undesirable in a crusher because it may lead to excessively dusty conditions in the work area.

Figure 12.19 illustrates a hammer crusher. Material is broken by direct impact of the hammers, by being thrown against the case or breaker bars, and by compression and shear when nipped between the hammers and the case. The hammers are mounted on a heavy rotor and/or the shaft is attached to a heavy flywheel, to give a high moment of inertia of the rotating mass. This type of crusher is best

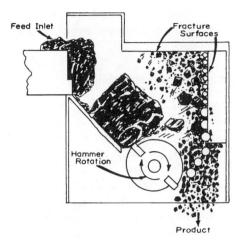

Figure 12.19. Heavy duty hammer crusher (Jeffrey Coal Buster).

suited for nonabrasive materials, although it is sometimes used for fairly abrasive rock because of its low capital cost: the user must then resurface the hammers at frequent intervals. Figure 12.20 shows a Cage–Pactor crusher, in which solid is crushed as it passes

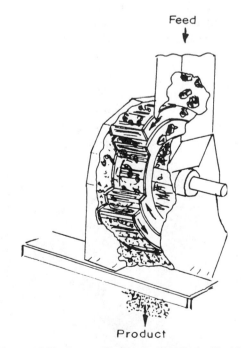

Figure 12.20. Cage mill (T. G. Gundlach Machine Company Cage–Pactor).

between rows of rotating bars. Again, this is best suited for relatively weak nonabrasive material such as coal.

12.3.2 Roll and Rod Mills

Figures 12.21, 12.22, and 12.23 show machines suitable for intermediate size reduction. The medium-duty hammer mill is especially suitable for sticky or tough materials that cannot be efficiently broken in rolls or rod mills, because of its ability to shear in addition to compress. The smooth roll crusher is widely used for size reduction in the laboratory, but it also has industrial application in preparing material with a top size of, say, 12 mesh (1.70 mm) and a minimum of fines (see Section 12.4). Because of abrasive wear the rolls have to be resurfaced at frequent intervals if the crushed material is strong and abrasive. Again, it is important to control the feed rate into the rolls to prevent the damaging forces arising from bed compression in place of steel– lump–steel crushing. The rod mill acts somewhat like a multiple set of rolls as the cylinder rotates; the bed of rods is carried up until it lies at an angle to the horizontal. It is then unstable and rods start to roll down the bed

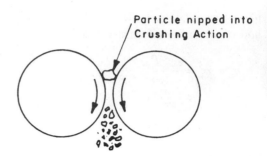

Figure 12.22. Smooth roll crusher.

surface, reenter the bed, and get carried up again. The rods rolling over one another act like sets of rolls, stressing particles in a similar manner. The power to the mill is used to lift the rods against gravity; the resulting potential energy of position is converted to kinetic energy as the rods fall, which in turn is converted to strain energy and, finally, to heat and sound.

However, there are two major differences between smooth roll crushers and rod mills. First, the rod mill is a *retention* device because fine fragments have to pass along the mill to overflow at the exit and can be rebroken again and again, so the mill is acting on a reservoir of powder. In this type of device, the *residence time distribution* (see below) of material in the mill is of importance, and more fines are produced. Second, there is obviously no con-

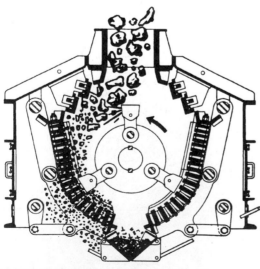

Figure 12.21. Medium-duty hammer mill (Jacobson Crusher Co.).

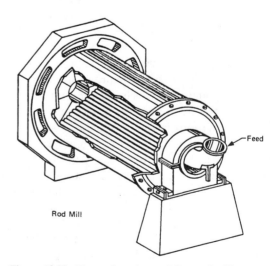

Figure 12.23. Illustration of a tumbling rod mill at rest.

trolled gap setting or controlled power to the turning rod, so it is not always possible to break large, strong lumps, which can then leave in the overflow. The force available for fracture is increased by making the steel rods heavier (larger diameter) and the mill diameter larger, but this is limited by excessive damage to the mill lining by the falling rods. Thus, the feed to a rod mill is normally less than about 25 mm in top size, depending on material strength, It is normally used for wet grinding. Abrasive wear on the rods means that worn-down rods must be removed and replaced with fresh rods at suitable intervals.

12.3.3 Tumbling Ball Mills

Figure 12.24 shows the tumbling ball mill, also a retention mill, which is very widely used for dry and wet grinding to relatively fine sizes. The principle is identical to that of the rod mill, but thc maximum force available to break large, strong lumps is even less, so the feed to the mill is rarely larger than 10 mm for strong rock. Because of its great industrial importance this type of mill has been widely investigated, and is discussed in detail below. Abrasive wear is easily handled by topping up the charge with fresh balls at frequent intervals and it is not necessary to stop the mill to add the balls. The mill shown has an overflow

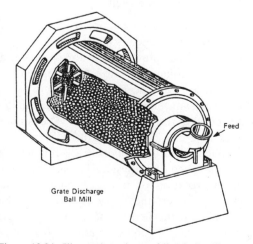

Grate Discharge
Ball Mill

Figure 12.24. Illustration of a tumbling ball mill at rest.

discharge for continuous wet grinding, while discharge through slots or grates that retain the balls is often used for continuous dry grinding. For grinding coal, the mill is swept with hot air to dry the coal and the fine coal removed in the exit air stream. Ball mills can be used for very fine dry grinding by air sweeping, with return of oversize particles to the mill feed from a high-efficiency (rotary) size classifier cutting at a small size to give a high circulating load.

12.3.4 Autogenous and Semi-Autogenous Mills

Autogenous tumbling mills are similar in principle to the tumbling ball mill, but use the material being broken as the breakage media. There are four major types. The first is essentially identical in construction to a ball mill, but the feed consists of two streams, a narrow size range of lumps of rock (e.g., 75 mm × 150 mm) and the normal fine crushed feed. The large rocks wear to round pebbles (hence, the name pebble mills) on tumbling and then act like steel balls on the rest of the feed. The feed rate of large rock is adjusted to keep a suitable load of pebbles in the mill. The second type has a large diameter-to-length ratio (typically 2 : 1) and takes a natural crushed feed containing rock typically up to 200 to 300 mm, with discharge through slots of typically 20 mm width. Since the feed rate has to be in balance with the rate at which the large lumps break themselves to less than 20 mm by their own tumbling action, it is not possible to vary the product size distribution over a wide range. In fact, the third type, semi-autogeneous mills, are identical but add some charge of large (4 in. = 100 mm) steel balls, typically a few percent of the mill volume, to increase output capacity. The Scandinavian countries and South Africa use a variant of this type with a smaller diameter-to-length ratio (typically 0.5 to 1), which behave like semi-autogeneous pebble mills.

Although very similar to tumbling ball mills, autogeneous and semi-autogeneous tumbling mills have some distinct features in their breakage action. Since rock has a lower density than steel, the power input per unit of mill volume is lower than in ball mills, so the equivalent ball milling action is reduced. However, a gradual decrease of the size of large lumps of rock is not a typical disintegrative breakage but has a major component of a chipping action in which pieces are broken off irregular feed shapes to give rounded material. The rounded lumps then abrade until the size is small enough to be broken by a larger lump. Both chipping and abrasion give small product fragments, so the mills give suitable qualities of finely ground material even when the product contains substantial amounts of very coarse particles.[11-14] Autogenous mills have lower capacity for a given mill volume than a ball mill and, hence, higher capital cost per unit of output, but they do not have the continuing cost of replacement steel balls. The use of semi-autogeneous mills allows the best economic balance to be reached between capital cost and cost of replacement steel.

The fourth type of autogeneous mill, the rotary breaker, is specific for coarse size reduction of coal. It has the added feature that the cylinder case is lightweight and contains many holes (typically 50 to 300 mm), so that material broken less than the desired top size falls through and forms the product. Coal is light enough and friable enough that self-breakage by tumbling gives high output without requiring a heavy shell to withstand pounding and abrasion.

12.3.5 Vibrating / Planetary / Centrifugal Ball Mills

There are two other variants of the ball mill. In the vibrating ball mill the cylinder is not rotated to cause tumbling but is packed almost full with balls and mounted on an eccentric that jerks it around the cylinder axis, thus causing the balls to vibrate in the cylinder. The mechanical stresses on the drive are high and the mill is not conveniently scaled to high continuous capacity. A small ball mill fitted

into a shaking mechanism is similar in principle and very useful for preparing laboratory samples of fine powders. The planetary or centrifugal mill[15,16] contains two or more rotating cylinders partially filled with balls, mounted at the periphery (parallel to the axis) of a bigger cylinder or frame that is also rotated. The respective speeds of rotation are set by gears to use the centrifugal force of the outer rotation to throw the balls across their cylinders as they rotate, thus replacing gravitational fall with much higher centrifugal force and also greatly increasing the number of balls moved per unit volume and time. A fairly recently developed mill[16] accomplishes the same purpose with a single horizontal mill shell mounted on an eccentric (with counterbalance weights), with the radius of gyration chosen to produce the effect of a centrifugal field moving around the mill with each gyration. This gives a high-force tumbling action of the ball charge but avoids the high force on the drive produced by the vibrating ball mill and is much simpler mechanically than planetary mills. The power input and capacity per unit volume of the mill is very high and it is suitable for underground treatment of ores in mining tunnels, thus saving millhouse construction costs. Abrasive wear is high and the mill is designed for rapid replacement of a removable lining in the mill.

12.3.6 Roller-Race Mills

Figure 12.25 gives an example of the class of mills known as vertical spindle mills or roller-race mills. The rotating table throws material through the roller-race and the pulverized material passes over the rim and is swept up by an air stream flowing through the annulus between the rim and the case. The stream passes to a classifier that returns oversize to the table, so that the rollers are acting upon a fairly thick bed of material. The basic action is that the rotation of the race pulls material under the roller, the roller is driven by this material, and the bed of material passing under the roller is nipped and crushed as it passes through the gap between the roller and the race. The rollers are loaded with massive

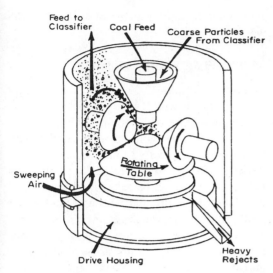

Figure 12.25. Illustration of roller-race mill (Krupp–Polysius Co.).

springs to give a high compressive force and the gap automatically adjusts to a height such that the mass flow of material pulled in equals the mass flow of compressed material passing through the gap. The bed compression produces breakage just like putting a bed of particles into a cylinder and applying high pressure. Fragments broken from the particles fit within spaces between the feed particles and the porosity of the bed decreases to a minimum at the gap where the pressure is highest (see later).

These mills are widely used for coal grinding, again with hot air drying and conveying, and for raw material grinding for cement manufacture. They give a produce size distribution of about 80% < 75 μm, from feeds with a top size of about 25 to 75 mm depending on the diameter of the rollers. The mills work well only for brittle feeds of a natural size distribution: if the feed consists entirely of large lumps, the rollers nip lumps rather than a feed bed. As a result, the rollers ride up, then fall, the mill runs roughly, and gives high abrasion of both rollers and bed plates. If the feed contains excessive fine material, mill operation is again unstable with gouging, and slip of the rollers over the bed of powder occurs because the bed develops fluid-like properties. How-

ever, with a proper size range of feed matched to roll diameter they are efficient mills with a lower energy consumption per ton of product than many other mill types. The compression of beds of powder by pulling material between rolls is used in three recent developments: the high-pressure grinding rolls machine (the Schönert mill), the Szego mill, and the Horomill. These are discussed in detail in the section on new developments (see later).

12.3.7 Hammer Mills

High-speed hammer mills, similar to the mill shown in Figure 12.21, are also used for relatively fine dry grinding of many nonabrasive materials, with much of the grinding action being by shear between the hammer tips and the case. The mills are air swept with built-in rotary size classifiers to retain coarser material in the mill.

12.3.8 Disc Mills

Figure 12.26 illustrates the disc mill, which also consists of surfaces rotating at high speed with respect to one another, but with the gap

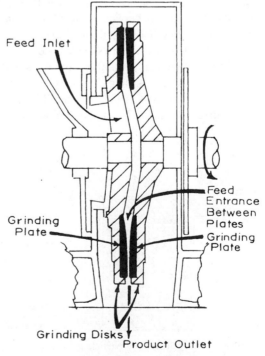

Figure 12.26. Disc mill.

between the discs readily adjustable during operation. The force application is by shear and compression as particles move into the narrower portions of the gap. There are several machines similar in principle but with different plate geometry.

12.3.9 Stirred Media Mills

Figure 12.27 shows a sand mill or Attritor, which consists of paddles turning in a bed of water and sand or small steel or ceramic balls. The large number of grinding particles give many breakage actions per unit time but the breakage action is mild, and the mill is most often used for comminution or deagglomeration of small, relatively weak particles or agglomerates, such as dyestuffs, pigments, clays, etc. A similar principle is used in the high-energy ball mill, with larger balls and high paddle speeds which give much higher forces and a high power input per unit of mill volume. These are used on a relatively small scale for preparing mechanical alloys by dry grinding of ductile metals. Larger versions are used for fine grinding of limestone and other fairly weak materials. In shear mills, slurry is flowing in a narrow annulus between a rotating drum and a stationary cylinder, with breakage caused by the high fluid shear forces across the annulus. They are generally suitable only for small weak particles or weak agglomerates. In some mills, a wider annulus is filled by small media. More intensive and uniform grinding action is then obtained, but at a cost of wall and media wear.

12.3.10 Fluid Energy Mills

Figures 12.28 and 12.29 shows types of fluid energy mill, in which small particles are suspended in high-velocity streams of air or steam obtained by expansion through nozzles with inlet pressures of 5 to 10 atmospheres. In the device illustrated in Figure 12.28, the tangential entry of high-velocity fluid creates a doughnut of swirling particles and fluid in the grinding chamber, which retains coarser particles by centrifugal action. The microturbulence of the gas stream causes high-speed impact of particle-on-particle, and the centrifugal size classification allows only fine sizes to leave the breakage zone. In Figure 12.29, the opposed jets cause high-speed collision of the particles, and a size classifier and fan

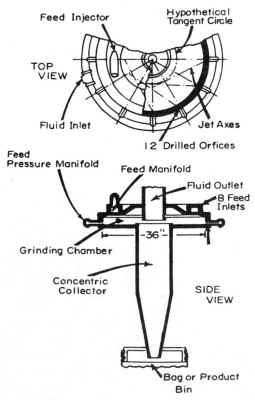

Figure 12.28. Fluid energy mill: Sturtevant Micronizer.

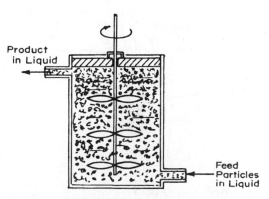

Figure 12.27. Stirred ball-particle mill: Attritor.

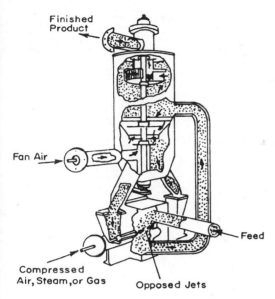

Figure 12.29. Fluid energy mill: Majac Jet Pulverizer.

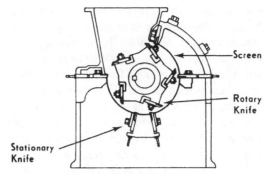

Figure 12.30. Illustration of rotary knife cutter mill.

system in the device returns larger sizes into the jet stream. The mills are designed to give fluid boundary layers on the containing surfaces, to reduce particle impact on the surfaces and the consequent abrasion. The specific energy consumption calculated from the energy required for air compression or steam-raising is high compared to mechanical grinders, but the mills are capable of producing very fine material (e.g., −5 μm) and are used primarily for specialty grinding of high-value materials or where cheap waste steam is available.

12.3.11 Shredders and Cutters

Figure 12.30 illustrates a whole class of mills designed specifically for size reduction of tough but nonabrasive materials such as polyvinyl chloride, Teflon, rubber, wood, etc. They rely on the cutting action, like scissors, between rotating and static sharp edges with narrow clearance. The efficiency of this type of mill is highly dependent on maintaining sharp cutting edges. Shredders, for example, for waste paper, and hogs for waste wood and bark fit into this category. A number of mechanical arrangements are used.

12.4 THE ANALYSIS OF SIZE REDUCTION PROCESSES

12.4.1 General Concepts

It is clear from the previous section that the multiplicity of mill types and breakage actions make it virtually impossible to formulate a general theory of the unit operation of size reduction. In most cases good mill design has evolved by trial-and-error starting from common-sense applications of the concepts of fracture. However, for devices that reduce large tonnages of material, using substantial electrical energy, there is considerable impetus for accurate process design rules and for techniques for optimization of the system. As in other unit operations, it is invaluable to construct mathematical models of the operation to aid in its understanding and optimization. In the last decades, considerable advances have been made in this respect using concepts very similar to those of chemical reactor theory.[17,18] The mill is considered equivalent to a reactor that accepts feed components (the set of feed sizes) and converts them to products (the set of product sizes), and a size-breakage rate (population) balance is performed on the reactor.

The rate at which a material breaks in a mill depends on its particle size as well as its strength characteristics. Normally, for any given mechanical action there will be particle sizes that are too big for efficient breakage because the action is not powerful enough, and particles that are too small for efficient breakage because the statistics of applying the action are not favorable. It is also apparent that the specific energy (kWh/ton) used for size reduction increases for breakage of finer and finer sizes because (1) it becomes more and more difficult to apply stress efficiently to millions of tiny particles and (2) the basic strength of brittle particles increases because large flaws (which can be stress-activated to fracture at low stress) become broken out as grinding proceeds to finer sizes. It is necessary, then, to analyze the breakage of each size range. It has been found convenient to use a $\sqrt{2}$ screen sequence to define the size ranges (e.g., "size" is defined as 16×20 mesh, 20×30 mesh, etc.) because material in one of these size intervals appears to behave like a uniform material, to a sufficient approximation. Since a geometric progression never reaches zero it is necessary to define a "sink" interval containing all material less than the smallest size measured. Thus, a feed size range can be split into n intervals, numbered 1 for the top size interval to n for the sink interval.

Using this basis, the size distribution from breaking a given "size" in one pass through the device is called the *progeny fragment distribution*, and is conveniently represented in the *cumulative* form "D_{ij} = weight fraction less than size x_i from breakage of larger size j," where x_i is the top size of interval i. Obviously, $D_{jj} = 1$ and $1 - D_{j+1, j}$ is the fraction of size j that remains of size j after passing; the fraction of size j transferred to size i is $d_{ij} = D_{ij} - D_{i+1, j}$. The set of numbers d_{ij} is called the *transfer number* matrix. For a *once-through* machine such as a roll crusher these values can be determined experimentally by crushing each size independently.

For a *retention* machine such as a ball mill, it is extremely valuable to define a *primary progeny fragment distribution*, B_{ij}, again cumulative, which is the mean set of product fragments produced from one breakage action, with the products then mixed back into the mill contents to wait to be selected for a second breakage, and so on. It has been found that the form of the primary B values is

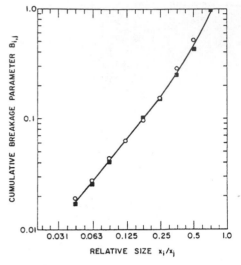

Figure 12.31. Typical cumulative primary progeny fragment distribution: ball milling of 20×30 mesh quartz. (■) dry; (○) wet.

similar to those of Figure 12.31 for many brittle materials and machine types: it is not difficult to see that this form is compatible with the tree of cracks illustrated in Figure 12.14. The slope γ of the finer end of the B plot is characteristic of the material and appears to be the same for all breaking sizes. In many cases, the B values are size-normalizable, that is, the curves of Figure 12.31 fall on top of one another for different breaking sizes. Thus, the "weight fraction less than a given fraction of the breaking size" is constant and

$$B_{ij} = B_{i+1, j+1} = B_{i+2, j+2}, \text{etc.}$$

For retention mills, the concept of *specific rate of breakage* S_i is applicable. Consider a mass W of powder in the mill, of which a weight fraction w_j is of size j. The specific rate of breakage S, for example, for size interval j, S_j is defined by:

Rate of breakage of size j to smaller sizes

$$= S_j w_j W \quad (12.24)$$

It has units of time^{-1} and is comparable to a first-order rate constant in chemical kinetics. A batch grinding test on a feed of size j is comparable to an homogeneous first-order "rate-of-reaction" experiment, and if S is constant, Eq. (12.24) goes to

$$d(w_jW)/dt = -S_jw_jW$$

and

$$w_j(t) = w_j(0)\exp(-S_jt) \quad (12.25)$$

Figure 12.32 shows a typical result. This first-order relation is observed so frequently that it can be called "normal" breakage, whereas non-first-order kinetics indicate some "abnormal" feature. Methods of estimating S and B values from experimental tests have been described,[18] but they are primarily useful for laboratory or pilot-scale test data, and it is at present frequently necessary to infer values for large devices by extrapolation from smaller scale results.

12.4.2 Mill Models

The function of a mill model is to describe the product size distribution. The model can then be used to assist in the analysis of the influence of design and operating variables on mill performance. For example, consider a simple once-through device such as the smooth roll crusher (see Fig. 12.22). It can be assumed as a first approximation that each size of the feed

lumps is stressed and fractured independently of the other sizes, as it reaches the region in the rolls where it is nipped and compressed. For a given gap setting, x_g say, breakage of each size will produce a mean set of progeny fragments denoted by D_{ij}. Then, for a feed consisting of weight fraction f_1 of size 1, f_2 of size 2, etc., the fraction p_L of product in size i is:

$$\left.\begin{array}{l} p_i = d_{i,1}f_1 + d_{i,2}f_2 + \cdots + d_{i,i}f_i \\ \\ P_i = D_{i,1}f_1 + D_{i,2}f_2 + \cdots + D_{i,i}f_i \end{array}\right\}$$
$$\text{or}$$
$$(12.26)$$

$$n \geq i \geq 1$$

where P_i is the cumulative fraction of material less than x_i in the crusher product. The logical analysis of how the D values vary with conditions is given later in this section as an example of the analysis of once-through devices.

Retention grinders such as the tumbling ball mill are very important industrially, and the mill model applicable to these is developed as follows. First, consider the simplest system of batch operation, with the powder getting finer and finer for longer and longer grinding times. Using the concept of primary progeny fragment distribution joined with Eq. (12.24) the "net rate of production of size i material is the sum of its production from all larger sizes minus its rate of breakage," or

$$W\frac{dw_i(t)}{dt} = -S_iw_i(t)W + W\sum_{\substack{j=1 \\ i>1}}^{i-1} b_{i,j}S_jw_j(t),$$

$$n \geq i \geq j \geq 1 \quad (12.27)$$

where W is the mass of powder in the mill and $b_{i,j}$ is the *primary* progeny fragment distribution in the interval form, $b_{ij} = B_{ij} - B_{i+1,j}$. This set of n equations is known as the batch grinding equation. If b_{ij} and S_i do not vary with time, it has the solution:[18,19]

$$w_i(t) = \sum_{j=1}^{i} d_{i,j}w_j(0), \qquad n \geq i \geq j \geq 1$$

$$(12.27a)$$

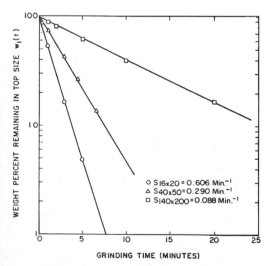

Figure 12.32. Typical first-order plot of batch grinding data, various sizes of cement clinker.

where the set of transfer numbers $d_{i,j}$ is computed from the algorithms

$$d_{i,j} = \begin{cases} 0 & i < j \\ e^{-S_i t} & i = j \\ \sum_{k=j}^{i-1} a_{i,k} a_{j,k} (e^{-S_k t} - e^{-S_i t}) & i > j \end{cases}$$

$$a_{i,j} = \begin{cases} -\sum_{k=i}^{j-1} a_{i,k} a_{j,k} & i < j \\ 1 & i = j \\ \dfrac{1}{(S_i - S_j)} \sum_{k=j}^{i-1} S_k b_{i,k} a_{k,j} & i > j \end{cases}$$

The equations are programmed[20] for computation on a PC, and the solution starts with $i = 1$, then $i = 2$, etc., using the feed size distribution $w_i(0)$. Figure 12.33 shows the computed solution compared to the smoothed experimental points for grinding of a narrow feed size, using experimentally determined values for S and B.

Second, consider a retention grinding machine where the powder flows uniformly, is ground, and is then fine enough to exit through an overflow or grate without preferential retention of larger sizes. If the flow through the

mill was plug flow, Eq. (12.27) would still apply with a grind time τ of $\tau = W/F$, F being the mass flow rate through the mill. However, retention mills will generally have a *residence time distribution* (RTD) defined by $\phi(t)\,dt =$ weight fraction of feed in at time 0 which leaves between time t and $t + dt$. This is due to mixing in the mill which brings some feed quickly to the discharge, while other material is back-mixed to the feed end and leaves later. Figure 12.34 gives an example determined by using a pulse of radiotraced powder in the mill feed and counting at the mill exit.[21] Then the steady product size distribution will be made up of material ground for all times over the RTD range, in a weighted sum:[18]

$$p_i = \int_0^{\infty} w_i(t) \phi(t)\,dt \qquad (12.28)$$

where $w_i(t)$ is the solution of Eq. (12.27) for the mill feed. For a fully mixed mill the mass-rate balance is "the rate of flow size i out = rate of flow size i in plus rate of production of size i by breakage of all larger sizes minus rate of breakage of size i." Thus,

$$Fp_i = Ff_i - S_i w_i W + W \sum_{\substack{j=1 \\ i>1}}^{i-1} b_{i,j} S_j w_j \qquad (12.29)$$

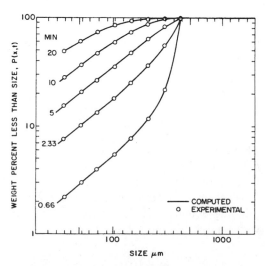

Figure 12.33. Comparison of computed to experimental size distributions for batch grinding.

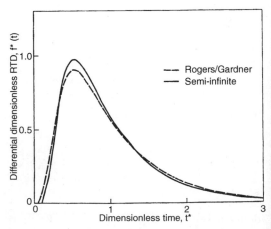

Figure 12.34. Residence time distribution for a 4.57 m diameter $\times$ 9.2 m long wet overflow discharge ball mill.

However, $p_j = w_j$ for a fully mixed system with no size classification at the mill exit. Using $\tau = W/F$

$$p_i = \frac{f_i + \tau \sum_{\substack{j=1 \\ i>1}}^{i-1} b_{i,j} S_j p_j}{1 + S_i \tau}, \quad n \geq i \geq j \geq 1$$

(12.29a)

This set of equations is readily computed sequentially starting at $i = 1$.

The variable used in the computations is the mean residence time τ, and any model can be computed for a range of τ values. Since $\tau = W/F$, the value of τ that gives the desired product size also specifies the mass W necessary to get a desired production rate F. Then the mill size needed to contain W is calculated. Of course, it is also necessary to have equations that give mill power, in order to determine the specific energy of grinding.

An important general conclusion can be reached by considering Eqs. (12.27) or (12.29) applied to a comparison of two milling systems operating on the same feed. Suppose that the B values are the same between the two systems, but that S values are different by a constant factor, $S_i' = kS_i$. Using Eq. (12.29a) as an example, applied to both mills,

$$p_i = \left(f_i + \tau \sum_{j=1}^{i-1} b_{i,j} S_j p_j \right) / (1 + S_i \tau)$$

(Mill 1)

$$p_i' = \left(f_i + \tau' \sum_{j=1}^{i-1} b_{i,j} S_j' p_j' \right) / (1 + S_i' \tau')$$

(Mill 2)

Substituting for S_i' in the second equation,

$$p_i' = \left(f_i + k\tau' \sum_{j=1}^{i-1} b_{i,j} S_j p_j' \right) / (1 + k\tau' S_i)$$

(Mill 2)

Obviously, $p_i' = p_i$ when $k\tau' = \tau$, that is, an identical set of size distributions is produced in mill 2 as in mill 1 *but with residence times decreased by the factor k*. If S values are dou-

bled ($k = 2$), the required residence time is halved. Thus, there can be *similitude* between a small mill and a large mill, with only a difference in time scale. The same result is obtained for batch or plug flow grinding, and for Eq. (12.28) providing the RTD is normalizable with respect to τ, that is, $\phi(t/\tau)$ is the same from one mill to another.

The use of these models is illustrated below. Experimental measurement of the variation of the values of S_j with mill conditions is the most explicit and logical means for describing mill operation and mill efficiency.

It is useful to have an approximate mill model that is simple enough for quick-hand calculations. The results of Figure 12.33 allow the deduction that Bond's "law"[22] applies to a reasonable approximation,

$$E = m_p t / W = E_I \left(\sqrt{\frac{100 \ \mu m}{x_{80P}}} - \sqrt{\frac{100 \ \mu m}{x_{80F}}} \right)$$

(12.30)

where m_p is the shaft mill power, x_{80P} is the size in micrometers at which 80% passes that size in the product, x_{80F} is the 80%-passing size of the feed, and the energy index E_I is determined from the data. E is the specific energy of grinding (kWh/ton) required to go from a specified feed of x_{80F} to a desired product of x_{80P}. This empirical equation enables rapid estimation of the grinding time or specific energy to go from any feed to any product, *assuming* that E_I is a constant. It does not give any information on the size *distribution* of the product nor does it take into account the size *distribution* of the feed. As might be expected, E_I is not closely constant from one mill to another, or for different mill conditions. As used in practice, E_I is determined for a given material from an experiment under standard conditions[23] using an empirical correlating equation that converts it to the value expected for an 8-ft diameter wet overflow ball mill operating in closed circuit. E_I is then known as the Bond Work Index W_I, which has the physical meaning of the hypothetical kWh/ton necessary to go from a very large feed to 80% passing 100 μm, in the 8-ft diameter mill. Empirical correction factors based on prior experience are used to allow for different conditions and mill diameter.[22]

12.4.3 Mill Circuits: Classification

In industrial practice, mills are frequently used in closed circuit, where the mill product is passed through a size classifier that gives two exit streams, a coarser stream returned to the mill feed and a finer stream, which is the final product. The operation of the classifier is best described by the set of classifier selectivity numbers, s_i, defined as the weight fraction of size i presented to the classifier that is sent to the coarse stream. These are readily calculated from experimentally measured size distributions of the three streams.[18] Figure 12.35 gives a typical example. It can be seen that a typical classifier is not ideal. It sends some coarse material to the product and returns some fine material back to the mill. The smaller the value of d_{50}, the bigger the overall fraction of the classifier feed that is directed into the recycle stream. The relation between the circuit feed and product and the mill feed and product is shown in Figure 12.36: defining the circulation ratio by $C = T/Q$, then

$$f_i(1 + C) = g_i + (1 + C)s_i p_i$$

and

$$q_i = (1 + C)(1 - s_i)p_i.$$

These are used in conjunction with the appropriate mill model to predict the circuit product size distribution from a mill circuit simulation.[18]

Figure 12.37 shows one interesting result from a simulation of a tumbling ball mill. If a mill circuit is designed to produce a size distribution passing through a control point ($\psi\%$ passing size x^*) from a given mill, then this specification can be met by a suitable feed rate through a classifier with set s_i values, or by a different feed rate with the classifier adjusted to cut at smaller sizes (and, hence, give more recycle and a larger C value). It is seen that there is a *permitted band* of size distributions through the control point, from $C = 0$ to $C = \infty$. Austin and Pérez[24] have shown that the limiting (steepest) size distribution obtained at high circulating load depends only on the primary progeny fragment distribution. Thus, it is a material characteristic and it is not possible for a customer to specify a steeper distribution. The higher circulating load also gives a higher circuit output rate Q tph (tons/h). The physical reason for these effects is that a high flow rate through the mill, $F = (1 + C)Q$, brings fine material rapidly to the classifier and removes it before it is overground. Thus, the mill contents contain on the average less fines and more coarser material, and coarser material breaks faster than fine material. *The general reason for closed circuit operation is to remove particles that are already fine enough, to prevent energy being wasted on grinding them even finer.*

The return of fine material back to the mill feed, due to the apparent bypass of the classifier as shown in Figure 12.35, decreases efficiency by leading to overgrinding. In principle, this can be compensated by higher circulation, but in practice (1) it may not be possible to pass enough mass through the mill to approach this limit without *overfilling* the mill leading to poor breakage action and (2) increased mass flow through a classifier may also increase the bypass fraction, thus defeating the action. For these reasons it is advantageous for a classifier to approach as closely as possible the ideal classification shown in Fig-

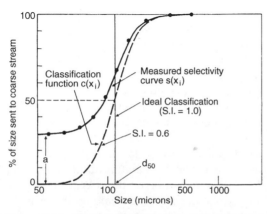

Figure 12.35. Illustration of selectivity values of a size classifier: a is an apparent bypass.

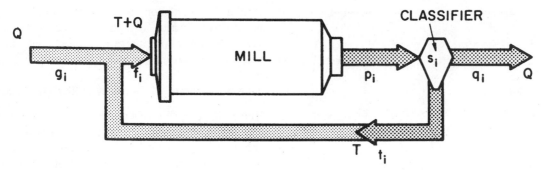

Figure 12.36. Normal closed circuit.

ure 12.35. *The function of efficient classification is to reduce the proportion of fine material by avoiding overgrinding of fines.* The concept of *indirect inefficiency* is that although a mill may be operating efficiently in transferring input energy to breakage it can be inefficient if that energy is used to break material that already meets specifications.

12.4.4 Non-First-Order Grinding and Slowing of Grinding Rate

It can be reasoned from fracture mechanics and the difficulty of efficiently stressing unit mass of very small particles that the specific rates of breakage are smaller for small particles than for larger ones. This has been confirmed for every type of mill investigated to date. However, there is an additional effect of

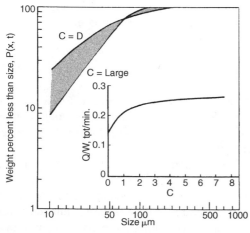

Figure 12.37. Permitted band of size distributions passing through a desired point, with varying circulating load.

small size in retention devices such as ball mills and roller-race mills. As fine material builds up in the bed of powder, the breakage of *all* sizes slows down. This appears to be partly due to coating of the grinding surfaces but principally due to a *cushioning* action. In dry grinding, it is argued[25] that the agglomerative forces between fine particles impart a fluid-like nature to the bed that can absorb impact without giving high stress to particles directly under the stressing surfaces. This can be likened to trying to grind particles suspended in a sponge; the energy of a falling ball or passing roller is spread over a large elastic mass instead of being concentrated on a small mass of solid. In addition, air trapped in such a bed cannot rapidly flow out of the bed in the path of the stressing surface because of the high drag forces, so it moves away carrying particles with it, much like a liquid parting to let a solid ball fall through.

It is sometimes possible to predict the correct product size distribution even in the presence of slowing-down effects, by performing the simulation with a *false* residence time θ that is less than the real residence time t. A slowing-down factor K can be defined by $K = \theta/t$, which then also represents the ratio of the actual mean value of S_i from time 0 to t to the first-order value S_i. Figure 12.38 shows values of K for four different materials, plotted against the fraction of fine material less than 10 μm in size. It is apparent that different materials develop the slowing-down process at different amounts of fines. The magnitude of the effect can be seen from Figure 12.39, where it takes 20 min to reach a size

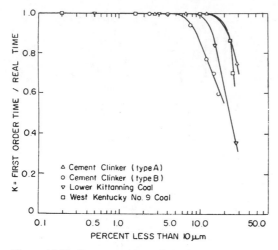

Figure 12.38. Representation of slowing down of rate of breakage with build-up of fines, K values, for dry grinding in a batch ball mill.

distribution which would have been obtained in 7 min if grinding had stayed first-order. Since the rate of energy input to the ball mill is almost constant, the slowing-down process produces greatly reduced grinding efficiency and leads to high specific grinding energy. A

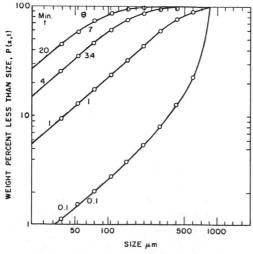

Figure 12.39. Comparison between computed and experimental size distributions of 20×30 mesh Lower Kittanning coal ground for different times in the ball mill. θ = first-order time, t = real time.

similar phenomenon is observed with wet grinding of slurries at high solids content. In very fine ball milling there may be changes in the primary progeny distribution as well.[26, 27]

12.4.5 Analysis of Smooth Roll Crushers

The unit operation of crushing does not usually give problems for brittle materials such as rocks and ores, so it has not yet received the amount of theoretical analysis given to fine grinding. However, the general concepts of the process engineering analysis of crushers can be illustrated by using smooth roll crushers as an example. There are five facets to the analysis: (1) the correct feed size, (2) the maximum force required, (3) the capacity in tons per hour, (4) the product size distribution, and (5) the maximum power required.

Gaudin[28] gives the relation between angle of nip Θ, coefficient of friction η, particle size x, gap x_g, and roll diameter d as

$$\left.\begin{aligned}
\tan(\Theta/2) &\leq \eta \\
\cos(\Theta/2) &= \frac{d + x_g}{d + x}
\end{aligned}\right\} \qquad (12.31)$$

See Figure 12.40. If x is too big the particle will not be nipped since Θ will give $\tan(\Theta/2) > \eta$. Austin et al.[29] have pointed out that there is little published information on the effective values of η between the crushed materials and the rolls as a function of material, roll speed, surface roughness, etc. When the gap is small compared to roll diameter, Eq. (12.31) gives

$$\frac{x}{d} \leq (1 + \eta^2)^{1/2} - 1 \qquad (12.32)$$

to assure nipping of feed size x. For a value of $\eta = 0.5$ this states that the maximum lump size should be less than $1/10$ of the roller diameter.

The maximum force tending to separate the rollers can be estimated by assuming the worst possible case, that is, simultaneous compression to failure of the maximum lumps of size x_m at all places along the rolls. Assuming that the lumps are small compared to the roll di-

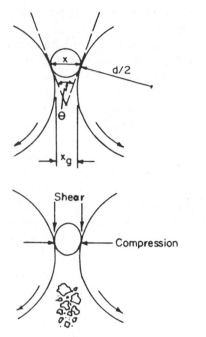

Figure 12.40. Illustration of nip angle in rolls and crushing-shear forces.

ameter, unit length of rolls contains $1/x_m$ large lumps each requiring the force P^* given by Eq. (12.23a). Thus,

$$\text{Maximum force} = x_m \sigma_c L/4 \quad (12.33)$$

where σ_c is the compressive resistance defined by $\sigma_c = P^*/(x_m/2)^2$, and L is the length of the rolls.

Gaudin[28] has described the concept of capacity calculation. Assuming that the feed is in sizes less than x_m the maximum capacity (at choke feeding) per unit length of rolls is given by the ribbon of solid which can be pulled through the rolls:

$$Q_{max} = u\rho(1 - \theta_g)x_g = u\rho(1 - \theta_c)x_c \quad (12.34)$$

where $1 - \theta_g$ is the volume fraction of solids at the gap, ρ the true solid density, and u the circumferential velocity of the rolls. Obviously $(1 - \theta_g)x_g = (1 - \theta_c)x_c$, where θ_c is the porosity of the feed picked up by the rolls at the critical angle of nip and x_c the value of x

at that location. Since the geometry of the system requires that $x_g < x_c$ it is clear that the material is consolidated as it passes towards the gap. To avoid reaching a highly consolidated bed that acts like a noncompressible solid, the feed rate is controlled (nonchoke feeding) to give a high porosity of the feed so that compression to the gap size gives porosities greater than about 0.3, thus $Q_{max} \approx 0.7\rho u x_g$. In practice, Q is even lower, as determined by experience. It depends on the forces required to consolidate the bed and typical values for coals[30] are 0.2 to 0.9 of Q_{max}, with lower values for strong materials and smaller rolls.

If rolls are operated at different speeds, the arithmetic mean is used for u. Different roll speeds are sometimes used to give an extra component of shear in addition to compression. If the feed contains lumps too big to be fully nipped, such lumps will build up in the mill inlet and will eventually abrade to a size which allows them to be nipped and pulled in completely. This, of course, reduces the feed rate. In circumstances where it is desirable to use too large lumps (to reduce the number of stages of crushing, for example), ribbed or toothed rolls are used to increase the ability of the rolls to pull in larger lumps, especially for relatively weak and friable materials.

A method of predicting the product size distribution obtained from any feed has been proposed by Austin et al.[29] They showed that the values of the transfer numbers d_{ij} depend on the ratio of the particle size x to the gap size x_g, providing the roll diameter is much bigger than x and x_g. Thus, values of d_{ij} measured experimentally for one gap setting can be converted to any other gap setting. Then Eq. (12.26) is used to calculate the product size distribution. However, this method requires the experimental measurement of the total d_{ij} matrix, which is very time consuming.

To reduce the description of the d_{ij} matrix to a few parameters, they assumed that the breakage processes could be described as shown in Figure 12.41. First, feed particles of size j have a probability s_j of breaking and a

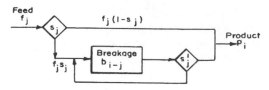

Figure 12.41. Equivalent circuit for a once-through roll crusher with multiple fracture actions.

probability $1 - s_j$ of falling through the rolls without fracture: clearly, the fraction of larger sizes that fall through the roll gap without breakage is zero, $s_j = 1$, and sizes much smaller than the gap do not break, $s_j = 0$. Experimental values are shown in Figure 12.42 and were found to fit the empirical relation

$$1 - s_j = \frac{1}{1 + \left(\dfrac{x_j/x_g}{d_{50}/x_g}\right)^{6.6}} \quad (12.35)$$

where d_{50}/x_g is characteristic of the material. Second, it is assumed that all sizes break into a normalized primary progeny fragment distribution b_{i-j}, where b_1 is the weight fraction of breakage products of one size that appears in the next lower size, b_2 appears in the size below that, etc. Third, it is assumed that a fragment of size i produced by fracture in the rolls has in turn a probability s_i' of being rebroken or $1 - s_i'$ of passing through the roll gap. Since this material results from fracture it is already in a favorable position to be nipped,

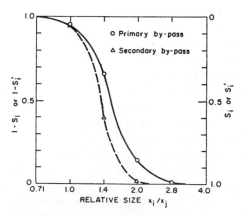

Figure 12.42. Measured primary bypass (fraction unbroken) and estimated secondary bypass for feeds of $\sqrt{2}$ screen intervals of Lower Freeport coal.

so it is expected that s_i' will be greater than or equal to s_i.

Considering the repeated fracture of size 1 material, $1 - s_1$ falls through the rolls to size 1 product, and s_1 breaks. The material resulting from the breakage follows two routes—material that passes the gap to give product and material retained to fracture again. Let $a_{ij} = b_{ij}(1 - s_i')$, which has the physical meaning "when size j breaks, a_{ij} is the fraction sent to product size i." Let $c_{ij} = b_{i,j}s_i'$ which is "when size j breaks, c_{ij} is the fraction sent to size i for a further breakage." Then, the broken quantity s_1 distributes itself as $a_{2,1}$ in size 2, $a_{3,1} + c_{2,1}a_{3,2}$ in size 3,

$$a_{4,1} + c_{2,1}a_{4,2} + c_{3,1}a_{4,3} + c_{2,1}c_{3,2}a_{4,3}$$

in size 4, and so on; $a_{4,1}$ is the product from breakage of size 1 to size 4, $c_{2,1}a_{4,2}$ is 1 breaking to 2 breaking to 4, $c_{3,1}a_{4,3}$ is 1 breaking to 3 breaking to 4, $c_{2,1}c_{3,2}a_{4,3}$ is 1 breaking to 2 breaking to 3 breaking to 4, and so on. Thus,

$$\begin{aligned}
d_{1,1} &= 1 - s_1 \\
d_{2,1} &= s_1 a_{2,1} \\
d_{3,1} &= s_1(a_{3,1} + c_{2,1}a_{3,2}) \quad (12.36) \\
d_{4,1} &= s_1(a_{4,1} + c_{2,1}a_{4,2} \\
&\quad + c_{3,1}a_{4,3} + c_{2,1}c_{3,2}a_{4,2})
\end{aligned}$$

etc., until c values become zero. The equation is readily converted to d_{ij} replacing 1 with j and 2 with $j + 1$, etc. Then the total size distribution from a feed of f_i is obtained from Eq. (12.26).

Austin et al.[29,31] treated the above problem somewhat differently by developing the mass balance equations for the equivalent circuit of Figure 12.41 as if s_i and s_i' were due to external classifiers and they developed a method for calculating b_{ij} values from the test data. They found that the values of b_{i-j} in the cumulative form fitted the empirical function (see Fig. 12.31):

$$B_{i-j} = \Phi\left(\frac{x_{i-1}}{x_j}\right)^{\gamma} + (1 - \Phi)\left(\frac{x_{i-1}}{x_j}\right)^{\beta} \quad (12.37)$$

where Φ, γ, β are characteristic parameters for the material, as shown in Table 12.1.

Table 12.1. Characteristic Breakage Parameters Determined from Smooth Roll Crusher Tests[31]

MATERIAL	Φ	γ	β	d_{50}/x_g
Rhyolite	0.29	0.83	3.6	1.45
Diabase	0.40	0.84	4.0	1.40
Coals				
Shamokin anthracite, PA	0.30	1.05	5.0	1.70
Illinois #6	0.36	0.81	3.0	1.66
Ohio #9	0.33	0.95	4.2	1.93
Western Kentucky #9	0.47	1.05	4.0	1.81
Belle Ayre, Wyoming	0.49	1.17	4.0	1.70
Pittsburgh E. Seam, PA.	0.32	0.81	3.0	1.66
Upper Freeport, PA.	0.39	0.96	4.0	1.56
Lower Freeport, PA.	0.50	1.05	4.5	1.54

By a trial-and-error matching of computed size distributions with experimental values they determined that s_i' could be estimated from s_i values by

$$s_i' = \begin{cases} s_{i-1} & i < i_g - 1 \\ (s_{i_g-1} + s_{i_g-2})/2 & i = i_g - 1 \\ s_i & i \geq i_g \end{cases} \quad (12.38)$$

where i_g is the interval number corresponding to the gap setting. Thus, a simulation model is constructed for smooth roll crushers which has the experimentally determined material characteristics of Φ, γ, β, and d_{50}/x_g. Figure 12.43 shows a typical match of computed versus experimental results.

They also simulated the effect of passing the product through a screen and recycling above-size material to the roll feed. The minimum production of fines was obtained when the gap and the screen were of the same size (see Figure 12.44) even though the circulating load was relatively small. Larger gap settings and the associated high circulating load produced very little change in the final product. This is because fine material is not acted on by the crusher as it passes through, so a high circulating load is no advantage.

The process of fracturing unit volume of feed to less than the gap setting requires stressing the original broken volume (s_1) of

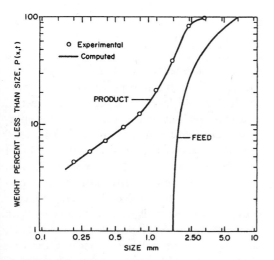

Figure 12.43. Crusher product size distribution from 3 × 12 mesh feed.

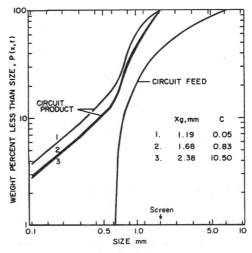

Figure 12.44. Simulated circuit product size distribution for 3 × 30 mesh Illinois #6 coal as a function of gap setting at ideal screening of 12 mesh.

particle size 1, stressing again the fraction of this volume that undergoes a second fracture, stressing again the fragments of these fragments that undergo a third fracture, and so on. The total stressed volume is readily calculated as s_1 plus the sum of all c terms, that is,

$$s_1(1 + c_{2,1} + c_{3,1} + c_{4,1}$$
$$+ \cdots + c_{2,1}c_{3,2} + c_{2,1}c_{4,2}$$
$$+ \cdots + c_{3,1}c_{4,3} + \cdots + c_{2,1}c_{3,2}c_{4,3} + \cdots).$$

If it is assumed that the strain energy per unit stressed volume required to produce fracture is a constant, which is known[32] as Kick's "law," the total stressed volume is proportional to the ideal specific energy required to grind size 1 to less than the gap setting. Defining a reduction ratio by x_1/x_g, Figure 12.45 shows the relation of the volume of repeated crushing to reduction ratio. In practice, it is usually found that a larger reduction ratio requires a bigger increase of specific energy than that predicted by Figure 12.45 because smaller lumps become relatively stronger (require higher stress to cause breakage).

If the crusher is run nearer to choke feeding then breakage owing to bed compression becomes an additional factor. As we will see later, fracture by bed compression in place of steel–particle–steel nipping fracture tends to produce size distributions with proportionally more fines than expected; also additional energy is used in the bed compression.

The capacity and product size distributions of other crushers can be analyzed in a similar fashion.[33-36] For example, a jaw crusher acts on a maximum solid volume rate of $A(1 - \theta_c)u$, where A is the throat area, θ_c is the feed porosity, and the velocity of flow u is determined by the fall of solid under gravity as the jaw opens. There is repeated breakage and fall as the material moves down the crusher until it passes the gap which is a mean of the open and closed side settings. The analysis is similar for gyratory crushers, although the rotational motion can aid the rate of material moving down.

12.4.6 Analysis of Tumbling Ball Milling

12.4.6.1 Influence of Mill Conditions

The tumbling ball mill is the most widely used device for fine grinding of brittle materials on an industrial scale. Because of its simplicity, it is mechanically reliable, which is very important in continuous process streams, and it is available in sizes ranging from small laboratory mills to industrial mills of 5 m diameter by 10 m long, or even larger. It is a retention device, where a bed of powder is acted upon by the tumbling balls and the mean residence time of solid in the bed is typically a few minutes to 30 min depending on the desired degree of size reduction. It has certain disadvantages. First, the mill power is almost independent of the level of filling by the powder, so a mill operated at lower than design capacity is inefficient because (1) if the powder level is held at a normal level, a low solid feed rate gives a long residence time ($\tau = W/F$), and the energy is used to grind finer than necessary and (2) if the level is dropped to keep τ constant, the energy is used to tumble balls on balls without enough powder between them, also giving excess ball wear. Second, the cost of replacing steel balls as they wear is substan-

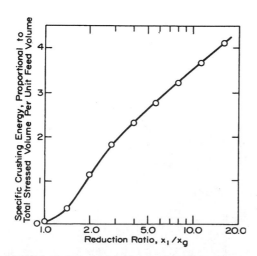

Figure 12.45. The total crushed volume per unit feed volume for roll crushing of a coal (Upper Freeport) through a smooth roll crusher, as a function of the particle size to gap size ratio.

tial, and the steel or rubber lining of the mill has also to be replaced every 2 or 3 years. Third, the "slowing-down" process comes into play for very fine dry grinding or fine wet grinding of viscous pulps. Thus, grinding can become inefficient and consume high energy.

The major fraction of the direct power required to turn the mill (excluding motor and drive losses) is used in the act of raising the balls. On the other hand, the more balls raised per unit time, the higher the rates of breakage of powder in the mill because the tumbling of the raised balls gives the breakage action. Thus, the variation of power input with mill conditions is likely to be a direct index of the best breakage conditions. Figure 12.46 shows typical power variation with rotational speed and with ball filling. The critical rotational speed is defined as the speed where balls on the case would start to centrifuge and is readily shown to be

$$\text{critical speed} = \frac{76.6}{\sqrt{D - d}, \text{ in feet}}$$

$$= \frac{42.2}{\sqrt{D - d}, \text{ in meters}} \text{ rpm}$$

$$(12.39)$$

where D is mill diameter, and d is ball diameter. Figure 12.46 shows that the power passes through a maximum at about 80% of critical speed. This varies somewhat with mill diameter and ball load because the force of a heavy ball charge acting on the case tends to prevent slip between the balls and the case (thus aiding the raising of balls) for larger mills. For large mills with steel balls the rotational speed is usually in the range 65% to 75% of critical speed to avoid cataracting of balls onto the mill case, which can damage the mill lining. The figure also shows that the maximum power is obtained at about 45% filling of the mill volume by the ball bed at rest (calculated assuming the ball bed has a porosity of 0.4), $J = (M/\rho_b V)(1/0.6)$, where V is mill volume, M mass of balls, ρ_b true density of ball material, and J is the fraction of mill volume filled by the bed of balls. Continuous overflow mills,

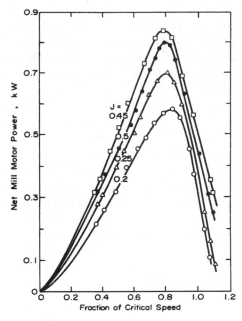

Figure 12.46. Variation of net mill motor power with critical speed as a function of ball loading: 2-ft diameter laboratory mill.

however, are normally run at $J < 0.4$ to prevent balls blocking the overflow or the feed entry.

The specific rates of breakage can be determined in laboratory or pilot-scale mills by batch tests with controlled powder and ball filling, and controlled pulp density if wet. Figure 12.47 shows a typical result for the variation of S_i with particle size. The rates of breakage are low for sizes that are relatively large with respect to the ball diameter because (1) the particles are so big that the force required to break them is achieved only by relatively few of the tumbles and (2) the particles are too big to be nipped by a ball–ball collision [see Eq. (12.31)]. Small sizes also break slowly because (1) their basic strength is higher due to removal of large flaws and (2) the mass of particles captured in a ball–ball collision becomes smaller and smaller as particle size decreases with respect to ball size. Large ball diameters are better for breaking large particles but small balls are better for breaking small particles because there are many more ball–ball collisions for a given mass of small balls than for the same mass of large balls. This means that there is an optimum mixture of ball sizes in the mill to go from any feed size distribution to any ball mill product.

The slope α shown in Fig. 12.47 is characteristic of the material. It is also found that the primary progeny distributions in the first-order breakage region, which occurs to the left of the maxima in the curves, can be fitted by Eq. (12.37), and the values of Φ, γ, and β are also characteristic of the material. Examples are given in Ref. 18. Especially, a material with a small value of γ will produce proportionately more fines on grinding.

12.4.6.2 Major Variables

The major variables involved in ball milling, in addition to these material characteristics are:

- the ball loading in the mill
- the distribution of ball sizes in the mill and the ball density (the balls must have a hard surface)
- the load of powder or suspended solid in the mill
- the rotational speed of the mill, as a fraction of critical speed, and the lifting action of mill lifters built into the mill lining
- the slurry density and viscosity in wet milling
- the dispersing action of chemicals used as grinding additives

plus, of course, the diameter and length of the mill. In addition, the degree of recycle and the efficiency of size classification or air (gas) sweeping to remove fines are also important factors to prevent overgrinding or the development of slowing-down effects. For example, tests show that a ball mill that is underfilled with solid is inefficient because the breakage zones where balls collide with balls or the case are not filled and energy is wasted by steel-on-steel collisions. On the other hand, overfilling by powder or slurry is also found to be

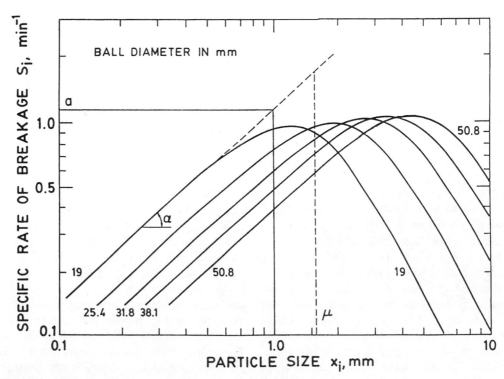

Figure 12.47. Predicted variation of S_i values with particle size for different ball diameters: copper ore ($\sqrt{2}$ intervals).

inefficient because it appears to cushion the breakage action. A general rule-of-thumb is that the solid should just occupy the interstices of the ball bed calculated with the bed at rest. Inefficiencies of this type, or the use of a mismatch of ball sizes to the particle size, or the use of too dilute or too concentrated slurry, etc., are examples of *direct inefficiency* as distinct from the indirect inefficiency of overgrinding. The reader is referred to Ref. 18 for more detailed discussions of the effects of the major variables in ball milling. Since this information is fairly up-to-date, it will not be repeated here. More recent work includes extended treatments of the optimization of the distribution of ball sizes in the mill,[37] the influence of slurry density in wet ball milling,[38,39] the mass transport of slurry through a ball mill[40,41] and predictive equations of mill power.[42,43]

Models for autogeneous and semiautogeneous grinding mills are not so well developed although the basic principles are very similar to those for other tumbling media mills. Recent work on constructing these models[44-46] includes the kinetics of chipping of large rock to form smaller pebbles, self-fracture of rock by its own tumbling action, mass transport through discharge grates, and mill power equations.

12.4.7 Analysis of Roller-Race Mills

The type of mill exemplified in Figure 12.25 is the second most important type of mill (after tumbling media mills) from the aspect of the tonnage of material ground annually. A recent analysis[47] has given a detailed account of the powder technology associated with this type of mill and the analysis is summarized here. The rotation of the table (race) brings material under the rollers, which ride up and rotate as the material passes underneath. Since the rollers are heavy and are loaded by massive springs, there is a vertical force acting down on the roller that generally depends on how high the roller is forced against the springs. Let this force per roller be denoted by Φ. The force is scaled with respect to mill size by

expressing it as a formal "grinding pressure" P defined by

$$P = \frac{\Phi}{Ld} \qquad (12.40)$$

where L is the length of the roller and d its diameter. Figure 12.48 illustrates the geometries involved, which shows that the system is like a choke-fed single roll crusher operating against a flat plate. It can be treated in a somewhat similar fashion to the double roll crusher system, except that the gap x_g is not set but is a natural consequence of the material pulled under the roller, that is, it is a *floating* roller.

Immediately, the mass flow under the roller is given by Eq. (12.34) modified to include roller length

$$Q = \rho u L (1 - \theta_g) = \rho u L (1 - \theta_c) x_c \qquad (12.41)$$

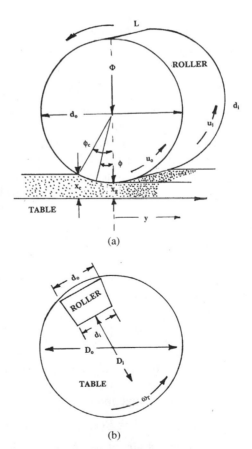

Figure 12.48. Illustration of roller geometry and notation for the analysis of roller-race mills (race is called table).

where θ_c and x_c are defined where slip ceases and material is pulled in without further slip and moves at the horizontal table velocity u. If there are no large lumps in the feed (to avoid chatter of the floating roller), the material is pulled in as a bed and crushed by compression of the bed. Until the bed is nipped for crushing there is very little work done on the material. The vertical compression pressure is essentially zero at the critical angle of nip ϕ_c, but it increases as the material moves toward the gap and reaches a maximum at the gap where the degree of compression is highest, $\phi = 0$. Let the resolved vertical pressure at ϕ be denoted by $P(\phi)$. Since the critical angle of nip for bed crushing is less than $12°$, $\sin \phi \approx \phi$ and $\cos \phi \approx 1.0$ and the total vertical force is

$$\Phi = \left(\frac{Ld}{2}\right) \int_0^{\phi_c} P(\phi)\, d\phi \quad (12.42)$$

From the definition of formal grinding pressure

$$P = (\tfrac{1}{2}) \int_0^{\phi_c} P(\phi)\, d\phi \quad (12.43)$$

Consider a thin vertical column of powder nipped at ϕ_c and moving at velocity u toward the gap. Define a linear strain ϵ by the fractional change in vertical dimension $\epsilon = (x_c - x)/x_c$. The relation between strain ϵ and porosity θ_c between ϕ_c and $\phi = 0$, with $\epsilon = 0$ at ϕ_c, is

$$\epsilon_g = \frac{\theta_c - \theta_g}{1 - \theta_g} \quad (12.44)$$

A simple geometric construction gives the relation between ϵ and ϕ as

$$\frac{\epsilon}{\epsilon_g} = 1 - \left(\frac{\phi}{\phi_c}\right)^2 \quad (12.45)$$

Let the relation between the vertical pressure $P(\phi)$ and the linear strain ϵ be the (unknown) function "stress = function of strain," that is,

$$P(\phi) = P(\epsilon) \quad (12.46)$$

where ϵ is the strain at ϕ. A hypothetical maximum strain ϵ_{max} is defined when the bed

is compressed to zero porosity, $\epsilon_{max} = \theta_c$ from Eq. (12.44). Then Eq. (12.43) becomes

$$P = \left(\frac{\phi_c}{4\epsilon_g}\right) \int_0^{\epsilon_g} \frac{P(\epsilon)\, d\epsilon}{\sqrt{1 - \epsilon/\epsilon_g}} \quad (12.47)$$

using Eq. (12.45) and its differentiation (ϕ_c in radians). Thus, the strain at the gap under a grinding pressure P is determined by the function $P(\epsilon)$ and the critical angle of nip,

$$P = \left(\frac{\phi_c}{4}\right) I_1(\epsilon_g) \quad (12.47a)$$

where I_1 is the integral of Eq. (12.47), which increases as ϵ_g increases.

Now consider the work done as the column of powder is compressed. By integrating force times the distance the force moves, from ϕ_c to $\phi = 0$, it is readily shown that

$$m_p = \left(\frac{uLd}{2}\right) \int_0^{\phi_c} \phi P(\phi)\, d\phi \quad (12.48)$$

where m_p is the net mill power per roller. With the same substitutions as before,

$$m_p = \left(\frac{uL\, d\phi_c^2}{4\epsilon_g}\right) \int_0^{\epsilon_g} P(\epsilon)\, d\epsilon \quad (12.49)$$

or

$$m_p = \frac{uL\, d\phi_c^2}{4} I_2(\epsilon_g) \quad (12.49a)$$

Thus, mill power is proportional to the velocity of the roller, its length and diameter, and the number of rollers. The compression characteristics of the material being ground enter via the critical angle of bed nip, the feed porosity, and the function of Eq. (12.46). The effect of grinding pressure also depends on the bed compression characteristics, as can be seen by substituting Eq. (12.47a) into Eq. (12.49a),

$$m_p = uL\, d\phi_c P I_2(\epsilon_g)/I_1(\epsilon_g) \quad (12.50)$$

It is convenient to put Eq. (12.50) in the form

$$m_p = \overline{\phi} uL\, dP \quad (12.51)$$

where $\overline{\phi}$ is a dimensionless factor called the *specific power* factor (per roller). Since a bed becomes more difficult to compress further once it is partially compressed, the value of $\bar{\phi}$ generally decreases as grinding pressure is increased. The factor is constant only at sufficiently low grinding pressures where the relation of Eq. (12.46) is linear, $P(\epsilon) = C\epsilon$, since then the integrals become $I_1 = (\frac{4}{3})C\epsilon_g$ and $I_2 = (\frac{1}{2})C\epsilon_g$ and $\bar{\phi} = (\frac{3}{8})\phi_c$. For a typical angle ϕ_c of $12° = 0.21$ radians, $\bar{\phi} = 0.079$ (per roller).

It must be realized that the formal grinding pressure defined by $P = \text{force}/Ld$ is much smaller than the actual maximum pressure at the gap. The average pressure over the region ϕ_c to 0 is $2P/\phi_c$, but this is the integral of a sharply rising stress–strain curve, so much higher stress exists at the gap. Bed compaction involves fracture of particles, with small product fragments fitting into the interstices of larger particles.

The flow rate under the roller will also generally decrease as grinding pressure is increased because x_c becomes smaller as x_g becomes smaller. Again, it is convenient to express the flow equation, Eq. (12.41), in the form:

$$Q = \rho u L d (1 - \theta_g)(x_g/d) = \dot{m}\rho u L d \quad (12.52)$$

where $\dot{m}$ is a dimensionless factor called the *specific capacity* factor. It is readily shown from the definition of strain and Eq. (12.44) that

$$\dot{m} = \left(\frac{1 - \epsilon_{\max}}{\epsilon_g}\right)\left(\frac{1 - \cos\phi_c}{2}\right) \quad (12.53)$$

Since the strain at the gap increases as grinding pressure increases via the relation of Eq. (12.47), the value of $\dot{m}$ decreases. It should be noted that the roller-race cannot be operated at very low pressure because there would be insufficient downward force to prevent slip between the roller and the powder bed, and the table would not transfer rotational velocity to the roller.

The specific grinding energy E for material flowing under the roller is m_p/Q and

$$E = \frac{\bar{\phi}P}{\dot{m}\rho} \quad (12.54)$$

Since both $\bar{\phi}$ and $\dot{m}$ decrease as grinding pressure is increased, the value of E can be approximately proportional to P over a limited range of P.

It must be understood that Q is the rate of material being crushed per roller, not the flow rate in and out of the mill. The centrifugal action of the table is constantly throwing material out of the race, where it is swept up in a high-velocity air stream. Larger particles fall back into the race as the gas velocity decreases above the annulus, and larger particles (and some fines) are returned to the race from the built-on classifier at the top of the mill. Thus the mill can be considered as a fully mixed retention mill where there is breakage action under the rollers and a reservoir of powder not under the rollers. Let p_i be the product size distribution out of this reservoir of weight W, f_i the size distribution of feed into the reservoir, and w_i the size distribution within the reservoir. A mass breakage rate balance on material entering and leaving the breakage zones and the reservoir gives

$$Fp_i = Ff_i - F_N(1 - d_{i,i})w_i + F_N \sum_{\substack{j=1 \\ i>1}}^{i-1} d_{i,j}w_j \quad (12.55)$$

where F is the feed rate in and out of the race and F_N is the rate in and out of N rollers in the race.

For the fully mixed assumption $w_i = p_i$, and rearranging Eq. (12.55) using $\tau = W/F$ gives

$$p_i = \frac{f_i + \tau \sum_{j=1}^{i-1} \bar{b}_{i,j}(F_N/W)(1 - d_{j,j})p_j}{1 + \tau(F_N/W)(1 - d_{i,i})} \quad (12.56)$$

where the apparent primary breakage distribution $\bar{b}_{i,j}$ is defined by the breakage products in one pass under the roller,

$$\bar{b}_{i,j} = \frac{d_{i,j}}{1 - d_{j,j}} \quad (12.56a)$$

Comparing with the usual equation, Eq. (12.29), it is seen that the specific rate of breakage is given by

$$S_i = (1 - d_{i,i})(F_N/W) \quad (12.57)$$

and

$$p_i = \frac{f_i + \tau \sum_{j=1}^{i-1} \bar{b}_{i,j} S_j p_j}{1 + \tau S_i} \qquad (12.58)$$

where F_N is given by NQ, Q being the flow rate under each roller at choke feeding, Eqs. (12.52) and (12.53). Thus, it is seen that the roller-race mill is one mill where it is possible to describe S_i and $b_{i,j}$ values in terms of a precisely known breakage zone.

The values of S_i and $\bar{b}_{i,j}$ have been determined in a laboratory scale roller-race mill and the equations enable these data to be scaled for pilot-scale and full-scale simulations, as follows. To start, because it is not easy to determine the mass of the reservoir of powder in an operating mill, it is convenient to replace S_i values with the *absolute rate of breakage* A_i defined by $A_i = S_i W$. A_i has the dimension mass/time, for example, kg/s, and is physically the instantaneous rate of breakage of size i (under specified conditions) if all of W were of size i. Equation (12.58) then becomes

$$p_i = \frac{f_i + (1/F)\sum_{j=1}^{i-1} \bar{b}_{i,j} A_j p_j}{1 + A_i/F} \qquad (12.58a)$$

Then, from Eqs. (12.52), (12.53), and (12.57)

$$A_i = NQ(1 - d_{i,i}) \qquad (12.59)$$

or

$$A_i = N\left(\frac{1 - \epsilon_{max}}{\epsilon_g}\right)\left(\frac{1 - \cos \phi_c}{2}\right)$$

$$\times (1 - d_{i,i})(\rho u L d) \qquad (12.59a)$$

At the same grinding pressure in the laboratory mill as in the full-scale mill it can be assumed that the bed compression ϵ_g, the hypothetical maximum strain ϵ_{max}, the critical angle of nip ϕ_c and the degree of breakage in one pass under the roller, $1 - d_{i,i}$, are the same since these values depend on grinding pressure, not the size of the roller. Thus A_i values are scaled by

$$A_i = A_{iT}\left(\frac{uLdN}{u_T L_T d_T N_T}\right) \qquad (12.59b)$$

where the suffix T refers to the laboratory test conditions. Equation (12.58a) combined with

classification and recycle in the usual manner[18] can then be solved for any value of F, and the value of F adjusted to give the desired final product size. If the total circulation ratio is C, the actual kg/s of final product, Q say, is given in the usual way by:

$$Q = F/(1 + C) \qquad (12.60)$$

where C is defined as the ratio of mass flow of final product to material returned to the race from internal or external size classification. Tests on a limited number of U.S.A. coals gave the following empirical relations for A_{iT} and $\bar{b}_{i,j}$ as a function of the Hardgrove Grindability Index and the grinding pressure:

$$A_{0T} = 0.172\left(1 + 1.08\frac{HGI}{100}\right)P \qquad (12.61)$$

where P is expressed in MPa and A_{0T} is the absolute rate of breakage of 18×25 mesh $(1 \times 0.841$ mm) coal in kg/min:

$$A_{iT} = A_{0T}\left(\frac{x_i}{x_0}\right)^\alpha \qquad (12.62)$$

where x_0 is the standard size of 1 mm and α is the material characteristic given by

$$\alpha = 0.58 - (2.4)(10^{-3})HGI \qquad (12.63)$$

and the characteristic breakage distribution parameters of Eq. (12.37) are given by

$$\left.\begin{array}{l} \beta = 5 \\ \gamma = 1.23 - (2.32)(10^{-3})HGI \\ \Phi = 0.58 + (2.6)(10^{-3})HGI \end{array}\right\} \qquad (12.64)$$

The values were determined for test conditions of $N_T = 2$, $u_T = 0.0565$ m/s, $d_T = 0.060$ m, $L_T = 0.016$ m, and a sufficient depth of bed to ensure choke-feeding to the two rollers.

Some comments can be made. First, the specific rates of breakage for coal ground in the laboratory mill ($d_T = 0.060$ m) are shown in Figure 12.49. It is seen that the simple power function of Equation (12.62) does not apply to larger sizes, where x_i/d is greater than about 1/25. The increased breakage rates above this size are due to the greater ability of a roller to nip single particles than to nip a bed of fine feed. The decrease at even larger sizes is due to the inability of the rollers to nip

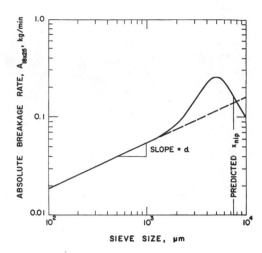

Figure 12.49. Absolute breakage rate of 18×25 mesh Elkhorn coal as a function of particle size.

particles that are larger still. However, feeds containing particles too large in reference to the roller diameter are avoided in practice because they give rise to chattering of the rollers. Second, the linear increase in specific breakage rates with increasing grinding pressure cannot be extrapolated to high grinding pressures because the coals (especially soft coals) will cake onto the rollers and cause slip, which leads to loss of energy as frictional heat instead of causing breakage. Third, the fraction of particles of a given size that do not break in one pass under the rollers are reincorporated into a new bed fed into the next pass and can break at the same specific breakage rate, thus preserving the first-order nature of the breakage kinetics. Every reapplication of grinding pressure will cause further breakage. A typical result is that the feed to the mill is rolled over about 10 times before it leaves the classifier as final product.

Fourth, in practice the rollers in an industrial mill are generally loaded with massive springs initially compressed to a preload, and any material passing under the roller is subjected to this minimum grinding pressure, P_0 say, plus the weight of the roller, M say. However, as the bed is pulled under the roller

the rise of the roller forces against the spring and the grinding pressure increases

$$P = P_0 + kx_g + Mg \qquad (12.65)$$

where k is the spring constant of the precompressed spring and g is the gravitational constant. For example, an industrial mill with steel rollers of 1.22 m (48 in.) diameter and 0.43 m (17 in.) length subjected to a preload per roller of 1.8×10^5 Newtons (40,000 lbf) will have a minimum grinding pressure of about 0.42 MPa. However, such a roller is expected to rise about 38 mm ($\approx$ 1.5 in.), and with a spring constant of 0.72×10^4 Newtons/mm (40,000 lbf/in.), this will give an extra grinding pressure of about 0.48 MPa, that is, the total grinding pressure per roller is about 0.9 MPa. Equation (12.58a) shows that a lower mill capacity F gives a finer product size distribution. However, the equation is valid only with almost constant A_i values as long as the reservoir W in the mill is sufficient to choke-feed the rollers. If the feed rate is made too small, the value of W will fall below this level as the rotating race throws material out, Q in Eq. (12.59) will change to a lower value and F and A_i each change by the same factor. Then the product size distribution will *not* get finer and, in fact, the smaller raise of the rollers will reduce the grinding pressure, cause less breakage and the product size distribution may get coarser, as demonstrated by Austin et al.[48-50] The mill power will fall as the rollers are underfed and, to get fine product, it is necessary to have a race designed to retain powder, plus efficient classification to give a high rate of recycle to the bed. Finally, the empirical equations for A_{0T}, α, β, γ, and Φ are based on limited data and it is advisable for values to be determined directly for any coal or other material under study.

12.5 NEW MILLS

12.5.1 High-Pressure Grinding Rolls Mill

New designs of mills are constantly being patented and constructed in small-scale versions, but most are variants on existing mill designs and operate with the same fundamental principles. However, there are several new mill designs that result in large part from the investigations of Professor Klaus Schönert in Germany. By studying the breakage of powder beds by compression in a piston-cylinder system, he showed[51] that the specific energy of size reduction was significantly less (20% to 30%) than that for tumbling ball mills, that is, the use of energy to cause breakage was more efficient in this type of system. This is because the particles are all subject to the stressing action and less energy is wasted (1) by impacts that are not sufficient to cause fracture, (2) by steel–steel collisions that do not trap particles for breakage, and (3) due to frictional losses from powder and media movement in unconfined systems. To apply this principle in practice, he invented a mill that is essentially a double roll crusher that has one roll free to move against a large applied force and that is choke-fed from a hopper above the gap. This type of mill, called *a high-pressure grinding rolls (HPGR)* mill, has been developed commercially by the Krupp–Polysius Company and (under license from Schönert and Krupp–Polysius) by the KHD–Humboldt Wedag Company, both of Germany, and by others (also under license). Figure 12.50 shows the principle:[52] both rolls are driven by electric motors connected by special couplings that permit the free roll to move in its containing tracks and the force is applied by an hydraulic pressure system which allows for easy control of the grinding pressure. The very high stresses at the gap require that the rolls be of strong and hard material to avoid surface cracking and reduce abrasion and the rolls must be thick enough to withstand the strain.

The mill has been very successful in the cement industry as a pregrinder to conventional long (tube) tumbling ball mills. When grinding cement clinker at formal grinding pressures of 2 to 6 MPa, the resulting compressed powder passes through the gap as a coherent strip that can then be deagglomerated with a hammer mill or in a following ball mill. Much harder materials do not briquette,

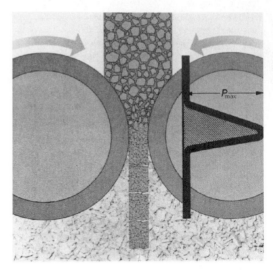

Figure 12.50. Principle of the high pressure grinding rolls (HPGR) mill (KHD–Humboldt Wedag).

while compacted softer materials such as coal tend to stick to the rolls and have to be removed with scalping blades. The mill has also been used for grinding diamond-bearing rock to liberate the diamonds since there is less breakage of the strong diamonds and more preferential fracture along a diamond–rock interface.[53]

Although very different in appearance, the basic action is very similar to that of the roller-race mill discussed previously. The feed material pulled in the rolls is nipped with a critical angle of nip, compressed (which causes breakage) to a maximum high pressure at the gap, and the gap automatically adjusts to pass the compressed cake. As we have already noted, the roller-race mill is more efficient than many other types of mill, and so is the HPGR mill: confined compression of beds of particles is generally more efficient than other grinding methods. The major differences between the roller-race mill and the HPGR mill are (1) that the critical angle of nip for the HPGR mill (two rolls of equal diameter) is half that for the roller-race (flat surface) mill, (2) the grinding pressures used in the HPGR mill are several times higher, and (3) the HPGR mill always has two rolls. Austin[47] has shown that allowing for these differences gives descriptive equations identical to those for the roller-race mill and that it is possible to take data from a laboratory HPGR mill and predict the mill power and capacity of roller-race mills. Usually a roller-race mill performs many repeated compressions, each at relatively low pressure, while the HPGR mill performs one compression at high pressure. The specific energy of grinding for a given duty is probably very similar whichever method is used. This is especially true when the HPGR mill is operated in closed circuit with a classifier.

Nearly all the comments made about roller-race mills also apply to HPGR mills. For example, both mills will chatter if feed sizes are too large for the roller or rolls diameter. Both machines give flat product size distributions with a relatively high proportion of fines because fine material produced in the initial compression is further broken as the material is pulled into the gap. Overgrinding of fines will be less in the roller-race mill if a high recirculation from an efficient classification is used, because of the lower pressures. The simpler mechanical design of the HPGR mill makes it easier to scale to high capacities and KHD offers sizes up to 1.7 m roll diameter and 2 m roll length with capacities up to 500 tons/h.[53] There are two major disadvantages to the mills. Unlike tumbling media mills where wear of steel media is readily compensated by frequent addition of fresh media without stopping the mill, wear on rollers, races and rolls has to be corrected by dismantling the equipment when wear has progressed too far. The Krupp–Polysius Company designs mills where this disassembly and roller replacement or resurfacing with segment sections can be done rapidly. KHD–Humboldt Wedag[54] have rollers fitted with hardened studs that hold compressed cake on the rolls, thus giving autogeneous surface protection (see Figure 12.51). Another disadvantage is that feeds containing a high proportion of fines give rise to erratic mill operation and, hence, machine vibration. This has been suggested to be caused by fluidization of the bed as it passes toward the

Figure 12.51. Studded roller surface for autogenous wear protection (KHD–Humboldt Wedag).

gap, owing to compression of the air contained in the bed which is escaping upward. It has also been suggested[47] that the fines impact a fluid-like property to the dry bed so that instead of moving into the rolls (or under a roller) as a locked bed, the bed can shear and collapse. Under these circumstances the presence of some moisture may be beneficial by providing capillary forces between particles, but the water content must be low enough to allow free bed compression.

12.5.2 The Horizontal Roller Mill (Horomill®)

Figure 12.52 shows another form of bed compression mill, recently introduced by the FCB company of France. The mill is specifically designed for dry grinding of cement clinker and consists of a horizontal stationary roller that rotates on its axis, inside of a horizontal mill cylinder that is driven to rotate on the mill axis. The grinding portion of the roller presses with a controlled force against a grinding track on this inside of the mill cylinder.[55] Like roller-race mills, the mill is a retention mill where the level of material in the mill is controlled to give choke feeding of the gap between the roller and the track; the roller

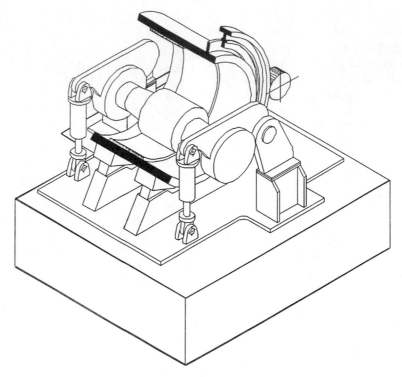

Figure 12.52. The Horomill® (FCB Groupe Fives–Lille).

will float to pass the material pulled into the gap, and the product passing under the roller will mix into the reservoir of material and be reground by repeated passes under the roller. Dry powder flows out of the mill and is lifted in a bucket elevator to a high-efficiency air classifier, with return of coarse material to the mill feed.

The grinding pressure is quoted as "moderate" and the mill is not air-swept like a conventional roller-race mill. The comments made on roller-race mills and high-pressure grinding rolls apply also to this mill and the mills will probably give similar specific grinding energies, although the power used for classification is probably higher for air-swept roller-race mills. It is easier to ensure choke-feeding in the Horomill® and in the HPGR mill as compared to roller-race mills where the rotating table both drives the rollers and throws material into the air stream, but the deagglomeration and rapid removal of fines is

an advantage for roller-race mills when used on softer materials such as coals which tend to form strong compacts under high pressure.

12.5.3 The Szego Mill

The original concept is due to the late L. L. Szego and the mill has been developed in Toronto, Ontario by *General Comminution, Inc.*, in close collaboration with University of Toronto researchers in the Department of Chemical Engineering. As a result, while industrial utilization of the mills is still modest, there is a great deal of published material available. The mill is a planetary ring-roller mill, consisting principally of a stationary grinding cylinder inside which a number of helically grooved rollers rotate, being flexibly suspended between flanges connected to a central drive shaft (see Fig. 12.53).

The material is fed by gravity, or pumped into a top feed cylinder if wet, and is discharged continuously at the bottom of the

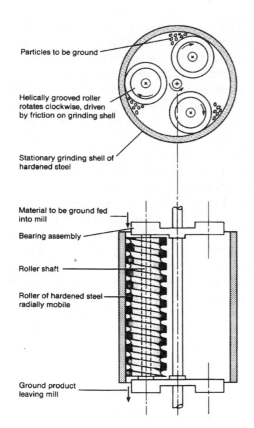

Particles to be ground

Helically grooved roller
rotates clockwise, driven
by friction on grinding shell

Stationary grinding shell of
hardened steel

Material to be ground fed
into mill

Bearing assembly

Roller shaft

Roller of hardened steel
radially mobile

Ground product
leaving mill

Figure 12.53. The Szego Mill (General Comminution Inc.).

mill. The feed particles are repeatedly crushed between the rollers and the stationary grinding surface. The crushing force is created mainly by the radial acceleration of the rollers; shearing action is induced by the high velocity gradients generated in the mill. Hence, the primary forces acting on the particles are the crushing and shearing forces produced by the circumferential motion of the rollers. The basic action of the mill is somewhat similar to that of the roller race mill and the HPGR mill. The rollers rotate about their own axes, pull material under the rollers with a critical angle of nip, and pass out compressed broken material. The rollers will float away from the stationary grinding cylinder to a gap that depends on the centrifugal force and the compression properties of the bed. The force on the rollers is controlled by the speed of rotation around the central axis in the mill cylinder, with a higher velocity giving a higher force per unit mass of roller.

An important feature is the ability of the roller grooves to aid the transport of material through the mill, thus providing a means to control the residence time, the number of times material is rolled over, and, hence, mill capacity and product size distribution. This transporting action is particularly important with materials that do not readily flow by gravity, such as pastes and sticky materials. The mill has several design variables that can be utilized to meet specific product requirements. The important variables are the number of rollers, their mass, diameter and length, and the shape, size, and number of starts of the helical grooves on the rollers. Increase in the number of starts gives a steeper angle for the helical grooves. As the number of rollers is increased, the product becomes finer. Heavier rollers and higher rotational speeds generate the greater crushing forces which may be needed for strong materials. The ridge/groove size ratio can be changed to increase or decrease the effective pressure acting on the particles. The common groove shapes are rectangular and tapered; the latter will decrease the chance of particles getting stuck in the grooves.

If several passes through the mill are required to get a sufficiently fine product, multiple-stage mills can be used that have several sets of rollers fixed onto the same rotor. This allows various design combinations of different roller sizes and ridge/groove size ratios in different stages for optimal mill performance. The operating variables for the mill are the material feed rate, its consistency (if wet), and the rotational speed of the rotor. Typically, the rotational speed is between 400 and 1200 rpm, depending on equipment size, which translates to roller velocities of 6 to 10 m/s.

Most work with the Szego Mill has been done on the grinding of coal in oil or water [56] for the preparation of coal-slurry fuels. Limestone, mica, talc, and other filled materials have been tested,[57,58] as have various waste materials, for example, hog fuel,[59] sawdust,[60] and waste paper,[61] the latter for use as a reinforcing filler in cellulose–plastic composites. Wet grinding of grains, as a preprocessing step for hydrolysis and fermentation to alcohol,[62] is another interesting application. The mill is characterized by high capacity per unit volume and modest power consumption. It is very versatile; in wet grinding it can also handle highly viscous materials such as thick pastes, that is, high solids concentrations, without extreme loss of efficiency.[63] Within reason, not only particle size distribution but also particle shape can be controlled, for example, from granular to flaky.[64]

Another group of applications involve grinding combined with other operations or processing. The simultaneous grinding and agglomeration (SGA) process,[65,66] as an example, combines grinding and selective oil agglomeration of coal with oil in water for coal beneficiation. In the conventional process, developed at the National Research Council of Canada, oil or a hydrocarbon solvent is added to finely ground coal in water. Intense mixing breaks the oil into fine droplets and allows the hydrophobic coal particles to collect onto the droplets, leaving the hydrophilic ash (noncombustable mineral matter) behind in the water.[67] A period of milder stirring allows the coal–oil particles to grow into larger spherical agglomerates for separation from the aqueous phase by screening or other means. The combined SGA process uses the Szego Mill to replace the grinding and high-shear mixing steps, with considerable equipment simplification and energy savings,[66] with results comparable to the conventional process. Other grinding mills such as ball or agitated media mills are not suitable, as the sticky agglomerates would coat the balls and either reduce the grinding efficiency greatly or block the mill, whereas the Szego Mill will operate owing to the positive transporting action of the roller grooves. The objective of those studies was to make beneficiated coal–oil–water slurry fuels as an oil replacement in industrial or utility boilers.

Other combined processes tested involve grinding and extraction, applied to oil extraction from rapeseed (canola);[68] and simultaneous grinding and reaction, in a coal liquefaction study.[69] When a thick slurry is being ground and a very fine product is required, a continuous recycle system without classification is used since classification is very difficult at high slurry or paste viscosity. The mill is then run long enough to give the product the desired fineness. Metals have been ground that way down to submicron flake thicknesses.[70]

A significant effort has been expended on mill modeling. This includes performance modeling using the population balance approach,[71,72] with breakage functions and grinding kinetics for single and multipass grinding for both wet and dry operation. A dynamic model[73] of fluid flow between a roller ridge and the stationary grinding cylinder has been made for wet grinding. The centrifugal forces are balanced by pressure development in the squeezed film of paste; the model allows, currently for a Newtonian fluid, computation of the total dynamic force field, velocities, shear stresses, etc., as well as the clearance between the roller-ridge and the grinding surface. Integration of these events, in combination with a confirmed mechanism of material transport through the mill, allows prediction of the residence time distribution and an upper limit to the product particle size distribution.[73]

Szego Mills are available in laboratory and pilot sizes as well as in small industrial sizes with throughputs of 1 to 10 tons/h. Compared to a ball mill, throughput per unit volume in the Szego Mill is some 30 times higher and the specific power consumption due to the high power density is typically 30% lower, as is characteristic of bed compression mills. While the Szego Mill is a compact and efficient grinder for many applications, very hard and abrasive materials excluded, its special niche is grinding wet at high solids loading; a toothpaste-like consistency appears to be the best. Special mills have been built for operation at high temperatures and pressures, further enhancing the range of applications of this mill.

12.5.4 The DESI Mill

This mill is another example of a mill that uses a principle similar to that of an existing type of mill but that incorporates changes allowing it to embrace also new applications. It has been developed in Estonia by the company Desintegraator and is in use in various parts of the former Soviet Union, with applications ranging from industrial minerals to fuels to biological materials. A great deal of work on the mill has also been done at the Tallinn Technical University, but there are relatively few publications, and most of these are in Russian. During privatization in the early 1990s, the original company was broken into smaller entities and information is available from the Desintegraator Association or from DESI-E Ltd., both in Tallinn, Estonia.

Invented by the late Dr. J. Hint some 40 years ago, the DESI mill was first used with the development of silicalcite, a strong building material made of sand and lime ground together. Mechanical activation imparted to the materials by the mill accounts for its high strength; the development of both silicalcite and the mill is described in a 600 page monograph by Hint.[74] The DESI is an impact mill comprising of two rotors moving at high speed in opposite directions. Thus the mill has the same principle as the Cage–Pactor mill shown in Figure 12.20 but it is specifically designed for fine grinding. The material fed to the cen-ter of the rotors passes through the working zone within a few hundredths of a second. The particles are disintegrated by collision with the multiple rows of grinding elements and by particle–particle attrition in the air stream. The grinding elements serve as targets for the colliding material and as accelerators for the next collision (see Fig. 12.54). The material typically undergoes two to eight collisions with the grinding elements.

Whereas many mills, including the HPGR mill, break particles by internal tension produced by compressive forces applied relatively slowly, in high-speed impact mills, the DESI included, breakage occurs by a different process of producing tension. The particles experience free, unrestricted impact at high velocity, typically in the 30 to 200 m/s range in the DESI. (It has been shown by Vervoon and Austin[75] that pellets moving at 30 m/s reach a maximum impact force within a few microseconds after impact when they strike a rigid target containing a force transducer). An intensive compression wave starts from the area of contact and surges through the particle at high velocity, with the stresses exceeding the normal compressive strength of the particle. When the compression wave reaches the opposite side of the particle, it is reflected as a tension wave of the same intensity. The particle then starts to break up. The multiple propagation of waves in the particle and its

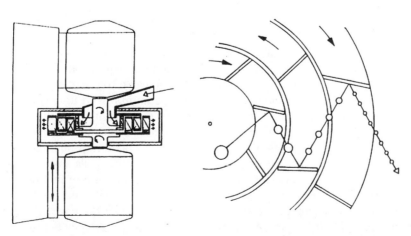

Figure 12.54. Operating principle of the DESI impact-roller mill (DESI-E Ltd.).

fragmentation are believed to activate the material chemically.[76] Hence, mechanochemical activation of the material occurs which may have beneficial effects on downstream processing, or even for simultaneous grinding and reaction. Such activation effects have been observed with chemical catalysts, building material (e.g., silicalcite), fertilizers, and in various biological systems. The DESI mill can be used for selective grinding of weaker components in a heterogeneous material by judicious selection of the speed of rotation to give impact forces between those required to break the respective materials.[76] Besides effective grinding, the fast rotation of the grinding elements in opposite directions allows excellent micromixing of solids or solids and liquids. The mill can also be used to treat sticky materials since the powerful centrifugal forces discourage adhesion.

For fine, and especially ultrafine grinding, the DESI mill is used with a built-in aerodynamic classifier, which recycles coarse material for regrinding. The fine product enters a collector and de-dusting system. DESI mills are available in a wide capacity range, from small laboratory units with capacities of 5 to 10 kg/h through to industrial units with capacities up to 100 t/h, the latter for limestone grinding in a DESI 31 M-8 mill. The total assembly weighs 14 t, with gross dimensions, m, of 4.5 length, 2.6 width, and 2.4 height, including motors, and a power rating of 500 to 1200 kW.

There are many DESI mills in industrial use covering a number of applications, with a range of quoted product particle sizes varying from 90 wt% < 5 μm to 90 wt% < 3 mm. Many more materials have been ground in laboratory settings down to the micrometer size. Apparently, most units are custom-designed, with the number of rows as well as size and inclination of the grinding elements being important variables in addition to the rotor diameter. The mill rotors are self-balancing and the grinding elements are reinforced with wear-resistant ceramics; chamber walls are also reinforced where required. An extensive amount of work has been done on wear, with many combinations of both target and abrasive particle materials as well as velocity, particle size, impact angle, etc.[77]

The main unique feature of this type of mill is the ability to mechanically activate many materials.[78,79] Such a claim is supported by extensive research; a more recent presentation[80] has summarized some of this work, including mechanical activation of polymers and biological systems in the disintegrator. Mill design and operating conditions were related to the resultant activation. Again, custom design is essential, for the desired objectives and the particular materials, in situ reactions or enhanced downstream processing. Of course, the same comments can be made about high-speed hammer mills, which operate at similar impact velocities.

12.5.5 The Nutating Mill

This mill is being developed by the Warmley company in Australia,[81,82] specifically for dry or wet grinding at high power density of brittle materials such as metalliferous ores. It has several similarities to the planetary and centrifugal mills[16] described previously since it is a mill that uses grinding balls at high g forces, but these forces are produced in a different way. The mill shell is in the form of an inverted cone, with feed from above into the narrow end of the cone. The shell is rotated about the center line of the cone, which is at an angle to the vertical. This axis is mechanically forced to rotate at the same time to form the surface of a narrow cone with the tip of the cone at a fixed point on the vertical (just like the earth rotating on its own axis but also moving in orbit with its axis not perpendicular to the plane containing the orbit path). This wobbling planetary action produces high g forces and rapid movement around and across the mill of the balls inside. The mill grinds very rapidly because of the high forces and the high power density and the feed discharges at the large end of the cone. The mill is capable of very fine grinding by adjusting the feed and discharge rate to give a long mean residence time while maintaining an appropriate hold-up of powder or slurry to avoid steel-on-steel collisions. As with all high power density mills using grinding media, the wear rate of media and shell liners is high and the energy efficiency is not going to be better than that of a more conventional tumbling media mill, but the mills are small for a high capacity. High power density machines are especially suited for very fine grinding, to avoid having to use a large machine to give a small amount of suitable product. The application of the concepts of mill modeling to the nutating mill is well advanced and it is possible to predict optimum conditions, capacities, and product size distributions from tests on a new material in a laboratory-scale mill.

12.6 FUTURE WORK

It is still true that much work remains to be done to raise the technical understanding of the unit operation of size reduction to that of the other (perhaps fundamentally simpler) unit operations such as heat transfer, distillation, absorption, etc. The mechanical stressing conditions inside mills are complex, and the fracture and disintegration of natural materials is a complex phenomenon. It must be emphasized that for size reduction we are concerned not only with the conditions at which fracture occurs but also the size distribution of the set of fragments resulting from the fracture.

The conversion of electrical energy via mechanical action to surface energy of fracture is thermodynamically very inefficient. However, based on the industrial requirements of cost, throughput, wear, and reliability of operation, it is difficult to see how to improve existing devices substantially or how to invent new ones with much greater efficiency. The material in this chapter has been limited to the powder technology relevant to crushers and mills that are in commercial operation with proven benefits for particular applications. Research on different methods of breakage and new types of mill is proceeding, of course, but until this research produces industrially important results it falls outside of the scope of this chapter.

The methodology of characterizing a size reduction operation by examining the specific rates of breakage and the primary progeny fragment distributions has proved very informative. Again, however, there are no precise descriptions of why the values of S_i and B_{ij} vary in the ways observed. The variations are often sensible from simple physical reasoning, but the quantitative relations involved are still essentially empirical.

The choice of a certain crusher-mill combination for a given job is generally made intuitively at present; the choice is not the logical result of a precise set of rules or calculations. Programming of the calculations for computation with current desktop computers and available software is not the problem: it is inadequate systemic, quantitative descriptions of how machines and materials behave that prevent full use of the techniques of mill and mill circuit simulation.

The mechanisms of the slowing down of size reduction that is observed as fines accumulate remain to be investigated in detail, and this branch of investigation will undoubtedly involve the nature of the cohesive interaction between particles, dry and in dense slurries, and the effect of grinding additives on these forces.

The better utilization of many ores, fuels, and other materials in the future may involve requirements of mechanical reduction to ultrafine sizes. This represents a branch of investigation that has come to the fore but that poses many problems in theory, experimental technique, and engineering design.

REFERENCES

1. A. Nadai, *Theory of Flow and Fracture of Solids*, McGraw Hill, New York, p. 89 (1950). See also *Developments in Fracture Mechanics*, Vol. 1, edited by G. G. Shell, Applied Science Publishers, London (1979).

2. A. A. Griffith, "Phenomena of Rupture and Flow in Solids," *Philos. Trans. R. Soc. Lond. 221A*:163 (1920).

3. A. A. Griffith, "The Theory of Rupture," *Proc. First Int. Conf. for Applied Mechanics*, Delft (1924).

4. G. R. Irwin, *Fracture Dynamics: Fracturing of Metals*, American Society of Metals (1948); Orowan, E., "Fracture and Strength of Solids," *Reports of Progress in Physics*, Physical Society, London, *12*:185 (1949).

5. R. P. von Rittinger, *Lehrbuch der Aufbereitungskunde*, Ernst v. Korn., Berlin (1857), quoted in many surveys of grinding theory.

6. H. E. Rose, private communication (1964).

7. I. J. Lin and S. Nadir, "Review of the Phase Transformations and Synthesis of Inorganic Solids by Mechanical Treatment," *Mat. Sci. Eng. 39*:193–209 (1979).

8. J. S. Benjamin, "Mechanical Alloying," *Sci. Am. 234*:41–48 (May 1976). See also: C. Suryanarajan, *Bibliography on Mechanical Alloying and Milling*, Cambridge Interscience Publ., 380 pp. (1995).

9. N. H. Macmillan, "Chemisorption Induced Variations in the Plasticity and Fracture of Non-metals," in *Surface Effects in Crystal Plasticity*, Nordhoff, Leyden, p. 629 (1977).

10. A. R. C. Westwood and J. J. Mills, "Application of Chemo-mechanical Effects to Fracture-dependent Industrial Processes," ibid., p. 835.

11. L. G. Austin, C. A. Barahona, and J. M. Menacho, "Fast and Slow Chipping Fracture and Abrasion in Autogenous Grinding," *Powder Technol. 46*(1):81–87 (1986).

12. L. G. Austin, N. P. Weymont, C. A. Barahona, and K. Suryanarayana, "An Improved Simulation Model for Semi-Autogenous Grinding," *Powder Technol. 47*(3):265–283 (1986).

13. L. G. Austin, C. A. Barahona, and J. M. Menacho, "Investigations of Autogenous and Semi-Autogenous Grinding in Tumbling Mills," preprinted for World Congress Particle Technology, Nuremburg, Federal Republic of Germany, April 1986; *Powder Technol. 51*:283–294 (1987).

14. L. G. Austin and S. Tangsriponkul, "A More General Treatment of Abrasion-Chipping Processes Applicable to FAG/SAG Milling," *Particle Particle Syst. Character. 11*:345–350 (1994).

15. A. A. Bradley, P. S. Lloyd, D. A. White, and P. W. Willows, "High-Speed Centrifugal Milling and Its Potential in the Milling Industry," *S. Afr. Mechan. Eng. 22*:129–134 (1972).

16. A. L. Hinde and F. B. Verardi, Studies on Design of Centrifugal Mill Grinding Circuits." *Proc. 3rd IFAC Symposium, Automation in Mining, Mineral and Metal Processing*, Montreal, Canada, p 283–294 (Aug., 1980). See also: L. P. Kitschen and P. J. Lloyd, "The Centrifugal Mill: Experience with a New Grinding System and its Applications." *Proc. 14th IMPC*, Toronto (1982).

17. L. G. Austin, "A Review Introduction to the Description of Grinding as a Rate Process," *Powder Technol. 5*:1–17 (1971/72).

18. L. G. Austin, R. R. Klimpel, and P. T. Luckie, *The Process Engineering of Size Reduction: Ball Milling*, AIME, New York, 561 p (1984).

19. K. Reid, "A Solution to the Batch Grinding Equation," *Chem. Eng. Sci. 20*:953 (1965).

20. T. Trimarchi and L. G. Austin, "A Ball Mill Circuit Simulator in Object-Oriented Programming," available from the Mineral Processing Section, Department of Mineral Engineering, The Pennsylvania State University, University Park, PA 16802.

21. R. S. C. Rogers and R. P. Gardner, "Use of a Finite-stage Transport Concept for Analyzing Residence Time Distributions of Continuous Processes," *AIChE J. 25*:229 (1979).

22. F. C. Bond, "Crushing and Grinding Calculations," *Brit. Chem. Eng. 6*:378 (1965).

23. C. A. Rowland, Jr. and M. M. Kjos, "Rod and Ball Mills," in *Mineral Processing Plant Design*, edited by A. L. Mular and R. B. Bhappu, AIME, New York, pp. 239–278 (1978).

24. L. G. Austin and J. W. Pérez, "A Note on Limiting Size Distributions from Closed Circuit Mills," *Powder Technol. 16*:291–293 (1977).

25. L. G. Austin and P. Bagga, "An Analysis of Fine Dry Grinding in Ball Mills," *Powder Technol. 28*:83–90 (1981).

26. L. G. Austin, M. Yekeler, and R. Hogg, "The Kinetics of Ultrafine Dry Grinding in a Laboratory Tumbling Ball Mill," *Proceedings of Second World Congress Particle Technology*, Kyoto, Japan, p 405–413 (September 1990).

27. L. G. Austin, M. Yekeler, T. F. Dumm, and R. Hogg, "Kinetics and Shape Factors of Ultrafine Grinding in a Laboratory Tumbling Ball Mill," *Particle Particle Syst. Character. 7*:242–247 (1990).

28. A. M. Gaudin, *Principles of Mineral Dressing*, McGraw-Hill, New York, p 41–43 (1939).

29. L. G. Austin, D. R. Van Orden, and J. W. Pérez, "A Preliminary Analysis of Smooth Roll Crushers," *Int. J. Miner. Proc. 6*:321–336 (1980).

30. L. G. Austin and J. D. McClung, "Size Reduction of Coal," in AIME Handbook, *Coal Preparation*, Harvey Mudd Series, edited by J. Leonard, p 189–219 (1991).

31. L. G. Austin, K. Shoji, D. R. Van Orden, B. McWilliams, and J. W. Pérez, "Breakage Parameters of Some Materials in Smooth Roll Crushers," *Powder Technol. 28*:245–251 (1981).

32. F. Kick, *Dinger Polytech. J. 247*:1 (1883); *250*:141 (1883).

33. W. J. Whiten, "Simulation of Crushing Plants with Models Developed Using Multiple Spline Regres-

sion," *J. S. Afr. Inst. Mining Metal.* 72:257–264 (1972).

34. W. J. Whiten, "Application of Computer Methods in Mineral Industries," *Proc. 10th Intl. Mining Processing Congress*; ibid. 73:317–323 (1973).

35. A. Kumar, "An Investigation of a General Mathematical Model for Predicting the Product Distribution from a Roll Crusher and a Cone Crusher." M. S. Thesis in Mineral Processing. The Pennsylvania State University, University Park, PA 16802 (1986).

36. V. Singhal, "An Investigation of the Applicability of a Crusher Model to Jaw Crushing" M.S. Thesis in Mineral Processing, The Pennsylvania State University, University Park, PA 16802 (1985).

37. F. Concha, R. Santelices, and L. G. Austin, "Optimization of the Ball Charge in a Tumbling Mill," *XVI International Mining Processing Congress*, Stockholm (June 1988).

38. C. Tangsathitkulchai and L. G. Austin, "The Effect of Slurry Density on Breakage Parameters of Quartz, Coal and Copper Ore in a Laboratory Ball Mill," *Powder Technol.* 42:287–296 (1985).

39. C. Tangsathitkulchai and L. G. Austin, "Slurry Density Effects on Ball Milling in a Laboratory Ball Mill," *Powder Technol.* 59(4):285–293 (1989).

40. R. C. Klimpel, L. G. Austin, and R. Hogg, "The Mass Transport of Slurry and Solid in a Laboratory Overflow Ball Mill," *Miner. Metal. Proc.* 6:73–78 (1989).

41. R. C. Klimpel and L. G. Austin, "An Investigation of Wet Grinding in a Laboratory Overflow Ball Mill," *Miner. Metal. Proc.* 6(1):7–14 (1988).

42. L. G. Austin, W. Hilton, and B. Hall, "Mill Power for Conical (Hardinge) Type Ball Mills," *Miner. Eng.* 5(2):183–192 (1992).

43. J. J. Cilliers, L. G. Austin, P. Leger, and A. Deneys, "A Method of Investigating Rod Motion in a Laboratory Rod Mill," *Miner. Eng.* 7:533–549 (1994).

44. L. G. Austin, J. M. Menacho, and F. Pearcy, "A General Model for Semi-Autogenous and Autogenous Milling," *Proc. 20th Int. Symp. on the Application of Mathematics and Computers in the Mineral Industries*, edited by R. P. King and I. J. Barker, Mintek, Johannesburg, South Africa, 2:107–126 (October 1987).

45. L. G. Austin, "State of the Art in Modeling and Design of Autogenous and SAG Mills," in *Challenges in Mineral Processing*, edited by K. V. S. Sastry and M. C. Fuerstenau, Society of Mining Engineering, Inc., Littleton, CO, p 173–193 (1989).

46. L. G. Austin, "A Mill Power Equation for SAG Mills," *Miner. Metal. Proc.* 7(1):57–62 (1990).

47. L. G. Austin, "The Theory of Roller-Race Mills," available from the Mineral Processing Section, Department of Mineral Engineering, The Pennsylva-

nia State University, University Park, PA 16802, submitted for publication.

48. L. G. Austin, J. Shah, J. Wang, E. Gallagher, and P. T. Luckie, "An Analysis of Ball-and-Race Milling: Part I, The Hardgrove Mill," *Powder Technol.* 29:263–275 (1981).

49. L. G. Austin, P. T. Luckie, and K. Shoji, "An Analysis of Ball-and-Race Milling: Part II, The Babcock E-17 Mill," *Powder Technol.* 33:113–125 (1982).

50. L. G. Austin, P. T. Luckie, and K. Shoji, "An Analysis of Ball-and-Race Milling: Part III, Scale-up to Industrial Mills," *Powder Technol.* 33:127–134 (1982).

51. K. Schönert, "Energetische Aspekte des Zerkleinerns spröder Stoffe," *Zement-Kalk-Gips, 32*(1):1–9 (1979).

52. F. Fischer-Helwig, "Current State of Roller Press Design," KHD Symposium '92 "Modern Roller Press Technology," KHD Humboldt-Wedag AG, Cologne, p 73–79 (1992).

53. H. Kellerwessel, "High-Pressure Particle-Bed Comminution: Principles, Application, Testing and Scale-up, Details of Equipment Design," KHD Humboldt-Wedag AG Paper, Cologne, 51 p (1993).

54. S. Strasser, "Current State of Roller Press Technology," KHD Symposium '92 "Modern Roller Press Technology," KHD Humboldt-Wedag AG, Cologne, p 11–21 (1992).

55. The Horomill, Objectif 93/9 A2B2, FCB, Division Cimenterie, Groupe Fives Lille, Lille, France.

56. E. A. J. Gandolfi, G. Papachristodoulou, and O. Trass, "Preparation of Coal-Slurry Fuels with the Szego Mill," *Powder Technol.* 40:269–282 (1984).

57. E. A. J. Gandolfi, V. R. Koka, and O. Trass, "Fine Grinding Applications with the Szego Mill," in *Proc. 12th Powder & Bulk Solids Conference / Exhibition*, Rosemount, IL, p 448–457 (1987).

58. O. Trass and E. A. J. Gandolfi, "Fine Grinding of Mica in the Szego Mill," *Powder Technol.* 60(3):273–279 (1990).

59. O. Trass and R. Gravelsins, "Fine Grinding of Wood Chips and Wood Wastes with the Szego Mill," in *Proc. 6th Bioenergy Seminar*, Vancouver, B.C., February 1987, p 198–204 (1988).

60. R. Gravelsins and O. Trass, "Wet Grinding of Wood with the Szego Mill," in *Proc. 7th Cdn. Bioenergy R & D Seminar*, edited by E. N. Hogan, Ottawa, Ontario, p 281–286 (April 1989).

61. T. Molder and O. Trass, "Grinding of Waste Paper and Rice Hulls with the Szego Mill for Use as Plastics Fillers," *Int. J. Miner. Proc.* (in press).

62. O. Trass, E. A. J. Gandolfi, and E. Daugulis, "Development of an Integrated Fine-Grinding, Hydrolysis, Ethanol Fermentation Process," in *Proceedings, "Energy from Biomass and Wastes XIV"*

Conference, Lake Buena Vista, Florida, 16 p (Jan./Feb. 1990).

63. O. Trass, E. Edusei, and E. A. J. Gandolfi, "Wet Grinding of Coal and Limestone with the Szego Mill at High Solids Concentrations," in *14th Intl. Conf. on Coal Slurry Technology*, Clearwater, FL, April 24–27, 1989; also *Proc. 15th Conf.*, p A115–128 (1990).

64. V. R. Koka, G. Papachristodoulou, and O. Trass, "Particle Shapes Produced by Comminution in the Szego Mill," *Particle Particle Syst. Character. 12*:158–165 (1995).

65. O. Trass and O. Bajor, "Modified Oil Agglomeration Process for Coal Beneficiation. II. Simultaneous Grinding and Oil Agglomeration," *Can. J. Chem. Eng. 66*:286–290 (1988).

66. O. Trass, P. D. Campbell, V. R. Koka, and E. R. Vasquez, "Modified Oil Agglomeration Process for Coal Beneficiation. IV. Pilot Plant Demonstration of the Simultaneous Grinding-Agglomeration Process," *Can. J. Chem. Eng. 72*:113–118 (1994).

67. C. E. Capes and R. G. Germain, "Selective Oil Agglomeration in Fine Coal Beneficiation," in *"Physical Cleaning of Coal, Present and Developing Methods,"* edited by Y. A Lin, Marcell-Dekker, New York, p 293–359 (1982).

68. L. L. Diosady, L. J. Rubin, and O. Trass, "Solvent Grinding and Extraction of Rapeseed," *Proc. 6th World Rapeseed Congress*, Paris, France, p 1460–1465 (May 1983).

69. O. Trass and E. R. Vasquez, "Liquifaction of Coal with Simultaneous Grinding," in *Proc. 15th Intl. Conf. on Coal Slurry Technology*, Clearwater, FL, p 337–349 (1990).

70. O. Trass and T. Lustvee, "Preparation of Aluminum Pastes with the Szego Mill," *Pacific Region Meeting*, Fine Particle Society, Honolulu, Hawaii (August 1983).

71. V. R. Koka and O. Trass, "Determination of Breakage Parameters and Modelling of Coal Breakage in the Szego Mill," *Powder Technol. 51*(2):201–214 (1987).

72. V. R. Koka and O. Trass, "Estimation of Breakage Parameters in Grinding Operations using a Direct Search Method," *Int. J. Miner. Proc. 23*:137–150 (1988).

73. O. Trass and G. L. Papachristodoulou, "Dynamic Modelling of Wet Grinding in the Szego Mill," in *Proceedings, 2nd World Congress Particle Technology*, Kyoto, Japan, Vol. II, p 471–179 (1990). See also: G. L. Papachristodoulou, "The Dynamic Modelling of the Szego Mill in Wet Grinding Operations," Ph.D. Thesis, University of Toronto (1982).

74. J. Hint, "Fundamentals of the Manufacture of Silicalcite Products," Gosstroiizdat, Leningrad, 601 p (in Russian) (1962).

75. P. M. M. Vervoorn and L. G. Austin, "The Analysis of Repeated Breakage Events as an Equivalent Rate Process," *Powder Technol. 63*:141–147 (1990).

76. A. Tymanok, "Grinding by Collision. Disintegrator and its Use in Technology: Review of Principles and Recent Results," Internal Report, Tallinn Technical University, Estonia, 8 p (1993).

77. H. Uuemois, H. Kangur, and I. Veerus, "Wear in the High-Speed Impact Mills," in *Proc. 8th European Symposium on Comminution*, Stockholm, Sweden, p 513–524 (May 1994).

78. J. Hint, "Uber der Wirkungsgrad der Mechanischen Aktivierung. Eininge Ergebnisse der Aktivierung von Feststoffen mittels grosser Mechanischer Energien," *Aufbereitungstechnik* (1971).

79. J. Hint, "About the Fourth Component of Technology," Valgus, Tallinn, Estonia, p 66–72 (in Russian) (1979).

80. B. Kipnis and L. Vanaselja, "Uber die Anvendung von Desintegratoren in Technologie der Mechanoaktivierung und Mechanochemie," Intl. Fachtagung *"Forstchritte in Theorie und Praxis der Aufbereitungstechnik,"* Freiberg, Germany, p 155–160 (1989).

81. J. M. Boyes, "High-Intensity Centrifugal Milling—A Practical Solution," *Int. J. Miner. Proc. 22*:413–430 (1988).

82. D. I. Hoyer and J. M. Boyes, "The High-Intensity Nutating Mill—A Batch Ball Milling Simulator," *Miner. Eng. 3*:35–51.

13
Sedimentation

Wu Chen and Keith J. Scott[†]

CONTENTS

13.1 INTRODUCTION

Gravity sedimentation is a widely used method of separating solids/liquid mixtures and includes diverse applications such as clarification of waste water, thickening of milled gold ore pulps, flotation of suspended sewage solids, and countercurrent washing of soluble metal from acid-leached suspensions. These operations typically are performed in relatively large single-compartment tanks such as shown in Figures 13.1, 13.2, and 13.3. The discussion in this chapter concentrates on sedimentation in liquids.

Suspensions of solids normally settle naturally, as long as there is a difference in density between solid and liquid. Given time a suspension separates into a clear liquid layer above, a supernatant, and a sediment below, which remains "saturated" with liquid. Such batch sedimentation can be carried out on a large scale in tanks, ponds, or lagoons.

To achieve continuous operation, it is necessary merely to supply a steady stream of fresh suspension, the feed, to the center (or end) of the sedimentation vessel and to remove continuously the lighter liquid phase,

---

[†] Deceased.

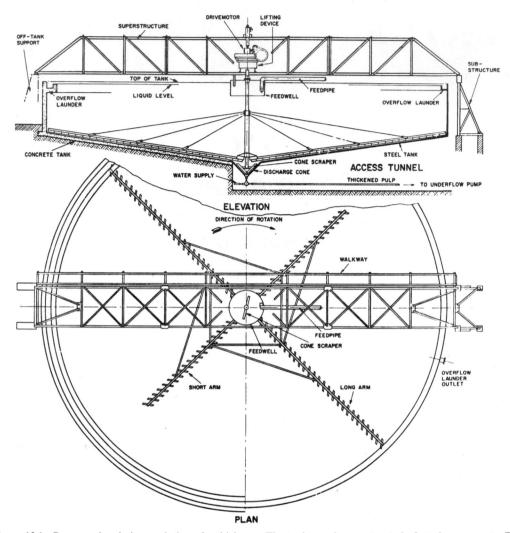

Figure 13.1. Cross-sectional view and plan of a thickener. The tank may be constructed of steel or concrete. The rake lifts vertically if it encounters an unusual resistance.

termed the overflow. Solids removal is normally achieved by continuous raking of the thickened sediment toward the center (or opposite end) of the tank, from where it is pumped out as the underflow stream.

The relative simplicity of both the process and the mechanical equipment involved makes gravity sedimentation the least costly of the available solids/liquid separation techniques.[2,3]

The process has the capability of treating high water flow rates with relatively little hardware[1]

while usually achieving a high degree of clarity in the overflow.

Other solids/liquid separation techniques, however, need to be considered as an alternative, or addition to, gravity sedimentation if:

1. The solids stream must have a low moisture content.
2. The loss of 10% to 15% of the liquid in the feed to the underflow is not acceptable.
3. The cost of the required floor space is excessive or space is not available.

Figure 13.2. View of an operating clarifier. The larger tank, the secondary clarifier, represents one of the final stages of producing clear drinking water from purified sewage in the Standard Water Reclamation Plant, a pilot plat at Daspoort, Pretoria. (Courtesy of National Institute for Water Research, SCIR.)

Figure 13.3. View of an empty thickener at a Transvaal gold mine. Such tanks handle up to 15,000 tons/day of solids and four to five times as much water.

4. The process must be carried out under pressure.

Table 13.1 shows the advantages of various solids/liquid separation methods.

Combinations of techniques may be used to improve the effectiveness of separation, such as a vacuum filter immediately following a thickener to dewater or wash the thickener underflow. In selecting a separation process it is essential, therefore, to consider wider aspects than just the pros and cons of individual techniques. Some guidelines are available in the literature for selecting equipment[2,5-7] but these should be supplemented by sufficient knowledge in this field. Discussions with specialists or equipment suppliers can help in formulating likely solutions for a given problem.

13.1.1 Objectives in Gravity Sedimentation

Sedimentation is distinguished into two primary functions. The first is *clarification*, in which absence of solids in the liquid overflow is the essential requirement and a relatively high proportion of liquid in the underflow can be tolerated. On the other hand, in *thickening*, the minimum quantity of liquid in the underflow is the main objective and the presence of up to a few percent of suspended solids in the overflow (often harmlessly recirculated) is of secondary concern. The distinction is therefore in the end result rather than the process; in thickening, the solids concentration in the feed stream is increased by sedimentation while in clarification the solids are removed by this process.

Each of the two functions can be optimized and controlled separately. The turbidity of a clarification tank overflow is related to slowly settling fine solids which may be flocculated to form larger faster settling units. The control of overflow clarity is therefore affected by the selection of flocculant, its dosage, and by control of the volumetric feed rate.

Underflow density of a thickener depends on the height of sediment in the tank, the degree of flocculation in the suspension (flocculated material tends to incorporate more liquid than dispersed particles), and on the underflow pumping rate. Because flocculation and feed rate affect both the overflow clarity

Table 13.1. A Qualitative Comparative Guide to the Particular Advantages of Various Solids/Liquid Separation Techniques.

SOLIDS / LIQUID SEPARATION TECHNIQUE	RATIO OF THROUGHPUT TO FLOOR AREA	CLARITY OF LIQUID	MOISTURE CONTENT OF SOLID STREAM	EASE OF WASHING SOLIDS	OVERALL COST CAPITAL PLUS OPERATING
Sedimentation					
Gravity	Low	Good	High	Require repeat operations	Low[a]
Centrifugal	High	Good to excellent	Medium	Possible	High
Cyclone	Very high	Very poor	High	Require repeat operations	Low
Filtration	High	Good	Low	Easy	High[a]
Screening	High	Very Poor	Medium	Easy	Medium
Drying	Medium	—	Extremely low	—	High

[a] The operation cost of vacuum filters in the S.A. gold mining industry ($\sim$ 75 million tons/yr) is six to eight times the cost of gravity thickening.[4]

and the underflow density it is seldom possible to optimize both clarification and thickening simultaneously.[8]

13.1.2 Applications of Gravity Sedimentation

Sedimentation processes are used extensively throughout the world in many industries, water purification, and waste water treatment. These large-tonnage operations are often carried out in remote locations or nonurban areas where land is available and relatively inexpensive, and hence the use of large tanks is not a serious disadvantage.

Sedimentation is also practiced on a smaller scale in a variety of processes. Increasing attention is being given to development of higher capacity thickeners, that is, those of reduced area per unit throughput or thickeners that can produce thick underflows equivalent to filter cakes. This interest is minimizing the disadvantages of sedimentation, rather than selecting alternative types of equipment, is an indication of the desirability of the positive features of gravity sedimentation as a means of separating solids/liquid mixtures.

13.2 THEORY OF SEDIMENTATION

This section covers the fundamental aspects of the sedimentation of particles, whether as single spheres in an unbounded liquid or in a mixture of many other particles in a finite suspension. As much of the published verification of the theory was carried out in laboratory measuring cylinders, this section also discusses the theory of batch settling tests. How the settling behavior of suspensions observed in batch tests is used in the design of sedimentation equipment is covered separately (p. 1002), while the nomenclature used is given at the end of the chapter.

The settling of a single sphere in an unbounded fluid represents the simplest case of solids/liquid sedimentation. This ideal condition is seldom encountered in practice,[1] how-ever. The complicating factors that arise in real situations, dealt with more fully in subsequent sections, are:

1. Nonspherical and irregularly shaped particles
2. The simultaneous presence of large number particles
3. The presence of mutual particle attraction in which individual particles lose their identity and are grouped into agglomerates (flocs) by chemical–physical forces
4. The method of measuring of settling rate for gravity settler design
5. Wall effects.

13.2.1 Sedimentation of a Sphere in an Infinite Fluid

When a single spherical particles is suspended at rest in a liquid it experiences two opposing forces, B_F and G_F, as shown in Figure 13.4. Provided the densities of the solid and the liquid are not equal, there will be an unbalanced force, the difference between G_F, the downward gravitational attractive force, and B_F, the Archimedean upward thrust or buoyancy force. This unbalanced force, $(G_F - B_F)$, equal to $V(\rho_s - \rho_L)g$, causes the particle to accelerate (downward if positive) and attain a velocity relative to the liquid. Skin friction, that is, the resistance offered by a fluid to the motion of a solid, then results in the development of a drag force, F_D, which opposes the motion and increases with increasing particle

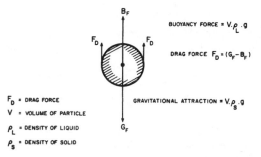

Figure 13.4. Forces acting on a spherical particle in a liquid.

velocity. The drag reduces the acceleration, and finally the value of the drag force becomes equal to the original driving force $(G_F - B_F)$ and there are no further unopposed forces acting on the particle, it continues to travel at a constant rate called its terminal settling velocity, u_∞. We may then write:

$$F_D = V(\rho_s - \rho_L)g \qquad (13.1)$$

This equation evaluates the magnitude of the drag force for any size particle but does not relate it to its unknown settling velocity.

This relationship has been formulated for a sphere in an infinite fluid[9] for slow flows but its general solution depends on the type and magnitude of flow around the particle as characterized by the dimensionless entity known as the Reynolds number.

13.2.1.1 Fluid Flow Around a Particle and the Reynolds Number

When the particle velocity is low, drag is due largely to the viscosity of the liquid and this flow is called viscous, laminar, or streamlined. At high velocities, the fluid streamlines do not continue completely around the particle but break up into vortices with the result that turbulent eddies and inertial forces also contribute to the drag. This finally develops into fully turbulent flow.

The criterion for distinguishing between flow conditions is the dimensionless particle Reynolds number:

$$\mathrm{Re} = \frac{d_p \cdot u_{sr} \cdot \rho_L}{\mu} \qquad (13.2)$$

where

d_p = particle diameter
u_{sr} = relative velocity between particle and liquid
ρ_L = liquid density
μ = liquid viscosity.

For particles,[12b] streamlined flow occurs below $\mathrm{Re} \sim 0.3$, turbulent flow above $\mathrm{Re} \sim 2 \times 10^5$, whereas in the intermediate region, in which inertial forces become increasingly

significant, the flow is called transitional. These limiting values for the particle Re are orders of magnitude lower than for flow in pipes in which the fluid streamlines are constrained by the boundary walls.

13.2.1.2 Laminar Flow

The analytical solution for the magnitude of the drag on a single sphere, settling under streamlined flow conditions in an unbounded liquid, is given by Stokes[9] as:

$$F_D = 3\pi\mu d_p u_\phi \qquad (13.3)$$

where u_ϕ = terminal velocity of the sphere in an infinite fluid in streamlined flow.

Even for this simplified condition, however, Eq. (13.3) is only a close approximation and, for greater accuracy, additional terms have been found to be necessary. Proudman and Pearson[10] for example, advocate the equation:

$$F_D = 3\pi\mu d_p u_\phi \left[1 + \tfrac{3}{16}\,\mathrm{Re} \right.$$
$$\left. + \tfrac{9}{160}(\mathrm{Re})^2 \ln\left\{\frac{\mathrm{Re}}{2}\right\} + \cdots \right] \quad (13.4)$$

It has become common practice[11] to express the forces exerted on moving bodies by the fluid in terms of a dimensionless drag coefficient C_D, obtained by dividing the drag force F_D by $\rho_L u_{sr}^2 / 2$ and by the area of the body projected onto the plane normal to u_{sr}. For a sphere, this area is $\pi d_p^2 / 4$; hence the coefficient is:

$$C_D = \frac{F_D}{\dfrac{\rho_L u_{sr}^2}{2} \cdot \dfrac{\pi d_p^2}{4}} \qquad (13.5)$$

which, together with Eqs. (13.2) and (13.4) and setting $u_{sr} = u_\phi$, becomes

$$C_D = \frac{24}{\mathrm{Re}} + \frac{9}{2} + \frac{27}{20}\,\mathrm{Re}\ln\left\{\frac{\mathrm{Re}}{2}\right\} + \cdots$$
$$(13.6)$$

The practical significance of terms following $24/\mathrm{Re}$ can be tested by considering the largest

Reynolds number likely to be encountered in real situations. The maximum overflow rate of operating sedimentation equipment quoted by Perry and Green[12a] is $u_{sr} \sim 0.8$ mm/s; the 95% upper limiting diameter in their particulate slurries is estimated as $d_p \sim 0.2$ mm, and, if we accept the most common sedimentation liquid as being water at 20°C, $\rho_L = 1 \times 10^3$ kg/m^3, $\mu = 1 \times 10^{-3}$ kg/ms, and therefore Re should lie largely in the range 0 to 0.16.

The accurate value of C_D for Re = 0.16 according to Eq. (13.6) is 153.95, while using the first term only, C_D = 150.0, representing a difference of 2.6%. This maximum error is quite acceptable in settler design as variations in ambient temperature of only 1 to 2°C can, by changing the liquid viscosity, result in greater variations in sedimentation rate. The second and third terms in Eq. (13.6) may therefore be safely neglected and the equation simplified to give:

$$C_D = \frac{24}{Re} \qquad (13.7)$$

which is an alternative expression of Eq. (13.3). As the particle is a sphere, V in Eq. (13.1) is $\pi d_p^3/6$ and from Eqs. (13.1) and (13.3) we can now write

$$u_\phi = \frac{d_p^2(\rho_s - \rho_L)g}{18\mu} \qquad (13.8)$$

This equation provides a means of calculating the terminal settling velocity of a single sphere of diameter d_p in an unbounded fluid, in streamlined flow, as determined by the physical properties of the sphere and fluid, that is, their densities and the fluid viscosity. Alternatively, it permits the estimation of the diameter of a particle by observing its settling velocity under these prescribed conditions. For a spherical quartz particle in water at 20°C, the maximum permissible diameter is $d_p \sim 85$ μm if the error in u_ϕ using Eq. (13.8) is not to exceed 5%.

13.2.1.3 Transitional and Turbulent Flow

Grit chambers, a special type of sedimentation basin used in sewage treatment, are designed for removal of coarse sands larger than 200 μm. Such large particles settle in transitional or turbulent flow but for the determination of their terminal velocities, no alternative simple expression similar to Eq. (13.8) exists. To calculate their terminal velocity, $u_{\phi t}$, the particle Reynolds number must first be known, but this cannot be known until the value of $u_{\phi t}$ is determined. A trial-and-error solution is one means of arriving at its value.[13]

The following more direct and accurate algorithm has, however, been found useful in avoiding both this repetitive procedure and the inaccurate graphical or cumbersome interpolation of the Re vs. $C_D Re^2$ values given in Perry and Green.[12b] The algorithm includes also the laminar flow region discussed previously.

Algorithm for Calculating $u_{\phi t}$ from d_p.

1. Calculate the entity $J = C_D Re^2$ which does not contain $u_{\phi t}$:

$$J = C_D Re^2$$
$$= \frac{4}{3} \cdot \frac{(\rho_s - \rho_L)\rho_L d_p^3}{\mu^2} \cdot g \qquad (13.9)$$

2. From J calculate the required values of a and b according to the data given in Table 13.2 using the appropriate range of J values.

Table 13.2. Values of a and b for Calculating Re from $J = C_D Re^2$ in Any Flow Regime.

FLOW REGIME	$\sim d_p$	$J = C_D Re^2$	a	b	$\sim$ Re
Laminar (Stokes' law)	1–75 μm	0–10	24	1	0–0.4
Transition region I	75–350 μm	$10–10^3$	$33.7J^{-0.19}$	$1.05J^{-0.05}$	0.4–20
Transition region II	0.35–2 mm	$10^3–2 \times 10^5$	$63.6J^{-0.29}$	$1.0J^{-0.046}$	20–600
Transition region III	2–15 mm	$2 \times 10^5–1 \times 10^8$	$91.7J^{-0.30}$	$0.96J^{-0.039}$	$600–1.5 \times 10^4$
Newton's Law	1.5×10 cm	$1 \times 10^8–1.7 \times 10^{10}$	0.67	0.5	$1.5 \times 10^4–2 \times 10^5$

3. Calculate

$$\mathrm{Re} = \frac{J^b}{a} \quad (13.10)$$

4. Finally, determine

$$u_{\phi t} = \frac{\mathrm{Re} \cdot \mu}{d_p \cdot \rho_L} \quad (13.11)$$

The values calculated in this way are accurate to within 5% to 6%. A subroutine for the trial-and-error solution mentioned earlier[13] is less accurate; for example, at Re ~ 70 the error is 18%. The algorithm covers a wide range of sphere diameters from 1 μm up to 10 cm and embraces the laminar, transitional, and Newton's flow regimes. When, however, the flow is known beforehand to be laminar (Re < 0.3), Eq. (13.8) provides exactly the same answer in fewer steps.

A theoretically derived[120] equation:

$$\mathrm{Re} = 20.52[(1 + 0.0921 J^{0.5})^{0.5} - 1]^2 \quad (13.12)$$

gives results that are within 7% of the experimental values for Re up to 7000.

A free-settling equation, valid not only for all particle sizes but also covering a wide range of naturally occurring shapes (see following section), is presented by Swanson;[14] his calculated values are within 20% of the measured values in most cases.

13.2.2 Nonspherical Particles in an Infinite Fluid

Spherical particles rarely occur in solids–liquid separation practice, some more common shapes being angular and flaky particles derived from crushing or weathering, amorphous fluffy precipitates, randomly formed agglomerates, and regular polyhedral or needle-shaped crystals.

The drag on a nonspherical particle depends on its shape and its orientation with respect to the direction of motion.[12b] If a body possesses spherical isotropy (such as a cube or octahedron) and is placed initially in any orientation in a liquid and allowed to fall without initial spin, it will not rotate to a different position but fall vertically in its original orientation. Most real particles are not symmetrical, however, and they experience not only drag forces parallel to the stream velocity, but also lateral (lift) forces at right angles to the stream. This may cause drift to one side during settling, rotation to a position of maximum resistance, steady rotation, or even a wobbling motion.

Even when such nontranslational motions are neglected, the calculation of the terminal settling velocity of nonspherical particles using an equation similar to Eq. (13.8), that is $u_{ns} = f(\rho_s, \rho_L, \mu, d_p)$ requires first that the shape is known or can be determined; second, that a representative "diameter" d_p can be assigned to this shape; and third, that a drag equation be available similar to Eq. (13.3) for a sphere.[15] The problem is thus complicated, and certainly no analytical solution exists for the irregular shaped particles encountered in practice.

However, empirical methods for dealing with shape are available. These are presented briefly with reference to some of the effects of increasing departure from spherical form shown in Table 13.3. This table compares particles of various shapes on the common basis of having the same volume (arbitrarily 1 mm³), that is, of each possessing an equal gravitational attractive force. Column 3 shows that as the shape departs increasingly from spherical, the surface area increases from 4.84 for a sphere to 8.35 for a cylindrical needle of the same volume, that is, an increase of 73%. The skin friction and hence drag for nonspherical particles will thus be greater than their terminal settling velocities correspondingly lower than for a sphere of the same volume.

Skin friction is related to surface area. If, in the drag Eq. (13.3), d_p is taken as the diameter of a sphere having the same surface area as a given shaped particle,[15] from Eqs. (13.1) and (13.3) it can be shown that:

$$\frac{u_{ns}}{u_{\phi t}} = \frac{d_p}{d_A} = K \quad (13.13)$$

Table 13.3. Various "Diameters" of Particles of Equal Volume (1 mm³) but Differing in Their Shape.

SHAPE	DESCRIPTION	CALCULATED SURFACE AREA (mm²)	DIAMETER OF A SPHERE (mm)			RATIO d_M/d_{scr}
			HAVING THE SAME SURFACE AREA AS PARTICLE d_A	SHOWING THE SAME PROJECTED AREA AS PARTICLE (SIZING BY MICROSCOPE) d_M	PASSING THE SAME MINIMUM SQUARE APERTURE (SCREEN ANALYSIS) d_{scr}	
Sphere	–	4.84	$1.24 = d_p$	1.24	1.24	1.00
Icosahedron	20 Equilateral triangles	5.15	1.28	1.32	1.23	1.07
Dodecahedron	12 Pentagons	5.31	1.30	1.35	1.27	1.06
Cube-octahedron	6 Octagons and 8 equilateral triangles	5.69	1.35	1.14	–	–
Octahedron	8 Equilateral triangles	5.72	1.35	1.36	1.24	1.10
Hexahedron (cube)	6 Squares	6.00	1.38	1.13	1.00	1.13
Tetrahedron	4 Equilateral triangles	7.20	1.51	1.51	1.44	1.05
Plate (5:5:1)	–	8.19	1.61	1.93	1.45	1.33
Needle (10:1)	Right circular cylinder	8.35	1.63	1.80	0.50	3.60

where

$u_{\phi t}$ = settling rate of sphere of diameter d_p having the same volume as the nonspherical particle

d_A = diameter of sphere having same surface area as the nonspherical particle

K_p = shape factor.

K_p is normally smaller than 1, indicating lower settling rate owing to the larger surface area in nonspherical particles. This treatment is not rigorous because Eq. (13.3) applies strictly to spheres. However, values of K_p calculated as above using the data in Table 13.3 compare reasonably well with the experimentally determined values[16] as shown in Table 13.4. The agreement, although not exact, serves to illustrate the principle of using shape correction factors, based on more than one characteristic diameter of a particle, to estimate its settling rate. The particle must, however, be of known shape.

The more usual situation is that neither the shape nor the size is known, that is, d_p and d_A both need to be determined experimentally.

Two commonly used methods of measuring particle size are microscopy and sieve analysis. For spheres both techniques give identical results but for nonspherical particles, the two methods, because they measure different properties, give different characteristic diameters for the same particle.

Thus, in microscopy the diameter of a particle, d_M, is the diameter of a circle with the same projected area as the particle which normally lies "flat," while in sieving the particle has a probability of being presented to the apertures in the most favorable position for passing, d_{scr}. Thus, for an elongated or platy particle, the two larger dimensions are recorded by microscopy while for sieving the two smaller dimensions are recorded.

Columns 5 to 7 in Table 13.3 compare the mean diameters of various shaped particles as measured by these two methods. It can be seen that d_M is, as expected, consistently larger than d_{scr}, except for a sphere where they are equal. The ratio d_M/d_{scr}, therefore, also provides a basis for formulating shape correction factors but it is less consistent than the ratio d_p/d_A. The latter can, however, be used only if d_p and the shape are known.

Much work has been done on characterizing the shapes of microscopic images alone,[16-18] but the weakness in applying such results to sedimentation[19] is that only two dimensions are considered. Thus, a cube and a square plate both of side d settle at quite different rates but are indistinguishable under the light microscope while a right circular cylinder of equal height and diameter may appear spherical to one observer and cubic to another depending on its position on the microscope slide. Different shape factors would therefore by applied to the same particle.

Both the experimental procedures for measuring shape factors as well as their interpretation are therefore involved but the main purpose is merely to apply a correction factor to Stokes' law so as to be able to calculate the settling velocity of any nonspherical particle. The necessity for this, however, is completely eliminated if for the measurement of d_p, we use an alternative, very commonly used method of particle sizing—the sedimentation method.

Table 13.4. Comparison of Calculated and Observed Shape Factors K_p.

	SHAPE FACTOR K_p	
SHAPE	CALCULATED, d_p/d_A	EXPERIMENTAL RESULT[16]
Sphere	1.0	1.0
Cube-octahedron	0.92	0.96–0.98
Octahedron	0.92	0.93–0.95
Cube	0.90	0.92–0.94
Tetrahedron	0.82	0.82–0.86

This technique supplies information on the settling velocity, u_{ns}, of any particle of unknown shape that when inserted into Eq. (13.8) gives the exact equivalent Stokes' diameter, d_{ns}. This is the diameter of a sphere of settling velocity identical to that of the nonspherical particle. It combines both the "true diameter" of the particle and its shape correction factor into a single term. For the transitional and turbulent flow regimes such direct observation is the only means of determining the settling velocities of nonspherical particles.[15] Investigations of settling of such particles have therefore been mainly experimental.[16, 20, 21] Most particles encountered in industrial practice are far from spherical to such an extent that not only the overall shape plays a role but also the microsurface topography.

The experimental observations of Richards and Locke[21] on the terminal settling velocities of various sized quartz particles u_{ns}, obtained by screening, may be used to determine the shape factor for this irregular material. The results are plotted in Figure 13.5.

It can be seen that the shape factor depends on the size d_p. Relating the shape factor to the corresponding Reynolds number Re < 1,

the data are scattered but average out at about 0.8; for Re > 2000 the shape factor is steady at 0.47, while in the transition region the factor shows a steady decrease with increasing Re.

The shape factor for a given nonspherical particle is therefore not even a constant for that particle but dependent also on the prevailing conditions.

Measurements of settling velocity of nonspherical particles are therefore simpler and of more use than prediction of these velocities from independent size and shape determinations.

13.2.3 Settling in the Presence of Other Particles

Many correlations have been presented to describe the effect of higher solids concentrations on the settling rate of uniformly dispersed particles.[23] Two effects have been observed. One is that some particles may loosely associate into a group, separated from each other by several diameters, and act as an entity descending at a higher rate than that corresponding to the expected terminal velocity of the individual particles. Such "clusters" are often transient and their occurrence has been observed[24-26] predominantly at low particle concentrations. Other particles in the same suspension remain single and may even show negative settling rates when being carried upward by the return flow from the rapidly descending clusters. Tory and Pickard[27] present a stochastic model that accounts for these wide variations in settling rate. They noted that in spite of variations of settling velocity between particles, the mean settling velocity as shown by their overall rate of descent remained remarkably constant.

The second effect is that as concentration increases each particle is subjected to increased drag owing to the higher volume of return flow fluid displaced by the sedimenting particles. Alternatively, the ideal fluid flow around each particle is disturbed by the presence of its neighbors.

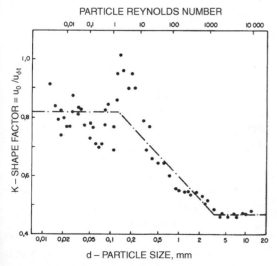

Figure 13.5. Variation of shape factor of quartz with particle size and Re.

13.2.3.1 Suspension of Uniform Particles

With closely sized particles uniformly distributed in a settling cylinder, a visible interface between suspension and liquid forms at commencement of settling even at low concentrations because of the constant means descent rate of each particle. In effect this interface is the one that exists between suspension and air before commencement of sedimentation. It is usually hazy, however, because particles are never exactly identical. At higher concentrations this interface becomes increasingly well defined and sharp, forming even for particles with a considerable range of sizes (see later). As such suspensions separate when dilute, because of the wide variation of settling velocities present, the formation of a distinct interface at higher concentrations indicates that interference between particles is such that particles of all sizes descend jointly, that is, they are in *hindered settling*.

Richardson and Zaki Equation. The settling velocity of the interface, u_s, was noted by Richardson and Zaki[28] to be related to the velocity of a single particle u_∞, and the concentration ϕ_s, by:

$$u_s = u_\infty(1 - \phi_s)^n \qquad (13.14)$$

where

u_s = mean settling rate of particles (particle–supernatant interface) in a container in the presence of many others

u_∞ = terminal velocity of a single representative particle, that is, u_s when $\phi_s = 0$, under otherwise similar conditions. It is a constant for a given solid–liquid system and equivalent to u_w in Eq. (13.31)

ϕ_s = volume fraction of particles (dimensionless) = C/ρ_s

C = mass concentration of particles, for example, kg/m^3

n = a constant = $f[d_p/D, \text{Re}]$

u_s/u_∞ = the hindrance factor = $(1 - \phi_s)^n$.

For streamlined flow, which has been shown to be the most common in thickening practice and hindering settling, n becomes independent of Re and was determined experimentally[28] to be

$$n = 4.65 + 19.5d_p/D \qquad (13.15)$$

and hence a second constant for a given system. An exact value of n in the range 4.65 to 5 is, however, seldom critical and, as shown later, a value of $n = 4.7$ is found to give a satisfactory correlation for a large number of real suspensions even where d_p is unknown, that is, when an exact value of n cannot be calculated from Eq. (13.15). A similar compromise value of $n = 4.7$ (n between 4.65 and 4.78) was arrived at also by Watanabe.[29]

Although Eq. (13.14) was derived empirically, various authors[15, 30, 31] have shown its general validity on theoretical grounds. The hindrance factor term is a simple one[15] that permits its modification to deal with sedimentation of irregularly shaped particles and particle aggregates which will be discussed later. Alternative hindering settling equations are much more complicated.[23] In some the concentration term appears in various forms up to five times in one equation rather than once as in Eq. (13.14).

Experimental Verification of the Richardson and Zaki Equation. As $(1 - \phi_s)$ is a fraction and n a positive number, Eq. (13.14) indicates a decrease in particle settling rate with increasing volume fraction of solids. Figure 13.6 compares the experimentally observed values of u_s for two suspensions of glass spheres with predicted values based on Eq. (13.14). The value of u_∞ in this equation was calculated from Eq. (13.8) and n calculated from Eq. (13.15). The agreement can be seen to be good. For an uncharacterized suspension for which d_p and hence the constants u_∞ and n are not known, it should be possible to estimate them from Eq. (13.14) by means of a plot of $\ln u_s$ versus $\ln(1 - \phi_s)$ and using the intercept and slope of the best fitting straight

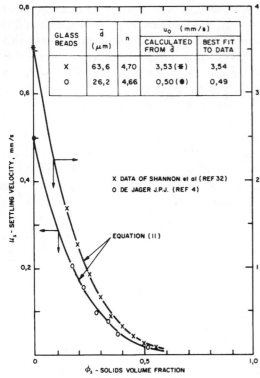

GLASS BEADS	$\bar{d}$ (μm)	n	u_0 (mm/s)	
			CALCULATED FROM $\bar{d}$	BEST FIT TO DATA
X	63,6	4,70	3,53 (✳)	3,54
O	26,2	4,66	0,50 (●)	0,49

X DATA OF SHANNON et al (REF 32)

O DE JAGER J.P.J. (REF 4)

EQUATION (II)

Figure 13.6. Settling velocity of glass bead suspension as a function of solids concentration.

line. The intercept is u_∞ and the slope is n. This is one basis for using hindered settling rates, u_s, to determine the mean particle size of a suspension of particles.

Two independent values of u_∞ estimated in this way are compared to Figure 13.6 and the excellent agreement found between them has been shown[33] to apply also to a variety of solids and liquids (d_p = 13 to 1740 μm; ρ_L = 890 to 1070 kg/m^3; and μ = 1 to 7 cp). The reliability of the Richardson and Zaki equation, together with their value of n, is therefore well established for relatively closely sized spherical particles up to the maximum attainable free-settling concentrations.

13.2.3.2 Suspensions Consisting of a Range of Particle Sizes, Shapes, and Densities

Suspensions consisting of uniform particles are rarely met within practice. Real slurries contain a range of particle sizes and often differ-

ent materials with various shapes and densities. Although such particles tend to segregate when dilute, at normal thickener feed concentration, mutual retardation of the particles in a batch test causes hindered of "zone-settling"[34] with a uniform particle settling velocity regardless of size. It is therefore less reliable to calculate u_∞ from Eq. (13.8) and, for reasons that are discussed in the next section, use is made of an alternative form of Eq. (13.14):

$$u_s^{1/n} = u_\infty^{1/n} - u_\infty^{1/n} \cdot \phi_s \qquad (13.16)$$

where n may be taken as equal to 4.7, that is, $1/n = 0.213$.

A plot of $u_s^{0.213}$ versus ϕ_s should therefore give a straight line from which either the slope or the intercept can now be used to determine u_∞. Although, because of differing d_p values, the values of u_∞ vary between different suspensions, the hindrance factor of any suspension $(1 - \phi_s)^{4.7}$, should be constant at a given concentration. The relative velocity, u_s/u_∞, should therefore be the same for each suspension at the same concentration and plots of $(u_s/u_\infty)^{0.213}$ versus ϕ_s should yield a single straight line of slope = -1 because from Eq. (13.16):

$$\left[\frac{u_s}{u_\infty}\right]^{0.213} = (1 - \phi_s) \qquad (13.17)$$

The data of Figure 13.6 are replotted on this basis in Figure 13.7 and they can be seen to fall closely on this theoretical line with intercepts at +1 and a negative slope of 45°. In addition, summarized data on spheres from an extensive survey of the literature[23] are included and the agreement with Eq. (13.17) is within 6%. Bearing in mind that no allowance was made in this work for fairly wide variations in d_p/D and its effect upon n and u_∞, this agreement is reasonable.

When, however, the sedimentation rates of angular quartz particles rather than spheres are compared on the same basis, Figure 13.7 shows that the settling rate now decreases

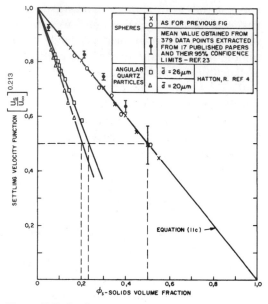

Figure 13.7. Settling velocity of spherical and angular particles plotted according to Eq. (13.17).

more rapidly with increasing concentration than predicted by Eq. (13.17). The spheres reach a hindrance factor $u_s/u_\infty = 0.0385$ at $\phi_s = 0.5$ while this same retardation is experienced by quartz particles at a concentration as low as 0.2. As the retardation in settling velocity of a particle in hindered settling is due to the interference offered to its ideal return fluid flow pattern by the presence of its neighbors, it must be concluded from hydraulic similarity considerations that quartz particles present a greater effective blockage to the return flow than can be expected from their volume. A unit volume of quartz must in fact have the retardation effect of $0.5/0.2 = 2.5$ volumes of an equivalent sphere. The plausible inference is that such angular particles carry with them attached water because of their roughness[14, 15] and this stagnant water behaves as if the volume of the particle were effectively increased. The net solids concentration is therefore greater than the volume of dry solids present.

By assuming the effective solids fraction for any degree of retardation to be similar to the volume fraction of spheres at the same retardation, we can calculate the proportion of

fixed water. Thus, if a quartz suspension at $\phi_s = 0.2$ settles according to Eq. (13.17) as if it experienced the same drag as a suspension of spheres at $\phi_s = 0.5$, its effective total solids volume must in fact be equal to 0.5, with the extra volume being made up of stagnant water than moves with the particle.

Because the stagnant water behaves as if solid, the lines for quartz in Figure 13.7 remain straight but because of the unknown quantity of water, they are of unknown slope. If we call the slope k_v, Eq. (13.17) can be rewritten

$$\left[\frac{u_s}{u_\infty}\right]^{0.213} = (1 - k_v\phi_s) \qquad (13.18)$$

or

$$u_s = u_\infty(1 - k_v\phi_s)^{4.7} \qquad (13.19)$$

Comparing with Eq. (13.14) it can be seen that the original concentration term ϕ_s is replaced by $k_v\phi_s$, which now represents the effective solids volume fraction. For spheres, where there is no stagnant liquid, $k_v = 1$, while for the two quartz suspensions $k_v \sim 2.5$ with the finer sample carrying relatively more water, that is, having a slightly higher value of k_v.

Correlation of the settling data for quartz according to Eq. (13.14), that is, $k_v = 1$ and n variable, gives values of n ranging from 11.8 to 14.6 depending on ϕ_s. The value of n is therefore not only much higher than expected from theory[30] (n lies between 1 and 8) but more seriously is not a constant for a given system. Many such correlations have been attempted,[35-40] with values of the exponent as high as 466.7, but as shown by Capes,[41, 42] these are reduced to expected levels if due allowance is made for the fixed water associated with particle agglomeration or irregular shape.

13.2.4 Aggregated Suspensions

Natural aggregation is frequently present in particle suspensions,[1, 43] especially at higher concentrations such as in thickener feeds, where the mutual proximity of the particles causes them to adhere and settle together as clumps rather than as single particles. This increase in "particle" size results in faster settling, and in thickening the effect is often exploited. In clarification, the solids concentration in the liquid is much lower, a natural aggregation is largely absent. Artificial flocculation is therefore always required. It can be brought about by reducing the mutually repellent charges on the particles by means of electrolytes (coagulation) or by bridging particles by the simultaneous adsorption of polymers.

In all cases, aggregates are produced, each consisting of a large number of varying size primary solid particles, associated together into a single relatively large sedimentation unit or floc. Such a floc includes not only this loosely held solids structure but also the interstitial stagnant water.[15] Flocs have a density lower than the solid particles, due to this water, but have a greatly increased diameter so that their settling rates are several orders of magnitude higher than those of the original individual particles. No distinction is made here between the terms flocs and aggregates or in their method of production, as only the sedimentation behavior of the final aggregates is of concern at this stage.

13.2.4.1 Types of Settling Behavior in Aggregated Suspensions

Previous sections dealt with discrete individual particles as the primary sedimentation units. Resuspension of these unaggregated pulps in a batch test to prepare a uniform suspension usually does not alter the size, shape, or settling characteristics of these units from test to test. In flocculated suspensions, however, the sedimentation units (aggregates) are freshly formed only after agitation ceases. The shape of the resultant sedimentation curve (height of interface H versus time t) depends on solids concentration,[44, 45] that is, on the number of primary particles present and their mutual proximity when agitation is stopped and the shearing force is removed.

Dilute Suspensions. When the suspension is dilute the aggregates (flocs) are formed independent of each other—they are widely spaced in the intervening liquid and descend through it as individual entities. After agitation ceases ($t = 0$) the floc formation time is fast[46] compared to the time over which sedimentation is observed in a batch test. For instance, in the presence of a coagulant, the silky appearance of dispersed micaceous clays noted during stirring disappears within seconds after agitation is stopped. Dilute suspensions of flocs therefore show a constant interface descent rate from zero time (curve A_1 in Figure 13.8).

Intermediate Concentrations. At higher concentrations, the particles have a better chance of forming larger flocs. At the start of a batch settling test, that is, after the cessation of agitation, the suspension appears to have an "induction" period (curve B_1, Figure 13.8) during which the relatively low initial sedimentation rate u_i increases with time either gradually,[44] or in discrete steps[45–51] and subsequently reaches a higher constant rate, u_s. The maximum steady value is accepted as the settling rate in a static batch test.[43, 52–54] Two phases in this acceleration process are shown in Figure 13.9A to C.

This rate is higher than would be expected from extrapolation of sedimentation data in the dilute rate (Fig. 13.10), indicating that the mode of sedimentation is now different.[58] The higher settling rates are attained not only by the formation of larger flocs but also by the reduction of resistance to relative movement of flocs and liquid. An anisotropic structure in the suspension with liquid channels of low flow resistance in an upward direction is formed during the induction period. A similar argument was used to explain the accelerated settling of intermediate concentration suspensions in the presence of particles of density lower than,[59] equal to,[60] or greater than[61] the density of the fluid.

At intermediate concentrations the flocs must be closer together than in dilute suspen-

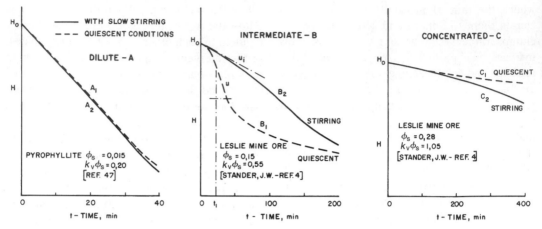

Figure 13.8. Mode of settling in aggregated suspensions depends on solids concentration and presence of mild agitation.

sions. Their probability of touching or bridging by particle growth is therefore much higher and this three-dimensional interaction between solids is likely to be involved in the formation of the channel structure with flow channels being developed between the flocs.[44] It is therefore not surprising that the induction period increases with concentration.[121] In the intermediate concentration range the maximum steady settling rates u_s decrease with concentration, but to a lesser extent than for pulps in the dilute range (Fig. 13.10). The decrease is not unexpected being at higher floc concentrations less voidage between them is available for channel flow and fewer and somewhat narrower channels may be formed.

Concentrated Suspensions. The solid particles are in a compression zone. The suspension does not attain any degree of "mobility" but subsides at a sluggish and ever decreasing rate. When agitation is stopped the particles are closer together and are able to form a three-dimensional structure like a packed bed. The lower layers can be further compacted by the weight of solids from above. The particles collapse inwardly toward each other and consequently liquid is expressed from these layers. This liquid moves upward through the bed and because of the tight packing, mainly between

the primary particles.[64] Some channeling may occur and often a few very large channel openings (volcanoes) are observed at the interface.

Settling curves such as those shown in Figure 13.8 A_1 to C_1 were obtained[66] also for sedimentation of the coal particles in oil, that is, a nonaqueous system, indicating their general nature.

Slow Agitation in Aggregated Suspensions. The effect of very mild stirring (0.1 to 2 rpm) depends on the concentration regime present. In dilute suspensions the formation and the subsequent sedimentation of the flocs is neither aided nor hindered and settling rates are therefore little affected (curve A_1 and A_2, Fig. 13.8). In intermediate suspensions, horizontal shear hinders the formation of short-circuit flow channels and materially decreases the maximum settling rate attained[47,62] (curve B_2 rather than B_1).

In concentrated slurries, mild mechanical disturbance promotes the shearing of the particle–particle links. Under quiescent conditions the three-dimensional structure that forms after cessation of agitation tends to resist collapse because of friction at the points of particle contact and the support from the base and the walls. The mass of solid above may not be sufficient to overcome the strength

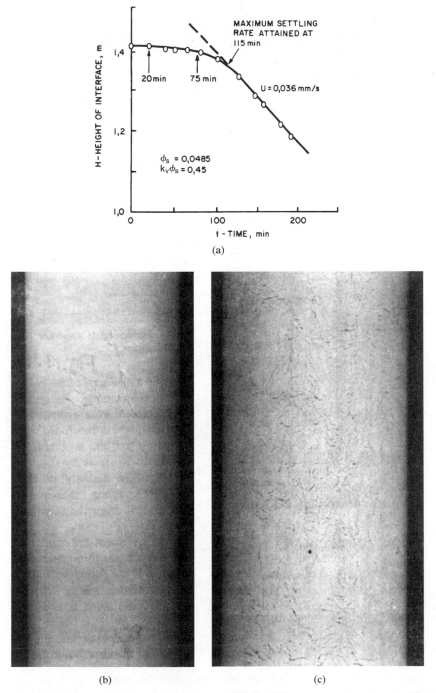

Figure 13.9. (A) Intermediate setting in a desanded mine pulp[47] showing the height of the pulp interface when photographs in B and C were taken. (B) Commencement of the break-up of the initially gelled mass; $t = 20$ min. (C) Agglomerated structure beginning to appear; $t = 75$ min.

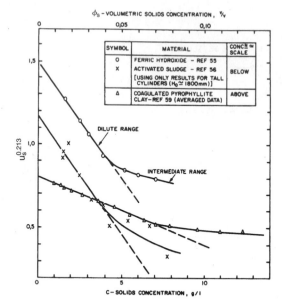

Figure 13.10. Settling velocities at low concentrations according to Eq. (13.20) of three widely different types of suspensions.

of the particle structure and hence slow shearing of the suspension results in bonds being broken, a realignment of the particles, release of water, and hence promotion of the subsidence of the interface, curve C_2 in Figure 13.8.

13.2.4.2 Calculating Settling Velocity of Aggregated Suspensions

The Richardson and Zaki equation [Eq. (13.19)] had been applied to predict the settling velocities of aggregated suspensions. It was quite successful in the dilute concentration range. Deviations from predictions were encountered for the intermediate suspensions.

Dilute Suspensions. With water in the voids between the particles of a single aggregate the volume occupied by the flocs, defined as the total volume of the envelopes surrounding each, is therefore greater than the volume fraction of actual solids present. The diameter, density, and concentration of the flocs should

therefore replace these corresponding terms for spheres in Eqs. (13.8) and (13.14).

Observation of the interface setting rate u_s of such suspensions over a range of concentrations ϕ_s can be expected to be correlated according to Eq. (13.19), which may be rewritten as:

$$u_s^{0.213} = u_F^{0.213}(1 - k_v \phi_s) \qquad (13.20)$$

where

u_F = terminal settling velocity of one representative floc
$k_v \phi_s$ = volume fraction of flocs
k_v = volume of floc per unit volume of contained solid.

Where, as in the case of hydroxide precipitates or organic suspensions, the volume fraction of the actual solids is not readily determined, k_v can be equally well expressed as k' = volume of floc per unit mass of contained solid and concentration C = the mass of solids per unit volume of suspensions. As $k' = k_v/\rho_s$ and $C = \phi_s \rho_s$, $k'C$ becomes identical to $k_v \phi_s$, that is, it is not necessary to know ρ_s.

As for dispersed particles in hindering settling, a plot of the experimental values $u_s^{0.213}$ and ϕ_s should give a straight line of intercept $I = u_F^{0.213}$, and of slope $I \cdot k_v$ over the concentration range pertaining to dilute suspensions. This has been found to apply to a large variety of suspensions and plots for a typical metal hydroxide precipitate, clay-mineral and activated sludge, are shown in Figure 13.10.

A similar settling pattern is demonstrated by each of the suspensions shown in Figure 13.10 despite the variation in their nature and composition. There is the expected linear plot at low concentrations followed by settling rates that deviate positively at higher values.

The data of Figure 13.10 are replotted, replacing the abscissa ϕ_s by the floc volume concentration $k\phi_s$, and normalizing the interface velocity with respect to u_F, in Figure 13.11. It is now the equivalent of Eq. (13.17) with (assumed) spherical flocs in the place of solid spheres and should therefore appear as the 45° line in Figure 13.7.

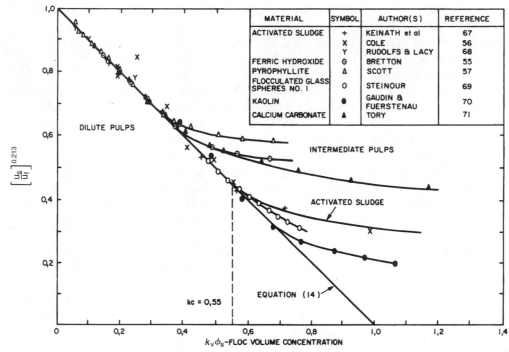

Figure 13.11. Dimensionless maximum interface sedimentation rates; data of Figure 13.10 and other sources plotted on normalized coordinates.

To demonstrate the general applicability of this plot, other aggregated suspensions have been included. For all suspensions in the dilute range, where the flocs behave as separate entities, there is a straight line of slope = −1 as expected, up to a characteristic value of $k_v \phi_s$, followed by increasing positive deviation from Eq. (13.20). This marks the onset of channel flow and an increased permeability of the suspension as a whole.

For a specific slurry the point of departure from the straight line indicates the upper limiting concentration of its dilute range. Similar materials, for example, activated sludges (Fig. 13.11) or red mud,[58] show similar limiting concentrations, that is, $k_v \phi_s$ at the point of departure from Eq. (13.20) does not differ significantly for the same material even though it may be derived from different sources.

Intermediate Concentrations. The slurry interface at concentrations higher than for the dilute range settles faster than expected from a suspension of flocs assumed to consist of individual rigid spheres. This onset of intermediate settling behavior is observed at $k_v \phi_s$ values from about 0.35 onward (Fig. 13.11). The onset would be expected to be related to the size and shape of the flocs and the interparticle and interfloc forces, which all play a role in the formation of a structure in the flocculated suspension. The settling rate is better determined by direct measurement.

13.2.5 Measuring Settling Rates

As discussed earlier, the Richardson and Zaki type equations correlate experimental data fairly well to a certain extent. A real world slurry contains a variety of particles with different shapes, sizes, and densities. The u_F in the Richardson and Zaki equation cannot be calculated directly but must be determined by actual settling tests (as in Fig. 13.10). As discussed earlier the settling rate deviates from the equation as the concentration becomes thicker and thicker. It is obvious that an actual settling test is still essential in studying sedimentation phenomena.

Sedimentation phenomena are usually studied by observing the behavior of suspensions placed in cylinders (frequently with a volume of 1 to 2 liters). Sedimentation in a cylinder consists of descent of particles and rise of sediment from the bottom. Two typical sedimentation curves are shown in Figure 13.12 for a slow settling clay (attapulgite) and a fast settling microbarite (principally BaSO$_4$ used for weighting purposes in drilling muds). The height of the descending interfaces along AB and AC and the rise of the sediment along OB and OD are shown. The slopes of the lines AB and AC yield the settling velocities relative to the container walls. For microbarite, $u_s = (420 - 70)/500 = 0.7$ mm/s and for attapulgite, line AC yields $u_s = 0.032$ mm/s.

As the solids settle, liquid is displaced upward. The downward flux $u_s \phi_s$ equals the upward liquid flux $(1 - \phi_s)u_L$ and

$$u_s \phi_s + (1 - \phi_s)u_L = 0 \qquad (13.21)$$

The velocity of the liquid is given by

$$u_L = -\frac{\phi_s}{\phi_L} u_s = -\frac{\phi_s}{1 - \phi_s} u_s \quad (13.22)$$

The velocity u_{sr} of the solids relative to the liquid is the most significant quantity rather than the observed velocity. It is given by

$$u_{sr} = u_s - u_L = \left(1 + \frac{\phi_s}{1 - \phi_s}\right)u_s = \frac{u_s}{1 - \phi_s}$$
$$(13.23)$$

The relative velocity is larger than the absolute velocity (u_s, the velocity relative to the wall of settling chamber, that is, the velocity measured during the settling test). As ϕ_s was essentially zero for the sedimentation of a single particle, the Stoke's velocity requires no correction.

In Figure 13.13 the various stages involved in a batch sedimentation of dilute to moderately concentrated suspension in a cylinder are illustrated. Along AB, the rate of sedimentation is constant and this rate is taken as the settling velocity at this initial concentration. From B to C, known as the first falling rate period, the slope decreases, indicating that the concentration is increasing. Simultaneously, the sediment is rising from the bottom as shown by the L versus t curve. When the upper descending boundary meets the ascending sediment at the compression point C, the compression period (also called the second falling rate period) begins. Further decrease in height is effected solely by flow of liquid out of the compaction zone because of the weight of the solid particles. When the final structure carries the entire weight of the sediment, liquid flow ceases.

For thickener design (to be discussed later), the relationship between solid settling flux and concentration is required. A series of batch

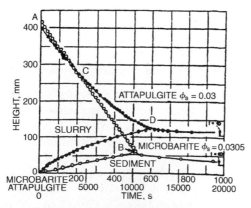

Figure 13.12. The sedimentation of microbarite and attapulgite. The initial slurry heights are 420 mm for microbarite and 405 mm for attapulgite.

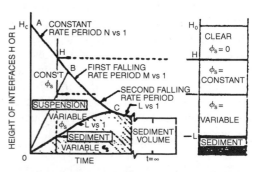

Figure 13.13. The various stages of sedimentation are illustrated. Conditions in the cylinder at a time corresponding to height H are shown.

settling tests at different concentrations lead to a relationship between u_{sr} and ϕ_s as shown in Figure 13.14. Particles tend to settle independently in dilute slurries, and consequently, there is no unique settling velocity for such slurries. As concentration of the slurry ϕ_s increases and settling of large particles is impeded by the presence of small particles, a point is reached were all particles presumably have identical velocities and settle as a "zone." Ultimately, as the concentration continues to increase, a point is reached where the solids form a cake capable of transmitting stresses through points of contact. The solids then enter into the matrix. As a crude approximation the null stress solid concentration ϵ_{s0} marks the beginning of the cake zone.

When uniform particles settle, a distinct interface is present even for dilute slurries; and the distinction between zone and dilute settling disappears. The extrapolated velocity corresponds to the Stokes velocity. Although there is not theoretical Stokes velocity when slurries with particles having a range of sizes are involved, an extrapolation to point A as shown in Figure 13.14A is employed to produce a pseudo-Stoke's velocity that can be used in empirical correlations.

In the zone settling region, it is generally assumed that the relative settling velocity is a unique function of concentration. If the size range does not include large, dense particles or submicron particles with high diffusion coefficients, settling will be predominantly in the zone mode. Fine, dispersed particles with diameters less than 0.1 micron will diffuse out of the descending slurry–liquid interface into the supernatant region.

In Figure 13.14A, the relative sedimentation curve is shown as terminating at a value of $\phi_s = \epsilon_{s0}$, where the particles enter into physical contact and form a cake. The velocity of the cake surface is no longer a unique function of the slurry concentration. It depends on the rate at which liquid is squeezed out of the cake by the weight of the cake. Nevertheless, many investigators have mathematically treated the compression zone in the same manner as the first falling-rate period. Ultimately as shown in Figure 13.12 by the point marked $t = \infty$, the sediment reaches a point at which there is no more compaction. At that point, the solid velocities are everywhere zero. Very few reliable data involving sedimentation velocities at concentrations near the cake region have appeared in the literature. Data are different to obtain and difficult to interpret.

13.2.5.1 Kynch Theory

Kynch[132] (1952) made an important improvement in the sedimentation theory. Instead of performing a series of batch tests to obtain the flux–concentration relation, Kynch developed a means to achieve that by a single batch sedimentation test.

The first falling rate period as shown in Figure 13.13 is the result of action that takes

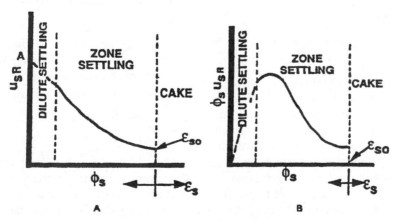

Figure 13.14. Relative settling velocities and relative flux as a function of concentration.

place at the bottom of the cylinder. When particles reach the bottom and start to form a cake, the liquid is squeezed out and flows upward. The upflowing liquid retards the settling of the particles above, resulting in a more concentrated slurry. The retarding effect propagates upward through the settling particles and can be treated as a signal carried by a characteristic wave of constant concentration. The constant rate period comes to an end when the first characteristic reaches the supernatant–slurry interface. Successive signals originating at the surface of the sediment lead to a decreasing sedimentation velocity throughout the suspension.

The equation of continuity in a settling column is given by

$$\frac{\partial \phi_s}{\partial t} + \frac{\partial (u_s \phi_s)}{\partial x} = 0 \qquad (13.24)$$

where

x = distance up from bottom of the settling column

t = time.

If the settling rate is a function of concentration only,

$$\partial (u_s \phi_s)/\partial x = d(u_s \phi_s)/d\phi_s \cdot \partial \phi_s/\partial x$$
$$= v(\phi_s)\partial \phi_s/\partial_x$$

Equation (13.24) can be rewritten as:

$$\frac{\partial \phi_s}{\partial t} + v(\phi_s)\frac{\partial \phi_s}{\partial x} = 0 \qquad (13.25)$$

The solution of Eq. (13.25) is

$$x = v(\phi_s)t + \text{constant} \qquad (13.26)$$

Equation (13.26) represents a straight line for a characteristic of constant concentration.

The method by which Kynch obtained the flux curve from a single settling plot is illustrated by Figure 13.15. If a characteristic with slope v emanating from the origin travels upward, it meets the supernatant–slurry interface at R. The total solids in the column will pass this characteristic during this period ($t = 0$ to t_R). Because the settling velocity is considered as function of concentration only, the

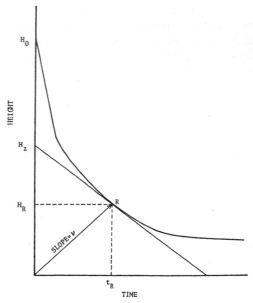

Figure 13.15. Kynch construction for a batch settling curve.

moment the particles pass the characteristic, they have the same concentration and same settling rate. A material balance over the characteristic takes the form

$$t_R \phi_{sR}\{v + (-u_s)\} = \phi_{s0}H_0 \qquad (13.27)$$

where

ϕ_{s0} = initial slurry concentration
ϕ_{sR} = slurry concentration at point R
v = characteristic velocity.

From Figure 13.15, it can be seen $v = H_R/t_R$, substituting into Eq. (13.27) gives

$$\phi_{sR} = \phi_{s0}H_0/\{H_R + (-u_s)t_R\} \qquad (13.28)$$

The settling velocity u_s is the slope of the settling curve at R, $(dH/dt)_{t=tR}$. Also the intercept of the tangent at R is

$$H_z = H_R + (-u_s)t_R \qquad (13.29)$$

Substituting Eq. (13.29) into Eq. (13.28) yields

$$\phi_{sR} = \frac{\phi_{s0}H_0}{H_z} \qquad (13.30)$$

Therefore, a relation between the solid concentration and sedimentation rate can be obtained along the first falling rate period of a settling curve.

Kynch ignored the sediment at the bottom of the settling chamber. Therefore, he argued that the constant in Eq. (13.26) is zero and all the characteristics emanate from the origin of Figure 13.15. Tiller[124] took into account the effect of the sediment rising from the bottom, revised Kynch's argument, and suggested that the characteristics come from the surface of the sediment. Fitch[125] considered the characteristic as a kind of concentration discontinuity that emanates either from the origin or from the cake surface depending on the initial concentration of the suspension and the shape of the flux curve. He states that the surface of the sediment was also a concentration discontinuity propagating upward. At the moment the characteristic leaves the cake surface, these two discontinuities should have the same velocity. Therefore, a characteristic should rise tangentially from the cake surface.

13.2.6 The Effect of Container Walls

When a particle sediments in a closed column rather than in an infinite liquid, it displaces its volume of liquid from a lower to higher level and the wall interferes with the ideal liquid flow pattern. This results in an additional drag on the particle and a reduction in the free settling velocity u_∞ by a factor W, which decreases with increasing ratio d_p/D, where D is the diameter of the container. The retardation effect may be expressed as follows:

$$u_w = u_\infty \cdot W \qquad (13.31)$$

and many expressions for W have been proposed.[1]

According to Francis[22]

$$W = \left[\frac{1 - 0.475(d/D)}{1 - (d/D)} \right]^{-4} \qquad (13.32)$$

for streamlined flow, while Garside and Al-Dinbouni[23] give a simpler equation: $W = [1 + 2.35(d/D)]^{-1}$, applicable for Reynolds numbers between 3 and 1200.

For a particle of $d_p = 200$ μm, settling in a one-liter graduate cylinder ($D \sim 60$ mm) or in a 25 m diameter tank, the calculated settling velocities, u_w, are 99.3% and 99.998% respectively of its velocity in an unbounded fluid.

Although wall effects in full-scale equipment may be safely ignored, design of such equipment is often based on settling rates observed in small laboratory glassware, for which corrections may therefore sometimes be required. In practice this correction is, however, usually neglected as the consequent error is both small and conservative.

13.3 THICKENING

Gravity thickening provides a means for economically removing a large fraction of the liquid in a slurry. The process is shown schematically in Figure 13.16 and the equipment used in Figures 13.1 and 13.3. In thickener technology, a slurry, sludge, pulp, or mud all describe a suspension of solid particles in a liquid. Schematically clarifiers and thickeners appear to be identical, and there is no sharp line between the two. In general, clarification involves suspensions in the dilute ppm (mg/liter) range whereas thickening tends to treat more concentrated slurries in the 1% and above range. However, it needs to be noted that the meaning of "dilute" or "concentrated" varies from industry to industry. For instance, the feed to a thickener used in mineral industries could be 5% by volume which could be equal or higher than the concentration of a cake produced in municipal waste water applications. Care must be exercised in interpretation of concentration limits

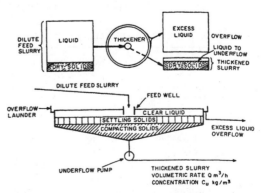

Figure 13.16. Schematic view of continuous thickening process.

suggested as *typical* for feed and outputs of solids–liquid separation processes.

Recovery and further processing of solids generally following thickening. Consequently, the concentration of the underflow is critical to subsequent operations. The density of the underflow from a clarifier is of less importance. As less solids are involved, the mechanical equipment for clarifiers is light compared to that for thickeners in which large volumes of dense materials generally require heavy raking systems.

In idealized free settling theory, the settling rate is considered to be a unique function of slurry concentration. Design methods based on this principle apply to cases when no sediments are present, and the underflow is simply a suspension with a higher concentration than the feed. When higher underflow concentrations are desired, sediments subject to compressive effects due to the unbuoyed weight of the solids are required. In the sediment, particles enter into contact, and the solid velocity is no longer a unique function of concentration. The liquid and solid fluxes are determined through the use of the Darcy–Shirato[126] equations relating the relative velocity of the solids to the liquid, pressure gradient, and the permeability.

These two distinct mechanisms of thickening process are discussed separately in this text.

13.3.1 Nomenclature

Different nomenclatures used in this field have been a source of confusion. A comparison of the symbols used in this text is listed in Table 13.5. For the most part in the industry, concentrations are given in mass/unit volume (kg/m^3 or lb/ft^3). They are represented by the letter C, which is generally employed by authors writing on free-settling theory. The same letter C has been used for both the suspension and the sediment. The new trend in the solids–liquid separation field is to use volume fractions that provide true concentration comparisons among different processes. It was also found advantageous to use different symbols for the free-settling (ϕ_s) and compression zones (ϵ_s).

13.3.2 Thickening in the Free Settling Region

Most of the existing design methods for continuous gravity thickeners fall into this category. The methodology provides a means of determining the area requirements of thickeners. The settling velocities and fluxes are required.

13.3.2.1 Design Procedures

The Coe and Clevenger Method. Coe and Clevenger[52] were the first authors to establish a rational method for the sizing of thickeners. They studied the settling of metallurgical pulps and correlated batch sedimentation phenomena with the design of continuous thickeners.

In a continuous thickener, the settling flux is taken relative to the bulk flow of the slurry. The slurry as a whole is also moving downward owing to continuous volumetric draw-off at

Table 13.5. Symbols for Thickening.

	MASS CONCENTRATION: SUSPENSION OR SEDIMENT	VOLUME CONCENTRATION	
		SUSPENSION	SEDIMENT
Variable concentration	C	ϕ_s	ϵ_s
Feed	C_F	ϕ_{sF}	—
Underflow	C_u	ϕ_{su}	ϵ_{su}
Critical concentration	C_{crit}	$\phi_{s,crit}$	
Settling velocity	u_s	u_s	u_s
Solid flux (Cu_s)	G	q_s	q_s
Solid flux at critical concentration	G_{min}	q_{smin}	—

the base. If the underflow pumping rate is Q m³/h and the area of cylindrical section of the thickener is A m² then the bulk slurry velocity is Q/A m/h and the solids flux due to underflow pumping alone, called the underflow flux, is $(Q/A)C$. The total flux G_T, the solids flux relative to the walls, is the sum of the settling flux and the underflow flux, that is,

$$G_T = u_s C + (Q/A)C = (u_s + Q/A)C$$
$$(13.33)$$

The concentration of the underflow under steady-state conditions must meet the materials balance: rate at which solids pass through the thickener equals the rate at which solids are discharged in the underflow, that is,

$$G_T = \frac{Q}{A} C_u \qquad (13.34)$$

Eliminating Q/A between Eqs. (13.33) and (13.34) yields

$$G_T = \frac{u_s}{\dfrac{1}{C} - \dfrac{1}{C_u}} \qquad (13.35)$$

Equation (13.35) is the design equation. To apply Coe and Clevenger's method, the batch settling rates u_s at concentrations ranging from that of the feed to the thickest "free-settling" slurries need to be measured. By inserting the corresponding values of u_s and C in Eq. (13.35), G_T can be calculated from the feed C_F to the underflow C_u. Thickening is a process of reducing the mean interparticle distance. Therefore, sometime during their passage through the thickener, feed particles have to traverse the range of interparticle distances

from the one representing the feed to the thickened underflow. All intermediate concentrations will therefore exist even if only as transients. The maximum solids throughput of a thickener is governed by the concentration layer that has the lowest solids flux. The minimum value G_{min} is then selected for designing the cross-sectional area for the thickener.

Example 13.1. An aqueous slurry of a mineral is to be thickened from 10 lb/ft³ to a concentrated underflow of 60 lb/ft³. The amount of solids to be recovered is 350 tons (dry basis). The density of the solids is 200 lb/ft³.

a. Calculate the overflow and underflow rates.

The volumetric concentrations of the feed (ϕ_{sF}) and underflow (ϕ_{su}) are
$$\phi_{sF} = 10/200 = 0.05$$
$$\phi_{su} = 60/200 = 0.30$$

The volumetric rate of the feed and the underflow are
$$Q_F = 350 \cdot 2000/10 = 70,000 \text{ ft}^3/\text{day}$$
$$Q_u = 350 \cdot 2000/60 = 11,667 \text{ ft}^3/\text{day}$$

The overflow rate is
$$Q_{over} = Q_F - Q_u = 70,000 - 11,667$$
$$= 58,333 \text{ ft}^3/\text{day}.$$

b. Determine the unit area and the total area.

Batch settling tests need to be conducted at different slurry concentrations to obtain the settling rate and flux information. The following batch sedimentation velocities were determined at 80°F.

C, lb/ft³	10	15	20	25	30	40	50	60
u_s, ft/h	6.13	4.13	2.66	1.65	1.10	0.65	0.40	0.26
G_T, lb/ft²·h	73.6	82.6	79.8	70.7	66.0	78.0	120.0	∞

The solid flux G_T is calculated according to Eq. (13.35):

$$G_T = 6.13/(\tfrac{1}{10} - \tfrac{1}{60}) = 73.6 \; \text{lb/ft}^2 \cdot \text{h}$$

The minimum value of G_T is 66 lb/ft$^2 \cdot$ h at $C = 30$ lb/ft^3. This represents the choke points where the thickener is expected to operate. The total area required can then be calculated

$$\text{area} = 350 \cdot 2000/24/66 = 411 \; \text{ft}^2$$

Applying a safety factor of 25% leads to 550 ft^2, corresponding to a diameter of 27 ft.

c. During winter, the average water temperature is 40°F. How should this factor affect the thickener design?

Without an actual settling test of the slurry at 40°F (which is probably the most reliable way), it can be assumed that the settling velocities are inversely proportional to the viscosity of water. Viscosities are

80°F	0.86 cp
40°F	1.55 cp

With the velocity reduced by the ratio $0.86/1.55 = 0.55$, the minimum flux becomes $0.55 \cdot 66 = 36.3$ lb/ft$^2 \cdot$ h. The area required is

$$\text{area} = 350 \cdot 2000/24/36.3 = 804 \; \text{ft}^2$$

If a 25% safety factor were applied, a 1000 ft^2 (36 ft in diameter) unit should be designed. Clearly, varying seasonal effects are significant and must be considered.

d. Assuming that the distance from the suspension surface to the overflow is 5 ft, determine the largest sized particle that would be carried over in the overflow for both summer (80°F) and winter (40°F) operation.

The up flow velocity of water is $58{,}333/1000/24/3600 = 0.000675$ ft/s $= 0.000206$ m/s. Using Stoke's settling law [Eq. (13.8)] $u_s = d^2(\rho_s - \rho_L)g/18\mu$, for summer

$$d = \left[\frac{18\mu u_s}{(\rho_s - \rho_L)g}\right]^{0.5}$$

$$= \left[\frac{18 \cdot 0.00086 \cdot 0.000206}{(3210 - 1000) \cdot 9.8}\right]^{0.5}$$

$$= 12 \; \mu\text{m}$$

For convenience, the variables were converted to SI units during the calculation. The result shows a particle larger than 12 μm would not be carried over in the overflow. For winter operation

$$d = \left[\frac{18\mu u_s}{(\rho_s - \rho_L)g}\right]^{0.5}$$

$$= \left[\frac{18 \cdot 0.00155 \cdot 0.000206}{(3210 - 1000) \cdot 9.8}\right]^{0.5}$$

$$= 16 \; \mu\text{m}$$

Larger particles can flow out. The thickener has poorer performance in the winter. Bear in mind that this is an oversimplified calculation as the flow patterns in a thickener involve circulation owing to the introduction of the feed and are complex.

The Hassett[65] Method. This is very similar to the Coe and Clevenger method. The graphical representation of the total flux curve [Eq. (13.33)] is shown in Figure 13.17. Disregarding the very dilute concentrations that normally apply only in clarification, the total flux curve can be seen to show a minimum value at M. The concentration of the minimum flux zone

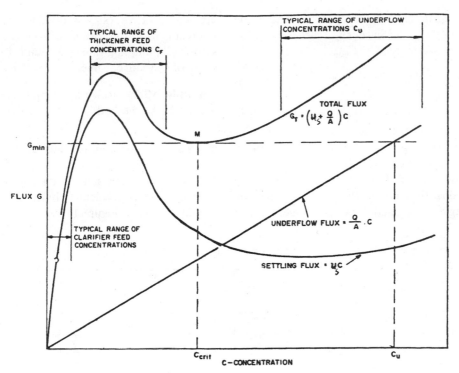

Figure 13.17. Underflow and settling fluxes may be summed graphically to give the total flux (Hassett method).

is (by definition) C_{crit} and its total flux (the minimum value of G_T) is

$$G_{min} = (u_{s,crit} + Q/A)C_{crit} \quad (13.36)$$

where U_{crit} = settling velocity at C_{crit}. Material balance [Eq. (13.34)] gives

$$G_{min} = \frac{Q}{A}C_u \quad (13.37)$$

The underflow flux line $(Q/A)G$ reaches the value G_{min} at the concentration $C = C_u$ (Fig. 13.17).

Determination of G_{min} according to Hassett's construction requires the drawing of a total flux curve for each value selected for the pumping rate Q and gives C_u only after G_{min} is determined. Although it has the advantage of clearly illustrating the minimum value of G_T, it is cumbersome as C_u is normally the primary thickening objective.

The Yoshioka Method. In 1957, Prof. N. Yoshioka of Kyoto University developed a procedure that has been looked on with favor by authors writing on thickening. The Hassett method suffers from the need to plot a separate curve for each underflow rate. In the Yoshioka procedure, only one graph is needed. For illustration, Figure 13.17 is replotted in 13.18. A line is drawn through point P (C_u on the abscissa) at an angle of which the tangent is $-Q/A$. Congruency considerations dictate that this line intercepts the ordinate at the value of G_{min}. The equation for this line is

$$G = G_{min} - \frac{Q}{A}C \quad (13.38)$$

which at C_{crit} attains the value $G_{min} - QC_{crit}/A$. At this concentration the equation for the settling flux, which is $u_sC = G_T -$

QC/A [Eq. (13.33)], also attains the value $G_{min} - QC_{crit}/A$. Since G_T reaches its minimum at M where $C = C_{crit}$

$$\frac{dG_T}{dC} = \frac{d(u_s C)}{dC} + \frac{Q}{A} = 0 \quad (13.39)$$

Thus, $d(u_s C)/dt = -Q/A$ at $C = C_{crit}$. The line drawn from G_{min} to P therefore coincides with the settling flux curve at C_{crit} and forms a tangent to this curve at N.

Yoshioka et al.[72] proposed determining G_{min} more directly than via the total flux curve by starting from C_u and drawing a tangent to the

$u_s C$ curve. Thus G_{min} is obtained from the intercept with the ordinate and the corresponding pumping rate Q, found from the slope of the tangent line.

Example 13.2. Rework Example 13.1 with the Yoshioka method.

b. Determine the unit area and the total area.

The same batch settling data in Example 13.2 are used to calculate batch settling flux and plotted in Figure 13.19.

C, lb/ft^3	10	15	20	25	30	40	50	60
U, ft/h	6.13	4.13	2.66	1.65	·1.10	0.65	0.40	0.26
UC, lb/ft$^2 \cdot$ h	61.3	62.0	53.2	41.3	33.0	26.0	20.0	15.6

The G_{min} determined graphically is 66 lb/ft$^2 \cdot$ h at $C = 30$ lb/ft^3, which is the same as obtained by the Coe and Clevenger method.

The Talamage and Firch Method. As only one concentration limits thickener throughput for a selected underflow concentration (or underflow pumping rate) it is an advantage if this value could be arrived at directly, thereby eliminating the need to measure the settling rates of many irrelevant concentration values. Talmage and Fitch[63] developed a procedure for obtaining the required minimum flux from a single batch settling test on the proposed thickener feed by observing not only the initial steady settling rate but also the complete $H - t$ settling curve.

The Kynch model discussed earlier was extended by Talmage and Fitch to design a continuous thickener. The solids flux that can be passed through the thickener is

$$G = C_0 H_0 / t_u \quad (13.40)$$

where t_u is the time required for settling solids to reach underflow concentration. The equa-

tion correlates the initial slurry height and concentration in a batch test to the continuous thickener underflow concentration C_u and its equivalent batch test solids height is

$$H_u = C_0 H_0 / C_u \quad (13.41)$$

To apply this method:

1. An attainable and acceptable C_u is selected.
2. H_u is calculated from Eq. (13.41).
3. An "underflow line" is drawn parallel to the time axis at $H = H_u$. If it intersects the settling curve above the compression point, then t_u is read directly from the settling curve (Fig. 13.20A). If the underflow line intersects the settling curve below the compression point, t_u is obtained as the intersection of the underflow line with a tangent drawn to the settling curve at the compression point (Fig. 13.20B). Once the value of t_u is obtained, the solids flux which corresponds to the required thickener area is calculated from Eq. (13.40).

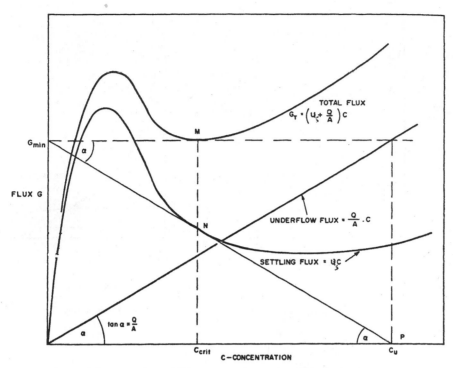

Figure 13.18. Yoshioka construction on a settling flux curve.

13.3.3 Thickening in the Compression Region

All the above design procedures employ flux theory which is based on the rate of sedimentation being uniquely dependent on the solid concentration. Such an assumption works reasonably well until particles enter into contact and form a sediment. In the sediment, Darcy's law applies, and the solid flux is no longer a unique function of the solids concentration. The solid flux is constant and independent of depth in the steady-state flow. However, concentration changes with depth and is a function of underflow rate and compressibility parameters.

Basic variables involved in gravity thickening are underflow concentration (ϵ_{su}), solid flux (q_s), and the height (L) of the compression zone. Other important quantities are permeability (K), local solidosity (volume fraction of solids, ϵ_s), liquid viscosity (μ), densities of solids (ρ_s) and liquid (ρ_L), and gravity (g). Flux theory omits most of these parameters and depends on the solution of the continuity equation [Eq. (13.24)]. Design of thickeners with a compression zone depends on the integration of the Darcy equation.[127-130]

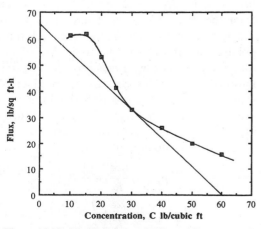

Figure 13.19. Yoshioka construction on Example 13.2.

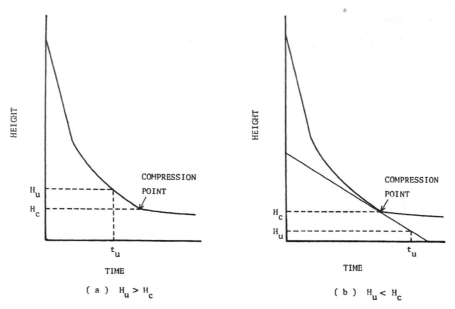

Figure 13.20. Locating underflow time in the Talmage and Fitch method.

13.3.3.1 Model for Continuous Thickener

The model is based on an idealized flat-bottomed, steady-state thickener. In Figure 13.21, the various regions are listed as overflow, feed-transition, thickening, and discharge. The increasing volume fraction of solids from ϵ_{s0} to ϵ_{su} is shown. Both solid and liquid fluxes are constant at any point x above the bottom of a thickener. Both solid and liquid have downward (positive) velocities, with the solids flowing down more rapidly. No liquid is squeezed upward. At the surface of the sediment, all the solids are accepted. A portion of the liquid is rejected and flows upward and out from the overflow. A material balance of the flow in the sediment compression zone is

$$q_F \phi_{SF} = q_s = q_{su} = \epsilon_s u_s = \epsilon_{sB} u_{sB} \quad (13.42)$$

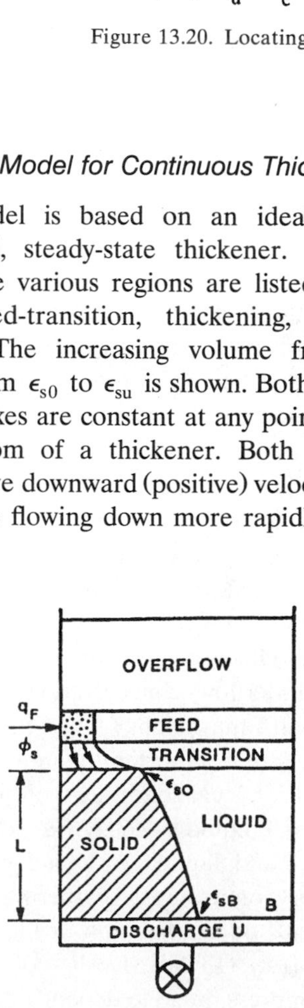

It needs to be noted that q_s is equivalent to G_T used in Eq. (13.33) for the Coe and Clevenger method. The flux of solids (q_s) is also called the "superficial solid velocity." The first term in Eq. (13.42) gives the solids flux from the feed. The second and third terms simply reaffirm that the flux q_s at an arbitrary point equals the flux in the underflow of the thickener. The last two items provide the same information with respect to the product of volume fraction of solids and the local true average velocity (u_s, u_{sB}). The pressure drop in a continuous thickener can be expressed by the Darcy–Shirato equation as:

$$\frac{dp_L}{dx} = -\rho_L g + \frac{\mu\epsilon}{K}(u - u_s)$$

$$= -\rho_L g + \frac{\mu\epsilon}{K}\left(\frac{q}{\epsilon} - \frac{q_s}{\epsilon_s}\right) \quad (13.43)$$

Figure 13.21. Idealized thickener.

where the relative velocities of solids and liquid are used. For compressible sediments, the structure parameters such as permeability (K), porosity (ϵ), and solidosity (ϵ_s) are considered to be functions of solids compressive pressure (p_s). The liquid pressure p_L in Eq. (13.43) must be replaced by p_s before integration is possible. Assuming point contact among particles, a force balance over a distance dx (Fig. 13.22) leads to

$$\frac{dp_L}{dx} + \frac{dp_s}{dx} + g(\rho_s \epsilon_s + \rho_L \epsilon) = 0 \quad (13.44)$$

where the effective pressure (p_s) represents the net vertical stress divided by the cross-sectional area. Combining Eqs. (13.43) and (13.44), the particulate structure equation with p_s results:

$$\frac{dp_s}{dx} = -(\rho_s - \rho_L)g\epsilon_s - \frac{\mu\epsilon}{K}\left(\frac{q}{\epsilon} - \frac{q_s}{\epsilon_s}\right)$$
$$(13.45)$$

The underflow concentration is given by $\epsilon_{su} = q_{su}/(q_{su} + q_u) = q_s(q_s + q)$, so q can be expressed as:

$$q = \left(\frac{1}{\epsilon_{su}} - 1\right)q_s \quad (13.46)$$

Substituting Eq. (13.46) into Eqs. (13.43) and (13.45) and eliminating q yields

$$\frac{dp_L}{dx} = -\rho_L g + \frac{\mu q_s}{K}\left(\frac{1}{\epsilon_s} - \frac{1}{\epsilon_{su}}\right) \quad (13.47)$$

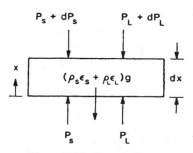

Figure 13.22. Force balance.

$$\frac{dp_s}{dx} = -(\rho_s - \rho_L)g\epsilon_s - \frac{\mu q_s}{K}\left(\frac{1}{\epsilon_s} - \frac{1}{\epsilon_{su}}\right)$$
$$(13.48)$$

The solids flux q_s becomes an input parameter as well as the underflow concentration ϵ_{su}. It is necessary to have constitutive equations relating K and ϵ_s to p_s. It is best to obtain these constitutive relations experimentally. The following expressions, which have been used for cake filtration, are adopted here:

$$\epsilon_s = \epsilon_{s0}\left(1 + \frac{p_s}{p_a}\right)^\beta ; \quad K = K_0\left(1 + \frac{p_s}{p_a}\right)^{-\delta}$$
$$(13.49)$$

where ϵ_{s0} and K_0 are solidosity and permeability under null stress. The degree of compressibility is related to the parameters β, δ, and p_a. All these parameters need to be determined experimentally.

13.3.3.2 Solution Methodology

For a continuous thickener operated under steady-state conditions, the solidosity at the bottom of the thickener is assumed to be the same as the underflow solids concentration. This also implies that the velocity of the solids equals the velocity of the liquid at the bottom.

To start the solution, selected values of ϵ_{su} and q_s are substituted into Eqs. (13.47) and (13.48). At the sediment surface $p_s = 0$. At the bottom p_s must equal a value that corresponds to ϵ_{su}. Equations (13.47) and (13.48) can be solved numerically as long as constitutive relations such as Eq. (13.49) are available.

13.3.3.3 Thickener Behavior

Equations (13.47) and (13.48) were solved for kaolin flat D (a type of clay). Figure 13.23 shows that at a given underflow concentration ϵ_{su}, increasing q_s results in higher values of L. It can also be noted that the plots are characterized by two asymptotes. The horizontal asymptote corresponds to a long detention time in which the Darcy term in Eq. (13.48) is negligible. A thickener operating far into this region would be oversized. The vertical asymptote represents the maximum flow rate that is possible with the required underflow concentration. Physically, the vertical asymptote corresponds to a condition in which the unbuoyed weight of the sediment is balanced by the Darcian drag. Therefore, no compressive pressure is available to thicken the solids and an infinite height of sediment is required to obtain the given underflow concentration. Operation should lie in a range in which flux varies from 25% to 75% of the limiting value of the flux.

13.3.3.4 Application of Deep Thickener

Deep thickener technology was developed and exploited in Alcan alumina plants.[131] The significantly higher underflow concentration is the most significant advantage over conventional thickeners. It was reported that they can achieve underflow concentration from 90% to 95% of that obtained with rotary vacuum filters. In addition, it had the advantages of lower capital and maintenance costs, lower area requirement, increased recovery of valuable chemicals, and even the production capacity. Table 13.6 shows the performance comparisons of a deep cone thickener with the conventional thickeners. As a result, Alcan had installed 20 new deep thickeners and converted 10 traditional multideck thickeners to deep thickeners.

13.4 CLARIFICATION

As implied in the name, the purpose of clarification is to remove turbidity or suspended solids from a murky liquid and render it crystal clear. It is used in a wide variety of industries, applied to raw materials, intermediates, and products and, increasingly in recent years, to waste streams.

The treatment of raw water supplied; clarification of solutions in the sugar, metal, and inorganic chemical industries; removal of fine catalyst particles from petroleum intermediates; polishing of beer in racking tanks after addition of finings; and the disposal of industrial waste water are but a few examples of the use of sedimentation for the commercial-scale clarification of liquids.

Sedimentation is, however, not the only means of achieving this,[5,6,74] and for a fuller coverage of clarification, the chapters dealing with filtration methods should also be consulted.

Clarification by gravity sedimentation is carried out in circular tanks similar to that shown in Figure 13.2, but of lighter construction than those shown in Figures 13.1 and 13.3, and also in rectangular tanks.[75] In the treatment of potable water, long rectangular basins are considered to be hydraulically more stable, with less short-circuiting between feed and overflow points, especially in larger plants.[76] In flocculated sewage treatment in the Toronto area,[77] a long, rectangular horizontal-flow settling tank was stated to be "much better" than a circular tank even when based on the same overflow rate and detention time.

In the food industry, the relatively prolonged residence time in normal gravity settling tanks sometimes leads to fermentation and deterioration. The processing time and the liquid inventory may both be reduced, however, by the use of centrifugal clarifiers. In these the gravitational force, g, is increased to 1000 to 10,000 times with a corresponding

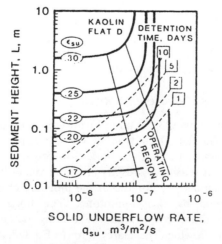

Figure 13.23. Height versus solid underflow rate at constant underflow concentration for kaolin flat D.

Table 13.6. Comparisons of Conventional with Deep Thickener.[131]

	CONVENTIONAL THICKENER	DEEP THICKENER
Diameter	120 ft	40ft
Height	15–20 ft	40–60 ft
Underflow concentration	30–35 wt%	45–55 wt%
Overflow clarity	< 200 mg/liter	< 100 mg/liter
Flocculant dosage	20–40 g/ton	50–80 g/ton
Upward velocity	0.5 m/h	3 m/h
Solids loading	1 to 2 mt solids/m^2 day	10–15 mt solids/m^2-day
Capital costs	very high	2–4 times lower

From Ref. 131

increase in the settling velocity of the solids. Examples are the treatment of olive oil to avoid rancidity and the separation of yeast cells from beer to cause a rapid termination of their growth not possible in normal gravity sedimentation.[12c] Detailed discussion of centrifugal sedimentation[5] is, however, beyond the scope of this chapter.

Clarification of gold-bearing solutions in Southern Africa is traditionally done by precoat filtration. A full-scale test[133] has shown, however, that a prior gravity sedimentation step can reduce the overall cost of clarification to 60% that of filtration alone. This specific example confirms that the relative lower cost of gravity sedimentation in general, indicated in Table 13.1, applies also to clarification.

13.4.1 Comparison of Clarifiers and Thickeners

The concentration of solids in feed to clarifiers ranges from 0.1 to 10 kg/m^3 which is between 20 and 1000 times more dilute than for thickeners. As the main purpose of clarification is the removal of the solid matter from the liquid, the concentration at which these solids (usually waste) are rejected from the clarifier is of reduced importance. The solids are usually finer than in thickener feeds and require flocculation for efficient settling. In a fully loaded thickener, the deep layers of sedimenting and compacting solids are a prominent feature of its depth-concentration profile whereas in clarifiers the major volume of the tank is occupied by relatively quiescent liquid.

As a consequence, the incoming feed, which may differ both in density and temperature from the contents of the tank, can readily upset the ideal flow pattern. The mode of feed entry and overflow removal is therefore more critical than in thickeners[76, 78] and model studies should be used to investigate the hydraulic effects of novel designs[79] including the effects of baffles, weirs, and distributor plates.[80]

In thickening, the solids settle by hindering settling with an interface between suspension and liquid so that size segregation of fines is minimal. Because of the lower solids concentration in clarifiers the flocs descend independently, with the larger particles or flocs reaching the sludge level faster than the slower settling fine material. If it were not for this range of settling rates, the normal settling flux curve (Figs. 13.17 and 13.18) could be used in design. Instead, a different flux curve would be required for each species of particle size and particle concentration present. However, as the maximum solids throughput is not of primary importance in clarifiers, a different approach is used in design.

More details of the conditions necessary for the separation of various sizes and densities of particles in a mixed suspension are given by Masliyah.[123]

13.4.2 Pretreatment for Sedimentation

Effective clarification often depends on flocculating agents for success. Even when sedimentation is feasible without such an aid,[1] pretreatment can result in both a reduction of the size of tanks required and an increased clarity of the product. Pretreatment also applied to thickener feed when throughputs have to be increased but this is considered to be a more expensive remedy than installing additional tanks for the long term.[73, 79]

Flocculants are additives that cause suspended solids to agglomerate into flocs which act like single large particles and therefore settle more rapidly than their smaller components. Floc formation is brought about by coagulation, by capture in hydrous precipitate, or by the formation of polymer bridges between particles.

Coagulation occurs when the mutual electrostatic repulsive forces between particles are sufficiently reduced, by the addition of ions of opposite sign, to permit the London–van der Waals attractive forces to cause aggregation of the particles. This requires either a pH change or the addition of preferably polyvalent ions or a combination of these actions. Lime and alum are common coagulation additives.

At a suitable pH, alum addition will lead to the formation of a hydrous aluminum hydroxide precipitate in which particles may be captured.

Polymer flocculation, whether by natural or synthetic neutral polymers or polyelectrolytes, may be considered[82] to take place in two stages:

1. adsorption of the polymer onto a particle surface, attributed to hydrogen bonding or ion adsorption, and
2. flocculation of the particles either as a direct result of the London–van der Waals attractive forces or due to physical polymer bridges formed between the particles.

These bridges may be formed by the two ends of one polymer molecule being attracted to two different particles,[83] or by the "loops" of polymer chains on one particle being attracted to the loops of another.[82] Polymer flocculation is extremely sensitive to the molecular weight of the polymer used.[84] Because of the differ-

ent mechanisms, coagulation is a reversible process whereas flocculation is not.

The combined use, first of an electrolyte to reduce repulsive charge, followed by a reduced quantity of the relatively more expensive polymer, often leads to a less costly pretreatment process than the use of either alone. There are of course restrictions in selecting flocculants for potable water and foodstuffs. Less efficient but edible natural products such as starches and gums find a useful application here.

The quantity of flocculant normally required to cause efficient flocculation is only a small fraction of that which can be adsorbed on the relatively large solid surface available, and polymer flocculant molecules are usually quickly and completely removed from solution. Contamination of the clarified liquid with residual free flocculants is therefore usually absent but not impossible and could lead to problems at another point in the circuit[1] or in the application of the product. A bigger problem at the flocculant addition stage is to ensure that the limited quantity of flocculant is equally distributed between all the particles in the suspension.

Although the principles of flocculation are reasonably well understood, selection and application of the best flocculant for a particular suspension is still an art.[1, 5, 83, 84] Thus negatively charged solids may be flocculated by cationic flocculants as expected but it is also possible that better results may be obtained with an anionic polymer after addition of a divalent cation.[82] Determination of the best conditions and selecting the best product from a range of similar type flocculants is therefore based on results of laboratory batch tests on the liquid to be clarified.[87] Such tests, if properly carried out, can indicate not only the chemicals to add, the required amounts and the order of their addition, but also the degree of stirring, method of application, and the wait period required either before the next addition or the commencement of sedimentation, that is, the point in time at which the flocculated suspension should be admitted to the clarifier.

A typical sequence is:

1. addition of lime solution to the feed under conditions of rapid mixing to ensure good dispersion and adsorption onto the solids,
2. addition of polymer solution, as dilute as possible (normally a few ppm) and at more than one point of application, also under conditions of rapidly mixing to ensure good distribution in the liquid, and finally
3. slower mixing or gentle shear of the clarifier feed so that the treated particles, now ready to adhere to each other, grow into flocs and then into larger aggregates to cause rapid sedimentation and to incorporate the finer particles which otherwise cause residual turbidity after settling.

It must, however, be borne in mind that in designing flocculation equipment the detailed mechanism of polymer flocculation are generally not as well understood as the theory, and the process cannot be applied without some empiricism and resort to pilot scale experiments, and even then the results cannot be scaled up with certainty.[88]

13.4.3 Floc Strength

Excess shear of the formed flocs should be avoided, especially those produced by irreversible polymer bridging. In one difficult case, dropping from the end of the launder into the feed well was sufficient to rupture the flocs. Removal of the launder from the feed well and lowering its end a few centimeters below the water surface of the clarifier provided a smoother entry for these fragile flocs and solids throughput was increased two to three times as a result. Smith and Kitchener[89] present three techniques whereby the strength of particle adhesion in flocculating media may be measured. They found, as a quick check, that an increased floc size was a good indication of an increased strength of adhesion.

The processes of coagulation, flocculation, and then settling of the flocs can be combined in a single vessel[90] on a large scale. It is claimed that well-formed flocs, produced in a central (feed well) zone, are smoothly transferred to the clarification zone with minimal floc disruption. This leads to a simpler process and reduced reagent consumption.

In repeated settling of the same flocculated solids such as in countercurrent decanting, floc shear in the transfer pumps is unavoidable. Although this can be minimized,[91] it is normally necessary to add a further quantity of flocculant at each stage, half or one third of the original dose being fairly typical. The actual quantity required can be adjusted by observing the performance of the clarifiers involved.

13.4.4 Final Settled Volume

It may appear at first to be paradoxical, but the greater the attraction between particles, brought about by pretreatment, the further they stay apart in the final sediment. This is related to the stability of an open structure of adhering particles compared to the same structure when the particles are dispersed. The latter, as they descent individually from suspension, cannot remain in contact with the first two or three particles they encounter in the build-up zone, because of the repulsive charge, and continue to descend by sliding or rolling until they reach a point of maximum stability where they fill the lowest remaining gap in the structure. Dispersed sediments are therefore compact while flocculated sediments are voluminous[49] and more readily break down on shear.

13.4.5 Theory and Design

Clarifiers are sized on the basis of the permitted (or desired) upper limit of solids in the overflow and the settling rate of these solids. If the size distribution of the solids in the feed is known, it can be converted to a settling velocity distribution according to Eq. (13.8). A typical plot is shown in Figure 13.24.

As the feed enters the clarifier the particles with settling velocities greater than the upflow rate settle to the bottom while the rest are

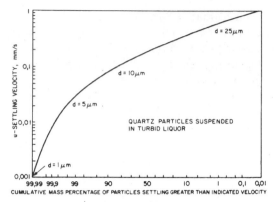

Figure 13.24. Settling velocity distribution of suspended quartz particles.

carried over. The proportion of feed particles appearing in the overflow can be determined from Figure 13.24 for any given overflow rate and their actual concentration then depends on the total quantity of solids in the feed. The settler area is governed by the maximum allowable overflow rate.

When the "particles" to be settled are flocs or agglomerates of the originally dispersed fine solids, their size characterization is not simple. It is best to directly measure in the laboratory the essential parameter, settling velocity distribution. This is illustrated in the examples below.

13.4.5.1 Two Examples of Estimating Clarifier Areas

(1) A liquid containing dispersed solids

A pregnant gold solution, obtained by rotary vacuum filtration of cyanide-leached ore, contains 550 mg/liter of 1 to 30 μm quartz fines and requires clarification to a limiting concentration of 20 mg/liter. What size sedimentation tank is required for a flow of 600 m^3/h?

The solids to be removed are $(550 - 20)$ mg/liter or 96.4% of the mass of incoming solids. As only solids that settle faster than the upward velocity of the overflow can be collected, the maximum overflow rate must be low enough to collect this percentage of the incoming solids. From Figure 13.24, based on

a size distribution of this material, 96.4% of the solids have settling velocities exceeding 0.052 mm/s. The maximum upflow rate, u_s, is therefore $0.052 * 3600/1000 = 0.187$ m/h.

As throughput $q_F = u_s *$ area of clarifier (A)

$$600 \text{ m}^3/\text{h} = 0.187 \text{ m/h} * A \text{ m}^2$$

$$A = 3200 \text{ m}^2$$

and diameter of circular tank = 64 m.

(2) Clarification of a liquid after flocculation

Assuming that 64 m diameter tank is too large for the site, what steps can be taken to reduce it? Flocculant tests in the laboratory indicated good settling behavior of the solids in the feed after addition of 5 mg/liter of ferric chloride coagulant followed by 0.2 mg/liter of a polyacrylamide flocculant.

For sizing the clarifier a 10-liter sample of the pretreated turbid gold solution was then gently added to a transparent settling tube ~ 65 mm diameter and 2 to 3 m deep. Provision was made for periodically sampling the liquid at known depths either by lowering a siphon tube or through suitably spaced side ports.[2, 92]

At time zero, the tube was immediately filled, the contents were sampled at the top, middle, and near the bottom to determine the original (feed) suspended solids concentration and check on even solids distribution. When the liquid started to clear, the time was noted and samples taken at all levels from top to bottom in that order, and analyzed for suspended solids. The sampling was repeated after four or five similar periods until all samples indicated a suspended solids concentration below the desired limit. Results of a typical test are shown in Table 13.7.

Sampling at H m below the surface after t h static settling is exactly equivalent to sampling the overflow liquid from a continuous clarifier operating at an upflow velocity $q_F/A = H/t$. It can be seen from the results, as is to be expected, that the suspended solids (SS) at any level decreases with time and at any time increases with depth. These data provide the maximum permissible upflow ve-

Table 13.7. Suspended Solids Values at Various Depths at Different Times.

DEPTH FROM SURFACE (m)	TIME OF SAMPLING (min)										
	0	5		10		15		20		25	
	SS (mg/liter)	u_8 (m/h)	SS (mg/liter)	u_8 (m/h)	SS (mg/liter)	u_8 (m/h)	SS (mg/liter)	u_8 (m/h)	SS (mg/liter)	u_8 (m/h)	SS (mg/liter)
0.25		3 (4.8)	10 (20)	1.5	3	1	2	0.75	1	0.6	2
0.5	550	6	30	3	7	2	3	1.5	2	1.2	Nil
1.0		12	140	6	23	4	7	3	3	2.4	3
1.5	560	18	380	9	65	6 (6.5)	15 (20)	4.5	5	3.6	4
2.0		24	520	12	140	8	35	6	9	4.8	5
2.5	540	30	550	15	250	10	65	7.5	15	6	7

locity for any desired suspended solid in the overflow. By interpolating the SS data after 5 min settling, it may be estimated that for a clarifier product containing no more than 20 mg/liter of suspended solids, the upflow must not exceed 4.8 m/h. For a volumetric throughput $q_F = 600$ m^3/h this means a tank of diameter = 12.6 m.

If the exercise is repeated with the 15 min results, however, the diameter is found unexpectedly to decrease to 10.8 m. This could not occur unless there were a change in the nature of the settling solids with time. For the various times shown in Table 13.7 it can be seen, however, that the SS value for a fixed upflow rate, say $u_s = 6$ m/h, decreases steadily with time from 30 mg/liter at 5 min to 7 mg/liter at 25 min. Therefore, both upflow rate and detention time are important in clarifier design. (Note: This aspect of detention time is quite different from the idea that it may be required to achieve maximum sludge thickening.)[90] Particles or flocs continue to grow during settling, either as a result of faster settling units overtaking and coalescing with slower ones or due to velocity gradients in the fluid.[143]

In a cylindrical tank of diameter D the detention time $t_0 = \pi D^2 H / 4q_F$ while the upflow rate is the detention time $t_0 \, 4q_F / \pi D^2$. Therefore the dimensions of the tank are

$$D = \left[\frac{4q_F}{\pi u_s} \right]^{1/2}$$

and

$$H = t_D * u_s$$

In the example $q_F = 600$ m^3/h and if we select a detention time

$t_D = 5$ min, u_s must be 4.8 m/h for SS
$\qquad < 20$ mg/liter
$$D = 12.6 \text{ m}$$
and
$$H = 0.40 \text{ m}$$

If a longer detention time is selected, say $t_D = 25$ min, then u_s becomes 10.8 m/h (by extrapolation):

$$D = 8.4 \text{ m}$$
and
$$H = 4.5 \text{ m}$$

which represents a narrower tank but much deeper. This apparent wide choice of possibilities is, however, limited by considering the standard sizes available as settling tanks of this diameter normally come with a standard depth of ~ 2.5 m, and therefore $u_s * t_D = 2.5$ m. The products $u_s * t_D$ for various times taken from Table 13.7 are shown in Table 13.8. From the final column the required value of 2.5 m occurs at about 19 min; therefore:

$$t_D = 19/60 = 0.315 \text{ h}$$
$$u_s = 2.5/0.315 = 7.9 \text{ m/h}$$

and

$$D = 9.8 \text{ m}$$

In actual practice a converted 10 m diameter pachuca tank was used[133] for this duty, in this case probably by reversing the procedure and adding the necessary quantity of flocculant to suit the size tank available. As for thickeners, the calculated value of D is rounded up to the next largest standard diameter tank manufactured and this has the double advantage of further lowering the upflow velocity and increasing the detention time, which is always useful for contingencies.

13.5 NONCONVENTIONAL SEDIMENTATION PROCESSES AND EQUIPMENT

Normal thickeners and clarifiers are large relatively empty tanks or basins that provide virtually quiescent conditions and sufficient surface area and volume for optimum settlement. This simple process has changed little since the turn of the century, and, based on the

Table 13.8. Values of $t_D * u_s$ Calculated from Table 13.7

TIME (min)	u_s FOR 20 mg/liter SUSPENDED SOLIDS	$t_D * u_s$ (m)
5	4.8	0.40
10	5.6	0.93
15	6.6	1.65
20	8.3	2.77
25	10.8	4.50

percentage tonnage currently handled, appears to be as well established as ever.

However, situations do arise where the conventional design is not ideal for special conditions. These include sedimentation of solids whose density is only marginally greater than that of the fluid; the clarification of high-temperature or very volatile liquids that must in consequence be kept under pressure during treatment, or when new or additional sedimentation equipment must be installed in existing premises, and floor space has become severely limited. In some cases the cost of creating the additional space may be many times greater than that of the equipment itself. In underground mining for instance, 10 to 20 m^3 of virgin rock has to be blasted out for every m^2 of settling area[95] required. Such situations have led to the introduction of a number of innovations, some of which have reached the commercial exploitation stage.

Their common purpose is to increase solids throughput per unit of superficial area. This can be done either through an improvement in actual particle sedimentation rates whereby more solids are settled per square meter of actual settling area or due to a multiple design where, for the same projected floor area, a larger net settling area is provided. Mechanical innovations such as improved designs of feedwell and rake are also presented.

13.5.1 Lamella Settlers

According to Hukki et al,[96] the use of inclined baffles for settling fine solids dates back to 1882 when a French patent was granted to Gaillet and Huet. The principle has since then "been rediscovered many times in many countries."[96] Figure 13.25 shows the angle and general arrangement of the baffles in the early Hukki design, the mode of settlement of the solids, the method of calculating the floor area required. The ratio of floor area to settlement area is

$$R' = \frac{1}{N} + \frac{d}{H \cos \alpha} \left[\frac{N-1}{N} \right]$$

and so to minimize R', the number of plates should be large, and they should be deep,

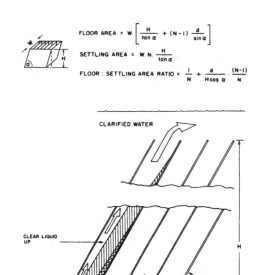

$$\text{FLOOR AREA} = W \left[\frac{H}{\tan \alpha} + (N-1) \frac{d}{\sin \alpha} \right]$$

$$\text{SETTLING AREA} = W.N. \frac{H}{\tan \alpha}$$

$$\text{FLOOR : SETTLING AREA RATIO} = \frac{1}{N} + \frac{d}{H \cos \alpha} \cdot \frac{(N-1)}{N}$$

Figure 13.25. Principles of the lammella settler.

closely spaced, and at as shallow an angle α as possible. The minimum value for α is prescribed by the flow properties of the sludge.

For plates of $H = 1.5$ m, at the usually accepted angle of $\alpha = 60°$ and spaced $d = 30$ mm apart, a 10-plate tank has an actual settling area about seven times greater than a conventional settler in the same space, whereas with 20 plates, this increases to 11. In the limit, as $N \to \infty$, the advantage becomes equal to $H(d/\cos \alpha)$, which in this case is 25. As the denominator $d/\cos \alpha$ represents the vertical distance between adjacent plates, the maximum advantage is, as would be expected, the number of plates that overlap in the vertical direction.

This theoretical figure is rarely achieved in practice, however, for, besides the obvious end effect, factors contributing to the lower efficiencies in the earlier designs are:

1. Manifolding problems of feeding a large number of plates

2. Entrainment of solids in the rapidly rising liquid stream due to internal mixing[97] giving rise to an increased overflow turbidity
3. Accumulation of solids on the plates.

These shortcomings, which are not always serious, have been recognized by various manufacturers and the following modified designs are available commercially:

1. Corrugated plates to avoid a continuous curtain of descending solids, and guard gutters to separate the flow of these solids from the incoming feed
2. Feeding the plates from the side to avoid feeding and discharging solids at the same point
3. Feeding from the top with overflow return pipes extending to clear water zones near the base
4. Stacking set of inclined plates, one above the other, each inclined in opposite directions to present a vertically zigzag profile for which settling solids should present the minimum disturbance to the clarified liquid
5. Continuous raking or low-frequency vibrators to assist both removal of solids from the plates and promote their compaction
6. Use of flexible textile materials or rubber for the "plates" to permit periodic dislodgement of solids or cleaning.

The lamella settler has been successfully used in many fields; in coal preparation plants[98, 99] for removal of fine mill scale from hot rolling mill wastes for reuse of the water[100] and also for separating metal hydroxides, fly ash, nickel, catalyst fines, cement dust, clarification of phosphoric acid, lime kiln scrubber water and paint booth water curtain.[100]

The *tubular settler*[101–103] and the rotating *spiral thickener*[104] operate on the same general principle of providing increased settling area per unit of floor area. In the latter, a "Swiss roll" provides a number of flow channels between curved walls through which the slurry flows and concentrates while the unit is slowly rotated.

Willis[103] reviews the practical design factors for tubular settlers including important points such as sludge collection, how to specify overflow rate, and presents various shapes of tube that may be used. Much of it applies also to conventional plate settlers but tubes are considered to overcome the hydraulic instability of "wide horizontal plates."

An alternative view,[97, 104] that lamella settlers offer the advantage of rapid sedimentation or additional clarifier capacity because of the decreased vertical fall height, can lead to confusion between a reduced throughput time and a real increase in total solids throughput. Only the former applies here as solids throughput is governed by available area and not settling height.

13.5.2 Upflow Solids Contact

In contrast to the lamella settler, which increases sedimentation "efficiency" by providing a multiplicity of settling planes, the upflow principle operates by improving the actual sedimentation characteristics of the feed. It is confined to flocculated suspensions or metal hydroxide precipitates in which the freshly formed loosely knit, voluminous flocs settle slowly owing to their smaller initial mean size, and their greater fragility, or their decreased density differential compared to compact agglomerates.

Such flocs may, however, mature with time resulting in larger, stable, denser, fast settling units, and when these aged flocs contact freshly formed material the settling characteristics of the latter are promoted.[80] The upflow principle exploits this phenomenon by adding the freshly flocculated feed, not into a normal feedwell above a body of clear liquid, but below the pulp interface of a bed of aged flocs. These flocs are thereby kept in a state of fluidization, and feed rates of up to 35 times higher than for conventional units are claimed. This is possible if the matured flocs have a settling rate 35 times faster than freshly formed flocs. This represents a special case of the benefits of detention time except that in this case the mass of the solid rather than the volume of the liquid is used to calculate detention.

Relatively long residence times are possible because of the high concentration of the solids in the fluidized zone. Escape by short-circuiting of the slower settling fresh flocs with the water from this zone is reduced by providing a zone sufficiently deep and uniform. There must therefore be some "filtering" action whereby the new flocs or their fragments collide with and are held by older flocs until they too mature and trap fresh material. The need to maintain a stock of fluidized solids means, however, a relatively long start-up period (while running at low efficiency) and the inability to recover quickly following flow reductions or shut-downs.

The original application appears to have been for water clarifications[105, 106] and a simplified design is shown in Figure 13.26. More recently it has been applied to conventional thickener feeds such as 20% uranium slurries[107] or zinc mine tailings[108] in units shown schematically in Figure 13.27.

A useful comparison of solids contact clarifiers, lamella separators, and inclined tube settlers is given by Mace and Laks[109] including cost data.

13.5.3 Flotation

Although froth flotation is used on a wide scale for separating between solids, particularly the flotation of mineral particles from

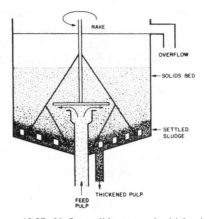

Figure 13.27. Upflow solids contact in thickening.

gangue, in this context it refers specifically to the removal of all solids suspended in a liquid to achieve complete clarification.

Where the density difference between solid and liquid is slight, as for example, the suspended organic solids in sugar juice and in sewage, sedimentation rates are low and sludge volumes usually high. Attachment of gas bubbles to the particles or flocs reduces their density still further to such an extent that the density different between "particles" and liquid is reversed and increased. Therefore, the driving force for separation is also increased.

Owing to the reversal in actual densities, the flocs now rise but at a rate more rapid than they descended previously and collect as a froth at the surface. They form a solids zone partially above the liquid surface so that the solids can drain to produce a "sludge" of considerably reduced liquid to solids ratio.[110]

The bubbles are normally generated by releasing the pressure of an air-saturated liquid stream[111] or by electrolysis.[112]

Further aspects of the process and detailed design of air flotation systems will be found elsewhere.[113]

13.5.4 Feedwell Design

Unconventional feedwell designs that have been proposed aim to improve the hydraulics of feed entry or the settling characteristics of

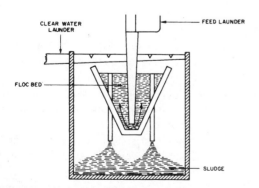

Figure 13.26. Upflow solids contact in water clarification.

the solids.[4, 73, 114] Tests of feedwell designs performed on model thickeners should be interpreted with caution and if possible repeated on a full-scale plant before making claims.[4] The dynamic effects are accentuated on models, and whether good[114] or bad[122] generally decrease on scale-up.

In clarifiers, internal circulation (caused by density currents) is to be avoided because it leads to short-circuiting and higher local up-flow velocities near the overflow. Various ways of overcoming this defect are discussed and a new type of feedwell presented by Firch and Lutz.[115] Within the normal cylindrical well are two adjacent, inwardly open, submerged channels between which the feed is split and introduced to each tangentially but in opposing directions. When these streams meet in "open water" near the center of the well their velocities are instantaneously dissipated as turbulence and residual velocity streams are thereby eliminated.

13.5.5 Rake Design

In common with other materials throughout the world, the gold ores mined on the East Rand in South Africa tend to be especially "sticky."[116] The clayey solids accumulate[117] on the arms of conventional rakes and often cause disastrous shock overloads when they eventually fall off in massive chunks. It is then also extremely difficult at the next half turn to rake the resultant "islands" to the discharge point.

The Klopper Dragrake[118] was developed to overcome this type of problem. It consists of a tubular steel arm to which are welded the raking blades. This arm is pivoted at the base in the center and suspended by means of cables or chains from a drag arm rotating above the liquid. Solids build-up on the arm is minimal and it has the ability to swing and lift automatically over any unusual loads. The great usefulness of this device is confirmed by its being used elsewhere[119] and by being taken up by international thickening manufacturers for the general treatment of thixotropic slurries and tacky muds.

ACKNOWLEDGEMENT

The author extends his gratitude to Professor Frank M. Tiller of the University of Houston for his review and recommendations during the preparation of this manuscript.

LIST OF SYMBOLS

A	Area of horizontal cross-section of thickener or clarifier
A_b	Area of batch test cylinder
a	Divisor for calculation of Re defined in Table 13.2
B_F	Buoyancy force or Archimedes' up thrust $= V \cdot \rho_L \cdot g$
b	Exponent for calculation of Re defined in Table 13.2
C	Mass concentration of particles, that is, mass of oven dry solids per unit volume of suspension
C_{crit}	Concentration of the minimum flux zone
C_D	Dimensionless drag coefficient defined in Eq. (13.5)
C_F	Concentration of thickener feed
C_{max}	Maximum concentration attained by random close packing of spheres, that is, in the absence of compression
C_0	Initial uniform concentration in a batch test
C_u	Concentration of thickener underflow
D	Diameter of settling vessel
d	Distance between plates in lamella settler (normal to plates)
d_p	Particle diameter
d_A	Diameter of a sphere having same surface area as a nonspherical particle
d_M	Diameter of circle with same projected area as image of particle in microscope
d_{ns}	Diameter of sphere with same terminal settling velocity as nonspherical particle
d_{scr}	Diameter of sphere just passing same square aperture as particle

F_D	Drag force on particle moving relative to liquid		lamella settler to vertically projected settling area of all its plates
G	Solids flux	Re	Reynolds number of a particle
G_{clar}	Solids flux in a clarifier		$(= d_p u \rho_L / \mu)$
G_F	Gravitational force of attraction $= V \cdot \rho_s \cdot g$	t	Time
		t_D	Detention time in a clarifier
G_{min}	Minimum value of total flux G_T	t_u	Time at which H_u intersects exten-
G_T	Total flux		sion of tangent to batch settling curve
g	Acceleration due to gravity		drawn before compression point
H	Height (1) to suspension–supernatant	t_1	Period during which initial settling
	interface in a batch test; (2) Vertical		rate u_i increases to a maximum steady
	height of lamella settler		value u_s; the induction period
H_0	Height of suspension in batch test at	u	Absolute velocity of liquid in a thick-
	zero time		ener
H_u	Underflow line $= H_0 C_0 / C_u$	u_F	u_0 for a suspension of flocs
J	Property of a particle–liquid system	u_i	Initial settling rate of an intermediate
	independent of $u_s (= C_D \mathrm{Re}^2)$		suspension
K	Permeability of a particular bed	u_{ns}	Terminal settling velocity of a non-
K_0	Permeability under null stress		spherical particle
K_p	Shape factor $= d_p / d_A$	u_0	Settling velocity of a single particle,
k_v	Effective volume of a sedimentation		representative of a suspension, that is,
	entity (single rough particle, cluster or		suspension settling velocity at $C = 0$,
	floc) per unit volume of contained		equivalent to u_w
	solids, that is, ratio volume of parti-	u_s	Settling velocity of a particle, a con-
	cles plus fixed water to volume of		centration zone, or a suspension–
	particles		supernatant interface relative to the
k'	Effective volume of a floc per unit		container
	mass of contained solids $(= k_v / \rho_s)$	u_{sB}	Absolute solid velocity at the bottom
N	Number of plates in a lamella settler		of thickener
n	Exponent in the Richardson and Zaki	$u_{s, crit}$	Settling velocity of minimum flux zone,
	equation $= 4.65 + 19.5 \ d/D$ in		C_{crit}
	streamlined flow, ~ 4.7 for most slur-	u_{sr}	Relative velocity between particle and
	ries in gravity sedimentation practice		liquid
p_a	Compressibility parameter in Eq.	u_w	Settling velocity in presence of con-
	(13.49)		tainer walls as opposed to settling in
p_L	Hydraulic pressure		an infinite liquid
p_s	Solid compressive pressure in the sed-	u_ϕ	Terminal settling velocity of a sphere
	iment		in an unbounded liquid in streamlined
Q	Volumetric underflow discharge rate		flow
	from a continuous thickener	$u_{\phi t}$	As for u_ϕ, but in transitional or tur-
q	Volumetric flow rate of liquid		bulent flow
q_F	Volumetric feed rate	V	(1) Volume of particle; (2) Volume of
q_s	Volumetric flow rate of solids in a		thickener
	thickener	W	(1) Retardation factor due to pres-
q_{smin}	Minimum value of total solid flux		ence of container walls, Eq. (13.31);
q_{su}	Volumetric flow rate of solid at the		(2) Width of lamella settler
	underflow	x	Distance up from bottom of the set-
R'	Ratio of floor area required to locate		tling column

α (1) Slope of "underflow flux line," $\tan \alpha = Q/A$. Equals rate of rise of C_{crit} in batch settling test; (2) Angle of plates in a lamella thickener to the horizontal

β Compressibility parameter in Eq. (13.49)

δ Compressibility parameter in Eq. (13.49)

ϵ Porosity

ϵ_s Solidosity, solid volume fraction in the thickener

ϵ_{s0} Solidosity under null stress

ϵ_{su} Solid volume fraction at the underflow

μ Liquid viscosity

ν Characteristic velocity; slope of settling flux curve

ρ_L Liquid density

ρ_s Particle or solids density

ϕ_s Volume fraction of particles in a suspension $= C/\rho_s$

$\phi_{s,\text{crit}}$ Solid volume fraction at a minimum flux

ϕ_{sF} Solid volume fraction of the feed

ϕ_{s0} Initial solid volume fraction in a settling column

ϕ_{su} Solid volume fraction at underflow

REFERENCES

1. M. J. Pearse, "Gravity Thickening Theories: A Review," Warren Spring Laboratory, Stevenage, England, No. LR 261 (MP) (1977).
2. R. W. Moore, Jr., "Liquid-Solids Separation," Presented at a symposium organized by Hunslet Taylor-Eimco Sales, Johannesburg (6 August 1971).
3. H. W. Campbell, R. J. Rush, and R. Tew, "Sludge Dewatering Design Manual," Res. Rep. No. 72. Res. Program Abatement Munic. Pollut. Provi. Canada-Ontario Agreement Great Lakes Water Quality (1978).
4. K. J. Scott, Unpublished work of co-worker (when mentioned). CERG-CSIR Internal Reports (1965–1970).
5. F. M. Tiller (ed), "How to Select Solid–Liquid Separation Equipment," *Chem. Eng.* 81:(9) 116–136 (April 29), (10) 98–104 (May 13, 1974).
6. B. Fitch, "Choosing a Separation Technique," *Chem. Eng. Prog.* 70(12):33–37 (1974).
7. D. B. Purchas, "Solid/Liquid Separation Equipment: A Preliminary Experimental Selection Programme," *Chem. Eng.* No. 328:47–49 (January, 1978).
8. B. S. Okunev, I. I. Morosova, Yu. A. Tisunov, and N. F. Kotov, "Automatic Control of Radial Thickener Operation," *Coke Chem. USSR* (12) 49–52 (1977).
9. C. G. Stokes, "On the Effect of the Internal Friction of Fluids on the Motion of Pendulums," *Trans. Cambridge Philos. Soc.* 9(II):8–106 (1851).
10. I. Proudman and J. R. A. Pearson, "Expansions at Small Reynolds Numbers for the Flow Past a Sphere and a Circular Cylinder," *J. Fluid Mech.* 2:237–262 (1957).
11. G. K. Batchelor, *An Introduction to Fluid Dynamics*, Cambridge University Press, London, p. 233 (1967).
12. R. H. Perry and D. Green, *Chemical Engineers' Handbook*, 6th edit., McGraw-Hill, New York. 12a, pp. 19–64, 12b, pp. 5–63, 64, 12c, pp. 1101 [Third Ed.-(1950)] (1984).
13. M. B. Sonnen, "Subroutine for Settling Velocities of Spheres," *Am. Soc. Civil Eng. J. Hydraulics Div.* 103 (HY9) 1097–1101 (1977).
14. V. F. Swanson, "A Free-Settling Equation Valid for All Particle Sizes," in *Conference: Proc. Tech. Program: Int. Powder Bulk Solids Handl. Pr.*, pp. 82–89 (Published 1978). (Available Ind. Sci. Conference Man., Inc., Chicago, IL)
15. P. G. W. Hawksley, "Survey of the Relative Motion of Particles and Fluids," *Br. J. App. Phys.* 5:S1–S5 (1954).
16. E. S. Pettyjohn and E. B. Christiansen, "Effect of Particle Shape on Free-Settling Rates of Isometric Particles," *Chem. Eng. Prog.* 44(2):157–172 (1948).
17. A. I. Medalia, "Dynamic Shape Factors of Particles," *Powder Technol.* 4:117–138 (1970/71).
18. J. Tsubaki and G. Jimbo, "A Proposed New Characterization of Particle Shape and Its Application," *Powder Technol.* 22:161–178 (1979).
19. S. T. Fong, J. K. Beddow, and A. F. Vetter, "A Refined Method of Particle Shape Representation," *Powder Technol.* 22:17–21 (1979).
20. J. M. Ziegler and B. Gills, Tables and Graphs for the Settling Velocity of Quartz in Water, Above the Range of Stokes' Law. Woods Hole Oceanographic Institution, Reference No. 59-36 (July, 1959).
21. R. H. Richards and C. E. Locke, *Textbook of Ore Dressing*, 3rd edit., McGraw-Hill, New York, Table 30, p. 129 (1940).
22. A. W. Francis, "Wall Effect in Falling Ball Method for Viscosity," *Physics* 4:403–406 (1933).

23. J. Garside and M. R. Al-Dibouni, "Velocity Voidage Relationships for Fluidization and Sedimentation in Solid–Liquid Systems," *Ind. Eng. Chem. Proc. Des. Dev.* 16:206–214 (1977).

24. B. H. Kaye and R. P. Boardman, "Cluster Formation in Dilute Suspensions," in *Proc. Symp. Interaction Between Fluids and Particles, Inst. Chem. Engr.* A17–21 (1962).

25. R. Johne, "Einfluss der Konzentration einer monodispersen Suspension auf die Sinkgeschwindigkeit ihrer Teilchen," *Chemie Ing Tech.* 38:428–430 (1966).

26. B. Koglin, "Dynamic Equilibrium of Settling Velocity Distribution in Dilute Suspensions of Spherical Irregularly Shaped Particles," in *Proceedings of the Conference on Particle Technology*, IIT Research Institute of Chicago, pp. 266–271, August 21–24 (1973).

27. E. M. Tory and D. K. Pickard, "A Three-Parameter Markov Model for Sedimentation," *Can. J. Chem. Eng.* 55:655–665 (1977).

28. J. F. Richardson and W. N. Zaki, "Sedimentation and Fluidization: Part 1," *Trans. Inst. Chem. Eng.* 32:35–53 (1954).

29. H. Watanabe, "Voidage Function in Particulate Fluid Systems," *Powder Technol.* 18:217–225 (1978).

30. A. D. Maude and R. L. Whitmore, "A Generalized Theory of Sedimentation," *Br. J. Appl. Phys.* 9:477–482 (1958).

31. W. C. Thacker and J. W. Lavelle, "Stability of Settling of Suspended Sediments," *Phys. Fluids* 21:291–292 (1978).

32. P. T. Shannon, E. Stroupe, and E. M. Tory, "Batch and Continuous Thickening. Basic Theory. Solids Flux for Rigid Spheres," *Ind. Eng. Chem. Fundam.* 2:203–211 (1963).

33. K. J. Scott and W. G. B. Mandersloot, "The Mean Particle Size in Hindered Settling of Multisized Particles," *Powder Technol.* 24:99–101 (1979).

34. B. Fitch, "Sedimentation Process Fundamentals," *Trans. AIME* 223:129–137 (1962).

35. J. P. Mogan, R. W. Taylor, and F. L. Booth, "The Value of the Exponent n in the Richardson and Zaki Equation, for Fine Solids Fluidized with Gases Under Pressure," *Powder Technol.* 4:286–289 (1970/71).

36. D. K. Vohra, "Sedimentation in a Viscous Suspending Medium," *Inst. Eng. (India) J. Chem. Eng.* 57:97–100 (1977).

37. A. K. Korol'kov, M. N. Kell, and A. A. Vasil'eva, "Dependence of the Rate of the Hindered Settling of Grains on Their Specific Surface," *Obogashch. Rud* 15:24–26 (1970). (Russ.) [Chem. Abstr. 74:89097 k (1971).]

38. L. Davies, D. Dollimore, and J. H. Sharp, "Sedimentation of Suspensions: Implications of Theories of Hindered Settling," *Powder Technol.* 13:123–132 (1976); 17:147–152 (1977).

39. G. B. Wallis, "A Simplified One-Dimensional Representation of Two-Component Vertical Flow and Its Applications to Batch Sedimentation," in *Proc. Symp. Interaction Between Fluids and Particles*, London, Institute of Chemical Engineers, pp. 9–16 (1963).

40. L. A. Adorj'an, "Determination of Thickener Dimensions from Sediment Compression and Permeability Test Results, *Trans. Inst. Min. Metal. C,* 85:C157–163 (1976).

41. C. E. Capes, "Particle Agglomeration and the Value of the Exponent n in the Richardson-Zaki Equation," *Powder Technol.* 10:303–306 (1974).

42. A. E. Fouda and C. E. Capes, "Hydrodynamic Particle Volume and Fluidized Bed Expansion," *Can. J. Chem. Eng.* 55:386–391 (1977).

43. L.-G. Eklund and A. Jernqvist, "Experimental Study of the Dynamics of a Vertical Continuous Thickener-I," *Chem. Eng. Sci.* 30:597–605 (1975).

44. A. S. Michaels and J. C. Bolger, "Settling Rates and Sediment Volumes of Flocculated Kaolin Suspensions," *Ind. Eng. Chem. Fundam.* 1:2433 (1962).

45. E. K. Obiakor and R. L. Whitmore, "Settling Phenomena in Flocculated Suspensions," *Rheol. Acta* 6:353–359 (1967).

46. P. G. Cooper, J. G. Rayner, and S. K. Nicol, "Flow Equation for Coagulated Suspensions," *J. Chem. Soc. Faraday Trans. I,* 74:785–794 (1978).

47. K. J. Scott, "Theory of Thickening: Factors Affecting Settling Rate of Solids in Flocculated Pulps," *Trans. Instn. Min. Metal* 77:C85–97 (1968); 78:C116–119, 244–245 (1969).

48. K. J. Scott, "Thickening of Calcium Carbonate Slurries. Comparison of Data with Results for Rigid Spheres," *Ind. Eng. Chem. Fundam.* 7:484–490 (1968).

49. K. J. Scott, "The Water Content of Flocs," *J. S. Afr. Inst. Min. Metal.* 65:357–367 (1965).

50. C. C. Harris, P. Somasundaran, and R. R. Jensen, "Sedimentation of Compressible Materials: Analysis of Batch Sedimentation Curve," *Powder Technol.* 11:75–84 (1975).

51. P. A. Vesilind, "Evaluation of Activated Sludge Thickening Theories," *J. San. Eng. Div. Proc. ASCE* 94:SA1, 185–191 (1968).

52. H. S. Coe and G. H. Clevenger, "Methods for Determining the Capacities of Slime-Settling Tanks," *Trans. AIME* 55:356–384 (1916).

53. B. Fitch, "Batch Tests Predict Thickener Performance," *Chem. Eng.* 78(19):83–88 (August 23, 1971).

54. B. Fitch, "Current Theory and Thickener Design," *Ind. Eng. Chem.* 58(10):18–28 (1966).

55. R. H. Bretton, "The Design of Continuous Thick-

eners," Ph.D. Thesis, Yale University (August, 1949).

56. J. A. Cole, "A Model for Activated Sludge Thickening," Ph.D. Thesis, University of Wisconsin (January, 1970).

57. K. J. Scott, "Continuous Thickening of Flocculated Suspensions," *Ind. Eng. Chem. Fundam.* 9:422–427 (1970).

58. G. Sarmiento and P. H. T. Uhlherr, "The Effect of Temperature on the Sedimentation Parameters of Flocculated Suspensions," *Powder Technol.* 22:139–142 (1979).

59. R. H. Weiland and R. R. McPherson, "Accelerated Settling by Addition of Buoyant Particles," *Ind. Eng. Chem. Fundam.* 18:45–49 (1979).

60. R. L. Whitmore, "The Sedimentation of Suspensions of Spheres," *Br. J. Appl. Phys.* 6:239–245 (1955).

61. P. Somasundaran, E. L. Smith, Jr., and C. C. Harris, "Effect of Coarser Particles on the Settling Characteristics of Phosphatic Slimes," in *Proceedings of the Conference on Particle Technology*, IIT Research Institute of Chicago, pp. 145–150, August 21–24 (1973).

62. J. B. McVaugh, Jr., Mathematical and Experimental Investigation of Nonsteady State Thickening of an Ideal Slurry," M.Sc. Thesis, Delaware University (May, 1975).

63. W. P. Talmage and E. B. Fitch, "Determining Thickener Unit Areas," *Ind. Eng. Chem.* 47:3841 (1955).

64. K. J. Scott, "Mathematical Models of Mechanisms of Thickening," *Ind. Eng. Chem. Fundam.* 5:109–113 (1966).

65. N. J. Hassett, "Mechanism of Thickening and Thickener Design," *Trans. Instn. Min. Metal.* 74:627–656 (1964/6S).

66. D. J. Slagle, Y. T. Shah, G. E. Klinzing, and J. G. Walters, "Settling of Coal in Coal–Oil Slurries," *Ind. Eng. Chem. Proc. Des. Dev.* 17:500–504 (1978).

67. T. M. Keinath, M. D. Ryckman, C. H. Dana, and D. A. Hofer, "A Unified Approach to the Design and Operation of the Activated Sludge System," Presented at the 21st Annual Industrial Wastes Conference, Purdue University (4 May, 1976).

68. W. Rudolfs and I. O. Lacy, "Settling and Compacting of Activated Sludge," *Sewage Works J.* 6:647–675 (1934).

69. H. H. Steinour, "Rate of Sedimentation: Concentration Flocculated Suspensions of Powders," *Ind. Eng. Chem.* 36:901–907 (1944).

70. A. M. Gaudin and M. C. Fuerstenau, "On the Mechanism of Thickening," in *International Mineral Processing Congress*, Institute of Mining and Metallurgy, London, pp. 115–127 (1960).

71. E. N. Tory, "Batch and Continuous Thickening of Slurries," Ph.D. Thesis, Purdue University (June, 1961).

72. N. Yoshioka, Y. Hotta, S. Tanaka, S. Naito, and S. Tsugami, "Continuous Thickening of Homogeneous Flocculated Slurries," *Chem. Eng. (Jpn)* 21:66–674 (1957).

73. E. H. D. Carman and D. P. Steyn, "Some Observations on Thickening," in *8th Commonwealth Min. Metall. Congress Australia* 6:443–454 (1965).

74. Anon., "Electrostatic Separation of Solids from Liquids," *Filtration Separation* 14:140–144 (March/April, 19–77).

75. Anon., "Rectangular versus Circular Settling Tanks," *Am. City* 88(10):98–99 (1973).

76. J. D. Walker, "Sedimentation Maximizes Clarity, Minimizes Turbidity in Potable Water Treatment," *Water Sewage Works.* Reference number 1978:R136-150 (11 pp.) (1978).

77. J. H. Tay, "Study of Settling Characteristics of Physical-Chemical Flocs in Sedimentation Tanks," *Diss. Abstr.* 38:1921 B (1978).

78. C. A. Lee, "Increasing Settling Tank Efficiency," *Plant. Eng.* 126–127, April 19 (1973).

79. C. E. Hubbell, "Hydraulic Characteristics of Various Circular Settling Tanks," *Am. Water Works Assoc. J.* 30:335–353 (1938).

80. Anon. "Sludge Settling Tank More Than Halves Retention Time," *Chem. Proc.* 13(5):4–7 (May, 1967).

81. C. G. Bruckmann, "Economic Aspects of Thickener Operations," Presented at Symposium S 19 on Thickener Design and Operation organized by the NCRL-CSIR, Pretoria (28 June, 1966).

82. G. H. Matheson and J. N. W. MacKenzie, "Flocculation and Thickening Coal-Washery Refuse Pulps," *Coal Age* 67:94–100 (December, 1962).

83. D. G. Hall, "The Role of Bridging in Colloid Flocculation," *Colloid Polymer Sci.* 252:241–243 (1974).

84. W. E. Walles, "Role of Flocculant Molecular Weight in the Coagulation of Suspensions," *J. Colloid Interface Sci.* 27:797–803 (1968).

85. Committee Report, "State of the Art of Coagulation Mechanism and Stoichiometry," *Am. Water Works Assoc. J.* 63:99–108 (1971).

86. S. L. Daniels, "A Survey of Flocculating Agents: Process Descriptions and Design Considerations," *AIChE Symp. Ser.* No. 136, 70:266–281 (1973).

87. F. M. Tiller, J. Wilensky, and P. J. Farrell, "Pretreatment of Slurries," *Chem. Eng.* 81(9):123–126 (April 29, 1974).

88. R. J. Akers, Flocculation. Publ. by I. Chem. E. Services for Inst. Chem. Eng., 15 Belgrave Square, London, England SWIX 8PT (1975).

89. D. K. W. Smith and J. A. Kitchener, "The Strength of Aggregates Formed in Flocculation," *Chem. Eng. Sci.* 33:1631–1636 (1978).

90. R. N. Kovalcik, "Single Waste-Treatment Vessel Both Flocculates and Clarifies," *Chem. Eng.* *85*(14):117–120 (June 19, 1978).

91. D. A. Dahlstrom and C. F. Cornell, "Thickening and Clarification," *Chem. Eng. Deskbook Issue* *78*(4):63–69 (February 15, 1971). "Sedimentation Systems from Laboratory Data," *Chem. Eng.* *68*(19):167–170 (September 18, 1961).

92. R. A. Conway and V. H. Edwards, "How to Design Sedimentation Systems from Laboratory Data," *Chem. Eng.* *68*(19):161–170 (September 18, 1961).

93. H. C. Bramer and R. D. Hoak, "Design Criteria for Sedimentation Basins," *I EC Proc. Des. Dev.* *1*:185–189 (1962).

94. E. B. Fitch, "The Significance of Detention in Sedimentation," *Sewage Ind. Wastes 29*:1123–1133 (1957).

95. W. H. Mitchell, "The Preparation and Treatment of Mine Water Prior to Pumping and Notes on Maintenance of Pumping Plant," *Inst. Certif. Eng. (S. Afr.) J. 26*:86–99 (1953).

96. R. T. Hukki, G. Diehl, and P. Vanninen, "Principles of Construction and Operation of the Channel and Syphon Thickener," in *7th Int. Mineral Processing Congress*, Vol 1, Gordon and Breach, New York, pp. 115–123 (1965).

97. R. F. Probstein and R. E. Hicks, "Lamella Settlers: A New Operating Mode for High Performance," *Ind. Water Eng. 15*:6–8 (1978).

98. R. L. Cook, "Compact Lamella Thickeners in Coal Preparation Plants," in *Second Symposium on Coal Prep.* (Washington, DC), at NCA/BCR (Natl. Coal Assoc./Bitum. Coal Res.) Coal Conf., and Expo. 3, Louisville, KY (October 19–21, 1976). (Available from NCA.)

99. R. L. Cook and J. J. Childless, "Performance of Lamella Thickeners in Coal Preparation Plants," *Min. Eng. 30*:566–571 (1978).

100. L. C. Meitzler and G. H. Weyermuller, "Compactness of Thickener Permits Installation in Limited Space," *Chem. Proc. 37*(5):18 (1974).

101. F. C. McMicheal, "Sedimentation in Inclined Tubes and Its Application for the Design of High Rate Sedimentation Devices," *J. Hydraul. Res. 10*:59–75 (1972).

102. D. Davis, S. Rogel, and K. Robe, "Increases Clarifier Capacity 85% at 1/20th Cost of New Unit," Chem. Process (U.S.) 36:10 (January, 1973).

103. R. M. Willis, "Tubular Settlers–A Technical Review," *J. Am. Water Works Assoc. 70*:331–335 (1978).

104. Axel Johnson Institute, "A New Separation Technique," Trade Brochure Box 13, Nynashamn, Sweden.

105. J. W. de Villiers, "An Investigation into the Design of Underground Settlers," *J. S. Afr. Inst. Min. Metal. 61*:501–521 (1961).

106. A. W. Bond, "Upflow Solids Contact Basin," *J. San. Eng. Div. Proc. ASCE* (SA6) 73–99 (1961).

107. Private communication from T. M. Stielau, Delkor Technik, Randburg, S.A.

108. Anon., "Revolutionary Thickener Design Tackles heaᴛly Flow of Zinc Mine Tailings," *Eng. Min. J. 179*:78–79 (April, 1978).

109. G. R. Mace and R. Laks, "Developments in Gravity Sedimentation," *Chem. Eng. Prog. 74*(7):77–83 (1978).

110. J. R. Bratby, "Aspects of Sludge Thickening by Dissolved-Air Flotation," *Water Pollut. Contr.* (Lond.) *77*:421–432 (1978).

111. M. T. Turner, "Use of Dissolved Air Flotation for the Thickening of Waste Activated Sludge," *Effluent Water Treat. J. 15*:243–251 (7 p.) (May, 1975).

112. E. R. Ramirez, "Dewatering Skimmings and Sludges with a Lectro-Thic Unit," Water and Wastewater Equip. Manuf. Assoc. Pollut. Control Conf., 3rd Annual Ind. Solutions '75: Air-Water-Noise-Solid Waste, Proc., pp. 467–78, ANN, April 1–4, 1975.

113. V. Gulas and R. Lindsey, "Factors Affecting the Design of Dissolved Air Flotation Systems." *J. Water Pollut. Control Fed.*

114. H. E. Cross, "A New Approach to the Design and Operation of Thickeners," *J. S. Afri. Inst. Min. Metall. 63*:271–289 (1963).

115. E. B. Fitch and W. A. Lutz, "Feedwells for Density Stabilization," *J. Water Pollut. Control Fed. 32*:147–156 (1960).

116. Anon., "Improved Thickener Rake," *S. Afr. Min. Eng. J.*, p. 415 (August 23, 1968).

117. F. R. Weber, "How to Select the Right Thickener," *Coal Min. Proc. 14*(5):98–100, 104, 116 (1977).

118. V. S. Dillon, "Special Features of the Kinross Mines, Ltd., Reduction Plant," in *Proc. 9th Commonwealth Mining Metall. Congress*, London, *Miner. Proc. Extr. Metal. 3*:485–508 (1969).

119. R. P. Plasket and D. A. Ireland, "Ancillary Smelter Operations and Sulphuric Acid Manufacture at Impala Platinum Limited," *J. S. Afr. Inst. Min. Metal.*, pp. 1–10 (August, 1976).

120. F. Concha and E. R. Almendra, "Settling Velocities of Particulate Systems. 1. Settling Velocities of Individual Spherical Particles," *Int. J. Miner. Proc. 5*:349–367 (1979).

121. K. J. Scott, "Experimental Study of Continuous Thickening of a Flocculated Silica Slurry," *Ind. Eng. Chem. Fundam. 7*:582–595 (1968).

122. L.-G. Ecklund, "Influence of Feed Conditions on Continuous Thickening," *Chem. Eng. Sci. 34*:1063–1066 (1979).

123. J. H. Masliyah, "Hindered Settling in a Multi-species Particle System," *Chem. Eng. Sci.* *34*:1166–1168 (1979).

124. F. M. Tiller, "Revision of Kynch Sedimentation Theory," *AIChE J. 27*:823–839 (1981).

125. B. Fitch, "Kynch Theory and Compression Zones," *AIChE. J. 29*:940–947 (1983).

126. F. M. Tiller and W. Chen, "Limiting Operating Conditions for Continuous Thickeners," *Chem. Eng. Sci. 43*(7):1695–1704 (1988).

127. J. L. Chandler, "Design of Deep Thickeners," Preprint, Institute of Chemical Engineers Symposium, Rugby, U.K. (1976).

128. P. Kos, "Fundamentals of Gravity Thickening," *Chem. Eng. Prog. 73*:99–105 (1977).

129. D. C. Dixon, "Effect of Sludge Funneling in Gravity Thickeners," *AIChE J. 26*:471–477 (1980).

130. W. Chen, "A Study of the Mechanism of Sedimentation," M.S. Thesis, University of Houston (1984).

131. R. D. Paradis, "Application of Alcan's Deep Thickener Technology for Thickening and Clarification," in American Filtration & Separation Society Annual Meeting, Chicago (1993).

132. G. J. Kynch, "A Theory of Sedimentation," *Trans. Faraday Soc. 48*:166–176 (1952).

133. L. E. Kun, R. O. Oelofsen, and E. J. J. Van Veuren, "Hopper Clarification of Gold Pregnant Solution at Vaal Reefs South," *J. S. Afr. Inst. Min. Metall. 78*:201–206 (1979).

14
Filtration of Solids from Liquid Streams

Larry Avery

CONTENTS

14.1 INTRODUCTION

In the modern filtration world, there are few industries that do not depend at least in part on the liquid filtration process. Current consumers demand products that are clear, free from contamination, attractive, and healthful. Filtration helps to accomplish these objectives.

Industry likewise demands commercial chemical and pharmaceutical products that are pure, meet stringent standards, and are of high quality. Pollution control regulations require clean water, sewage treatment, waste reduction, and sludge dewatering before disposal, all of which are controlled by various liquid–solid filtration separation process operations.

The diversity and rather little known ubiquity of the many common products made partly by filtration operations was the subject of a most interesting article entitled "Do We Need Filtration?" by Carl Jahreis, a research filtration engineer of the Shriver Co. He described more than 100 common products that one encounters daily that would not exist if it were not for liquid filtration operations.[1]

In light of competitive market conditions, increased energy cost, and government regulations, the continuing profitable manufacture of products requires a knowledge of how to filter particles suspended in the liquid phase. Also needed is a knowledge of how best to select a suitable filtering device, how to program and operate it efficiently, and, finally, how to perform needed postfiltration steps to provide the final quality product.

It is the aim of this chapter to explore this complex subject of particle removal from liquid streams, and to make understandable the mechanics involved in the practical modern filter devices used to accomplish this goal.

14.1.1 Definition

Liquid filtration is a two-phase physical separation of particulate solid matter from the liquid in which it is suspended. The means is a filter medium properly selected to retain the solid particles. The driving force may be gravity, vacuum, pressure, or centrifugal force.

14.1.2 Purposes

The general purposes of filtration can be broadly stated according to a frequently used classification as follows:

1. To clarify, purify, or sterilize a valuable liquid end product where the contaminants are discarded usually with the filter medium
2. To collect a solid valuable product with the liquid discarded, reused, or recycled
3. To save both phases as useful products
4. To dispose of as waste both the solids filtered out as dewatered sludge, and also the liquid portion as a waste material.

From the above it is apparent that there are two major divisions of filtration: clarification and solids or cake filtration.

14.1.3 Clarification

Clarification involves hundreds of liquid food products, juices, household products, water, coolant liquids, chemicals, hydraulic oils, gasoline, and even molten metals. Generally, the liquid has a very low percentage of solids, usually 0.1% or less. Typical media are in the shape of cartridges, bags, sheets, pads, and nonwovens. Operation is by pressure, and it is usually a small-scale batch process.

14.1.4 Cake Filtration

Cake filtration provides pigments, dyestuffs, minerals, chemicals, food such as yeast, pharmaceuticals, and catalyst. Filtration is a most important process in pollution control for dewatering waste water streams. It provides an economical way to solidify waste for regulated landfill and hazardous waste disposal facilities. Cake filtration may use pressure, vacuum, or centrifugal force. Equipment used includes filter presses, rotary vacuum filters, and centrifuges. Large-scale chemical, mineral, and waste sludges are usually cake filtrations.

14.1.5 Concept

The concept of filtration seems simple as shown in Figure 14.1. Ideally, all of the particles would be removed and the filtered effluent, or filtrate, would be perfectly clear. In practice, this never happens because the filter medium always permits some particles to pass through. Also, some liquid is always retained in the collected cake, which is the greater problem of the two.

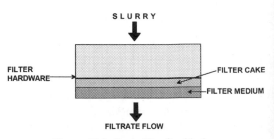

Figure 14.1. Basic filtration device.

14.1.6 The Four Basic Components

Important considerations of the four basic components are given in the following subsections.

Liquid. The liquid contains the suspended particles, and is called the feed, suspension, or slurry. The types of particles—the size characteristics, the density, the settling rate, the shape, softness, quantity, and chemical nature—and the viscosity of the liquid determine the filterability of the feed or slurry.

Medium. The medium is the porous material for collection of the particles. It determines the efficiency of the filtration, the mechanism involved, and the suitable operation of the filter itself.

Solids. Mostly, it is desirable for the solids to have as low a residual moisture content as possible, and to be free from the mother liquor; hence the need to wash the filter cake.

Filtrate, or Effluent. A high degree of clarity or purity is required for liquid products. Wastewater streams should be low in TSS (total suspended solids).

Filters can operate in either a batch or a continuous mode. Most batch filters operate on a small scale. However, some batch filter presses, for example, can handle solids loads up to 300 ft^3 per batch or about 9 tons depending on the density of the wet cake. Continuous belt or rotary vacuum filters can process up to 120 tons per day, again depending on the percent solids in the feed stream, and the density of the solids. Of course, batch filters can be operated in multiple parallel stages to produce essentially a continuous production output.

14.2 PHYSICAL MECHANISMS OF FILTRATION

14.2.1 Surface Filtration

The two almost classic concepts of liquid filtration are called surface filtration and depth filtration. They refer to the filter medium and the mechanism of the particle collection as regards the specific media. In the first case, as shown in Figure 14.2, the particles are retained from the suspension exactly on the face of the medium as the particles approach the medium at right angles. The principle is that the pores or openings in that medium are smaller than the particles contacting it, thus preventing them from passing through. The medium must be physically and mechanically strong enough to resist any pressure deformation preventing the pores from enlarging. The particles must be rigid or firm enough so as not to compress or squeeze through the openings. If all these conditions are met, we would have complete or absolute retention of the particles. With metal screens, perforated metals, porous ceramics, and some membrane filters, this condition can approach reality. In many cases it does not have to be perfect because the filtrate can be recirculated and trapped by the solids already built up on the medium. This cannot be done, however, in critical microbe filtrations of pharmaceutical products where a single pass must be complete to what is called a log-reduction value of 7, which indicates that there were 10 microbes found in the filtrate for a filtration efficiency of 0.9999999 (seven nines).

This same surface filter medium is also the desired type for cake filtrations, where the solids built up to 1-in. thickness or more based on the type of filter. Media with a smooth slick surface and a pore size in the 1 to 10 micron

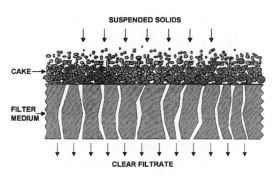

Figure 14.2. Surface filtration sketch.

range will accomplish this. Over time, there will be some penetration of the medium known as progressive plugging, but yet these filter cloths can last for hundreds of filtration cycles before they have to be replaced.

14.2.2 Depth Filtration

The other basic separation mechanism is known as depth filtration, and as the name implies, the solids are captured under the surface, and within the depth of the filter medium. This concept is shown in Figure 14.3. This can apply to membrane media that may be only 50 microns thick or fabrics and filter sheets that can be as much as 0.125-in. thick. This is not to say that very large particles may all collect on the surface so that it performs like a surface medium. This can happen with string wound filter cartridges, for example, which are layered to provide a porosity gradient from the outside feed side to the internal core for filtrate discharge.

The advantage of depth filters is that they can trap particles smaller than the average pore size in the medium. This is done by electrostatic forces, molecular forces, direct impingement on fibers, and attachment to the sidewalls of the interstices within the medium. This entrapment of particles within the depth leads to an important property of filter cartridges called dirt-load capacity. Even though cartridges are used for feed streams of under 0.1% solids, the higher the dirt-load capacity,

the longer the cartridge life and lower related filtration costs.

14.2.3 Pore Blocking

Another mechanism closely related to the above is pore blocking and particle bridging of the pores. The first is undesirable because it stops the flow. It occurs most often with relatively small particles, high viscosity, and low solids concentration.

Particle bridging results from particles collecting around the pore openings, and gradually closing over the opening. An increase in the suspension's particle concentration favors this mechanism. Once this occurs, cake filtration can take place.

14.3 FILTRATION THEORY

Filtration has long been considered more of a practical art still being developed than as an engineering science. Likewise, the theory of filtration operations has itself been the continuing subject of much study in the academic field. Many of the basic approaches for the last 75 years have been most important in developing fundamental theoretical relationships. The real beginning was the work of Darcy on capillary and pressure relationships in 1856.

His work was recently translated not without some difficulty from the French to English by J. B. Crump and critiqued by Tiller as related to our current theory.[43]

The equations expressing relationships between filtration variables have been applied to certain designs of equipment, but mostly they are helpful in interpreting pilot and laboratory tests and determining the specific cake resistance which is unique to each slurry. This specific cake resistance is affected by the basic factors plus the porosity and the specific surface of the particles in the suspension to be filtered.

The fundamental theory begins with the basic Darcy equation relating the flow rate Q

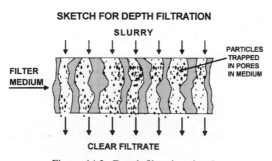

SKETCH FOR DEPTH FILTRATION

SLURRY

FILTER MEDIUM

PARTICLES TRAPPED IN PORES IN MEDIUM

CLEAR FILTRATE

Figure 14.3. Depth filtration sketch.

of viscosity μ through a bed of thickness L and area A and driving force p:

$$Q = K \frac{A \Delta p}{\mu L} \qquad (14.1)$$

where K is a constant referred to as the permeability of the filter bed. This equation is often written in the form

$$Q = \frac{A \Delta p}{\mu R} \qquad (14.2)$$

where R is called the medium resistance and is equal to L/K.

If the suspension were a clean liquid, the parameters in Eq. (14.1) would be constant, and the relationship between the flow and the pressure drop would produce a cumulative filtrate volume that would increase linearly with time. When the suspension contains particles, the resultant cake formation takes up more pressure so the flow decreases with time.

With cake forming, there are two resistances to flow, the cake and the filter medium as per the following equation:

$$Q = \frac{A \Delta p}{\mu (R + R_c)} \qquad (14.3)$$

This assumes the filter medium resistance to be constant, which in practice is not always precisely true because of particle impingement on the medium surface, and also progressive plugging of the media. Assuming the cake (if incompressible) is proportional to the amount of cake deposited, it follows that

$$R_c = \alpha w \qquad (14.4)$$

where w is the mass of cake deposited per unit area and α is the specific cake resistance. Substituting Eq. (14.4) in (14.3) gives

$$Q = \frac{\Delta p A}{\alpha \mu w + \mu R} \qquad (14.5)$$

This relates the flow rate Q to the pressure drop Δp, the mass of cake deposited w, and other parameters, some of which can be assumed to be constant. However, Δp may be constant or variable with time. The face area

may increase where cake builds on tubular or rotary drum surfaces. The viscosity stays constant if the temperature is likewise constant and the liquid is Newtonian.

The specific cake resistance α should be constant for incompressible cakes, but could vary slightly because of possible cake consolidation or feed approach velocity. However, most cakes are compressible, so the specific cake resistance changes with Δp_c. Then the average specific cake resistance α_{av} should replace α in Eq. (14.5). It can be determined by

$$\frac{1}{\alpha_{av}} = \frac{1}{\Delta p_c} \int_0^{\Delta p_c} \frac{d(\Delta p_c)}{\alpha} \qquad (14.6)$$

if the function $\alpha = f(\Delta p_c)$ is known from test data. If not, an experimental empirical relationship can be used over a limited pressure range:

$$\alpha = \alpha_0 (\Delta p_c)^n \qquad (14.7)$$

where α_0 is the resistance at unit applied Δp and n is a compressibility index (equal to zero for incompressible substance).

Using Eq. (14.7), the average cake resistance α_{av} can be shown to be:

$$\alpha_{av} = (1 - n) \alpha_0 (\Delta p_c)^n \qquad (14.8)$$

The mass of cake deposited per unit area w is a function of time in batch filtrations, and it can be related to the cumulative volume V in time t by

$$wA = cV \qquad (14.9)$$

where c is the concentration of solids in the suspension.

From the above initial analysis, basic equations for filtration operations for incompressible cakes using constant pressure and constant rate filtrations have been developed. From pilot tests, the specific cake resistance can be determined. Likewise, equations for compressible cake filtrations and relationships between the specific cake resistance and porosity and specific surface of the particles have been made known. These are expressed as the classic Kozeny–Carman equation.

The above is only a very basic outline of simple theory as based on an excellent presentation by L. Svarvosky on filtration fundamentals in his recent book. All of the basic equations mentioned above are included in detail in Ref. 44.

As research workers explore troublesome assumptions in the classic theory, new considerations are presented. Work by Tiller, Wakeman, Rushton, Willis, and others is adding to the field. Studies on formulas for constant pressure filtration and compaction of filter cakes were presented at the recent American Filtration meeting in Hershey, PA.[4]

14.4 FILTER MEDIA

Filter media are available in many different forms, and being the essential element of a filter, they should have as many of the following characteristics as possible. These pertain mostly to woven fabrics, but can apply to some nonwovens such as felt as well. They:

1. Should have particle retention suitable for the application, generally no more that actually required because of increased cost.
2. Should have low flow resistance.
3. Should be resistant to chemical degradation and any subsequent cleaning chemicals.
4. Should have enough physical strength to adapt to the type of filter equipment used and avoid problems from creasing.
5. Should not change form, stretch, or shrink during filtration or be susceptible to bacterial growth.
6. Should offer resistance to the maximum temperature of liquid to be filtered or subsequent washing or steam cleaning of media.
7. For cake filtration, should have a smooth and slick surface to facilitate unaided release of the filter cake.
8. Should not have loose fibers that shed into the cake or liquid being filtered.
9. Should be capable of being fabricated, sewed, fused, or adaptable to other types of converting operations.
10. Should have an economic service life.

Not all the above will be found in a single medium so that certain compromises will have to be made regarding cost, medium life, and performance.

14.4.1 Types of Filter Media

The most common types of media are woven fabrics, papers, and felts. Yarn types for woven media are shown in Fig. 14.4. Physical and chemical characteristics of the most frequently used fibers are shown in Table 14.1. In recent years, there has been an increasing interest in nonwoven textiles and also membranes, laminates, finemesh woven metal wire, and photoetched metals.

Also considered as media are screens, wedge wire, see Figure 14.5 grids, sand, perforated steel plates, porous ceramic, see Fig. 14.6 plastic, and carbon sheets and tubes. Thus, it can be seen that some media are flexible, some rigid, and some even granular. Pore size and porosity can vary considerably. Selecting the right filter fabric was covered by Clark.[45]

Figure 14.4. Yarn types. (Zurich Bolting Cloth Mfg. Co. Ltd.)

Table 14.1. Typical Characteristics for Common Fibers Used for Filter Cloths for Liquid Filtration. Temperatures are Approximate. Resistances Depend on Strength and Temperature of Acid or Alkali.

	MAXIMUM OPERATING TEMPERATURE					
FIBER	°F	°C	ACID RESISTANCE	ALKALI RESISTANCE	WET HEAT RESISTANCE	FLEX AND ABRASION
Acrylic	275	135	Excellent	Fair	Good	Good
Aramid	400	205	Fair	Good	Excellent	Very Good
Cotton	210	99	Poor	Good	Fair	Good
Nylon	250	121	Fair	Very Good	Good	Excellent
Polyester	284	140	Good	Fair	Good	Very Good
Polypropylene	200	94	Excellent	Excellent	Fair	Very Good

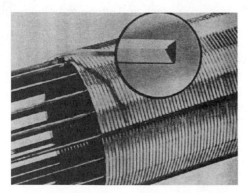

Figure 14.5. Wedgewire media. Wound to form tubular filter element. (Johnson Screens)

Ceramic media for severe corrosive environments are discussed by Sheppard.[46]

14.4.2 Selection of Filter Media

Generally the type of filter equipment selected will determine the appropriate filter medium based on years of experience by the filter manufacturers and users. For example, filter presses use filter clothes that are mostly synthetics such as polypropylene and polyester; pressure leaf filters use fine mesh stainless steel woven wire cloth; rotary vacuum filters use lighter weight, 5 to 6 oz/yd,² more open

SEM image shows cross-section of support and membrane layers that make up a 0.2 micron P19-40 element.

Figure 14.6. Porous ceramic. U.S. Filter Membralox® (Trade Mark)

cloths than filter presses; and belt filters use heavy-duty, 22 to 25 oz/yd^2 rugged synthetic woven cloths. Many clarifying filter presses use filter paper and sheets. Cartridges and filter bags are also used widely for clarifying filtrations. Once a medium has been established in practice, and there is a change in the process, or if a problem develops, such as insufficient particle retention, improvements can be made by gradually selecting a similar filter medium for test on a pilot scale. Then the new medium can be tried on plant equipment, usually with good results.

Where an untried application develops, the selection has to be made with a more critical look at all the desired characteristics. Here, lab tests will be required to determine a final choice.

Filtration and separation media characteristics along with advantages and typical uses are shown in Table 14.2.[47]

14.4.3 Nonwoven Media

A newer type of media showing increasing use is the nonwoven or bonded material. It has a web structure of entangled fibers made by a mechanical, thermal, or chemical bonding process. The filtering properties of these media are controlled, such as strength and uniformity of fiber orientation. A recent article explains the advantages and applications of the four basic technologies for bonding nonwoven webs, which are chemical, ultrasonic, needle punching, and adhesive melt.[48] The various types are designated by the method of formation such as card webs, air laid, wet formed, spunbonded, and melt blown. This nonwoven technology is explained by Shoemaker.[49] In addition, bonding mechanisms are given by Pangrazi.[48]

Major uses for nonwovens are for membrane supports and cartridge filters, especially for swimming pool water and other liquid filtrations. Advantages are pleatability, resistance to damage, good retention values, and flow rates. In roll form from 18 in. to 45 in. widths, they are widely used in filtering machine tool coolants in deep bed filters. An-

other major use is on continuous horizontal plate filters for coolants in the metal rolling industry and D & I can manufacturing operations.

In a study comparing pleated nonwoven media in a filter cartridge against a conventional wound cartridge design, it was found that the nonwoven media had a greater efficiency in particle removal than the wound yarn design.[65] The media used was a polyester material. Other materials are the cellulose, rayon, and nylon used in early nonwovens. More recently, aramids such as Nomex and Kevlar and fluorocarbons such as Teflon are being used.

14.5 MEMBRANES

A most important field of liquid filtration is the one that benefits from membrane filters. These are very thin microporous polymeric film sheet media from 10 to 100 microns thick. The range of separations is shown in Figure 14.7. The four basic types of membrane processes are:

1. Reverse osmosis (RO), with an osmotic pressure driving force separating a solvent, usually water, from a dissolved monovalent salt
2. Nanofiltration (NF), which rejects divalent salt, sugars, and disassociated acids
3. Ultrafiltration (UF), which separates or fractionates dissolved molecules by molecular weight and size
4. Microfiltration (MF), which is actually particle removal of very fine or colloidal particles.

There is some overlapping of the separation range, and since we are concerned only with particle filtration, we will discuss MF and the related range of UF. From their limited initial use 50 years ago in removing microorganisms in drinking water, they have had rapid growth to sales of over $900 million annually. They meet critical applications in gas, liquid, and solvent separations. Major uses are in desali-

Table 14.2. Filtration and Separation Media Characteristics.

TYPE	RELATIVE MEDIA COST RANGE	GENERAL EFFICIENCY RATINGS RANGE	RELATIVE MEDIA PERMEABILITY RANGE	CURRENT MARKET PENETRATION TREND AND USE	ADVANTAGE	DISADVANTAGE	COMMON USE
Ceramic	8–10	5–10	1–4	+	Chemical compatibility and high temperature capabilities.	Expensive and brittle.	Biotechnology-pharmaceutical. reusable.
Filter aid	1–2	2–7*	8–10	Same	Inexpensive, excellent filter cake base.	Disposal, mostly limited to pressure filtration.	Precoat for large volume pressure filtration e.g leaf pressure filters.
Glass	1–2	2–8	4–8	Same to –	High temperature, chemical compatibility, low stretch, low cost.	Limited media processing capabilities, yarn weakness.	Baghouse filtration, laboratory filters, HEPA filters.
Membranes	6–9	9–10	1–2	+	Narrow pore size distribution below one micron, many polymer choices.	High cost, low flow rates. Somewhat hard to process.	Pharmaceuticals, semiconductors, medical devices, ultrapure water.
Metal	3–8	4–9	3–10	Same to –	Reusable, high temperature, diverse properties, narrow pore size distribution.	Expensive, high cleaning costs.	Vibratory sifting, aerospace, polymer filters, reusable applications.

Table 14.2. Filtration and Separation Media Characteristics. Continued

TYPE	RELATIVE MEDIA COST RANGE	GENERAL EFFICIENCY RATINGS RANGE	RELATIVE MEDIA PERMEABILITY RANGE	CURRENT MARKET PENETRATION TREND AND USE	ADVANTAGE	DISADVANTAGE	COMMON USE
Nonwoven fabrics	1–3	1–8*	4–8	++	Low cost, dirt holding capability, diverse constructions.	Random pores, particle unloading, fiber migration disposal of media.	Chemical, medical, water, baghouse filters, strainer bags.
Paper	1–2	1–6	4–8	–	Dirt holding capabilities, diverse polymers, moldable.	Fiber release possible, poor wet strength, particle unloading, disposal of media.	Automotive, laboratory, air, and general process industries.
Porous plastics	3–7	4–8	2–5	Same	Dirt holding capabilities, diverse polymers, moldable.	Restricted to rigid forms, limited uses.	Medical, battery vents, water.
Precision woven synthetic screen fabrics	4–7	7–9	4–10	+	High flow with minimal resistance, precision pores, wide choice.	Especially expensive at lower size pore ratings (5–30 micron).	Dewatering, medical, aerospace, automotive, process filtration including belts.
Woven fabrics	3–5	2–6	2–7	Same	High wet/dry strength. Lower cost. Dirt holding capabilities. Wide choice.	Lower flow rates, random size pores, particle unloading.	Filter presses, RO channel separators, vacuum belts

*High range includes special conditions/processing and circumstances.
Range Ratings: 1—Represents lower cost or performance. 10—Represents higher cost or performance.

RANGE OF SEPARATION BY MEMBRANE PROCESSES

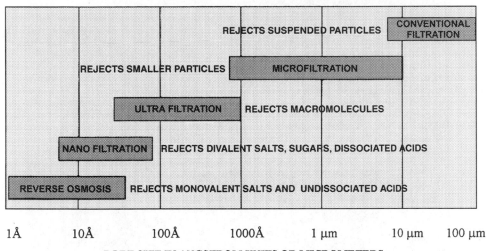

PORE SIZE IN ANGSTROM UNITS OR MICROMETERS

Figure 14.7. Range of separation by membrane processes.

nation, fluid sterilization, and waste water treatment, such as separating emulsified oily wastes. Microfiltration uses include removal of suspended particles from effluent waters, clarification of fruit juices and vinegars, and harvesting bacterial cells.

Cellulose esters were first used for making microfiltration membranes, but now various polymers including nylon, polyvinyl chloride (PVC), polypropylene, polysulfones, and polytetrafluorethylene (PTFE) are used. The latter can be made by stretching a thin sheet of the polymer carefully and bonding it to a porous substrate.[50] See Fig. 14.8. The membrane film can be from 100 to 250 microns thick. Other membrane preparation methods are track-etching and phase inversion casting. Details of these processes are given by Porter.[51]

Most of these membranes have physical and temperature limitations and may be subject to chemical and solvent attack. Recently ceramic membranes have become commercially available. Originally developed by the French nuclear industry, and now declassified, they are being used as tubular membrane filters with the membrane surface on the inside. Retention values down to less than 0.2 micron are available as shown in Figure 14.6. Another

ceramic application is for large discs in a rotary vacuum disc filter used in the mining industry. These filters can have several hundred square feet of filter area.

Membrane filters can be designed with flat sheets in a plate and frame support. Cartridges are made in a spiral wound or pleated membrane configuration. Hollow fiber membranes are in a tubular design.

Flat sheet media or pleated cartridges have a special application in the pharmaceutical industry for the purpose of sterilization of certain liquid batch products. The membrane media and its holder or housing must be sterilized, and tested for integrity. Since it is critical that no microbes or contaminants pass through the filter assembly, the pores in the membrane must not be larger than the microbes or particles to be retained. To verify this, a bubble point test must be done. The apparatus for performing this test is described by ASTM F136. The factors involved are absolute filtration, average size pore, and filtration efficiency. The variables and interpretation of this subject are discussed in detail by Johnston.[52]

One aspect of membrane filtration, and different from sterilizing filtrations, is that the flow patterns are not at right angles to the

1000X

Figure 14.8. Illustration of stretched polymer membrane media at 1000 × magnification. Courtesy W. G. Gore and Associates.

membrane, see Fig. 14.9, but tangent to the media, which is called crossflow filtration. (See Fig. 14.10.) This concept has been used in conventional cake filtration as a means to limit cake growth and increase output. However, in membrane filtration, it is essential that the crossflow be of sufficient velocity to offset a phenomenon known as "concentration polarization" in which the solute builds up on the surface of the membrane in concentrations

CONVENTIONAL FILTER

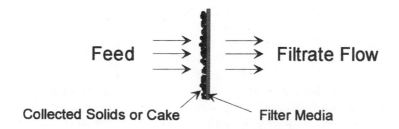

Feed ⟶ Filtrate Flow

Collected Solids or Cake — Filter Media

DIRECT-FLOW (PERPENDICULAR TO MEDIA)

Figure 14.9. Direct flow to media.

MEMBRANE FILTER

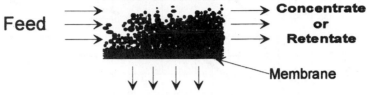

Feed

Concentrate
or
Retentate

Membrane

Permeate Flow

CROSS-FLOW (TANGENTIAL TO MEMBRANE)

Figure 14.10. Cross flow filtration.

much higher than in the bulk flow of the feed stream. There are ways of overcoming this problem, and, among others, Van den Berg and Smolden developed mathematical models to study it. They concluded that besides cross-flow filtration, reducing scaling of membranes, chemical treatment of the membrane surface, using corrugated membranes, and using appropriate pretreatment methods to increase the mass transfer coefficient are helpful.[53]

In the field of biological membrane separations, Gyure discusses in qualitative terms the many practical considerations in using cross-flow filtration. Continuous versus batch systems are compared and methods for effective cleaning of membranes are given.[54]

14.6 FILTER AIDS

Filter aids are loose powders, such as diatomaceous earth (DE) and expanded perlite, that are used to facilitate and improve the filtration of difficult to filter products, such as gels, hydroxides, and very fine particles. Their rigidity and high porosity make them suitable for this purpose. They are added to the slurry, thus forming a more permeable filter cake. Occurring as natural minerals, they are processed into about 10 different grades with ranges of particle sizes from 40 microns down to under 2 microns. The distribution of parti-

cle sizes determines the grade and the practical application. Flow rates of different grades are shown in Figure 14.11. Diatomite filtration systems can remove particles under 1 micron and at flow rates of from 0.2 to 2.0 gal/min/ft^2 on rotary vacuum filters.[56] This type of use is called precoat filtration; a 5- or 6-in. layer of filter aid is formed on the filter drum, and is gradually scraped off with a sharp knife edge along with a thin layer of the filtered solids. Precoat can also be used on sheet media and filter cloths as a porous layer, and also serves to facilitate cake removal from the medium.

Another common method of using filter aids, called admix or body-feed, is to add them to the suspension being filtered. The amount and grade used can be determined empirically, but generally it must be equal to or more than the weight of the suspended solids, and it can exceed this by up to 10%. If this optimum amount is not maintained, it is apparent that the formed filter cake will plug to end the cycle.

Although filter aids are inert, and up to 95% silica, they do have impurities such as iron, copper, etc. and the possible effect on the filtrate should be considered in their selection. Also, the amount of filter cake produced is greater, and this could add to disposal costs.

The permeability of diatomite filter aids is specified in Darcies, which is defined at unity if a liquid material passes 1 cc/cm^2 per sec-

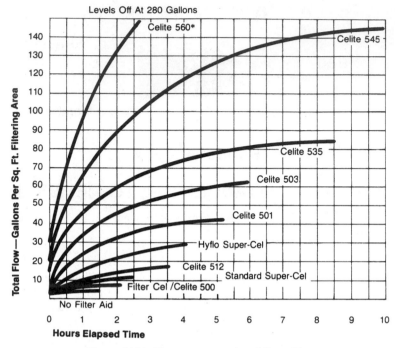

Figure 14.11. Flow rate properties of filter aids.

ond through a layer of 1 cm thick, the viscosity being 1 cp and the pressure drop 1 atm.

Another material used as a filter aid is cellulose fiber. Besides serving as a filter aid, particularly over screens on pressure leaf filters, it is combustible in case the filter cake has to be incinerated for disposal or product recovery.

A recent addition to filter aid materials is rice hull ash (RHA), which is from 92% to 97% amorphous silica dioxide. These calcined curved rigid particles have a porosity that makes them suitable for body feed filtrations. Examples and filtration characteristics are described by Reiber.[57]

Major uses of diatomite filter aids include food and chemical processing, brewing, pharmaceutical, metalworking, and electric power industries. Recently, they have been used more in the municipal water treatment field and also for clarifying drinking water supplies. More than 170 plants using DE have been installed since 1949. They are also effective in controlling the waterborne disease giardiasis.[58]

14.7 STAGES OF THE FILTER CYCLE

14.7.1 Pretreatment

Before the filtration actually starts, the condition of the slurry can be modified to a certain extent for the purpose of making the separation easier by increasing the size of the particles to be filtered out. Larger particles settle faster, and also make more porous and less resistant filter cakes. Both chemical and physical methods can be used.

Pretreatment, also called conditioning in water treatment, is done in several ways. For clarifying operations, with cartridge filters, for example, the use of a coarser filter before the final filter is a common approach. Choices as to the relative retention values need to be determined by tests for the most economical results.

In the case of cake filtration the most frequent method used is to thicken the solids content in the slurry. This has a great effect on the performance of cake filters. It affects the capacity and cake resistance. For example, for the same cycle time, if the concentration is increased by a factor of four, the production capacity is doubled. Or alternately, filtration area can be cut in half for the same capacity.[44]

Physical pretreatment can include heating the slurry to lower the viscosity and improve the flow rate. It may be cooled to chill and solidify waxes, for example, so they will filter out. Other physical means are ultrasonic and mechanical vibrations, magnetic treatment, and ionizing radiation.

By far, the most frequently used methods are addition of chemicals to coagulate or flocculate the particles by changing the particle charges. This is particularly helpful in filtering colloidal suspensions, usually considered as containing particles from 0.001 to 1.0 microns in size. Natural electrolytes such as alum, lime, ferrous sulfate, and ferric chloride decrease the surface charge and are called coagulants.

Flocculants can be either natural or synthetic chemicals, which cause dispersed particles to form relatively stable aggregates of particles. These settle and filter more easily. Higher molecular weight long-chain organic chemicals called polyelectrolytes are widely used in this process. They are available commercially in liquid, powder, or emulsion forms, and also anionic nonionic and cationic types. The science of selecting them has been highly developed.[59] Some modern filters such as belt pressure filters for sewage sludge filtrations would not be cost effective without the use of suitable polyelectrolytes.

Although there has been some confusion about the terms coagulation and flocculation, they are better thought of in terms of function. A good explanation is given by the publication by Zeta-Meter, Inc.[60] Coagulation takes place when the energy barrier between particles is lowered so that the net interaction is always attractive. This is also referred to as destabilization. Flocculation refers to the successful collisions that occur when the destabilized particles come together and form agglomerates and then visible floc masses. The

electrokinetic force that controls this process is called the zeta potential.

14.7.2 Filtration

After the pretreatment step, the slurry is fed to the filter by gravity, pressure pumps, or vacuum sources. For pressure filters, slurries are usually fed by diaphragm pumps. They are easy to control by compressed air, and when the filter is loaded with filter cake, they reach maximum pressure and stop. Gear pumps are often used for small clarifying filters. Higher pressure progressive-cavity pumps are used for sludge filtering up to 225 psi. Pumps frequently are automatically controlled to increase pressure gradually as the cake resistance increases.

14.7.3 Cake Washing

For cake products, such as pigments, the mother liquor must be removed. Formerly simple displacement washes were used that were inefficient owing to large volumes of wash fluid required. Also cracks were formed in the filter cake, causing bypassing of wash liquids. Recently, the membrane or diaphragm filter press has prevented this problem by squeezing the filter cake before single- or multiple-wash cycles.

Washing on continuous belt or vacuum filters is done by spray washes over the collected solids either in a single pass or in a countercurrent mode. Multiple washes are possible where needed and are effective.

14.7.4 Solids Discharge

In small polish and clarifying filters, the collected contamination is disposed of with the spent cartridge or filled filter bag. If hazardous, the volume of either can require compacting to save space and reduce disposal costs.

In filter presses, cakes are removed manually in small units. Larger filters have plate shifting devices that separate the plates, permitting the solids to fall into a receptacle below the filter. Conveyors can also be used under the filter to transfer the cakes to disposal containers, or to a downstream process such as a dryer. Rotary and belt continuous filters discharge over a roll or from a scraper.

14.7.5 Drying of Filter Cake

Drying of filter cake in filter presses can be done by compressing the cake to remove moisture, in many cases up to 75%. If additional drying is needed, air can be blown through the filter cake through the wash plates in the press. On rotary vacuum filters, air or dry steam can be used for drying. Some filters also have compression mechanisms on the top side of the filter drum.

14.7.6 Downstream Drying

A number of different cake dryers are available for waste sludge drying. Typical are countercurrent hot gas dryers, and paddle or heated blade type dryers. Product dryers utilize conventional spray drying equipment suitable for the crystals or solids collected. Resultant fine-dried powders are then packaged as completed product.

14.8 LITERATURE AND INFORMATION REVIEW

At the time the first edition of this handbook was published, there was a paucity of information in the general field of liquid filtration. There is a journal called *Filtration and Separation* started in 1964, and the Filtration Society was organized in England the following year. A series of Filtech conferences began in 1967 and have continued. Even so, at the time, concerned filtration engineers and academe were calling for more basic teaching and courses in filtration and separation.

However, in the last 10 years, a great deal more information has been published and many conferences were held. The new American Filtration and Separations Society publishes the *Fluid /Particle Separation Journal* devoted to all phases of the subject. Pioneered by Dr. Frank Tiller of the University of Houston, it has gained wide acceptance in our

industry. Many conferences have been held both in North America and in Europe and more than 1000 articles have been presented. It is encouraging to note that many younger engineers are becoming more active in the field, especially in research and development programs, many funded by the U.S. Department of Energy, the National Science Foundation, and the U.S. Environmental Protection Agency.

The research covers many areas such as cake compressibility, expression of solids, permeability studies, and recently an entire conference was devoted to the pore and porosity upon which all relationships in filtration ultimately depend. This was the Hershey Conference held in May 1991. The proceedings were not published but some papers were summarized.[3]

Having been able to survey most all of the work done, I will be selective and subjective while trying not to omit any important papers. Of course, the development of new theories to add to the already extensive literature continues. From the Pore Conference, Tiller gave a tutorial on the parameters of pipes and pores. A mathematical analogy was used in which pipe flow equations for friction factors and Reynolds numbers are modified for flow in porous media. Hypothetical pores are analyzed showing how the void ratio times the specific surface relates to a channel in a porous bed. Permeability and equivalent pore diameter are shown as a function of the fractional distance in both moderately compactable cakes and also highly compressible ones.[4]

Another study tried to resolve theoretical and experimental problems relating filtrate volume to time in constant-pressure filtrations. Problems arise in interpreting theoretical derivation and experimental techniques such as nonuniform cake deposition, variable slurry concentration, degradation of flocs, and clogging of cake and supporting media due to migrating fine particles. Reviewed are basic planar filtration theory, simplified equations for constant pressure filtration, parabolic data analysis, and determining instantaneous rates.[5]

One of the important areas of cake filtration is the compression or expression of filter cakes by mechanical means after the filtration part of the cycle is completed. Prof. M. Shirato from Nagoya University in Japan has done much work in this field. One of his recent research reports focussed on compression filters using hydraulic expression with a perforated membrane.[6] He has recently retired, but his successor, Prof. Murase, is continuing the work. He recently explored the problem of the filter cake expression being stopped before reaching equilibrium state, causing the cake stress to decrease as the material relaxes. The study analyzes this condition. It was found that the cake stress does not depend on either constant pressure or constant rate filtration.[7]

Willis looked at the mechanics of non-Newtonian fluids on nonstationary particles to determine the applicability of Darcy's law. They identified the physical significance and the limitations of this law under these circumstances.[8]

Willis and Chase considered multiphase processes in filtration. They proposed a general strategy for developing a fundamental framework and a systematic approach for evaluating any multiphase porous media process. Concepts of scale, analogy, and averaging, along with the characteristics of basic principles and scientific analysis are used.[9]

One of the most interesting pursuits of Prof. Frank Tiller, who at this writing is 76 years old, is a historical review of papers on filtration theory that were presented at technical meetings some 50 to 75 years ago. These early filtration researchers frequently raised pertinent questions that could not be answered at that time because of lack of instruments to make as precise determinations as we have now. However, Tiller reviews and comments on their questions and provides current theory in explanation of these early investigations. This is a very valuable contribution for new students of filtration and even experienced engineers involved in filtration process development.[10]

Tiller also presented two papers, one relating to relative liquid removal in filtration and expression detailing experimental techniques,[11] and the other concerning improved formulas for constant pressure filtration and compaction of filter cakes.[12] Dick et al. wrote about how capillary forces are related in compressible filter cake filtrations.[13]

Because rating of filter cartridges is a timely and sometimes controversial subject, many articles have been written on it. Johnston says the micron rating of a membrane or filter cartridge can frequently be misleading to the user. Because filtration is not a pure sieving process, its efficiency can be affected by the medium thickness, the nature of the fluid, and the fluid flow rate. He emphasizes that no single factor can characterize a filter medium—at least five are necessary: porosity, permeability, thickness, material of construction, and whether or not pores on one face are larger than on the opposite face.[14]

Many studies on cartridge filters address filter test methods rather than theory because critical applications depend on filtrate or product analysis with emphasis on final particle count. Williams wrote on testing performance of spool-wound cartridges.[15] Verdegan et al. covered recent advances in oil filter test methods for cartridge filters.[16] The effects of temperature and volume on filter integrity tests were studied by Scheer et al.[17] Another study, by Bentley and Lloyd,[18] concerned interpretation of ratings of cartridge filters.

Chiang wrote on the interfacial phenomena in fluid–particle separation. This article gives a complete and detailed study of the most important area of surface–interface relations. The degree and rate of separation are influenced by this behavior. The four basic selected points covered are: surface of solid particles, fluid–solid particle interface, application of interfacial surface tensions, and experimental techniques and instruments.[19]

New filter media were the subjects of many articles. Gregor updated media selection resulting from more demanding environmental regulations. Finer filter media and specialized media are also covered in this article along with options, cost, advantages, and efficiencies of existing and new media.[20] Mayer explains the use of spun-bonded polyolefin nonwovens for micro-filtration.[21]

Bergmann details a new growing market of filter media for blood and medical applications.[22] New uses of new nonwoven filters made by the melt blown process are presented by Manns. This new method produces microfine fibrous webs with fibers as fine as 1 to 10 microns in diameter. The material is made directly from thermoplastic resins and has a number of uses such as media for pleated cartridges.[23]

Membrane filters, one of the fastest growing segments of the entire separation spectrum, recently estimated in Ref. 97 at 10% per year and reaching $2 billion per year in 1996, was the subject of many articles. In fact, the annual membrane conference has had its tenth meeting. Membrane fouling in RO systems was discussed by Kronmiller; use of high-purity water for the semiconductor industry was described by Parekh; crossflow filtration in food applications such as fruit juices was evaluated by Short; and pervaporation future markets were outlined by Bartels.[24]

The market for microfiltration membranes for environmental purposes was covered by Cartwright for system design in pollution control with emphasis on the crossflow technique.[25] An article by Duran explained a new water treatment technology involving nanofiltration membranes in a spiral-wound configuration that function at 75 to 130 psi. This method replaces conventional lime treatment.[26] In the food industry, a method of using BASIC computer programs to solve problems of the effects of transmembrane pressure on orange juice concentration was described by Toledo.[27] The development of a special asymmetric membrane for hazardous waste removal in waste water in the electronics industry was discussed by Sternberg.[28] Also in the microelectronics field, where liquid purity is critical, a method of point-of-use treatment of chemical baths was given by Carr.

Reducing levels of impure chemicals that cause yield losses can be controlled by submicron filtration along with molecular sieve drying and fractional distillation as a high-purity solvent reprocessor. The hazardous waste resulting is minimized to about 3% of the original volume.[29] A new membrane process called nanofiltration is expected to widen the use of membranes in liquid-phase separations in the chemical process industries. An article compares properties and performance characteristics of commercially available NF membranes.[92] A current review of membrane separation technologies for wastewater treatment is presented by Cartwright and includes options and comparisons for selecting the best method.[95]

Pretreatment of slurries by chemical polyelectroytes is essential in many filtrations, and selecting the proper chemical is a task that frequently has to be done by testing. A good overview of the use of polymers and inorganic coagulants is presented by Mangravite.[31] Scheiner discusses the removal of toxic metals from waste water by testing 21 different flocculents. The testing procedures determined the optimum flocculating agents used to achieve allowable levels of cadmium and lead where hydroxide treatment did not work.[30]

Probably more articles and papers were devoted to equipment design, performance, and applications than anything else. We will mention only a few that are new or cover important improvements in existing equipment. Filtration has been combined recently with drying and other processing operations. An interesting review of this area, in which filtration is used with as many as 16 different processes relating to heat and mass transfer operations, was made by Yelshin. Robotic principles, automation, and a unique concept of using rotational machines and conveyor systems in the filtration process is presented.[32]

A new type of water screen filter is described that is self-cleaning by using a pressure senser to activate a back-flushing action. No shut-down is required and particles can be removed down to the 10 to 15 micron range. It can be used for any water that needs to be cleaned or recycled. Individual units can handle up to 5000 gal/min.[33]

Continuous belt and belt filter presses continue to increase in usage as manufacturers make improvements. Besides mineral and chemical processes, many applications are in waste water treatment. Schonstein discusses a vacuum belt press for paper dewatering use.[34] Mau shows how a vertical automatic pressure filter equipped with horizontal filter plates with squeeze diaphragms can improve sludge filtration.[35] Deutsch explains the operation, features available, and options for selecting belt filter presses.[36]

In waste water treatment, centrifuges have a unique advantage over other conventional filters in that they are enclosed, odor-free, safe, and require only minimum labor. One drawback, that of lower solids output, has been addressed by manufacturers and considerable improvements made. Leung describes a high solids decanter centrifuge that gives cake solids above 30% in dewatering mixed primary/secondary sludge.[37] Albertson also writes about improved designs for high cake solids and also use of centrifuges for mechanical dewatering processes in general.[38, 39] Morgenthaler assesses decanter centrifuges for environmental applications using feed rate, polymer addition, and concentration and the suspended solids in the feed, cake, and effluent. Equations are given for calculating polymer consumption, recovery of suspended solid particles, and determining the specific gravity based on density and weight percents.[93] De Loggio reviews recent design innovations in centrifuges for the chemical process industries. For example, new vertical decanter models can handle process streams up to 700° F and 150 psig. A good selection table of different types of centrifuges is presented.[94]

West discusses the disc-bowl centrifuge including centrifugal settlers, solid-bowl nozzle types, and conventional nozzle types. He also explains sigma theory in regard to the relationship between geometry and centrifugal ac-

celeration. Selection data and applications are also given.[40]

Ekberg describes a vacuum disc filter, long used in the minerals field, that uses a new sintered alumina disc medium with average pore sizes of 1.5 to 2.0 microns. See Figure 14.35. When the pores in the medium are filled with liquid, they prevent air passage in the vacuum cake drying part of the filter cycle because of the pressures created in the pores due to capillary action. Thus the filter discs are easier to back-flush, and do not require filter cloths.[41]

An entirely different type of filter is the tube press, which was invented 20 years ago for clay filtrations. It has recently been improved, with larger filter modules. It is now being used in the mining, chemical, and other fields as surveyed by Johns.[42]

14.9 TYPES AND DESCRIPTION OF LIQUID FILTER EQUIPMENT

Starting with batch equipment, then continuous, the various types of current filters in use will be described. Emphasis will be on most recent developments in design and application while still considering the older types, many of which are still widely used in industry. Parameters such as cake washing capabilities, driving forces, settling rates, types of discharge, and cake compression will be added where they relate to the particular filter.

14.9.1 Batch Filters

14.9.1.1 Filter Presses

The filter press is the most common type of pressure batch filter and the oldest, originating in the early 19th century. Its development into a modern efficient, versatile, and flexible filter has kept pace with technological improvements. As shown in Figure 14.12, it is a series of plates and frames, or recessed plates mounted on side bars and supported by a suitable structure. The plates are held together during filtration by hydraulic or mechanical pressure. The slurry is fed into a

Figure 14.12. Filter press showing plates and frames. (Avery Filter Co. Laboratory filter press)

specific port in parallel, and flows through the filter medium covering the drainage area of the plate. The filtered liquid is discharged through another suitable port, and the solids are collected within the filter chamber. They are released at the end of the filter cycle by separating the plates, either manually for small filter presses, or automatically for larger units.

Sizes can range from laboratory filters up to large production units of 6000 ft^2 of filter area and 350 ft^3 of cake capacity. Plates can be made from metals or polypropylene. Figure 14.13 shows typical polypropylene plates. The feed can be from the center or a corner of the plate. Plates and frames are mostly used for clarifying filtrations, where filter paper can be used over the filter cloth for easy removal of the solids, and clean papers can be inexpensively used for each batch.

More filter presses are used for cake operations, and the filter cloth frequently lasts many cycles, as many as 500 to 600 times repeat usage. Cake filters use recessed plates in which the solids are collected between plates in the recess on each adjacent plate, producing final cake thicknesses of from 25 mm to 50 mm, although 32 mm is most common. The major advantage of recessed plates is that when the filter press is opened for discharge, there is no frame to retain part of the cake, and it falls free by gravity from the open chamber. Automatic plate shifters can thus be used to facili-

Figure 14.13. Typical polypropylene plates. (Klinkau GmbH and Co.)

tate and automate the cake discharge. If necessary, all of the functions of the press cycle can be controlled by programmable computer systems including feed flow rate, mass solids in the feed, closing and operating pressures, change of feed pump pressures during the filter cycle, opening to discharge the filter cake, and closing the filter to start another cycle. Typical flow rates are from 0.1 to 1.0 gal/min per ft^2 of filter area. Cake solids content usually range from 25% to 40% depending on pressures and nature of the solids being filtered out.

14.9.1.2 Sheet Filter Presses

Sheet filter presses are so called because they use cellulose filter sheets of about 0.125 in. thick. Frequently they contain a charged powder to effect an attraction for submicron particles. These are depth media and are used in beverage, cosmetic, pharmaceutical, plating solution, electric discharge machining (EDM), and transformer oil filtrations. The filters are of stainless steel structural construction, and

maximum operating pressures may be 50 psi. A typical unit is shown in Figure 14.14.

14.9.1.3 Membrane Filter Presses

Also called diaphragm presses, membrane filter presses (Fig. 14.15) utilize a special plate with an impermeable flexible drainage area on the filter surface of the filter plate (Fig. 14.16). This is separated from the body of the plate, and can be inflated by air or water after the end of the cake-forming part of the filter press cycle. This compresses or squeezes the cake to remove more liquid. This is the most important development of the filter press in the last decade. The improvement in the performance is shown in Figure 14.17. A diagram showing operation of the membrane filter press is shown in Figure 14.18. Generally, solids content of up to 75% may be achieved with savings of downstream drying and sludge cake disposal costs. Examples are given by Mayer[61] for the waste disposal area as compared to rotary vacuum filters and centrifuges, and by Avery for the food and chemical industry.[62]

Figure 14.14. Sheet filter presses installed in a beverage plant. (Seitz Werke)

Figure 14.15. Membrane filter press showing overhead manifold and compressed air connections to membrane filter plates. (Avery Filter Co.)

Automation may also be applied to the cycle times; the squeeze function including time and pressure; blowdown, wash, and discharge actions; and cooling or heating the filter plates. Membrane plates are polypropylene, but some have steel bodies and replaceable neoprene diaphragms.

14.9.1.4 Vertical Automatic Pressure Filter

This filter has a plate stack similar to a filter press except it is in a vertical position as shown in Figure 14.19. The plates are horizontal, with a membrane on the upper side of the plate only. This limits the capacity of the filter, but it can have areas up to 11 m². Cake volumes per cycle can be up to 30 ft³. Because of short 10- to 25-min cycles, the overall output can be large. The filter cloth is a continuous belt passing in between the plates, and capable of being washed after every filter cycle (Fig. 14.20). The cake is removed from the 3 in. diameter discharge rollers by knife scrapers. The filters are automatically operated and are used extensively in the mining,

sugar, and chemical industries. Good washing of filter cakes is made easier because the cakes are in a horizontal position, thus permitting the wash liquid to uniformly flow through the filter cake. The wash water is readily removed by the squeeze action of the membrane, and a second wash can be done if needed.

14.9.1.5 Batch Filters Using Closed Pressure Vessels

These filters have in common a closed tank or housing containing the filter leaves, plates, bags, tubes, or cartridges. Pressures usually do not exceed 100 psi and the size can range from single-element cartridge filters to large horizontal tank leaf filters of up to 3000 ft² in filter area. The vessels can be made from stainless steel or other suitable materials including plastic for smaller sizes.

14.9.1.5.1 Pressure Leaf Filters. Pressure leaf filters can be vertical or horizontal tank designs, the latter capable of larger areas. A

Figure 14.16. Center feed membrane filter press plate cross-section showing separation of membrane from body of filter plate. (Lenser America, Inc.)

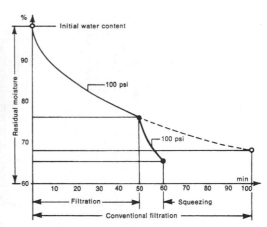

Figure 14.17. Conventional filtration versus membrane filter press operation showing reduction in cycle time with cake squeezing. (Lenser America, Inc.)

typical vertical unit is shown in Figure 14.21 and a horizontal one in Figure 14.22. A cutaway view of a filter leaf is shown in Figure 14.23. The center coarse drainage member is covered by a fine mesh, usually 60 mesh, or a 24 × 110 dutch twill weave wire cloth, and the frame riveted or bolted together. The leaves discharge the filtrate through a connection to a manifold pipe at the bottom of the filter.

Filter cloth can be used over the leaves, but most often the filters operate with a diatomaceous earth filter aid, as most of the applications are for clarifying liquids. The solids collected on the leaves can be sluiced off with water, or can be removed as a dry filter cake manually or by vibrators. Major uses are for fruit juices, beer, sugar solutions, wines, and chemicals.

14.9.1.5.2 Horizontal Plate Batch Filters.
The horizontal plate batch filter consists of multiple round filter plates of metal or plastic

enclosed in a vertical tank. The plates are the same size and are stacked vertically, the number relating to the size required. Generally the maximum area is about 200 ft.^{2} The slurry flow is either up the center tube, or in from the side, then through the filter medium, usually filter paper. Pressures are moderate, rarely exceeding 60 psi. A typical unit is shown in Figure 14.24.

This is primarily a clarifying filter, frequently using activated carbon decolorize particles, and a filter aid to assist the filtration. One advantage is that the filter cake formed on the horizontal surface is stabilized so that it is uniform and not affected by intermittent operation. Because of stacking of plates, the filter is compact, taking up little floor space. Inexpensive filter paper can be used and replaced for each batch, providing uniform flow conditions for each successive cycle because the filter medium resistance is the same each time.

14.9.1.5.3 Horizontal Pressure Plate Filters.
Horizontal pressure plate filters with centrifugal cake discharge permit multiprocessing stages such as filtering, washing, drying, and discharging of the filter cake automatically. A typical filter is shown in Figure 14.25. The filter leaves have a drainage member, with a

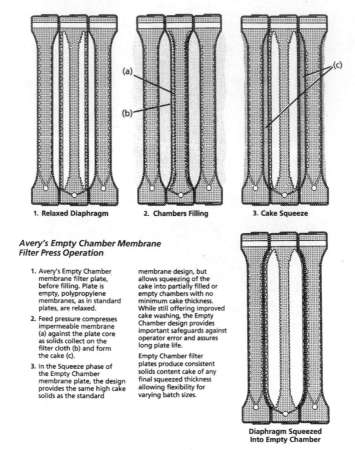

1. Relaxed Diaphragm   **2. Chambers Filling**   **3. Cake Squeeze**

Avery's Empty Chamber Membrane Filter Press Operation

1. Avery's Empty Chamber membrane filter plate, before filling. Plate is empty, polypropylene membranes, as in standard plates, are relaxed.

2. Feed pressure compresses impermeable membrane (a) against the plate core as solids collect on the filter cloth (b) and form the cake (c).

3. In the Squeeze phase of the Empty Chamber membrane plate, the design provides the same high cake solids as the standard

membrane design, but allows squeezing of the cake into partially filled or empty chambers with no minimum cake thickness. While still offering improved cake washing, the Empty Chamber design provides important safeguards against operator error and assures long plate life.

Empty Chamber filter plates produce consistent solids content cake of any final squeezed thickness allowing flexibility for varying batch sizes.

Diaphragm Squeezed Into Empty Chamber

Figure 14.18. Membrane filter press operation with new empty chamber membrane filter plate. (Avery Filter Co., Inc.)

fine stainless wire cloth over it. They may also be fitted with filter cloth. Once the filter cycle is complete, and the solids are on the plates, the feed is stopped, and the stack of plates is rotated at speeds up to 300 rpm to dislodge the filter cake. It can be done dry, or sluiced with a liquid. A different design is used for each type of discharge.

The closed filter vessel is safe in a hazardous environment and protects workers when they are filtering toxic chemicals. In addition, product purity is maintained and automation improves production and reduces labor costs. Typical uses include recovery of previous metal catalyst, gold precipitate recovery, separation of antibiotics, and removal of

catalyst and bleaching earths from edible oils and fatty acids. Units are available of up to a 1000 ft^2 area. Feed with medium to slow settling rates are typical.

14.9.1.5.4 Single Plate Pressure Nutsche Filters. Single plate pressure nutsche filters can have diameters up to 15 ft, areas of 135 ft^2, and with a 12 in. thick filter cake, a volume of 135 ft^3. In some cases, filter cakes can be thicker, giving more cake capacity. These filters serve the need for filtrations that can isolate the final product to maintain purity and avoid toxic exposure to the environment. Such needs are prevalent in the chemical and pharmaceutical industries. Thus there are several

Figure 14.19. Automatic vertical membrane filter with 203.4 ft^2 of filter area. (Larox)

manufacturers supplying this highly specialized filter. A typical unit with agitation is shown in Figure 14.26.

Being totally enclosed, they can operate under inert gas pressure, and even vacuum if needed. Their versatility comes from their ability not only to filter and wash the filter cake by displacement washes, but also to reslurry the cake, refilter it, smooth and compress the cake, dry it, and discharge it from the filter without opening it. All of these steps can be done automatically. They are made mostly from stainless steel, but other alloys can be used; there is even a glass-lined unit with agitation (Fig. 14.27).

14.9.1.5.5 Pressure Tubular Filters. Used for clarifying or solids thickening operations, pressure tubular filters are closed pressure vessels operating up to 100 psi, and feature a tubular filter element. There are a number of types used—wire wound perforated tubes, wedge wire tubes, porous ceramic, plastic, metal, and plastic tubes; and membrane-laminated felt covers over perforated tubes. A typical pressure tubular filter is shown in Figure 14.28. They are sometimes called candle filters because of the vertical tubes. They are placed in vertical vessels, in multiples usually on a tube sheet. The feed can be to the inside or outside of the filter tube. In either case, once the filter cycle is complete, a reverse hydraulic pulse of air, gas, or liquid dislodges the solids in about 5 min. In catalyst recovery and recycling, 100% can be captured and recycled. In some cases, tubes are pre-

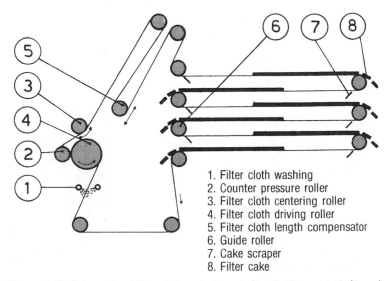

1. Filter cloth washing
2. Counter pressure roller
3. Filter cloth centering roller
4. Filter cloth driving roller
5. Filter cloth length compensator
6. Guide roller
7. Cake scraper
8. Filter cake

Figure 14.20. Schematic of filter cloth path through filter in Figure 14.19. (Larox)

Figure 14.21. Pressure leaf vertical tank filter. (Duriron, Filtration Systems Div.)

Figure 14.22. Pressure leaf horizontal tank filter, 950 ft^2. (Duriron, Filtration Systems Div.)

coated with a filter aid for polishing liquids such as water, solvents, beer, wine, reagents, boiler condensates, or acids.

Whether feeding to the inside or outside of the filter tube, the time of the backwash determines the overall filter capacity. A new design using a membrane-laminated medium over the support tube can back-pulse the tube as often as every 2 to 3 min and remove particles down to 0.5 microns. The discharged solids settle into the cone-shaped bottom of the filter vessel to be concentrated, and periodically removed (Fig. 14.29).

14.9.1.5.6 Cartridge Filters. Cartridge filters are a widely used and important class of liquid batch pressure filters. They maintain or improve cleanliness of many liquids including water, hydraulic fluids, food and beverage products, oils, paints, and many other products. Generally they remove small amounts of particulate contaminants, usually in amounts of less than 100 ppm, from relatively large

volumes of liquids. Many cartridge filters consist of a suitable metal or plastic housing that holds 1 to 16 cartridges for flow rates up to 10 gpm per 10 in. cartridge length. Larger pressure vessels may hold over 100 cartridges, and operate up to 150 psi. High-pressure metal cartridges can operate up to 500 psi. A group of typical cartridges is shown in Figure 14.30.

The original string-wound cartridge depth type was developed about 75 years ago, and since then many other designs have appeared including pleated, grooved, sintered, resin bonded, thermally bonded polypropylene microfiber, controlled gradient density, and solid porous cylinders. There is also a new compound radial pleat that increases loading capacity. The dirt holding capacity is one of the basic selection factors. Performance and particle size retention are the other most important parameters to consider. Multiple layers of media may hold activated carbon or diatomites

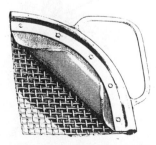

Figure 14.23. Cutaway view of typical filter leaf. Liquid-Solids Separation Corp.

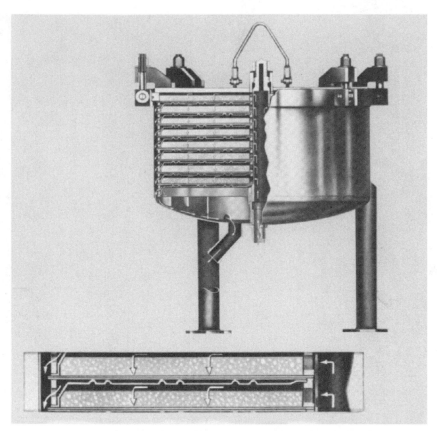

Figure 14.24. Typical horizontal pressure tank filter. (Ketema Inc.)

between them for organic material removal, special decolorizing, or finer filtering.

Although some cartridges are interchangeable with other manufacturers' housings, there are no current standards that relate ratings and performance among different suppliers in the chemical and related industries. On the other hand, critical hydraulic filter systems have standards set by ASME and the API.

Much work has been done on comparing different cartridges. For example, Sandstedt and Weisenberger[65] report on the confusion about micron ratings, saying there is frequent failure to specify the level of efficiency as part of the performance rating, that there is no acceptance of a single, standard test procedure to predict performance in many applications, and that the difference between clarification and classification is often overlooked. Their

tests showed that pleated cartridges, after cake formation begins, achieve 98 + % efficiency equal to about 1 micron performance regardless of the medium.

Other investigators have attempted to clarify time-dependent variations in cartridge performance. Juhasz questioned Beta, Beta Prime, and Epsilon rating procedures and suggests that three components of downstream level contamination should be considered, namely, instantaneous efficiency, unloading, and leakage.[66]

14.9.1.5.7 Bag Filters.   The concept here is a sewn filter bag of fabric, felt, or mesh, of a specific retention value that is made to fit into a filter housing with proper seals to prevent bypassing of the liquid to be filtered. Typically, a No. 1 standard filter bag size has 2.5 ft^2 of

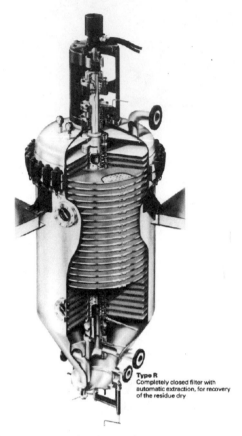

Type R
Completely closed filter with
automatic extraction, for recovery
of the residue dry

Figure 14.25. Horizontal plate filter with centrifugal cake discharge feature. (Steri-Tech, Inc. Funda)

Figure 14.26. Pressure nutsche filter 0.6 m² for pharmaceutical pilot plant. Filter is open to show single plate which supports the filter medium (Rosenmund, Inc.).

filter area, and a No. 2 bag has 5 ft². Filter bags and housings are shown in Figure 14.31. Generally, the percent solids in the feed stream are low, but may be as much as 1%, which is much greater than for cartridge filters. Multiple bag units are available and flow rates can reach 1000 gal/min. The original idea was developed by Wrotnowski, who found that the felt media did an excellent job of classifying paints and inks.[67] Subsequent developments have made the bag filter a popular clarifying filter. New and more retentive materials have increased the acceptance of this filter. An approach to increase the filter area and keep a small size was reported by Johnson,[68] using multiple layers with bypass openings. This design of filter bag has five times the filter area of a standard bag, and 15 times the dirt load

capacity of a standard filter cartridge. This means fewer changeovers and lower disposal costs. This is an advantage over cartridges because, for equal filter areas, the disposal value is less for bags than for cartridges.

Generally, bag filters offer higher flow rates with lower pressure drops. Cartridges with depth or extended surface area offer greater reliability and efficiency of particle removal.

14.9.2 Vacuum Filters

Using a vacuum as the driving force for filtration is common in rotary vacuum filters which have the advantage of continuous cake discharge. Forces up to 0.8 bar can be enough to produce improved filtration rates. Vacuum sources are simple, and the filters from small nutsches to large-scale equipment have been well developed since the first rotary vacuum

Figure 14.27. Special pressure nutsche with glass-lined vessel and agitator. (Zschokke Wartmann, Ltd.)

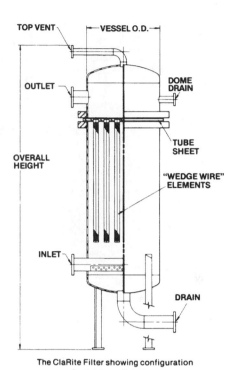

The ClaRite Filter showing configuration for "air bump" design.

Figure 14.28. Pressure tubular filter. (Croll-Reynolds)

drum filter was introduced in England over 100 years ago. Some disadvantages are that the production rate per unit area is low, so that larger equipment is needed for large volume. Some rotary vacuum filters can be 16 ft in diameter and 33 ft long. Solids content can be relatively low as compared to those obtained using pressure filters, and limits apply to volatile liquids.

14.9.2.1 Drum Filters

Vacuum filters are used for both solids recovery and fine clarifications that utilize diatomite and are called precoat filters. The vacuum drum filter is widely used in industry as shown in Figures 14.32 and 14.33. The drum has sections controlled by a rotary valve on which

the filtration, washing–drying, and discharge steps take place. Cake discharge is done by different methods such as scraper, air blow, rollers with strings, and belt. Filter cloths, mostly synthetic but sometimes metal, are used with air flows in the range of 25 to 100 cfm on the Frazier scale. Rotation speed is about 1 rpm. Maximum size can be 1700 ft^2, although from 200 to 600 ft^2 is much more common. Outputs on the very largest units can be up to 120 tons per day.

14.9.2.2 Precoat Drum Filter.

The precoat rotary vacuum filter is completely dependent on the use of a filter aid, which in almost all cases is diatomate. A typical filter system is shown in Figure 14.34. The diatomate slurry, from 4% to 8% by weight, from the precoat tank is applied to the filter drum to form a thickness of up to 6 in. After successful precoating, the filtration begins, using admix, also called body feed, which is fed along with the

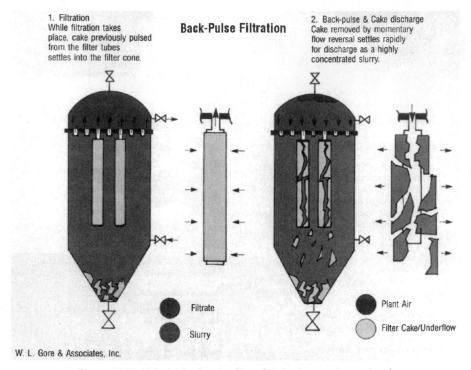

1. Filtration
While filtration takes place, cake previously pulsed from the filter tubes settles into the filter cone.

Back-Pulse Filtration

2. Back-pulse & Cake discharge
Cake removed by momentary flow reversal settles rapidly for discharge as a highly concentrated slurry.

Filtrate

Slurry

Plant Air

Filter Cake/Underflow

W. L. Gore & Associates, Inc.

Figure 14.29. Tubular back-pulse filter. (W. L. Gore and Associates)

influent liquid to the filter bowl. As the solids collect on the precoat, an advancing knife shaves off as little as 0.001 in. of solids for each revolution of the filter drum. Filtering factors are given by Smith[69] as:

Grade of precoat	Precoat cutting efficiency
Vacuum	Cake filterability
Drum speed	Cake drying time
Knife advance time	Liquid viscosity
Drum submergence	Precoat thickness
Solids in precoat	Continuous addition of filter aid

The most important objective is to optimize production output and to control cost by minimizing filter aid consumption. Determining the right combination of the above factors will require leaf test and careful observation of plant runs. Test filter leaves of 0.1 ft^2 of filter area are used with different grades of filter aids, cake thickness, and vacuum. Evaluating data indicate initial optimum conditions. Plant runs can then be observed to see if test results are effective. The angle and the desired rate of knife advance can often be determined only by trial and observation.

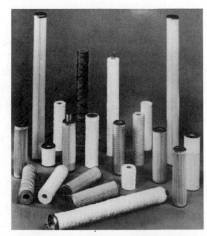

Figure 14.30. Cartridge filters—various types. (Parker-Hannifin Corp.)

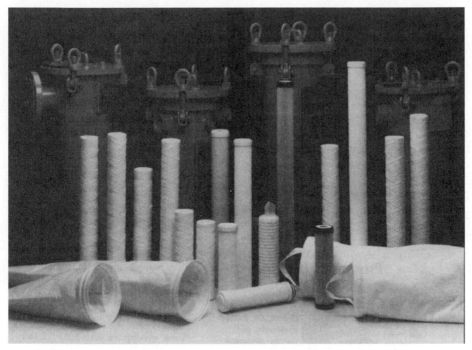

Figure 14.31. Bag and cartridge filters. (Commercial Filter Div. Parker-Hannifin Corp.)

14.9.2.3 Vacuum Disc Filters

Vacuum disc filters are continuous rotary filters with circular vertical filtering discs mounted on a horizontal hollow central shaft. (See Fig. 14.35.) The slurry fills a trough into which the filter discs are submerged. The discs are partitioned into sectors with suitable drainage members to permit the filtrate to be fed to the rotating control valve. The filter leaves are usually covered with a filter cloth. As the leaves rotate and become submerged, the automatic rotary valve applies vacuum and the solids form on the filter disc. Cake is removed by scraper or roller and it drops between the tank divisions. A recent development in porous ceramic technology has made possible filter discs that do not require filter cloth as covers. The disc surface has a very fine pore size, permitting fine filtration and easy cake release. Several of these new designs are reported to be operating successfully in Australia for zinc, lead, and copper concentrate filtrations.[70]

14.9.2.4 Horizontal Continuous Vacuum Filters

The filter surface is formed as a table, a belt, or multiple moving pans in a line or a circular arrangement. As with other filters with horizontal surfaces, they maintain cake stability and thickness, and permit easy cake washing including countercurrent. On the other hand, they take up more space, and cost more than drum filters. Their principal use is filtration of gypsum and phosphate rock residues, metallurgical sludges, pulp washing, and solvent extraction of oil seeds.

The horizontal belt vacuum filter is made in two basic designs, one with a heavy rubber underbelt carrying the filter cloth. This endless belt rides on a vacuum or suction trough and has lateral drainage grooves for the filtrate. The slurry is fed at one end, and filtering and washing take place at the end of the filter, where the solids are dropped from the moving belt.

The second variation of the belt filter uses

Figure 14.32. Vacuum continuous drum filter. (Kromline-Sanderson)

Figure 14.33. Vacuum precoat drum filter. (Witco, Kenite Div.)

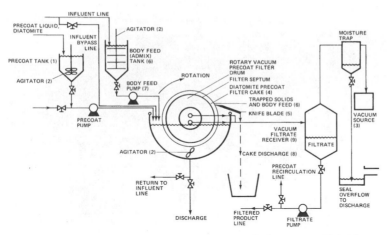

Figure 14.34. Complete precoat filtration system. (Arthur Basso, Ref. 56)

vacuum pans underneath the filter cloth with no rubber supporting belt. The vacuum boxes move forward intermittently as the vacuum is applied as the cloth and pan move together. This type is generally of lighter weight construction than the first design, and is used more on chemical, pharmaceuticals, and food products. Widths can be up to 2 meters and lengths to 40 meters.

14.9.3 Continuous Compression Belt Filter

Originally developed in Germany in the 1960s for dewatering pulps in the paper and food industries, the continuous compression belt filter quickly became adaptable for waste sludges. It was called a sewage sludge concentrator, although the common name now is a

Figure 14.35. Vacuum disc filter with ceramic filter discs. (Outokumpu Mintec USA Inc.)

Figure 14.36. Continuous belt filter press. (Komline-Sanderson)

belt filter press. A typical unit is shown in Figure 14.36. It is a heavy-duty mechanical machine that dewaters sludges that have been properly conditioned with polymers. There are various methods to do this, one of which is shown in Figure 14.37. This shows a modern controlled system that tends to optimize the polymer-to-sludge ratio to reduce polymer costs. A study of compressible sludge properties in belt presses was done by Wells.[71]

The process in this filter takes place in three steps:

1. Gravity settling, in which the free water drains from the treated sludge. Some me-chanical plows or rollers may be used here to assist in the drainage.

2. The wedge or low-pressure zone. The sludge flows onto a carrying filter belt, becoming sandwiched between this and another over filter belt. By converging they apply gradual increases in pressure to the sludge.

3. The dewatering continues as the two belts enter into a high-pressure or shear zone around pressure rolls. These high shear forces maximize the cake dryness. From here, the dewatered cake is continuously removed by a doctor blade on a discharge roller. The general configuration of the press is shown in Figure 14.38.

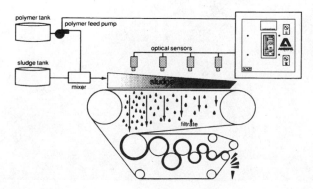

Figure 14.37. Typical polymer control systems for belt filter presses. (Andritz Ruthner, Inc.)

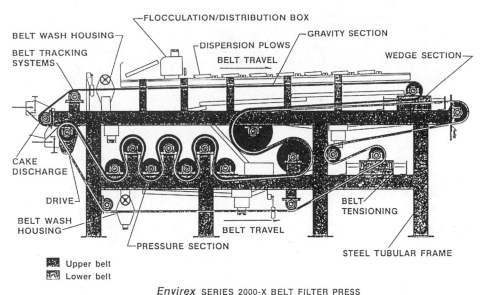

Figure 14.38. Schematic of a belt filter press. (Envirex Corp.)

Improving performance is detailed in an article by Lecey in which the mechanical variable, the sludge characteristics, and optimum operating conditions are discussed.[72] Recent developments have provided a number of new features and they are described by Deutsch.[73] Some installations in municipal treatment plants involve multiple units such as the seven units installed in a plant in Camden, NJ (Fig. 14.39).

14.9.4 Screw Presses

Screw presses provide another way of continuously compressing or squeezing a sludge material, particularly of coarse type materials such as organic waste, pulps, and fibrous materials.

Figure 14.39. Belt presses installed in a muncipal sewage treatment plant. (Enviroquip. Inc.)

It too benefits from the use of polymeric treatment chemicals to agglomerate the particles to be separated from the liquid. A schematic of a typical unit is shown in Figure 14.40. The screw is an extruder type with a tapered center shaft that compresses the product gradually as it moves toward the discharge end. Feed can be as low as 0.5% dry solids, and depending on feed composition, the solids discharged can range from 15% to 70% dry solids content. Throughput can reach 2 tons of dry solids per hour in the largest unit available.

14.9.5 Continuous Pressure Filters

Along with continuous vacuum and belt filters, there are several continuous pressure filters that are quite unique in design, providing special applications that make their relatively high cost acceptable.

14.9.5.1 BHS Fest Filter

The first of these is the BHS Fest filter, developed in Germany in the late 1930s. It is an entirely enclosed low-pressure (up to 50 psi) rotary drum filter in which the slurry is fed into a filter chamber which is a segmented part of the drum. Subsequent chambers complete the filtration, wash the cake, then dry it and prepare for discharge. In this self-contained environment protected unit, toxic, hazardous, or solvent materials can be pro-cessed, and solvent cake washing performed in pharmaceutical operations, dewaxing paraffin from oil–water mixtures, or removing extraction agents from food processes. A sketch is shown in Figure 14.41. Sizes are available from 0.12 to 7.68 m^2.

14.9.5.2 KDF Filter

Another continuous pressure filter is the KDF from Amafilter (Fig. 14.42) in Holland. Its design is a horizontal tank in which filter axles are mounted on a rotating main shaft, each with a particular number of elements attached. Both the elements and the main shaft rotate, using constant air pressure at 6 bar to effect the filtration. The air pressure gradient provides the driving force and is also used for displacement dewatering of the cake. A chain type conveyor is used for cake discharge. With 50 m^2 of filter area, it can produce filter cakes of very low moisture content at capacities up to 1750 kg/m^3 per hour. Developed in the early 1980s, the principles are detailed by Kleizen and Dosoudil.[74] Applications have been for coal fines, cement slurries, and coal flotation concentrates.

14.9.5.3 Ingersoll–Rand Filter

The Ingersoll–Rand continuous pressure filter is a third commercial device of this category as shown in Figure 14.43. It is used not only for

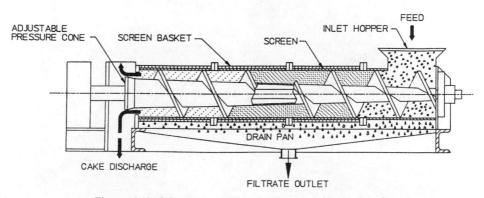

Figure 14.40. Schematic of a typical screw press. (Bepex Corp.)

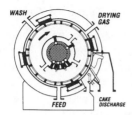

Figure 14.41. BHS Fest continuous pressure filter. (Komline-Sanderson)

filtering, but also as a slurry thickener. It utilizes the concept of limiting the growth of the filter cake by rotating filter cloth covered discs adjacent to stationary filter plates on a horizontal shaft inside a horizontal vessel. The cake thickness may be reduced to 1.0 mm with 3 mm clearance. This feature added remarkable flexibility to this continuous filter. High pressure in the range of 300 psi and thin cakes combine to produce high filter rates.[75]

This filter is also called an Artisan dynamic filter, a rotary filter press,[76] a crossflow filter with rotating elements,[77] and an axial filter, developed at Oak Ridge National Laboratory.[78] An ultrafiltration module has also been described based on this principle.[79]

The filter offers automated, continuous operation, compact design, a totally enclosed system, clear filtrates, and low operating costs.

One drawback is that the close tolerances cause wear with highly abrasive substances.

14.10 CENTRIFUGES

14.10.1 Use of Centrifuges

The use of centrifuges for liquid–particle separation is widespread in the chemical, food processing, mining, and pharmaceutical industries. More recently, they are being used more in pollution control, especially in municipal waste water treatment plants. They utilize the strong G-forces caused by high-speed rotation up to 10,000 rpm. In general, the power needed is proportional to the square of the operating speed, and the maintenance may even relate to the cube of the speed. Larger machines with higher capacities running at slower speeds can thus show power and maintenance savings; whereas smaller machines with lower capacity can effect higher G-forces for separating more difficult-to-separate materials. Particle separations can range from 50 to 1000 micron sizes for perforated basket types, and from 0.5 to 10 microns for disc types. Some decanter solid bowl models are capable of separating particles from 1 micron up to 1/4 in. in size. Flow capacities can be up to 1000 gal/min with solids loading up to 100 tons/h.

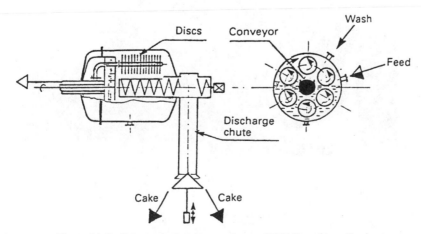

Figure 14.42. Schematic showing continuous KDF filter. (Amafilter)

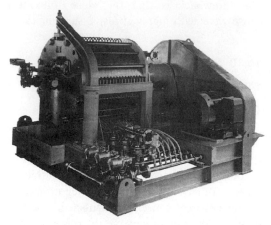

Figure 14.43. Automatic continuous rotary disc filter. (Ingersoll–Rand, Inc.)

14.10.2 Basic Types of Centrifuges

There are two basic types of centrifuges—filtering and sedimentation. The first type, as shown in Figure 14.44, uses a filter cloth or a screen element for fine particle separation in a perforated basket, with either vertical or horizontal configuration. The filtrate has low suspended solids and the trapped solids can be removed manually, or by mechanical devices. A sedimentation centrifuge, so called because it greatly accelerates the normal settling rate of particles by subjecting them to high centrifugal forces, is shown in Figure 14.45. There is no filter cloth, and the solids are forced up against an imperforate bowl, allowing the liquid to decant off the top. These types are also

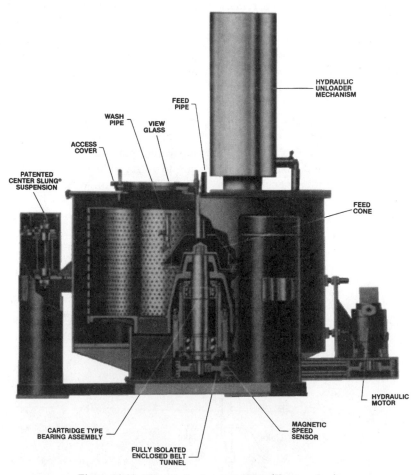

Figure 14.44. Filtering basket centrifuge. (Ketama, Inc.)

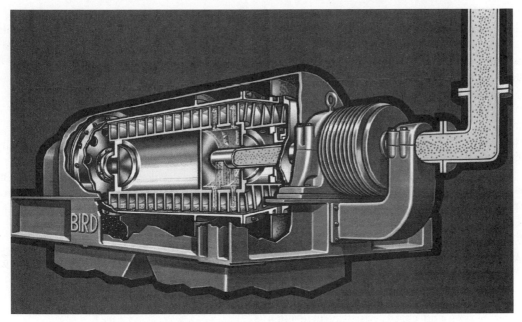

Figure 14.45. Typical solid bowl centrifuge. (Bird Machine Corp.)

called solid bowl, screen bowl, decanters, and disc machines.

14.10.2.1 Filtering Centrifuges

Filtering centrifuges, also called basket centrifuges, are commonly used in batch feed mode in the fine chemical and pharmaceutical industries for filtering and washing organic crystals, inorganic salts, and fine particles. They are available in sizes up to 40 ft^3 cake capacity, and with top or bottom drive. They can be automated and solids can be removed mechanically by plow or peeler devices. Stainless steel sanitary and vapor-tight designs are available. A new design offers ASME code for 35 psi steam for sterilization. A recent innovation has the basket mounted in a horizontal position with the filter cloth fastened at both ends of a movable drum. At the end of the filtering cycle, the drum insert moves axially and hydraulically into a discharge chamber, carefully turning the filter cloth inside out so that the solids are then on the outside of the cloth, and can be discharged by the continuing rotation of the drum. No residual product remains on the cloth, there is no manual operation, and the centrifuge is not opened during discharge. Solids can be loaded into a suitable container without being exposed to the environment. The machine comes in four sizes from drum diameters of 300 to 1000 mm. Throughputs vary from 100 to 300 kg/h. Up to 90 psig gas pressure can also be added to the bowl, maintaining liquid head and increasing filtration rates.

14.10.2.2 Solid Bowl Centrifuges

The increasing use of solid bowl decanter centrifuges in waste water treatment plants is due to their good solids dewatering capability of up to 35% cake solids for a mixed feed of primary and secondary sludge. They operate continuously, and because of their enclosed operation, reduce or eliminate odor problems. They tend to be favored for very large scale plants. For example, at the second largest municipal waste water dewatering facility in the United States, seven of these high-solids centrifuges have been installed. They are dewatering 350 dry tons per day. The centrifuges were chosen over filter presses and belt filters in the selection process described by Lipke.[80]

At another large municipal plant in Los Angeles, the sludge dewatering process has been optimized each year to reduce operating cost and yield dryer cakes. The three process variables changed were solids retention time, hydraulic loading rate, and polymer injection rate. How this was done is explained by Zschach.[81]

However, a smaller unit recently introduced is a modular centrifuge especially designed for treatment plants processing up to 5 MGD. The unit is compact and can also serve to thicken and dewater waste streams separately by being converted from one mode to the other in minutes.[82]

Norton discusses applications of decanter centrifuges in the oil drilling industry to recover barites and control viscosity in drilling fluids. Recent design changes permit increases in clarification, solids retention time, and general performance.[83]

14.10.2.3 Sizing Sedimentation Centrifuges

The key factors for sizing sedimentation centrifuges is the minimum required settling rate for the solids material if it is not to leave with the overflow. This can be expressed by the equation:

$$V_{s(req.)} = (h/2)/t = \tfrac{1}{2}(h/L)(Q/A) \quad (14.10)$$

where $h/2$ is the distance that an average particle must travel radially while settling, t is the residence time, L is the distance between the feed inlet and the overflow, Q is the volumetric throughput, and A is the cross-sectional area of the annulus, the liquid pool adjacent to the bowl wall. This suggests that the required settling rate for the average particle is the throughput divided by the settling surface area, a very familiar result in sedimentation. To determine the rate available, by using Stokes' law, the settling velocity V_s can be obtained from the equation

$$V_s = (\Delta \rho / 18\mu) g d^2 (G/g) \quad (14.11)$$

where d is the particle diameter in meters, μ is in (kg/m · s), Δp is the difference in density between the particle and fluid, g is the gravitation constant (9.81) m/s², and the ratio G/g is defined by the equation

$$G/g = \Omega_b^2 R_b / g \quad (14.12)$$

where Ω_b^2 is the rotation speed of the bowl in radians per second and R_b is the bowl radius in meters. This ratio measures the centrifugal acceleration developed in units of gravity. The required rate from Eq. (14.10) can be equated with the available rate from Eq. (14.11) and rearranged to give

$$Q = 2V_{s(1g)}(\Omega_b^2 R_{av}/g)(LA/h) \quad (14.13)$$

where $V_{s(1g)}$ is the settling velocity (Stokes velocity) and R_{av} is the average radius of the bowl and the pool. This equation shows that the throughput Q increases with the Stokes settling velocity, the intensity of centrifugation G/g, and the surface area for settling. This approach to sedimentation centrifuging is from Bershad et al.[84]

Further analysis of batch filtering centrifuges is given in this chapter considering the following mechanisms of compaction of the solids cake:

- Centrifugal force acting on the solids (minus the effect of buoyancy)
- Viscous drag on the solids due to liquid flow
- Resistance mechanisms due to the solids stress developed as the cake deforms
- An arching effect due to the radial geometry.

For a compressible cake, both permeability and porosity of the cake are functions of the solids stresses, These can be measured in the laboratory by using a hydraulic press[85] or a compression-permeability cell.[86]

14.10.2.4 High-Capacity Oscillating and Tumbler Centrifuges

High-capacity oscillating and tumbler centrifuges (up to 250 t/h solids) have relative low clarity of overflow. They are used on rapidly filtered products such as fire-coal, ore, sand, and coarse salts. To improve their operation, an increase in the residence time would be beneficial. A recent article explored the concept of using a step drum in a tumbler centrifuge. It was demonstrated that this leads to much greater improvement over conical basket machines. The drum should be designed with at least three steps for best results.[87]

14.11 FILTER EQUIPMENT SELECTION

With such a wide choice of many various kinds of filtration equipment, it would at first appear that choosing the optimum for the specific application would be confusing and frustrating. This is not usually the case, however. Many guidelines exist for the initial category of choice, and then more specific and well-defined parameters exist for narrowing the choice to a very few appropriate filters. In fact a recent article uses a best and worst choice of factors related on a scale from -2 to $+2$, with the final best choice indicated. There are, however, some warnings on borderline cases.[88]

More recently, a complete software program has been developed based upon the above system, but with the added input of practical or heuristic values so that the outcome becomes a real workable basis for making a very good first selection.[89]

The study of particle settling rates has been the classic approach to initial filter selection. A number of tables and guides have been published using this technique. A review of the most important guides was made by Mayer.[90]

The magnitude of the planned operation easily eliminates many small and batch filters and indicates continuous belt and rotary vacuum filters. Automatic vertical plate filters with short cycles and decanter type centrifuges provide volume production. Vacuum pan and table filters are economical only with large-scale operations. At the other end of the scale are

polishing and solids contaminant removing filters, most often small batch quantities which are best done with cartridge and bag filters.

Where laboratory facilities are available, much information can be gained by simple Buchner funnel and vacuum leaf tests. The basic lab test for coagulation clarification is the jar test, which permits testing a water or waste water with various coagulant chemicals. The CST (capillary suction time) test is used to evaluate filterability of waste sewage sludge. All these are described in detail with procedures in several texts, the most exacting of which is Purchas.[91] Methods using test results for scaling filters up to production size are also given.

Pilot plant and in-plant test are more complete and often more decisive than lab tests. In my estimation, the optimum program involves a test filter placed in the production plant and set up to filter a side stream from the existing process. This eliminates any possible variations in the liquid slurry that can be caused by shipping to a test facility, time factors, or chemical changes in the material. If the product has not yet been made in production, the pilot plant approach is desired. One advantage of the pilot plant is that the test runs can be made on a 24-h basis, giving more positive test data than laboratory testing.

After all the above have been done, a careful evaluation of last minute considerations must be made. Not yet mentioned, but obviously of major concern, is the relative cost of capital equipment and installation, of multiple choices if such exists. In some demanding and critical choices I have seen a very high price a secondary factor. In the final analysis, the ultimate desired quality of the final product is decisive.

REFERENCES

1. C. Jahreis, "Do We Need Filtration?" *Filtration Separation*. Cited in *Fluid Particle Sep. J* 5(1):53–54 (March 1992).
2. N. P. Cheremisinoff and D. S. Azbel, *Liquid Filtra-*

tion, pp. 93–97, Ann Arbor Science, Woburn, MA (1983).

3. *Fluid Particle Sep. J.*, The Hershey Conference on the Pore, pp. 12S–13S, 4-3 (Sept. 1991).

4. F. M. Tiller, "From Pipe to Porous Media Parameters," in Conference on the Pore, Hershey, PA, May, 1991. *Fluid Particle Sep. J.*, pp. 139–146, 4-2 (June 1991).

5. F. M. Tiller, "Tutorial: Interpretation of Filtration Data I & II," *Fluid Particle Sep. J.*, pp. 85–94, 3-2 (June 1990) and pp. 157–164, 3-3 (Sept. 1990).

6. M. Shirato, T. Murase, E. Iritani, and M. Iwata, "Advanced Studies in Solid/Liquid Separation, Filtration, Sedimentation and Expression," *Fluid Particle Sep. J.*, pp. 131–135, 2-3 (Sept. 1989).

7. T. Murase, M. Iwata, T. Adachi, and M. Wakita, "Stress Relaxation of Expressed Cake," *J. Chem. Eng. Jpn. 22*:655–659 (1989). Cited in *Fluid Particle Sep. J.*, p. 17D, 4-4 (Dec. 1991).

8. M. S. Willis, F. N. Desai, G. G. Chase, and I. Tosun, "A Continuum Mechanics Analysis of Darcy's Law," *Fluid Particle Sep. J.*, pp. 137–142, 3-3 (Sept. 1990).

9. M. S. Willis and G. G. Chase, "A Strategy for Multiphase Processes," *Fluid Particle Sep. J.*, pp. 55–59, 5-2 (June 1992).

10. F. M. Tiller, "Out of the Past." *Fluid Particle Sep. J.* Various issues since 2-1, March 1989.

11. F. M. Tiller and C. S. Yeh, "Relative Liquid Removal in Filtration and Expression," in *Proceedings American Filtration Society Meeting*, AFS, Kingwood, TX (March 1988).

12. F. M. Tiller and S. Hsyung, "Improved Formulas for Constant Pressure Filtration" and "Compaction of Filter Cakes," in *American Filtration Society Meeting*, Atlanta, GA, AFS Kingwood, TX (October 1987).

13. R. I. Dick, S. A. Wells, and B. R. Bierck, "Role of Capillary Forces in Compressible Cake Filtration," *Fluid Particle Sep. J.*, pp. 32–34, 1-1 (Sept. 1988).

14. P. R. Johnston, "The Micron Rating of a Filter Medium: A Discussion of the Performance of Filter Media," *Fluid Particle Sep. J.*, pp. 157–161, 2-3 (Sept. 1989).

15. C. J. Williams, "Testing Performance of Spool-Wound Cartridges," *Filtration Separation*, pp. 167 ff (Mar./Apr. 1992).

16. B. M. Verdegan, K. McBroom, and L. Liebmann, "Recent Developments in Oil Filter Test Methods," *Filtration Separation*, pp. 327–331, 29-4 (Jul./Aug. 1992).

17. L. A. Scheer, W. C. Steere, and C. M. Geisz, "Temperature and Volume Effect on Filter Integrity Tests," *Pharmaceut. Technol.*, pp. 22–32 (Feb. 1993).

18. J. M. Bentley and P. J. Lloyd, "Interpreting the Rating of Cartridge Filters," *Filtration Separation*, pp. 333–325, 29-4 (July/Aug. 1992).

19. S.-H. Chiang, "Interfacial Phenomena in Fluid Particle Separation," *Fluid Particle Sep. J.*, pp. 168–181, 5-4 (Dec. 1992).

20. E. C. Gregor, C. Wait, and J. R. Mollet, "Considerations in the Selection of Media of Process Applications," *Fluid Particle Sep. J.*, pp. 163–167, 5-4 (Dec. 1992).

21. E. Mayer and H. S. Lim, "Tyvek for Microfiltration Media," *Fluid Particle Sep. J.*, pp. 17–21 2-1 (March 1989).

22. L. Bergmann, "Filter Media for Blood Products," *Tech. Text. Int.*, pp. 14–16 (Sept. 1992).

23. J. Manns, "The Melt Blown Process: Its Role in Filtration," *Nonwovens Industry*, pp. 36–40, 22-10 (Oct. 1991).

24. D. Kronmiller, "Fouling Control in RO Systems;" Parekh, B. et al., "New Developments in the Filtration of High Purity DI Water for Semiconductor Manufacturing;" Short, J., "Cross Flow Filtration: The Solution is Clear;" and Bartels, C. R., "The Future for Pervaporation: Technology and Applications," in *10th Annual Membrane Conference*, Business Communications Co., Norwalk, CT (1992).

25. P. Cartwright, "Membrane Technology for Pollution Control, A System Design Primer," *FILTECH 91, Proceedings*, pp. 105–125, Filtration Society, W. Sussex, England (1991).

26. F. E. Duran, "Florida Treatment Plant Uses Low-Pressure Membranes," *Water Eng. Mgmt.*, pp. 26–28 (Jan. 1933).

27. R. F. Toledo, *Fundamentals of Food Process Engineering, 2nd edit.*, pp. 533–538. Van Nostrand Reinhold, New York (1991).

28. S. Sternberg, "Membrane Utilization in Hazardous Waste Removal from Wastewater in the Electronics Industry," *Environ. Prog.*, pp. 139–144, 6-3 (Aug. 1987).

29. G. Carr, "Continuous Purification of Process Chemicals," *Natl. Environ. J.*, pp. 20–24 (Sept./Oct. 1992).

30. O. C. Carter and B. J. Scheiner, "Removal of Toxic Metals from an Industrial Wastewater Using Flocculants," *Fluid Particle Sep. J.*, pp. 193–196, 4-4 (1991).

31. E. Mangravite, "Overview of the Use of Polymeric Flocculants and Inorganic Coagulants in Filtration," *Fluid Particle Sep. J.*, pp. 95–99, 2-2 (1989).

32. A. I. Yelshin, "The Development of Filtering Equipment For Hazardous Materials," *Fluid Particle Sep. J.* pp. 126–129, 5-3 (Sept. 1992).

33. Anon., "Self-Cleaning Filters Help Keep Cooling Tower Free of Corrosion, Scale Buildup," *Plant Eng.* 45–16 (Sept. 1991).

34. P. J. Schonstein, "Horizontal Vacuum Belt Filter for Chem. & Mining Industries," in Conference, Brisbane, Australia 1989. *Filtration Separation*, 28-2 (Mar./Apr. 1991).

35. R. W. Mau and G. S. Miller, "Improved Waste Water and Sludge Filtration in Fully Automated Pressure Filters," *Fluid Particle Sep. J.*, pp. 158–162, 5-4 (Dec. 1992).

36. N. D. Deutsch, "Options in Belt Filter Presses," *Water Eng. Mgmt.*, pp. 34–37 (Sept. 1987).

37. W-F. Leung and R. Havrin, "High Solids Decanter Centrifuge," *Fluid Particle Sep. J.*, pp. 44–48, 5-1 (March 1992).

38. O. E. Albertson, "Improved Centrifuge Design for High Cake Solids," *Fluid Particle Sep. J.*, pp. 28–35, 3-1 (1990).

39. O. E. Albertson, "Mechanical Dewatering Processes," *Fluid Particle Sep. J.*, pp. 56–70, 3-2 (June 1990).

40. J. West, "Disc-Bowl Centrifuges," *Chem. Eng.*, pp. 69–73 (Jan. 7, 1985).

41. B. Ekberg and J. Haarti, "Capillary Filtration," *Fluid Particle Sep. J.*, pp. 116–120, 5-3 (Sept. 1992).

42. F. Johns, "Tube Press," in FILTECH 1991 Conference, Karlsruhe, *Proceedings of Filtration Society*, pp. 183–195, W. Sussex, England (1991).

43. H. Darcy, "Determination of the Laws of Flow of Water through Sand," Translated by J. R. Crump, *Fluid Particle Sep. J.*, pp. 33–35, 2-1 (March 1989).

44. L. Svarovsky, *Solid–Liquid Separation*, 3rd edit. Chap. 9, Butterworth-Heinemann, Oxford, England (1991).

45. J. G. Clark, "Select the Right Fabric for Liquid-Solid Separation," *Chem. Eng. Prog.*, pp. 45–50 (Nov. 1990).

46. L. M. Sheppard, "Corrosion Resistant Ceramics for Severe Environments," *Am. Ceram. Soc. Bull.* pp. 1146–1158, 70-7 (1991).

47. E. C. Gregor, C. Wait, and J. R. Mollet, "Filtration and Separation Considerations in the Selection of Media for Process Applications," *Fluid Particle Sep. J.*, pp. 163–167, 5-4 (Dec. 1992).

48. A. Pangrazi, "Nonwoven Bonding Technologies," *Nonwovens Industry*, p. 32 ff, 23-10 (Oct. 1992).

49. W. Shoemaker, "Nonwoven Technology in Filter Media," *Nonwovens Industry*, p. 34 ff (Oct. 1984).

50. G. R. S. Smith, "Membrane Filter Cloth," in *American Filtration Society Conference Proceedings*, pp. 665–669, AFS, Kingwood, TX (1988).

51. M. Porter, "Membrane Filtration," Sect. 2.1, pp. 2–103, in *Handbook of Separation Techniques for Chemical Engineers*, edited by P. A. Schweitzer, McGraw-Hill, New York (1979).

52. P. R. Johnston, *Fluid Sterilization by Filtration*, Interpharm Press, Buffalo Grove, IL (1992).

53. G. B. Van den Berg and C. A. Smolden, "Flux Decline in Membrane Filters," *Filtration Separation*, pp. 115–121, 25-2 (Mar./Apr. 1988).

54. D. C. Gyure, "Set Realistic Goals for Crossflow Filtration," *Chem. Eng. Prog.*, pp. 60–66, 88-11 (Nov. 1992).

55. B. Culkin, "Vibratory Shear Enhanced Processing: An Answer to Membrane Fouling," *Chem. Proc.*, pp. 42–44 (Jan. 1991).

56. A. Basso, "Vacuum Filtration Using Filter Aids," *Chem. Eng.* (April 19, 1982).

57. R. Reiber, "RHA: A Surprisingly Effective Filter Aid," in *Proceedings of the American Filtration Society*, Vol. 6, Chicago, pp. 227–232, AFS, Kingwood, TX (1992).

58. R. Rees, "Diatomites Cut Filtration Costs," *Pollut. Eng.*, pp. 67–70 (April 1990).

59. L. Svarovsky, *Solid–Liquid Separation*, 3rd edit., pp. 334–335. Butterworth-Heinemann, Oxford, England (1991).

60. Anon., *Everything You Want to Know about Coagulation and Flocculation*, 3rd edit. Zeta-Meter, Long Island City, New York (1991).

61. E. Mayer, "Membrane Press Sludge Dewatering," in *American Filtration Society Proceedings*, Pittsburgh Conference. AFS, Kingwood, TX (1989).

62. Q. D. Avery, "Membrane Filter Presses," in *American Filtration Society Meeting*, Chicago, Vol. 6, AFS, Kingwood, TX (1992).

63. Q. D. Avery, "Automated System Design for Membrane Filter Presses," in *American Filtration Society Meeting*, Chicago, Vol. 7. AFS, Kingwood, TX (May 1993).

64. G. R. S. Smith and C. Rinschler, "Back-Pulse Liquid Filtration Enhances Tubular Filter Role," *Chem. Proc.*, pp. 48–53 (Jan. 1991).

65. H. Sandstedt and J. Weisenberger, "Cartridge Filter Performance and Micron Rating," *Filtration Separation* (Mar./Apr. 1985).

66. C. Juhasz, "Total Filter Performance," in *Proceedings American Filtration Society*, Ocean City, MD, AFS, Kingwood, TX (March 1988).

67. A. C. Wrotnowski, "Final Filtration with Felt Bag Strainers," *Chem. Eng. Prog.*, p. 89 ff (Oct. 1978).

68. T. W. Johnson, "Large Surface Area in Small Package," in *American Filtration Society Proceedings*, Ocean City, MD, pp. 671–678. AFS, Kingwood, TX (1988).

69. G. R. S. Smith, "How to Use Rotary, Vacuum, Precoat Filters," *Chem. Eng. 83*:84–90 (Feb. 16, 1976).

70. Anon., Case Study, *Chem. Equip.*, p. 29 (June 1992).

71. S. A. Wells, "Two-Dimensional, Steady-State Modeling of Compressible Cake Filtration in a Laterally Unconfined Domain," *Fluid Particle Sep. J.*, pp. 107–116, 4-2 (June 1991).

72. R. W. Lecey and K. A. Pietila, "Improving Belt Filter-Press Performance," *Chem. Eng.* (Nov. 28, 1983).

73. N. D. Deutsch, "Options in Belt Filter Presses," *Water Eng. Mgmt.*, pp. 34–37 (Sept. 1987).

74. H. H. Kleizen and M. Dosoudil, "Continuous Pressure Filtration: From Theory to KDF," in *Proceedings of the American Filtration Society Conference*, Ocean City, MD. AFS, Kingwood, TX (1988).

75. A. Bagdasarian and F. M. Tiller, "Operational Features of Staged, High-Pressure, Thin-Cake Filters," *Filtration Separation* p. 594–598 (Nov./Dec. 1978).

76. T. Toda, "Recent Advances in the Application of the Rotary Filter Press," *Filtration Separation*, pp. 118–122 (Mar./Apr. 1981).

77. L. Svarovsky, *Solid–Liquid Separation*, 3rd edit., p. 582, Butterworth-Heinemann, Oxford, England (1991).

78. I. I. Irizarry and D. B. Anthony, Ornl-Mit-129, Oak Ridge National Laboratory, 28 (April, 1971).

79. B. Hallstrom and M. Lopez-Leiva, "Description of a Rotating Ultra-Filtration Module," *Desalination* 24:273–279 (1978).

80. S. Lipke, "High-Solids Centrifuges Turn Out to be Surprise Dewatering Choice," pp. 22–24 *Water Eng. Mgmt.* (June 1990).

81. A. Zschach et al., "Hyperion's Recipe for Dry Cake," *Operations Forum*, pp. 16–19, 9-8 (Aug. 1992).

82. G. S. Sadowski, "Modular Centrifuges," Paper presented to the Texas Water Pollution Control Association (June 5, 1992).

83. V. K. Norton, "Centrifuges for Solids Control," *Fluid Particle Sep. J.*, pp. 180–181, 3-4 (Dec. 1990).

84. B. C. Bershad, R. M. Chaffiotte, and W-F. Leung, "Making Centrifugation Work for You," *Chem. Eng.*, pp. 84–89 (August, 1990).

85. F. M. Tiller and C. S. Yeh, "Relative Liquid Removal in Filtration and Expression," *Filtration Separation*, 17-2 (1990).

86. H. P. Grace, "Resistance and Compressibility of Filter Cakes," *Chem. Eng. Prog.* Parts 1 and 2 (1953).

87. F. Deshun and R. J. Wakeman, "Effects of Step Drums on Solids Residence Times in Conical Basket and Tumbler Centrifuges," *Filtration Separation*, 29-2 (Mar./Apr. 1992).

88. M. Ernst, R. M. Talcott, H. C. Romans, and G. R. S. Smith, "Tackle Solid–Liquid Separation Problems," *Chem. Eng. Prog.*, pp. 22–28, 87-7 (July 1991).

89. R. J. Wakeman and E. S. Tarlton, "Solid/Liquid Separation Equipment Simulation and Design—An Expert Systems Approach," *Filtration Separation*, 28-4 (May/June 1991).

90. E. Mayer, "Solid/Liquid Separation—Selection Techniques," *Fluid Particle Sep. J.*, pp. 129–139, 1-2 (Dec. 1988).

91. D. Purchas (ed.), *Solid/Liquid Separation Equipment Scale-Up*, Uplands Press, Croydon, U.K. (1986).

92. L. P. Raman, M. Cheryan, and N. Rajagopalan, "Consider Nanofiltration for Membrane Separations," *Chem. Eng. Prog.*, pp. 68–74, 90/3 (March 1994).

93. M. Morgenthaler, "Understanding Decanting Centrifuges and their Environmental Applications," in *American Filtration Society Workshop*, University of Houston, Houston, TX (Jan. 4, 1994).

94. T. De Loggio and A. Letki, "New Directions in Centrifuging," *Chem. Eng.*, pp. 70–76 (Jan. 1994).

95. P. Cartwright, "Membranes Meet New Environmental Challenges," *Chem. Eng.*, pp. 84–87 (Sept. 1994).

96. F. W. Schenck and R. E. Hebeda, "Starch Hydrolysis Products" p. 495, VCH Publishers, New York, 1992.

97. Survey Report #282. Membrane Separation Equipment, 1995. Future Technology Surveys Inc. Norcross, GA (see pg. 700 of this document).

15
Cyclones

David Leith and Donna Lee Jones

CONTENTS

15.1 INTRODUCTION

A cyclone is a device without moving parts that spins a gas stream to remove entrained particles by centrifugal force. Cyclones are simple and inexpensive to make, relatively economical to operate, and are adaptable to a wide range of operating conditions. Cyclones have been used throughout industry since the 1880s for the removal of dust from gases.[1] By the turn of the 20th century, they were used to collect sawdust and shavings in woodworking shops. Ten years later, cyclones began to control dust from cement kilns. Shortly thereafter, they were first used to remove fly ash from flue gas.

Cyclones can be made to withstand extreme temperatures and pressures, can accommodate high dust loadings, and can handle large gas flows. Although standard cyclone designs are inefficient for collecting particles smaller than about 5 microns, high-efficiency cyclones used alone or in series can collect particles between 2 and 5 microns. The standard design cyclones are probably the most frequently used dust collectors in industry.

The first published efforts to predict cyclone performance did not appear until about 1930. Extensive studies of the gas flow pattern in cyclones made during the 1940s led to the development of many models for predicting cyclone pressure drop and dust collection efficiency; efforts at modeling cyclone performance have continued to the present. Although our knowledge of what goes on inside a cyclone has increased over the years, the basic cyclone design shown in a 1885 German patent (No. 39,219) looks a good deal like a cyclone that might be used today.

15.2 PERFORMANCE CHARACTERISTICS

15.2.1 Types

Over the years, many different types of cyclones have been built. However, the "reverse-flow," or cone-under-cylinder design shown in Figure 15.1, is the type used most often for industrial gas cleaning. In this design, aerosol enters the cyclone at the cylinder top, where the shape of the entry causes the gas to spin. Tangential, scroll, and swirl vane entries have been used as shown in Figure 15.2; tangential entries are most common. After entering the cyclone, the gas forms a vortex with a high tangential velocity which gives particles entrained in the gas a high centrifugal force, throwing them to the cyclone wall for collection. Below the bottom of the gas exit duct, the spinning gas gradually migrates inward, to a "central core" along the cyclone

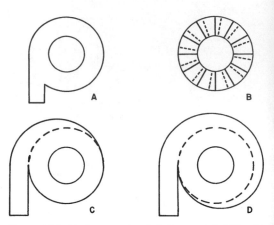

Figure 15.2. Cyclone entries. (*A*) Tangential; (*B*) swirl vane; (*C*) half scroll; (*D*) full scroll.

axis, and from here up, out the gas exit duct. Collected dust descends the cyclone walls to the dust outlet at the bottom of the cone, primarily due to the downward component of the gas velocity at the cyclone wall rather than due to gravity.

Figure 15.3 shows a "straight-through" cyclone. Here, dusty gas enters at one end while cleaned gas and separated dust exit separately at the opposite end. Again, the entry shape causes the inlet gas to spin. Swirl vane entries are used most often on straight-through cyclones. Within the cyclone, centrifugal force pushes particles to the wall. Cleaned gas leaves from a central exit duct while separated particles flow out with a small amount of purge gas through the annular dust discharge. The remainder of this chapter is devoted to the reverse-flow cyclone, as it is the type used by far the most often for industrial dust control. Particle collection theory for straight-through cyclones[3] is not as well developed as that for conventional reverse-flow cyclones.

15.2.2 Standard Designs

Figure 15.4 shows the eight dimensions of a reverse-flow cyclone. It is convenient to express cyclone dimensions as multiples of diameter, D. The "diameter ratios," a/D, b/D, De/D, S/D, h/D, H/D, and B/D, allow comparing the shape of two or more cyclones

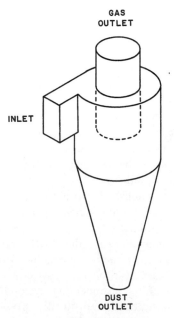

GAS OUTLET

INLET

DUST OUTLET

Figure 15.1. Reverse-flow cyclone.

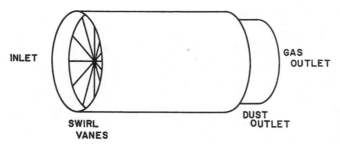

Figure 15.3. Straight-through cyclone.

that might differ in size. Several sets of dimension ratios, or "standard designs," are given in Table 15.1.

A comparison of the designs in Table 15.1 reveals that cyclone shape varies with recommended duty. A high-efficiency cyclone has a smaller inlet area (a/D and b/D) and exit area (De/D) than does a high-throughput cyclone. Gas outlet length (S/D) is less in the high-efficiency designs, probably because inlet height (a/D) is less. Outlet length should be greater than inlet height to be sure that a stable vortex is formed within the cyclone body.

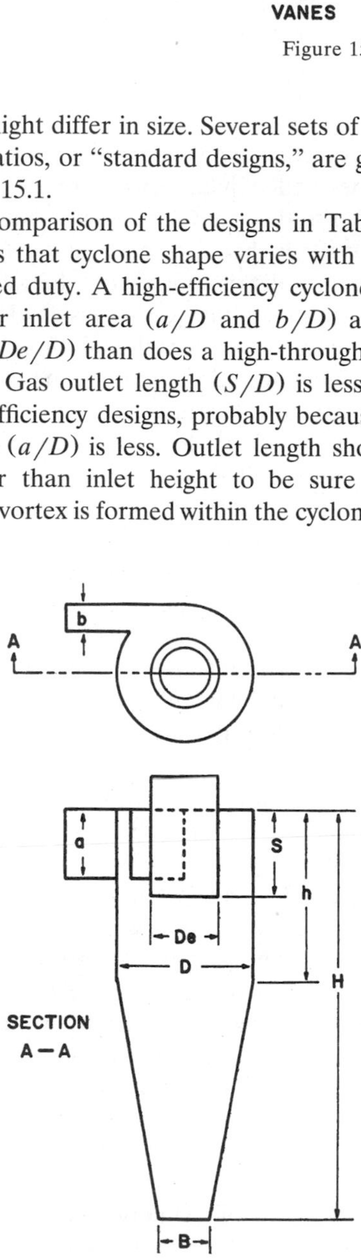

The high-efficiency and general-purpose standard designs have tangential gas entries whereas the high-throughput designs have scroll entries. High efficiency can be traded against high throughput for cyclones operating at the same pressure drop.

Because cyclone design changes with recommended duty, no single optimum cyclone design exists that will work best for all possible applications. Design of a cyclone appropriate for a particular task involves compromises among a number of cyclones, throughput per cyclone, pressure drop and efficiency.

15.2.3 Application Areas

Standard design cyclones collect particles larger than 5 to 10 microns in diameter with reasonable efficiency. For smaller particles, efficiency declines rapidly. With customized high-efficiency cyclones, particles less than 5 microns can be collected, although collection is limited to 1 micron particles or greater. For dust streams with particles larger than several hundred microns, a settling chamber can often by used with lower installation and operating costs than a cyclone system.

Cyclone pressure drop is similar to that found in other particle collection devices, except for high-energy scrubbers which can require much higher pressure drops. For cyclones, there generally is a trade-off between efficiency and pressure drop, with higher pressure drops associated with higher efficiency and vice versa. Values for cyclone pressure drop range from about 0.1 to 2 kPa (0.5 to 8 in. of water).

Figure 15.4. Dimensions of a reverse-flow cyclone.

Table 15.1. Standard Designs for Reverse-Flow Cyclones.

SOURCE	RECOMMENDED DUTY	D	a/D	b/D	De/D	S/D	h/D	H/D	B/D	ΔH	Q/D^2 (m/h)
Stairmand[4]	High-efficiency	1	0.5	0.2	0.5	0.5	1.5	4.0	0.375	6.4	5500
Swift[5]	High-efficiency	1	0.44	0.21	0.4	0.5	1.4	3.9	0.4	9.2	4940
Lapple[6,7]	General-purpose	1	0.5	0.25	0.5	0.625	2.0	4.0	0.25	8.0	6860
Swift[5]	General-purpose	1	0.5	0.25	0.5	0.6	1.75	3.75	0.4	7.6	6680
Stairmand[4]	High-throughput[a]	1	0.75	0.375	0.75	0.875	1.5	4.0	0.375	7.2	16,500
Swift[5]	High-throughput[a]	1	0.8	0.35	0.75	0.85	1.7	3.7	0.4	7.0	12,500

[a] Scroll type gas entry used.

A properly designed cyclone can process effectively dusts in very high concentrations, and in practice loadings of over 2000 g/m^3 (1000 gr/ft^3) have been accommodated.[4] Cyclones have the fortunate ability simultaneously to increase efficiency[8-11] and decrease pressure drop[12-14] with an increase in dust loading. This may come about due to the increased number of particles that move radially outward through the cyclone vortex when dust loading increases. This movement might hinder the formation of the vortex and thereby decrease pressure drop. Despite vortex suppression, efficiency might increase owing to the increased opportunity for larger particles to strike and collect smaller particles while they move toward the cyclone wall.

Cyclones are available in many sizes, and can be made from materials able to withstand extreme operating conditions. They are commercially available in sizes to process 50 to 50,000 m^3/h. Although smaller diameter units generally are more efficient,[4,15,16] a manifold must be used to connect many small cyclones together to process a large gas flow. Refractory lined cyclones have been operated at temperatures of 1000°C, while other units have run at pressures of several hundred atmospheres.[17] However, special materials of construction chosen to allow operation under extreme conditions may not always have good resistance to erosion of the cyclone walls by collected dust. Sticky, hygroscopic dusts may not discharge readily through the dust outlet[18] and these dusts may be better suited to collection in a scrubber. Small-diameter cyclones

housed together and operating in parallel, sometimes called multiclones, are frequently connected to the same dust bin without valves on the dust discharge of each unit. Unequal inlet pressure distribution across the inlet and exhaust manifolds may cause gas to flow out the exit duct of some cyclones, through the dust bin, then up and out through the dust exit and gas exit of other units. This flow pattern will adversely affect the performance of the cyclone system. The performance of multiclones is almost never as good as that of each small cyclone operating individually. However, multiclone performance should be better than that of a single large-diameter cyclone operating at the same pressure drop and handling the same gas flowrate as the manifolded design. The small-diameter cyclone, manifolded design does offer the advantage of compact installation.

Industrial processes use cyclones for unloading material from process gas streams, and for controlling particulate emissions to the atmosphere. The dry product collected in a cyclone can often be recycled to the plant for further processing. Among the processes using cyclones are coal driers, grain elevators, grain driers and milling operations, sawmills and wood-working shops, asphalt plant rotary rock driers, and detergent manufacturing processes. Teams of cyclones operating in series are used under high temperature and pressure conditions to collect catalyst dust from catalytic cracking units at oil refineries and to collect fly ash generated from coal combustion in a pressurized fluidized bed.

15.3 PERFORMANCE MODELING

15.3.1 Flow Pattern

To understand pressure loss and particle collection in a cyclone, it is important to understand the cyclone gas flow pattern. Cyclone flow pattern has been studied in some detail; the overall trend of gas motion has been generally confirmed by all workers. However, no generalized model of the flow pattern is available that will allow the prediction of all velocity components at any point in the cyclone. Each experimenter has concentrated on at most several cyclones of similar design, none of which is particularly similar to the units studied by others. Although each set of results is internally consistent, ambiguities arise when comparing the trends reported in several studies. The most complete review of the gas flow patterns in cyclones is probably that by Jackson.[1]

15.3.1.1. Tangential Gas Velocity

After entering the cyclone, the gas stream forms a confined vortex such that the tangential gas velocity, v_t, is related to the distance, r, from the cyclone axis by Eq. (15.1):

$$v_t r^n = \text{constant} \tag{15.1}$$

The vortex exponent, n, is +1 for an ideal fluid, 0 for velocity which is constant regardless of radial position, and −1 for rotation as a solid body. While theoretical descriptions of cyclone flow patterns[16,19,20] employ values of n from +1 to −1, data[13,21–23] indicate that the usual range is from 0.5 to 0.9, with most values around 0.5. Tangential gas velocity, therefore, increases from a minimum at the cyclone wall to higher values approaching the cyclone axis. Tangential velocities may be lower than the gas inlet velocity at the cyclone wall, but can exceed the inlet velocity by several times at some distance from the wall. Alexander[24] has given an empirical expression for dependence of vortex exponent on cyclone diameter in meters, D_m, at temperature of 283 K.

$$n = 0.67 D_m^{0.14} \tag{15.2}$$

He gives another expression to allow adjustment of the vortex exponent for gas temperature variations.

$$\frac{1 - n_1}{1 - n_2} = \left(\frac{T_1}{T_2}\right)^{0.3} \tag{15.3}$$

Figure 15.5 shows the relationship between vortex exponent, n, cyclone diameter, D, and gas temperature as given by Eqs. (15.2) and (15.3).

If Equation (15.1) is applied throughout the cyclone, tangential gas velocity would increase with decreasing radius to infinity at the cyclone axis. Actually, within the portion of the cyclone above the bottom of the gas outlet duct, tangential velocity is limited by the gas outlet wall. At the cyclone body wall, and at the gas exit wall, the tangential velocity falls off rapidly. If the gas exit is large relative to body diameter, wall effects may hinder the formation of the vortex to the point that tangential velocity becomes almost constant rather than increasing with decreasing radius as is normally the case.[13]

Below the gas exit, tangential velocity increases with decreasing radius, up to a maximum at a point defined as the radius of the cyclone central core. Within this core, tangential gas velocity decreases with decreasing radius, and falls ultimately to a value near zero at the cyclone axis. Within the core, Eq. (15.1) still describes the tangential gas velocity but with a revised value for the vortex exponent ranging from about −0.5 to less than −2.[23] The diameter of the central core ranges from about 0.3 to 1.0 times the gas outlet diameter.[4,25]

Jackson[1] believed that the core diameter decreases from a value close to that of the gas exit diameter directly below the gas exit to a value of about a third of the gas exit diameter near the bottom of the cyclone. Iozia (Jones) and Leith[26] found that the core diameter is relatively constant throughout the cyclone, giving a cylindrical shape to the core region.

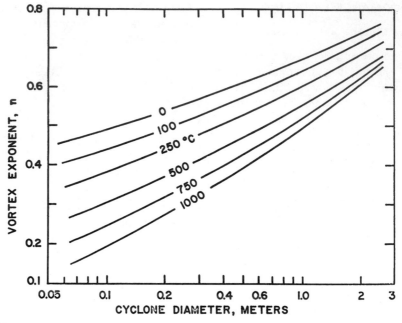

Figure 15.5. Vortex exponent n as a function of cyclone diameter and gas temperature, according to Eqs. (15.2) and (15.3).

Iozia (Jones) and Leith[26] found that the core diameter can be estimated from the cyclone inlet and outlet dimensions. The core length is calculated from geometry using the core diameter and other cyclone dimensions, but is limited by the size of the core diameter relative to the dust outlet diameter.[26] Figure 15.6 presents their anemometer measurements of tangential velocity at different positions in a reverse-flow cyclone with tangential entry.[26]

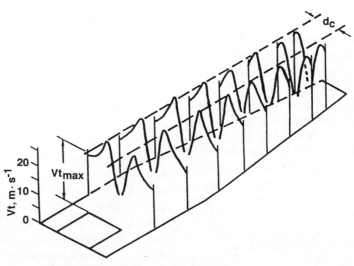

Figure 15.6. Tangential gas velocity in reverse-flow cyclone.

15.3.1.2 Vertical and Radial Gas Velocity

In general, the gas within the cyclone flows downward near the cyclone wall and upward near the cyclone axis; these vertical velocities, both downward and upward, are much less than tangential gas velocities. The radial position at which vertical velocity changes from down to up is relatively closer to the cyclone wall at the top of the unit than at the base of the cone. At all vertical locations, the velocity changeover point appears to be outside the central core. However, once within the core the upward gas velocity increases substantially. Figure 15.7 shows measurements of vertical gas velocity made by ter Linden at different positions in a reverse-flow cyclone with scroll entry.[22]

The radial component of gas velocity has not been measured as extensively as have the tangential and vertical components. Data show that inward radial velocity is low, constant with radial position, and approximately equal at all vertical positions within the cyclone below the gas outlet duct. However, these data are not self-consistent, as radial velocity must be greater near the central core from conservation of mass principles. Radial gas velocity is the most difficult velocity component to measure experimentally. Still, knowledge of this component is essential for determining particle collection efficiency through the "static particle" approach described below. Lack of data on this point has led to unproven speculation on the variability of the radial velocity which is used to explain the inadequacies of this efficiency theory. Figure 15.8 shows ter Linden's measurements of radial gas velocity at different vertical positions in a cyclone.[22]

15.3.1.3 Pressure Distribution

The total pressure at any point in a cyclone is the sum of the static pressure and velocity pressure at that point. Total pressure slowly decreases from a maximum value at the cyclone wall to a minimum value near the cyclone axis. With the high tangential gas velocities present in a cyclone, velocity pressures can be so high that static pressure becomes negative relative to the atmosphere. The static

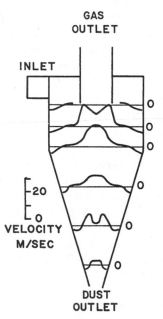

Figure 15.7. Vertical gas velocities in a reverse-flow cyclone.

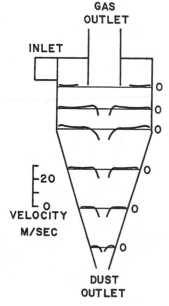

Figure 15.8. Radial gas velocities in a reverse-flow cyclone.

pressure within the central core can be negative, even when the cyclone is installed on the discharge side of a fan. The zone of negative static pressure can extend from the core through the dust outlet and if a suitable valve is not used at the dust outlet, into the dust collection bin. If no valve or a leaky valve is fitted and the dust collection bin is not airtight, dusty air from the bin will be drawn into the central core, up and out of the cyclone. For this reason the cyclone dust hopper should always be airtight. ter Linden's measurements of static and total pressure[22] in a cyclone are shown in Figure 15.9.

15.3.1.4 Overall Gas Flow Pattern

As gas enters the cyclone, it forms a vortex in the annulus above the gas outlet duct. Below the gas outlet, spinning gas gradually migrates into the central core. Near the cyclone walls, gas flows downward, whereas gas closer to and within the central core flows upward toward the gas outlet duct. At the narrow end of the cyclone, all the gas flows into the central core.

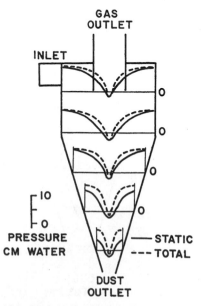

Figure 15.9. Static and total pressures in a reverse-flow cyclone.

First[23] showed that after the gas has entered the cyclone and makes one full revolution, it is not entirely displaced downward by gas entering subsequently. Some older gas is forced toward the cyclone axis in an inward spiral, a phenomenon First calls "lapping." As the newer gas squeezes the older gas toward the cyclone axis, the tangential velocity of the older gas increases through conservation of momentum.

15.3.2 Pressure Drop

Factors[13] that contribute to cyclone pressure drop, static pressure differential across the cyclone, include:

1. Gas expansion as it enters the cyclone
2. Formation of the vortex
3. Wall friction
4. Regain of the rotational kinetic energy as pressure energy.

The first three factors are probably the most important. Controversy exists over the importance of wall friction on pressure drop, as Iinoya[14] has shown that sand glued onto the cyclone walls, increasing wall roughness, actually decreases the pressure drop. If this is correct, then energy consumption due to vortex formation plays a greater role in pressure drop than does wall friction. First [23] also found that wall friction makes an insignificant contribution to overall pressure drop.

Devices such as an inlet vane, an extension of the inner wall of the tangential gas entry within the cyclone body up to a position close to the gas exit duct, and a cross baffle in the gas outlet duct will lower pressure drop. However, these devices probably suppress vortex formation,[1, 18, 27] and so decrease efficiency as well as pressure drop. Because a cyclone is a device for vortex generation, it is not logical to put attachments within it that inhibit vortex formation. Cyclones can be designed for low pressure drop without resorting to internal attachments that may impair efficiency.

Many investigators have developed expressions to predict pressure drop; some are empirical, some theoretical, and most a mixture of both. Despite the complexity of some pressure drop relationships, no single expression has been developed that will give a reliable estimate of pressure drop for all cyclones operating under all conditions.

Cyclone pressure loss is expressed most conveniently as a number of inlet velocity heads, ΔH. Velocity heads can be converted to loss in pressure units, ΔP, by Eq. (15.4):

$$\Delta P = \Delta H(\tfrac{1}{2}\rho_G v_i^2) \qquad (15.4)$$

The number of inlet velocity heads, ΔH, will be constant for any cyclone design although the pressure loss, ΔP, varies with different operating conditions. Pressure drop for a cyclone can best be established by determining ΔH experimentally for a particular cyclone design. The static pressure loss, ΔP, for geometrically similar cyclones can then be found from Eq. (15.4) for different operating conditions. Values of ΔH are listed in Table 15.1 for the standard design cyclones listed there.

Many analytical expressions for determining ΔH from cyclone geometry have been presented in the literature. Several are listed in Table 15.2. One review[30] found that the Barth,[25] Stairmand,[29] and Shepherd and Lapple[28] equations work better than those by Alexander[24] and First.[23] The Barth and Stairmand approaches are complex and require knowledge of all cyclone dimensions. The Shepherd and Lapple approach, Eq. (15.5), is simpler to use, and while it does not include all cyclone dimensions it nevertheless gives results about as good as those produced by the more complex calculation methods.

$$\Delta H = 16 \frac{ab}{De^2} \qquad (15.5)$$

Values of cyclone pressure drop calculated from theory may give results in error by 50% or more. There is currently no alternative to experimental testing when cyclone pressure drop must be known accurately.

15.3.3 Efficiency

Collection efficiency, η, is defined as that fraction of particles of a certain size that are collected by the cyclone. Experience in dealing with cyclones has shown that collection efficiency increases with:

1. Increasing particle diameter and density
2. Increasing gas inlet velocity
3. Decreasing cyclone diameter
4. Increasing cyclone length
5. Drawing some of the gas from the cyclone through the dust exit duct
6. Wetting the cyclone walls

A plot of collection efficiency against particle diameter is called a fractional efficiency curve or grade efficiency curve. A typical fractional efficiency curve for a cyclone is shown in Figure 15.10. Fractional efficiency rises rapidly at first, then flattens out and approaches unity for very large particles.

Particles are separated from the gas stream in a cyclone by spinning to the cyclone wall through centrifugal force. Figure 15.11 shows the forces acting on a particle rotating with tangential velocity u_t at radial position r. The particle moves radially outward with velocity u_r. The tangential velocity of the gas and that of the particle will be assumed equal, $v_t = u_t$. This is probably a reasonable assumption for small particles, for which efficiency is most difficult to determine.

The centrifugal force, F_c, acting on the particle is given in Eq. (15.18):

$$F_c = \frac{\pi d^3 \rho_p v_t^2}{6r} \qquad (15.18)$$

The drag force, F_d, acting on the particle as it moves rapidly outward can be given by Stokes' law; for larger particles with higher radial velocities Stokes' law becomes a progressively poorer approximation.

$$F_d = 3\pi \mu d(u_r - v_r) \qquad (15.19)$$

Equation (15.1), which describes gas tangential velocity as a function of radial position, gives tangential velocity at position r as a function

Table 15.2. Equations for Predicting Pressure Loss at Number of Inlet Velocity Heads, ΔH.

SOURCE	PRESSURE LOSS EQUATION	
Shepherd and Lapple[28]	$$\Delta H = 16 \frac{ab}{De^2}$$	(15.5)
First[23]	$$\Delta H = \frac{ab}{De^2} \left(\frac{24}{\left(\frac{h(H-h)}{D^2} \right)^{1/3}} \right)$$	(15.6)
Alexander[24]	$$\Delta H = 4.62 \frac{ab}{DDe} \left(\left(\left(\frac{D}{De} \right)^{2n} - 1 \right) \left(\frac{1-n}{n} \right) + f \left(\frac{D}{De} \right)^{2n} \right)$$	(15.7)
	$n = 0.67 D_{m}^{0.14}$ at 283 K	(15.2)
	$$\frac{1 - n_1}{1 - n_2} = \left(\frac{T_1}{T_2} \right)^{0.3}$$	(15.3)
	$$f = 0.8 \left(\frac{1}{n(1-n)} \left(\frac{4 - 2^{2n}}{3} \right) - \left(\frac{1-n}{n} \right) \right)$$	
	$$+ 0.2 \left((2^{2n} - 1) \left(\frac{1-n}{n} \right) + 1.5(2^{2n}) \right)$$	(15.8)
Stairmand[29]	$$\Delta H = 1 + 2\phi^2 \left(\frac{2(D-b)}{De} - 1 \right) + 2 \left(\frac{4ab}{\pi De^2} \right)^2$$	(15.9)
	$$\phi = \frac{- \left(\frac{De}{2(D-b)} \right)^{1/2} + \left(\frac{De}{2(D-b)} + \frac{4G^*A}{ab} \right)^{1/2}}{\frac{2G^*A}{ab}}$$	(15.10)
	$$A = \frac{\pi}{4}(D^2 - De^2) + \pi Dh + \pi DeS$$	
	$$+ \frac{\pi}{2}(D + B) \left((H - h)^2 + \left(\frac{D-B}{2} \right)^2 \right)^{1/2}$$	(15.11)
	$G^* = 0.005$	(15.12)
Barth[25]	$$\Delta H = \left(\frac{4ab\theta}{\pi De^2} \right)^2 \epsilon$$	(15.13)
	$$\epsilon = \frac{De}{D} \left(\frac{1}{(1 - 2\theta(H - S)(\lambda/De))^2} - 1 \right) + 4.4\theta^{-2/3} + 1$$	(15.14)
	$$\theta = \frac{\pi De(D - b)}{4ab\alpha^* + 2(H - S)(D - b)\pi\lambda}$$	(15.15)
	$$\alpha^* = 1 - \frac{1.2b}{D}$$	(15.16)
	$\lambda = 0.02$	(15.17)

of cyclone wall radius, r_w, and the tangential velocity at the wall, v_{tw}.

$$V_t r^n = \text{constant} = v_{tw} r_w^n \qquad (15.1)$$

Strictly speaking, the gas tangential velocity at the cyclone wall must be zero. However, the boundary layer at the cyclone wall is thin, as can be seen in Figure 15.6; and Eq. (15.1)

describes the tangential velocity near the wall with little error. The sum of the centrifugal and drag forces acting on the particle will equal its mass times its acceleration.

$$\left[\frac{\pi d^3 \rho_p v_t^2}{6r} - 3\pi\mu d(u_r - v_r) \right] = \frac{\pi d^3}{6} \rho_p \frac{d^2 r}{dt^2}$$
$$(15.20)$$

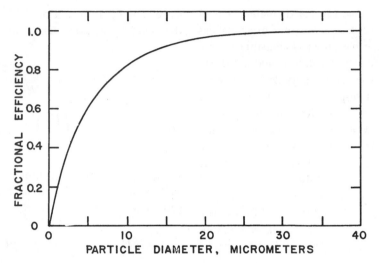

Figure 15.10. Typical cyclone fractional efficiency curve.

Simplifying and making the substitutions $u_r = dr/dt$ and $v_t = v_{tw}r_w^n/r^n$ yields Eq. (15.21):

$$\frac{d^2r}{dt^2} + \frac{18\mu}{d^2\rho_p}\frac{dr}{dt} - \left(\frac{v_{tw}^2 r_w^{2n}}{r^{(2n+1)}} + \frac{18\mu v_r}{d^2\rho_p}\right) = 0$$

(15.21)

Equation (15.21) describes radial particle motion within a vortex and underlies many approaches used to calculate cyclone collection efficiency. Unfortunately, Eq. (15.21) has not been solved analytically. Approximate solutions can be found by postulating various flow conditions within the cyclone, allowing deletion of some terms in the equation. All these approximations are open to criticism.

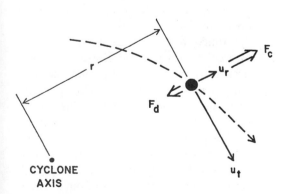

CYCLONE
AXIS

Figure 15.11. Forces on a particle in a cyclone.

The relative importance of each term will change with each cyclone design and particle diameter. It is unlikely that any approximation will yield good results for all applications.

The theoretical efficiency of a cyclone can be characterized in terms of a "critical particle" diameter, d_{100}. The critical particle is that which, according to theory, is collected with 100% efficiency. Since collection efficiency increases gradually with increasing particle diameter and approaches 100% only as a limit, the critical particle is not observed experimentally.

A more easily verified theoretical construct is the "cut diameter" or d_{50}, the particle size that is collected with 50% efficiency. Calculations of critical or cut diameter can be used to generate the cyclone fractional efficiency curve shown in Figure 15.10.

15.3.3.1 Critical Diameter: The Static Particle Approach

Gas at the edge of the central core will have maximum tangential velocity, v_{max}. Gas flowing radially inward to the cyclone central core will flow past the core edge with an average inward radial velocity given by Eq. (15.22):

$$v_r = \frac{-Q}{2\pi r_{core}(H - S)}$$

(15.22)

For particles of the critical diameter, the inward drag force caused by the inrushing gas will just balance the outward centrifugal force caused by their rapid rotation about the cyclone axis. These "static particles" will theoretically remain suspended at the edge of the central core. Larger particles should spin out to the cyclone wall and become collected, and smaller particles should flow past the static particles into the central core and out the cyclone. As they are stationary, the critical particles will have no radial acceleration or velocity $(d^2r/dt^2 = dr/dt = 0)$. From Eq. (15.1), $v_{tw}^2 r_w^{2n} = v_{max}^2 r_{core}^{2n}$, which when substituted into Eq. (15.21) yields the critical particle diameter.

$$d_{100} = \left(\frac{9Q\mu}{\pi(H-S)\rho_p v_{max}^2} \right)^{1/2} \quad (15.23)$$

According to static particle theory, "fractional degree of dust separation in the case of a (critically sized) particle suddenly increases from 0 to 100%."[25] The shape of experimental fractional efficiency curves, which show a gradual increase in efficiency with increasing particle diameter, is explained as being due to variations in the gas inward drift velocity, v_r, along the cyclone axis.

Barth[25] and Stairmand[4] used different assumptions in the static particle approach to develop equations for critical diameter. Iozia (Jones) and Leith[26] used experimental data to develop an equation to predict cut diameter using the static particle approach. These equations are shown in Table 15.3.

Iozia (Jones) and Leith[26] predicted *maximum tangential velocity* from inlet velocity and a dimensionless geometry parameter.

$$C_I = \left(\frac{ab}{D^2} \right)^{0.61} \left(\frac{De}{D} \right)^{-0.74} \left(\frac{H}{D} \right)^{-0.33} \quad (15.26)$$

$$V_{max} = 6.1V_iC_I \quad (15.27)$$

Core length, used in Eq. (15.26), depends on the value of the core diameter d_c.

$$d_c = 0.52D \left(\frac{ab}{D^2} \right)^{-0.25}$$

$$\times \left(\frac{De}{D} \right)^{1.53} \quad (15.28)$$

When $d_c > B$, the core intercepts the cyclone walls and the core length is calculated from geometry.

$$z_c = (H - S) - ((H - h)/(D/B - 1))$$

$$\times ((d_c/B) - 1) \quad (15.29)$$

When $d_c < B$, the core length extends to the bottom of the cyclone.

$$z_c = H - S \quad (15.30)$$

15.3.3.2 Critical Diameter: The Timed Flight Approach

The timed flight approach is another way to calculate critical diameter, and involves a different set of assumptions for solving Eq. (15.21). Let r_i be the innermost radial position at which particles enter the cyclone. Particles entering at this point must cross the distance from r_i to the cyclone wall to be collected, and if a particle is not in the cyclone long enough to travel this distance it will escape collection. In the timed flight approach, the particle's radial acceleration and the gas radial velocity are arbitrarily set equal to zero and neglected $(d^2r/dt^2 = v_r = 0)$. The gas velocity in the entrance duct is assumed equal to the velocity at the cyclone wall, $v_r = v_{tw}$. These assumptions allow solution of Eq. (15.21) for particle radial position as a function of particle residence time within the cyclone, t. The particle with critical diameter will travel from its initial

Table 15.3. Equations Derived from Eq. (15.20) for Predicting Collection Characteristics.

$$\frac{d^2r}{dt^2} + \frac{18\mu}{d^2\rho_p}\frac{dr}{dt} - \left(\frac{v_{tw}^2 r_w^{2n}}{r^{(2n+1)}} + \frac{18\mu v_r}{d^2\rho_p}\right) = 0 \qquad (15.21)$$

STATIC PARTICLE APPROACH	$\dfrac{d^2r}{dt^2}$	$\dfrac{dr}{dt}$	ASSUMPTIONS r_{core}	v_{max}	RESULTANT EQUATION	
Barth[25]	0	0	$De/2$	$\dfrac{4Q\theta}{\pi De^2}$	$d_{100} = \left(\dfrac{9Q\mu}{\pi(H-S)\rho_p v_{max}^2}\right)^{1/2}$	(15.23)
Stairmand[4]	0	0	$De/4$	$\dfrac{Q\phi}{ab}\left(\dfrac{2\left(D-\dfrac{b}{2}\right)}{De}\right)^{1/2}$	$d_{100} = \left(\dfrac{9Q\mu}{2\pi(H-S)\rho_p v_{max}^2}\right)^{1/2}$	(15.24)
Iozia (Jones)[26] and Leith	0	0	$d_c/2$	$6.1\,V_i\cdot C_I$	$d_{50} = \left(\dfrac{9Q\mu}{\pi z_c \rho_p V_{max}^2}\right)^{1/2}$	(15.25)

TIMED FLIGHT APPROACH	$\dfrac{d^2r}{dt^2}$	v_r	r_i	ASSUMPTIONS n	t	RESULTANT EQUATION	
Rosin et al.[20]	0	0	$\dfrac{D}{2}-b$	0	$\dfrac{\pi DN}{v_i}$	$d_{100} = \left(\dfrac{9\mu b}{\pi\rho_p v_i N}\left(1-\dfrac{b}{D}\right)\right)^{1/2}$	(15.32)
Lapple and Shepherd[32]	0	0	$\dfrac{D-De}{2}$	0	$\dfrac{\pi DN}{v_i}\left(1-\dfrac{De}{2D}\right)$	$d_{100} = \left(\dfrac{9\mu De}{2\pi\rho_p v_i N}\right)^{1/2}$	(15.33)
Davies[33]	0	0	$\dfrac{De}{2}$	1	$\dfrac{H}{v_i}$	$d_{100} = \left(\dfrac{9\mu D^2\left(1-\left(\dfrac{De}{D}\right)^4\right)}{8\rho_p v_i H}\right)^{1/2}$	(15.34)
Lapple[6]	0	0	$\dfrac{D-b}{2}$	0	$\dfrac{\pi DN}{v_i}\left(1-\dfrac{De}{2D}\right)$	$d_{50} = \left(\dfrac{9\mu b}{2\pi\rho_p v_i N}\right)^{1/2}$	(15.35)
Leith and Licht[34]	0	0	0	n	depends on geometry and throughput	$\eta = 1 - \exp(-2(C_L\psi)^{1/(2n+2)})$	(15.36)

position at r_i and just reach the cyclone wall in time t.

$$d_{100} = \left[\frac{9\mu D^2\left(1-\dfrac{2r_i}{D}\right)^{2n+2}}{4(n+1)v_i^2\rho_p t}\right]^{1/2} \qquad (15.31)$$

Investigators have made assumptions about the initial particle radial position, r_i, and the value for vortex exponent, n. Residence time, t, is sometimes defined in terms of an empirical "number of turns," N, that the gas stream makes within the cyclone. The value for N reportedly varies from 0.3 to 10, with a mean value of about 5.[31] Table 15.3 gives several sets of assumptions for r_i, n, and t along with the resultant equations for either critical or cut diameter. In the timed flight approach, particles the size of the cut diameter theoretically enter the cyclone at the midpoint of gas entry.

15.3.3.3 The Fractional Efficiency Curve

Critical particle diameter is useful only as a rough estimator of cyclone efficiency. For more precise work, as when estimating overall cyclone efficiency on a dust with a range of particle sizes, the entire fractional efficiency curve is necessary. Lapple[6] and Barth[25] have developed generalized plots of efficiency versus a dimensionless particle parameter. Lapple's parameter is defined as particle diameter over the cut diameter calculated from Eq. (15.35). This plot is given in Figure 15.12 and is valid for cyclones of the Lapple design listed in Table 15.1. No figures for cyclones of other design are available.

Iozia (Jones) and Leith[26,35] developed an equation to predict fractional efficiency from the dimensionless particle parameter of cut diameter calculated from Eq. (15.25) over particle diameter. The fractional efficiency curve is defined by using the particle parameter in a "logistic" equation.

$$\eta_d = \frac{1}{1 + (d_{50}/d)^\beta} \quad (15.37)$$

The logistic slope parameter, β, is estimated from cut diameter (in centimeters) and a dimensionless geometry parameter, C_β, that depends on inlet dimensions:

$$C_\beta = ab/D^2 \quad (15.38)$$

$$\ln \beta = 0.62 - 0.87 \ln(d_{50}\text{-cm})$$
$$+ 5.21 \ln C_\beta$$
$$+ 1.05(\ln C_\beta)^2 \quad (15.39)$$

Efficiency data from the literature[36] were used to compare the prediction of efficiency using Eqs. (15.25) and (15.37) through (15.39) against other theories.[17,25,34,37] Equations (15.25) and (15.39) were found to predict efficiency significantly better than the other theories.[35]

Leith and Licht[34] combined an approximate solution to Eq. (15.21) with the assumption that uncollected dust is remixed within the cyclone gas stream due to gas stream turbulence. The assumptions they made for solving Eq. (15.21) are listed in Table 15.3. The resultant equation predicts the fractional efficiency curve:

$$\eta = 1 - \exp\left(-2(C_L \psi)^{1/(2n+2)}\right) \quad (15.36)$$

Here, the vortex exponent, n, can be calculated from Eq. (15.2) and (15.3), or found from

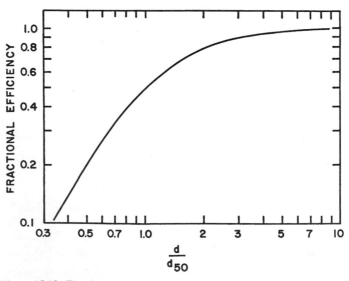

Figure 15.12. Fractional efficiency versus d/d_{50} for Lapple design cyclone.

Figure 15.5. The influence of particle and gas properties are combined into ψ, a dimensionless inertia parameter or Stokes' number:

$$\psi = \frac{d^2 \rho_p v_i (n + 1)}{18 \mu D} \qquad (15.40)$$

The effect of cyclone geometry is consolidated in C_L, a dimensionless geometry parameter. The geometry parameter depends only on the cyclone dimension ratios, and is independent of size

$$C_L = \frac{\pi D^2}{ab} \left(2 \left(1 - \left(\frac{De}{D} \right)^2 \right) \left(\frac{S}{D} - \frac{a}{2D} \right) \right.$$
$$+ \frac{1}{3} \left(\frac{S + z_c - h}{D} \right) \left(1 + \frac{d_c}{D} + \left(\frac{d_c}{D} \right)^2 \right)$$
$$\left. + \frac{h}{D} - \left(\frac{De}{D} \right)^2 \frac{z_c}{D} - \frac{S}{D} \right) \qquad (15.41)$$

Core length, z_c, is found from an equation developed by Alexander:[24]

$$z_c = 2.3 \, De \left(\frac{D^2}{ab} \right)^{1/3} \qquad (15.42)$$

The diameter of the core, d_c, can be determined from Eq. (15.43).

$$d_c = D - (D - B) \left(\frac{S + z_c - h}{H - h} \right) \qquad (15.43)$$

Equation (15.36) implies that a cyclone with a high value of geometry parameter, C_L, should have a higher efficiency than a unit with a low value of C_L for particles of all sizes and for all operating conditions. The efficiency capabilities of alternative cyclone designs can be evaluated by comparing their values of C_L in the same way that pressure drop requirements are evaluated by comparing values of ΔH. Equation (15.36) was tested against experimental data from the literature[38] and was found to predict the data reasonably well.

The equations discussed in Table 15.3 may be useful for determining the efficiency of industrial-sized cyclones, a few meters or less in diameter. The gas flow assumptions used in

the theories discussed here may not apply to smaller cyclones. For these small cyclones, alternative collection efficiency expressions may be more appropriate.[39-42]

Although theoretical calculations of critical particle diameter and fractional efficiency are useful, they, like theoretical pressure drop calculations, may predict performance substantially in error from that experienced in the field. All efficiency theories discussed have been for tangential entry, reverse-flow cyclones as shown in Figure 15.1. Their applicability is unknown to other cyclone designs, such as those with either scroll or swirl vane entries, or to straight-through cyclones of the type in Figure 15.2. The best way to determine cyclone fractional efficiency characteristics is to test the cyclone in the laboratory or in a pilot test program.

Once an experimental fractional efficiency curve has been developed for a cyclone operating under known conditions, the fractional efficiency curve can be determined for a cyclone of the same design under different operating conditions by adjusting the efficiency curve of the test cyclone. According to one theory,[34] two cyclones will have the same efficiency when their Stokes numbers are the same. If the test cyclone has known efficiency on particles of size d_1, a similar cyclone will have the same efficiency on particles d_2, where:

$$d_2 = d_1 \left(\frac{Q_1}{Q_2} \frac{\rho_{p1}}{\rho_{p2}} \frac{\mu_2}{\mu_1} \frac{D_2}{D_1} \right)^{1/2} \qquad (15.44)$$

This analysis assumes that the diameters of the two cyclones are close enough that the value of the vortex exponent, n, does not change appreciably. The fractional efficiency curve for the similar cyclone can be constructed from the curve for the tested cyclone by picking a series of coordinates from the experimentally derived efficiency curve and calculating the analogous coordinates for the similar cyclone from Eq. (15.44).

The accuracy of this procedure decreases as each of the ratios in Eq. (15.44), Q_1/Q_2, etc., departs more and more from unity. The procedure is especially suspect when predicting the performance of cyclones with much greater diameter and throughput than the test model. Also, when adapting results based on an experimental dust to a different dust, particle shape may change as well as density. Nevertheless, a fractional efficiency curve calculated using this procedure is strongly preferred over one determined strictly from theory.

15.3.4 Other Variables Affecting Performance

Although cyclone performance theories express the effect of many variables on cyclone performance, several variables known to influence pressure drop and efficiency are not considered.

Increasing inlet dust concentration, c_i, simultaneously increases collection efficiency and decreases pressure drop. Briggs quantified the influence of dust loading on pressure drop.

$$\Delta P_{dusty} = \frac{\Delta P_{clean}}{1 + 0.0086(c_i)^{1/2}} \quad (15.45)$$

Here, c_i has the dimensions of grams per cubic meter. The effect on efficiency of changing inlet loading from c_{i1} to c_{i2} can be found[11] from:

$$\frac{100 - \eta_1}{100 - \eta_2} = \left(\frac{c_{i2}}{c_{i1}}\right)^{0.182} \quad (15.46)$$

Presumably the values of efficiency and concentration in Eq. (15.46) are for polydisperse dusts and the relationship applies to overall dust concentration and efficiency rather than to values for any one particle size.

If the tangential velocity of the gas near the cyclone wall is too high, saltation will occur; particles will bounce along the cyclone wall and not be separated effectively from the gas stream. Kalen and Zenz[43] have examined this phenomenon, and its implications for cyclone design are discussed by Koch and Lict.[44] An empirical equation (15.47), which gives the cyclone inlet velocity above which *saltation* occurs, v_{is}, is:

$$v_{is} = 2400 \frac{\mu \rho_p}{\rho_G^2} D^{0.2} \frac{(b/D)^{1.2}}{1 - b/D} \quad (15.47)$$

SI units (m, kg, s) must be used in this equation. Cyclone efficiency increases with inlet velocity up to about 1.25 v_{is}; further increases in inlet velocity cause a decrease in efficiency as saltation and reentrainment of collected dust become more important.

Stairmand[4] showed that the overall efficiency of a well-designed cyclone increases from its normal value of 92% to an increased value of 93.6% when about 10% of the gas flow is drawn through the dust outlet. A similar "base purge" increased the efficiency of a poorer cyclone design from 89.1% to 92.2%. Stairmand believes this efficiency increase is due to a reduced reentrainment of separated dust in the dust outlet region. The disadvantages of this practice are that it requires the use of otherwise unnecessary auxiliary fans and ducts to draw off the purge, and that if the purge is recycled to the cyclone inlet, the cyclone must be sized to handle the purge air as well as the process air. In practice, base purge is seldom used.

Stairmand[4] also reported that efficiency increases from a normal value of 92% to 93.7% for the well-designed cyclone and 89.1% to 93.2% for the poorer design when these cyclones operate with wetted walls. The wetted walls may reduce reentrainment of collected dust throughout the cyclone. Disadvantages of this practice are that water piping is required and that the collected dust is in a slurry.

15.3.5 Overall Efficiency on Polydisperse Dusts

Industrial dusts contain particles of many sizes. To calculate the overall cyclone collection efficiency, $\eta_{overall}$, on such a dust one must multiply efficiency for each particle size by the fraction of particles in the dust that are of that size. The sum of these products is the overall fractional efficiency for the cyclone. Table 15.4

Table 15.4. Overall Collection Efficiency Calculation Using Numerical Integration of Eq. (15.48).

(1) SIZE RANGE (MICROMETERS)	(2) MEAN SIZE (MICROMETERS)	(3) FRACTION IN RANGE	(4) EFFICIENCY ON MEAN SIZE	(5) FRACTIONAL EFFICIENCY COLUMNS (3) × (4)
0–2	1	0.10	0.03	0.00
2–5	3.5	0.10	0.38	0.04
5–10	7.5	0.10	0.81	0.08
10–20	15	0.15	0.96	0.14
20–30	25	0.10	0.99	0.10
30–40	35	0.10	1.00	0.10
40–60	50	0.15	1.00	0.15
60–76	68	0.10	1.00	0.10
76–104	90	0.07	1.00	0.07
104–150	127	0.03	1.00	0.03
Total		1.00		0.81

illustrates this process for the cyclone whose fractional efficiency curve is shown in Figure 15.10. The dust size distribution is plotted in Figure 15.13. Equation (15.48) is the formal mathematical statement of this process.

$$\eta_{\text{overall}} = \int_0^1 \eta_d \, dG \qquad (15.48)$$

Here, n_d is the efficiency on particles of a certain size, d, and dG is the fraction of all particles of that size in the dust. The overall efficiency for this cyclone on this dust is found to be about 85%.

15.4 CYCLONE DESIGN

15.4.1 Necessary Design Information

Before attempting to design a cyclone or cyclone system it is important to consider cyclone limitations, to be sure that an alternative control device might not work better. Cyclones may be unsuited for collection of particles less than about 5 microns in diameter, as efficiency falls off rapidly for particles smaller than this. Other types of collectors such as fabric filters, electrostatic precipitators, and some kinds of scrubbers will be able to collect these small particles more efficiently. If the inlet dust loading is high and the desired outlet concentration is low, it may be necessary to use a higher efficiency collector such as a fabric filter either instead of, or in conjunction with (usually after) a cyclone system. Sticky or hygroscopic dusts may stick to the cyclone walls, and not discharge into the collection hopper. For dusts of this type, a scrubber may be a better collector choice than a cyclone.

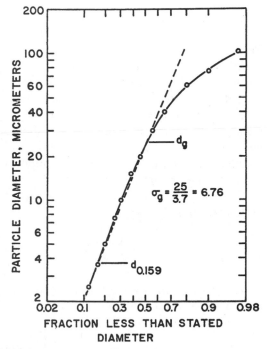

Figure 15.13. Particle size distribution from Table 15.4.

To design a cyclone or any collection device, the inlet dust concentration and size distribution must be known. Although preliminary estimates of expected dust properties are available from the literature,[45-47] this information should always be obtained by stack sampling when possible. Of course, when designing control equipment for a plant that has yet not been constructed, stack testing is impossible and in this case the design will have to be based on data obtained from similar plants in conjunction with the design plans for the process to be controlled.

Design criteria such as gas flow rate, temperature, and particle density—material density, not apparent or bulk density—special conditions of corrosivity, particle abrasiveness, and fluctuations in gas flow should be noted. These data requirements are summarized in Table 15.5. All the data necessary for design of a cyclone system can be obtained from a stack test performed on the gas stream to be cleaned.

15.4.2 Cyclone Specification

Usually cyclones are not custom designed. Rather an accepted standard design is selected, such as one listed in Table 15.1 or a manufacturer's proprietary design. Cyclone diameter can be determined from gas flowrate Q, using the value for Q/D^2 tabulated for each standard design given in Table 15.1. Once diameter is known the remaining seven dimensions can be determined from the dimension ratios of the standard design selected. For volumetric gas flows larger than about 20,000 m^3/h it is often better to use several smaller cyclones in parallel rather than one large cyclone. This is because collection efficiency decreases with increasing cyclone diameter and also because of possible problems with space or headroom requirements for very large cyclones.

A fractional efficiency curve for the selected design can be determined by one of the methods discussed above. The overall collection efficiency for the selected cyclone design, inlet dust size distribution and concentration to be processed, and outlet dust concentration desired can then be determined from the methods describe previously.

A cyclone can be *custom* designed to perform a specific dust collection job.[48] This approach will give a cyclone with a greater collection efficiency, smaller size, or lower pressure drop than a cyclone with a standard design. The "optimized" cyclone design procedure requires trial and error calculations that are better suited for a microcomputer or programmable calculator than by hand.

First, determine a preliminary cyclone diameter from Eq. (15.49):

$$D_{\mathrm{m}} = \frac{(\rho_{\mathrm{p}} Q_{\mathrm{cyclone}})^{1/3}}{275} \qquad (15.49)$$

Particle density and flow must be in units of m–kg–h in this equation. If the diameter calculated from Eq. (15.49) is greater than 2 m, then the flow should be divided to accommodate at least two cyclones from the start. In most situations, two or more cyclones should be used to allow flexibility in operation and maintenance, and to avoid a system shutdown if one cyclone becomes plugged.

The flow going to each cyclone is calculated by Eq. (15.50):

$$Q_{\mathrm{cyclone}} = \frac{Q_{\mathrm{system}}}{N_{\mathrm{C}}} \qquad (15.50)$$

Next, pick a target value for outlet concentration or overall cyclone efficiency, which is determined by Eq. (15.51):

$$c_0 = c_i(1 - \eta_{\mathrm{overall}}) \qquad (15.51)$$

Table 15.5. Data Necessary for Cyclone Design.

Particle size distribution
Inlet dust loading (g/m^3)
Particle density (kg/m^3)
Gas flowrate (m^3/h)
Gas temperature $(\degree C)$
Special conditions of corrosivity, abrasiveness,
    fluctuations in gas flow, etc.

Using the design parameters in Table 15.6 calculate the overall cyclone collection efficiency of the dust stream with three different cyclone designs that correspond to design parameter K values of 1.5, 3, and 4.4. The cyclone diameter calculated from Eq. (15.49) and the cyclone flow calculated from Eq. (15.50) are needed in the efficiency calculations. The highest efficiency cyclone design will correspond to K equal to 1.5 and the lowest efficiency cyclone design will correspond to K equal to 4.4. The overall efficiency for collection of polydisperse dust is found from the fractional efficiency curve generated using Eq. (15.37) and numerical integration of

Eq. (15.48) These calculations were shown previously in Table 15.4.

Plot the K values against the predicted overall efficiency. From the line joining the three points, determine the closest K value from Table 15.6 that corresponds to the target efficiency. The design in Table 15.6 that corresponds to this K value is the optimized cyclone design. At this point the cyclone design is fixed.

The pressure drop for the system will be determined from the number of cyclones and flow going to each cyclone. The cyclone pressure drop is calculated from Eq. (15.4) using the ΔH values from Table 15.6 and the inlet

Table 15.6. Optimized Designs.

DESIGN PARAMETER K	a/D	b/D	De/D	H/D	h/D	S/D	B/D	ΔH
1.5	0.16	0.30	0.26	6	1.5	0.16	0.26	11.4
1.6	0.18	0.30	0.28	6	1.5	0.18	0.28	11.0
1.7	0.20	0.30	0.30	6	1.5	0.20	0.30	10.7
1.8	0.22	0.30	0.31	6	1.5	0.22	0.31	11.0
1.9	0.25	0.30	0.32	6	1.5	0.25	0.32	11.7
2.0	0.27	0.30	0.33	6	1.5	0.27	0.33	11.9
2.1	0.29	0.30	0.34	6	1.5	0.29	0.34	12.0
2.2	0.31	0.30	0.35	6	1.5	0.31	0.35	12.1
2.3	0.34	0.30	0.36	6	1.5	0.34	0.36	12.6
2.4	0.38	0.30	0.37	6	1.5	0.38	0.37	13.3
2.5	0.40	0.30	0.38	6	1.5	0.40	0.38	13.3
2.6	0.43	0.30	0.39	6	1.5	0.43	0.39	13.6
2.7	0.48	0.30	0.41	6	1.5	0.48	0.41	13.7
2.8	0.51	0.28	0.42	6	1.5	0.51	0.42	12.7
2.9	0.54	0.26	0.43	6	1.5	0.54	0.43	12.1
3.0	0.57	0.25	0.44	6	1.5	0.57	0.44	11.8
3.1	0.60	0.25	0.44	6	1.5	0.60	0.44	12.4
3.2	0.63	0.25	0.45	6	1.5	0.63	0.45	12.4
3.3	0.66	0.25	0.46	6	1.5	0.66	0.46	12.5
3.4	0.69	0.25	0.47	6	1.5	0.69	0.47	12.5
3.5	0.73	0.25	0.48	6	1.5	0.73	0.48	12.7
3.6	0.76	0.25	0.48	6	1.5	0.76	0.48	13.2
3.7	0.79	0.25	0.49	6	1.5	0.79	0.49	13.2
3.8	0.82	0.25	0.50	6	1.5	0.82	0.50	13.1
3.9	0.85	0.25	0.50	6	1.5	0.85	0.50	13.6
4.0	0.89	0.25	0.51	6	1.5	0.89	0.51	13.7
4.1	0.93	0.25	0.51	6	1.5	0.93	0.51	14.3
4.2	0.96	0.25	0.52	6	1.5	0.96	0.52	14.2
4.3	0.99	0.25	0.52	6	1,5	0.99	0.52	14.6
4.4	1.00	0.25	0.52	6	1.5	1.00	0.52	14.8

velocity of the cyclone with flow equal to $Q_{cyclone}$ calculated from Eq. (15.50). If the calculated pressure drop is too high, then the number of cyclones should be increased in Eq. (15.50) until an acceptable pressure drop is obtained. As discussed earlier, pressure drop calculated from theory can be considerably higher or lower than actual. The limitations of the system being designed should be considered before a final decision is made.

Once the number of cyclones is fixed, the design diameter of the cyclone is calculated from Eq. (15.49) with Q equal to $Q_{cyclone}$.

In some cases, no cyclone system will provide adequate collection efficiency or suitable pressure drop; in this case, teams of cyclones in series or alternate control devices should be considered.

15.4.3 Design Example

The design procedures described in the previous section can be illustrated by the following design example. Suppose 40,000 m³/h of air containing rock dust comes from a rotary dryer at 100°C. It is desired to collect as much dust as possible for recycle to the process. The effluent from the cyclone system will go to a scrubber for final control before release to the atmosphere. The maximum loading to the scrubber should be 10 g/m³; however, 8 g/m³ or less is preferable. A stack test finds the dust loading from the dryer is 50 g/m³. The pressure drop for the cyclone must be less than 2 kPa.

15.4.3.1 Example Using Standard Designs

Any of the standard cyclone designs shown in Table 15.1 may be used. Example calculations using the Stairmand design will be shown here. The diameter of the cyclone can be found from cyclone flow and the value for Q/D^2 in Table 15.1 as in Eq. (15.52).

$$D_M = \frac{(40{,}000(m^3/h)/2)^{1/2}}{(5500)^{1/2}} \quad (15.52)$$

$$D_M = 1.91 \text{ m}$$

Since the cyclone diameter calculated with Eq. (15.52) and a flow of 40,000 m³/h is greater than 2 m, two cyclones should be used in the preliminary design calculations. The flow to each cyclone then, calculated from Eq. (15.50), is 40,000 divided by 2 or 20,000 m³/h.

The other dimensions of the cyclone can be calculated from the dimension ratios given in Table 15.1 and are given in Table 15.7. Inlet gas velocity will be found from Eq. (15.53).

$$V_i = \frac{(20{,}000 \text{ m}^3/\text{h})(1 \text{ h}/3600 \text{ s})}{(0.95 \text{ m})(0.38 \text{ m})}$$

$$= 15.3 \text{ m/s} \quad (15.53)$$

The pressure loss for the system can be calculated from ΔH for the Stairmand design from Table 15.1 and Equation (15.4).

$$\Delta P = 6.4(1/2)(0.944 \text{ kg/m}^3)(15.3 \text{ m/s})^2$$

$$\Delta P = 710 \text{ kPa} \quad (15.54)$$

The expected efficiency and outlet dust loading concentration can be calculated using the method outlined previously. The maximum tangential velocity is calculated from Eqs. (15.26) and (15.27).

$$C_I = \left[(0.95 \text{ m})(0.38 \text{ m})/(1.91 \text{ m})^2 \right]^{0.61}$$

$$\times [(0.95 \text{ m})/(1.91 \text{ m})]^{-0.74}$$

$$\times [(7.63 \text{ m})/(1.91 \text{ m})]^{-0.33}$$

$$C_I = 0.26 \quad (15.55)$$

$$V_{max} = 6.1(15.3 \text{ m/s})(0.26)$$
$$V_{max} = 24.2 \text{ m/s} \quad (15.56)$$

Core diameter is calculated using Eq. (15.28).

$$d_c = (0.52)(1.91)(0.1)^{-0.25}(0.5)^{1.53}$$
$$d_c = 0.61 \text{ m} \quad (15.57)$$

Since core diameter is less than the outlet diameter (0.72 m), core length is calculated with Eq. (15.30) rather than Eq. (15.29)

$$z_c = 7.63 - 0.95 = 6.68 \text{ m} \quad (15.58)$$

Table 15.7. Cyclone Specifications for Design Example.

DESIGN PARAMETERS	STAIRMAND STANDARD DESIGN		OPTIMIZED DESIGN
	TWO CYCLONES	FOUR CYCLONES	TWO CYCLONES
Diameter	1.91	1.35	1.34
a	0.95	0.67	0.92
b	0.38	0.27	0.33
De	0.95	0.67	0.63
H	7.63	5.39	8.03
h	2.86	2.02	2.01
S	0.95	0.67	0.67
B	0.72	0.51	0.63
ΔH	6.4	6.4	12.5
Inlet velocity (V_i), m/s	15	15	18
Maximum tangential velocity (V_{max}), m/s	24	24	36
Maximum inlet velocity for no saltation, m/s	30	28	39
Core diameter, m	0.61	0.43	0.34
Core length, m	6.7	4.7	7.4
d_{50}, microns	5.9	5.0	3.8
Logistic slope parameter (β)	1.9	2.2	4.8
Overall efficiency (η)$_{overall}$	0.81	0.84	0.84
Pressure drop, kPa	0.71	0.71	1.91
Outlet loading, g/m^3	9.4	8.1	8.1

Next, the value for d_{50} is calculated from Eq. (15.25).

$$d_{50} = \{[9(20,000 \text{ m}^3/\text{h})(1 \text{ h}/3600 \text{ s})$$

$$\times (2.17 \times 10^{-5} \text{ kg/m} \cdot \text{s})]/\pi (6.68 \text{ m})$$

$$\times (2500 \text{ kg/m}^3)(24.2 \text{ m/s})^2\}^{1/2}$$

$$(15.59)$$

$$d_{50} = 5.9 \times 10^{-6} \text{ m}$$

The value for the logistic slope parameter (β) is calculated from Eqs. (15.38) and (15.39).

$$C_\beta = (0.95 \text{ m})(0.38 \text{ m})/(1.91 \text{ m})^2 \quad (15.60)$$

$$C_\beta = 0.1$$

$$\ln \beta = 0.62 - 0.87 \ln(5.9 \times 10^{-4} \text{ cm})$$

$$+ 5.21 \ln(0.1) + 1.05(\ln 0.1)^2 \quad (15.61)$$

$$\ln \beta = 0.659$$

$$\beta = 1.93 \quad (15.62)$$

Cyclone efficiency can now be found for the Stairmand design using Eq. (15.37) for particles of any size, d.

$$\eta_d = \frac{1}{1 + [(5.9 \times 10^{-6} \text{ m})/d]^{1.93}} \quad (15.63)$$

The relationship between particle diameter and collection efficiency for the Stairmand design given by Eq. (15.37) is plotted in Figure 15.14. The overall efficiency for this cyclone on particles with the distribution given in Figure 15.13 is determined through calculations shown previously in Table 15.4. Overall efficiency was found to be 81% for this design; outlet dust concentration found with Eq. (15.51) then will be 9.4 g/m^3. Although this concentration meets the minimum requirements, it is higher than the target efficiency. Therefore, the number of cyclones is increased until the target efficiency is reached. The calculations performed above for the

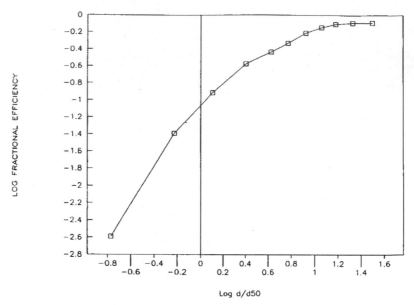

Figure 15.14. Fractional efficiency versus d/d_{50} for Stairmand design cyclone.

two-cyclone system are repeated for a three- and four-cyclone system. The four-cyclone system reaches the goal of 8 g/m³ and 84% control. The results of the calculations for a four-cyclone system are also shown in Table 15.7. The four-cyclone system also meets the design objective in terms of pressure drop.

15.4.3.2 Example Using Customized Design

As in the standard procedure, the system will initially consist of two cyclones operating in parallel. The diameter of each cyclone is found with Eq. (15.49) and cyclone flow calculated with Eq. (15.50).

$$D_{\mathrm{m}} = \frac{[(2500 \text{ kg/m}^3)(20{,}000 \text{ m}^3/\text{h})]^{1/3}}{275}$$

$$D_{\mathrm{m}} = 1.34 \text{ m} \qquad (15.64)$$

Next a plot of design parameter K versus collection efficiency is used to find the optimized cyclone design for the situation. Figure 15.15 shows the plot of design parameter, K, of 1.5, 3, and 4.4 versus collection efficiency for the particle distribution given in Table

15.13 and the two-cyclone system. The method to calculate overall collection efficiency for this system is analogous to the calculations shown in Eqs. (15.53) through (15.63) for the Stairmand design.

From the plot in Figure 15.15 a K value of 3.4 is found that corresponds closest to 84% collection efficiency, the target efficiency of the system. The optimized design dimensions corresponding to this K value and cyclone diameter calculated from Eq. (15.49) are shown in Table 15.7.

The pressure drop for the optimized design is calculated from a ΔH value of 12.5 given in Table 15.6 for the optimized design ($K = 3.4$) using Eq. (15.4). Since the efficiency and pressure drop of the optimized design meet the design objectives, only two cyclones will be necessary.

15.4.4 Other Design Considerations

After the shape and size of a cyclone or cyclone system have been decided, it is important to consider the additional design criteria that will ensure long, trouble-free operation of the system.

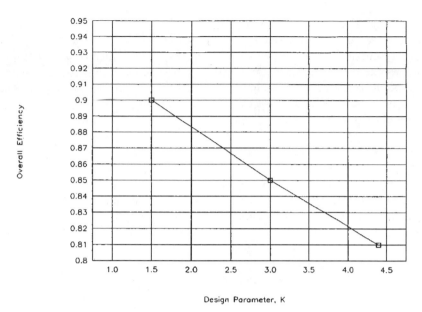

Figure 15.15. Design parameter, K, versus collection efficiency for example design problem.

When small cyclones, less than 300 mm or so in diameter, are used wall erosion may pose a serious problem[8] if the dust is abrasive. Larger dust particles strike the cyclone wall more forcefully and have more effect than smaller particles. Abrasion can be especially troublesome around welded seams, and occurs whether the seams are horizontal or vertical. The seam itself may not be as susceptible as the cyclone wall around the seam, which may have been softened through annealing during the welding process.[8] To minimize the effect of wall erosion, several steps can be taken. Often, commercially available small-diameter cyclones are cast rather than fabricated from sheet metal. Casting eliminates the problem with erosion around weldments, and may provide a thicker wall. Replacement wear plates are sometimes installed on the cyclone wall opposite to the tangential gas inlet. When installing a wear plate, it is important that the plate be fitted to maintain a smooth interior wall. Failure to maintain a smooth wall will hasten the erosion of the wear plate or the wall around the edges of the plate, and may also adversely affect cyclone efficiency. Wear plates and entire cyclone interior walls have been rubber coated to reduce erosion.

It is essential that air not be allowed to leak into the cyclone through the dust outlet. Leakage at this point can keep the cyclone from discharging dust to the dust bin, and if sufficiently severe, can lower collection efficiency to zero.[1] Leakage through the dust exit can even occur with the cyclone on the pressure side of a fan owing to the low static pressure in the cyclone central core, although the problem is more pronounced if the cyclone is on the suction side. When possible, cyclones are mounted on the upstream or suction side of a fan to minimize wear of the fan impeller from the dust in an uncleaned gas stream.

If the cyclone operates on the effluent from a batch process, the unit is often directly connected to a dust bin below the dust exit without an intervening valve. When using this arrangement, it is essential that the dust bin be emptied before it fills and blocks the cyclone dust exit. In this case, the gas flowing to the cyclone must be diverted before the dust bin can be emptied. The dust bin must be airtight to prevent leakage from the bin entering the cyclone dust exit.

A better solution to dust exit sealing is through use of a valve between the exit and dust bin. The valve must allow for the continuous discharge of collected dust, but not permit backflow of air. A valved dust discharge arrangement is essential for a continuous process as it allows the collection bin to be emptied at any time. Rotary values are often used[8] when the negative pressure at the dust exit is less than about 1 kPa (100 mm of water column).

Cyclone gas inlet velocities are frequently of the order of 15 m/s while duct velocities are usually lower than this. To minimize pressure drop through the cyclone system it is important to provide a good transition between the inlet ductwork and the cyclone inlet.

Attempts have been made to regain some of the rotational energy in the outlet gas stream by modifying the shape of the gas outlet. A thorough review of these devices is provided by Stern et al.[18] Reverse scrolls mounted above the gas outlet duct, and curved or straight vanes within the gas outlet duct have been used. These devices usually provide a reduction in pressure drop in the 10% range, but despite careful design, collection efficiency may be adversely affected. Pressure recovery devices are not generally used.

Build-up of collected dust on the cyclone walls can be a problem, especially where soft small-diameter, hygroscopic particles are collected. When build-up occurs, it can sometimes be scoured out by feeding some large-diameter, hard particles as an abrasive. Wall deposition of hygroscopic dusts is aggravated by condensation of moisture from the gas stream on the cyclone walls, when the cyclone is mounted outdoors in winter. If the problem occurs only on start-up, preheating the cyclone by either warming the inlet gas stream or running gas through without dust may help. Cyclones whose inlet walls are smooth and that operate at inlet velocities in excess of 15 m/s will be less prone to wall build-up.

LIST OF SYMBOLS

a	Gas entry height
A	Inside surface area of cyclone
b	Gas entry width
B	Dust outlet diameter
c_i	Inlet dust concentration
c_0	Outlet dust concentration
C_I	Cyclone geometry parameter, Iozia (Jones) and Leith
C_L	Cyclone geometry parameter, Leith and Licht
C_β	Logistic cyclone geometry parameter
d	Particle diameter
d_c	Diameter of cyclone core
d_{50}	Cut particle diameter, theoretically collected with 50% efficiency
d_{50}-cm	Cut diameter, in centimeters
d_{100}	Critical particle diameter, theoretically collected with 100% efficiency
D	Cyclone cylinder diameter
D_m	Cyclone cylinder diameter, meters
De	Gas outlet diameter
f	Factor in Eq. (15.8)
F_c	Centrifugal force acting on particle
F_d	Drag force acting on particle
g	Acceleration of gravity, 9.81 m/s^2
G	Dust cumulative size distribution
G^*	Friction factor, 0.005
h	Cyclone cylinder height
H	Cyclone overall height
K	Optimum design parameter
n	Vortex exponent
N	Number of turns gas makes within cyclone
N_c	Number of cyclones
Q	Volumetric gas flowrate
$Q_{cyclone}$	Flow going to one cyclone
$Q_{overall}$	Total flow of the system
r	Radial distance from cyclone axis
r_{core}	Radial distance from cyclone axis to edge of central core
r_i	Radial distance from cyclone axis to innermost particle at entry
r_w	Radial distance from cyclone axis to cyclone wall, $D/2$
S	Gas outlet height
t	Time
T	Absolute temperature, K
u	Particle velocity

u_r	Radial component of particle velocity dr/dt
u_t	Tangential component of particle velocity
v	Gas velocity
v_i	Gas inlet velocity, Q/ab
v_{is}	Gas inlet velocity above which saltation occurs
v_{max}	Maximum gas tangential velocity
v_r	Radial component of gas velocity
v_t	Tangential component of gas velocity
v_{tw}	Gas tangential velocity at cyclone outer wall
z_c	Length of the core
α^*	Factor in Eq. (15.16)
β	Logistic slope parameter
ϵ	Loss factor
η_d	Fractional collection efficiency of particles of one size, d
$\eta_{overall}$	Overall collection efficiency for polydisperse dust
ΔH	Pressure drop expressed as number of inlet velocity heads
ΔP	Pressure drop expressed as static pressure head
θ	Ratio of maximum tangential gas velocity to gas velocity in gas outlet, see Eq. (15.15)
λ	Friction factor, 0.02
μ	Gas viscosity
ρ_G	Gas density
ρ_p	Particle density
ϕ	Ratio of maximum tangential gas velocity to velocity within gas entry
ψ	Inertia parameter

REFERENCES

1. R. Jackson, *Mechanical Equipment for Removing Grit and Dust from Gases*, Cheney and Sons, Banbury, England (1963).
2. W. Barth, *Staub 21*:382 (1961).
3. J. I. T. Stenhouse and M. Trow, in *Proceedings of Second World Filtration Congress*, 1 Katharine St., Croydon CR9 1LB, England (1979).
4. C. J. Stairmand, *Trans. Inst. Chem. Eng. 29*:356 (1951).
5. P. Swift, *Steam and Heating Engineer 38*:453 (1969).
6. C. Lapple, *Chem. Eng. 58*:144 (1951).
7. R. H. Perry and C. H. Chilton, *Chemical Engineer's Handbook*, 5th edit., McGraw-Hill, New York (1973).
8. H. J. van Ebbenhorst Tengbergen, *De Ingenieur*. 77th Year of Publication, W1 (1965).
9. H. J. van Ebbenhorst Tengbergen, *Staub 25*:44 (1965).
10. W. A. Baxter, in *Source Control by Centrifugal Force and Gravity*. K. J. Caplan, in *Air Pollution*, Vol. 3, 2nd edit., edited by A. C. Stern, Academic, New York (1968).
11. L. C. Whiton, *Chem. Met. Eng. 39*:150 (1932).
12. L. W. Briggs, *Trans. Am. Inst. Chem. Eng. 42*:511 (1946).
13. C. B. Shepherd and C. E. Lapple, *Ind. Eng. Chem. 31*:972 (1939).
14. K. Iinoya, *Mem. Fac. Eng. Nagoya Univ. 5* (Sept. 1953).
15. E. Anderson, *Chem. Met. Eng. 40*:525 (1933).
16. M. A. Lissman, *Chem. Met. Eng. 37*:630 (1930).
17. C. E. Lapple, *Amer. Ind. Hyg. Assoc. Quart. 11*:40 (1950).
18. A. C. Stern, K. J. Caplan, and P. D. Bush, *Cyclone Dust Collectors*, American Petroleum Institute, New York (1956).
19. M. Seillan, *Chal. Ind. 10*:233 (1929).
20. P. Rosin, E. Rammler, and E. Intelmann, *V.D.I. (Ver. Deut. Ing.) Z. 76*:433 (1932).
21. F. Procket, *Glasers Ann. 107*:43 (1930).
22. A. J. ter Linden, *Proc. Inst. Mech. Engrs. (London) 160*:233 (1949).
23. M. W. First, Sc.D. thesis. Harvard University, Boston (1950).
24. R. McK. Alexander, *Proc. Australas, Inst. Mining Met. N.S. 152–153*:203 (1949).
25. W. Barth, *Brennst.-Waerme-Kraft 8*:1 (1956).
26. D. L. Iozia (Jones) and D. Leith, *Aerosol Sci. Technol. 10*:491 (1989).
27. K. J. Caplan, in *Air Pollution*, Vol. 4, 3rd edit., edited by A. C. Stern, Academic Press, New York (1977).
28. C. B. Shepherd and C. E. Lapple, *Ing. Eng. Chem. 32*:1246 (1940).
29. C. J. Stairmand, *Engineering (London) 168*:409 (1949).
30. D. Leith and D. Mehta, *Atmos. Environ. 7*:527 (1973).
31. S. K. Friedlander, L. Silverman, P. Drinker, and M. W. First, *Handbook on Air Cleaning*. U.S.A. E.C., AECD-3361, NYO-1572, Washington (1952).
32. C. E. Lapple and C. B. Shepherd, *Ind. Eng. Chem. 32*:605 (1940).
33. C. N. Davies, *Proc. Inst. Mech. Engrs. (London) 10*:185 (1952).

34. D. Leith and W. Licht, *A.I.Ch.E. Symposium Scr.* *68*:196 (1972).

35. D. L. Iozia (Jones) and D. Leith, *Aerosol Sc. and Technol. 12*:598 (1990).

36. J. A. Dirgo and D. Leith, *Filtration Separation 22*:119 (1985).

37. P. W. Dietz, *Assoc. Ind. Chem. Eng. J. 27*:288 (1981).

38. N. A. Fuchs, *The Mechanics of Aerosols*, Pergamon, New York (1964).

39. T. Chan and M. Lippman, *Environ. Sci. Technol. 11*:377 (1977).

40. W. Licht, T. Chan, and M. Lippman, *Environ. Sci. Technol. 11*:1021 (1977).

41. W. B. Smith, D. L. Iozia, and D. B. Harris, *J. Aerosol Sci. 14*:402 (1983).

42. W. B. Smith, R. R. Wilson, D. B. Harris, *Environ. Sci. Technol. 13* (1979).

43. B. Kalen and F. A. Zenz, *A.I.Ch.E. Symposium Ser. 70*(137):388 (1974).

44. W. Koch and W. Licht, *Chem. Eng. 84*:80 (Nov. 4, 1977).

45. Midwest Research Institute. *Handbook of Emissions, Effluents and Control Practices for Stationary Particulate Pollution Sources*. NAPCA contract CPA 22-69-104, NTIS Publication No. PB 203-522, Springfield, VA (1970).

46. U.S. Environmental Protection Agency. *Compilation of Air Pollution Emission Factors*, 2nd edit., Publication No. AP 42 (April 1973).

47. J. A. Danielson, *Air Pollution Engineering Manual*, 2nd edit., EPA. Publication No. AP 40 (May 1973).

48. D. L. Iozia (Jones) and D. Leith, *Filtration Separation 24*:272 (1989).

16

The Electrostatic Precipitator: Application and Concepts

Jacob Katz

CONTENTS

16.1 INTRODUCTION

16.1.1 General Comments

The electrostatic precipitator uses electrical forces to capture either liquid or solid particulate matter from a flue gas system. We tend to classify the precipitator as a high-efficiency collector, comparable to the fabric bag-house or high-pressure venturi scrubber. As such, collection efficiencies of 99.5% plus are within a design range for most applications. A prime characteristic separating the various methods for high-efficiency collection is that the precipitator concentrates its primary energy forces on the particle, rather than on the carrier gas stream. However, the gas stream or process characteristics will generally determine whether the particle will be easily collected or prove difficult to contain by electrical forces.

An interesting facet of precipitation is that even after many years of application, simple fundamental knowledge has eluded personnel involved with this equipment. That is why we continually have to face the same field problems and why performance of the precipitators often varies greatly from its design criteria. This lack of understanding cannot be overcome with more theoretical coverage, but rather there is a need for practical concepts and field information to be clearly identified and distributed.

For that reason, this chapter attempts to provide a brief description for some of the key areas of precipitation without regard for detailed theory. It actually consists of excerpts from the book *The Art of Electrostatic Precipitation* written for the practitioner. The bibliography at the end of this chapter also includes sources of literature that can be used to upgrade the theoretical knowledge of the precipitator.

16.1.2 Performance Comments

Each industrial process presents its own specific problems or application factors when precipitators are considered. Subtle changes in the process, of raw materials and equipment, can often produce a wide band of performance characteristics in a specific precipitator. These subtle changes probably produce the greatest number of installations that fail to meet expected collection efficiencies because of reduced electrical power input. A secondary contribution to subperformance levels is a combination of design shortcomings including, among others, gas distribution, gas sneakage, and insufficient sectionalization. A third group of factors that keeps the collector from performing consistently on a satisfactory basis is the balance of reliability, either in the evacuation of material from hoppers or in the failure of precipitator components.

16.1.3 Applications

The use of precipitators has been applied in all the basic as well as some exotic industries over the years. Collection of particulate matter in a dry-type precipitator with flue gas temperatures between 250 to 700°F has been the most popular application. However, specific process characteristics will usually determine the design and type of precipitator utilized. There are process situations where the effectiveness of the electrical collector is questionable because the material is difficult to

handle. Some typical industrial processes that have successfully employed the precipitator include:

PROCESSES	PRINCIPAL MATERIAL COLLECTED
Utility	Fly ash (SiO_2, Al_2O_3, Fe_2O_3)
Industrial boiler houses	Fly ash
Oxygen steelmaking furnaces	Iron oxide (Fe_3O_4)
Cement kilns	Calcium oxide, silicon oxide
Pulp and paper	Sodium sulfate

A large number of precipitators have also cleaned process gases from blast furnaces, sinter plants, open hearths, coke ovens, gypsum plants, catalytic cracking, smelters, sulfuric acid, and phosphoric acid plants. Other processes that use precipitators include electric-arc furnaces, scarfing machines, incinerators, and the carbon black industry. Other precipitators have also recovered valuable metals in special situations.

16.1.4 Precipitator Arrangements

A convenient and logical method of comparing precipitators is the total collecting surface area and the amount of mechanical and electrical sectionalization. The basic terminology used to describe sectionalization follows and an illustration is shown in Figure 16.1.

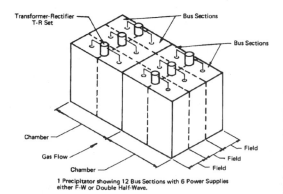

1 Precipitator showing 12 Bus Sections with 6 Power Supplies either F-W or Double Half-Wave.

Figure 16.1. Typical precipitator arrangements showing terminology and method of applying power supplies.

Precipitator. A single precipitator is an arrangement of collecting surfaces and discharge electrodes contained within an independent housing.

Bus Section. The smallest portion of the precipitator that can be independently deenergized.

High-Voltage Power Supply. The power supply unit to produce the high voltage required for precipitation, consisting of a transformer–rectifier combination and assorted controls. Numerous bus sections can be energized by one power supply.

Field. A field of a precipitator is an arrangement of bus sections in the direction of gas flow that is energized by one or more power supplies situated laterally across the gas flow.

Collecting Surfaces. The individually ground components that make up the collecting system and that collectively provide the total area of the precipitator for the deposition of particulate.

Collecting Surface Area. The total flat projected area of collecting surface exposed to the electrostatic field (effective length × effective height × number of sides).

Effective Height. Total height of collecting surface measured from top to bottom.

Effective Width. Total number of gas passages multiplied by the center to center spacing of the collecting surfaces. (Disregard shape of collection surface.)

Effective Cross-Sectional Area. Effective width times effective height.

Gas Passage. Formed by two adjacent rows of collecting surfaces.

Discharge Electrode. The component that is installed in the high-voltage system to provide the function of ionizing the gas and creating the electric field.

Collecting Surface Rapper. A device for imparting vibration or shock to the collecting surface to dislodge the deposited particulate.

Aspect Ratio. The length of the precipitator divided by its height.

16.1.5 Basic Concepts of Precipitation

Probably the best way to gain an insight into the process of precipitation is to study a relationship generally known as the Deutsch–Anderson equation. This equation and adaptations of it are well covered in several books.[1,2] It describes the factors involved in the collection efficiency of the precipitator as shown in its simplest form:

Collection efficiency

$$N = 1 - e^{-(A/V)W}$$

where

A = effective collecting electrode area of the precipitator (m^2)
V = gas flow rate through the precipitator (acm/s)
W = migration velocity (m/s).

This equation has been used extensively in the above form in past years. Unfortunately, while the relationship is scientifically valid, there are a number of operating parameters that can cause the exponent to be in error by as much as a factor of two or more. It is well to remember that the basic D-A equation can be used as an indicator or tool, but has limitations more often than not unless equated with some practical and empirical considerations by the designer. Values used can either be in the English or metric systems.

The exponent term W, known as the migration velocity, actually represents the speed of movement of the particle toward the collector surface under the influence of an electrical field. While we would consider it more an indicator than actual velocity, it does have a finite value that can be used for comparison purposes. This migration velocity is comprised of:

$$W = \frac{aE_0 E_\rho}{2\pi\theta}$$

where

a = particle radius, microns

E_0 = strength of field in which particles are charged, statvolts/cm (represented by the peak voltage)

E_ρ = strength of field in which particles are collected, statvolts/cm (normally the field close to the collecting plates)

θ = viscosity of frictional resistance coefficient of the gas.

High levels of voltage and useful corona power in the precipitator, all other conditions being equal, are synonymous with high collection efficiencies. Figure 16.2 shows a typical performance curve of the effect on efficiency by changes in the peak voltage of a precipitator. This simple curve can represent only one situation because each precipitator will have its own characteristic curve based on many factors. The important point to remember is

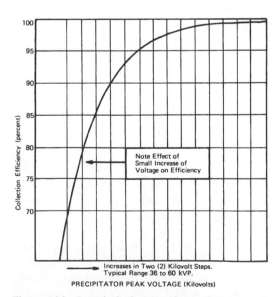

Figure 16.2. A typical electrostatic precipitator peak voltage versus dust collection efficiency curve shows how efficiency increases with voltage.

that small changes can produce substantial changes in power, and hence in the efficiency of the collector. This is especially true at the lower levels of power input. It is therefore important to understand the factors that affect the electrical characteristics of the precipitator.

16.1.6 Main Factors Affecting Electrical Characteristics

Optimum power input to the precipitator varies among processes and even changes on a minute to minute basis for certain applications. There are seven basic factors that directly affect the electrical characteristics. These are:

1. Design of power supply
2. Physical design of precipitator
3. Design of electrode system
4. Characteristics of gas stream
5. Effect of process changes
6. Characteristics of particulate
7. Maintenance factors.

The power supply must be matched correctly for the precipitator section or service expected, or several difficulties can arise:

1. The impedance of the power supply, including a ballast resistance or reactor in the primary winding of the transformer, may not be sufficient to dampen the severity of electrical breakdowns in the precipitator. This condition is especially likely if the power supply rating is much larger than the actual operating level.
2. If the physical size of the precipitator is too large compared to the size of the power supply, then lower than desirable precipitator voltages may exist because the current rating of the supply becomes the limiting factor.
3. The gas and particulate matter conditions can drastically alter the voltage–current relationship and produce lower voltage fields than expected because a small power supply becomes current limited.

The average precipitator can be sensitive to process changes in the following ways:

1. Changes to gas temperature (effect on density)
2. Changes in gas pressure (effect on density)
3. Changes in gas flow rate
4. Changes in gaseous composition
5. Changes in particulate chemical characteristics
6. Changes in particulate concentration or loading
7. Changes in the size distribution of the particulate
8. Changes in the electrical conducting characteristics of the particulate.

It is difficult to separate the effect of one process change on another. If the rate of process change is rapid, the readings can change almost instantaneously. On the other hand, rapid changes of temperature may not be seen readily on the meters because of the heat sink effect of the precipitator. Some changes in the process will cause large variations of voltage-current readings, while others will cause subtle effects.

The size distribution of the particulate matter can have a bearing on electrical readings. For example, iron oxide fume from a basic oxygen vessel contains a predominance of submicron particles that will react like a space-charge in a vacuum tube. This can actually impede the flow of precipitator current and thereby elevate the voltage potential across the space. This condition can become serious enough to completely nullify the precipitator process depending on the electrode geometry and the concentration level of the submicron particles.

16.2 FACTORS AND EFFECTS

16.2.1 Gas Flow Rate

While the true measurement of the gas flow rate commands an important place in specifications and performance tests, in practice, it is often less critical than other factors. This comment is especially true with the larger designs that exist today. Even with the 1.8 m/s (6 ft/s) or more velocity designs, it is often the quality of gas that weighs most importantly.

Whether or not the relationship of higher gas flow rate to reduced efficiency becomes critical is dependent in large part on the characteristics of the particle. Certainly, large porous particles such as combustible grit found in fly ash applications will be sensitive to increased velocities. On the other hand, fine-sized particulate matter that tends to agglomerate in the deposited layer of the collecting surface will resist easy reentrainment into the gas stream. With low levels of power input and low aspect ratios, high gas flow rates can often be observed in reduced performances of the precipitator.

16.2.2 Gas Flow Distribution

Gas distribution problems are of concern from the standpoint of velocity, temperature, and concentration of material as well as particle size. If one area of precipitation has become worse in recent years, it is in gas distribution. The trend toward larger collectors has meant greater difficulty in transferring the gas leaving the inlet nozzle to an acceptable pattern at the face of the precipitator. Granted optimum gas distribution is not as critical in the larger units with all fields serviceable, but the margin can be quickly lost with outages of equipment.

Probably one fallacy in gas distribution is placing too much emphasis on the results of model studies. The model cannot foresee the fallout of material during periods of reduced operation that will often distort the actual flow pattern.

16.2.3 Gas Temperature

The level of gas temperature in the precipitator opens up many areas of interest, especially the effect on the viscosity of the gas stream. But the major effects of temperature lie in the modification of the electrical characteristics and the reactions of the particles as they deposit on surfaces. The effect on metal corrosion by changes in flue gas temperature must also be considered.

Practically all of the particulate matter handled in precipitators will go through a wide spectrum of electrical characteristics for the temperature range of 200 to 750°F. Much of this has to do with condensation effects and surface leakage at the lower range and conductivity changes in the bulk material at the higher temperatures. The true effects at any given temperature will depend on the moisture level and chemical composition of the particles. Of greater interest would be whether the precipitator is operating in critical temperature zones for that particulate material. For example:

1. High sulfur coal for pulverized coal-fired precipitators would be critical in the 250 to 280°F zone.
2. Lower sulfur coal for this same precipitator might find its most critical zone between 310 to 360°F.
3. Cement precipitators might find its most critical range in the 350 to 400°F.

A variation in electrical readings may occur with as little as 10 to 15°F movements in the process gases. In some fly ash installations a 15°F change has meant a three to fourfold increase in emissions.

The ability to change flue gas temperatures from critical zones is as important to successful precipitator performance as any other design feature. As with variations in gas flow, short-term variations in flue gas temperature should be controlled in order to minimize losses from the collector. In fact, it is usually better to operate at a less than optimum uniform temperature rather than experience variations. The heat sink effect of the internal structure will tend to mask effects of the temperature cycle if it is less than 10 minutes in duration.

16.2.4 Rappers and Reentrainment

Nowhere in the original Deutsch–Anderson equation is an allowance made for the losses that occur in transferring the collected material from electrode to hoppers. The interplay between the electrical forces holding the material on the collecting surface and the rapping device attempting to remove it provides a real challenge for effective precipitation. But this challenge does not merit the priority some people have placed on high rapping forces. This statement is valid as long as the rapper mechanism is sufficient to impart at least 10 to 25 Gs to the support structure holding a group of collector plates. With many process conditions, even a substandard rapper system will not effect performance adversely. But when the build-up on the collecting surface reaches over 1.9 cm ($\frac{3}{4}$ in.) it is prudent to assess whether the rapper system is sufficient. Great emphasis should be placed on the reliability of the rapper and control circuitry components. This has become more important as collection efficiency levels have increased.

The effect of rapping on precipitator performance is whether puffing losses are observed or measured since this can denote a significant reentrainment of material from the collector surfaces caused by the rapper operation. While the reentrainment puff is usually a mechanical occurrence, the operation of the rapping device can sometimes effect the electrical characteristics at the same time, thus aggravating the magnitude of the problem. The vibration of the high voltage frames could produce an electrical disturbance dependent on the structural integrity of the discharge electrode system.

16.2.5 Power Supply Characteristics

As precipitators have grown in size so have the power supplies grown in kva ratings. This trend to larger transformer–rectifier capacities has introduced some difficulties in stability, and yes, even in the performance of the precipitator if a gross mismatch occurs between the size of the power supply and the field to be energized.

It is well to understand a basic concept of precipitation that each field of any installation will only effectively absorb the amount of power that the existing gas, dust, and internal structure integrity allows. Therefore, the actual voltage-current requirements of a precipitator field may be drastically different than shown by the full load rating of the power supply.

16.2.6 Operation and Maintenance Factors

Many success stories of the high collection performance of precipitators are well documented. But to many users, a constant battle is waged to maintain these performance levels. A major reason for this situation lies in the basic design of components for the overall system that produces sensitivity for breakdowns. A concerted effort must be made by the user to understand all the inputs to the potential problems of maintenance.

Obviously, it is exceedingly difficult to predict where some of the maintenance troubles may occur, but there are eight key areas that can be emphasized:

1. Raw material and operation forecasts— original design
2. Design concepts
3. Construction phase
4. Initial check and training
5. Personnel assignment
6. Control of process
7. Record keeping
8. The actual maintenance program.

16.2.7 Construction Phase

The best precipitator design can be adversely affected in the fabricating and erection phases. Just how the quality of welding is controlled, or the shaping of the component is finally accomplished in the shop, could have a significant effect on the final operating characteristics of the precipitator.

How the material is handled on the job site can be important to prevent a tendency to distort long electrode elements. Weather protection is of primary concern if long storage time is required.

There are advantages for the user to assign an inspector during the actual erection phase. Cross-checks of the actual erection procedures are important.

16.2.8 Personnel Assignment

Just who to assign to the precipitator system should be given much thought. The value of the initial check-out and contacts with the manufacturer can be lost if the user representative is moved to another assignment.

A person who can be assigned long term to oversee the precipitator system and monitor the process as to how it affects the collector is probably the best investment a company can make. Recent years have shown the advantage of close supervision for the large precipitator installations.

16.3 RESISTIVITY

16.3.1 Introduction

Much emphasis has already been placed on the fact that effective precipitation coincides with the occurrence of optimum amounts of electrical power input in the corona process. While power input is sometimes limited by structure or individual component defects, the performance of limited power installations occurs under conditions of excessive-electrical resistivity of the collected material, usually expressed in ohm-centimeters.

All finely divided particles that are generated in the basic industrial processes have critical temperature zones that can affect the electrical operation of the precipitator. The chemical composition of the bulk of dust particles contain common constituents even if the make-up varies somewhat in weight fractions. Given similar gas conditions, it might be hard to differentiate electrically whether it was fly ash, cement dust, or iron oxide being handled in the collector. Of course, that is a simplistic statement because in practice there are an infinite number of process conditions where differences in raw material can alter the electrical characteristics of the particles.

Fortunately, high moisture contents in the flue gas stream (such as found in wet process cement applications or other applications where spray water is used to cool the gases) will usually nullify the subtle chemical particle composition and provide ample power inputs. You can call water vapor a primary conditioning agent that will control resistivity problems if the quantity of water used is effectively matched to the gas temperature levels of the flue gas entering the collector.

When moisture levels in the flue gas are low—usually below 10% by volume—the chemical make-up of the particle becomes a dominant factor in controlling electrical characteristics.

The classification of this characteristic of the particle is simply related to its ability to conduct or resist the passage of electric current. This ability is not critical for the individual particle as it drifts in the gas stream, but becomes important after it deposits on the collecting surface.

One of several power input limits can occur:

1. The voltage limit of the T–R set can be reached before any other limitation.
2. Either the primary or secondary winding current limit could be reached.
3. Or spark-over can occur within the field limiting the available power from the T–R set.

These three limits should be well understood. First, the voltage limit is rarely observed on normal precipitator applications, but it can occur with an appreciable mismatch of the T–R set to the load requirement. That is, it can occur when a large capacity of supply is connected to a small surface area field and is combined with high concentrations of finer sized particles.

Second, the current limit is observed more often. Adjusting the conduction angle of the secondary current to approximately 86% by the correct application of linear reactors will generally produce a match of the primary and secondary currents. As the conduction angle decreases from 86%, the primary current will trend toward higher readings relative to the secondary current reading. The rated secondary current will sometimes be achieved before the primary current limit if conduction angles rise past the 86% point when higher levels of impedance are applied in the primary circuit.

This brings us to the third limit, which is a spark-over between the discharge electrode and collecting surface. When this occurs, the power supply voltage must be reduced to keep the breakdowns within a reasonable level. This level could range from a nominal 150 sparks/min for the inlet field to the occasional spark for the outlet field.

However, the designer predicated his precipitator performance on a power parameter that now may not be attained because of the limitation imposed by spark-over. Basically, precipitation spark-over can occur by one of two mechanisms:

1. The impressed voltage is greater than the spacing and the physical contour or conditions between electrode surfaces will allow, regardless of the characteristic of the particulate matter. This condition is often observed with electrode misalignment where one or more of the discharge electrodes has moved too close to the collecting surface.

Another important factor that can cause premature breakdowns is the presence of severe discontinuities on the collector surface opposite the corona-emitting zones of the discharge electrode. This type of breakdown tends to provide a greater electrical disturbance compared to the spark-over caused by high resistivity.

2. Spark-over caused by high resistivity levels is the most common reason for low power inputs to the precipitator. The resistance of the layer of collected material on the collect-

ing surface is the prime reason spark-over will occur. This layer will develop a voltage drop based on three factors: the resistivity value × the layer thickness × the current density. If the voltage drop is greater than the dust layer can withstand, then breakdown within the layer occurs.

This phenomenon is not unlike the breakdown of a capacitor. Reduction of the layer resistivity can be achieved by a number of methods including flue gas additives and process modifications. Sufficient reduction of the layer thickness is often difficult to obtain because as resistivity increases, so does the tenacity of the particles to stick together and adhere to the collecting surface.

The third component of the voltage drop is the current density. This means that the amount of corona current attempting to pass through the dust layer must be reduced if either the resistivity value or layer thickness increases. Otherwise, spark-over can occur. For example, 0.43 ma/m^2 (40 ma/1000 ft^2) may occur with a material resistivity of 10^{10} ohm-cm. But if the actual resistivity was 10^{11} ohm-cm the current density might have to decrease to 0.27 ma/m^2 (25 ma/1000 ft^2) to keep spark-over at a reasonable level. In other words, the higher the resistivity level, the lower the current density must be to keep the precipitation process functional. It is not uncommon to see current densities below 0.054 ma/m^2 (5 ma/1000 ft^2) on certain fly ash applications.

16.3.2 Effects of Resistivity on Power Levels

What this means in a practical sense is that an infinite number of voltage and current readings can occur in the precipitator that will not in any way match the name-plate data of the T–R sets. The important thing to remember is that higher resistivity conditions will decrease power inputs because of the spark-over limitation. Superimposed on a resistivity problem is the possible condition of the internal structure causing the spark-over to occur at a much lower power level. Once the dust resistivity reaches a critical point, its deposition on a sharp edge, or for that matter any kind of discontinuity on the collecting surface, will cause a localized electrical stress build-up point that will draw the spark. This is why uniformity of alignment and elimination of all internal irregularities becomes more important as the resistivity moves up from the moderate range.

It is difficult to specify various resistivity levels as denoting good or bad operation. The poor physical design of the precipitator components from a high-voltage standpoint can alter spark-over levels. It is advantageous to group resistivity into three basic zones; low, moderate, and high. The moderate range would generally encompass a resistivity from 10^9 to 10^{11} ohm-cm and is considered the best zone for effective precipitation. A finer grouping might show the following:

COMMENTS	RESISTIVITY RANGE
10^4 to 10^7 ohm-cm	Usually high conductive material—hard to retain— low-voltage fields present.
10^8 to 10^9 ohm-cm	Sensitive stage where lack of resistive characteristics can sometimes hurt— especially in fly ash cases.
10^{10} to 10^{11} ohm-cm	Appears to be the best range to shoot for— should show some spark-over in precipitator.
10^{12} to 10^{13} ohm-cm	Range usually associated with low sulfur coals— reduced power in all fields can exist.
Over 10^{13} ohm-cm	Not commonly observed in basic industries with normal moisture contents. Can produce severe electrical disturbances.

The description of spark-over can be defined as an electrical breakdown through an isolated gas path between the negative and positive electrodes. The case of the threshold resistive spark-over where the discharge streamers occur from the deposited dust layer is considered the start of back corona. This situation is not unlike that which occurs in atmospheric lightning when positive streamers from earth actually draw the localized stroke from the negatively charged clouds. But in the case of very high resistivity, a severe back-corona condition can occur characterized by greatly reduced voltage and high current densities without spark-over.

When low resistivity exists with low-voltage conditions, it is difficult to achieve high collection because of dust reentrainment losses. Power consumption is high because of the high current flow through the dust layer caused by a practically nil electrical resistance. Internal inspections usually show collecting surfaces devoid of any buildup. During this low resistivity, it will be difficult to achieve the guaranteed efficiency at even half the design gas velocity.

Moderate resistivity will allow dust particles to bond together in the dust layer by forming a charged dipole relationship not unlike those found in a magnet. The opposite polarities provide good adhesiveness at the tangent contact points of adjacent particles and even aid in holding these particles together as they become dislodged by the rapping procedure. Reentrainment losses are at a minimum level.

When the resistivity increases into the high range, the bonding can become severe and layer dislodgment by rapping is difficult. However, at some point the reentrainment portion of the total precipitator losses can be higher than in the moderate range and are caused by the reduced power levels. A sizeable part of the precipitator dust loss in high-resistivity cases can also occur during high spark-over conditions because each spark blows out a small volume of dust from the layer. The Deutsch equation showed that the migration velocity contained the product of two precipitator voltage gradients without regard to the corona current component. The magnitude of the current flow will partially depend on the resistivity conditions for any given voltage. For a fly ash example showing 280 V on a primary voltmeter, a current may indicate 750 ma with low resistivity, 400 ma in the moderate range, or 150 ma in a higher resistivity range. In this example, the 400 ma condition would probably provide the better collection performance on a higher velocity precipitator. It is always advantageous to work toward higher voltage gradients and take whatever corona current results. The only exceptions would be current suppression caused by discharge electrode build-up or excessive space charge caused by high concentration levels of fine-sized particles.

One important concept is that each process will produce a particulate matter whose resistivity will usually decrease rapidly on the low temperature side of a peak, while decreasing at a lesser rate on the high temperature side. Figure 16.3 shows a typical plot of resistivity obtained by laboratory analysis for dust entering a cement precipitator. A typical fly ash from an eastern bituminous coal source with

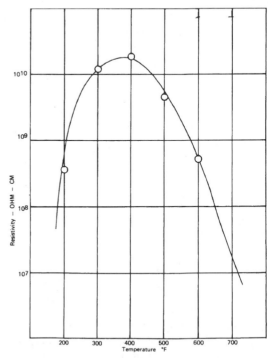

Figure 16.3 Resistivity curve for dust at inlet to cement precipitator: Lab measurement at 4000 V and 25% H_2O.

2% to 3% sulfur might show a similar pattern at approximately 6% moisture by volume.

16.4 OPERATION AND MAINTENANCE

This section intends to provide many practical items and advice for personnel who continually work with the precipitator. All precipitators have common trouble areas, yet all precipitator designs will perform well given certain operating conditions. However, it is the recurring trouble areas that nullify good performance and often cause costly production losses because of the outages required to correct these difficulties.

16.4.1 Internal Inspection

The internal inspection of the precipitator is an important operating and maintenance tool that can provide many benefits if done thoroughly. This is where a person who is trained to do the job on a periodic basis can build up a knowledge of observations and data that will help define causes of difficulties and even catch areas of impending trouble. The internal condition of the small precipitator with only one field may require this careful inspection more than the multifield unit. Internal defects in any one field of the bigger collectors can often be detected electrically during operation, while it might be difficult in a small unit to ascertain whether a change in the voltage-current characteristics is due to the process or to an internal defect.

16.4.2 Alignment of Electrode Systems

The net effect on the electrical readings by the characteristics of the gas and particulate matter will often be contingent on the proper spacing or alignment between the electrodes. Meter readings may indicate a resistivity problem, while close spacing or even a specific electrode design may be causing the spark-sensitive precipitator. There is no substitute for careful measurements and inspections.

The degree of misalignment allowable in a precipitator is dependent on a number of factors. In highly resistive conditions, deviations of 6 mm ($\frac{1}{4}$ in.) in some discharge electrodes from the center of a passage could provide sensitive electrical breakdowns at slightly reduced voltages. Other installations can stand electrode misalignment of 25 mm (1 in.) and more, and the effect on performance cannot be easily observed. Nevertheless, significant misalignment between the high voltage and collecting electrodes, whether with weighted wire or rigid frame, should be cause for alarm.

16.4.3 Particulate Removal from Hoppers

Difficulties with the evacuation of material from hoppers can provide a continuous headache for plant personnel. The large-sized precipitator can have a good potential performance nullified by hopper troubles. This does not have to occur. Unfortunately, in the quest to produce high-performance units in order to meet stringent regulations, the emphasis on good hopper design was less than the concerns with SCA or power input. Hopper difficulties are often fought for years because key modifications are not applied, especially to correct the underlying causes. This lack of commitment to eliminate hopper troubles has produced a sad reputation for precipitators. The problems are many, the solutions often hard to come by, but do not hesitate to expend time and money to obtain a satisfactory performance of the hopper system.

Process characteristics, the type of material being handled, and moisture and gas temperature levels are important inputs in understanding the reasons for hopper problems. Without a reliable evacuation system from the hopper flange to the storage area, the burden on the hopper components may be too much. Based on many field experiences, here are several main factors.

16.4.4 Insufficient Heat and Insulation

Lack of proper insulation and heat input, especially at the bottom apex of the hopper, has probably accounted for most of the troubles experienced. Regardless of the gas and dust characteristics, the ability to keep the wall surface temperature of the lower hopper no less than 250°F is most important.

16.4.5 Air Inleakage into Hoppers

Entry of outside air into any part of the hopper system is considered poor practice. Aside from the effect on performance, excessive air can cool down the inside wall surfaces, or condense moisture in some of the high water-vapor installations. Unfortunately, most of the screw conveyor installations coupled with process precipitators are conducive to this condition.

16.4.6 Level Detectors

While it is always better to place time and money in the prevention of hopper difficulties, the detection of build-ups by some method is desirable in most applications. These devices can utilize gamma radiation, sound, capacitance, pressure differential, temperature, or even paddle-wheel methodology for the detection of excessive build-up. Any method that does not require components within the hopper appears the most desirable.

Several comments: Use detectors as maintenance tools rather than to identify full hoppers. For example, if an automatic batch cycle allows a maximum 90 cm (3 ft) of build-up in the inlet hoppers, it is well to locate the detectors no more than 150 cm (5 ft) above the apex flange. The object is to alert the operator before a major hopper fill-up exists, yet minimize frequent detector alarms. A rule of thumb would allow the lapsed time from normal dust height to alarm level to equal the same length of time it takes the dust to rise from the apex to the normal height. Remember that the pyramid design allows for a greater volume of material to accumulate in each foot of hopper height. One problem arises in the uneven build-up that occurs by the slope and corner

effect of the hoppers. The detector should recognize this effect and be placed a little lower at the center of the wall or a little higher near the corners.

16.4.7 Outage Clean-Down of Electrodes

If good shut-down procedures are followed on most installations, the degree of build-up on electrode surfaces will usually require no further cleaning during the outage. That does not mean than 6 to 10 mm ($\frac{1}{4}$ to $\frac{3}{8}$ in.) mounds of deposit will not exist, but the build-up will be spotty with most of the surface holding less than 3–6 mm ($\frac{1}{8}$ to $\frac{1}{4}$ in.) thick mounds. There are exceptions, especially caused with high-resistivity materials or other operating characteristics.

Whether any manual cleaning is implemented during the outage depends on several factors. If it is an annual outage with certain planned work on the electrode system, then a water wash of the unit might be considered. It is not recommended to use this type of cleaning unless it is necessary to perform major work on the system. Depending on the time of year and thoroughness of the washing, some rusting and corrosion pockets can be accelerated.

16.4.8 Important Troubleshooting Approaches

Because of the high-voltage danger, familiarity with all the safety aspects of the system cannot be overstressed. Even portions of equipment inside each control cabinet will be at a 480 V potential, so care must be taken in any measurement procedure. The manufacturer's manual should be well studied for guidelines in the handling of certain control difficulties.

If any troubles occur initially with control circuit components, fuses, or any other low voltage trouble source, correct these problems post-haste, since the high-voltage portion of the precipitator tends to supply enough potential difficulties of its own.

16.4.9 Normal Versus Abnormal Power Characteristics

An understanding of the electrical readings of the precipitator must be a starting point in coping with this collector. It was already stressed that the name-plate electrical values will not be observed on many fields of the precipitator, so the patterns of meter readings become an important tool of evaluation.

The key word is "uniformity" of patterns of each precipitator because gross power values used to compare one unit to another sometimes provide questionable evaluation results. It is recommended that comparison of the voltage to current flow value of each field be ascertained under normal conditions as well as process variations.

By now, you should well understand that the control panel readings are a reflection of everything that is occurring in the precipitator. The magnitude as well as the trends of the readings will generally tell a story of normal precipitator performance as well as abnormal conditions. A sound knowledge of the effective voltage-current characteristics can allow one to judge emissions on reproducible process operations almost as well as actual stack tests.

The primary winding voltage and current readings should provide valid reflections on what is occurring in the secondary circuit of the high-voltage transformer. However, the presence of secondary as well as primary meters does provide added monitor capability. In all difficulties within the precipitator, the two voltmeters and the two ammeters will work in unison for specific characteristics. That is, when the primary voltage is low, the secondary voltage should also be low, while the ammeters could both be showing relatively high values (see Fig. 16.4).

Probably 80% to 90% of the problems that occur in precipitators will tend to reduce volt-

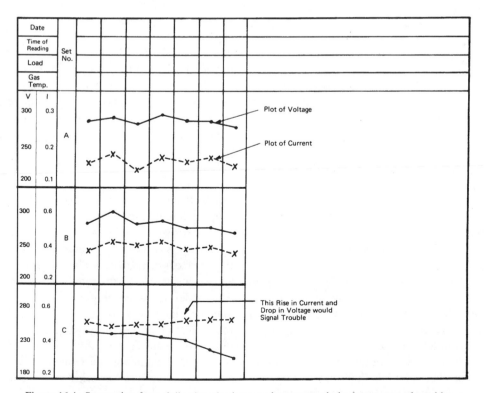

Figure 16.4. Suggestion for a daily plot of voltage and current to help detect start of troubles.

ages and raise currents at the same time. These problems are usually associated with a multitude of difficulties with electrode failure, dust build-ups in hoppers, and electrical leakage over insulator surfaces.

As each field of a multifield precipitator does its work, the reduction of suspended particulate matter in the flue gas will alter the voltage-current relationship from inlet to outlet. This phenomenon is best observed in the moderate resistivity range. A change in the resistivity of the material in each field can alter the patterns, but the slope of the pattern is mostly space charge oriented. For example:

	PRI. VOLTS	SEC. MA	SPARKS / MIN
Inlet	360	400	50
Center	330	550	20
Outlet	300	750	occ.

In other words, for identically sized fields, we are generally looking for a stepped decrease in voltage and stepped increase in current from inlet to outlet. Some key concepts in meter observations include:

1. Patterns in voltage and current readings from inlet to outlet should form some type of recognizable pattern unless internal defects cloud the issue.
2. The high-resistivity range can produce a relatively low flat voltage and current pattern, but only in the very high resistivity zones (10^{12} and above). Generally you should see an increasing pattern of current flow in the direction of gas flow.
3. The moderate range of resistivity would show the greatest magnitude of change from inlet to outlet.
4. As the resistivity becomes low enough so that all sparkover ceases from this cause, the voltage and current readings tend to flatten out again from inlet to outlet, but at a much higher power level.
5. It is only in the moderate to high resistivity ranges that internal electrode defects will show major distortions in the voltage-

current patterns between fields and adjacent cells.
6. Effects of resistivity can completely nulify the effects of the space charge on the inlet to outlet patterns.

16.4.10 Reentrainment

Some additional comments in the area of reentrainment of material are warranted. The effects of resistivity on reentrainment have already been mentioned, but the subject is much more complex. How far the material moves out into the gas system, where it redeposits in relation to its original position, and whether it changed its physical character are but some of the unknown factors in the reentrainment syndrome.

A common description of the dust layer sliding down the collecting surface does not usually occur. The shock or tremor imparted by the rapper appears to more often dislodge some percentage of material from its resting place. If the particles have had a chance to agglomerate, the adverse effects of reentrainment are minimized. It is when the particles bounce back into the gas stream in the same condition as they were collected that troubles begin to mount. This is where proper resistivity and the timing between raps can play an important part in this interesting phase of precipitation.

Some key concepts include:

1. Always use the internal inspection and other visual means to help ascertain the lowest rapping intensity possibly commensurate with other performance observations.

2. Always attempt to match the rapping to the dust characteristics or resistivity. For example, a low resistivity requires soft rapping, the moderate range requires a harder blow, and the high-resistivity zone means real trouble. Remember that hard rapping with high-resistivity materials usually exhibits limited success and changing the resistivity is usually a much better way to achieve a satisfactory performance.

3. Do not feel that the inlet field must be overrapped because it handles the bulk of material. Be aware that puffs out of the stack or other signs of reentrainment are not unique to the outlet fields. Actually, the material collected in the inlet field of some precipitators is often easier to dislodge, and excessive carry-over adds to the reentrainment potential of the following fields.

4. Rapping loss is not usually uniform across the precipitator, even discounting the effects of resistivity gradients. When reentrainment losses are observed, investigations into possible problems with gas distribution is highly recommended.

5. When reentrainment losses are severe, lengthening the time between raps on the collecting surfaces in the direction of inlet to outlet is usually recommended. As a first adjustment it might be advisable to double the rap time on succeeding fields. For example, if the inlet field rapped every 5 min, then the second field would be rapped every 10 min, the third field every 20 min, and so on.

6. The more the power characteristics are improved, the better the chance for reentrainment losses to diminish. Rappers should always operate across one field before the cycle moves on to the next field. As discussed in the text, excessive dust disturbances on the collected layer can lead to adverse electrical field activity in certain cases. Allowing the surface contour to smooth out slightly between raps can relieve localized stress points and reduce the spark-over potential. Usually 5 to 10 min time duration is needed to observe this phenomenon where it will occur.

16.4.11 Gas Distribution

Whether gas distribution is effecting the precipitator performance adversely can be related to many factors. Efforts are usually worthwhile in exploring improvements in the gas distribution pattern if one or more of the following conditions exist:

1. Aspect ratios of 1.0 or less.
2. Average calculated gas velocity of more than 1.8 m/s (6 ft/s) through the precipitator.
3. Reentrainment dust losses above 50% of the total ESP emission losses.
4. Low-resistivity characteristics are apparent with an absence of spark-over.
5. High-resistivity characteristics are apparent with less than an average of 0.11 ma/m^2 (10 ma/1000 ft^2) of collecting surface.

Field observations have pointed out a number of concepts:

1. The gas flow vectors in a dynamic system will tend to keep going in the direction pointed, until striking another obstruction. This concept is fundamental to the understanding of why some installations have problems.

2. The velocity of the gases entering an expansion plenum will determine the final patterns at the face of the precipitator. If there is a poor vector pattern at the entry of the nozzle, then higher flow rates will usually aggravate the distribution by the time the gases reach the precipitator.

3. The 40% to 50% open area diffuser plates will provide little correction of a poor gas pattern if the pressure drop across the plate is less than 13 mm (0.5 in.) H$_2$O. However, these plates will reduce the rolling action of the gas, and most of the kinetic energy will be transferred into smaller jets. Generally, below the 3.0 to 4.6 m/s (10 to 15 ft/s) range, only minimal benefits will accrue in the gas spreading effects of the low pressure drop diffuser.

4. With a 40% to 50% open area diffuser plate, which is commonly used, any gas vectors striking the plate at 45° or more from the perpendicular will have a sizeable fraction of that gas flow slide across the plate.

5. Any flue expansion with more than about an 8° slope will generally have some separation of gas from the surface. The common practice of 30 to 45° plenum expansions tends to present distribution problems for that reason.

6. Any process whose flue gases contain particles over 30 microns in diameter could

get into distribution troubles by the settling of dust in the expansion plenum. This condition becomes worse during long periods of reduced process operation with its attendent low gas velocities.

16.5 GAS CONDITIONING

The preferred method to improve the performance of existing precipitators involves the use of higher power inputs. Poor resistivity levels can be overcome by the modification of the flue gas characteristics. The term *gas conditioning* normally refers to the various methods used for injection of chemical constituents into the flue gas stream, primarily to help alter resistivity levels in the precipitator. This term should include any method, whether or not it is inherent in the process or supplied from an external source.

16.5.1 Concepts

The process should be first explored to determine if inherent changes in operation equipment can modify resistivity levels. Some of these techniques include:

1. Substitute, blend, or prepare some of the bad actors in the raw materials or fuel in a manner conducive to precipitation. This may require some modification of the handling equipment.
2. More efforts on the maintenance of moisture levels in the flue gas is important. Just the elimination of inleakage air will have this net effect.
3. Awareness of the temperature effect on resistivity must be uppermost for any process change. Even the elimination of high to low gas temperature zones may help moderate a poor performance to one that is acceptable.

Probably the use of additional moisture in the flue gas by way of water sprays or steam injection can be considered a primary method of gas conditioning. Water addition to many process gas streams is often part of the operation. Water forms part of the raw material in some cases, or in others, is primarily used to control gas temperature levels at the discharge of the process. Fuel supplies another source of moisture. As previously discussed, moisture contents over 20% by volume tend to nullify resistivity problems depending on the gas temperature range at the precipitator. The use of steam is much less utilized because of the cost factor, but it is useful on a short-term basis where water may present condensation problems.

The use of chemical additives offers a secondary approach if the time factor or economics dictates that any modification of the process is not a satisfactory route. Fly ash collection has been the greatest area of implementation for this method in recent years. There are a number of companies and techniques available in the gas conditioning field and success has been achieved on difficult installations.

16.6 DESIGN AND PERFORMANCE CONCEPTS

The simple Deutsch equation is a valid way to understand how the various critical inputs can affect the performance of the precipitator. As mentioned earlier, the exponent can be low, as much as a factor of two, because of a number of problems that the designer did not foresee. Excessive reentrainment and poor gas distribution were two of the prime reasons for the disparity between the theoretical and actual results.

Recent designs have taken the migration exponent to another $\frac{1}{2}$ power or less to correct for previous problem areas and provide additional margin for the fine-sized particles existing in the latter fields of the precipitator. What this means is that a doubling of physical precipitator is indicated compared to what was considered a standard design of the early 1970s. Whether this is warranted is based on two factors: whether or not good design concepts are applied and, second, the confidence of the user that he can exercise some control over the process gas and dust characteristics. In retrospect, the design of 8 to 12 years ago could meet its guarantees if conditions were optimized. Designs of 40% to 50% greater surface area over those of the past now appear quite reasonable if you weigh all the factors of today's environment. This still means carefully addressing the process characteristics and applying a commitment to proper operating and maintenance techniques.

Each precipitator field should be considered a separate collector unit, and for that matter, each gas passage of a field must perform well to attain the best bottom line of the overall system. For high collection efficiencies to be achieved, the inlet field must perform near design levels usually in the nominal 80% range. Theoretically, that means about 80% of the particulate matter would deposit in the front hopper. This is why the inlet field looms important in any upgrading program.

I would stress a few points:

1. Each succeeding field works on the residual of the preceding field, but the potential collection efficiency tends to decrease in the direction of flow. Part of the reason is that collection values are harder to achieve as the magnitude of particles decrease.

2. Another reason is that the particles that are left in the gas stream in the latter half of the precipitator are more difficult to collect since they usually consist of the finer sized segment. Unless current densities above 0.22 ma/m^2 (20 ma/1000 ft^2) are observed in these latter fields, their collection efficiencies can deviate substantially from design.

16.7 EFFECT OF PARTICLE SIZE

The effect of the particle size on precipitation is seen in the component relationship that represents the migration velocity of the Deutsch equation. This exponent indicates that the smaller sized particle is more difficult to

remove from the flue gas stream. Although there is some validity to this concept, too much is made of this point in the practical application of the precipitator. It is difficult to analyze effectively because the particle size, shape, and chemical make-up interact in many diverse ways.

Each basic industry tends to produce particulate matter from some form of a grinding, combustion, or condensation process. Normally, the discrete larger particle of material found in a flue gas will be more irregular in shape and will be more chemically associated to that of the process raw material. Particles formed by condensation in the process tend to be submicrometer in size and more spherical in shape, while often deviating from the chemical characteristics of the larger particle found in the gas stream of the same process.

The effect of the particle size on the electrical precipitation can be identified in a number of ways:

1. The larger the particle the more electrical charge can be accumulated on its surface, and this condition provides an increased velocity of the particle toward the collector surfaces of the precipitator.

2. Electrical precipitation probably performs the least on a particle size about one-half micron in diameter. Collection of particle sizes less than one-half micron improves with benefits of Brownian motion in the vicinity of the collection site, while the larger sized particles benefit from the greater levels of charging.

3. However, it appears that a large number of the smaller particles tends to adhere to the larger particles, so that it is difficult to separate the practical effect of the sizing segments on the overall efficiency of collection.

4. A population of particles that is more homogeneous in sizing will often make the deposited layer of material on the collector surfaces more difficult to dislodge by rapping forces. As a rule, the larger particles of material, because of the effect of greater

porosity in the layer, will allow for easier removal by the mapping mechanism.

5. Both size and chemical segregation of particles will tend to occur throughout the length of the precipitator. The outlet electrical fields will often contain a greater percentage of the finer sized particles as well as the more chemically active material, such as condensed alkali and acidic ingredients.

REFERENCES

Theoretical Background

1. H. J. White, *Industrial Electrostatic Precipitation*, Addison-Wesley, Reading, MA (1963).

2. M. Robinson, "Electrostatic Precipitation," in *Air Pollution Control I*, edited by W. Strauss, Wiley-Interscience, New York, NY (1971).

3. J. D. Cobine, *Gaseous Conductors*, Dover Publications, New York, NY (1958).

Operation and Maintenance

J. Katz, *The Art of Electrostatic Precipitation*, Scholium International Inc. Port Washington, NY (3rd printing 1989).

Manuals and Publications on ESP's

5. The National Technical Information Service, Springfield, VA.

6. The McIlvaine Company, Northbrook, IL.

7. Air Pollution Control Association, Pittsburgh, PA.

8. Industrial Gas Cleaning Institute, Alexandria, VA.

17
Granular Bed Filters

PART 1. THE THEORY

Gabriel I. Tardos

CONTENTS

17.1.1 INTRODUCTION

The separation of airborne dust in granular beds takes place in either a "cake" or a "noncake" (deep-bed) filtration mode depending on the region in the filter in which particle deposition actually occurs. During cake filtration, as the name implies, initially deposited dust layers serve as collection media for subsequent filtration and the granules in the bed serve only as a support for the separated dust. The main mechanism of particle separation is sieving: incoming dust particles are retained on the already deposited dust. This results in a significant increase in thickness of the deposited layer as filtration proceeds and is accompanied by a large increase in pressure drop. This in turn causes a compression of the deposited layer and hence results in a higher filtration efficiency as more and more dust accumulates at the surface of the cake. The efficiency of the filter in cake filtration is overwhelmingly a function of the deposited layer's pore size and increases dramatically with pressure drop. If the dust particle size is larger than the pore size, dust is filtered and the efficiency is very high (practically 100%). However, if the dust is smaller than the open pore size, a cake is not formed and deep-bed filtration takes place.

During noncake or "deep-bed" filtration, dust particles are captured on each and every one of the granules or collectors of the filter. As filtration progresses, deposits of dust slowly fill the interstices of the granular bed starting with the contact points between granules without drastically altering the geometry of the filter or the pressure drop through the bed. The filtration in this case is overwhelmingly influenced by the size of dust particles and by the thickness of the granular filter in the direction of the flow.

The theoretical calculations presented in this section pertain only to the case of deep-bed (noncake) filtration in granular packed, moving, or fluidized beds. In these cases dust is collected either inside the filter or distinct collectors or particles deposit on each other without significantly altering the geometry of the filter as dust collection proceeds. The pressure drop in the filter, $\Delta p/L$, under these conditions can be calculated from the well-known Ergun correlation,[1,2] which in dimensionless form is given as:

$$f_0[\epsilon^3/(1 - \epsilon)] = 180(1 - \epsilon)/\mathrm{Re}_0 + 1.8 \tag{17.1.1}$$

The actual pressure drop per unit thickness of filter is then evaluated from the equation:

$$\Delta p/L = f_0 \rho U_0^2/2a \tag{17.1.2}$$

where the Reynolds number is expressed as $\mathrm{Re}_0 = 2aU_0 \rho/\mu$, L is the thickness of the filter in the direction of the flow, a is some average granule radius, and U_0 is the superficial gas velocity in the filter.

17.1.2 TOTAL BED EFFICIENCY

The efficiency with which dust is collected in a filter, η, can simply be calculated from the concentration of airborne dust entering n_{in}, and leaving the filter n_{out} as:

$$\eta = 1 - n_{in}/n_{out} \tag{17.1.3}$$

Extensive studies of deep-bed filtration in both granular and fabric filters have revealed that the total efficiency is an exponential function of the filter thickness and this can be expressed by the equation:

$$\eta = 1 - \exp[-1.5(1 - \epsilon)(K/2a)E] \tag{17.1.4}$$

where $L/2a$ is the number of collector layers in the filter, ϵ is the relative void volume, and E is the so-called single collector efficiency. The quantity E is defined as the ratio of the number of all airborne (dust) particles captured by a single collector in the bed to the total number of dust particles flowing toward it in a circular tube of cross-sectional area πa^2. The implicit assumption in Eq. (17.1.4) is that all collectors act as if they were independent within the filter as shown in Figure 17.1.1 and hence experience similar filtration phenomena. Equation (17.1.4) can be used in a predictive way provided the single collector efficiency E can be calculated from first principles.

A somewhat different but in principle equivalent way of computing the total bed efficiency is to use the concept of the unit cell efficiency, e, so that:

$$\eta = 1 - [1 - e]^n \tag{17.1.5}$$

The quantity e is defined as the ratio of number of airborne dust captured by a collector (granule) to the total number of dust particles flowing toward it in a square duct of cross-sectional area, l^2, where the length l is given by:

$$l = 2[\pi/6(1 - \epsilon)]^{1/3} a \tag{17.1.6}$$

The quantity n is the number of layers of unit cells in the filter, $n = L/l$. Comparing Eqs. (17.1.4), (17.1.5), and (17.1.6), the ratio of the

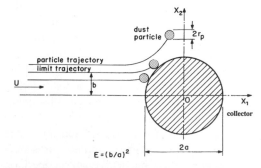

Figure 17.1.1. Schematic representation of dust particle deposition on a sphere. (Copyright *AIChE*. Vol. 31, No. 7, p. 1095, July (1985). (Reproduced with permission.)

single collector and the unit cell efficiencies is given by:

$$e/E = 1.2(1 - \epsilon)^{2/3} \qquad (17.1.7)$$

Whereas the definition of the single collector efficiency is somewhat arbitrary and its value can exceed unity in some cases (this may be difficult to justify on purely mechanistic grounds) the unit cell efficiency has a clear physical meaning. For a detailed discussion of the different efficiencies and their definitions, the reader is directed to the exhaustive monograph on granular filtration by Tien.[3] The remainder of this section is dedicated to ways of calculating the single collector efficiency E which hence allows the prediction of the total efficiency, η.

17.1.3 COLLECTION MECHANISMS IN DEEP-BED FILTRATION

Collection of small airborne dust by granules (collectors) in a packed or fluidized bed is due to external forces that cause the dust to deviate from the fluid stream lines and thereby to impact and stick to the collector. The forces that are most frequently associated with filtration in granular beds are inertia, diffusion, gravity, and electrical effects. While inertial and gravitational forces are characteristic of large particles of the order of microns and tens of microns, diffusion becomes important only for very fine particles in the submicron range;[4] electrical forces, if present, are effective in the whole range of particle sizes. For relatively small particles and in the absence of electrostatics, the so-called interception effect becomes important. This is a purely geometric "mechanism" and is due to the finite size of the dust particles, that is, even if the particles follow the fluid stream lines exactly some stream lines will approach the collector to a distance smaller than the radius, r_p, of the dust particle, as can be seen in Figure 17.1.1, thereby causing deposition.

Table 17.1.1 presents a summary of the important mechanisms that cause deposition in a granular bed; each of the mechanisms is governed by a characteristic dimensionless number that is defined in the second column of the table. Because electrical effects are caused by

Table 17.1.1. Collection Mechanisms in Granular Beds.

MECHANISM	CHARACTERISTIC DIMENSIONLESS NUMBER	EQUATION	REMARKS
Interception	$R_p = r_p/a$ Interception parameter	$E_R \cong 1.5g^3(\epsilon)R_p^2$ $E_R \cong (3/\epsilon)R_p$	$Re_0 < 1$ $Re_0 < 30$ Geometric effect[b]
Diffusion	$Pe = 2aU_0/D_B$ Peclet number	$E_D = 4g(\epsilon)Pe^{2/3}$ $E_D = 4.52/(\epsilon Pe)^{1/2}$	$Re_0 < 1$[b] $Re_0 < 30$
Gravity	$Ga = ag/U_0^2$ Galileo number	$E_G \cong GaSt$	Independent of flow to a first approximation
Inertia	$St = 2Cp_pU_0r_p^2/9\mu a$ Stokes number	$E_I = 2St'^{3.9}$ $(4.34^{-6} + St'^{3.9})$ $0.1 < St' < 0.03$	$St' = St[1 + 1.75\,Re_0/$ $150(1 - \epsilon)]^c$
Electrical effects	K_e Electrical number[a]	$E_{el} = -4K_e$ $E_{el} = K_{ex}/(1 + K_{ex})$	Coulombic force only External electric field only

[a] See definitions in Table 17.1.2
[b] See expressions for $g(\epsilon)$ in Table 17.1.3.
[c] See other expressions in Table 17.1.4.

a combination of charges present on the particle, the collector, or both, the characteristic electrical number K_e is given separately in Table 17.1.2. Table 17.1.3 is a summary of expressions for the correction factor $g(\epsilon)$ that appears in the equations describing interceptional and diffusional efficiencies, while Table 17.1.4 contains theoretical and experimental relations to calculate the efficiency due to inertial effects. As seen in Table 17.1.1, the Reynolds number $\text{Re}_0 = 2aU_0/\nu$ enters explicitly only in the expression of the inertial disposition; one has to note, however, that expressions for interception and diffusion are different for low and high Reynolds number flows as shown in Table 17.1.1.

Although it is quite simple to predict filtration efficiencies if only one mechanism is active by using the expressions in Table 17.1.1, in reality a combination of effects almost always exists. A general practice in this case is to add the predicted values for each individual mechanism by using the equation:

$$E = 1 - (1 - E_R)(1 - E_D)(1 - E_G)$$
$$\times (1 - E_I)(1 - E_{el}) \qquad (17.1.8)$$

which, if all efficiencies are small compared to unity, simply becomes:

$$E \cong E_R + E_D + E_G + E_I + E_{el} \quad (17.1.9)$$

The assumption behind Eq. (17.1.8) is that different mechanisms act independently; this was demonstrated to be true for the case of diffusion, interception, and inertia;[24] interception and gravity; and interception, diffusion,

Table 17.1.2. Electrical Forces Between Particles and Characteristic Parameters, K_e.

	FORCE, F_e	DESCRIPTION	PARAMETER K_e
c	Coulombic force	Both collector and particle are charged.	$K_c = CQ_eQ_p/24\pi^2\epsilon_f r_p a^2\mu U_0$
ic	Charged-particle image force	Particle only is charged. Charge separation induced in collector.	$K_{ic} = \gamma_c CQ_p^2/24\pi^2\epsilon_f r_p a^2\mu U_0$
ip	Charged-collector image force	Collector only is charged. Charge separation induced in particle.	$K_{ip} = \gamma_p CQ_c^2 r_p^3/12\pi^2\epsilon_f a^5\mu U_0$
ex	External electric field force	Particle only is charged. Charge separation in collector induced by external electric field.	$K_{ex} = CQ_p E_0/6\pi r_p\mu U_0$
icp	Electric dipole interaction force	Neither body is charged. Charge separation in both bodies induced by external electric field.	$K_{icp} = 2\gamma_c\gamma_p\epsilon_f Cr_p^2 E_0^2/a\mu U_0$

ϵ_f, Dielectric constant of fluid;
ϵ_p, dielectric constant of particle;
ϵ_c, dielectric constant of collector.
$\gamma_c = (\epsilon_c - \epsilon_f)/(\epsilon_c + 2\epsilon_f)$.
$\gamma_p = (\epsilon_p - \epsilon_f)/(\epsilon_p + 2\epsilon_f)$.

Table 17.1.3. Values of Correction Factor $g(\epsilon)$.

AUTHOR	$g(\epsilon)$	RANGE
Pfeffer[5]	$\{2[1 - (1 - \epsilon)^{5/3}]/[2 - 3(1 - \epsilon)^{1/3} + 3(1 - \epsilon)^{5/3} - 2(1 - \epsilon)^2]\}^{1/3}$	$Re_0 < 0.01$ $Pe \geq 1000$
Tardos et al.[6]	$\{\epsilon/[2 - \epsilon - (9/5)(1 - \epsilon)^{1/3} - (1/5)(1 - \epsilon)^2]\}^{1/3}$	$Re_0 < 0.01$ $Pe \geq 1000$
Sirkar[7,8]	$\{[2 + 1.5(1 - \epsilon) + 1.5[8(1 - \epsilon) - 3(1 - \epsilon)^2]^{1/2}]/\epsilon[2 - 3(1 - \epsilon)]\}^{1/3}$	$Re_0 < 1$ $\varepsilon > 0.33$ $Pe \geq 1000$
Tardos et al.[9]	$1.31/\epsilon$	$0.3 < \epsilon < 0.7$ $Re_0 < 0.01$ $Pe \geq 1000$
Tan et al.[10]	$1.1/\epsilon$	$Re_0 < 1$ $0.35 < \epsilon < 0.7$
Wilson and Geankoplis[11]	$1.09/\epsilon$	$Re_0 < 10$ $0.35 < \epsilon < 0.7$
Thoenes and Kramers[12]	$1.448/\epsilon$	$Re_0 < 10$ $\epsilon = 0.746$
Karabellas et al.[13]	$1.19/\epsilon$	$Re_0 < 10$ $\epsilon = 0.26$
Sorensen and Stewart[14]	$1.104/\epsilon$ $1.17/\epsilon$	$\epsilon = 0.476$ $\epsilon = 0.26$

gravity, and weak electric effects.[25] Strong electric effects due to Coulombic attraction and strong external electric fields (see Table 17.1.1) cannot be combined with inertial effects and have to be considered separately.[26-28]

To complete the picture of collection of small airborne dust by a granule (collector) in a granular bed, the phenomenon of bounce-off has to be mentioned. It was observed by many researchers that at relatively high gas velocities or large particle sizes, while inertial effects ensure that dust particles collide with collectors following their tortuous way through the filter, the dust is in fact not collected and instead bounces off on contact and is, in the end, not retained by the filter. This behavior

Table 17.1.4. Empirical Correlations for Single-Sphere Efficiency Due to Inertial Effects.

AUTHOR	E	RANGE
Paretsky[15]	$2 \times St^{1.13}$	$St < 0.01$
Meisen and Mathur[16]	$0.00075 + 2.6 \times St$	$St < 0.01$
Doganoglu[17]	$2.89 \times St$	$d_c \leq 100$ micron
	$0.0583 \times Re \times St$	$d_c \leq 600$ micron
Thambimuthu et al.[18]	$10^5 \times St^3$	$0.001 < St < 0.01$
Schmidt et al.[19]	$3.75 \times St$	$St < 0.05$
Goren[10]	$1270 \times St^{9/4}$	$0.001 < St < 0.02$
Pendse and Tien[a21]	$(1 + 0.04\,Re)[St]$	
D'Ottavio and Goren[c21]	$St_{eff}^{3.55}/(1.67 + St_{eff}^{3.55})$	$0.33 < \epsilon < 0.38$
Gal, Tardos and Pfeffer (1985)[b23]	$2St'^{3.9}/(4.3 \times 10^{-6} + St'^{3.9})$	$0.01 < St' < 0.02$

[a] Interception neglected.
[b] $St' = St[1 + 1.75\,Re_0/150(1 - \epsilon)]$.
[c] $St_{eff} = f(Re, \epsilon)St$.
$f(Re, \epsilon) = (1 - h^{5/3})/(1 - 1.5h^{1/3} + 1.5h^{5/3} - h^2) + 1.14\,Re_0^{1/2}/\epsilon^{2/3}$ where $h = 1 - \epsilon$.

results in a reduced efficiency at particle Stokes numbers larger than about St > 0.01. Tien[3] introduces the coefficient of adhesion probability Υ given by

$$\Upsilon = 0.00318 \, St^{-1.248} \qquad (17.1.10)$$

to account for this effect. For practical calculations, the efficiency E obtained from Eq. (17.1.8) has to be multiplied by the factor Υ if the Stokes number exceeds the value St = 0.01 even if the deposition is overwhelmingly influenced by electrostatic effects.

17.1.4 EXPERIMENTAL VERIFICATION

A schematic representation of the experimental apparatus to test a granular bed filter is depicted in Figure 17.1.2, a schematic of the test section is also shown. In the case of an electrically enhanced filter, a wire mesh electrode is added to the top of the bed where the electric field is applied and a radioactive source, used to neutralize the generated aerosols (dust particles), is followed by a particle charger (not shown in the figure). The complexity of the set-up is required by the need to very carefully control the dusty gas flow, the particle and granule electric charge (or lack of it), and the granule–wall interaction in the bed. Additional problems are also generated by the sensitivity of the particle counter (Royco counter in the figure). Filtration experiments usually require the generation of a dilute stream of test aerosols (usually latex particles of known size) which are subsequently passed through the filter at known flow rate and the concentration in and out of the bed is carefully measured. These experiments are repeated with a whole range of specially manufactured test dusts or aerosols of different sizes and sometimes composition and electrical properties. To control electric charges, the test particles are first neutralized and then electrically charged to the appropriate level before entering the bed. Experiments are performed at different gas flow rates and at different electric fields if electric effects are present.

Figures 17.1.3 and 17.1.4 show measured[28, 29] and calculated filter efficiencies using the equations give in Table 17.1.1. The dust particles used in these experiments are of the latex aerosol type, which are commonly used in industry to test filters as mentioned previously. Figure 17.1.3 shows filtration efficiencies, E, as a function of gas superficial velocity in a sand bed of grain average size of 450 μm. The calculated values are for large Reynolds numbers (upper line in the figure) and very low Reynolds numbers (viscous flow) using the equations of Table 17.1.1 for diffusion, interception, gravity, and inertia. The experimental values of the Reynolds number are depicted with arrows on the lower side of the figure. As seen, the data follow the calculations as expected: for Reynolds numbers below about $Re_0 = 3$, the data fit viscous flow calculations well, whereas for values of the Reynolds number of the order of $Re_0 = 30$ and higher, the measured data follow the calculations for potential flow. One can clearly see the effect of bounce-off at superficial gas velocities larger than about 2 m/s. One has to note here that the data presented above are an exceedingly exaggerated case in which the limits of the theoretical calculations are being checked. Granular filters are usually operated at gas velocities of the order of 2 to 30 cm/s, where it is clearly seen that calculated values fall quite close to the measured ones.

Figure 17.1.4 shows results for an electrically enhanced filter operated with an external electric field. The shape of the efficiency curves (total efficiencies η in this case) are typical of granular filters: efficiencies are high for small dust particles below 0.1 μm and large dust particles above 1 μm in diameter and are lower between these two limits. Increasing the applied electric field results in a significant improvement in efficiency even at the high gas velocity of $U_0 = 0.5$ m/s as shown in the figure.

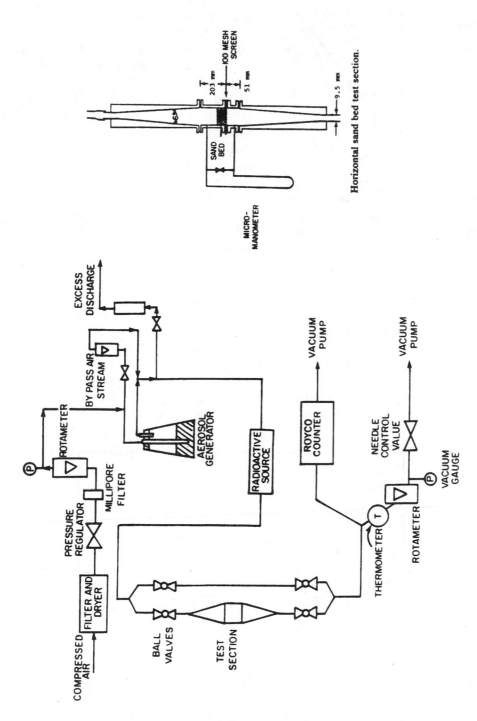

Figure 17.1.2. Schematic of experimental apparatus. (Copyright Academic Press, Inc. *Journal of Celloid and Interface Science*, Vol. 71, No. 3, October (1979). (Reproduced with permission.)

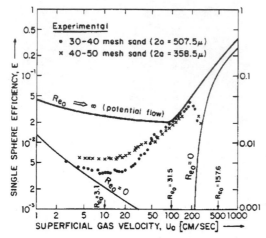

Figure 17.1.3. Single-sphere efficiency versus superficial gas velocity. Filtration of 1.1 μm latex aerosols. Theoretical values computed for: bed porosity $\epsilon = 0.4$, granule diameter $2a = 0.45$ mm, and particle density $\rho_p = 2 g/cm^3$. (Copyright Academic Press, Inc. *Journal of Celloid and Interface Science*, Vol. 71, No. 3, October (1979). (Reproduced with permission.)

17.1.5 CONCLUDING REMARKS

A theoretical approach was outlined to predict both the single collector efficiency E and the total bed efficiency η of a deep-bed granular filter from first principles. The method is based on the assumption that on the average each granule in the bed plays the role of a collector and that overall, the effects of all collectors can be integrated to yield an exponential decay in concentration along the filter. The filtration process is then divided into individual mechanisms by which airborne dust deviates from the fluid streamlines and can, at least in principle, collide with a granule and stick to it. Some experimental evidence is given to show that this model is at least somewhat realistic and that carefully measured efficiencies can in fact be predicted theoretically with some degree of confidence.

Dust filtration in a cake clearly does not fit the above model and hence the reader is referred to the pertinent literature[30-32] for further information. Fortunately, deep-bed filtration is almost always the important mode of separation of small particles whereas cake filtration becomes important when the granular filter is overwhelmingly clogged with dust and under those conditions the efficiency becomes very high. Because of pressure drop considerations the operation of granular bed filters in the clogged regime is not economically and technically attractive.

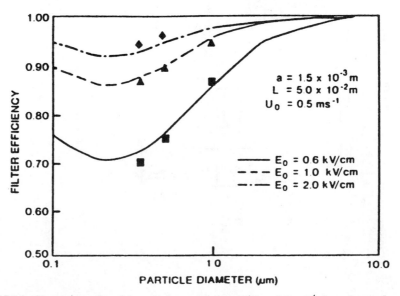

Figure 17.1.4. Comparison of model predictions with theory ($U_0 = 0.5$ ms^{-1}, $2a = 3$ mm, $L = 5$ cm).

LIST OF SYMBOLS

a	Collector (granule) efficiency
C	Cunningham correction factor
$D_B = KT/6\pi\mu r_p C$	Dust particle diffusion coefficient
$d_c = 2a$	Collector (sphere) diameter)
e	Unit cell efficiency
E	Single-sphere (collector) efficiency
E_D	Single-sphere efficiency due to diffusion
E_{el}	Single-sphere efficiency due to electrical effect
E_G	Single-sphere efficiency due to gravity
E_I	Single-sphere efficiency due to inertia
E_0	Applied electric field
E_R	Single-sphere efficiency due to interception
f_0	dimensionless pressure drop defined in Eq. (17.1.2)
g	Acceleration of gravity
$\mathrm{Ga} = ag/U_0^2$	Galileo number
$q(\epsilon)$	Porosity-dependent function given in Table 17.1.3
K	Boltzman constant
K_c	Electrical number defined in Table 17.1.2
l	Unit cell size defined in Eq. (17.1.6)
L	Filter bed height
n_{in}	Inlet aerosol concentration
n_{out}	Outlet aerosol concentration
$n = L/l$	Number of unit cell layers
$\mathrm{Pe} = 2aU_0/D_B$	Peclet number
Q_c	Collector (sphere) electric charge
Q_p	Particle (dust) electric charge
r_p	Dust particle radius
$R_p = r_p/a$	Interception parameter
$\mathrm{Re}_0 = 2aU_0/\nu$	Reynolds number
$\mathrm{St} = 2C\rho_p U_0 r_p^2/9\mu a$	Stokes number
T	Absolute gas temperature
U_0	Superficial flow velocity
Δp	Pressure drop through the packed bed

Greek letters

ϵ	Bed porosity
ρ_p	Dust particle density
μ	Gas viscosity
$\nu = \mu/\rho$	Gas kinematic viscosity
η	Total filtration efficiency
ρ	Gas density
γ	Adhesion probability coefficient defined in Eq. (17.1.9)

REFERENCES

1. S. Ergun, "Fluid Flow Through Packed Columns," *Chem. Eng. Prog. 48*:89–94 (1952).
2. I. F. Macdonald, M. S. El Sayed, K. Mow, and F. A. L. Dullien, "Flow Through Porous Media— The Ergun Equation Revisited," *Ind. Eng. Chem. Fund. 18*:199–208 (1979).
3. C. Tien, *Granular Filtration of Aerosols and Hydrosols*, Butterworths, Boston and London (1989).
4. G. I. Tardos, N. Abuaf, and C. Gutfinger, "Diffusional Filtration of Dust in a Fluidized Bed," *Atmos. Environ. 10*:389–394, April (1978).
5. R. Pfeffer, "Heat and Mass Transfer in Multi-Particle Systems," *IEC Fund. 3*:380 (1964).
6. G. I. Tardos, C. Gutfinger, and N. Abuaf, "High Peclat Number Mass Transfer to a Sphere in a Fixed or Fluidized Bed," *AIChE J. 22*:1146–1149 (1976).
7. K. K. Sirkar, "Creeping Flow Mass Transfer to a Single Active Sphere in a Random Spherical Inactive Particle Cloud at High Schmidt Numbers," *Chem. Eng. Sci. 29*:863 (1974).
8. K. K. Sirkar, "Transport in Packed Beds at Intermediate Reynolds Numbers," *Ind. Eng. Chem. Fund. 14*:73 (1975).
9. G. I. Tardos, N. Abuaf, and C. Gutfinger, "Dust Deposition in Granular Bed Filters—Theories and Experiments," *JAPCA 28*(4):354–363 (1978).
10. A. Y. Tan, B. D. Prasher, and J. A. Guin, "Mass Transfer in Nonuniform Packing," *AIChE J. 21*(2):396 (March 1975).
11. E. J. Wilson and C. J. Geankopis, "Liquid Mass Transfer at Very Low Reynolds Numbers in Packed Beds," *Ind. Eng. Chem. Fund. 5*:9 (1966).
12. D. Theones and H. Kramers, "Mass Transfer from a Sphere in Various Regular Packings to a Flowing Fluid," *Chem. Eng. Sci. 8*:271 (1958).
13. A. J. Karabellas, T. H. Wegner, and T. J. Hanratty, "Use of Asymptotic Relations to Correlate Mass Transfer Data in Packed Beds," *Chem. Eng. Sci. 26*:1581 (1971).
14. J. P. Sorenson and W. E. Stewart, "Computation of Forced Convection in Slow Flow through Ducts and Packed Beds—I, II, III, IV," *Chem. Eng. Sci. 29*:819 (1974).
15. L. C. Paretsky et al., "Panel Bed Filter for Simultaneous Removal of Fly Ash and Sulfur Dioxide," *J. APCA 21*:204 (1971).
16. A. Meisen and K. B. Mathur, "Multi-Phase Flow Systems," *Inst. Chem. Eng. Symp. Ser. 38*, Paper K3 (1974).
17. Y. Doganoglu, Ph.D. Dissertation, McGill University (1975).
18. K. V. Thambimuthu et al., Symp. Deposition and Filtration of Particles from Gases and Liquids, *Soc. Chem. Ind. (London)* p. 107 (1978).
19. E. W. Schmidt et al., "Filtration of Aerosols in a Granular Bed," *J. APCA 28*(2):143 (1978).
20. L. S. Goren, "Aerosol Filtration by Granular Beds," *EPA Symp. in Transfer and Utilization of Particulate Control Technology (Rpt.) EPA-600-7-79-004* (1978).
21. H. Pendse and C. Tien, "General Correlation of the Initial Collection Efficiency of Granular Filter Beds," *AIChE J. 28*(4):677 (1982).
22. T. D'Ottavio and L. S. Goren, "Aerosol Capture in Granular Beds in the Impaction Dominated Regime," *Aerosol Sci. Tech. 2*:91 (1983).
23. E. Gal, G. I. Tardos, and R. Pfeffer, "Inertial Effects in Granular Bed Filtration," *AIChE J. 31*:1093 (1985).
24. C. Gutfinger and G. I. Tardos, "Theoretical and Experimental Investigation of Granular Bed Dust Filters," *Atmos. Environ. 13*(6):853 (1979).
25. R. Pfeffer, G. I. Tardos, and L. Pismen, "Capture of Aerosols on a Sphere in the Presence of Weak Electrostatic Forces," *IEC Fund. 20* (1981).
26. K. A. Nielsen and J. C. Hill, "Collection of Inertialess Particles on Spheres with Electrical Forces," *IEC Fund. 15*:149 (1976).
27. K. A. Nielsen and J. C. Hill, "Capture of Particles on Spheres by Inertial and Electrical Forces," *Chem. Eng. Commun. 12*:1/1 (1981).
28. G. I. Tardos and R. W. L. Snaddon, "Separation of Charged Aerosols in Granular Beds with Imposed Electric Fields," *AIChE Symp. Ser. 235*(80):60 (1984).
29. G. I. Tardos, C. Gutfinger, and R. Pfeffer, "Experiments on Aerosol Filtration in Granular Sand Beds,"*J. Col. Int. Sci. 71*(3):616 (1979)
30. R. P. Donovan, *Fabric Filtration for Combustion Sources*, Marcel Dekker, New York (1985).
31. R. C. Flagan and H. H. Seinfeld, *Fundamentals of Air Pollution Engineering*, Prentice-Hall, Englewood Cliffs, NJ (1988).
32. F. Loeffler, "Collection of Particles by Fiber Filters," in *Air Pollution Control Part I*, edited by W. Strauss, John Wiley & Sons, New York.

PART 2. APPLICATION and DESIGN

Frederick A. Zenz

CONTENTS

17.2.1 INTRODUCTION

The suggestion that extraneous solids could be removed from gaseous streams by passage through a bed of particles is to this day regarded by many as novel, if not improbable, despite the fact that the average engineer is well aware of the technology of treating liquid streams by passage through beds of particles for purposes of removing undesirable elements via ion exchange and more simply for removing any suspended solids. The suggestion becomes more acceptable and credible by considering the analogy between the well accepted bag filter, which represents a tortuous path through the interstices in a layer, or bed, of interlaced cylinders (fibers), as opposed to the granular bed which represents a tortuous path through the interstices in a layer of interlaced spherical or angular shapes.

The novelty of the concept, in principle, vanishes rapidly when a search of the patent and technical literature reveals that shallow bed, granular, filtration devices in various forms have been proposed, explored, tested, and even marketed over the period from at least 1896 to the present day.

17.2.2 PURPOSES AND APPLICATIONS

The principal interest in development of granular bed filters today rests in removal of the particulates from hot pressurized process gases which could then be utilized to drive turbines and other equipment for efficient recovery of otherwise lost energy. A secondary interest lies in ensuring that exhaust gases meet local and national emission standards for particulate matter, and a third in the simple incentive for an efficient filter not subject to high maintenance costs.

Although no universally accepted standards have been specified or agreed upon for the allowable particulate loadings of gases driving turbines, it is generally agreed that the lifetime and performance efficiency of such turbines can justifiably be increased by reducing the particulate content of the feed gas stream via a reasonable cost filter. As an example, some typical particulate loading specifications are summarized in Table 17.2.1. Electrostatic precipitators, scrubbers, and fabric filters have been successfully used for removing particulates from exhaust gases at moderate levels of temperature and pressure but not under high temperature (e.g., 1000° to 2000°F) and pressure conditions. Lowering the temperature and pressure of the turbine inlet gas stream for the purpose of facilitating particle removal by using proven techniques would result in large losses of energy. The increased prospective utilization of combined cycle gas/steam turbine/electrical generating systems, coal gasification, fluidized bed combustion, and synthetic gaseous fuels has intensified the interest in granular bed filters. Unfortunately the effect on turbine blades appears to be a function of the physical properties of the impinging particles. In catalytic cracking of petroleum fractions, carryover silica-alumina catalyst fines remaining in the exhaust gases after passing through three stages of cyclones have shown power recovery turbine blade erosion to a degree necessitating reblading only at 5-year intervals.[1] The severity with coal combustion or gasification fines is reportedly far greater.

Other than the incentives of power recovery the increasingly stringent EPA emissions standards in the U.S.A. create definitive advantages for dry collection to avoid the handling of slurries from wet scrubbers. With cyclones not likely to meet the standards, with electrostatic precipitators representing high installation and maintenance costs, and with bag houses at 2 to 10 CFM/ft^2 representing large installations requiring considerable space, the potential of a 40 to 100 CFM/ft^2 high-efficiency, low-maintenance, granular bed filter is an inevitable attraction even in conventional electrical utility applications.

Other than power recovery and meeting emission standards there exist a host of industrial applications in which a "sand" medium would offer considerable process advantage over a fabric. Condensibles in a gas stream can cause severe bag failures, particularly in systems carrying cement kiln and similar calcined effluents, which can solidify on the bags. Sand media can be effectively dried and even washed in situ, and light accretions removed by the grinding action of the media. Where collected particulates can represent a fire hazard as in the collection of carbon black, cellulosic solids, and similar materials, a noncombustible filter medium would have immeasurable advantage.

In the popular literature are found such terms as "gravel bed filters," "panel bed filters," "expandable bed filters," "moving bed filters," "sand filters," "loose-surface filters," "porous bed filters," "MB filters," and a host of others all of which pertain to versions of what are generically referred to as granular bed filters.

Table 17.2.1. Typical Turbine Specifications.[a]

SOURCE	PARTICULATE LOADING (MAXIMUM ALLOWABLE)
United Aircraft	0.8 lb/million SCF low-BTU fuel gas (or approximately 12 ppm)
Westinghouse	0.03% (in fuel oil) (or 300 ppm by weight)
	0.002 grains/SCF all smaller than 6 μm (for fluidized bed coal combustors)
General Electric (for aircraft-type turbines)	30 ppm by weight in fuel gas (10 μm maximum)
Brown Boveri	1–2 ppm by weight (in gases entering turbine)
DOE (estimates)	0.75 grains/SCF (or approx. 2.6 ppm by weight) in 0–2 μm range
	0.001 grains/SCF (or approx. 35 ppb by weight) in 2–6 μm range
Exxon (estimates)	45–1 mg/m^3 (0.02 to 0.0004 grains/SCF)

[a] NTIS-BP 266 231 Feb. 77.

All potential commercializations of the concept rest on four basic interrelated factors:

1. Collection efficiency
2. Cleaning or regeneration capability
3. Capacity
4. Competitive cost.

Their practicality lies simply in the operational details of their technology which might best be illustrated by a relatively chronological review of the principal attempts at commercial development over the past 30 years and the status of such work today.

17.2.3 POROUS SINTERED GRANULE BEDS

Particle filtration via porous membranes rests primarily on the formation of a filter cake, removable by a reverse flow of fluid when the resistance of the cake (or the pressure drop) exceeds any desired level. The porous membrane may take the form of a bed of particles or for that matter a mat of fibers such as is the case with cloth bag filters. The thinner the membrane the lower the overall resistance with or without a filter cake and hence by analogy to the near monolayer of fibers in a cloth collector one could construct a thin bed or sheet of granules by sintering a shallow layer of metallic granules in a high temperature furnace. Such sheets of sintered metal granules in various forms are sold commercially for filtration purposes. Their principal application is in liquid systems, such as the maintenance of dirt-free fuel lines in aircraft engines. They have had limited application in the recovery of carryover from fluid bed reactors and similar fine-particle processing, but are not broadly acceptable. Their commercialization stemmed from the work of Dr. David Pall whose interest in the early 1940s lay in the development of a gaseous diffusion barrier for isotope separation in connection with the Manhattan District project during World War II.

In the late 1940s the Pall Corporation (now Pall Trinity Corp.) supplied such porous metal in tubular form with the suggestion that their application would be analogous to a cloth bag filter in which the cloth is replaced by a rigid thin bed of sintered metal particles.[2] As illustrated in Figure 17.2.1, in normal operation valve 1 would be open, allowing dusty inlet gas to flow through the elements, depositing a filter cake on their outer upstream surfaces and leaving through the exhaust plenum. When the cake has grown to a thickness exhibiting high pressure drop, valve 1 is closed, valve 2 is opened, and a short, high-pressure pulse of air is admitted in reverse flow through the porous elements to dislodge the filter cake which falls to the bottom of the containing vessel, to be eventually withdrawn through valve 3. In continuous use, the valves operate on a timed cycle and the containing vessel is provided with a multiplicity of porous elements and separate plenums for localized reverse cleaning.

There are a number of inherent disadvantages to this form of filter. In order to obtain structural strength, the granules making up the porous element must be small in size to

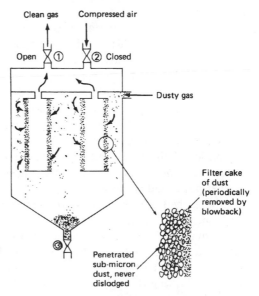

Figure 17.2.1. Porous sintered granule filter.

present sufficient bonding surfaces. This results in high-pressure drop or low gas capacity despite the only approximately $\frac{1}{8}$ in. wall thickness of the elements. Simultaneously, the thin-wall tubes are very subject to cracking as a result of repeated thermal shocks between normal high-temperature operation and relatively colder reverse-flow cleaning blasts. In addition, over extended operation, the submicron fines penetrate the interstices of the elements. These fines become trapped and are unable to be blown out with the reverse cleaning blast. Thus, these elements build up a residual pressure drop that further limits their capacity. They are not recommended where only very fine particles are to be collected; they operate most satisfactorily in handling streams containing sufficient coarse particles, in effect, to build up their own precoat. Such instances are rare, since where they occur the particle loading is usually so high that the cost of the elements is prohibitive even if they were never to fail by thermal shock.

17.2.4 CONTINUOUS MOVING-BED FILTERS

As early as 1924, Cramp[3,4] proposed a "dust curtain" system for filtration of blast furnace effluent gases. The dust curtain was a vertical layer of the collected dust itself, held in place by an arrangement of grids and louvers packed under its own weight. Dusty gas passed through this curtain horizontally, adding more dust to it. The excess dust gradually overflowed the louvers and fell to the bottom. The curtain was renewed periodically by dropping dust out of the bottom of the grids and adding more at the top. It is not clear whether this concept was ever put to industrial use.

In the early 1950s the late Morton Dorfan (Mechanical Industries, Inc.) introduced the Dorfan Impingo Filter,[5,6] which consisted basically of a downflowing bed of gravel retained between louvered walls. Dust-laden gas was filtered by blowing through the bed normal to the direction of gravel flow, and filtered dust,

hopefully carried downward with the gravel, was trapped within its interstices, as illustrated in Figure 17.2.2.

Four units, each of 17,000 CFM air capacity and consisting of two cells each, were installed in a plant to collect asbestos rock dust from a stream of flue gas coming from a direct-fired dryer in which the rock was dried prior to milling. The dust was 100% finer than 100-mesh and 60% finer than 10 μm. The concentration entering the collector was approximately 6 grains/ft^3 and that leaving about 0.2 grains/ft.^{3}

Dorfan's filter circumvented the thermal shock failure of Pall's sintered beds by avoiding reverse-flow cleaning and instead circulating the bed granules, plus collected dust, through a vibrating screen. However, this consequently required continual replenishment of the bed and hence an enormous and costly gravel circulation system. In order to provide reasonable gas throughput capacity, the bed granules had to be relatively large so as not to be blown off the retaining louvers; in practice "gravel" of plus $\frac{1}{4}$ to minus $1\frac{1}{2}$ in. was recom-

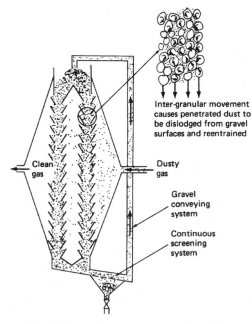

Inter-granular movement causes penetrated dust to be dislodged from gravel surfaces and reentrained

Clean gas

Dusty gas

Gravel conveying system

Continuous screening system

Figure 17.2.2. Continuous moving-bed filter.

mended. The commensurately large size of the bed interstices resulted in lower collection efficiency, and required staging beds in series, as shown in Figure 17.2.2, to achieve reasonable overall collection efficiencies. Such staging only added to the enormity of the required gravel circulation. Basically, micron-size particle collection efficiencies of the order of 99 + % could never be attained with such a moving bed system because intergranule movement in a downflowing mass physically dislodges collected particles and makes them subject to reentrainment.

Simultaneous with Dorfan's development a similar device was reported in the Russian literature by Zhitkevich[7] for the removal of peat dust from air by passing it through a vertical moving bed of pieces of peat of 5 mm in diameter. The rate of downward movement of the bed was controlled by the speed of a screw conveyor removing the dusty peat at the base of the unit in much the same fashion as rotary valves (not shown in Figure 17.2.2) at the bottom of each louvered column in Dorfan's arrangement. It is not known to what degree Zhitkevich's device found industrial application.

In 1954 Egleson et al.[8] published information on pilot tests of a downwardly flowing coke bed proposed as a possible means of filtering coal dust from a stream of synthesis gas produced in the gasification of pulverized coal. An 8.5 ft deep bed of 0.125 to 0.40 in. diameter coke particles moved vertically downward by gravity flow through a 12 in. I.D. column counter to an upwardly flowing stream of dusty air. At inlet loadings of 2 to 8 grains/ft^3 of 200-mesh dust, collection efficiencies were reported as high as 99.9% with superficial air velocities of 33 CFM/ft^2 giving pressure drops of only 0.3 in. of water per foot of bed depth. The dust-laden coke leaving the bottom of the column was washed with water and continuously recirculated to the top of the column. Though achieving, by means of counterflow, a higher efficiency than the Dorfan or similar crossflow devices, it became obvious that the enormity of the granular media han-

dling equipment would again make industrial application impractical.

17.2.5 INTERMITTENT MOVING-BED FILTERS

In the late 1950s it occurred to A. M. Squires (presently associated with the Chemical Engineering Department at the Virginia Polytechnic Institute) that the Dorfan filter could be made far more efficient by using a finer medium and arresting the bed motion during a filtration period, after which an accumulated cake of filtered fines would be removed by moving just the dusty sand. The desirable arrangement illustrated in Figure 17.2.3 would circumvent the fouling action of unremovable penetrated fines (as experienced with Pall type filters) and hopefully lessen the burden of circulating the immense quantities of sand associated with continuously moving beds (as exemplified by Dorfan type units).

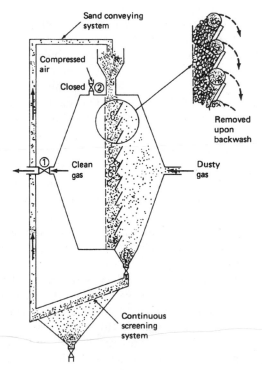

Figure 17.2.3. Intermittent moving-bed filter.

In the arrangement pictured in Figure 17.2.3 the dusty gas passes through a thin bed of stationary sand held between a panel of louvers and a fine-mesh screen. The filtered fines build up a cake on the exposed bed surfaces and to some extent penetrate the interstices. When the resistance of the cake has reached an undesirable level, the clean gas outlet valve 1 is closed and a short pulse of compressed air is blasted in reverse flow through the sand bed by opening valve 2. In continuous use, the valves operate on a timed cycle and the containing vessel is provided with a multiplicity of panels exhausting to a partitioned plenum, permitting localized reverse flow cleaning of individual panels.

The backwash pulse is sufficient to physically lift the sand beds as a mass, with minimum interparticle movement, so that a surface layer of sand between each pair of louvers is physically ejected from the panel and falls to the bottom of the filter vessel along with the collected filter cake. The expelled sand is immediately replaced by downward movement of fresh sand from the overhead hoppers. This only intermittent downward movement, coupled with a fine size of "gravel," permits high-efficiency collection and avoids the build-up of resistance due to penetrating submicron fines because these are expelled along with the surface layer of their surrounding sand. Incomplete removal of the sand containing penetrated fines would lead to their eventual accumulation near the bottom of the panel, by gravity flow, after repeated cleaning cycles, causing a gas flow maldistribution and excessive pressure drop. In order to minimize penetration and hence minimize sand circulation, the gas throughput was limited to approximately 10 to 12 $ACFM/ft^2$ of exposed bed surface.

Referred to as the "Loose-Surface" or LS filter, several models were built and tested in the period of 1959 to 1961 at the laboratories of F. A. Zenz then in Roslyn Harbor, NY. Figure 17.2.4 illustrates the largest of these. Typical experimental results using hand-sieved bank sand as the filter medium, and bagged flyash from a Consolidated Edison power plant as the redispersed dust, are given in Table 17.2.2.

The simplicity of the laboratory unit of Figure 17.2.4 appeared sufficiently attractive to warrant its purchase in 1960 by the Fuller Company (of Catasauqua, PA, a subsidiary of General American Transportation). After filing their patent[9] in Squires' name and spending 3 years in an attempt to build equally successful test models in their own laboratories, Fuller abandoned the commercial development of Squires' filter and currently maintain solely a nonexclusive interest in the original patent. Fuller's decision to halt their development was based on (1) an inability to restrict the amount of sand lost on each cleaning cycle to levels commensurate with economical recirculation costs, (2) the realization that the cost of steel and support structures was far greater than Squires' original estimates, and (3) that at 10 CFM/ft^2 the capacity was not competitive with bag filters and hence a limited market would possibly exist only in a few very high-temperature applications. At the time of its purchase by Fuller the LS filter was considered a potential competitor of electrostatic precipitators for power plant flyash collection.

In view of subsequent developments it is interesting to note that during the early period of laboratory development of the LS filter it was suggested that the circulation of sand could be entirely avoided by installing a retaining screen on the clean side of the panel, thus trapping the sand in a number of superimposed beds that could be cleaned by a reverse fluidizing flow of gas to elutriate out the collected fines. Squires reasoned that this would not permit high filtration efficiency because the mixing action accompanying fluidization would result in backmixing of collected fines so that eventually the sand bed would be dusty throughout and therefore some amount of collected dust would find its way into the clean side of the filter. Exploration of this suggestion was therefore not carried further.

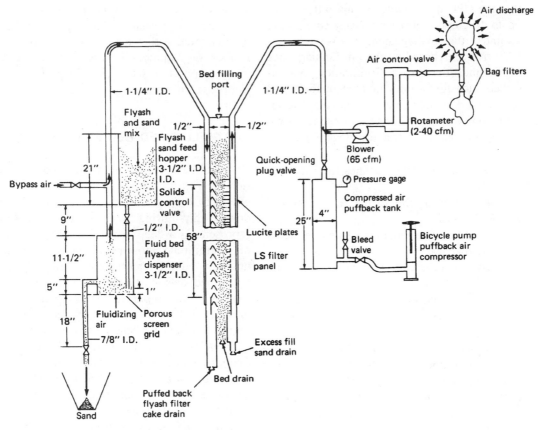

Figure 17.2.4. Flow diagram of test unit LS number 2.

Upon joining the faculty of City College in 1966 Squires obtained substantial financial support from EPA and similar government agencies to essentially repeat the LS development with, in some instances, minor modifications in bed depth or louver configuration.[10-12]

Inspired by Squires' modification, the principles of the Dorfan filter was investigated anew at the U.S. Bureau of Mines.[13, 14] In this instance the test apparatus, designed by Combustion Power Corporation (CPC), utilized downward moving beds of sand grains. The CPC design comprises essentially the panel form of the Dorfan filter rotated into a cylindrical arrangement and operated with intermittent media downflow. With the expectation

Table 17.2.2. Typical Loose Surface, Panel Bed Filter, Test Results.

SAND MEDIUM PARTICLE SIZE	16- TO 30-MESH	30- TO 100-MESH
Flyash loading in feed gas, grains/ft^3	2–6	7–9
Face velocity, ACFM/ft^2	8–12	6
Initial (clean) bed pressure drop, H$_2$O	0.1	0.38
Maximum pressure drop (prior to blowback), H$_2$O	0.6	0.6
Flyash laydown between cleanings, oz	14	2–3
Collection efficiency	99.7%	99.8%

that the CPC filter would live up to the claims of its developers on economically competitive grounds, the management of Weyerhauser Industries purchased all the outstanding stock of Combustion Power Corporation and installed such filters on a number of their own bark boilers. Unfortunately such complete installations were found to require the following component steps illustrated in Figure 17.2.5:

1. Sieving of the collected fines from the bed media
2. Conveyance of the media to a superimposed vessel
3. Elutriation of residual fines from the media by fluidization
4. Recovery of the elutriated fines in a "small" bag filter
5. Redistribution of the granular media to the filter annulus.

A complete installation therefore takes on many undesirable and costly complexities quite apart from the otherwise simple principle. In addition to an inadequate overall collection efficiency (including the periods during media movement) a complete installation reaches

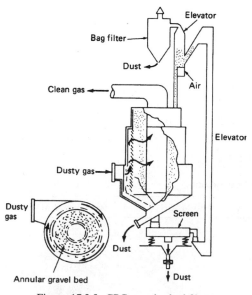

Figure 17.2.5. CPC annular bed filter.

levels of cost that encounter a severe sales resistance. Weyerhauser has reportedly been disenchanted with their own installations and further development is presently only minimally supported with U.S. Department of Energy (DOE) funds.

17.2.6 FLUIDIZED BED FILTERS

It should be obvious from the foregoing experiences with crossflow moving beds that high-efficiency collection is not possible unless the downwardly moving "wall" is very thick (e.g., representing a deep bed). The rubbing action of intergranule movement and the continuously changing interstitial configurations allow entrapped dust particles to be blown by the gas stream to deeper and deeper penetrations of the filter bed. Therefore very deep beds are required to achieve only reasonable collection efficiencies. It would, therefore, be expected that gas fluidized beds would similarly yield relatively low collection efficiencies. This conclusion is generally borne out by several such reported investigations.[15-17]

Under U.S. Army Chemical Corps Contract DA-18-064-CML-2758, a mechanically (vibrated) fluidized bed of sand was investigated for its possible filtration potential. The point of this investigation was to determine whether the poorer efficiency of gas fluidized beds was solely attributable to intergranule movement (as in moving bed filters) or significantly affected by the passage of unfiltered gas in the bubbles rising through the bed. The vibratory fluidized bed was operated with gas downflow, thus avoiding bubble formation and the accompanying bypassing of dusty gas. The results showed insignificant improvement over a bubbling fluidized bed, thus leading again to the conclusion that high filtration efficiency is compatible only with stationary media.

It is, however, of interest to note that a shallow moving fluidized bed of raw bauxite feed has been used to filter the carryover of particulates from the off-gases of electrolytic cells used in the production of aluminum metal. Such a unit has been operated by Aluminum Co. of America following the description in their patent.[18] The filtered particulates are primarily pitch or asphalt employed as a binder in the cell anodes. Though they appear to be dry they are sufficiently tacky to adhere to most surfaces on contact. Kaiser Aluminum installed an electrostatic precipitator at their Chalmette, Louisiana plant which operated for only a short time until it became completely coated with a layer resembling the undercoating on an automobile. The sticky nature of the collected particulates might be the principal reason for the practical application of a fluid bed filter in this instance, particularly because the filter medium is itself the raw material continuously fed to the cells, thus requiring no recirculation equipment and efficiently returning the asphalt fines to the cells as further fuel. The filtered off gases leaving the fluid bed are in any case subsequently passed through a bag house for cleanup of nonsticky fines to achieve the level of overall efficiency necessary to pass EPA standards. A filter of the Dorfan type was also proposed for such aluminum plant operations.[19]

17.2.7 GRANULAR BED FILTERS MECHANICALLY CLEANED

As opposed to the LS filter's attempt to compromise between the principles illustrated in Figures 17.2.1 and 17.2.2 (e.g., fixed bed filtration but minimal sand circulation) Max and Wolfgang Berz[20] in 1957 proposed an arrangement requiring no sand circulation. The cleaning would be carried out by flowing a gas in reverse direction through the bed while subjecting it to mechanical vibration of sufficient magnitude to cause the intergranule movement necessary for removal of entrapped fines. The principle of their proposed filter is illustrated in Figure 17.2.6.

In normal operation dusty feed gas would flow upward through a stationary fixed bed of sand held on a horizontal screen of a mesh size smaller than the bed granules. A filter

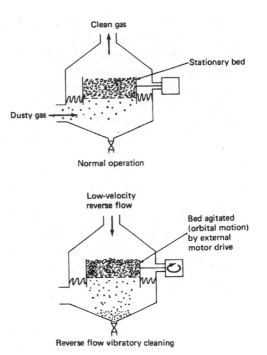

Figure 17.2.6. Mechanically cleaned granular bed filter.

cake might develop and adhere to the lower surface of the screen and some amount of fines penetration would occur within the interstices of the bed. Because they employed a granule size many times larger than the approximately 30-mesh sand investigated by Fuller, they experienced a greater degree of fines penetration (as opposed to cake build-up), required deeper beds, and operated at considerably higher throughput per unit of bed surface area.

When resistance to flow reached a level requiring bed cleaning, the dusty feed gas was diverted to another unit and a reverse flow of gas at low velocity passed down through the bed, while it was simultaneously vibrated in a manner such as to impart an orbital rotation to the particle mass. Thus, the mechanically induced interparticle motion dislodged the penetrated fines so that they could be swept out of the bed by the downflowing breeze. In continuous use, the reverse flow and bed vibration were automated on a timed cycle and the containing vessel again provided with multiple units cleaned individually in sequence.

The Berz' MB filter was sold commercially by Lurgi Apparatebau G.M.bH. of Frankfurt, Germany, until about the end of 1969 when it was abruptly withdrawn from the market. Its shortcoming lies in the strain imposed on the necessary flexing membranes associated with the vibrating technique. It is conceivable that at low temperatures where rubber membranes and spring-supported bed mounts are feasible, such a filter could operate with a reasonable life. However, at low temperatures it could never economically compete with the simplicity of bag houses. At high temperatures only metal bellows would suffice as the flexing membrane, and their life expectancy in the hot and dusty environment is impractically short.

Undaunted by their mechanical failure, Berz devised an alternate arrangement now marketed through Gesellshaft fur Entstau-

bungsanlagen m.b.H. of Munich, Germany. In this version, illustrated in Figure 17.2.7, the gravel medium is agitated during the cleaning cycle by the stirring action of a rotating rake whose fingers are imbedded in the gravel medium. Though the long-term mechanical integrity of the raking system as well as its possible introduction of flow paths of lower efficiency might be objectively questioned, it appeared that this filter might receive reasonable acceptance in at least the cement industry. Users, however, were not as completely satisfied as might be implied in Berz' optimistic presentation to the 1972 IEEE cement industry technical conference held in May, 1972. The cleaning period requires several minutes and the module undergoing cleaning is isolated from the dusty gas by a valve system during the cleaning period. Carrier Corp. (Rexnord Division) installed such filters at over a dozen cement plants;[21] however, rake fail-

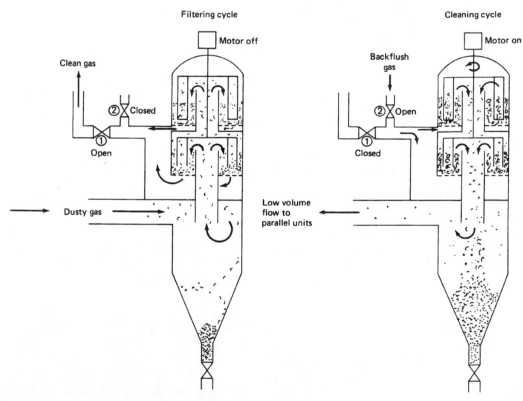

Figure 17.2.7. Alternate form of mechanically cleaned granular bed filter.

ures began, as anticipated, within about a year. Present efforts are concentrating on possible incorporation of a means of periodically removing and replacing the bed medium.

17.2.8 GRANULAR BED FILTERS PNEUMATICALLY CLEANED

In 1970 the Ducon Co. introduced the "Expandable Bed" filter,[22-24] in which a number of superimposed beds of sand, trapped between retaining screens, would be cleaned by a reverse flow of fluidizing gas. Its principles are illustrated in Figure 17.2.8. In normal opera-

tion dusty gas passes through vertical arrays of parallel shallow granular beds held within metal-walled compartments sealed at top and bottom by perforated meshes finer in aperture than the size of the bed granules. The dusty gas enters the beds at relatively high velocities, in the range of 40 to 100 CFM/ft^2 of bed surface area. When flow resistance reaches a level requiring the bed to be cleaned, a sufficient momentary reverse flow of compressed gas is admitted to fluidize the bed granules, or "expand" the bed, and thus by entrainment from these agitated beds expel the particles collected and agglomerated in the bed inter-

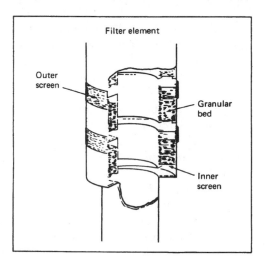

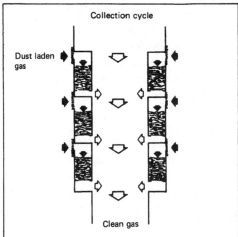

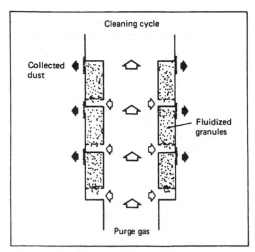

Figure 17.2.8. Pneumatically cleaned granular bed filter.

stices. No loss of sand can occur because any granules that might reach the end of a compartment cannot escape through the perforated retaining mesh.

The fluidized expansion of the granule bed and the accompanying intergranule movement, which allows efficient cleaning, are directly analogous to the action in bag filters. Upon cyclically reversing the flow of gas, particulates trapped between the fibers are expelled as the cloth flexes or expands. This movement of the fibers in the expansion of a bag allows such complete cleaning (depending upon the number of flexings during each cycle) that bags can always be returned to near their original pressure drop characteristics. By sufficient and proper duration of backwash flows, the trapped fines in the beds of Figure 17.2.8 can also be nearly completely expelled by the analogous intergranular bed flexing, bed expansion, or bed fluidization. Thus fluidization affords an economical nonmechanical cleaning mechanism compatible with high-temperature filter operation, requiring no costly recycle conveying of sand, no complex redistribution hoppers, no subsequent need to separate collected fines from filter element sand, and hence no attrition of bed granules which normally accompanies repeated solids handling operations.

In practice it was discovered that the intermingling of collected fines within the granular bed, which inevitably occurs as a result of the fluidized agitation during cleaning, resulted in an eventual build-up to an equilibrium fines content which then increased the collection efficiency, presumably simply because it presented a more tortuous interstitial path for the dusty gas flow. It was also found that the economic optimum operation lay in high gas capacity with minor sacrifice in efficiency. In this connection it is interesting to note that, in a paper dealing with the panel bed filter, Pfeffer[12] reported also observing higher efficiencies when using dusty as opposed to fresh sand. However, it is also obvious that following a reverse flow cleaning there could exist a small amount of the collected dust adhering to

"sand" media lying directly atop the bed supporting grid. This dust might well be blown off into the clean side of the filter when it is again placed in filtering operation and so contribute to a less than 100% filtration efficiency depending on cleaning frequency.

17.2.9 TECHNOLOGICAL STATUS OF SYSTEMS UNDER DEVELOPMENT AND UNDER COMMERCIALIZATION

The technological success and eventual commercialization, of any one or more of the many granular bed filter concepts, lies in

1. Satisfactory dust collection efficiency
2. Reproducible regeneration or filter media cleaning
3. Competitive total installation cost.

Unfortunately these three aspects are rather intimately interrelated and in all likelihood only eventual full-scale testing will narrow the field; this becomes partially evident in even a cursory review of the major commercial contenders.

The Fuller-Squires Panel Bed concept though abandoned by Fuller has undergone a series of studies related principally to improvements in louver design. There is no doubt that this concept affords excellent collection efficiency and that following each blowback the dusty gas faces a reproducibly cleaned filter bed. The questions that remain as yet incompletely answered are whether sieving the dust-laden sand drawn from the filter will produce sufficient cleaning. Will some residual dust particles cling to the media and escape into the cleaned gas stream upon return to the louvers? Would an additional fluidized bed elutriation column provide sufficient cleaning; how would this fluidizing gas stream then be cleaned and what would such additional equipment add to the total cost? How many louvers can be uniformly cleansed upon each blast of blowback gas? What is the investment increment for the media circulation and distribution system and what degree of media attrition might be anticipated. No tests of the Fuller–Squires panels have ever been conducted with continuous media recirculation or in an industrial environment.

The *Combustion Power Corporation Annular Bed* as illustrated in Figure 17.2.5 has undergone full scale operation on bark boiler effluent giving results such as shown in Table 17.2.3. It is unknown whether the reported media efficiencies include periods during which the media were in downward movement or only periods when the media were stationary. The media pressure drops may also refer to clean conditions prior to dust buildup. In any event the media efficiencies are substantially below the 99 + % levels normally sought in granular bed filtration. Efficiencies in the 90% range could more effectively be attained with multiclone centrifugal separators. Reliable cost figures for a complete installation including all the elements in Figure 17.2.5 are unavailable.

CPC's principal development experience centered around a pilot unit installed at the Snoqualmie, Washington plant of Weyerhaeuser Lumber Co. to clean up effluent from a hog mill waste conductor. The pebble bed filter was in this instance a single downflowing sand annulus $8\frac{1}{2}$ ft O.D. and about 6 ft I.D. with an effective filtering height of 16 ft. The face velocity and effective inlet area are quoted by CPC on the basis of the mean diameter of the annulus and in this instance correspond to 100 to 163 ft/min (or ACFM/ft^2) and to 365 ft^2, respectively. The pebbles were angular in shape and ranged from $\frac{1}{8}$ to $\frac{1}{4}$ in. in average diameter. They move downward through the annulus in gravity flow at a bulk velocity of 2 to 4 ft/h. The pebble inventory is of the order of 40 tons though only about 20 tons are in the region exposed to gas flow. At 35,000 ACFM gas flow (~ 100 ft/min) with inlet loadings of 0.17 to 0.3 grains/dry ft^3 and outlet loadings of 0.053 grains/dry ft^3 the pressure drop across the downflowing pebble bed was of the order of 12 in. of water. The unit was operated at 300° to 350°F. Collection efficiencies in all

tests ranged from 75% to 97%. The feed particulates showed at best about 27% smaller than 0.7 μm and 17% larger than 20 μm. The loss analysis showed 57% smaller than 0.7 μm and 11% larger than 20 μm.

Relative to public utility flue gas clean-up requirements the above performance figures fall far short of the necessary goals. Nevertheless CPC appears interested in promoting their filter for such application based on the premise that the Weyerhaeuser unit was not intended for high efficiency and that a finer size of pebble combined with removal of the coarser feed solids by tangential feed gas entry into the containing vessel shell might yield the necessary overall efficiency for utility purposes.

CPC's conceptual design for a 1000 MW electric utility burning 12,000 tons of coal/day with 3,300,000 ACFM of 400°F flue gas, would consist of three banks of scrubbers each having eight cylindrical filter vessels and each vessel containing four annular elements. They estimated a pressure drop of 8 to 10 in. of water across these filters and in 1974 a turnkey installed cost including fans and motors of $2.00 to $2.50 per ACFM. This would involve the sieving and circulation of 225 to 450 tons of pebbles per hour or as much as 10 times the ash rate assuming the flyash loading to the filters is of the order of 0.2 grains/ft^3. Higher inlet loadings would require proportionately higher pebble circulation rates. The conceptual design is based on recirculating the flyash-freed pebbles pneumatically to the top of the filters where the pebbles are collected via cyclones and the conveying gas vented through bag filters. The pebbles are again freed of the collected flyash via screens and elutriation; the latter would involve more bag house area but less screen wear. It was not clear whether the $2 to $2.50/ACFM cost estimate included the solids handling and/or separation equipment.

In terms of a test unit for installation in a public utility's flue gas system operating on a slip stream, CPC felt that a 100,000 ACFM

Table 17.2.3. CPC Annular Bed Filter Test Data[a] (Power Boiler Effluent—Waste Wood and Fuel Oil Fired).

MEDIA SIZE (IN.)	MEDIA FACE VELOCITY (ft/min)	MEDIA PRESSURE DROP (H_2O)	CYCLONE PRESSURE DROP (H_2O)	LOADING, Gr/dry SCF			COLLECTION EFFICIENCY (%)		
				CYCLONE INLET	MEDIA INLET	MEDIA OUTLET	CYCLONE	MEDIA	TOTAL
0.125–0.250	125	6	1.2	2.768	0.875	0.075	68.4	91.4	97.3
C.125–0.250	150	9.3	2.0	1.486	0.609	0.080	59	86.9	94.6
0.065–0.130	150	11.8	1.4	2.542	0.80	0.07	68.5	91.3	97.3
0.065–0.130	125	9.7	1.0	4.719	0.618	0.026	86.9	95.7	99.4

[a]Weyerhaeuser New Bern N.C. Pulp Mill, January, 1975.

unit would be the minimum size for obtaining reliable and scalable data.

The Canmet-Prasco Hitec System[25, 26] represents a development very similar to the CPC design as illustrated in Figure 17.2.9. Initially, the packed-bed region is filled with appropriately sized granular material, as shown by the shaded portion of the drawing. The hot "dirty" gases enter at the location marked "1" and are directed by pipes to pass horizontally through the packed bed, which is held in place by louvers. As the dust is removed, the pores of the filter material begin to plug up, causing an increase of pressure. When this happens, some filter material drops from a chute onto a conveyor belt at "5," and fresh, properly sized material moves down the column "3" from above the louvered section to replace it. The dirty filter material goes to the top of the unit in a bucket elevator and is dumped onto an inclined screen "6" where most of the dust "7" is sifted out for disposal. The remainder of the filter material is then fed back into the unit near the top of the vertical column into what is called an elutriation column. The gases to be cleaned enter at the bottom of this elutriation column and come in contact with the falling filter material. By controlling the speed of these gases, the finer particles of dust still in the filter material that were not removed by the screen are carried up with the gas and follow the path shown by the arrows marked "2," leading into the packed bed. After cleaning, the gases are vented to the atmosphere at "4." The granular filter material is too heavy to be carried up by the gas stream and thus falls down into the storage area above the louvers for reuse. In this manner the problem of the plugging of the filter material has been overcome.

Tests carried out in CANMET laboratories have shown the system to be capable of capturing dust particles down to 1 μm (0.00004 in.) in diameter at efficiencies of over 99.9%. Under appropriate operating conditions, the results obtained on the experimental unit showed the exhaust gases to contain less than 0.05 grain per ft^3. A commercial-size unit capable of treating 7500 ft^3 of "dirty" gas per minute has been constructed and installed in Winnipeg on a furnace used to melt cast iron. This unit is undergoing testing to determine its operational capabilities and whether it will meet the Manitoba Government regulations for pollution abatement devices.

The concurrent sizing of the filtration bed material by the gas being treated plus the baffle above the bed, which results in a desired size gradation across the packed bed, are the novelties claimed to be essential to the achievement of the performance goals. Typical laboratory scale tests with ambient air are summarized in Table 17.2.4. Subsequent tests on an Ancast Industries Ltd. cupola in Winnipeg, Manitoba showed average collection efficiencies of 99.4% for particles larger than 25 μm, 97.3% for particles larger than 1 μm, and 50.3% for particles smaller than 1 μm. The resulting overall efficiency of 87.9% was insufficient to bring the effluent loading to

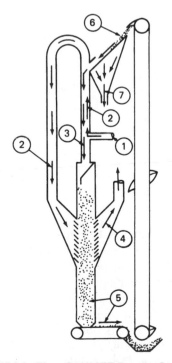

Figure 17.2.9. The CANMET-Prasco Ltd. Hitec System.

Table 17.2.4. Laboratory Scale Filtration Data for the CANMET-HITEC System.

PACKED BED FILTER							DUSTY
MATERIAL			DUST CHARACTERISTICS			FILTER	AIR
MEAN DIAM. (μm)	RATE (lb/min)	PRESSURE DROP (IN. OF H_2O)	TYPE[b]	INLET CONC. (gr/SCF)	OUTLET CONC. (gr/SCF)	COLLECTION EFFICIENCY (%)	FLOW RATE (SCFM)
1100	0.97	2.5–7.5	MCD	135	0.050	99.96	10
1100	0.62	5.0	MCD	134	0.036	99.97	10
1100	0.44	7.0	MCD	134	0.054	99.96	10
1100	0.96	5.0	MCD	45.5	0.175	99.62	15
1100	0.98	5.0	MCD	87.8	0.196	99.78	15
1100	0.98	3.5–6.5	MCD	81.6	0.156	99.81	15
600	0.62	5.0	MCD	135	0.011	> 99.99	10
600	0.36	5.0	MCD	132	0.023	99.98	10
600	0.96	5.0	MCD	82.4	0.056	99.93	15
600	1.21	5.0	MCD	85.1	0.196	99.77	15
600	0	1.4	GB	56.6	0	100	10
600	0	1.6	GB	65.8	0.14	99.79	10
600	0	1.9	GB	137	1.28	99.07	10
600	0	2.0	GB	150	3.02	97.98	10
600	0.50	2.2	GB	148	4.22	97.15	10
600	1.22	2.2	GB	118	1.18	99.00	10
600	1.74	2.2	GB	153	0.880	99.42	10
600	2.20	2.2	GB	130	0.420	99.67	10
250	1.44	5.0[a]	MCD	30.5	1.91	93.74	10
250	1.44	5.0[a]	MCD	31.4	1.59	94.94	10
250	1.68	5.0[a]	MCD	35.7	0.637	98.22	10
800	0.19	3.5	MCD	141	0.061	99.96	10
800	0.38	3.5	MCD	144	0.082	99.94	10
800	0.76	3.5	MCD	153	0.109	99.93	10
900	0.28	3.5	MCD	144	0.010	> 99.99	10
900	0.53	3.5	MCD	142	0.016	99.99	10
900	0.89	5.0	MCD	82.6	0.091	99.89	15
1500	0.04	3.5	MCD	38.0	0.023	99.94	10
1500	0.23	3.0	MCD	81.2	0.051	99.94	10
1500	0.44	3.0	MCD	118	0.068	99.94	10

[a] Pressure varied widely; figure reported is average pressure drop.
[b] MCD—mixed cupola dust (8% < 10μm, 10% < 30 μm, 50% < 150 μm, 100% < 1000 μm) GB—glass beads (1 to 25 μm).
[c] Filter face area: 6 $\times$ 6 in.
Filter bed depth: 3 in.

below the Manitoba Department of the Environment's limit of 0.25 grains per standard cubic foot. Failure to reach the efficiencies anticipated from laboratory tests was attributed to poor cupola management which led to an unusual amount of submicron material in its effluent. License for fabrication and sale of the Hitec system has been granted to Prasco Ltd. of Winnipeg, Manitoba.

The Kawasaki louvered bed development is again patterned on the Dorfan moving bed principle. Little is known of this work other than tests carried out at 85 to 295 CFM passing through a 10 in. wide $\times$ 40 in. high louvered panel holding 8 in. thick beds of downflowing silica sand either 1250 to 2500 μm or 2500 to 5000 μm in diameter. Figure 17.2.10 summarizes results obtained with an oil-fired utility flyash as the test dust.

The Ducon Fluidizing Reverse Flow filter described in principle in Figure 17.2.8 has undergone several stages of detailed development as

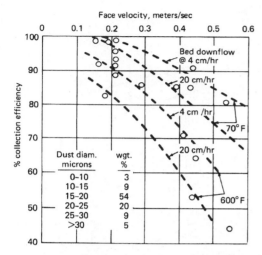

Figure 17.2.10. Reported filtration efficiencies of 8 in. deep beds of 1,250 to 5,000 μm diameter sands.

earlier models displayed performance failures in piloted industrial environments for a variety of reasons. The modular design presently offered[28] is illustrated in Figure 17.2.11, its operating principle is still the same as portrayed in Figure 17.2.8, but its design details are now regarded with extreme care. For example:

1. Bed depth has been optimized at 1.5 in. because shallower beds impose unrealistic fabrication and erection tolerances and deeper beds add unnecessary and undesirable pressure drop.
2. Filter media sand particle size has been optimized at 250 to 600 μm because coarser materials require excessive fluidizing velocity (blowback gas consumption) and entail reduced collection efficiency whereas finer particles would entrain too readily upon fluidization and would require too fragile a bed support.
3. Reverse flow fluidization at 120 CFM/ft² for at most 6 to 8 s appears an optimum for cleaning to maintain an equilibrium low concentration fines inventory without excessive pressure drop; the degree of cleaning is calculable from fluidized bed entrainment correlations.
4. The bed supports are designed as screened perforated plates with a multiplicity of tiny

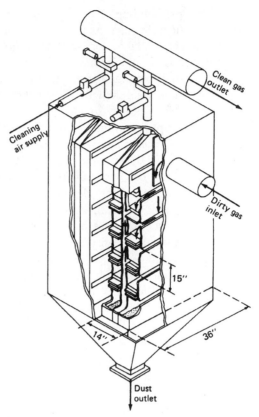

Figure 17.2.11. The Ducon fluidizing reverse flow granular bed filter. 900 ACFM nominal capacity of each 14 × 36 × 90 in. element consisting of 12 beds in 2 stacks of 6 each, at 50 ACFM/ft²; 5 in. of water "clean" pressure drop with ambient air through $1\frac{1}{2}$ in. deep beds of 30- to 60-mesh sand at rated capacity.

holes to distribute the fluidizing gas bubbles over the entire bed in close proximity and with sufficient pressure drop to distribute equally to each bed in the stacked module.
5. The disengaging height above each bed is set at no less than $10\frac{1}{2}$ in. in order to avoid filter media loss (without an inlet screen) compatible with the sand size, density, and reverse flow velocity.

No tests have as yet been carried out in an industrial environment with a unit bearing all of these design considerations and limited to face velocities of about 40 CFM/ft² with a maximum 1 psi build-up of pressure drop before reverse flow cleaning. As in the case of all granular bed filters, laboratory scale tests show excellent filtration efficiencies as illustrated for example in Table 17.2.5 and Figure 17.2.12 from the results published by Westinghouse researchers.[29] The ultimate technical feasibility of this concept lies in the ability to achieve cleaning of the beds in a practical manner.

For purposes of illustration Figure 17.2.13 represents a filter composed of four modules each containing two filter elements. In the normal continuous mode of operation valves D-1, D-2, etc. would always be open so that the dusty gas would be distributed from the feed manifold equally into all elements of all modules continuously. To maintain any desired pressure difference between the feed and product manifolds, with minimal fluctuations with time as the filter elements accumulated dust, the elements would be blown back individually on a timed cycle by sequentially opening and closing valve B-1, then B-2, then B-3, etc. Such sequential blowback cleaning of

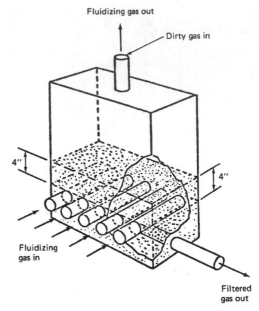

Figure 17.2.12. Typical GBF filtration.[29]

individual elements follows conventional bag house practice. Consider for the moment the effect of such flow reversal on the other elements within the filter. During the time that valve B-1 is open no dusty gas can enter element E-1; therefore, elements E-2 through E-8 are subject to a dusty gas feed rate increased by a factor of $\frac{8}{7}$ or 14%. However, in addition to this relatively minor increase, these seven elements must also absorb the blowback gas from element E-1. Though dependent on such factors as bed particle size, frequency of blowback, size and density of collected fines, etc., it is current practice to specify a blowback rate that creates a superficial fluidizing velocity of 2ft/s or 120 ACFM/ft^2 of filter bed surface. If the elements were operating at a design face velocity of 50 ACFM/ft^2 then obviously blowing back E-1 would increase the flow rate to the remaining seven elements from 50 ACFM/ft^2 to:

$$50(\tfrac{8}{7}) + (120/7) = 74 \text{ ACFM/sq. ft.}$$

Table 17.2.5. Typical GBF Filtration Efficiencies: Gas Velocity 50 fpm $\pm$ 10%. Bed Material 16- to 20-mesh Ottawa sand. Bed Depth 4 in.[29]

PARTICLE SIZE (μm)	INLET DUST (mg/m^3)	OUTLET DUST (mg/m^3)	COLLECTION EFFICIENCY (%)
0–0.3	32	0.39	98.8
0.3–0.45	20	0.08	99.6
0.45–0.75	33	0.12	99.6
0.75–1.5	53	0.08	99.8
1.5–2.3	47	0.04	99.9
2.3–3.3	32	0.04	99.9
3.3–5.0	53	0.04	99.9
5.0–8.0	46	—	100
8.0	108	—	100
	513	—	100

RUN NO.	INLET LOADING (g/m^3)	OVERALL EFFICIENCY (%)
6–24	5.8	99.96 (99.92)
6–25	0.94	99.92 (99.8)
6–26	1.17	99.90 (99.7)

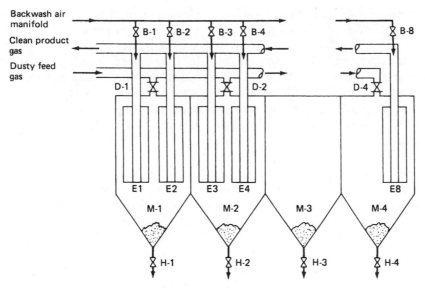

Figure 17.2.13. Illustrative arrangement of GBF modules.

during the blowback period. This can be expressed analytically in the form:

$$\%OL = 100\left[\frac{T_E + (12/5)}{T_E - 1}\right] - 1$$

where

$\%OL = \%$ increase or overload in dusty gas face velocity through elements not being cleaned

T_E = total number of elements in filter and only one element backwashed at any moment.

Represented graphically in Figure 17.2.14 it becomes obvious that any such installation whether on a pilot or industrial scale should preferably be provided with no less than 10 to 20 elements no matter what their size. Lower gas capacity for pilot tests should be provided with elements containing a smaller number of stacked beds rather than a smaller total number of elements.

Note in relation to Figure 17.2.13 it is assumed the reverse blowback from E-1 would be equally distributed to all seven remaining

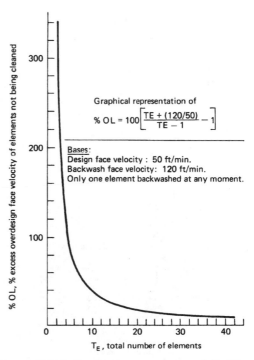

Figure 17.2.14. Graphical representation of % OL = 100 [TE + (120/50)/TE − 1] − 1. Bases: design face velocity; 50 ft/min; backwash face velocity: 120 ft/min. Only one element backwashed at any one moment.

elements; however, this is probably unrealistic since it requires an absolutely zero frictional resistance in valves B-1, B-2, etc., as well as in the entire dusty gas manifold, and in addition, ignores the inertia transients. In practice, element E-2 would be considerably more overloaded. Some reasonable limits must be placed on this overload since it can result in dust breakthrough by the force of the high velocity interstitial gas circumventing particle impact by heightened particle pick-up. For example, 100 ACFM/ft^2 through a bed of 40% voids amounts to an interstitial velocity a little over 4 ft/s which will exceed the saltation velocity of a variety of particulates depending on their properties. If the bed interstices were filled to some equilibrium content with collected fines, this effective interstitial velocity could easily reach 8 ft/s and cause reentrainment. The permissible overload cannot be specified without detailed knowledge of the system characteristics and properties.

The PEMMCO Restricted Circulation Filter illustrated in Figure 17.2.15 is one[30] of two concepts under development designed to minimize the filter media losses experienced in the Fuller–Squires type of unit or to permit cleaning,[31] without any reverse flowing gas. Only the shallow layers facing the dusty gas in Figure 17.2.15 are blown off by the reverse flow

cleaning gas; the gaps at the bases of the louvers prevent immediate media replacement and therefore excessive media blow off. A basic advantage of such granular bed filters, in which the medium is removed and replenished, lies in their adaptability to simultaneously act as gaseous adsorbers or chemical reactors as in the RESOX process for flue gas desulfurization and conversion to elemental sulfur.[32, 33]

The Melcher Electrofluidized Bed Filter consisting typically of millimeter sized particles is stressed by an imposed electric field that effectively polarizes the particles. They then act as collection sites for previously charged fine particulates entrained in the dusty fluidizing feed gases. The advantages claimed for the electrofluidized bed derive from the greatly reduced residence time for effective cleaning of the gas, realized by virtue of the bed's large collection surface area per unit volume, and the ease of handling the bed particles for removal of the collected fines.[35, 36]

Collection efficiencies in excess of 98% have been reported[37] for submicron asphaltic particles in 10 cm deep beds of 2 mm sand. Superficial velocities ranged from 1.5 to 2 m/s with bed pressure drops typically about 12 cm of water. Electrical energization required less than 80 W/1000 CFM. Though these conditions are a vast improvement over even multistaged[38] nonelectrified fluidized beds they still fall short of such typical fixed bed results as illustrated in Figure 17.2.12 despite their substantially greater face velocity.

The application of an electrical potential gradient to a bed of ceramic beads for filtration of petroleum fluid catalytic cracking fractionator liquid bottoms has been reported[39] as achieving exit solids contents as low as 0.01% by weight. Offered by Gulf Science & Technology Co. it again suggests the possibility of application to fixed bed gas filters though this concept has been considered beyond the limits of economic feasibility.

It appears to be the general concensus that some form of granular bed filter will eventually emerge as a significant industrial tool very

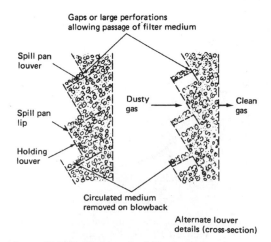

Figure 17.2.15. Design principle of restricted circulation granular bed filter-adsorber.

likely some day supplanting the electrostatic precipitator. The time will be determined by economic pressures, developmental details and further air quality regulations.

REFERENCES

1. A. P. Krueding, *Chem. Eng. Prog.* 71(10):56–61 (1975).
2. D. B. Pall, *Ind. Eng. Chem.* 45:1197 (1953).
3. G. B. Cramp, *Chem. Met. Eng.* 30:400–401 (March 10, 1924).
4. G. B. Cramp, *Blast Furnace Steel Plant,* 12:101–103 (February 1924).
5. *The Dorfan Impingo Filter*, Bull. no. 4, Mechanical Industries, Inc., 541 Wood Street, Pittsburgh 22, Pa. (1954).
6. M. I. Dorfan, *Electric Furnace Steel Proc.* 10:41–60 (1952).
7. L. K. Zhitkevich, *Trudy Inst. Energ. Akad. Nauk. Belorus, SSR* 1:150–160 (1954).
8. G. C. Egleson, H. P. Simons, L. J. Kane, and A. E. Sands, *Ind. Eng. Chem.* 46:1157–1162 (1954).
9. A. M. Squires, U.S. Pat. No. 3,296,775; filed 10/16/62, issued January 10, 1967.
10. L. Paretsky et al., *J.A.P.C.A.* 21:204–209 (April 1971).
11. A. M. Squires and R. Pfeffer, *J.A.P.C.A.* 20:534–538 (August 1970).
12. K. C. Lee, A. M. Squires, and R. Pfeffer, "Filtration of Fly Ash and Puffback in a Panel Bed Filter," paper No. 25a; 67th annual A.I.Ch.E. meeting, Washington, D.C. (December 3, 1974).
13. Bureau of Mines, Report of Investigations No. 7276 (July 1969).
14. E.P.R.I. Report 243-1 (November 1974).
15. H. P. Meisner and H. S. Mickley, *Ind. Eng. Chem.* 41:1238–1242 (1949).
16. J. P. Pilney and E. E. Erickson, *J.A.P.C.A.* 18(10):64–685 (October 1968).
17. C. H. Black and R. W. Boubel, *IEC Proc. Des. Dev.* 8(4):573–578 (October 1969).
18. L. L. Knapp and C. C. Cook, U.S. Pat. No. 3,503,184; filed 3/7/68, issued March 31, 1970.
19. A. F. Johnson, U.S. Patent No. 3,470,075; filed 2/6/67, reissued as Re 27, 383, May 30, 1972.
20. M. Berz and W. Berz, U.S. Pat. No. 3,090,180; filed May 19, 1960; see *Staub*, 24(11):449–452 (1954).
21. Rexnord Corp., *Environmental Sci. and Tech.* 8(7):600–601 (July 1974).
22. Bulletin No. F-9671, The Ducon Co., 147 E. Second Street, Mileola, L.I., N.Y. (1971).
23. U.S. Patent No. 3,410,055; filed 10/26/66; issued November 12, 1968.
24. B. Kalen, U.S. Patent No. 3,798,882; filed 9/28/71; issued March 26, 1974.
25. R. K. Buhr and E. Darke, CANMET Phys. Met. Res. Labs. Reports MRP/PMRL (CF) 76-4 (R) and 75-10 (FT).
26. R. K. Buhr and R. D. Warda, CANMET Report MRP/PMRL-75-2 (R) (January 14, 1975).
27. Private communication (1977) Kawasaki Corp.
28. U.S. Patent No. 4,067,704; filed 10/18/76; issued January 10, 1978.
29. D. F. Ciliberti, D. L. Keairns, and D. H. Archer, "Particulate Control for Pressurized Fluidized-Bed Combustion Processes," presented at the 5th International Conference on Fluidized Bed Combustion (December 13, 1977).
30. U.S. Patent No. 3,770,388; filed 5/24/71; issued November 6, 1973.
31. U.S. Patent No. 3,800,508; filed 10/26/70; issued April 2, 1974.
32. W. F. Bischoff, Jr., and P. Steiner, *Chem Eng.* pp. 74, 75 (January 6, 1975).
33. G. O. Layman, *Environ. Sci. Tech.* 9(8):712–713 (August, 1975).
34. K. Zahedi and J. R. Melcher, *J.A.P.C.A.* 26:345 (1976).
35. J. C. Alexander and J. R. Melcher, *IEC Fund.* 16(3):311–317 (1977).
36. *Popular Science*, p. 10 (August 1975).
37. P. B. Zieve, K. Kahedi, J. R. Melcher, and J. F. Denton; *Envir. Sci. Tech.* 12(1):96–99 (January 1978).
38. R. G. Patterson and M. L. Jackson, *A.I.Ch.E. Symp. Ser.* no. 161, vol. 73, pp. 64–73 (1977).
39. Gulftronic Separator Systems, *Chem. Proc.*, p. 20, (mid-April 1978).

BIBLIOGRAPHY

Including some references to liquid filtration, where theory or equipment arrangements are pertinent to granular bed filtration development.

"Backing Germany's Magnesium Bid" (Knapsack-Griesheim Flowing Coke Bed Filter), *Chem. Week*, pp. 71–72 (June 17, 1961).
"Big Dryer," *Chem. Eng. News*, p. 131 (October 23, 1961).
J. S. M. Botterill and E. Aynsley, "The Collection of Airborne Dusts Parts 1 and 2," *Br. Chem. Eng.*, 12(10):1593–1598 (October, 1967); ibid., 12(12): 1899–1903 (December, 1967).
"Braided-Wire Tubes Increase Filtering Efficiency," *Chem. Eng.*, p. 116 (June 20, 1966).
Br. Chem. Eng. 15(4):549 (April 1970).

Chem. Eng., pp. 90–91 (September 27, 1965).

Chem. Week, pp. 71–73 (June 17, 1961).

"Coal Filters for Waste Treatment," *Chem. Eng.,* p. 122 (March 14, 1966).

C. N. Cochran, W. C. Sleppy, and W. B. Frank, "Chemistry of Evolution and Recovery of Fumes in Aluminum Smelting," Paper no. A70-22, Metallurgical Society of the AIME meeting held February 16-19, 1970.

J. T. Cookson, "Removal of Submicron Particles in Packed Beds," *Environ. Sci. Tech. 4*(2):128–134 (February 1970).

E. D. Ermenc, *Chem. Eng.,* pp. 87–94 (May 29, 1961).

N. Fuchs and A. Kirsch, "The Effect of Condensation of a Vapor on the Grains and of Evaporation from their Surfaces on the Deposition of Aerosols in Granular Beds," *Chem. Eng. Sci. 20*:181–185 (1965).

G. Funke, "Pollution and Nuisance Control Activities," Zement-Kalks-Gips, no. 5, pp. 209–219 (1968).

L. Goldman, "Is the Gravel Layer Suitable as an Air Filter," *Wasser, Luft Betrieb 6*(7):233–236 (May, 1962).

J. P. Herzig, D. M. Le Clerc, and P. Le Goff, "Flow of Suspensions through Porous Media-Application to Deep Filtration," *IEC 62*(5):8–35 (May 1970).

S. Jackson and S. Calvert, "Entrained Particle Collection in Packed Beds," *A.I.Ch.E. J. 12*(6):1075–1078 (November 1966).

B. Kalen and F. A. Zenz, "Filtering Effluent from a Cat Cracker," *Chem. Eng. Prog. 49*:67–71 (June 1973).

Y. V. Krasovitskii and V. A. Zhuzhikov, "Separation of Dust from a Gas Stream by Filtration at Constant Velocity," *Khim. Prom., 49*(2):49–52 (1963); Translated by E. K. Wilip as ANL-Trans-572 (February 1968).

W. D. Lovett and F. T. Cuniff, "Air Pollution Control by Activated Carbon" (Moving Bed Adsorber Panel), *Chem. Eng. Prog. 70*(5):43–47 (May 1974).

A. Maroudas and P. Eisenklam, "Clarification of Suspensions, A Study of Particle Deposition in Granular Media, Part I—Some Observations on Particle Deposition," *Chem. Eng. Sci. 20*:867–873 (1964).

A. Maroudas and P. Eisenklam, "Part II—A Theory of Clarification," *Chem. Eng. Sci. 20*:875–888 (1965).

R. E. Pasceri and S. K. Freidlander, "The Efficiency of Fibrous Aerosol Filters: Deposition by Diffusion of Particles of Finite Diameter," *Can. J. Chem. Eng. 38*(6):212–213 (December 1960).

R. D. Rea, "Plume-Free Stacks Achieved in Sulfuric Acid Production," *Chem. Proc.,* pp. 13–14 (January 1971).

"Sand Filter Saves Space," *Chem. Eng.,* p. 112 (September 21, 1970).

E. W. Schmidt, J. A. Giesche, P. Gelfand, T. W. Lugar, and D. A. Furlong, "Filtration Theory for Granular Beds," *J. A.P.C.A. 28*(2):143–146 (1978).

"Simultaneous Sulfur Dioxide and Fly Ash Removal," *Environ. Sci. Tech. 5*(1):18–19 (January 1971).

L. Spielman and S. L. Goren, "Model for Predicting Pressure Drop and Filtration Efficiency in Fibrous Media," *Environ. Sci. Tech. 2*(4):270–287 (April 1968).

L. A. Spielman and S. L. Goren, "Capture of Small Particles by London Forces from Low-Speed Liquid Flows," *Environ. Sci. Tech. 4*(2):135–140 (February 1970).

D. I. Tardos, N. Abuaf, and C. Gutfinger, "Dust Deposition in Granular Bed Filters: Theories and Experiments," *J. A.P.C.A. 28*(4):354–363 (1978).

R. M. Werner and L. A. Clarenburg, "Aerosol Filters," *IEC Proc. Des. Dev. 4*(3):288–299 (1965).

R. L. Zahradnik, J. Anyigbo, R. A. Steinberg, and H. L. Toor, "Simultaneous Removal of Fly Ash and Sulfur Dioxide from Gas Streams by a Shaft-Filter-Sorber," *Environ Sci. Tech 4*:663 (1970).

F. A. Zenz and H. Krockta, "The Shallow Expandable Bed—A Versatile Processing Tool," *A.I.Ch.E. Sym. Ser. 67*(116):245–250 (1971).

F. A. Zenz and H. Krockta, "The Evolution of Granular Beds for Gas Filtration and Adsorption," *Br. Chem. Eng. Proc. Tech. 17*(3):224–227 (March 1972).

18
Wet Scrubber Particulate Collection

Douglas W. Cooper

CONTENTS

18.1 INTRODUCTION

18.1.1 Emission Control Goals

Activities involving powders can result in the generation of airborne particulate material, aerosols, which may need to be controlled because of concern about health, because of laws and regulations, or because of an economic incentive for process material recovery. The principal alternatives for fine particle control are cyclones, filters, scrubbers, and electrostatic precipitators. Alternatives are generally compared with respect to effectiveness and cost.

Scrubbers are air pollution control devices that use liquid to collect particles or gases or both from a gas stream. Usually, the liquid used is water, occasionally with a surface-active agent added. The use of scrubbers to remove gases is not discussed here.

18.1.2 Control System Options

The considerations for the selection among the types of air pollution control equipment are summarized in Figure 18.1. Scrubbers are

used not only for air pollution control, however, but also to recover valuable materials, to cool gas streams, and to add liquid or vapor to gas streams.

A succinct comparison of wet collectors (scrubbers) and dry collectors (cyclones, filters, electrostatic precipitators) was presented by Strauss[2] and is shown in Table 18.1. Economic comparisons are given at the end of this chap-

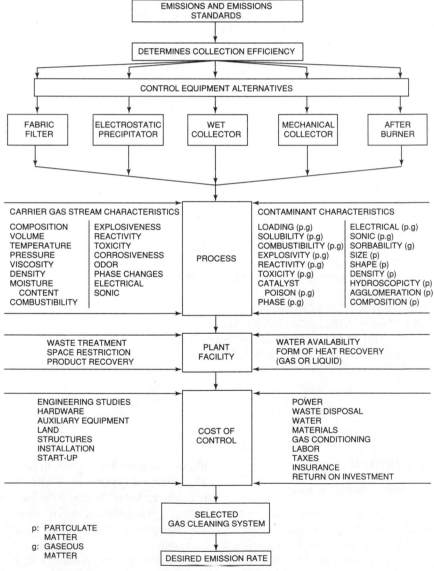

Figure 18.1. Process for selection of gas cleaning equipment.[1]

ter, but as a rough guideline: scrubbers have higher efficiencies and higher costs than cyclones, they can be made to have efficiencies comparable to those for filters and electrostatic precipitators but at higher operating costs and, generally, lower equipment costs. Further information on various air and gas cleaning devices is presented in Figure 18.2.

Figure 18.3 gives approximate collection efficiency as a function of particle size (fractional efficiency) for each of the major collector types. It is useful for qualitative comparisons only, as each of these devices has collection efficiency characteristics more complicated than the relationship shown.

18.1.3 Types of Scrubbers

Scrubbers capture particles on droplets, liquid surfaces, or liquid-coated surfaces. The droplets may be formed independently of the gas flow, or they may be atomized by the flow. Scrubbers using preformed sprays include: spray towers, cyclone spray scrubbers, water-jet scrubbers, and mechanical scrubbers. Venturi and orifice scrubbers are usually designed to produce a spray by gas atomization of the scrubbing liquid. Impingement scrubbers and sieve plates involve flow into or through a volume of liquid. Some of these scrubber types are shown in Figure 18.4. Particles are captured primarily on liquid-coated surfaces in packed-bed scrubbers, fluidized-bed scrubbers, and fibrous-bed scrubbers. Many of these different types use about the same amount of power to achieve the same degree of particle collection. The choice of scrubber type is therefore dictated by space constraints, the availability of certain kinds of power (e.g., waste heat) and equipment (such as pumps, fans, ducting, piping), and the aspects of the

Table 18.1. Advantages and Disadvantages of Wet and Dry Collectors.[2]

WET COLLECTORS	DRY COLLECTORS
Advantages	Advantages
1. Can collect gases and particles at the same time.	1. Recovery of dry material may give final product without further treatment.
2. Recovers soluble material, and the material can be pumped to another plant for further treatment.	2. Freedom from corrosion in most cases.
3. High-temperature gases cooled and washed.	3. Less storage capacity required for product.
4. Corrosive gases and mists can be recovered and neutralized.	4. Combustible filters may be used for radioactive wastes.
5. No fire or explosion hazard if suitable scrubbing liquid used (usually water).	5. Particles greater than 0.05 μm may be collected with long equipment life and high collection efficiency.
6. Plant generally small in size compared to dry collectors such as bag houses or electrostatic precipitators.	
Disadvantages	Disadvantages
1. Soluble materials must be recrystalized.	1. Hygroscopic materials may form solid cake and be difficult to shake off.
2. Insoluble materials require settling in filtration plant.	2. Maintenance of plant and disposal of dry dust may be dangerous to operatives.
3. Waste liquids require disposal, which may be difficult.	3. High temperatures may limit means of collection.
4. Mists and vapors may be entrained in effluent gas streams.	4. Limitation of use for corrosive mists for some plants (e.g., bag houses).
5. Washed air will be saturated with liquid vapor have high humidity and low dew point.	5. Creation of secondary dust problem during disposal of dust.
6. Very small particles (submicrometer sizes) are difficult to wet, and so will pass through plant.	
7. Corrosion problems.	
8. Liquid may freeze in cold weather.	

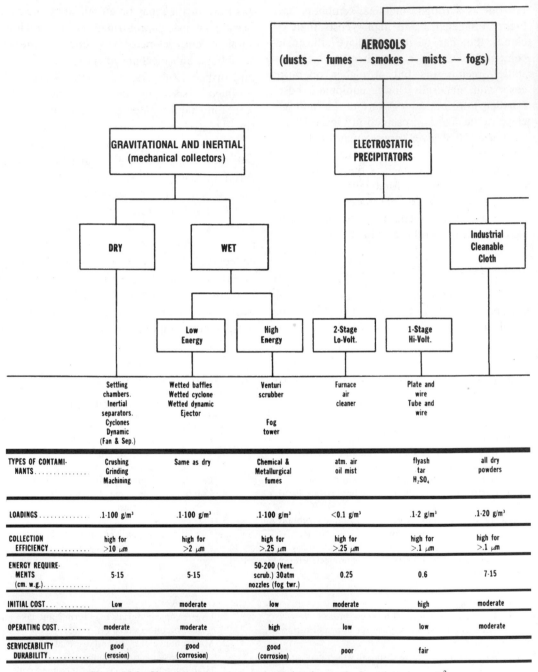

Figure 18.2. Characteristics of air and gas cleaning methods and equipment.[3]

dust/scrubber combination that affect plugging, corrosion, and the handling of liquid and solid waste.

In a *spray tower*, the particle-laden gas stream flows upward through a spray falling downward. In a *spray chamber*, the gas flow is generally horizontal and the spray is often in a cross-current orientation. In a *cyclone spray scrubber*, a spray is introduced near the cyclone entrance. The relative motion of spray

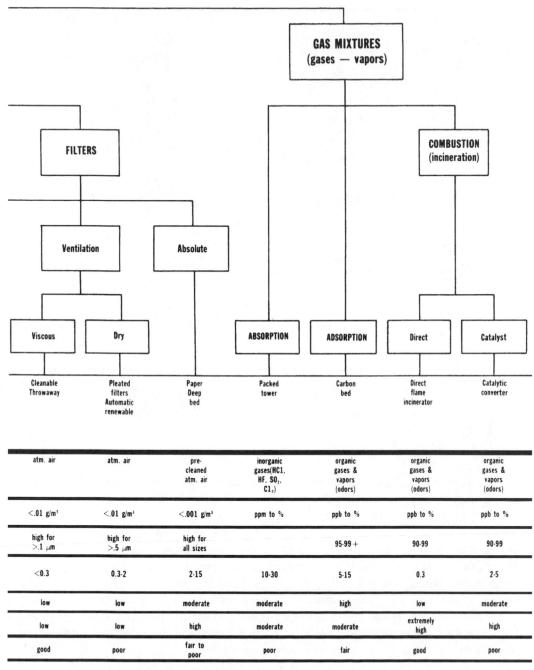

atm. air	atm. air	pre-cleaned atm. air	inorganic gases(HCl. HF. SO₂, Cl₂)	organic gases & vapors (odors)	organic gases & vapors (odors)	organic gases & vapors (odors)
<.01 g/m³	<.01 g/m³	<.001 g/m³	ppm to %	ppb to %	ppb to %	ppb to %
high for >.1 μm	high for >.5 μm	high for all sizes		95-99 +	90-99	90-99
<0.3	0.3-2	2-15	10-30	5-15	0.3	2-5
low	low	moderate	moderate	high	low	moderate
low	low	high	moderate	moderate	extremely high	high
good	poor	fair to poor	poor	fair	good	poor

Figure 18.2. Continued

and particles induced by the cyclones aids collection of the particles and also of the drops containing the captured particulate material. *Mechanical scrubbers* use sprays and moving baffling (fans, etc.) to induce particle capture. A *venturi scrubber* accelerates the gas in a converging channel, introduces a spray near the throat section, then decelerates the gas through a very gradually tapered diverging section. *Orifice scrubbers* work in a similar

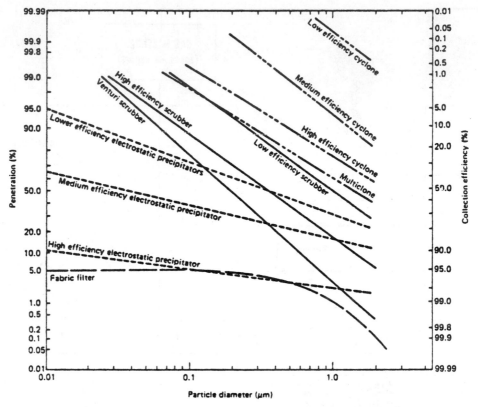

Figure 18.3. Extrapolated fractional efficiency of control equipment.[4]

fashion, except that the throat is an orifice, an abrupt change in duct cross-sectional area, of negligible length, followed by an abrupt expansion.

Impingement scrubbers direct the gas at the surface of a liquid, causing intimate liquid/particle mixing, using a variety of geometries. *Tray towers (sieve plates)* have a series of multiply perforated plates arranged vertically in such a way that water introduced at the top of the array flows downward from plate to plate, and the particle-laden gas is passed through the plate perforations and through the liquid in essentially cross-current flow at each stage; for the array, the flow is counter-current. *Packed* and *fluidized beds* may have sprays providing scrubbing liquid either cocurrently or in a counter-current

fashion; a great variety of packing materials are employed; often these are designed to remove gases as well as particles from the flow, for which their relatively large surface-to-volume ratios and residence times are advantageous.

Pressure drop and collection efficiency equations are given in this chapter for many of these scrubber types. The *Scrubber Handbook*[6] and manufacturer publications provide fuller descriptions.

Typically, a scrubber system will include not only the scrubber but also a demister for removing the droplets separated in the scrubbing process, and a clarifier for concentrating the solids and removing them from the liquid effluent. For pollution control, both the demister and the clarifier are important. Demisters

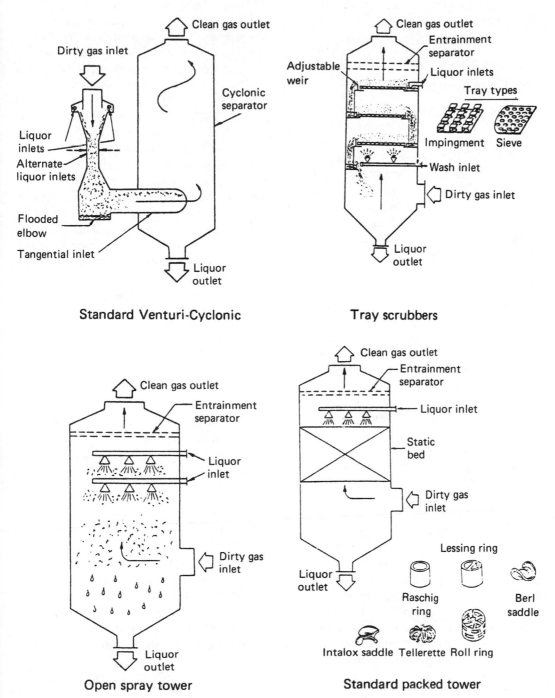

Figure 18.4. Examples of scrubbers.[6]

are covered in this chapter. The removal of solids from liquid streams by filtration is covered in Chapter 15.

18.2 POWER CONSUMPTION

18.2.1 Introduction

The difference in gas pressure at the inlet and outlet of the scrubber, due to the resistance to gas flow of the scrubber, is the pressure drop, ΔP (N/m^2 = Pa), the energy consumption per unit volume of air. For scrubbers having appreciable collection efficiency for submicron particulates, the energy costs can outweigh all the other costs, so minimizing pressure drop, while maintaining adequate collection efficiency, is important.

18.2.2 Definition

Pressure represents potential energy per unit volume, and the product of the pressure drop and the volume rate of flow of the gas or liquid represents power consumption. The metric units for pressure drop are N/m^2 and for the product of pressure drop and volume flow rate is $(N/m^2)(m^3/s)$ = N-m/s or watts (W). Frequently, pressure differences are measured by manometers and are given as inches of water (in. WC or in. WG); 6.3 in. of water pressure drop is equivalent to one horsepower per 1000 ft^3/min; 1.0 in. WG is 249 N/m^2. The electrical power consumed by a fan moving the gas through this pressure drop will be the product of the pressure drop and the gas volume flow rate, divided by the fan/motor efficiency (typically about 0.6), $Q\Delta P/E_f$. The pressure drop across spray nozzles, the volume flow rate of spray, and the electrical energy consumed by the pump have the analogous relationship. As energy costs rise, pressure drops across fans and pumps become more important as a design consideration.

18.2.3 Contacting Power

For many years, some scrubber experts thought that the collection efficiency of *any* type of scrubber would be the same for a given aerosol if the power consumption were the same.[7] Although this is a useful rule of thumb, it is not strictly correct. Testing an orifice scrubber, a multiple orifice contactor, and a variety of spray type scrubbers, Semrau et al.[8] found for test aerosols that were primarily submicrometer, that the various spray scrubbers gave the same collection efficiency for the same aerosol at given levels of power consumption, but the orifice scrubber did somewhat better. Calvert[9] also found that at conditions appropriate for scrubbing submicrometer particles efficiently, power consumption differences among scrubber types became appreciable. The difference becomes evident for high-energy scrubbing, which is precisely where it is most important. As total pressure drop is increased for a venturi or orifice scrubber, it becomes advantageous, in terms of collection efficiency, to divide the pressure drop equally between two or more scrubbers in series rather than concentrate the power consumption in a single-stage scrubber.[10, 11]

18.2.4 Other Types Of Power Consumption

Besides the power that is used for maintaining the pressure drop across the scrubber and across spray nozzles (where used), power may be required in the following:

1. Monitoring scrubber performance
2. Keeping scrubber elements above freezing temperatures
3. Filtering the scrubbing liquid
4. Drawing air to the scrubber and forcing it through a demister and to a stack
5. Heating scrubber outlet to decrease or prevent condensation in stack or plume
6. Electrostatic augmentation, if charged droplet scrubbing is used
7. Rotating a mechanical element within the scrubber to enhance droplet disintegration and particle capture
8. Generating steam for subsequent condensation to enhance scrubbing

9. Cooling the gas in association with condensation scrubbing
10. Handling and disposing of the solid and liquid wastes.

Most of these aspects are discussed at greater length below.

18.3   COLLECTION EFFICIENCY

18.3.1   Introduction

Scrubbers are designed to achieve adequate efficiency (or acceptable penetration) at minimum cost, and for high-energy scrubbers ($\Delta P > 3 \times 10^3$ N/m^2 or 12 in. WG), this means at nearly the minimum power consumption.[12]

18.3.2   Collection Efficiency and Penetration

The total mass collection efficiency (often called "total efficiency") is the difference between the inlet mass flux ($\dot{M}_0$) and the outlet mass flux ($\dot{M}$), divided by the inlet mass flux:

$$\overline{E} = \left(\dot{M}_0 - \dot{M} \right)/\dot{M}_0 \qquad (18.1)$$

The total penetration is just $\overline{Pt} = \dot{M}/\dot{M}_0$; it is the fraction of the mass that penetrates the device. The efficiency for a particular particle size (or narrow size range) is called the "fractional efficiency" and is, for particles in size range i:

$$E_i = \left(\dot{M}_0 - \dot{M} \right)_i/\dot{M}_{0i} \qquad (18.2)$$

To compare gas–particle separation devices, one generally needs their fractional efficiencies over the particle size range of interest. To get total mass efficiency from fractional efficiencies, one multiplies the fractional efficiencies, size interval by size interval, with the fractions of aerosol mass in each size interval and sums the products, closely related to the numerical integration of Eq. (18.4). For the scrubbers discussed below, equations will be given for determining collection efficiency as a

function of particle aerodynamic diameter, because impaction is almost always the predominant collection mechanism for scrubbers, especially for particles larger than about 1 μm, and impaction is a function of the particle aerodynamic diameter. The particle aerodynamic diameter, d_{pa}, is the diameter of a solid particle having the density of water that would have the same terminal settling velocity due to gravity as does the particle in question. From Stokes's law (with the Cunningham correction), this means that $(1 + 2.5\lambda/d_p)\rho_p d_p^2 = (1 + 2.5\lambda/d_{pa})\rho_w d_{pa}^2$. λ is the mean free path of the gas molecules, 0.065 μm at standard temperature and pressure; ρ_p is the particle density, ρ_w is the density of water.

Figure 18.5 gives the aerodynamic diameters of spherical particles of the densities indicated, as functions of particle physical (geometric) diameter.[13] For particles for which the Cunningham correction is negligible ($d_p \gg 10\lambda$), the aerodynamic diameter is $d_{pa} = (\rho_p/\rho_w)^{1/2}d_p$.

Generally, rather than using efficiency, one works with penetration, Pt, 1 minus the frac-

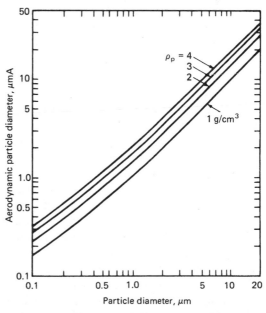

Figure 18.5. Relation between physical and aerodynamic diameter.[13]

tion collected. The aerosol aerodynamic diameter mass size distribution $m(d_{pa})$ is defined so that $m(d_{pa})$ represents the fraction (of the total mass concentration of the aerosol) having aerodynamic diameters between d_{pa} and $d_{pa} + dd_{pa}$. Thus, the distribution is normalized to unity:

$$\int_0^\infty m(d_{pa})dd_{pa} = 1 \qquad (18.3)$$

and the fraction (by mass) of particles that will penetrate the scrubber is given by

$$\overline{Pt} = \int_0^\infty Pt(d_{pa})m(d_{pa})dd_{pa} \qquad (18.4)$$

This is sometimes called the "integrated penetration" or the "total penetration." The product of the total penetration and the inlet mass concentration gives the outlet mass concentration, often the quantity of interest. The outlet particle size distribution becomes $Pt(d_{pa})m(d_{pa})/\overline{Pt}$.

Often it is convenient and sufficiently accurate to approximate the size distribution of an aerosol with a log-normal distribution. (This is the same as saying the logarithms of the particle diameters are distributed normally.) The two parameters describing a log-normal distribution are its median (d_g), which for a log-normal distribution equals its geometric mean, and its geometric standard deviation (σ_g). Of the aerosol mass, 68% is due to particles having diameters between d_g/σ_g and $d_g\sigma_g$; 95% of the mass is due to particles of diameters between d_g/σ_g^2 and $d_g\sigma_g^2$. The log-normal distribution is used in venturi scrubber design algorithms: Calvert[9] presented several figures that are convenient to use for scrubber design for particles having log-normal size distributions, once the cut diameter (d_{pc}), the diameter for which $E(d_{pc}) = 0.50$, is determined. Others have used the log-normal assumption for closed-form evaluations of Eq. (18.4) by approximating the fractional efficiency curve as a cumulative log-normal curve.[15, 17, 106]

18.3.3 Single Obstacle Efficiency

Formulas for scrubber collection efficiency require the single collector (obstacle) efficiency, η, which involves a number of physical mechanisms. It is defined for a single collector in an unbounded stream as:

$$\eta = \frac{\text{flow area cleaned}}{\text{collector cross-sectional area}} \qquad (18.5)$$

The calculation of η depends in part on the flow past the collector. Two flow models are commonly in use: viscous flow and potential flow. Viscous flow is an appropriate model when the obstacle Reynolds number is small; that is, when:

$$Re_c = \rho_G(U_G - U_c)D_c/\mu_G \ll 1 \qquad (18.6)$$

Although the motion of the dust particles in the gas stream often meets this Reynolds number criterion, the flow around the collectors usually does not. (A flow of air at 0.1 m/s past a fiber or droplet 100 μm. in diameter gives $Re_c \sim 1$.) The model of potential flow is derived for $Re_c \gg 1$, but even in this regime it is appropriate only up to where the flow separates and forms a wake that trails behind the obstacle.

The single collector efficiency can be calculated for various collection mechanisms separately and then combined as though the mechanisms acted independently.[18] It is more accurate and more difficult to solve for the particle trajectories in the appropriate flow field, including the collection forces and mechanisms.[19-21]

18.3.4 Collection Mechanisms

When a dust particle strikes the collection surface because of its inertia, the collection is said to be due to *impaction*. The impaction process can be characterized by the impaction parameter, ψ:

$$\psi = \frac{U_G C(d_{ae})\rho_w d_{ae}^2}{18\mu_G D_c} \qquad (18.7)$$

where D_c is the collector diameter and μ_G is the gas velocity.

The following expression approximates the single sphere efficiency for impaction:[6]

$$\eta_I = \psi^2/(\psi + 0.35)^2 \qquad (18.8)$$

Impaction is usually the most important collection mechanism for scrubbers, for particles larger than 0.1 μm. Figure 18.6 gives the impaction efficiency for several collector geometries versus the impaction parameter.[22]

"Interception" occurs when a particle strikes a collector even though the particle center would not have. Incorporating it correctly with other collection mechanisms really means altering the boundary conditions for the problem. Define N_R as the ratio of (spherical) particle radius to (spherical or cylindrical) collector radius. The incremental efficiency due to "interception" (above that of impaction, if operative) is between $2N_R$ and $3N_R$ for potential flow around a spherical collector and between N_R and $2N_R$ for potential flow around a cylinder, for inertialess and highly massive particles, respectively.[23, 107]

Capture by *diffusion* occurs because of the Brownian motion of the particles. It becomes appreciable only as the Peclet number (the gas velocity times the collector diameter divided by the particle diffusivity) becomes much less

than 1, which is rarely the case for $d_p >$ 0.1 μm.

Electrostatic forces have been employed to augment the collection efficiency of scrubbers. The case in which the particles and the collectors are charged (Coulombic interaction) typically produces a much greater effect than those cases in which either the collectors or the particles are charged, but not both (image force interactions). The "migration velocity" is the terminal velocity of a particle at the surface of the collector, due to the electrical forces. For electrostatic interaction to be important, the migration velocity should not be very much smaller than the product of the relative velocity and the collection efficiency due to all other mechanisms, $\eta(U_G - U_c)$. Figure 18.7 gives the migration velocities calculated by assuming that the particles were charged to saturation in a 10 kV/cm field (or they are uncharged) and the same field is produced by the collectors (or they are uncharged), for particles of the size indicated and a 100-μm spherical collector.[24] Electrostatic collection is intrinsically energy-efficient because the collection force can be applied directly to the particles, rather than indirectly to the particles through moving the gas.

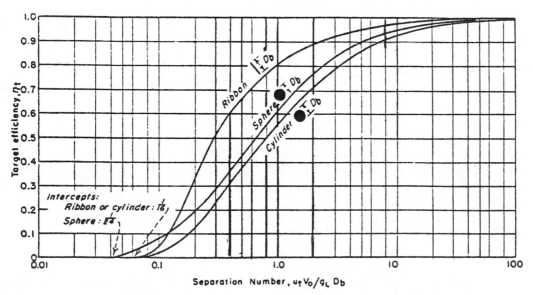

Figure 18.6. Target efficiency of spheres, cylinders, and ribbons.[22]

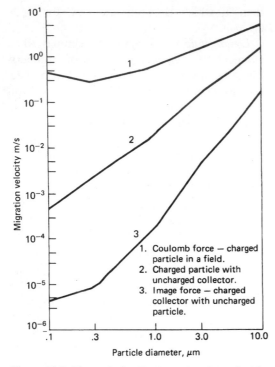

Figure 18.7. Theoretical collection migration velocities for three electrostatic mechanisms.[24]

The collection efficiency for a scrubber can be obtained from the collection efficiency (η) of its obstacles (droplets, beads, fibers, etc.) as follows. The number of particles collected in time, dt, as the particles flow (parallel to the x axis) through an infinitesimal volume, $dV = A\,dx$, is:

$$dN_p = -\eta n_p A_c (U_G - U_c)\,dt \quad (18.9)$$

where dN_p is the number of particles collected, n_p is the particle number concentration, A is the scrubber cross-sectional area, A_c is the obstacles (collectors) cross-sectional area, perpendicular to the flow, and U_G and U_c are the velocities of the gas and the collectors (if moving).

If the concentrations and velocities of collectors and particles are uniform perpendicular to the flow, then (using $dt = dx/U_G$):

$$\frac{dn_p}{n_p} = -\eta \left(\frac{U_G - U_c}{U_G}\right)\frac{A_c\,dx}{V} \quad (18.10)$$

From this point, the derivation can be done with various degrees of sophistication:

1. The calculation of the single droplet efficiency η can include some or all of the following mechanisms: impaction, interception, diffusion, electrostatic interactions, diffusiophoresis, thermophoresis.
2. The velocities U_G and U_c can be calculated in detail, including their dependence on position.
3. Any spatial variation of the collectors can be taken into account.
4. Various averages of collector areas, A_c, can be used, or a functional form for their distribution employed.

If the various quantities in the right-hand side of Eq. (18.10) are uniform, then

$$Pt = \frac{n_p}{n_{p0}} = \exp\left(-\eta\left[\frac{U_G - U_c}{U_G}\right]\frac{A_c L}{V}\right)$$

$$(18.11)$$

where n_{p0} = concentration at $x = 0$, and n_p is the concentration at $x = L$.

For stationary collectors, this becomes:

$$Pt = \exp(-\eta A_c L/V) \quad (18.12)$$

18.3.5 Predicting Total Efficiency

In sections to follow, fractional efficiency equations will be given for various scrubber types. In general, they come from assuming particle and collector concentrations to be uniform perpendicular to the mean flow and inertial impaction to be the dominant collection mechanism. Once the fractional efficiencies are known, the total efficiency is determined from:

$$\bar{E} = 1 - \overline{Pt} = \int_0^\infty E(d_{pa})m(d_{pa})dd_{pa} \quad (18.13)$$

or

$$\bar{E} = \sum_{i=1}^m E_i(d_{pa_i})\left(\dot{M}_{0i}/\dot{M}_0\right) \quad (18.14)$$

18.4 SCRUBBER SELECTION

18.4.1 Introduction

Generally, power consumption costs and other operating costs for scrubbers operating at high pressure drops (greater than about 2.5 kPa) are greater than or comparable to the annualized equipment costs. Further, the collection efficiency of one type of scrubber compared with another, on a given aerosol, will be about the same for a given pressure drop (thus a given energy consumption). Thus, the choice among scrubber types may depend on factors other than collection efficiency and power consumption. Krockta and Lucas[25] presented a detailed list of the factors to be considered in selecting a scrubber for a particular application, a list prepared by a committee of the Air Pollution Control Association. Among these factors are: economic aspects, including capital expenditures, operating and maintenance costs; environmental factors, such as climate and the resources for power and waste treatment; engineering factors, including such particle characteristics as size distribution, concentration, solubility in scrubbing liquid, chemical reactivity, abrasiveness; gas characteristics, such as temperature, humidity, pressure, and chemical composition; and such scrubbing liquid characteristics as viscosity, density, surface tension, and solids concentration.

Although a scrubber type might be operable at pressure drops outside its conventional design range, the information in Table 18.2 is useful in suggesting what scrubbers are appropriate for achieving at least 80% efficiency on the particle sizes indicated.[26]

The following special considerations apply.[26]

1. *Gas absorption.* For collecting gases and vapors as well as particles, counter-current flow is to be preferred, with steps taken to maximize surface area and contact time, using, for example, a packed bed if plugging can be avoided.
2. *Plugging.* Fibrous beds and packed beds are susceptible to plugging, and generally one should use the more open types of scrubbers (venturi, orifice, preformed spray) for heavy aerosol and concentrations (≥ 10 g/m^3); high recirculation rates may also lead to plugging of spray nozzles; it may be advantageous to have a low-energy scrubber or a cyclone upstream of a high-energy scrubber to help prevent plugging.
3. *Reentrainment.* Once the scrubbing liquid has captured the particles, the liquid must be retained; scrubbing droplets must be captured by an efficient demister; failure to do so for dyes and pigments, for example, can be serious.
4. *Stack-condensate fallout.* The condensation of scrubbing liquid within the exhaust stack can cause spray to be generated from the stack walls if stack velocities become too high.
5. *Freezing.* Cold-weather operation must include provisions for preventing freezing during operation and for preventing dam-

Table 18.2. "Minimum" Particle Size for Various Types of Scrubbers.[a]

	PRESSURE DROP (in. water)	PRESSURE DROP (kPa)	MIN. PARTICLE SIZE (μm)[b]
Spray towers	0.5–1.5	0.12–0.38	10
Cyclone spray scrubbers	2–10	0.5–2.5	2–10
Impingement scrubbers	2–50	0.5–12	1–5
Packed- and fluidized-bed scrubbers	2–50	0.5–12	1–10
Orifice scrubbers	5–100	1.2–25	1
Venturi scrubbers	5–100	1.2–25	0.8
Fibrous-bed scrubbers	5–110	1.2–28	0.5

[a] Adapted from Ref. 26.
[b] Smallest particle size for which the scrubber has at least 80% collection efficiency.

age due to freezing when the scrubber is not in operation.

A limited survey of scrubber applications was carried out by Calvert et al.,[6] the results of which are shown in Table 18.3.[27] (It was noted that the number of sources surveyed was small.) The following comments were among those made:[27]

1. Packed bed and fibrous scrubbers were used in applications requiring the collection of gases, liquids, and those particles (soluble or nonadhering) that tend not to plug.
2. Preformed (hydraulic) sprays were mainly used to capture gases.
3. Mechanically aided scrubbers were rarely found.

From this table, it appears that centrifugal scrubbers were preferred for the coarse dusts (≥ 10 μm) from crushing operations, but for the fine dusts from smelting operations (much of it ≤ 1 μm), gas-atomized scrubbers were dominant, and hydraulic ("preformed") spray scrubbers were of secondary importance.

18.4.2 Summary

For control of coarse aerosols, such as powders formed by disintegration of bulk material, low-energy, open-structure scrubbers such as those with preformed sprays or impingement scrubbers, are often applicable with little risk of plugging. For fine aerosols, such as from powders made from condensation processes (including gas-phase reactions), high-energy scrubbers, such as venturi scrubbers, can be applied successfully, with attention to the prevention of plugging and reentrainment.

18.5 ATOMIZED SPRAY SCRUBBERS (VENTURI, ORIFICE, IMPINGEMENT)

18.5.1 Introduction

Directing a high-velocity flow of gas across a liquid surface forms drops, which can then be used as collectors of particles in the gas stream. A variety of atomizing scrubbers work this way. Three different examples are shown in Figure 18.8. In atomizing scrubbers the air flow controls both the droplet size distribution and the ratio of droplet volume to gas volume, the liquid-to-gas flow ratio (Q_L/Q_G), but in hydraulic spray scrubbers the droplet size distribution can be changed independently of the liquid-to-gas ratio, and in various column-type scrubbers, the liquid flow rate can also be changed without affecting the collector size,

Table 18.3. Results of Survey of Scrubber Applications.[27]

						SCRUBBER TYPE				
					GAS-					
		MASSIVE	FIBER	PREFORMED	ATOMIZED			IMPINGE-	MECH.	MOVING
PROCESS	PLATE[a]	PACKING	BED	SPRAY	SPRAY	CENTRIFUGAL	BAFFLE	MENT	AIDED	BED
Calcining	6	2	—	13	21	—	—	43	—	—
	(1)[b]	(1)	(0)	(5)	(23)	(0)	(0)	(3)	(0)	(0)
Combustion	17	—	—	5	2	2	—	29	—	9
	(3)	(0)	(0)	(2)	(2)	(1)	(0)	(2)	(0)	(2)
Crushing	6	—	—	—	—	26	—	14	—	5
	(1)	(0)	(0)	(0)	(0)	(11)	(0)	(1)	(0)	(1)
Drying	39	—	—	10	18	70	100	—	25	64
	(7)	(0)	(0)	(4)	(19)	(30)	(1)	(0)	(1)	(14)
Gas removal	17	72	40	45	9	2	—	14	50	5
	(3)	(33)	(2)	(18)	(10)	(1)	(0)	(1)	(2)	(1)
Liquid-mist	0	24	60	7	—	—	—	—	—	—
recovery	(0)	(11)	(3)	(3)	(0)	(0)	(0)	(0)	(0)	(0)
Smelting	17	2	—	20	50	—	—	—	25	18
	(3)	(1)	(0)	(8)	(54)	(0)	(0)	(0)	(1)	(4)

[a] Read vertically. Example: 39% of all plate-type scrubbers are used to control discharges from drying processes.
[b] Numbers in parenthesis refer to number of operators reporting information to the survey.

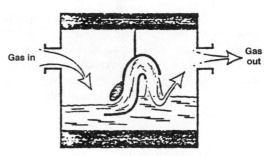

Figure 18.9. An impingement scrubber.[27] Excerpted by special permission from *Chemical Engineering* (Aug. 29, 1977), copyright © 1977 by McGraw-Hill, Inc., New York, NY 10020.

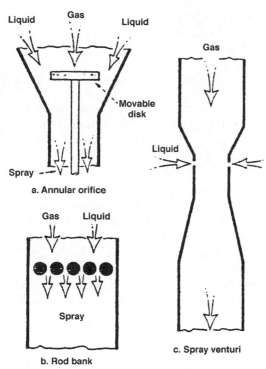

Figure 18.8. Three atomizing scrubber types.[27] Excerpted by special permission from *Chemical Engineering* (Aug. 29, 1977), copyright © 1977 by McGraw-Hill, Inc., New York, NY 10020.

which in turn affects the collector efficiency. A *venturi scrubber* has a converging section, a throat, and a diverging section. It accelerates the gas in the converging channel, introduces liquid (often as a spray) near the throat, where most of the particle collection occurs, then decelerates the gas and droplets in the diverging sections, generally quite gradually tapered. It is very widely used and receives special attention here. *Orifice scrubbers* use much the same principles. They are generally made from a single opening put in a place in the duct, with the plate being wetted by a flow of liquid which is then atomized at the plate edge. *Impingement scrubbers* direct a flow of gas at the surface of a liquid, using a variety of geometries, causing intimate mixing of liquid and particles due to atomization and turbulence. An example is shown in Figure 18.9.

18.5.2 Venturi Scrubber

Venturi scrubbers are quite popular, especially in applications (such as metallurgical emissions control) where efficiencies of 90% or more are required for particles of 1 μm diameter or smaller. Such applications may require pressure drops of 10 kPa or more. Venturis are relatively simple to build, using geometries whose cross-sections are either circular or rectangular. Figure 18.8a shows an adjustable-throat venturi scrubber. Here liquid is introduced near the top of the converging section, to be atomized by the high-velocity gas at the throat. The diverging section is often followed by a flooded elbow, and the material not caught at the elbow is captured in a mist eliminator, such as a cyclone. Adjustable throats are needed where the gas volume flow is variable. For rectangular throats, the area can be changed by adjusting the throat width; for circular throats, usually a disk will be inserted to form an annular throat, which can be adjusted conveniently by moving the disc to various positions in the converging section.

18.5.2.1 Power Consumption

The pressure drop is the main contributor to power consumption. Without liquid flow, a venturi would have a pressure drop of about one-tenth the gas "velocity pressure," the latter being $0.5\rho_G U^2$. This is small in comparison with the energy consumed in accelerating the droplets to the gas velocity, some of which energy is regained in the expanding section. After reviewing several correlations for pressure drop, Yung et al.[28] recommended the following equation (a misprint has been corrected):

$$\Delta P = -2\rho_L U_G^2 \left(\frac{Q_L}{Q_G} \right)$$

$$\times [1 - X^2 + (X^4 - X^2)^{0.5}] \quad (18.15a)$$

where Q_L and Q_G are the liquid and gas volume flow rates, U_G is the gas velocity in the throat, and X is the dimensionless throat length:

$$X = 1 + 3L C_{D1} \rho_G / 16 D_d \rho_L \quad (18.15b)$$

in which L is the length of the throat and C_{D1} is the drag coefficient for the droplets at the throat:

$$C_D = 0.22 + (24/Re_T)(1 + 0.15 Re_T^{0.6}) \quad (18.16)$$

$$Re_T = \rho_G U_G D_T / \mu_G \quad (18.17)$$

D_T is the throat diameter and D_d is the diameter that characterizes the droplets. For air and water, the expression for atomized droplet D_d of Nukiyama and Tanasawa becomes[29] (SI units):

$$D_d = 0.0050/U_G + 0.92(Q_L/Q_G)^{1.5} \quad (18.18)$$

with D_d in m, U_G in m/s, and the flows in m³/s. (See Table 18.4.) Equation (18.15a) for pressure drop compared well with data for liquid-to-gas ratios of 10^{-4} to 10^{-3}. It was assumed in deriving the equation that all the drops are accelerated in the venturi throat and that none of the momentum thus imparted is recovered as pressure gain when the drops decelerate in the diffuser, that there is no initial axial component of velocity for the droplets, that the flow is one-dimensional, incompressible, and adiabatic, that at any cross-section the liquid fraction is small, and that the new pressure difference of wall friction minus pressure recovery in the diffuser is negligible. If the throat length is long enough to accelerate the droplets to the velocity of the gas, the term in brackets becomes 0.5 and Eq. (18.15a) reduces to that presented earlier by Calvert:[30]

$$\Delta P = \beta \rho_L (Q_L/Q_G) U_G^2 \quad (18.19)$$

Calvert's original value for β was 1.00,[30] but it has been found that $\beta = 0.85$ agrees better with experimental data.[31]

Table 18.4. Droplet Sizes Predicted by Nukiyama – Tanasawa Equation for Various Gas Velocities for Air and Water

U_G (m/s)	$0.0050/U_G$ (mm)	Q_L/Q_G (10^{-3})	$0.92(Q_L/Q_G)^{1.5}$ (mm)		D_d (mm)
1	5.0	1	0.029		5.0
		10		0.92	5.9
3	1.7	1	0.029		1.7
		10		0.92	2.6
10	0.50	1	0.029		0.53
		10		0.92	1.42
30	0.17	1	0.029		0.20
		10		0.92	1.09
100	0.050	1	0.029		0.079
		10		0.92	0.97
300	0.017	1	0.029		0.046
		10		0.92	0.94

18.5.2.2 Collection Efficiency

Mathematical models developed to predict penetration and pressure drop for venturi scrubbers have limitations due to the assumptions that go into their derivations, and conclusions based on model results must be interpreted cautiously. However, good models can help in obtaining improvements in scrubber performance.

Models for venturi scrubber performance have been developed and reported by Johnstone et al.,[32] Calvert,[30,33] Boll,[34] Taheri and Shieh,[35] Goel and Hollands,[36] and by Yung et al.[28,29] Assumptions made in these derivations differ; they use various relationships between impaction parameter and single droplet collection efficiency, make a variety of assumptions regarding drop velocity at the time atomization occurs, and assume particle collection occurs in various parts of the venturi. All the models assume monodisperse droplets and complete liquid utilization, except that Taheri and Shieh[35] incorporated particle and droplet concentration distribution.

The model most frequently used for penetration and pressure drop in a venturi scrubber is probably the model presented by Calvert[30,33] and Calvert et al.[16] These equations are used in this section. This approach considers the same processes described in all venturi models, but its equations are more tractable. Calvert et al.[16] showed that agreement between the theoretical predictions and data is generally good, although this agreement is helped by an adjustable constant, f, in the equation for the penetration. Drop velocity at atomization is assumed to be "f" times the gas velocity in the venturi throat, where f is between 0 and 1. With proper selection of this constant, the theory and data can be made to agree. The utility of the Calvert penetration model largely depends on the extent to which f remains constant for all venturi scrubbers and for all aerosols. Values of f from 0.25 to 0.5 were reported by Calvert,[30,31] the larger values being appropriate for more hydrophilic particles and larger gas flow rates.

The Calvert[6,16] equation for penetration through a venturi is:

$$-\ln Pt = \frac{2Q_L \rho_L D_d U_G}{55 Q_G \mu_G K}$$

$$\times \left[0.7 + fK - 1.4 \ln\left(\frac{0.7 + fK}{0.7} \right) \right.$$

$$\left. - \frac{0.49}{0.7 + fK} \right]. \qquad (18.20)$$

The parameter K is just 2ψ [Eq. (18.7)].

This equation can fairly readily be programmed on a computer or even a programmable calculator.[38] Using dimensional analysis, we identified a group that helps describe scrubber performance, the performance number, N_p:[12]

$$N_p = C\rho_p d_p^2 \Delta P / 18\mu_G^2 = \tau \Delta P / \mu_G \quad (18.21)$$

where τ is the particle aerodynamic relaxation time.[23] Let the pressure drop be given by Eq. (18.19); then this model predicts the following minimum penetration (Pt^*) at a given pressure drop:[39]

$$Pt^* = \exp\left[-0.124(f^2/\beta)N_p \right] \quad (18.22)$$

Figure 18.10 shows the number of transfer units

$$N_{tu} = -\ln Pt \qquad (18.23)$$

versus performance number for experiments with small venturis and for the minimum penetration conditions, as calculated with Eq. (18.22). Ideally, all liquid will be fully atomized to droplets immediately upon injection, and all droplets will accelerate to the full gas throat velocity. In this case, which represents full liquid utilization with no pressure regain due to droplet deceleration, the values of constants f and β are unity and f^2/β becomes unity as well. Smaller values of f^2/β represent less complete liquid utilization and little or no gain.

The dependence of f or β on variables under the control of the venturi designer has not yet been quantified. Calvert[31] suggested f and β may depend upon the size of the ven-

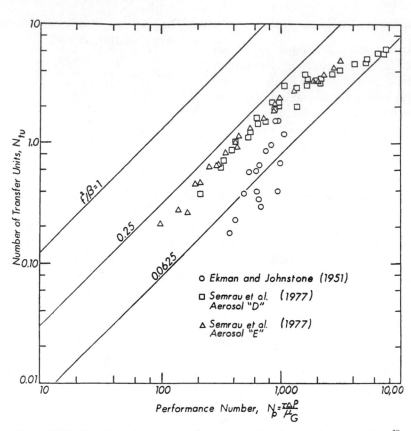

Figure 18.10. Transfer units versus performance number for atomizing scrubbers.[39]

turi, the method of liquid injection, or other factors. In practice, f and β are seldom if ever known with certainty; in designing a venturi they are estimated on the basis of past experience.

Two more venturi scrubber models have recently appeared in the literature.[111, 112] The results of the first indicated "dispersity of the droplet size distribution only slightly affects collection efficiency over the operating range normally encountered."[111] The other author concluded that polydispersity makes a difference and that "calculations based on the assumption that droplets are monodisperse result in an underestimation of the efficiency."[112]

18.5.2.3 Optimization of Design

The factors to be decided upon in the design of the scrubber include: angles of convergence and divergence, throat cross-sectional area and length, and liquid-to-gas ratio. The angles of convergence and divergence are thought not to be critical within the range of conventional designs (20 to 25° and 5 to 7°). The cross-sectional area will be determined by the gas volume flow rate and the desired throat velocity. The throat length criterion has been proposed as:[29]

$$4/3 \leq C_{D1} L \rho_G / D_d \rho_L \leq 2 \quad (18.24)$$

It represents a compromise between increased particle collection and increased frictional flow resistance as throat length is increased. The liquid-to-gas ratio (typically around 10^{-3}) affects both the pressure drop and the collection efficiency; in general, increased values of the ratio Q_L/Q_G improve collection efficiency at a given pressure drop but also increase the amount of water to be handled for recirculation and disposal.

For a fixed value of scrubber performance number, N_p, and predetermined value of f^2/β, Eq. (18.23) can be shown to depend only upon the fK product.[39] The value of fK for which penetration is minimized is:

$$fK = 1.10 \qquad (18.25)$$

according to the Calvert model, assuming $X \gg 1$. Smaller values of throat length lead to larger values of fK being optimal. Reasons for an optimum value for fK can be discussed in terms of an optimum droplet diameter. A drop larger than the optimum will sweep through a larger volume of particle-laden gas and have a larger surface area. However, a larger droplet will also have a smaller single droplet collection efficiency, owing to impaction, contribute more to pressure drop, and, for a given amount of liquid used, fewer such drops will be produced. The optimum droplet diameter reflects the best compromise among these factors. Although Semrau et al.[8] did not find, experimentally, an independent effect of liquid-to-gas ratio and pressure drop, Ekman and Johnstone[40] and Muir et al.[41] found that venturi efficiency improved at a given pressure drop when the liquid-to-gas ratio increased, thus when thc droplet size increased.

For any selected value of pressure drop, for particles of specified d_{pa} and with f and β fixed, it is possible to predict the gas velocity and liquid-to-gas ratio that should produce drops of optimum size and allow operation at theoretically maximum efficiency, using Eqs. (18.21), (18.22), and (18.24). Three equations can be written, using as unknowns the gas velocity, liquid-to-gas ratio, and droplet diameter at optimum conditions. The solutions to these equations show the operating conditions necessary to produce theoretically optimum performance. The simultaneous solution of these equations allows determination of optimum gas velocity, U_G^*, and optimum liquid-to-gas ratio, $(Q_L/Q_G)^*$:[39]

$$U_G^* = \sqrt{\frac{0.005 + (2.5 \times 10^{-5} + 6.69 f\tau (\Delta P/\rho_L)^{3/2})^{1/2}}{3.65 f\tau}}$$

$$(18.26)$$

The diameter of the "optimum drops" that correspond to these U_G^* and $(Q_L/Q_G)^*$ conditions can be found from Eq. (18.18). (U_G must be in units of m/s and $\Delta P/\rho_L$ in N·m/kg, however.)

Nukiyama and Tanasawa found Equation (18.18) empirically for the following conditions: 70 m/s $< U_G <$ 230 m/s, $8 \times 10^{-5} <$ $Q_L/Q_G < 1 \times 10^{-3}$, corresponding to atomized drops 20 μm $< d_d <$ 100 μm in diameter.[39] The present analysis indicates that for particles larger than about 0.5 μm in diameter, for venturis and long throats, optimum sized atomized drops are larger than those for which Eq. (18.18) can be used with confidence. Other relationships for the diameter of atomized drops might be used with greater confidence for larger atomized drop diameters.[34]

This analysis is for collection of monodisperse particles by monodisperse droplets. The extension of this work to polydisperse aerosols requires further investigation. As a first approximation, one might use the simple expression for penetration:[6,9,27]

$$Pt \doteq \exp\left(-A d_{pa}^B\right) \qquad (18.27)$$

($B \doteq 2$) and use the curves presented in these references for the total mass (integrated) penetration for log-normal aerosols of various values of σ_g as functions of d_{pc}/d_{pg}. For this approximation, performance is optimized by determining the particle cut diameter needed to achieve the integrated penetration desired, then finding the optimal conditions as described previously to give 50% efficiency for a particle of that cut diameter (see Fig. 18.11).

18.5.2.4 Design Example[39]

Consider the design of a venturi scrubber to collect particles 0.5 μm in diameter with 90% efficiency from 5 m³/s of gas, using the minimum possible pressure drop. Assume $\rho_p =$

2000 kg/m^3, $\rho_G \doteq 1.2$ kg/m^3, and $\mu_G = 1.8 \times 10^{-5}$ kg/m · s, respectively. Further, follow Calvert's suggestion that often $f = 0.5$ and $\beta = 0.85$.[31]

First, determine the number of transfer units required:

$$N_{tu} = -\ln Pt = -\ln(0.10) = 2.30 \quad (18.28)$$

Next, find the minimum scrubber performance number:

$$N_p = \frac{N_{tu}}{0.0124f^2/\beta} = \frac{2.30}{0.0124(0.5)^2/0.85}$$

$$= 631 \quad (18.29)$$

Particle relaxation time can be found from:

$$\tau = \frac{d_p^2 \rho_p C}{18\mu_G} = \frac{(5 \times 10^{-7})^2(2000)(1.33)}{18(1.8 \times 10^{-5})}$$

$$= 2.1 \times 10^{-6} \text{ s} \quad (18.30)$$

The lowest pressure drop theoretically necessary under these conditions can now be obtained from Eq. (18.21):

$$\Delta P = \frac{N_p \mu_G}{\tau} = \frac{(631)(1.8 \times 10^{-5})}{2.1 \times 10^{-6}}$$

$$= 5410 \text{ Pa} \quad (18.31)$$

The gas velocity in the venturi throat (assumed very long) required to generate optimum-sized drops at this pressure drop is given by Eq. (18.26):

$$U_G = \sqrt{\frac{0.005 + \left(2.5 \times 10^{-5} + 6.69(0.5)(2.1 \times 10^{-6})\left(\frac{5410}{1000}\right)^{3/2}\right)^{1/2}}{3.65(0.5)(2.1 \times 10^{-6})}} = 64 \text{ m/s} \quad (18.34)$$

The cross-sectional area of the venturi throat to provide this gas velocity is:

$$A = Q_G/U_G = (5 \text{ m}^3/\text{s})/(64 \text{ m/s})$$

$$= 0.078 \text{ m}^2 \quad (18.32)$$

A circular throat 0.32 m in diameter will serve.

The liquid flow rate required can now be found from Eq. (18.27):

$$Q_L = \frac{Q_G \Delta P}{\rho_L \beta U_G^2} = \frac{(5)(5410)}{(1000)(0.85)(64)^2}$$

$$\times 7.8 \times 10^{-3} \text{ m}^3/\text{s} \quad (18.33)$$

With the diameter of the venturi throat and liquid flow rate fixed, the essential design of the venturi is complete.

18.5.3 Orifice Scrubbers

Orifice scrubbers are often made by inserting a plate with a hole or slit into a vertical run of ducting, irrigating the plate so that scrubbing droplets are formed at the aperture. They operate quite similarly to venturi scrubbers and are effectively venturis with zero throat length and 180° angle of convergence and divergence. The pressure drop is given by Eq. (18.19) for venturi scrubbers. In their summary of pressure drop and efficiency equations useful for various scrubbers, Yung and Calvert[42] used the same equations, (18.19) (with $\beta = 0.85$) and (18.20). Venturi and orifice scrubbers behave almost identically.[8, 21]

A wetted butterfly valve was tested by Taheri et al.[43] and found to be an inexpensive variable orifice scrubber, convenient for use on variable gas flows. Pressure drop and collection efficiency data were given in their article, and although they did not compare the performance with that of other orifice scrubbers, it should be much the same.

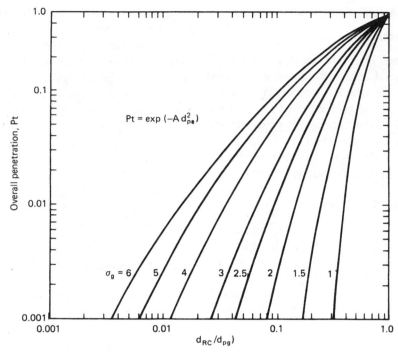

Figure 18.11. Integrated (overall) penetration as a function of cut diameter and particle parameters.[13]

18.5.4 Impingement Scrubbers

There are a variety of designs for relatively low-energy impingement scrubbers, such as that shown in Figure 18.9. These gas-atomizing scrubbers have collection efficiencies similar to those of venturi or orifice scrubbers operating at the same pressure drop.[27, 42] Equation (18.19) for pressure drop and (18.20) for efficiency apply, with $\beta = 1.0$.[42]

18.5.5 Multiple-Stage Scrubbing

In 1976, we proposed that for high-energy scrubbing there were situations in which multiple-stage devices, such as two venturis in series, would be more efficient for a given level of energy consumption than would a single-stage device, even without particle growth or other condensation effects.[10] This contradicted the contacting power theory and conventional beliefs. In 1979, Muir and Meheisi demonstrated the truth of this hypothesis by experiments with venturi scrubbers.[11] For venturi scrubbers, the power advantage of multiple-stage scrubbing becomes appreciable when the scrubber performance number, $N_p = \tau \Delta P / \mu_G$, becomes greater than or on the order of 10^3.[12] It is likely the same will hold true for other atomization scrubbers. Added improvement due to particle growth and water vapor flux forces make multiple-stage scrubbing potentially even more attractive.

18.5.6 Summary

Venturi and other gas-atomized scrubbers have similar pressure drop and efficiency characteristics. They can be used over a wide range of operating conditions, and are simple, rugged, and resistant to plugging. The formation of fine spray means demisters are essential to their successful operation.

18.6 HYDRAULIC SPRAY SCRUBBERS

18.6.1 Introduction

The scrubbers discussed in this section all have preformed sprays, produced by nozzles and having their droplet size distributions determined by nozzle geometry and the properties of the liquid but not by the properties of the gas or its flow rate. The types of scrubbers covered here are: spray chambers, ejectors, and centrifugal (cyclone) spray scrubbers.

18.6.2 Spray Chambers

Spray chambers are conceptually quite simple. Gas and particles flow through a chamber with sprays directed co-current, cross-current, or counter-current to the flow, the latter being advantageous if gases are also to be removed in the scrubbing process. A demister is generally used as well.

18.6.2.1 Power Consumption

The power consumed in a spray chamber is the product of the gas volume flow rate and gas pressure drop (usually negligible) plus the power consumed by the nozzles, the sum of the products of their volume flow rates and pressure drops. Similarly, the demister consumes power due to the added gas flow resistance. The particle size for which collection efficiency becomes negligible is that for which the impaction parameter is substantially less than 1; therefore, droplet size and velocity are important. Throughout the chamber, the droplet velocity relative to the gas will be within the range of the initial droplet velocity and the droplet terminal settling velocity, the larger limiting how small a particle can be captured. The initial velocity of the droplet as it leaves the nozzle will be, in the potential flow approximation:

$$U_d = (2\Delta P_L/\rho_L)^{1/2} \qquad (18.35)$$

Since the impaction parameter depends on $d_{pa}^2 U_d$, the pressure on the nozzles would have to be increased by about $2^4 = 16$ to get the same single droplet efficiency on a particle half the size. Actually, the situation is more complicated: the liquid and gas flow rates, the trajectory of the droplet and its stopping distance, and the geometry of the scrubber should go into such an analysis. It has been found that spray scrubbers are not quite as efficient in collecting particles as are venturi scrubbers for the same power consumption.[8] Preformed spray scrubbers use rather high liquid-to-gas ratios, 4 to 12×10^{-3}, and this high water usage is often coupled with problems of corrosion, erosion, and plugging of the spray nozzles.[27]

18.6.2.2 Collection Efficiency

The penetration of a cross-current spray chamber can be approximated by:[31]

$$Pt = \exp[-\eta_I 3Q_L h/2Q_G D_d] \quad (18.36)$$

where η_I is the impaction collection efficiency [Eq. (18.8)] and h is the dimension of the scrubber traversed by the drops. For D_d it is consistent with the derivation of this equation to use the ratio of the mean cubed diameter to the mean squared diameter, $\overline{D_d^3}/\overline{D_d^2}$, a ratio also known as the "Sauter mean diameter." For a counter-current spray operated vertically and having height z, Pt can be estimated from Eq. (18.36) by replacing h with $zU_d/(U_d - U_G)$.[31] The choice of velocity at which to evaluate Eq. (18.36) can be problematic, and we emphasize the equation is quite approximate.

18.6.3 Ejectors

The motion of droplets ejected from a nozzle spraying co-currently with the gas flow can be used to collect particles and to move the gas which bears them. This eliminates the need for fans in corrosive and erosive atmospheres.[44] Ideally, the momentum transferred from the nozzle would go entirely to the mixture of gas and droplets. Some units need $Q_L/Q_G \sim 10^{-2}$ to generate drafts of even 250 Pa, however, Harris[44] analyzed theoretically such a scrubber to predict collecting efficiencies for particles, vapors, and gases.

Where the pressure for the nozzles can be obtained from utilizing waste heat, such scrubbers may be economically advantageous.[45] One ejector was found to have collection efficiency quite similar to a venturi scrubber for the same power consumption.[46]

18.6.4 Centrifugal Scrubbers

By impacting a rotary motion to a gas, a cyclone can remove particles by a mechanism similar to impaction. Introducing a spray at the inlet of a cyclone can enhance particle collection by both capturing particles on the droplets and by preventing reentrainment of captured particles from the walls. The cyclone, or other centrifugal scrubber, may serve as its own demister and is resistant to plugging.

The pressure drop across the cyclone will be somewhat greater than what it would be without the spray. The penetration is approximately what would be predicted without the spray times $\exp[-3Q_L h \eta_I/2Q_G D_d]$, where h is the difference between the inner and outer radius of the cyclone.[41]

A centrifugal scrubber was tested[42] and found to have collection efficiency equal to that predicted for a venturi scrubber, Eq. (18.20), with $f = 0.4$, which is within the range of performance found for venturis ($f = 0.25$ to 0.50).

18.7 WETTED PACKED BEDS AND FIBROUS MATS

18.7.1 Introduction

Wetted packed beds and fibrous mats can be advantageously used for the collection of mists, gases, and vapors. They tend to plug, however, when used to capture insoluble particulate material, so they may not find much use in powder technology. The tendency to plug depends on the particle size distribution, the gas and liquid flow rates, the particle concentrations in the gas and in the liquid, and the dimensions of the interstices in the bed, so that there are situations in which such scrubbers might be employed.

18.7.2 Packed Beds

Packed beds have been used for the separation of one gaseous constituent from another and to a lesser extent for the separation of particulates from gases. For a collector having packing with a mean surface-to-volume diameter D_{sv} (equal to six times the total solid volume of the packing material divided by total surface area) the Ergun equation holds for the pressure drop when operated dry:[48]

$$\Delta p = \frac{150 \mu_G U_G L (1 - e)^2}{D_{sv}^2 e^3}$$
$$+ \frac{1.74 \rho_G U_G^2 (1 - e) L}{D_{sv} e^3} \quad (18.37)$$

where e is the volume void fraction (dimensionless) and L is the length of the bed. This equation is just the sum of the Blake–Kozeny equation for laminar flow plus the Burke–Plummer equation for turbulent flow. For $Re = \rho_G U_G D_{sv}/\mu_G > 10^3$, the first term is negligible.

The pressure drop for the dry bed will be less than that for the wet bed, but the calculations for predicting the latter are beyond our scope here, for which the reader should consult Ref. 26.

Calvert[31] presented this equation for the penetration of a packed column for particles caught due to inertia:

$$Pt = \exp[-3.5 \psi L/e D_c] \quad (18.38)$$

where ψ is the impaction parameter (18.7) and D_c is the diameter of the collectors making up the bed. This relationship can also be written as:[31]

$$Pt = \exp\left[-A d_{pa}^2\right] \quad (18.39)$$

with

$$A = 0.69/d_{pc}^2 \qquad (18.40)$$

where d_{pc} is the cut diameter, $Pt(d_{pc}) = 0.5$:

$$d_{pc} = D_c(3.6e\mu_G/LU_G\rho_p)^{1/2} \qquad (18.41)$$

Figure 18.11 can be used to determine the overall mass penetration for an aerosol which has a log-normal distribution with a mass median = geometric mean diameter d_{pg} and a geometric standard deviation σ_p.

18.7.3 Fluidized Beds

A fluidized bed results when the upward flow of gas through packing that is unconstrained at its top becomes sufficient to support the weight of the bed. At this velocity, the packing material moves freely. At greater velocities the packing material may be carried off in the gas stream, but for wetted fluidized beds, unacceptable levels of liquid entrainment would likely occur before this velocity was reached.

As gas velocity is increased, pressure drop across such a bed increases as it would for any packed bed. The bed becomes fluidized when the pressure drop equals the weight per unit area of the bed and its associated liquid; further increases in velocity give much less added pressure drop increase per unit velocity increase.

Dry fluidized beds are receiving much attention for coal desulfurization, but their tendency to form channels and bubbles has limited their use in particle collection. Initial tests of a wetted fluidized bed were unusually promising. The results of subsequent tests have not shown that they have an energy advantage over most other scrubbers in collecting particulate matter.[27] Where mass-transfer as well as particulate collection is important, the wetted fluidized bed may be advantageous.

18.7.4 Wetted Fibrous Mats

Dry fiber mats are covered more extensively in the chapter on filtration. Wetted fibrous filter mats are attractive for scrubbing, in that the collection due to interception on fine fibers offers the hope of doing somewhat better in terms of efficiency versus pressure drop than do other scrubbers, which rely on impaction.[31, 49]

The pressure drop across a fibrous filter can be estimated from the traditional Kozeny–Carman equation for pressure drop in laminar flow:[50]

$$\Delta P = k'S^2\mu_G U_G L(1 - e)^2/e^3 \qquad (18.42)$$

which for circular cylinders becomes:

$$\Delta P = 16k'\mu_G U_G L(1 - e)^2/D_c^2 e^3 \qquad (18.43)$$

where k' is the Kozeny constant, equal to about 5 for porosities between 0.2 and 0.8. (The surface-to-volume ratio of the fibers is $S = 4/D$ for cylinders of diameter D.) For fibers oriented transverse to the flow, k' is 6.0, and for fibers parallel to the flow it is 3.1.[2]

Pressure drop depends strongly on porosity. In Table 18.5 $(1 - e)^2/e^3$ is given for $e = 0.2$, $0.3, \ldots, 0.8$. Over that range of porosities, the pressure drop changes a factor of 1000.

The Kozeny–Carman equation is for $e < 0.8$. Davies[51] cited his prior research with filter pads of different materials, having porosities from 0.7 to 0.994 as support for the equation:

$$\Delta P = 64\mu_G U_G L(1 - e)^{1.5}$$
$$\times (1 + 56(1 - e)^3)/D_c^2 \qquad (18.44)$$

Clearly, the wetted mat will have greater flow resistance than when dry, however.

The collection efficiency of a clean fibrous bed is approximately:[52]

$$E = 1 - \exp[-4(1 - e)L\eta_c'/\pi eD_c] \qquad (18.45)$$

Table 18.5. Values of $(1 - e)^2/e^3$ for e from 0.2 to 0.6.

POROSITY e	$(1 - e)^2/e^3$
0.2	80.
0.3	18.1
0.4	5.6
0.5	2.0
0.6	0.74
0.7	0.26
0.8	0.078

where η_c is the collection efficiency of a single fiber transverse to the flow, and n'_c is the collection efficiency of that fiber as part of a mat. If the collection is due to impaction, then:[53]

$$\eta'_c = \eta_c[1 + a(1 - e)] \qquad (18.46)$$

where a is about 5 to 20. For impaction, the collection efficiency of a single fiber is approximately (from fitting data presented by May and Clifford[54]):

$$\eta_c = \psi^2/(\psi^2 + 0.64) \qquad (18.47)$$

where ψ is the impaction parameter.

A wetted fiber filter was tested in the laboratory and modeled mathematically in a fashion quite similar to the analysis above.[49] It produced a somewhat higher efficiency than would be predicted for a venturi scrubber operating at the same pressure drop (about 7.5 kPa), which was attributed in part to the interception mechanism. (The fiber diameters were approximately 50 μm and the mat porosity was 0.97.) The model correctly predicted a sharp decline in penetration as particle aerodynamic diameter became greater than 0.5 μm.

18.8 TRAY TOWERS

Tray towers have one or more perforated plate trays that are irrigated with water and through which gas travels and is scrubbed. Often a series of such plates will be used, with the liquid introduced at the top of the scrubber to travel from tray to tray via "downcomers" or by trickling through the holes in the plate (see Fig. 18.4). If the holes have (submerged) baffles or targets connected to them at which the jet of gas and liquid are directed, one has an "impingement plate" scrubber; if there are holes but no impingement targets, one has a "perforated plate."

18.8.1 Sieve Plates

A "sieve plate" is a common type of plate scrubber adapted from gas–liquid contacting uses. At liquid or gas flow rates that are too high for the design, flooding will occur, marked by a sharp decrease in liquid throughput and an increase in pressure drop. Avoiding this condition is one important goal of the design. If the gas flow becomes too low, liquid can seep through the perforations, decreasing contacting efficacy. Design equations are available to prevent either of these malfunctions.[26]

The pressure drop across the plates is due to the resistance to gas flow due to the geometrical arrangement itself, as when dry, and the added resistance of the flow through the scrubbing liquid. For each of the dry plates, the gas flow can be apportioned among the holes, and this equation for pressure drop used:[13]

$$\Delta P_{dry} = 1.14\left[0.4(1.25 - f_h)\right.$$
$$\left. + (1 - f_h)^2\right](\rho_G U_G^2/2) \quad (18.48)$$

in which f_h is the fraction of the plate area represented by the holes and U_G is the gas velocity through the holes. The other major contribution to pressure drop for each plate comes from the hydrostatic pressure represented by the height of the liquid on each plate (H_{weir}) as determined by the weir (often about 5 cm in height):

$$\Delta P_{weir} = \rho_L g H_{weir} \qquad (18.49)$$

The pressure drop across each plate is approximately $\Delta P_{dry} + \Delta P_{weir}$. More exact formulas and more design details are available elsewhere.[26]

For hydrophilic (wettable) aerosols, Taheri and Calvert found the following relationship for the penetration through a sieve plate scrubber as a function of particle size:[55]

$$Pt = \exp(-80F_L^2\psi) \qquad (18.50a)$$

for

$$0.38 \le F_L \le 0.65 \qquad (18.50b)$$

where F_L is the volume of clear liquid per volume of froth, m^3/m^3, and ψ is an impaction parameter, Eq. (18.7), based on the hole diameter and the velocity of the gas through the hole. Taheri and Calvert found that hydrophobic aerosols were collected less effectively than hydrophilic, and that the addition of wetting agents lessened collection efficiency, by creating a less dense froth.[55]

18.8.2 Impingement Plates

The pressure drop for impingement plates can be estimated by the equations given above for perforated plates.

Impingement plate penetration is predicted to be:[31]

$$Pt = \exp\left(-0.693 d_{pa}^2/d_{pc}^2\right) \quad (18.51)$$

with

$$d_{pc} = (1.37 \mu_G n_h D_h^3/Q_G)^{1/2} \quad (18.52)$$

in which n_h is the number of holes per unit area and D_h is the hole diameter; the source of this design equation noted a lack of reliable experimental data to support it. Calvert also estimated the cut diameters of two-plate and three-plate systems as 88% and 83% of the one plate system.[31] Note that increasing the number of trays from one to three often will not greatly increase collection efficiency for particles though it may for gases.[27]

Equation (18.52) can be used with Figure 18.11 to estimate total penetration for aerosols with log-normal distributions.

18.8.3 Foam Scrubbers

The formation of low density foam ($F_L \ll 1$) from a perforated plate has been the basis of several foam scrubber designs.[56-58] Unlike most scrubbers, impaction may not be the predominating mechanism. The longer residence times characteristic of such scrubbers and the small dimensions of the foam bubbles give sedimentation and diffusion more importance than usual, augmented by the interception effect. An important design problem is the breaking up of the foam and the capture of the fine particle-bearing droplets from the breaking up. One evaluation made in 1977 had this conclusion: "The operating cost of foam scrubbing with 99% surfactant recycle is an order of magnitude higher than that of the most expensive conventional method."[59] Further residence times were $\geq 10^1$ s, suggesting substantial construction costs for high-volume-flow operations.

18.9 CONDENSATION SCRUBBING

18.9.1 Theory and Experiment

Decades ago, Schauer[60] and Lapple and Kamack[61] found that the addition of steam to the gas to be scrubbed could bring about marked improvements in scrubber collection efficiency. Samrau[62] noted anomalously high collection efficiencies reported for scrubbers in which condensation occurred. (An extensive literature review of the work done before 1973 is available in the report by Calvert and co-workers.[63]) Several factors act:

Condensation of water vapor on spray scrubber droplets, caused by the droplets being at temperatures below the saturation temperature of the gas, can enhance particulate capture due to diffusiophoresis, the principal component of which is the net flow of water molecules toward the droplets; this is accompanied by a more subtle force due to the concentration gradient of the water molecules. Diffusiophoresis is accompanied by thermal forces tending to oppose it, however.

The diffusiophoresis "flux force" mechanism was discussed in detail by Waldmann and Schmitt. The existence of this mechanism is evident from the experimental results of Lapple and Kamack[61] and Semrau et al.[65] The latter, for example, noted a large difference in efficiency between wet scrubbers operating with hot versus cold water sprays. They suggested the differences could be caused by evaporation from the hot water droplets, which would produce a diffusiophoretic force away from the drop surface and therefore would result in reduced efficiency.

Sparks and Pilat[66] calculated particle collection efficiencies by droplets, assuming that (1) condensation, or (2) evaporation, or (3) neither, occurred. The collection mechanisms studied were inertial impaction and diffusiophoresis. Condensation was shown to enhance, whereas evaporation was shown to diminish, the collection by inertial impaction, the effects being more pronounced for the smaller particles.

Condensation of water vapor on droplets will also cause a temperature gradient. The latent heat of vaporization must be conducted away from the droplet. This may offset the effects of diffusiophoresis.[67]

Calculations were made for collection by droplets of 100, 500, and 1000 μm diameters in a spray tower.[68] The gas was assumed saturated and the droplets were taken to be cooler, warmer, or at the same temperature of the gas, so that particle growth due to condensation was not a factor. Single droplet collection efficiencies for the condensing case were quite insensitive to particle size, the mechanisms considered being impaction, diffusion, diffusiophoresis, and thermophoresis. The condensation/evaporation effects were greater for the larger droplets due to the longer maintenance of the temperature gradients.

Whitmore[69] found that the fraction of particles collected due to the flow of water vapor to scrubber surfaces and droplets was approximately equal to the fraction of the gas that condensed.[70]

Condensation of water vapor on particles can lead to enhanced capture due to the increase in particle aerodynamic diameter. Soluble particles will become droplets at humidities greater than their "transition" humidities,[71] the humidities a solution made from the bulk material would produce in air in a closed vessel. (For NaCl, for example, this is 75% relative humidity.) A hydrophilic liquid such as sulfuric acid does not have a transition humidity; such a droplet changes size to be in equilibrium with any ambient humidity. Aerosols made of hygroscopic liquids and solids change their volumes approximately in proportion to $1/(1 - H)$, where H is the fractional humidity.[72, 73] Hydrophobic particles will

not grow until the gas is supersaturated, often to a multiple of the saturation vapor concentration; this condition is hard to create because soluble condensation nuclei, almost always present, will compete for the water vapor, making it hard to achieve supersaturation. Typically, about 75% of the condensing vapor goes to the cold surfaces of the scrubber and 25% of the particles.[70] Lancaster and Strauss[74] concluded that diffusiophoresis was less important than particle growth in conventional scrubbers in which steam is injected. Calvert et al.[63] and Calvert and Jhaveri[75] showed that condensation scrubber efficiency is insensitive to particle size. (Therefore, condensation scrubbing would be potentially competitive with high-energy scrubbers when high collection efficiencies for submicron particles are required.) They found also that condensation collection increases as the concentration of particles decreases.[63, 75] The available moisture is shared by fewer particles, which thereby grow larger and are collected more easily than otherwise. Thermophoresis was shown by them to be of minor importance compared with diffusiophoresis and the effect of particle enlargement by condensation.[63, 75]

In experiments using hydrophobic oil drop aerosols with diameters of roughly 2 μm, Jacko and Holcomb[77] determined that the penetration of a multiple-tray sieve plate scrubber decreased from 0.44 to 0.03 as steam injection was added; the steam injection ratio, the mass of water per mass of dry gas, was 0.43. Lowering the scrubber water temperature from 57°C to 15°C decreased the penetration from 0.11 to 0.05 at an injection ratio of 0.25.

18.9.2 Application

Humidification of a gas, by addition of steam for example, consumes energy. The use of condensation scrubbing, therefore, is more likely to be economically attractive in those applications where waste heat is available. A summary of condensation scrubbing was prepared by Calvert and Parker,[42] from which Table 18.6 is taken, showing the major indus-

Table 18.6. Major Industrial Particulate Sources for Which Condensation Scrubbing is Attractive.[42]

INDUSTRY	SOURCE
Iron and steel	Sinter plants
	Coke manufacture
	Blast furnaces
	Steel furnaces
	Scarfing
Forest products	Wigwam burners
	Pulp mills
Lime	Rotary kilns
	Vertical kilns
Primary nonferrous	
Aluminum	Calcining
	Reduction cells
Copper	Roasting
	Reverberatory furnaces
	Converters
Zinc	Roasting
	Sintering
	Distillation
Lead	Sintering
	Blast furnaces
	Dross reverberatory furnaces
Asphalt	Paving material
	Roofing materials
Ferroalloys	Blast furnaces
	Electric furnaces
Iron foundry	Furnaces
Secondary nonferrous metals	
Copper	Material preparation
	Smelting and refining
Aluminum	Sweating furnaces
	Refining furnaces
	Chlorine fluxing
Lead	Pot furnaces
	Blast furnaces
	Reverberatory furnaces
Zinc	Sweating furnaces
	Distillation furnaces

trial sources of particulate emissions for which this technique is attractive. Figure 18.12, from the same report, shows a conceptual design. An analysis is presented in that report that concludes that condensation scrubbing would be economically superior (in both capital and operating costs) to a conventional high-energy scrubber for a gray iron cupola.

The feasibility of a "flux force/condensation" system was demonstrated in the control of emissions from a secondary metals recovery furnace, controlling a flow rate of 3.3 m^3/s maximum, using a quencher, a sieve plate column, and an entrainment separator: "The system was generally capable of 90% to 95% efficiency on particles with a mass median aerodynamic diameter of 0.75 μm."[76] The pressure drop was 7 kPa (27 in. WG) and it was estimated that a conventional high energy scrubber would have required 3 to 7 times as much pressure drop to achieve the same range of efficiencies.

The use of condensation scrubbing seems likely to increase. In some cases it would be a relatively low-cost modification to upgrade a scrubber already in operation.

18.10 ELECTROSTATIC AUGMENTATION

18.10.1 Introduction

Collecting particles by impaction requires accelerating the gas in which the particles are suspended to cause particle deposition due to particle inertia, an inherently inefficient approach, considering that particle mass concentrations are roughly one-thousandth or less of gas densities. Charging the scrubber surfaces or charging the particles produces electrostatic forces operating on the particles directly, not using the gas as an intermediary, and is inherently more energy efficient.

18.10.2 Theory

The two major types of electrostatic force that are significant in the collection of particles in scrubbers are the Coulomb force, which occurs when a charged particle is subjected to an electric field (such as from a charged droplet), and the induced charge ("dipole," "image") force, which is caused by the presence of an inhomogeneous field. Two other electrostatic forces can sometimes be significant: the image

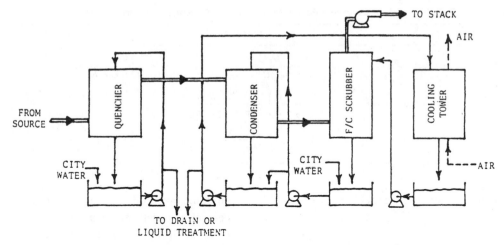

Figure 18.12. Conceptual design for a condensation scrubber.[42]

force due to the interaction between a charged particle and an uncharged collector and the mutual repulsion (or attraction) of the aerosol particles themselves.[23]

The Coulomb force (F_{Qq}) exerted on a particle of charge Q_p is:

$$F_{Qq} = Q_p E \qquad (18.53)$$

in which E is the electrical field created by the collector. The force due to induced polarization in the particle in an inhomogeneous field is:[23]

$$F_{q0} = \frac{1}{4\pi\epsilon_0} \chi_E V_p \, \text{grad}(E^2) \quad (18.54)$$

in which

$$\chi_E = (3/8\pi)(\epsilon_p - 1)/(\epsilon_p + 2) \quad (18.55)$$

for spherical particles, where ϵ_p is the dielectric constant of the particle relative to the dielectric constant of a vacuum and V_p is the volume of the particle. The gradient of homogeneous electric fields is zero, so this force occurs only in inhomogeneous fields.

The collection efficiency (η) of an obstacle is defined as the area of the oncoming gas it cleans divided by the cross-sectional area it presents to the flow. Where both the particles and the collector are charged, the collection efficiency of a collector of any shape for particles assumed to have negligible inertia can be shown to be:[78, 79]

$$\eta_E = 4K \qquad (18.56)$$

in which K is the ratio of the particle terminal velocity, calculated for the force as evaluated at the surface of the collector, to the velocity of the free stream (U_G, the superficial or mean gas velocity). See Table 18.7 for definitions of K for spherical and cylindrical collectors.[79]

The parameter K can be used to sum up experimental and theoretical results concerning collection efficiencies for various conditions, as done in Table 18.8.[78, 79] In Table 18.8 are listed the collector geometry, the force type, its radial dependence (the forms for F_{0q} are approximate), the range of K for which the efficiency expression is correct, and the efficiency, η_E. The expression "$O(K^{2/5})$" means that the efficiency is roughly $K^{2/5}$, with a correction factor of order unity which will be somewhat different depending upon the flow field. Calculation of K (Table 18.7) and the use of the material in Table 18.8 provide a simple method for estimating the electrostatic contribution to collection efficiency for these cases.

To increase the effect of electrostatics, one can charge the particles. The saturation charge due to charging a particle ($d_p \gtrsim 1\ \mu$m) in the

Table 18.7. Definitions of the Electrical Force Parameter (K) for Spherical Particles.[21,78,79]

	SPHERE	CYLCINDER
Coulombic force (F_{Qq})	$\dfrac{CQ_cQ_p}{24\pi^2\epsilon_0 r_p R_c^2 \mu_G U_G}$	$\dfrac{C(Q_c/L)Q_p}{12\pi^2\epsilon_0 r_p R_c \mu_G U_G}$
Charged-collector image force (F_{Q0})	$\left(\dfrac{\epsilon_p - 1}{\epsilon_p + 2}\right)\dfrac{CQ_c^2 r_p^2}{12\pi^2\epsilon_0 R_c^5 \mu_G U_G}$	$\left(\dfrac{\epsilon_p - 1}{\epsilon_p + 2}\right)\dfrac{C(Q_c/L)^2 r_p^2}{6\pi^2\epsilon_0 R_c^3 \mu_G U_G}$
Charged-particle image force (F_{0q})	$\left(\dfrac{\epsilon_c - 1}{\epsilon_c + 2}\right)\dfrac{CQ_p^2}{24\pi^2\epsilon_0 r_p R_c^2 \mu_G U_G}$	$\left(\dfrac{\epsilon_c - 1}{\epsilon_c + 1}\right)\dfrac{CQ_p^2}{96\pi^2\epsilon_0 r_p R_c^2 \mu_G U_G}$

presence of an electric field of strength E_0 is:[80]

$$q_s = 3\left[\epsilon_p/(\epsilon_p + 2)\right]d_p^2\pi\epsilon_0 E_0 \quad (18.57)$$

Particles much smaller than 1 μm can be charged by diffusion of ions at relatively high concentrations. (See the chapter on electrostatic precipitation.)

18.10.3 Applications

Three electrostatic scrubbers that have received attention are:

1. A scrubber developed at the University of Washington that uses particles charged to one polarity and droplets from spray nozzles kept at a high voltage of the opposite polarity.[81,82]
2. A scrubber developed by TRW, Inc, that uses the geometry of an electrostatic precipitator but replaces the corona-producing central wire with an array of electrohydrodynamic sprays to produce droplets that both transfer charge to the particles, so they will be captured by the grounded collector plates, and that capture the particles by impaction and electrostatic interactions.[83,84]
3. A scrubber developed by Air Pollution Systems, Inc. that uses a novel particle charging geometry[85] to produce high levels of charge on the particles, which are then collected by inertial and electrostatic forces in a venturi scrubber.

Table 18.9 (adapted from one presented in Ref. 86) gives information determined by experiments done during development programs, so these results are not definitive. The power which a conventional venturi would use to produce 90% efficiency at 0.5 μm aerody-

Table 18.8. Summary of Experimental, Theoretical Results for Collection Efficiencies for Electrostatic Interactions and Inertialess Particles.[21,78,79]

COLLECTOR	FORCE	R^n	K RANGE	EFFICIENCY
Sphere	F_{Qq}	R^{-2}	all	$4K$
	F_{Q0}	R^{-5}	$K \gg 1$	$0(K^{2/5})$
			$K \ll 1$	$4K$
	F_{0q}	$\sim R^{-5}$	$K \gg 1$	$0(K^{2/5})$
			$K \ll 1$	$0(K^{1/2})$
Cylinder	F_{Qq}	R^{-1}	all	πK
	F_{Q0}	R^{-3}	$K \gg 1$	$0(K^{1/3})$
			$K \ll 1$	πK
	F_{0q}	$\sim R^{-2}$	$K \gg 1$	$0(K^{1/2})$
			$K \ll 1$	$0(K^{1/2})$

Table 18.9. Comparisons of Electrostatic Droplet Scrubbers Based on Developmental Units.

ELECTROSTATIC SCRUBBER	POWER USED [W/(m³/s)]	LIQUID-TO-GAS RATIO (10^{-3})	COLLECTION EFFICIENCY AT (AERODYNAMIC DIAMETER):	
			0.5 μm	1.0 μm
University of Washington	0.22	0.8–2.3	0.99	0.97
APS, Inc.	1.3	1.4	0.96	0.90
TRW, Inc.	0.5–0.7	0.12	0.90	0.85–0.95

Adapted from Ref. 86.

namic diameter is about 5 kW/(m³/s), much greater than the power used in these devices.[86]

Compared with electrostatic precipitators of similar efficiency, high-efficiency venturi scrubbers are typically smaller and less expensive in capital costs but use more power and have higher operating costs. Electrostatic scrubbers are likely to show capital and operating costs that are between those for scrubbers and those for electrostatic precipitators and should be judged by their annualized costs rather than by their power consumptions alone. (Cost comparison methodology is treated briefly at the end of this chapter.) Corrosion problems and electrical isolation problems can also be significant.

18.11 DEMISTERS AND ENTRAINMENT SEPARATORS

18.11.1 Introduction

A scrubber uses liquid surfaces to rid the gas stream of particles. Inevitably, the scrubber produces droplets containing solid and dissolved material which must be captured before the gas is emitted to the atmosphere, either to meet emissions limits or to prevent damage to fans, ducting, etc. (Wetted surfaces produce droplets due to atomization or to the liquids falling from the surfaces.) The droplet size produced will be a function of scrubber type, geometry, power consumption, and flow velocity; for example, for a packed bed cross-flow scrubber, Bell and Strauss[87] found the number mean droplet size to decrease from 400 to 100

μm as superficial velocity in the scrubber increased 50% (from 3 to 5 m/s). As liquid usage is increased, so is droplet entrainment; as scrubber energy input is increased, through increased pressure drop in the gas or increased spray nozzle pressure, the entrained liquid droplets can be expected to become smaller, their mass concentration greater. Droplets may contain captured particulate matter; even without captured matter, they dry to become fine solid particles due to dissolved minerals ("hardness") in the water. To prevent emission of material due to droplet reentrainment, the scrubber should be followed by a demister (also called an "entrainment separator" or a "mist eliminator").

The design goals for demisters were summed up by Bell and Strauss:

In general, mist eliminators should have the following characteristics: low cost, ease of manufacture and installation, low pressure drop, and high efficiencies over a wide range of superficial gas velocities and mist loadings. The units should be self-draining and self-cleaning, with low operating and maintenance charges, able to operate for long periods without attention.[87]

Entrainment separator design can be improved by using guidelines recently published.[109, 110] For fibrous beds or packed beds, optimal efficiency at a fixed pressure drop (or minimum pressure drop at a fixed efficiency) can be obtained by choosing a collector element size and collector face velocity such that the impaction parameter is approximately 1 for the droplet size of interest.

18.11.2 Types

Droplets formed from atomization from scrubber surfaces have number mean diameters $\sim 10^2$ to 10^3 μm. Generally, the higher the gas velocity, the smaller the droplets. Droplets from hydraulic spray scrubbers will be similar to the spray. In cases where a mist forms from condensation of water vapor, the droplets will be ≤ 1 μm; these are much more difficult to collect and are not discussed further here. (This condition should be avoided.) For the larger droplets, the collection mechanisms which come into use are: gravity settling, centrifugal collection, and impaction.[88]

Centrifugal collection or inertial impaction are really the same collection mechanism: the gas stream has its direction (and sometimes speed) changed, and the droplet's inertia gives it a velocity component toward the collector wall, perpendicular to the mean gas flow. Devices operating on this principle include cyclones, baffled chambers (using chevrons, corrugated sheets, etc.) and packed beds, with packing material of many geometries and wide range of characteristic collector dimensions and volume void fractions.

18.11.3 Pressure Drop

The pressure drop across the mist eliminator can be identified as friction drag and form drag, proportional to velocity and velocity squared, respectively. The pressure drop as a function of superficial velocity (gas volume flow divided by scrubber cross-sectional area before any baffles or obstacles are introduced) will be of the form:

$$\Delta P = aU_G + bU_G^2 \qquad (18.58)$$

As the gas flow Reynolds number in the scrubber increases the U_G^2 term will predominate. Packed bed pressure drop can be estimated using the Ergun Eq. (18.37) and predictive correlations are available for cyclones.[87] The pressure drop required will be determined by the collection efficiency needed, so the rela-

tionship between pressure drop and efficiency is discussed next.

18.11.4 Collection Efficiency

"Primary" collection efficiency is the fraction of the droplets that are caught on collection surfaces. Total collection efficiency is the fraction of droplets that are retained by the mist eliminator. The difference is due to reentrainment of the captured droplets.

Inertial collection of a spray droplet is correlated with the droplet impaction parameter. Thus, demister collection efficiency can be expected to change as a function of droplet size. Calculation of the total efficiency requires integrating the collection efficiency as a function of droplet size over the droplet size distribution. The method Calvert[9] described for obtaining the total collection efficiency for droplets (or aerosols) assumed to be lognormally distributed in droplet size with known mass median diameter and geometric standard deviation has been presented above (see Fig. 18.11). Figure 18.13 shows the droplet cut diameter as a function of entrainment separator pressure drop for several types of separators (note: 1 cm WC = 98 Pa).[90] The power advantage of the wire mesh is apparent, although this may be offset by cleaning/plugging problems.

Porous fibrous structures or wire meshes are often used as mist eliminators. Strauss[91] presented Table 18.10 (from Griwatz et al.[92]). The eliminator "type" descriptions were:

1. 20-μm diameter Teflon (DuPont) fibers combined with 152-μm wire;
2. American Air Filter Type T bonded fiberglass;
3. Mine Safety Appliances bonded fiberglass;
4. 9-μm diameter fiberglass mixed and knitted with 127-μm wire (Mine Safety Appliances);
5. knitted wire mesh (Farr Type 68-44 MHZ).

18.11.5. Reentrainment of Droplets

Increasing the velocity increases the pressure drop and may lead to increased collection efficiency. Beyond some velocity, however,

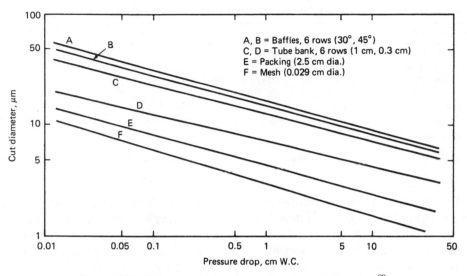

Figure 18.13. Entrainment separator performance cut diameters.[90]

flooding of the separator or reentrainment of droplets from the separator surfaces will produce an increase in emissions, thus an apparent decrease in efficiency. Approximate values of superficial velocity at which entrainment begins are given below.[88]

SEPARATOR	GAS VELOCITY (m/s)
Zigzag with upward gas flow and horizontal baffles	3.7–4.6
Zigzag with horizontal gas flow and vertical baffles	4.6–6.1
Cyclone (gas inlet velocity)	30.5–39.6
Knitted mesh with vertical gas flow	3.1–4.6
Knitted mesh with horizontal gas flow	4.6–7.0
Tube bank with vertical gas flow	3.7–4.9
Tube bank with horizontal gas flow	5.5–7.0

As Calvert concluded, "Liquid drainage is best when the gas flow is horizontal and collection surfaces are near vertical; also, with this configuration, reentrainment occurs at higher flow rates than for horizontal elements."[88]

18.11.6 Plugging

Captured solids and solute material that precipitates from solution can build up on scrubber and entrainment separator surfaces, leading to increased flow resistance. Recirculation of the scrubbing liquid will aggravate this condition. Some actions that may help reduce this problem are:

1. Reduce slurry concentrations.
2. Design collection elements to have nearly vertical surfaces.
3. Provide for washing of the collection surface.
4. Avoid drying of the surfaces; if scrubber is shut off, clean before reusing.
5. Design using geometries having larger rather than smaller minimum flow path dimensions (i.e., choose a chevron over a knitted mesh, other things being equal).

18.11.7 Summary

Mist eliminators ("demisters") are needed for almost all scrubbers. They capture droplets by

Table 18.10. Operating Characteristics and Efficiencies for Fiber Mist Eliminators.[91,92]

TYPE (SEE TEXT)	1	2	3	4	5
Bed depth (mm)	67	61	125	125	100
Flow velocity (m/s)	2.0	1.44	2.0	2.0	2.26
Pressure drop (Pa)	322–555	200–300	250–475	250–425	65–87
Efficiency (%)					
at 100 μm	26	100	100	100	100
at 10 μm	36	100	100	100	90
at 0.6 μm	31	7	20	22	1
at 0.3 μm	7	5	4	—	0

inertial mechanisms, generally with superficial velocities (except for cyclones) less than 7 m/s, to prevent reentrainment. Horizontal gas flow is preferable to vertical flow, but plugging can be a problem for either, especially where dissolved solids are being used to scrub gases from the effluent stream. For more detailed information, see the work by Strauss.[91]

18.12 SUNDRY DESIGN CONSIDERATIONS

18.12.1 Introduction

Covered here are several factors which should be taken into consideration in design but that did not fit conveniently into other sections of this chapter.

18.12.2 Corrosion

Corrosion problems are specific to the particular source type under control.

Case histories of scrubber applications and problems in the metallurgical industry were presented by Steiner and Thompson:[93] Abrasion of mild steel piping used to carry slurry occurred in a venturi used to control gaseous and particulate emissions from a boiler; the problem was cured by using rubber-lined piping and valving. In a sinter plant application, corrosion of carbon steel in the liquid flow lines was a problem, perhaps due to inadequate pH control; no such problem occurred in the air flow passages, where type 304 stainless steel was used. In a third situation, control

of an open-hearth furnace, nonstainless steel components were corroded severely; lime neutralization led to problems of scaling due to calcium sulfate, later mitigated by switching to caustic for neutralization.

Hoxie and Tuffnell summarized extensive tests in scrubbers used for flue gas desulfurization:[94] carbon steel and type 304L stainless steel were inadequate in the wet areas, and type 316L steel was occasionally attacked, specifically by certain combinations of pH and chloride concentration. They presented detailed information for more than a dozen steels.

Three options for corrosion protection were identified by Busch et al.:[95] liners, different materials of construction, thicker materials. They presented cost comparisons for various steels and a steel and rubber liner combination.

Further information on corrosion control may be obtained from the National Association of Corrosion Engineers, 2400 West Loop South, Houston, TX 77027, U.S.A.

18.12.3 Wetting Agents

There is some belief among pollution control engineers that the addition of wetting agents to the scrubbing liquid can improve scrubber performance.[108] In certain cases, it is certainly possible that the droplet size distribution of the hydraulic or atomized spray will become somewhat better suited to scrubbing the aerosol, but it is as likely that the size distribution will become less suited. It seems preferable to change the nozzles, the flow rates, or the pressures than to introduce wetting agents to the liquid, which will represent added material expense and perhaps added water pollution control costs.

In some cases, the scrubbing improvement noted in scrubbing wettable versus nonwettable particles may have been due to hygroscopic growth of the former. Experiments have shown that wettable and nonwettable particles are caught with equal efficiency by drops (at a given impaction parameter value), except in the rare instances where nonwetting particles coat the droplet to the degree that other particles strike them and are not retained.[96]

18.12.4 Scale-Up

Even some companies with extensive experience with scrubbers have made it a policy to use pilot scale scrubbers to help design the full-scale scrubber to be used in a particular application.[97] Even so, some assumptions must be made in scaling up the results of such test.

Two sets of investigators[98,99] found improved performance in larger scrubbers at a given pressure drop, perhaps due to increased turbulence at the higher Reynolds numbers. On the other hand, Behie and Beeckmans[100] reviewed many previous investigations and concluded there were no appreciable effects due to scaling up a scrubber.

18.12.5 Water Pollution

As water quality standards and water pollution control requirements become more stringent, scrubber design must increasingly take water treatment into account, influencing water usage rates, recycling rates, construction materials, and additive selection (such as for pH control). This is well beyond our scope, however.

18.12.6 Mechanical Aids

The use of wetted fans or other blade-type mechanical methods for disintegrating droplets

and capturing particles is not particularly attractive; collection efficiency is not quite as good for such scrubbers as for venturi scrubbers at the same power consumption and problems of corrosion, erosion, and vibration are inherent in such designs.[27]

18.13 COSTS

A great many factors contribute to the total cost of a scrubber. Figure 18.14 shows a generalized cost evaluation scheme.[1] The source and its operating characteristics will influence the choice of control type, its capacity, efficiency, construction materials, and thus the costs of control. Handling the collected materials is costly, though there may be salvage value. Note that the cost of the control hardware is only a part of the total cost, especially for high-energy scrubbers.

One approach for cost comparisons of various particulate control options is that described by Edmisten and Bunyard.[101] The goal is to develop a single cost parameter, here the *total annualized cost*, with which to compare different air pollution control devices. This is quite useful because, for example, electrostatic precipitators have relatively high initial costs and relatively low operating costs in comparison to scrubbers of similar collection efficiency.

The costs can be divided into three categories:[101]

1. *Capital investment cost*. This includes the control hardware cost, the cost of auxiliary equipment and the cost of installation, including initial studies.
2. *Maintenance and operating costs*. These are taken on a yearly basis, averaged over the life of the equipment.
3. *Capital charges*. These are what it costs to borrow the money equivalent to the capital investment, plus taxes and insurance.

To convert these various costs into a single number, the total annualized cost, one sums

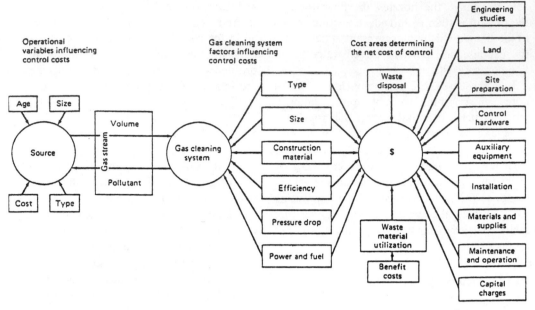

Figure 18.14. Diagram of a cost evaluation scheme for a pollutant control system.[1]

the annual capital investment depreciation, the operating and maintenance costs, and the capital charges.

The usual method of depreciation in such contexts is to assume straight-line depreciation: Estimate the life of the equipment, Edmisten and Bunyard[101] suggested 15 years, and figure the yearly depreciation as the capital investment divided by the life expectancy. Thus, the total annualized cost is given by the sum of capital investment divided by the lifetime plus yearly maintenance and operating costs and capital charges, Generally, mainte-

nance and capital charges will be nearly proportional to the capital investment. Table 18.11 is based on a survey done as part of the preparation of the *Scrubber Handbook*[6] and allows one to make a rough estimate of the installed cost of a scrubber, based on the current Marshall and Stevens Index. The fixed capital investment is about three times the installed cost.[31] Table 18.12 shows the conditions that can affect the installed costs of control devices, factors that are reflected in the ranges attributed to costs in Table 18.11.

For more details on conventional control

Table 18.11. Reported Costs of Complete, Installed Scrubber Systems.[31]

	MEAN COST/acfm[a,b] (AT acfm LISTED)					
SCRUBBER TYPE	1000	10,000	50,000	100,000	HIGH / MEAN	LOW / MEAN
Venturi	$14.00	$5.50	$3.00	$2.20	3	$\frac{1}{3}$
Packed bed	$14.00	$3.00	$0.80	—	3	$\frac{1}{3}$
Spray	$50.00	$5.00	$1.00	$0.70	2	$\frac{1}{2}$
Centrifugal	$3.00	$1.30	$0.70	—	2	$\frac{1}{2}$
Impingement and entrainment	$8.00	$3.50	$2.00	$1.50	1.5	0.7
Mobile bed	—	$3.00	$2.00	—	1.5	0.7

[a] Costs are for Marshall and Stevens Index of about 280.

[b] acfm: Actual cubic feet per minute. (1000 acfm = 0.47 m^3/s.)

Table 18.12. Conditions Affecting Cost of Control Devices Installed.[105]

COST CATEGORY	LOW COST	HIGH COST
Equipment transportation	Minimum distance; simple loading and unloading procedures	Long distance; complex procedure for loading and unloading
Plant age	Hardware designed as an integral part of new plant	Hardware installed into confines of old plant requiring structural or process modification or alteration
Available space	Vacant area for location of control system	Little vacant space, requires extensive steel support construction and site preparation
Corrosiveness of gas	Noncorrosive gas	Acidic emissions requiring high alloy accessory equipment using special handling and construction techniques
Complexity of start-up	Simple start-up, no extensive adjustment required	Requires extensive adjustments; testing; considerable downtime
Instrumentation	Little required	Complex instrumentation required to assure reliability of control or constant monitoring of gas stream
Guarantee on performance	None needed	Required to ensure designed control efficiency
Degree of assembly	Control hardware shipped completely assembled	Control hardware to be assembled and erected in the field
Degree of engineering design	Autonomous "package" control system	Control system requiring extensive integration into process, insulation to correct temperature problem, noise abatement
Utilities	Electricity, water, waste disposal facilities readily available	Electrical and waste treatment facilities must be expanded, water supply must be developed or expanded
Collected waste material handling	No special treatment facilities or handling required	Special treatment facilities or handling required
Labor	Low wages in geographical area	Overtime and/or high wages in geographical area

device costs, see the article by Edmisten and Bunyard[101] and the articles by Hanf and MacDonald[102] and by Fraser and Eaton[103] and Neveril et al.,[104] who presented graphs and equations for estimating the prices for electrostatic precipitators, venturi scrubbers, fabric filters, incinerators, and absorbers, as well as the costs of auxiliary equipment, ductwork and dampers, and such other costs as operating, maintenance, and installation.

Although much of the necessary information on costs will have to be obtained from manufacturers for a specific application, one can readily estimate power costs. Power consumption figures are often given in terms of

kW per m^3/s flow rate or horsepower per 1000 actual cubic feet per minute flow rate. (Note: 1 hp = 0.746 kW; 1000 acfm = 0.472 m^3/s). When the power is given as hydraulic power (pressure drop times volume flow rate), a pump/fan/motor efficiency factor must be used (as a divisor) to convert to actual electrical power; this efficiency factor is generally about 0.6, whether fans are moving gas or pumps are moving liquid. The power cost is given by the product of: volume rate of gas flow, power consumption per unit flow of gas, cost per unit of energy, and operating time.

Certain forms of power may be nearly free: the recovery of waste heat is free with regard to operating costs, although it will add to the capital investment and to the costs associated with capital investment. As with other costs, the power costs will vary considerably from situation to situation.

Specific circumstances will also greatly affect waste disposal costs. Scrubbers produce waste water that must be handled properly; waste water treatment produces solid wastes that must be used or disposed. Generally the final phase is to convert the captured material into a solid for such uses as landfill or to recycle some or all of the captured material. Solid waste disposal cost can be broken down into costs of hauling and cost of disposal. Hauling costs are dependent on the type of equipment, length of hauls, type of route, and traffic encountered, and the number of employees necessary. Cost of disposal usually means the cost of a sanitary landfill. Components of sanitary landfill cost are cost of site, degree of compaction, and cost of developing such things as access roads, water supply, fences, landscaping, water runoff diversion facilities, etc.

LIST OF SYMBOLS

a	Subscript: aerodynamic; coefficient in Equation (19.62), N-s/m^3
b	Coefficient in Eq. (18.58) N-s^2/m^4
c	Subscript: collector, cut
d	Diameter, m; subscript: droplet
e	Void volume fraction
f	Subscript: fan; empirical parameter for venturi efficiency
g	Subscript: geometric mean; gravitational acceleration, 9.8 m/s^2
h	Length, m; subscript: hole
i	Index number; subscript: particle size interval
k	Kozeny constant
m	Particle size distribution by mass, m^{-1}
n	Number concentration, m^{-3}; number per area, m^{-2}
0	Subscript: initial, vacuum

p	Subscript: particle
q	Subscript: charge on particle
r	Radius, m
t	Time, s
W	Subscript: water
x	Coordinate axis, m
y	Coordinate axis, m
z	Coordinate axis, m
A	Area, m^2; coefficient in Eq. 18.27, m^{-B}
B	Exponent in Eq. 18.27
$C(d_p)$	Cunningham correction (approximately $1 + 2.5\lambda/d_p$)
C_D	Drag coefficient
D	Diameter, m
E	Efficiency
F	Volume fraction; force, N
G	Subscript: gas
H	Relative humidity (fraction)
I	Subscript: impaction
K	Ratio of particle terminal velocity to gas free stream velocity; an impaction parameter $= 2\psi$
L	Length, m; subscript: liquid
M	Mass, kg
M	Mass flux, kg/s
N	Number, ratio
P	Pressure, N/m^2
Q	Volume flow rate, m^3/s; electrical charge, coul; subscript: charge on collector
R	Radius, m; subscript: interception
S	Surface-to-volume ratio, m^{-1}
T	Temperature, K; subscript: throat
U	Velocity, m/s
V	Volume, m^3
Pt	Penetration
Re	Reynolds number
Stk	Impaction number
β	Coefficient in pressure drop, Eq. (18.19)
ϵ	Dielectric constant
η	Efficiency
λ	Mean free path, m
μ	Viscosity, N-s/m^2
ρ	Density, kg/m^3
σ	Standard deviation

ψ Impaction parameter

Δ Difference or charge

REFERENCES

1. A. E. Vandergrift, L. J. Shannon, E. W. Lawless, P. G. Gorman, E. E. Sallee, and M. Reichel, "Particulate Systems Study," Vol III, *Handbook of Emission Properties*. APTD-0745 (NTIS PB 203 522), US EPA (1971).

2. W. Strauss, *Industrial Gas Cleaning*, Pergamon, New York (1966).

3. M. W. First, Harvard School of Public Health, Boston, MA (1979).

4. L. J. Shannon, P. G. Gorman, and M. Reichel, "Particulate Pollutant Systems Study," Vol. II, *Fine Particle Emissions*. APTD-0744 (NTIS PB 203 522), US EPA (1971).

5. Courtesy of the Industrial Gas Cleaning Institute, Alexandria, VA.

6. S. Calvert, J. Goldschmid, D. Leith, and D. Mehta, *Scrubber Handbook*, US EPA, NTIS PB 213 016 (1972).

7. K. Semrau and C. L. Witham, *Wet Scrubber Liquid Utilization*. Stanford Research Institute, Menlo Park, CA., EPA-650/2-74-108, US EPA (October 1974).

8. K. T. Semrau, C. L. Witham, and W. W. Kerlin, *Energy Utilization by Wet Scrubbers*. EPA-600/2-77-234, US EPA (1977).

9. S. Calvert, "Engineering Design of Wet Scrubbers," *J. Air Pollut. Contr. Assn,* 24:929–934 (1974).

10. D. W. Cooper, "Theoretical Comparison of Efficiency and Power for Single-Stage and Multiple-Stage Particulate Scrubbing," *Atmos. Environ.* 10:1001–1004 (1976).

11. D. M. Muir and Y. Miheisi, "Comparison of the Performance of a Single- and Two-stage Variable-throat Venturi Scrubber," *Atmos. Environ.* 13:1187–1196 (1979).

12. D. W. Cooper, "Optimizing Venturi Scrubber Performance Through Modeling." Presented at 2nd Symposium on Transfer and Utilization of Particulate Control Technology, Denver, CO, 23–27 July 1979, sponsored by US EPA (1979).

13. S. C. Yung and S. Calvert, *Particulate Control Highlights: Performance and Design Model for Scrubbers*. EPA-600/8-78-005b, US EPA (1978).

14. J. Aitchison and J. A. C. Brown, *The Log-Normal Distribution*, Cambridge University Press, Cambridge (1957).

15. G. Kubie, "A Note on the Treatment of Impactor Data for Some Aerosols," *J. Aerosol Sci.* 2:23–30 (1971).

16. S. Calvert, D. Lundgren, and D. S. Mehta, "Venturi Scrubber Performance," *J. Air Pollut. Contr. Assn.* 22:529–532 (1972).

17. J. L. Held and D. W. Cooper, "Theoretical Investigation of the Effects of Relative Humidity on Aerosol Respirable Fraction," *Atmos. Environ.* 13:1419–1425 (1979).

18. P. Knettig and J. M. Beeckmans, "Inertial Capture of Aerosol Particles by Swarms of Accelerating Spheres," *J. Aerosol Sci.* 5:225–233 (1974).

19. L. E. Sparks, "The Effect of Scrubber Operating and Design Parameters on the Collection of Particulate Air Pollutants," Ph.D. dissertation (Civil Engineering), University of Washington, Seattle (1971).

20. H. F. George and G. W. Poehlein, "Capture of Aerosol Particles by Spherical Collectors: Electrostatic, Inertial, Interception, and Viscous Effects," *Env. Sci. Technol.* 8:46–49 (1974).

21. K. A. Nielsen and J. C. Hill, "Collection of Inertialess Particles on Spheres With Electrical Forces," *Ind. Eng. Chem., Fundam.* 15:149–157 (1976). "Capture of Particles on Spheres by Inertial and Electrical Forces," *Ind. Eng. Chem., Fundam.* 15:157–163 (1976).

22. I. Langmuir and K. B. Blodgett, "A Mathematical Investigation of Water Droplet Trajectories," U.S. Army Air Forces Tech. Report. 5418 (NTIS: PB 27565) (February 1946).

23. N. A. Fuchs, *Mechanics of Aerosols*, Pergamon, New York (1964).

24. D. W. Cooper, "Fine Particle Control by Electrostatic Augmentation of Existing Methods," Paper 75-0.2.1 presented at the 68th Annual Meeting of the Air Pollution Control Assoc., Boston, MA (1975).

25. H. Krockta and R. L. Lucas, "Information Required for the Selection and Performance Evaluation of Wet Scrubbers," *J. Air Pollut. Contr. Assn.* 22:459–462 (1972).

26. R. H. Perry and C. H. Chilton, *Chemical Engineers' Handbook*, 5th ed., McGraw-Hill, New York (1973).

27. S. Calvert, "How to Choose a Particulate Scrubber," *Chem. Eng.*, pp. 54–68 (August 29, 1977).

28. S. C. Yung, H. F. Barbarika, and S. Calvert, "Pressure Loss in Venturi Scrubbers," *J. Air Pollut. Contr. Assn.* 27:348–351 (1977).

29. S. Yung, S. Calvert, and H. F. Barbarika, "Venturi Scrubber Performance Model," EPA-600/2-77-172, US EPA (1977).

30. S. Calvert, "Source of Control by Liquid Scrubbing," in *Air Pollution*, edited by A. C. Stern, Academic Press, New York (1968).

31. S. Calvert, "Scrubbing," in *Air Pollution*, edited by A. C. Stern, Academic Press, New York (1977).

32. H. F. Johnstone, R. B. Field, and M. C. Tassler, "Gas Absorption and Aerosol Collection in Venturi Scrubber," *Ind. Eng. Chem. 46*:1602–1608 (1954).

33. S. Calvert, "Venturi and Other Atomizing Scrubber Efficiency and Pressure Drop," *A.I.Ch.E. J. 16*:392–396 (1970).

34. R. H. Boll, "Particle Collection and Pressure Drop in Venturi Scrubbers," *Ind. Eng. Chem. Fund. 12*:40–50 (1973).

35. M. Taheri and C. M. Shieh, "Mathematical Modeling of Atomizing Scrubbers," *A.I.Ch.E. J. 21*(1):153–157 (1975).

36. K. C. Goel and K. G. T. Hollands, "Optimum Design of Venturi Scrubbers," *Atmos. Environ. 11*:837–845 (1977).

37. S. Calvert, D. Lundgren, and D. S. Mehta, "Venturi Scrubber Performance," *J. Air Pollut. Control Assn. 22*:529–532 (1972).

38. L. E. Sparks, "SR-52 Programmable Calculator Programs for Venturi Scrubbers and Electrostatic Precipitators," EPA 600/7-78-026, Office of Research and Development, US EPA (March 1978).

39. D. Leith and D. W. Cooper, "Venturi Scrubber Optimization," *Atmos. Environ. 14*:657–664 (1980).

40. F. O. Ekman and H. F. Johnstone, "Collection of Aerosols in a Venturi Scrubber," *Ind. Eng. Chem. 43*:1358–1370 (1951).

41. D. M. Muir, C. D. Grant, and Y. Miheisi, "Relationship between Collection Efficiency and Energy Consumption of Wet Dust Collectors," *Filtration Separation 15*:332–340 (1978).

42. S. Calvert and R. Parker, "Particulate Control Highlights: Flux Force/Condensation Wet Scrubbing," EPA-600/8-78-005c, US EPA (June 1978).

43. M. Taheri, S. A. Beg, and M. Beizie, "Gas Cleaning in a Wetted Butterfly Valve," *J. Air Pollut. Contr. Assn. 22*:794–798 (1972).

44. L. S. Harris, "Fume Scrubbing with the Ejector Venturi System," *Chem. Eng. Prog. 62*:55–59 (1966).

45. H. E. Gardenier, "Submicron Particulate Scrubbing with a Two Phase Jet Scrubber," *J. Air Pollut. Contr. Assn. 24*:954–957 (1974).

46. D. W. Cooper and D. P. Anderson, "Dynactor Scrubber Evaluation," EPA-650/2-74-083a, US EPA (June 1975).

47. S. Calvert, N. C. Jhaveri, and S. Yung, "Fine Particle Scrubber Performance Tests," EPA-650/2-74-093, US EPA (October 1974).

48. R. B. Bird, W. E. Stewart, and E. N. Lightfoot, *Transport Phenomena*, Wiley & Sons, New York (1960).

49. J. D. Brady, D. W. Cooper, and M. T. Rei, "A Wet Collector of Fine Particles," *Chem. Eng. Prog. 73*(8):45–53 (1977).

50. J. Happel and H. Brenner, *Low Reynolds Number Hydrodynamics*, Prentice-Hall, Englewood Cliffs (1965).

51. C. N. Davies, *Air Filtration*, Academic Press, New York (1973).

52. K. Iinoya and C. Orr, Jr., "Filtration," in *Air Pollution*, edited by A. C. Stern, Academic Press, New York (1977).

53. C. Y. Chen, "Filtration of Aerosols by Fibrous Media," *Chem. Rev. 55*:595–623 (1955).

54. K. R. May and R. Clifford, "The Impaction of Aerosol Particles on Cylinders, Spheres, Ribbons, and Discs," *Ann. Occup. Hyg. 10*:83–95 (1967).

55. M. Teheri and S. Calvert, "Removal of Small Particles from Air by Foam in a Sieve-plate Column," *J. Air Pollut. Contr. Assn. 18*:240–245 (1968).

56. B. S. Javorsky, "Gas Cleaning with the Foam Scrubber," *Filtration Separation 9*:173 (1972).

57. B. Javorsky, "Fume Control and Gas Cleaning with an Industrial Scale Foam Bed Scrubber," *Filtration Separation 10*:21 (1973).

58. T. E. Ctvrtnicek, H. H. S. Yu, C. M. Moscowitz, and G. H. Ramsey, "Fine Particulate Control Using Foam Scrubbing," in *Novel Concepts and Advanced Technology in Particulate-Gas Separation*, edited by T. Ariman, University of Notre Dame, Notre Dame, Ind. (1978).

59. G. Ramsey, "Evaluation of Foam Scrubbing as a Method for Collecting Fine Particulate," EPA-600/2-77-197, US EPA (September 1977).

60. P. J. Schauer, "Removal of Submicron Aerosol Particles from a Moving Gas Stream," *Ind. Eng. Chem. 43*(9):1532–1538 (July 1951).

61. C. E. Lapple and H. J. Kamack, "Performance of Wet Dust Scrubbers," *Chem. Eng. Prog. 51*:110–121 (1955).

62. K. T. Semrau, "Dust Scrubber Design—A Critique on the State of the Art," *J. Air Pollut. Contr. Assn. 13*:587–594 (1963).

63. S. Calvert, J. Goldschmid, D. Leith, and N. C. Jhaveri, "Feasibility of Flux Force/Condensation Scrubbing for Fine Particulate Collection," APT. Inc., Riverside, CA, EPA-650/5-73-076, US EPA (1973).

64. L. Waldmann and K. H. Schmitt, "Thermophoresis and Diffusiophoresis of Aerosols," in *Aerosol Science*, edited by C. N. Davies, Academic Press, New York (1966).

65. K. T. Semrau, C. W. Marynowski, K. E. Lunde, and C. E. Lapple, "Influence of Power Input on Efficiency of Dust Scrubber," *Ind. Eng. Chem. 50*:1615–1620 (1958).

66. L. E. Sparks and M. J. Pilat, "Effect of Diffusiophoresis on Particle Collection by Wet Scrubbers," *Atmos. Environ. 4*:651–660 (1970).

67. W. G. N. Slinn and J. M. Hales, "A Re-evaluation of the Role of Thermophoresis as a Mechanism of

In- and Below-cloud Scavenging," *J. Atmos. Sci.* 28:1465–1471 (1971).

68. M. J. Pilat and A. Prem, "Effect of Diffusiophoresis and Thermophoresis on the Overall Particle Collection Efficiency of Spray Droplet Scrubbers," *J. Air Pollut. Contr. Assn.* 27:982–988 (1977).

69. P. J. Whitmore, "Diffusiophoretic Particle Collection Under Turbulent Conditions," Ph.D. thesis, University of British Columbia, Canada (1976).

70. S. Calvert and R. Parker, "Particulate Control Highlights: Fine Particle Scrubber Research," EPA-600/8-78-005a, US EPA (June 1978).

71. C. Orr, Jr., F. K. Hurd, and W. J. Corbett, "Aerosol Size and Relative Humidity," *J. Coll. Sci.* 13:472–482 (1958).

72. M. Neiburger and M. G. Wurtele, "On the Nature and Size of Particles in Haze, Fog and Stratus of the Los Angeles Region," *Chem. Rev.* 44:321–335 (1949).

73. D. W. Cooper, D. W. Underhill, and M. J. Ellenbecker, "A Critique of the U.S. Standard for Industrial Exposure to Sodium Hydroxide Aerosols," *Am. Indus. Hyg. Assn. J.* 40:365–371 (1979).

74. B. W. Lancaster and W. Strauss, "A Study of Stream Injection into Wet Scrubbers," *Ind. Eng. Chem. Fund.* 10:362–369 (1971).

75. S. Calvert and N. C. Jhaveri, "Flux Force/Condensation Scrubbing," *J. Air Pollut. Contr. Assn.* 24:947–952 (1974).

76. S. Calvert, S. Gandhi, D. L. Harmon, and L. E. Sparks, "FF/C Scrubber Demonstration on a Secondary Metals Recovery Furnace," *J. Air Pollut. Contr. Assn.* 27:1076–1080 (1977).

77. R. B. Jacko and M. L. Holcomb, "A Parametric Study of Flux Force/Condensation Scrubber for the Removal of Fine Hydrophobic Particles." Paper 78-17.2 presented at the 71st Annual Meeting of the Air Pollution Control Association, Houston, TX (June 1978).

78. D. W. Cooper, "Approximate Equations for Predicting Electrostatic Particle Collection." in *Novel Concepts and Advanced Technology in Particulate-Gas Separation*, edited by T. Ariman, University of Notre Dame, Notre Dame, Ind. (1978).

79. K. A. Nielsen, "Written Discussion," in *Novel Concepts and Advanced Technology in Particulate-Gas Separation*, edited by T. Ariman, University of Notre Dame, Notre Dame, Ind. (1978).

80. S. Oglesby, Jr. and G. B. Nichols, "Electrostatic Precipitation," in *Air Pollution*, edited by A. C. Stern, Academic Press, New York (1977).

81. M. J. Pilat, S. A. Jaasund, and L. E. Sparks, "Collection of Aerosol Particles by Electrostatic Droplet Spray Scrubbers," *Env. Sci. Technol.* 4:360–362 (1974).

82. M. J. Pilat, "Collection of Aerosol Particles by Electrostatic Droplet Spray Scrubbers," *J. Air Pollut. Contr. Assn* 25:176–178 (1975).

83. C. W. Lear, W. F. Krieve, and E. Cohen, "Charged Droplet Scrubbing for Fine Particle Control," *J. Air Pollut. Contr. Assn.* 25:184–189 (1975).

84. S. Calvert, S. C. Yung, H. Barbarika, and R. G. Patterson, "Evaluation of Four Novel Fine Particulate Collection Devices," EPA-600/2-78-062, US EPA, March (1978).

85. M. T. Kearns, "High Intensity Ionization Applied to Venturi Scrubbing," *J. Air Pollut. Contr. Assn.* 29:383–385 (1979).

86. D. C. Drehmel, "Advanced Electrostatic Collection Concepts," *J. Air Pollut. Contr. Assn.* 27:1090–1092 (1977).

87. C. G. Bell and W. Strauss, "Effectiveness of Vertical Mist Eliminators in a Cross Flow Scrubber," *J. Air Pollut. Contr. Assn.* 23:967–969 (1973).

88. S. Calvert, "Guidelines for Selecting Mist Eliminators," *Chem. Eng.*, 109–112 (February 27, 1978).

89. D. Leith and D. Mehta, "Cyclone Performance and Design," *Atmos. Environ.* 7:527–549 (1973).

90. S. Calvert and R. Parker, "Particulate Control Highlights: Fine Particle Scrubber Research," EPA-600/8-78-005a, US EPA (June 1978).

91. W. Strauss, "Mist Eliminators," in *Air Pollution*, edited by A. C. Stern, Academic Press, New York (1977).

92. G. H. Griwatz, J. V. Friel, and J. L. Creehouse, Report 71-45, U.S. Atomic Energy Commission, Mine Safety Applications Research Corp., Evans City, PA (1971).

93. B. A. Steiner and R. J. Thompson, "Wet Scrubbing Experience for Steel Mill Applications," *J. Air Pollut. Contr. Assn.* 27:1069–1075 (1977).

94. E. C. Hoxie and G. W. Tuffnell, "A Summary of INCO Corrosion Tests in Power Plant Flue Gas Scrubbing Processes," in *Resolving Corrosion Problems in Air Pollution Equipment*. National Association of Corrosion Engrs., Houston, TX (1976).

95. J. S. Busch, W. E. MacMath, and M. S. Lin, "Design and Cost of High Energy Scrubbers: 1. The Basic Scrubber," *Pollut. Engrg.*, pp. 28–32 (January 1973).

96. L. D. Stulov, F. I. Murashkevich, and N. A. Fuchs, "The Efficiency of Collision of Solid Aerosol Particles with Water Surfaces," *J. Aerosol Sci.* 9:1–6 (1978).

97. R. W. McIlvaine, "When to Pilot and When to Use Theoretical Predictions of Required Venturi Pressure Drop." Paper 77-17.1 presented at the 70th Annual Meeting of the Air Pollution Control Association, Toronto, Canada (1977).

98. M. Taheri, S. A. Beg, and M. Beizie, "The Effect of Scale-up on the Performance of High Energy

Scrubbers," *J. Air Pollut. Contr. Assn. 23*:963–966 (1973).

99. N. S. Balakreshnan and G. H. S. Cheng, "Scale-up Effect of Venturi Scrubber." Paper 78-17.3 presented at the 71st Annual Meeting of the Air Pollution Control Association, Houston, TX (June 1978).

100. S. W. Behie and J. M. Beeckmans, "Effects of Water Injection Arrangement on the Performance of a Venturi Scrubber," *J. Air Pollut. Contr. Assn. 24*:943–945 (1974).

101. N. G. Edmisten and F. L. Bunyard, "A Systematic Procedure for Determining the Cost of Controlling Particulate Emissions from Industrial Sources," *J. Air Pollut. Contr. Assn. 20*:446–452 (1970).

102. E. M. Hanf and J. W. MacDonald, "Economic Evaluation of Wet Scrubbers," *Chem. Eng. Prog. 7*(3):48–52 (1975).

103. M. D. Fraser and D. R. Eaton, "Cost Models for Venturi Scrubber System." Presented at 68th Annual Meeting of the Air Pollution Control Association, Boston (1975).

104. R. B. Neveril, J. U. Price, and K. L. Engdahl, "Capital and Operating Costs of Selected Air Pollution Control Systems-I.-V." *J. Air Pollut. Contr.*

Assn. 28:829–836, 963–968, 1069–1072, 1171–1174, 1253–1256 (1978).

105. A. C. Stern, H. C. Wohlers, R. W. Boubel, and W. P. Lowry, *Fundamentals of Air Pollution*, Academic Press, New York (1973).

106. D. W. Cooper, "On the Products of Lognormal and Cumulative Lognormal Particle Size Distributions," *J. Aerosol Sci. 13*:111–120 (1982).

107. K. W. Lee and J. A. Gieseke, "A Note on the Approximations of Interceptional Collection Efficiencies," *J. Aerosol Sci. 11*:335–341 (1980).

108. D. S. F. Atkinson and W. Strauss, "Droplet Size and Surface Tension in Venturi Scrubbers," *J. Air Pollut. Contr. Assn. 28*:1114–1118 (1978).

109. D. W. Cooper, "Filter Beds: Energy-Efficient Packing Diameter," *J. Air Pollut. Contr. Assn. 32*:205–208 (1982).

110. D. W. Cooper, "Optimizing Filter Fiber Diameter," *Atmos. Environ. 16*:1529–1533 (1982).

111. T. D. Placek and L. K. Peters, "Analysis of Particulate Removal in Venturi Scrubbers—Effect of Operating Variables on Performance," *AIChE J. 27*:984–993 (1981).

112. L. P. Bayvel, "The Effect of the Polydispersity of Drops on the Efficiency of a Venturi Scrubber," *TransIChemE, 60*:31–34 (1982).

19
Fire and Explosion Hazards in Powder Handling and Processing

Stanley S. Grossel

CONTENTS

19.1 INTRODUCTION

When storing, transferring, or processing bulk solids and powders consideration must be given to the proper design of the equipment and systems to prevent dust explosions and fire, or to mitigating their effects, if they occur.

The subject of dust explosions is too large and complicated to cover in depth in this chapter, but certain aspects are discussed to present some fundamentals and background material. For further reading on the subject, consult the technical publications and books by NFPA,[1,2] Bartknecht,[3] and Eckhoff,[4] to name a few recent ones.

A dust explosion is in reality a dust deflagration, that is, a combustion phenomenon in which the propagation of the combustion zone occurs at a velocity that is less than the speed of sound in the unreacted dust. However, for conformity with common usage, it is referred to as a dust explosion in this chapter.

Dust explosions and fires are the principal hazards associated with dust handling systems. Other hazards that may occur include:

1. The development of electrostatic charges on the conveyed material or system components which might ignite vapors or dusts in associated processes
2. Unexpected electrical shocks from static charges on ungrounded components, causing involuntary reaction
3. In the case of toxic dusts, health hazards associated with even small leaks or with maintenance work on the system.

In the following sections we discuss principles of dust explosions, factors affecting dust explosions, ignition sources, basic system design considerations, dust explosion prevention and protection methods, and application to industrial processes and equipment.

19.2 PRINCIPLES OF DUST EXPLOSIONS

19.2.1 Introduction

A dust explosion results when finely divided combustible matter is dispersed into an atmosphere containing sufficient oxygen to permit combustion and a source of ignition of appropriate energy is present. Dust explosions have certain similarities to gas explosions, especially with regard to the chemical processes involved, and in cases where the particle size of the dust is less than 5 μm. However, there are significant differences that make dust explosions more difficult to achieve. For a dust explosion to occur, a degree of turbulence must be present, if only to disperse the dust into a suspension. Gas explosions can occur when the gas is in a quiescent state, the mixture being homogeneous and consisting of molecular-size particles. The suspensions of dusts encountered in dust explosions are, however, unlikely to be homogeneous, normally containing a range of concentrations of particles that are many orders of magnitude larger

and heavier than gas molecules and that settle out of suspension owing to gravity. The processes of a dust explosion involve such a high rate of combustion that individual particles and agglomerates are either consumed or oxidized. The combustion of carbon present in organic materials will produce gaseous products that in themselves take up more space than the solids of the parent material. In addition, an expanding flame front will result from the ignition of flammable gases produced by the decomposition of the dust. A dust explosion therefore produces a system requiring more space owing to expansion of the hot gaseous products. In industrial plants, the heat released during a dust explosion is likely to exceed the natural rate of cooling and consequently an explosion would be accompanied by significant, and in some cases uncontrolled expansion effects. In an unconfined situation, a dust explosion would result in mainly localized flames and pressure effects. However, in confined situations, such as those commonly found in plants handling particulate matter, the expansion effects are likely to be sufficient to rupture the plant equipment or piping unless they are suppressed or vented.

A number of conditions must be satisfied simultaneously for a dust explosion to occur:

1. The dust must be combustible.
2. The dust must be a suspension in the atmosphere, which must contain sufficient oxygen to support combustion.
3. The dust must have a particle size distribution that will propagate a flame.
4. The dust concentration in the suspension must be within the explosible range.
5. The dust suspension must be in contact with an ignition source of sufficient energy.

If these conditions are satisfied, the hazard from a dust explosion depends on the explosibility of the dust, the volume and characteristics of the vessel or chamber containing the dust suspension, the dispersion and concentration of the dust suspension, and the degree of turbulence in the vessel.

The explosibility of a dust can be determined by tests that are described by Eckhoff[4] and Field.[5]

19.2.2 Lower Explosive Limit

Dusts, like gases, have lower and upper explosive limits. The lower explosive limits (also called minimum explosive concentration) for many dusts are available in the open technical literature. They are usually expressed as grams per cubic meter or sometimes as grams per liter. Extensive tables are given in the books by Eckhoff[4] and Palmer.[6] Data are meager for upper explosive limits as they are difficult to experimentally determine because of problems in achieving adequate suspension of the dust during testing. The value of the lower explosive limit depends on a number of factors such as the composition of the dust, its particle size distribution, and to some extent, the strength of the ignition source.

19.2.3 Oxidant

The oxidant in a dust explosion is normally the oxygen in air. However, other oxidants, such as the halogens, can also lead to an explosion, and should be considered. There is a limiting oxygen concentration (LOC), also called maximum safe oxygen concentration (MSOC), below which combustion will not occur. The LOC for dusts depends on the composition and particle size distribution of the solids. Values of LOC for most organic chemical dusts lie in the range of 10 to 16 volume percent. Palmer[6] lists LOC data for many dusts, as does NFPA 69.[2]

19.2.4 Maximum Explosion Pressure and Maximum Rate of Pressure Rise

When a dust explosion occurs, two of the factors influencing the security of the explosion are the maximum explosion pressure (P_{max}) and the maximum rate of pressure rise $(dP/dt)_{max}$. These two quantities determine the pressure build-up to which equipment is subjected, and are needed to calculate vent areas. Experimental data for these two quantities should be obtained using a 20-liter test vessel as a minimum size.[1] Older data obtained in the Hartmann bomb (U.S. Bureau of Mines) should not be used for sizing deflagration vents by the methods given in NFPA 68. Data on P_{max} and $(dP/dt)_{max}$ are available for many dusts.[1,4]

19.2.5 Minimum Ignition Temperature

The minimum ignition temperature of a dust suspension is the lowest temperature at which it will ignite spontaneously and propagate the flame. It depends on the size and shape of the apparatus used to measure it as well as the rate of rise in temperature of the dust, the particle size, and moisture content of the dust. Therefore, minimum ignition temperatures have to be determined in a standardized type of apparatus to enable meaningful comparisons between dusts.[4,5] Minimum ignition temperatures are used to establish a maximum safe operating temperature for processes such as drying. Refer to the books by Field[5] and Palmer[6] for data on minimum ignition temperatures.

19.2.6 Minimum Ignition Energy (MIE)

Minimum ignition energies are measured to provide data on the possibility of ignition of dust clouds by electrostatic sparks. Powders that have low ignition energies, for example, below 15 mJ, are often regarded as particularly hazardous because of the possibility of ignition by operators who have become accidentally charged electrostatically. The MIE of a dust cloud depends on the dust concentration, particle size, moisture content, etc. The lowest value of the MIE is found at a certain optimum mixture. It is this value (at this optimum mixture) that is usually quoted as the MIE. Values of MIE for dusts vary from 10 to hundreds of millijoules. Values of MIE for many dusts can be found in the books by Eckoff,[4] Field,[5] and Palmer.[6]

19.2.7 Flame Propagation

The rate of propagation of a flame, that is, flame speed, in a dust explosion cannot be readily predicted as in the case of gas explosions. In the case of gases, the flame speed reaches a maximum at or near the stoichiometric mixture, that is, that mixture in which all the gas just reacts with the available oxygen. The flame speed in a dust explosion reaches a maximum when there is an excess of dust and reduces significantly only when the dust concentration is several times the stoichiometric mixture. Dusts can produce more serious explosions than gases because there is a tendency to a slower flame speed resulting in a longer residence time and a greater total impulse. The flame speed is not constant and depends on a number of variables, the most significant probably being the chemical composition, particle size, concentration, and moisture content of the dust, and the nature and turbulence of the gas in which the dust is dispersed. The flame speed increases with increase in turbulence and with decrease in particle size, provided that the dust is evenly dispersed. In industrial situations turbulence should be expected, but it is unlikely that the dust dispersion will be completely homogeneous.

19.2.8 Explosibility Rating

As mentioned in Section 19.2.4, key characteristics of a closed-vessel deflagration are the maximum pressure attained, P_{max}, and the maximum rate of pressure rise, $(dP/dt)_{max}$, developed during the event. The most widely used measure of the explosibility of a combustible material is computed from the maximum rate of pressure rise attained by combustion in a closed vessel. The index of explosibility, as developed by Bartknecht,[3] is defined as:

$$K_{st} = (dP/dt)_{max}V^{1/3}$$

where V is the volume of the test vessel and $(dP/dt)_{max}$ is the maximum rate of pressure rise attained over the range of fuel/air ratios

tested. The value of $(dP/dt)_{max}$ will be a maximum for a particular fuel concentration, referred to as the "optimum" concentration, and is characteristic of the particular combustible. The K_{st} value has been found to be nearly invariant with $V^{1/3}$ only for measurements of $(dP/dt)_{max}$ made in vessels 20 liters or larger in size. For this reason it is important that K_{st} values be determined according to an approved standard that employs a vessel of at least 20 liters volume.

Another classification of explosibility of dusts uses the concept of dust class, which is related to K_{st} values, as follows:

DUST CLASS	K_{st} (bar-m/s)
St-0	Nonexplosible
St-1	< 201
St-2	201 to 300
St-3	> 300

K_{st} values of various materials have been tabulated in NFPA 68[1] and Eckhoff's book.[4] These values should be used only as first-order guidelines. The design of protection equipment for a particular process should be based on the measured combustion properties of the actual product being handled.

Both K_{st} values and dust class are used for sizing deflagration vents.[1]

19.2.9 Primary and Secondary Explosions

Dust explosions can be divided into two types: primary and secondary explosions. A primary explosion occurs in equipment when dust is airborne in an atmosphere containing sufficient oxidant (usually oxygen) for combustion and is subjected to an ignition source of sufficient energy. Secondary explosions result when the flame ball emitted from equipment experiencing a primary explosion ignites combustible dust in the immediate vicinity. This exterior dust is usually from fugitive dust that has been allowed to settle and accumulate on horizontal surfaces. The secondary explosion often can be much more violent than the primary explosion because the pressure from the secondary explosion can be transmitted throughout a plant building, resulting in structural collapse. In addition to these pressure effects, the flames of a dust explosion can propagate significant distance and spread fire to areas not in the immediate vicinity of the primary explosion.

19.3 FACTORS AFFECTING DUST EXPLOSIONS

The following chemical and physical factors influence the initiation and propagation of a dust explosion: chemical reactivity, moisture content, particle size/specific surface area, dust concentration, oxygen content of oxidizer gas, turbulence, initial temperature of dust clouds, initial pressure of dust clouds, effect of inert gas or dust, and combustible gas or vapor mixed with the dust cloud (hybrid mixtures). These are discussed briefly below. For more extensive discussion of these factors, refer to the books by Eckhoff[4] and Field.[5]

19.3.1 Chemical Reactivity

Increasing chemical reactivity of dusts, similar to gases and vapor, leads to increasing explosion severity. Examples of highly reactive powders are metals (e.g., Al, Mg, Ti, Zr, etc.) that possess very high heats of oxidation. For example, the maximum reaction temperature of a metal powder explosion may reach well above 3000 K, whereas the maximum temperature reached in an explosion of an organic powder will usually be 2000 K to 3000 K (about the same as a gas explosion). Also, whereas the maximum pressure reached in an explosion of an organic dust is in the range of 7 to 10 bars, some metal dust explosions may generate maximum pressures in excess of 10 bars.

The presence of specific chemical groups in organic material can give an indication of the explosion risk, for example, COOH, OH, NH_2, NO_2, $C = N$, $C \equiv N$, and $N = N$ tend to increase the explosion hazard, whereas the incorporation of the halogens Cl, Br, and F generally results in a reduced explosion hazard.

19.3.2 Moisture Content

Many powders contain moisture, the amount depending on the presence of moisture from the previous processing steps, the hydrophilic nature of the powder, and the relative humidity of the surrounding atmosphere. In general, the presence of moisture is beneficial as it tends to decrease the explosibility of a dust in two different, but synergistic ways. First, as the moisture content increases, the dust particles generally become more cohesive and form agglomerates that are more difficult to disperse. Second, any heat applied to a suspension of moist dust will first be used to vaporize the moisture (water and solvent) and will therefore not be used in the combustion process.

Moisture in a dust reduces both ignition sensitivity and explosion violence of dust clouds. Figure 19.1 illustrates the influence of moisture content on the minimum electric spark ignition energy (MIE), and Figure 19.2 shows how the maximum pressure rise is reduced with increasing moisture content. The ignition delay characterizes the state of turbulence of the dust cloud at the moment of ignition in the sense that the turbulence intensity decreases as the ignition delay increases. However, it is not possible to predict, a priori, a moisture content that would be sufficient to prevent an explosion from occurring as this varies with other factors as well, such as the nature and particle size of the dust. As a general rule, in normal industrial operations a dust explosion is probably unlikely to occur if the dust being processed has a moisture content in excess of 30%.[5] The only sure way of determining the moisture content needed to prevent an explosion is by experimental tests.

19.3.3 Particle Size / Specific Surface Area

One of the most important physical properties of a powder that affects dust explosions is the particle size distribution. This is illustrated in

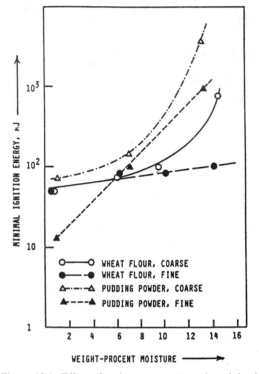

Figure 19.1. Effect of moisture content on the minimal ignition energy (MIE) of two powders.

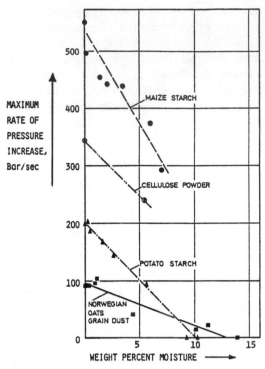

Figure 19.2. Effect of moisture content on the explosion severity of some agricultural dusts.

Table 19.1, which shows that for a given mass of dust the smaller the particle diameter, the greater the amount of surface area available for reaction. It is for this reason that explosion severity (the maximum pressure and rate of pressure rise) increases with decreasing particle size (see Fig. 19.3).

As the particle size decreases, particle volume and mass decrease sharply (see Table 19.1) so that it requires a smaller amount of energy to bring finer particles to their ignition temperature than larger particles. For this reason, explosion sensitivity will increase (e.g., lower MIEs) as particle size decreases (see Fig. 19.4). Also, the lower explosion limit

Table 19.1. Relation of Particle Size (Length) to Particle (Specific) Surface Area and Volume (Particles in the Form of Cubes; Density = 1000 kg / m³).

PARTICLE LENGTH (μm)	PARTICLE SURFACE AREA (m²)	PARTICLE VOLUME (m³)	PARTICLE MASS (kg)	PARTICLE SPECIFIC SURFACE AREA (m²/kg)	NUMBER OF PARTICLES PER KG (kg⁻¹)
0.1	6×10^{-14}	10^{-21}	10^{-18}	60,000	10^{18}
1.0	6×10^{-12}	10^{-18}	10^{-15}	6,000	10^{15}
10.0	6×10^{-10}	10^{-15}	10^{-12}	600	10^{12}
100.0	6×10^{-8}	10^{-12}	10^{-9}	60	10^{9}

$(1\ \mu\text{m} = 10^{-6}\ \text{m})$

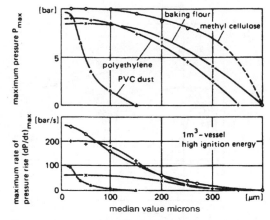

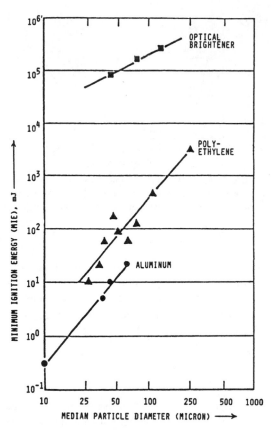

Figure 19.3. Effect of average particle diameter of dusts on the maximum pressure and the maximum rate of pressure rise developed by a deflagration in a 1 m³ vessel.

(minimum explosible concentration) decreases as the particle size decreases (see Fig. 19.5).

As a general rule, combustible dust clouds containing particles normally less than 420 μm may deflagrate more readily than larger particles.[1] However, tests should be conducted to determine the effect of particle size on explosibility of powders.

19.3.4 Dust Concentration

Unlike gases and vapors, the most severe explosion behavior for dusts is not found at the stoichiometric composition, but at concentrations considerably higher. This is because a dust explosion is a surface phenomenon. Thus, a powder in a stoichiometric concentration expressed in terms of weight is actually far under the stoichiometric composition in terms of surface area.

Explosion rate, $(dp/dt)_{max}$, and minimum ignition energy vary with dust concentration, as shown in Figure 19.6, where C is the minimum explosible concentration, C_{stoich} the stoichiometric concentration, and C_u the maximum explosible concentration.

19.3.5 Oxygen Content of Oxidizer Gas

As one would expect, both explosion violence and ignition sensitivity increase with increasing oxygen concentration, as shown in Figure

Figure 19.4. Effect of particle size on minimum ignition energy (MIE).

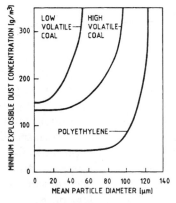

Figure 19.5. Influence of mean particle diameter on minimum explosible concentration for three different dusts in 20-liter USBM vessel.

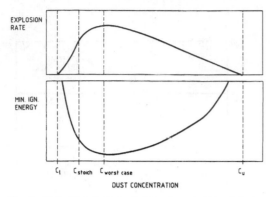

Figure 19.6. Illustration of typical variation of explosion rate and minimum electric spark ignition energy (MIE) with dust concentration within the explosible range.

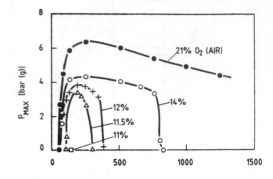

19.7. Furthermore, as shown in this figure, the explosible dust concentration range was narrowed, in particular on the fuel-rich side, as the oxygen content decreased. Figure 19.8 shows the influence of oxygen content on the MIE of three organic powders.

19.3.6 Turbulence

Turbulence is usually present in industrial dust–air systems, especially in pneumatic conveying systems. At the onset of a dust explosion a degree of turbulence will already exist that will be increased as the flame front moves through the dust. It is extremely difficult to quantify turbulence in dust explosions because it is likely to be nonuniform and the normal flow of a given process will be grossly distorted. The turbulent dispersion of combustible dusts results in an increased explosion hazard because the access of oxygen to the active surfaces of the dust is greatly improved. This results in faster reaction rates at the solid–gas interface and a corresponding enhancement in heat-transfer processes. Turbulence is also likely to cause the flame front to fragment, producing sites from which combustion can develop simultaneously, and resulting in greater explosion pressures.

Initial turbulence in closed vessels results in both higher maximum pressure and higher

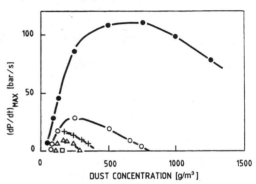

Figure 19.7. Influence of oxygen content in the gas on the maximum explosion pressure and maximum rate of pressure rise of brown coal dust for various concentrations. Nitrogen as inert gas. 1 m^3 ISO standard explosion vessel.

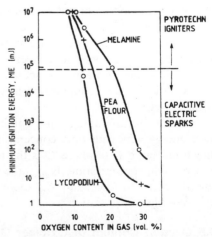

Figure 19.8. Influence of oxygen content in gas on minimum ignition energy of dust clouds.

maximum rates of pressure rise than would be obtained if the fuel–oxidant mixture were at initially quiescent conditions. This is shown in Figure 19.9.

While increased turbulence strongly increases explosion severity, its effect on explosion sensitivity (MIE) is usually the opposite. The MIE will increase as turbulence increases. This can be explained on the following basis. For dust particles to be ignited they must be exposed to an energy source for a sufficient period of time to allow them to heat up and react. An energy source is usually located in a specific place, so that rapid air movement induced by turbulence shortens the length of time that particles are present within a given volume; thus the particles have less time available to be activated, and therefore require more energy.

19.3.7 Initial Temperature of Dust Clouds

As the initial temperature of a dust cloud increases, the minimum explosible dust concentration (LEL) decreases (see Fig. 19.10). Also, as the initial temperature increases, the

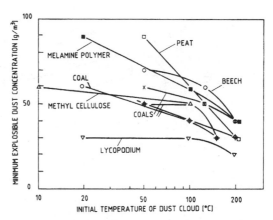

Figure 19.10. Influence of initial temperature of dust clouds on minimum explosible dust concentration in air at 1 bar (abs.).

minimum ignition energy (MIE) decreases, as shown in Figure 19.11. The influence of increasing temperature on P_{max} and $(dp/dt)_{max}$ is shown in Figure 19.12.

19.3.8 Initial Pressure of Dust Clouds

Increasing the initial pressure results in an increase in both P_{max} and $(dp/dt)_{max}$ as shown in Figure 19.13. The influence of increasing pressure on minimum explosible concentration is illustrated in Figure 19.14.

19.3.9 Effect of Inert Gas on Dust

Increasing the concentration of gaseous inerts in air decreases the oxygen concentration and has the effects discussed in Section 19.3.5. The

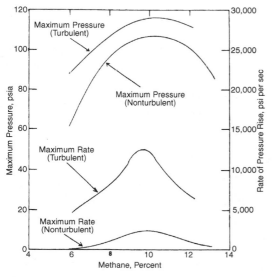

Figure 19.9. Maximum pressure and rate of pressure rise for turbulent and nonturbulent methane/air mixtures in a 1 ft^3 closed vessel.

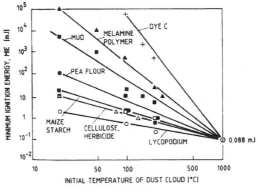

Figure 19.11. Influence of initial temperature of dust cloud on minimum electric spark ignition energy.

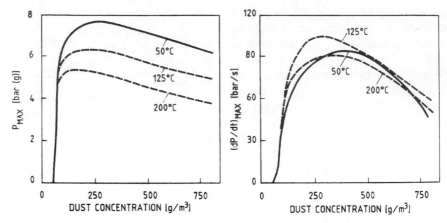

Figure 19.12. Influence of initial temperature of dust cloud on explosion development in 1 m³ closed vessel. Bituminous coal dust in air.

oxygen is normally replaced by nitrogen or carbon dioxide, although argon, flue gas, or steam may be used in some circumstances. This process is called inerting and is discussed in Section 19.6.

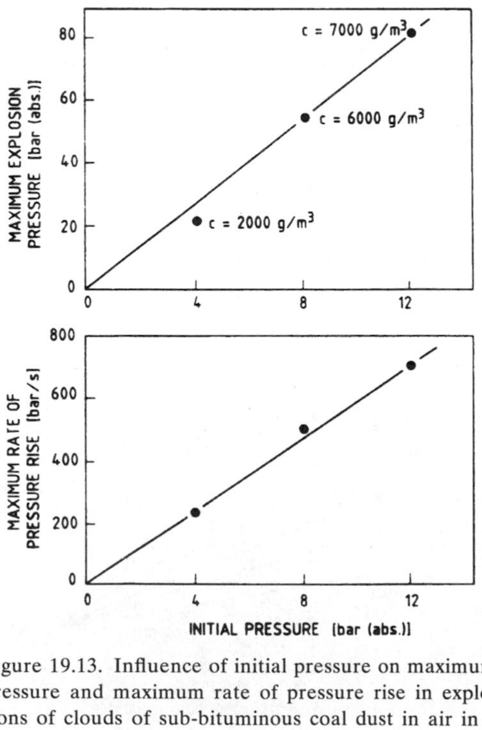

Inert dust added to combustible dust–air mixtures also acts as an explosion inhibitor by interfering with the diffusion process of the oxygen to the active surfaces of the combustible dust and by acting as a significant heat sink. The rates of reaction and heat transfer are considerably lowered, resulting in a reduced explosion hazard. This technique is used primarily in the coal mining industry, and some recent research work on this subject was presented by Amyotte and Pegg.[7]

19.3.10 Hybrid Mixtures

Hybrid mixtures are those containing a combustible gas with either a combustible dust or

Figure 19.13. Influence of initial pressure on maximum pressure and maximum rate of pressure rise in explosions of clouds of sub-bituminous coal dust in air in a 15-liter closed bomb: median particle size by mass 100 μm.

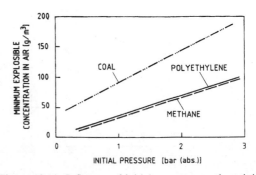

Figure 19.14. Influence of initial pressure on the minimum explosible concentration of two dusts and methane in air.

a combustible mist, and are often encountered in drying operations. The presence of the combustible gas has a strong influence on the burring characteristics of the dust, the severity depending on the nature and concentration of the gas. In essence, hybrid mixtures represent an increased explosion hazard compared with that already presented by the combustible dust alone. The effects are as follows:

1. A hybrid mixture will explode more violently than a dust–air mixture alone, even if the gas concentration is below its LEL.
2. The ignition energy and ignition temperature of hybrid mixtures will be lower than that of dust–air mixtures alone.
3. The minimum explosible concentration of hybrid mixtures is lower than that of the dust itself in air, even if the concentration of flammable gas is below its LEL.
4. For hybrid mixtures the maximum pressure and rate of pressure rise during a deflagration may increase considerably in comparison to a dust–air mixture alone.

19.4 IGNITION SOURCES

As mentioned in Section 19.2 a dust explosion requires an ignition source of appropriate energy. In general, the most important characteristics of the ignition source are:

1. The type of ignition.
2. The amount of energy expended (Joules).
3. The power of the ignition source, that is, the rate at which the energy is expanded over a time (Joules/s).
4. The temperature of the ignition source.
5. The surface area and form of the ignition source.
6. The place where ignition occurs.

A good discussion of these ignition sources is presented by Eckhoff[4] and Field.[5] The main types of ignition sources are:

- Electric sparks
- Electrostatic discharge sparks
- Flames (open fire)
- Friction heating or sparks
- Hot surfaces
- Impact sparks
- Incandescent material
- Spontaneous heating
- Welding or cutting operations

Electrostatic discharge sparks are one of the most commonly occurring ignition sources, and have been the cause of many dust explosions.

Electrostatic charges can develop on bulk solids and powders being conveyed or processed, especially organic ones. These charges occur because of the contacts made between surfaces during the movement of particles. The charges on a powder particle are governed by three factors: (1) the charge production rate, (2) the charge leakage rate when the particle is in contact with a ground, and (3) the electrical breakdown of air initiated by the high field around the charged particle.

An electrostatic spark occurs when an isolated object that has been allowed to accumulate charge is suddenly grounded. The accumulation of static electricity on an object produces an electric field around it and a spark will occur if the field strength exceeds the breakdown value of the surrounding atmosphere. For air, this is approximately 3000 kV/m.

A number of good books are available that discuss electrostatic spark hazards and methods of preventing or mitigating them.[8–11]

19.5 GENERAL PLANT DESIGN CONSIDERATIONS

In designing a plant handling or processing powders and bulk solids some general design principles should be followed in order to prevent or minimize the potential for dust explosions. These are:

1. Where possible select less dusty alternatives for materials and minimize attrition.

2. Minimize handling of dusty materials and design handling systems to minimize dust generation and the size of dust clouds.
3. Avoid the accumulation of dust (which can be disturbed to form a dust cloud) by the detailed design of equipment, building, and working practices.
4. Anticipate possible ignition sources and eliminate them, as far as is reasonably practicable, by appropriate equipment design, bonding, grounding, maintenance, and working practices.
5. Take appropriate additional measures, where practicable, such as inerting, containment, venting, or suppression.
6. Isolate vulnerable plant equipment as appropriate. For example, dust collectors should be located outdoors or on roofs, if feasible.

19.6 DUST EXPLOSION PREVENTION AND PROTECTION METHODS

19.6.1 Introduction

To prevent dust explosions or mitigate their effects two groups of methods are used in industry, that is, prevention and protection. Prevention methods include:

1. Removal of ignition sources.
2. Prevention or minimization of dust cloud formation.
3. Oxidant concentration reduction (inerting).
4. Combustible concentration reduction (ventilation or air dilution).

Protection methods include:

1. Deflagration venting
2. Deflagration suppression
3. Deflagration pressure containment
4. Deflagration isolation systems.

These methods are discussed in detail in several books and association publications.[1-6,12-14] A brief review of some of these methods is presented below.

19.6.2 Removal of Ignition Sources

Various methods for removing or controlling the ignition sources listed in Section 19.4 are presented by Schofield and Abbott.[13] They present ignition prevention techniques for size reduction equipment, pneumatic conveyors, mechanical conveyors, dryers, storage bins and silos, and dust filters (bag houses). In addition to these techniques, fans and blowers can be specified to have spark-proof construction.

19.6.3 Inerting

Inerting is probably the most commonly used prevention technique. It is of particular use for very strongly explosible dusts ($K_{st} > 600$ bar/s) and where hybrid mixtures are present. Inerting is often used for grinding or drying operations that otherwise would result in frequent explosions.

Nitrogen is the most commonly used gas for inerting. However, carbon dioxide, argon, helium, and flue gases may also be used. Table 19.2 shows the relative merits of these gases. In choosing an inerting gas, the reactivity of the dust and gas must be considered, as some metal dusts, for example, can react violently with carbon dioxide and some can even burn in nitrogen, Schofield and Abbott[13] and NFPA 69[2] present a thorough discussion of the design and application of inerting gas systems.

19.6.4 Deflagration Venting

Protection of process vessels and enclosures can be accomplished quite frequently by deflagration venting, which is the most widely used and least expensive protection method. A deflagration vent is an opening, normally provided with a cover, in a vessel or enclosure that allows combustion-generated gases to expand and flow. Its purpose is to limit the deflagration pressure so that damage to the vessel or enclosure is limited to an acceptable level. Flames and burning powder will be ejected from the vent so that the positioning of the vent must take into consideration the location of nearby equipment, buildings, and

Table 19.2. Relative Merits of Inert Gases.

GAS	ADVANTAGES	DISADVANTAGES
Carbon dioxide	Readily available in compressed form, from proprietary inert gas generators, and in some cases as a waste gas from on-site processes	Some metal dusts react violently with carbon dioxide (e.g., aluminum)
	Effective—higher oxygen levels (per cent by volume) are permissible compared with nitrogen Moderate cost	Flow of carbon dioxide can generate considerable electrostatic charge
Nitrogen	Readily available in compressed or cryogenic form, and in some cases as a waste gas from on-site processes	Less effective in volume/volume terms than carbon dioxide
	Moderate cost	Some metal dusts react with nitrogen (e.g., magnesium) at high temperature
Flue gases	Often readily available as a waste gas from on-site processes or from inert gas generators	Requires additional equipment to: Cool the gas, Remove contaminants, Monitor or remove combustible vapors, Remove incandescent material May react with dusts
	Often available at low cost	Storage of flue gas may not be practical. so that adequate quantities may not always be available, for example during a furnace shutdown
Argon or helium	Unlikely to contaminate products or react with them	Expensive

roads where operating people may pass. If toxic or other very hazardous materials are processed in the equipment to be protected, then venting as a protective measure should not be used. Recoil forces on the vented vessel or equipment may cause failure of supports if they are not taken into account. If vessels or equipment provided with deflagration vents are located inside a room, vent ducts should be installed to discharge the flames, combustion products, and pressure to outside of the room. Vent ducts increase the pressure on the discharge side of the vent, owing to frictional pressure drop, so that the reduced explosion pressure in the vessel can increase significantly in comparison to the situation in which there is no vent duct.

The sizing of deflagration vents is based on research done primarily in Germany, Switzerland, and Norway[3,4] and is summarized in NFPA 68[1] and the books by Lunn,[12,14] Bartknecht,[3] and Eckhoff.[4] The sizing method depends on whether the equipment to be protected is a low-strength or high-strength enclosure. Low-strength enclosures are those that cannot withstand internal pressure greater than 1.5 psig (0.1 bar ga.), such as rooms, buildings, and certain equipment such as bag houses. All structural elements must be considered in making a strength assessment, including, walls, ceilings, doors, seals, etc. Equipment capable of withstanding an internal pressure greater than 1.5 psig is considered a high-strength enclosure.

Vent areas for low-strength structures can be calculated by the following equation:

$$A_v = CA_s/(P_{red})^{1/2}$$

where

A_v = vent area (ft^2 or m^2)
C = combustible-dependent constant (see Table 19.3)
A_s = internal surface area of enclosure, to include walls, floor, and ceiling (ft^2 or m^2)
P_{red} = maximum overpressure tolerable by weakest structural element, psi or kPa. P_{red} is defined as a pressure two-thirds of the ultimate strength of the weakest part of the enclosure

Vent areas for high-strength enclosures can be sized either by equations or nomographs[1, 12] based on values of K_{st} or dust class (see Section 19.2.8). The equations and nomographs from which they were derived can be applied within the following constraints:

1. Initially quiescent dust mixture
2. No internal obstructions that may enhance turbulence development during deflagration
3. A maximum ignition energy of 10 J
4. Initial pressure of 101.3 kPa (14.7 psia)
5. Enclosure length-to-diameter ratio $(L/d) < 5$
6. $1 < V < 1000$ m^3
7. $20 < P_{red} < 200$ kPa g
8. $10 < P_{stat} < 50$ kPa g.

Table 19.3. Combustible-Dependent Constant for Low-Strength Enclosures.

COMBUSTIBLE	C (psi)$^{1/2}$	C (kPa)$^{1/2}$
Anhydrous ammonia	0.05	0.13
Methane	0.14	0.37
Gases with $S_u < 0.6$ m/s	0.17	0.45
St-1 dusts	0.10	0.26
St-2 dusts	0.12	0.30
St-3 dusts	0.20	0.51

S_u is the fundamental burning velocity. See Table B-1 of NFPA 68 for values of S_u for a number of gases.

Figure 19.15 shows one venting nomograph using K_{st} as a parameter and Figure 19.16 shows one of the nomographs using the dust class as a parameter. The equations given in NFPA 68 were derived from the nomographs.

The thrust force resulting from the recoil of the vented vessel can be calculated by the following formula[1]:

$$F_r = 1.2A (P_{red})$$

where

F_r = reaction force resulting from venting (lb)
A = vent area (in.^{2})
P_{red} = maximum pressure developed during venting (psig).

19.6.5 Deflagration Suppression

A deflagration is not an instantaneous phenomenon, but takes some time to build up destructive pressure in a vessel. Typically it takes 30 to 100 ms before destructive pressures are achieved. Therefore, it is possible to suppress an explosion utilizing equipment that detects an incipient explosion very soon after ignition occurs and injects a sufficient amount of a chemical agent at a fast enough rate to extinguish all flame before a destructive overpressure develops. Suppression is most often used when it is not possible to vent the contents of a vessel to a safe place, for example, where a toxic dust would be emitted or the fireball would impinge on people or adjacent equipment.

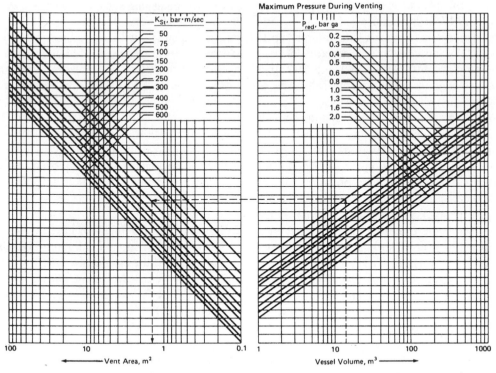

Figure 19.15. Venting nomograph for dusts—P_{max} = 0.1 bar ga.

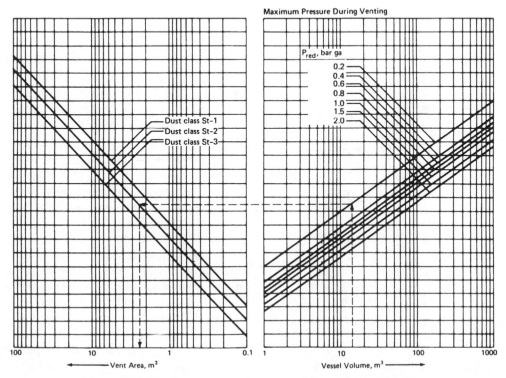

Figure 19.16. Venting nomograph for classes of dusts—P_{max} = 0.1 bar ga.

The principles of suppression are shown in Figure 19.17. It is technically feasible to suppress explosions in vessels with volumes up to 1000 m³.[15] Suppression systems are normally used only for dust classes St-1 and St-2. It is possible only in some exceptional cases to suppress dust class St-3 explosions.

A deflagration suppression system consists of three basic subsystems for (1) detection, (2) extinguishment, and (3) control and supervision. Incipient deflagrations are detected using pressure detectors, rate of pressure rise, or "rate" detectors, or optical flame detectors. Optical detectors, employing ultraviolet radiation sensors, are preferred in unenclosed environments with nonabsorbing ultraviolet atmospheres. Examples of such environments are solvent storage and pump rooms and aerosol filling rooms. Pressure detectors are employed in closed process equipment and particularly where dusty atmospheres prevail. Rate detectors find use in processes that operate at pressures significantly above or below atmospheric.

The extinguishing subsystem consists of one or more high rate discharge (HRD) extinguishers charged with agent and propellant. Normally dry nitrogen is used to propel the agent. The propellant overpressure is normally in the range of 2 to 6 MPa (300 to 900 psig), depending on the supplier. Explosively opened valves, usually 70 to 125 mm in diameter, ensure rapid agent delivery which is critical to effective suppression. One of several types of extinguishing agents are employed, usually selected from among the following:

1. Water
2. Dry chemical formulations based on sodium bicarbonate or monoammonium phosphate
3. Halon substitutes (halons, which were used for many years, are being phased out because of their deleterious effect on the ozone layer).

The extinguishing mechanisms whereby each agent works is a combination of thermal quenching (100% in the case of water) and chemical inhibition, a discussion of which is beyond the scope of this chapter. The selection of agent is usually based on several considerations such as effectiveness, toxicity, product compatibility, residual inerting, and volatility. The halons are particularly versatile agents but are now subject to production phase-out owing to their adverse effect on stratospheric ozone. Alternative environmentally safe chemicals are being developed by several chemical manufacturers but these remain to be proven effective in explosion protection applications. As such, dry chemical agents are more commonly specified in suppression applications.

Control of these systems is achieved using an electronic power supply having battery back-up power. This unit supervises the suppression system circuitry to ensure integrity of the system and supplies the current to discharge the explosive actuators employed to open the HRD extinguishers. Normally the process being protected by the suppression system is automatically shut down on detec-

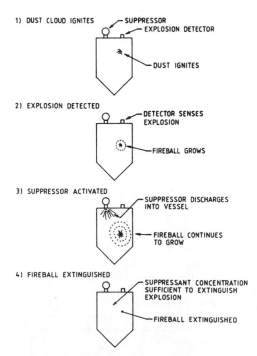

Figure 19.17. Principle of suppression.

tion of an incipient deflagration. Figure 19.18 is a schematic diagram of a suppression system. For further details on deflagration suppression systems refer to NFPA 69[2] and Schofield and Abbott.[13]

19.6.6 Deflagration Pressure Containment

One recently developed method of explosion protection is to design the equipment in which the deflagration may take place to contain the pressure developed. Two approaches are available:

1. Pressure resistance: the vessel or process equipment is designed to prevent permanent deformation on rupture.
2. Pressure shock resistance: the vessel or process equipment is designed to withstand the explosion pressure without rupture, but is subject to permanent deformation in the event of an explosion occurring.

NFTA 69[2] presents equations for calculating the design pressure for these two cases, based on an article by Noronha et al.[16]

The design pressure shall be calculated according to the following equations:

$$P_r = \frac{1.5[R(P_i + 14.7) - 14.7]}{F_u}$$

$$P_d = \frac{1.5[R(P_i + 14.7) - 14.7]}{F_y}$$

where

P_r = the design pressure to prevent rupture due to internal deflagration (psig)

P_d = the design pressure to prevent deformation due to internal deflagration (psig)

P_i = the maximum initial pressure at which the combustible atmosphere exists (psig)

R = the ratio of the maximum deflagration pressure to the maximum initial pressure, as described below

F_u = the ratio of the ultimate stress of the vessel to the allowable stress of the vessel

F_y = the ratio of the yield stress of the vessel to the allowable stress of the vessel.

For vessels fabricated of low-carbon steel and low-alloy stainless steel $F_u = 4.0$ and $F_y = 2.0$. The dimensionless ratio R is the ratio of the maximum deflagration pressure, in absolute pressure units, to the maximum initial pressure, in consistent absolute pressure units. As a practical design basis (because optimum conditions seldom exist in industrial equipment) for most gas–air mixtures R shall be taken as 9; for organic dust–air mixtures R shall be taken as 10. For St-3 dust–air mixtures R shall be taken as 13. An exception exists in that a different value of R shall be permitted to be used if appropriate test data or calculations are available to confirm its suitability.

For operating temperatures below 25°C (77°F), the value of R shall be adjusted according to the following formula:

$$R' = R\left(\frac{298}{273 + T_i}\right)$$

where R is either 9.0 or 10.0 and T_i is the operating temperature in °C.

19.6.7 Deflagration Isolation Systems

Deflagration isolation systems for dust explosions can be of the following types:

● Automatic fast acting closing valves
● Suppressant barriers

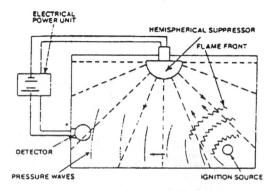

Figure 19.18. Schematic diagram of an explosion suppression system.

- Material chokes
- Flame front diverters.

These are discussed briefly below.

19.6.7.1 Automatic Fast-Acting Closing Valves

Fast-acting closing valves are available in several designs, including flap and slide valves. They are activated by an explosion detector that triggers an explosive charge that releases compressed air or nitrogen from a cylinder, which in turn closes the valve. Such a system is shown in Figure 19.19. The required closing time depends on the distance between the remote pressure or flame sensor and the valve, and the type of dust. Typical closing times for such valves are between 25 and 50 ms. The valve is usually installed 5 to 10 m from the detectors. Both pressure detectors (with threshold detection levels around 0.1 bar) and optical/radiation detectors are used. Pressure detectors are favored in most dusty applications because of the possibility of blinding an optical detector.

Rapid-action valves have to be tested under explosion conditions similar to that expected in actual operation to determine their effectiveness as a flame barrier and their pressure ratings before actual use in practice. Bartknecht[3] and Schofield and Abbott[13] discuss these in more detail.

19.6.7.2 Suppressant Barriers

Suppressant barriers are similar to suppression systems used in equipment but are used in pipelines. They can stop fully developed dust explosions at a predetermined pipe location and limit the course of the explosion to a defined pipe section.

For a given explosion velocity the quantity of suppressant required per unit area of pipe cross-section is constant and does not depend on pipe diameter. The quantity of suppressant required is typically 20 to 100 kg/m^2 of pipe cross-section.[13] Suppressant barriers have been used effectively for pipelines up to 2500 mm in diameter. Bartknecht[3] presents a thorough discussion of these barrier systems. Figure 19.20 shows such a system.

19.6.7.3 Material Chokes

Explosion isolation can also be achieved by the judicious selection and design of mechanical conveying equipment such as screw conveyors and rotary valves (air locks). These types of equipment provide a "choke" of material (powders or bulk solids) to prevent the propagation of an explosion. However, some burning material can be swept through such choke devices if they are not stopped immediately after an explosion is detected, and to prevent such an occurrence, an inerting concentration of suppressant is often injected into the connecting piping.

Bartknecht[3] gives some criteria for the design of rotary valves to enable them to protect against explosion propagation.

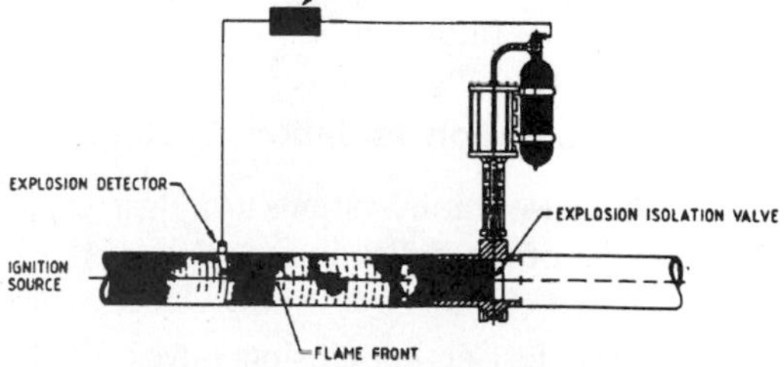

Figure 19.19. Rapid action valve.

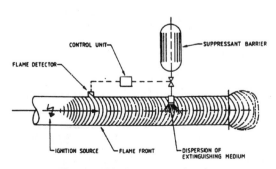

Figure 19.20. Suppressant barrier.

19.6.7.4 Flame Front Diverters

A fairly recent device for providing deflagration isolation is the flame front diverter. It consists of pipelines that are interconnected by a special pipe section, which is closed from the atmosphere by a cover or rupture disk (see Fig. 19.21). The basic principle is that the explosion is vented at a point where the flow direction is changed by 180°. Owing to the inertia of the fast flow caused by the explosion the flow will maintain its direction upward rather than making a 180° turn as during normal flow.

BURSTING DISC
OR OTHER VENT COVER

Figure 19.21. Section through device for interrupting dust explosions in ducts by combining change of flow direction and venting. Flow direction may also be opposite to that indicated by arrows.

19.7 APPLICATIONS TO INDUSTRIAL PROCESSES AND EQUIPMENT

The following section discusses the design of various powder and bulk solids handling and processing equipment to minimize dust explosions, and the application of preventive and protective measures to this equipment. Four groups of processing equipment are covered:

- Crushing and milling equipment
- Dryers
- Powders mixers
- Conveyors and dust removal equipment.

19.7.1 Crushing and Milling Equipment

The type of crusher or mill has an effect on the propensity for a dust explosion. In crushers and roll mills, the dust concentration is mostly below the LEL because of the nature of the comminution process itself. In the case of screen mills and in jet mills, the probability of ignition sources is usually very low. For fluid jet mills, nitrogen can be used in lieu of air which will inert the operation. Mills are available in shock-resistant construction so that they can withstand an internal dust explosion.

Whenever possible, one should use mill types that minimize dust cloud formation and generation of ignition services by high-speed impact (i.e., mills with low-speed rotors).

Table 19.4 lists appropriate means for preventing and mitigating dust explosions in crushers and mills.[17] In this table "X" indicates the most appropriate means of protection, and "(X)" implies the use of the means indicated is possible, but that these methods are not used as frequently as those indicated by an "X." For example, Table 19.4 indicates that adding an inert dust to explosible dust in some mills is a means of preventing a dust explosion, but this method is not usually feasible as the product would be contaminated by the inert dust.

It is sometimes more feasible to isolate a crusher or mill from other equipment by locating it in an enclosed room with deflagration vent panels in an outside wall.

19.7.2 Dryers

Table 19.5 lists methods for preventing and mitigating dust explosions in a number of dryer types.

Spray dryers and fluid bed dryers usually operate at dust concentrations significantly below the LEL, which adds to their safety. However, dust deposits are often generated on walls, etc., so that smoldering spots may develop, depending on the temperature and oxygen concentration. A number of dryers can be designed with a closed-loop nitrogen system,

Table 19.4. Appropriate Means for Preventing and Mitigating Dust Explosions in Chemical Process Plant.

MEANS OF EXPLOSION PREVENTION / MITIGATION — CRUSHING AND MILLING EQUIPMENT	DUST CONCENTR. < MIN. EXPL. CONCENTR.	INERTING BY ADDING INERT GAS	INTRINSIC INERTING	EVACUATION OF PROCESS EQUIPMENT	ADDITION OF INERT SOLIDS	ELIMINATION OF IGNITION SOURCES	EXPLOSION RESISTANT EQUIPMENT	EXPLOSION VENTING	AUTOMATIC EXPLOSION SUPPRESSION	EXPLOSION ISOLATION
Ball mills		X		(X)	(X)		X	(X)		
Vibratory mills		(X)			(X)		X	(X)		
Crushers	X					(X)	X	(X)	(X)	
Roll mills	X					(X)				
Screen mills		(X)					X	(X)		
Air jet mills		(X)					X	(X)		
Pin mills		(X)			(X)		X	(X)	(X)	
Impact mills					(X)		X	(X)	(X)	
Rotary knife cutters	(X)						X	(X)	(X)	
Hammer mills		(X)			(X)		X	(X)	(X)	

From Noha, 1989.[17]

for example, plate and belt dryers. They can also be designed in dust-tight and gas-tight construction. Two good references on dryer safety are the book by Abbott[18] and the article by Gibson et al.[19]

19.7.3 Powder Mixers

Powder mixing can be accomplished in both batch and continuous mixers, which are available in a variety of designs. Among these are tumbling mixers (V-type and double-cone), orbiting screw, U-trough, and fluidized bed. Those with rotating mixing elements (orbiting screw, U-trough) can cause friction sparks if the elements come in contact with the wall of the vessel. Table 19.6 lists protection methods for preventing and mitigating dust explosions in powder mixers. As can be seen from the table, elimination of ignition sources by proper design is the most commonly used method, but inerting and even venting is frequently used.

19.7.4 Conveyors and Dust Removal Equipment

Conveyors for powders and bulk solids are available as mechanical conveyors or pneumatic conveyors.

Pneumatic conveying systems normally have the greatest proclivity for dust explosions and fires among conveyors, for the following reasons:

1. Generation of static electricity by contact between particles themselves and between particles and the pipewall.

Table 19.5. Appropriate Means for Preventing and Mitigating Dust Explosions in Chemical Process Plant.

POWDER DRYERS (MEANS OF EXPLOSION PREVENTION / MITIGATION)	DUST CONCENTR. < MIN. EXPL. CONCENTR.	INERTING BY ADDING INERT GAS	INTRINSIC INERTING	EVACUATION OF PROCESS EQUIPMENT	ADDITION OF INERT SOLIDS	ELIMINATION OF IGNITION SOURCES	EXPLOSION RESISTANT EQUIPMENT	EXPLOSION VENTING	AUTOMATIC EXPLOSION SUPPRESSION	EXPLOSION ISOLATION
Spray dryers (nozzle)	X	(X)	X			X		(X)	(X)	
Spray dryers (disc)	X	(X)	X					(X)	(X)	
Fluidized bed dryers		(X)	(X)			X	(X)	(X)	(X)	
Stream dryers	(X)	(X)				X	(X)	(X)		
Spin-flash dryers						(X)	X	(X)	(X)	
Belt dryers	X					(X)				
Plate dryers	X					(X)				
Paddle dryers	X	X		(X)	(X)	X	(X)	(X)	(X)	

From Noha, 1989.[17]

2. The possibility of dust concentrations within the explosible range at the delivery point where the dust is separated from the air (silos, cyclones, bag houses).

3. The possibility that heated particles created during grinding or drying may be carried in a pneumatic transport system and fanned to a glow by the high air velocity. These can then cause an ignition in the storage or collection system at the end of the pneumatic conveyor. Tramp metal in pneumatic systems may also cause frictional heating or sparks as it is passed through the system.

Mechanical conveyors are less prone to fires and explosions than pneumatic conveyors, but they also can experience them if adequate design and operational precautions are not taken into account. Grossel[20] discusses safety considerations in conveying of bulk solids and powders, including recommendations about protective techniques. NFPA 650[21] also discusses safety aspects of pneumatic conveying systems. Dust collectors and cyclones have experienced fires and explosions in many processes, and protective techniques must be provided for them. Palmer[6] pays specific attention to dust explosions in cyclones and dust collectors. Factory Mutual Engineering Corporation (FMEC) also presents information on protecting dust collectors.[22] Venting and suppression are commonly used for dust collector protection. Also, some manufacturers of cylindrical dust collectors can design them for 50 psig which will contain a deflagration. Table 19.7 lists appropriate techniques for preventing and mitigating dust explosions in conveying and dust removal equipment.

Table 19.6. Appropriate Means for Preventing and Mitigating Dust Explosions in Chemical Process Plant.

POWDER MIXERS (MEANS OF EXPLOSION PREVENTION / MITIGATION)	DUST CONCENTR. < MIN. EXPL. CONCENTR.	INERTING BY ADDING INERT GAS	INTRINSIC INERTING	EVACUATION OF PROCESS EQUIPMENT	ADDITION OF INERT SOLIDS	ELIMINATION OF IGNITION SOURCES	EXPLOSION RESISTANT EQUIPMENT	EXPLOSION VENTING	AUTOMATIC EXPLOSION SUPPRESSION	EXPLOSION ISOLATION
With mixing tools:										
High-speed	(X)	X		(X)			X	(X)	(X)	
Low-speed	(X)	(X)		(X)		X	(X)	(X)	(X)	
Without mixing tools										
Drum mixers		(X)		(X)		X	(X)			
Tumbling mixers		(X)		(X)		X	(X)			
Double cone mixers		(X)		(X)		X	(X)			
Air flow mixers:										
Fluidized bed mixers						X	(X)	(X)		
Air mixers						X	(X)	(X)		

From Noha, 1989.[17]

Additional protective measures for dust collectors should include the following:

1. Water deluge spray headers on the clean side above the bags or cartridges to extinguish a fire. The water supply piping to the deluge header may be hardpiped if the bag house is indoors or in a warm climate, or a dry-pipe system should be used if the bag house is outdoors in a cold climate where freeze-up may occur.
2. High-temperature sensor and alarm to warn of a possible fire inside the bag house. This may be interlocked with an automated block valve in the water supply piping to the deluge spray header.
3. Proper grounding of the bag house to dissipate electrostatic charges.
4. A broken bag detector with an alarm to alert operating personnel that unfiltered dust may be emitting into the atmosphere. This is especially important if the dust is toxic.

19.7.5 General Recommendations

The discussion in the previous sections and the recommended preventative and mitigating methods listed in Tables 19.4, 19.5, 19.6, and 19.7 should be regarded as only a starting point for further investigation rather than a final answer. The protection technique finally chosen will be the result of detailed analysis of many relevant factors for each specific type of equipment. These will include economics, impact of the protective measures on nearby equipment and people, and the fact that some protective measures are not suitable for cer-

Table 19.7. Appropriate Means for Preventing and Mitigating Dust Explosions in Chemical Process Plant.

MEANS OF EXPLOSION PREVENTION / MITIGATION — POWDER / DUST CONVEYORS AND DUST REMOVAL EQUIPMENT	DUST CONCENTR. < MIN. EXPL. CONCENTR.	INERTING BY ADDING INERT GAS	INTRINSIC INERTING	EVACUATION OF PROCESS EQUIPMENT	ADDITION OF INERT SOLIDS	ELIMINATION OF IGNITION SOURCES	EXPLOSION RESISTANT EQUIPMENT	EXPLOSION VENTING	AUTOMATIC EXPLOSION SUPPRESSION	EXPLOSION ISOLATION
Screw conveyers	(X)	(X)				(X)	X	(X)		
Chain conveyors	(X)						X	(X)	(X)	
Bucket elevators		(X)				(X)	X	X	(X)	
Conveyor belts	X									
Shaker loaders							X	(X)		
Rotary locks		(X)				X	X	(X)		
Pneumatic transport equipment	(X)	(X)				X	(X)			(X)
Dust filters and cyclones		(X)				X	X	(X)	(X)	
Industrial vacuum cleaning installations	X					X				

From Noha, 1989.[17]

tain types of equipment because of their construction or design.

REFERENCES

1. NFPA 68, *Venting of Deflagrations*, National Fire Protection Association, Quincy, MA (1994).
2. NFPA 69, *Explosion Prevention Systems*, National Fire Protection Association, Quincy, MA (1992).
3. W. Bartknecht, *Dust Explosions-Course, Prevention, Protection*, Springer-Verlag, Berlin, Germany, and New York (1989) (English Translation).
4. R. K. Eckhoff, *Dust Explosions in the Process Industries*, Butterworth-Heinemann Ltd., Oxford, UK and Boston, MA (1991).
5. P. Field, *Dust Explosions* (*Handbook of Powder Technology*, Vol, 4), Elsevier, Amsterdam, The Netherlands (1982).
6. K. N. Palmer, *Dust Explosions and Fires*, Chapman Hall, London, UK (1973).
7. P. R. Amyotte and M. J. Pegg, Proceedings of the 26th Annual AIChE Loss Prevention Symposium (1992).
8. J. Cross, *Electrostatics: Principles, Problems, and Applications*, Adam Hilger (IOC Publishing Ltd.), Bristol, UK (1987).
9. H. Haase, *Electrostatic Hazards: Their Evaluation and Control*, Verlag Chemie, Weinheim, West Germany and New York (1977) (English translation by M. Wald).
10. M. Glor, *Electrostatic Hazards in Powder Handling*, John Wiley & Sons, New York (1988).
11. G. Luttgens and M. Glor, *Understanding and Controlling Static Electricity*, Expert Verlag, Ehningen bei Boblingen, Germany (1989).
12. G. Lunn, *Dust Explosion Prevention and Protection, Part 1—Venting*, 2nd edit., Institution of Chemical Engineers, Rugby, England (1992).
13. C. Schofield and J. A. Abbott, *Guide to Dust Explosion Prevention and Protection, Part 2—Ignition Prevention, Containment, Inerting, Suppression and Isolation*, Institution of Chemical Engineers, Rugby, UK (1988).

14. G. A. Lunn, *Guide to Dust Explosion Prevention and Protection, Part 3—Venting of Weak Explosions and the Effect of Vent Ducts*, Institution of Chemical Engineers, Rugby, UK (1988).

15. P. E. Moore and W. Bartknecht, Proceedings of the International Loss Prevention Symposium, Cannes, France (September 1986).

16. J. A. Noronha, M. T. Merry, and W. C. Reid, *Plant/Operat. Prog. 1*(1) (January 1982).

17. K. Noha, *VDI—Berichte*, No. 701, pp. 681–693 (1989).

18. J. Abbott (ed.), *Prevention of Fires and Explosions in Dryers: A User Guide*, Institution of Chemical Engineers, Rugby, UK (1990).

19. N. Gibson, D. J. Harper, and R. L. Rogers, *Plant/Operat. Prog. 4*:181–189 (1985).

20. S. S. Grossel, *J. Loss Prevent. Proc. Indust. 1*:62–74 (April 1988).

21. NFPA 650, *Pneumatic Conveying Systems*, National Fire Protection Association, Quincy, MA (1989).

22. FMEC, Loss Prevention Data Sheet 7-73, *Dust Collectors*, Factory Mutual Engineering Corporation, Norwood, MA (1991).

20
Respirable Dust Hazards

B. H. Kaye

CONTENTS

20.1 INTRODUCTION

Damage to the human lung from breathing a dusty atmosphere is not new. Scientists who have studied Egyptian mummies have found cases of silicosis, a disease caused by damage to the lung from inhaling very fine particles of silica. These Egyptian incidents of silicosis probably were caused by the fact that it was common practice to create vases and hollow vessels by grinding sandstone with a harder stone by rotating the material under the drill of hard material. This work was often carried out in poorly ventilated buildings. For a discussion of silicosis among early miners in the 1600s in the silver mines of South America see Ref. 1. In a seminal book published in 1955 Donald Hunter reviewed the history of lung diseases created by dust from industrial activity. In the book he describes the often shocking conditions in which people were expected to work.[2] In particular he quotes from a book

written in 1843 about the conditions among workers in the Sheffield, England cutlery trade.

"Thus fork grinding is always performed on a dry stone and in this consists a peculiarly destructive character of the industry. In the room in which it is carried on there are generally 8 to 10 individuals who work and the dust which is created composed of the fine particles of stone and metal rises in clouds and pervades the atmosphere to which they are confined."

The 1843 book describes how a study of the records of 61 fork grinders who died between 1825 and 1840 showed that 35 of these were under 30 years of age.

Although the gross excesses of dusty environments such as those of 1843 abated with the development of factory safety conditions and laws, there was still a very high death rate among workers such as coal miners and asbestos workers well into the 1950s. In the

United States alone it was estimated that over 500,000 workers and dependents received compensations for coal miner's lung (a disease also called pneumoconiosis and black lung).[3]

Good industrial housekeeping in Western industrial countries has reduced a problem of dust-initiated diseases to the problems of long-term exposure to low levels of dangerous dusts.[4,5] In this brief review of dust diseases it is not possible to do more than present the basic concepts involved in the hazards posed by respirable dust and to review some of the more widely spread diseases along with references to further studies of such diseases.

In occupational health and hygiene the term "respirable dust" has a specific meaning. To understand what is meant by respirable dust in occupational health and hygiene studies consider the drawing of the lung shown in Figure 20.1.[6]

When looking at dust hazards, scientists do not always know the density of the various types of fineparticles present in a dust. For this reason many occupational health studies make use of a parameter known as the aerodynamic diameter of a fineparticle. The aerodynamic diameter is defined as the size of a sphere of unit density that would have the same falling speed as the dust fineparticle.

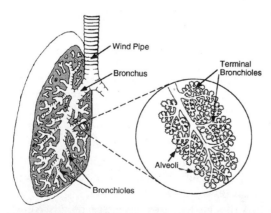

Figure 20.1. Respirable dust, in occupational health and hygiene, refers strictly to particles having an aerodynamic diameter of 5 μm or less. This is the size that can penetrate to the alveoli where there are no cilia to clean the dust from the lung.

In occupational hygiene studies respirable dust is defined as dust having an aerodynamic diameter of less than 5 μm. The exact value of this limiting upper size varies slightly from one country to another. Thus the value is 7 μm in Great Britain. The significance of this upper size limit is that in general dust fineparticles below this size can reach down into the alveoli of the lung where there are fewer clearance mechanisms to defend the lung.[3]

In the early days of occupational hygiene, when one was concerned with the removal of gross amounts of dust, the aerodynamic diameter of a fineparticle was a sufficient measure. Today, however, as we are concerned with more exotic dusts such as fumes from nuclear reactors and the detailed properties of such pollutants as diesel exhausts, the aerodynamic diameter is only one of several parameters that must be measured to adequately characterize the relevant properties of a dangerous dust. Thus, in Figure 20.2, three sets of isoaerodynamic diameter dust fineparticles of different types, as prepared using the Stöber centrifuge, are shown.[7,8]

In this diagram, circles depicting the aerodynamic diameter of the fineparticles and the Stokes diameter are shown. The Stokes diameter is defined as the diameter of the sphere having the same density as the fineparticles that has the same falling speed as the dust fineparticles. It can be seen that the aerodynamic and Stokes diameters of the coal fineparticles are smaller than the physical size of the dust.[9] This is because coal has micropores, making it of lighter density than that of the nominal material. It can be seen that if the particles are relatively compact then the dust particles are almost the same size as the aerodynamic size, but as they get more jagged they are considerably larger than the aerodynamic size. For such profiles it can be shown that the fractal dimension, a measure of the ruggedness of the structure, is a useful parameter for characterizing the significant parameters of the dust fineparticles. The fractal dimensions of the isoaerodynamic dust particles are shown below each profile.[10-12] The fineparticles

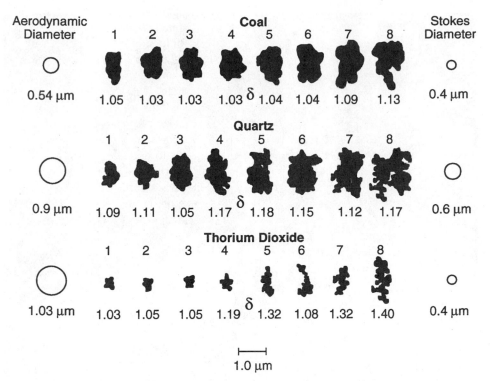

Figure 20.2. Within groups of isoaerodynamic fineparticles, fineparticles of the same aerodynamic diameter, increasing physical size within a group appears to accompany increasing fractal dimension.[7-9, 12]

shown in Figure 20.2 are essentially silhouettes. If one looks at the actual profile of the highly rugged thorium dioxide, as depicted in the original publication, it can be seen to be very porous. If one were trying to measure the burden of adsorbed cancer-causing chemicals carried by such dust into the lung, the use of the simplified aerodynamics as a characteristic parameter would lead to a gross underestimate of the hazard. Also the surface reactivity of such rugged profiles would be far greater than anticipated for the measured aerodynamic diameter. To characterize such hazardous dusts fully one needs not only the aerodynamic diameter but also the physical size, which would govern the ability of the dust fineparticle to lodge in the wall of the lung. The fractal dimension of rugged fineparticles would help in the assessment of the hazard burden or reactivity of the fineparticle.[12]

In the higher parts of the lung, leading to the alveoli, there are hairlike organs called cilia. These cilia, with a whiplike action, move the dust up into the trachea where they are either swallowed, moving into the digestive system, or they can be spat out of the mouth. One particular type of dust, which is very open structured, is diesel exhaust fumes. Thus in Figure 20.3 a set of diesel oil combustion soot products are shown at high magnification.[13] Such fineparticles have very low aerodynamic diameter and move with the inflow of breath. However, because of the large, real size they are easily captured on the walls of the tubes feeding the alveoli. There is some indication that workers exposed to diesel exhaust fumes can suffer from cancer of the bladder, which would indicate that the diesel exhaust fineparticles are not penetrating the lung but are being expelled into the digestive system by the

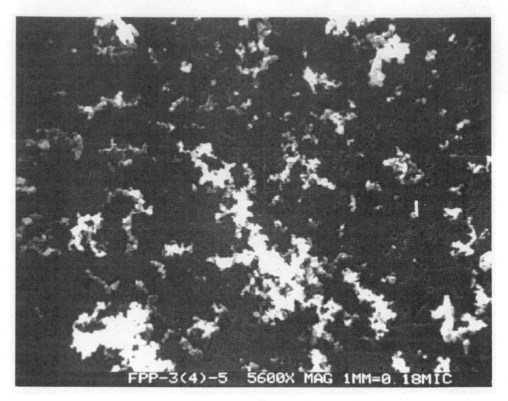

Figure 20.3. Soot fineparticles from free-burning diesel fuel are open-structured and fluffy with enormous surface area capable of carrying large loads of adsorbed carcinogenic combustion chemicals into the body. This type of dangerous dust has small aerodynamic size but large physical size and is easily captured in the higher regions of the lung before reaching the alveoli.[12, 13]

cilia and subsequent swallowing of the soot bearing mucus causes problems in the bladder.[14]

Although respirable dust is considered the major candidate for causing lung diseases such as pneumoconiosis and silicosis, lung cancer often starts higher in the lung on the walls of the bronchioles and the bronchus. It is thought that this is due to the fact that factors in some individual lifestyles damage the cilia, interfering with the cleaning mechanisms. The subsequent irritation of such sites by the inhaled dust initiates the development of a cancer. For example, it is believed that the nicotine in cigarette smoke paralyzes the cilia interfering with their ability to clear dust from the lungs.

The interaction of a lifestyle factor with the physical danger from a respirable dust is described as a synergistic interaction. In Figure 20.4 the synergistic interaction of cigarette smoking and the exposure to asbestos dust in shipyard workers is illustrated.[15-17]

A possible aggravation of the lung by cigarette smoking is a contentious issue between the mining industry and the unions at the time of this writing.[17, 18]

Many different instruments are used to monitor dust levels in the working environment and it is possible in this chapter only to give an indication of two or three of the modern monitoring technologies.[3, 19-21] The physical design of one of the instruments that splits dust to be characterized into respirable and coarse dust fractions is the dichotomous sampler shown in Figure 20.5.[11, 19]

The fractionation achieved in the dichotomous aerosol sampler is based on the principle of impaction used in many different aerosol

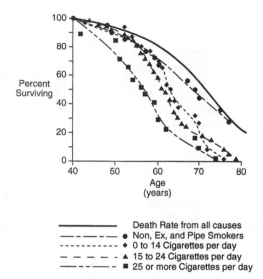

Death Rate from all causes
● Non, Ex, and Pipe Smokers
◆ 0 to 14 Cigarettes per day
▲ 15 to 24 Cigarettes per day
■ 25 or more Cigarettes per day

Figure 20.4. Epidemiological studies of the death rates of shipyard workers in Belfast, involved in the removal of asbestos insulation from ships, indicate that synergistic interaction of lifestyle factors and respirable dust can greatly increase the death rate as compared to the effect of the respirable dust alone.

sampling devices. The basic principle used in an impactor is illustrated in Figure 20.5a. A jet of dusty air is made to impinge on a flat surface. The presence of this flat surface diverts the jet in a circular path. This creates centrifugal forces on the dust particles in the air stream. Larger particles, above a certain cutoff size, are thrown downwards onto the surface by the centrifugal action. The cut size of the impactor depends on the flow velocity of the air stream and the distance between the exit orifice of the jet and the collecting surface.[11] This simple type of impactor device suffers from the problem that hard fineparticles, such as quartz dust fineparticles, tend to bounce when they hit the surface. On rebound they are reentrained in the moving air system. To avoid the problems of bounce and possible reentrainment of the fineparticles by the moving air stream a system known as the virtual surface is employed. The basic principles of this system are shown in Figure 20.5b. A static air reservoir is placed beneath the small orifice, intercepting the flow of dusty air. Just as

in the case of the solid surface impactor, the dust fineparticles are centrifuged out to the central point beneath the air jet as the air stream turns. Now, however, the fineparticles thrown out of the airstream fall into the air reservoir where they can be collected at a later time. The first virtual impactor surfaces developed were found to suffer from the fact that the reservoir below the air jet oscillated. In more modern systems a small amount of air is sucked down through the reservoir constituting the virtual surface to suppress this oscillation as shown in Figure 20.5c. To suppress the oscillation of the surface of the virtual impactor reservoir 1/49th of the total air supply is sucked through the filter in the reservoir used to collect coarser fineparticles.[19] Obviously one must monitor the air flow so that as soon as the oversize collector filter carries a certain load one must change both filters of the device. In the dichotomous sampler shown in Figure 20.5c, instead of a jet being used to direct the dust fineparticles into the reservoir an orifice in a plate is used. It is found that this orifice acts as a half centrifuge turn as distinct from a quarter centrifuge that is operative in the simple jet impactor of Figure 20.5a. The flow through the orifice and the distance between the orifice and the surface of the virtual impactor reservoir is adjusted so that fineparticles having aerodynamic diameters greater than 5 μm are sent into the virtual impactor reservoir whereas the respirable dust fraction carries on through the system to the fines collector filter shown in the diagram.

Another device widely used in monitoring the air in a working environment is the cyclone shown in Figure 20.6 The air to be inspected is directed tangentially into a conical body. As the air spirals down this conical body the larger fineparticles are thrown out to the walls of the cyclone.[3, 11] These larger fineparticles fall down the walls of the cyclone and collect at the base. The vortex flow of the cyclone eventually moves up through the center of the device leaving through a central pipe. This air flow is then directed to a high-

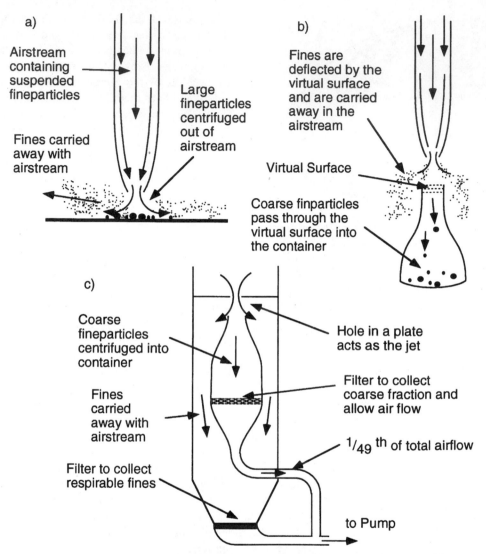

Figure 20.5. Impactors are often used to separate coarse fineparticles from a stream of dusty air. (*a*) A simple jet impactor deposits coarse fineparticles on a surface by centrifugal action on the airstream. (*b*) A virtual impactor addresses some of the problems associated with a simple impactor by using a reservoir of trapped air to capture coarse fineparticles. (*c*) The dichotomous sampler allows a small airflow through the collection chamber to prevent resonance vibration of the virtual surface.

efficiency membrane filter to collect the respirable dust. The personal cyclones used to monitor working air in places such as mines have flow rates and dimensions so that only dust with aerodynamic diameters smaller than 5μm can pass onto the filter. At the end of a working shift the filter is removed and the weight of powder deposited recorded.[3, 20]

In Figure 20.7 a new system based on an instrument known as TEOM being used by the Bureau of Mines and other scientists to evaluate respirable dust is shown. The term TEOM stands for Tapered Element Oscillating Microbalance. The name describes the essential element of Figure 20.7a shown separately in Figure 20.7b.

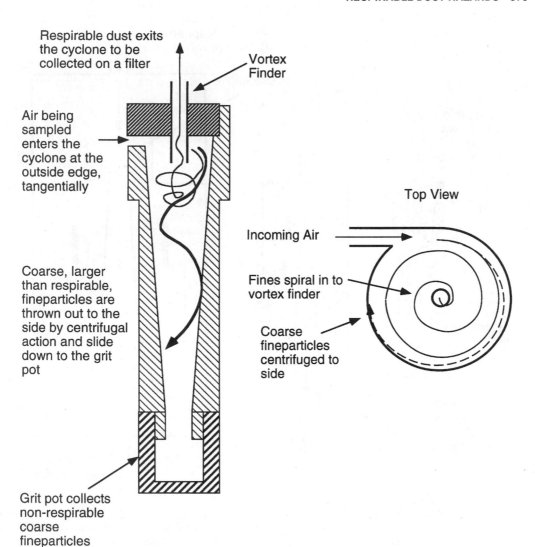

Respirable dust exits
the cyclone to be
collected on a filter

Vortex
Finder

Air being
sampled
enters the
cyclone at the
outside edge,
tangentially

Top View

Incoming Air

Coarse, larger
than respirable,
fineparticles are
thrown out to the
side by centrifugal
action and slide
down to the grit
pot

Fines spiral in to
vortex finder

Coarse
fineparticles
centrifuged to
side

Grit pot collects
non-respirable
coarse
fineparticles

Figure 20.6. A simple cyclone can be used to separate coarse fineparticles from respirable dust to be characterized by using centrifugal action to send the coarse fineparticles to the wall of the cyclone so they fall into the grit pot.

It is interesting to note that the TEOM monitor evolved from space research projects aimed at measuring the mass of dust grains encountered in the tails of comets. In space one cannot weigh objects because they do not have weight in the absence of a large gravitational field. The TEOM device measures the mass of a fineparticle from the change in the oscillating behavior of the equipment as the dust accumulates on the filter at the top of the tapered element shown in Figure 20.7b.

Because of the way in which it works, the orientation of the device is immaterial; it can be used upside-down or on its side depending on the available space for mounting the device. When measuring dust in the work environment the device is equipped with the prestage of a cyclone that removes anything other than respirable dust from the air stream. Fineparticles having respirable diameters, that is, smaller than 5 μm aerodynamic diameter, are deposited on the filter and after the end of

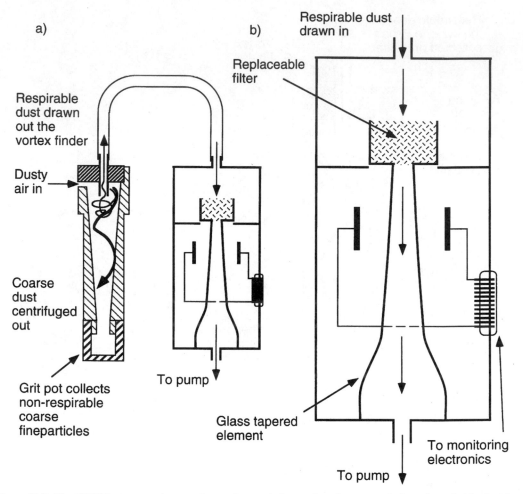

Figure 20.7. The TEOM mass monitor can be used to actively monitor the accumulation of respirable dust in a working environment. (*a*) A cyclone is used in series with the TEOM monitor in order to remove coarse fineparticles from the airstream. (*b*) Respirable dust is captured by the filter element of the TEOM, increasing the mass and changing the oscillation frequency of the system.

a shift, the miner brings the TEOM element to a central point where the deposited mass from operation during a working shift is measured. The system is shown in Figure 20.8.[21-24]

20.2 SPECIFIC RESPIRABLE DUST HAZARDS IN INDUSTRY

Historically, one of the major areas of disease from industrial dust was in the mining industry, where deposits of coal dust in the lung

gave rise to an illness known as pneumoconiosis, also known as black lung. This caused emphysema, and in severe cases, lung cancer. The disease was particularly prevalent among the hard coal miners of South Wales, where they mine the very dense coal known as anthracite. It was not always clear where the health hazard came from. Some workers believe that it was the presence of silica in the coal, or in adjacent seams to where the coal was being mined, that gave rise to the health hazards. It is hard to realize how working

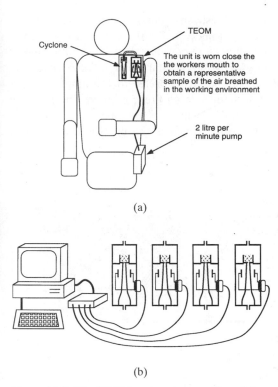

(a)

(b)

Figure 20.8. The TEOM system in operation. When monitoring the quality of air in the workplace, the equipment must be kept compact enough for the worker to carry comfortably. (*a*) The TEOM, cyclone, and pump being worn by a miner. Note that the intake is as near to the mouth and nose as is practical. (*b*) Several TEOM units being prepared for data collection after the working day.

conditions have changed in the mining industry. Back in the 1950s, the father of one of my friends worked in the south Yorkshire coal field lying on his side in an 18-in. seam of coal, swinging his pick, and pushing out the coal with his feet. Today the coal industry in Western industrial nations is largely mechanized and large machines are used to cut the coal. However, the struggle to abate coal dust in the working areas continues, using sprays to suppress dust and providing respirators in particularly dangerous working areas.

Hard-rock miners, in such industries as gold mining and nickel mining, were at risk from silica dust. However, it must be pointed out that dangerous silica dusts have to be freshly shattered quartz dusts. Aged dust tends to be less dangerous than the freshly shattered material. This is probably a function of a chemical activity of the freshly generated dust surfaces. Sand blasters in foundries and in the ceramic industry can also be exposed to dangerous levels of silica dust.[25-27]

A controversial practice in the mines of Canada involved the breathing of aluminum dust at the beginning of a shift in the belief that the aluminum dust in the lung could prevent silicosis. The practice developed from very limited data based on the study of the health of seven rabbits exposed to alumina and silica dusts. There is a possibility that Alzheimer's disease may be associated with aluminum in the lifestyle of an individual. (Alzheimer's disease is a progressive form of senile dementia. There is no doubt that some of the early-onset Alzheimer's cases are genetically linked. The possibility that aluminum could be a factor is relevant to later age onset cases.[29-31])

A newsitem by Raphals reviews the work of Rifat, who did an epidemiological study of Alzheimer's disease among miners who breathed aluminum dust as a prophylactic against silicosis.[32] This controversial study indicates that there is a higher level of Alzheimer's disease among miners who were made to breathe aluminum dust. Because this would involve large amounts of compensation payments the study is being challenged and is considered controversial.

Fiberglass, made from glass that is essentially a silicate, is a controversial topic in industrial hygiene. Some workers believe that because the silicate is in an amorphous chemical state, there is no hazard to a lung from fiberglass if inhaled directly into the lung other than an irritation factor. Other people believe that it causes a health hazard.[28]

The problem in assessing the health hazard of the fiberglass is again partly linked to the problem of lifestyle involving cigarette smoking and also the fact that people working in the industry may have been exposed to dangerous dusts in other industries.

A major dust health hazard is posed by the handling of asbestos. Unfortunately the term asbestos is a generic term referring to various forms of mixed metal–oxide–silicates.[33] The physical appearance and chemical names of the two main groups of asbestos compounds are shown in Figure 20.9. There is considerable controversy over what constitutes a safe level of asbestos dust. The main type of asbestos, mined in South Africa, is one of the amphibole materials called crocidolite that is known by the popular name of "blue asbestos." Industry used to prefer to use amphibole asbestos for making fireproof pipes and building materials owing to its long straight fibers. Chrysotile, which is the main asbestos mined

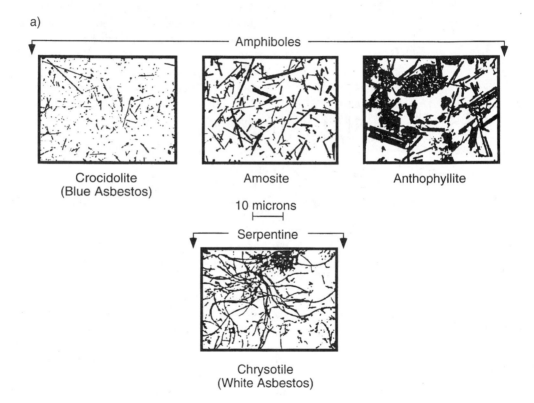

b)		Amphiboles		Serpentine
	Crocidolite	**Amosite**	**Anthophyllite**	**Chrysotile**
Composition	$Na_2O\ Fe_2O_4$	5 5FeO	$7MgO\ 8SiO_2$	$3MgO\ 2SiO_2$
	$3FeO\ 8SiO_2$	1 5MgO	H_2O	$2H_2O$
	H_2O	$8SiO_2\ H_2O$		
Specific Gravity	3.00-3.45	2.60-3.00	2.85-3.50	2.36-2.50
Crystal Structure	monoclinic	monoclinic	orthorhombic	monoclinic
Refractive Index	1.69-1.71	1.66-1.70	1.60-1.66	1.49-1.57

Figure 20.9. Asbestos can be divided into two main families of minerals known as amphiboles and serpentines. (*a*) Physical appearance of the two families of asbestos. (*b*) Chemical properties of the various types of asbestos.

in Canada, has curly fibers, which is preferred for use in the making of fireproof blankets and clothing for industrial workers. Chrysotile is not as dangerous as amphibole asbestos because its curliness prevents penetration into the lung when inhaled. Also the material is more soluble in body fluids than blue asbestos.[34, 35]

Blue asbestos becomes more dangerous as it is handled because the fibers break down to smaller, more easily respirable sizes. Thus it is least dangerous to the miners and has proven to be very dangerous for workers who remove fire insulation from old ships that are being stripped down into useful materials. For the same reason there is some controversy as to whether it is safe to remove asbestos used as fire insulation in buildings. Some people argue that it should simply be sprayed with a sealant and left in place because it is more dangerous to actually remove the material.[35]

At one time there was a great deal of debate over the safety of chrysotile asbestos but the debate was clouded by the fact that chrysotile, as mined in Canada, contains a small amount of tremolite. Industrial processes are being developed to remove the tremolite to increase the safety of asbestos.

Several diseases are attributed to the inhalation of asbestos fibers. The simplest is known as white lung in which the lung suffers from a burden of deposited asbestos fibers that create emphysema and eventually lung cancer. The most dangerous disease caused by inhalation of asbestos fiber is known by the term mesotheloma, which is a cancer of the lining of the lung cavity. It is believed that mesotheloma is caused by fibers less than 1.5 μm in diameter and greater than 8 μm in length. It is believed that such fibers, when they are trapped in the lung, work their way through the lung wall, as they move during the act of breathing, and that they then pierce the walls of the cells of the lung lining, damaging the genetic structure of the cell and resulting in the start of a cancer. Cancer in general is a disease caused by malfunctioning of the genetic information in the nucleus of living cells.

Such disturbance to the genetic structure of the cell can either be chemical (giving us the term carcinogenic chemicals) or physical, such as direct damage caused by penetration of a long spearlike fiber into the center of the living cell. We have already shown in Figure 20.4 that deaths of asbestos workers in the Belfast shipyards can be greatly increased by the synergistic effects of cigarette smoking. It is believed that asbestos fiber damage, when cigarette smoke is present, probably arises from the fact that carcinogenic chemicals in cigarette smoke are adsorbed onto the fibers and that the chemical hazard is greatly increased by the fact that the adsorption process increases the chemical activity of the adsorbed chemicals. Thus some chemicals appear to be 15 times more active when adsorbed onto fibers of asbestos than when present as a simple chemical spray. Some scientists, who believe that the major problem with asbestos is the fibrous nature of the dust, urge great caution be taken in replacing asbestos with ceramic fibers which may cause the same problem.[36-40]

At one time, talcum powder contained asbestos and although it has been removed from modern products, in North America one should be aware that sometimes unauthorized importation of cosmetic material from the third world may result in the individual being exposed to dangerous levels of asbestos.

Strict industrial procedures for handing asbestos fibers have been introduced and the regulations of various countries should be studied for detailed information.

One of the problems when working with various types of dust is that changes in industrial practice have made previously safe dusts a possible problem. Thus, for many years carpenters and furniture workers have worked with low-speed tools on natural woods. The switch to bonded plywoods and chip boards, in which there is synthetic glue, and the working of such woods with high-speed tools can cause the chemical breakdown of the glue by means of the heat generated during working processes. The dust in such an environment is potentially carcinogenic because of the glue byproducts deposited on the dust. This is a possible explanation of the fact that recently it has been found that there is a high incidence of nasal cancer in modern industrial woodworking environments.[41-43]

Histoplasmosis is a lung disease caused by infection with a fungus of the fungal family genohistoplasma. It is marked by the benign involvement of lymph nodes of the trachea and bronchi. Usually the condition is one of emphysema but it can proceed to a dangerous level. Cases have been known among workers who work with musty books in poorly ventilated secondhand book stores and among people who knock down musty swallows' nests in old agricultural buildings. Agricultural workers generally can suffer health problems caused by inhaling fungal spores from things such as moldy hay. Also dusts prevalent in granaries can cause health problems.[44-46]

Such health hazards are not necessarily confined to the farm. The writer knows of a case where a player suffered an asthma attack caused by the fungal spores leaving a moldy straw broom when playing the game of curling. In curling, the player vigorously sweeps the ice in the front of the moving stone, known as "a rock," to help the rock go farther. The cloud of fungal spores released from the moldy broom during such a game initiated a severe allergy attack that required hospital treatment.

Byssinosis is a disease that affects cotton workers who breathe in many small fragments of the fibers used to make the cotton. The term byssinosis comes from the Hebrew word *Bysisus*, meaning fine white linen. It is essentially a disease of textile workers who work with many different natural fibers. Medical experts do not class byssinosis as a true pneumoconiosis because fibrosis of the lung does not occur in this disease. In the textile industry byssinosis is often known as brown lung.

Bagassocis is another respiratory illness caused by inhaling fungal spores and fibrous dust produced by storing the waste products of the sugar cane processing industry. Bagasse is the name given to the fibrous residue left from the processing of sugar cane. It is a Spanish word that has the same meaning as the English word dregs. It is the residue, or dregs, of the sugar cane harvest.

One should always be careful of dust generated in a poorly ventilated atmosphere. Thus recently an industrial disease has been detected among hairdressers who work with cosmetic sprays in a poorly ventilated atmosphere. This disease has been given the name thesarosis. Unless the worker is protected with a proper respirator, welding fumes can cause problems.[47] Older hazards that are now basically controlled are such problems as police officers being subjected to lead poisoning by breathing the lead aerosols produced when firing guns using lead bullets in the confined space of a firing range. Dentists started to suffer from a form of silicosis from the debris from high-speed drills using in dental work before it was appreciated that it was necessary to wear masks to protect the dental worker against such problems. Artists are not always aware of the fact that making such items as stained glass windows, which involves the soldering of lead strips, can also give the workers lead poisoning from the aerosol generated during the act of soldering.

REFERENCES

1. M. J. Allison, "Paleo-Pathology in Peru," *Natural History* February 1979, pp. 74–82.
2. D. Hunter, *The Diseases of Occupations*, 6th edit. (1978); first published 1955, Hodder & Stoughton. See especially Chapter 14, "The Pneumoconioses."
3. F. P. Perera and A. Karim Ahmed, *Respirable Particles; The Impact of Airborne Fineparticles on Health and the Environment*, Ballinger Publishing Company, Cambridge, MA, a subsidiary of Harper & Row (1979).
4. H. Gavaghan, "Healthy Miners but Fewer Jobs," *New Scientist*, March 15, 1984, p. 22.
5. "Evaluation of Coal Mining Technology," Publications Officer, The Technical Chain Center, 114 Cromwell Road, London, SW7 4ES. This article contains information on dust diseases in coal miners.
6. Bloor, *Science Spectrum*.

7. Reproduced from B. H. Kaye, *A Randomwalk Through Fractal Dimensions*. VCH Publishers, Weinheim, Germany (1989).

8. W. Stöber and H. Flachsbart, *Environ. Sci. Technol.* 3:1280 (1969).

9. P. Kotrappa, "Shape Factors for Aerosols of Coal, Uranium Dioxide in the Respirable Size Range," in *Assessment of Airborne Particles*, edited by T. Mercer, E. Morrow and W. Stober, Charles C. Thomas, Springfield, IL, Ch. 16 (1973).

10. B. H. Kaye, "The Physical Significance of the Fractal Structure of Some Respirable Dusts," in preparation.

11. For a discussion of the concepts of aerodynamic diameters and Stokes diameter and the design of equipment for measuring aerosol size distribution in the working environment see B. H. Kaye, *Direct Characterization of Fineparticles*, John Wiley & Sons, New York (1981). See also *Characterizing Powders and Mists*, to be published by VCH Publishers in Weinheim, Germany. The anticipated publication date is June 1997.

12. For a discussion of the fractal structure of dust fineparticles and the techniques used for measuring the boundary fractals of respirable dust see B. H. Kaye, *A Randomwalk Through Fractal Dimensions*, 2nd edit., VCH Publishers, Weinheim (1994).

13. R. G. Pinnick, T. Fernandez, B. D. Hinds, C. W. Bruce, R. W. Schaefer, and J. D. Pendleton, "Dust Generated by Vehicular Traffic on Unpaved Roadways: Sizes and Infrared Extinction Characteristics," *Aerosol Sci. Technol.* 9:99–121 (1985).

14. See newsitem "Cancer Fears for Pastry Cooks," *New Scientist*, p. 28, June 19, 1986.

15. P. C. Elmes, "Health Risks from Inhaled Dusts and Fibers," *R. Soc. Health J.*, June 1977.

16. P. C. Elmes and Simpson, *B. J. Med.* 33–174.

17. For a discussion of the synergistic effects of smoking and asbestos fibers see discussion in B. H. Kaye, *Science and the Detective; Selected Readings in Forensic Science*, VCH Publishers, Weinheim, pp. 251–259 (1995).

18. See article by W. List, "Panel Makes Connection Between Hardrock Mining and Cancer," *Can. Occup. Safety*, November/December 1994.

19. T. G. Dzubay, R. K. Stevens, and C. M. Peterson, "X-ray Fluorescence Analysis of Environmental Samples in Applications of the Dichotomos Sampler to the Characterization of Ambient Aerosols," edited by T. Dzubay, Ann Arbor Science Publishers, Ann Arbor, MI (1978).

20. J. H. Vincent, *Aerosol Science for Industrial Hygienists*, Pergamon-Elsevier, Oxford, England, and Tarrytown, New York (1995).

21. K. L. Williams and R. P. Vincent, "Evaluation of the TEOM Dust Monitor," Bureau of Mines Information Circular, 1986, United States Department of the Interior.

22. H. Patashinck and G. Ruppercht, "Microweighing Goes on Line in Real Time," *Research and Development*, Technical Publishing, June 1986.

23. Commercial information available from Ruppercht and Patashinck Inc., 17 Maple Road, P.O. Box 330, Voorheesville, NY 12186.

24. H. Patashnick and G. Rupprecht, "Advances in Microweighing Technology," Reprinted from *Am. Lab.*, July 1986, pp.

25. G. R. Yourt, "Gravimetric Sampler Assesses Risk of Silicosis," Canadian Mining Journal, October 1972, pp. 46, 48 and 49.

26. C. J. Williams and R. E. Hallee, "An Industrial Hazard—Silica Dust," *Am. Lab.*, pp. 17–27.

27. H. W. Glindmeyer and Y. H. Hammad, "Contributing Factors of Sand Blasters Silicosis: Inadequate Respiratory Equipment and Standards," *J. Occup. Med.* 30(12):917–921 (1988).

28. See M. Hamer, "Fiberglass Linked to Lung Disease," *New Scientist*, October 24, 1992, p. 4.

29. "The Case Against Aluminum," *Can. Res.*, pp. 32–35, March 1988.

30. W. Glenn, "Aluminum: Can It Damage the Brain?" *Occup. Health Safety Can.* 2(6), 1986.

31. L. Tataryn, "Some Miners Are Dying for a Living," *Toronto Star*, Tuesday, September 18, 1979, p. A10.

32. P. Raphals, "Study of Miners Heightens Aluminum Fears," *New Scientist*, 18:17 (August 1990).

33. L. McGenty, "A Ban on Asbestos," *New Scientist*, July 14, 1977, pp. 96–97.

34. News Story, "An Overblown Asbestos Scare. The Dangers Are Minimum in Most Buildings Says a New Study," *Time*, January 29, 1990.

35. J. Zuckerbrot, "Risky Business, Debating the Use of Asbestos in Canada," *Occup. Health Safety Can.* 4(5): Number 32–94 (1988).

36. Newsitem, "Germans Deem Glass and Ceramic Fibers Carcinogenic," in *Chem. Eng.*, October 1993, p. 27.

37. "Asbestos Users Step Up Search for Substitutes," *Chem. Eng.*, October 27, 1986, pp. 18–26.

38. R. Burger, "Getting Rid of Asbestos," *Chem. Eng.*, June 22, 1987, pp. 167–168 and 170.

39. Regulations Respecting Asbestos Made Under the Occupational Health and Safety Act revised Statutes of Ontario, 1980, Chapter 321, Issued August 1982, Ontario Ministry of Labour, Occupational Health and Safety Division, 400 University Avenue, Toronto, Ontario, M7A 1T7.

40. For a review of some of the legal problems posed by asbestos injury lawsuits see the discussion "The Synergistic Killers" in B. H. Kaye, *Science and the Detective; Selected Readings in Forensic Science*, pp. 251–259, VCH Publishers, Weinheim (1995).

41. W. Glenn, "Wood Dust—Tree Bites Man," *Occup. Health Safety Can.* 4(2):18–21 (1988).

42. "Carcinogenic Hazard of Wood Dust," *Toxicity Review* 15, Health and Safety Executive, London, England, October 1984.

43. B. Woods and C. D. Calnan, "Toxic Woods," *Br. J. Dermatol.* 1995, Supplement 13, 1976.

44. J. Mannon and E. Johnson, "Fungi Down on the Farm," *New Scientist*, February 28, 1995, pp. 12–16.

45. See, for example, News Story "Moldy Birds Nests Give Seven Men Respiratory Disease," *Toronto Star*, March 27, 1979.

46. R. Drennon Watson, "Trouble in Store" (A discussion of problems such as moldy hay hazards), *New Scientist*, April 22, 1976, pp. 170–172.

47. See "Welding Fumes," W. Glenn, *Occup. Health Safety Can.* 4(6):18–21.

Index